Encyclopedia of Natural Resources

Volume I—Land

Encyclopedia of Natural Resources

Volume I—Land

Edited by
Yeqiao Wang

CRC Press is an imprint of the
Taylor & Francis Group, an **informa** business

CRC Press
Taylor & Francis Group
6000 Broken Sound Parkway NW, Suite 300
Boca Raton, FL 33487-2742

Printed on acid-free paper
Version Date: 20140501

International Standard Book Number-13: 978-1-4398-5258-3 (Hardback)

Visit the Taylor & Francis Web site at
http://www.taylorandfrancis.com

and the CRC Press Web site at
http://www.crcpress.com

Land

Water

Air

Editorial Advisory Board

Contributors

Jazuri Abdullah / *Department of Civil and Environmental Engineering, Engineering Research Center, Colorado State University, Fort Collins, Colorado, U.S.A.*

Gayatri Acharya / *Environmental Economist, World Bank Institute (WBI), Washington, District of Columbia, U.S.A.*

Vicenç Acuña / *Catalan Institute for Water Research, Girona, Spain*

Gordon N. Ajonina / *CWCS Coastal Forests and Mangrove Programme, Cameroon Wildlife Conservation Society, and Institute of Fisheries and Aquatic Sciences, University of Douala, Yabassi, Douala, Cameroon*

Erhan Akça / *Adiyaman University, Adiyaman, Turkey*

Jose A. Amador / *Laboratory of Soil Ecology and Microbiology, University of Rhode Island, Kingston, Rhode Island, U.S.A.*

Kathryn M. Anderson / *Department of Zoology, University of British Columbia, British Columbia, Vancouver, Canada*

Kim A. Anderson / *Environmental and Molecular Toxicology, Oregon State University, Corvallis, Oregon, U.S.A.*

Konstantinos M. Andreadis / *Jet Propulsion Laboratory, California Institute of Technology, Pasadena, California, U.S.A.*

Frank Asche / *Department of Industrial Economics, University of Stavanger, Stavanger, Norway*

Richard Aspinall / *Honorary Research Fellow, James Hutton Institute, Aberdeen, U.K.*

Peter V. August / *Department of Natural Resources Science, University of Rhode Island, Kingston, Rhode Island, U.S.A.*

James L. Baker / *Department of Agricultural and Biosystems Engineering, Iowa State University, Ames, Iowa, U.S.A.*

Andrew H. Baldwin / *Department of Environmental Science and Technology, University of Maryland, College Park, Maryland, U.S.A.*

Robert D. Ballard / *Graduate School of Oceanography, University of Rhode Island, Narragansett, Rhode Island, U.S.A.*

Roger G. Barry / *Cooperative Institute for Research in Environmental Sciences, National Snow and Ice Data Center, Boulder, Colorado, U.S.A.*

Austin Becker / *Emmett Interdisciplinary Program in Environment and Resources (E-IPER), Stanford University, Stanford, California, U.S.A.*

Jürgen Bender / *Federal Research Institute for Rural Areas, Forestry and Fisheries, Thünen Institute of Biodiversity, Braunschweig, Germany*

Jagtar S. Bhatti / *Canadian Forest Service, Northern Forestry Centre, Edmonton, Alberta, Canada*

John Boardman / *Department of Geographical and Environmental Science, University of Cape Town, Cape Town, South Africa and Environmental Change Institute, University of Oxford, Oxford, U.K.*

Derek B. Booth / *Center for Water and Watershed Studies, University of Washington, Seattle, Washington, U.S.A.*

John Borrelli / *Department of Civil Engineering, Texas Tech University, Lubbock, Texas, U.S.A.*

Virginie Bouchard / *School of Natural Resources, The Ohio State University, Columbus, Ohio, U.S.A.*

Thomas Boving / *Department of Geosciences/Department of Civil and Environmental Engineering, University of Rhode Island, Kingston, Rhode Island, U.S.A.*

James Boyd / *Resources for the Future, Washington, District of Columbia, U.S.A.*

John Van Brahana / *Division of Geology, Department of Geosciences, University of Arkansas, Fayetteville, Arkansas, U.S.A.*

Helen Bramley / *The UWA Institute of Agriculture, University of Western Australia, Crawley, Western Australia, Australia*

Michael L. Brennan / *Center for Ocean Exploration, University of Rhode Island, Narragansett, Rhode Island, U.S.A.*

Sylvie M. Brouder / *Department of Agronomy, Purdue University, West Lafayette, Indiana, U.S.A.*

Jaclyn N. Brown / *Wealth from Oceans National Research Flagship, CSIRO Marine and Atmospheric Research, Hobart, Tasmania, Australia*

Bill Buffum / *Department of Natural Resources Science, University of Rhode Island, Kingston, Rhode Island, U.S.A.*

Thomas J. Burbey / *Department of Geosciences, Virginia Tech University, Blacksburg, Virginia, U.S.A.*

Mary T. Burke / *UC Davis Arboreturm, University of California, Davis, California, U.S.A.*

Robert H. Burris / *Department of Biochemistry, College of Agriculture and Life Sciences, University of Wisconsin, Madison, Wisconsin, U.S.A.*

Richard Burroughs / *Department of Marine Affairs, University of Rhode Island, Kingston, Rhode Island, U.S.A.*

Tim P. Burt / *Department of Geography, University of Durham, Durham, U.K.*

Thomas J. Butler / *Cary Institute of Ecosystem Studies, Millbrook, and Cornell University, Ithaca, New York, U.S.A.*

Frederick H. Buttel / *Department of Rural Sociology, University of Wisconsin, Madison, Madison, Wisconsin, U.S.A.*

Carrie J. Byron / *Gulf of Maine Research Institute, Portland, Maine, U.S.A.*

Steven X. Cadrin / *Department of Fisheries Oceanography, University of Massachusetts, Fairhaven, Massachusetts, U.S.A.*

Rebecca L. Caldwell / *Department of Geological Sciences, Indiana University, Bloomington, Indiana, U.S.A.*

Daniel E. Campbell / *Atlantic Ecology Division (AED), National Health and Environmental Effects Research Laboratory (NHEERL), U.S. Environmental Protection Agency (EPA), Narragansett, Rhode Island, U.S.A.*

Lucila Candela / *Department of Geotechnical Engineering and Geo-Sciences, Technical University of Catalonia (UPC), Barcelona, Spain*

Edward Capone / *Northeast River Forecast Center, National Oceanic and Atmospheric Administration (NOAA), Taunton, Massachusetts, U.S.A.*

Jennifer Caselle / *Marine Science Institute, University of California, Santa Barbara, California, U.S.A.*

Don P. Chambers / *College of Marine Science, University of South Florida, St. Petersburg, Florida, U.S.A.*

Thomas N. Chase / *Department of Civil, Environmental and Architectural Engineering and Cooperative Institute for Research in the Environmental Sciences (CIRES), University of Colorado at Boulder, Boulder, Colorado, U.S.A.*

Gargi Chaudhuri / *Department of Geography and Earth Science, University of Wisconsin, La Crosse, Wisconsin, U.S.A.*

Long S. Chiu / *Department of Atmospheric, Oceanic and Earth Sciences, George Mason University, Fairfax, Virginia, U.S.A.*

L. M. Chu / *School of Life Sciences, Chinese University of Hong Kong, Hong Kong, China*

Daniel L. Civco / *Department of Natural Resources and the Environment, University of Connecticut, Storrs, Connecticut, U.S.A.*

John C. Clausen / *Department of Natural Resources and the Environment, University of Connecticut, Storrs, Connecticut, U.S.A.*

Sharon A. Clay / *Plant Science Department, South Dakota State University, Brookings, South Dakota, U.S.A.*

Michael T. Clegg / *Department of Ecology and Evolutionary Biology, University of California, Irvine, Irvine, California, U.S.A.*

Gopalasamy Reuben Clements / *School of Marine and Tropical Biology, James Cook University, Cairns, Queensland, Australia*

David A. Cleveland / *Environmental Studies Program, University of California, Santa Barbara, California, U.S.A.*

Janet Coit / *Rhode Island Department of Environmental Management, Providence, Rhode Island, U.S.A.*

Jill S. M. Coleman / *Department of Geography, Ball State University, Muncie, Indiana, U.S.A.*

Sean D. Connell / *Southern Seas Ecology Laboratories, School of Earth and Environmental Science, University of Adelaide, Adelaide, South Australia, Australia*

William H. Conner / *Baruch Institute of Coastal Ecology and Forest Science, Clemson University, Georgetown, South Carolina, U.S.A.*

Jeffrey D. Corbin / *Department of Biological Sciences, Union College, Schenectady, New York, U.S.A.*

Richard T. Corlett / *Xishuangbanna Tropical Botanical Garden, Chinese Academy of Sciences, Yunnan, China*

Ray Correll / *Commonwealth Scientific and Industrial Research Organisation (CSIRO), Adelaide, South Australia, Australia*

Robert Costanza / *Institute for Sustainable Solutions, Portland State University, Portland, Oregon, U.S.A.*

Roland C. de Gouvenain / *Department of Biology, Rhode Island College, Providence, Rhode Island, U.S.A.*

Marinés de la Peña-Domene / *Department of Biological Sciences, University of Illinois at Chicago, Chicago, Illinois, U.S.A.*

Eddy De Pauw / *International Center for Agricultural Research in the Dry Areas (ICARDA), Aleppo, Syria*

Sherri DeFauw / *New England Plant, Soil and Water Laboratory, Agricultural Research Service, U.S. Department of Agriculture (USDA-ARS), University of Maine, Orono, Maine, U.S.A.*

Ahmed M. Degu / *Department of Civil and Environmental Engineering, Tennessee Technological University, Cookeville, Tennessee, U.S.A.*

Heidi M. Dierssen / *Department of Biology, Norwegian University of Science and Technology, Trondheim, Norway, and Department of Marine Sciences/Geography, University of Connecticut, Groton, Connecticut, U.S.A.*

Michael E. Dietz / *Center for Land Use Education and Research (CLEAR), University of Connecticut, Storrs, Connecticut, U.S.A.*

Peter Dillon / *Commonwealth Scientific and Industrial Research Organisation (CSIRO), Adelaide, South Australia, Australia*

Endre Dobos / *Department of Physical Geography and Environmental Sciences, University of Miskolc, Miskolc-Egyetemváros, Hungary*

Lindsay M. Dreiss / *Department of Natural Resources and the Environment, University of Connecticut, Storrs, Connecticut, U.S.A.*

Douglas A. Edmonds / *Department of Geological Sciences, Indiana University, Bloomington, Indiana, U.S.A.*

David Ehrenfeld / *Department of Ecology, Evolution, and Natural Resources, Rutgers University, New Brunswick, New Jersey, U.S.A.*

Florent Engelmann / *Institute of Research for Development (IRD), Montpellier, France*

Hari Eswaran / *National Resources Conservation Service, U.S. Department of Agriculture (USDA-NRCS), Washington, District of Columbia, U.S.A.*

M. Tomedi Eyango / *Institute of Fisheries and Aquatic Sciences, University of Douala, Yabassi, Douala, Cameroon*

Daniel B. Fagre / *Northern Rocky Mountain Science Center, U.S. Geological Survey (USGS), West Glacier, Montana, U.S.A.*

Souleymane Fall / *College of Agriculture Environment and Nutrition Science, and College of Engineering, Tuskegee University, Tuskegee, Alabama, U.S.A.*

Lynn Fandrich / *Oregon State University, Corvallis, Oregon, U.S.A.*

Thomas E. Fenton / *Soil Morphology and Genesis, Agronomy Department, Iowas State University, Ames, Iowa, U.S.A.*

Charles W. Fetter, Jr. / *C. W. Fetter, Jr. Associates, Oshkosh, Wisconsin, U.S.A.*

Joseph Fiksel / *Center for Resilience, The Ohio State University, Columbus, and Office of Research and Development, U.S. Environmental Protection Agency (EPA), Cincinnati, Ohio, U.S.A.*

Joshua B. Fisher / *Jet Propulsion Laboratory, California Institute of Technology, Pasadena, California, U.S.A.*

Patrick J. Fitzpatrick / *Mississippi State University, Stennis Space Center, Mississippi, U.S.A.*

Graham E. Forrester / *Department of Natural Resources Science, University of Rhode Island, Kingston, Rhode Island, U.S.A.*

Neil W. Foster / *Canadian Forest Service, Ontario, Sault Ste. Marie, Ontario, Canada*

Thomas G. Franti / *Department of Biological Systems Engineering, University of Nebraska, Lincoln, Nebraska, U.S.A.*

Alan J. Franzluebbers / *Agricultural Research Service, U.S. Department of Agriculture (USDA-ARS), Watkinsville, Georgia, U.S.A.*

Lisa Freudenberger / *Center for Development Research (ZEF), University of Bonn, Bonn, Germany*

Congbin Fu / *Institute of Atmospheric Physics, Chinese Academy of Sciences, Beijing, China*

Scott Glenn / *Coastal Ocean Observation Laboratory, Institute of Marine and Coastal Sciences, School of Environmental and Biological Sciences, Rutgers University, New Brunswick, New Jersey, U.S.A.*

Brij Gopal / *Centre for Inland Waters in South Asia, Jaipur, India*
L. R. Gopinath / *M.S. Swaminathan Research Foundation, Chennai, India*
Josef H. Görres / *Department of Plant and Soil Science, University of Vermont, Burlington, Vermont, U.S.A.*
Felicity S. Graham / *Wealth from Oceans National Research Flagship, CSIRO Marine and Atmospheric Research, and Institute for Marine and Antarctic Studies, University of Tasmania, Hobart, Tasmania, Australia*
David Gregg / *Rhode Island Natural History Survey, Kingston, Rhode Island, U.S.A.*
Burak Güneralp / *Department of Geography, Texas A&M University, College Station, Texas, U.S.A.*
Sally D. Hacker / *Department of Integrative Biology, Oregon State University, Corvallis, Oregon, U.S.A.*
C. Michael Hall / *Department of Management, University of Canterbury, Christchurch, New Zealand, Centre for Tourism, University of Eastern Finland, Savonlinna, and Department of Geography, University of Oulu, Oulu, Finland*
Pertti Hari / *Department of Forestry, University of Helsinki, Helsinki, Finland*
Christopher D. G. Harley / *Department of Zoology, University of British Columbia, British Columbia, Vancouver, Canada*
Mark E. Harmon / *Oregon State University, Corvallis, Oregon, U.S.A.*
John L. Havlin / *Department of Soil Science, North Carolina State University, Raleigh, North Carolina, U.S.A.*
Michael J. Hayes / *National Drought Mitigation Center, Lincoln, Nebraska, U.S.A.*
Hong S. He / *School of Natural Resources, University of Missouri, Columbia, Missouri, U.S.A.*
Richard W. Healy / *U.S. Geological Survey (USGS), Lakewood, Colorado, U.S.A.*
Alan D. Hecht / *Office of Research and Development, U.S. Environmental Protection Agency (EPA), Washington, District of Columbia, U.S.A.*
Robert W. Hill / *Biological and Irrigation Engineering Department, Utah State University, Logan, Utah, U.S.A.*
Curtis H. Hinman / *School of Agricultural, Human and Natural Resource Sciences, Washington State University, Tacoma, Washington, U.S.A.*
Kyle D. Hoagland / *School of Natural Resources, University of Nebraska, Lincoln, Nebraska, U.S.A.*
Faisal Hossain / *Department of Civil and Environmental Engineering, Tennessee Technological University, Cookeville, Tennessee, U.S.A.*
Jinrong Hu / *Yuen Yuen Research Centre for Satellite Remote Sensing, Institute of Space and Earth Information Science, Chinese University of Hong Kong, Hong Kong, and Laboratory of Coastal Zone Studies, Shenzhen Research Institute, Shenzhen, China*
T. G. Huntington / *U.S. Geological Survey (USGS), Augusta, Maine, U.S.A.*
Thomas P. Husband / *Department of Natural Resources Science, University of Rhode Island, Kingston, Rhode Island, U.S.A.*
Stephen Hutchinson / *Baruch Institute of Coastal Ecology and Forest Science, Clemson University, Georgetown, South Carolina, U.S.A.*
Kristen C. Hychka / *Atlantic Ecology Division, Office of Research and Development, U.S. Environmental Protection Agency (EPA), Narragansett, Rhode Island, U.S.A.*
C. Rhett Jackson / *Daniel B. Warnell School of Forest Resources, University of Georgia, Athens, Georgia, U.S.A.*

Jennifer P. Jorve / *Department of Zoology, University of British Columbia, British Columbia, Vancouver, Canada*

Pierre Y. Julien / *Department of Civil and Environmental Engineering, Engineering Research Center, Colorado State University, Fort Collins, Colorado, U.S.A.*

Manjit S. Kang / *Department of Plant Pathology, Kansas State University, Manhattan, Kansas, U.S.A.*

Selim Kapur / *University of Çukurova, Adana, Turkey*

Darryl J. Keith / *Atlantic Ecology Division, U.S. Environmental Protection Agency (EPA), Narragansett, Rhode Island, U.S.A.*

Olivia Kellner / *Department of Earth, Atmospheric, and Planetary Sciences, Purdue University, West Lafayette, Indiana, U.S.A.*

D. Q. Kellogg / *Department of Natural Resources Science, University of Rhode Island, Kingston, Rhode Island, U.S.A.*

Jaehoon Kim / *Department of Civil and Environmental Engineering, Engineering Research Center, Colorado State University, Fort Collins, Colorado, U.S.A.*

Jinwoo Kim / *The Ohio State University, Columbus, Ohio, U.S.A.*

Marcos Kogan / *Integrated Plant Protection Center, Oregon State University, Corvallis, Oregon, U.S.A.*

Josh Kohut / *Coastal Ocean Observation Laboratory, Institute of Marine and Coastal Sciences, School of Environmental and Biological Sciences, Rutgers University, New Brunswick, New Jersey, U.S.A.*

Rai Kookana / *Commonwealth Scientific and Industrial Research Organisation (CSIRO), Adelaide, South Australia, Australia*

Rebecca L. Kordas / *Department of Zoology, University of British Columbia, British Columbia, Vancouver, Canada*

Ken W. Krauss / *National Wetlands Research Center, U.S. Geological Survey (USGS), Lafayette, Louisiana, U.S.A.*

Betty J. Kreakie / *Atlantic Ecology Division, Office of Research and Development, U.S. Environmental Protection Agency (EPA), Narragansett, Rhode Island, U.S.A.*

Jürgen Kreuzwieser / *Institute of Forest Botany and Tree Physiology, University of Freiburg, Freiburg, Germany*

Anne Kuhn / *Atlantic Ecology Division, U.S. Environmental Protection Agency (EPA), Narragansett, Rhode Island, U.S.A.*

William P. Kustas / *Hydrology and Remote Sensing Lab, Agricultural Research Service, U.S. Department of Agriculture (USDA-ARS), Beltsville, Maryland, U.S.A.*

R. Lal / *Carbon Management and Sequestration Center, The Ohio State University, Columbus, Ohio, U.S.A.*

Matthias Langensiepen / *Department of Modeling Plant Systems, Humboldt University of Berlin, Berlin, Germany*

Jean-Claude Lefeuvre / *Laboratory of the Evolution of Natural and Modified Systems, University of Rennes, Rennes, France*

Gene E. Likens / *Cary Institute of Ecosystem Studies, Millbrook, New York, and University of Connecticut, Storrs, Connecticut, U.S.A.*

Yuling Liu / *Cooperative Institute for Climate and Satellites, University of Maryland, College Park, Maryland, U.S.A.*

Zhong Lu / *Cascades Volcano Observatory, U.S. Geological Survey (USGS), Vancouver, Washington, U.S.A.*

Ariel E. Lugo / *International Institute of Tropical Forestry, Forest Service, U.S. Department of Agriculture (USDA-FS), Río Piedras, Puerto Rico*

Alison MacNeil / *Northeast River Forecast Center, National Oceanic and Atmospheric Administration (NOAA), Taunton, Massachusetts, U.S.A.*

Carol Mallory-Smith / *Oregon State University, Corvallis, Oregon, U.S.A.*

Jeffrey A. Markert / *Biology Department, Providence College, Providence, Rhode Island, U.S.A.*

Toshihisa Matsui / *NASA Goddard Space Flight Center, National Aeronautics and Space Administration (NASA), Greenbelt, and ESSIC, University of Maryland, College Park, Maryland, U.S.A.*

Thomas Joseph McGreevy, Jr. / *Biology Department, Boston University, Boston, Massachusetts, U.S.A.*

Gregory McIsaac / *Natural Resources and Environmental Sciences, University of Illinois, Urbana, Illinois, U.S.A.*

Ernesto Medina / *International Institute of Tropical Forestry, Forest Service, U.S. Department of Agriculture (USDA-FS), Río Piedras, Puerto Rico, and Center for Ecology, Venezuelan Institute for Scientific Research, Caracas, Venezuela*

Miguel A. Medina, Jr. / *Department of Civil and Environmental Engineering, Duke University, Durham, North Carolina, U.S.A.*

Mallavarapu Megharaj / *Commonwealth Scientific and Industrial Research Organisation (CSIRO), Adelaide, South Australia, Australia*

Cevza Melek Kazezyılmaz-Alhan / *Department of Civil Engineering, Istanbul University, Istanbul, Turkey*

Forrest M. Mims III / *Geronimo Creek Observatory, Seguin, Texas, U.S.A.*

Emily S. Minor / *Department of Biological Sciences and Institute for Environmental Science and Policy, University of Illinois at Chicago, Chicago, Illinois, U.S.A.*

Niti B. Mishra / *University of Texas at Austin, Austin, Texas, U.S.A.*

Vasubandhu Misra / *Department of Earth, Ocean and Atmospheric Science and Center for Ocean-Atmospheric Prediction Studies, Florida State University, Tallahassee, Florida, U.S.A.*

David M. Mocko / *NASA Goddard Space Flight Center, National Aeronautics and Space Administration (NASA), Greenbelt, and SAIC, Beltsville, Maryland, U.S.A.*

Adam H. Monahan / *School of Earth and Ocean Sciences, University of Victoria, Victoria, British Columbia, Canada*

Miranda Y. Mortlock / *School of Agriculture and Food Sciences, University of Queensland, Brisbane, Queensland, Australia*

Ravendra Naidu / *Commonwealth Scientific and Industrial Research Organisation (CSIRO), Adelaide, South Australia, Australia*

V. Arivudai Nambi / *M.S. Swaminathan Research Foundation, Chennai, India*

Jocelyn C. Nelson / *Department of Zoology, University of British Columbia, British Columbia, Vancouver, Canada*

John Nimmo / *U.S. Geological Survey (USGS), Menlo Park, California, U.S.A.*

Dev Niyogi / *Department of Agronomy and Department of Earth, Atmospheric, and Planetary Sciences, Purdue University, West Lafayette, Indiana, U.S.A.*

Barry R. Noon / *Department of Fish, Wildlife, and Conservation Biology, Colorado State University, Fort Collins, Colorado, U.S.A.*

David Noone / *Department of Civil, Environmental and Architectural Engineering and Cooperative Institute for Research in the Environmental Sciences (CIRES), University of Colorado at Boulder, Boulder, Colorado, U.S.A.*

Jesse Norris / *Centre for Atmospheric Science, School of Earth, Atmospheric and Environmental Sciences, University of Manchester, Manchester, U.K.*

Alyssa Novak / *University of New Hampshire, Durham, New Hampshire, U.S.A.*

Brittany L. Oakes / *Department of Biological Sciences, Union College, Schenectady, New York, U.S.A.*

Micheal D. K. Owen / *Department of Agronomy, Iowa State University, Ames, Iowa, U.S.A.*

Antônio R. Panizzi / *Embrapa Trigo, Passo Fundo, Brazil*

Peter W. C. Paton / *Department of Natural Resources Science, University of Rhode Island, Kingston, Rhode Island, U.S.A.*

Jan Pergl / *Department of Invasion Ecology, Institute of Botany, Academy of Sciences of the Czech Republic (CAS), Průhonice, Czech Republic*

Debra P. C. Peters / *Jornada Experimental Range, Agricultural Research Service, U.S. Department of Agriculture (USDA-ARS), Las Cruces, New Mexico, U.S.A.*

Manon Picard / *Department of Zoology, University of British Columbia, British Columbia, Vancouver, Canada*

Roger A. Pielke, Sr. / *Cooperative Institute for Research in Environmental Sciences (CIRES), University of Colorado at Boulder, Boulder, Colorado, U.S.A.*

Finn C. Pillsbury / *Jornada Experimental Range, Agricultural Research Service, U.S. Department of Agriculture (USDA-ARS), Las Cruces, New Mexico, U.S.A.*

David Pimentel / *Department of Entomology, Cornell University, Ithaca, New York, U.S.A.*

Håkan Pleijel / *Applied Environmental Science, Göteborg University, Göteborg, Sweden*

Tony Prato / *Department of Agricultural and Applied Economics, University of Missouri, Columbia, Missouri, U.S.A.*

Stephen A. Prior / *National Soil Dynamics Laboratory, Agricultural Research Service, U.S. Department of Agriculture (USDA-ARS), Auburn, Alabama, U.S.A.*

Seth G. Pritchard / *Department of Biology, College of Charleston, Charleston, South Carolina, U.S.A.*

Jeffrey P. Privette / *National Climate Data Center, National Environmental Satellite Data Information Service, National Oceanic and Atmospheric Administration (NOAA), Asheville, North Carolina, U.S.A.*

Petr Pyšek / *Department of Invasion Ecology, Institute of Botany, Academy of Sciences of the Czech Republic (CAS), Průhonice, Czech Republic*

Nageswararao C. Rachaputi / *Queensland Department of Primary Industries, Kingaroy, Queensland, Australia*

Deeksha Rastogi / *Department of Atmospheric Sciences, University of Illinois, Urbana, Illinois, U.S.A.*

John P. Reganold / *Department of Crop and Soil Sciences, Washington State University, Pullman, Washington, U.S.A.*

Paul Reich / *National Resources Conservation Service, U.S. Department of Agriculture (USDA-NRCS), Washington, District of Columbia, U.S.A.*

Heinz Rennenberg / *Intitute of Forest Botany and Tree Physiology, University of Freiburg, Freiburg, Germany*

John S. Roberts / *New Technology Department, Research and Development, Rich Products Corporation, Buffalo, New York, U.S.A.*

Gilbert L. Rochon / *Tuskegee University, Tuskegee, Alabama, U.S.A.*

Cathy A. Roheim / *Department of Agricultural Economics and Rural Sociology, University of Idaho, Moscow, Idaho, U.S.A.*

Somnath Baidya Roy / *Department of Atmospheric Sciences, University of Illinois, Urbana, Illinois, U.S.A.*

G. Brett Runion / *National Soil Dynamics Laboratory, Agricultural Research Service, U.S. Department of Agriculture (USDA-ARS), Auburn, Alabama, U.S.A.*

Bayden D. Russell / *Southern Seas Ecology Laboratories, School of Earth and Environmental Science, University of Adelaide, Adelaide, South Australia, Australia*

Grace Saba / *Coastal Ocean Observation Laboratory, Institute of Marine and Coastal Sciences, School of Environmental and Biological Sciences, Rutgers University, New Brunswick, New Jersey, U.S.A.*

Sergi Sabater / *Catalan Institute for Water Research, Girona, Spain*

Uriel N. Safriel / *Center of Environmental Conventions, Jacob Blaustein Institutes for Desert Research, and Department of Ecology, Evolution and Behavior, Hebrew University of Jerusalem, Jerusalem, Israel*

William Saunders / *Northeast River Forecast Center, National Oceanic and Atmospheric Administration (NOAA), Taunton, Massachusetts, U.S.A.*

Mary C. Savin / *Department of Crop, Soil, and Environmental Sciences, University of Arkansas, Fayetteville, Arkansas, U.S.A.*

Jennifer L. Schaeffer / *Food Safety and Environmental Stewardship Program, Oregon State University, Corvallis, Oregon, U.S.A.*

Neil T. Schock / *Institute for Great Lakes Research, CMU Biological Station, and Department of Biology, Central Michigan University, Mt. Pleasant, Michigan, U.S.A.*

Oscar Schofield / *Coastal Ocean Observation Laboratory, Institute of Marine and Coastal Sciences, School of Environmental and Biological Sciences, Rutgers University, New Brunswick, New Jersey, U.S.A.*

David M. Schultz / *Centre for Atmospheric Science, School of Earth, Atmospheric and Environmental Sciences, University of Manchester, Manchester, U.K.*

Yongwei Sheng / *Department of Geography, University of California, Los Angeles, Los Angeles, California, U.S.A.*

Xun Shi / *Department of Geography, Dartmouth College, Hanover, New Hampshire, U.S.A.*

Frederick T. Short / *University of New Hampshire, Durham, New Hampshire, U.S.A.*

Paul Short / *Canadian Sphagnum Peat Moss Association, St. Albert, Alberta, Canada*

C. K. Shum / *Division of Geodetic Science, School of Earth Sciences, The Ohio State University, Columbus, Ohio, U.S.A.*

Kadambot H. M. Siddique / *The UWA Institute of Agriculture, University of Western Australia, Crawley, Western Australia, Australia*

Thomas R. Sinclair / *Crop Genetics and Environmental Research Unit, U.S. Department of Agriculture (USDA), University of Florida, Gainesville, Florida, U.S.A.*

Michael J. Singer / *Department of Land, Air, and Water Resources, University of California, Davis, California, U.S.A.*

Ajay Singh / *Department of Biology, University of Waterloo, Waterloo, Ontario, Canada*

Martin D. Smith / *Nicholas School of the Environment and Department of Economics, Duke University, Durham, North Carolina, U.S.A.*

Daniela Soleri / *Geography Department, University of California, Santa Barbara, California, U.S.A.*

Bretton Somers / *Department of Geography and Anthropology, Louisiana State University, Baton Rouge, Louisiana, U.S.A.*

Jan Hanning Sommer / *Center for Development Research (ZEF), University of Bonn, Bonn, Germany*

Roy F. Spalding / *Water Science Laboratory, University of Nebraska, Lincoln, Nebraska, U.S.A.*

Jean L. Steiner / *Agricultural Research Service, U.S. Department of Agriculture (USDA-ARS), El Reno, Oklahoma, U.S.A.*

Deborah Stinner / *Ohio Agriculture Research and Development Center, The Ohio State University, Wooster, Ohio, U.S.A.*

Mark H. Stolt / *Department of Natural Resources Science, University of Rhode Island, Kingston, Rhode Island, U.S.A.*

David A. Stonestrom / *U.S. Geological Survey (USGS), Menlo Park, California, U.S.A.*

Nigel E. Stork / *Environment Futures Centre, Griffith School of Environment, Griffith University, Brisbane, Queensland, Australia*

Matthew L. Stutz / *Department of Chemistry, Physics, and Geoscience, Meredith College, Raleigh, North Carolina, U.S.A.*

Donglian Sun / *Department of Geography and Geoinformation Science, George Mason University, Fairfax, Virginia, U.S.A.*

Philip Sura / *Department of Earth, Ocean and Atmospheric Science and Center for Ocean-Atmospheric Prediction Studies, Florida State University, Tallahassee, Florida, U.S.A.*

Takehiko Takano / *Tohoku Gakuin University, Sendai, Japan*

Albert E. Theberge, Jr. / *Central Library, National Oceanic and Atmospheric Administration (NOAA), Silver Spring, Maryland, U.S.A.*

Ralph W. Tiner / *National Wetlands Inventory (NWI), U.S. Fish and Wildlife Service, Hadley, Massachusetts, U.S.A.*

Ronald F. Turco / *Department of Agronomy, Purdue University, West Lafayette, Indiana, U.S.A.*

Donald G. Uzarski / *Institute for Great Lakes Research, CMU Biological Station, and Department of Biology, Central Michigan University, Mt. Pleasant, Michigan, U.S.A.*

Fernando Valladares / *Centro de Ciencias Medioambientales, Madrid, Spain*

Arnold G. van der Valk / *Ecology, Evolution and Organismal Biology, Iowa State University, Ames, Iowa, U.S.A.*

George F. Vance / *Department of Ecosystem Sciences and Management, University of Wyoming, Laramie, Wyoming, U.S.A.*

Kathleen J. Vigness-Raposa / *Marine Acoustics, Inc., Middletown, Rhode Island, U.S.A.*

Dale H. Vitt / *Department of Plant Biology and Center for Ecology, Southern Illinois University, Carbondale, Illinois, U.S.A.*

John C. Volin / *Department of Natural Resources and the Environment, University of Connecticut, Storrs, Connecticut, U.S.A.*

Daniel von Schiller / *Catalan Institute for Water Research, Girona, Spain*

James A. Voogt / *Department of Geography, University of Western Ontario, London, Ontario, Canada*

Jonathan M. Wachter / *Department of Crop and Soil Sciences, Washington State University, Pullman, Washington, U.S.A.*

H. Jesse Walker / *Department of Geography and Anthropology, Louisiana State University, Baton Rouge, Louisiana, U.S.A.*

Ivan A. Walter / *Ivan's Engineering, Inc., Denver, Colorado, U.S.A.*

Pao K. Wang / *Atmospheric and Oceanic Sciences, University of Wisconsin, Madison, Wisconsin, U.S.A.*

Yeqiao Wang / *Department of Natural Resources Science, University of Rhode Island, Kingston, Rhode Island, U.S.A.*

Zhuo Wang / *Department of Atmospheric Sciences, University of Illinois at Urbana-Champaign, Urbana, Illinois, U.S.A.*
Owen P. Ward / *Department of Biology, University of Waterloo, Waterloo, Ontario, Canada*
Elizabeth R. Waters / *Department of Biology, San Diego State University, San Diego, California, U.S.A.*
Hans-Joachim Weigel / *Federal Research Institute for Rural Areas, Forestry and Fisheries, Thünen Institute of Biodiversity, Braunschweig, Germany*
W. W. Wenzel / *Institute of Soil Research, University of Natural Resources and Life Sciences, Vienna, Austria*
French Wetmore / *French & Associates, Ltd., Park Forest, Illinois, U.S.A.*
Ryan L. Wheeler / *Institute for Great Lakes Research, CMU Biological Station, and Department of Biology, Central Michigan University, Mt. Pleasant, Michigan, U.S.A.*
Penny E. Widdison / *Department of Geography, University of Durham, Durham, U.K.*
Robert L. Wilby / *Department of Geography, University of Loughborough, Loughborough, U.K.*
Donald A. Wilhite / *National Drought Mitigation Center, Lincoln, Nebraska, U.S.A.*
John Wilkin / *Coastal Ocean Observation Laboratory, Institute of Marine and Coastal Sciences, School of Environmental and Biological Sciences, Rutgers University, New Brunswick, New Jersey, U.S.A.*
Victoria A. Wojcik / *Pollinator Partnership, San Francisco, California, U.S.A.*
Graeme C. Wright / *Department of Primary Industries, Queensland Department of Primary Industries, Kingaroy, Queensland, Australia*
Marcia Glaze Wyatt / *Department of Geology, University of Colorado at Boulder, Boulder, Colorado, U.S.A.*
Jason Yang / *Department of Geography, Ball State University, Muncie, Indiana, U.S.A.*
Jian Yang / *State Key Laboratory of Forest and Soil Ecology, Institute of Applied Ecology, Chinese Academy of Sciences, Shenyang, China*
Qingsheng Yang / *Department of Resources and Environment, Guangdong University of Business Studies, Guangzhou, China*
Xiaojun Yang / *Department of Geography, Florida State University, Tallahassee, Florida, U.S.A.*
Xu Yi / *Coastal Ocean Observation Laboratory, Institute of Marine and Coastal Sciences, School of Environmental and Biological Sciences, Rutgers University, New Brunswick, New Jersey, U.S.A.*
Kenneth R. Young / *University of Texas at Austin, Austin, Texas, U.S.A.*
Yunyue Yu / *Center for Satellite Applications and Research, National Environmental Satellite Data Information Service, National Oceanic and Atmospheric Administration (NOAA), College Park, Maryland, U.S.A.*
Baiping Zhang / *State Key Lab for Resources and Environment Information System, Institute of Geographic Sciences and Natural Resources Research, Chinese Academy of Sciences, Beijing, China*
Yuanzhi Zhang / *Yuen Yuen Research Centre for Satellite Remote Sensing, Institute of Space and Earth Information Science, Chinese University of Hong Kong, Hong Kong, and Laboratory of Coastal Zone Studies, Shenzhen Research Institute, Shenzhen, China*
Fang Zhao / *State Key Lab for Resources and Environment Information System, Institute of Geographic Sciences and Natural Resources Research, Chinese Academy of Sciences, Beijing, China*
Jinping Zhao / *Ocean University of China, Qingdao, China*

Yuyu Zhou / *Joint Global Change Research Institute, Pacific Northwest National Laboratory, College Park, Maryland, U.S.A.*
Carl L. Zimmerman / *Department of Urban and Environmental Policy and Planning, Tufts University, Medford, Massachusetts, U.S.A.*
Michael A. Zoebisch / *German Agency for International Cooperation, Tashkent, Uzbekistan*

Contents

Land

Land (cont'd.)

Water

Water (cont'd.)

Air

Topical Table of Contents

Land

Biodiversity

Genetics Resources

Coastal Issues

Benthic Resources

Land (cont'd.)

Coastal Issues (cont'd.)

Benthic Resources (cont'd.)

Coastal Resources

Ecosystems

Ecosystem Services

Forests

Hydrology

Landscape

Landscape Analysis

Management

Land (cont'd.)

Management (sont'd.)

Land Capability

Soil

Sustainability

Land (cont'd.)

Wetlands

Coastal Wetlands

Water

Coastal Issues

Benthic Resources

Coastal Resources

Freshwater

Hydrology

Marine Issues

Coastal Environment

Water (cont'd.)

Marine Issues (cont'd.)

Coastal Environment (cont'd.)

Marine Resources

Air

Atmosphere

Climate

Meteorology

Preface

With unprecedented attention on global change, one of the focuses of debate is about the availability and sustainability of natural resources and about how to balance the equilibrium between what society demands from natural environments and what the natural resource base can provide. It is critical to gain a full understanding about the consequences of the changing resource bases to the degradation of ecological integrity and the sustainability of life. Natural resources represent such a broad scope of complex and challenging topics; in this field, a reference book must cover a vast number of topics in order to be titled an encyclopedia.

A total of 275 scholars from 25 countries including Australia, Austria, Brazil, China, Cameroon, Canada, Czech Republic, Finland, France, Germany, Hungary, India, Israel, Japan, New Zealand, Norway, Puerto Rico, Spain, Sweden, Syria, Turkey, United Kingdom, United States, Uzbekistan, and Venezuela contributed 188 entries to the two volumes of this printed set. Volume I (*Land*) includes 98 entries that cover the topical areas of renewable and nonrenewable natural resources such as forest and vegetative; soil; terrestrial coastal and inland wetlands; landscape structure and function and change; biological diversity; ecosystem services, protected areas and management; natural resource economics; and resource security and sustainability. In Volume II, Water includes 59 entries and Air includes 31 entries. The *Water* entries cover topical areas such as fresh water, groundwater, water quality and watersheds, ice and snow, coastal environments, and marine resources and economics. The *Air* entries cover topical areas such as atmosphere, meteorology and weather, climate resources, and effects of climate change.

This *ENR* is a reference series that is under continuous development. In addition to the entries included in this printed set, continuously developed entries will be published through the online version of the *ENR* and will be included in the future edition of the printed copy. I look forward to working with a broader scope of contributors and reviewers for the further development of this series of publication.

Yeqiao Wang
University of Rhode Island

Acknowledgments

I am honored to have had this opportunity and privilege to work on this important publication. Many people who helped tremendously during the process deserve special acknowledgment. First and foremost, I thank the 275 contributors from 25 countries and all the reviewers from different countries around the world. Their expertise, insights, dedication, hardwork, and professionalism ensure the quality of this encyclopedia.

I appreciate the guidance and contribution of the members of the Editorial Advisory Board: Drs. Peter V. August of University of Rhode Island; Kate Moran of the Ocean Networks Canada at the University of Victoria; Roger A. Pielke, Sr. of Colorado State University and University of Colorado at Boulder; Russell G. Congalton of University of New Hampshire; Giles Foody of the University of Nottingham; Jefferson M. Fox of the East-West Center; Xiaowen Li of the Chinese Academy of Science; Hui Lin of the Chinese University of Hong Kong; Jing Ming Chen of University of Toronto; and Darryl J. Keith of the U.S. Environmental Protection Agency.

I wish to express my gratitude to my respected colleagues and scholars, in particular Drs. Peter V. August, Peter W.C. Paton, Thomas R. Husband, Jose A. Amador, Mark H. Stolt, Graham E. Forrester, Cathy A. Roheim, Thomas Boving, Richard Burroughs, Bill Buffum, Even Preisser, Serena M. Moseman-Valtierra, Nancy Karraker, Arthur J. Gold, Scott McWilliam, Laura Meyerson, David Abedon, Daniel L. Civco, John C. Volin, John C. Clausen, Jiansheng Yang, and Yuyu Zhou, for their encouragement, support, contribution, and service during the preparation of this publication.

The preparation for the development of this publication was started in spring 2010. I appreciate the initial contact from Irma Shagla, Editor for Environmental Sciences and Engineering of the Taylor & Francis Group/CRC Press, for the conceptual development of this publication. Claire Miller and Molly Pohlig from the Encyclopedia Program of the Taylor & Francis Group managed the preparation of this publication with leadership, insights, and the highest standard of professionalism. A special tribute goes to Susan Lee, a former editor of the encyclopedia program of the Taylor & Francis Group. Susan was dedicated to the development of this encyclopedia for 3.5 years from the very beginning of the project till the very last days of her very young life. It would have been impossible to achieve what we have accomplished today without Susan's diligence, gentle diplomacy, and superior organizational skills. I am so sorry that Susan could not be with us to see the publication of this encyclopedia.

The inspiration for working on this reference series came from my 30 years of research and teaching experiences in different stages of my professional career. I am grateful for the opportunity of working with many top-notch scholars, staff members, administrators, and enthusiastic students throughout the time.

As always, the most special appreciation is due to my wife and daughters for their love, patience, understanding, and encouragement during the preparation of this publication. I wish my parents and previous professors of soil ecology and of climatology from the School of Geography of the Northeast Normal University could see this set of publications.

Aims and Scope

Land, water, and air are the most precious natural resources that sustain life and civilization. Maintenance of clean air and water and preservation of land resources and native biological diversity are among the challenges that we are facing for the sustainability the well-being of all on the planet Earth. Natural and anthropogenic forces have affected constantly land, water, and air resources through interactive processes such as shifting climate patterns, disturbing hydrological regimes and alternating landscape configurations and compositions. Improvements in understanding of the complexity of land, water, and air systems and their interactions with human activities represent priorities in scientific research, technology development, education programs, and administrative actions for conservation and management of natural resources.

The *Encyclopedia of Natural Resources* (ENR) consists of two volumes (Volume I: *Land* and Volume II: *Water* and *Air*). The *ENR* is designed to provide an authoritative and organized informative reference covering a broad spectrum of topics such as the forcing factors and habitats of life, their histories, current status, future trends, and their societal connections, economic values, and management. The topical entries are authored by world-class scientists and scholars. The contents represent state-of-the-art science and technology development and perspectives of resource management. Efforts were made for each topical entries to be written at a level that allows a broad scope of audience to understand. Public and private libraries, educational and research institutions, scientists, scholars, and students will be the primary audience of this encyclopedia.

About the Editor-in-Chief

Dr. Yeqiao Wang is a professor at the Department of Natural Resources Science, College of the Environment and Life Sciences, University of Rhode Island. He earned his BS from the Northeast Normal University in 1982 and his MS degree in remote sensing and mapping from the Chinese Academy of Sciences in 1987. He earned an MS and a PhD in natural resources management & engineering from the University of Connecticut in 1992 and 1995, respectively. From 1995 to 1999, he held the position of assistant professor in the Department of Geography and Department of Anthropology, University of Illinois at Chicago. He has been on the faculty of the University of Rhode Island since 1999. In addition to his tenured position, he held an adjunct research associate position at the Field Museum of Natural History in Chicago. He has also served as a guest professor and an adjunct professor at universities in the United States and in China. Among his awards and recognitions, Dr. Wang was a recipient of the prestigious Presidential Early Career Award for Scientists and Engineers (PECASE) by former U.S. president William J. Clinton in 2000.

Dr. Wang's specialties are in terrestrial remote sensing and applications of geoinformatics in natural resources analysis and mapping. His research projects have been funded by different agencies that supported his scientific studies in various regions of the United States, in East and West Africa, and in regions in China. Besides peer-reviewed journal publications, Dr. Wang edited *Remote Sensing of Coastal Environments* and *Remote Sensing of Protected Lands* published by the CRC Press in 2009 and 2010, respectively. He has also authored and edited over 10 science books in Chinese.

Encyclopedia of Natural Resources

Land

Volume I

Agriculture–Wetlands
Pages 1–588

Agriculture: Organic

John P. Reganold
Jonathan M. Wachter
Department of Crop and Soil Sciences, Washington State University, Pullman, Washington, U.S.A.

Abstract
Organic agriculture is a quickly expanding global market force that seeks socioeconomic and environmental sustainability on local and global scales. It relies on the integration of a diversity of farm components, cycling of nutrients and other resources, and stewardship of soil and environment. This entry discusses organic agricultural practices, the history and extent of organic agriculture, organic certification, and the sustainability of organic agriculture. The performance of organic farming systems in the context of sustainability metrics indicates that organic agriculture, now occupying about 1% of global agricultural land, can play a much larger role in feeding the world in an unpredictable future. No one farming system alone will safely feed the planet. Rather, a blend of organic and other innovative farming systems, such as agroforestry, integrated farming, conservation agriculture, and mixed crop/livestock systems, will be needed for future global food and ecosystem security.

INTRODUCTION

During the past two decades, organic farming has become one of the fastest growing segments of agriculture all over the world.[1] Increasing demands for and expanding supermarket involvement with organic foods in Europe and North America have created a demand for more types of organic food, including tropical and out-of-season foods, and opened the market for growing and exporting organic foods in Africa, Latin America, and Oceania.[2] Today, the US$59.1 billion global market for organic foods is dominated by Europe (US$28 billion) and North America (US$29 billion).[1] As a result of organic agriculture being a quickly expanding global market force, more arable land, research funding, and research test sites are being devoted to organic farming worldwide.

Organic farming systems do not represent a return to the past. Organic agriculture, sometimes called biological or ecological agriculture, combines traditional conservation-minded farming methods with modern farming technologies. It virtually excludes such conventional inputs as synthetic fertilizers, pesticides, and pharmaceuticals. Instead, it uses naturally derived chemicals or products as defined by organic certification programs, and it emphasizes improving the soil with compost additions and animal and green manures, controlling pests naturally, rotating crops, and diversifying crops and livestock.[3] Organic farmers use modern equipment, certified seed, soil and water conservation practices, improved crop varieties, and the latest innovations in feeding and handling livestock. Organic farming systems range from strict closed-cycle systems that go beyond organic certification guidelines by limiting external inputs as much as possible to more standard systems that simply follow organic certification guidelines.

Organic agriculture is an alternative to conventional agriculture that seeks socioeconomic and environmental sustainability on local and global scales. Organic farming systems have been shown to be energy efficient, environmentally sound, productive, stable, and tending toward long-term sustainability.[4,5] This entry discusses organic agricultural practices, the history and extent of organic agriculture, organic certification, and the sustainability of organic agriculture.

ORGANIC AGRICULTURAL PRACTICES

Organic farming systems rely on ecologically based practices. A central component of organic farming systems is the rotation of crops—a planned succession of various crops grown on one field. Rotating crops often results in increased yields when compared with monocultures (growing the same crop on the same field year after year).[3] In most cases, monocultures can be perpetuated only by adding large amounts of fertilizers and pesticides. Rotating crops provides better weed and insect control, less disease buildup, more efficient nutrient cycling, and other benefits. Alternating two crops, such as corn and soybeans, is considered a simple rotation. More complex rotations require three or more crops and often a four- to seven-year (or more) cycle to complete. In growing a more diversified group of crops in rotation, a farmer is less affected by price fluctuations of one or two crops, and may experience more year-to-year financial stability.

Encyclopedia of Natural Resources DOI: 10.1081/E-ENRL-120049206

There are disadvantages, too, however. They include the need for more equipment to grow a number of different crops, the reduction of acreage planted with government-supported crops, and the need for more time and information to manage a greater number of crops.

Organic agriculture not only has a diverse assortment of crops in rotation, but also maintains diversity from mixing species and varieties of crops and from systematically integrating crops, trees, and livestock. When most of North Dakota experienced a severe drought during the 1988 growing season, for example, many monocropping wheat farmers had no grain to harvest. Organic farmers with more diversified systems, however, had sales of their livestock to fall back on, or were able to harvest their late-seeded crops or drought-tolerant varieties.[3]

Maintaining healthy soils by regularly adding crop residues, manures, and other organic materials to the soil is another central feature of organic farming. Organic matter improves soil structure, increases its water storage capacity, enhances fertility, promotes soil biological activity, and contributes to the tilth, or physical condition, of the soil. The better the tilth, the more easily the soil can be tilled or direct-seeded, and the easier it is for seedlings to emerge and for roots to extend downward. Water readily infiltrates soils with good tilth, thereby minimizing surface runoff and soil erosion. Organic materials also feed earthworms and soil microbes, which contribute to soil aeration and structure, and plant health.

The main sources of plant nutrients in organic farming systems are nitrogen-fixing green manures, composted animal manures and plant materials, and plant residues. A green manure crop is typically a grass or legume that is plowed into the soil or surface-mulched at the end of a growing season to enhance soil productivity and tilth. Green manures help control weeds, insect pests, and soil erosion, while also providing forage for livestock and cover for wildlife. Fertility on organic farms is usually biologically mediated, depending on macro- and microorganisms to break down organic materials and make nutrients available to plants.

Weed management is often cited as a primary challenge for organic farmers. Crop rotations are designed to diversify resource competition against weeds, and to diversify timing and type of operations in which weeds can be controlled. Dense green manures, cover crops, and pasture phases of a rotation can compete against weeds, as well as providing a heavy mulch to physically suppress emergence. Direct methods of weed control, such as shallow tillage, solarization, and hoeing, are also common.

Organic farmers use disease-resistant crop varieties and biological controls (such as natural predators or parasites that keep pest populations below injurious levels). They also select tillage methods, planting times, crop rotations, and plant-residue management practices to optimize the environment for beneficial insects that control pest species, or to deprive pests of a habitat. Organically certified pesticides, usually used as a last resort to control insects, diseases, and weeds, are applied when pests are most vulnerable or when any beneficial species and natural predators are least likely to be harmed.

HISTORY AND EXTENT OF ORGANIC AGRICULTURE

European organic agriculture emerged in 1924 when Rudolf Steiner held his course on biodynamic agriculture. In the 1930s and 1940s, organic agriculture was developed in Britain by Lady Eve Balfour and Sir Albert Howard, in Switzerland by Hans Müller, in the United States by J. I. Rodale, and in Japan by Masanobu Fukuoka. Since the beginning of the 1990s, development of organic agriculture in Europe has been supported by government subsidies. In many other countries of the world, organic agriculture was established because of the growing demand for organic products in Europe, the United States, and Japan. Today, organic farming is gaining increasing acceptance by the public at large.

Organic agriculture is practiced in almost all countries of the world, and its share of agricultural land and farms is growing everywhere. In 2010, roughly 37.1 million ha of lands were managed organically worldwide.[1] The major part of this area is located in Australia (12 million ha), Argentina (4.18 million ha) and the United States (1.95 million ha). Oceania holds 35.1% of the world's organic land, followed by Europe (23.4%), and Latin America (23.1%). In North America, lands spanning more than 2.5 million ha are managed organically. In most Asian and African countries, the area under organic management is still low.

ORGANIC CERTIFICATION

Growers are turning to certified organic farming systems as a potential way to lower input costs, decrease reliance on nonrenewable resources, capture high-value markets and premium prices, and boost farm income. Organic certification provides verification that products are indeed produced according to certain standards. For consumers who want to buy organic foods, the organic certification standards ensure that they can be confident in knowing what they are buying. For farmers, these standards create clear guidelines on how to take advantage of the increasing demand for organic products. For the organic products industry, these standards provide an important marketing tool to help boost trade in organic products.

"Certified organic" means that agricultural products have been grown and processed according to the specific standards of various national, state, or private certification organizations. Certifying agents review applications from farmers and processors for certification eligibility; qualified inspectors conduct annual onsite inspections of their operations. Inspectors talk with producers and observe their production or processing practices to determine if they are in compliance with organic standards. Organic certification standards detail the methods, practices, and substances that can be used in producing and handling organic crops and livestock, as well as processed products. For example, the U.S. Department of Agriculture's organic standards went into effect in 2002, establishing national standards that a food labeled "organic" must meet, whether it is grown in the United States or imported from other countries.[6] The European Union developed a regulatory framework to oversee organic practices across all of Europe beginning in 1991. The International Federation of Organic Agriculture Movements (IFOAM) established the IFOAM Accreditation Programme to provide international equivalency of certification bodies worldwide, put into effect in 2000.[7]

While requirements vary slightly between agencies, they all prohibit the use of synthetic pesticides and fertilizers, the use of growth hormones and antibiotics in livestock (with some exceptions for the sake of humane treatment), and the use of genetically modified organisms. Critics of organic certification argue that it encourages farmers to follow the minimum standards for organic certification and undervalue the agroecological elements of organic farming, that the cost of certification gives favor to large farms over small operations, and that national and global certification schemes do not account for important local and regional differences in farming systems.[4,8]

THE SUSTAINABILITY OF ORGANIC AGRICULTURE

Organic agriculture addresses many serious problems afflicting world food production: high energy costs, groundwater contamination, soil erosion, loss of productivity, depletion of fossil resources, low farm incomes, and risks to human health and wildlife habitats. However, just because a farm is organic does not mean that it is sustainable. For any farm to be sustainable, whether it is organic or conventional, it must (i) produce adequate yields of good quality, (ii) enhance the natural resource base and environment, (iii) be profitable, and (iv) contribute to the well-being of farmers, farm workers, and rural communities.[9] So, if an organic farm produces high yields of nutritious food and is environmentally friendly and energy efficient, but is not profitable—then it is not sustainable. Likewise, a conventional farm that meets all the sustainability criteria but pollutes a nearby river with sediment because of soil erosion is not sustainable.

The performance of organic farming systems in the context of various sustainability metrics (Fig. 1) indicates that organic agriculture can play a role in producing food in a world facing potential resource limits and an unpredictable climate. Although meta-analyses of hundreds of organic/conventional yield comparison show that yields from organic farms are on average 20% lower than those from conventional farms, they also show that adoption of organic agriculture under agroecological conditions where it performs best may close the yield gap between organic and conventional systems.[10,11] For example, de Ponti et al.[10] found that the best yielding organically grown crops are rice (6% lower than conventional), soybeans (8% lower), corn (11% lower), and grass clover (11% lower). In comparison, the highest-yielding organic crops identified by Seufert et al.[11] were organic fruits (3% lower yield than conventional), rain-fed legumes, such as soybeans (5% lower), and oilseed crops (11% lower).

Research has found that organic foods, compared with their conventional counterparts, have significantly fewer pesticide residues in produce and a lower incidence of antibiotic-resistant bacteria in chicken and pork.[12,13] Research has also found organic food to be more nutritious (generally higher antioxidant and vitamin C levels) in some fruits and vegetables than conventionally produced foods.[14,15]

Reviews of comparison studies on environmental quality generally support the perception that organic farming systems are more environmentally friendly than conventional farming systems.[4,16–19] Environmental benefits of organic agriculture include enhanced soil quality and less soil erosion.[17,19,20] For example, in a meta-analysis of the database for comparison studies of organic and conventional farming systems, Gattinger et al.[21] found higher soil organic carbon concentrations and stocks in topsoils under organic farming. In a specific comparison study of commercial organic and conventional strawberry farms, Reganold et al.[22] found that organically farmed soils were of higher quality and exhibited greater numbers of endemic (unique) genes and greater functional gene abundance and diversity for several biogeochemical processes, such as nitrogen fixation and pesticide degradation (Fig. 2).

Studies have shown that organic farms generally have more plant diversity, greater faunal diversity (insects, soil fauna and microbes, birds), and often more habitat and landscape diversity than conventional farms.[4,17–19] As organic agriculture uses virtually no synthetic pesticides, there is little to no risk of synthetic pesticide pollution of ground and surface waters.[16] Concerning the impact of organic farming systems on nitrate and phosphorous leaching and greenhouse gas emissions, Mondelaers et al.[20] found that organic farming systems score better than conventional farming for these items when expressed in terms of per production area; however, given the lower land-use efficiency of organic farming in developed countries, this positive effect is less pronounced or not present at all when expressed in per unit product. Organic systems have been shown to usually be more energy efficient—energy consumed per hectare and per product—than their conventional counterparts.[16,18,19,23]

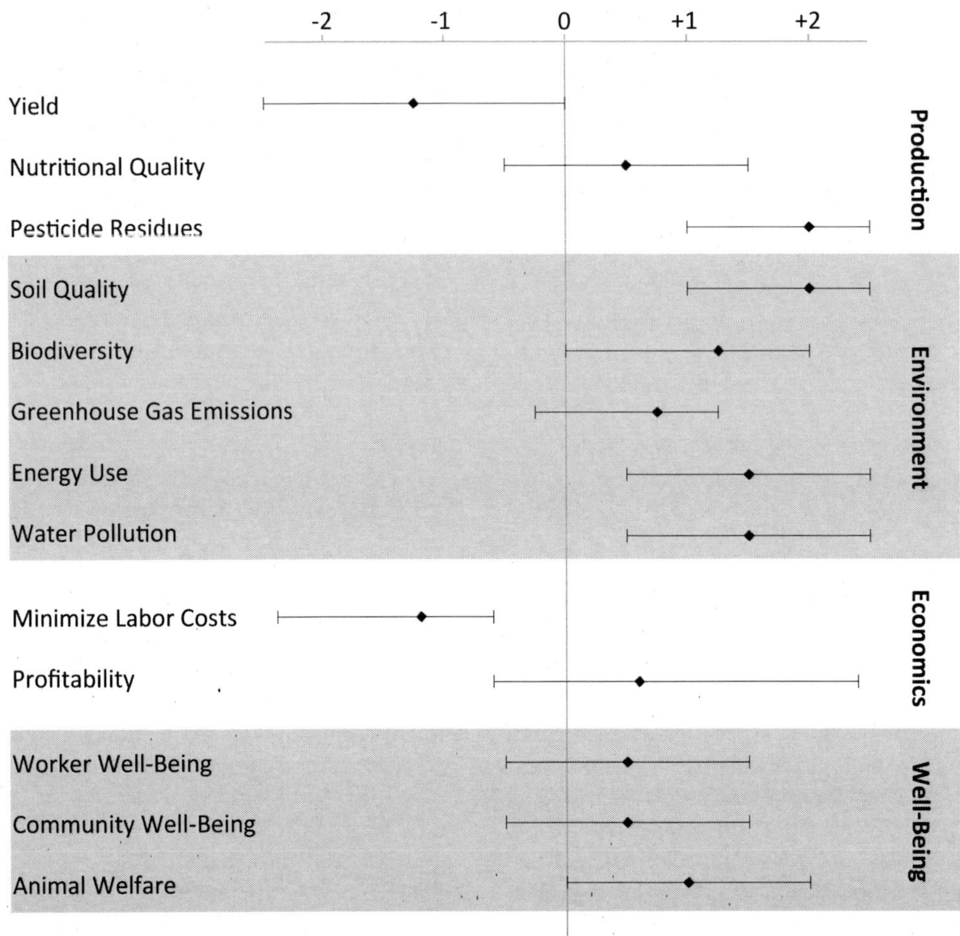

Fig. 1 Assessment of organic farming relative to conventional farming in the four areas of sustainability (+2 much better, +1 better, 0 not different, −1 worse, −2 much worse).

Evidence in the literature is growing that adoption of organic farming systems produces financial benefits. However, transitioning to organic production (usually three years for crops) can be economically challenging and more information intensive. This transition period is, therefore, financially the most difficult time for organic farmers. After transition, the net return per hectare for organic compared with conventional farms is often equal or higher because of good yields and price premiums, as shown by numerous studies.[4,24–27] Studies generally do not account for externalities—whether the crop or product has damaged, preserved, or enhanced the environment (natural capital) in its production process. Upon incorporation of such externalities, especially costs, into economic assessments of farming systems, we may find that many currently profitable farming systems are uneconomical and therefore unsustainable. At a minimum, the profitability of some organic systems with greater environmental benefits compared to their conventional counterparts would increase. The challenge facing policymakers is to incorporate the value of ecosystem processes into the traditional marketplace, thereby supporting food producers in their attempts to employ sustainable practices.

Organic agriculture aims to produce sufficient quantities of humanely raised healthy food while treating and paying all workers fairly. Organic farmers often produce foods for their community or region. Guidelines of IFOAM stipulate that organic farmers should be able to support themselves and other workers with fair incomes, while maintaining safe and dignified working conditions. Furthermore, organic certified animals must be raised humanely under conditions that allow for the expression of their natural behaviors and needs.[7] For example, USDA organic regulations require that livestock have access to the outdoors year-round and that sick animals be treated as needed, even with the use of antibiotics if required.[6]

Research studies comparing organic and conventional farming systems for the social aspects of agricultural sustainability are scarce despite their importance in providing farmers with knowledge to design systems that balance different social sustainability goals to improve overall sustainability.[9] MacRae et al.[28] found some evidence in North America that when a community has many sustainable (including organic) producers, there are positive shifts in community economic development and social

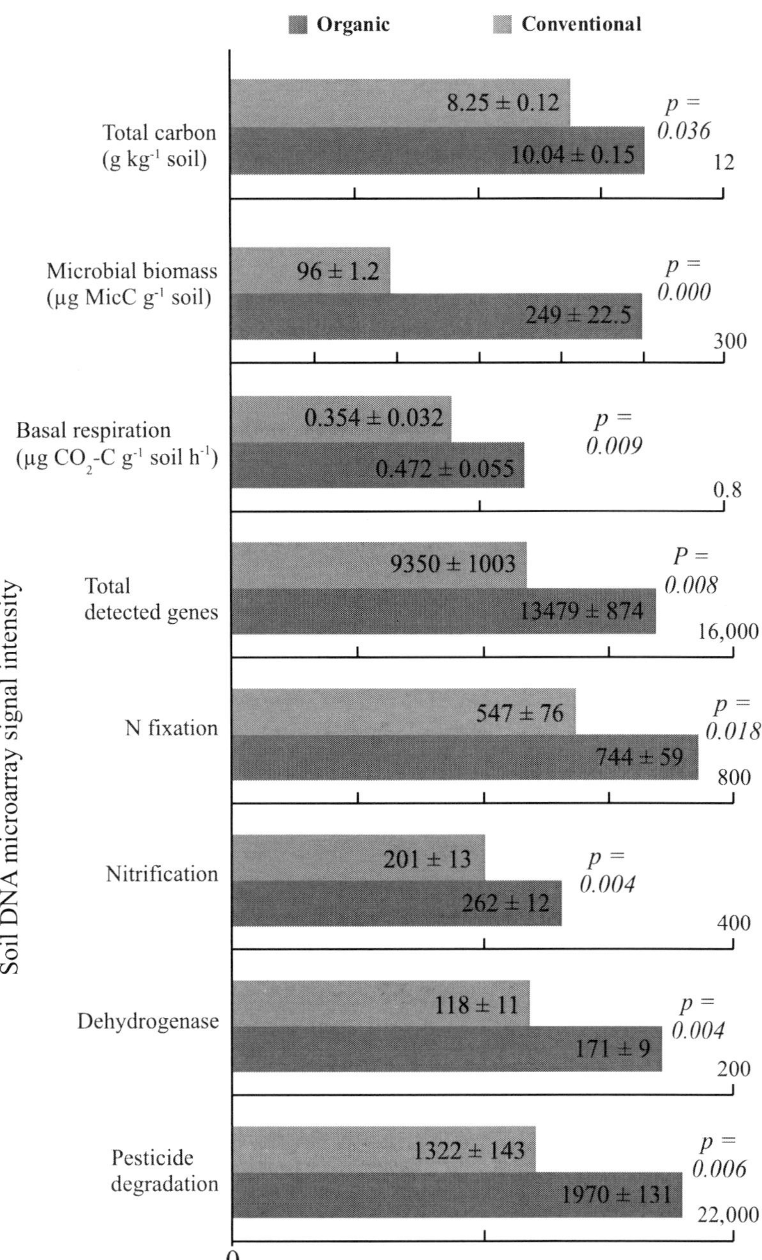

Fig. 2 Comparison of soil quality and biology from organic and conventional strawberry farms.
Source: Data selected from Reganold et al.[22]

interaction; the reasons appear to be related to the need to hire more labor, the increased demand for local goods and services, and a greater commitment to participation in civic institutions.

CONCLUSIONS

If we want to feed a growing world population, producing adequate yields is vital. But, as described above, sufficient productivity is only one of four main goals that must be met for agriculture to be sustainable. The other three are enhancing the natural-resource base and environment, making farming financially viable, and contributing to the well-being of farmers and their communities. Mainstream conventional farming systems have provided growing supplies of food and other products, but often at the expense of the other three sustainability goals. Although organic farming may generally have lower yields, a U.S. National Academy of Sciences report identified organic farming as one of several innovative, more diverse systems that better integrate production, environmental, and socioeconomic objectives (Fig. 1).[9] Other innovative systems include agroforestry, integrated (hybrid organic/conventional), conservation agriculture, grass-fed livestock production, and mixed crop/livestock systems.

No one agricultural system alone, however, will safely feed the planet. Rather, a blend of innovative farming systems will be needed for future global food and ecosystem security. The performance of organic farming systems in the context of sustainability metrics indicates that organic agriculture, now occupying about 1% of global agricultural land, can play a much larger role in feeding the world.

REFERENCES

1. Willer, H.; Kilcher, L., Eds. *The World of Organic Agriculture: Statistics and Emerging Trends 2012*; International Federation of Organic Agriculture Movements: Bonn, Germany, 2012.
2. Aschemann, J.; Hamm, U.; Naspetti, S.; Zanoli, R. The organic market. In *Organic Agriculture: A Global Perspective*; Kristiansen, P., Taji, A., and Reganold, J., Eds.; Comstock Publishing Associates: Ithaca, New York, 2007; 123–151.
3. Chiras, D.D.; Reganold, J.P. *Natural Resource Conservation: Management for a Sustainable Future, 10th Ed.*; Prentice Hall: Upper Saddle River, New Jersey, 2010.
4. Lotter, D.W. Organic agriculture. J. Sustain. Agr. **2003**, *21* (4), 59–128.
5. Halberg, N.; Alrøe, H.F.; Knudsen, M.T.; Kristensen, E.S. Eds. *Global Development of Organic Agriculture: Challenges and Prospects*; CABI: Wallingford, UK, 2007.
6. http://www.ams.usda.gov/nop/ (accessed October 2012).
7. http://www.ifoam.org (accessed August 2012).
8. Schmid, O. Development of standards for organic farming. In *Organic Farming: An International History*; Lockeretz, W., Ed.; CABI Publishing: Oxfordshire, UK, 2007; 152–174.
9. National Research Council. *Toward Sustainable Agricultural Systems in the 21st Century*; The National Academies Press: Washington, D.C., 2010.
10. de Ponti, T.; Rijk, B.; van Ittersum, M.K. The crop yield gap between organic and conventional agriculture. Agr. Syst. **2012**, *108*, 1–9.
11. Seufert, V.; Ramankutty, N.; Foley, J.A. Comparing the yields of organic and conventional agriculture. Nature **2012**, *485* (7397), 229–232.
12. Baker, B.P.; Benbrook, C.M.; Groth III, E.; Benbrook, K.L. Pesticide residues in conventional, integrated pest management (IPM)-grown and organic foods: insights from three U.S. data sets. Food Addit. Contam. **2002**, *19* (5), 427–446.
13. Smith-Spangler, C.; Brandeau, M.L.; Hunter, G.E.; Bavinger, J.C.; Pearson, M.; Eschbach, P.J.; Sundaram, V.; Liu, H.; Schirmer, P.; Stave, C.; Olkin, I.; Bravata, D.M. Are organic foods safer or healthier than conventional alternatives? Ann. Intern. Med. **2012**, *157* (5), 348–366.
14. Brandt, K.; Leifert, C.; Sanderson, R.; Seal, C.J. Agroecosystem management and nutritional quality of plant foods: the case of organic fruits and vegetables. Critical Rev. Plant Sci. **2011**, *30* (1–2), 177–197.
15. Lairon, D. Nutritional quality and safety of organic food: A review. Agron. Sustain. Dev. **2010**, *30* (1), 33–41.
16. Alföldi, T.; Fliessbach, A.; Geier, U.; Kilcher, L.; Niggli, U.; Pfiffner, L.; Stolze, M.; Willer, H. Organic agriculture and the environment. In *Organic Agriculture, Environment, and Food Security*; Scialabba, N.E.-H., Hattam, C., Eds.; Food and Agriculture Organization: Rome, 2002; Chapter 2. Available online at http://www.fao.org/docrep/005/y4137e/y4137e00.htm (accessed September 2012).
17. Gomiero, T.; Pimentel, D.; Paoletti, M.G. Environmental impact of different agricultural management practices: conventional vs. organic agriculture. Crit. Rev. Plant Sci. **2011**, *30* (1–2), 95–124.
18. Kaspercyk, N.; Knickel, K. Environmental impacts of organic farming. In *Organic Agriculture: A Global Perspective*; Kristiansen, P., Taji, A., Reganold, J., Eds; Comstock Publishing Associates: Ithaca, 2006; 259–294.
19. Tuomisto, H.L.; Hodge, I.D.; Riordan, P.; Macdonald, D.W. Does organic farming reduce environmental impacts?–A meta-analysis of European research. J. Environ. Manage. **2012**, *112*, 309–320.
20. Mondelaers, K.; Aertsens, J.; Van Huylenbroeck, G. A meta-analysis of the differences in environmental impacts between organic and conventional farming. Brit. Food J. **2009**, *111* (10), 1098–1119.
21. Gattinger, A.; Muller, A.; Haeni, M.; Skinner, C.; Fliessbach, A.; Buchmann, N.; Mäder, P.; Stolzea, M.; Smith, P.; Scialabb, N.E.-H.; Niggli, U. Enhanced top soil carbon stocks under organic farming. P. Natl. Acad. Sci. USA **2012**, *109* (44), 18226–18231.
22. Reganold, J.P.; Andrews, P.K.; Reeve, J.R.; Carpenter-Boggs, L.; Schadt, C.W.; Alldredge, J.R.; Ross, C.F.; Davies, N.M.; Zhou, J. Fruit and soil quality of organic and conventional strawberry agroecosystems. PLoS ONE **2010**, *5* (9), e12346, DOI: 10.1371/journal.pone.0012346.
23. Lockeretz, W.; Shearer, G.; Kohl, D.H. Organic farming in the Corn Belt. Science **1981**, *211* (4482), 540–546.
24. Padel, S.; Lampkin, N.H. Farm-level performance of organic farming systems: an overview. In *The Economics of Organic Farming: An International Perspective*; Lampkin, N.H., Ed.; CABI Publishing: Oxfordshire, UK, 1994; 201–219.
25. Pimentel, D.; Hepperly, P.; Hanson, J.; Douds, D.; Seidel, R. Environmental, energetic, and economic comparisons of organic and conventional farming systems. BioScience **2005**, *55* (7), 573–582.
26. Reganold, J.P.; Glover, J.D.; Andrews, P.K.; Hinman, H.R. Sustainability of three apple production systems. Nature **2001**, *410* (6831), 926–930.
27. Smolik, J.D.; Dobbs, T.L.; Rickerl, D.H. The relative sustainability of alternative, conventional, and reduced-till farming systems. Am. J. Alt. Agr. **1995**, *10* (1), 25–35.
28. MacRae, R.J.; Frick, B.; Martin, R.C. Economic and social impacts of organic production systems. Can. J. Plant Sci. **2007**, *87* (5), 1037–1044.

Albedo

Endre Dobos
Department of Physical Geography and Environmental Sciences, University of Miskolc, Miskolc-Egyetemváros, Hungary

Abstract
The fraction of the incident radiation that is reflected from the surface is called the albedo. Albedo plays a major role in the energy balance of the Earth's surface, as it defines the rate of the absorbed portion of the incident solar radiation. Soil albedo is a complex feature, which is determined by many soil dependent and independent (environmental) characteristics. The process that results in the reflected radiation is called reflectance, whereby the energy of radiation is reradiated by the chemical constituents (e.g., atoms or molecules) of the surface layer approximately half the thickness of wavelength.

INTRODUCTION

The portion of solar radiation not reflected by the Earth's surface is absorbed by the soil or the vegetation, which interacts with the incident radiation. The absorbed energy can increase the soil temperature or the rate of evapotranspiration from the surface of the soil–vegetation system. Some of the energy that is absorbed and transformed into heat is reradiated at a longer wavelength than the incoming radiation. That is why the peak terrestrial radiation occurs in the infrared spectrum while the peak incident radiation occurs in the blue–green portion of the visible spectrum.

The albedo value ranges from 0 to 1. The value of 0 refers to a blackbody, a theoretical media that absorbs 100% of the incident radiation. Albedo ranging from 0.1–0.2 refers to dark-colored, rough soil surfaces, while the values around 0.4–0.5 represent smooth, light-colored soil surfaces. The albedo of snow cover, especially fresh, deep snow, can reach as high as 0.9. The value of 1 refers to an ideal reflector surface (an absolute white surface) in which all the energy falling on the surface is reflected. The mean albedo of the Earth system is 0.36 ± 0.06 (Table 1).[1]

FACTORS AFFECTING ALBEDO

Albedo varies diurnally and seasonally due to the changing sun angle.[2],[3] In general, the lower the sun angle the higher the albedo. Besides the sun angle, many of the surface characteristics have large impact on the albedo. The most significant factors affecting the soil albedo are the type and condition of the vegetation covering the soil surface, soil moisture content, organic matter content, particle size, iron-oxides, mineral composition, soluble salts, and parent material.[4]

The type and the condition of the vegetation have a strong impact on the surface albedo. Forest vegetation with multilevel canopy has a low albedo because the incident radiation can penetrate deeply into the forest canopy where it bounces back and forth between the branches and leaves and get trapped by the canopy.[5] The albedo for grassland and cropland ranges between 0.1 and 0.25.[6-9]

Changes in soil moisture content change the absorbance and reflectance characteristics of the soil. Increase in soil moisture content increases the portion of the incident solar radiation absorbed by the soil system. This relationship is well known and used for soil color differentiation when the Munsell color chart is used. The colors of dry and moist soil samples are always different. The higher the soil moisture content, the darker the color and lower the albedo. However, this relationship is valid only for soil moisture contents up to the field capacity. Beyond field capacity, the increase in soil moisture content does not darken the color any more, but starts building up a water sheet on the aggregate surface, creating a shiny and better reflecting surface, which increases the reflectance and thus the albedo. This phenomenon is the major reason for differences in the albedo among soils of different textural classes. Clayey soils can maintain high moisture content in the presence of water supply, while the sandy textured soils drain and dry out much more rapidly. Due to the differences in the resulting soil moisture content between the texture classes, there are differences in the reflectance and absorbance characteristics and so in the albedo (Fig. 1).

Surface roughness defines the type of reflection. Shiny, smooth surfaces, like water body, plant leaves, or wet soil surfaces may be near-perfect, specular reflectors, which may reflect well and show relatively high albedo for lower sun angles. Rough surfaces represent lower albedo values, especially when sun angle is low and the shading effect

Encyclopedia of Natural Resources DOI: 10.1081/E-ENRL-120014334

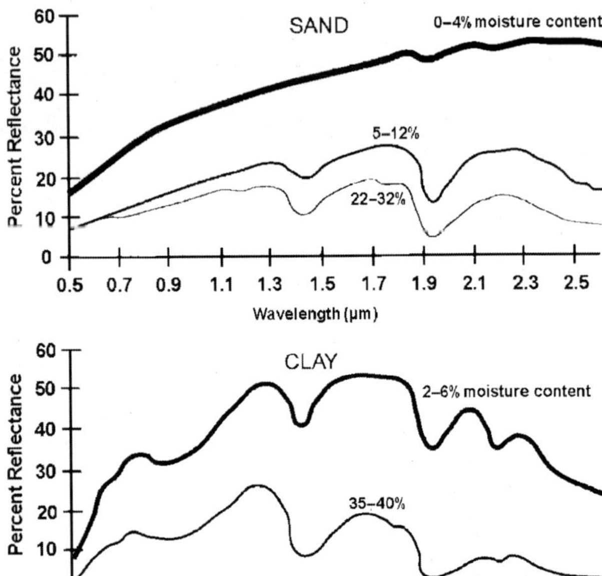

Fig. 1 The higher the moisture content the lower the reflectance throughout the visible and near-infrared region, especially along the water absorption bands at 1.4 μm and 1.7 μm. Notice the differences in the reflectance characteristics between the clayey and sandy soils.
Source: Adapted from Hoffer.[10]

Table 1 The approximated ranges of albedo of natural surfaces

Natural surface types	Approximated albedo
Blackbody	0
Forest	0.05–0.2
Grassland and cropland	0.1–0.25
Dark-colored soil surfaces	0.1–0.2
Dry sandy soil	0.25–0.45
Dry clay soil	0.15–0.35
Sand	0.2–0.4
Mean albedo of the Earth	0.36
Granite	0.3–0.35
Glacial ice	0.3–0.4
Light-colored soil surfaces	0.4–0.5
Dry salt cover	0.5
Fresh, deep snow	0.9
Water	0.1–1
Absolute white surface	1

lowers the reflection. There are measurable differences in the surface roughness among soil textural classes. Fine-textured, dry soils with small particle size produce high albedo due to relatively smooth surface. However, clayey soils are often wet, and soil moisture absorbs the incident radiation and decreases albedo. Conversely, dry, coarse textured soils with relatively large particles (sand grains) reflect larger portions of the incident radiation than clayey soils.

Surface color is determined by the interaction of the surface material with the visible spectra of the incident solar radiation. Soil color is a differentiating factor in all the soil classification systems. It reflects many of the most important soil physical and chemical characteristics. One of the most significant coloring agents of the soils is the soil organic matter content. Soil organic matter content increases the absorbance of the soil. Thus the higher the organic matter content, the lower the albedo. Iron oxides increase the reflectance in the red portion of the spectrum while causing a decrease in the blue–green and infrared portion. Salt crust on the surface increases the albedo dramatically. That is why mapping of salt-affected area with remotely sensed images is a very powerful tool for soil surveyors.

MEASUREMENT OF ALBEDO

The theoretical concept of measuring albedo is simple. A radiation sensor (pyranometer) is pointed upwards to measure the incident radiation and then quickly flipped downwards to measure the reflected radiation. For deriving the albedo, the quantity of the reflected radiation has to be divided by the one for the incident radiation. In fact, the actual measurement of surface albedo under natural condition is rather complex. The problem is threefold. First, the incident radiation does not only come from the radiation source directly, but also from diffused light from other directions. Secondly, the reflector surfaces do not reflect equally in all directions and thirdly, the sensors gather light only from a small range of angles. Thus, our measurements of reflectance are only samples of the bidirectional reflectance distribution function (BRDF). Albedo is often defined as an overall average reflection coefficient of an object. More precisely the terms of spectral and total albedo are differentiated. The spectral albedo refers to the reflectance in a given wavelength, while the albedo is calculated as an integral of the spectral reflectivity times the radiation, over all wavelengths in the visible spectrum. A good estimation of the surface albedo can be done using clear-sky satellite measurements.[11]

CONCLUSIONS

Albedo measures the overall reflectance of the surface. It provides lots of useful information about the soil system and helps to better understand the soil energy balance. But different wavelengths of sunlight are normally not equally reflected, which gives rise to a variable color of surfaces and differences in reflectance of certain wavelengths due to differences in physical or chemical characteristics of the soil surface. Differences in soil albedo can be measured with radiometers.

REFERENCES

1. Weast, R.C. *Handbook of Chemistry and Physics*; CRC Press: Boca Raton, FL, 1982.
2. Matthews, E. Vegetation, land-use and seasonal albedo data sets. In *Global Change Data Base Africa Documentation*; Appendix D; NOAA/NGDC, 1984.
3. Kotoda, K. *Estimation of River Basin Evapotranspiration. Environmental Research Center Papers*; University of Tsukuba: 1986; Vol. 8.
4. Baumgardner, M. F.; Sylva, L.F.; Biehl, L.L.; Stoner, E.R. Reflectance properties of soils. Adv. Agron. **1985**, 38, 1–44.
5. Geiger, R. *The Climate Near the Ground*; Harvard University Press: Cambridge, MA, 1965.
6. Jones, H.G. *Plants and Microclimate: A Quantitative Approach to Environmental Plant Physiology*, 2nd Ed.; The Cambridge University Press: Cambridge, U.K., 1992.
7. Jensen, M.E.; Burman, R.D.; Allen, R.G. Evapotranspiration and irrigation water requirements. In *ASCE Man. Rep. Pract. 70*; American Society of Civil Engineers: New York, NY, 1990.
8. Oke, T.R. *Boundary Layer Climates*; Methuen: New York, NY, 1978.
9. Van Wijk, W.R.; Scholte Ubing, D.W. Radiation. In *Physics of Plant Environment*; Van Wijk, W.R., Ed.; North-Holland Publishing Co.: Amsterdam, the Netherlands, 1963; 62–101.
10. Hoffer, R.M. Biological and physical considerations in applying computer-aided analysis techniques in remote sensing data. In *Remote Sensing: The Quantitative Approach*; Swain, P.H., Davis, S.M., Eds.; McGraw Hill: San Francisco, 1978; 227–289.
11. Li, Z.; Garand, L. Estimation of surface albedo from space: A parameterization for global application. J. Geophys. Res. **1994**, *99*, 8335–8350.

Altitudinal Belts: Global Mountains, Patterns and Mechanisms

Baiping Zhang
Fang Zhao
State Key Lab for Resources and Environment Information System, Institute of Geographic Sciences and Natural Resources Research, Chinese Academy of Sciences, Beijing, China

Abstract
Climatically sharp altitudinal gradient leads to the diversity and complexity of altitudinal (vegetation) belts and their vertical combinations (spectra) in mountains of the world. A total of 80 altitudinal belts could be identified globally. Almost all altitudinal belts have a trend toward increasing in elevation from high to low latitude, from oceanic to continental climatic regions, from the rim to the interior of immense mountain masses, and from shady slope to sunny slope. Many factors interplay to give rise to their global and local patterns of distribution, among which the most significant geographical factors include latitude, mass elevation effect, and continentality; the most effective ecological factors involve mean temperature of the warmest month, mean temperature of the coldest month, continentality, and annual precipitation.

INTRODUCTION

This entry is to show the diversity and complexity of global altitudinal belts, their basic patterns, and the main mechanisms. The mountains of the world are characterized by numerous and colorful altitudinal vegetation belts and their local-specific combination from mountain base upwards to mountain top (i.e., spectra of altitudinal belts, Fig. 1) thanks to the extremely complex and varied environment on different altitudes and slopes. On one hand, these belts follow a general trend of ascending from polar to tropical regions, from island to continental inland, and from the periphery to central parts of large mountain massifs; on the other hand, local factors (geomorphic, human, and soil factors; and even ocean currents, species competition, beetle infestations, invasive pathogens, and herbivores) could distort the general trend of global-scale distribution and modify the altitudinal limits of these belts to varying degrees. In other words, global and local factors interplay to form the actual distribution of altitudinal belts. Many factors have been used to depict the altitudinal distribution of alpine timberline, for example, warmest-month temperature (10°C), growing season temperature (5.5–7.5°C), soil temperature, etc. But they are often separately or singly used.

It is the integrated effect of these factors that really matters. Our latest study from a geographical perspective shows that, if latitude, mass elevation effect (MEE), and continentality are used as independent factors to linearly model the altitudinal distribution of timberline (N = 516), the coefficient of determination (R^2) can be as high as 0.904 for the Northern Hemisphere. This nearly perfectly explains the altitudinal distribution of timberline globally.

Some 14 ecological factors have been thought to relate with altitudinal limits of montane vegetation belts. Our most recent study shows that four factors (mean temperature of the warmest month, mean temperature of the coldest month, continentality, and annual precipitation) could be used to model the global-scale distribution of timberline ($R^2 \approx 0.94$).

GLOBAL ALTITUDINAL BELTS: DIVERSITY AND COMPLEXITY

Environmental gradient is about 1000 times as much in vertical direction as in horizontal, giving rise to the development of distinct altitudinal (vegetation) belts appearing successively with increased elevation in mountains, especially in high mountains. Their types and altitudinal positions are quite different from mountain to mountain and from place to place, thanks to the intricate interaction of zonal, nonzonal, and even human factors causing altitudinal belts. As a result, mountains display extreme heterogeneity and ecological variety, and are characterized by the most diverse landscapes and biological resources in the world. In total, 80 types of altitudinal belts have been identified globally,[1] and their types and altitudinal limits can combine in unlimited ways in different mountains and even on varied slopes, making a kaleidoscopic world of altitudinal belts (Fig. 1). In the southeastern Himalayas nine altitudinal belts can be identified from piedmont plains to snow-coved peaks, which combine to form an altitudinal spectrum with the largest number of vegetation belts in the world, whereas in the Ural Mountains, only two to three altitudinal belts can be found. The upper timberline can reach up to 4800 m

Encyclopedia of Natural Resources DOI: 10.1081/E-ENRL-120049856

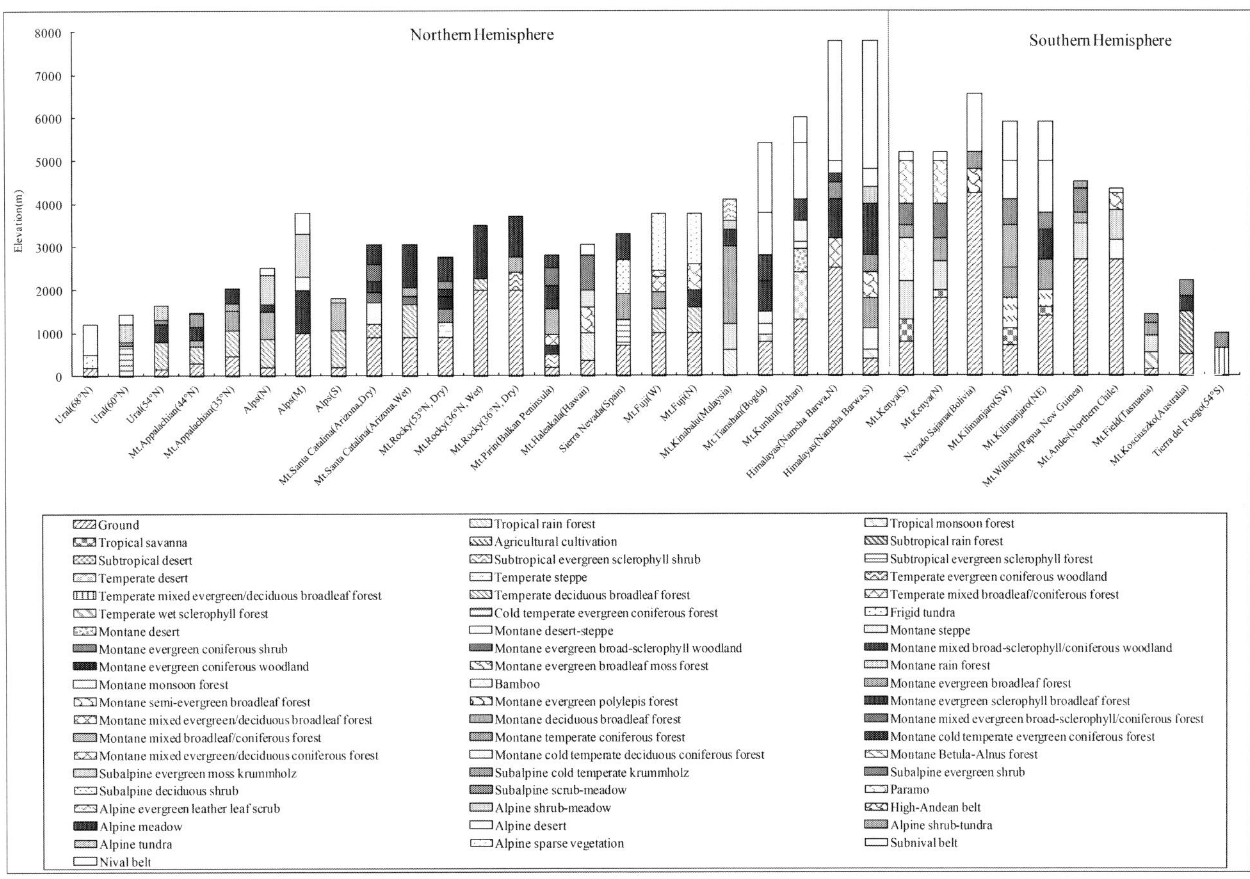

Fig. 1 The representative spectra of altitudinal belts of the main mountain ranges of the world.

above sea level in the southeastern Tibetan Plateau and even higher in the central Andes, compared to only about 2000 m or even lower in similar-latitude mountains, especially those on islands. The number, elevation, and combination of altitudinal belts show quite differently in different flanks/slopes in arid/semiarid temperate mountains, but rather moderately in humid tropical and subtropical mountains.

PATTERNS OF ALTITUDINAL BELTS

Almost all altitudinal belts have a trend toward increasing in elevation from high to low latitudes, from oceanic to continental climatic regions, from the rim to the interior of immense mountain masses, and from shady slopes to sunny slopes. Especially, the natural high-elevation tree line shows a close association with a physics-driven boundary, the snow line, suggesting a common climatic driver.[2] The upper timberline (forest line) or tree line, as a global phenomenon and the most studied limit of altitudinal belts, is a bioclimatic reference line against which other vertical bioclimatic zones can be defined.

From High to Low Latitudes

Typically, snowline and timberline, just as other altitudinal belt limits, ascend from high to low latitudes, but reach their summit at 30° northern latitude and 20° southern latitude, and between these latitudes are relatively low value areas, forming a saddle-shaped global distribution pattern. So, there is no uniform relationship between tree line altitudinal position and latitude. Usually, tree lines decline/ascend steeply between temperate and boreal latitudes but are relatively constant between ~30°N and 20°S. And the pattern is not symmetrical around the equator.[3,4]

From Coastal to Inland Areas of Continents

From coastal to inland areas of any continent, the limits of altitudinal belts ascend gradually, usually by 1000–2000 m or even larger. The 2000 m amplitude seen within a narrow latitudinal belt in the temperate zone illustrates that latitude is a very imprecise predictor of tree-line elevation,[2] or the inland position and relief are as significant factors as latitude for the elevation of altitudinal belts. In addition, compared to mountains at similar latitudes in continental areas, tree line on oceanic islands is substantially lower than the continental tree line,[5] and the forest line on the islands is found at 1000–2000 m lower elevations.[6]

From the Outer Margins to the Interior of Mountain Massifs

More than 100 years ago, a tendency was observed for temperature-related parameters, such as tree line and snowline, to occur at higher elevations in the central Alps than on their outer margins.[7] This is more evident in the Tibetan Plateau, where the timberline ascends from 3500 m in the southeastern rim to 4700 m in the interior and the snowline from 5000 m to 6200 m. In the European Pyrenees, the timberline is at 1500 m in the rim and 2300–2400 m in the interior. In the comparatively high and dry western cordillera of Bolivia, for example, the upper limit of the occurrences of *Polylepis tarapacana (P. tomentella)* is about 1000 m higher than the upper limit of *P. tarapacana* in the humid and lower eastern cordillera.[8] *Polylepis* climbs even higher, to 5200 m, and even to 5400 m, making the highest tree line in the world.[9]

Difference between Shady and Sunny Slopes and between Windward and Leeward Slopes

Usually, a certain type of altitudinal belt appears at a relatively higher position in sunny slopes than in shady slopes. In the Catalina Mts. of Arizona, the fir forest, ponderosa pine forest, oak pine woodland, and dry woodland and Chaparral belts are all at a higher elevation in the south-facing slope than in the north-facing slope.[10] In Mt. Kenya of Africa, the orotropical cloud forest lies between 3100 and 3200 m in the humid south slope and 2650 and 3000 m in the arid north slope.[11] Extremes are not present on the sunny slopes of certain belts (e.g., montane forest), as are present in shady slopes, as in the northern Tien Shan Mountains of northwest arid China. Some east–west extending ranges often serve as significant geographic demarcation (e.g., the Qinling Mts. in central China, as the transition from subtropical to warm-temperate zone), and different spectra of altitudinal belts develop in the southern and northern flanks. On the other hand, altitudinal belts show differently in windward and leeward slopes, often with a higher position in the leeward slope. If the sunny slope is the windward slope at the same time, a complicated situation occurs—the altitudinal belt in question may be at a higher position in the shady slope, depending on the degree of aridity resulting from the interplay of so many climatic factors.

The Upward Shifting of Altitudinal Belts with Global Warming

Theoretically, an increase of mean temperature by 6°C would mean an upward shift of timberline by 1200 m.[12] However, the fluctuation of altitudinal belts are related with many other factors (e.g., topography, tree species, etc.), also including the importance of site history to the present timberline dynamics.[13,14] With global warming, some altitudinal belts may disappear. In arid and semiarid areas, the future dynamics of the present montane forest stands at timberline might be controlled by moisture supply rather than by higher temperatures. In addition, fluctuations of climates and timberline were not synchronous and occurred also at different regional and local intensities. In case of a sudden rise of temperature as has been predicted by many models, mountain timberlines would not respond spontaneously, but rather with a time lag of several decades or even centuries. The general trend is being modified by regional, local, and temporal variations and thus is different as to its extent, intensity, and process of change. Consequently, it seems difficult to make generally acceptable statements without restricting them at the same time in view of the many regional and local peculiarities.

MECHANISM FOR THE ELEVATION OF ALTITUDINAL BELTS

Many factors together give rise to the great diversity, complexity, and varying elevation of mountain altitudinal belts. Globally, the altitudinal position of all altitudinal belts follows a similar trend to timberline and permanent snow line, suggesting a common physical driver (e.g., temperature). On smaller scales, more factors (topography, microclimate, soil conditions, etc.) participate in the making of altitudinal belts. As a result, altitudinal belts' heterogeneity increases from the global to the regional, landscape, and local scales. In addition, factors and processes at one scale may not be as important as they are at another scale. Generally, these factors could be classified as geographical, ecological, physiological, and even human interference.

Geographical Factors

Latitude

In the very early stage of altitudinal belt research, especially on upper timberline, it was noted that the elevation of tree line is closely related with the latitude in which it appears, and many regional tree-line latitude-elevation models were developed. Some authors related the altitude of the climatic timberline to geographical latitude and calculated "timberline gradients." For example, the gradient between the northern border of Oregon and northern California is 184 m, in the Appalachians 83 m,[15] in the Ural Mountains 71 m, in Middle Siberia 89 m, and in East Siberia 76 m per degree northern latitude.[16] In the southern Andes between 35°S and 55°S, timberline declines south for about 80 m per degree of latitude.[17] It is clear that the empirical gradients are quite different because of the geographical location of the longitudinal (north–south oriented) transects to which they are related.

Mass elevation effect

Larger mountain massifs serve as a heating surface absorbing solar radiation and transforming it to long-wave energy. In other words, the climate becomes increasingly continental from rim to the central parts of the mountain areas, with lower precipitation and higher percentages of sunshine in the inner parts compared to the outer mountain ranges. Consequently, temperature is higher than in the free atmosphere at any given elevation, which makes the identical altitudinal belt at a higher position in the interior than in the outer margins of mountain massifs. This is the so-called "Massenerhebung effect" or "mass elevation effect,"[7] sometimes called "mountain mass effect." This can effectively explain the large difference in elevation of altitudinal belts in similar latitudes. Some extremely high timberlines (in the central Andes and the Tibetan Plateau) are mainly the result of MEE. Without MEE, the highest timberline should not surpass 3500 m above sea level in any mountains. The magnitude of MEE was considered to be closely related with the mean elevation of a mountain massif; however, what is really significant is the base elevation of the local basins or valleys in the inner parts of the massif. To a large extent, the local base elevation could represent MEE, and it has been proved to contribute a lot to the elevation of snowline and timberline.[18,19]

Continentality

Gradual ascending of altitudinal belts from coastal to continental inland areas is due to general decreased precipitation and increased degree of continentality in the same direction. In reality, continentality and MEE often combine to shape the spatial pattern of altitudinal belts on a large scale, namely, making it possible for the altitudinal belts to ascend to a very high position, especially in the case of immense plateaus and mountain ranges in the interior of continents. In addition, the forest line on the islands is found between 1000 and 2000 m lower elevations compared to mountains at similar latitudes in immense continental areas,[6] and four factors that are thought to contribute to the forest line depression include: (a) drought on trade-wind exposed island peaks with stable temperature inversions, (b) the absence of well-adapted high-altitude tree species on isolated islands, (c) immaturity of volcanic soils, and (d) an only small mountain mass effect that influences the vertical temperature gradient.

Slope-facing effect

Differences in radiation and aridity on different facing slopes may lead to difference in elevation and area of altitudinal belts. The contrast becomes more evident from tropical to temperate zones, and from coastal humid to inland arid areas. But this is only a local driver for altitudinal belts.

Of course, it is the integrated action of these factors that really matters. The latest study shows that, if latitude, MEE, and continentality are used as independent factors to linearly model the altitudinal distribution of timberline (N = 516), the coefficient of determination (R^2) can be as high as 0.904 for the northern hemisphere.[20] This nearly perfectly explains the global distribution of alpine timberline.

Ecological Factors

Many temperature and aridity factors have been found to closely relate to the elevation of altitudinal belts, including the 10°C isotherm for the warmest-month, mean annual temperature, annual range of temperature, growing-season temperature, soil temperature, aridity, mean annual precipitation, the existence of well-adapted tree-line species, competition between communities or species, etc. In total, 14 ecological factors have been used to separately explain the altitudinal belts. In exploring the "ultimate cause" of altitudinal tree line, it was found that low mean temperatures in the rooting zone during the growing season or all-year-round (at tropical tree lines) is the critical factor controlling directly worldwide tree line position.[21–24] The latest research shows that the current natural high-elevation tree line is associated with a growing season that is at least three months long and during which the mean air temperature reaches at least 6.4°C.[2] With it, in addition to some necessary fine tuning that accounts for regional aridity and snow pack effects, the global pattern of natural climate-driven tree-line positions can be modeled with great confidence. The obtained error term of ± 0.7 K corresponds to a mean uncertainty of ±130 m of elevation, sufficiently robust at a global scale to encourage a discussion of the likely causes of the global tree-line isotherm. Our most recent study shows that four factors (mean temperature of the warmest month, mean temperature of coldest month, continentality, and annual precipitation) together could almost perfectly explain the global-scale distribution of timberline ($R^2 \approx 0.94$).

Physiological Factors

UV-B (280–315 nm) is damaging and mutagenic to living organisms.[25] It is known to be particularly high in the alpine zone of tropical mountains. This involves the occurrence of dwarf forest at anomalously low altitudes, especially for isolated peaks in tropical mountains. It has also been suggested that temperature could operate via the soil,[26] lower temperatures leading to a greater accumulation of organic matter and to changes in nutrient status. The increasingly poor supply of nitrogen and phosphorus depends on decrease in temperature and the increase in frequency of fog is likely to be important not only in governing the distribution of forest types but also in influencing their biomass and physiognomy. This also leads to relatively lower elevation of forest belts in tropical mountains and islands.

CONCLUSIONS

The diversity and complexity of altitudinal belts of the world mountains result from many factors, global and local, zonal and nonzonal, uplifting (e.g., MEE) and depressing (e.g., summit syndrome and wet climate), etc.

As a result, global altitudinal belts have common general patterns and varied local patterns. A difference of 2000–3000 m in elevation of alpine timberline exists both in north–south direction and in similar latitudes. The significance of the traditional latitude-elevation models for altitudinal belts is rather limited. Any single factor, no matter how effective, could not satisfactorily explain the limits of altitudinal belts. On the other hand, it is also not adequate to use too many factors to explain the altitudinal belts, because that means we say nothing about them.

Latitude, MEE, and continentality are the most important geographical factors for altitudinal belts. With them as independent variables, a simple regression equation could almost perfectly explain the elevation of altitudinal belts on global scale. From ecological perspective, mean temperature of the warmest month, mean temperature of the coldest month, continentality (also an ecological factor), and annual precipitation are four most significant factors, which together could also rather satisfactorily explain the global distribution of altitudinal belts.

REFERENCES

1. Zhang, B.; Tan, J.; Yao, Y. *Digital Integration and Patterns of Mountain Altitudinal Belts*; Environmental Science Press: Beijing, China, 2009.
2. Körner, C. *Alpine Treelines: Functional Ecology of the Global High Elevation Tree Limits;* Springer Basel: Heidelberg, New York, Dordrecht, London, 2012; 26.
3. Troll, C. *Geoecology of the High Mountain Regions of Eurasia*; Franz Steiner Verlag Gmbh: Wiesbaden, 1972; 1–16.
4. Körner, C. A re-assessment of high elevation treeline positions and their explanation. Oecologia **1998**, *115* (4), 445–459.
5. Berdanier, A.B. Global treeline position. Nat. Educ. Knowledge **2010**, *1* (11), 11.
6. Leuschner, C. Timberline and alpine vegetation on the tropical and warm-temperate oceanic islands of the world—Elevation, structure and floristics. Vegetatio **1996**, *123* (2), 193–206.
7. Quervain, A.D. Die Hebung der atmosphärischen lsothermenin der Schweizer Alpen und ihre Beziehung zu deren Höhengrenzen. Gerlands Beitr. Geophys **1904**, *6*, 481–533.
8. Kessler, M. *Polylepis-Wälder Boliviens: Taxa, Ökologie, Verbreitung und Geschichte*; Charmer: Berlin, Stuttgart, 1995; 246.
9. Holtmeier, F.K. Mountain timberlines: ecology, patchiness, and dynamics. In *Advances in Global Change Research*, Springer, 2009; 36, 1–437.
10. http://www.geo.arizona.edu/Antevs/biomes/azlifzon.html (accessed August 2012).
11. Bussmann, R.W. Vegetation zonation and nomenclature of African mountains - an overview. Lyonia **2006**, *11* (1), 41–66.
12. Zimmermann, N.E., Bolliger, J., Gehrig-Fasel, J., Guisan, A., Kienast, F., Lischke, H., Rickebusch, S. and Wohlgemut, T. Wo wachsen die Bäume in 100 Jahren? Forum für Wissen **2006**, 63–71.
13. Payette, S., Filion, L., Delwaide, A. and Begin, C. Reconstruction of the treeline vegetation response to long-term climate change. Nature **1989**, *341*, 429–432.
14. Lescop-Sinclair, K. and Payette, S. Recent advance of the arctic treeline along the eastern coast of Hudson Bay. J. Ecol. **1995**, *83*, 929–936.
15. Cogbill, C.V. and White, P.S. The latitude-elevation relationship for spruce-fir forest and treeline along the Appalachian mountain chain. Vegetatio **1991**, *94*, 153–175.
16. Malyshev, L. Levels of the upper forest boundary in northern Asia. Vegetatio, **1993**, *109* (2), 175–186.
17. Crawford, R.M.M. *Studies in Plant Survival: Ecological Case Histories of Plant Adaptation to Adversity;* Blackwell Scientific Publications: London, Edinburgh, Boston, MA, Melbourne, Paris, Berlin, Wien, 1989.
18. Han, F.; Zhang, B.; Yao, Y.; Zhu, Y.; and Pang, Y. Mass elevation effect and its contribution to the altitude of snowline in the Tibetan plateau and surrounding areas. Arct. Antarct. Alp. Res., **2011**, *43* (2), 207–212.
19. Han, F.; Yao, Y.; Dai, S.; Wang, C.; Sun, R.; and Zhang, B. Mass elevation effect and its forcing on timberline altitude. J. Geographical Sci., **2012**, *22* (4), 609–616.
20. Zhao, F.; Zhang, B.; Pang, Y.; Yao, Y.; Han, F.; Zhang, S.; and Qi, W. Mass elevation effect and its contribution to the altitude of timberline in the northern hemisphere. Acta Geogr. Sinica **2012**, *67* (11), 1556–1564.
21. Paulsen, J., Weber, U.M. and Körner, C. Tree growth near treeline: abrupt or gradual reduction with altitude. Arct. Antarct. Alp. Res. **2000**, *32* (1), 14–20.
22. Hoch, G. and Körner, C. The carbon charging of pines at the climatic treeline: a global comparison. Oecologia **2003**, *135*, 10–21.
23. Hoch, G and Körner, C. A test of treeline theory on a montane permafrost island. Arct. Antarct. Alp. Res. **2006**, *38* (1), 113–119.
24. Körner, C. and Paulsen, J. A world-wide study of high altitude treeline temperatures. J. Biogeogr. **2004**, *31*, 713–732.
25. Flenley, J. Ultraviolet insolation and the tropical rainforest: altitudinal variations, Quaternary and recent change, extinctions, and biodiversity. In *Tropical Rainforest Responses to Climatic Change*; Praxis Publishing: Chichester, UK, 2007; 219–235.
26. Grubb, J. Interpretation of the "Massenerhebung" effect on the tropical mountains. Nature **1971**, *229* (5279), 44–45.

BIBLIOGRAPHY

1. Barbour, M.G.; Billings, W.D. *North American Terrestrial vegetation*; Cambridge University Press: New York, 2000; 107 pp.
2. Bear, R.; Hill, W.; Pickering, C.M. Distribution and diversity of exotic plant species in montane to alpine areas of Kosciuszko National Park. Cunninghamia **2006**, *9* (4), 559–570.
3. Ellenberg, H. *Vegetation Ecology of Central Europe*; Cambridge University Press: New York, 1988; 731 pp.
4. Hemp, A. Continuum or zonation? Altitudinal gradients in the forest vegetation of Mt. Kilimanjaro. Plant Ecol. **2006**, *184* (1), 27–42.
5. Henning, I. Horizontal and veritcal arrangement of vegetation in the Ural-System. In *Geoecology of the High Mountain*

Regions of Eurasia; Troll, C., Ed.; Franz Steiner Verlag Gmbh: Wiesbaden, 1972; 264–275.

6. Hoch, G.; Körner, C. Growth, demography and carbon relations of Polylepis trees at the world's highest treeline. Funct. Ecol. **2005**, *19*, 941–951.
7. Hope, G.S. The vegetational history of Mt. Wilhelm, Papua New Guinea. J. Ecol. **1976**, *64*, 627–661.
8. Kitayama, K. An altitudinal transect study of the vegetation on Mount Kinabalu, Borneo. Plant Ecol. **1992**, *102* (2), 149–171.
9. Kitayama, K.; Mueller-Dombois, D. An altitudinal transect analysis of the windward vegetation on Haleakalā, a Hawaiian island mountain: (2) vegetation zonation. Phytocoenologia **1994**, *24*, 135–154.
10. Li, B. The Vertical Spectra of Vegetation in the Mt.Namjagbarwa. Reg Mt. Res. **1984**, *3*, 174–181.
11. Li, S.; Zhang, X. Principles of Classification and Characteristics of vegetation altitudinal belts in Xinjiang. Chin. J. Plant Ecol **1966**, *4* (1), 132–141.
12. Massaccesi, G.; Roig, F.A.; Martı´nez Pastur, G.J.; Barrera, M.D. Growth patterns of Nothofagus pumilio trees along altitudinal gradients in Tierra del Fuego, Argentina. Trees Struct. Funct. **2008**, *22* (2), 245–255.
13. Messerli, B.; Winiger, M. Climate, environmental change, and resources of the African mountains from the Mediterranean to the equator. Mt. Res. Dev. **1992**, *12* (4), 315–336.
14. Nagy, L; Thompson, D; Grabherr, G; Körner, C. *Alpine Biodiversity in Europe*; Springer Press: Berlin, 2003; 167 pp.
15. Niemelä, T.; Pellikka, P.; Museum, B. Zonation and characteristics of the vegetation of Mt. Kenya. In *Taita Hills and Kenya. Expedition Reports of the Department of Geography*; Pellikka, P.; Ylhäisi, J; Clark, B., Eds.; University of Helsinki: Helsinki, 2004; 14–20.
16. Ogden, J.; Powell, J.A.A. Quantitative description of the forest vegetation on an altitudinal gradient in the Mount Field National Park, Tasmania, and a discussion of its history and dynamics. Aust. J. Ecol. **1979**, *4* (3), 293–325.
17. Ohsawa, M. An interpretation of latitudinal patterns of forest limits in south and east Asian mountains. J. Ecol. **1990**, *78* (2), 326–339.
18. Ohsawa, M. Differentiation of vegetation zones and species strategies in the subalpine region of Mt. Fuji. Vegetatio **1984**, *57*, 15–52.
19. Peng, B. Preliminary discussion on the vertical zonation in the Mt.Namjagbarwa region. Mt. Res. **1984**, *2* (3), 182–189.
20. Smith, J.M. Vegetation of the summit zone of Mount-Kinabalu. New Phytol **1980**, *84* (3), 547–573.
21. Stefanova, I.; Ammann, B. Lateglacial and Holocene vegetation belts in the Pirin Mountains (southwestern Bulgaria). The Holocene **2003**, *13* (1), 97–107.
22. Villagrfin, C.; Armesto, J.J.; Arroyo, M.T.K. Vegetation in a high Andean transect between Turi and Cerro Le6n in Northern Chile. Vegetatio **1981**, *48*, 3–16.
23. Walter, H. *Vegetation of the Earth and Ecological Systems of the Geo-Biosphere*; Springer Verlag: Berlin, 1979; 107 pp.
24. Whittaker, R.H. Vegetation of the great smoky mountains. Ecol. Monogr. **1956**, *26* (1), 1–80.
25. Zhang, B.; Zhou, C.; Chen, S. The Geo-info-spectrum of Montane Altitudinal Belts in China. Acta Geographica Sinica **2003**, *2*, 163–170.

Aquaforests and Aquaforestry: Africa

Gordon N. Ajonina
CWCS Coastal Forests and Mangrove Programme, Cameroon Wildlife Conservation Society, and Institute of Fisheries and Aquatic Sciences, University of Douala, Yabassi, Douala, Cameroon

M. Tomedi Eyango
Institute of Fisheries and Aquatic Sciences, University of Douala, Yabassi, Douala, Cameroon

Abstract

As water covers over 70% of the Earth's surface, all over the world from the mountain to the sea and particularly visible in tropical savanna, urban and periurban areas, the natural vegetation associated with waterways which is mostly represented by forests at various levels of inundation from permanent (swamp forests) to dry riparian forests is credited to be among the most species-rich ecosystems. Aquatic forest ecosystems provide critical services for man and wildlife. They are important both to flora and to fauna living in water as well as those living on land. These forests are degraded or lost at an alarming rate due to factors mostly anthropogenic. This entry sets to raise the status of such forests and to demonstrate their capacity and potentials to satisfy mankind needs if sustainably managed under an emerging management regime of "aquaforestry." Divided into two major parts: part 1—aquaforests deals with the definitions, characteristics, typology, distribution, structure, and functions of aquaforests including major benefits, threats, and drivers of degradation and loss of aquaforests; part 2—aquaforestry covers the essentials of strategies for sustainable management of aquaforests within integrated river basin approach for sustained production of environmental goods and services to satisfy mankind under the new management regime of "aquaforestry" or "aquatic forestry" with special reference to African systems.

INTRODUCTION

As water covers over 70% of the Earth's surface, all over the world from the mountain to the sea and particularly visible in tropical savanna, urban and periurban areas, the natural vegetation associated with waterways which is mostly represented by forests at various levels of inundation from permanent (swamp forests) to dry riparian forests is credited to be among the most species-rich ecosystems.[1] These forests which we now group as "aquaforests," "aquatic forests," "blue forests," or forested wetlands are important areas for global sources and quality and stream environment.[2] These forests are widely found in many large landscape in the world especially peat forests in the temperate zones, the Amazonian swamp forests, swamp forests in Malaysia, and the peat swamps of Indonesia—to name a few examples. In Africa, these forests abound with the most species-rich forests at least for woody plants in the water areas of West Central Africa (from Mt. Cameroon South East into Gabon) and in Northeast Madagascar.[3] In proportion to their area within a watershed, they perform more ecological and productive functions than do adjacent uplands.[4] They harbor a diversified flora and physical structure.[5] Ecologists have also examined the value of these forests, especially riparian forests, as habitats for many animals and recognized them as a priority area for conservation of terrestrial mammals,[6,7] fisheries as well as birdlife.[5,8,9] They are marginal among the wooded vegetation and because of their advanced state of degradation. Despite the fact that there is growing recognition of the ecological, hydrologic, biogeochemical, sociocultural, economic, and aesthetic importance of these forests in Africa, they have remained insufficiently studied and managed as key vegetation formation for biodiversity protection and sustained livelihoods for surrounding communities. Aquaforests have often been ignored or excluded from vegetation studies in favor of upland forests.

This entry sets to raise the status of such forests especially in Africa and to demonstrate their capacity and potentials to satisfy mankind's needs if sustainably managed under an emerging management regime of "aquaforestry." Hence, the entry is divided into two major parts: part 1—aquaforests deals with the definitions, characteristics, typology, distribution, structure, and functions of aquaforests including major benefits, threats, and drivers of degradation and loss of aquaforests; part 2—aquaforestry covers the essentials of strategies for sustainable management of aquaforests within integrated river basin approach for sustained production of environmental goods and services to satisfy mankind under the new management regime of "aquaforestry" or "aquatic forestry."

Encyclopedia of Natural Resources DOI: 10.1081/E-ENRL-120047504

AQUA-FORESTS

Definitions

"Aquaforests," "aquatic forests," "blue forests," or "forested wetlands" is an umbrella term that we use to refer to diverse range of ecosystems with vegetation (trees, shrubs) and marshes that is permanently or temporally covered by either fresh or salt water or bordering water courses. Riverine landscapes are heterogeneous, dynamic, and biologically and spatially complex.[10] Within riverine landscapes, riparian forests are transitional between terrestrial and aquatic ecosystems and are distinguished by gradients in biophysical conditions, ecological processes, and biota. They are portions of terrestrial ecosystems that influence exchanges of energy and matter between aquatic and uplands ecosystems[4] and contribute to the diversity and function of both terrestrial and aquatic ecosystems.[11] At all latitudes, swamps, riparian, or streamside forests are recognized as distinct terrestrial ecosystems.[12] They occur in areas of intense land–water interaction and are known to support concentrated and diverse assemblages of wildlife and plant species landscape mosaic, spatial heterogeneity, complex environmental gradients, and unique natural disturbance regimes[13] which instigate a wide variety and abundance of resources and substrate.[11] These forests are adapted to life on banks of river estuaries and lagoons where salt-water flooding follows a daily cycle or to freshwater flooding of a permanent or seasonal nature in forest areas.[14]

General Characteristics of Aquaforests

Certain general characteristics of these forest habitats can be identified:

- *Critical role of hydrology in system sustenance and integrity*: As pointed out by Brinson (1990),[15] these forests own their dynamics, structure, and composition (species and habitat diversity and food webs) to river processes of inundation, sediments dynamics (transport of sediments), and biogeochemistry and nutrient cycling hence the erosive forces of water. The characteristic plant species, plant communities, and associated aquatic or semiaquatic animal species are intrinsically linked to the role of water as both an agent of natural disturbance and as a critical requirement of biota survival. Alternating environmental stress[16] such as periodic flooding is a form of disturbance to which many of the taxa occurring in riparian communities appear well adapted by sprouting from remaining root or trunk (e.g. *Pterocarpus santalinoides*, *Cola laurifolia*, *Szygium guineense*, *Cynometra megalophylla*).[17] According to Acker et al. (2003),[11] flooding apparently promotes complexity at the life form, species composition, and stand structural levels, thus the significance of channel constraint for severity of floods may be different in mountainous versus relatively flat terrain. Frequent floods are known to maintain the native riparian vegetation in a mosaic of different successional stages.[18,19] Sediments deposited by floods can create initial seedbeds for germination but also injure seedlings.[20] Research on the effects of flooding have shown that flooding increases wood and leaf decomposition rates and sites that are flooded generally have a lower accumulation of organic matter.[21] As a result, these forests tend to be very diverse in species composition and physical structure.[22]
- *Fragile and vulnerable ecosystems.* They resemble island species in their vulnerability to environmental stresses. They have a limited range (often a single watershed, lake, or river system) and low population numbers.
- *Transition ecosystems.* These ecosystems are seemingly transitional ecosystems (ecotones) between terrestrial systems and open water systems but most heavily influenced by terrestrial ones. Where they occur, they attract species from both systems and are often very productive regions with high species richness.[23]
- *High litter production.* Mangroves and other aquaforests produce considerable litter such as leaves, fruits, and deadwood. While some of this detritus stay in the forests, the rest end up in estuaries. There, biological degradation by bacteria, fungi, and larger organisms results in detrital complexes that appear to be the most important source of energy for maintaining estuarine fisheries and supporting tropical marine coastal ecosystems.
- *High levels of primary productivity.* The high levels of primary production coupled with their characteristic habitat complexity provide the support base for some of the world's highest levels of biodiversity. Many support important populations of wildlife and plants including endangered species, terrestrial mammals such as primates, and often a spectacular concentration of birds. Numerous aquaforest areas form part of international flyways for migratory birds.
- *Aquaforests as biodiversity vegetation hotspots.* Hotspots of biodiversity are areas particularly rich in species, rare, endemic, threatened species, or some combination of these attributes.[24] These forests can be considered as hotspots of biodiversity for several reasons:
    - They are among the most vulnerable forest formations at all latitudes, yet of high ecological importance.
    - Their high species richness is relatively in small areas (patches and narrow corridors). Natta (2003)[17] collected one-third of the estimated species in the Benin flora in just 20 ha of riparian forests.
    - Their numerous species are rare, threatened, or with superior adaptability to a specific habitat.
    - They are habitat for an endemic plant species in a fire-prone environment.
    - They are a vital ecosystem for numerous wild and birdlife species.
    - They protect many water courses all over the landscape.

Typology and Distribution of Aquaforests

Aquaforests are associated with natural or artificial water bodies or courses (streams, rivers, lakes, tidewaters, etc.) from mountain (limit of tree-line) to the sea. They can be classified based on landscape units; dominant vegetation, hydroperiod, and salinity broadly into (see Figs. 1 and 2):

- *Inland aquaforests*
  - Upland watersheds and catchment forests
  - Swamp forests and marshes
  - Riverine/riparian/fringing/gallery forests
  - Periodically inundated/floodplain forests/flooded forests
- *Coastal and marine aquaforests*
  - Transitional swamps
  - Intertidal forests (mangrove forests)
- *Artificial aquaforests*
  - Planted forests around rivers or lakes, e.g., *Raphia* forests
  - Planted mangroves

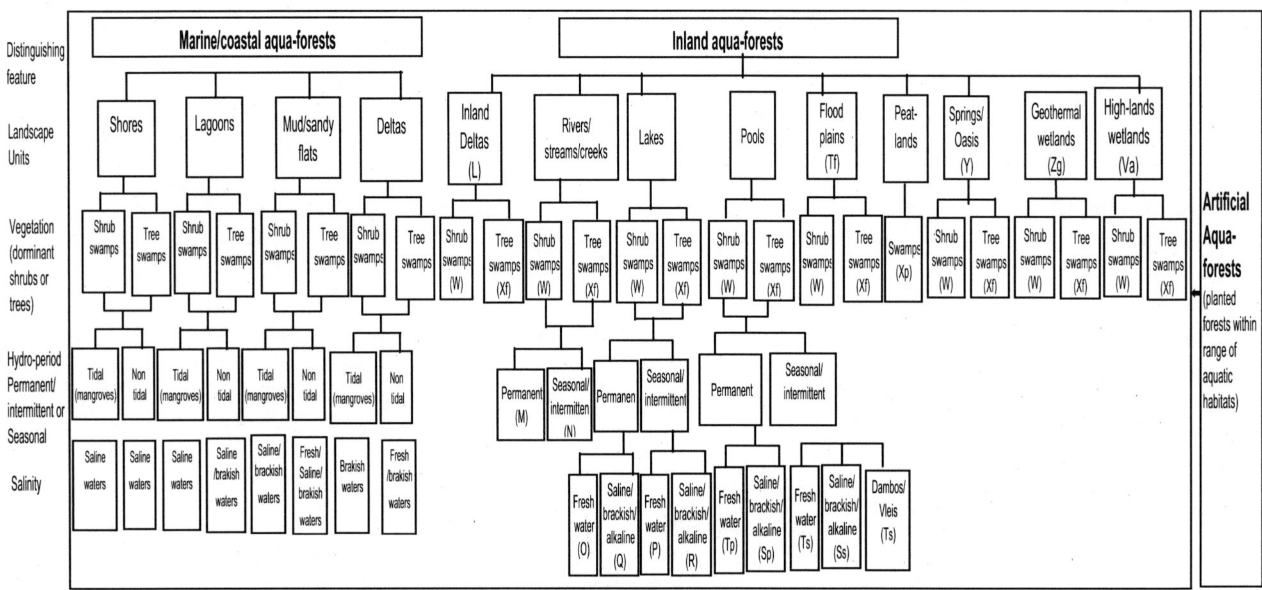

Fig. 1 Attempted classification of aquaforests following Ramsar classification of wetlands habitats and codes.

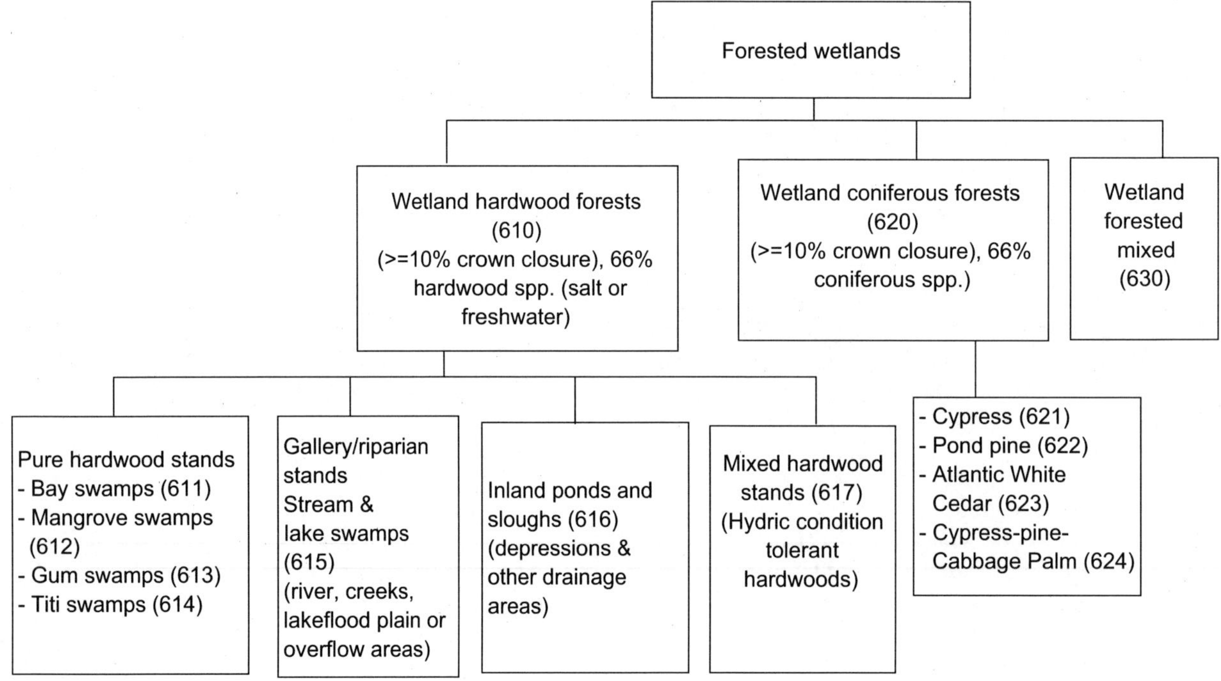

Fig. 2 Classification following The Florida Land Use Cover and Forms Classification System (FLUCCS) and codes.

Swamp forests are found in still waters around lake margins and certain parts of floodplains. These typically consist of *Ficus* species, borassus palms, and *Syzgium*.[25] Extensive areas are found in the Congo basin and the Niger Delta. The Guineo-Congolian swamp forests have a diversified endemic flora. Cameroon's swamp forests along the Nyong River are unusual and important representatives of this type of habitat.

Mangrove is a broad name used to describe over 50 species from five families worldwide, which have been adapted to live in sheltered low-wave action intertidal areas within the tropical and subtropical areas. They are more intensive in the deltas of large rivers, but also occur along small bays and lagoons. The major element of mangrove community is generally comprised of a pure stand of a particular mangrove species; each stand may have a unique associated fauna. Table 1 shows the global distribution of mangrove forests now occupying less than 140,000 km^2, of which most (33.4%) is in Southeast Asia and 15% in South America. In Africa, mangroves are found discontinuously from Senegal to Angola on west coast and from Somalia to South Africa on the east coast. Extensive stands are also found in Madagascar.

Structure and Function of Aquaforest Ecosystems

A cross-section through any aquatic forest habitat will display both its horizontal and vertical structure (see Fig. 3).

Horizontal structure

A horizontal structure will display a water body within the land mass with different substrates due to various levels of inundation from submerged (swamp forest) within open water through temporarily inundated to dry forests on firm ground.

Vertical structure

The vertical structure of any aquaforest will show the characteristic three tree-layered strata of any matured forest depending on the ecological region on land boundary (the upper tree stratum: 30–45 m, the middle tree stratum:

Table 1 World distribution of mangrove areas per regions

Region	Area (km^2)	Proportion of global (%)
East and South Africa	7.917	5.2
Middle East	624	0.4
South Asia	10.344	6.8
Southeast Asia	51.049	33.4
East Asia	215	0.1
Australia	10.171	6.7
Pacific Ocean	57.17	3.8
North and Central America	22.402	14.7
South America	23.882	15.7
West and Central Africa	20.04	13.2
Total	**134.325**	**100.0**

Source: Adapted from Spalding et al. (2010).

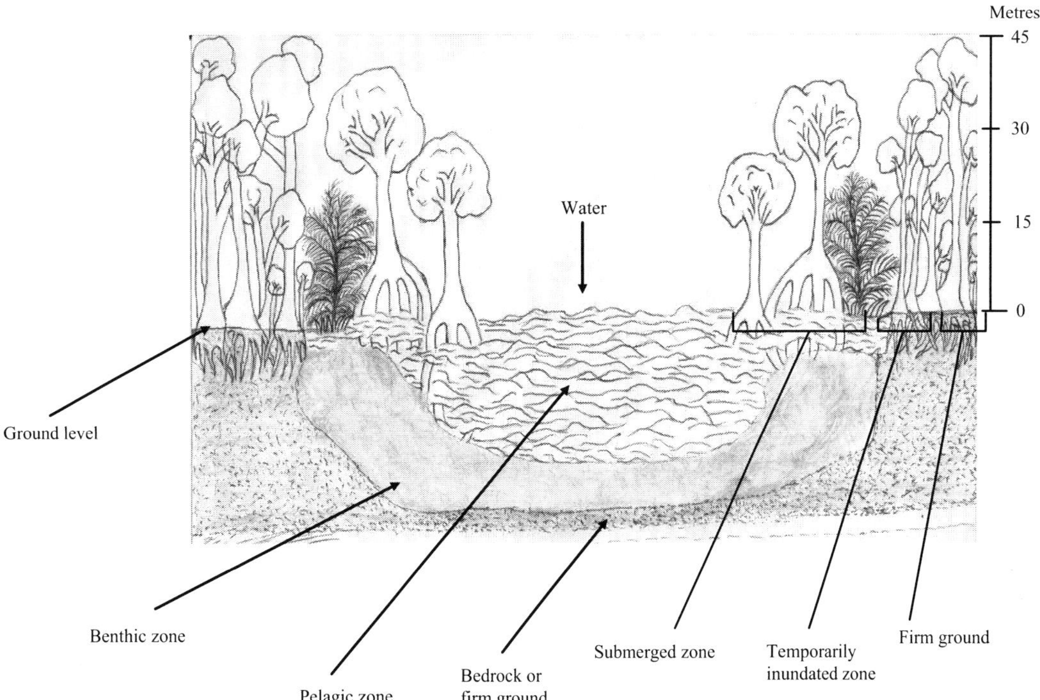

Fig. 3 Cross-section through aquatic forest habitat showing horizontal and vertical layers.

23–30 m and lower tree stratum: 9–23 m), and a shrub layer: <9 m also consisting of young trees with a herb layer also with tree seedlings with an admixture of herbs which may attain 1 meter.[26]

For the deep water part such as any water body, a vertical section through it will reveal two distinct zones: the benthic zone which consisting of the layers closer to the bottom of the water body above the bottom sediment varying from a few cm to several 1000s of meters as in the open ocean. The biotas that dwell in this zone (termed benthos) consist of organisms that have adapted to tolerate low temperatures and high pressure, as well as low oxygen levels found in this zone. Many of them have bottom-dwelling adaptations. Since light cannot penetrate this depth, this zone lacks the ability to photosynthesis as its energy source. The main energy source of this zone consists of organic materials that drift down the upper layers, and this region is dominated by detritivores and scavengers. The second layer is the pelagic zone (Greek meaning: "open sea") which includes the free water column that interacts with the surface layers of a water body directly with the atmosphere. Physical and chemical properties of this zone vary greatly because of the vastness of this area, which extends from the uppermost waters down to the deeper layers near the benthic zone of a water column. As the depth increases, favorable life-sustaining features of the pelagic zone reduce, resulting in a decrease in the biota, as well.

A riverine or aquaforest is an ecosystem with dynamics, structure, and composition (species and habitat diversity and food webs) that depend on river processes of erosion, sediment transport, flooding inundation, and alluvial deposition.[15] These processes are briefly described as follows:

- *Nutrient flow.* Rivers and streams move nutrients and minerals from watershed and lowland areas to the sea and all of the points between. For example, floodplains are extraordinarily productive agricultural areas as rivers transport organic material and sediments over the floodplain soil. The flooding also releases terrestrial nutrients into the aquatic system, generating increased aquatic plant production and food for spawning fish and their young.[27] These also play a role in sustaining dryland gazing systems.
- *Species distribution.* The distribution of both terrestrial and aquatic species can depend on one another. Quite a few African amphibians, reptilian, bird, and mammal species are found near riverine or floodplain habitats, such as buffaloes, certain antelopes, snakes, and cane rats.[28] Similarly, terrestrial wildlife depends on ephemeral rivers in arid areas, for both water and the associated plant life. The distributions of some floodplain tree species depend on animal activity such as hippopotamus that influence the distribution of *Acacia albida* along the Luanga and Zambezi river valleys in Zambia. The hippo and other wildlife eat the seeds, enhancing germination and fostering dispersion.
- *Food webs.* Riparian habitats provide leaves, fruits, fruits, seeds, flowers, and branches to river and streams. A number of freshwater fishes and invertebrates depend on this terrestrial plant material for shelter and food. This is important as Lowel-McConnell[29] remarked that certain lakes (Ntomba and Mai-Ndombe in the Congo basin and the Great Lakes—Lakes Tanganyika, Malawi, and Victoria) are particularly important for the evolution of Africa's endemic fishes, interestingly the richness of the fauna especially in Lake Ntomba appears to be due to food and nutrients from the surrounding forest, the water in the lakes being impoverished in nutrients.

Aquaforests Relationships with Upland Communities

Forested wetlands occur in various climatic, geomorphic, edaphic, hydroperiod, and geographic settings and have similar as well as contrasting features with surrounding upland plant communities with regard to species composition, structure, and ecosystem diversity.

In most savanna regions, narrow bands of vegetation in the vicinity of the river contrast strongly with open forests and savanna woodlands which otherwise dominate the landscape.[5] Numerous studies confirm not only the great richness of these forests over upland forests.[30] It is now accepted that these forests harbor more species of birds[8] and mammals[6] than upland habitats. Roché[31] found bird species richness in riverine flood plains to be twice that in adjacent uplands while[32] documented that the majority of bird species in arid regions of Great Basin (USA) are associated with riparian habitats. Also in the intensive natural landscape of Australia's tropical savanna, species richness and abundance of birds were significantly greater in riparian forests than in the matched riparian forests areas and riparian vegetations allow many species to extend their distributions into lower rainfall areas.[5] The physical and vegetation characteristics of the streamside area differ from those upslope because of frequent inundation, soil saturation, and physical disturbance of streamside vegetation due to flood flows, mass soil movement, etc.[23,33]

On the contrary, along the Helena River and its tributaries and in the Grampians (Western Australia), riparian forests were generally less species rich than adjacent upland communities.[16,34] According to these authors, their results were corroborated by other studies in other parts of Australia and South Africa,[35] but not in Europe. Riparian forests in the Oregon Coast Range are said to have lower tree density than upland stands.[36,37] Reasons for such richness pattern have not been fully investigated.

Light penetration that depends upon topographic position, slope, and vegetation characteristics plays a major role in the ecology of the riparian forest and protection of the light environment is generally seen as an important function of buffers that aims to protect riparian habitat values.[38] The ecotone between the riparian and upland vegetation may be very narrow or gradual depending on the fact that gradients are sharp or not. In the latter case, it could be the site of highest species richness because of the presence of species from both communities.[16,39] In Benin, field observations by Natta (2003)[17] show that riparian forests tree flora is not only more diverse but also has marked differences in structure, abundance, and composition with surrounding upland plant communities (e.g., in Yarpao, Pénéssoulou, Idadjo, etc.).

Benefits of Aquaforests

A large body of work has demonstrated that such forests, especially riparian forests, play a critical role in regulating interactions between terrestrial and aquatic components of temperate zones landscapes;[23,40] however, there have been relatively few studies dedicated to them in the tropics[41] in general. Aquatic ecosystems provide innumerable ecological and economic benefits (see Table 2), for example:

- *Provide habitat corridors.* As tropical forests become more fragmented due to deforestation, riparian forests play a crucial role in providing habitat corridors between forest patches to increase landscape connectivity.[42]
- *Protection.* Aquaforests serve as "green-shields" helping to stabilize shores, prevent erosion, and protect many coastal areas from storms through their damping of wind and wave action and their retention of the bottom sediments by plant roots. Trees lining the sides of rivers help to prevent erosion of the banks, while those lining the shores of lakes play a critical role in the ecological balance of the lakes. Wetlands also control floods by absorbing floodwaters and releasing runoff evenly.
- *Wastewater treatment and reduction of sediment loading.* Forested wetlands provide wastewater treatment playing a central role in water purification (are therefore ecofilters) preventing eutrophication by absorbing nutrients and retaining sediments and toxicants and also aiding in flood control and protection of coastal areas. If they are destroyed, these functions must be achieved artificially at a significant cost!
- *Perenniating water supply.* Aqua forests help to ensure a year-round water supply by their role in groundwater recharging (i.e., the movement of water from the wetlands into the underground aquifers) and discharging (i.e., when underground water moves upward into the wetland). It has been shown that where wetlands have been lost the hydrological cycle is affected and water supply during the dry season has been lost.[25]
- *As fisheries and other aquatic fauna habitats enhance fishery production.* Forested wetlands including mangrove forests serve as feeding, breeding, hatching, and nursery grounds for numerous commercially important fin and shell fish species. African river basins with associated wetlands are highly productive ecosystems with an estimated 40% of fish in Africa coming from riverine and floodplain fisheries.[43] Fish also represent an irreplaceable source of protein (about 25%) for a continent in which there is often a widespread and chronic shortage of animal protein. Most of African lakes and rivers and a number of African wetlands support tribes whose livelihood has historically depended upon fishing.
- *Aquaforest products and resources.* Wetlands are the source of a variety of products which can be harvested sustainably such as fish, fodder, timber, nontimber forest products (e.g., resins and medicines), agricultural crops, and wildlife resources (e.g., meat, skins, honey, and both bird and turtle eggs). Worldwide, people use the products of mangroves and the mangrove environment for subsistence.
- *Energy generation.* Most aquaforests provide vast sources of energy, especially mangroves provide fuel wood for cooking and fish preservation.
- *Tourisms and other cultural values.* Aquaforests and wetlands offer great potential for tourism because they are important to both terrestrial and aquatic life. In one spot, tourists can view both large mammals and an astonishing abundance of waterfowl.
- *Climate change indicators and mitigation.* Wetlands often provide visible indicators of climate change when water courses (streams, rivers, and lakes) begin to dry off. They are also considerable areas for carbon capture and storage. The mangrove ecosystems have the capacity to store great amounts of organic carbon; significant enrichments of organic carbon have been reported even at several meters of depth.[44,45]

Threats to Aquaforests, Drivers, and Consequences of Degradation and Loss

Despite the growing recognition of the ecological, hydrologic, biogeochemical, sociocultural, economic, and aesthetic importance of these forests, they are degraded or lost at an alarming rate due to factors mostly anthropogenic:

- Pollution

Pollution from farms, cities, and factories can affect adjacent aquatic areas as well as areas further downstream. Agricultural threats include both pesticide and fertilizer runoff. The chemicals in pesticides runoff become more concentrated and toxic as they work their way up the food chain. They accumulate in the bodies of fish and other higher-level organisms. Fertilizer runoff increases the nutrient loading in waters, thereby causing eutrophication and algal blooms. This impacts fisheries and drinking water and reduces biological diversity.[46] Municipal pollution

Table 2 Ecosystem services in aqua-forest habitats

Provisioning	
Food	Production of fish, wild game, fruits, and grains
Fresh water	Storage and retention of water for domestic, industrial, and agricultural use
Fibre & fuel	Production of timber and fuel wood, peat, fodder
Biochemical	Extraction of pharmaceuticals, natural products including salts, medicines, and other materials from biota
Genetic materials	Genes for resistance to plant pathogens, ornamental species, etc.
Others	Extraction of minerals and other natural products, space for ports/transportation
Regulating	
Climate regulation	Source of and sink for greenhouse gases; influence local and regional temperature, precipitation, and other climatic processes
Water regulation (hydrological flows)	Groundwater recharge/discharge
Water purification & waste treatement	Retention, recovery and removal/filtration of excess nutrients, other pollutants, and waste water
Erosion regulation	Retention of soils and sediments
Natural hazard regulation	Flood control, storm protection
Pollination	Habitats for pollinators
Others	Waste disposal
Cultural	
Spiritual & inspirational	Source of inspiration; many religions attach spiritual and religious values to aspects of wetland ecosystems
Recreational	Opportunities for recreational and touristic activities
Aesthetic	Many people find beauty or aesthetic value in aspects of wetland ecosystems
Educational	Opportunities for formal and informal education and training
Supporting	
Soil formation	Sediment retention and accumulation of organic matter
Nutrient cycling	Storage, recycling, processing and acquisition of nutrients
Option use values	
	Future unknown and speculative benefits. Currently unknown potential future uses of aquaforest biodiversity

Source: Adapted from Millennium Ecosystem Assessment Ecosystems and Human Well Being (2005) and Forest Trends (2010).

also increases eutrophication as the greater the number of people living along a river, the greater the amount of nitrates in the river.[47] Industrial pollution mainly threatens coastal waters as the oceans are utilized by most nations worldwide as a vast dumping ground for waste. It is estimated that more than 90% of all chemical refuse and other material entering coastal waters remain there in sediments, wetlands, fringing reefs, and other coastal ecosystems. The most widespread and serious chemical pollutants are hydrogenated hydrocarbons (such as pesticides, herbicides, and PDBs), heavy metals, petroleum products, and fertilizers.[48] These substances can cause tumors and disease in estuarine and coastal fish. In addition, plastic and other debris, such as pieces of nets, entangle and kill a variety of aquatic animals.

- Aquaforest site conversion

For agricultural production

The rapidly increasing human population in Africa and other developing nations puts heavy demand on increased food production, leading to agricultural conversion of critical wetlands.

Conversion not only damages the environment but seldom works mainly due to the development of acid sulfate soil conditions. Many mangrove areas have soil containing large amounts of pyrate sulfur that when exposed to air, oxidize to release sulphuric acid. The soil then becomes extremely acidic and vey high in soluble salts, a condition which often leads to nutrient and fertilizer problems resulting in the impairment and even failure of most crops especially rice, coconut, etc. Even if drained land does produce reasonable yields for several years, long-term sustainability on the site is virtually impossible because of the resulting environmental degradation. Wetland drainage can have additional effects of disrupting the water supply of nearby towns affecting the quality of water and altering the microclimate.

For aquaculture production

Improperly sited aquaculture projects can greatly damage freshwater ecosystems. If cultured stocks escape to the wild, they can affect native gene pools. In addition, the potential spread of pests and diseases poses a serious risk to wild stocks since the spread of farmed species happens all too often. From experience, Bartley and Minchin (1995)[49] noted that complete containment of exotic species in aquaculture facilities is nearly impossible. Unfortunately, economic analyses of aquaculture often neglect the economic cost of conversion. For example, for mangrove conversion to aquaculture ponds, Dixon (1989)[50] noted that usually only the marketable mangrove forestry products are considered, ignoring the economic benefit mangroves provide to fisheries both within mangroves and in nearly coastal and estuarine ecosystems. Further, this neglects the fact that some products or services such as nutrient flows to estuaries do not have market prices. Finally, the decision as to whether or not mangroves should be converted is made by comparing the minimum partial estimate with the total expected benefit from conversion. No wonder mangroves are being lost as such a rapid pace. In fact, large areas of mangrove have been cleared worldwide for the establishment of shrimp mariculture. In addition to the destruction of mangroves, shrimp farming reduces the area of estuaries due to the diking of sand and mudflats to create ponds and water intake channels and changes estuarine flow by channelization and by controlling the flow of water into the ponds. Waterfowl numbers decline due to habitat degradation and in some cases through routine shooting of those birds believed to eat shrimp.

- Deforestation from within and adjoining areas

Due to grazing, dams, embankments or other urban infrastructural development projects

Deforestation damages both freshwater and marine ecosystems. The deforestation of riparian areas due to grazing, dams, embankments, or other urban infrastructural development projects can cause stream temperatures to fluctuate widely, reduce dissolved oxygen concentrations and reduce the contribution of terrestrial nutrients, including leaf letter to aquatic food webs. The erosion that results from deforestation can change the shape of a stream or river harming the quality of spawning grounds and habitat. Sedimentation can also smother spawning grounds and reduce habitat complexity of rocky substrates. Sediments harm freshwater food webs in a variety of ways. They reduce the rate of light penetration and hence rate of photosynthesis. They cover benthic algae, reducing foraging efficiency in herbivorous fish. They reduce the nutritional value of detritus (organic waste).[51] They also interfere with the feeding apparatus of filter-feeding organisms. Sedimentary particles themselves can abrade the bodies of aquatic organisms. Fortunately, maintaining srip of riparian forest can do much to minimize these effects.

Wood bioenergy issues

Vast areas of aquaforests are also devastated to meet rising energy demands in both rural and urban areas especially in developing countries where wood is a major source of energy. Throughout the coasts of West and Central Africa, mangrove forests are largely cut to preserve fish catch through smoking.[52–54]

Fire

Riparian plants are not dependent upon fire for renewal, but fire can influence the composition and structure of riparian ecosystems,[55] in particular the understory in combination or not with other factors such as flooding.[56] In Central New Mexico (USA), the suppression of flooding along RFs has also increased forest floor litter and woody debris which may have contributed to the increased frequency and severity of fires.[21]

- Water diversion

Dams and water diversions are known to modify surface flow rates, flood periodicity, and sediment and nutrient transport often to the detriment of riparian plants.[20] When channeling occurs within riparian systems, removal of sediments and nutrients from surface runoff is less effective.[57] Large dams are constructed for various reasons such as to increase agricultural production from large-scale irrigation, generate hydroelectricity, control floods, and assure water supplies. The benefits of large dams are usually not as large as expected and their adverse effects are often severely underestimated. These include the following:

- *Habitat degradation and reduced biodiversity.* Dams directly impact riverine channel characteristics and habitat quality and greatly modify the river's influence downstream. They alter the timing and volume of river flow which reduces the sediment and nutrient transport downstream, alters temperature and salinity and at times even increases the river's acidity due to plant decomposition. Mangrove forest systems are influenced by reduction of freshwater inflows. The loss of such inflows affects the three most important factors in maintaining mangrove ecosystems: a sufficient amount of water, a sufficient supply of nutrients, and stability of the substrate. Mangrove forests degrade without periodic pulses of freshwater. They can also be degraded by saline fronts caused by dams. Channel mouth closure due to sedimentation is a growing threat to lagoonal mangroves in West and Central Africa especially in Ghana.[58]
- *Declining fisheries and long-established indigenous patterns of floodplain agriculture including livestock production.*

  Studies have demonstrated that fish yield is directly related to river and catchment size with rivers with extensive floodplains having a higher yield than those with smaller floodplains.[59] In addition to degrading riverine habitat, dams directly block fish migration, both of marine fish that spawn in rivers and riverine fish that spawn in marine realms. Dams can also disrupt the hydrological and chemical cues needed to introduce migratory and spawning behavior in fish. Finally, the profound changes caused by dams exacerbate the spread of exotic fish species. On the impact of dams on floodplain fisheries, construction of the Kainji Dam in Nigeria resulted in the loss of 50% of the fish catches in lands extending to 200 km downstream of the dam.[60] Dams also impact lacustrine and coastal fisheries due to irregular freshwater input.

- Encroachments from urbanization

Many aquaforests areas have been encroached by increased urbanization from burgeoning population pressures and so are lost at a rapid rate. Urbanization on lands adjacent to intact riparian woodland has substantial impacts on riparian bird species richness, density, and community composition.[61,62] Likewise, invasion of the naturalized shrubs potentially alters competitive hierarchies and disturbance regimes in riparian systems.[56]

- Climate change

Wetlands are one of the most vulnerable ecosystems to global warming. Effects are more visible when water courses (streams, rivers, and lakes) begin to dry off. Although considerable areas for carbon capture and storage can also rapidly emit carbon dioxide when degraded or destroyed.

- *Policy deficiencies, overlapping jurisdiction, and inadequate planning.* Above all else, aquatic systems require integrated planning. Too often, however, governmental agencies consider aquatic systems only for a single purpose rather than for multiple uses benefiting a variety of sectors. A number of government agencies (fisheries, tourism, urban sanitation and water, agriculture, and forestry) can have responsibilities over aquatic areas and resources leading to confusion, inertia, and ineffective planning. Overlapping jurisdiction can occur horizontally (within the same level of government) and vertically (across different levels of government from local authorities to regional). In the case of wetlands, for example, despite increasing efforts to conserve these areas, many wetlands are lost because of competing government priorities. Governments can have a stated commitment to wetlands conservation while at same time their national agricultural policy favors wetlands drainage!

  Subsidies according to Shumway (1999)[46] often lead to waste and misuse of aquatic resources. These include agricultural subsidies for pesticides, fertilizers, and water, forestry subsidies which can exacerbate deforestation and erosion in marginal and vulnerable areas and tax relief for overseas investors with little incentives to sustainably manage a given resource.
- *Introduction of exotic species.* According to Shumway (1999)[46] citing other sources, invasion of exotics is believed to be the second greatest threat to global biodiversity after habitat loss. Islands, coastal estuaries and lakes are at particular risk. Some are deliberately introduced with good intentions of filling apparently

"empty" niches and producing more big fish, to control unwanted organisms to improve sport fishing and to control fish production in ponds through the introduction of predators and also with the introduction of the now extensive West-Central African invasive mangrove palm (*Nypa fruticans*) from Asia to Nigeria in 1906. Others have been introduced accidentally through translocation especially as most African river systems are connected by tunnels, pipes, and canals for the sake of development infrastructure. In Africa, two South American species have caused considerable havoc to wetlands systems: water hyacinth (*Eichhornia crassipes*) and *Salvinia molesta*. Both species rapidly colonize water bodies, forming a dense floating mat of interlocking plants.[23] These species crowd out native vegetation, reduce light penetration, limit water column mixing, and increase detrital inputs. Kaufman and Ochumba (1993)[63] report that a single water hyacinth plant can produce 140 million daughter plants per year enough to cover 140 ha with a weight of 28,000 tons, the seeds can surviving for 30 years in mud, making the plants almost impossible to eradicate completely.

- *Over exploitation of aquaforest resources.* Overfishing and postharvest losses due to lack of infrastructure for handling/processing facilities which causes waste. In most African countries between 15–25% the catch is lost due to spoilage, wastage at the time of capture, insect infestation, or improper handling and storage. Some countries have postharvest losses of up to 50% particularly those where smoking and drying are the normal preservation methods.[46]

MANAGING AQUAFORESTS (AQUAFORESTRY OR AQUATIC FORESTRY)

Definitions and Scope of Aquaforestry

Given the aforementioned ecological, hydrologic, biogeochemical, sociocultural, economic, and aesthetic importance of aquaforests and their threats to degradation and loss, there is a need to manage these forests sustainably to perpetuate their goods and services to mankind through desired land-use management practices of aquaforestry. We therefore use the term "aquaforestry" to include all sustainable land-use management practices aimed at perpetuating the ecological services of aquaforests within the framework of the integrated basin approach. This is not much different from the traditional forestry practices but needs to be highly ecological, integrative, and participative in all its dimensions to make it produce the desired land-use management change for aquaforests. The challenging task of sustainable aquaforest management is being confronted by the need to assess these resources (aquaforest inventory and mensuration) the outcome of which can produce management practices that may range from conservation, sustainable utilization to regeneration/restoration of degraded habitats.

Aquaforest Inventory and Mensuration

Effective management strategies must be devised to preserve and restore riparian corridors[64] and shores[65] which rely on accurate understanding of the structure and dynamics of aquaforests communities.[66] Forest inventories and mensuration is a means to obtain reliable information on the stock and dynamics of forest resources through various remote sensing and ground survey methods. Because of diversity of landscapes and tree forms in aquaforests, the traditional forest inventory and mensuration procedures[67,68] need to be adapted to aquaforests and especially for mangrove vegetation[69–71] with characteristic stilt roots.

Aquaforest Conservation

Cost-effective conservation of biodiversity requires that maximum biodiversity should be protected in a minimum area.[72] In proportion to their area within a watershed, aquaforests perform more ecological and productive functions than do adjacent uplands.[4] The vital ecological, hydrological, and biogeochemical vital functions of these forests bordering waterways can be protected through conservation that can preserve a large range of plants, animals, water sources, soils, and watersheds. In Amazon and the Congo basins, numerous streams and rivers provide huge potentials for increasing the conservation value of deforested and fragmented landscape through the protection of linear remnants along waterways.[73] Not only do they constitute a natural habitat or the last refuge for many species, but they also lodge many endemic species and extinction-menaced species[74] and usually act as routes for movement of terrestrial plants and animals across the landscape.[42] It should be noted that simply protecting aquaforests in a buffer zone may not be adequate to ensure their existence in the long term.[75] Instead, the management of these forests must be a component of good watershed or landscape management. Therefore, awareness has to be raised to various stakeholders of such forests as unique physical and natural systems in their own right and warranting special management and protection. Integrated management of these forests that optimize their values as habitat for native plants and animals requires planning and acting with all stakeholders at both site-specific and watershed levels. This can also be achieved within the framework of the emerging engagement of private sector and other land-use developers in various payments for ecosystem services arrangements.

Sustainable Utilization of Aquaforests

It is important that all human interactions within and outside aquaforests should embrace the sustainable exploitation concept. This range from consumptive uses (harvests of timber and nontimber forest resources) to nonconsumptive uses (cultural, tourisms, etc.). Appropriate technologies should be employed, especially the reduced impact logging approach.[76] In the Congo basin logging practice,[77] aquaforests are preserved in logging concessions and classified under sensitive habitats barred from logging activities while efficient wood energy utilization technology is becoming widespread in Africa to reduce the traditional open-cooking ovens. In the mangrove zones of central Africa such improved fish smoking ovens or cook stoves could reduce as much as 60% wood used.[78]

Regenerating and Restoring Degraded Aquaforests

Aquasilviculture

Where aquaforests are degraded or lost altogether, they may be regenerated by natural regeneration methods and/or artificial regeneration methods through replanting (reforestation) or planting (afforestation) where they did not exist before. The silvicultural techniques will differ from one type of forest to the other and will require appropriate adaptation to the edaphic and hydrological peculiarities of aquaforests. For example, a *Rhizophora* dominated mangrove can be regenerated only through viviparous propagules not through cuttings as this widely spread mangrove genus in West-Central Africa does not coppice. Lagoon, estuaries, or coastal shores are opened to facilitate tidal movements to facilitate the establishment of mangrove stands.

Wetlands farming

The natural rise and fall of rivers and streams can be used to advantage in farming. IUCN notes that if properly managed, natural wetland agriculture (i.e., agriculture that takes advantage of the normal flooding cycle rather than draining the site) can yield substantial benefits to rural communities. Natural wetlands farming occurs in the inland Delta in Mali,[79] the flood plains of the larger rivers in the Sudan and Sahel Savanna zones and estuarine swamps of the Upper Guinea coast. In the large floodplains, complex systems of flood-advance and flood-retreat agriculture have been developed.[25]

CONCLUSION

Aquatic forest ecosystems provide critical services for man and wildlife. They are important both to flora and to fauna living in water as well as those living on land. Aquatic biodiversity though threatened is still abundant especially in Africa with one of the world's richest treasures and equally importantly to African nations and peoples. We have the chance to conserve some of this wealth if action is taken now and alternatives are provided. It is not too late. Integrated approaches to conservation and development are necessary. The goal of aquaforestry or aquatic forestry is to seek to perpetuate aquaforests for sustained production of ecological services derived from this special and fragile forested habitats.

ACKNOWLEDGMENTS

The authors thanks the students and lecturers of the Institute of Fisheries and Aquatic Sciences of the University of Douala at Yabassi, Cameroon where aquaforests and aquaforestry have been taught as a series of courses and who greatly contributed to shaping the ideas presented in this entry. The authors also acknowledge support of the Cameroon Wildlife Conservation Society (CWCS) in collaboration with the Cameroon Ministry of Forestry & Wildlife (MINFOF) and Ministry of Environment, Nature Protection and Sustainable Development (MINEPDED) and funding partners through its Coastal forests and Mangrove Conservation Programme, Cameroon. We also thank Mr. Lissouck Benard and Ms. Jeanette Wiwa of Government High School Mouanko for the drawings in Fig. 3.

Life forms (LF) following Raunkaier (1934), Schnell (1971) and Keay & Hepper (1954–1972) are

- Phanerophytes (Ph): megaphanerophytes (MPh: >30m), mesophanerophytes (mPh: 8–29m), microphanerophytes (mph: 2–7 m), and nanophanerophytes (nph <2m)
- Therophytes (Th: plant survive as seeds)
- Hemicryptophytes (Hc: perennating buds at soil surface)
- Chamaephytes (Ch: perennating buds near soil surface)
- Lianas (L)
- Cryptophytes (Ch): Geophytes (Ge: perennating buds underground), Helophytes (Hel: perennating buds underground in marshes), hydrophytes (hyd: perennating buds in water)
- Epiphytes (Ep)
- Parasites (Par)

Phytogeographic types (PT) following White (1986) and Keay & Hepper (1954–1972) are

- Species widely distributed in the tropics (Cosmopolitan – cosmo, pantropical – pan, Afro-American – AA and Paleotropical – Paleo)
- Species widely distributed in Africa (Tropical Africa – TA, Pluri Regional in Africa – PRA)
- Regional species in Africa (Sudanian – S, Guinean – G, Sudano-Guinean – SG, Sudano-Zambesian – SZ, Guineo Congolian – GC)

Appendix 1 Typical aquaforest higher plant species of West and Central Africa

						Habitat					
						Forest zone			Savanna zone		
No.	Species	Family	Habit	Life forms*	Geo affinity*	River banks/ fringing forests	Fresh water swamps	Tidal swamps/mangroves	River banks/ fringing forests	Fresh water swamps	Tidal swamps/ mangroves
1	*Acacia ataxacantha*	Mimosaceae	tree	mph	S				x		
2	*Aeschynomene elaphroxylon*	Papilionaceae	tree	mph	SG	x	x		x	x	
3	*Afzelia africana*	Caesalpiniaceae	tree	mPh	SG	x			x		
4	*Albizia malacophylla*	Mimosaceae	tree	mPh	SG				x		
5	*Albizia zygia*	Mimosaceae	tree	mPh	SG				x		
6	*Alchornea cordifolia*	Euphorbiaceae	tree	mph	TA	x	x		x	x	
7	*Allanblackia floribunda*	Guittiferae	tree	mPh	GC	x					
8	*Allophylus africanus*	Sapindaceae	shrub	nph	Pan	x	x	x	x	x	x
9	*Alstonia boonei*	Apocynaceae	tree	mPh	GC	x	x				
10	*Alstonia congensis*	Apocynaceae	tree	MPh	GC	x	x				
11	*Ancistrophyllum* sp.	Palmae	liana	mph	GC		x				
12	*Ancylobotrys amoena*	Apocynaceae	liana	nph	SG				x		
13	*Andira inermis*	Papilionaceae	tree	mph	SG				x	x	
14	*Annona senegalensis*	Annonaceae	shrub	nph	TA	x			x		
15	*Anogeissus leiocarpus*	Combretaceae	tree	mph	SG	x			x		
16	*Anthocleista nobilis*	Loganiaceae	tree	mPh	GC		x				
17	*Anthocleista schweinfurthii*	Loganiaceae	tree	mPh	GC		x				
18	*Anthocleista vogelii*	Loganiaceae	tree	mPh	SG				x	x	
19	*Antiaris africana*	Moraceae	tree	mPh	SG				x		
20	*Antidesma venosum*	Euphorbiaceae	tree	mph	TA	x	x		x	x	
21	*Aphania sengalensis*	Sapindaceae	tree	mph	TA	x	x	x	x	x	x
22	*Avicennia germinans*	Avicenniaceae	tree	mPh	GC			x			x
23	*Berlinia grandiflora*	Caesalpiniaceae	tree	mPh	GC	x			x		
24	*Berlinia grandiflora*	Caesalpiniaceae	tree	mPh	SG				x		
25	*Borassus aethiopum*	Palmae	palm	mPh	TA				x		
26	*Bosqueia angolense*	Moraceae	tree	MPh	GC	x			x		
27	*Brachystegia eurycoma*	Caesalpiniaceae	tree	MPh	GC	x					
28	*Breonadia salicina*	Rubiaceae	tree	mph	SG				x		

(*Continued*)

Appendix 1 (*Continued*) Typical aquaforest higher plant species of West and Central Africa

						Habitat					
						Forest zone			Savanna zone		
No.	Species	Family	Habit	Life forms*	Geo affinity*	River banks/fringing forests	Fresh water swamps	Tidal swamps/mangroves	River banks/fringing forests	Fresh water swamps	Tidal swamps/mangroves
29	*Bridelia micrantha*	Euphorbiaceae	tree	mPh	SG				x		
30	*Calamus* sp.	Palmae	liana	mph	GC		x				
31	*Carapa procera*	Meliaceae	tree	mPh	GC	x					
32	*Cassia sieberiana*	Caesalpiniaceae	tree	mph	SG				x		
33	*Ceiba pentandra*	Bombacaceae	tree	mPh	SG				x		
34	*Celtis integrifolia*	Ulmaceae	tree	mPh	SG				x		
35	*Clausena ainisata*	Rutaceae	shrub	nph	SG				x		
36	*Cleistopholis glauca*	Annonaceae	tree	mPh	GC	x					
37	*Cleistopholis patens*	Annonaceae	tree	mPh	GC	x	x				
38	*Clerodendrum thyrsoideum*	Verbenaceae	liana	mph	SG						
39	*Cola laurifolia*	Sterculiaceae	tree	mPh	GC	x			x		
40	*Combretum paniculatum*	Combretaceae	shrub	nph	TA	x			x		
41	*Combretum tomentosum*	Combretaceae	tree	mPh	SG	x			x		
42	*Cordia myxa*	Boraginaceae	tree	mph	SG				x		
43	*Cordia sinensis*	Boraginaceae	tree	mph	Pan	x	x		x	x	
44	*Crataeva adansonii*	Capparaceae	tree	nph	TA	x	x		x	x	
45	*Cynometra megalophylla*	Caesalpiniaceae	tree	mPh	GC	x	x				
46	*Cynometra sanagaensis*	Caesalpiniaceae	tree	mPh	GC	x	x				
47	*Cynometra vogelii*	Caesalpiniaceae	tree	mph	SG				x		
48	*Dalbergia melanoxylon*	Papilionaceae	tree	mph	Pan	x			x		
49	*Dalbergia sissoo*	Papilionaceae	tree	mph	Pan	x			x		
50	*Detarium senegalense*	Caesalpiniaceae	tree	mph	SG				x		
51	*Dialium dingklagei*	Caesalpiniaceae	tree	mPh	GC	x	x				
52	*Dialium guineense*	Caesalpiniaceae	tree	mph	SG	x			x		
53	*Diospyros mespiliformis*	Ebenaceae	tree	mPh	SG				x		
54	*Dracaena arborea*	Agavaceae	tree	mPh	GC	x	x				
55	*Endodesmia calophylloides*	Hypericaceae	tree	mPh	GC		x				
56	*Entada manii*	Mimosaceae	liana	mph	GC	x					
57	*Erythrophleum suaveolens*	Caesalpiniaceae	tree	mPh	SG	x			x		

58	*Eugenia nigerina*	Myrtaceae	shrub	nph	SG				x		
59	*Ficus abutilifolia*	Moraceae	shrub	nph	GC				x		
60	*Ficus asperifolia*	Moraceae	shrub	nph	GC	x					
61	*Ficus capreifolia*	Moraceae	shrub	nph	GC				x		
62	*Ficus dicranostyla*	Moraceae	shrub	nph	GC				x		
63	*Ficus exasperata*	Moraceae	shrub	nph	GC				x		
64	*Ficus glumosa*	Moraceae	shrub	nph	GC				x		
65	*Ficus natalensis*	Moraceae	shrub	nph	GC				x		
66	*Ficus ovata*	Moraceae	shrub	nph	GC				x		
67	*Ficus polita*	Moraceae	shrub	nph	GC				x		
68	*Ficus sur*	Moraceae	shrub	nph	GC				x		
69	*Ficus sycomorus*	Moraceae	shrub	nph	GC				x		
70	*Ficus thonningii*	Moraceae	shrub	nph	GC				x		
71	*Ficus trichopoda*	Moraceae	tree	mPh	GC				x	x	
72	*Ficus vallis-choudae*	Moraceae	tree	mPh	GC				x		
73	*Ficus vogelii*	Moraceae	tree	mPh	GC				x		
74	*Flacourtia indica*	Flacourtiaceae	tree	mph	SG				x		
75	*Gambeya africana*	Sapotaceae	tree	MPh	GC	x					
76	*Garcinia livingstonei*	Guittiferae	tree	mph	SG				x		
77	*Garcinia ovalifolia*	Guittiferae	tree	mph	SG				x		
78	*Gardenia imperialis*	Rubiaceae	tree	mph	GC	x	x				
79	*Guibourtia capallifera*	Caesalpiniaceae	tree	mph	SG				x		
80	*Guibourtia demeusei*	Caesalpiniaceae	tree	MPh	GC	x	x				
81	*Hippocratea africana*	Hippocrateaceae	liana	mph	SG				x		
82	*Holarrhena floribunda*	Apocynaceae	tree	mph	GC	x			x		
83	*Ipomoea carnea*	Convolvulaceae	shrub	nph	Pan	x	x		x	x	
84	*Irvingia smithii*	Simaroubaceae	tree	mPh	PRA	x					
85	*Jasminum dichotomum*	Oleaceae	liana	mph	SG				x		
86	*Kaya senegalensis*	Meliaceae	tree	MPh	SG				x		
87	*Keayodendron bridelioides*	Euphorbiaceae	tree	mPh	GC	x	x				
88	*Keetia cornelia*	Rubiaceae	shrub	nph	SG				x		
89	*Laguncularia racemosa*	Combretaceae	tree	mPh	G			x			x

(Continued)

Appendix 1 (*Continued*) Typical aquaforest higher plant species of West and Central Africa

						Habitat					
						Forest zone			Savanna zone		
No.	Species	Family	Habit	Life forms*	Geo affinity*	River banks/ fringing forests	Fresh water swamps	Tidal swamps/mangroves	River banks/ fringing forests	Fresh water swamps	Tidal swamps/ mangroves
90	*Lecaniodiscus cupanioides*	Sapindaceae	shrub	nph	SG				x		
91	*Leptoderris brachyptera*	Papilionaceae	liana	LmPh	GC	x					
92	*Loeseneriella africana*	Hippocrateaceae	liana	mPh	SG				x		
93	*Lonchocarpus cyanescens*	Papilionaceae	shrub	mph	SG				x		
94	*Lonchocarpus griffonianus*	Papilionaceae	tree	mPh	GC		x				
95	*Lonchocarpus sericeus*	Papilionaceae	tree	mPh	SG				x		
96	*Lophira alata*	Ochnaceae	tree	MPh	SG		x				
97	*Macrosphyra longistyla*	Rubiaceae	shrub	nph	SG				x		
98	*Malacantha alnifolia*	Sapotaceae	tree	nph	TA				x		
99	*Manilkara multinervis*	Sapotaceae	tree	mPh	TA	x					
100	*Militia excelsa*	Moraceae	tree	mPh	SG				x		
101	*Mimosa pigra*	Mimosaceae	shrub	nph	Pan	x					
102	*Mitragyna ciliata*	Rubiaceae	tree	MPh	GC	x	x				
103	*Mitragyna inermis*	Rubiaceae	shrub	nph	SG				x		
104	*Mitragyna stipulosa*	Rubiaceae	tree	mPh	SG		x				
105	*Morelia senegalensis*	Rubiaceae	shrub	mph	SG	x	x		x		
106	*Napoleonaea vogelii*	Lecythidaceae	tree	mph	G	x					
107	*Nauclea pobeguinii*	Rubiaceae	tree	mph	G		x				
108	*Neocarya macrophylla*	Chrysobalanaceae	tree	nph	SG				x		
109	*Olax subsorpioides*	Olacacea	shrub	nph	SG				x		
110	*Oncoba spinosa*	Flacourtiaceae	tree	mph	SG				x		
111	*Ouratea glaberrima*	Ochnaceae	shrub	nph	GC	x					
112	*Oxytenanthera abyssinica*	Poaceae	bamboo	mPh	SG				x		
113	*Pachystela pobeguiniana*	Sapotaceae	shrub	nph	TA				x		
114	*Parinari congensis*	Chrysobalanaceae	tree	MPh	GC	x					
115	*Paullinia pinnata*	Sapindaceae	liana	nph	TA				x		
116	*Pericopsis laxiflora*	Papilionaceae	tree	mph	SG				x		
117	*Phoenix reclinata*	Palmae	palm	mPh	SG	x	x		x	x	
118	*Phyllanthus muellerianus*	Euphorbiaceae	shrub	nph	SG				x		

119	*Phyllanthus reticulatus*	Euphorbiaceae	liana	nph	SG	x			x		
120	*Phyllanthus welwitschianus*	Euphorbiaceae	shrub	nph	SG	x			x		
121	*Plagiosiphon longitubus*	Caesalpiniaceae	tree	mPh	GC	x	x				
122	*Psychotria calva*	Rubiaceae	shrub	nph	GC	x					
123	*Pterocarpus santalinoides*	Papilionaceae	tree	mPh	PRA	x					
124	*Raphia hockeri/vinifera*	Palmae	palm	mPh	SG	x	x		x	x	
125	*Raphia sudanica*	Palmae	palm	mPh	SG				x	x	
126	*Rhizophora harrisonii*	Rhizophoraceae	tree	mPh	TA			x			x
127	*Rhizophora mangle*	Rhizophoraceae	shrub	nph	AA			x			x
128	*Rhizophora racemosa*	Rhizophoraceae	tree	mPh	TA			x			x
129	*Ricinodendron heudelotii*	Euphorbiaceae	tree	mPh	GC	x					
130	*Rinorea* spp.	Violaceae	liana	mPh	SG				x		
131	*Rotula aquatica*	Boraginaceae	shrub	nph	Pan	x	x	x	x	x	x
132	*Rourea thomsonii*	Connaraceae	shrub	mph	SG				x	x	
133	*Rytigynia senegalensis*	Rubiaceae	shrub	nph	SG				x		
134	*Saba comorensis*	Apocynaceae	liana	mPh	SG	x			x		
135	*Saba senegalensis*	Apocynaceae	liana	mPh	SG	x					
136	*Sacoglottis gabonensis*	Humiriaceae	tree	mPh	GC		x				
137	*Salix subserrata*	Salicaceae	shrub	nph	Pan	x	x	x	x	x	x
138	*Santaloides afzelii*	Connaraceae	shrub	nph	SG				x		
139	*Scyphocephalium mannii*	Myristericacee	tree	mPh	GC		x				
140	*Sesbania sesban*	Papilionaceae	shrub	nph	Pan	x	x		x	x	
141	*Sorindeia grandifolia*	Anacardiaceae	tree	mPh	GC	x					
142	*Spathodea campanulata*	Bignoniaceae	tree	mPh	GC	x					
143	*Spathodea tomentosa*	Bignoniaceae	tree	mPh	SG				x		
144	*Spondianthus preussii*	Euphorbiaceae	tree	mPh	GC	x	x				
145	*Spondias mombin*	Anacardiaceae	tree	mPh	GC	x			x		

(*Continued*)

Appendix 1 (*Continued*) Typical aquaforest higher plant species of West and Central Africa

						Habitat					
						Forest zone			Savanna zone		
No.	Species	Family	Habit	Life forms*	Geo affinity*	River banks/ fringing forests	Fresh water swamps	Tidal swamps/mangroves	River banks/ fringing forests	Fresh water swamps	Tidal swamps/ mangroves
146	*Sterculia tragacantha*	Sterculiaceae	tree	mPh	SG	x	x		x	x	
147	*Strombosia grandifolia*	Olacacea	tree	mPh	GC	x					
148	*Strophanthus sarmentosus*	Apocynaceae	liana	mPh	SG	x			x		
149	*Symphonia globulifera*	Guittiferae	tree	mPh	GC		x				
150	*Synsepalum brevipes*	Sapotaceae	tree	mph	GC	x					
151	*Syzygium guineense*	Myrtaceae	tree	mPh	TA	x	x				
152	*Taccazea apiculata*	Aslepiadaceae	liana	mPh	GC	x					
153	*Talbotiella batesii*	Caesalpiniaceae	tree	mPh	GC	x	x				
154	*Terminalia glaucescens*	Combretaceae	tree	mPh	SG				x		
155	*Terminalia laxiflora*	Combretaceae	tree	mPh	SG				x		
156	*Tetracera alnifolia*	Dilleniaceae	liana	mph	SG				x		
157	*Uapaca guineensis*	Euphorbiaceae	tree	mPh	GC	x					
158	*Uapaca heudelotii*	Euphorbiaceae	tree	mPh	GC	x					
159	*Uapaca togoensis*	Euphorbiaceae	tree	mPh	SG				x		
160	*Uvaria chamae*	Annonaceae	shrub	nph	SG				x		
161	*Vitex chrysocarpa*	Verbenaceae	tree	mPh	SG				x		
162	*Vitex doniana*	Verbenaceae	tree	mPh	SG				x		
163	*Vitex madiensis*	Verbenaceae	tree	mPh	SG				x		
164	*Voacanga africana*	Apocynaceae	tree	mPh	SG				x		
165	*Xylopia aethiopica*	Annonaceae	tree	mPh	GC	x			x		
166	*Xylopia aurantiiodora*	Annonaceae	tree	mph	GC	x	x				
167	*Zanthoxylum zanthoxyloides*	Rutaceae	shrub	nph	SG				x		
168	*Ziziphus spina-christi*	Rhamnaceae	shrub	nph	SG				x		
	Total					**75**	**43**	**9**	**109**	**19**	**9**
	As % of total					**44.6**	**25.6**	**5.4**	**64.9**	**11.3**	**5.4**

Source: Compiled from Arbornnier, 2004; Thikakul, 1985; Gledhill, 1972; Etukudo, et al., 1994.

Appendix 2 Typical aquaforest higher fauna species of West and Central Africa

No.	Scientific name	Family	Common name	Type of aquaforests		
				Fringing forest	Fresh water	Mangroves/ Coastal
1	*Albula vulpes*	Albulidae	Sea banana	0	1	1
2	*Alectis alexandrinus*	Carangidae	Jack fish	0	0	1
3	*Alestes* sp.	Alestidae	Alestes	0	1	1
4	*Alopochen aegyptiacus*	Anantidae	Ouette of Egypt	0	1	0
5	*Antennarius striatus*	Antenarridae	Frog fish	0	1	1
6	*Aplocheilichthys spilauchen*	Poeciliidae	Poeciliids	1	0	0
7	*Aristichthys nobilis*	Cyprinidae	Big carp head	0	1	0
8	*Arius* spp.	Bagridae	Cat fish	0	1	1
9	*Barbus* sp.	Gobeiidae	River barber	1	1	1
10	*Batrachoides liberiensis*	Batrechoididae	Toadfish	0	0	1
11	*Brachydeuterus auritus*	Haemulidae	Grunds	0	0	1
12	*Brycinus longipinnus*	Alestidae	African tetras	1	1	1
13	*Bufo regularis*	Bufonidae	Toad	1	0	0
14	*Carassius auratus*	Cyprinidae	Gilded (bronzed) carasin	0	1	0
15	*Carcharhinus leucas*	Carcharhinidae	Shark	0	0	1
16	*Chrysichtys nigrodigitatus*	Bagridae	Cat fish	0	1	0
17	*Cirrhina molitorella*	Cyprinidae	Mud carp	0	1	0
18	*Clarias gariepinus*	Claridae	Cat fish	1	1	0
19	*Clupea harengus*	Clupeidae	Herring	0	0	1
20	*Coryphaena hippurus*	Coryphaenidae	Sea bream	0	0	1
21	*Crocodilus fuscus*	Crocodylidae	Caiman	1	1	0
22	*Crocodylus* sp.	Crocodylidae	Crocodile	0	1	1
23	*Ctenopharyngodon idella*	Cyprinidae	Herbivorous carp	0	1	0
24	*Cynoglosis* spp.	Soleidae	Sole	0	1	1
25	*Cynoglossus senegalensis*	Soleidae	Sole	0	0	1
26	*Cyprinus carpio*	Cyprinidae	Common carp	0	1	0
27	*Dasyatis* spp.	Rhinobatidae	Skate	0	1	1
28	*Dermocherys coriacea*	Dermochelyidae	Tortoise	0	1	1
29	*Dicentractchus labrax*	Serranidae	Common bar	0	0	1
30	*Dorminator lebretonis*	Eleotridae	Sleeper	0	1	1
31	*Dreissena polymorpha*	Dreissenidae	Streaked mussel	0	1	0
32	*Drepana africana*	Drepaneidae	Disk	0	0	1
33	*Elops lacerta*	Elopidae	Ladyfish	0	1	1
34	*Engraulis encrasicolus*	Engraulidae	Anchovies	0	0	1
35	*Ephippion guttifer*	Tetraodontidae	Puffer	1	0	0
36	*Epinephelus* sp.	Serranidae	Grouper	0	0	1
37	*Ethmalosa fimbriata*	Clupeidae	Wadding	0	1	1
38	*Fluviatilis* sp.	Fluviatilis	Ecrevisse	0	1	1

(*Continued*)

Appendix 2 (*Continued*) Typical aquaforest higher fauna species of West and Central Africa

				Type of aquaforests		
No.	Scientific name	Family	Common name	Fringing forest	Fresh water	Mangroves/ Coastal
39	*Fodiator acutes*	Exocotidae	Wheel fish	0	0	1
40	*Galeoides decadactylus*	Polynemidae	Threadfins	0	0	1
41	*Gymnarcus niloticus*	Gymnarchidae	Aba – aba, Frankfish	0	0	1
42	*Gymnura micrura*	Gymnuridae	Butterfly-rays	0	0	1
43	*Hemichromis faciatus*	Cichlidae	Panther fish	0	1	0
44	*Hemiramphus balao*	Hemiramphidae	Halfbeaks	0	0	1
45	*Hepsetus odoe*	Hepsetidae	Pikes	0	1	1
46	*Hetérobrancus longifilis*	Claridae	Cat fish	1	1	0
47	*Heterotis niloticus*	Osteoglossidae	Kanga	0	1	0
48	*Hippoglossus Hippoglossus*	Pleuronectidae	Halibut	0	0	1
49	*Hippoglossus stenolepis*	Pleuronectidae	Halibut	0	0	1
50	*Hydrochoeris hydrochoeris*	Hydrochoeridae	Water pig	0	1	0
51	*Hydrocynus forskalii*	Alestidae	Tiger fish	0	0	1
52	*Hypleurochilus langi*	Blennidae	Blennie	0	1	1
53	*Hypophtalmichthys molitrix*	Cyprinidae	Silver carp	0	1	0
54	*Inia geoffrensis*	Platanistidae	Fresh	0	0	1
55	*Labeo* sp.	Cyprinidae	Labeo	0	1	0
56	*Lates niloticus*	Centropomidae	Captain	0	1	0
57	*Lepomis gibbosus*	Centrachidae	Perche sun	0	1	0
58	*Lisha africana*	Clupeidae	Clupeids	0	1	1
59	*Litjanus agennes*	Lutjaidae	Snapper	1	1	1
60	*Liza* sp.	Mugilidae	Mullet	0	1	1
61	*Malapterus electricus*	Malapteruriade	Electricial fish	0	1	1
62	*Mugil cephalus*	Mugilidae	Mullet	0	1	1
63	*Mylopharyngodon piceus*	Cichlidae	Black carp	0	0	1
64	*Myocastor coypus*	Myocastoridae	Coypu	0	0	1
65	*Myrophis plumbeus*	Ophichthyidae	Snake eel	0	1	1
66	*Oncorhynchus mykiss*	Salmonidae	Rainbow trout	0	1	0
67	*Ondatra zibethicus*	Muridae	Muskrat	0	1	0
68	*Oreochromis mosambicus*	Cichlidae	Tilapia	0	1	0
69	*Oreochromis niloticus*	Cichlidae	Nile Tilapia	0	1	0
70	*Oxyura jamaicensis*	Anantidae	Fuzz Erismature	0	1	0
71	*Pangasius gigas*	Pangasiides	Mekong's fish	1	1	1
72	*Parachanna abscura*	Channidae	Viper fish	1	1	0
73	*Paraconger arisona*	Ophichthyidae	Conger	0	1	1
74	*Penaeus* sp.	Peneidae	Shrimp	0	1	1
75	*Periphtalmus* sp.	Gobeiidae	Periophtalm	0	0	0
76	*Poecilia latipina*	Poeciliidae	Molly	0	0	1

(*Continued*)

Appendix 2 (*Continued*) Typical aquaforest higher fauna species of West and Central Africa

				Type of aquaforests		
No.	**Scientific name**	**Family**	**Common name**	**Fringing forest**	**Fresh water**	**Mangroves/ Coastal**
77	*Poecilia reticulata*	Poeciliidae	Guppy	0	0	1
78	*Polydacrylus* sp.	Polynemidae	Small African threadfin	0	0	1
79	*Polypterus senegalus*	Polyteridae	Senegal bichir	0	0	1
80	*Pomedasys jubilini*	Sparidae	Sompat grunt	0	0	1
81	*Procyon lotor*	Procyonidae	Racoon	0	1	0
82	*Psettodes belcheri*	Psettodiae	Psettodis	0	0	1
83	*Psettus sebas*	Drepaneidae	African moony	0	0	1
84	*Pseudotolithus brachygnathus*	Sciaenidae	Drums	0	1	1
85	*Pseudotolithus elongatus*	Sciaenidae	Bobo croaker	0	1	1
86	*Pseudotolithus senegalensis*	Sciaenidae	Bar	0	0	1
87	*Pteroscion peli*	Sciaenidae	Fried fish	0	1	1
88	*Rana* sp.	Ranidae	Frog	0	1	0
89	*Rhinobatos rhinobatos*	Rhinobatidae	Guitar skate	0	0	1
90	*Salmo salar*	Salmonidae	Salmon	0	0	1
91	*Sardinella maderensis*	Clupeidae	Clupeids	0	1	1
92	*Sarotherodon galilaeus*	Cichlidae	Tilapia	0	1	0
93	*Schilber mystus*	Schilbedae	Cat fish	0	1	0
94	*Scomber scombrus*	Scombridae	Mackerel	0	0	1
95	*Sphyraena barracuda*	Sphyraenidae	Barracuda	0	0	1
96	*Sphyraena dubia*	Hepsetidae	Brochet	0	0	1
97	*Sphyraena piscatorium*	Spyraenidae	Guinean barracuda	0	0	1
98	*Strongylura senegalensis*	Belonidae	Orphies	1	1	1
99	*Synodontis* sp.	Mochokidae	Synodontis	0	1	0
100	*Tilapia guineensis*	Cichlidae	Tilapia	0	1	0
101	*Trichechus senegalensis*	Trichechidae	Manatee	1	1	1
102	*Trichiurus lepturus*	Trichiuridae	Cutlassfish	0	0	1

Source: Compiled from Stiassny et al., 2007.

REFERENCES

1. Nilsson, C.; Jansson, R.; Zinko, U. Long term responses of river-margin vegetation to water-level regulation. Science **1997**, *276*, 798–800.
2. Trimbe, S.W. Decreased rates of alluvial sediment storage in the Coon Creek basin, Wiscosin, 1975–1993. Science **1999**, *285*, 1244–1246.
3. Gentry, A.H. Tropical forest biodiversity: distributional patterns and their conservational significance. Oikos **1992**, *63*, 19–28.
4. National Research Council (NRC). *Riparian Areas: Functions and Strategies for Management. Summary. WSTB CRZFSM*; National Academy Press: Washington, 2002.
5. Woinarski, J.C.Z.; Brock, C.; Armstrong, M.; Hempel, C.; Cheal, D.; Brennan, K. Bird distribution in riparian vegetation in the extensive natural landscape of Australia's tropical savanna: a broad-scale survey and analysis of a distributional data base. J. Biogeography **2000**, *27*, 843–868.
6. Doyle, A.T. Use of riparian and upland habitats by small mammals. J. Mammal. **1990**, *71*, 14–23.
7. Darveau, M.; Labbé, P.; Beauchesne, P.; Bélanger, L.; Huot, J. The use of riparian forest strips by small mammals in a boreal balsam fir forest. Forest Ecol. Manag. **2001**, *143*, 95–104.
8. Larue, P.; Bélanger, L.; Huot, J. Riparian edge effects on boreal balsam fir bird communities. Canadian J. Forest Resources **1995**, *25*, 555–566.

9. Saab, V. Importance of spatial scale to habitat use by breeding birds in riparian forests: a hierarchical analysis. Ecol. Appl. **1999**, *9*, 135–151.
10. Ward, J.V.; Tockner, K.; Arscott, D.B.; Claret, C. Riverine landscape diversity. Freshwater Biol. **2002**, *47*, 517–539.
11. Acker, A.S.; Lienkaemper, G.; McKee, W.A.; Swanson, F.J.; MIller, S.D. Composition, complexity, and tree mortality in riparian forests in the central Western Cascades of Oregon. Forest Ecol. Manag. **2003**, *173*, 293–308.
12. Cordes, L.D.; Hughes, F.M.R.; Getty, M. Factors affecting the regeneration and distribution of riparian woodlands along a northern praire river: the Red Deer River Alberta, Canada. J. Biogeography **1997**, *24*, 675–695.
13. Robinson, C.T.; Tockner, K.; Ward, J.V. The fauna of dynamic riverine landscape. Freshwater Biol. **2002**, *47*, 661–677.
14. Gledhill, D. *West African trees*. Longman. 1972.
15. Brinson, M.M. Riparian forests. In *Forested Wetlands Ecosystems of the World 15*; Lugo, A.E., Brinson, M., Brown, S. Eds.; Elsevier, Amsterdam **1990**, 87–141
16. Hancock, C.N.; Ladd P.G.; Froend, R.H. Biodiversity and management of riparian vegetation in Western Australia. Forest Ecol. Manag. **1996**, *85*, 239–250.
17. Natta, A.K. *Ecological Assessment of Riparian Forests in Benin: Phytodiversity, Phytosociology and Spatial Distribution of Tree Species*: PhD thesis Wageningen University: Netherlands, 2003.
18. Salo, J. Kalliola, R.; Häkkinen, I.; Mäkinen, Y.; Puhakka, M.; Coley, P.D. River dynamics and the diversity of Amazon lowlands forests. Nature **1986**, *322*, 254–258.
19. Naiman, R.J.; Décamps, H.; Polleck, M. The role of riparian corridors in maintaining regional biodiversity. Ecol. Appl. **1993**, *3*, 209–212.
20. Levine, C.M.; Stromberg, J.C. Effects of flooding on native and exotic plant seedlings: implications for restoring southwestern riparian forests by manipulating water and sediment flows. J. Arid Environ. **2001**, *49*, 111–131.
21. Ellis, L.M.; Molles, M.C., Jr.; Crawford, C.S. Influence of experimental flooding on litter dynamics in a Rio Grande riparian of central New Mexico. Restoration Ecol. **1999**, *7*, 1–13.
22. Gregory, S.V.; Swanson, F.J.; MsKee, W.A.; Cummins, K.W. An ecosystem perspective of riparaian zones: focus on links between land and water. BioScience **1991**, *41*, 540–551.
23. Harper, D.M.; Mavuti, K.M. Freshwater wetlands and marshes. *In*: *East African Ecosystems and their Conservation*; McClanahan, T.R., Young, Y.P., Eds.; Oxford University Press: New York, NY, 1996; 217–239 pp.
24. Reid, W.V. Biodiversity hotspots. Tree **1998**, *13* (7), 275–280.
25. IUCN. *Wetland Conservation: A Review of Current Issues and Required Action*; Dugan, P.J., Ed.; Gland: Switzerland, 1990.
26. Hopkins, B. *Forest and Savanna. An Introduction to Terrestrial Ecology with Special Reference to West Africa*; Heinemann: London, 1974.
27. Scudder, T. The need and justification for maintaining transboundary flood regimes: The Africa Case. Natural Resources Journal: University of New Mexico School of Law **1991**, *31*, 75–107.
28. Cooper, S.D. Rivers and streams. In *East African Ecosystems and their Conservation*; In: McClanahan, T.R., Young, T.P., Eds.; Oxford University Press: New York, NY, 1996; 133–170 pp.
29. Lowel-McConnell, R.H. Ecological studies in tropical fish communities, Cambridge University Press: Cambridge, 1987.
30. Melick, D.R.; Ashton, D.H. The effects of natural disturbances on warm temperate rainforests in South-eastern Australia. Aust. J. Ecol. **1991**, *39*, 1 30.
31. Roché, J. The use of historical data in the ecological zonation of rivers: the case of the 'tern zone'. Vie et Milieu **1993**, *43*, 27–41.
32. Wakentin, I.G.; Reed, J.M. Effects of habitat type and degradation on avian species richness in great Basin riparian habitats. Great Basin Nat. **1999**, *59*, 205–212.
33. Fetherston, K.L.; Naiman, R.J.; Billy, R.E. Large woody debris, physical process and riparian forest development in montane river networks of the Pacific Northwest. Geomorphology **1995**, *13*, 133–144.
34. Enright, N.J.; Miller, B.P.; Crawford, D. Environmental correlates of vegetation paterns and species richness in the northern Grampians, Victoria. Aust. J. Ecol. **1994**, *19*, 159–168.
35. Cowling, R.M.; Holmes, P.M. Flora and vegetation. In: *The Ecology of Fynbos: Nutrient, Fire and Diversity;* Cowling, R.M., Ed.; Oxford University Press: Cape Town, 1992.
36. Gregory, S.V.; Beschta, R.; Swanson, F.J.; Sedell, J.; Reeves, G.; Everest F. *Abundance of Conifers in Oregon's Riparian Forests*, COPE Report, Vol 3. No. 2. College of Forestry, Oregon State University: Corvallis, OR, 1990, 5–6 pp.
37. Nierenberg, T.; Hibbs, D.E. Characterization of unmanaged riparian areas in the central coast range of Western Oregon. Forest Ecol. Manag. **2000**, *129*, 195–206.
38. Dignan, P.; Bren, L. A study of logging on the understorey light environment in riparian buffer strips in a south-east Australian forests. Forest Ecol. Manag. **2003**, *172*, 161–172.
39. Tilman, D. *Plant Strategies and the Dynamics and Structure of Plant Communities*. Princeton University Press: Princeton, 1988.
40. Gilliam, J.W. Riparian Wetlands and Water Quality. J. Environ. Qual. **1994**, *25*, 896–900.
41. Groffman, P.M.; McDowell, W.H.; Myers, J.C.; Merriam, J.L. Soil microbial biomass and activity in tropical riparian forests. Soil Biol. Biochem. **2001**, *33*, 1339–1348.
42. Forman, R.T.T.; Godron, M. *Landscape Ecology*. John Wiley and Sons: New York, 1986.
43. FAO. *Review of the State of World Fishery Resources*; FAO Fisheries Circular no. 710. (Rev. 2). FAO: Rome, 1981.
44. Bouillon, S.; Dahdouh-Guebas, F.; Rao, A.V.V.S.; Koedam, N.; Dehairs, F. Sources of organic carbon in mangrove sediments: variability and possible ecological implications. Hydrobiology **2003**, *495*, 33–39.
45. Donato, D.C.; Kauffman, J.B.; Murdiyarso, D.; Sofyan Kurnianto, S.; Melanie Stidham, M.; MarkkuKanninen, M. Mangroves among the most carbon-rich forests in the tropics. Nat. Geosci. **2011**, *4*, doi: 10.1038/NGEO1123.
46. Shumway, C.A. *Forgotten waters: Freshwater and marine ecosystems in Africa. Strategies for Biodiversity Conservation and Sustainable Development*; Boston University: USA, 1999.

47. Cole, J.J.; Peierls, B.I.; Caraco, N.E.; Pace, M.I. Nitrogen loading of rivers as a human-driven process. In *Humans as Components of Ecosystems*; McDonnell, M., Pickett, S., Eds., Springer-Verlag, New York, 1993; 141–157 pp.
48. National Research Council (NRC). Understanding Riverine biodiversity: A Research Agenda for the Nation; Academy of Sciences: Washington, D.C, 1995.
49. Bartley, D.; Minchin, D. Precautionary approach to introductions and transfers of aquatic organisms. In Precautionary approach to fisheries, Part 2, FAO Fish. Tech. Pap. **1995**, *350* (2), 159–189.
50. Dixon, J.A. *Valuation of mangroves*. Tropical Coastal Area Management. ICLARM Newsl. **1989**, *4* (3), 1–6.
51. Graham, A. Siltation of stone-surface periphyton in rivers by clay-sized particles from low concentrations in suspension. Hydrobiologia **1990**, *199*, 107–115.
52. Ajonina, G.N.; Usongo, L. Preliminary quantitative impact assessment of wood extraction on the mangroves of Douala-Edea Forest Reserve, Cameroon. Trop. Biodivers. **2001,** *7* (2–3), 137–149.
53. Ajonina, P.U.; Ajonina, G.N.; Jin, E.; Mekongo, F.; Ayissi, I.; Usongo, L. Gender roles and economics of exploitation, processing and marketing of bivalves and impacts on forest resources in the Douala-Edaa Wildlife Reserve, Cameroon. Int. J. Sustainable Develop. World Ecol. **2005,** *12*, 161–172.
54. Feka, N.Z.; Ajonina, G.N. Drivers causing decline of mangrove in West-Central Africa, a review. Int. J. Biodivers. Sci. Ecosystem Serv. Manag. 2011, doi:10.1080/21513732.2011.634436.
55. Kellman, M.; Meave, J. Fire in the tropical gallery forests of Belize. J. Biogeography **1997**, *24* (1), 23–34.
56. Ellis, L.M. Short-term response of woody plants to fire in a Rio Grande riparian forest of Central New Mexico, USA. Biol. Conserv. **2001**, *97*, 159–170.
57. Norris, V.O.L. The use of buffer zones to protect water quality: a review. Water Resources Manag. **1993**, *7*, 257–272.
58. Ajonina, G. *Rapid Assessment of Mangrove status and Conditions for use to Assess Potential for Marine Payment for Ecosystem Services in Amanzuri and Surrounding Areas in the Western Coastal Region of Ghana, West Africa*, Consultancy report to Coastal Resource Center, 2011.
59. Ruwa, R.K. Intertidal wetlands. In *East African Ecosystems and their Conservation*; McClanahan, T.R., Young, Y.P., Eds.; Oxford Univ. Press: New York, N.Y, 1996; 101–130 pp.
60. Scudder, T. Recent experiences with river basin development in the tropics and subtropics. Nat. Resources Forum **1994**, *18*, 101–113.
61. Katibath, E.F. A brief history of riparian forests in the Central Valley of California. In *California Riparian Systems: Ecology, Conservation and Productive Management*; Warner R.E., Hendrix, K.M., Eds.; University of California Press: Berkeley, 1984; 23–29 pp.
62. Rottenborn, S.C. Predicting the impacts of urbanization on riparian bird communities. Biol. Conserv. **1999**, *88*, 289–299.
63. Kaufman, L.; Ochumba, P. Evolutionary and conservation biology of ciclid fishes as revealed by faunal remnants in northern Lake Victoria. Conserv. Biol. **1993**, *7*, 719–730.
64. Taylor, J.R.; Cardonne, M.A.; Mitsch, W.J. Bottomland hardwood forests: their functions and values. In *Cumulative Impacts: Illustrated by Bottomland Hardwood Wetlands Ecosytems*; Gosselink, J.P., Lee, L.C., Muir, T.A., Eds.; Lewis Publishers: Chelsea MI, 1990, 13–86 pp.
65. Spalding, M.; Kainyuma, M.; Collins, L. *World Atlas of Mangroves*; Earthscan, London, 2010.
66. Sagers, C.L.; Lyon, J. Gradient analysis in a riparian landscape: contrasts in amongst forests layers. Forest Ecol. Manag. **1997**, *96*, 13–26.
67. Loetsch, I. Zohier, F.; Haller, K. *Forest Inventory Vol. 2*; Second Edition BIV: Germany, 1973.
68. Husch, B.; Beers, T.W.; Keuhaw, J.A., Jr. *Forest Mensuration*, 4th Ed.; John Wiley & Sons, 2003. pp.
69. Pool, D.G.; Snedaker, S.C.; Lugo, A.E. Structure of mangrove forests in Florida, Puerto Rico, Mexico and Costa Rica. Biotropica **1977**, *9*, 195–212.
70. Cintron, G.; Schaeffer-Novelli, Y. Methods of studying mangrove structure. In *The Mangrove Ecosystem: Research Methods*; Snedaker, S., Snedaker, J.G., Eds.; UNESCO Publication, 1984.
71. Kjerfve, B.; Macintosh, D.J. The impact of climatic change on mangrove ecosystems. In *Mangrove Ecosystems Studies in Latin America and Africa;* Kjerfve, B.B., Lacerda, L.D., Diop S. H., Eds.; UNESCO: Paris, **1997**; 1–7 pp.
72. Williams, P.H.; Humpries, C.J.; Vane–Wright, R.I. Measuring biodiversity: Taxonomic relatedness for conservation priorities. Aust. Syst. Bot. **1991**, *4*, 665–679.
73. De Lima, M.G.; Gascon, C. The conservation value of linear forest remnants in central Amazonia. Biol. Conserv. **1999**, *91*, 241–247.
74. Roggeri, H. *Zones Humides Tropicales D'eau Douce. Guide de Connaissances Actuelles Et De La Gestion Durable.* WIW, Union Europeene, Universitë de Leiden. Kluwer Academic Publishers: Pays–Bas, 1995.
75. Hibbs, D.E.; Bower, A.L. Riparian forests in the Oregon Coast Range. Forest Ecol. Manag. **2001**, *154*, 201–213.
76. Blas, D.E.; Perez, M.R. Prospects for reduced impact logging in Central African logging concessions. Forest Ecology and Management **2008**, *256*, 1509–1516.
77. Cerutti, P.O.; Nasi, R.; Tacconi, L. Sustainable forest management in Cameroon needs more than approved forest management plans. Ecol. Soc. **2008**, *13* (2), 36.
78. Feka, N.Z.; Chuyong, G.B.; Ajonina, G.N. Sustainable utilization of mangroves using improved fish smoking systems: A management perspective from the Douala-Edea Wildlife Reserve, Cameroon. Trop. Conserv. Sci. **2009**, *4*, 450–468.
79. Wetlands International. *Planting Trees to Eat Fish: Field Experiences in Wetlands and Poverty Reduction*; Wetlands International, Wageningen: the Netherlands, 2009.

BIBLIOGRAPHY

1. Arbornnier, M. *Trees, Shrubs and Lianas of West African Dry Zones*; CIRAD, MNHN, Paris, 2004.
2. Etukudo, I.G.; Akpan-Ebe, I.N.; Udofia, A.; Attah, V.I. *Elements of Forestry*; Usanga & Sons. UYO: Nigeria, 1994.
3. Forest Trends. Payments for Ecosystem Services: Getting Started in Marine and Coastal Ecosystems A Primer. 2010.

4. Keay, R.W.J.; Hepper, F.N. *Flora of West tropical Africa*, 2nd Ed.; Millbank: London, 1954–1972; Vol. I-III pp.
5. Millennium Ecosystem Assessment. *Ecosystems and Human Well-Being: Synthesis*; World Resources Institute: Washington, D.C. 2005.
6. Raunkaier, C. *The Life Forms of Plants and Statistical Plant Geography*. Clarendon Press: London, 1934.
7. Schnell, R. *Introduction à la Phytogéographie des Pays Tropicaux*; Volume II. Milieux, les groupements végétaux. Edition Gauthier–Villars : Paris, France, 1971; 503–951 pp.
8. Stiassny, M.L.J.; Teugels, G.G.; Hopkins, C.D. *Poissons d'eaux douces et saumâtres de basse Guinée, ouest de l'Afrique centrale/ The Fresh and Fishes of Lower Guinea, West-Central Africa*. Volume 1. IRD, Publications scientifiques du Museum, MRAC: Paris, 2007.
9. Swanson, F.J.; Johnson, S.L; Gregory, S.V.; Acker, A.S. Flood disturbance in a forested Mountain Landscape. Bioscience **1998**, *48*, 681–689.
10. Thikakul, S. *Manual of Dendrology-Cameroon*. Groupe Poulin Quebec: Canada, 1985.
11. White, F. *La végétation de l'Afrique. Mémoire accompagnant la carte de végétation de l'Afrique*. UNESCO/AETFAT/UNSO, 1986.

Biodegradation and Bioremediation

Owen P. Ward
Ajay Singh
Department of Biology, University of Waterloo, Waterloo, Ontario, Canada

Abstract
Biodegradation processes mediate recycling of organic materials in the environment. In bioremediation processes, micro-organisms and plants participate in the biodegradation and removal of hazardous contaminants to restore the polluted environment. Although micro-organisms thrive in aqueous environments, the majority of organic chemicals and environmental contaminants are hydrophobic and sorb to soil particles. To access these contaminants, microbes may interact directly with the contaminant or soil particulates or both or secrete biosurfactants to mobilize the contaminant into the aqueous phase. A variety of in situ and ex situ bioremediation strategies are employed, aimed at promoting biodegradation of the chemical pollutants present in the contaminated medium.

INTRODUCTION

In their most general sense, biodegradation processes are central to all biological systems. In these processes, cells transform and recycle biological and synthetic chemicals and materials into smaller molecules to generate energy and/or metabolic intermediates needed for cell functioning including maintenance, biosynthesis, and growth. Indeed, these processes also contribute to the unique adaptation properties of organisms, for example, enabling microbes to adjust to changing environments by making the requisite biochemical and physiological changes mediated by biodegradative and biosynthetic metabolic processes. Bioremediation represents an important applied segment among these biodegradation processes in which technologies are generally engineered and optimized to exploit the same biodegradation capabilities especially of microbes, for the removal of organic contaminants and sometimes, metal contaminants from the environment. The principal objective of this entry is to provide an overview of bioremediation processes and systems as well as a perspective on recent developments and future prospects. General principles of biodegradation processes are first considered to provide a context for addressing bioremediation technologies.

BIODEGRADATION OVERVIEW

On a macro scale, biodegradation processes often mediate recycling of organic materials, including animal, plant and microbial materials, present in the environment. Enzymes secreted by microbes and other species attack and depolymerize the principal biological polymeric structures, namely, complexes containing carbohydrates, proteins, nucleic acids, lipids, and their derivatives into oligomeric or monomeric components, which may be taken up and metabolized by cells. These processes are promoted in the digestive systems of humans, animals, and other species mediated by endogenous enzymes from these organisms or supported by the microbial populations present or both. These processes also occur everywhere in the environment where microbes can thrive, most notably in soil and in aqueous media. Processes in soil are enhanced around plant roots where nutrients released from the roots support the growth of plant growth-promoting bacteria. Such processes are also actively promoted in the practices of agriculture and horticulture and in myriad solid and liquid waste treatment systems. Some examples of general environmental processes and practices for the biodegradation of natural organic materials include processes for silage making, composting, and wastewater treatment (Table 1). The reader is referred to references[1–3] for further discussion of these topics.

BIOREMEDIATION OVERVIEW

Industrial-scale processes for the recovery of petroleum and the manufacturing, distribution, and use of chemicals invariably lead to the transfer of chemical contaminants into the environment, most notably into soils, groundwater, shorelines, and water bodies. These contaminants range from alkanes and mono- and polycyclic aromatic hydrocarbon compounds, naturally present in crude oil or generated through combustion, to high-volume synthetic chemicals including chlorine-, nitrogen-, and phosphate-containing

Encyclopedia of Natural Resources DOI: 10.1081/E-ENRL-120047482

Table 1 Example processes for biodegradation of natural organic materials

- **Silage-making processes**, whereby fibrous plant materials are rendered more digestible for animal feeding by conversion of sugars to organic acids and some enzymatic cellulose/hemicellulose degradation, mediated by microbial anaerobic digestion.
- **Composting processes**, used to aerobically transform plant and other biological and organic materials, most frequently derived from agricultural, horticultural, food, and municipal wastes. A variety of microbial species, most notably fungi and bacteria including mesophiles and thermophiles, participate in these exothermic biodegradative processes to generate beneficial organic fertilizers devoid of pathogens, which are killed as thermophilic conditions are achieved.
- **Biological wastewater treatment processes**, where the objective is to remove contaminants from domestic, municipal, or industrial waste streams. Micro-organisms are primarily responsible for aerobic and anaerobic degradation and assimilation of the organic components together with some removal/concentration of heavy metals. The resulting accumulated biosolids may be further processed and transformed into beneficial organic fertilizers for use in agriculture.

organic chemicals and many others. In addition, soils become contaminated with inorganic chemicals. Just as microbes have the ability to concentrate and accumulate metals in wastewater treatment processes, specific microbes can accumulate or transform some of these inorganic species or both.

The extent of the threat to the environment posed by petroleum and organic chemicals may be appreciated by reference to the production volumes of these products. Global crude oil production and world refining capacity is approaching 80 million barrels per day. Estimates for global production volumes of synthetic polymers and plastics approximate to 140 million tons per annum.[4] In the United States alone, annual production volumes of each of ethylene, propylene, vinyl chloride, and benzene range from 5 to 20 million tons, with volumes of many other organic chemicals ranging from 1 to 5 million tons.[5]

Some of the most widely distributed organic environmental contaminants are aliphatic and aromatic hydrocarbons, including alkanes, mono- and polycyclic aromatics, present in crude oils or in refined oil fuels or petrochemicals; halogenated aliphatic, especially chlorinated species including chlorinated methanes, ethanes, and ethenes such as trichloroethene (TCE), dichloroethene (DCE), perchloroethene (PCE), and vinyl chloride (VC); mono- and polyhalogenated phenols and benzoates, including pentachlorophenol (PCP); polychlorinated biphenyls (PCBs); chloroaromatics including 2,4-dichlorophenoxyacetate (2,4-D), dichlorodiphenyltrichloroethane (DDT), and related compounds; nitroaromatics including nitrobenzene and 2,4,6 trinitrotoluene (TNT); organophosphates; nitrate esters (including glycerol trinitrate); nitrogen heterocycles (including the explosive compound hexahydro-1,3,5-trinitrotriazine or RDX); and polyesters, polylactates, polyurethanes, and nylon.[6,7]

The recognition that microbes in soil could degrade, transform, or accumulate many of the chemicals polluting the soil provided momentum for the undertaking of research to exploit these activities to clean up contaminated soil and other media, using processes now termed as bioremediation. Indeed, the observed effectiveness of bioremediation as a process for the degradation of oils contaminating the shorelines at Prince William Sound, Alaska due to the *Exxon Valdez* oil spill led to the development of a variety of strategies for the application of bioremediation processes for soil remediation.[8,9] Bioremediation is generally considered to be the low-cost option in comparison to physical and chemical alternatives.[10]

Bioremediation technologies may be adapted to a variety of different contaminants and environmental conditions, and in particular, conditions can be developed for the selection and enrichment of microbial populations having the requisite and unique degradation abilities to remediate the specific contaminants present at a particular site.[11] Bioremediation technologies are also advantageous where contaminants have become diluted or widely dispersed over time.

A vast array of physical environments exist where biodegradation and bioremediation processes may thrive. Broadly speaking, these environments will all contain water or at least be moist because the presence of water is a prerequisite for microbial growth and biodegradation activity. The nature of the microbial type or population which develops will be influenced by a wide range of factors including the physical properties of the medium; salt concentration; solid content, surface properties and particle size; nature and concentration(s) of contaminant(s); nature and concentration of nutrients present to support microbial growth; relative humidity; presence of oxygen or other electron acceptors; pH; and temperature.

CONTAMINANT BIODEGRADABILITY

Although many organic chemicals are biodegradable by microbes present in the environment, other natural or synthetic chemicals structures exhibit resistance or recalcitrance to biodegradation and tend to persist in the environment for prolonged periods of time. Microbial transformation of organic pollutants is mediated primarily by enzyme-catalyzed reactions but often with the support of other cell components, associated with cell surfaces, surfactants, chelators, vesicles, and organelles. The bio-transforming enzymes may be located within the cell protoplasm or be membrane associated (or a combination of these) or indeed in a few cases may be extracellular. A good example of a combined intracellular/membrane-bound system is the *alk* gene system for alkane degradation, whereas extracellular fungal peroxidases and laccases are non-specific mechanisms for the attack of organic pollutants, mediated by the generation of free radicals.[11] Microbial enzyme transformation of contaminants may lead to complete contaminant mineralization or to the production of breakdown products which may be assimilated into the cells' central pathways for energy or biomass production or both. Alternatively, contaminants may be partially degraded or transformed into metabolites, which the cell cannot further utilize, whereby such metabolites may accumulate in the cell or be excreted. These metabolites may be further transformed by other cell types, among the multitude of microbes present in the ecosystem, which may have different metabolic capacities. Alternatively, the metabolites may be generally recalcitrant to biodegradation and may bioaccumulate. Contaminants in well-aerated environments will predominantly be degraded by aerobic metabolic processes, whereas contaminants in a non-aerated subsurface environment may be anaerobically degraded.

A review of aerobic pathways for the biodegradation of organic contaminants including petroleum hydrocarbons, chlorinated hydrocarbons, nitroaromatics and nitrogen heterocycles, organophosphate derivatives, and plastics reveals the following reaction types: dioxygenases, monooxygenases, hydroxylases, ligninases, and reductases among others. Anaerobic biodegradations typically involve initial activation mechanisms such as carboxylation, methylation, hydroxylation, and dehydrogenation followed by addition reactions, anaerobic reductions, and hydrolytic reactions.[12]

Microbes have the potential to remediate metals in contaminated media by exploiting a combination of mechanisms including precipitation, sorption, biosurfactant complexation, biofilm entrapment, and metabolic uptake. Some microbes can also transform metals including radionuclides, for example, by reducing mercury into a volatile form, which can be removed. In metal bioremediation applications, there is particular interest in exploiting microbes, which have been isolated from mine tailings and heavy metal-contaminated sites, as these strains exhibit superior metal-scavenging capacities, greater resistance to metal toxicity (sometimes mediated by metal efflux mechanisms), and ionizing radiation.[13] Examples of interesting strains for these applications include certain *Pseudomonas* species, including recombinant strains, *Ralstonia eutropha*, *Thiobacillus ferrooxidans*, and *Deinococcus radiodurans*.

CONTAMINANT TOXICITY

The principal concern regarding chemical contamination of the environment relates to contaminant toxicity. Many of the chemical contaminants, but especially the more recalcitrant chemicals, are toxic to higher life-forms or are known carcinogens or both. As such, their presence represents a threat to human and animal health. Many of these chemicals are classified as hazardous by national regulatory agencies, including the United States Environmental Protection Agency (USEPA). For example, 16 polycyclic aromatic hydrocarbons (PAHs) are considered to be priority pollutants by the USEPA and 7 of these are classified as probable human carcinogens.[14]

Toxicity concerns as a result of chemical contamination of the environment do not just relate to the toxicities of the original chemical compounds transferred. There are many examples where the "parent" contaminant may get chemically or biologically transformed into reaction products or catabolites, which are as toxic as or even much more toxic than the parent compound. For example, surfactants present in detergents, such as nonylphenol ethoxylates, are often biotransformed into nonylphenol end products with endocrine-disrupting activity as a result of microbial transformation.[15] The presence of strong chemical oxidizing or reducing agents in the contaminated medium can give rise to non-specific conversions of the contaminants into more reactive and more toxic products.[16]

In heavily contaminated media, the toxicity of the medium may hinder the implementation of bioremediation strategies as a result of the contaminant, at a higher concentration, being toxic to the microbial population which may have the metabolic capability to biotransform the contaminant. In a manner akin to antibiotic resistance, some microbial cells have developed resistance to specific toxic organic compounds such as toluene, by using efflux pumps to extrude toxic contaminants from the cell cytoplasm or periplasmic space.[17] It is also thought that, where toxic contaminants are only partially transformed by one species, efflux pumps may have the potential to excrete the metabolites such that they may be made available to other members of a consortium with the capacity for degradation of the excreted metabolites. In metal bioremediation applications, there is particular interest in exploiting microbes, which have been isolated from mine tailings and heavy metal-contaminated sites, as these strains exhibit greater resistance to metal toxicity (sometimes mediated by metal efflux mechanisms) and ionizing radiation.[13]

The extent of toxicity of a known toxic contaminant cannot be simply related to its concentration in the contaminated medium. Actual toxicity may be diminished by sorption of the contaminant to soil particles or organic matter. In addition, biological species present in or added to the contaminated medium may adsorb the contaminant or secrete biological molecules including chelators or surfactants, which effectively sequester the contaminant, thereby reducing its toxicity. Arguably, the presence of other components in the contaminated medium or indeed produced by the biological organisms present could enhance toxicity. For example, a hydrophobic contaminant, with a tendency to sorb to soil, might become mobilized and solubilized by the presence of lipids or surfactants and possibly become more toxic to the organisms present.

Scientists have developed toxicity-testing methods, whereby contaminant-sensitive organisms, including microbes, algae, plants, fish, macroinvertebrates, and cultured cell lines, are used as biomonitors of pollutant accumulation in soil and aquatic systems and also for following the reduction in the toxicity of the contaminated media as a result of the implemented remediation processes. These biomonitoring tests rely on observing whole-organism biological impacts, related, for example, to growth, behavior, reproduction, or mortality. Alternatively, toxicity tests may involve the measurement of intracellular biomarkers, for example, heat shock proteins; enzymes including antioxidant enzymes, DNA repair enzymes, acetylcholinesterase, cytochrome P450 monooxygenase, and glutathione S transferase; and DNA mutations or chromosomal defects.[18]

CONTAMINANT BIOAVAILABILITY

Microbes generally thrive and grow in aqueous phases and are most efficient at accessing nutrients that are water soluble. However, petroleum components, as well as most other chemical contaminants, are water insoluble or hydrophobic and hence less bioavailable to microbes. In addition, soil is the most dominant medium for these contaminants, and hydrophobic contaminants tend to bind strongly to soil particles, thereby further reducing their bioavailability to microbes capable of biodegrading the contaminants and rendering them more persistent in the environment. Soil bioremediation processes require adequate amounts of water present in the soil to support microbial activity, and the contaminant-biodegrading organisms have to have the means to access the hydrophobic or soil particle-bound contaminants or both. Two general natural microbial mechanisms are known to facilitate this contaminant access. First, cells may sorb and uptake the contaminant by direct interaction between the microbes and the hydrophobic substance/particulate matter. Alternatively, the microbes may secrete biosurfactants into the surrounding environment, which may pseudosolubilize and mobilize the contaminant through micelle formation, which facilitates contaminant contact with the microbes and promotes cellular uptake of the hydrophobic substrate. As amphipathic molecules, biosurfactants contain both hydrophilic and hydrophobic ends. In the presence of hydrophobic contaminants in an aqueous environment, micelle formation results when the biosurfactant molecules associate with and solubilize the hydrophobic contaminant in the center of the micelle, while the hydrophilic ends of the biosurfactant molecules remain on the outside of the micelle in contact with the aqueous phase.[19,20] These contrasting processes were well illustrated by culturing oil-degrading strains of a *Pseudomonas* sp. and *Rhodococcus* sp. separately and together in aqueous media supplemented with crude oil. The *Pseudomonas* was located predominantly in the aqueous phase and did not associate with oil droplets, whereas the *Rhodococcus* cells were concentrated at the oil–water interface.[21] The greatest degradation occurred where both mechanisms were in play, namely, when a coculture of the strains was used.

In bioremediation processes, the natural mechanisms of contaminant mobilization and pseudosolubilization, promoted by biosurfactants, may be replaced or supplemented by the use of added chemical surfactants. The chemical, nonylphenol ethoxylate surfactant, substantially increased crude oil biodegradation by a consortium of oil-degrading organisms.[22] However, biosurfactants or chemical surfactants may also inhibit degradation of contaminants, whereby isolating the contaminant in the center of the micelles effectively prevent uptake by the degrading microbes. For example, Billingsley et al.[23] observed that at chemical surfactant concentrations above the critical micelle value, the presence of an anionic surfactant promoted, whereas nonionic surfactants inhibited, biotransformation as compared to no surfactant controls. These and related topics are discussed in more detail by Ward.[24]

GENERAL MICROBIAL APPROACHES TO REMEDIATION OF CONTAMINANTS

Two general strategies for the implementation of bioremediation processes were identified around the time of the *Exxon Valdez* oil spill and cleanup, biostimulation, and bioaugmentation. The objective of biostimulation processes is to promote the growth of the segment of microbes present in the contaminated environment, which can degrade the contaminants. If the contaminated environmental medium does not contain microbes, which can degrade the contaminant, bioaugmentation with a selected inoculum already acclimated to grow on the contaminant may be needed.

Biostimulation

Most organic contaminants in the environment are carbon rich and the growth of organisms that can use this carbon source is usually limited by the availability of other nutrients, supplements, and electron acceptors. Nutrients that are most widely used in aerobic bioremediation processes are nitrogen and phosphate, and getting oxygen to the medium is often the most challenging engineering component. In surface soil bioremediation processes, oxygen is supplied from the atmosphere, and atmospheric oxygen is also exploited in reactor-based processes, engineered soil piles, and in situ processes by sparging air to the contaminated soil or slurry.[25] Other strategies involve the supply of pure oxygen gas, ozone, or manganese peroxide. In anaerobic bioremediation, delivery of adequate amounts of suitable electron acceptors to the medium is a priority strategy. Examples of electron acceptors are sulfate, nitrate, and ferric iron.

There are also other more specialized biostimulation approaches. Especially with hydrophobic contaminants, which are less accessible to microbes, surfactants, biosurfactants, or indeed, easily biodegraded solvents and vegetable oils may be added to solubilize these contaminants, releasing them from particulate matter and rendering them more amenable to biodegradation.[19,20,26] Where contaminants are typically degraded by cometabolic processes requiring the presence of primary substrates, remediation of the contaminant may be stimulated by the addition of a primary substrate. A further example of biostimulation, in the case of aerobic bioremediation of clays having low porosities that resist aeration/oxygen delivery systems, is to add low-density bulking agents such as plant materials, sawdust, and wood chips to increase the porosity of the medium and allow easy delivery of air.

Bioaugmentation

Examples where a bioaugmentation strategy has been beneficial include bioremediation of very recalcitrant contaminants such as TNT, phthalate esters, and carbon tetrachloride.[27,28] This strategy is also necessary where the contaminants are not degraded by natural organisms but where recombinant organisms have been engineered, which are capable of degrading the contaminant. Bioaugmentation is also applicable where highly accelerated biodegradation processes are required, for example, for accelerated reactor-based processes for the remediation of hydrocarbon sludges.[29] In many cases, bioaugmentation has been shown not to enhance biodegradation rates beyond what may be achieved through biostimulation, as was most notably illustrated in the case of studies during the *Exxon Valdez* oil spill cleanup.

BIOREMEDIATION TECHNOLOGIES

There are a wide range of possible approaches to the bioremediation of contaminated environments. They include soil bioremediation processes involving natural attenuation, in situ subsurface treatments and use of biopiles and composting approaches, and landfarming and bioreactor treatments. Crude oil spills at sea may be addressed through shorelines or deep-sea bioremediation strategies.

Monitored Natural Attenuation or Intrinsic Remediation

This process is the simplest and least-invasive bioremediation approach.[30] The process presumes that the environment contains sufficient nutrients to support the indigenous microbial population in the biodegradation of contaminants. Oxygen availability is often a rate-limiting factor although anaerobic biodegradation may also be important. Intrinsic remediation processes may proceed for years, if not decades, as was shown with remediation of PCBs in the Hudson riverbed.

In Situ Subsurface Bioremediation

This approach involves designing a subsurface configuration to allow growth and contaminant biodegradation by the microbes present through the efficient supply of nutrients, oxygen, or other electron acceptors through injector wells and air-sparging systems into the unsaturated or saturated zone, generally below the contaminated plume.[31] Processes may include a soil vapor extraction or other vapor containment system to prevent the release of volatile contaminants from the site to the atmosphere.

Engineered Biopiles

Engineered soil biopiles are typically fitted with a series of aeration pipes to distribute air to the contaminated soil via positive or negative pressure.[31] With soils contaminated with volatile hydrocarbons, the air pumps generally draw a vacuum such that air enters at the surface of the pile drawing any undegraded volatiles into the pipework system where they may be directed to a biofilter or other volatile organic carbon (VOC) treatment or collection system. Addition of bulking agents to the soil may be required to facilitate airflow through the soil, and care must be taken to control soil humidity such that growth and biodegradation processes are facilitated.

Soil Composting

In the application of composting strategies to soil bioremediation, biomaterials including wood chips or other plant materials are mixed with the contaminated soil in approximately equal amounts. As in general composting, aerobic biodegradation processes result in heat generation. Contaminants are degraded by the bacteria and fungi associated with the compost, and it is known that contaminants may also react irreversibly with plant and humic substance components.[31]

Landfarming

This bioremediation approach involves surface atmospheric aeration processes enhanced by frequent tilling and maintenance of moisture levels at 40–60% saturation via sprinklers or other means. The process is applicable to in situ soils contaminated near the surface or to spreading and treating excavated soils at a designated landfarming location.[31] Weathering, sorption, evaporation or volatilization, leaching, and photo-oxidation processes may cause the removal of certain hydrocarbon compounds during bioremediation, resulting in overestimation of the extent of biodegradation.[7]

Soil Bioreactors

Soil slurry bioreactors have advantages for bioremediation processes similar to the general fermentation and other bioprocesses, namely, those conditions in the mixture can approach homogeneity, and processes can be optimally controlled to promote microbial growth and contaminant biodegradation.[26] Although capital and energy costs are high in these processes, high rates and extents of degradation may be achieved with high levels of process dependability and reliability.[29]

Bioremediation of Shorelines

Marine shoreline bioremediation processes are challenging due to the potential for changing tides to wash away nutrients and microbes from the contaminated shore. The capacities of microbes to remediate contaminated soil, coastlines, and water bodies was first demonstrated on a large scale in the cleanup of the crude oil, which spilled from the Exxon-Valdez tanker which ran aground off the coast of Alaska, near Prince William Sound. Remediation of the hydrocarbons was promoted by the addition of specific oleophilic nitrogen fertilizers, which bound to the oil contaminants to support microbial growth and biodegradation.[8,9] In contrast, addition of microbial inoculants was ineffective.

Deep-Sea Bioremediation

The capacity of microbes to degrade crude oil at sea was illustrated during one of the world's largest environmental disasters, when the explosion at British Petroleum's Deep Water Horizon oil rig resulted in an estimated 4.9 million barrels of crude oil spilling into the Gulf of Mexico. On August 4, 2010, just three weeks after the well was capped, the U.S. government estimated that 50% of the oil was already gone from the Gulf, and the rest was being rapidly degraded. Dr. Kenneth Lee put all of this in a scientific perspective: "The consensus now is that microbial degradation was a major process responsible for cleaning up the Deepwater Horizon spill." (http://www.dfo-mpo.gc.ca/science/publications/article/2012/01-25-12-eng.html).

PHYTOREMEDIATION

In phytoremediation processes, removal of inorganic or organic contaminants from media such as soil is promoted by plants or associations between plant roots and microbes in soil and aqueous media.[32,33] In these processes, specific contaminants may be extracted, degraded, or immobilized. The mechanisms involved include the following: phytodegradation, whereby root exudates from the plant enhance microbial growth and contaminant biodegradation in the root surroundings, rhizofiltration where soluble contaminants are sorbed onto or into the roots, phytoextraction involving uptake of metals by roots and their translocation in the plants. The plants may be harvested for disposal, and phytovolatilization, where contaminants taken up by the plant, are possibly, but not always, modified and transpired into the atmosphere.

AIR BIOFILTRATION

Contaminants in an air stream, for example, VOCs, may be remediated through air biofiltration, whereby the contaminated air passes through a porous medium, which supports growth of microbes as attached biofilms.[34] The contaminants and oxygen are transferred from the air stream to the biofilm where biodegradation occurs, thereby decontaminating the air stream. Biofilter media include wood branches, bark or chips, peat, compost, synthetic media, coated ceramic or other particulates, and various combinations of these constituents. For effective performance, the biofilter environment must support efficient bacterial growth and facilitate effective transfer and sorption of contaminants from the air stream into the biofilm and access to the biodegrading microbes. The air stream or filtration medium must be appropriately humidified to insure that adequate water activity is maintained to support microbial growth and biodegradation while not oversaturating the medium such that air flow rate is retarded. The flow rate of the mobile phase (air in this case) is controlled to achieve a retention time in the biofilter sufficient to remove and biodegrade the contaminants in the air stream to a desired specification.

CONCLUSION

The purpose of surface or groundwater remediation is not only to enhance the biodegradation, transformation, or detoxification of pollutants, but also to protect the quality of soil to maintain environmental quality and sustain biological productivity within the ecosystem boundaries. Although biological processes are typically implemented at a relatively low cost, implementation of bioremediation technologies requires knowledge of interdisciplinary sciences involving microbiology, chemistry, engineering, ecology, and hydrogeology. Bioremediation methods have been successfully used to treat polluted soils, oily sludges, and groundwater contaminated by industrial chemicals and solvents, petroleum hydrocarbons, pesticides, and heavy metals.

REFERENCES

1. Bitton, G. *Wastewater Microbiology*, 4th Ed.; Wiley-Blackwell: New Jersey, 2011.
2. Kutzner, H.J. Microbiology of composting. In *Biotechnology*, 2nd Ed.; Rehm, H.-J., Reed, G., Eds.; Wiley-VCH: Weinheim, 2001; Vol. 3C, 35–94.
3. Storm, I.M.; Kristensen, N.B.; Raun B.M.; Smedsgaard J.; Thrane, U. Dynamics in the microbiology of maize silage during whole-season storage. J. Appl. Microbiol. **2010**, *109*, 1017–1026.
4. Shimao, M. Biodegradation of plastics. Curr. Opin. Biotechnol. **2001**, *12*, 242–247.
5. Singh, A.; Ward, O.P. Soil bioremediation and phytoremediation – an overview. In *Applied Bioremediation and Phytoremediation*; Singh, A., Ward, O.P., Eds.; Soil Biology Series; Springer-Verlag: Germany, 2004; Vol. 1, 1–12.
6. Megharaj, M.; Ramakrishnan, B.; Venkateswarlu, K.; Sethunathan, N.; Naidu, R. Bioremediation approaches for organic pollutants: A critical perspective. Environ. Intern. **2011**, *37*, 1362–1375.
7. Singh, A.; Ward, O.P., Eds.; *Biodegradation and Bioremediation, Soil Biology Series*; Springer-Verlag: Germany; 2004; Vol. 2.
8. Prince, R.C. Bioremediation of marine oil spills. Trends Biotechnol. **1997**, *15*, 158–160.
9. Head, I.M.; Martin Jones, D.; Röling, W.F.M. Marine microorganisms make a meal of oil. Nature Rev. Microbiol. **2006**, *4*, 173–182.
10. USEPA. Cleaning up the nations waste sites: markets and technology trends. Technology Innovation and Field Services Division, Office of Solid Waste and Emergency Response, EPA 542-R-04-015; 2004.
11. Van Hamme, J.; Singh, A.; Ward, O.P. Recent advances in petroleum microbiology. Microbiol. Mol. Biol. Rev. **2003**, *67*, 503–549.
12. Van Hamme, J.D. Bioavailability and biodegradation of organic pollutants – a microbial perspective. In *Biodegradation and Bioremediation*; Singh, A., Ward, O.P., Eds.; Soil Biology Series; Springer-Verlag: Germany, 2004; Vol. 2, 37–56.
13. Marques, A.P.G.C.; Rangel, A.O.S.S.; Castro, P.M.L. Remediation of heavy metal contaminated soils: an overview of site remediation techniques. Crit. Rev. Environ. Sci. Technol. **2011**, *41*, 879–914.
14. U.S. Environmental Protection Agency. *Provisional Guidance for Quantitative Risk Assessment of Polycyclic Aromatic Hydrocarbons, EPA/600/R-93/089*; USEPA: Washington, D.C., 1993.
15. Vazquez-Duhalt, R.; Marquez-Rocha, F.; Ponce, E.; Licea, A.F.; Viana, M.T. Nonylphenol, an integrated vision of a pollutant. Appl. Ecol. Environ. Res. **2005**, *4*, 1–25.
16. Crawford, R.L.; Hess, T.F.; Paszczynski, A. Combined biological and abiological degradation of xenobiotic compounds. In *Biodegradation and Bioremediation*; Singh, A., Ward, O.P., Eds.; Soil Biology Series, Springer-Verlag: Germany, 2004; Vol. 2, 251–278.
17. Ramos, J.L.; Duque, E.; Godoy, P.; Segura, A. Efflux pumps involved in toluene tolerance in *Pseudomonas putida* DOT-T1E. J. Bacteriol. **1998**, *180*, 3323–3329.
18. Singh, A.; Kuhad, R.C.; Shareefdeen, Z.; Ward, O.P. Methods for monitoring and assessment of bioremediation processes. In *Biodegradation and Bioremediation*; Singh, A., Ward, O.P., Eds.; Soil Biology Series, Springer-Verlag, Germany, 2004; Vol. 2, 279–304.
19. Van Hamme, J.D.; Singh, A.; Ward, O.P. Surfactants in microbiology and biotechnology: Part 1. Physiological aspects. Biotechnol. Adv. **2006**, *24*, 604–620.
20. Singh, A.; Van Hamme, J.D.; Ward, O.P. Surfactants in microbiology and biotechnology: Part 2. Application aspects. Biotechnol. Adv. **2007**, *25*, 99–121.
21. VanHamme, J.D.; Ward, O.P. Physical and metabolic interactions of *Pseudomonas* sp. Strain-B45 and *Rhodococcus* sp. F9-D79 during growth on crude oil and the effect of chemical surfactants thereon. Appl. Environ. Microbiol. **2001**, *67*, 4874–4879.
22. VanHamme, J.; Ward, O.P. Influence of chemical surfactants on the biodegradation of crude oil by a mixed culture. Can. J. Microbiol. **1999**, *45*, 130–137.
23. Billingsley, K.A.; Backus, S.M.; Wilson, S.; Singh, A.; Ward, O.P. Remediation of PCBs in soil by surfactant washing and biodegradation in the wash by *Pseudomonas* sp. LB400. Biotechnol. Lett. **2002**, *24*, 1827–1832.
24. Ward, O.P. Biosurfactants in biodegradation and bioremediation. Adv. Exper. Med. Biol. **2010**, *672*, 65–74.
25. Ward, O.P.; Singh, A.; Van Hamme, J.D.; Voordouw, G. Petroleum microbiology. In *Encyclopedia of Microbiology*, 3rd Edn.; Schaechter, M. Ed.; Elsevier: Amsterdam, 2009; 443–456.
26. Yapa, C.L.; Gana, S.; Ng, H.K. Application of vegetable oils in the treatment of polycyclic aromatic hydrocarbons-contaminated soils. J. Hazard. Mater. **2010**, *177*, 28–41.
27. Tyagi, M.; Fonseca, M.M.R.; de Carvalho, C.C.C.R. Bioaugmentation and biostimulation strategies to improve the effectiveness of bioremediation processes. Biodegradation **2011**, *22*, 231–241.
28. Ye, J.; Singh, A.; Ward, O.P. Biodegradation of nitroaromatics and other nitrogen-containing xenobiotics. World J. Microbiol. Biotechnol. **2004**, *20*, 117–135.
29. Ward, O.P.; Singh, A.; Van Hamme, J. Accelerated biodegradation of petroleum hydrocarbon waste. J. Ind. Microbiol. Biotechnol. **2003**, *30*, 260–270.
30. Delisle, S.; Greer, C.W. Remediation of organic pollutants through natural attenuation. In *Applied Bioremediation and Phytoremediation*; Singh, A., Ward, O.P., Eds.; Soil Biology Series, Springer-Verlag: Germany, 2004; Vol. 1, 159–186.
31. Ward, O.P.; Singh, A. Evaluation of current soil bioremediation technologies. In *Applied Bioremediation and*

Phytoremediation. Singh, A., Ward, O.P., Eds.; Soil Biology Series; Springer-Verlag: Germany, 2004; Vol. 1, 187–214.

32. Macek, T.; Kotrba, P., Svatos, A.; Novakova, M.; Demnerova, K.; Mackova, M. Novel roles for genetically modified plants in environmental protection. Trends Biotechnol. **2008**, *26*, 146–152.
33. Gerhardt, K.E.; Xiao-Dong Huang, X.-D.; Glick, B.R.; Bruce M. Greenberg, B.M. Phytoremediation and rhizoremediation of organic soil contaminants: potential and challenges. Plant Sci. **2009**, *176*, 20–30.
34. Shareefdeen, Z.; Singh, A., Eds.; *Biotechnology for Odour and Air Pollution Control*; Springer-Verlag: Germany, 2005.

Biodiversity: Agriculture

Deborah Stinner
Ohio Agriculture Research and Development Center, The Ohio State University, Wooster, Ohio, U.S.A.

Abstract
No other human activity has greater impact on the Earth's biodiversity than agriculture. Expansion of human agricultural activity around the globe historically has resulted in significant impacts on global biodiversity. Although agriculture and biodiversity often are inversely related, biodiversity enhancement can be a key organizing principle in sustainable agroecosystems.

INTRODUCTION

No other human activity has greater impact on the Earth's biodiversity than agriculture. From its origins some 12,000 years ago, the goal of agriculturists has been to enhance production of desired species over competing species. Expansion of human agricultural activity around the globe historically has resulted in significant impacts on global biodiversity in four major ways: 1) loss of wild biodiversity and species shifts resulting from conversion of native ecosystems by agroecosystems; 2) influence of agroecosystems' structure and function on agrobiodiversity; 3) offsite impacts of agricultural practices; and 4) loss of genetic diversity among and within agricultural species.[1] Although agriculture and biodiversity often are inversely related, biodiversity enhancement can be a key organizing principle in sustainable agroecosystems.[2–4]

AGRICULTURE'S IMPACT ON WILD BIODIVERSITY GLOBALLY

Historically, the earliest subsistence farmers and pastoralists had low population densities and limited technology and their small-scale patchworks of fields, pastures, and home gardens had little net effect on global biodiversity.[2] In some ecosystems, agricultural activity may have actually increased biodiversity because more diverse habitats and ecotones were created—a pattern that may still exist in some areas.[2] However, as surplus agricultural production allowed human populations to increase and with the development of civilizations, the impacts of agriculture on wild biodiversity increased, even to the point that biodiversity loss may have contributed to the decline of some ancient civilizations.[2] Since 1650, there has been at least a 600% increase in the worldwide deforestation of native ecosystems for agriculture and wood extraction that have resulted in radical changes to wild biodiversity globally.[2]

Wild biodiversity is more threatened now than at any time since the extinction of the dinosaurs, with nearly 24% of all mammals, 12% of birds, and almost 14% of plants threatened with extinction.[2] If current trends continue, it is estimated that at least 25% of the Earth's species could become extinct or drastically reduced by the middle of this century.[2] Conversion of natural ecosystems to agroecosystems is a primary cause of these alarming trends.[2,4,5] At least 28% of the Earth's land area currently is devoted to agriculture to some degree.[1] Intensive agriculture dominates 10% of the earth's total land area and is part of the landscape mosaic on another 17%, while extensive grazing covers an additional 10–20%.[2] Nearly half of the global temperate broadleaf and mixed-forest and tropical and subtropical dry and monsoon broadleaf forest ecosystems are converted to agricultural use (45.8% and 43.4%, respectively).[1,2] However, agriculture's greatest impact has been on grassland ecosystems, including temperate grasslands, savannas, and shrublands (34.2%); flooded grasslands and savannas (20.2%); and montane grasslands and shrublands (9.8%).[1] Combined, 64.2% of the Earth's grassland ecosystems have been converted to agriculture, primarily for production of cereal grasses—maize, rice and wheat.[1,2] In the past 20 years, net expansion of agricultural land has claimed approximately 130,000 km^2/yr globally, mostly at the expense of forest and grassland ecosystems, but also from wetlands and deserts.[1]

The native ecosystems that agriculture has replaced typically had high biodiversity. A hectare of tropical rain forest may contain over 100 species of trees and at least 10–30 animal species for every plant species, leading to estimates of 200,000 or more total species.[6] In contrast, the world's agroecosystems are dominated by some 12 species of grains, 23 vegetable crops, and about 35 fruit and nut crop species.[1] Furthermore, conversion of native ecosystems to agriculture causes dramatic shifts in ecosystem structure and function that affect ecosystem processes above and below ground including energy flow, nutrient cycling,

Encyclopedia of Natural Resources DOI: 10.1081/E-ENRL-120010502

water cycling, food web dynamics, and biodiversity at all trophic levels.[7] The amount of wild biodiversity loss depends on the degree of fragmentation of the native landscape. Whereas some species require vast continuous areas of native habitat, many can survive as long as the appropriate size and number of patches with connecting corridors of native habitat are left intact and provided that barriers to species movement—such as road and irrigation networks—are limited. However, when conversion leads to critical levels of native landscape fragmentation, chain reactions of biodiversity loss have been observed as interdependent species lose the resources they need to survive.[2] Loss of wild biodiversity at this level leads to loss of numerous ecosystem benefits that are essential to agriculture, e.g., 1) drought and flood mitigation; 2) soil erosion control and soil quality regeneration; 3) pollination of crops and natural vegetation; 4) nutrient cycling; and 5) control of most agricultural pests.[5]

STRUCTURE AND FUNCTION OF AGROECOSYSTEMS AND BIODIVERISTY

The structure and function of agroecosystems are largely determined by local context, including interaction of ecological conditions (including bio-, geo-, and chemical) with social factors, including farmers' economic needs, cultural and spiritual values, and social structure and technology. Two types of agrobiodiversity have been defined:[8] planned biodiversity is the specific crops and/or livestock that are planted and managed and associated biodiversity is nonagricultural species that find the environment created by the production system compatible (e.g., weeds, insect and disease pests, predators and parasites of pest organisms, and symbiotic and mutualistic species).[8] Planned and associated biodiversity can enhance stability and predictability of agroecosystems.[5] Traditional forms of agriculture—such as home gardens and shade coffee farms in the New and Old World tropics[4,6,8] and traditional Amish dairy farms in North America[9]—have a complex and diverse spatial and vertical structure and high planned and associated biodiversity. For example, traditional neotropical agroforestry systems commonly contain over 100 annual and perennial plant species per field.[4] Traditional agroecosystems create landscape patterns of small-scale diverse patches with many edges, habitat patches, and corridors for wild biodiversity.

In contrast to traditional agroecosystems, the vertical and horizontal structure of modern industrial agroecosystems is simplified into monocultures on a large scale that create landscape patterns of widespread extreme genetic uniformity with few edges, habitat patches, and corridors for dispersal.[2] For example, in the United States 60–70% of the total soybean area is planted with two to three varieties, 72% of the potato area with four varieties, and 53% of the cotton area with three varieties.[6] The structure and function of industrial livestock agriculture impose similar negative impacts on biodiversity worldwide.[10] Livestock operations for all major species—particularly swine, poultry, beef, and dairy—are becoming increasingly concentrated, with feed produced in monocultures and brought to the animals in feedlots. Even in more extensive grazing operations, although good management can increase plant biodiversity,[3,10] these systems replace native forests and/or grasslands that once supported highly diverse complexes of coadapted plants and migratory grazing and browsing ungulates and their predators.[10]

OFFSITE IMPACTS OF AGRICULTURAL ACTIVITIES

The third major way that agriculture impacts global biodiversity is through the direct and indirect offsite effects of the various managements used to maintain their structure and function. Growing annual species in large monocultures goes against ecological forces of plant community succession; therefore, a great deal of intervention is required to maintain high levels of production. Fertilizers applied to maximize production of crop plants create favorable habitat for other plant species that are adapted to nutrient-enriched conditions, including alien invasive species. Tillage, herbicides, and genetic engineering may prevent competition between crop plants and annual and perennial weeds. Widespread monocultures of nutrient enriched plants create an easily exploited resource for insect pests and disease organisms. Insecticides, fungicides, and genetic engineering may protect crops from these competitors. Furthermore, conventional cropping agroecosystems are notoriously leaky (i.e., the sheer volume of external inputs being applied in combination with soil disturbance and decreased soil quality often exceeds the capacity of the agroecosystem to absorb and process the inputs).[1] Concentrated livestock agriculture also can be a major source of chemical and biological pollution. As a result of these many factors, sediment, excess fertilizer, manure, and pesticides run off into streams and down into groundwater. Hydrological alterations to land and natural streams in combination with chemical and biological pollution cause considerable reductions in aquatic biodiversity that can extend throughout whole watershed systems.[2] The Hypoxia in the Gulf of Mexico is a dead zone that covers 18,000 km^2 where aquatic biodiversity has been drastically reduced by impacts of agriculture in the Mississippi River watershed.[2] A new concern regarding offsite impacts of modern agriculture on biodiversity is the genetic pollution that can result as genetically modified crops expand worldwide.[1] Possible transfer of genes for resistance to weeds, insects, fungi, and viruses could overwhelm wild populations and communities.[2]

LOSS OF DIVERSITY WITHIN AGRICULTURAL SPECIES

Of 7000 crop species, less than 2% are currently important, only 30 of which provide an estimated 90% of the world's calorie intake—with wheat, rice, and maize alone providing more than half of plant-derived calories.[1] Some 30–40 animal species have been used extensively for agriculture worldwide, but fewer than 14 account for over 90% of global livestock production, whereas some 30% of international domesticated breeds are threatened with extinction.[1] There are additional trends of decreased varietal and landrace diversity within crop species as more farmers adopt modern high-yielding varieties.[1,4] These alarming trends have prompted government policy recommendations whose purpose is to 1) ensure that current agricultural genetic diversity in plants is preserved in seed banks and plant and germplasm collections (ex situ) or as growing crops (in situ), particularly wild relatives of major crops and livestock breeds in their centers of origin; and 2) ensure that wild crop and livestock relatives are conserved in carefully identified natural systems1.[5]

CONCLUSION

Biodiversity as a Principle of Agroecosystem Management

Although industrial agriculture is generally inversely related to biodiversity, there are promising examples of alternative agroecosystems that protect and enhance biodiversity and are also highly productive.[2–4] Some of these include: 1) organic agriculture; 2) sustainable agriculture; 3) permaculture; 4) natural system agriculture; 5) holistic management; and 6) ecoagriculture.[2,3] These models are based on ecological principles and the assumption that biodiversity can contribute significantly to sustainable agricultural production. Within ecoagriculture, the following strategies are proposed to protect and enhance wild biodiversity: 1) create biodiversity reserves that also benefit local farming communities; 2) develop habitat networks in nonfarmed areas; 3) reduce (or reverse) conversion of wild lands to agriculture by increasing farm productivity; 4) minimize agricultural pollution; 5) modify management of soil, water, and vegetation resources; and 6) modify farming systems to mimic natural ecosystems.[2] Examples of specific management practices that sustain or enhance biodiversity include: 1) hedgerows; 2) dykes with wild herbage; 3) polyculture; 4) agroforestry; 5) rotation with legumes; 6) dead and living mulches; 7) strip crops, ribbon cropping, and alley cropping; 8) minimum tillage, no-tillage, and ridge tillage; 9) mosaic landscape porosity; 10) organic farming; 11) biological pest control and integrated pest management; 12) plant resistance; and 13) germplasm diversity.[11] New research, particularly if conducted in a participatory mode with farmers, should lead to many more ways to protect and enhance biodiversity, including: rotational grazing of high-diversity grasslands for dairy and beef cattle production, timber and pulp production systems that use perennial plants, high-diversity mixtures of single annual crops and/or rotational diversity, and precision agriculture that closely matches small-scale soil conditions with optimal crop genotypes.[5] Although specific management practices are helpful, also needed are whole-farm planning approaches and decision-making processes that encompass farmers' values and economic needs, in addition to environmental concerns for biodiversity[2] (e.g., holistic management).[3] More research and education and policies that encourage farmers and consumers to appreciate the ecological, economic, and quality-of-life values of biodiversity are needed to thwart current threats to global

Fig. 1 Ecoagricultural strategies in practice.

biodiversity and agrobiodiversity, and to address the many challenges and opportunities of global sustainable food security.[4]

REFERENCES

1. Wood, S.; Sebastian, K.; Scherr, S.J. *Pilot Analysis of Global Ecosystems: Agroecosystems*; International Food Policy Research Institute and World Resources Institute: Washington, DC, 2000.
2. McNeely, J.A.; Scherr, S.J. *Ecoagriculture: Strategies to Feed the World and Save Wild Biodiversity*; Island Press: Washington, 2003.
3. Stinner, D.H.; Stinner, B.R.; Martsolf, E. Biodiversity as an organizing principle in agroecosystem management: Case studies of holistic resource management practitioners in the USA. Agric. Ecosys. Environ. **1997**, *62* (2), 199–213.
4. http://www.wri.org/wri/sustag/lab-01.html (accessed February 2003).
5. http://www.cast-science.org/bid.pdf (assessed December 2002).
6. Altieri, M.A. The ecological role of biodiversity in agroecosystems. Agric. Ecosys. Environ. **1999**, *74* (1–3), 19–31.
7. Swift, M.J.; Anderson, J.M. Biodiversity and Ecosystem Function in Agricultural Systems. In *Biodiversity and Ecosystem Function*; Schuleze, E.D., Mooney, H.A., Eds.; Springer-Verlag: New York, 1993; 15–41.
8. Vandermeer, J. Biodiversity loss in and around agroecosystems. In *Biodiversity and Human Health*; Grifo, F., Rosenthal, J., Eds.; Island Press: Washington, DC, 1997; 111–127.
9. Moore, R.H.; Stinner, D.H.; Kline, D.; Kline, E. Honoring creation and tending the garden: Amish views of biodiversity. In *Cultural and Spiritual Values of Biodiversity*; Posey, D.A., Ed.; United Nations Environment Programme, Intermediate Technology Publications: London, 1999; 305–309.
10. Blackburn, H.W.; de Haan, C. Livestock and biodiversity. In *Biodiversity in Agroecosystems*; Collins, W.W., Qualset, C.O., Eds.; CRC: New York, 1999; 85–99.
11. Paoletti, M.G.; Pimental, D.; Stinner, B.R.; Stinner, D. Agroecosystem biodiversity: Matching production and conservation biology. In *Biotic Diversity in Agroecosystems*; Paoletti, M.G., Pimental, D., Eds.; Elsevier: New York, 1992; 3–23.

Biodiversity: Climate Change

Lisa Freudenberger
Jan Hanning Sommer
Center for Development Research (ZEF), University of Bonn, Bonn, Germany

Abstract

Climate and biodiversity are interacting in multiple and complex ways. Climate parameters such as temperature and precipitation influence the distribution of species and ecosystems, while in turn biodiversity can affect climate. Therefore, climate change is expected to have a dramatic effect on biodiversity at all levels including species and their habitats, species interactions, ecosystems, and genetic diversity. As biodiversity is the basis for many ecosystem services humanity depends on, climate change may also affect the constant provision of these services. Different concepts and methods have been developed to explain the relationship between climate change and biodiversity and to predict future biodiversity changes under climate change scenarios. This task represents a major challenge as it is associated with a high degree of complexity and uncertainty, but provides valuable output for science and policy. The most prominent strategies to tackle climate change are either to mitigate its incidence or to adapt to its implications. Climate change mitigation refers to the avoidance of climate-relevant emissions while climate change adaptation describes adjustment measures of the natural or social system to cope with new climatic conditions. Both, climate change mitigation and adaptation strategies can be based on pure technological solutions or the natural functions of biodiversity making use of ecosystem-based or ecosystem engineering strategies. In the past, biodiversity conservation and climate change policy have often operated in isolation while joint actions would have been indicated. Recently, a closer collaboration was started and international initiatives were established to facilitate the utilization of synergies and cross-cutting issues.

INTRODUCTION

Biodiversity, the diversity of life on Earth and its evolution, is closely linked to climate conditions, dynamics, and climate history.[1–3] From the poles to the tropics, very strong climate gradients exist that in turn influence the distribution of species and the patterns of biodiversity.[4,5] Temperature and precipitation are the main factors in this regard. There is strong empirical evidence that broad-scale gradients of species richness can be best predicted by global climatic conditions in combination with topographical parameters such as altitude and topographical diversity.[6–8] The habitat of each species is delimited, amongst others, by its physiological level of tolerance to deal with climatic extremes, such as drought or frost. Moreover, climate affects the characteristics of soils such as the availability of nutrients and the water cycle. Finally, not only are temperature and precipitation important but also the length and intensity of solar radiation. The sun represents the global source of energy and hence facilitates primary productivity, the production of plant biomass by photosynthesis, and influences metabolism rates in general. The cycling of energy within an ecosystem is positively correlated with its complexity and functionality and thus with its species richness, genetic diversity, and biomass productivity. This is leading to higher energy dissipation causing a positive feedback loop.[9,10]

Climate change means a mid- to long-term change in one or more climate parameters, such as an increase of temperature, or changes in the magnitude and frequency of extreme events such as droughts and heavy rainfall events. It is to a large extent driven by a rise of the concentration of greenhouse gasses in the atmosphere, such as carbon dioxide.[11] Carbon dioxide, on the other hand, also has a fertilizing effect as it supports plant growth and biomass production.[12]

Climate and biodiversity are interdependent, as biodiversity also influences the climate on the micro- to mesoscale. Biodiversity facilitates the provisioning of ecosystem services including climatically relevant functions such as temperature regulation and water cycling. It has been shown, for example, that higher amounts of biomass have a cooling effect (e.g., the climate in cities can be improved by parks) and dead wood in forests is leading to the stabilization of microclimatic conditions.[13] As a result, climatic conditions are not only affecting the distribution and evolution of biodiversity but also vice versa.[14,15]

Encyclopedia of Natural Resources DOI: 10.1081/E-ENRL-120047624

THE IMPACT OF CLIMATE CHANGE ON BIODIVERSITY

Climate change is affecting biodiversity at all its levels and at different spatial and temporal scales. It has a potential impact not only on the individual and species level but also on the ecosystem level affecting ecosystem functionality and services. For a systematic classification, see Geyer et al.[16]

Impact on Individuals, Populations, and Habitats

Probably the most studied impacts of climate change can be found on the individual and population levels of animals and plants. These include changes in their behavior and physiology comprising their metabolism, growth rates, life cycling, immune functions, and also their population dynamics such as changes in gene pools, dispersal abilities, population growth rates, or sex ratios.[17] Climate change is also often having an impact on species distributions through changes in habitat quality. This may result in the spatial shift of species distributions if new suitable areas are available and if the species is able to reach them. However, in many cases it is more likely that species will not be able to adapt, and populations or even whole species may become extinct as conditions turn unsuitable.[18] For example, corals are suffering from increasing water temperatures (see Table 1).

Species in some ecosystem types are more vulnerable than others as, e.g., at mountain tops where no cooler refuges are available when temperature rise drives species uphill.[19] But also other types of physical boundaries such as roads, dams, or fragmented habitats in general may restrict species to escape unsuitable conditions.[20] Therefore, habitat connectivity and landscape permeability are critical to mitigate those negative effects. Fragmentation may also cause the loss of genetic diversity, as species populations are becoming increasingly isolated and genetic exchange is delimited. This leads to a decrease in genetic or intraspecific diversity.[21] Reduced levels of intraspecific diversity are also relevant for the adaptive capacity of species under changing environmental conditions since higher genetic diversity increases the probability that some individuals are able to survive and reproduce under the changing conditions.[22] Lower genetic diversity on the other hand is likely to negatively affect the persistence of species under changing conditions.

Impacts on the Interactions between Species and the Community

Climate change impacts species interactions and community structure in many different ways. Often, this is a consequence of temporal or spatial decoupling between the interacting species, resulting in changes or loss of trophical relationships and other interactions.[24] A popular example is the temporal decoupling between the earlier flowering times of plants well before the awakening from hibernation of bees as their pollinators.[25] At the same time, new interactions may become apparent under climate change, e.g., through the spread of invasive species.[26] "Winners" of climate change may extend their distribution ranges into areas where there are fewer competitors or predators and can reach high densities. All these changes may lead to general alterations in species composition and community structure having an impact in turn on other biotic and abiotic factors of whole ecosystems.

Impact on Ecosystems and Ecosystem Services

Climate change has a direct impact on abiotic conditions such as humidity, evaporation, wind patterns, microclimatic conditions, snow loads, water bodies, and the condition of soils. Impacts on ecosystem structure, processes, and dynamics include changes in photosynthetic activities, and the frequency and intensity of disturbances, e.g., from floods or fire events, nutrient cycling, and succession.[27] Finally, not only species are likely to shift their distribution under altered climatic conditions but also ecosystem types.[28,29] The melting of the permafrost soil leading to a spatial increase of the Taiga vegetation or the drying of the Mediterranean region are just two

Table 1 Coral bleaching

Coral reefs are described as the rainforests of the sea as they represent the most diverse marine habitat. They are, however, very vulnerable to climate change since water temperature increase is causing large-scale diebacks of coral reefs through a process called coral bleaching. Corals live in symbiosis with algae (zooxanthellae) that grow in corals, give them their specific color, and provide them with nutrients. When the temperature increases, zooxanthellae start to produce toxins leading to the release of the algae from the coral and the loss of the corals' color. Corals are able to survive a limited time without nutrients from the symbiotic algae but a general increase of water temperature and temperature fluctuations during the last years coupled with other stressing factors such as overfishing, water pollution, acidification, and mechanical destruction has already caused the loss of large proportions of coral reefs. About 19% of the original coral reefs have already been lost and about 35% are threatened. Predictions estimate that about 50% of the world's human population will live along coasts by 2015.[23] Altogether, these stressors are likely to lead to ever increasing pressures on coral reefs and the loss of essential ecosystem services they provide.

examples of climate change impacts on ecosystem-type distributions.

Changes in ecosystem functions, processes, and distributions will also affect ecosystem services.[30] In some areas of the world, extreme events such as droughts and floods will have tremendous effects on the availability of food, fiber, timber, and water. Climate change will also influence several supporting services, such as water cycling, primary production, and nutrient cycling as well as different regulating services like pest regulation, seed dispersal, and water purification. Finally, climate change is having a general impact on the appearance of ecosystems and therefore on their spiritual, religious, or aesthetical value. Altogether, these impacts on ecosystems may directly or indirectly influence human society and feed conflicts about scarce resources, food, and clean water.

System Resilience to Climate Change

Climate change impacts are not uniformly distributed and different biodiversity elements will suffer from climate change to varying degrees. The vulnerability of biodiversity, e.g., an ecosystem or a certain species, is a function of its exposure and sensitivity toward climate change as well as its ability to adapt to these changes.[11,31] This means that an element of biodiversity that has a high exposure towards climate change can still be less vulnerable than another element with low exposure as long as it is either less sensitive or has the ability to adapt to these changes, e.g., through the spatial shift of a species distribution. This is connected to the related concept of resilience. Traditionally, resilience has been described as the ability of a system or a species to return to its original state after disturbance. However, recent trends in ecological studies have highlighted the dynamic behavior of ecosystems, therefore giving rise to a different definition where resilience means merely the ability of a system to return to a stable state that could be different from its original one.[32] According to this concept, systems can have different stable states and it is possible that an ecosystem can transform to another without losing its functionality. The potential nonlinear impact of climate change on ecosystems and biodiversity has also been described with the concept of tipping points (see Figure 1 and Table 2). These are "points of no return" beyond which it is impossible to stop or even reverse the negative impacts from climate change.[33] Different definitions for all of these concepts can be found in literature and a discussion on the relation between them and their differentiation is still ongoing.

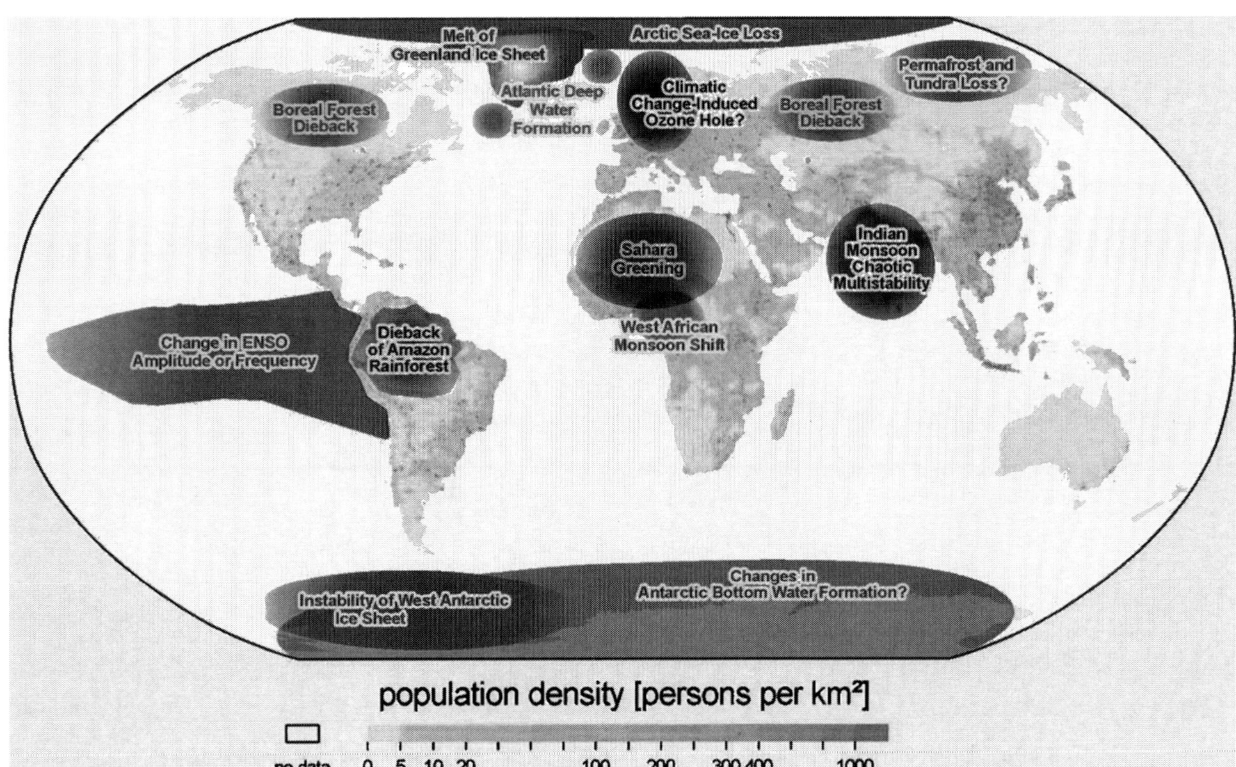

Fig. 1 Map of global tipping elements.
Source: Adapted from Schellnhuber & Held,[55] Lenton et al.,[33] by permission of Oxford University Press.

Table 2 Tipping points

Tipping points describe points in time when a system does not develop any more in a way that can be predicted by previous linear relationships but reaches a point of no return beyond which a trend has either significantly changed its pace or its direction. Tipping elements in the climate system describe global subsystems that are likely to shift to a completely different state if climate change continues and tipping points are the corresponding points in time (see Figure 1).[33] These elements have been self-stabilizing over a long period of time, and may most likely not recover once they are destroyed. Climate change impacts become much more difficult to predict after tipping points have been reached or complex feedback mechanisms are getting more important.

METHODOLOGICAL AND CONCEPTUAL APPROACHES TO ASSESS THE IMPACTS OF CLIMATE CHANGE

The impacts of climate change on biodiversity are often difficult to assess since predictions of the future are difficult to verify with recent or historical data, especially if the expected future conditions have never existed in the past. Gaining evidence by conducting field experiments is very time consuming and may provide results only at a very late temporal stage, and lab experiments are not able to tackle the response of biodiversity and ecosystems in their full complexity. Additionally, climate change–induced impacts are often overshadowed and camouflaged by other stressors such as landscape fragmentation, land degradation, landuse intensification, pollution, deforestation, or over-exploitation in general. Nevertheless, different methodologies have been developed to assess the impacts of climate change on biodiversity.[34–36]

Experimental Approaches

Classical empirical research is designed as a systematic experiment that can take place either in the field or within the laboratory. Species are exposed to different environmental conditions, e.g., higher temperature or lower precipitation rates, alone or in combination to test how climate change may affect species and ecosystems. While lab experiments have the advantage of enabling the control of external factors, field experiments capture a higher level of complexity. Different studies have been conducted monitoring, e.g., growth rates of plants under increased temperature and higher carbon dioxide levels.[37] Empirical evidence is of high importance to understand the underlying processes of climate change and biodiversity interactions. However, their practical value often remains limited as they are usually focused on a relatively narrow research question, a specific location, a particular ecosystem type, and a specific set of species. Nevertheless, empirical research is providing the real world data necessary to extrapolate and estimate the future impact of climate change.

Modeling Approaches and Scenario Development

The impacts of climate change on biodiversity have often been assessed through modeling approaches. Modeling approaches can be categorized as either statistical or mechanistic, although hybrid forms of both also exist. Statistical models use empirical data to find correlations between, e.g., the number of species and recent climatic conditions that may be used to assess the impact of future climate conditions on species richness. For example, biodiversity parameters for an area of interest can be compared to those of other areas in the world that today already have climate conditions similar to the expected future ones of the area of interest (space for time substitution approach).[38,39] Mechanistic or process-based models incorporate generally known relationships with the aim to reproduce the processes relevant for the ecosystem or species studied. Common examples for the application of mechanistic models are vulnerability assessments of ecosystems under climate change scenarios.[28] Models provide valuable information for research, nature conservation, and decision making but are often associated with a high degree of uncertainty. Climate change scenarios are also often combined with different specifications to assess, e.g., the impact of different policy options.[40,41] As climate change is modeled based on different emission scenarios, modeling outputs can be highly dependent on the socio-economic assumptions being made under the specific scenario. Furthermore, there are different uncertainties that are resulting from complex climate–biodiversity–society interactions.[42,43] When interpreting the results of model outputs, it is therefore important to take all model restrictions into consideration and to understand the results as one out of many potential futures.

Vulnerability Indices and Relative Assessments

While modeling approaches usually rely to some extent on quantitative empirical data from the past, there are other approaches that are utilizing heuristics to assess potential future impacts of climate change. These assessments are often only relative measurements or indices and can only be interpreted when comparing different species, ecosystems, or areas. A common concept to assess the impacts of climate change on biodiversity is, e.g., the vulnerability framework.[31] Different indicators or expert knowledge are used to assess the sensitivity, adaptive capacity, and exposure to climate change and are combined in a vulnerability index. These relative assessments have the advantage of combining various different proxies often with different units of measurement while being easy to interpret.[44] However, they do not provide true quantitative data that can be directly validated by field measurements or interpreted on its own without comparing the index value to different areas or times.

Complexities of Climate Change Impacts on Biodiversity and Non-Knowledge Management

Scenario development and modeling are very important tools to deal with the uncertainties and knowledge gaps of climate change and its impacts. However, it should be noted that dealing with ecosystems and climate change means dealing with uncertainties and non-knowledge.[45] The precautionary concept that is promoted by the United Nations Convention on Biodiversity (CBD), which was signed at the Rio Earth Summit in 1992, becomes particularly relevant in terms of biodiversity conservation under climate change.[46] It endorses a proactive policy approach, where future risks are minimized and actions are not necessarily prevented by knowledge gaps. The impacts of climate change on biodiversity are characterized by a high level of complexity, feedback mechanisms, and non-linear dependencies. It is difficult to capture this complexity through simple statistical relationships and to assess future impacts of climate change on biodiversity. Different concepts to describe these complex relationships have been introduced, all of which deal with the high level of uncertainty. Although we can develop many different models and scenarios for future climate change, each of them corresponds only to one of many possible futures and it is very unlikely that the real future will exactly match a model that is being developed today. Nevertheless, we can use these models or even non-quantitative scenarios and heuristics to identify trends and paint pictures of how the future might look.

CLIMATE CHANGE MITIGATION AND ADAPTATION

Conservation biology used to be a relatively isolated political field motivated mainly by ethical or even religious obligations to protect nature. In the past 10 years, a more anthropocentric perspective has advanced where biodiversity is not only appreciated for its inherent value but also for the services that are provided to human society. From this perspective, biodiversity conservation is indispensable for the provision of ecosystem services that humanity is depending on, such as the production of food or the filtration of water.[47] Another development has been the consideration of the interconnections between climate change and biodiversity. It has been realized by both, the biodiversity conservation as well as the climate change community, that synergies as well as conflicts between both policy fields exist that need to be dealt with.[48–51] There are two categories, mitigation and adaptation, that conceptualize how nature conservation can contribute to the management of climate change impacts. While mitigation corresponds to actions undertaken to alleviate future climate change, adaptation measures are aiming at enabling biodiversity to persist under climate change. Mitigation approaches mainly refer to strategies targeting the reduction of greenhouse gases in the atmosphere through either the minimization of emissions, e.g., the protection of carbon sources, or the maximization of greenhouse gas uptake, e.g., the protection or restoration of carbon sinks. On the other hand, adaptation measurements assume that at least some intensity of climate change will take place and that biodiversity can only be protected effectively if adaptive strategies are employed. Despite the straightforward definition of mitigation and adaptation, both are often not isolated but interconnected. In many cases, adaptation measures are only applicable if climate change is not exceeding critical thresholds and, vice versa, ecosystems are more likely to contribute effectively to climate change mitigation if they are in a healthy condition.

During the last few years, a new concept has become increasingly important, namely, ecosystem-based adaptation and mitigation.[52,53] Ecosystem-based adaptation refers to the utilization and strengthening of natural functions of ecosystems to make societies less vulnerable to the negative impacts of climate change. One example for ecosystem-based adaptation would be the abandonment of alluvial areas for construction in order to maintain space for flood protection, but other examples also exist where natural solutions can provide more effective and efficient options

Table 3 Examples for strategies aiming at the mitigation of climate change and adaptation toward its impacts

	Mitigation of climate change through higher carbon sequestration	**Adaptation** to increased frequency of floods through climate change
Technological solution	Use of carbon capture and storage technology (CCS)	Building of artificial dams for flood protection
Ecological engineering	Rewetting of peatlands to increase carbon capture	Restoration and control of alluvial areas for flood protection
Ecosystem based	Protection of peatlands to maintain natural carbon capture functionality	Abandoning alluvial areas for natural flood protection

than technological advances for adaptation to climate change. Ecosystem-based mitigation follows the same approach but with the aim of increasing the capacity of ecosystems to store and capture carbon, e.g., through the conservation of natural carbon sinks and sources. Ecological engineering, on the other hand, focuses on actions coupling both technological and ecological solutions to benefit society.[54] These often relate to the restoration or enhancement of natural ecosystem functions or the use of these functions within technological solutions. One of the reasons that these two concepts, ecosystem-based actions and ecological engineering, have gained in relevance is their potential to explicitly combine aspects of climate change policy with biodiversity conservation. Table 3 gives examples for mitigation and adaptation strategies for all three options. While nature conservation and climate change policy have often acted independently from each other, it has been realized that collaborative approaches often provide synergies and beneficial effects for both policy areas. Furthermore, it turned out that pure technological solutions aiming at the mitigation and adaptation to climate change often entail substantial costs and efforts. Ecosystem-based actions and ecological engineering have the potential to provide cost-effective strategies to cope with future climate change.

POLITICS OF CLIMATE CHANGE AND BIODIVERSITY

Different international initiatives exist to tackle the challenge of climate change and its impact on biodiversity. While both the United Nations Framework Convention on Climate Change (UN-FCCC) and the Convention on Biological Diversity (CBD) were founded at the Earth Summit 1992 in Rio de Janeíro, they remained relatively isolated in the following 20 years. It is only recently that a better integration of climate change policy and biodiversity conservation goals is getting higher priority on the political agenda. The UN-REDD (Reducing Emissions from Deforestation and Forest Degradation) program, which is part of the Clean Development Mechanism of the Kyoto Protocol[11] and currently probably the most powerful instrument for international climate change mitigation, aims at the reduction of greenhouse gases through the creation of financial incentives paid for the mitigation of deforestation and forest degradation. To avoid potential negative effects on biodiversity and society, it has been developed towards the REDD+ mechanism, with a stronger focus on multiple benefits including biodiversity conservation. Climate change has also been officially included as a main cross-cutting issue of the CBD since 2004 after the establishment of an Ad Hoc Technical Expert Group (AHTEG) on climate change and biodiversity in 2001. During the last few years, the Rio Conventions Pavilion has been founded at the 10th Conference of Parties of the CBD in Nagoya (Japan, 2010) as a platform aiming at the identification of synergies between the different conventions that were established in 1992 in Rio. In 2012, as a pendant to the Intergovernmental Panel on Climate Change, a new Intergovernmental Platform on Biodiversity and Ecosystems Services (IPBES) has been established, an independent intergovernmental body to strengthen the science policy interface in the field of biodiversity and ecosystem services. In consideration of the various interactions between climate change and biodiversity, a collaboration and coordination between both policy areas is likely to become increasingly necessary.

CONCLUSION

Climate change and biodiversity represent both very complex issues, and many answers to the question of how they are interacting are still elusive. However, there is strong evidence that climate change will have negative impacts on biodiversity, ecosystem service supply, and thereby on human society. The paradigm of natural sciences to base research on empirical data can only be partly fulfilled in the context of climate change research, as research questions include predictions about possible future developments not experienced in the past. Nevertheless, the expected pace and magnitude of climate change suggests that prompt and flexible actions are needed to prevent unbearable and irreversible negative impacts. Both climate change mitigation and adaptation strategies need to be implemented, incorporating concepts of conservationists and climate researchers in order to gain the best possible achievements.

ACKNOWLEDGMENTS

The authors thank the Center for Development Research, University of Bonn, the Nees Institute for the Biodiversity of Plants, University of Bonn, the Centre for Econics and Ecosystem Services, and the University for Sustainable Development Eberswalde for their support in conducting research on climate change and biodiversity and for the development of this entry.

REFERENCES

1. O'Brien, E. Water-energy dynamics, climate, and prediction of woody plant species richness: An interim general model. J. Biogeogr. **1998**, *25* (2), 379–398.
2. Hawkins, B.A.; Field, R.; Cornell, H.V.; Currie, D.J.; Guégan, J.F.; Kaufman, D.M.; Kerr, J.T.; Mittelbach, G.G.; Oberdorff, T.; O'Brien, E.M.; Porter, E.E.; Turner, J.R.G. Energy, water and broad-scale geographic patterns of species richness. Ecology **2003**, *84* (12), 3105–3117.
3. Currie, D.J.; Mittelbach, G.G.; Cornell, H.V.; Field, R.; Guégan, J.F.; Hawkins, B.A.; Kaufman, D.M.; Kerr, J.T.; Oberdorff, T.; O'Brien, E.; Turner, J.R.G. Predictions

and tests of climate-based hypotheses of broad-scale variation in taxonomic richness. Ecol Lett. **2004**, *7* (12), 1121–1134.
4. Barthlott, W.; Mutke, J.; Rafiqpoor, M.D.; Kier, G.; Kreft, H. Global centres of vascular plant diversity. Nova Acta Leopoldina, *92*, 1–83.
5. Kier, G.; Mutke, J.; Dinerstein, E.; Ricketts, T.H.; Küper, W.; Kreft, H.; Barthlott, W. Global patterns of plant diversity and floristic knowledge. J. Biogeogr. **2005**, *32* (7), 1107–1116.
6. Field, R.; Hawkins, B.A.; Cornell, H.V.; Currie, D.J.; Diniz-Filho, J.A.F.; Guégan, J.F.; Kaufman, D.M.; Kerr, J.T.; Mittelbach, G.G.; Oberdorff, T.; O'Brien, E.M.; Turner, J.R.G. Spatial species-richness gradients across scales: A meta-analysis. J. Biogeogr. **2009**, *36* (1), 132–147.
7. Sommer, J.H.; Kreft, H.; Kier, G.; Jetz, W.; Mutke, J.; Barthlott, W. Projected impacts of climate change on regional capacities for global plant species richness. Proc. R. Soc. B Biol. Sci. **2010**, *277* (1692), 2271–2280.
8. Kreft, H.; Jetz, W. Global patterns and determinants of vascular plant diversity. Proc. Natl. Acad. Sci. **2007**, *104* (14), 5925–5930.
9. Fath, B.D.; Jørgensen, S.E.; Patten, B.C.; Straskraba, M. Ecosystem growth and development. Biosystems **2004**, *77* (1–3), 213–228.
10. Jørgensen, S.E.; Patten, B.C.; Straškraba, M. Ecosystems emerging. Ecol. Model. **2000**, *126* (2–3), 249–284.
11. IPCC. *Climate Change 2007: Impacts, Adaptation and Vulnerability*, Working Group II Contribution to the Fourth Assessment Report of the Intergovernmental Panel on Climate Change; IPCC Secretariat: Geneva, 2008.
12. Ainsworth, E.A.; Rogers, A. The response of photosynthesis and stomatal conductance to rising [CO_2]: Mechanisms and environmental interactions. Plant Cell Environ. **2007**, *30* (3), 258–270.
13. Norris, C.; Hobson, P.; Ibisch, P.L. Microclimate and vegetation function as indicators of forest thermodynamic efficiency. J. Appl. Ecol. **2012**, *49* (3), 562–570.
14. Battersby, S. Call in the clouds. New Sci. **2013**, *218* (2923), 32–35.
15. Bonan, G.B. Forests and climate change: Forcings, feedbacks, and the climate benefits of forests. Science **2008**, *320* (5882), 1444–1449.
16. Geyer, J.; Kiefer, I.; Kreft, S.; Chavez, V; Salafsky, N.; Jeltsch, F.; Ibisch, P.L. Classification of climate-change-induced stresses on biological diversity. Conserv. Biol. **2011**, *25* (4), 708–715.
17. Walther, G.-R.; Post, E.; Convey, P.; Menzel, A.; Parmesan, C.; Beebee, T.J.C.; Fromentin, J.-M.; Guldberg, O.H.; Bairlein, F. Ecological responses to recent climate change. Nature **2002**, *416* (6879), 389–395.
18. Thomas, C.D.; Cameron, A.; Green, R.E.; Bakkenes, M.; Beaumont, L.J.; Collingham, Y.C.; Erasmus, B.F.N.; de Siqueira, M.F.; Grainger, A.; Hannah, L.; Hughes, L.; Huntley, B.; van Jaarsveld, A.S.; Midgley, G.F.; Miles, L.; Huerta, M.A.O.; Peterson, A.T.; Phillips, O.L.; Williams, S.E. Extinction risk from climate change. Nature **2004**, *427* (6970), 145–148.
19. Thuiller, W. Climate change threats to plant diversity in Europe. Proc. Natl. Acad. Sci. **2005**, *102* (23), 8245–8250.
20. Opdam, P.; Wascher, D. Climate change meets habitat fragmentation: linking landscape and biogeographical scale levels in research and conservation. Biol. Conserv. **2004**, *117* (3), 285–297.
21. Keller, I.; Largiader, C.R. Recent habitat fragmentation caused by major roads leads to reduction of gene flow and loss of genetic variability in ground beetles. Proc. R. Soc. B Biol. Sci. **2003**, *270* (1513), 417–423.
22. Jump, A.S.; Marchant, R.; Peñuelas, J. Environmental change and the option value of genetic diversity. Trends Plant Sci. **2009**, *14* (1), 51–58.
23. Wilkinson, C., Ed. *Status of Coral Reefs of the World: 2008*; Townsville, Australia, 2008.
24. Winder, M.; Schindler, D.E. Climate change uncouples trophic itneractions in an aquatic ecosystem. Ecology **2004**, *85* (8), 2100–2106.
25. Kudo, G.; Ida, T.Y. Early onset of spring increases the phenological mismatch between plants and pollinators. Ecology **2013**, *94* (10), 2311–2320.
26. Mainka, S.A.; Howard, G.W. Climate change and invasive species: Double jeopardy. Integr. Zoolog. **2010**, *5* (2), 102–111.
27. Littell, J.S.; Oneil, E.E.; McKenzie, D.; Hicke, J.A.; Lutz, J.A.; Norheim, R.A.; Elsner, M.M. Forest ecosystems, disturbance, and climatic change in Washington State, USA. Climatic Change **2010**, *102* (1–2), 129–158.
28. Scholze, M.; Knorr, W.; Arnell, N.W.; Prentice, I.C. A climate-change risk analysis for world ecosystems. Proc. Natl. Acad. Sci. **2006**, *103* (35), 13116–13120.
29. Gonzalez, P.; Neilson, R.P.; Lenihan, J.M.; Drapek, R.J. Global patterns in the vulnerability of ecosystems to vegetation shifts due to climate change. Global Ecol. Biogeogr. **2010**, *19* (6), 755–768.
30. Mooney, H.; Larigauderie, A.; Cesario, M.; Elmquist, T.; Guldberg, O.H.; Lavorel, S.; Mace, G.M.; Palmer, M.; Scholes, R.; Yahara, T. Biodiversity, climate change, and ecosystem services. Curr. Opin. Environ. Sustainability **2009**, *1* (1), 46–54.
31. Schröter, D. Ecosystem service supply and vulnerability to global change in Europe. Science **2005**, *310* (5752), 1333–1337.
32. Walker, B.H.; Gunderson, L.H.; Kinzig, A.P.; Folke, C.; Carpenter, S.; Schultz, L. A handful of heuristics and some propositions for understanding resilience in social-ecological systems. Ecol. Soc. **2006**, *11* (1), 13.
33. Lenton, T.M.; Held, H.; Kriegler, E.; Hall, J.W.; Lucht, W.; Rahmstorf, S.; Schellnhuber, H.J. Tipping elements in the Earth's climate system. Proc. Natl. Acad. Sci. U.S.A. **2008**, *105* (6), 1786–1793.
34. Guisan, A.; Thuiller, W. Predicting species distribution: offering more than simple habitat models. Ecol Lett. **2005**, *8* (9), 993–1009.
35. Araujo, M.B.; Rahbek, C. How does climate change affect biodiversity? Science **2006**, *313* (5792), 1396–1397.
36. Algar, A.C.; Kharouba, H.M.; Young, E.R.; Kerr, J.T. Predicting the future of species diversity: macroecological theory, climate change, and direct tests of alternative forecasting methods. Ecography **2009**, *32* (1), 22–33.
37. Ghannoum, O.; Phillips, N.G.; Conroy, J.P.; Smith, R.A.; Attard, R.D.; Woodfield, R.; Logan, B.A.; Lewis, J.D.; Tissue, D.T. Exposure to preindustrial, current and future atmospheric CO_2 and temperature differentially affects growth and photosynthesis in *Eucalyptus*. Global Change Biol. **2010**, *16* (1), 303–319.

38. Hijmans, R.J.; Graham, C.H. The ability of climate envelope models to predict the effect of climate change on species distributions. Global Change Biol. **2006**, *12* (12), 2272–2281.
39. Dolgener, N.; Freudenberger, L.; Schneeweiss, N.; Ibisch, P.L.; Tiedemann, R. Projecting current and potential future distribution of the Fire-bellied toad *Bombina bombina* under climate change in north-eastern Germany. Reg. Environ. Change **2013**, doi:10.1007/s10113-013-0468-9.
40. Moss, R.H.; Edmonds, J.A.; Hibbard, K.A.; Manning, M.R.; Rose, S.K.; van Vuuren, D.P.; Carter, T.R.; Emori, S.; Kainuma, M.; Kram, T.; Meehl, G.A.; Mitchell, J.F.B.; Nakicenovic, N.; Riahi, K.; Smith, S.J.; Stouffer, R.J.; Thomson, A.M.; Weyant, J.P.; Wilbanks, T.J. The next generation of scenarios for climate change research and assessment. Nature **2010**, *463* (7282), 747–756.
41. Parry, M.L.; Rosenzweig, C.; Iglesias, A.; Livermored, M.; Fischere, G. Effects of climate change on global food production under SRES emissions and socio-economic scenarios. Global Environ. Change **2004**, *14* (1), 53–67.
42. Hansen, A.J.; Neilson, R.P.; Dale, V.H.; Flather, C.H.; Iverson, L.R.; Currie, D.J.; Shafer, S.; Cook, R.; Bartlein, P.J. Global change in forests: Responses of species, communities, and biomes. BioScience **2001**, *51* (9), 765.
43. Freudenberger, L.; Schluck, M.; Hobson, P.; Sommer, H.; Cramer, W.; Barthlott, W.; Ibisch, P.L. A view on global patterns and interlinkages of biodiversity and human development. In *Interdependence of Biodiversity and Development Under Global Change*; Ibisch, P.L., Vega, A.E., Herrmann, T.M., Eds.; Secretariat of the Convention on Biological Diversity: Montreal, Quebec, Canada, 2010; Vol. 54, pp. 37–56.
44. Saltelli, A.; Nardo, M.; Saisana, M.; Tarantola, S. Composite indicators–The controversy and the way forward. In *Statistics, Knowledge and Policy: Key Indicators to Inform Decision Making*; OECD, Ed., 2005, pp. 359–373.
45. Ibisch, P.L.; Hobson, P. Blindspots and sustainability under global change: Non-knowledge illiteracy as a key challenge to a knowledge society. In *Global Change Management: Knowledge Gaps, Blindspots and Unknowables*, 1st Ed.; Ibisch, P.L., Geiger, L., Cybulla, F., Eds.; Nomos: Baden-Baden, 2012; vol. 1, pp. 15–54.
46. CBD, Ed. Preamble of the convention on biological diversity. In *Handbook of the Convention on Biological Diversity*; Earthscan Publications: London, Sterling, VA, 1992.
47. Reid, W.V. *Ecosystems and Human Well-Being: General Synthesis: A Report of the Millennium Ecosystem Assessment*; Island Press: Washington, DC, 2005.
48. Heller, N.E.; Zavaleta, E.S. Biodiversity management in the face of climate change: A review of 22 years of recommendations. Biol. Conserv. **2009**, *142* (1), 14–32.
49. Poiani, K.A.; Goldman, R.L.; Hobson, J.; Hoekstra, J.M.; Nelson, K.S. Redesigning biodiversity conservation projects for climate change: examples from the field. Biodivers. Conserv. **2011**, *20* (1), 185–201.
50. Freudenberger, L.; Hobson, P.R.; Schluck, M.; Ibisch, P.L. A global map of the functionality of terrestrial ecosystems. Ecol. Complexity **2012**, *12*, 13–22.
51. Freudenberger, L.; Hobson, P.; Schluck, M.; Kreft, S.; Vohland, K.; Sommer, H.; Reichle, S.; Nowicki, C.; Barthlott, W.; Ibisch, P.L. Nature conservation: Priority-setting needs a global change. Biodivers. Conserv. **2013**, *22* (5), 1255–1281.
52. Munang, R.; Thiaw, I.; Alverson, K.; Mumba, M.; Liu, J.; Rivington, M. Climate change and Ecosystem-based Adaptation: A new pragmatic approach to buffering climate change impacts. Curr. Opin. Environ. Sustainability **2013**, *5* (1), 67–71.
53. Doswald, N.; Osti, M. *Ecosystem-Based Approaches to Adaptation and Mitigation - Good Practice Examples and Lessons Learned in Europe*; BfN: Bonn, 2011.
54. Mitsch, W.J.; Jørgensen, S.E. Ecological engineering: A field whose time has come. Ecol. Eng. **2003**, *20* (5), 363–377.
55. Schellnhuber, H.J.; Held, H. How fragile is the Earth system. In *Managing the Earth: The Eleventh Linacre Lectures*; Briden, J.C., Downing, T.E., Eds.; Oxford University Press: Oxford, UK, 2002; 5–34.

BIBLIOGRAPHY

1. Blair, T.; Schellnhuber, H.-J.; Cramer, W.P., Eds. *Avoiding Dangerous Climate Change*; Cambridge University Press: Cambridge, England, 2006.
2. Hilty, J.A.; Chester, C.C.; Cross, M.S., Eds. *Climate and Conservation: Landscape and Seascape Science, Planning, and Action*; Island Press: Washington, DC, 2012.
3. Kapos, V.; Campbell, A. *Review of the Literature on the Links between Biodiversity and Climate Change: Impacts, Adaptation, and Mitigation*; Secretariat of the Convention on Biological Diversity: Montreal, Quebec, Canada, 2009.
4. Lovejoy, T.E.; Hannah, L.J., Eds. *Climate Change and Biodiversity*; Yale University Press: New Haven, 2005.
5. Maes, F.; Cliquet, A.; Du Plessis, W.; McLeod-Kilmurray, H. Biodiversity and climate change. *Linkages at International, National and Local Levels* (The Iucn Academy of Environmental Law); Edward Elgar Publishing Ltd, 2013.
6. Pearson, R. *Driven to Extinction: The Impact of Climate Change on Biodiversity*; Natural History Museum: London, UK, 2011.
7. UNEP-WCMC, Ed. *Carbon and Biodiversity: A Demonstration Atlas*; UNEP-WCMC: Cambridge, UK, 2008.

Biodiversity: Conservation

Nigel E. Stork
Environment Futures Centre, Griffith School of Environment, Griffith University, Brisbane, Queensland, Australia

Abstract
Biodiversity, the diversity of life from genes to species to ecosystems, is found on almost all parts of Earth's surface but is greatest in species terms in the tropics. Arguably, there are 5 ± 3 million species on Earth, only about 1.5 million of which have been named, with arthropods being the largest group. Tropical rain forests and coral reefs are the centers of biodiversity for the terrestrial and marine environments. Conservation, the protection of organisms and their habitats, is essential because of the increasing threats to biodiversity from large-scale environmental change and more recently climate change. Conservation biology is the science developed to protect biodiversity and ecosystems. The impact of environmental change is becoming so severe that many consider that a large proportion of the world's species are threatened with extinction. Current extinction rates are increasing, and extinctions, previously mostly on islands, are becoming more prevalent in continental landmasses.

INTRODUCTION

The term "biodiversity," the contraction of "biological diversity," came into common usage following the signing of the Convention on Biological Diversity at the United Nations Conference on Environment (commonly known as the "Earth Summit") in Rio de Janeiro in 1992. This Convention was signed by 193 nations (ratified by 168) and is one of the most widely recognized international treaties. Prior to the Convention, biodiversity was a rapidly developing concept among scientists. It was first featured in a book edited by Wilson and Peters [1] and signified a rising concern for the world's fauna and flora. This also followed several decades of concern for the world's environment as the conservation movement also began to grow.

But what is biodiversity? The Convention defines biodiversity as "the variability among living organisms from all sources including terrestrial, marine and other aquatic ecosystems and ecological complexes of which they are a part; this includes diversity within species, between species and ecosystems." However, biodiversity means different things to different people and is more complex than this simple definition. It encompasses not just hierarchies of taxonomic and ecological scales and other more or less independent scales such as geographic scales and scaling in body size of organisms. A search on Google for the definition of biodiversity comes up with an interesting word cloud (Fig.1).

THE DIMENSIONS OF BIODIVERSITY: HOW MUCH DO WE KNOW?

Phyletic and Genetic Diversity

Classical taxonomic systems had traditionally grouped most organisms into two Kingdoms, Animalia and Plantae. In these systems, a wide array of micro-organisms had been loosely grouped together in a third Kingdom, Protista. Recognizing this group was a "grab bag," in the last few decades, other Kingdoms have been recognized, and current classifications now replace Protista with separate Kingdoms of Bacteria, Protozoa, Chromista, and Fungi.[2,3] Alternative classifications, also based on ribososmal (r)RNA sequences, group all life into three Domains: Bacteria, Archaea, and Eucarya, which are above Kingdoms in the taxonomic classification system[4] (Fig. 2). In this classification where the lines represent genetic distance, it is clear that all the multicellular groups are relatively closely related genetically compared to some of the former Protista. It also shows that animals, plants, and fungi share a great deal of genetic material.

Darwin observed in his *Origin of Species* that "systematists are far from being pleased in finding variability in important characters," and yet it is this genetic component that is the basis for evolution, and without it, evolution and adaptation cannot take place. From the origins of first life on Earth, genetic variation, adaptation, and evolution have led to the enormous diversity of life that we see today and in past extinct life-forms. The distribution of genetic variation within populations of individuals can also tell us a lot about migration between populations, past processes of colonization, and the effects of habitat disruption and loss. How variation is distributed within individuals of a population can also tell us about breeding structures (inbreeding versus outbreeding), reproductive success of different individuals, and a whole range of other patterns. Such genetic variation also leads to morphological variation in individuals—the kind of variation we see everyday when we look at ourselves as humans and other animals. In addition, the environment an individual lives in can also have an impact and lead to additional environmental variation. For example, the amount of food a larva of a Rhinoceros beetle has available

Encyclopedia of Natural Resources DOI: 10.1081/E-ENRL-120047425

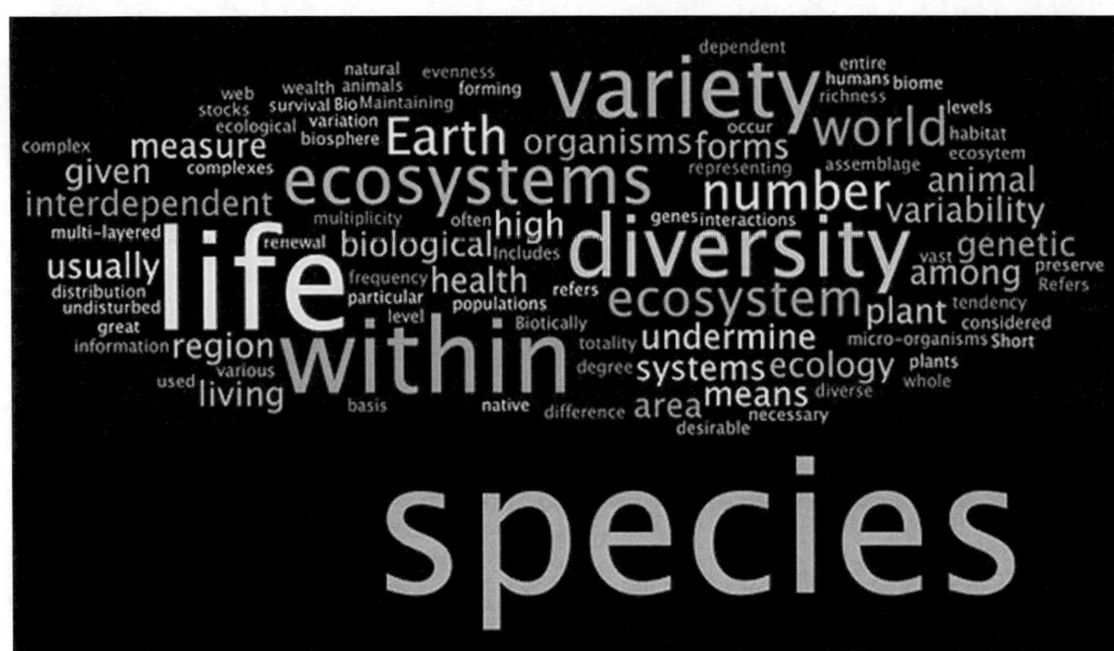

Fig. 1 This word cloud results from a Google search for the definition of biodiversity. (http://thepimmgroup.org/1022/defining-biodiversity-toward-a-consensus/).

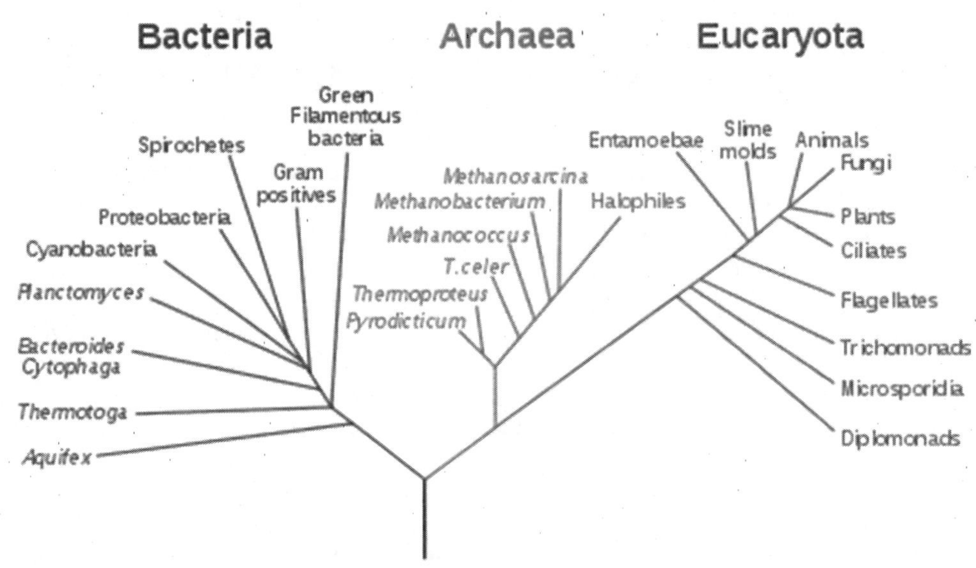

Fig. 2 Phylogenetic tree based on onribososmal (r)RNA sequences groups all life into three Domains Bacteria, Archaea, and Eucarya, which are above Kingdoms in the taxonomic classification system.
Source: Adapted from Woese, et al.[4]

to it will determine not just the size of the adult beetle but also the shape. Larger males will have larger horns and hence may be more competitive when courting females.

Species Diversity: How Many Species Are There and How Many Have Been Described?

Taxonomists have been naming and describing species since the Swedish biologist Carolus Linnaeus (later known as Carl von Linné) created the universal system of naming species (his 10th edition of the Systema Natura published in 1758 is generally considered the start of modern taxonomy).[5] However, determining just how many species have been described is not a simple task as taxonomists often disagree on what constitutes a species because individuals can vary so much. The proportions of known synonyms range from over 80% and 90% in some alga genera, 7% to 80% (32% overall) for insects, 33% to 88% for groups of seed plants, 38% for world molluscs, 50% for marine fish, and 40% for all marine species.[6] Considering past rates of synonymy, it is probable that overall, at least 20% of the currently estimated described species are yet to be discovered synonyms. There are only a few groups of organisms where it is likely that all or virtually all species on Earth have been described, such as the birds and large mammals. For others, such as the insects, nematodes, and fungi, the task is still far from complete. Estimates suggest that 1.5–1.8 million species have been named and described so far, and about two-thirds of these have been cataloged in global databases.

The famous English evolutionary biologist J.B.S. Haldane was asked in the 1930s what he could deduce about "the Creator" from his creation and he replied "an inordinate fondness for beetles." Not surprisingly, about 400,000 species of beetles have been described (about 25% of all named species), and with other insects and other arthropods, this brings up a total of around 1 million species. But just how many species there are including the undescribed ones remains a mystery and is difficult to estimate. In the past, some have suggested that there might be as few as 2 million species, whereas others have suggested as high as 100 million species. Three different methods now suggest that estimates of around 5 ± 3 million species are more likely. Two of these use the new and extensive (but as yet incomplete) catalogs of described species to provide estimates for all species. The first method used extrapolation of the rate of discovery of selected higher taxa to predict 8.7 million (±1.3 million SE) eukaryotic species globally, of which 25% (2.2 million ± 0.18 million SE) were marine.[7] The second method modeled global species richness based on the rates of species description and predicted that 0.3 million marine and 1.7 million terrestrial species may exist.[8] A third model used uncertainty analysis of the specificity of arthropods on tropical trees to suggest there may be 6.1–7.8 million terrestrial arthropod species.[9,10] This latter method showed that the key variable for global insect species richness is the level of host specificity of insects to individual tree species. It also has 90% confidence intervals of 3.6–13.7 million species, suggesting that higher global estimates are extremely unlikely. The finding that marine organisms comprise only about 25% of all species on Earth is surprising, given that first, oceans cover 70% of the surface and second, that at the higher taxonomic level of Phylum, most phyla are found in the marine environment with the terrestrial fauna being less rich. However, the marine environment has few barriers to dispersal, whereas land and freshwater habitats are highly fragmented. The largest group in terms of species, after the arthropods, is the plants with an estimated 400,000 species, and most of these are terrestrial. The explosion of diversity that arose through the evolution of the flowering plants and the associated insects that live and feed on them may account for the apparent disproportionate proportions of these two groups from the total.

The International Census of Marine Microbes (ICoMM) estimates that 3.6 x 10^{30} microbial cells account for more than 90% of the total biomass in the oceans, and there is the possibility that viral particles may be up to 100 times greater than this number. Archaea, one of the two prokaryotic Domains of life, dominate oceanic mid-waters. For prokaryotes, normal species definitions do not apply, and the term phylotype is used. Great progress has been made in recent years through the use of genomics to characterize the diversity of phylotypes in different habitats, but achieving global phylotype estimates for all prokaryotes seems to be some way off.

Where Is Biodiversity Found?

In the terrestrial environment, it is an almost universal principle across all taxa that the diversity of species is highest close to the equator and decreases toward the poles. At the same time, diversity is highest in lowlands and decreases with altitude. Similarly, areas with higher rainfall have the highest diversity at a given latitude or altitude; hence, tropical rainforests are renowned for their high diversity. Consequently, countries with large areas of tropical rain forest feature highly in the list of those countries with the largest numbers of species of plants, birds, and mammals (Table 1). There is some evidence that in tropical forests there is a midaltitude peak in species richness, but this result may well be spurious because of sampling problems. Table 2 shows the diversity of plant species found in small areas of different forests around the world where typically researchers have measured and identified to species, every tree stem above 1 cm diameter. The highest diversity of plants is found in lowland tropical rain forest plots, particularly those with high annual rainfall and a less distinct seasonal pattern such as the plots in Lambir Hills (Malaysia) and Yasuni (Ecuador). With this high plant richness is an associated high insect and other arthropod species richness. For the marine realm, patterns of species richness do not always follow the same diversity patterns. Here, the diversity of organisms associated with coral reefs gives a high species richness in the tropics for some groups, but the deep-sea welling's of nutrients in higher latitudes result in the reversal of the normally low diversity in lower latitudes for some groups that are dependent on these nutrients.

There has been some debate about whether within the tropical forests there are more species in the forest canopy than in the ground layer (including soil), with the focus largely being on arthropods. For this reason, some have called the forest canopy "the last biological frontier." However, evidence from a long-term study using equal sampling of the forest and canopy layers shows that of the 1500 beetle species collected, roughly 25% are canopy specialists, 25% are ground specialists, and 50% are shared between these strata.[12] Beetles are probably a good indicator for other arthropods because they comprise 40% of described arthropods and they are biologically very diverse. So, although the canopy of trees is a very important stratum for biodiversity, because that is where most photosynthesis, flowering, and fruiting take place, it seems that the soil and the ground

Table 1 World ranking of mega-biodiversity countries as assessed by estimates of the numbers of species of mammals, birds, and flowering plants found in these countries

Country	Mammals	Birds	Flowering plants
Mexico	450	1026	25,000
Indonesia	436	1531	27,500
Zaire	415	1,096	11,000
Brazil	394	1,635	55,000
China	394	1,244	30,000
Colombia	359	1,695	50,000
Peru	344	1,678	17,121
India	316	1,219	15,000
Venezuela	305	1,296	20,000
Ecuador	302	1,559	18,250
Cameroon	297	874	8,000
Malaysia	286	736	15,000
Australia	252	751	15,000
South Africa	247	790	23,000
Panama	218	926	9,000
Papua New Guinea	214	708	10,000
Vietnam	213	761	7,000
Costa Rica	205	850	11,000
Philippines	153	556	8,000
Madagascar	105	253	9,000

Source: Adapted from Paine & Status.[11]

Table 2 The number of individuals and number of species richness of trees in a selected list of surveyed 50- and 25-hectare plots around the world

CTFS plot location	Latitude	Plot size (ha)	No. spp	No. trees
Wind River, USA	45.8	25.6	25	31,162
Wabikon Lake Forest, USA	45.6	25.6	36	57,388
Haliburton Forest, Canada	45.3	13.53	30	46,339
Changbaishan, China	42.4	25	52	38,902
Smithsonian Cons. Biol. Inst. USA	38.9	25.6	65	41,031
Yosemite National Park, USA	37.8	25.6	23	34,458
Gutianshan, China	29.3	24	159	140,700
Xishuangbanna, China	21.6	20	468	95,834
Huai Kha Khaeng, Thailand	15.6	50	251	72,500
Mo Singto, Thailand	14.4	30.5	262	no data
Mudumalai, India	11.6	50	72	25,500
Barro Colorado Island, Panama	9.2	50	299	208,400
Khao Chong, Thailand	7.5	24	593	121,500
Sinharaja, Sri Lanka	6.4	25	204	193,400
Korup, Cameroon	5.1	50	494	329,000
Lambir, Malaysia	4.2	52	1,182	359,600
Pasoh, Malaysia	3.0	50	814	335,400
Yasuni, Ecuador	-0.7	50	1,114	151,300

Source: Data from the Center for Forest Science, Smithsonian Institution.

Abbreviations: CTFS, The Center for Tropical Forest Science; ha, Hectare; Smithsonian Cons. Biol. Inst., Smithsonian Conservation Biology Institute; Spp., Species.

Note: These data are for all trees on the plots that are greater than 10 cm in diameter (as measured on tree trunks at around 1.5 meters above ground).

layer are equally important for biodiversity as that is where the roots of trees are found, where seeds develop into young plants, and where decomposition takes place.

The 17th-century poet Jonathan Swift wrote "So nat'ralists observe, a flea hath smaller fleas that on him prey, and these have smaller fleas that bite 'em, and so proceed ad infinitum." This observation that larger organisms have smaller parasitic ones is true about life. Some groups of parasites can be very host specific, such as lice, whereas others, such as fleas, are less host specific and may be found on a wide range of host animals. In general, the larger, more abundant, and more geographically widespread species of organisms, such as the mammals or large fish, have the richest parasitic load. The obvious conclusion from Swift's observation is that most species should be parasites, but this has yet to be statistically tested.

CONSERVATION

Conservation, the protection of animals, plants, and other organisms and their habitats probably first appeared in Assyria in 700 BC where game reserves were set aside for royal hunting. In the United Kingdom, some of the best-protected and most species-rich forests, such as Windsor Great Park and Epping Forest, were protected to conserve deer for hunting by royalty from the 15th and 16th centuries. This had followed a sharp decline in the area of forests, particularly oak, as large areas had been cleared for timber to build warships in the 1500s. This kind of conservation for sustainable use—in this case of deer—is different from what some conservationists would consider conservation—which is to protect species. The so-called conservation movement is often more concerned with the latter kind of conservation.

Protected Areas

Across the world, an average of 10–15% of the terrestrial environment is protected in some way or other.[13] In contrast, there is far less marine protection, with only 1.2% protected.[14] Much of the oceans are considered international waters with very limited international environmental protection.

The International Union for Conservation of Nature (IUCN), originally called the International Union for the Preservation of Nature (IUPN) and for a short time also called the World Conservation Union, is the world's oldest and largest global environmental organization. Its mission is "to influence, encourage and assist societies throughout the world to conserve the integrity and diversity of nature and to ensure that any use of natural resources is equitable and ecologically sustainable." IUCN is the largest global environmental network with more than 1000 government and non-governmental member organizations, and almost 11,000 volunteer scientists in more than 160 countries. Its program of work is discussed and agreed every 4 years at IUCN's World Conservation Congress. IUCN recognizes six categories of protected management areas: Ia) Strict Nature Reserve and Ib) Wilderness Area, II) National Park, III) Natural Monument or Feature, IV) Habitat/Species Management Area, V) Protected Landscape/Seascape, and VI) Protected Area with sustainable use of natural resources. Each of these gives different levels of protection or sustainable use or both.

Threats to Biodiversity and the Environment

In the last century, there has been a very rapid increase in the world's human population and as a result, increasing demands on the environment for food, water, and other commodities. Efforts to improve agricultural productivity have increased the levels of pest organisms, and as a response chemicals have been developed to manage weeds, fungi, and pest insects and other animals. In the 1960s, concerns were first raised about environmental pollution through the use of toxic chemicals as insecticides and herbicides. Much of the concern was about health risks to humans and other animals but also because of the harm to the environment. Rachel Carson's *Silent Spring*[15] was at the vanguard of this movement. At the same time, the industrial revolution resulted in problems of industrial waste and chemical pollution of water systems. Other sometimes more insidious forms of environmental pollution include acid rain caused by the release of nitrous oxide and sulfur dioxide from the burning of fossil fuels and subsequently deposited through rain as nitric acid and sulfuric acid during rainfall events. Often, the acid rain, which causes dieback and often death of trees, was affecting areas of forests away from the area of initial production and pollution.

Perhaps as much as 70–80% of all terrestrial biodiversity is found in tropical forests. Conservation biologists talk about the importance of old growth or primary forests, particularly tropical rainforests as these contain the greatest diversity of life. The term "virgin" forests is nowadays largely ignored as virtually all forests have been impacted directly or indirectly by man in some way. It was in 1979 that Norman Myers identified the loss of tropical forests as a serious environmental issue.[16] At the time, he suggested that shifting cultivators, provided with new road access, was the major cause of forest loss. Now, selective logging of forest for timber or clear felling to provide new areas for agriculture are seen as the major threats. Subsequently, many suggested that the loss of species associated with cleared primary forest might cause a manmade extinction crisis. However, the large-scale losses have not been observed as yet perhaps due to targeted conservation efforts and the fact that in many places logged forests have remained as secondary forests and that the extinction of many species threatened by habitat loss has been delayed, thus forming an extinction debt that may occur sometime in the future. As of 2010, about 50% of the world's tropical forests fell into the broad secondary-forest category.[17] Although many species that are normally found in primary forests appear to be able to survive in some kinds of secondary forests, particularly selectively logged forests, it is clear that for others there is no substitute for primary forests.[18]

Climate change is already having effects on biodiversity and, given the Intergovernmental Panel on Climate Change (IPCC) scenarios for increasing levels of carbon dioxide (CO_2) in the atmosphere, rising temperatures, and changing rainfall patterns, both the distribution and the existence of many species will be affected.[19] Modeling of the impact of rising temperatures on biodiversity suggests that species that are restricted in distribution to mountaintops will move up the mountains and many will disappear, particularly in the tropics. A 5°C rise in temperature could mean a roughly 1000-meter rise in elevation for species if they are to continue in the same climate envelope. For lowland species, a similar temperature increase could mean they would need to move many thousands of kilometers poleward if there are no higher elevations available. In the marine environment, increasing CO_2 levels will increase acidification of seas, and this could lead to the demise of many corals. Rising temperatures, changing rainfall patterns, and ocean acidification are likely to act synergistically with other threats such as hunting, overharvesting, and habitat loss to increase biodiversity loss and extinctions. Precisely how they may act together and what this may mean for biodiversity is still largely unknown.

Conservation Biology

The science concerned with the management of biodiversity is usually referred to as conservation biology. The management and conservation of biodiversity draw not only on the ecology of species but also on a range of disciplines including animal behavior, chemistry, genetics, geography, demography, medicine, political science, and philosophy. It is a truly multidisciplinary field. Strategies for conservation will often combine theoretical disciplines such as population biology with landscape management options such as establishing protected areas of one kind or another. Others will include the removal and protection of endangered species with reintroductions back into the wild, once viable populations have been created. Other mechanisms will involve the removal or management of threats such as hunting, overfishing, or overharvesting and the reduction or removal of pest animals and weeds. Other activities include controlling the movement or activities of people that are likely to be detrimental to the protected species or protected areas. Such actions usually require government legislation and policy development in addition to financial resources and dedicated staff.

Global Action

A number of approaches have been adopted to tackle the problem of conservation of biodiversity, and these differ in the way that the problems and goals have been defined. Conservation International, for example, adopted an idea first identified by Norman Myers[20] and targeted the protection of biodiversity through the identification of global biodiversity hotspots. They identify these areas as being places where there are more than 1500 endemic (not found anywhere else) species and 70% habitat change or loss. So far, they have identified 36 biodiversity hotspots, which they claim cover 2.3% of the terrestrial land surface and yet include 50% and 40% of the world's plants and vertebrates, respectively.[21] BirdLife International focuses, as the name suggests, on the world's birds and because this is the best known of all organisms, their Important Bird Areas is based on the occurrence of key bird species that are vulnerable to global extinction or whose populations are otherwise irreplaceable.

EXTINCTION RATES

Many scientists believe that the extent and rate of environmental change is so great that the Earth is heading toward a catastrophic "Holocene" mass extinction event like the five mass extinctions that occurred in the geological past and separately wiped out 65–95% of the species that existed at the time. For one of the best-known groups, birds, it appears that extinctions are already at between 50 to 100 times the so-called "background rate" and could be a 1000 times that rate in the next 50 to 100 years.[22] The background rate is a very rubbery figure of how many species extinctions per year there may have been per million existing species—for the last 600 million years, some have suggested that this might be around one extinction per year per million species.[23,24] IUCN has worked with biologists around the world to develop a list of extinct, threatened, and endangered species called the Red List of Threatened Species. In 2012, the list of species that are recognized as having gone extinct in the last 200 years stood at about 1500 species of plants and animals. The status of about 45,000–60,000 species is monitored, particularly birds, mammals, and other vertebrates, and so many species of other groups lack close observation. Of course, recognizing when a species has disappeared, particularly for species that are difficult to observe, is difficult, and so many more species may have gone extinct without anyone recognizing the fact. At the same time, some species, which were thought to have gone extinct are miraculously rediscovered. One example of this is the Lord Howe Island stick insect, which had disappeared soon after the introduction of the black rat on this small (16 km^2) island in 1918. It was rediscovered on a very small island off the coast of Lord Howe Island in 2001, and a plan to reintroduce it was put in place alongside a plan to eradicate the black rat.

A large proportion of the species that are known to have gone extinct in the last 200 years are from islands and are species that had a geographically very small range. Their extinction was caused predominantly by habitat loss, by hunting, or by invasive organisms. There is strong evidence that now most extinctions are of perhaps more widespread species in continental areas, and their demise is due to habitat loss and other threats.

CONCLUSION

The term "biodiversity" arose in the 1980s when concern for our environment was increasing and led eventually to global agreements to increase the conservation of animals and plants. In spite of these agreements, environmental degradation and climate change are likely to continue to impact severely on the world's biota. In coming decades, extinction rates will rise, and many iconic species as well as many largely unknown species will disappear.

REFERENCES

1. Wilson, E.O.; Peters, F.M. *Biodiversity*. National Academy Press: Washington, D.C., 1988; 326–332 pp.
2. Cavalier-Smith, T. Only six kingdoms of life. Proc. R. Soc. Lond. B **2004**, *271* (1545), 1251–1262.
3. Simpson, A.G.B.; Andrew, R.J. The real 'kingdoms' of eukaryotes. Curr. Biol. **2004**, *14* (17), R165–R167.
4. Woese, C.R.; Kandler, O.; Wheelis, M.L. Towards a natural system of organisms: Proposal for the domain Archaea, Bacteria, and Eucarya (Euryarchaeota/Crenarchaeota/kingdom/evolution). PNAS **1990**, *87*, 4576–4579.
5. Linnaeus, C. (in Latin). Systema naturae per regna trianaturae: Secundum classes, ordines, genera, species, cum characteribus, differentiis, synonymis, locis (10th edition ed.). Stockholm: Laurentius Salvius. 1758.
6. Costello, M.J.; May, R.M.; Stork, N. E. Can we name Earth's species before they go extinct? Science. **2013**, *339*, 413–416, DOI: 10.1126/science.1230318
7. Mora, C.; Tittensor, D.P.; Adl, S.; Simpson, A.G.B.; Worm, B. How many species are there on Earth and in the Ocean? PLoSBiol. **2011**, *9* (8), e1001127, doi:10.1371/journal.pbio.1001127.
8. Costello, M.J.; Wilson, S.P.; Houlding, B. Predicting total global species richness using rates of species description and estimates of taxonomic effort. Syst. Biol. **2012**, *61*, 871–883.
9. Hamilton, A.J.; Basset, Y.; Benke, K.K.; Grimbacher, P.S.; Miller, S.E.; Novotný, V.; Samuelson, G.A.; Stork, N.E.; Weiblen, G.D.; Yen, J.D.L. Quantifying uncertainty of tropical arthropod species richness. Amer. Nat. **2010**, *176*, 90–95.
10. Hamilton, A.J.; Basset, Y.; Benke, K.K.; Grimbacher, P.S.; Miller, S.E.; Novotný, V.; Samuelson, G.A.; Stork, N.E.; Weiblen,G.D.; Yen, J.D.L. Correction. Amer. Nat. **2011**, *177*, 544–545.
11. Stork, N.E.; Grimbacher, P.S. Beetle assemblages from an Australian tropical rainforest show that the canopy and the ground strata contribute equally to biodiversity. Proc. Roy. Soc. Series B. **2006**, *273*, 1965–1975.
12. Soutullo, A. Extent of the Global Network of Terrestrial Protected Areas. Cons. Biol. **2010**, *24*, 362–363.
13. Global Ocean Protection: *Present Status and Future Possibilities*: IUCN, Gland. 2010; 96 pp.
14. Carson, R. *Silent Spring*: Houghton, Mifflin, Boston, USA. 1962.
15. Myers, N. *The Sinking Ark: A New Look at the Problem of Disappearing Species*: Oxford, U.K., Pergamon, 1979; 307 pp.
16. Chazdon, R.L.; Peres, C.A.; Dent, D; Sheil, D; Lugo, A.E; Lamb, D; Stork, N.E; Miller, S. The potential for species conservation in tropical secondary forests. Conserv. Biol. **2009**, *23*, 1406–1417
17. Gibson, L.; Lee, T.M.; Koh, L.P.; Brook, B.W.; Gardner, T.A.; Barlow, J.; Peres, C.A.; Bradshaw, C.J.A.; Laurance, W.F.; Lovejoy, T.E.; Sodhi. N.S. Primary forests are irreplaceable for sustaining tropical biodiversity. Nature **2011**, *478*, 378–381, doi:10.1038/nature10425.
18. IPCC Fourth Assessment Report: *Climate Change*: 2007; 104 pp.
19. Myers, N. Threatened biotas: 'hotspots' in tropical forests. Environmentalist **1988**, *8*, 1–20.
20. Mittermeier, R.A.; Turner, W.R.; Larsen, F.W.; Brooks, T.M.; Gascon, C. Global biodiversity Conservation: The Critical role of Hotspots. In *Biodiversity Hotspots. Distribution and Protection of Conservation Areas*. Frank, E.Z., and Jan, C.H, Eds.; Springer: Heidelberg 2011; 3–20 pp.
21. Pimm, S.L.; Raven, P.H. Extinction by numbers. Nature **2000**, *403*, 843–845.
22. May, R.M.; Lawton, J.H.; Stork, N.E. Assessing extinction rates. In *Extinction Rates*: John, H., Robert, M., Eds.; Oxford University Press: Oxford, 1995; 1–24 pp.
23. Pimm, S.K.; Russell, G.J.; Gittleman, J.L.; Brooks, T.M. The future of biodiversity. Science **1995**, *269*, 347–349.
24. Giam, A.; Brook, B.W.; Gregory, S.D.; Koh, L.P.; Gibson, L.; Stork, N.E.; Fordham, D.A.; Bradshaw, C.J.A. A global assessment of threats to biodiversity across continents and islands. Nature (in review).

BIBLIOGRAPHY

1. Lindenmayer, D.; Burgman, M. *Practical Conservation Biology*; CSIRO Publishing: Collingwood, Australia, 2005; 433–445 pp.
2. Norse, E.A., Crowder, L.B. (Eds.) *Marine Conservation Biology*; The Science of Maintaining the Sea's Biodiversity, Island Press: ISBN 978-1-55963-662-9, 2005.
3. Primack, R.B. *Essentials of Conservation Biology*, 2nd Ed.; Sunderland, MA: Sinauer, 1998.

Biodiversity: Habitat Suitability

Betty J. Kreakie
Atlantic Ecology Division, Office of Research and Development, U.S. Environmental Protection Agency (EPA), Narragansett, Rhode Island, U.S.A.

Abstract

Habitat suitability quantifies the relationship between species and habitat and is evaluated according to the species' fitness (i.e., proportion of birth rate to death rate). Even though selecting only optimal habitat might maximize evolutionary success, species are not always in a habitat that optimizes fitness. In fact, species live in habitats that result in varying fitness levels. By studying habitat suitability range and making assumptions about how fitness relates to occupancy, we can make predictions about the probability of occurrence across a landscape. Spatial predictions of occurrence have a wide range of applied uses. Habitat modeling helps make predictions about how species may respond to climate change and other habitat changes. Conservation biologists use models for endangered species protection (e.g., identification of introduction sites and critical habitat). Models can also be used to identify areas that are vulnerable to invasive species. Habitat suitability models simplify a large amount of ecological information and therefore require fastidious interpretation. Despite simplifying assumptions, the results provide insight into large ecological problems.

INTRODUCTION

Habitat suitability is the capacity of a habitat to support a particular species.[1] To fully understand habitat suitability and how it is used in applied research, it is necessary to understand concepts of fitness, habitat quality, and basic niche theory. The remaining portion of the introduction will cover these concepts. The subsequent section discusses how these theories are used to build spatial models of species distribution. The final main section explores some of the areas of conservation and land management that use these types of habitat suitability models.

Species Fitness and Habitat Quality

In general, the fitness of a population is the number of births per generation (or birth rate) minus the number of deaths per generation (death rate).[2] Furthermore, a population is in suitable habitat only if the environment can support the population (e.g., the population sustains itself and is able to maintain its population size). The fitness of the population must, therefore, be greater than (birth rate greater than death rate) or equal to zero. If we were to look at a species existing in a habitat of optimal quality, we would expect to observe two possible fitness responses.[3] First, if the population is at equilibrium—perhaps this population has existed for several generations in this optimal habitat—the birth rate would be equal to the death rate. This population is said to be at carrying capacity, the largest population size the habitat can support with given levels of resources. Alternatively, if the population has not reached carrying capacity and the birth rate is higher than the death rate, the population size would be increasing. Species, however, do not exist only in optimal habitats (for reasons we will discuss later). Species can be found in a range of habitats of varying quality, which would result in varying levels of fitness.[4] To define habitat suitability, we need to quantify the relationship between fitness and the habitat of a particular area.

A logical assumption might be that a species would only exist in habitats where it could maximize its fitness. Natural selection preferentially selects populations that contribute the largest number of offspring to the next generation. Accordingly, we might expect to see populations existing only in optimal habitats because those populations are being preferentially selected. However, existence in an optimal habitat is not always the case.[5] The possible reasons for living in a range of habitat are nearly infinite. However, numerous reasons are due to the dynamic nature of our world; most characteristics of the environment are not static and therefore populations are unlikely to reach equilibrium at a given location.[6] Climates change, as does land cover. Often, populations are limited in dispersal capabilities and cannot move or migrate to follow these changes in geographic space.[7] If they are able to migrate, it is not certain if this new area will provide their optimal habitat or even be sufficient to support the population. Additionally, they might not be able to compete with an established competitor population in a new location. Despite any changes in climate or land cover, the established population might still be comfortable in their habitat range and feel no pressure to acquiesce the habitat to the displaced population.

Encyclopedia of Natural Resources DOI: 10.1081/E-ENRL-120047431

Habitat Suitability and Niche Theory

We cannot talk about defining habitat suitability without at least a cursory discussion of the niche theory.[8] The niche theory has a long history and a complex literature, which is well beyond the scope of this text. However, for a thorough review of the niche theory history and a modern synthesis, see Chase and Leibold.[9] For our purposes of defining habitat suitability, the broad Chase and Leibold definition of "niche" is appropriate:

> *"[T]he joint description of the environmental conditions that allow a species to satisfy its minimum requirements so that birth rate of a local population is equal to or greater than its death rate along with the set of per capita effects of that species on these environmental conditions."*

As you can see, the first part of the broad niche definition is nearly the definition of habitat suitability presented earlier. In fact, through history the "per capita effects of that species on these environmental conditions" portion of the niche definition was often ignored or omitted. Some will refer to this more concise or limited definition of niche as the Grinnellian Niche, based on the work of Joseph Grinnell.[10]

The "n-Dimensional Hypervolume"

G. Evelyn Hutchinson was perhaps the first, and definitely the most famous, to propose defining a niche by what he called the "n-dimensional hypervolume".[11] His concept of defining the niche by its "n-dimensional hypervolume" is still essentially the core of habitat suitability quantification. The concept is as follows. For a given species, elements of the environment are measured at each known species occurrence location. In other words, each time that species is observed, environmental characteristics are measured at that location. The list of environmental characteristics depends on the type of species (i.e., if the species is a plant, terrestrial animal, fish, fungus, etc.), the availability or ability to measure a particular variable, and the researchers' assumptions about the species. Typically, the measured parameters might include temperature, precipitation, elevation, pH (soil or water), habitat type, and so forth. From these parameters, researchers would be able to define the niche as the multidimensional space where that species "lives." The niche is not defined as latitude and longitude extents on a landscape but instead in terms of physiological tolerances to environmental parameters. It might even be possible to assume that where (in that multidimensional space) we observe the highest population numbers is where the habitat is optimal. Reciprocally, areas with low or diminishing population numbers might be classified as sub-optimal (or fringe) habitats. This method of defining a niche's n-dimensional hypervolume is also a means to explore the range of habitat suitability.[12]

HABITAT SUITABILITY MODELING

Besides pure ecological curiosity, understanding the range extent and spatial distribution of a species' suitable habitat is often necessary for effective land management and species conservation (the following section will concentrate on exploring some specific applications). Based only on observational data, our ability to understand species' habitat suitability throughout an entire range is limited. These limitations are mainly due to sampling. To understand comprehensively the behavior of species habitat suitability throughout the entire range, we would need to sample the entire range. Additionally, we would need to sample areas that are not habitats. This allows for predictions across the entire landscape and not just in areas of known occurrence. Furthermore, we would need to conduct this sampling of habitats and non-habitats throughout the year and over many years. This would allow us to understand how the species moves across the landscape throughout the year and how this spatial distribution changes between years. In others words, we would need to sample everywhere all of the time. Even with high sample intensity, not observing your species during a survey does not necessarily indicate a true absence. Some species are cryptic; they may be present in a location but maybe simply undetected by the researcher. The compounded effects of these sampling limitations necessitate habitat modeling. There are numerous statistical approaches used to carry out this type of research (see Guisan and Zimmermann[13] for a review).

The previous discussion of habitat suitability involved relating environmental parameters to a species' fitness. Habitat suitability modeling takes the previous definition subtly in a new direction. Habitat modeling incorporates environmental information from sampled habitats and non-habitats and makes predictions about the likelihood of species occurrence in an unsampled location.[12,14] The more similar a novel unsampled location is to the optimal in the species' habitat multidimensional space (again returning to the concept "n-dimensional hyperspace"), we would predict a higher probability of occurrence.[8] This, of course, is contingent on such factors as the ability of the species to reach this unsampled location. This type of modeling and subtle variations of it has been called many names: species distribution modeling, habitat suitability index, niche modeling, ecological niche factor analysis, climate envelopment modeling, and numerous others. However, they are all generally based on the same simple concept; there is some means to predict the likelihood of occurrence (or predicted suitability) in an unsampled location based on what we know about the habitat at a known occurrence. Figure 1 provides an illustration of typical habitat suitability results.[15]

Data Requirements

Habitat suitability modeling is employed because we are simply not capable of sampling the entire range of a species.

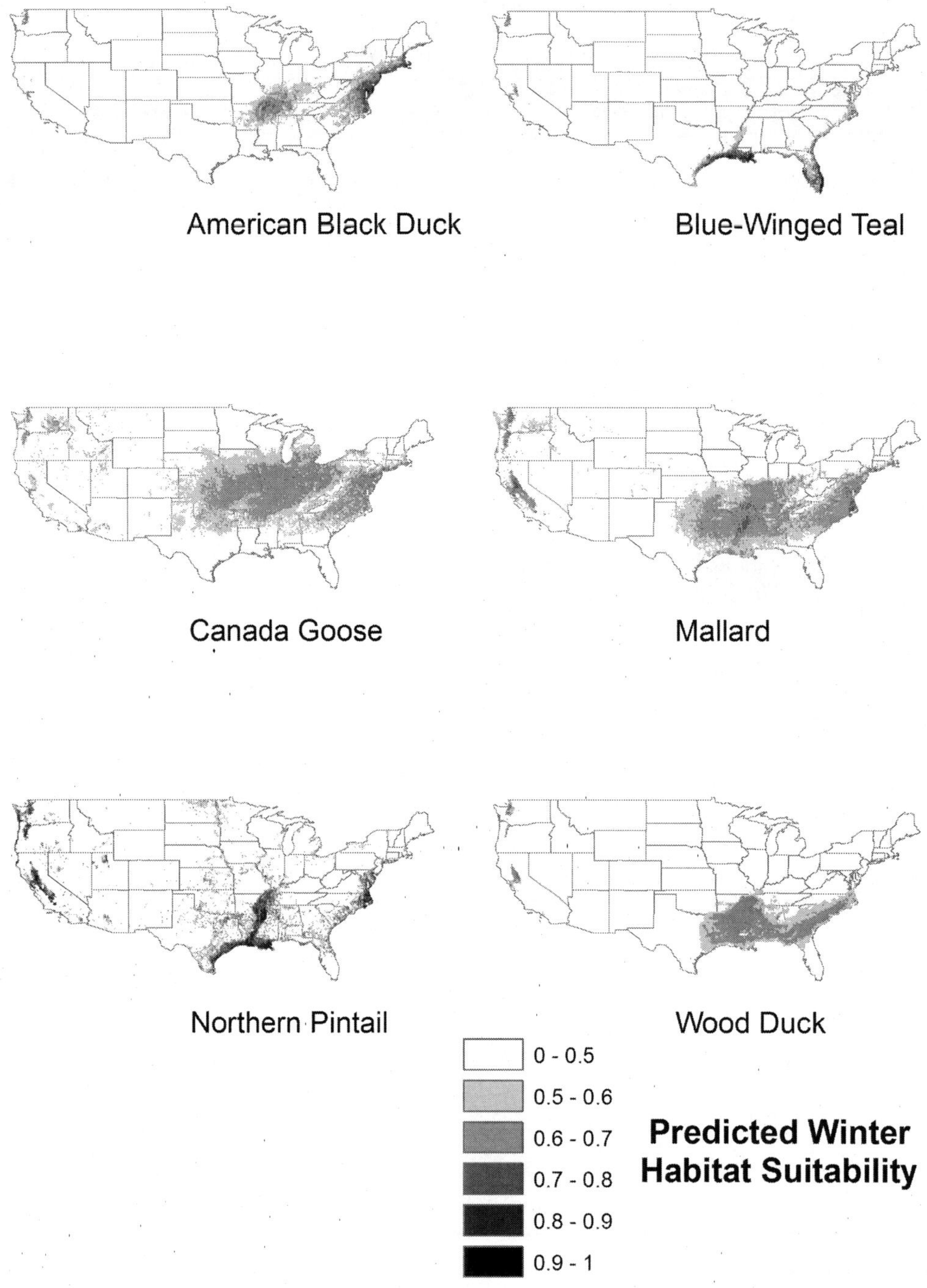

Fig. 1 Map of winter-predicted suitability for six waterfowl species in the continental United States. Predictions were created using MaxEnt[16] with data from the U.S. Geological Survey's Banding Bird Laboratory 1990–1999 data.[17] Temperature, precipitation, elevation, and water table depth were the predicted variables used to model the predicted distribution of these species' habiat.[15]
Source: Adapted from Kreakie et al.[15]

Even though habitat suitability modeling approaches are employed to compensate for insufficient observational data, there are still considerable data requirements. We discussed how most modeling approaches require information about both habitats and non-habitats; in other words, descriptions of where the species was observed as well as where it was not observed. As there are large amounts of presence-only data available (e.g., herbarium records, bird banding), modeling approaches have been developed to utilize this type of presence-only data.[16,18] For this reason and others, a model approach should be selected based on type of data available. As mentioned in the preceding text, some species may be more cryptic than others; therefore it is important to understand the detection probability of the particular species.[19,20] It is also important to understand any sampling bias in the data. For example, if a bird species was only sampled in the summer, it is unlikely that sampling data can be used to make predictions about winter locations. Or if one area was sampled with a much higher frequency than any other portion of the range, this will skew the overall occurrence predictions. It is likely that predictions will be high for areas that are similar to intensively sampled locations (regardless of true relationship to optimal habitat) unless the sample bias is statistically corrected.

Limitations Due to Data and Approach

The predictions from habitat suitability modeling must be interpreted carefully with a clear understanding of the limitations associated with the data and modeling method.[21,22] Species data that are spatially sparse or temporally sporadic may impact the uncertainty we have when interpreting the results.[22–24] In addition to understanding the limitations of the species data, the same level of scrutiny is needed for the environmental parameters. Variables that are included and omitted might lead to drastically different predictions of species occurrence probability. Often it is possible to create predictions that seem to perform very well; yet in reality, these results are meaningless. Due to issues of overfitting the model predictions to the sample data, the results may be irrelevant to species. To alleviate some concerns about model selection and performance, the general reader of the research results does not need to have an in-depth understanding of the mathematics behind all modeling approaches and methods of performance evaluation. Researchers should demonstrate to readers that they recognize these issues and have taken steps to alleviate some of the ramifications.

Limitations Due to Biological Assumptions

The shortcomings of habitat suitability modeling are more than methodological. In using habitat models, we make assumptions about the biology of a species that should be considered when interpreting the results.[25] The model predictions are based on the assumption that the species can successfully disperse to all locations. This might not be the case. There may be barriers in the landscapes that the species simply cannot cross (e.g., roads or rivers). It might simply be too great a distance for the species to travel. Model predictions are based on sightings of a species, which we assume is living in a suitable habitat and is static in time.[26] However, it is possible for a species to be found in an area where it has reduced fitness (what ecologists call sink habitat).[27,28] Therefore, model predictions could be based on a habitat to which the species is actually maladapted. Also, in general, model predictions are based on a single species, when in reality all species live in communities where they interact with different species.[29,30]

These shortcomings of habitat suitability modeling do not mean that we should abandon this area of research. The following section will present several areas of research where these approaches are very useful. However, understanding the shortcomings will help to increase the effectiveness of their use.[31]

A Brief Introduction to Mechanistic Approaches

Our discussion of habitat suitability modeling has focused entirely on methods that are considered correlative approaches. These methods attempt to elucidate a relationship between species location data and environmental parameters. However, there is an area of research that makes an attempt to define habitat range from a completely different angle. Mechanistic models estimate occurrence likelihood based on resource requirements and physiological constraints of a species.[32,33] This approach requires a detailed understanding of the physiological tolerances of a species, which may require extensive laboratory work. Once a researcher determines these thresholds, they determine the spatial distribution of the appropriate habitat. Kearney et al. have significantly advanced this research avenue with their detailed work on reptiles and amphibians.[34,35]

APPLICATIONS OF HABITAT SUITABILITY MODELS

There are numerous uses for the predictions created from habitat suitability modeling, and these results have been applied in a wide range of disciplines. The following provides a concise summary of a few key examples in ecology. In addition, here are some further references for more insight into the applications of habitat suitability modeling.[36–39]

Applications for Habitat Change

One of the main uses of habitat suitability modeling is to predict how species may respond to changes in habitat.[40] Researchers gain information about the multidimensional habitat space in which the species is currently living and assume the species will continue to be found in the same multidimensional space in the future. Then data is gathered about how the landscape is predicted to look in the future. These future landscapes are used to predict the spatial distribution of the species' future habitat. This type of future scenario modeling has been used to predict how species may respond to converting a habitat from one type to another. For example, Broton et al. calculated the predicted effect of different irrigation scenarios on bird habitat.[41] Using this type of scenario modeling, Sánchez-Cordero forecasted how endemic mammal species of Mexico might respond to deforestation.[42] This approach is commonly used to predict new species range maps for various climate change scenarios. Early work in this area by Iverson et al. made predictions about different tree species in eastern United States.[43,44] Iverson's work illustrates how some species might expand their range limits while others may potentially contract.

Applications for Invasive Species Management

Habitat suitability modeling can be a useful tool in managing invasive species. Once researchers understand the native habitat of a potentially invasive species, they can use the information to predict which areas of the introduced habitat might be at a higher risk of invasion.[45,46] This work helps establish areas that might need to be monitored or regulated more closely.[47] For example, Ward used this approach to make predictions about the potential distribution of six ant species in New Zealand, finding a propensity for lowlands and coastal areas.[48] This also included information about the time duration of the invasion, illustrating a positive correlation between the time spent in the introduced habitat and its range. Zambrano et al. used these methods to make predictions about fish (carp and tilapia) invasions in stream reaches.[49] This study illustrated that each species presents different patterns of risk across the landscape. The majority of work assumes that the habitat requirements do not shift between native and introduced habitats, which may not always be true.[50]

Applications for Endangered Species Management

Habitat suitability has a wide range of applications for the management of endangered species. For species in the United States that are officially listed as endangered or threatened, legislation mandates that key habitats be identified. To accomplish this goal, the U.S. Fish and Wildlife Service (USFWS) developed methods to calculate an area's habitat suitability index (see Schamberger et al.[51] for more detail on methods and application). Additionally, endangered species management may require introduction into novel areas. These introductions could be from captivity-bred population (e.g., zoo breeding programs) or from struggling populations that are in isolated pockets of habitats. Habitat suitability modeling is used to identify areas that might optimize results. This application is really an attempt to identify areas that will maximize fitness (i.e., maximizing birth rate while minimizing the death rate). Understanding an area's potential reproductive fitness for a species is critical for this work. Often, the number of individuals introduced is low, and potential stochastic effects can be high.

CONCLUSION

Ecologists have taken the principles of niche theory and successfully developed tools for conservation biology and land management. By understanding where on the landscape a given species is presently flourishing (i.e., where it has the highest fitness), we can make inferences about how a species may respond in unsampled areas and novel habitat or to scenarios of climate or land change. This area of research continues to improve and expand. Remotely sensed environmental data is increasing the spatial extent of the modeling endeavors.[52,53] As a result, regional and global distribution modeling are becoming more accessible. Also, large biological databases make it possible for researchers to share their survey data and receive range-wide information about their target species without sampling the entire range.[54] The Global Biodiversity Information Facility (GBIF) collects and manages biodiversity data and provides it freely to researchers via the internet (http://www.gbif.org). Improved technology and methods will expand the spatial extent of study areas, allow for finer resolution predictions, and reduce our uncertainty in results. These improvements should enhance our ability to effectively mitigate ecological problems.

ACKNOWLEDGMENTS

I would like to thank all the reviewers for their useful comments and suggestions. I would also like to thank Dwaine Kreakie for all his diligent editorial work. The U.S. EPA's Office of Research and Development (ORD) tracking number for this entry is ORD-003218.

REFERENCES

1. Bernstein, C.; Krebs, J.R.; Kacelnik, A. *Distribution of Birds Amongst Habitats: Theory and Relevance to Conservation. Bird Population Studies: Relevance to Conservation and Management*; Oxford University Press: Oxford, UK, 1991; 317–345.
2. Pianka, E.R. *Evolutionary Ecology*, 7th Ed. eBook; Eric R.P. Ed.; 2011.
3. Futuyma, D.J.; Moreno, G. The evolution of ecological specialization. Annu. Rev. Ecol. Syst. **1988**, 207–233.
4. Johnson, M.D. Measuring habitat quality: A review. Condor **2007**, *109*, 489–504.
5. Terborgh, J. Habitat selection in Amazonian birds. In *Habitat Selection in Birds*; Cody, M., Ed.; Academic Press: New York, 1985; 311–338.
6. Franklin, J. Moving beyond static species distribution models in support of conservation biogeography. Diversity and Distributions **2010**, *16*, 321–330.
7. Araujo, M.B.; Guisan, A. Five (or so) challenges for species distribution modelling. J. Biogeogr. **2006**, *33*, 1677–1688.

8. Hirzel, A.H.; Le Lay, G. Habitat suitability modelling and niche theory. J. Appl. Ecol. **2008**, *45*, 1372–1381.
9. Chase, J.M.; Leibold, M.A. *Ecological Niches: Linking Classical and Contemporary Approaches*; University of Chicago Press: Chicago, IL, 2003; 212 p.
10. Grinnell, J. The niche-relationships of the California Thrasher. The Auk. **1917**, *34*, 427–433.
11. Hutchinson, G.E. Cold Spring Harbor Symposium on Quantitative Biology. Concluding Remarks **1957**, *22*, 415–427.
12. Pulliam, H.R. On the relationship between niche and distribution. Ecol. Lett. **2000**, *3*, 349–361.
13. Guisan, A.; Zimmermann, N.E. Predictive habitat distribution models in ecology. Ecol. Model. **2000**, *135*, 147–186.
14. Augustin, N.; Mugglestone, M.; Buckland, S. An autologistic model for the spatial distribution of wildlife. J.Appl. Ecol. **1996**, *33*, 339–347.
15. Kreakie, B.J.; Fan, Y.; Keitt, T.H. Enhanced migratory waterfowl distribution modeling by inclusion of depth to water table data. PloS One **2012**, *7*, e30142.
16. Phillips, S.J.; Dudík, M. Modeling of species distributions with Maxent: New extensions and a comprehensive evaluation. Ecography **2008**, *31*, 161–175.
17. Buckley, P.; Francis, C.M.; Blancher, P.; DeSante, D.F.; Robbins, C.S.; Smith, G.; Cannell, P. The North American Bird Banding Program: Into the 21st Century (El Programa de Anillaje de Aves de Norte América: Hacia el Siglo 21). J. Field Ornithol. **1998**, *69* (4), 511–529.
18. Hirzel, A.H.; Helfer, V.; Metral, F. Assessing habitat-suitability models with a virtual species. Ecol. Model. **2001**, *145*, 111–121.
19. Royle, J.A.; Nichols, J.D.; Kéry, M. Modelling occurrence and abundance of species when detection is imperfect. Oikos **2005**, *110*, 353–359.
20. Mackenzie, D.I. Modeling the probability of resource use: The effect of, and dealing with, detecting a species imperfectly. J. Wildlife Manag. **2006**, *70*, 367–374.
21. Hirzel, A.; Guisan, A. Which is the optimal sampling strategy for habitat suitability modelling. Ecol. Model. **2002**, *157*, 331–341.
22. Barry, S.; Elith, J. Error and uncertainty in habitat models. J. Appl. Ecol. **2006**, *43*, 413–423.
23. Stockwell, D.R.B.; Peterson, A.T. Effects of sample size on accuracy of species distribution models. Ecol. Model. **2002**, *148*, 1–13.
24. Wisz, M.S.; Hijmans, R.; Li, J.; Peterson, A.T.; Graham, C.; Guisan, A. Effects of sample size on the performance of species distribution models. Divers. Distributions **2008**, *14*, 763–773.
25. Cassini, M.H. Ecological principles of species distribution models: The habitat matching rule. J. Biogeogr. **2011**, *38*, 2057–2065.
26. Guisan, A.; Theurillat, J.P.; Deil, U.; Loidi, J. Equilibrium modeling of alpine plant distribution: How far can we go?, Vegetation and climate. A selection of contributions presented at the 42nd Symposium of the International Association of Vegetation Science, Bilbao, Spain, 26–30 July 1999. Gebrüder Borntraeger Verlagsbuchhandlung **2000**, 353–384.
27. Dias, P.C. Sources and sinks in population biology. Trends Ecol. Evol. **1996**, *11*, 326–330.
28. Hanski, I.; Hanski, I.A. *Metapopulation Ecology*; Oxford University Press: Oxford, UK, 1999; 313 p.
29. Austin, M.P. Spatial prediction of species distribution: An interface between ecological theory and statistical modelling. Ecol. Model. **2002**, *157*, 101–118.
30. Elith, J.; Leathwick, J.R. Species distribution models: Ecological explanation and prediction across space and time. Annu. Rev. Ecol. Evol. Syst. **2009**, *40*, 677–697.
31. Hernandez, P.A.; Graham, C.H.; Master, L.L.; Albert, D.L. The effect of sample size and species characteristics on performance of different species distribution modeling methods. Ecography **2006**, *29*, 773–785.
32. Kearney, M.; Porter, W. Mechanistic niche modelling: Combining physiological and spatial data to predict species' ranges. Ecol. Lett. **2009**, *12*, 334–350.
33. Guisan, A.; Thuiller, W. Predicting species distribution: Offering more than simple habitat models. Ecol. Lett. **2005**, *8*, 993–1009.
34. Kearney, M.; Porter, W.P. Mapping the fundamental niche: Physiology, climate, and the distribution of a nocturnal lizard. Ecology **2004**, *85*, 3119–3131.
35. Kearney, M.; Phillips, B.L.; Tracy, C.R.; Christian, K.A.; Betts, G.; Porter, W.P. Modelling species distributions without using species distributions: the cane toad in Australia under current and future climates. Ecography **2008**, *31*, 423–434.
36. Brooks, T.M.; Mittermeier, R.A.; da Fonseca, G.A.B.; Gerlach, J.; Hoffmann, M.; Lamoreux, J.F.; Mittermeier, C.G.; Pilgrim, J.D.; Rodrigues, A.S.L. Global biodiversity conservation priorities. Science **2006**, *313*, 58–61.
37. García, A. Using ecological niche modelling to identify diversity hotspots for the herpetofauna of Pacific lowlands and adjacent interior valleys of Mexico. Biol. Conserv. **2006**, *130*, 25–46.
38. Naidoo, R.; Balmford, A.; Costanza, R.; Fisher, B.; Green, R.E.; Lehner, B.; Malcolm, T.; Ricketts, T.H. Global mapping of ecosystem services and conservation priorities. Proc. Nat. Acad. Sci. **2008**, *105*, 9495–9500.
39. Costa, G.C.; Nogueira, C.; Machado, R.B.; Colli, G.R. Sampling bias and the use of ecological niche modeling in conservation planning: A field evaluation in a biodiversity hotspot. Biodivers. Conserv. **2010**, *19*, 883–899.
40. Pearson, R.G.; Dawson, T.P. Predicting the impacts of climate change on the distribution of species: Are bioclimate envelope models useful? Global Ecol. Biogeogr. **2003**, *12*, 361–371.
41. Brotons, L.; Mañosa, S.; Estrada, J. Modelling the effects of irrigation schemes on the distribution of steppe birds in Mediterranean farmland. Biodivers. Conserv. **2004**, *13*, 1039–1058.
42. Sánchez-Cordero, V.; Illoldi-Rangel, P.; Linaje, M.; Sarkar, S.; Peterson, A.T. Deforestation and extant distributions of Mexican endemic mammals. Biol. Conserv. **2005**, *126*, 465–473.
43. Iverson, L.R.; Prasad, A.M. Predicting abundance of 80 tree species following climate change in the eastern United States. Ecol. Monogr. **1998**, *68*, 465–485.
44. Iverson, L.R.; Prasad, A.; Schwartz, M.W. Modeling potential future individual tree-species distributions in the eastern United States under a climate change scenario: A case study with *Pinus virginiana*. Ecol. Model. **1999**, *115*, 77–93.
45. Peterson, A.T. Predicting the geography of species' invasions via ecological niche modeling. Q. Rev. Biol. **2003**, *78*, 419–433.

46. Thuiller, W.; Richardson, D.M.; PYŠEK, P.; Midgley, G.F.; Hughes, G.O; Rouget, M. Niche-based modelling as a tool for predicting the risk of alien plant invasions at a global scale. Global Change Biol. **2005**, *11*, 2234–2250.
47. Herborg, L.M.; Jerde, C.L.; Lodge, D.M.; Ruiz, G.M.; MacIsaac, H.J. Predicting invasion risk using measures of introduction effort and environmental niche models. Ecol. Appl. **2007**, *17*, 663–674.
48. Ward, D.F. Modelling the potential geographic distribution of invasive ant species in New Zealand. Biol. Invasions **2007**, *9*, 723–735.
49. Zambrano, L.; Martínez-Meyer, E.; Menezes, N.; Peterson, A.T. Invasive potential of common carp (*Cyprinus carpio*) and Nile tilapia (*Oreochromis niloticus*) in American freshwater systems. Can. J. Fisheries Aqua. Sci. **2006**, *63*, 1903–1910.
50. Broennimann, O.; Treier, U.A.; Müller-Schärer, H.; Thuiller, W.; Peterson, A.; Guisan, A. Evidence of climatic niche shift during biological invasion. Ecol. Lett. **2007**, *10*, 701–709.
51. Schamberger, M.; Farmer, A.H.; Terrell, T.J. Habitat Suitability Index (HSI) Models: Introduction. 1982, U.S. Fish and Wildlife Service Special Report FWS-OBS-82-10.
52. Kerr, J.T.; Ostrovsky, M. From space to species: Ecological applications for remote sensing. Trends Ecol. Evol. **2003**, *18*, 299–305.
53. Bradley, B.A.; Fleishman, E. Can remote sensing of land cover improve species distribution modelling? J. Biogeogr. **2008**, *35*, 1158–1159.
54. Corsi, F.; de Leeuw, J.; Skidmore, A. *Modeling Species Distribution with GIS. Research Techniques in Animal Ecology*; Columbia University Press: New York, 2000; 389–434.

Biodiversity: Tropical Agroforestry

Thomas P. Husband
Department of Natural Resources Science, University of Rhode Island, Kingston, Rhode Island, U.S.A.

Abstract
In recent years, agroforestry has shown promise as being a valuable way to combine the goals of biological diversity preservation and sustainable farming. Although ideally biodiversity would be sustained through the preservation of natural landscapes, in many parts of the world this is not feasible with the pressures to grow food and competition with other land uses. Agroforestry systems can support a diversity of native flora and fauna, which is particularly important in fragmented landscapes. Lands on which agroforestry is practiced also may serve as important buffer zones to more natural habitats and aid in creating contiguous natural corridors for the dispersal of native species. A growing body of scientific data demonstrates that certain agroforestry approaches can enhance the diversity of many native taxonomic groups, including birds, mammals, insects, amphibians, and plants. Agroforestry may be a critical tool that natural resource managers will use to create "forest-like" habitats while meeting the ever-increasing human needs for food and other resources.

INTRODUCTION

The goal of this entry is to introduce the reader to basic concepts of agroforestry and how they relate to landscape ecology and the enhancement of biodiversity in the tropics. In recent years, the term *biodiversity* has become a household word throughout much of the world. According to the Food and Agriculture Organization of the United Nations (FAO), "'Biological diversity' encompasses the variety of existing life forms, the ecological roles they perform and the genetic diversity they contain. In forest ecosystems, biological diversity allows species to adapt continuously to dynamically evolving environmental conditions, to maintain the potential for tree breeding and improvement (to meet human needs for goods and services and changing end-use requirements), and to support their ecosystem functions."[1] The role of agroforestry in maintaining or enhancing biodiversity, particularly in the tropics, is the primary focus of this encyclopedia entry.

AGROFORESTRY

Agroforestry Defined

According to Young,[2] "Agroforestry is a collective name for land-use systems in which woody perennials (e.g., trees, shrubs, etc.) are grown in association with herbaceous plants (e.g., crops, pastures) and/or livestock in a spatial arrangement, a rotation or both, and in which there are both ecological and economic interactions between the tree and non-tree components of the system." Gascon et al.[3] simply defined agroforestry as "the practice of integrating trees and other large woody perennials on farms and throughout the agricultural landscape." Agroforestry practices include[2] trees on cropland, plantation crop combinations, multistory-tree gardens, hedgerow intercropping, boundary planting, trees on erosion-control structures, windbreaks and shelterbelts, biomass transfer, trees on rangeland or pastures, plantation crops with pastures, live fences, fodder banks, woodlots with multipurpose management, and reclamation forestry leading to multiple use. Agroforestry involves mixing trees or shrubs or both with non-woody crops and sometimes livestock for the purpose of crop diversification, improved profitability, and improved environmental stewardship. Agroforestry may involve the introduction of long-term crops (trees and shrubs) into open agricultural landscapes, or conversely, the introduction of agricultural husbandry of a wide range of crops and livestock into existing woodlands. Agroforestry crops include wood fiber; both native and exotic tree species; non-wood crops from trees; traditional agricultural crops such as horticultural crops, field crops, forages, special crops, and livestock; and non-timber forest crops. Agroforestry practices involve growing woody perennials, usually trees, along with more than one farm crop growing on the same land area at the same time, which can cause both positive and negative competition between crops for light, water, and nutrients. Successful implementation of agroforestry practices requires good planning and management.

Encyclopedia of Natural Resources DOI: 10.1081/E-ENRL-120048077

The Value of Agroforestry

According to Garret,[4] "When properly designed and integrated, agroforestry can protect crops and improve crop yields, shelter livestock, reduce animal stress while improving weight gain, and enhance resource stewardship and land conservation. From creating consumer markets and reducing nonpoint source pollution to protecting waterfowl habitat and developing new farm crops, agroforestry is moving to the forefront in meeting the needs of landowners and natural resource professionals while keeping the family farm economically viable and the environment in which we live healthy." Gold and Garret[5] point out that, in the tropics, agroforestry has been long practiced and widely accepted by farmers as an alternative to traditional slash-and-burn agriculture and to conventional agriculture where slopes are steep and on marginal land. As a result, ecologically based agroforestry has helped to restore and maintain biodiversity, stabilize the ecology of farms and their watersheds, help sustain the production of basic human needs, and create marketplace opportunities for millions of rural poor.[6–8]

Much important advancement in tropical agroforestry has been made through the work of scientists from various institutions around the world. Among those institutions is the World Agroforestry Centre (ICRAF), an independent not-for-profit research organization. ICRAF is part of an alliance of the Consultative Group on International Agricultural Research (CGIAR) centers that are "dedicated to generating and applying the best available knowledge to stimulate agricultural growth, raise farmers' incomes, and protect the environment." ICRAF's vision is a rural transformation in the developing world as smallholder households strategically increase their use of trees in agricultural landscapes to improve their food security, nutrition, income, health, shelter, energy resources, and environmental sustainability. The Centre's mission is to generate science-based knowledge about the diverse roles that trees play in agricultural landscapes, and use its research to advance policies and practices that benefit the poor and the environment.[9] ICRAF lists the various benefits of agroforestry, including the following:

- Enriching the asset base of poor households with farm-grown trees.
- Enhancing soil fertility and livestock productivity on farms.
- Linking poor households to markets for high-value fruits, oils, cash crops, and medicines.
- Balancing improved productivity with the sustainable management of natural resources.
- Maintaining or enhancing the supply of environmental services in agricultural landscapes for water, soil health, carbon sequestration, and biodiversity.

Agroforestry is increasingly recognized as a valuable strategy with the potential to diversify production of agricultural and forest products for greater social, economic, and environmental benefits.[3] Relative to agroforestry, ICRAF has outlined seven key challenges inter-related to the Millennium Development Goals (MDG) and the Water, Energy, Health, Agriculture, and Biodiversity (WEHAB) initiative.[6,10] They include the following: i) Help eradicate hunger through basic, propoor food production systems in disadvantaged areas based on agroforestry methods; ii) lift more rural poor from poverty through market-driven, locally led tree cultivation systems that generate income and build assets; iii) advance health and nutrition of the rural poor through agroforestry systems; iv) conserve biodiversity through integrated conservation development solutions based on agroforestry technologies, innovative institutions, and better policies; v) protect watershed services through agroforestry-based solutions that enable the poor to be rewarded for their provision of these services; vi) assist the rural poor to better adapt to climate change, and to benefit from emerging carbon markets, through tree cultivation; and vii) build human and institutional capacity in agroforestry research and development. Agroforestry has had a significant role in increasing productivity of landscapes for the benefit of human populations, thus contributing to many of these goals.

Many of the MDGs focus on environmental stability and the integration of the principles of sustainability into policies and programs to reverse the loss of environmental services. According to Garrity,[6] the environmental services that are of greatest relevance are watershed protection, biodiversity conservation, and climate change mitigation and adaptation. A major emerging trend in agroforestry research and development is in the area of enhancing environmental services while meeting farmers' needs for food and income. The latter is critical, as about 90% of the biodiversity resources in the tropics are located in human-dominated or working landscapes.[6] The intensification of agroforestry systems can reduce the exploitation of protected areas,[11,12] increase biodiversity in agricultural landscapes, and actually increase the species and within-species diversity of trees in farming systems.[4,6,13]

BIODIVERSITY

Biodiversity Defined

Biodiversity comprises the complexity of all life forms. The Convention on Biological Diversity defines its area of concern as "... the variability among living organisms from all sources, including, *inter alia*, terrestrial, marine and other aquatic ecosystems and the ecological complexes of which they are part; this includes diversity within species, between species, and of ecosystems."[14] Myers[15] states, "Biodiversity embraces the totality of life forms,

Table 1 The species diversity of two different hypothetical communities

Species	Community 1	Community 2
A	90	34
B	5	33
C	5	33
Total Individuals	100	100
Species Diversity (H')	0.39	1.09

from the planetary spectrum of species to subunits of species (races, populations) together with ecosystems and their ecological processes."

Sometimes thought of simply as the number of species in a given area, the concept of biodiversity is much more complex. In addition to a count of species, or *species richness*, it includes elements of *species diversity*, *genetic diversity*, *functional diversity*, and *ecological diversity*. Species diversity includes both the number of species in an area and the measure of their relative abundance or spatial distribution. For example, in Table 1, each of the two hypothetical communities has three species with a total of 100 individuals. The species richness is the same; however, as calculated by the Shannon Index, [16] the species diversity is higher in Community 2. As with most calculations of species diversity, the Shannon Index takes into account evenness, or a measure of the relative proportions represented by each species. Community 2 has a higher degree of evenness, and thus higher species diversity.

Genetic diversity is now being considered more frequently as a critical element of biodiversity.[17] This includes various measures of the variety of genetic material within a species or a population. Low *genetic diversity* can lead to short-, medium-, and long-term problems, such as inbreeding depression that shifts mean phenotypes in a direction that causes a reduction in fitness. *Functional diversity*, another important component, includes the biological and chemical processes such as energy flow and matter recycling needed for the survival of species, communities, and ecosystems.[18] Functional diversity also comprises the many ecological interactions among species (e.g., competition, predation, parasitism, mutualism, etc.), as well as the natural disturbances (e.g., fire, rain events, etc.) that many species and communities require. Ecological diversity considers the variety of terrestrial and aquatic ecosystems that are found in an area of the Earth.[18] Together, these four components comprise biodiversity.

The Value of Biodiversity

The values gained from biodiversity are associated with different scales and the individual components outlined above, including landscapes, ecosystems, species, populations, genes, and various ecosystem functions. Usually, the value of biodiversity must be judged on the basis of the scale or component critical to the situation; thus no single measure of biodiversity value can be employed universally. Generally, four different types of values can be recognized with regard to biodiversity.[19,20] First, biodiversity could be considered to have its own value, often called its *intrinsic* value. Cultural, social, aesthetic, and ethical benefits are all components of intrinsic value. Second is the *utilitarian* or *material goods* value of biodiversity, which includes all of the subsistence and commercial benefits of species or their genes by various sectors of society.[15] Third, biodiversity may have *serependic* value, that is, the belief that parts of an ecosystem may have future, but as of yet unknown, values to humans. Last, biodiversity has a *functional* or *ecosystems services* value, that is, it contributes to ecosystem life support functions and the preservation of ecological structure and integrity. At some level, all of these types of values may be considered ethnocentric and depend upon the context of the human culture in which they are viewed. In general, biodiversity is considered valuable to humans because it provides us with natural resources (e.g., food, water, wood paper products, energy, medicines), natural services (e.g., air and water purification, soil fertility, waste disposal, pest control), and aesthetic pleasure (e.g., wildlife, national parks and nature reserves, natural scenic beauty). Tropical ecosystems, where most agroforestry systems are employed, are particularly recognized for their significant biodiversity because they contain over half of the terrestrial species on our planet.[21]

TRENDS IN THE WORLD FORESTS

One important role of agroforestry with respect to environmental services is that of ameliorating the downward trends in the world's forests. Based upon 2010 statistics,[1] the total forest in the world was estimated to be slightly more than 4 billion hectares or about 31% of the total land area of the Earth (Fig. 1). Tropical forests make up about 47% of this total.[1] According to the FAO,[1] deforestation continues at an alarming rate, mostly the result of conversion to agricultural land. In recent years, this loss has been slowed significantly around the world by the planting of new forest trees, landscape restoration, and natural expansion.[1] Overall, net global forest loss from 2000 to 2010 was 13 million hectares per year, down from 16 million hectares per year for the period 1990–2000.[22] The greatest losses from 2000 to 2010 were incurred in South America where over 3.9 million hectares were lost each year, followed by Africa, which lost about 3.4 million hectares per year (Fig. 2). Clearly, the forests of the tropics have been the most impacted in the world.

For a variety of reasons, trees are being planted at increasing rates around the world. Even so, forest

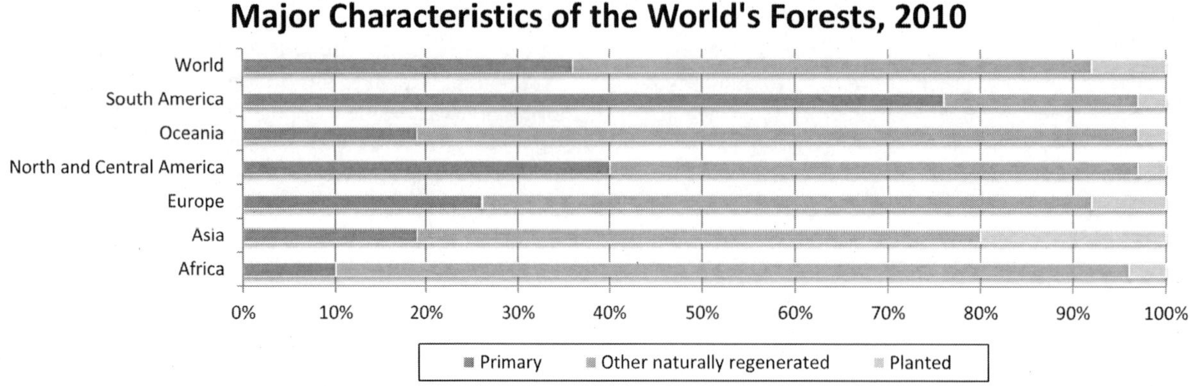

Fig. 1 Major characteristics of the world's forests, 2010.
Source: From http://www.fao.org/forestry/26026-0f6e61f1755a602cea05a15f9a1e7c450.jpg (accessed October 2013); FAO Global Forest Resource Assessment 2010. FAO Forestry Paper No. 163. http://www.fao.org/docrep/013/i1757e/i1757e.pdf. (accessed October 2013).[1]

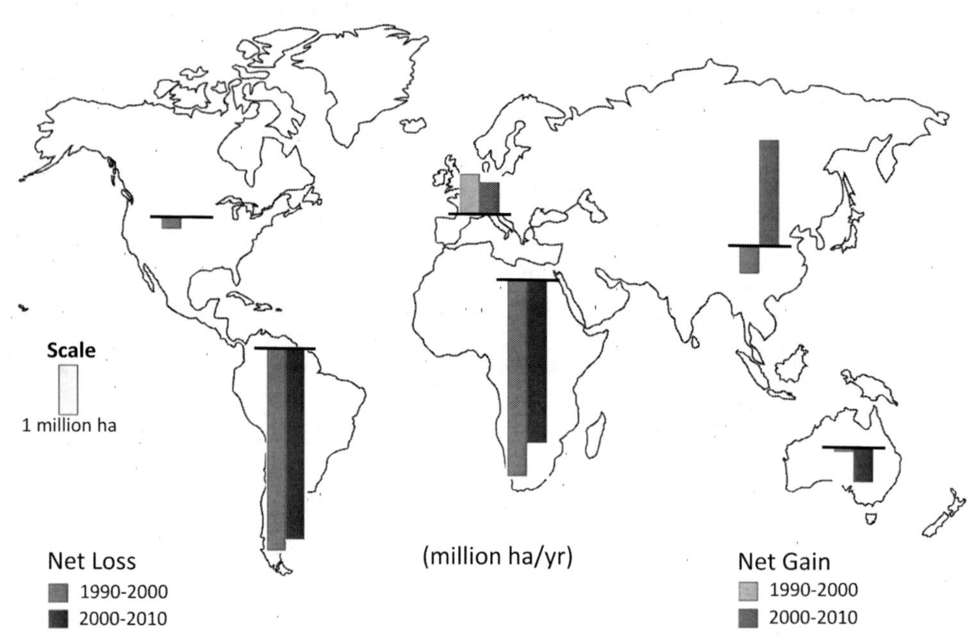

Fig. 2 Changes in forest area by region of the world, 1990–2010.
Source: From FAO Global Forest Resource Assessment 2010. FAO Forestry Paper No. 163. http://www.fao.org/docrep/013/i1757e/i1757e.pdf. (accessed October 2013).[1]

plantations comprise only about 4–5% of the total forest area.[1] Of all forests, less than 12% are designated as having conservation of biodiversity as their primary function.[1] Although deforestation, including conversion to agricultural land, continues at an alarmingly high rate, agroforestry likely can play an important role in lessening the impact of this trend.

AGROFORESTRY AND BIODIVERSITY PRESERVATION

The two major approaches to protecting biodiversity, particularly in forested environments, are i) creating protected parks and ii) establishing agroforestry areas. Parks have proven to be an effective way of protecting biodiversity, although land-use pressures compromise many parks, and their management is often underfunded.[23] In addition, parks are often subject to practices such as illegal poaching and overuse by visitors. Additionally, in some countries, it is politically unpopular to "lock up" needed resources in protected parks. An alternative approach is to use agroforestry to enhance the biodiversity of existing habitats. Agroforestry also may be a useful tool for reducing the impact of deforestation and edge effects on tropical parks and biodiversity corridors that link protected habitats.[24,25] At the same time, agroforestry can produce sustainable harvests of food crops and wood and perhaps contribute to the mitigation of greenhouse gases that contribute to climate change.[6,26]

STUDIES OF AGROFORESTRY AND BIODIVERSITY

The study of the relationship between agroforestry and biodiversity has steadily advanced in the past decade.[24] The impacts of agroforestry on biodiversity have been studied in coffee,[24,25,27–36] cacao,[37,38] tea,[3,39] banana and plantain,[36,40] rubber,[41] and several other crops. Among the agroforestry crops, biodiversity of coffee-dominated landscapes has perhaps been the most studied. Studies in traditional shaded coffee plantations have shown that agroecosystems can serve as a refuge for biodiversity.[28–30] However, most of the initial work on coffee and biodiversity was conducted on birds, insects, and arthropods.[38,42–54]

The relationship between agroforestry systems and the natural biodiversity of an area has been found to be very complicated. Much more research is needed to acquire an understanding of the ability of agroforestry systems to support the natural ecosystem structure and function. To date, some generalities have been supported: complex agroforestry systems are better than monocrop systems; shade coffee is better than sun coffee; and systems using native plants tend to be generally more biologically diverse. As landscape changes in the world seem to be inevitable, scientists and managers alike must examine whether ecosystem management strategies, such as those employing agroforestry, can limit ecological losses while satisfying economic and societal gains. This examination of trade-offs has been conducted for several agroforestry crops in a number of countries, but there is still far too little known about the human dimension of land-use change and the consequences for ecosystem functioning and biodiversity retention.[55]

PLANNED BIODIVERSITY

Biodiversity in the tropics has benefited through the implementation of the so-called planned biodiversity. Planned biodiversity refers to the crops and livestock introduced and sustained in an agroecosystem by a farmer, whereas unplanned biodiversity includes all other organisms (e.g., soil flora and fauna, herbivores, carnivores, decomposers) that exist in or colonize the agroecosystem. A number of agroforestry practices appear to promote planned biodiversity, namely, home gardens, annual systems, and perennial systems.[56] Perhaps the most notable is the practice of creating home gardens.[13] Home gardens, as defined by Perfecto and Vandermeer [56] are "land-use systems involving multipurpose trees and shrubs in association with annual and perennial agricultural crops as well as livestock (usually small animals) that are intensively managed by family labor within the compound of individual houses." Practiced in the humid tropics for centuries,[57] home garden systems are often very diverse, have several strata in the canopy, and provide multiple ecological and agricultural functions.[58–63] Although most of the research in home gardens is qualitative and descriptive,[56] more recent research that has employed cluster and other multivariate quantitative analyses[60,62,64,65] supports earlier descriptive work in evaluating the role of home gardens in the conservation of biodiversity. Kumar and Nair[66] documented that most studies of home gardens, whether descriptive or quantitative and experimental, demonstrate that these agroforestry systems have a high planned biodiversity.

CERTIFICATIONS AND BIODIVERSITY

Certifications associated with biodiversity have had an increasingly role in promoting biodiversity in tropical agroforestry systems. For example, research on avifauna has led to the formulation of certification criteria by the Smithsonian Migratory Bird Center for the so-called Certified Bird Friendly® farms.[67] Although the criteria are well thought out and represent a significant step toward growing agroforestry crops while supporting biodiversity, the number of farms certified in the program is relatively low. For example, only 28 coffee farms or co-operatives are now certified, representing about 1,402 individual growers in 7 countries. The total production of certified farms is only slightly over 9.6 million pounds of coffee per year, with the total land area equal to 7,332 hectares. To give some perspective to these numbers, there were 133,308,000 bags of coffee produced in 2010 by over 26 million people in some 70 countries.[68] If a figure of 60 kilograms per bag is assumed (bag weights vary somewhat from country to country), this equates to approximately 8 billion kilograms of coffee grown in the world. The total area on which coffee is grown peaked in 2000 at 10,719,591 and was last estimated in 2009 to be about 9,841,317 hectares worldwide.[69] Thus, Certified Bird Friendly® farms represent, relative to the pounds of coffee grown worldwide and the area on which it is grown, only 0.05% and 0.07%, respectively.

Other programs that are contributing to promoting biodiversity in agroforestry landscapes include Rainforest Alliance and UTZ certifications.[70–72] These programs offer third-party certification, verification, and validation for about 100 different crops, including coffee, tea, bananas, chocolate, and even roses. For example, to qualify for Rainforest Alliance certification, farms must meet the standards of the Sustainable Agriculturc Network.[73] In 2003, various large- and medium-sized companies started selling Rainforest Alliance Certified coffee in North America, Europe, and Asia. By 2006, global retail sales of Rainforest

Alliance Certified coffee, bananas, and chocolate exceeded U.S. $1 billion with more than 10,000 certified farms in 15 countries. In subsequent years, certification has grown steadily and expanded to many other crops such as flowers, pineapple, and tea. These types of certifications appear to be making significant contributions to sustainability and the preservation of biodiversity in tropical countries around the world.

CONCLUSION

Agroforestry systems in the tropics sometimes resemble natural rain forests, particularly in terms of their structure. Through the incorporation of trees on farmlands and bringing degraded lands under agroforestry systems, many areas now support the ecological functions and biodiversity as do more natural tropical forests. Agroforestry has been thought of as a promising biodiversity-friendly land-use strategy that conserves a significant proportion of tropical rain forest diversity while providing significant economic returns.[56] There exists a profound need for more quantitative information on the potential trade-offs between biodiversity loss and agricultural intensification in the tropics.

ACKNOWLEDGMENTS

I wish to thank CIRAD (the French research center working with developing countries to tackle international agricultural and development issues) for providing me with the facilities and technical support necessary to complete this task. I also wish to thank the University of Rhode Island for supporting this work through a sabbatical leave. I especially wish to thank Dr. Philippe Vaast for reviewing the manuscript and offering constructive criticism.

REFERENCES

1. FAO. *Plant genetic resources: their conservation in situ for human use. Prepared in collaboration with the United Nations Educational, Scientific and Cultural Organization (UNESCO)*; UNEP and IUCN: Rome, 1989.
2. Young, A. *Agroforestry For Soil Conservation*; ICRAF Science and Practice of Agroforestry: BPCC Wheatons Ltd, Exeter, UK. 1986.
3. Gascon, C.; de Fonseca, G.A.B.; Sechrest, W.; Billmark, K.A.; Sanderson, J. Chapter 1: Biodiversity conservation in deforested and fragmented tropical landscapes: an overview. In *Agroforestry and Biodiversity Conservation in Tropical Landscapes*; Schroth, G., de Fonseca, G. A.B., Harvey, C.A., Gascon, C., Vasconcelos, H.L., Izac, A-M.N., Eds.; Island Press: Washington, D.C., 2004; 15–32.
4. Garrett, H.E., Ed.; *North American Agroforestry: An Integrated Science and Practice*; 2nd Ed. Agronomy Society of America: Madison, Wisconsin, USA, 2009.
5. Gold, M.A.; Garrett, H.E. Agroforestry nomenclature, concepts, and practices. In *North American Agroforestry: An Integrated Science and Practice*; 2nd Ed.; Garrett, H.E., Ed.; Agronomy Society of America: Madison, Wisconsin, USA, 2009; 45–55.
6. Garrity, D.P. Agroforestry and the achievement of the Millennium Development Goals. Agrofor. Syst. **2004**, *61*, 5–17.
7. Russell, D.; Franzel, S. Trees of prosperity: Agroforestry, markets and the African smallholder. Agrofor. Syst. **2004**, *61*, 345–355.
8. Nair, P.K.R.; Allen, S.C.; Bannister, M.E. Agroforestry today: An analysis of the 750 presentations to the 1st World Congress of Agroforestry 2004. J. For. **2005**, *103*, 417–421.
9. http://www.worldagroforestry.org/about_us. (accessed September 2011).
10. Jose, S. Agroforestry for ecosystem services and environmental benefits: an overview. Agrofor. Syst. **2009**, *76*, 1–10.
11. Murniati, G.; Garrity D.P.; Gintings, A.N. The contribution of agroforestry systems to reducing farmers' dependence on the resources of adjacent national parks: a case study from Sumatra Indonesia. Agrofor. Syst. **2001**, *52*, 171–184.
12. Garrity, D.P.; Amoroso, V.B., Koffa, S.; Catacutan, D.; Buenavista, G.; Fay, P.; Dar, W.D. Landcare on the poverty-protection interface in an Asian watershed. In *Integrated Natural Resource Management: Linking Productivity, the Environment, and Development*. Campbell, B.M., Sayer, J.A., Eds.; CABI Publishing: Cambridge, MA, USA. 2003, 195–210 pp.
13. Nair, P.K.R. *An Introduction to Agroforestry*. Springer: Dordrecht, the Netherlands, 1993.
14. Heywood, V.H.; Bates, I. Introduction. In *Global Biodiversity Assessment*; Heywood, V.H., Watson, R.T., Eds.; Cambridge University Press: Cambridge, England, UNEP, 1995; 1–21 pp.
15. Myers, N. Environmental services of biodiversity. Proc. Natl. Acad. Sci. **1996**, *93*, 2764–2769.
16. Weaver, W.; Shannon, C.E. *The Mathematical Theory of Communication*; University of Illinois: Urbana, Illinois, 1949.
17. Atta-Krah, K.; Kindt, R.; Skilton, J.N.; Amaral, W. Managing biological and genetic diversity in tropical agroforestry. Agrofor. Syst. **2004**, *61*, 183–194.
18. Miller, G.T.; Spoolman, S. *Living in the Environment; 18th Ed., Cengage: Stamford, CT, USA,2012.*
19. Nunes, P.A.D.; van der Bergh, J.C.J.M. Economic valuation of biodiversity: sense or nonsense? Ecol. Econ. **2001**, *39*, 203–222.
20. Swift, M.J.; Izac, A.-M.N.; van Noordwijk, M. Biodiversity and ecosystem services in agricultural landscapes—are we asking the right questions? Agric. Ecosyst. Environ. **2004**, *104* (1), 113–134.
21. Wiens, J.J.; Graham, C.H.; Moen, D.S.; Smith, S.A.; Reeder, T.W. Evolutionary and ecological causes of the latitudinal diversity gradient in hylid frogs: treefrog trees unearth the roots of high tropical diversity. Am. Nat. **2006**, *168*, 579–596.
22. Food and Agriculture Organization of the United Nations. *State of the World's Forests*; Rome, Italy, 2011.

23. Bruner, A.G.; Gullison, R.E.; Rice, R.E.; Fonseca, G.A. Effectiveness of parks in protecting tropical biodiversity. Science **2001**, *291*, 125–128.
24. Husband, T.P.; Abedon, D.H.; Paton, P.W.C. Surveying Mammal Biodiversity in Coffee-Dominated Landscapes in Costa Rica (2005–2007). Abstract of presentation at the Second International Symposium for the Multistrata Agroforestry Systems with Perennial Crops: Making Ecosystem Services Count for Farmers, Consumers, and the Environment: Turrialba, Costa Rica, Sept. 17–21, 2007.
25. Husband, T.P.; Abedon, D.H.; Donelan, E.; Paton, P.W.C. Do in Coffee-Dominated Landscapes Support Mammal Biodiversity: Costa Rica 2005–2008. Abstract of presentation at the Second World Agroforestry Congress: Nairobi, Kenya, Aug. 23–29, 2009.
26. Albrecht, A.; Kandji, S.T. Review: Carbon sequestration in tropical agroforestry systems. Agriculture, Ecosyst. Environ. **2003**, *99*, 15–27.
27. Philpott, S.M.; Arendt, W.J.; Armbrecht, I.; Bichier, P.; Diestch, T.V.; Gordon, C.; Greenberg, R.; Perfecto, I.; Reynoso-Santos, R.; Soto-Pinto, L.; Tejeda-Cruz, C.; Williams-Linera, G.; Valenzuela, J.; Zolotoff; J. M. Biodiversity Loss in Latin American Coffee Landscapes: Review of the Evidence on Ants, Birds, and Trees. Conserv. Biol. **2008**, *22* (5), 1093–1105.
28. Escalante, E.E. Coffee and agroforestry in Venezuela. Agrofor. Today **1995**, *7* (3–4), 5–7.
29. Perfecto, I.; Snelling, R. Biodiversity and the transformation of a tropical agroecosystem: ants in coffee plantations. Ecological Applications **1995**, *5*, 1084–1097.
30. Perfecto, I.; Rice, R.A.; Greenberg, R.; Van der Voort, M.E. Shade coffee: a disappearing refuge for biodiversity. BioSci. **1996**, *46*, 598–608.
31. Perfecto I.; Mas, A.; Dietsch, T.; Vandermeer, J. Conservation of biodiversity in coffee agroecosystems: a tri-taxa comparison in southern Mexico. Biodiver. Conserv. **2003**, *12*, 1239–1252.
32. Moguel, P.; Toledo, V.M. Biodiversity conservation in traditional coffee systems in Mexico. Conserv. Biol. **1999**, *12* (1), 1–11.
33. Klein, A.-M.; Steffan-Dewenter, I.; Buchori, D.; Tscharntke, T. Effects of land-use intensity in tropical agroforestry systems on coffee flower-visiting and trap-nesting bees and wasps. Conserv. Biol. **2002**, *16* (4), 1003–1014.
34. Mas, A. H.; Dietsch, T. V. An index of management intensity for coffee agroecosystems to evaluate butterfly species richness. Ecol. Appl. **2003**, *13*, 1491–1501.
35. McNeely, J.A. Nature vs. nurture: managing relationships between forests, agroforestry and wild biodiversity. Agrofor. Syst. **2004**, *61–62* (1–3), 155–165.
36. Bhagwat, S.A.; Willis, K.J.; Birks, H.J.B.; Whittaker, R.J. Agroforestry: a refuge for tropical biodiversity? Trends Ecol. Evol. **2008**, *23* (5), 261–267.
37. Bentley, J.W.; Boa, E.; Stonehouse, J. Neighbor trees: shade, intercropping, and cacao in Ecuador. Hum. Ecol. **2004**, *32* (2), 241–270.
38. Asare, R. A review on cocoa agroforestry as a means for biodiversity conservation. Paper presented at World Cocoa Foundation Partnership Conference, Brussels, May 2006.
39. van Biervliet, O.; Wiśniewski, K.; Daniels, J.; Vonesh, J. R. Effects of tea plantations on stream invertebrates in a global biodiversity hotspot in Africa. Biotropica **2009**, *41* (4), 469–475.
40. Harvey, C.A.; Gonzalez, J.; Somarriba, E. Dung beetle and terrestrial mammal diversity in forests, indigenous agroforestry systems and plantain monocultures in Talamanca, Costa Rica. Biodivers. Conserv. **2006**, *15* (2), 555–585.
41. Penot, E. Prospects for conservation of biodiversity within productive rubber agroforests in Indonesia (CIRAD-Tera, Montpellier (France)), Conference International Workshop on Management of secondary and logged-over forests in Indonesia, CIFOR Bogor, Indonesia, Nov. 17–19, 1997.
42. Wunderle, J.; Latta S. Avian abundance in sun and shade coffee plantations and remnant pine forest in the Cordillera Central, Dominican Republic. Ornitologia Neotrop. **1996**, *7*, 19–34.
43. Estrada, A.; Coates-Estrada, S.; Merrit, D., Jr. Anthropogenic landscape changes and avian diversity at Los Tuxtlas, Mexico. Biodiver. Conserv. **1997**, *6*, 19–43.
44. Greenberg, R.; Bichier P.; Cruz Angon A.; Reitsma R. Bird populations in shade and sun coffee plantations in Central Guatemala. Conserv. Biol. **1997a**, *11*, 448–459.
45. Greenberg, R.; Bichier P.; Sterling J. Bird populations in rustic and planted shade coffee plantations of Eastern Chiapas, Mexico. Biotropica **1997b**, *29*, 501–514.
46. Calvo, L.; Blake, J. Bird diversity and abundance on two different shade coffee plantations in Guatemala. Birds Conserv. Int. **1998**, *8*, 297–308.
47. Wunderle, J.M., Jr. Avian distribution in Dominican shade coffee plantations: area and habitat relationships. J. Field Ornithol. **1999**, *70*, 58–70.
48. Johnson, M.D. Effects of shade tree species and crop structure on the arthropod and bird communities in a Jamaican coffee plantation. Biotropica **2000**, *32*, 133–145.
49. Nestel, D.; Dickschen, F. The foraging kinetics of ground ant communities in different coffee agroecosystems. Oecologia **1990**, *84*, 58–63.
50. Ibarra-Núñez, G.; Garcıa Ballinas, J.A.; Moreno-Próspero, M.A. *La comunidad de artrópodos de dos cafetales con diferentes prácticas agrıcolas: el caso de los Hymenópteros Resúmenes, XXVIII Congreso Nacional de Entomologıa, Sociedad Mexicana de Entomología*;. Universidad de las Américas: Cholula, Puebla, Mexico, 1993.
51. Perfecto, I.; Vandermeer, J.H. Understanding biodiversity loss in agroecosystems: reduction of ant diversity resulting from transformation of the coffee ecosystem in Costa Rica. Entomology. Trends Agric. Sci. **1994**, *2*, 7–13.
52. Perfecto, I.; Vandermeer, J.H. The quality of the agroecological matrix in a tropical montane landscape: ants in coffee plantations in southern Mexico. Conserv. Biol. **2002**, *16*, 174–182.
53. Perfecto, I.; Vandermeer J.H.; Hanson, P.; Cartín V. Arthropod biodiversity loss and the transformation of a tropical agro-ecosystem. Biodivers. Conserv. **1997**, *6*, 935–945.
54. Ibarra-Núñez, G.; Garcia-Ballinas, J.A. Diversidad de tres familias de arañas tejedoras (Araneae: Araneidae, Tetragnathidae, Theridiiae) en cafetales del soconusco, Chiapas, Mexico. Folia Entomol. Mexicana **1998**, *102*, 11–20.

55. Steffen-Dewenter, I.; Kessler, M.; Barkmann, J.; Bos, M.M.; Buchori, D. Tradeoffs between income, biodiversity, and ecosystem functioning during tropical rainforest conversion and agroforestry intensification. Proc. Natl. Acad. Sci. USA. **2007**, *104*, 4973–4978.
56. Perfecto, I.; Vandermeer, J. Biodiversity Conservation in Tropical Agroecosystems. Ann. N. Y. Acad. Sci. **2008**, *1134*, 173–200.
57. Montagnini, F. Homegardens of Mesoamerica: biodiversity, food security, and nutrient management. In *Tropical Homegardens: A Time-Tested Example of Sustainable Agroforestry*. Kumar, B.M., Nair, P.K.R., Eds.; Springer: the Netherlands, 2006; 1–23.
58. Fernandes, E.C.M.; Nair, P.K.R. An evaluation of the structure and function of tropical homegardens. Agric. Syst. **1986**, *21*, 279–310.
59. Coomes, O.T.; Ban, N. Cultivated plant species diversity in home gardens of an Amazonian peasant village in northeastern Peru. Econ. Bot. **2004**, *58*, 420–434.
60. Kehlenbeck, K.; Maass, B.L. Crop diversity and classification of homegardens in Central Sulawesi, Indonesia. Agro. Syst. **2004**, *63*, 53–62.
61. Major, J.; Clement C.R.; DiTommaso, A. Influence of market orientation on food plant diversity of farms located on Amazonian Dark Earth in the region of Manaus, Amazonas, Brazil. Econ. Bot. **2005**, *59*, 77–86.
62. Peyre, A.; Guidal, A.; Wiersum, K.F.; Bongers, F. Dynamics of homegarden structure and function in Karala, India. Agrofor. Syst. **2006**, *66*, 101–115.
63. Pandey, C.B.; Rai, R.B.; Singh, L.; Singh, A.K. Homegardens of Andaman and Nicobar, Indian. Agric. Syst. **2007**, *92*, 1–22.
64. Blanckaert, I.; Swennen, R.L.; Paredes Flores, M.; Rosas Lopez, R.; Lira Saade, R. Floristic composition, plant uses and management practices in homegardens of San Rafael Coxcatlan, Valley of Tehuacan-Cuicatlan, Mexico. J. Arid Environ. **2004**, *57*, 39–62.
65. Albuquerque, U.P.; Andrade, L.H.C.; Caballero, J. Structure and floristics of homegardens in Northeastern Brazil. J. Arid Environ. **2005**, *62*, 491–506.
66. Kumar, B.M.; Nair, P.K.R. *Tropical Homegardens: A Time-Tested Example of Sustainable Agroforestry*; Springer: the Netherlands, 2006.
67. http://nationalzoo.si.edu/scbi/migratorybirds/coffee/farmer.cfm. (accessed September 2011).
68. http://www.ico.org/prices/po.htm. (accessed September 2011).
69. http://faostat.fao.org/. (accessed October 2011).
70. Perfecto, I.; Vandermeer, J.; Mas A.; Soto Pinto, L. Biodiversity, yield, and shade coffee certification. Ecol. Econ. **2005**, *54*, 435–446.
71. http://www.rainforest-alliance.org/agriculture/certification. (accessed 5 October 2011).
72. http://www.utzcertified.org/en/aboututzcertified. (accessed October 2011).
73. http://sanstandards.org/sitio/. (accessed October 2011).

Biodiversity: Values

David Ehrenfeld
Department of Ecology, Evolution, and Natural Resources, Rutgers University, New Brunswick, New Jersey, U.S.A.

Abstract

Biodiversity has been valued in many ways, depending on the perspective of the person who is doing the valuing. Those who understand biodiversity as "nature" or "wilderness" may find different values than others who see biodiversity as "resources" or yet others who define it scientifically in terms of "genetic or ecological richness." The values that emerge from these very different viewpoints can be grouped in two broad categories: intrinsic and instrumental, which can overlap. Intrinsic values are those that are thought to be inherent in the species or species assemblages themselves. They are derived from a higher authority, either God or an ethical system (such as animal rights), and do not depend on any usefulness of the species to humans. Instrumental values are those that are derived from usefulness. Species are prized for providing food, shelter, clothing, medicine, recreation, transport, energy, or in the case of species assemblages, ecosystem resilience. Since the start of the conservation movement, and with the rise of the environmental sciences, instrumental values, especially the ones concerning ecosystem resilience, have gained in relative importance, and their proponents have sometimes battled with those who advocate the protection of biodiversity for its own sake. Ultimately, however, the two value systems do not necessarily have to clash.

INTRODUCTION

The idea that biodiversity has values predates the word "biodiversity," which only came into common use as a contraction of "biological diversity" in the 1980s. Biodiversity can be defined as the variety of life-forms in a given place, with "life-forms" ranging from genotypes to species and other taxonomic categories to species assemblages to entire ecosystems. Historically, values have been accorded to "wilderness," "natural resources," "species diversity," "genetic diversity," "community and ecosystem diversity," and the more general term "nature," which subsumes them all. Each of these and the other terms related to biodiversity has had its own set of values associated with it, but these values can be broadly divided into two categories, *intrinsic* and *instrumental*.

Intrinsic values are those believed to be inherent in biodiversity itself, rather than in any usefulness to humans that the biodiversity may confer. Thus, species considered useless for human purposes may yet be accorded intrinsic values. The values ascribed to wilderness and the values invoked by the animal rights movement are frequently intrinsic.

Instrumental values, conversely, are associated with components of biodiversity or entire natural systems that have a known, assumed, or suspected usefulness in the lives of people. The values assigned to natural resources and to biodiversity as a source of human health and ecosystem stability are usually instrumental. Since the growth of the physical environmental sciences and ecology, these instrumental values have gained steadily in importance, often overshadowing the older intrinsic values.

Any aspect of biodiversity may have both intrinsic and instrumental values associated with it. There is considerable overlap, and the lines separating the two systems can sometimes be blurred or missing.

In the modern world, the subject of biodiversity's values is frequently raised as a justification for environmental protection or conservation. When this happens, there can be sharp polarization between those who value the preservation or restoration of biodiversity primarily because it serves specific needs of humanity, including ecosystem health, and those who justify conservation because it maintains the order of nature without reference to human needs. Nevertheless, there is not always a contradiction between instrumental and intrinsic values, and a coexistence of the two positions can occur.

INTRINSIC VALUES

The study of biodiversity in the West owes much to the early influence of Aristotle, whose *Generation of Animals* inspired both Islamic and western Christian scientists. In other parts of the world, there have also been long-established traditions of knowing and valuing biodiversity. A comprehensive and systematic discussion of the place and value of nature in each of the world's major religions and regions (including Shamanism, Shintoism, Taoism, Confucianism, Zoroastrianism, Buddhism, Hinduism, Judaism, Christianity, Islam,

Encyclopedia of Natural Resources DOI: 10.1081/E-ENRL-120047424

and others) can be found in the book *Religion and the Order of Nature*, by Seyyed Hossein Nasr.[1]

In 9th-century Baghdad, the biologist, Islamic philosopher, and grandson of an African slave Abu Uthman Amr ibn Bahr al-Kinani al-Fuqaimi al-Basri, known as al-Jahiz, published his multivolume work, the *Book of Animals*, some of which survives. al-Jahiz, who was the first to provide a scientific classification of biodiversity (animals), described approximately 350 animal types, and, 1000 years before Darwin and Wallace, gave a concise explanation of evolution by means of natural selection. To al-Jahiz, the value of biodiversity was not primarily in the usefulness of any of its particular elements but in demonstrating the wonders of God's creation: "For this purpose the small and slight carries as much weight as the great and vast."[2] In other words, God's imprint on every species or ecological system is what establishes its value—the value is inherent and not related to the particular properties of one species or another. On the other hand, al-Jahiz does see a special instrumental value to biodiversity because he believes that contemplation of it will lead people to a better appreciation of God.

Nearly 900 years later, in a different age, a different part of the world, and a different culture, the pioneering botanist John Ray, often described as the father of English natural history, found similar values in biodiversity. Ray, an Anglican priest, began the modern plant classification system that was later continued and developed by Linnaeus, and was the first to give a biological definition of the term "species."[3] Ray's book, *The Wisdom of God Manifested in the Works of the Creation* (1691), was enormously popular and was published in numerous editions, continuing long after his death in 1705. In this book, he considered biodiversity to include species of plants, animals, and microbes, as well as anatomical structures and physiological systems.[4]

To Ray, the value of biodiversity was clear, and strikingly similar to that of al-Jahiz. From the start, he makes plain what that value is: "If the Number of Creatures be so exceeding great, how great, nay immense, must need be the Power and Wisdom of him who formed them all!" Unlike al-Jahiz, however, Ray takes great pains to distance himself from a utilitarian valuing of biodiversity: "How can Man give Thanks and Praise to God for the Use of his Limbs and Senses, and [for] those his good Creatures which serve for his Sustenance, when he cannot be sure they were made in any Respect for him; nay, when 'tis as likely they were not, and that he doth abuse them to serve Ends for which they were never intended?" The value of biodiversity, says Ray, is entirely to God and for God. Quoting the book of Proverbs, he wrote: "God made all things for himself. . . . He hath no Need of them: he hath no use of them. No, he made them for the manifestation of his Power, Wisdom, and Goodness, and that he might receive from the Creatures that were able to take Notice thereof, his Tribute of Praise."

Later in his book, Ray qualifies this argument somewhat by cataloging a long list of practical instrumental values of plant and animal species to humans, including values that have yet to be discovered and the value of intellectual stimulation. He notes that even apparently worthless or noxious creatures (especially insects) may provide sustenance to other species that we do value. Nevertheless, the ultimate value of biodiversity that Ray always returns to is the intrinsic value relating to God.[4]

The celebration of nature for its own sake was established in 18th-century North America by the naturalists and artists Mark Catesby and John James Audubon, and by the father-and-son team of John and William Bartram. John Bartram, who traveled primarily in Virginia, Pennsylvania, New York, and the southeastern states, collected seeds and live specimens of every species of new plant he could find, bringing them back to his garden in Philadelphia [still preserved as a 46-acre (19-ha) U.S. National Historic Landmark]. A Quaker farmer with a rudimentary education, John corresponded with many of the leading European botanists, including Linnaeus, who referred to him as the greatest natural botanist in the world. Some Bartram plant specimens remain in the collection of the British Museum (Natural History).[4]

William Bartram is primarily known for his travels in the southeastern United States, especially Florida. His book, commonly titled *Bartram's Travels* (1791), gained him international fame.[5,6] Coleridge's poem *Xanadu* was inspired by the book's descriptions of Florida springs and sinkholes. William's written descriptions and paintings of plants, insects, birds, reptiles, and fish were widely circulated, as were the specimens he sent back to his father in Philadelphia. John and William's most celebrated discovery, in 1765, was a small, flowering tree found by the Altamaha River in Georgia, which they named *Franklinia alatamaha*, after John's friend Benjamin Franklin. Soon extinct in the wild, *Franklinia* survives in cultivation as descendants of specimens in John Bartram's garden.

The Bartrams, especially John, were practical people, always looking for new species, areas with fertile soils, and mineral deposits that might be put to use. Yet they also devoted enormous effort to finding and describing species (like *Franklinia*) that seemed of no practical use whatsoever, whose value was, as far as they knew, entirely intrinsic; and their enthusiasm at discovering these species remains vivid to this day.

The 19th-century Prussian explorer and naturalist Alexander von Humboldt carried out daring explorations, especially in South America, founded the science of physical geography (the Humboldt Current, etc.), also established new frontiers in the study of linguistics and education, and described many new species of animals and plants, some of which were named after him. von Humboldt's discoveries of species and ecosystem processes reflected his delight in nature for its own sake, irrespective of utilitarian values. In his celebrated work *Views of Nature*, translated into English in 1850, Humboldt wrote: "When the active spirit of man is directed to the investigation of nature, or when in imagination he scans the vast fields of organic creation, among the varied emotions excited in his mind there is none more profound or vivid than that awakened by the universal profusion of life."[7] von Humboldt's feelings were mirrored in the writings of the many 18th-, 19th-, and 20th-century naturalist explorers who spread out around the globe searching for and collecting new species, many of which ended up in great private and museum collections.

From the days of Ray to the present, the theme of the intrinsic value of biodiversity has reappeared many times in various ways. Henry David Thoreau, writing in *Walden; or, Life in the Woods* (1854), described and celebrated the diversity of life in his native Massachusetts. He often distanced himself from instrumental values, as, for example, when he extols vegetarianism: "I have no doubt that it is part of the destiny of the human race, in its gradual improvement, to leave off eating animals," and he continues, "All nature is your congratulation. . . . The greatest gains and values are farthest from being appreciated."

In consideration of the intrinsic values of biodiversity, the concept of animal rights should be mentioned, because in this case, the value of animals, including domesticated animals, is completely disconnected from their utility to humans. This formulation of intrinsic worth has, historically, led to contentious debates with utilitarians, even to ridicule, as in Samuel Butler's satiric chapter on the "Rights of Vegetables," in his novel *Erewhon* (1872). More recently, in support of intrinsic rights, the law scholar Christopher D. Stone advanced the argument that trees should have legal standing in U.S. courts, in his 1972 book *Should Trees Have Standing? Toward Legal Rights for Natural Objects*.[8] This notion was echoed in the same year by U.S. Supreme Court Justice William O. Douglas, who wrote, "the voice of the inanimate (non-human) object should not be stilled" in a famous legal dissent. He was referring to groves of trees, alpine meadows, aquatic insects, otters, elk, and other parts of biodiversity, stating that they had a right to be represented in court by those humans who had an "intimate relation" to them.[9]

The English ecologist Charles S. Elton wrote: "The first [reason for conservation], which is not usually put first, is really religious. There are some millions of people in the world who think that animals have a right to exist and be left alone, or at any rate that they should not be persecuted or made extinct as species. Some people will believe this even when it is quite dangerous to themselves." Although he was well aware of its instrumental values, the first great value of biodiversity, Elton asserted, was embodied in "the right relation between man and living things."[10]

Biologist David Ehrenfeld has pointed out that it is easy to exaggerate the utilitarian values of biodiversity: "When everything is called a resource, the word loses all meaning"; and he gave examples, including contemporary ecological debates over ecosystem diversity and stability. Concluding a discussion of intrinsic values, he cited the biblical account of Noah's ark. Noah is described as taking at least one male and female of each animal species on board the ark, presumably including the worthless and noxious ones. Ehrenfeld called this strategy the "Noah Principle": species should be conserved (valued) simply "because they exist." In his discussion of the Noah Principle, he cited contemporary arguments for the conservation of pathogens and vermin—even the smallpox virus.[11]

INSTRUMENTAL VALUES

Arguments for the instrumental values of biodiversity are also long established. An oft-quoted passage from the Hebrew Bible (Deuteronomy 20:19,20) states: "When in your war against a city you have to besiege it a long time in order to capture it, you must not destroy its trees, wielding the axe against them. You may eat of them, but you must not cut them down. Are trees of the field human to withdraw before you into the besieged city? Only trees that you know do not yield food may be destroyed."[12] This passage forms the principal basis for the Jewish commandment of *bal tashhit* (do not destroy), an ancient series of Jewish environmental regulations, which were rabbinically extended to a wide range of transgressions, including the overgrazing of the countryside, unjustified killing of animals or feeding them toxic foods, sport hunting, extinction of species, destruction of cultivated varieties of plants, air and water pollution, overconsumption of anything, and the waste of resources.[13,14] The original source of the commandment and much of the Jewish system of biodiversity values that followed were certainly utilitarian or instrumental in nature. There are, however, some partial exceptions. The general Jewish prohibition against cruelty to animals, *ts'ar baalei hayyim* (pain of living things) is based on the following passage in Deuteronomy 22:6: "If, along the road, you chance on a bird's nest . . . with fledglings or eggs and the mother sitting over [them], do not take the mother together with her young. Let the mother go, and take only the young, in order that you may fare well and have a long life."[12] This value seems largely intrinsic, although a modern interpretation would include the ecological utility of leaving the female to reproduce in subsequent seasons. Whether this is part of the biblical intent is not clear.

The English naturalist Gilbert White, whose book *The Natural History and Antiquities of Selbourne* (1789) inspired study and appreciation of the intrinsic value of nature around the world, was, nevertheless, one of the first to quantify a species' value. In his Letter XXVI to Daines Barrington, White, in an early example of ecological economics, carefully calculated the value, per pound, of the common rush, *Juncus effusus*. Rushes, when dried and dipped in bacon fat or other grease, made an inexpensive substitute for candles, giving "a good clear light. . . . Thus a poor family will enjoy 5 1/2 hours of comfortable light for a farthing." White estimated that one-and-a-half pounds of dried rushes would last a family as much as a year.[15]

The modern drive to assign instrumental values to biodiversity dates to 1864, when the American linguist, lawyer, legislator, writer, and a founder of environmental science George Perkins Marsh, following in the intellectual footsteps of von Humboldt, published his book *Man and Nature; or Physical Geography as Modified by Human Action*. Marsh, who traveled in Europe, the Middle East, and North Africa, and who read easily and widely in 20 languages, was the first to provide a truly comprehensive description of many of the instrumental values of biodiversity. He demonstrated that forests promote even-stream flow, mitigate flooding, protect against avalanches, stabilize climate, prevent soil erosion, and halt the movement of sand dunes, besides providing valuable raw materials. He showed the importance of birds in limiting outbreaks of destructive insects and in sowing seeds. He documented in great detail the economic benefits and liabilities of exotic plant species, the impact of earthworms on soil porosity and fertility, the usefulness of insects in destroying dead and decaying animal and vegetable matter, and the importance of micro-organisms in the formation of deposits of lime and in the process of fermentation. Above all, he sounded a very modern warning that we destroy biodiversity at our own peril: "If man is destined to inhabit the Earth much longer. . . he will learn to put a wiser estimate on the [value of the] works of creation."[16]

Marsh's warnings about the critical values to people of biodiversity have been echoed and amplified in the 20th and 21st centuries. The terms most associated with these values are "natural capital" and "ecosystem services," both of which indicate their instrumental nature. In a major paper entitled "Biodiversity loss and its impact on humanity," Bradley Cardinale and 16 other authors noted that "ecosystem services are the suite of benefits that ecosystems provide to humanity." Those pertaining to the role of biodiversity include food, wood, fresh water, climate regulation, pollination, and pest and disease control. The authors state, "There is now unequivocal evidence that biodiversity loss reduces the efficiency by which ecological communities capture biologically essential resources, produce biomass, [and] decompose and recycle biologically essential nutrients. . . . There is mounting evidence that biodiversity increases the stability of ecosystem functions through time. . . . Diverse communities are more productive because they contain key species that have a large influence on productivity, and differences in functional traits among organisms increase total resource capture." The authors note that "there are (a small number of) instances where increased biodiversity may be deleterious," and give examples, cautioning "against making sweeping statements that biodiversity always brings benefits to society."[17]

In a section called "Valuing biodiversity," the paper continues: "Economists have developed a wide array of tools to estimate the value of natural and managed ecosystems and the market and non-marketed services that they provide. Although there are good estimates of society's willingness to pay for a number of non-marketed ecosystem services, we still know little about the marginal value of biodiversity (that is, value associated with changes in the variation of genes, species and functional traits) in the production of those services."[17] The value of natural capital has been further explored in a related paper by Ehrlich, Kareiva, and Daily, who describe software-based models that "help decision makers by quantifying and mapping the generation, distribution and economic value of ecosystem services under alternative scenarios."[18]

The idea that the maintenance of and exposure to biodiversity, from microbes to ecosystems, improve human health and well-being has gained many adherents. This improvement comes about in a number of ways. First, many medicinals are obtained directly or indirectly from biological species. Examples are statin drugs from *Aspergillus*, *Pleurotus*, and *Monascus* fungi, an antidiabetic from gila monster lizard (*Heloderma suspectum*) saliva, a powerful painkiller from the venom of marine cone snails (*Conus*), the anticancer drug taxol from the Pacific yew (*Taxus brevifolia*), and the potent antimalarial artemisinin from wormwood (*Artemisia annua*).[19]

Second, mounting evidence has shown that childhood exposure to diverse species of bacteria and fungi, and probably to intestinal parasites, protects against the development of asthma and allergies. Children who grow up in environments, such as traditional farms, that provide a wide range of microbial exposure develop less asthma and allergy than children in more sanitary environments. Similarly, it has been proposed that exposure to certain bacteria may prevent or treat particular cancers. For example, the bacterium *Helicobacter pylori*, although causing gastric ulcers and gastric cancer, may reduce the risk of esophageal cancer. The diversity of the microbiome, especially that of the skin and internal organs, is increasingly linked to human health.[20–24]

There is a third, more sweeping association between biodiversity and health. In the words of the author and childhood development expert Richard Louv, "A growing body of research links our mental, physical, and spiritual health directly to our association with nature." Louv has coined the term "nature deficit disorder" to describe a constellation of maladies that afflict the increasing number of children who spend most of their time indoors, out of contact with the natural world. For many children in industrial nations, their relationship with nature is intellectual, not experiential. The consequences, it is claimed, include attention deficit disorder, obesity, depression, and eventually hypertension and heart disease.[25] Similar maladies afflict adults. The modern popularization of this idea can be traced to biologist E. O. Wilson's use, beginning in 1984, of the older term "biophilia," in which he claimed that human beings have a deep, evolved relationship with other species, which often confers a mutual survival value.[26,27]

Even earlier, the botanist Hugh Iltis ascribed a similar importance to biodiversity.[28]

THE VALUES BATTLEGROUND

The proponents of the intrinsic values of biodiversity have at times struggled with those who favor instrumental values. One of the first and most long-lasting examples of this conflict was the controversy, beginning in 1908, about whether to dam California's Tuolumne River in Yosemite National Park's Hetch Hetchy Valley to create a reservoir for the City of San Francisco. Gifford Pinchot, early advocate of the conservation of natural resources, and eventually the first Chief Forester of the U.S. Forest Service, was strongly in favor of the dam. He wrote: "The first principle of conservation is development, the use of the natural resources now existing on this continent for the benefit of the people who live here now." On the other side was John Muir, naturalist, writer, inventor, and founder of the Sierra Club. Muir, whose writing had earlier prompted the creation of Yosemite Park, wrote: "God has cared for these trees, saved them from drought, disease, avalanches, and a thousand . . . leveling tempests and floods; but He cannot save them from fools." Both men appealed to their friend President Theodore Roosevelt, but the issue was not decided until after Roosevelt had left office. In 1914, the U.S. Congress legislated the damming of the Hetch Hetchy Valley. John Muir died several months later.[29] In 1987, the U.S. Secretary of the Interior proposed restoration of the valley, and a year later, the U.S. Bureau of Reclamation concurred. The Sierra Club immediately endorsed breaching the dam. The battle continues, with fierce arguments on both sides. In a 2012 editorial entitled "Hetch Hetchy's Past and Future," *The New York Times* again supported breaching the dam: "Ninety-nine years ago, this newspaper argued against the construction of the dam. We lost, but the much bigger loss was nature's."[30]

Since Hetch-Hetchy, there have been countless struggles in nearly every country on Earth between those who seek to develop natural resources for some practical purpose and those who wish to preserve the species and ecosystems that would be lost to development. In virtually all of these cases, the arguments against or for the preservation of biodiversity echo the sorts of values expressed by Pinchot and Muir.

Much of the fight about intrinsic versus instrumental values centers on the attempt by some economists to quantify the worth to humanity of biodiversity. A United Nations Environmental Program initiative, "The Economics of Ecosystems and Biodiversity," puts the annual value of ecosystems services at $5 trillion, with the loss of forests worldwide amounting to a loss of ecosystems services of $2–4.5 trillion a year. Ecological economist Robert Costanza argues that we need to understand the economic value of biodiversity and should create new institutions for managing natural capital if we are to promote the public good. "We've used up all the frontier," says Costanza. "People are recognizing this, but our institutions haven't caught up." Ignoring the value of natural capital has serious economic consequences, Costanza and others point out.[31]

In opposition, biologist Douglas J. McCauley claims that although putting a value on their ecosystem services can occasionally be useful in protecting specific parts of biodiversity, "We will make more progress in the long run by appealing to people's hearts rather than to their wallets. If we oversell the message that ecosystems are important because they provide services, we will have effectively sold out on nature." McCauley provides a detailed list of reasons why he believes that placing a dollar value on nature is a poor idea.[32]

Some scientists have stated that putting a dollar value on nature ignores both the very approximate and incomplete nature of the economic calculations involved and the vast extent of our ignorance about ecological systems. It has been pointed out that when we give a specific value to the ecosystem services of biodiversity, any destructive exploitation of that biodiversity that yields higher-dollar profits will appear to outweigh the value of the biodiversity itself, justifying its destruction. Moreover, it is claimed, an emphasis on the quantification of instrumental values leads, by analogy, to the problematic notion that the parts of the world with the highest number of species, mostly tropical, are more deserving of conservation funding than less biodiverse regions, which are mostly in temperate or colder latitudes. Wealth of species is surely important, it is noted, but there are more ways to value the biodiversity of a region than by counting its species.[33]

CONFLICT RESOLUTION

There has been controversy over the values of biodiversity, but there has also been effective coexistence and integration. One of the best examples can be seen in the writings of the wildlife ecologist Aldo Leopold. Leopold is considered the founder of the science of game management, which regulates the abundance of animal species that are hunted, trapped, or fished, maximizing their instrumental value. Beyond simple game management, however, Leopold was one of the first to see the necessity of ecosystem management, showing that game species could only be maintained if understood in their full ecological context.[34,35]

The richness of Leopold's ecological wisdom promoted a continuous interaction, seen in his work and writings, between the different ways that biodiversity can be valued. The dedication of Leopold's book *Game Management* (1933) was to his father, "Pioneer in Sportsmanship." The young Leopold was an avid hunter; in the preface to *Game Management*, he states, "Control [of wildlife] comes from the co-ordination of science and use."[36] Yet Leopold changed as he grew older; his awareness of less human-centered values grew, as did his application of ecological science to game management and conservation. In his essay "Thinking Like a Mountain," from his book *A Sand County Almanac* (1949), he described the day when he began to rethink his earlier premises and prejudices. He had just shot a female wolf who was playing with her pups by a river. Wolves eat deer; as a deer hunter he felt that wolves were fair game. Then he climbed down to the riverbank to look at his kill. "We reached the old wolf in time to watch a fierce green fire dying in her eyes. I realized then, and have known ever since, that there was something new to me in those eyes—something known only to her and to the mountain. I was young then, and full of trigger-itch; I thought that because fewer wolves meant more deer, that no wolves would mean hunters' paradise. But after seeing the green fire die, I sensed that neither the wolf nor the mountain agreed with such a view. . . . Perhaps this is behind Thoreau's dictum: In wildness is the salvation of the world. Perhaps this is the hidden meaning in the howl of the wolf, long known among mountains, but seldom perceived among men."[37]

The interplay between intrinsic and instrumental valuing of nature is complex here. Leopold came to realize, as a wildlife manager, that too many deer would overgraze and ruin the mountain ecosystem, ultimately resulting in fewer deer, a decrease in the ecosystem's instrumental value to hunters. A healthy deer population, he believed, needs wolves. The other part of his thinking, however, was also clear: the wolf had an intrinsic value of her own, a value that deserved his respect. And beyond the wolf was the mountain, the functional ecological health of the land community.

In the essays incorporated in the posthumously published *A Sand County Almanac*, Leopold showed that the instrumental and intrinsic values of biodiversity can interact, forming a seamless whole. He wrote: "Most members of the land community have no economic value. . . . Yet these creatures are members of the biotic community, and if (as I believe) its stability depends on its integrity, they are entitled to continuance." In this statement, he claims that an overall instrumental value of biodiversity, ecosystem resilience, is made up of the collective values of its species, many of which have no individual economic worth. In his essay "A Biotic View of Land,"[38] Leopold had written: "The emergence of ecology has placed the economic biologist in a peculiar dilemma: with one hand he points out the accumulated findings of his search for utility, or lack of utility, in this or that species, with the other he lifts the veil from a biota so complex that no man can say where utility begins or ends. . . . The old categories of 'useful' and 'harmful' have validity only as conditioned by time, place, and circumstance. The only sure conclusion is that the biota as a whole is useful, [including] not only plants and animals, but soils and waters as well."

Does this mean that ultimately the highest value of biodiversity is its instrumental ecosystem stability? Not necessarily. Leopold goes on to write, using the word "land" as we would use biodiversity, "It is inconceivable to me that an ethical relation to land can exist without love, respect, and admiration for land, and a high regard for its value. By value, I of course mean something far broader than mere economic value, I mean value in the philosophical sense."[37]

The values of biodiversity are a complex mix of the economic and the religious or philosophical. In the long history of thought about the values of nature, these two strands have been both separated in conflict and fused in coexistence, according to circumstances. This pattern seems likely to continue in the foreseeable future.

REFERENCES

1. Nasr, S.H. *Religion and the Order of Nature: The 1994 Cadbury Lectures at the University of Birmingham*; Oxford University Press: USA, 1996; 310 pp.
2. Lunde, P. Science: the Islamic legacy: the book of animals. Saudi Aramco World **1982**, *33* (3), 1–4.
3. Mayr, E. *The Growth of Biological Thought: Diversity, Evolution, and Inheritance*; Harvard University Press: Cambridge, Mass, 1982; 974 pp.
4. Ray, J. *The Wisdom of God Manifested in the Works of the Creation*, 10th ed.; Innys, W., Manby, R., Eds.; London, 1735; Vol. 18, 1–17.
5. Cruickshank, H.G., ed. *John and William Bartram's America*; Devin-Adair: New York, 1957.
6. Van Doren, M. *Travels of William Bartram*; Dover: New York, 1928.
7. Von Humboldt, A.; Otté, E.C.; Bohn, H.G. *Views of Nature: or Contemplations on the Sublime Phenomena of Nature;* Henry G.B., Eds.; London, 1850; Vol. 210, 459.
8. Stone, C.D. *Should Trees have Standing? Toward Legal Rights for Natural Objects*, rev. Ed.; Avon: New York, 1974; 102.
9. Sierra Club v. Morton - 405 U.S. 727 (1972).
10. Elton, C.S. *The Ecology of Invasions by Animals and Plants;* Methuen: London, 1958; 659–666.
11. Ehrenfeld, D. *The Arrogance of Humanism*; Oxford University Press: Oxford, 1978; 286 pp.
12. JPS. *Hebrew-English Tanakh*, 2nd Ed.; Jewish Publication Society: Philadelphia, 1999.
13. Freudenstein, E. *Ecology and the Jewish Tradition*; Atid: New York, 1970.
14. Helfand, J. Ecology and the Jewish tradition: a postscript. Judaism **1971**, *20*, 330–335.
15. White, G. *The Natural History of Selbourne;* Penguin Books: London, 1977; 320 pp.
16. Marsh, G.P. *Man and Nature*, Lowenthal, O., Ed.; Harvard University Press: Cambridge, Massachusetts, 1965.
17. Cardinale, B.J.; Duffy, E.; Gonzalez, A.; Hooper, D.U.; Perrings, C.; Venail, P.; Narwani, A.; Mace, G.M.; Tilman, D.; Wardle, D.A.; Kinzig, A.P.; Daily, G.C.; Loreau, M.; Grace, J.B.; Larigauderie, A.; Srivastava, D.S.; Naeem, S. Biodiversity loss and its impact on humanity. Nature **2012**, *486*, 59–67.

18. Ehrlich, P.R.; Kareiva, P.M.; Daily, G.C. Securing natural capital and expanding equity to rescale civilization. Nature **2012**, *486*, 68–73.
19. Conniff, R. A bitter pill. Conservation **2012**, *13* (1), 18–23.
20. Leonardi-Bee, J.; Pritchard, D.; Britton, J. Asthma and current intestinal parasite infection: systematic review and meta-analysis. Am. J. Respir. Crit. Care Med. **2006**, *174* (5), 514–523.
21. Yazdanbakhsh, M.; Kremsner, P.G.; van Ree, R. Allergy, parasites, and the hygiene hypothesis. Science **2002**, *296*, 490–494.
22. Ege, M.J.; Mayer, M.; Normand, A.-C.; Genuneit, J.; Cookson, WO.; Braun-Fahrländer, C.; Heederik, D.; Piarroux, R.; von Mutius, E. Exposure to environmental microorganisms and childhood asthma. N. Eng. J. Med. 2011, *364*, 701–709.
23. Mager, D.L. Bacteria and cancer: cause, coincidence or cure? A review. J. Transl. Med. **2006**, *4*, 14.
24. http://www.translational-medicine.com/content/4/1/14.
25. Dunn, R. *The Wild Life of our Bodies: Predators, Parasites, and Partners that Shape Who We are Today*; Harper Collins: New York, 2011; 290.
26. Louv, R. *Last Child in the Woods: Saving our Children from Nature-Deficit Disorder;* Algonquin Books of Chapel Hill: Chapel Hill, NC, 2005; 335.
27. Wilson, E.O. *Biophilia: The Human Bond with Other Species;* Harvard University Press: Cambridge, MA, 1984; 157.
28. Kellert, S.R.; Wilson, E.O. *The Biophilia Hypothesis;* Island Press: Washington, DC, 1993; 234–240.
29. Iltis, H.H. To the taxonomist and ecologist: whose fight is the preservation of nature? BioScience **1967**, 17, 886–890.
30. Ehrenfeld, D. *Conserving Life on Earth;* Oxford University Press: New York, 1972; 264.
31. Editorial. *Hetch-Hetchy's Past and Future*. The New York Times, Feb. 17, 2012; A26.
32. Schwartz, J.D. Should we put a dollar value on nature? http://www.time.com/time/business/article/0,8559,1970173,00.html (accessed March 2010).
33. McCauley, D.J. Selling out on nature. Nature **2006**, *443*, 27–28.
34. Ehrenfeld, D. *Becoming Good Ancestors: How We Balance Nature, Community, and Technology*; Oxford University Press: New York, 2009; 97–127.
35. Meine, C. *Correction Lines: Essays on Land, Leopold, and Conservation*; Island Press: Washington, DC, 2004; 296.
36. Meine, C. and Knight, R.L. *The Essential Aldo Leopold: Quotations and Commentaries*; University of Wisconsin Press: Madison, WI, 1999; 384 pp.
37. Leopold, A. *Game Management*; Charles Scribner's Sons: New York, 1933; 481 pp.
38. Leopold, A.A. *Sand County Almanac: With Other Essays on Conservation from Round River*; Oxford University Press: New York, 1966; 242.
39. Leopold, A. A biotic view of land. J. For. **1939**, *37*, 727–730.

BIBLIOGRAPHY

1. Rolston, H., III. *A New Environmental Ethics: The Next Millennium for Life on Earth*; Routledge: New York, 2012; 244.
2. Rolston, H., III. *Environmental Ethics*; Temple University Press: Philadelphia, 1988; 400.

Biomes

Paul Reich
Hari Eswaran
National Resources Conservation Service, U.S. Department of Agriculture (USDA-NRCS), Washington, District of Columbia, U.S.A.

Abstract
The term *biome* is generally used to identify the natural habitat conditions around the world. Two major determinants of biome type are precipitation and air temperature. A third variable that affects the habitat type is the soil. This entry presents a general overview of the soils characterizing the major biomes of the world.

INTRODUCTION

A biome is defined as "a community of organisms interacting with one another and with the chemical and physical factors making up their environments."[1] For most purposes, the term biome is used to identify the natural habitat conditions around the world. Depending on the purpose, the global ecosystem is divided into units each characterized by a specific combination of climatic factors. Two major determinants of biome type are precipitation (total and its distribution) and air temperature. These two elements of climate have been commonly used to define the major biomes of the world. A third variable that affects the habitat type is the soil. This section presents a general overview of the soils characterizing the major biomes of the world.

Detailed maps showing the biomes of the world are not available due to conceptual differences of definitions and reliable global databases. There are many excellent and detailed studies of specific habitats around the world, and using these and the global soils and climate database of the world,[2] a map showing the distribution of the major biomes was drawn. The terms used to describe the biomes are common in use, but their subdivisions are based on important differentiating factors.

MAJOR BIOMES

The five major biomes of the world and their subdivisions are listed in Table 1 and their distribution is shown in Fig. 1 Some geographers have used elevation as a differentiating factor and recognized a "montane biome." This biome's small extent precludes its description in this section. Few geographers recognize the Mediterranean biome as presented here as a subdivision of the Temperate biome. Within each biome, some major distinction is made. In the Polar biome, an attempt is made to differentiate areas with permafrost and warmer areas with intermittent permafrost. The latter is termed "interfrost." The Boreal Forest biome has a humid and a semiarid counterpart similar to the temperate and tropical zones. The semiarid vegetation is generally grass or shrubs. The desert areas are divided into hot, cool, and cold equivalents.

Deserts occupy about 30% of the global ice-free landmass. Potential evapotranspiration exceeds precipitation during most days of the year, and the biome shows the greatest contrast in temperature, both diurnal and seasonal. Desert vegetation is very sparse with a few isolated shrubs dominated by succulents and annuals. Grasses and forbs become more dominant at the margins. The tropics occupy about 27% of the landmass and are characterized by only a small variation in temperature during the year. The soil moisture conditions range from semiarid to humid. The availability of soil moisture determines the biota. The temperate grassland and forest biome occupies about 15%. The cold biomes, the boreal and polar, are mainly in the northern latitudes with small areas in South America. The low temperatures control the flora and fauna, and vegetation is scarce to nonexistent in the extreme cold polar regions. Each of these areas has specific kinds of soils that have developed as a response to the specific soil moisture and temperature conditions. The diversity in vegetation has its equivalent in fauna with animal species adapting to the specific bio-climatic conditions. Within each of the biomes, there are specific conditions that promote unique flora and fauna. Examples of such localized systems are volcanic hot springs, wetlands, and oases.

Soils of the Major Biomes

Soil temperature and moisture, with their seasonal and annual variations, are an integral part of the soil classification

Encyclopedia of Natural Resources DOI: 10.1081/E-ENRL-120042695

Table 1 Major global biomes with subdivisions and their areas

			Global land area		
Biome	**Subdivision**	**Remarks**	**In thousand km^2**	**In%**	**In%**
Polar	Permafrost	Mosses and lichens	10,547	8.06	15.46
	Interfrost	Shrubs and stunted trees	9,685	7.40	
Boreal	Semi arid	Shrub	3,552	2.72	9.94
	Humid	Forest	9,446	7.22	
Temperate	Semiarid	Grassland	7,348	5.62	15.14
	Humid	Forest	12,453	9.52	
	Mediterranean warm	Shrubs and fores	3,624	2.77	3.37
	Mediterranean cold	Forbs and shrubs	791	0.60	
Desert	Hot	Barren	4,418	3.38	29.47
	Cool	Barren	28,521	21.81	
	Cold	Barren	5,599	4.28	
Tropical	Semiarid	Grassland, savanna	20,299	15.52	26.62
	Humid	Forests	14,514	11.10	
Total			130,796	100.00	

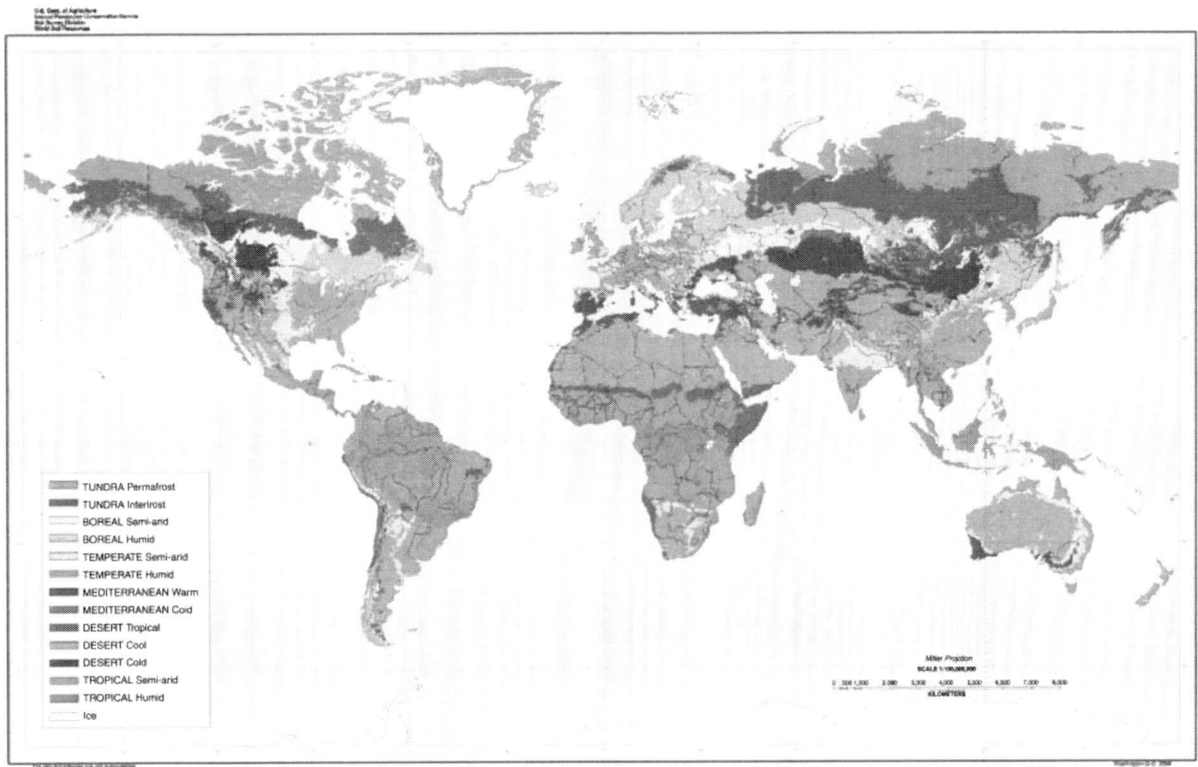

Fig. 1 Major biomes.

system called Soil Taxonomy[3] which is used here. As Fig. 1 and Table 1 are based on soil moisture and temperature conditions, there is an implied link between the biomes and their soil resource endowments. The purpose of this introductory section is to present the general geographic distribution of the biomes and the major soils characterizing each. The major soil orders are listed in Table 2. The subsequent sections elaborate on the soil resources of each of the six biomes listed in Table 1.

Polar Biome

Characterized largely by a mean annual soil temperature of less than 0°C, the Polar biome occupies a large land

Table 2 Dominant soils in the major biomes (Area in thousand km^2)

Soil order	Ice-free land	Polar		Boreal		Temperate				Desert			Tropics	
		Permafrost	Interfrost	Semiarid	Humid	Semiarid	Humid	Mediterranean warm	Mediterranean cold	Warm	Cool	Cold	Semiarid	Humid
Gelisols	11,869	10,476	1,393	—	—	—	—	—	—	—	—	—	—	—
Histosols	1,526	—	—	299	686	17	82	1	8	—	23	93	37	281
Spodosols	4,596	55	1,058	227	2,566	105	450	39	6	—	—	28	—	61
Andisols	975	6	56	47	196	26	135	32	—	17	10	14	198	235
Oxisols	9,811	—	—	—	—	9	184	—	—	27	4	—	3,387	6,199
Vertisols	3,160	—	—	2	15	595	303	99	0	239	651	0	1,170	87
Aridisols	15,629	1	—	60	76	353	18	203	23	1,824	11,014	2,013	44	1
Ultisols	11,053	—	—	1	14	266	3,122	20	18	—	—	—	4,371	3,240
Mollisols	9,161	9	147	1,059	511	1,280	1,232	874	323	49	1,110	2,255	136	176
Alfisols	12,621	—	—	1,002	1,816	1,867	2,149	860	125	—	—	—	4,165	638
Inceptisols	21,805	—	6,751	167	3,333	1,636	3,078	683	89	—	—	—	3,566	2,502
Entisols	21,467	—	280	172	232	1,173	1,697	807	45	1,708	10,693	342	3,225	1,093
Miscellaneous	7,122	—	—	516	1	20	4	6	153	553	5,016	853	—	—
Total	130,796	10,547	9,685	3,552	9,446	7,348	12,453	3,624	791	4,418	28,521	5,599	20,299	14,514

mass adjacent to the Arctic circle to about 60°N latitude. This is generally the ice-free land of the northern latitudes. The sub-zero soil temperatures that prevail during most of the year are conducive to the formation of permafrost. The dominant soils of the region are Gelisols, occupying 59% of the Polar biome. Turbels freeze and thaw once or more during a year which triggers cryoturbation or physical mixing of the soil material. The Orthels are generally shallower or drier soils and have little or no cryoturbation. The low temperatures and periodic moisture saturation promote the accumulation of organic matter. The Histels are characterized by the high organic matter and occupy about 5% of the Polar biome. These organic soils form the largest contiguous extent of such soils in the world and serve as an important sink of CO_2. Other soils that occur in the Tundra zone are Inceptisols 33%, Spodosols 6%, while Mollisols and Entisols are each 1% of the area.[4]

Boreal Forest Biome

Bordering the southern flank of the Polar biome is the Boreal Forest biome in the northern hemisphere. The high volcanic area between Chile and Argentina has similar climatic regimes but the nature of the soil and physiography may result in a different set of habitat conditions. There are about 243,000 km^2 of Andisols or volcanic ash soils characterizing the Boreal Forest biome in the southern hemisphere. Such soils only occur as a small area in the Kamchatca Peninsula in the Northern Hemisphere. In Northern Europe and Siberia, the Boreal Forest biome has a humid part and is bordered in its southern periphery by a semiarid to arid part. In Canada, the semiarid part is in the middle of the continent. A range of soils are present and the most extensive are the Spodosols, which occupy 2.7 million km^2 (20.6%). The Spodosols are mostly in the humid part of the Boreal Forest biome and form under acid vegetation. Histosols are present in the depressions in this cold region.

Temperate Grassland and Forest Biome

The Temperate Grassland and Forest biome extends from about 25 to 55°N with counterparts in the Southern Hemisphere. Large areas in this belt are also deserts. Due to the favorable climate and soil endowments of this biome, much of the land is used for agriculture and native habitats are local and sporadic. About 50% of the zone (12.1 million km^2) is occupied by Mollisols and Alfisols (grasslands), and Ultisols in the forests. Their general good fertility and tilth have made them in great demand for grain production. Due to the long history of civilization in the Temperate Grassland and Forest biome of Europe and China, pristine ecosystems are rare. In large areas, there have been successive replacements and changes in the floral and faunal composition. Within the Temperate biome, are areas characterized by moist winters and dry summers. These Mediterranean conditions are conducive to unique ecosystems. These areas occur around the Mediterranean Sea and small areas in the western U.S. and southern Australia.

Desert Biome

Deserts occupy about 38.6 million km^2 and may be distinguished as warm (or tropical), cool (or temperate) and cold (or boreal) kinds of habitats. Though the biome is characterized by lack of moisture for normal vegetative growth of most plants, the ambient temperature conditions further distinguish the habitat conditions. About 38.6% of the Desert biome is occupied by Aridisols, which by definition have some kind of subsurface horizon. Shifting sands or moving dunes occupy about 13.7% and Entisols, which are very recent deposits, occupying about 33% of the Desert biome. The harshness of the environment has resulted in plants and animals with special adaptive features. Many of the Aridisols are increasingly used for agriculture when irrigation facilities are made available. In this fragile ecosystem, both soil and habitat conditions are drastically altered when irrigation is introduced.

Tropics Biome

An absence of winter and summer temperature extremes characterizes the Tropics biome, which occupy about 34.8 million km^2. Availability of moisture separates the semiarid from the humid tropics. As a corollary to the desert, in the humid tropics there are the perhumid areas where the potential evapotranspiration never exceeds the precipitation during any month of the year. This is a biomic condition that deserves much greater detailed studies. The Tropics biome is the home to the Oxisols (9.6 million km^2), which are unique to this biome. These are highly weathered soils where the original vegetation is closed or open forests. Most of the plant nutrients are concentrated in the top 5 cm of the soil and recycled through plant uptake and leaf fall. Ultisols also occur in this biome. Apart from these soils, there are small areas of most of the other soil orders. The wetlands of the tropics present a separate and unique habitat condition and those along the coast have some special soils.

REFERENCES

1. Tootil, E., Ed.; *Dictionary of Biology*; Intercontinental Kook Productions, Ltd.: Maidenhead, Berkshire, England, 1980.
2. Eswaran, H.; Beinroth, F.H.; Kimble, J.; Cook, T. Soil diversity in the tropics: implications for agricultural

development. In *Myths and Science of Soils of the Tropics*; Lal, R., Sanchez, P.A., Eds.; Soil Sci. Soc. Am. Spec. Publ. **1992;** *29*, 1–16.

3. Soil survey staff. In *Soil Taxonomy: A Basic System of Soil Classification for Making and Interpreting Soil Surveys*, 2nd Ed.; Natural Resources Conservation Service. U.S. Department of Agriculture, Handbook 436; U.S. Government Printing Office: Washington, DC, 1999.
4. *Keys to Soil Taxonomy*, 9th Ed.; U.S. Government Printing Office: 2003.

Boreal Forests: Climate Change

Pertti Hari
Department of Forestry, University of Helsinki, Helsinki, Finland

Biomes—
Ecosystems

Abstract

Solar radiation is the source of energy of phenomena taking place in the atmosphere and in forest ecosystems. Absorption either in the atmosphere or at the Earth's surface converts radiation energy into heat and reflection conveys solar radiation into the space. The properties of the atmosphere have been stable during the last 10,000 years, and consequently, the climate has also been stable after the ice age. Expansion of agriculture into the forests, use of fossil fuels, and industrialization have increased the flow of carbon dioxide and several other gases into the atmosphere, resulting in changes in atmospheric concentrations. The increasing concentrations of greenhouse gases (e.g., CO_2, CH_4, N_2O) have reduced the thermal radiation of Earth into space resulting in accumulation of energy in the atmosphere. Greenhouse gas concentrations will increase during the twenty-first century and consequently exacerbate the climate change. The metabolism of vegetation and microbes in the soil has reacted and will react to climate change. The increasing atmospheric CO_2 concentration has enhanced and will enhance photosynthesis and accelerated decomposition of proteins in the soil due to higher temperatures, which provides an additional source of nitrogen for trees and for ground vegetation. Temperature increase has prolonged and will prolong the photosynthetically active period, increasing the availability of sugars for growth and metabolism. The wood growth responds to the changes in the metabolism and the wood mass in boreal forests is strongly increasing. The changes in the size of carbon pools in the soil of boreal forests are of minor importance. The biomass in the boreal forests will increase and the boreal forests are able to bind about 6% of the anthropogenic CO_2 emissions.

INTRODUCTION

Coniferous forests at high latitudes are called boreal forests. They are mostly located between the latitudes 50°N and 70°N and they extend on the Eurasian and American continents from ocean to ocean. The stands are often rather evenly aged and the number of species is low, often one species dominating the stands. The mean biomass of trees is about 4.2×10^7 g ha^{-1}.[1] Dwarf shrubs dominate the field layer, but grasses and herbs also exist, especially in fertile soils. The carbon pool is often larger in the soil than in the living biomass.

Clear summer and winter seasons are characteristic of boreal forests. The mean July temperature is around 15–20°C and the mean January temperature is as low as −10 to −20°C. The annual variation is smaller near the oceans than in the middle of the continents. The climate in the boreal forests is rather humid because the rather low temperatures hinder evaporation, and the rainfall is usually small.

Human activities have increased the flow of several compounds into the atmosphere and the atmospheric composition reacts to these flows. The concentrations of several compounds are increasing in the air. The atmosphere hinders thermal radiation into the space more than previously, resulting in climate change. The boreal forests are reacting and will react to the changes in the environmental factors, especially atmospheric CO_2 concentration and temperature.

CLIMATE CHANGE

The air bubbles in ice in Antarctica and in Greenland have stored the climatic history on Earth for 700,000 years. Slow, in the time scale of 1000 years, and rather large changes characterize temperature and CO_2 concentration (Fig. 1). These oscillations in temperature are evidently generated by small periodic changes in the spinning of Earth and in the circulation of Earth around the Sun. The climate is evidently rather stable in the time scale of 10,000 years and the temperature has varied very little after the ice age.

The human population on Earth and its need for food and energy before 1800 were so small that the human influence on the material and energy fluxes had only minor effects on the atmospheric concentrations. The rapid population growth and development of technology changed the situation, and human activities started to change the atmospheric concentrations and climate. Exponential growth has been characteristic for human-induced fluxes; thus, the changes in the atmospheric concentrations and climate have also been strongest during the twentieth century.

The need of new agricultural land for crops dominated the anthropogenic CO_2 fluxes into the atmosphere in the

Encyclopedia of Natural Resources DOI: 10.1081/E-ENRL-120047444

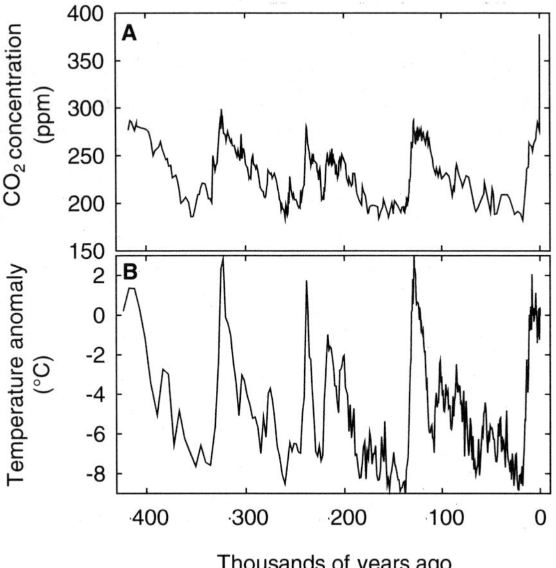

Fig. 1 (**A**) Concentration of CO_2 and (**B**) change in temperature during the last 420,000 years. The CO_2 concentration data are compiled from Vostock,[2] Siple station ice core data,[3] and atmospheric measurements.[4] The temperatures are obtained from Vostok ice core isotopes as difference from present values.[5–8]
Source: Copyright by Springer.

nineteenth century and the release of CO_2 from burning fossil fuels grew slowly in importance and now it covers about 90% of the anthropogenic emissions. Over two centuries, the atmospheric CO_2 concentration has increased from 280 ppm to nearly 400 ppm in the present.

Solar radiation is the source of energy for processes and transport phenomena taking place in the atmosphere. Several phenomena convert energy into other forms in the atmosphere. Gases and aerosols in the atmosphere absorb solar radiation and convert radiation energy into heat. Aerosols (clouds) reflect radiation into the space. Latent heat flux, i.e., flux of water vapor, is important for the atmospheric dynamics since evaporation consumes and condensation releases large amounts of energy.

All radiations have their spectrum, i.e., the radiation energy flux at each wavelength of the radiation. All bodies on earth emit thermal radiation. The spectrum of thermal radiation and the radiation flux depend on the temperature of the emitting body. Solar radiation occurs mainly within the range of 300–2000 nm, and half of its energy is within the visible light from 400 nm to 700 nm. The bodies on Earth emit thermal radiation at wavelengths of about 10,000 nm.

Gas molecules can absorb solar and thermal radiation. The ability to absorb, the absorbance, depends strongly on the wavelength of the radiation. The so-called greenhouse gases, i.e., CO_2, CH_4, and O_3, are transparent at visible light (0.3–0.7 nm), but they absorb the thermal radiation of several wavelengths. The greenhouse gases are able to partially absorb the thermal radiation from earth at several wavelengths.

Solar radiation generates flows of material and energy in the atmosphere, resulting in the climate at each location, warm and humid near the equator and cool and humid in the boreal region. The solar radiation has a very strong annual pattern in boreal areas due to the tilting position of the spinning axis of the Earth. This annual pattern is reflected in the climate, quite warm summers and very cold winters.

The physical phenomena, such as absorption and emission of radiation, take place in the atmosphere at a molecular level, and we have to convert our knowledge to deal with the global climate. Numeric simulation models enable the conversion of the knowledge dealing with different processes and transport phenomena in the atmosphere and in the oceans at the global scale. Several versions of global climatic models (GCMs), often also called global circulation models, have been constructed. The basic structure of the models is similar in these models.

The atmosphere is treated with volume elements, voxels, which are 200–300 km wide and the thickness varies from a few hundred meters to several kilometers at the top of the atmosphere. The temperature, water vapor concentration, three-dimensional wind velocities, etc., are treated in each voxel. The mass and energy fluxes change temperature and concentration in the voxels. The in and out fluxes of energy and material are combined with the conservation principle of mass, energy, and momentum, resulting in differential equations describing temperature, gas concentrations, and wind velocities.

The GCMs are able to simulate the climate at different locations rather well, utilizing physical knowledge and conservation principles. They are able to model the climate of boreal forests, rather warm or cool summer and cold winter, especially in the central parts of the continents. Although the GCMs simulate the global climate rather well, they are still under development.

Analysis and simulations of the present climate change are the most important applications of the GCMs. The absorption of the thermal radiation from earth in the atmosphere is increasing. This effect is built in the GCMs since they deal with all relevant aspects of energy flows in the atmosphere. There are two facts that generate the response of our climate to increasing greenhouse gas concentrations: 1) the atmosphere is quite transparent for the solar radiation, increasing greenhouse gas concentrations generate only a minor increase in the absorption of solar radiation, and 2) the greenhouse gases only partially absorb the thermal radiation of several wavelengths from Earth and any increase in concentrations of greenhouse gases results in enhanced absorption in the atmosphere. This fundamental physical basis of the present climate change is well understood and widely tested with measurements in the laboratory and in the field.

The development of the greenhouse concentrations in the atmosphere is well known during the last two centuries. After the ice age, the concentrations have been quite stable for about 10,000 years and the strong increase has occurred during the last 200 years. The temperature has been stable for the same 10,000 years, although some small variations have taken place (Fig. 1).

The temperature measurements started in the eighteenth century and systematic monitoring began during the nineteenth century. The measurements in the boreal region indicate rather irregular fluctuations and a clear increasing trend of temperature starting around 1970. The phonological observations and dates of formation of ice in the Tornionjoki River show similar patterns.

The development of the atmospheric greenhouse gas concentrations during the twenty-first century is not clear. However, the coal resources are so large that the availability of coal will not limit the use of fossil fuels. Several countries have tried to reduce CO_2 emissions according to the Kyoto agreement with some success in the industrialized countries. However, the rapid development in China, India, and some other developing countries has overridden the reduction of CO_2 emissions and the emissions are growing very fast. The atmospheric carbon budget has been intensively studied and alternative developments of CO_2 concentration in the atmosphere can be found at http://www.globalcarbonproject.org/carbonbudget.

The expected concentration of the atmospheric CO_2 at the year 2100 varies greatly between scenarios from stabilization at 500 ppm to a strong increasing trend at 1000 ppm. The development of CO_2 concentration generates the largest uncertainty in the simulated global temperatures in the year 2100. The simulations of the climate change during the twentieth century show a clear increasing trend in temperature, especially in the boreal region.

The evidence of rapid temperature increase is piling up. The most important one is systematic temperature measurements during the last century. They indicate a clear increasing global trend and an especially strong increase at high latitudes (50–70°N). There has also been additional information from observations made in nature. The flowering of birch and bird cherry trees is taking place earlier than previously. The melting of ice cover on Tornionjoki River is occurring earlier than previously in the long time series covering the years 1693–2013. The polar ice on the Arctic Sea has been recorded low in the year 2012 and the glaciers are drawing back.

BOREAL FORESTS

Boreal forest stands are often born after a catastrophe, forest fire, storm, or clear cut. The trees grow and the ecosystem develops until the next catastrophe and thereafter the stand starts again. The ground vegetation is rich after a catastrophe until the shading of the growing trees begins to reduce the growth of dwarf shrubs and herbs. The litter fall and root exudates from trees and ground vegetation feed the microbes in the soil.

Living organisms have long chains of enzymes, membrane pumps, and pigments for metabolic and specialized tissues for transport tasks, especially for transport of water and sugars within the individuals. The enzymes, membrane pumps, and pigment complexes are proteins. Proteins are amino acid polymers and their nitrogen content is as high as 15–17%.

Photosynthesis converts the solar radiation energy into chemical form as sugars, which are the main source of raw material of growth and the only source of energy for metabolism. The sugars are used for the synthesis of macromolecules, such as cellulose, lignin, and lipids, in the growing cells in needles, in woody components, and in fine roots. The macromolecules enter as litter into the organic material pool in the soil. The microbes decompose the macromolecules with extracellular enzymes and utilize the resulting small molecules in their metabolism. The metabolism of vegetation and microbes requires energy that is obtained from small carbon molecules in the process called respiration, then adenosine diphosphate is converted to energy-rich adenosine triphosphate (ATP) and CO_2 is released into the atmosphere. The flow of carbon compounds through a forest ecosystem is visualized in Fig. 2.

Processes consume or produce materials giving rise to concentration and pressure differences that generate flows within the system. The carbon flows through the system via photosynthesis and respiration by either vegetation or microbes. The nitrogen fluxes in and out of the system are very small, and the nitrogen circulates effectively within the system. The effects of global change on processes are highlighted with dashed arrows in Fig. 2.

The metabolism of vegetation and microbes is based on the action of enzymes, membrane pumps, and pigments. These proteins generate the important role of nitrogen as a nutrient, and the metabolism of a cell is impossible without proteins. We can say that photosynthetic products provide the structure and nitrogen uptake, together with photosynthesis, the tools in the structure.

Roots take up nitrogen as ions or amino acids. The ions are utilized in the synthesis of amino acids that are combined in the synthesis of proteins. When a cell dies, about half of the proteins in the cell are decomposed into amino acids for reuse in growing tissues and the other half enters an organic matter pool in the soil. The microbes decompose the proteins in the litter with extracellular enzymes to amino acids that can be used in the metabolism of microbes and vegetation. Microbes emit ammonium ions that are picked up by vegetation. Very small fluxes, i.e., deposition, biological nitrogen fixing, and leaching, connect the forest ecosystem with its surroundings, and thus the nitrogen circulates in a rather closed loop in the forest ecosystem.

The transpiration of vegetation is about hundredfold when compared with photosynthesis. Thus, trees need an effective transport system for water. We consider the woody structures in branches, stem, and coarse roots as water

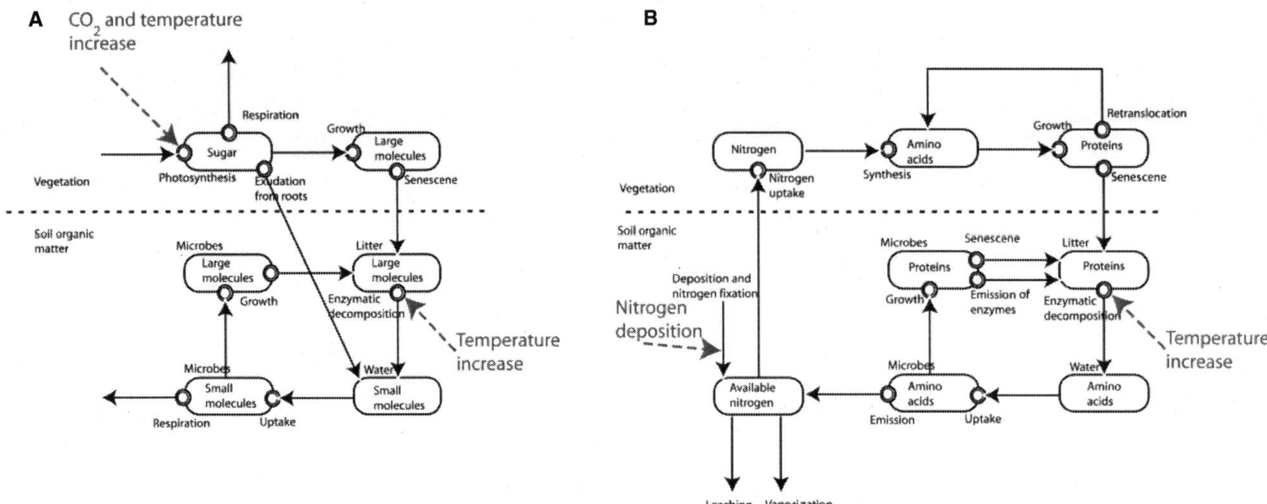

Fig. 2 (**A**) Carbon and (**B**) nitrogen fluxes in a forest ecosystem. Boxes, arrows, and double circles denote amounts, flows, and processes, respectively.

transport systems in the tree. We use sapwood area as a measure of the water transport capacity of wood. We assume that there is a balance between the water transport in branches, stem, and coarse roots and the needle mass in a whorl.

A coniferous tree annually forms a whorl on the top of the stem. Thereafter, the whorl grows and finally dies. The trees have to solve the allocation problem, i.e., the use of available sugars and nitrogen for the growth of needles, water pipes, and fine roots. We assume that the trees solve the allocation problem annually at the level of whorls. The annual amount of sugars and nitrogen, i.e., the amount of sugars and nitrogen flown into the whorl during a year, is used for the growth of the whorl in such a way that: 1) all available sugars and nitrogen are used for growth and to obtain ATP for metabolism; 2) the roots can provide the nitrogen needed for the synthesis of proteins in the new tissues; and 3) the capacity of the water transport system meets the need of water in needles. The above three requirements result in carbon and nitrogen balance equations and the growths are obtained as the solution of the equations. These balance equations are very similar to those describing the behavior of heat and concentrations in a voxel in the GCMs.

When we describe all the carbon and energy fluxes in Fig. 2, we obtain the forest ecosystem model MicroForest. The model describes cellulose, lignin, lipid, starch, and protein pools in the ecosystem in trees, in ground vegetation, and in soil. We have tested the model with measurements of tree diameter and height from very young to rather mature stands. We measured five stands in Estonia. The model estimated with measurement at station for measuring ecosystem - atmosphere relations II is able to predict the tree growths in Estonia without any changes in the parameter values. We demonstrate the fit between measured and predicted growths for the stands with best and weakest fit and the means of the five stands (Fig. 3). The prediction of the mean growths is especially close to the measured one, indicating that random noise dominates the discrepancies at stand level. We have determined only the initial state of soil fertility from measurements in Estonia.

The Effect of Climate Change on Boreal Forests

The atmospheric CO_2 concentration is increasing and the climate is getting warmer in the boreal regions. In addition, the nitrogen deposition, although it is rather low, is high when compared to the flux a century ago and other fluxes between forest ecosystems and their surroundings. Boreal forests are reacting to the climate change, i.e., changes in carbon dioxide and nitrogen ion concentrations and in temperature.

The response of boreal forest to climate change takes place at the cell level in vegetation and in the soil. These small-scale reactions have to be converted to the annual ecosystem level. We had the same problem when several small-scale processes were combined to obtain the changes in the behavior of the atmosphere. We used GCMs for the transition to the global level based on the material and energy flows in the atmosphere. Similarly, we can use MicroForest as a tool for the transition from cell level to ecosystems.

Availability of CO_2 limits photosynthesis, especially at high light intensities. Thus, the increase in the atmospheric CO_2 concentration has increased the formation of sugars, i.e., the flow of radiation energy into chemical energy. In addition, the temperature increase expands the photosynthetically active period and in this way also enhances the formation of sugars in photosynthesis.

The protein pool in the soils of boreal forests is large, about 0.5 kg m^{-2}. Microbes release ammonium ions from the protein pool in decomposition with extracellular enzymes. The annual release of NH_4 is small when

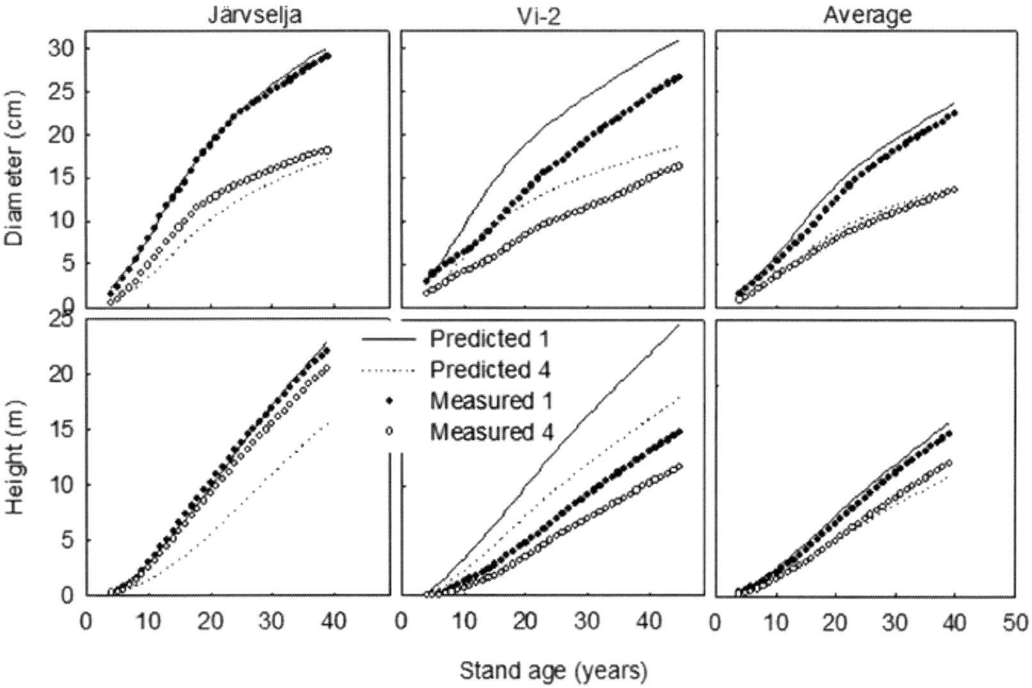

Fig. 3 Predicted and measured diameters and heights at two sites in Estonia [Järvselja the best fit and Vihterpalu 2 (Vi-2) the poorest fit] and the average of all the sites in Estonia.
Source: Copyright by Springer.

compared with the amount of proteins, only about 2% of the protein pool in the soil. The rate of enzymatic reactions depends strongly on temperature; thus, the climate change accelerates the decomposition of proteins in the boreal forest soils, resulting in increased availability of nitrogen for vegetation.

The use of fossil fuels and nitrogen fertilizers in agriculture has increased the flux of reactive nitrogen into the atmosphere. The additional reactive nitrogen changes its chemical form in the reactions in the atmosphere and finally falls down as either wet or dry deposition within some thousand kilometers from the location of the emissions. The nitrogen deposition increased rapidly during the later part of the twentieth century to nearly 10-fold when compared with the pre-industrial values, providing an additional source of nitrogen for boreal forests.

Climate change has increased and will increase photosynthesis and has accelerated and will accelerate the decomposition of proteins in the soil, and nitrogen deposition has provided and will provide additional reactive nitrogen for vegetation. Thus, the metabolism of trees and microbes and the material fluxes are reacting to the climate change in boreal forests. These changes in the material fluxes are described in the ecosystem model called MicroForest.[9] The simulations with MicroForest indicate considerable changes in boreal forest and in photosynthesis, needle mass, and stem volume as a response to the changing climate (Fig. 4).

Feedback from Boreal Forests to Climate Change

The responses of boreal forests to climate change have generated and will generate feedback to the climate change. The carbon pool in the boreal forests will change, absorption of solar radiation into coniferous trees will enhance, and aerosol formation and growth will increase, resulting in increasing reflection of solar radiation.

There are several changes in the carbon pools of a boreal forest due to the climatic change as seen in Fig. 4. We can convert these changes to the level of boreal forests with MicroForest simulations. We introduce three additional assumptions: 1) even aged stands form the boreal forests; 2) the rotation time is 100 years; and 3) the age distribution of the stands is even. We divided the boreal forests into six areas, three on each continent, and we applied an area-specific development of temperature and nitrogen deposition during the period 1750–2100 in the simulations of carbon pools in the boreal forests.[10] The responses varied to some extent between areas, high values for densely populated areas of high nitrogen deposition and low ones for extremely remote areas. The obtained changes in the carbon pools in the boreal forests are rather large. The annual carbon uptake by boreal forests is about 0.5 Pg yr^{-1}, which is about 6% of the annual anthropogenic CO_2 emissions in the year 2000 (Fig. 5).

The snow reflects nearly all solar radiation and, in contrast, the coniferous needles effectively absorb visible

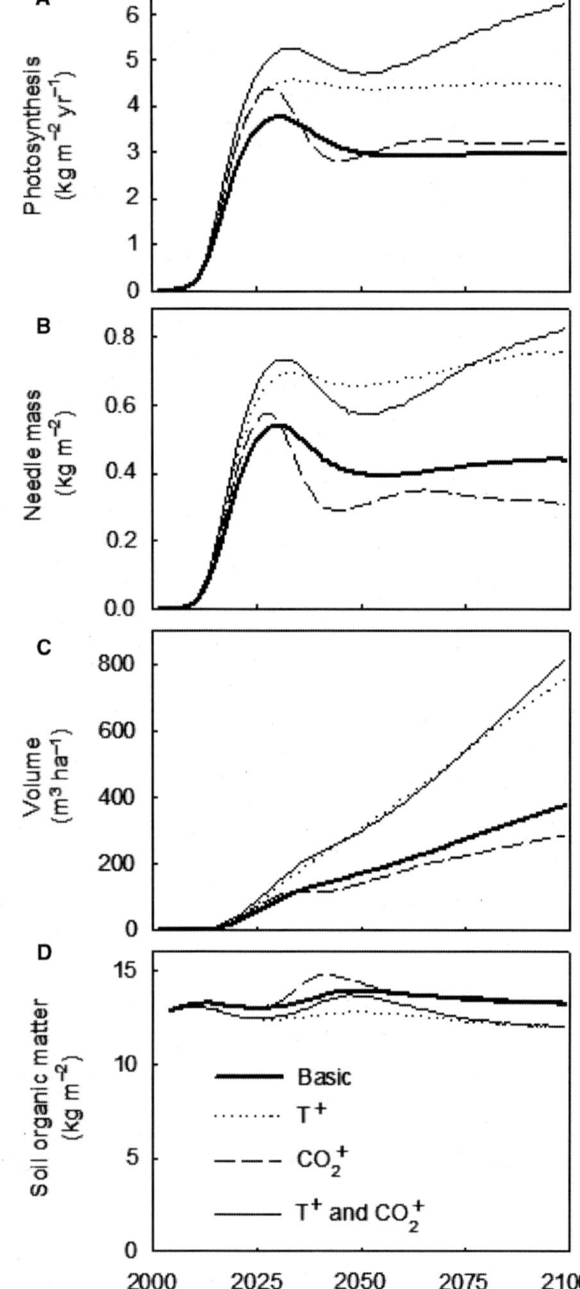

Fig. 4 (**A**) Simulated photosynthesis (expressed in units of sugars), (**B**) needle mass, (**C**) volume, and (**D**) mass of soil organic matter (dry weight) by MicroForest for the years 2000–2100 with different scenarios: no change in temperature or atmospheric CO_2 concentration, increase in temperature (0.4°C/decade), increase in CO_2 concentration (1.5 ppm yr^{-1}), and increase in both temperature and CO_2 concentration.
Source: Copyright by Springer.

light. The boreal forests will expand to tundra and their needle mass will increase. The expansion of the area of boreal forests is a rather slow phenomenon, but the increase in needle mass takes place in the annual time scale. We simulated with GCM the effect of twofold needle mass on the monthly mean temperatures in boreal forests. The

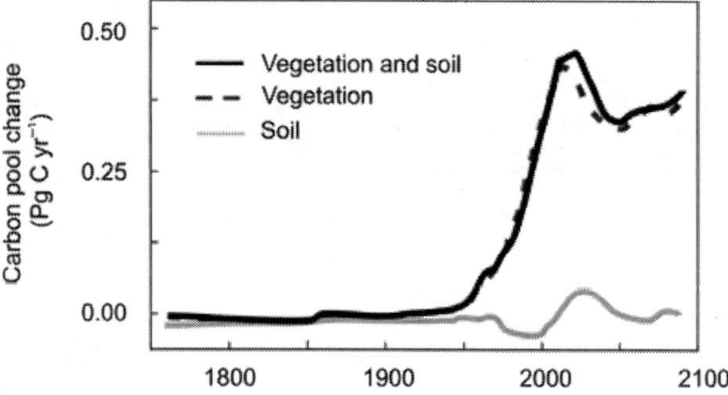

Fig. 5 Simulated changes in carbon pools of vegetation, soils, and ecosystems in boreal forests during 1750–2100.

simulations indicated a clear increase in the spring, even 1.5°C in Central Asia and 1°C in Central North America, and in near coasts the increase was clearly smaller.[11] In midsummer the effects were clearly smaller.

The aerosols scatter solar radiation in the atmosphere and reduce the solar energy flux to the Earth's surface and in this way cool the climate. The vegetation, especially trees, emit volatile organic compounds (VOCs) that react in the atmosphere and lose their volatility. Thereafter, they condense on aerosol particles and the particles grow bigger. Riipinen et al.[12] roughly estimated the effects of doubling VOC emissions on the radiation energy reaching the earth's surface. This simulation procedure involves large uncertainties and the effect of aerosols in the changing climate is weakly known.

CONCLUSION

The physical basis of climate change is well understood and the simulation models (GCMs) are based on the physical knowledge of the phenomena involved and on conservation principles. The simulated changes are in a reasonable agreement with measurements and observations in nature. In addition, the responses of boreal forests are quite well understood: the photosynthesis is enhanced, decomposition of proteins in the soil is accelerated, and nitrogen deposition provides additional nitrogen for vegetation. The biomass in the boreal forests will increase and the boreal forests are able to bind about 6% of the anthropogenic emissions. In addition, the increasing needle mass is able to enhance the absorption of solar radiation in the spring. The role of changing aerosol concentrations increases the uncertainty in the analysis of climate change.

REFERENCES

1. Botkin, D.B.; Simpson, L. The first statistically valid estimate of biomass for a large region. Biogeochemistry **1990**, *9* (9), 161–174.
2. Barnola, J.-M.; Raynaud, D.; Lorius, C.; Barkov, N.I. Historical CO_2 record from the Vostok ice core. In *Trends: A Compendium of Data on Global Change*; Carbon Dioxide

Information Analysis Center, Oak Ridge National Laboratory, U.S. Department of Energy: Oak Ridge, Tennessee, USA, 2003.
3. Friedli, H.; Lötscher, H.; Oesschger, H.; Siegenthaler, U.; Stauffer, B. Ice core record of $^{13}C/^{12}C$ ratio of atmospheric CO_2 in the past two centuries. Nature **1986**, *324* (6094), 237–238.
4. Keeling, C.D.; Whorf, T.P. Atmospheric CO_2 records from sites in the SIO air sampling network. In *Trends: A Compendium of Data on Global Change*; Carbon Dioxide Information Analysis Center, Oak Ridge National Laboratory, U.S. Department of Energy: Oak Ridge, Tennessee, USA, 2005.
5. Jouzel, J.; Lorius, C.; Petit, J.R.; Genthon, C.; Barkov, N.I.; Kotlyakov, V.M.; Petrov, V.M. Vostok ice core: A continuous isotope temperature record over the last climatic cycle (160,000 years). Nature **1987**, *329*, 403–408.
6. Jouzel, J.; Barkov, N.I.; Barnola, J.M.; Bender, M.; Chappellaz, J.; Genthon, C.; Kotlyakov, V.M.; Lipenkov, V.; Lorius, C.; Petit, J.R.; Raynaud, D.; Raisbeck, G.; Ritz, C.; Sowers, T.; Stievenard, M.; Yiou, F.; Yiou, P. Extending the Vostok ice-core record of palaeoclimate to the penultimate glacial period. Nature **1993**, *364* (6436), 407–412.
7. Jouzel, J.; Waelbroeck, C.; Malaize, B.; Bender, M.; Petit, J.R.; Stievenard, M.; Barkov, N.I.; Barnola, J.M.; King, T.; Kotlyakov, V.M.; Lipenkov, V.; Lorius, C.; Raynaud, D.; Ritz, C.; Sowers, T. Climatic interpretation of the recently extended Vostok ice records. Clim. Dyn. **1996**, *12* (8), 513–521.
8. Petit, J.R.; Jouzel, J.; Raynaud, D.; Barkov, N.I.; Barnola, J.-M.; Basile, I.; Bender, M.; Chappellaz, J.; Davis, M.; Delayque, G.; Delmotte, M.; Kotlyakov, V.M.; Legrand, M.; Lipenkov, V.Y.; Lorius, C.; Pepin, L.; Ritz, C.; Saltzman, E.; Stievenard, M. Climate and atmospheric history of the past 420,000 years from the Vostok ice core, Antarctica. Nature **1999**, *399* (6735), 429–436.
9. Hari, P; Nikinmaa, E. Ecosystem responses to climate change. In *Boreal Forest and Climate Change*; Springer Dordrecht Heidelberg: New York, London, 2008; 499–503.
10. Hari, P.; Bäck, J.; Nikinmaa, E. Responses of forest ecosystems to climate change. In *Physical and Physiological Forest Ecology*; Springer Dordrecht Heidelberg: New York, London, 2013; 435–439.
11. Räisänen, J.; Smolander, S. Climatic effects of increased leaf area: Reduced surface albedo and increased transpiration. In *Physical and Physiological Forest Ecology*; Springer Dordrecht Heidelberg: New York, London, 2013; 446–454.
12. Riipinen, I.; Petäjä, T.; Hari, P.; dal Maso, M.; Bäck, J.; Kulmala, M. Forests, aerosols and climate change. In *Physical and Physiological Forest Ecology*; Springer Dordrecht Heidelberg: New York, London, 2013; 454–469.

Coastal Environments: Remote Sensing

Yeqiao Wang
Department of Natural Resources Science, University of Rhode Island, Kingston, Rhode Island, U.S.A.

Abstract

As coastal environments around the world continue to face unprecedented natural and anthropogenic threats, recent developments in remote sensing science and technologies have profoundly enhanced monitoring capabilities in coastal studies. Providing systematic treatment of key developments, this entry presents a comprehensive review of the state-of-the-art technology in this technically complex arena. The text illustrates the latest developments in active remote sensing, hyperspectral remote sensing, high spatial resolution remote sensing, and integration of remote sensing and in situ data, and provides a snapshot of a wide range of remote sensing applications in coastal issues.

INTRODUCTION

Coastal zone includes areas of continental shelves, islands or partially enclosed seas, estuaries, bays, lagoons, beaches, and terrestrial and aquatic ecosystems within watersheds that drain into coastal waters. Coastal zone is the most dynamic interface between land and sea and represents the most challenging frontier between human civilization and environmental conservation. Worldwide, over 38% of the human population live in the coastal zones.[1] In the United States, about 53% of the human population lives in the coastal counties.[2] An increasing proportion of the global population lives within the coastal zones of all major continents that require increasing attention to agricultural, industrial, and other human-related effects on coastal habitats and water quality, and their impacts on ecological dynamics, ecosystem health, and biological diversity.

Coastal environments contain a wide range of natural habitats such as sand dunes, barrier islands, tidal wetlands and marshes, mangrove forests, coral reefs, and submerged aquatic vegetation (SAV) that provide foods, shelters, and breeding grounds for terrestrial and marine species. Coastal habitats also provide irreplaceable services such as filtering pollutants and retaining nutrients, maintaining water quality, protecting shoreline, and absorbing flood waters. As coastal habitats are facing intensified natural and anthropogenic disturbances by direct impacts such as hurricane, tsunami, harmful algae bloom, and cumulative and secondary impacts, such as climate change, sea level rise, oil spill, and urban development, inventory and monitoring of coastal environments become one of the most challenging tasks of the society in resource management and humanity administration. Remote sensing science and technologies have profoundly changed the practice in the monitoring and understanding of the dynamics of coastal environments.

Remote sensing refers to art, science, and technology for Earth system data acquisition through nonphysical contact sensors or sensor systems mounted on space-borne, airborne, and other types of platforms; data processing and interpretation from automated and visual analysis; information generation under computerized and conventional mapping facilities; and applications of generated data and information for societal benefits and needs. Remote sensing data can be collected passively or actively through selected bandwidths of spectral ranges across the spectrum of the electromagnetic radiation with different spatial resolutions. The multispectral capabilities of remote sensing allow observation and measurement of biophysical characteristics of the Earth system components, while the multitemporal and multisensor capabilities allow tracking of changes in the characteristics over time.[3] Remote sensing has been broadly applied in physical world (e.g., the atmosphere, water, soil, rock), its living inhabitants (e.g., flora, fauna), and the processes at work (e.g., erosion, deforestation, urban sprawl, effects of climate and environmental change, etc.).

Coarse spatial resolution remote sensing data have been applied for coastal studies. For example, high concentrations of suspended particulate matter in coastal waters directly affect water column and benthic processes such as phytoplankton productivity,[4,5] coral growth,[6–9] productivity of SAV,[10] and nutrient dynamics.[11]

The Sea-viewing Wide Field-of-view Sensor (SeaWiFS) system acquires radiance data from the Earth in eight spectral bands with a maximum spatial resolution of 1 km at nadir. The global 9 km^2 spatial resolution Level 3 data provide daily, 8-day, monthly, and annual standard data products. Acker et al.[12] employed monthly chlorophyll-a data from the 8-year SeaWiFS mission and data from moderate-resolution imaging spectroradiometer (MODIS) to analyze the spatial pattern of chlorophyll concentrations and seasonal cycle. The data indicate that large coral reef

Encyclopedia of Natural Resources DOI: 10.1081/E-ENRL-120049150

complexes may be sources of either nutrients or chlorophyll-rich detritus and sediment, enhancing chlorophyll-a concentration in waters adjacent to the reefs.

Landsat data have been used in coastal applications for decades.[13–15] Landsat and similar types of imagery data have been applied in inventory mapping, change detection, and management of mangrove forests through visual interpretation,[16,17] vegetation index,[18] and to map seagrass coverage,[19,20] among others.[21–24] As archived Landsat images have been made available at no cost to user communities since early 2009,[25] coastal applications can take advantages from such type of data.

ACTIVE REMOTE SENSING OF COASTAL ENVIRONMENTS

Active remote sensors are the data acquisition systems that transmit electromagnetic pulses with a specific wavelength (λ) and measure the time of return (translated to distance) of the signal reflected from a target. For example, light detection and ranging (LiDAR) systems generally transmit pulses at wavelength around the visible domain (λ in nanometers), whereas radar pulses are in the microwave domain (λ in centimeters). LiDAR and interferometric synthetic aperture radar (InSAR) are among new developments in active remote sensing that are the particular interests in coastal studies.

Active radar sensors are well known for their all-weather and day-and-night imaging capabilities, which are effective for mapping coastal habitats over cloud-prone tropical and subtropical regions. The synthetic aperture radar (SAR) backscattering signal is composed of intensity and phase components. The intensity component of the signal is sensitive to terrain slope, surface roughness, and dielectric constant. Studies have demonstrated that SAR intensity images can map and monitor forested and nonforested wetlands occupying a range of coastal and inland settings.[26,27] SAR intensity data have been used to monitor floods and dry conditions, temporal variations in the hydrological conditions of wetlands, including classification of wetland vegetation at various geographic settings.[28–32] When the phase components of two SAR images of the same area acquired from similar vantage points at different times are combined through InSAR processing, an interferogram can be constructed to depict range changes between the radar and the ground, and can be further processed with a digital elevation model (DEM) to produce an image with centimeter to subcentimeter vertical precision. InSAR has been extensively utilized to study ground surface deformation associated with volcanic, earthquake, landslide, and land subsidence processes. Alsdorf et al.[33] found that an interferometric analysis of L-band (wavelength of 24 cm) Shuttle Imaging Radar-C (SIR-C) and Japanese Earth Resources Satellite (JERS-1) SAR imagery can yield centimeter-scale measurements of water level changes throughout inundated floodplain vegetation.

LiDAR technology includes small- and large-footprint laser scanners.[34] LiDAR has shown promising results for assessing coastal habitats.[35–37] Due to the rapid laser firing, LiDAR pulses can penetrate vegetation cover, which makes LiDAR technology well suited to measure topography in coastal salt marsh areas.[28] Airborne LiDAR data have been tested for identifying coral colonies on patch reefs.[39] Shuttle Radar Topography Mission (SRTM) data have been applied to explore 3D modeling of mangrove forests. Although the SRTM was designed to produce a global DEM of the Earth's surface, the SRTM InSAR measurement of height z is the sum of the ground elevation and the canopy height contribution. Therefore, SRTM DEM can be used to measure vegetation height as well as ground topography.[40]

Reported studies integrate LiDAR data, digital historical maps, and orthophotos to measure long-term coastal change rates, to map erosion hazard areas, and to understand coastal processes on a variety of timescales from storms to seasons to decades and longer. The studies in barrier islands along the U.S. Gulf coast and in rocky coastal environments of the U.S. west coast show how similar types of data can be used to map coastal hazards in a variety of geographic settings and at a variety of spatial scales. Datasets that have commonly been used for coastal zone assessments can be integrated with newer data sources to modernize and update analyses. Such assessments will continue to be critical to coastal management and planning, especially with the currently predicted rates of sea level rise through the 21st century.[41]

HYPERSPECTRAL REMOTE SENSING OF COASTAL ENVIRONMENTS

Hyperspectral technology brings new insights about remote sensing of coastal environments. Hyperion sensor on board of EO-1 satellite is a representative space-borne hyperspectral system. The Hyperion imaging spectrometer collects data in 30-m ground sample distance over a 7.5-km swath and provides 10-nm (sampling interval) 220 contiguous spectral bands of the solar reflected spectrum from 400 nm to 2500 nm. Airborne Visible Infrared Imaging Spectrometer (AVIRIS) is a representative airborne hyperspectral sensor system. The AVIRIS whiskbroom scanner collects data in the same spectral interval and range as the Hyperion system but with 20-m spatial resolution. Hyperspectral remote sensing has advantages in coastal wetlands characterization due to its large number of narrow, contiguous spectral bands as well as high horizontal resolution from airborne platforms. It has been used to map habitat heterogeneity[42] to determine plant cover distribution in salt marshes.[43] Hyperspectral images have been used to separate vigor types by detecting slight differences in coloration due to stress factors, infestation, or displacement by invading species.[44] A study mapped the onset and progression of coastal *Spartina alterniflora* marsh dieback using hyperspectral image data at the plant-leaf, canopy, and satellite levels without *a priori* information on where, when, or how long the dieback had proceeded.[45] Hyperspectral data offer an enhanced ability to determine dieback onset and track progression.

Increased population and urban development have contributed significantly to environmental pressures along many areas of the U.S. coastal zone. The pressures have resulted in substantial physical changes to beaches, loss of coastal wetlands, and decline in ambient water and sediment quality, and the addition of higher volumes of nutrients (primarily nitrogen and phosphorus) from an urban, nonpoint source runoff. Algal growth is stimulated when nutrient concentrations, from sources such as stream and river discharges, wastewater sewage facilities, and agricultural runoff, are increased beyond the natural background levels of estuaries and other coastal receiving waters. These excess nutrients and the associated increased algal growth can also lead to a series of events that can decrease water clarity, cause benthic degradation, and result in low concentrations of dissolved oxygen. Recent research interests include quantification of the effects of sampling design and measurement accuracy, frequency, and resolution on the ability to improve our quantitative knowledge of coastal water quality. A study determined the ecological condition of numerous individual embayments and estuaries along the southern New England coast, as well as the adjoining coastal ocean, over an annual cycle using airborne hyperspectral remote sensing data and the criteria for assessing chlorophyll *a* concentrations.[46] The assessments would have been sample intensive and expensive if conducted over an annual period using traditional field-based monitoring.

HIGH SPATIAL RESOLUTION REMOTE SENSING OF COASTAL ENVIRONMENTS

High spatial resolution remote sensing data provide much needed spatial details and variations at the submeter level for mapping dynamic coastal habitats. For example, besides changing in areas, degradations of mangrove forests due to changing environment and selective harvesting have significant effects on ecosystem integrity and functions. Accurate and effective mapping of mangrove forests is essential for monitoring changes in spatial distribution and species composition. Advancement of high spatial resolution, multispectral remote sensing data makes such an inventory and monitoring possible.[47] High spatial resolution multispectral satellite and airborne digital remote sensing data have been employed to evaluate benthic habitats.[20,48–51]

Assessment of the quantity of impervious surface areas (ISAs) in landscapes has become increasingly important with growing concern of its impact on the environment.[52] This is particularly true for coastal areas due to the impacts of ISA on aquatic systems and its role in transportation and concentration of pollutants. Urban runoff, mostly over impervious surface, is the leading source of pollution in U.S. estuaries, lakes, and rivers.[53] Precise data of ISA in spatial coverage and distribution patterns in association with landscape characterizations are critical for providing the key baseline information for effective coastal management and science-based decision making. An example of studies extracted information on urban ISA from true-color digital orthophotography data to reveal the intensity of urban development surrounding the Narragansett Bay, Rhode Island.[54]

Remote sensing applications in disaster management have become critically important to support preparation through response to natural and human-induced hazards and events affecting human populations in coastal zones. Increased exposure and density of human settlements in coastal regions increase the potential loss of life, property, and commodities that are at risk from intense coastal hazards. Remote sensing has a long history of being used to capture towns, harbors, and coastal lines affected by disasters and played a key role in recovery efforts post disasters. The history can be traced back from the earthquake and tsunami that struck Alaska on March 1964, to the five hurricanes of Dennis, Katrina, Ophelia, Rita, and Wilma, which directly devastated the U.S. coastal regions in 2005.[55] High-spatial resolution images with timely coverage were proved to be efficient and useful to make rescue and recovery plans.

INTEGRATION OF REMOTE SENSING AND IN SITU MEASUREMENT DATA

Integration of multispectral, multitemporal, multisensor airborne and space-borne remote sensing data with global positioning system (GPS) guided in situ observations becomes necessary for effective and timely inventory and monitoring of the coastal environments. For example, a significant amount of the coastal wetlands along the Long Island Sound in the northeastern United States has been lost over the past century due to urban development, filling, and dredging, or damaged due to human disturbance and modification. Beyond the physical loss of marshes, the species composition of marsh communities is changing. With the mounting pressures on coastal wetland areas, it is becoming increasingly important to identify and inventory the current extent and condition of coastal marshes located on the Long Island Sound estuary, implement a cost-effective way to track changes in wetlands over time, and monitor the effects of habitat restoration and management. Identification of distribution and health of individual marsh plant species like *Phragmites australis* using remote sensing is challenging because vegetation spectra are generally similar to one another throughout the visible to near-infrared (VNIR) spectrum. The reflectance of a single species may vary throughout the growing season due to variations in the amount and ratios of plant pigments, leaf moisture content, plant height, canopy effects, leaf angle distribution, and other structural characteristics.

A study addressed the use of multitemporal field spectral data, satellite imagery, and LiDAR top-of-canopy data to classify and map common salt marsh plant communities.[56] VNIR reflectance spectra were measured in the field to assess the phenological variability of the dominant species of *Spartina patens*, *Phragmites australis,* and *Typha* spp. The field spectra and single-date LiDAR canopy height data were used to define an object-oriented classification methodology for the plant communities in multitemporal QuickBird imagery. The classification was validated using an extensive field inventory of marsh species.

Mapping of SAV is inherently difficult, as most species are subtidal. Traditionally, mapping of SAV has been accomplished using a combination of aerial photographs and in situ sampling by divers using scuba. A reported study created a video reference database as field reference of benthic habitats.[57] During the study, an underwater video camera was attached to a sled and the sled was towed to a vessel. The position, heading, and speed of the vessel were outputs from a GPS receiver located on the vessel and overlaid on the live video image. To increase the spatial accuracy of the coordinates the study placed a GPS receiver in a kayak and positioned directly over the sled. A VCR/TV combo installed on the towing vessel recorded SAV conditions and geographic coordinates on video files. The recorded video files were related to the coordinates captured on the towed GPS unit through the time stamp. Based on speed, the amount of video equaling the required 30 m was recorded. The corresponding video files to each line segment were linked to the geographic information system database. Each video file was reviewed for the percent of SAV within each transect. The percent was estimated by dividing the number of seconds in which SAV was observed and the dominant SAV species in each video transect was noted. The line coverage was then overlaid on top of the satellite images to aid our classification or conduct accuracy assessment.

CONCLUSIONS

The complexity of coastal constituents imposes significant challenges in application of remote sensing technologies due to the dynamic spatial and temporal nature of coastal habitats. Coastal environments are the boundary between the land and the ocean. Remote sensing of coastal environments straddles the technical boundary of land and ocean satellite remote sensing. Satellite radiometers optimized for remote sensing of terrestrial environments, as well as those for the open ocean, are not well suited for quantitative remote sensing of phytoplankton biomass, sediment, and other constituents of coastal waters. Remote sensing of coastal waters requires high spatial resolution to resolve characteristically small features as well as daily or high-frequency coverage to understand many of the important coastal processes of these dynamic regions. The ideal satellite radiometer, for measurements of coastal waters would, for example, have a finer spatial resolution and higher frequency of revisit time. The future promises many exciting advances in new and improved sensors for use as well as improved geospatial processing tools and computing technology to study coastal ecosystems.

REFERENCES

1. Crossett, K.M.; Culliton, T.J.; Wiley, P.C.; Goodspeed, T.R. *Population trends along the coastal United States: 1980–2008*, Coastal Trends Report Series; National Oceanic and Atmospheric Administration, National Ocean Service Management and Budget Office: Silver Spring, MD, 2004; 54 pp.
2. Small, C.; Cohen, J.E. Continental Physiography, Climate, and the Global Distribution of Human Population. Curr. Anthropol. **2004**, *45* (2), 269–277.
3. Wang, Y.; Moskovits, D.K. Tracking fragmentation of natural communities and changes in land cover: Applications of landsat data for conservation in an urban landscape (Chicago Wilderness) Conserv. Biol. **2001**, *15* (4), 835–843.
4. Cole, B.E.; Cloern, J.E. An empirical model for estimating phytoplankton productivity in estuaries. Marine Ecology. Prog. Ser. **1987**, *36*, 299–305.
5. Cloern, J.E. Turbidity as a control on phytoplankton biomass and productivity in estuaries. Continental Shelf Res. **1987**, *7* (11), 1367– 1381.
6. Dodge, R.E.; Aller, R.; Thompson, J. Coral growth related to resuspension of bottom sediments. Nature **1974**, *247*, 574–577.
7. Miller, R.L.; Cruise, J.F., Effects of suspended sediments on coral growth: Evidence from remote sensing and hydrologic modeling. Remote Sensing Environ. **1995**, *53*, 177–187.
8. Torres, J.L.; Morelock, J. Effect of terrigenous sediment influx on coral cover and linear extension rates of three Caribbean massive coral species. Caribbean J. Sci. **2002**, *38* (3–4), 222–229.
9. McLaughlin, C.J.; Smith, C.A.; Buddemeier, R.W.; Bartley, J.D.; Maxwell, B.A. Rivers, runoff and reefs. Global Planetary Change **2003**, *39* (1–2), 191–199.
10. Dennison, W.C.; Orth, R.J.; Moore, K.A.; Stevenson, J.C.; Carter, V.; Kollar, S. Assessing water quality with submersed aquatic vegetation. Bioscience **1993**, *43*, 86–94.
11. Mayer, L.M.; Keil, R.G.; Macko, S.A.; Joye, S.B.; Ruttenberg, K.C.; Aller, R.C. The importance of suspended particulates in riverine delivery of bioavailable nitrogen to coastal zones. Global Biogeochemical Cycles **1998**, *12*, 573– 579.
12. Acker, J.; Leptoukh, J.; Shen, S.; Zhu T.; Kempler, S. Remotely-sensed chlorophyll a observations of the northern Red Sea indicate seasonal variability and influence of coastal reefs, J. Mar. Syst. **2008**, *69*, 191–204.

13. Munday, J.C., Jr.; Alfoldi, T.T. Landsat test of diffuse reflectance models for aquatic suspended solids measurements. Remote Sensing Environ. **1979**, *8*, 169–183.
14. Bukata, R.P.; Jerome, J.H.; Bruton, J.E. Particulate concentrations in Lake St. Clair as recorded by a shipborne multispectral optical monitoring system. Remote Sensing Environ. **1988**, *25*, 201–229.
15. Ritchie, J.C.; Cooper, C.M.; Shiebe, F.R. The relationship of MSS and TM digital data with suspended sediments, chlorophyll and temperature in Moon Lake, Mississipi. Remote Sensing Environ. **1990**, *33*, 137–148.
16. Gang, P.O.; Agatsiva, J.L. The current status of mangroves along the Kenyan coast, a case study of Mida Creek mangroves based on remote sensing. Hydrobiologia **1992**, *247*, 29–36.
17. Wang, Y.; Bonynge, G.; Nugranad, J.; Traber, M.; Ngusaru, A.; Tobey, J.; Hale, L.; Bowen, R.; Makota, V. Remote sensing of mangrove change along the Tanzania coast. Mar. Geodesy **2003**, *26* (1–2), 35–48.
18. Jensen, J.R.; Ramset, E.; Davis, B.A.; and Thoemke, C.W. The measurement of mangrove characteristics in south-west Florida using SPOT multispectral data. Geocarto Int. **1991**, *2*, 13–21.
19. Armstrong, R.A. Remote sensing of submerged vegetation canopies for biomass estimation. Int. J. Remote Sensing **1993**, *14* (3), 621–627.
20. Lathrop, R.G.; Montesano, P.; Haag, S. A multi-scale segmentation approach to mapping seagrass habitats using airborne digital camera imagery. Photogrammetric Eng. Remote Sensing **2006**, *72* (6), 665–675.
21. Gao, J. A hybrid method toward accurate mapping of mangroves in a marginal habitat from SPOT multispectral data. Int. J. Remote Sensing **1998**, *19*, 1887–189.
22. Green, E.P.; Mumby, P.J.; Edwards, A.J.; Clark, C.D.; Ellis, A.C. The assessment of mangrove areas using high resolution multispectral airborne imagery. J. Coastal Res. **1998**, *14*, 433–443.
23. Rasolofoharinoro, M.; Blasco, F.; Bellan, M.F.; Aizpuru, M.; Gauquelin, T.; Denis, J. A remote sensing based methodology for mangrove studies in Madagascar. Int. J. Remote Sensing **1998**, *19*, 1873–1886.
24. Pasqualini, V.; Iltis, J.; Dessay, N.; Lointier, M.; Gurlorget, O.; Polidori, C. Mangrove mapping in North-Western Madagascar using SPOT-XS and SIR-C radar data, Hydrobiologia **1999**, *413*, 127–133.
25. Woodcock, C.E.; Allen, R.; Anderson, M.; Belward, A.; Bindschadler, R.; Cohen, W.; Gao, F.; Goward, S.N.; Helder, D.; Helmer, E.; Nemani, R.; Oreopoulos, L.; Schott, J.; Thenkabail, P.S.; Vermote, E.F.; Vogelmann, J.; Wulder, M.A.; Wynne, R. Free access to landsat imagery. Science **2008**, *320*, 1011.
26. Ramsey, E.W., III.; Monitoring flooding in coastal wetlands by using radar imagery and ground-based measurements. Int. J. Remote Sensing **1995**, *16* (13), 2495–2502.
27. Ramsey, E., III.; Rangoonwala, A. Canopy reflectance related to marsh dieback onset and progression in coastal Louisiana. Photogrammetric Eng. Remote Sensing **2006**, *72* (6), 641–652.
28. Hess, L.L.; Melack, J.M. Mapping wetland hydrology and vegetation with synthetic aperture radar. Int. J. Ecol. Environ. Sci. **1994**, *20*, 197–205.
29. Hess, L.L.; Melack, J.M.; Filoso, S.; Wang, Y. Delineation of inundated area and vegetation along the Amazon floodplain with the SIR-C synthetic aperture radar. IEEE Trans. Geosci. Remote Sensing **1995**, *33* (4), 896–904.
30. Kasischke, E.S.; Bourgeau-Chavez, L.L. Monitoring south Florida wetlands using ERS-1 SAR imagery. Photogrammetric Eng. Remote Sensing **1997**, *63*, 281–291.
31. Simard, M.; De Grandi, G.; Saatchi, S.; Mayaux, P. Mapping tropical coastal vegetation using JERS-1 and ERS-1 radar data with a decision tree classifier. Int. J. Remote Sensing **2002**, *23* (7), 1461–1474.
32. Kiage, L.M.; Walker, N.D.; Balasubramanian, S.; Babin, A.; Barras, J. Applications of RADARSAT-1 synthetic aperture radar imagery to assess hurricane-related flooding of coastal Louisiana. Int. J. Remote Sensing **2005**, *26* (24), 5359–5380.
33. Alsdorf, D.; Melack, J.; Dunne, T.; Mertes, L.; Hess, L.; Smith, L. Interferometric radar measurements of water level changes on the Amazon floodplain. Nature **2000**, *404*, 174–177, 2000.
34. Lefsky, M.A.; Cohen, W.B.; Parker, G.G.; Harding, D.J. LiDAR remote sensing for ecosystem studies. Bioscience **2002**, *52*, 19–30.
35. Brock, J.C.; Sallenger, A.H.; Krabill, W.B.; Swift, R.N.; Wright, C.W. Recognition of fiducial surfaces in lidar surveys of coastal topography. Photogrammetric Eng. Remote Sensing **2001**, *67*, 1245–1258.
36. Brock, J.C.; Wright, C.W.; Sallenger, A.H.; Krabill, W.B.; Swift, R.N. Basis and methods of NASA Airborne Topographic Mapper Lidar surveys for coastal studies. J. Coastal Res. **2002**, *18*, 1–13.
37. Nayegandhl, A.; Brock, J.C.; Wright, C.W.; O'Connell, J. Evaluating a small footprint, waveform-resolving LiDAR over coastal vegetation communities. Photogrammetric Eng. Remote Sensing **2006**, *72* (12), 1407–1417.
38. Montane, J.M.; Torres, R., Accuracy assessment of LiDAR saltmarsh topographic data using RTK GPS. Photogrammetric Eng. Remote Sensing **2006**, *72* (8), 961–967.
39. Brock, J.; Wright, C.W.; Hernandez, R.; Thompson, P. Airborne lidar sensing of massive stony coral colonies on patch reefs in the northern Florida reef tract. Remote Sensing Environ. **2006**, *104* (2006), 31–42.
40. Simard, M.; Zhang, K.; Rivera-Monroy, V.H.; Ross, M.S.; Ruiz, P.L.; Castañeda-Moya, E.; Twilley, R.R.; Rodriguez, E. Mapping height and biomass of mangrove forests in Everglades National Park with SRTM elevation data. Photogrammetric Eng. Remote Sensing **2006**, *72* (3), 299–311.
41. Hapke, C.J. Integration of LiDAR and historical maps to measure coastal change on a variety of time and spatial scales (Ch. 4). In *Remote Sensing of Coastal Environments*; Wang, Y., Ed.; CRC Press: Boca Raton, Florida, 2009; 61–78 pp.
42. Artigas, F.J.; Yang, J. Hyperspectral remote sensing of habitat heterogeneity between tide-restricted and tide-open areas in New Jersey Meadowlands. Urban Habitat **2004**, *2*, 1.
43. Belluco, E.; Camuffo,M.; Ferrari, S.; Modenese, L.; Silvestri, S.; Marani, A.; Marani, M. Mapping salt marsh vegetation by multispectral and hyperspectral remote sensing. Remote Sensing Environ. **2006**, *105*, 54.

44. Artigas, F.J.; Yang, J. Hyperspectral remote sensing of marsh species and plant vigor gradient in the New Jersey Meadowland. Int. J. Remote Sensing **2005**, *26*, 5209.
45. Ramsey, E. III.; Rangoonwala, A. Mapping the onset and progression of marsh dieback (Ch. 5). In *Remote Sensing of Coastal Environments*; Wang, Y., Ed.; CRC Press: Boca Raton, Florida, 2009; 123–149 pp.
46. Keith, D.J. Estimating chlorophyll conditions in Southern New England Coastal Waters from hyperspectral aircraft remote sensing (Ch. 7). In *Remote Sensing of Coastal Environments*; Wang, Y., Ed.; CRC Press: Boca Raton, Florida, 2009; 151–172 pp.
47. Wulder, M.A.; Hall, R.J.; Coops, N.C.; Franklin, S.E. High spatial resolution remotely sensed data for ecosystem characterization. Bioscience **2004**, *6*, 511–521.
48. Su, H.; Karna, D.; Fraim, E.; Fitzgerald, M.; Dominguez, R.; Myers, J.S.; Coffland, B.; Handley, L.R.; Mace, T. Evaluation of eelgrass beds mapping using a high-resolution airborne multispectral scanner. Photogrammetric Eng. Remote Sensing **2006**, *72* (7), 789–797.
49. Deepak, M.; Narumalani, S.; Rundquist, D.; Lawson, M. Benthic habitat mapping in tropical marine environments using QuickBird multispectral data. Photogrammetric Eng. Remote Sensing **2006**, *72* (9), 1037–1048.
50. Wang, L.; Sousa, W.P.; Gong, P.; Biging, G.S. Comparison of IKONOS and QuickBird images for mapping mangrove species on the Caribbean coast of Panama. Remote Sensing Environ. **2004**, *91*, 432–440.
51. Wang, L.; Silvan-Cardenas, J.L.; Sousa, W.P. Neural network classification of mangrove species from multi-seasonal IKONOS imagery. Photogrammetric Eng. Remote Sensing **2008**, *74* (7), 921–927.
52. Civco, D.L.; Hurd, J.D.; Wilson, E.H.; Arnold, C.L.; Prisloe, S. Quantifying and describing urbanizing landscapes in the Northeast United States. Photogrammetric Eng. Remote Sensing **2002**, *68*, 1083–1090.
53. Booth, D.B.; Jackson, C.R. Urbanization of aquatic systems: degradation thresholds, stormwater detection, and the limits of mitigation. J. Am. Water Resources Assoc. **1997,** *35*, 1077–1090.
54. Zhou, Y.; Wang, Y.; Gold, A.J.; August, P.V. Modeling watershed rainfall - runoff using impervious surface-area data with high spatial resolution. Hydrogeol. J. **2010**, *18* (6), 1413–1423.
55. White, S.; Aslaken, M. NOAA's use of direct georeferencing to support emergency response. PERS **2006**, *72* (6), 623–627.
56. Gilmore, M.S.; Wilson, E.H.; Barrett, N.; Civco, D.L.; Prisloe, S.; Hurd, J.D.; Chadwick, C. Integrating multitemporal spectral and structural information to map wetland vegetation in a lower Connecticut River tidal marsh. Remote Sensing Environ. **2008**, *112*, 4048–4060.
57. Wang, Y.; Traber, M.; Milstead, B.; Stevens, S. Terrestrial and submerged aquatic vegetation mapping in fire Island National seashore using high spatial resolution remote sensing data. Mar. Geodesy **2006**, *30* (1), 77–95.

Coastal Erosion and Shoreline Change

Matthew L. Stutz
Department of Chemistry, Physics, and Geoscience, Meredith College, Raleigh, North Carolina, U.S.A.

Abstract

This entry broadly addresses the causes of and responses to coastal erosion and shoreline change. Three major causes of shoreline change are examined in detail: sea level change, storms and waves, and sediment supply and transport. The history of both postglacial and recent global sea level change is discussed, and factors that influence relative sea level change are also fully considered. Short-term and long-term impacts of storms, including dune recession, cliff and bluff erosion, and overwash are described. Lastly, sediment transport related to longshore drift and tidal dynamics are included. Three major responses to coastal erosion and their impacts are then described: hard stabilization, soft stabilization, and relocation. Sea walls, jetties, and groins are described as the most common forms of hard coastal stabilization. Beach nourishment is covered as the major form of soft stabilization. Finally, examples of structural and community-scale relocation efforts are included. The entry concludes with a projection of what shoreline change may look like in the future, specifically in the context of climate change.

INTRODUCTION

The shoreline defines the boundary between land and sea. The position of the shoreline is in constant flux, existing in a dynamic equilibrium between shifting coastal sediment, fluctuating waves and currents, and sea level. Shorelines in most places worldwide are experiencing erosion (i.e., landward movement) and may erode even more rapidly in the next century if global sea level continues to rise at its present pace. Coastal societies have a long history of attempting to hold back the sea; however, future attempts are likely to be more widespread, more extreme, more expensive, and in some cases more futile.

This synopsis of shoreline change and coastal erosion first addresses the underlying causes of shoreline change and its coastal impacts—sea level change, storms, and sediment supply. It also describes some of the common responses to coastal erosion—hard stabilization, beach nourishment, and relocation. The conclusion considers the future implications of global climate change and rising sea level on shoreline change and coastal societies.

Herein, the general term "shoreline change" will be used to apply to any movement of the shoreline landward or seaward. "Coastal erosion" is a term that denotes a landward movement of the shoreline due to loss of sediment, increased wave activity, and/or a rise in sea level. "Accretion" is used to describe a seaward movement of the shoreline.

CAUSES OF SHORELINE EROSION

The shoreline exists in a balance between three elements of the coastal environment—sea level, waves and currents, and sediment. A change or shift in any of these factors results in shoreline change. The direction of the change—landward or seaward—depends upon the specific nature of the change in the coastal environment. In general, an influx of new sediment, reduction in wave intensity, or fall in sea level produces shoreline accretion. Removal of sediment, increase in wave intensity, or rise in sea level usually results in shoreline erosion.

Sea Level

The *effective sea level* for any given coastal location is influenced by both *eustatic* and *relative* sea level. Eustatic sea level is governed by the volume of water in the oceans, and is "global." Relative sea level is governed by the height of the land itself, and is generally "local." A rise in either eustatic or relative sea level results in the inundation of land and shoreline erosion.

Eustatic sea level

Eustatic sea level rise results from an increase in ocean volume. Ocean volume increases due to the addition of water to the ocean, principally through the melting of land-based glacial ice. At the end of the Pleistocene Ice Age when vast ice sheets covered North America and Europe, eustatic sea level was 130 m lower (Fig. 1),[1] and continental shorelines were located at the edge of today's continental shelf—in some cases more than 100 km seaward. Eustatic sea level rose a total of 0.2 m during the last century and is presently rising by 3 mm each year according to satellite altimeter measurements and coastal tide gauges (Fig. 2).

Encyclopedia of Natural Resources DOI: 10.1081/E-ENRL-120047497

The Intergovernmental Panel on Climate Change (IPCC) attributes about half of the present eustatic sea level rise to glacial melting and the other half to the thermal expansion of ocean water.[2] The present rate of sea level rise is widely projected to accelerate, rising another 0.5 to 2 m in the coming century. This estimate is highly uncertain due to the complexity of ice sheet behavior in Antarctica and Greenland, the two largest stores of glacial ice. The Antarctic and Greenland ice sheets presently hold enough ice to add about 70 m and 7 m, respectively, to sea level if completely melted.

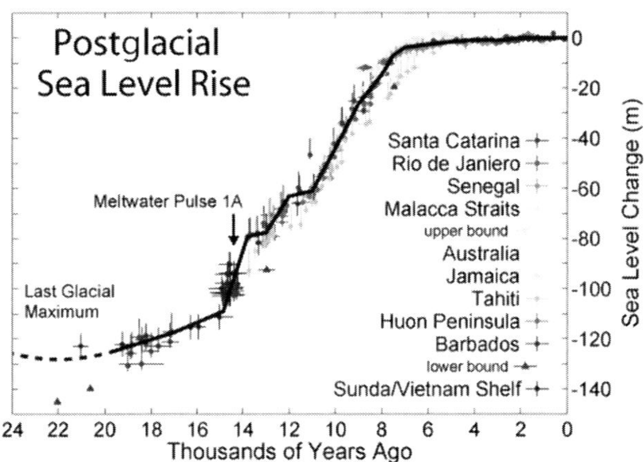

Fig. 1 Global eustatic sea level rose rapidly in the late Pleistocene and early Holocene Epochs but has been nearly stable for the past 6000 years.
Source: Adapted from Rohde, Wikimedia Commons, http://commons.wikimedia.org/wiki/File:Post-Glacial_Sea_Level.png (accessed March 2012).

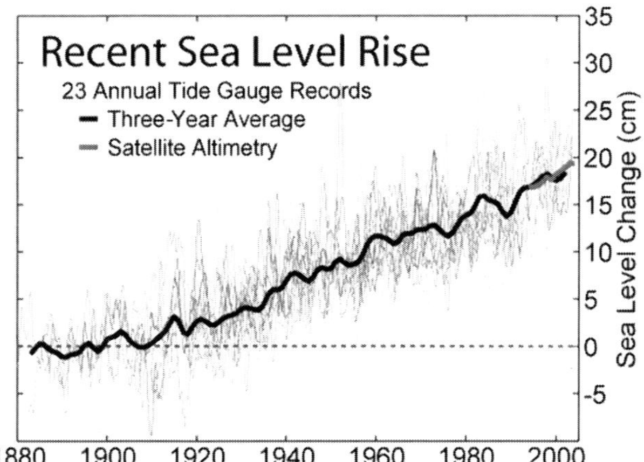

Fig. 2 Modern tide gauge records show a rate of eustatic sea level rise of 2 mm/yr between 1900 and 2000. Satellite altimetry measured since 1990 indicates a rate of 3 mm/yr.
Source: Adapted from Rohde, Wikimedia Commons, http://commons.wikimedia.org/wiki/File:Recent_Sea_Level_Rise.png (accessed March 2012).

Relative sea level

Relative sea level rise results from localized vertical land movement. Tectonic land movement is a significant source of relative sea level rise. The March 2011 Japanese earthquake caused some coastal areas to instantaneously sink several feet, flooding some coastal villages now during normal daily high tides.[3] Relative sea level rise also affects deltas because as their thick sediment deposits compact, the delta surface gradually subsides. Compaction is naturally offset by continuous new sediment deposition. However, upstream dams and levees starve many deltas today, such as the Nile Delta, of essential new sediment buildup. Wetland drainage and petroleum extraction—often accompanying urban or agricultural development of deltas—accelerates compaction and is a significant cause of subsidence on the Mississippi Delta. The Mississippi Delta is sinking as much as 9 mm each year and loses about one football field of area *every hour* to erosion.[4] The city of New Orleans is as much as 3 m (10 ft) below sea level due to this long-term subsidence. The impact of Hurricane Katrina was more extreme as a result of the long-term subsidence. One last cause of relative sea level rise is *isostasy*, the uplift or subsidence of the earth's crust in response to the addition or removal of overlying weight. The massive Pleistocene ice sheets covering North America and Eurasia depressed the crust underneath and pushed up a bulge in the crust around the edges. When the ice sheets melted, the isostatic effect reversed and is still ongoing. The land around Hudson Bay, Canada, and Scandinavia is actually rising, causing relative sea level to fall. Coasts throughout much of the United States and western Europe, located on the former ice age bulge, are now gradually sinking.[1,5] The addition of more than 100 m of sea water over the continental shelf also contributes to subsidence through *hyrdroisostasy*.[1]

Waves and Currents

Waves and currents are chiefly influenced by storm activity and daily tides. The impacts of storms on shorelines can range from minor and temporary to catastrophic and permanent. The magnitude and duration of storms largely determine what type of impact the storm has. The height of tides at a given coastal location is relatively consistent and predictable; however, tidal currents in and around coastal inlets vary greatly due to the natural shifting of inlet shoals and to human modification to inlets.

Storms

Storms are meteorological events that cause temporarily elevated tide levels, or *storm surge*, and wave heights to impact the shoreline. The following storm impact classification illustrates the range of coastal impacts caused by storms.[6] Storm impacts begin with a narrowing of the dry beach as larger waves begin surging higher up the beach (the *run-up regime*) and may soon begin to scour the base of dunes (the *collision regime*). Most storms fall into these two categories, and their impacts are usually temporary. The sand eroded by beach and dune erosion is swept offshore and built up into offshore bars. The return of calm waves brings this sand gradually back to the beach over a period of days to months. A major storm may break through low, narrow sections of the dunes, washing water and sand inland (the *overwash regime*). A catastrophic storm can completely drown and wash over the dunes, causing wholesale landward shoreline retreat (the *inundation regime*). Shoreline erosion caused by overwash is generally permanent, because the sand is moved inland and cannot return to the beach by waves. At the inundation level, new inlets may cut completely through barrier islands or, in rare cases, cause barrier islands to disappear permanently (Fig. 3).

Storm impacts on the shoreline are also closely related to the coastal topography. The Chandaleur Islands in Louisiana were almost completely destroyed by Hurricane Katrina. Whereas the Chandaleur Islands were uniformly low lying, narrow, and easily inundated, wider islands in the Gulf of Mexico with higher dunes survived the immense storm surge and waves of Katrina, experiencing some overwash but not complete destruction.[7]

Along rocky coasts, storm waves may pound the base of steep coastal bluffs, slowly weakening the bluffs and eventually causing a collapse of the bluff. Bluff erosion is more extreme in California during years with a strong *El Nino*, which brings more frequent storms and higher water levels to the Pacific U.S. coast. Sea level rise can also accelerate bluff erosion. All bluff erosion is permanent although the collapsed debris may temporarily protect the base of the bluff from further erosion. Future storms or sea level rise will eventually remove the debris and renew bluff erosion.

Fig. 3 Hurricane Irene generated significant erosion and overwash on the North Carolina Outer Banks, north of Cape Hatteras. Storm surge completely breached the island here and destroyed a section of Highway 12. Note that the highway makes a broad landward curve at this location, due to relocation of the highway following damage from previous storms.
Source: Adapted from United States Geological Survey Coastal and Marine Geology, August 31, 2011, http://coastal.er.usgs.gov/hurricanes/irene/post-storm-photos/20110830/20110831_150835d.jpg (accessed March 2012).

Tides and currents

The daily rise and fall of ocean tides can generate strong tidal currents, particularly in coastal bays, estuaries, and lagoons. These environments are similar in that they are narrow passages that intensify the flow of tidal currents. In barrier island systems, *tidal inlets* are the narrow passes that separate two islands and connect the ocean with a quiescent barrier lagoon. Inlets are typically narrow unstable channels that allow ocean water to circulate into and out of lagoons as the tide rises and falls, respectively. Tidal currents capture and transport enormous amounts of sediment, forming large shoals just inside and outside tidal inlets. The rising (flood) tide flowing into the lagoon deposits a flood tidal delta just inside the lagoon. The falling (ebb) tide flowing out of the lagoon deposits an ebb tidal delta just seaward of the inlet.

Tidal inlet channels and their tidal deltas are in constant flux, but they are often cyclical and therefore predictable on a decadal timescale. Whenever the channel bends toward one side of the inlet, the ebb tidal delta also shifts its sand to the barrier island on that side of the inlet and the island begins accreting seaward. The island on the opposite side of the inlet will conversely erode rapidly. When the channel shifts back, the pattern reverses. Some tidal inlets migrate alongshore as opposed to shifting back and forth. The barrier island that the inlet is migrating toward is steadily consumed, while the updrift island is lengthened at the same time.

Sediment Supply

The action of waves, wind, and tidal currents causes a continuous movement of coastal sediment. Net loss of sediment along any shoreline reach always results in coastal erosion. A net gain of sediment results in shoreline accretion or a seaward advance of the shoreline. Storms or other events that cause periods of larger waves and higher water levels cause net sediment loss by sweeping sand from the beach and dunes offshore to form offshore bars. Fair-weather waves nudge that sand back to the beach slowly, resulting in net sediment gain and widening of the beach.

Longshore transport

Most beaches experience a *longshore current*, which for swimmers means drifting slowly down the beach away from their beach towel as they swim in the surf. The longshore current is caused by waves approaching from any direction other than directly ashore—the wave crests are not parallel to the beach. Also, if waves are higher along one stretch of beach compared to an adjacent beach, the longshore current will flow along the beach from where the waves are high to where they are lower. Sand eroded from one beach can be transported along the beach by the resulting longshore drift.

Evidence of longshore transport is prevalent along *spits*, which are curved accumulations of sediment that extend across shallow bays. Along rocky coasts, where bays are too deep for spits to cross, sediment is nonetheless swept alongshore where it accumulates at the head of the bay. For many years, the accepted geologic wisdom was that all shorelines ultimately straighten out as waves erode sediment from headlands and fills in bays. The exception to this rule is *cuspate forelands* (e.g., Cape Hatteras, NC), which actually build seaward due to a large net gain of sediment from longshore transport. Because cuspate forelands jut out seaward, each side is sheltered from waves approaching from the opposite side. Thus, all waves and longshore transport converge toward the foreland, continuing its growth.[8]

Because beach sediment is often eroded from headlands and transported alongshore, the construction of seawalls along eroding headlands can also result in erosion downdrift because new sediment is blocked from entering the longshore drift. Delta shorelines are also starved of sediment if dams capture sediment upstream or flood-control levees prevent sediment from spreading over the delta. Heavy-mineral sand has long been mined directly from beaches and dunes where these dark minerals are concentrated in conspicuous layers (much like gold is concentrated in alluvial channels), resulting in a direct loss of sediment and subsequent erosion. All of these examples reveal that virtually all human intervention in the coastal zone is followed by reduced sediment supply and increased erosion somewhere.

RESPONSES TO SHORELINE EROSION

Shorelines are transient features due to the always-shifting nature of waves, currents, and wind along the coast. Beaches are always eroding or accreting. No harm is done if erosion is occurring in an unpopulated area. The "problem" of erosion only exists where it threatens human interests such as buildings or roads. Three major categories of erosion-control measures have been tried around the world—(1) hard stabilization, including seawalls, jetties, and groins; (2) soft stabilization, primarily beach nourishment; and (3) relocation.[9]

Hard Stabilization

Sea walls

Sea walls are shore-parallel walls designed to stop further landward erosion. Waves still impact beaches in front of

Fig. 4 View of the Galveston sea wall that was built after the 1900 hurricane that killed more than 6000 residents. One of the drawbacks of sea walls is seen here as wave run-up reaches the base of the wall, leaving no visible beach for recreation. **Source:** Adapted from Morton et al.[4]

the seawall, which almost inevitably results in the narrowing or complete disappearance of the usable beach (Fig. 4). Sea walls can be constructed from a wide variety of materials, including boulder revetments, concrete, steel, and wooden bulkheads, gabions, and sandbags as a few examples. For centuries, sea walls were the most preferred strategy to slow coastal erosion. Their negative effects on the beach environment—especially the loss of the dry "visible" beach—has led to a shift in strategy toward beach nourishment.[10] Sea walls are no longer permitted in many locations due to these concerns.

Jetties and groins

Jetties and groins are engineered structures that extend offshore allowing them to block the longshore transport of sand. Jetties are employed at tidal inlets to improve navigation by blocking sediment from entering the channel (Fig. 5). They also alter the flow of tidal currents and the size and position of the tidal deltas. They are frequently accompanied by dredging, which affects sediment transport in inlets with or without jetties. Instead of flowing to the next island, sediment will be trapped in the "hole" created by the dredging. Groins are built along beaches rather than inlets.

The primary problem with groins and jetties is that they interrupt the natural longshore transport of sediment. By trapping sediment on the updrift side, they cause sediment loss and erosion on the downdrift side. This may lead to the proliferation of more and more groins further downdrift. The jetties built at Ocean City, Maryland, in the 1930s and the groins built at Westhampton Beach, New York, in the 1960s demonstrate the impact of jetties and groins on the erosion of downdrift beaches.

Fig. 5 View of jetties at Humboldt Bay, California. The jetties interrupt longshore sediment transport, causing a net buildup of sediment on the north jetty (left) and a net loss to the south.
Source: Adapted from Campbell, R. U.S. Army Corps of Engineers Digital Visual Library. Retrieved from Wikimedia Commons, http://commons.wikimedia.org/wiki/File:Humboldt_Bay_and_Eureka_aerial_view.jpg (accessed March 2012).

Fig. 6 View of a nourished beach construction in Tybee Island near Savannah, Georgia. Sediment dredged from offshore is discharged from the pipeline and repositioned by the bulldozers.
Source: Adapted from Jordan, J. U.S. Army Corps of Engineers Digital Visual Library, http://eportal.usace.army.mil/sites/DVL/DVL%20Images/Forms/DispForm.aspx?ID=1146 (accessed March 2012).

Beach Nourishment

Beach nourishment is the most commonly employed response to erosion today, largely due to the negative aspects of hard stabilization. Beach nourishment simply involves the emplacement of sand to widen the beach. Although nourished beaches keep the natural ability to adjust to the dynamic ocean, their drawback is that they must be renourished frequently to maintain their width—as often as every 3 to 4 years. Carolina Beach, North Carolina, has been renourished more than 20 times since its first beach nourishment project.[11]

For beach nourishment, sand is typically dredged and pumped from a nearby tidal inlet, shipping channel, or offshore location, or else trucked in a land source. Grading equipment is subsequently used to design it into a predetermined shape (Fig. 6). In some instances, it is shaped into a wide berm a few feet above normal high tide. It may also be shaped into taller artificial dunes. Such an artificial dune was built along the North Carolina Outer Banks in the 1930s and 1940s, and these have been built extensively throughout the Netherlands today.

The limiting factor in beach nourishment is a sufficient supply of quality sand, which is not as readily available as one would think. Nourishment sand is usually of different quality than the natural beach sand, commonly being coarser or finer than the natural sand. It may also contain more mud, shells, or even rock depending on where the sand was obtained. Detailed geologic mapping of the continental shelf allows the identification of good-quality sand; however, the best sand is not always close to the beach, and lesser-quality sand is frequently used due to convenience and cost. Seafloor dredging may also alter wave patterns and currents, causing accelerated erosion of the nourished beach or adjacent beaches. In the Netherlands, sand cannot be dredged less than 2 km offshore to minimize the loss of beach sand offshore. Dredging sand from the seafloor also causes considerable damage to benthic habitats.

Relocation

Relocation is the movement of threatened structures to a more inland location. The most famous move in recent times occurred when the Cape Hatteras Lighthouse in North Carolina was moved in 1999 a half mile inland (Fig. 7). It was not the first move for the lighthouse—it had moved to its former location in the late 1800s from its original site. During the lighthouse's lifetime, the shoreline has eroded almost 1500 ft. To the north of Cape Hatteras, North Carolina Highway 12 also feels the impact of persistent, annual erosion. Numerous sections of the highway are buried by overwash each year as a result of frequent winter storms and occasional hurricanes. Severe storms, such as Hurricanes Irene (2011) and Isabel (2003) often destroy the road completely and cut new inlets through the island. In these instances, the road is commonly moved landward when it is rebuilt. The length of the road and the frequency of storms in North Carolina would make beach nourishment too expensive and impractical here.

More dramatic relocation efforts are already occurring worldwide and will likely become more prevalent in the future as sea level continues to rise. The Alaskan village of Shishmaref, home to about 500 mostly native Alaskans, will soon be moved to the mainland of Seward Peninsula in the Bering Strait.[12] The village has tried other measures of

Fig. 7 The Cape Hatteras Lighthouse was moved in 1999 from its former location (top of the photo) along the path through the center of the photo to the site cleared in the foreground. It is now 1600 ft from the shoreline.
Source: Adapted from United States National Park Service, http://www.nps.gov/caha/historyculture/images/LH-move-path-ariel.jpg (accessed March 2012).

erosion control for several decades but none provided a permanent solution. In the south Pacific, the 2000 residents of the Carteret Islands (officially part of Papua New Guinea) have already abandoned their coral atoll. Other low-lying Pacific atolls, including Kiribati, may also soon follow suit.

THE FUTURE OF SHORELINE EROSION

World eustatic sea level will be 30 cm or 1 ft higher a century from now if the present rise of 3 mm/yr continues unchanged. There is a growing scientific consensus based on improved observational methods that the Greenland ice sheet—which would raise sea level by 7 m (more than 20 ft) if completely melted—is melting at an accelerating rate. It is unlikely that the entire ice sheet will melt within the next century; however, continued acceleration of the ice sheet's shrinkage could easily raise sea level by 1 to 2 m (3–6 ft) by 2100.[12] This scenario would expose all coastlines to the rate of submergence now seen only in Louisiana and a handful of other places where local relative sea level rise has been dramatically accelerated by humans.

What does this mean for coastal erosion rates? It is a certainty that due to the local variability of topography, rock types, and wave and storm patterns that coastal erosion rates will vary enormously as the rate of sea level rise increases. But it is also a certainty that all of the world's coastlines and a significant portion of the world's population will deal with increased environmental, economic, and political stresses due to sea level rise. All of recorded civilization has developed and expanded during a time of stable sea level and coastlines. The last century has seen global population grow by 5 billion people. In 2011, the world population officially eclipsed 7 billion, roughly half of whom live in low-lying coastal areas. In addition to the thousands of Shishmarefs of the world, sea level rise will impact the Shanghais of the world as well—making it more difficult for hundreds of millions to get fresh water, grow food on fertile low-lying deltas, and protect homes from floods.

The last period of Earth's history that witnessed a rate of sea level rise of more than 1 m/century occurred when the last remnants of the Pleistocene ice sheets melted between 11,000 and 6000 years ago (Fig. 1). Sea levels rose more than 60 m (200 ft) over that period as coastlines worldwide retreated rapidly across the continental shelf, stopping roughly where they stand today. Based on the location of the 60 m depth contour on the continental shelf today, Atlantic coast shorelines would have eroded 15 times greater than average erosion rates today,[13] comparable to the extreme erosion rates in Louisiana.[4] In such a scenario, we would see an increasing number of vulnerable barrier island systems such as the North Carolina Outer Banks disintegrate much like the Chandeleur Islands.[14] Coastal marsh and wetland loss would greatly accelerate in places such as Chesapeake Bay and in already threatened and crowded deltas such as the Ganges, Mekong, Nile, Rhine, and Yangtze.

CONCLUSION

Although there is much uncertainty regarding future sea level rise and coastal erosion, the possibility of a catastrophic sea level rise is growing larger as we continue to see warmer temperatures in the Arctic and more rapid glacial melting. We may not yet grasp the full implications for coastal regions because we have no historical parallel with which to compare such a scenario. The traditional responses to coastal erosion such as sea walls and beach nourishment may not hold up to accelerating shoreline change. A long and methodical withdrawal from vulnerable shorelines is the most prudent approach for the future; however, this would require a level of long-term planning and cooperation—both political and economic and local, national, and global—that has only been rarely achieved in modern human society.

REFERENCES

1. Woodroffe, C.J. Ed.; *Coasts: Form, Process, and Evolution*; Cambridge University Press: Cambridge, UK, 2003.
2. Solomon, S.; Qin, D.; Manning, M.; Chen, Z.; Marquis, M.; Averyt, K.B.; Tignor, M.; Miller, H.L. Eds.; *Climate Change 2007: The Physical Science Basis. Contribution of Working Group I to the Fourth Assessment Report of the Intergovernmental Panel on Climate Change*; Cambridge University Press: Cambridge, UK, 2007.
3. Gibbons, H. Japan Lashed by Powerful Earthquake, Devastating Tsunami, Sound Waves, United States Geological Survey, March 2011, http://soundwaves.usgs.gov/2011/03/ (accessed March 2012).
4. Morton, R.A.; Miller, T.M.; Moore, L.J. *National Assessment of Shoreline Change: Part 1, Historical Shoreline Changes and Associated Coastal Land Loss along the U.S. Gulf of Mexico. Open file report 2004–1043*; United States Geological Survey: St. Petersburg, FL, 2004, http://pubs.usgs.gov/of/2004/1043 (accessed March 2012).
5. Pirazzoli, P.A. *World Atlas of Holocene Sea-Level Changes*; Elsevier Oceanography Series; Elsevier: Amsterdam, 1991; Vol. 58.
6. Sallenger, A.H., Jr. Storm impact scale for barrier islands. J. Coastal Res. **2000**, *16* (3), 890–895.
7. Sallenger, A.H., Jr.; Wright, C.W.; Lillycrop, W.J.; Howd, P.A.; Stockdon, H.F.; Guy, K.K.; Morgan, K.L.M. Extreme changes to barrier islands along the central Gulf of Mexico coast during Hurricane Katrina. *In Science and the Storms—The USGS Response to the Hurricanes of 2005*. Farris, G.S., Smith, G.J., Crane, M.P., Demas, C.R., Robbins, L.L., Lavoie, D.L., Eds.; United States Geological Survey Circular 1306, 2007; 113–118, http://pubs.usgs.gov/circ/1306/ (accessed March 2012).
8. Ashton, A.; Murray, A.B.; Arnoult, O. Formation of coastline features by large-scale instabilities induced by high-angle waves. Nature **2001**, *414*, 296–300.
9. Kaufman, W.; Pilkey, O.H. *The Beaches Are Moving: The Drowning of America's Shoreline*; Duke University Press: Durham, North Carolina, 1979.
10. Pilkey, O.H.; Wright, H.L., III. Seawalls versus beaches. J. Coastal Res. **1988**, *4*, 41–64.
11. Valverde, H.L.; Trembanis, A.C.; Pilkey, O.H. Summary of beach nourishment episodes on the U.S. East Coast Barrier Islands. J. Coastal Res. **1999**, *15* (4), 1100–1118.
12. Pilkey, O.H.; Young, R. *The Rising Sea*; Island Press: Washington, D.C., 2009.
13. CCSP. *Coastal Sensitivity to Sea-Level Rise: A Focus on the Mid-Atlantic Region.* A report by the U.S. Climate Change Science Program and the Subcommittee on Global Change Research. Titus, J.G. (Coordinating Lead Author), Anderson, K.E., Cahoon, D.R., Gesch, D.B., Gill, S.K., Gutierrez, B.T., Thieler, E.R., Williams, S.J. (Lead Authors); United States Environmental Protection Agency: Washington, D.C., U.S.A., 2009.
14. Riggs, S.R.; Ames, D.V.; Culver, S.J.; Mallinson, D.J. *The Battle for North Carolina's Coast: Evolutionary History, Present Crisis, and Vision for the Future*; University of North Carolina Press: Chapel Hill, NC, 2011.

Community Forestry: Sustainability and Equity Issues

Bill Buffum
Department of Natural Resources Science, University of Rhode Island, Kingston, Rhode Island, U.S.A.

Abstract
Governments around the world are increasingly handing over the authority for the management of national forests to the local communities. By 2002, governments had given communities legal rights to manage 380 million hectares of forest, most of which had been transferred during the previous 15 years. This approach has been particularly widespread in South Asia, where Nepal and India have the largest and oldest programs. More than 13,000 community forest user groups in Nepal are managing 25% of the national forests, while in India 53,000 forest protection committees are managing 18% of national forests. The experience of Nepal and India in the 1980s and 1990s has encouraged the development of community forestry programs in many other countries. Overall, community forestry programs have been highly successful at providing a wide range of economic and social benefits to the participating user groups. However, concerns have been raised about the ability of user groups to manage their community forests (CF) on a sustainable basis and share products in an equitable manner. This entry reviews the literature related to the sustainability and equity of forest management in CF.

INTRODUCTION

In recent years, many governments around the world have given local communities the authority to manage blocks of national forest.[1–3] As of 2002, governments had given communities legal rights to manage 380 million hectares of forest, 57% of which had been transferred during the previous 15 years.[4] This approach has been particularly widespread in South Asia. Nepal and India have the oldest and largest programs in the region: More than 13,000 community forest user groups in Nepal are managing 25% of the country's total national forests,[5] while in India 53,000 forest protection committees are managing 18% of the country's national forests.[6] The experience of Nepal and India in the 1980s and 1990s influenced the development of forestry programs in many other countries across the world.[7] Overall, community forestry programs have been highly successful at providing a range of economic and social benefits to the participating user groups.[1–3] However, concerns have been raised about the ability of user groups to manage their community forests (CFs) on a sustainable basis and share the CF products in an equitable manner.

OPERATION OF COMMUNITY FORESTS

Community forestry has recently been defined as "forestry practices which directly involve direct forest users in common decision making processes and implementation of forestry activities."[8] There are many variations on how CFs operate in different countries. The key feature is that a group of traditional users of a national forest is given legal rights to manage a specific block of national forests under the guidance of the national forestry department (FD). Typically, the users must prepare a community forestry management plan for approval by the FD. Then a CF management committee elected by the members assumes responsibility for the management of the CF. The CF management committee organizes regular meetings of the users to implement the CF management plan. An important activity in all CFs is patrolling to ensure that illegal harvesting of non-timber forest products or forest grazing is not taking place. In some CFs, one or more forest guards are hired, while in other CFs the members take turns patrolling the forest on a voluntary basis. Besides patrolling, other management activities include nursery management, tree-planting in degraded areas, thinning and tending operations, harvesting, as well as grazing and fire protection. When members require wood products, they request the CF management committee to issue a permit. The management committee verifies whether the member has already received his/her allocation of products according to the guidelines in the management plan. The CFs generate income from royalty payments from the harvesting of wood products and contributions from visitors. The funds are used for a variety of purposes, including providing loans and emergency support to members, constructing CF offices, providing meals during work days and meetings, and contributing to local development activities. In some countries, the CF user group is obligated to share some of the revenue with the FD.

Encyclopedia of Natural Resources DOI: 10.1081/E-ENRL-120051664

SUSTAINABILITY OF SILVICULTURE PRACTICES IN CFs

Community forestry programs have had a positive impact on forest cover and production of wood products. CFs have been found to exhibit greater increases in forest cover than government forests, according to studies based on analyses of satellite imagery and aerial photography[9–11] and forest inventories.[12,13] Community forestry has been linked to reduced rates of deforestation in countries such as Mexico.[14] Community forestry may support carbon sequestration as a function of increased timber stocks: a study of CFs in Nepal estimated that annual forest growth resulted in the sequestration of 4700 tCO_2e per year in 630 hectares.[15] Community forestry also supports the efforts of governments in forest protection by increasing community participation in forest management: Illegal logging on public lands has been reported to cost governments US$10–15 billion per year.[4]

However, there are concerns that some CF user groups may lack the technical skills and discipline to harvest their CFs in a sustainable manner.[3,4,16,17] Not all CFs are successfully managed: "members of some groups fail to perceive the growing scarcity of their local forests, fail to create effective rules to counteract the incentives to overharvest, and fail to enforce their own rules."[18] In Nepal, it was reported that some CF user groups did not understand the forest inventory process, and the lack of simple approaches to regulating the harvests led to over-harvesting, especially in CFs containing high-value timber.[19] Community ownership has been reported to have resulted in over-harvesting and increased rates of deforestation in Papua New Guinea.[4] On the other hand, a study in Bhutan found that user groups were managing their CFs in accordance with the management plans without any indications of overharvesting or overgrazing.[20,21]

There is concern that user group harvesting approaches, which are usually based on single tree selection, may not promote the natural regeneration of preferred species and maintain forest diversity. This issue has rarely been studied in CFs, but in other forests single tree selection was reported to reduce diversity and favor shade-tolerant species.[22,23] Another study reported that regeneration and stem density of important species had not recovered two decades after selective logging in India.[24] However, other studies have reported that single tree selection caused less damage to forests than harvesting systems that create large gaps in the forest canopy,[25] maintained tree diversity,[26–29] and was operationally more efficient than group selection.[30] Decisions about whether single tree selection is appropriate for a particular CF will have to be based on the silvics of the concerned tree species.

Additionally, there is concern that user groups may high-grade the CFs through excessive removal of the straightest stems of economically valuable species, resulting in residual stands containing crooked and otherwise deformed overstory trees and a genetic selection of inferior genotypes.[31–33] This process of "high-grading," which occurs when forest managers maximize current profits rather than long-term productivity, has been well documented in North America,[34,35] but is widespread in many other regions.

Surprisingly, the sustainability of harvesting approaches in CFs has rarely been addressed by researchers: A recent review of participatory forest management (PFM) concluded that "the most conspicuous absence from the PFM literature is a thorough analysis of forestry itself, the technology of intervention in a forest ecosystem to provide a reliable supply of the products and benefits required by the users."[1] A number of authors have noted that professional foresters have not yet been able to provide silvicultural guidance to meet the diverse needs of community forestry user groups.[1,3,36,37] Victor[37] summed up this issue as follows: "Unfortunately, the development and spread of alternative forest management practices (barefoot silviculture) for CF management has yet to really begin. Most foresters' knowledge is limited to mensuration and silviculture developed for timber production at an industrial scale. As a result, it is difficult for government foresters to readily provide appropriate advice to users and to recognize good and bad community forestry practice or to suggest feasible improvements."

With the rapid expansion of community forestry in developing countries, it is important to assess the sustainability of CF harvesting practices, which are generally based on indigenous knowledge and traditional practices of selection cutting. If these harvesting practices are not sustainable, there is an urgent need for technical assistance and training.

EQUITY IN COMMUNITY FORESTS

Despite the many positive achievements of CFs, there is increasing evidence that community forestry programs do not always provide benefits equally to all social groups, and in some cases CFs may actually increase inequality.[7,38] The emerging equity issues are part of what the donor and practitioner literature refers to as "second-generation" issues of community forestry.[1] Two aspects of equity apply to community forestry; political equity, which involves participation in decision making and the ability of stakeholders to express their ideas and concerns, and economic equity, which involves the distribution of benefits.[39] In the context of community forestry, equity is usually understood to refer to "fairness" rather than "equality": The management of a CF is considered to be equitable if the members view their share of use and decision-making rights as being fair.[40]

Several studies have investigated the lack of political equity (participation in decision making) in CFs. Menzies[2] noted that "traditional societies are rarely equitable and participatory … there may be aspects of traditional social capital that do not favor 'just and wise' management of forest resources." Management committees in many countries were dominated by political, social, or economic elites and thus resulted in "elite capture" of the CF management process by men and high caste groups.[16,41] Cultural norms have discouraged women from participating in CF meetings,[42] and women were excluded from decision making even in seemingly participatory user groups.[43] However, there are also examples of positive political equity in CFs: In contrast to the above studies, some authors have reported the positive impacts of increased understanding and collaboration in CFs between dominant and disadvantaged ethnic groups, and between men and women,[44–46] and other studies have noted that the involvement of elites in community development activities does not necessarily lead to elite capture.[47,48]

Other studies have documented a lack of economic equity (distribution of benefits) in some CFs. For example, some management committees in Nepal practiced favoritism in distributing products;[42,49] CF rules were biased toward meeting the needs of richer households;[50] and influential members were allowed to graze their animals in CFs even when the bylaws of the CF prohibited forest grazing.[51] In other CFs, the poorest members of the user groups had less access to CF products than the richer and higher caste members,[52–56] and benefited less from loans from CF user group funds.[57]

Equity has environmental consequences in addition to social consequences: A study of CFs in the Indian Himalayas found that improved forest condition was associated with reduced levels of conflict among users and greater involvement of women in decision making. The CFs that did not have internal conflicts were better able to make effective decisions about forest protection, and women were able to make sound management decisions due to their strong involvement in the collection of forest products.[58] Thus, national governments have many reasons for adopting community forestry programs that ensure equity in CFs. The linkage between equity and conservation is reflected in the objectives of the UN Convention on Biological Diversity, which include the conservation of biological diversity, sustainable use of its components, and fair and equitable benefit sharing.[59]

ATTRIBUTES OF CFs AND NATIONAL FORESTRY POLICIES THAT RESULT IN THE SUCCESSFUL MANAGEMENT OF CFs

Several authors have conducted multi-country studies to identify the attributes of successful user groups in an attempt to understand why some forestry user groups are better able than others to manage their forests sustainably and equitably.[40,60–62] These studies are complicated by different views of what constitutes success in a CF. The most common indicator of successful forest management is improved forest condition, which was used in recent analyses of forest governance in the Indian Himalaya[58] and Nepal.[63] However, other indicators of success, such as equity of benefit sharing or fulfillment of local needs, may be equally important.[61] For example, some CFs in Nepal exhibited significant increases in forest cover but produced few benefits for the user groups, and the CF products were unfairly distributed among the members, especially to poor users who lacked access to forest products on private land.[64]

The set of positive attributes prepared by Ostrom[60] (Table 1) is more detailed than other recent analyses of forestry user groups.[58,61,62,65] Ostrom's list includes attributes of user groups that are conducive to a community adopting or changing rules, limiting access to their forests.[60] These attributes are relevant because CF user groups in many countries must initiate the process of forming a CF even though the government subsequently plays a significant role in supporting the user group. Ostrom's list also includes several attributes of CFs, including that the forest should not be so degraded that it is useless to organize,

Table 1 Summary of Ostrom's attributes of successful user groups

The forest should not be so degraded that it is useless to organize, or so underutilized that there is little advantage from organizing.
Reliable and valid information about the general condition of the resource should be available at reasonable costs.
The availability of resource units should be relatively predictable.
The resource should be sufficiently small, given the transportation and communication technology in use, so that users can develop accurate knowledge of external boundaries and internal microenvironments.
Users should be dependent on the resource for a major portion of their livelihood or other variables of importance to them.
Users should have a shared image of the resource and how their actions affect each other and the resource.
Users should have a sufficiently low discount rate in relation to future benefits to be achieved from the resource.
Users with higher economic and political assets should be similarly affected by a current pattern of use.
Users should trust each other to keep promises and relate to one another with reciprocity.
Users should be able to determine access and harvesting rules without external authorities countermanding them.
Users should have learned at least minimal skills of organization through participation in other local associations or learning from neighboring groups.

or so underutilized that there is little advantage from organizing. Governments in many countries have been unwilling to hand over well-stocked forests for community management.[2,3] For example, forest policies in Nepal since 2000 have restricted the handover of well-stocked forests in the Terai/Churia regions while continuing to promote the handover of relatively low-value forests in the mid-hills.[6,44] The Forest Department in India reportedly retained control of the most productive forest land and allocated fragmented and degraded patches for community management.[6] In Bhutan, however, after initially only handing over degraded forests to communities, the government decided to give villagers a stronger incentive to participate by stipulating that CFs should be approximately 50% well-stocked.[66,67] This provision was probably influenced by the abundance of forest resources in Bhutan, but appears to reflect a genuine willingness to give up control of some national forests in order to promote economic development at the village level.[68]

CONCLUSIONS

Community forestry programs are expanding throughout the world. The experience to date has largely been positive: CFs in many countries have been able to provide a range of social and economic benefits, and many CF user groups have demonstrated that they are capable of effectively managing the forest resources in their CFs. However, there still are questions about the sustainability of harvesting and management systems in CFs, and the lack of political and economic equity in some CFs is troubling. It is crucial that additional research be conducted on these issues to provide guidance for the development of national community forestry policies and programs so that appropriate technical support is provided to CF user groups to maximize their ability to manage CFs in a sustainable and equitable manner.

REFERENCES

1. Lawrence, A. Beyond the second generation: Towards adaptiveness in participatory forest management. Vet. Sci. Nutr. Nat. Resour. **2007**, *2* (28), 1–15.
2. Menzies, N.K. *'Global Gleanings' Lessons from Six Studies of Community Based Forest Management*, Report for Ford Foundation's Environment and Development Affinity Group (EDAG); Ford Foundation: Berkley, 2002, p 31.
3. Carter, J. *Recent Experience in Collaborative Management - CIFOR Occasional Paper No. 43*; Centre for International Forestry Research: Bogor, 2005, p 48.
4. White, A.; Martin, A. *Who Owns the World's Forests? Forest Tenure and Public Forests in Transition*; Center for International Environmental Law: Washington, 2002; p 30.
5. Pokharel, B.K.; Paudel, D. Impacts of armed conflicts on community forest user groups in Nepal: Can community forestry survive and contribute to peace building at local level? Eur. Trop. For. Res. Netw. **2005**, *43/44* (Winter 2005/2005), 83–86.
6. Agrawal, A.; Ostrom, E. Collective action, property rights, and decentralization in resource use in India and Nepal. Polit. Soc. **2001**, *29* (4), 485–514.
7. Hobley, M. Building state-people relationships in forestry. In *Forest Policy and Environment Programme: Grey Literature*; Forest Policy and Environment Programme, Overseas Development Institute: London, 2005; p 7.
8. Maryudi, A.; Devkotab, R.R.; Schusserb, C.; Yufanyib, C.; Sallab, M.; Aurenhammerb, H.; Rotchanaphatharawitb, R.; Krott, M. Back to basics: Considerations in evaluating the outcomes of community forestry. For. Policy Econ. **2012**, *14* (1), 1–5.
9. Gautam, A.P.; Shivakoti, G.P.; Webb, E.L. Forest cover change, physiography, local economy, and institutions in a mountain watershed in Nepal. Environ. Manag. **2004**, *33* (1), 48–61.
10. Gautam, A.P.; Webb, E.L.; Eiumnoh, A. GIS assessment of land use/land cover changes associated with community forestry implementation in the Middle Hills of Nepal. Mt. Res. Dev. **2002**, *22* (1), 63–69.
11. Sakurai, T.; Rayamajhi, S.; Pokharel, R.K.; Otsuka, K. Efficiency of timber production in community and private forestry in Nepal. Environ. Dev. Econ. **2004**, *9* (4), 539–561.
12. Branney, P.; Yadav, K.P. *Changes in Community Forest Condition and Management 1994–1998: Analysis of Information from the Forest Resource Assessment Study and Socio-economic Study in the Koshi Hills, Nepal*, Project Report No. G/NUCFP/32; UK Community Forestry Project: Kathmandu, 1998; p 64.
13. Yadav, N.P.; Dev, O.P.; Springate-Baginski, O.; Soussan, J. Forest management and utilization under community forestry. J. For. Livelihood **2003**, *3* (1), 37–50.
14. Barsimantov, J.; Kendall, J. Community forestry, common property, and deforestation in eight Mexican states. J. Environ. Dev. **2012**, *21* (4), 414–437.
15. Bhattarai, T.; Skutsch, M.; Midmore, D.J.; Rana, E.B. The carbon sequestration potential of community based forest management in Nepal. Int. J. Clim. Change **2012**, *3* (2), 233–254.
16. Kellert, S.R.; Mehta, J.N.; Ebbin, S.A.; Lichtenfeld, L.L. Community natural resource management: Promise, rhetoric, and reality. Soc. Nat. Resour. **2000**, *13* (8), 705–715.
17. Roberts, E.H.; Gautam, M.K. *Community Forestry Lessons for Australia: A Review of International Case Studies*; The Australian National University: Canberra, 2003; p 16.
18. Gibson, C.C.; McKean, M.A.; Ostrom, E., Eds. Some initial theoretical lessons. In *People and Forests: Communities, Institutions, and Governance*; MIT Press: Cambridge, Mass, 2000; 227–242.
19. Maharjan, M.R. Yield regulation techniques for community forest management in Nepal: Opportunities and constraints. In *Cultivating Forests - Alternative Forest Management Practices and Techniques for Community Forestry*; Victor, M., Barash, A., Eds.; RECOFTC: Bangkok, 2001; 189–197.
20. Buffum, B.; Gratzer, G.; Tenzin, Y. The sustainability of selection cutting in a late successional broadleaved community forest in Bhutan. For. Ecol. Manag. **2008**, *256* (12), 2084–2091.
21. Buffum, B.; Gratzer, G.; Tenzin, Y. Forest grazing and natural regeneration in a late successional broadleaved community forest in Bhutan. Mt. Res. Dev. **2009**, *29* (1), 30–35.

22. Jenkins, M.A.; Parker, G.R. Composition and diversity of woody vegetation in silvicultural openings of southern Indiana forests. For. Ecol. Manag. **1998**, *109* (1), 57–74.
23. Schuler, T.M. Fifty years of partial harvesting in a mixed mesophytic forest: Composition and productivity. Can. J. For. Res. **2004**, *34* (5), 985–997.
24. Ganesan, R.; Davidar, P. Effect of logging on the structure and regeneration of important fruit bearing trees in a wet evergreen forest, southern Western Ghats, India. J. Trop. For. Sci. **2003**, *15* (1), 12 25.
25. Rice, R.E.; Gullison, R.E.; Reid, J.W. Can sustainable management save tropical forests? Sci. Am. **1997**, *276* (4), 44–49.
26. Cannon, C.H.; Peart, D.R.; Leighton, M. Tree species diversity in commercially logged Bornean rainforest. Science **1998**, *281* (5381), 1366–1368.
27. Dekker, M.; de Graaf, N.R. Pioneer and climax tree regeneration following selective logging with silviculture in Suriname. For. Ecol. Manag. **2003**, *172* (2–3), 183–190.
28. Montagnini, F.; Eibl, B.; Szczipanski, L.; Ríos, R. Tree regeneration and species diversity following conventional and uniform spacing methods of selective cutting in a subtropical humid forest reserve. Biotropica **1998**, *30* (3), 349–361.
29. Webb, E.L.; Peralta, R. Tree community diversity of lowland swamp forest in Northeast Costa Rica, and changes associated with controlled selective logging. Biodiv. Conserv. **1998**, *7* (5), 565–583.
30. Suadicani, K.; Fjeld, D. Single-tree and group selection in montane Norway spruce stands: Factors influencing operational efficiency. Scand. J. For. Res. **2001**, *16* (1), 79–87.
31. Clatterbuck, W.K. *Treatments for Improving Degraded Hardwood Stands. Publication FOR-104*; University of Kentucky: Knoxville, 2006.
32. Nyland, R.D. Rehabilitating cutover stands: Some ideas to ponder. In *Proceedings of the Conference on Diameter-Limit Cutting in Northeastern Forests*; Northeastern Research Station, U.S. Forest Service: Amherst, MA, 2006; 47–51.
33. Oliver, C.D.; Larson, B.C. *Forest Stand Dynamics*, Update Ed.; Wiley: New York, 1996; xviii + 520 p.
34. Kenefic, L.S.; Nyland, R.D. *Diameter-Limit Cutting and Silviculture in Northeastern Forests*; Forest Service, United States Department of Agriculture: Newtown Square, 2005.
35. Kenefic, L.S.; Nyland, R.D. Facilitating a dialogue about diameter-limit cutting. In *Proceedings of the Conference on Diameter-Limit Cutting in Northeastern Forests*; Northeastern Research Station, U.S. Forest Service: Amherst, MA, 2006; 1–3.
36. Donovan, D.G. Where's the forestry in community forestry? In *Cultivating Forests: Alternative Forest Management Practices and Techniques for Community Forestry*; RECOFTC: Bangkok, Thailand, 2001; 3–23.
37. Victor, M. The search for a barefoot silviculture: Too little, Too early? In *Cultivating Forests: Alternative Forest Management Practices and Techniques for Community Forestry*; RECOFTC: Bangkok, Thailand, 2001; vi–xviii.
38. Schreckenberg, K.; Luttrell, C.; Moss, C. Participatory forest management: An overview. In *Forest Policy and Environment Programme: Grey Literature*; Forest Policy and Environment Programme, Overseas Development Institute: London, 2006; p 7.
39. Mahanty, S.; Fox, J.; Nurse, M.; Stephen, P.; McLees, L., Eds. Equity in community-based resource management. In *Hanging in the Balance: Equity in Community-Based Natural Resource Management in Asia*; RECOFTC and East-West Center: Bangkok, 2006; 1–13.
40. McKean, M.A. Common property: What is it, What is it good for, and what makes it work? In *People and Forests: Communities, Institutions, and Governance*; Gibson, C.C., McKean, M.A., Ostrom, E., Eds.; MIT Press: Cambridge, MA, 2000; 27–55.
41. Uprety, D.R. Conflicts in natural resource management – Examples from community forestry. Jahrbuch der Österreichischen Gesellschaft für Agrarökonomie **2006**, *15*, 143–155.
42. Springate-Baginski, O.; Dev, O.P.; Yadav, N.P.; Soussan, J. Community forest management in the middle hills of Nepal: The changing context. J. For. Livelihood **2003**, *3* (1), 5–10.
43. Agarwal, B. Participatory exclusions, community forestry, and gender: An analysis for South Asia and a conceptual framework. World Dev. **2001**, *29* (10), 1623–1648.
44. Bhattarai, B. Widening the gap between terai and hill farmers in Nepal: The implications of the New Forest Policy 2000. In *Hanging in the Balance: Equity in Community-Based Natural Resource Management in Asia*; Mahanty, S., Fox, J., Nurse, M., Stephen, P., McLees, L., Eds.; RECOFTC and East-West Center: Bangkok, 2006; 143–161.
45. Buffum, B.; Lawrence, A.; Temphel, K.J. Equity in community forests in Bhutan. Int. For. Rev. **2010**, *12* (3), 187–199.
46. Luintel, H. Do civil society organizations promote equity in community forestry? A reflection from Nepal's experiences. In *Hanging in the Balance: Equity in Community-Based Natural Resource Management in Asia*; Mahanty, S., Fox, J., Nurse, M., McLees, L., Eds.; RECOFTC and East-West Center: Bangkok, 2006; 122–142.
47. Dasgupta, A.; Beard, V.A. Community driven development, collective action and elite capture in Indonesia. Dev. Change **2007**, *38* (2), 229–249.
48. Mansuri, G.; Rao, V. Community-based and -driven development: A critical review. World Bank Res. Obs. **2004**, *19* (1), 1–39.
49. Varalakshmi, V. Joint forest management and conflict in Haryana, India. In *Community-Based Forest Resource Conflict Management*; Means, K., Josayma, C., Eds.; Food and Agriculture Organization, United Nations: Rome, 2002; p 321.
50. Adhikari, B. Poverty, property rights and collective action: Understanding the distributive aspects of common property resource management. Environ. Dev. Econ. **2005**, *10* (1), 7–31.
51. Pandit, B.H.; Thapa, G.B. Poverty and resource degradation under different common forest resource management systems in the mountains of Nepal. Soc. Nat. Resour. **2004**, *17* (1), 1–16.
52. Adhikari, B.; Di Falco, S.; Lovett, J.C. Household characteristics and forest dependency: Evidence from common property forest management in Nepal. Ecol. Econ. **2004**, *48* (2), 245–257.
53. Chhetry, B.; Francis, P.; Gurung, M.; Iversen, V.; Kafle, G.; Pain, A.; Seeley, J. A framework for the analysis of community forestry performance in the terai. J. For. Livelihood **2005**, *4* (2), 1–16.
54. Malla, Y.; Neupane, H.R.; Branney, P.J. Why aren't poor people benefiting more from community forestry? J. For. Livelihood **2003**, *3* (1), 78–90.

55. Nightingale, A. Nature-society and development: Social, cultural and ecological change in Nepal. Geoforum **2003**, *34* (4), 525–540.
56. Sarin, M. *Who is Gaining? Who is Losing? Gender and Equity Concerns in Joint Forest Management*; Society for the Promotion of Wastelands Development: New Delhi, 1998; p 81.
57. Pokharel, B.; Nurse, M. Forest and people's livelihood: Benefiting the poor from community forestry. J. For. Livelihood **2004**, *4* (1), 19–30.
58. Agrawal, A.; Chhatre, A. Explaining success on the commons: Community forest governance in the Indian Himalaya. World Dev. **2006**, *34* (1), 149–166.
59. Knapp, S. Dynamic diversity. Nature **2003**, *422*, p 475.
60. Ostrom, E. Self-Governance and Forest Resources, CIFOR Occasional Paper No. 20; CIFOR: Jakarta, 1999; p 15.
61. Pagdee, A.; Kim, Y.-S.; Daugherty, P.J. What makes community forest management successful: A meta-study from community forests throughout the world. Soc. Nat. Resour. **2006**, *19* (1), 33–52.
62. Persha, L.; Agrawal, A.; Chhatre, A. Social and ecological synergy: Local rulemaking, forest livelihoods, and biodiversity conservation. Science **2011**, *331* (6024), 1606–1608.
63. Nagendra, H. Tenure and forest conditions: Community forestry in the Nepal Terai. Environ. Conser. **2002**, *29* (4), 530–539.
64. Shrestha, K.K.; McManus, P. The politics of community participation in natural resource management: Lessons from community forestry in Nepal. Aust. For. **2008**, *71* (2), 135–146.
65. Nagendra, H. Drivers of reforestation in human-dominated forests. Proc. Natl. Acad. Sci. **2007**, *104* (39), 15218–15223.
66. Chhetri, B.B. Review of experiences of social forestry and forestry extension in Bhutan. In *Social Forestry and Forestry Extension - Report of the National Workshop on Social Forestry and Forestry Extension*, UNDP/FAO Forest Resources Management and Institutional Development Project; Ministry of Agriculture, Royal Government of Bhutan: Lingmethang, 1992; 14–29.
67. Upadhyay, K. Review of community forest/community protected forest rules. In *Social Forestry and Forestry Extension - Report of the National Workshop on Social Forestry and Forestry Extension*, UNDP/FAO Forest Resources Management and Institutional Development Project; Department of Forestry, Ministry of Agriculture, Royal Government of Bhutan: Lingmethang, 1992; 92–103.
68. Buffum, B. Why is there no tragedy in these commons? An analysis of forest user groups and forest policy in Bhutan. Sustainability **2012**, *4* (7), 1448–1465.

Conserved Lands: Stewardship

Peter V. August
Department of Natural Resources Science, University of Rhode Island, Kingston, Rhode Island, U.S.A.
Janet Coit
Rhode Island Department of Environmental Management, Providence, Rhode Island, U.S.A.
David Gregg
Rhode Island Natural History Survey, Kingston, Rhode Island, U.S.A.

Biomes—Ecosystems

Abstract
Land that has been conserved for natural resource protection requires careful and ongoing management. There are many factors that can degrade protected lands. These include changes in native plant or animal communities caused by pests, pathogens, or invasive species; poaching; illegal resource harvesting; disturbance to plants and animals from intensive human use; malicious destruction of resources; and ecological changes in nearby landscapes. Every protected property, regardless of size, must have an explicit conservation goal, a management plan, stewardship consistent with the management plan, monitoring to determine if management goals are being achieved, and adaptive adjustment of management plans if goals are not being met or new threats to the integrity of the protected land emerge.

INTRODUCTION

Acquiring ownership of, or development rights to land is one of the most effective ways of conserving natural and cultural resources at local, regional, and national scales. Once acquired, protected land requires constant stewardship. Land protection is done for many reasons: to protect the fauna and flora of a site, to protect cultural resources, to ensure that ecosystem services are maintained, to allow sustainable use of natural resources of the site, and to afford public access for recreational purposes.[1,2] Land conservation occurs at multiple scales, in all ecosystems (terrestrial, aquatic, and marine), and is accomplished by many kinds of institutions and public agencies. National governments protect watersheds, forests, and rangelands for resource protection and in many cases, for sustainable resource harvesting. National governments also conserve large national park systems for the benefit of biota and present and future generations of citizens. Conservation occurs at smaller jurisdictions as well; for example, states, counties, and towns. Large nonprofit organizations, such as The Nature Conservancy and the Audubon Society in the United States, are very effective in protecting land, as are small nonprofit organizations such as local land trusts and conservancies. Regardless of the size of the conservation organization or the property preserved, all protected lands require ongoing management and stewardship.

The consequences of not stewarding protected lands can jeopardize the very resources that are meant to be preserved. Developing and implementing a management plan for protected land is one good way of cataloging and prioritizing stewardship responsibilities and we review steps in management planning here. Threats to protected lands can rapidly change; therefore, monitoring protected lands is an important part of the process. Finally, the technical knowledge within a conservation organization, the availability of staff and equipment, and funding for protected land management are frequently inadequate for the magnitude of the task, thus, innovative collaborations and efficient implementation of management plans are necessary.

Protected land stewardship is a challenging endeavor and has many components which will be reviewed here. Large government agencies or other organizations that own and manage large areas of protected lands typically have dedicated programs for stewardship activities. For example, the National Park Service (NPS) oversees a large and complex network of protected areas in the United States. Every park in the NPS system has a General Management Plan, which articulates the management needs of a park and how the park will meet those needs.[3] Furthermore, the NPS has a sophisticated program—the NPS Inventory and Monitoring Program—to systematically monitor environmental conditions in parks to know if management and stewardship activities are having the desired results and to be vigilant to new or unforeseen threats to the ecological integrity of a park.[4] Other conservation agencies that oversee large areas of land have similar programs such as the U.S. Fish and Wildlife Services,[5] U.S. Bureau of Land Management,[6] and U.S. Forest Service.[7] Small conservation organizations, such as local conservancies and land trusts, typically do not have dedicated staff or program resources for protected

Encyclopedia of Natural Resources DOI: 10.1081/E-ENRL-120047473

land stewardship, yet they own significant areas of land. In the United States, for example, there are 1700 different land trusts that control over 150,000 km^2 of land.[5] The focus of this entry is to describe the conservation land stewardship process that would be followed by a small conservation organization.

LAND PROTECTION AND STEWARDSHIP

There are a variety of reasons to protect land and many stewardship strategies that can be used to meet conservation goals (Table 1). Conservation lands are typically protected by securing fee simple ownership, control over future development rights, or establishment of permitted and restricted uses of the land through zoning controls.[2] However a property is controlled, the development and implementation of an ongoing management plan for the protected property are essential. There are many threats to the ecological integrity of protected lands (Table 2, Fig. 1). Pests, pathogens, and invasive species can impact native fauna and flora of a site. Illegal poaching or harvesting natural resources can diminish the biota. Soil compaction, erosion, and vegetation disturbance in fragile ecosystems caused by motor vehicle riding (e.g., all-terrain vehicles, motorcycles) can result in serious environmental damage. Malicious acts, such as garbage dumping, littering, theft of cultural resources, and vandalism, can diminish the esthetic value of a site. A carefully designed management plan will be attentive to these threats. Protected land management should happen in a systematic manner (Fig. 2). The basic steps are discussed in detail in the coming sections.

Site Assessment and Baseline Inventory

An essential first step in conservation land management is a site (the parcel) and landscape (what is around it) survey to inventory current conditions and catalog important habitats. The purpose of the site assessment is to evaluate the presence and condition of important natural resources and to identify threats to the focal resources and the ecosystem as a whole. Target resources can be species, habitats, rivers, landforms, viewscapes, farms, cultural resources, groundwater, or ecosystem services and ecological processes.[8,9] Site assessment can be a complex activity and require personnel familiar with local ecological conditions. Assessment involves field survey and consolidation of relevant geospatial information such as data from Geographic Information Systems (wetlands, rare species occurrences, land use, soils, landform, etc.), digital imagery, and the results of previous reconnaissance of the area if available. The initial site assessment establishes a baseline condition for easement monitoring and tracking changes in the condition of the property. Landscape analysis provides a regional context for conservation and provides insight in gains and losses of dispersal corridors, up-watershed threats that would jeopardize site-level habitats or species, and land use changes on nearby properties that might enhance or diminish the conservation value of a particular parcel.[10] There is a growing number of protocols that have been advanced to perform a baseline inventory.[11,12]

Development of Management Goals and Plan

Development of management goals is an essential step and the goals of the management plan will reflect the values of

Table 1 Examples of common conservation goals and management activities that achieve them

Goal	Management and stewardship actions
Preserve biodiversity	Protect large tracts of land. Connect separate refuges with corridors. Increase the size of refuges. Enforce policies against poaching wildlife or harvesting plants. Ensure adjacent land uses are not a source of invasive species, pests, or pathogens. Monitor for invasive species, pests, and pathogens. Monitor population levels of key or indicator species of plants and animals.
Preserve cultural resources	Protect sites or regions of interest; limit public access to sensitive sites.
Preserve ecosystem services	Ensure that conservation land boundaries encompass whole watersheds, maintain diversity of habitats, and allow public access and nondestructive forms of recreation so that patrons can benefit from esthetic values.
Preserve aesthetic or recreational values	Provide trails and interpretive services for public access; encourage hiking, hunting, and fishing if appropriate for the site; manage viewscapes and soundscapes to preserve natural conditions.
Ensure sustainable resource use	If a site permits, allow controlled harvest of sustainable resources, such as wood products, game and fish, plants, fruits, berries, and fungi.
Agricultural preservation	Obtain development rights and easements to working lands and waters to ensure they will remain in agricultural land use.

Table 2 Examples of common threats to resources on conserved lands

Resource	Threat
Biodiversity	Habitat destruction, poaching; illegal harvest of plants or animals; spread of invasive species, pests, and pathogens; loss of fitness due to small population sizes; illegal motor vehicle access in sensitive habitats.
Cultural resources	Vandalism, theft, and alteration of landscape context (e.g., viewscapes, soundscapes).
Ecosystem services	Groundwater withdrawal outside refuge, habitat destruction on borders of refuge, and high-impact land uses outside refuge in watershed.
Aesthetic or recreational values	Trespassing, illegal motor vehicle access, dumping, degradation of natural viewscapes or soundscapes.

Fig. 1 Examples of threats to protected lands: (**A**) Vandalism (destruction of signage); (**B**) illegal dumping of refuse; (**C**) all-terrain vehicle track damage to wetland habitat.
Source: Photo courtesy, The Rhode Island Chapter of The Nature Conservancy.

the conservation institution and the purpose for acquiring the conservation land.[13] Common management goals include stewarding the land for water (ground and/or surface) protection, conservation of biodiversity, forestry, farming, esthetic values, fish and wildlife management, and recreation (hiking, paddling, fishing, hunting, etc.). Management goals of one property can affect the viability of nearby conservation properties owned by others; hence there is a need for coordinated management when protected lands are in a mosaic of small parcels managed by different institutions. It is important that a management plan establish a basis or rationale for prioritizing resources

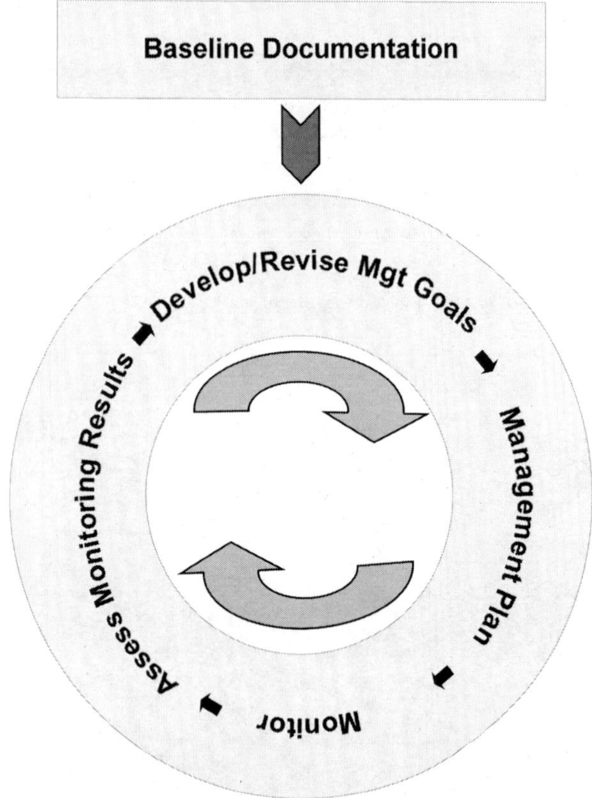

Fig. 2 Management cycle of protected lands: Steps involved in managing protected lands.

and stewardship actions, so that scarce resources are applied over time can still achieve large, long-term goals. Once goals for a property are established, a management plan can be developed and clear measures of success must be identified. These measures are the basis for ongoing monitoring and are the indicators of success or failure of the management plan.

Implementation of Management Plan

Management plans can vary in complexity; simple management plans for conservation lands may require little activity beyond periodic monitoring. Other stewardship activities, however, could require considerable work, expertise, and investment in personnel, supplies, equipment, and time in the field. Examples of expensive and complex management tasks include habitat restoration, creating and maintaining trail systems, forest management (selective cutting to maintain the age and target species composition of the forest), and removal or control of invasive species.[14,15] The implementation of a management plan can be a challenge for conservation organizations. Large land owners do not always have the staff or resources to manage extensive properties. Similarly, small, local conservation organizations, such as land trusts, rarely have the technical wherewithal or financial ability to take on complex management activities. Partnerships and collaborative projects are one way to perform complex stewardship tasks.

Monitoring

Vigilant monitoring is required of conservation properties and the land surrounding them. Easements must be monitored on a regular basis to ensure that fee owners are managing properties in a manner that is consistent with the easement that is held by the conservation organization. Violations in easements will have legal and policy ramifications which will need to be addressed.

All properties must be visited on a regular basis to protect against adverse possession claims and to identify inappropriate human activities occurring on or near them. Vandalism, dumping, erosion caused by motor vehicles, illegal hunting, and wood cutting are, unfortunately, common on conservation lands (Fig. 1). Monitoring for disturbances such as these is relatively straightforward for small properties but requires constant attention. Monitoring over large, expansive conservation lands can be logistically difficult, especially if access is difficult.[16] Ecological monitoring must be done on and around conservation lands to ensure that invasive plants and animals have not become established, pests or pathogens are not present, stewardship activities are yielding desired outcomes, target species are still present, and ecosystem health remains high. Regional changes in the creation of impervious surfaces, which increase storm water runoff, water withdrawal in the watershed, land use conversion, and habitat fragmentation can degrade the condition and viability inside conservation lands.

Synthesis, Reflection, and Adaptive Stewardship

Monitoring provides the data from which decisions about the efficacy of the management plan can be made. A careful analysis of monitoring data determines if the management plan is working and the ecological condition of the property is changing. If the desired results are not occurring, the management plan should be modified to meet the conservation goals of the property. This is a critical step and follows the logic of the adaptive management paradigm.[17] New, unanticipated stewardship challenges can emerge rapidly. Management plans must be dynamic documents and capable of changing as knowledge or needs require.

CONCLUSIONS

Acquiring land is an effective way to protect natural resources and careful stewardship of conserved lands ensures that the condition of the natural resources is preserved. Stewardship is an ongoing process and a long-term commitment. Many factors can diminish the value of protected land and these must be monitored, and when present, mitigated. The process of protected land management has a number of steps and begins with a careful resource inventory of the protected property and establishment of a suite of management goals that will direct stewardship activities. Ongoing monitoring to ensure that management goals are being met is an essential component of the process.

Large land owners, such as U.S. NPS or U.S. Forest Service, have very complex management programs to ensure that the property in their care retains the environmental and cultural resources the lands were obtained to protect. Small land owners, such as land trusts and local conservancies, frequently do not have dedicated staff resources, knowledge, or budgets to steward their land, but innovative partnerships are one way the resources of many institutions can be leveraged to achieve effective land stewardship even by small land owners. One model is the Rhode Island Conservation Stewardship Collaborative (RICSC),[18] an alliance of federal, state, municipal, and nonprofit conservation organizations who partner to (from its mission statement) "... *advance long-term protection and stewardship of terrestrial, aquatic, coastal, estuarine, and marine areas in Rhode Island that have been conserved by fee, easement, or other means.*" The RICSC tackles systemic, state-wide, impediments to good conservation land stewardship and provides training materials, protocols, and technical capacity to assist conservation land owners in their stewardship challenges.

ACKNOWLEDGMENTS

The following individuals have provided us thoughtful guidance and support for our work in conservation land management and stewardship: Julie Sharpe, Peggy Sharpe, Rupert Friday, Cathy Sparks, Scott Ruhren, Sharon Marino, Larry Taft, Jennifer Pereira, Charles Vandemoer, and Kathleen Wainwright. The Rhode Island Conservation Stewardship Collaborative has challenged us to develop procedures, protocols, and guidance for conservation agencies, especially local land trusts, in supporting their protected land management. The University of Rhode Island Cooperative Extension program, URI Department of Natural Resources Science, and USDA Renewable Resources Extension Act have steadily supported our work in conservation land management and stewardship.

REFERENCES

1. Daily, G.C.; Ehrlich, P.R.; Goulder, L.H.; Alexander, S.; Lubchenco, J.; Matson, P.A.; Mooney, H.A.; Postel, S.; Schneider, S.H.; Tilman, D.; Woodwell, G.M. Ecosystem services: benefits supplied to human societies by natural ecosystems. Issues Ecol. **1997**, *2* (1), 1–16.
2. Duerksen, C.; Snyder, C. *Nature Friendly Communities: Habitat Protection and Land Use Planning*; Island Press: Washington D.C., 2005; 421 pp.
3. National Park Service. General Management Plan Dynamic Sourcebook. http://planning.nps.gov/GMPSourcebook/GMPHome.htm (accessed February 2012).
4. Oakley, K.L.; Thomas, L.P.; Fancy, S.G. Guidelines for long-term monitoring protocols. Wildlife Soc. Bull. **2003**, *31* (4), 1000–1003.
5. United States Fish and Wildlife Service. Comprehensive Conservation Planning Process. http://www.fws.gov/policy/602fw3.html (accessed February 2012).
6. United States Bureau of Land Management. Land Use Planning. http://www.blm.gov/planning/index.html (accessed February 2012).
7. United States Forest Service. Strategic Planning and Resource Assessment. http://www.fs.fed.us/plan (accessed February 2012).
8. Turner, R.K.; Daily, G.C. The ecosystem services framework and natural capital conservation. Environ. Resour. Econ. **2008**, *39* (1), 25–35.
9. Lokocz, E.; Ryan, R.L.; Sadler, A.J. Motivations for land protection and stewardship: Exploring place attachment and rural landscape character in Massachusetts. Landscape Urban Plan. **2011**, *99* (2), 65–76.
10. Theobald, D.M. Targeting conservation action through assessment of protection and exurban threats. Conserv. Biol. **2003**, *17* (6), 1624–1637.
11. Land Trust Standards and Practice, 2004. Land Trust Alliance. http://www.lta.org. (accessed October 2011).
12. Ruhren, S. Baseline documentation and inventory protocol. Rhode Island Conservation Stewardship Collaborative, 2011. http://www.ricsc.org/Projects/Docs_2009/CSC_BDIP.pdf (accessed October 2011)
13. Carwardine, J.; Wilson, K.A.; Watts, M.; Etter, A.; Klein, C.J.; Possingham, H.P. Avoiding costly conservation mistakes: The importance of defining actions and costs in spatial priority setting. PLoS ONE **2008**, *3* (7), e2586. doi:10.1371/journal.pone.0002586 (accessed October 2011).
14. Anderson, P. Ecological restoration and creation: a review, Biol. J. Linnean Soc. **1995**, *56* (S1), 187–211.
15. Mack, R.N.; Simberloff, D.; Lonsdale, W.M.; Evans, H.; Clout, M.; Bazzaz, F.A. Biotic invasions: causes, epidemiology, global consequences and control. Ecol. Appl. **2000**, *10* (3), 689–710.
16. Upgren, A.; Bernard, C.; Clay, R.P.; de Silva, N.; Foster, M.N.; James, R.; Kasecker, T.; Knox, D.; Rial, A.; Roxburgh, L.; Storey, R.J.; Williams, K.J. Key biodiversity areas in wilderness. International Journal of Wilderness, **2009**, *15* (2), 14–17.
17. Walters, C. Adaptive Management of Renewable Resources; MacMillan Publishers: New York, 1986; 374 pp.
18. RICSC. Rhode Island Conservation Stewardship Collaborative. http://www.ricsc.org (accessed October 2011)

Desertification

Uriel N. Safriel
Center of Environmental Conventions, Jacob Blaustein Institutes for Desert Research, and Department of Ecology, Evolution and Behavior, Hebrew University of Jerusalem, Jerusalem, Israel

Abstract
Desertification is a persistent reduction of land's biological productivity of economic value—a terminal state of land degradation process. Land degradation and desertification (LDD) can occur anywhere but the focus is on drylands, whose natural low productivity is water limited. LDD is a syndrome of interactive biophysical processes whose direct drivers are land resource overexploitation by land users. Underlying causes of LDD are economic, demographic, social, cultural, and political, operating across scales. About a quarter of global land is degraded and active LDD is ongoing. LDD is of concern: it is believed to contribute to poverty, migrations, refugees, and conflicts; it undermines global food security and the emissions from degraded lands exacerbate global climate change, which is projected to increase the contribution of drought to LDD, and has already increased drying and a geographical expansion of some drylands. Some dryland communities demonstrate resilience to the underlying causes and take measures for avoiding LDD and the United Nations Convention to Combat Desertification (UNCCD) is instrumental in assisting developing countries addressing their LDD.

INTRODUCTION

"Desertification" is a human-driven process, which impinges on a significant Earth's natural capital, land, by reducing its productivity. Desertification is probably as old as the pastoral and farming livelihoods, but has attracted attention only in the 20th century, such that as it turned into the new millennium, desertification has been ranked among the greatest contemporary environmental problems.[1] Though literally the term links with deserts, desertification hardly occurs in deserts. Rather, desertification affects non-desert areas, all around the globe, and is of major concern, as it is a driver of poverty, mainly in the rural areas of the developing world, but at the same time it puts at risk food security at the global scale. This entry explains what desertification is, and describes how it occurs, where, by whom, and why. It then elaborates on its implications to people across scales, from local to global. This is followed by scanning through the measures taken by people and the international community to address this threat to human survival and global sustainability. The entry closes with the effect of global climate change on desertification, currently and in the future.

WHAT IS DESERTIFICATION?

Deserts—areas of poor vegetation cover and of the driest climate on earth—were molded by climatic processes, such as scarce rainfall and intense evaporation. These jointly constrain plant productivity such that much of the desert appears lifeless. As the world's deserts were formed by natural climatic processes over long periods during which they have grown and shrunk, the term "desertification" can grammatically describe desert formation. However, the term was apparently coined first in a 1949 publication,[2] not for describing a climate-driven outward spreading of deserts, but for a human-driven decline of land productivity to levels reminiscent to those of deserts, while climate remained a non-desert one. The term came into widespread international use only as of the 1970s, then being defined and redefined ever since. All definitions, however, relate to a land attribute critical to all people, everywhere and always—the marketed biological products, food and fiber, which are derived from the land's biological productivity. Thus, desertification is best expressed by a persistent reduction, relative to the potential, of the land's biological productivity of an economic value.[1,3]

DESERTIFICATION AND LAND DEGRADATION, DRYLANDS AND NON-DRYLANDS

The decline in biological productivity of economic value is driven by biophysical processes of "land degradation," which can operate on lands used by people anywhere on earth, but mostly affect lands whose productivity is constrained by water. Water is one of the raw materials required for the biological production process, but atmospheric CO_2 and soil minerals, as well as light are required too. Each of these can limit productivity, even if all the others are

Encyclopedia of Natural Resources DOI: 10.1081/E-ENRL-120047475

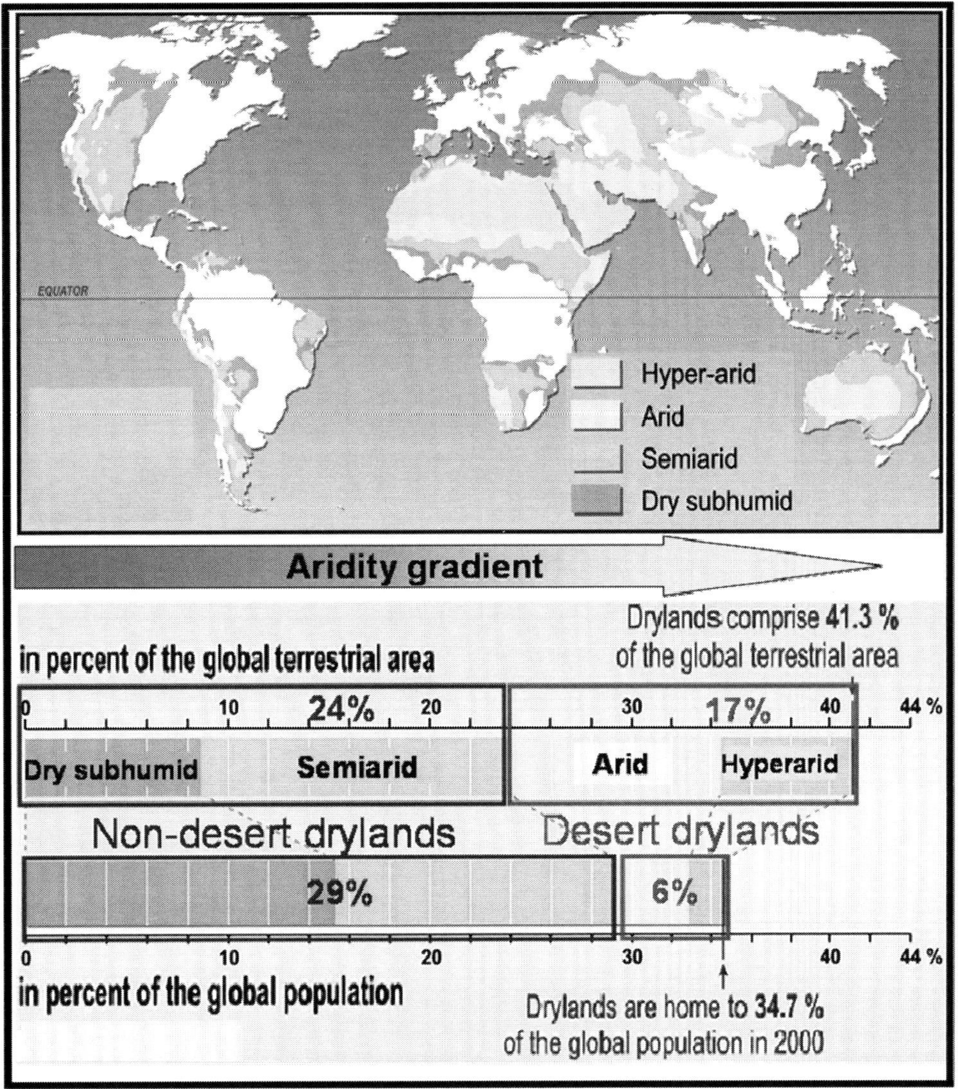

Fig. 1 The global drylands. Geographic layout of the four sections of the aridity gradient, their percent share of global land, their human populations, and their percent share of global population.
Source: Adapted from Adeel et al.[1]

available in abundance. Land degradation occurs in areas of low water and high evaporation, thus making soil water the limiting factor. This water scarcity exists in all lands exposed to a potential annual evaporative loss that is at least about one and one-half times greater than the mean annual gain through rainfall. These are "drylands," and depending on their degree of aridity, they are classified into desert drylands (hyper arid and arid drylands) and non-desert ones (semiarid and dry subhumid drylands) (Fig. 1).

The United Nations Convention to Combat Desertification (UNCCD) confined its mandate (to "combat desertification") to degradation of drylands only, excluding the most arid ones, the hyper arid desert drylands. Also, the stakeholders of the UNCCD habitually use the two terms, "land degradation" and "desertification" interchangeably or jointly. However, there is an increasing recent tendency to address "land degradation" as a process, while regarding "desertification" as a terminal, extreme state of this process; and, to use both terms for processes impinging on biological productivity in non-dryland areas too.[3,4] These tendencies have driven an increasing usage of an acronym for Land Degradation and Desertification, namely, LDD.

HOW LDD IS CAUSED?

Biological productivity depends on the land's soil, which provides trees, forage and crop plants physical anchor, and raw materials for the production process. The loose and aerated texture of the soil is due to its organic matter that binds the soil's mineral particles together. The binding creates aggregates with air spaces between them that retain water in which the plant mineral nutrients are dissolved. This particulate texture of soil, however, makes it vulnerable to the mechanical force of wind, that blows away the soil's fine components what often leaves sandy areas and dunes, and of rainwater that carries soils away. These soil erosion processes reduce fertility as most nutrients are confined to the uppermost soil layers, which are directly exposed to the eroding forces. Since the rate of soil formation is extremely slow, for all practical purposes soil is nonrenewable and hence topsoil loss is one of the major causes of persistent productivity decline. It is plant cover that provides for soil protection, by mitigating wind force and reducing the impact of raindrops that would otherwise destroy the delicate soil texture. Breakage of the aggregates results in soil crusting, leading to land surface sealing that reduces soil water absorption. Rather than infiltrating and enriching soil moisture, rainwater turn into surface runoff that erodes the soil or leaching away dissolved soil minerals and its organic compounds. These losses directly reduce fertility and indirectly the water holding capacity of the soil, what also slows down the soil microbial activity essential in nutrient recycling. Thus, due to soil erosion plants may lose their physical support as well as their production resources. Another major driver of declining productivity is soil salinization. Almost all types of water, including rainwater, contain some dissolved salts, which are left in the top soils after soil moisture evaporates or is transpired by the plants. This accumulated salinity reduces plants' ability to suck moisture from the soil, and the resulting soil sodicity (predominance of sodium ion in soil moisture) limits water absorption by soils. To conclude, while climatically driven water scarcity makes dryland productivity low compared to most other lands,[1] drylands are vulnerable to soil erosion and salinization (Fig. 2), the major physical and chemical processes, respectively, leading to LDD, whose ultimate expression is productivity loss.

WHO CAUSES LDD?

Depending on whether the biological products of economic value are derived directly from wild plants (e.g., woody plants used for timber and firewood, and other edible wild species); or indirectly from free-ranging livestock feeding on wild plants; or from cultivated crop plants; lands and their productivity are managed by their users through interventions expressed in natural resource exploitation rates, livestock stocking rates, and soil fertility levels, respectively. These interventions sustain peoples, societies, and their prosperity, yet they often pose potential threats to the land's productivity, which could ultimately impinge on the land users themselves, and on others. This is because while all productivity is derived from plants, their exploitative removal, either through grazing or harvesting, exposes the land to degradation processes once a threshold of plant cover is removed and/or one of regeneration relative to exploitation rate is crossed.

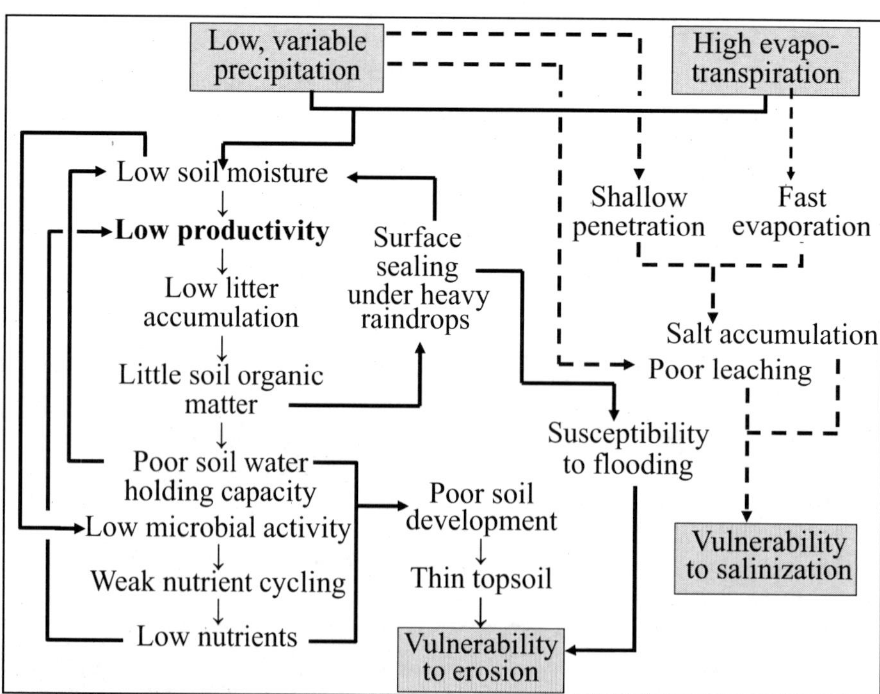

Fig. 2 The dryland syndrome—climate-driven natural processes driving soil biophysical processes leading to inherent low productivity, to erosion vulnerability (solid arrows, two negative feedback loops), and to salinization vulnerability (broken arrows).
Source: Adapted from Fig. 2 in p. 232, Safriel.[31]

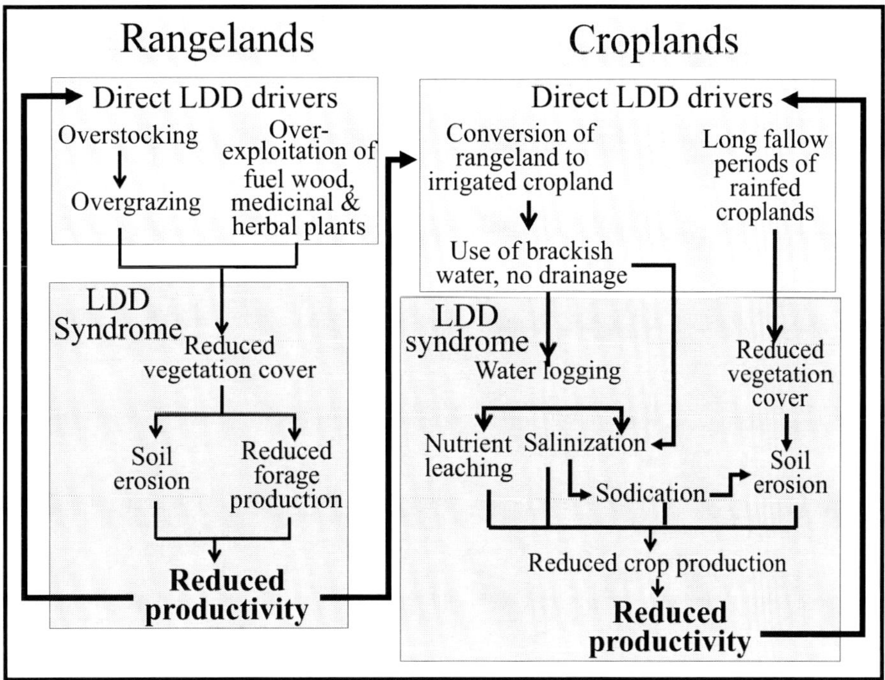

Fig. 3 Actions of land users that constitute the proximate, direct drivers (top boxes) of the land degradation and desertification (LDD) syndrome (bottom boxes) and their feedbacks and linkage (thick arrows), in each of the two prevailing dryland livelihoods. For further explanation see text.

Regarding the pastoral livelihood, high livestock densities maintained throughout a long period reduce the overall soil protective cover; exploit vegetation faster than it can regenerate, and export nutrients (through selling livestock and their products) faster than their natural renewability. These encourage non-forage plants to dominate the range, promote encroachment of bushes that obstruct livestock foraging movement, and reduce soil physical resilience to livestock trampling, which compacts soil surface, reduces rainfall infiltration thus increasing risks of powerful surface runoff that erodes the soil and changes the local topography. Individually or combined, these processes lead to a persistent rangeland productivity loss (Fig. 3). As to the farming livelihood, long fallow periods expose the croplands to erosion forces, but in short fallow periods, a high number of crops cultivated each year and a large amount of nutrients exported in the harvested agricultural products, individually or combined are of a potential to cross the threshold of land degradation. This is through nutrient loss that is faster than its renewability and a decline in soil organic matter, which reduces soil permeability, and the water holding capacity leading to increased erosion vulnerability. Irrigation, used mainly in drylands for increasing the productivity of otherwise rainfed agriculture, has dramatically increased not only the economic productivity but also soil salinity. Dryland water is scarce and hence not used for leaching soil salinity and when used, nutrients are leached too. Furthermore, much of dryland water is brackish, and irrigation systems devoid of drainage cause water logging, leading to further soil salinization. The irrigation-driven increased economic productivity amplifies nutrient export through harvest, which needs to be balanced by fertilizers. This combination of irrigation and fertilization could generate immediate economic gains, often succeeded by LDD (Fig. 3).

Practitioners of both livelihoods, but more often of the pastoral one, also exploit the woody plants of natural or planted woodlands and of rangelands, for timber and fuel wood, or remove all vegetation cover of woodlands or rangelands, to be replaced by expanding cultivation. These processes, too, expose large swaths of land to erosion processes and eventual LDD. To conclude, human land uses directed at increasing the inherently low dryland productivity of economic value are of a potential that is often materialized to bring about LDD. The persistent decline in productivity, rather than its aspired increase, attests to a preceding shift from grazing, cropping, and exploitation, to overgrazing, overcropping, and overexploitation—activities of land users that constitute the direct, biophysical drivers of the LDD processes.[5]

WHY LAND USERS CAUSE LDD?

It can be expected that land users are well adapted to their local environment and hence possess the knowledge for using their lands sustainably, i.e., without degrading it. Since land degradation does occur in areas used by experienced users, factors beyond their control may be involved. Indeed, indirect drivers of change, or underlying causes of

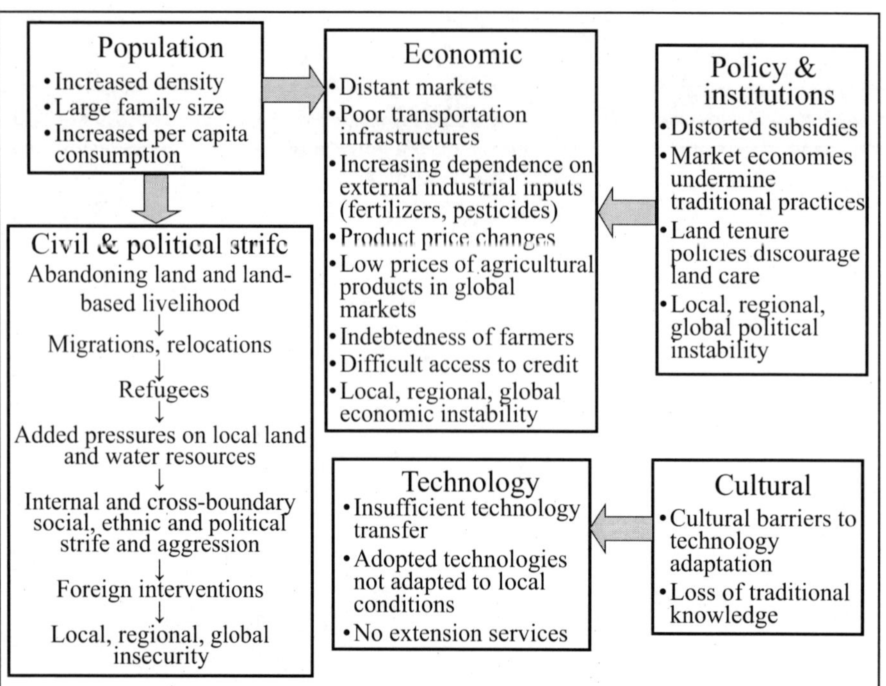

Fig. 4 Underlying socioeconomic and policy indirect land degradation and desertification drivers and their interactions.
Source: Adapted from Fig. 3, p. 173 in Reynolds et al.[7]

LDD that affect the direct, proximate causes that change non-degraded to a degraded land can be invoked (e.g., Fig. 3). These can be population (or demographic), economic, technological, policy and institutional, cultural, and political drivers.[6] Each of these (Fig. 4) can affect the direct LDD biophysical drivers alone, jointly, linearly or interactively, across different temporal and spatial scales.[7] However, in spite of an increasing research and political interest in these LDD drivers, there is not enough field evidence, hence also agreement, regarding the role andrelative significance of each of the variables that structure the network of pathways to LDD, and therefore the LDD "syndrome" is currently illustrated by paradigms that can be perceived as structured programs for further research. An "LDD paradigm" (Fig. 5) integrates elements of the Desertification Paradigm,[8] the Dryland Livelihood Paradigm[9] and the Dryland Development Paradigm.[10] Under this paradigm LDD is driven by mutually interactive biophysical processes within the land's "Ecological System, and by mutually interactive socio-economic and policy processes within a "Human System," integrated into an interactive "Human-Environment System" that oscillates between land's sustainable use and LDD. The LDD paradigm attends the effect of LDD on human well-being at different spatial scales—effects that often further exacerbate rather than mitigate the LDD state. These are mainly poverty at the local scale, and migrations, refugees, conflicts and foreign interventions at the regional and even global scales. In short, land users are driven to overexploit land, mostly by economic factors that in themselves can be driven by demographic variables, social and cultural factors, and by economic and social policies and political processes molded at diverse regional and global processes on which they have no control.

WHERE AND HOW MUCH LDD THERE IS?

Since biological productivity expressed in market-traded products of economic value is regularly monitored a detected decline in the production of crops or in livestock-derived products can be used to delineate LDD-affected areas and quantify their spatial extent. However, the quantities of marketable agricultural products per unit area do not necessarily reflect land productivity, as these much depend on the land users' management decisions, driven mainly by economic considerations. Therefore, instead of monitoring marketed land products scientists turned to monitoring land and soil attributes serving as indicators of productivity. But since different studies use different indicators and measurement methods, it is not surprising that five global assessments during the last four decades resulted in LDD estimates ranging from 15% to 63% of global land and 4% to 74% of its subset of global drylands.[11] A critical review of these studies suggested (with "medium certainty") that 10–20% of the global drylands have been degraded by year 2000,[8] while an earlier study qualified 12% of non-dryland as degraded.[12] The latter study qualified 5.7% and 3.8% of Asian and African lands as degraded, respectively, and 1.9%, 1.7%, 1.2%, and 0.8% of lands in South America, Europe, North America, and Australasia as degraded, respectively; which combined to classify 15.1% of global land as degraded.[13]

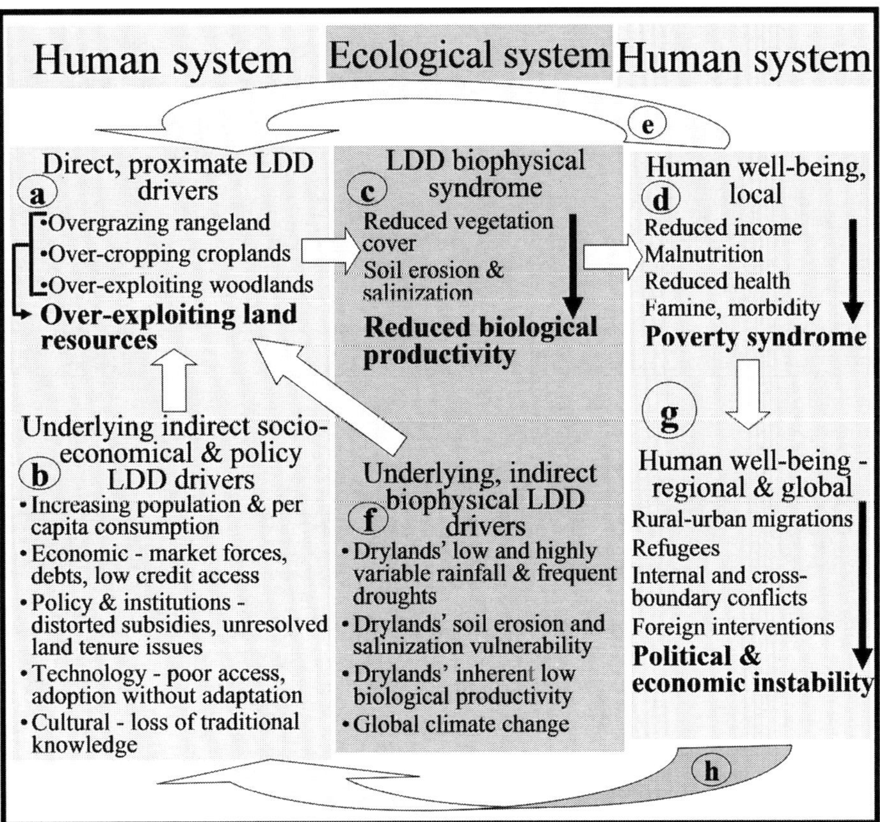

Fig. 5 The land degradation and desertification (LDD) paradigm—interactions and feedback loops: Direct proximate LDD drivers (a) (detailed in Fig. 3), driven by underlying socioeconomic and policy drivers (b) (detailed in Fig. 4) generate the LDD syndrome (c) (detailed in Fig. 3); the reduced human well-being impinges on the direct drivers thus further exacerbating LDD (e); underlying indirect biophysical drivers also drive people to further exacerbate LDD (f) (detailed in Fig. 2); the locally affected human well-being (d) also has regional and global repercussions (g), and human well-being at all levels impinges on the underlying LDD drivers (h). Vertical arrows indicate the linear effects of variables from top to bottom of lists, while bullets indicate the independent effects of each variable.

The development of frequent and accessible time series of Earth monitoring from space enabled assessing not just the amount of degraded land but the more recent rates of ongoing LDD. Satellite-born sensors of radiation reflected and emitted from the Earth's surface, and especially from its vegetation cover, are used for following the course of biological productivity changes within given time periods. Most studies attend selected regions but one[14] addresses the dynamics of biological productivity of vegetation cover at a global scale. This 23-year (1981–2003) analysis of remote sensing images revealed persistently declining overall productivity (i.e., not just the subset of direct economic value) at 24% of the global land, on which 1.5 billion people reside. It mainly occurred in Africa south of the equator, Southeast Asia and south China, north-central Australia, the Argentinean Pampas and swaths of the Siberian and North American taiga. The study also found that land degradation in China is more severe in forests than in croplands. However, measures taken to control the productivity–rainfall interaction need to be refined in making such assessment fully reliable.[15] Nevertheless, the available studies demonstrate that as much as a quarter of the global land is already degraded, and active LDD is still in force, which points at the vulnerability of non-degraded lands around the globe, especially in drylands.

WHY LDD IS OF CONCERN?

LDD reduces the capacity of earth to provide food; hence, it is of concern not only to the direct land users but to humanity at large. Furthermore, even the LDD's local effect on the well-being of the direct land users is implicated with generating externalities of a global concern, as the local LDD-driven poverty leads to within-country, cross-boundary, and overseas migrations of refugees, which often trigger conflicts of variable severity and foreign and international interventions (Figs. 4,5). This component of the LDD paradigm is widely alluded to, though hard data on the causation and the dimension of each of these processes at the global scale are scarce. Regarding poverty, countries' statistics of poor rural land users do not distinguish between LDD-driven poverty and poverty not driven by LDD; it adds the poverty of land users with that of rural people who do not live off the land, as well as that of the country's dryland and

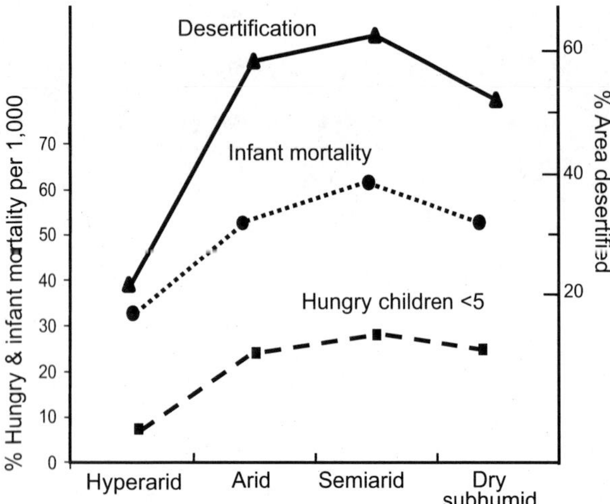

Fig. 6 Trends of land degradation and desertification and poverty indicators across the aridity gradient of the global drylands. Land degradation and desertification (LDD) data are of GLASOD,[12] with the "low degradation" category excluded; data of infant mortality per 1000 births and percentage of hungry children under the age of 5, drylands of OECD countries excluded, are from the Millennium Ecosystem Assessment.[8]
Source: Adapted from Fig. 4 of Safriel.[32]

non-dryland inhabitants. Thus, only localized statistics (e.g., 68–84% poverty in the drylands in northern Kenya that is higher than in other, non-dryland parts of the country) suggest a prevalence of poverty in drylands of developing countries, and also has led to a rough estimate of about nearly one billion people in poor rural drylands.[16] A direct, quantifiable linkage between poverty and LDD at the global scale is so far provided only by the observation that the poverty indicators—infant mortality and malnutrition of under-5-year-old children track the trajectory of desertification severity across the global aridity gradient (Fig. 6). Similar to the LDD-poverty nexus, there is no hard data on the linkage among poverty, migrations, refugees, and conflicts, and evidence for LDD being instrumental is circumstantial. Most cited are migrations following severe droughts, while only a few cases present a direct linkage, for example, the Tambacuanda region of Senegal's 1960s migrations driven by soil erosion-triggered decline of crop production—to the capital, other cities, other African countries, and Europe.[17] Social scientists also disagree on the role of environmental degradation (LDD included) in generating conflicts and violence, and concluded that empirical evidence is still insufficient to demonstrate a tangible effect.[18] However, comprehensive research of the Darfur conflict[19] asserted that LDD in both croplands and rangelands has been one of the major factors in that crisis but it has been also intertwined with a range of other socioeconomic and political issues. Even if LDD does not always generate local poverty of regional and global repercussions, any local loss of land productivity undermines food security not only locally, but also at the global scale. It has been estimated that for feeding the global population by year 2050, 8.9 million km^2 of good cultivable land would be required, on top of the 15 million km^2 already under cultivation.[20] However, only 4.4 million km^2 of cultivable land will be available.[21] Therefore, LDD which practically converts cultivable to non-cultivable land could dramatically aggravate the already fragile global food security. LDD is of concern not only due to loss of crop and forage production but also due to loss of other benefits that people derive from the land, such as water regulation and purification, local climate amelioration, pollination, biodiversity conservation, and others. These benefits, called "ecosystem services," are an expression of the land's functionality as an ecological system. Adopting a holistic approach, LDD is increasingly described as a persistent, substantial reduction in a bundle of services provided to humans by cultivated, range, or natural ecosystems.[1,22] The persistent decline in the provision of one of these dryland ecosystem services, carbon sequestration (the uptake of atmospheric CO_2 by plants, and transforming it to plant biomass and litter) is of global concern, too. All dryland vegetation and all dryland soils store 14% and 20% of the global organic carbon, respectively.[8] LDD-associated soil erosion and loss of plant cover and plant productivity deplete this land's storage of organic carbon. This locally removed organic carbon is eventually oxidized and emitted back to the atmosphere, thus increasing atmospheric CO_2 concentration and exacerbating climate change, globally. Indeed, drylands' LDD-driven emissions are estimated to comprise about 4% of the total global emissions from all sources combined (in year 2000),[1] and added LDD would further amplify global climate change.

WHAT HAS BEEN DONE AND WHAT CAN BE DONE?

That most lands used in drylands are not yet affected by LDD suggests that drylands can be used sustainably. Many cases have been documented, mostly in east and west African drylands, in which rural communities adapted to rapid population growth, globalization, market development, and technological change and avoided LDD. These apparently detrimental changes triggered innovative actions: investments in more intensive yet sustainable land use, which helped reaping benefits of the enlarged market opportunity; exploitation of local bioclimatic and social comparative advantages; access to technologies that increased land and labor productivity faster than population growth; and improved access to growing markets.[23] Avoiding LDD can be also achieved by adding income-generating "alternative" land-independent livelihoods like tourism and aquaculture,[24] or generating supplementary income from seasonal migration to urban centers.[17]

Nevertheless, it is developing countries that share most of the global drylands,[8] as well as probably the most rural poverty and LDD risks. Noting these facts and being aware of the global threats of dryland LDD, the international community negotiated and adopted the UNCCD, with the intention of coordinating and promoting assistance of industrial (developed) to developing countries in addressing their LDD problems. Entering into force in 1996, the 195 country parties of the UNCCD together with the Secretariat and the subsidiary bodies of the Convention succeeded in raising awareness, encouraging knowledge and research, and promoting capacity-building and partnerships in projects on the ground, as well as installing mechanisms for assessing success. A major thrust of the international community is to drive the political process of setting a sustainable development goal—striving for Land Degradation Neutral World—and its associated target of achieving Zero Net Land Degradation (ZNLD) by 2030, initiated at the United Nations Conference on Sustainable Development convened in June 2012 in Rio De Janeiro. The rationale behind this target is that completely halting further decline of the productive, non-degraded land within a relatively short time is unattainable. However, the amount of non-degraded land that becomes degraded within a time period can be offset by a similar amount of already degraded land, whose productivity is restored within the same time period. Once this offset is attained, the net rate of LDD is zero. Setting such goals and targets can help shape expectations and create the conditions for all stakeholders to assess progress and take appropriate action in addressing LDD.

As to reducing LDD for addressing the ZNLD target, "sustainable land management" (SLM) is widely prescribed. In practice, any land that supports the user does not lead LDD to qualify as SLM. More specifically, practices of land use need to adaptively maintain the resilience of the used ecosystems to all LDD drivers, such that the thresholds in ecosystem functions will not be crossed and the service provision will be sustained.[25] Avoiding degradation may be easier and cheaper than restoring already degraded land but, as land is often degraded inadvertently while restoration is planned, its success prospects are relatively high. The first stage is removal of the LDD causes, to be followed by selecting from the rich literature of ecological restoration[26] the toolbox appropriate for the local state and the set restoration target.

DESERTIFICATION AND CLIMATE, AND CLIMATE CHANGE

Drylands are not only exposed to climate-driven water scarcity but also to high between-year variability. This includes relatively short periods (<5 years) during which rainfall and hence productivity remain far below the long-term average. These periods are termed "droughts," and since their associated productivity decline is reversible, this productivity decline does not qualify as desertification. It can be expected that a drought-driven productivity decline would persist after the drought terminate, provided that it has been severe and exceptionally long. Such events, if ever occurred, have not been reliably documented. On the other hand, drylands in which human-induced LDD processes have been already initiated are likely to be pushed by a drought to cross the desertification threshold, such that productivity decline would persist after the drought terminates. Such cases are rare, too. Even the reduced productivity during the Sahel droughts of 1968–1973 and 1982–1984 of the 20th century, which took a heavy toll on properties and lives and aroused the world's concern that culminated in adopting the UNCCD, was restored once the droughts terminated.[27] Nevertheless, though droughts alone do not cause LDD, they do exacerbate LDD and its effect on human well-being, and hence droughts are listed among the underlying, indirect biophysical LDD drivers (Fig. 5). The role of drought in desertification is laid out in the full name of the UNCCD—a convention *"to combat desertification, in those countries experiencing serious droughts ..."* and in its objective, that is not only to" *combat desertification"* but also to *"mitigate the effects of droughts."*[28] In spite of the difference between "combating" and "mitigating," an emerging and widely used acronym for the UNCCD's subject matter is Desertification, Land Degradation, and Drought, or DLDD.

Coining the DLDD acronym is partly justified given that the frequency of droughts and their severity are projected to increase, driven by the ongoing global warming and climate change. Indeed, climate change (and poverty) is projected to intensify desertification by 2050 in all the four Millennium Ecosystem Assessment's scenarios of global governance.[1] Climate change's effect on global aridity is already evident. Comparison of the climatic conditions, precipitation, and temperatures that prevailed in 1931–1960 and used for setting the drylands' boundaries, with those prevailing in the period of 1961–1990, demonstrated that the spatial extent of the African hyper-arid and arid drylands increased by 51 and 3 million ha, respectively, which comprise 1.7% and 0.1% of their area extent in 1931–1960, respectively. More significant is the "loss" of 25 million ha of African non-drylands of 1931–1960 that turned into dry subhumid drylands during 1961–1990.[29] Furthermore, rainfall decreases and evapotranspiration increases are projected to intensify this trend in many drylands by mid-century (IPCC 2008).[30] To conclude, though current LDD is driven by humans at the local scale, climate change, apparently human-driven, too, also leads to local productivity persistent decline, that is LDD, but one in which the local land users are not directly involved.

CONCLUSIONS

Desertification is a persistent reduction in land's biological productivity of an economic value, as well as in other services provided by the land's ecosystems. It is an extreme, terminal state of the land degradation process, and LDD can occur anywhere but the focus is on drylands; 40% of global land whose inherent low productivity is constrained by water. LDD comprises a syndrome of dynamically interactive biophysical processes, directly driven by land management practices of land users, themselves driven by underlying socioeconomic and policy interlinked processes that cross temporal and spatial scales, as well as by the inherent vulnerability of dryland soils to LDD, and by the effects of droughts.

Though the cumulative area of degraded land is only about a quarter of the globe's productive land, the process is still ongoing apparently at an increasing rate. LDD is therefore of global concern—the poverty it inflicts on the local land users is a source of migrations, refugees, and conflicts across boundaries and scales; and the local loss of productive land risks the already fragile global food security. In addition, LDD and global climate changes are interlinked; climate change currently increases dryness and hence LDD vulnerability, and LDD reduces the land's carbon storage locally thus increasing emissions and warming, globally. The good news is that the UNCCD succeeded in raising awareness to the LDD plight, such that the international community seems ready to commit itself to significantly reducing further degradation and investing in restoring already degraded lands. Knowledge, institutions and instruments are available, and the political will and resources need to be secured.

REFERENCES

1. Adeel, Z.; Safriel, U.; Niemeijer, D.; White, R. *Ecosystems and Human Well-Being: Desertification Synthesis*; The Millennium Ecosystem Assessment; World Resources Institute: Washington, DC, 2005.
2. Aubreville, A. Climats, Forest, et Desertification de l'Afrique Tropicale. Societe de Editions 3/ Geographiques, Maritime et Coloniales, Paris, 1949.
3. Vogt, J.V.; Safriel, U.; Von Maltitz, G.; Sokona, Y.; Zougmore, R.; Bastin, G.; Hill, J. Monitoring and assessment of land degradation and desertification: towards new conceptual and integrated approaches. Land Degrad. Dev. **2011**, *22*, 150–165.
4. Safriel, U.N. Global Land Degradation: State, Risks and Prospects Under Global Change. In *Soil, Society and Global Change*; Bigas, H., Gudbransson, G.I., Monanarella, L.; Arnalds, O., Eds.; European Communities: Italy, 2009; 46–52.
5. http://dsd-consortium.jrc.ec.europa.eu/documents/WG1_White-Paper_Draft-2_20090818.pdf (accessed March 2012). White Paper of DSD Working Group 1, August 2009, Integrated Methods for Monitoring and Assessing Desertification/Land Degradation Processes and Drivers.
6. Geist, H.J.; Lambin, E.F. Dynamic causal patterns of desertification. Bioscience **2004**, *54*, 817–829.
7. Reynolds, J.F.; Grainger, A.; Stafford Smith, D.M.; Bastin, G.; Garcia-Barrios, L.; Fernandez, R.J.; Janssen, M.A.; Jurgens, N.; Scholes, R.J.; Veldkamp, A.; Verstraete, M.M.; Von Maltitz, G.; Zdruli, P. Scientific concepts for an integrated analysis of desertification. Land Degrad. Dev. **2011**, *22*, 166–183.
8. Safriel, U.N.; Adeel, Z. Dryalnd Systems. In *Ecosystems and Human Well-Being: Current State and Trends*; Hassan, R., Scholes, R., Ash, N., Eds.; Island Press: Washington, 2005; 623–662.
9. Safriel, U.N. Alternative livelihoods for attaining sustainability and security in drylands. In *Coping with Global Environmental Change, Disasters and Security*; Brauch, H.G., Oswald Spring, U., Mesjasz, C., Grin, J., Kameri-Mbote, P., Chourou, B., Dunay, P., Birkmann, J. Eds.; Vol 5, Hexagon Series on Human and Environmental Security and Peace, edited by J. Springer: Berlin, 2011; 835–852.
10. Reynolds, J.F.; Stafford Smith, D.M.; Lambin, E.F.; Turner, B.L. II,; Mortimore, M.; Batterbury, S.P.J.; Downing, T.E.; Dowlatabadi, H.; Fernandez, R.J., Herrick, J.E., Huber-Sannwald, E.; Jiang, H.; Leemans, R,; Lynam, T.; Maestre, F.T.; Ayarza, M., Walker, B. Global desertification: building a science for dryland development. Science **2007**, *316*, 847–851.
11. Safriel, U.N. The assessment of global trends in land degradation. In *Climate and Land Degradation*; Sivakumar M.V.K., Ndiaugui, N., Eds.; Springer: Berlin, 2007; 1–38.
12. http://www.isric.org/projects/global-assessment-human-induced-soil-degradation-glasod (accessed March 2012) GLASOD, Global Assessment of Soil Degradation, International Soil Reference and Information Centre, Wageningen, Netherlands, and United Nations Environment Programme, Nairobi, Kenya; 1990.
13. Middleton, N.; Thomas, D. Eds., *World Atlas of Desertification*; Arnold: London, 1997; 18 p.
14. Bai, Z.; Dent, D.; Olsson, L.; Schaepman, M.E. Proxy global assessment of land degradation. Soil Use Manag. **2008**, *24*, 223–243.
15. Wessels, K. Comments on proxy global assessment of land degradation, by Bai et al. 2008. Soil Use Manag. **2009**, *25*, 91–92.
16. Dobie, P. Poverty and the drylands. UNDP, Nairobi. September 2001. http://arabstates.undp.org/contents/file/Poverty-and-the-Drylands-Challenge-Paper.pdf (accessed March 2012).
17. Leighton, M. Desertification and migration. In *Governing Global Desertification: Linking Environmental Degradation, Poverty and Participation;* Johnson, P.-M., Mayrand, K., Paquin, M., Eds.; Ashgate Publishing: Aldershot, England, 2006; 43–58.
18. Safriel, U.; Adeel, Z. Development paths of drylands—is sustainability achievable? Sustain. Sci. J. **2008**, *3* (1), 117–123.
19. Sudan Post-Conflict Environmental Assessment: United Nations Environment Programme; Nairobi, Kenya, 2007. http://postconflict.unep.ch/publications/UNEP_Sudan.pdf (accessed March 2012).

20. Fischer, G.; Shah, M.; van Velthuizen, H.' Nachtergaele, F. O. Global agroecological assessment for agriculture in the 21st century. 2001, http://www.iiasa.ac.at/Research/LUC/SAEZ/index.html (accessed March 2012).
21. Tilman, D.; Fargione, J.; Wolff, B.; D'Antonio, C.; Dobson, A.; Howarth, R.; Schindler, D.; Schlesinger, W. H.; Simberloff, D.; Swackhamers, D. Forecasting agriculturally driven global environmental change. Science **2001**, *292*, 281–284.
22. Verstratete, M. M.; Scholes, R.J.; Stafford Smith, M. Climate and desertification: looking at an old problem through new lenses. Front. Ecol. Environ. **2009**, *7* (8), 421–428.
23. Nkonia, E.; Winslow, M.; Reed, S.; Mortimore, M.; Mirzagaev, A. Monitoring and assessing the influence of social, economic and policy factors on sustainable land management in drylands. Land Degrad. Dev. **2011**, *22*, 240–247.
24. Adeel, Z.; Safriel, U. Achieving sustainability by introducing alternative livelihoods. Sustain. Sci. J. **2008**, *3* (1), 125–133.
25. Cowie, A.L.; Penman, T.D.; Gorssen, L.; Winslow, M.D.; Lehmann, J.; Tyrrell, T.D.; Twomlow, S.; Wilkes, A.; Lal, R.; Jones, J.W.; Paulsch, A.; Kellner, K.; Akhtar-Schuster, M. Towards sustainable land management in the drylands: scientific connections in monitoring and assessing dryland degradation, climate change and biodiversity. Land Degrad. Dev. **2011**, *22*, 248–260.
26. Bainbridge, D.A. A. Guide for Desert and Dryland Restoration; Island Press: Washington, 2007.
27. Prince, S.D.; Brown De Colstoun, E, Kravitz, L.L. Evidence from rain use efficiencies does not indicate extensive Sahelian desertification. Clim. Change **1988**, *4*, 359–458.
28. United Nations Convention to Combat Desertification (UNCCD) http://www.unccd.int.
29. Hulme, M.; Marsh, R; Jones, P.D. Global change in a humidity index between 1931–1960 and 1961–1990. Clim. Res. **1992**, *2*, 1–22.
30. Parry, M.L.; Canziani, O.F.; Palutikof J.P., van der Linden, P.J.; Hanson, C.E. Eds.; Climate Change 2007: Impacts, Adaptation and Vulnerability. Contribution of Working Group II to the Fourth Assessment Report of the Intergovernmental Panel on Climate Change. Cambridge University Press: Cambridge, UK and New York, NY, USA., 2007.
31. Safriel, U.N. Dryland development, desertification and security in the Mediterranean. In *Desertification in the Mediterranean Region. A Security Issue*. Kepner, W.G., Rubio, J.L., Mouat, D.A., Pedrazzini, F., Eds.; NATO Security through Science Series, Volume 3, Springer Publishers: Germany, 2006; 227–250.
32. Safriel, U.N. Deserts and Desertification: Challenges but also Opportunities. Land Degrad. Dev. **2009**, *20*, 353–366.

BIBLIOGRAPHY

1. Allington, G.R.H.; Valone, T.J. Reversal of desertification: the role of physical and chemical soil properties. J. Arid Environ. **2010**, *74*, 973–977.
2. Briassoulis, H. *Policy Integration for Complex Environmental Problems. The Example of Mediterranean Desertification*; Ashgate: Burlington, USA, 2005.
3. Corell, E. *The Negotiable Desert. Expert Knowledge in the Negotiations of the Convention to Combat Desertification*; Linkoping University, Lindoping: Sweden, 1999; 286 pp.
4. Fantechi, R.; Peter, D.; Balabanis, P.; Rubio, J.L. *Desertification in a European Context: Physical and Socio-Economic Aspects*; European Commission: Luxembourg, 1955.
5. Grainger, A. *The Threatening Desert: Controlling Desertification*; Earthscan and UNEP: Nairobi, 1990.
6. Safriel, U.N. The concept of sustainability in dryland ecosystems. In *Arid Lands Management – Toward Ecological Sustainability;* Hoekstra, T. W., Shachak, M., Eds.; University of Illinois Press: Urbana, 1999; 117–140.
7. Schlesinger, W.H.; Reynolds, J.F.; Cunningham, G.L.; Huenneke, L.F.; Jarrell, W.M.; Virginia O.A.; Whitford, W.G. Biological feedbacks in global desertification. Science **1990**, *247*, 1043–1048.
8. Swift, J.; Leach, M.; Mearns, R. *Desertification. Narratives, Winners & Losers*. James Currey: Oxford, UK, 1996.
9. Tiffen, M.; Mortimore, M.; Gichuki, F. *More People, Less Erosion: Environmental Recovery in Kenya*; J. Wiley: Chichester, England, 1994.
10. Thomas, D.S.G.; Middleton, N. *Desertification: Exploring the Myth*; J. Wiley: Chichester, 1994.
11. Williams, M.A.J.; Balling, R.C. *Interactions of Desertification and Climate*; Arnold, Ahalsted Press: 1996; 372.
12. Xu, D.Y.; Kang, X.W.; Zhuang, D.F; Pan, J.J. Multi-scale quantitative assessment of the relative roles of climate change and human activities in desertification – A case study of the Ordos Plateau, China. J. Arid Environ. **2010**, *74*, 498–507.

Diversity: Species

Jeffrey D. Corbin
Brittany L. Oakes
Department of Biological Sciences, Union College, Schenectady, New York, U.S.A.

Biomes—Ecosystems

Abstract

Species diversity is a measure of the number of species, either with or without consideration of relative abundances. Statistical analysis estimates that there are approximately 8.7 million (±1.3 million SE) eukaryotic species on Earth. Several global patterns of species diversity have been described, including a peak in species diversity in tropical latitudes and fewer taxa as latitude decreases toward the poles. The mechanisms of species coexistence, and therefore maintenance of species diversity, at local scales is an active field of ecological theory. Models of coexistence include both equilibrium and non-equilibrium explanations. Human activities including land-use change, overexploitation, and, in the future, climate change have significantly influenced species numbers locally and globally by increasing the rates of extinction. Efforts to preserve species and limit rates of extinction have been justified both in terms of the benefits that species provide in economic and social terms (e.g., "ecosystem services") and in ethical terms.

INTRODUCTION

Species diversity is one of the three scales at which diversity can be measured, intermediate between genetic diversity and ecosystem (or habitat) diversity. The concept of biodiversity is often considered to encompass all three of these scales of diversity. In practice the two terms—species diversity and biodiversity—are often used interchangeably, particularly in popular usage or in the context of environmental conservation.

Mora and colleagues used statistical analysis to estimate that there are 8.7 million (±1.3 million SE) eukaryotic species on Earth; thus far, only 1.2 million have been identified and catalogued.[1] Prokaryotic diversity is less well-known but likely to add significantly to this estimate. Invertebrate species are the largest contributors to current estimates, with insect diversity reaching one million species and arachnids around 100,000 species.[2] Plant species rank second to invertebrate richness, owing to the notably high diversity of flowering plants (~280,000 species).[2]

QUANTIFYING SPECIES DIVERSITY

Species diversity is most simply quantified by counting the number of species, also often called species richness. Quantitative diversity indices include species' abundances along with the number of species. Among the most commonly used quantitative diversity indices are the inverse Simpson's index

$$\lambda = 1/\Sigma p_i^2$$

and the Shannon–Wiener index

$$H' = -\Sigma p_i \log p_i$$

where p_i = proportion of species i in the total sample of individuals. Each index is differentially sensitive to the presence of rare species, so often more than one index is calculated in order to give a more complete picture.[3]

PATTERNS OF SPECIES DIVERSITY

Tropical habitats, including mesic (or wet) forest and coral reefs, are renowned for their high species diversity. Tropical rainforests may have as many as 10,000 species per hectare; at smaller spatial scales, nutrient poor grasslands, such as those in the Coastal Plains of southeastern North America, are notable for having as many as 50 species per m^2. By contrast, habitats at northern latitudes, including boreal forest and tundra, and salt marshes contain relatively few species at both large (e.g., km^2) and small spatial scales.

At large scales, the number of species of a variety of taxa peaks at the lower (i.e., tropical) latitudes and decreases as latitude increases toward the poles.[4] Several hypotheses have been proposed to explain the latitudinal gradient in species richness. These hypotheses include the evolutionary age of the habitats, habitat heterogeneity, climatic stability, available environmental energy, higher rates of evolution in the tropics, and a sampling effect due to the arrangement of landmasses near the equator (i.e., the "mid-domain" effect).[5]

Encyclopedia of Natural Resources DOI: 10.1081/E-ENRL-120047440

Endemic species, or those restricted to a specific geographic location and found nowhere else on the Earth, are an important component of species diversity both within a specific habitat and as a contributor to global species diversity. Myers and colleagues overlapped areas with a significant number of endemic species (at least 0.5% of the world's total vascular plants, or 1,500 plant species, as endemics) with areas experiencing high rates of habitat loss (at least 70% of its original vegetation lost) to develop the concept of biodiversity hotspots[6,7] (Fig. 1). The 34 biodiversity hotspots that are currently recognized cover 2.3% of the Earth's surface; most are located in tropical zones. Benefits of identifying such hotspots include the potential to prioritize conservation funding toward areas that are both threatened by habitat loss and that would result in a significant loss of unique species.[6] This approach has been criticized, however, in that it does not take a variety of other components of diversity into account, including smaller-scale diversity or diversity of taxa besides plants.[8]

At a local scale, Whittaker differentiated between alpha (within-habitat), beta (between-habitat), and gamma (regional or between-ecosystem) diversity.[9] The number of species typically increases as the area sampled increases.[10,11] This species–area relationship has been described as a power function:

$$S = cA^z$$

where S is species richness, A is area sampled, and c and z are constants fit to the data. Graphically, it is often portrayed in logarithmic space, with the logarithm of species number having a linear relationship with the logarithm of area (Fig. 2).

MAINTENANCE OF SPECIES DIVERSITY

Understanding factors that maintain species diversity—that is, how multiple species can coexist in the same habitat rather than one species "winning" and competitively excluding others—has been a central focus of ecological theory since at least G. E. Hutchison's famous paper "Homage to Santa Rosalia, or why are there so many animals?"[12] Models of species coexistence and, therefore, maintenance of species diversity can be roughly divided into equilibrium and non-equilibrium models.[13] Equilibrium models of species coexistence assume that competitive exclusion of one species over another does not take place because direct competition is avoided, for example, because of niche differentiation (e.g., species are utilizing different resources) or because of spatial or temporal heterogeneity (e.g., resources and competitive advantages vary in space or time). Non-equilibrium models assume that periodic events such as predation (e.g., keystone predators)[14] or disturbances[13] prevent equilibrium and, therefore, competitive exclusion.[15]

THREATS TO SPECIES DIVERSITY

Periods of mass species extinctions have occurred at least five times in the Earth's history, but indications are that human activities have ushered a sixth episode. The impacts of human population on a variety of measures including land use, global climate, and biodiversity have led many to define the current epoch as the "Anthropocene."[16] Human activities may have led to the extinction of 5–20% of the

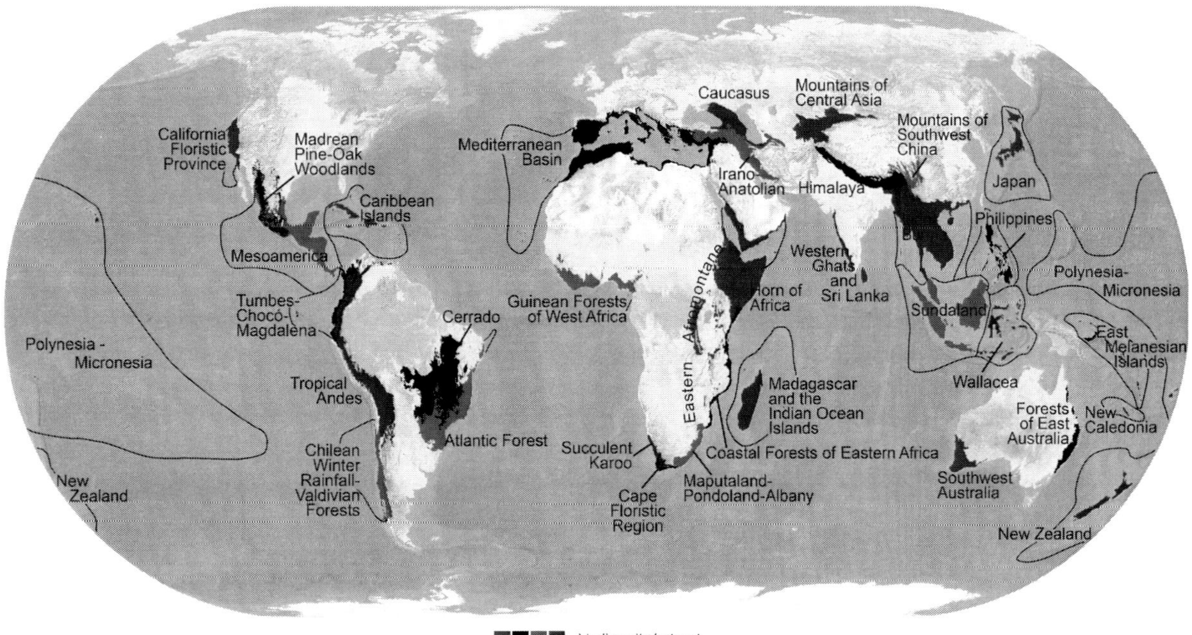

Fig. 1 Map of global biodiversity hotspots.
Source: Courtesy of Conservation International.[7]

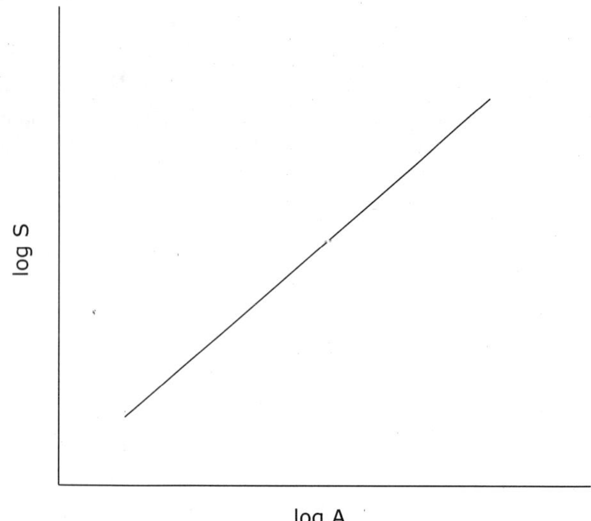

Fig. 2 The linear transformation (log S = log c + z log A) of the species–area power function, $S = cA^z$, where S is species richness, A represents the area sampled, and c and z are constants fit to the data.

species in some species groups,[17] well above background levels of extinctions that would naturally occur.[18] Change in land use, including habitat loss, conversion of rainforest habitat, and fragmentation, is considered the leading driver of modern extinctions. Other causes include overexploitation and invasion of non-native species. Forecasts of future drivers of species extinction suggest that climate change will be a significant factor (Fig. 3).[19,20]

BENEFITS OF SPECIES DIVERSITY

One of the most often-cited motivations for conserving species diversity is the benefit that biodiversity and nature provide to humans. Such benefits have been quantified as "ecosystem services" and include provisioning of food, fiber, pharmaceutical products, fresh water, etc.; regulation of flood waters, agricultural pests, or atmospheric carbon; cultural uses including recreation, enjoyment of the aesthetics, or spiritual values; and indirect benefits such as maintenance of nutrient cycling, pollination, or primary productivity.[21] In this framework, biodiversity provides the foundation for all other services even if it provides relatively little direct benefit to humans.

Species diversity is also believed to influence an ecosystem's stability. Both theory and empirical evidence suggest that ecosystems with more species have a greater ability to resist and recover from perturbations.[22] It is unclear, however, whether diversity is the driver of this pattern or merely a correlation with other ecological factors. Instead, stability seems to arise from the greater likelihood that a diverse community will include highly productive species or functional groups, or those that can respond differentially to perturbations and buffer ecological changes.[22]

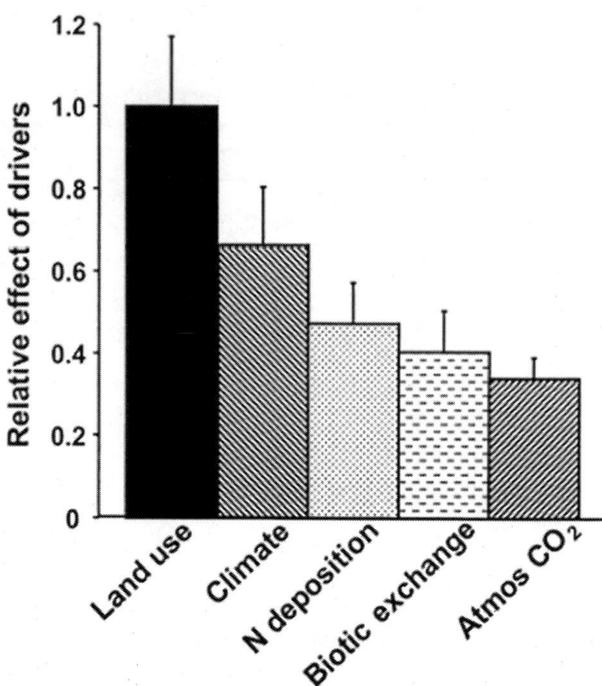

Fig. 3 Relative effect of major drivers of changes on biodiversity: Land use change, climate change, atmospheric nitrogen deposition and acid rain, biotic exchange including invasive species, and elevated atmospheric CO_2. Expected biodiversity change for each biome for the year 2100 was calculated as the product of the expected change in drivers times the impact of each driver on biodiversity for each biome. Values are averages of the estimates for each biome and they are made relative to the maximum change, which resulted from change in land use. Thin bars are standard errors and represent variability among biomes.
Source: Adapted from Sala et al.[19]

Finally, a principal motivation for conserving nature and species diversity derives from the appreciation that they have value in their own right. To say that a species, or nature in general, has a value of its own regardless of the benefits that humans may derive from them is to say that it has *intrinsic value*.[23] The writings of Aldo Leopold and John Muir gave rise to such an environmental ethic that places a value on species whether or not they provide any benefits to humans.

REFERENCES

1. Mora, C.; Tittensor, D.P.; Adl, S.; Simpson, A.G.B.; Worm, B. How many species are there on Earth and in the ocean? PLoS Biol. **2011**, *9* (8), 1–8.
2. The World Conservation Union. Summary statistics for globally threatened species. Table 1: Numbers of threatened species by major groups of organisms (1996–2010). In *IUCN Red List of Threatened Species*; IUCN; UK, 2010.
3. Peet, R.K. Relative diversity indices. Ecology **1975**, *56* (2), 496–498.
4. Hillebrand, H. On the generality of the latitudinal diversity gradient. Am. Nat. **2004**, *163* (2), 192–211.

5. Pianka, E.R. Latitudinal gradients in species diversity: A review of concepts. Am. Nat. **1966**, *100* (910), 33–46.
6. Myers, N.; Mittermeier, N.A; Mittermeier, C.G.; da Fonseca, G.A.B.; Kent, J. Biodiversity hotspots for conservation priority. Nature **2000**, *403* (6772), 853–858.
7. Conservation International, 2013. Biodiversity Hotspots Map. http://www.conservation.org/where/priority_areas/hotspots/Documents/CI_Biodiversity-Hotspots_2013_Map.pdf (accessed January 2013).
8. Kareiva, P.; Marvier, M. Conserving biodiversity coldspots. Am. Sci. **2003**, *91* (4), 344–351.
9. Whittaker, R.H. Evolution and measurement of species diversity. Taxon **1972**, *21* (2/3), 213–251.
10. Williams, C.B. Patterns in the Balance of Nature and Related Problems in Quantitative Ecology; Academic Press: New York, 1964.
11. Rosenzweig, M.L. *Species Diversity in Space and Time*; Cambridge University Press: Cambridge, 1995.
12. Hutchinson, G.E. Homage to Santa Rosalina or why are there so many kinds of animals? Am. Nat. **1959**, *93* (870), 145–159.
13. Connell, J.H. Diversity in tropical rainforests and coral reefs. Science **1978**, *199* (4335), 1302–1310.
14. Paine, R.T. Food web complexity and species diversity. Am. Nat. **1966**, *100* (910), 65–75.
15. Huston, M. A general hypothesis of species diversity. Am. Nat. **1979**, *113* (1), 81–101.
16. Vince, G. An epoch debate. Science **2011**, *334* (6052), 32–37.
17. Pimm, S.L.; Russell, G.J.; Gittleman, J.L.; Brooks, T.M. The future of biodiversity. Science **1995**, *269* (5222), 347–350.
18. Lawton, J.H.; May, R.M. *Extinction Rates*; Oxford University Press: Oxford, 1995.
19. Sala, O.; Chapin, F.S., III; Armesto, J.J.; Berlow, E.; Bloomfield, J.; Dirzo, R.; Huber-Sanwald, E.; Huenneke, L.F.; Jackson, R.B.; Kinzig, A.; Leemans, R.; Lodge, D.M.; Mooney, H.A.; Oesterheld, M.; Poff, N.L.; Sykes, M.T.; Walker, B.H.; Walker, M.; Wall, D.H. Global biodiversity scenarios for the year 2100. Science **2000**, *287* (5459), 1770–1774.
20. Thomas, C.D.; Cameron, A.; Green, R.E.; Bakkenes, M.; Beaumont, L.J.; Collingham, Y.C.; Erasmus, B.F.N.; de Siqueira, M.F.; Grainger, A.; Hannah, L.; Hughes, L.; Huntley, B.; van Jaarsveld, A.S.; Midgley, G.F.; Miles, L.; Ortega-Huerta, M.A.; Peterson, A.T.; Phillips, O.L.; Williams, S.E. Extinction risk from climate change. Nature **2004**, *427* (6970), 145–148.
21. Millennium Ecosystem Assessment. *Our Human Planet: Summary for Decision-Makers*; Island Press: Washington, D.C., 2005.
22. McCann, K.S. The diversity-stability debate. Nature **2000**, *405* (6783), 228–233.
23. Vilkka, L. *The Intrinsic Value of Nature*; Rodopi: Amsterdam, the Netherlands, 1997.

Ecological and Evolutionary Processes

Roland C. de Gouvenain
Department of Biology, Rhode Island College, Providence, Rhode Island, U.S.A.
Gopalasamy Reuben Clements
School of Marine and Tropical Biology, James Cook University, Cairns, Queensland, Australia

Abstract
Because many of the natural resources we harvest are the products of ecosystems, overexploitation of these resources can degrade the ecological and evolutionary processes that sustain these ecosystems. Loss of species and genetic diversity from unsustainable resource extraction removes the natural variation upon which natural selection operates to allow evolutionary change, and without which the earth's biota may no longer be able to adapt to human-induced or natural environmental changes. Sustainable resource extraction ensures not only that future human generations will enjoy these resources, but also that the ecosystems that generate these resources will maintain the capacity to do so as Earth's environments change.

INTRODUCTION

Natural resources, whether renewable like forests or fisheries, or non-renewable like crude oil or minerals, have provided generations of human beings with food, shelter, and spiritual or aesthetical enjoyment. Unfortunately, these resources are too often being destroyed or extracted at unsustainable rates. Only when we begin responsibly managing our resources within an ecosystem framework that considers long-term social good will we achieve some equilibrium between extraction and conservation.

NATURAL RESOURCES ARE COMPONENTS OF ECOSYSTEMS

Biotic (living) natural resources such as forests, wildlife, and fisheries are vital components of the different ecosystems on our planet; for instance, sphagnum moss is a component of the arctic tundra or boreal ecosystems, and gray reef sharks are components of tropical coral reef ecosystems. Abiotic (non-living) natural resources (for instance, the atmosphere, water, and soils) influence the health of those ecosystems. Thus, ecosystems are the source of many of our natural resources.[1,2] Soils, which took thousands of years to develop under the activity of microbes, fungi, and invertebrates, are also products and components of ecosystems; they sustained the advent of agriculture 10,000 years ago and support today's agricultural productivity.[1,3] Located deep underground where no life is found today, abiotic natural resources such as fossil fuels are nonetheless products of ancient forest or marine ecosystems now fossilized.[3]

ECOSYSTEMS AND THEIR RESOURCES ARE SHAPED BY EVOLUTIONARY CHANGES

All living natural resources such as forests, fisheries, and wildlife are not only shaped by ecosystems, they also influence how these ecosystems function and evolve. From the beginning of life on Earth 3.5 million years ago, ecosystems and their biotic and abiotic natural resources have been shaped by evolutionary changes in the world's biota (all the living organisms that inhabit the Earth). In fact, evolutionary forces were already in motion between 3 and 2 billion years ago, when immense colonies of single-celled *Cyanobacteria* were slowly transforming the early atmosphere of the Earth from a reducing one (low in oxygen gas) to an oxidizing one (high in oxygen gas) as a result of their photosynthetic activity.[4] This atmospheric transformation itself would, about 2.5 billion years ago, foster the evolution of more complex organisms that relied on aerobic respiration to harvest the energy contained in their food. These primitive eukaryotes would eventually (around 550 million years ago) evolve into animals, plants, and fungi.[4,5] Approximately 350 million years ago, during the tropical climate of the Carboniferous period, the Earth's biota would similarly shape its ecosystems when newly evolved tree-sized vascular terrestrial plants produced vast *Carboniferous forests*. These ancient forests were then fossilized over millions of years into thick coal deposits that

Encyclopedia of Natural Resources DOI: 10.1081/E-ENRL-120047426

fueled the industrial revolution and still powers today's electricity production.[3,5] Just as an oxygen-rich atmosphere was the product of photosynthesis by Cyanobacteria colonies, and just as today's coal deposits are the product of the Carboniferous forests, today's ecosystems are the product of their component organisms that have themselves evolved over millions of years in response to abiotic and biotic sets of natural selection factors relevant to each ecosystem.

NATURAL RESOURCE OVEREXPLOITATION JEOPARDIZES ECOSYSTEM HEALTH

The current natural resource management is driven by the exponential growth of the human population (around 7 billion people as of 2011) and socioeconomic goals that seek to maximize resource extraction with little attention to the perils of overexploitation.[1,6] Technological advances have accelerated the rate at which resources are being harvested, and this has resulted in irreversible alterations to ecosystems, especially those that are not managed according to sustainability guidelines, such as forest concessions that do not carry out reduced-impact logging. By increasing the amount and rate of extraction of biotic and abiotic natural resources, human populations change (often irreversibly) the species composition and ecology of natural ecosystems. This in turn influences the natural selection processes that shape these ecosystems, in addition to impacting the amount and location of natural resources available for future generations.

From the cold-adapted micro-organisms in the permafrost soils of Siberia to the salt-adapted trees of tropical mangroves, many plant, animal, and fungi species are pushed to the limit of their ecological tolerance as a result of the loss or transformation of their natural environment by humans. Perhaps the most well-known example is that of polar bears faced with an ever-shrinking Arctic sea ice under the influence of global climate change, but many more examples abound, including the loss of Asian tigers impacted by the logging of Asian tropical dipterocarp forests and their replacement with oil palm or rubber plantations, the disappearance of migratory bird species due to loss of stop-over wetlands, the collapse of oceanic fisheries due to overfishing, and the poaching of charismatic large mammals in increasingly logged rain forests.[7–13] Although some cases of rapid evolutionary adaptive response of plants and insects to environmental degradation or to climate change have been documented,[14,15] most studies report negative impacts (including species extinction) from the overharvesting of biotic resources or from the impacts of abiotic resource extraction.[11,13,15]

By reducing the biodiversity (the diversity of species) of the earth, we are not only losing plant, animal, and fungi species that could potentially provide future generations with critical resources, including yet undiscovered medicinal compounds, we are also losing the ecological services these ecosystems provide us with (such as the storage of clean water in forest watersheds). Furthermore, we are destroying the very species and ecological processes that have made these ecosystems productive and resilient to natural or human-induced environmental changes.[15,16] Losing walleye pollock (a prey species for predatory marine fish, birds, and mammals) because of climate change affects the other marine species in the Bering Sea food web, and degrades the health and productivity of the oceanic ecosystems that support economically valuable fisheries.[17]

As species diversity is lost, the raw material upon which natural selection operates is lost as well. Loss of intertidal species (for instance mangrove trees and seagrass beds) to development reduces the ability of coastal ecosystems to filter polluted effluents or to protect shorelines from storm flooding,[13] which can hinder future evolutionary adaptation of these coastal ecosystems to climate and sea-level changes. Worldwide, nearly 30% of the currently fished species are considered collapsed (>90% decline), and this decline occurred faster in species-poor ecosystems than in species-rich ones, suggesting that maintaining ecosystem biodiversity is the key to future ecosystem resilience in the face of human impact.[13,16] Monoculture plantations of cash crops, such as oil palm and rubber, are rapidly transforming tropical forests into biological deserts.[7,18] Southeast Asia has the highest rate of tropical deforestation, losing around 1.0% of its forests per year,[19] and some countries like the Philippines and Singapore have nearly no forest cover left. Indonesia's forest ecosystems are impacted by even higher rates of deforestation (2.6% per year), and two-thirds of its original forests are now replaced with oil palm plantations. In large part as a result of this rapid deforestation, more than 25% of the mammals, more than 20% of the birds, and more than 15% of the native plants species of Southeast Asia are now extinct.[11]

ECOLOGICALLY SUSTAINABLE NATURAL RESOURCE MANAGEMENT CAN PROTECT EVOLUTIONARY PROCESSES

To be ecologically sustainable, management of natural resources must protect the ecological integrity of ecosystems, including their biodiversity, so that the evolutionary processes that generated these ecosystems can be maintained in the future. This will increase the likelihood that these ecosystems will be able to evolve in response to changing global conditions,[16,20] especially because it is difficult to predict with certainty the characteristics of future terrestrial and aquatic environments.[21] For example, maintaining habitat connectivity (through wildlife corridors, for instance) can help ensure a beneficial exchange of genetic material among populations isolated by habitat conversion.[9,22,23] Conserving the health of soil ecosystems by protecting them from erosion and maintaining a diversity of nutrient-cycling microbes can ensure that future generations will enjoy sustained agricultural productivity.[24]

Perhaps the single-most important indicator of ecosystem health is biodiversity, and the genetic diversity it manifests is what natural selection can operate on to allow the natural processes of evolution to keep these ecosystems healthy, resilient, and productive.[16,20] Not only can productive fisheries not exist in species-poor oceans, or economically viable wood supplies not be maintained in unsustainably logged landscapes, but species-poor ecosystems are also much more likely to collapse in the face of natural or human-induced environmental change than species-rich ones.[16] Terrestrial and marine species diversity enhances ecosystem processes such as nutrient cycling and primary production and increases the resilience of ecosystems to disturbance and therefore the capacity of these ecosystems to provide services to future human generations.[13] Sustainable resource extraction is not only common sense for long-term economic benefits to humans, it is also a management practice that can give the ecosystems of the world a chance to adapt to and persist.[20] For instance, sustainably logged Malaysian tropical forests can still support reasonable densities of tiger populations,[25] and as long as forest structure and biodiversity are maintained, even old logging roads are traveled regularly by the big cats (Fig. 1).

To be sustainable, natural resource management should be both ecologically responsible, that is, it should respect the "economy of nature" as defined by the German biologist Ernst Haeckel, and socially ethical, as suggested by the American ecologist Aldo Leopold.[26] Societal decisions regarding both the exploitation and the conservation of natural resources should be based on ecological knowledge and research, because ecological sustainability is the key to long-term success,[23,27] and on social justice, because extracting and conserving resources involve trade-offs that have ecological and social costs.[28–30] This is especially urgent, given the fact that most of the world's biodiversity "hotspots" (areas with both high biodiversity and high rates of natural habitat loss)[10] are also areas of relatively high human population density and growth.[31]

To assess these trade-offs and make socially fair decisions regarding the extraction and conservation of natural resources, other points of view besides the dominant market-oriented western perspective should be solicited and appraised, including the traditional ecological knowledge of the world's rural cultures.[32] Giving resource management more legitimacy by involving local human communities in the decision process concerning the extraction and conservation of natural resources is likely to yield management actions that are not only ecologically but socially sustainable as well.[28] For instance, conservation of the Tampolo coastal rain forest in Madagascar has enjoyed support from local communities by mixing traditional covenant ceremonials, agricultural development, ecotourism, native tree species planting, and environmental education (Fig. 2). The Tampolo Forest, although only 800 hectares in size, is home to 90 species of ants, 56 bird species, 16 species of amphibians, 31 reptiles species, and 6 species of lemur.[33,34] Elsewhere, in Thailand, a successful community-based conservation approach has even resulted in the recovery of ungulate populations that were previously subjected to poaching pressure.[35] Community-based conservation approaches may involve negotiating trade-offs and compromises that are complex and thus require more time to achieve than the traditional "top down" conservation

Fig. 1 Tiger on abandoned logging road photographed by photo trap, Tembat Forest Reserve, Terengganu, Peninsular Malaysia.
Source: Photo by Rimba/Reuben Clements.

Fig. 2 Forest reserve and local inhabitants at a native tree and vegetable nursery, Tampolo Madagascar.
Source: Photos by Roland de Gouvenain.

approach. However, the process itself may not only empower local communities to own and manage their resources as caretakers, it may also enhance the educational benefit for all parties involved, and allow stakeholders to embrace this necessary commitment to long-term resource conservation for the sake of future human generations.

CONCLUSION

Although unsustainable extraction of natural resources and pollution from the use of these resources has modified nearly the entire set of ecosystems found on Earth, as long as ecological processes are not irremediably impacted and species diversity is maintained, intrinsic properties of these ecosystems can allow natural processes to restore their health and productivity.[9] It behooves us, as de facto caretakers of the Earth's environment, to keep its biota diverse and healthy so that natural evolutionary processes can provide future generations with an ecologically and socially livable world.

REFERENCES

1. Holechek, J.L.; Cole, R.A.; Fisher, J.T.; Valdez, R. *Natural Resources: Ecology, Economics, and Policy*; Pearson Prentice Hall: Upper Saddle River, NJ, 2003; 716–718 pp.
2. Weddell, B.J. *Conserving Living Natural Resources in the Context of a Changing World*; Cambridge University Press: Cambridge, 2002; 426 pp.
3. Chiras, D.D.; Reganold, J.P. *Natural Resource Conservation: Management for a Sustainable Future*; Pearson Prentice Hall: Upper Saddle River, 2005; 644 pp.
4. Knoll, A. *Life on a Young Planet*; Princeton University Press: Princeton, 2003; 296 pp.
5. Futuyma, D.J. *Evolution*, 2nd Ed.; Sinauer: Sunderland, 2009.
6. Rutherford, M.B. Resource management. In *Encyclopedia of Environmental Ethics and Philosophy*; Callicott, J.B.; Frodeman, R., Eds.; Macmillan – Gale: Farmington Hills, 2009; 198–201.
7. Aziz, S.A.; Laurance, W.F.; Clements, R. Forests reserved for rubber. Front. Ecol. Env. **2010**, *8*, 178.
8. Clements, R.; Rayan, D.M.; Ahmad Zafir, A.W.; Venkataraman, A.; Alfred, R.; Payne, J.; Ambu, L.N.; Sharma, D.S.K. Trio under threat: can we secure the future of rhinos, elephants and tigers in Malaysia? Biodivers. Conserv. **2010**, *19*, 1115–1136.
9. Gardner, T.A.; Barlow, J.; Chazdon, R.; Ewers, R.M.; Harvey, C.A.; Peres, C.A.; Sodhi, N.S. Prospects for tropical forest biodiversity in a human-modified world. Ecol. Letters **2009**, *12*, 561–582.
10. Myers, N.; Mittermeier, R.A.; Mittermeier, C.G.; da Fonseca, G.A.B.; Kent, J. Biodiversity hotspots for conservation priorities. Nature **2000**, *403*, 853–858.
11. Sodhi, N.S.; Koh, L.P.; Brook, B.W.; Ng, P.K.L. Southeast Asian biodiversity: an impending disaster. Trends Ecol. Evol. **2004**, *19*, 654–660.
12. Venter, O.; Meijaard, E.; Possingham, H.; Dennis, R.; Sheil, D.; Wich, S.; Hovani, L.; Wilson, K. Carbon payments as a safeguard for threatened tropical mammals. Conserv. Lett. **2009**, *2*, 123–129.
13. Worm, B.; Barbier, E.B.; Beaumont, N.; Duffy, J.E.; Folke, C.; Halpern, B.S.; Jackson, J.B.C.; Lotze, H.K.; Micheli, F.;

Palumbi, S.R.; Sala, E.; Selkoe, K.A.; Stachowicz, J.J.; Watson, R. Impacts of Biodiversity Loss on Ocean Ecosystem Services. Science **2006**, *314*, 787–790.
14. Bone, E.; Farres, A. Trends and rates of microevolution in plants. Genetica **2001**, *112–113*, 165–182.
15. Parmesan, C. Ecological and evolutionary responses to recent climate change. Ann. Rev. Ecol. Evol. Syst. **2006**, *37*, 637–669.
16. Elmqvist, T.; Folke, C.; Nyström, M.; Peterson, G.; Bengtsson, J.; Walker, B.; Norberg, J. Response Diversity, Ecosystem Change, and Resilience. Front. Ecol. Env. **2003**, *1*, 488–494.
17. Walther, G.-R.; Post, E.; Convey, P.; Menzel, A.; Parmesan, C.; Beebee, T.J.C.; Fromentin, J.-M.; Hoegh-Guldberg, O.; Bairlein, F. Ecological responses to recent climate change. Nature **2002**, *416*, 389–395.
18. Koh, L.P.; Wilcove, D.S. Is oil palm agriculture really destroying tropical biodiversity? Conserv. Lett. **2008**, *1*, 60–64.
19. Miettinen, J.; Shi, C.; Liew, S.C. Deforestation rates in insular Southeast Asia between 2000 and 2010. Glob. Change Biol. **2011**, *17*, 2261–2270.
20. Fazey, I.; Gamarra, J.G.P.; Fischer, J.; Reed, M.S.; Stringer, L.C.; Christie, M. Adaptation strategies for reducing vulnerability to future environmental change. Front. Ecol. Env. **2010**, *8*, 414–422.
21. Lawler, J.J.; Tear, T.H.; Pyke, C.; Shaw, M.R.; Gonzalez, P.; Kareiva, P.; Hansen, L.; Hannah, L.; Klausmeyer, K.; Aldous, A.; Bienz, C.; Pearsall, S. Resource management in a changing and uncertain climate. Front. Ecol. Env. **2010**, *8*, 35–43.
22. Chave, J.; Wiegand, K.; Levin, S. Spatial and biological aspects of reserve design. Environ. Model. Assess. **2002**, *7*, 115–122.
23. Sodhi, N.S.; Koh, L.P.; Clements, R.; Wanger, T.C.; Hill, J.K.; Hamer, K.C.; Clough, Y.; Tscharntke, T.; Posa, M.R.C.; Lee, T.M. Conserving Southeast Asian forest biodiversity in human-modified landscapes. Biol. Conserv. **2010**, *143*, 2375–2384.
24. Stringer, L. Can the UN Convention to Combat Desertification guide sustainable use of the world's soils? Front. Ecol. Env. **2008**, *3*, 138–144.
25. Rayan, D.M.; Mohamad, S.W. The importance of selectively logged forests for tiger *Panthera tigris* conservation: A population density estimate in Peninsular Malaysia. Oryx **2009**, *43*, 48–51.
26. Rozzi, R. The reciprocal links between evolutionary-ecological sciences and environmental ethics. BioScience **1999**, *49*, 911–921.
27. Palmer, M.A.; Bernhardt, E.S.; Chornesky, E.A.; Collins, S.L.; Dobson, A.P.; Duke, C.S.; Gold, B.D.; Jacobson, R.B.; Kingsland, S.E.; Kranz, R.H.; Mappin, M.J.; Martinez, M.L.; Micheli, F.; Morse, J.L.; Pace, M.L.; Pascual, M.; Palumbi, S.S.; Reichman, O.J.; Townsend, A.R.; Turner; M.G. Ecological science and sustainability for the 21st century. Front. Ecol. Env. **2005**, *3*, 4–11.
28. Brechin, S.R.; Wilshusen, P.R.; Fortwangler, C.L, West, P.C. *Contested Nature: Promoting International Biodiversity with Social Justice in the Twenty-first Century*; State University of New York Press: Albany, 2003; 321 pp.
29. Lugo, A.E. Management of Tropical Biodiversity. Ecol. Appl. **1995**, *5*, 956–961.
30. Peluso, N.L. *Rich Forests, Poor People: Resource Control and Resistance in Java*; University of California Press: Berkeley, 1992; 321 pp.
31. Cincotta, R.P.; Wisnewski, J.; Engelman, R. Human population in the biodiversity hotspots. Nature **2000**, *404*, 990–992.
32. Menzies, C.R. *Traditional Ecological Knowledge and Natural Resource Management*; University of Nebraska Press: Lincoln, 2006; 274 pp.
33. de Gouvenain, R.C.; Silander, J.A. Jr. Littoral forest. In *The Natural History of Madagascar*; Goodman, S.M.; Benstead, J., Eds.; Field Museum of Natural History: Chicago, 2003; 103–111.
34. Ratsirarson, J.; Goodman, S.M. Inventaire biologique de la forêt littorale de Tampolo (Fenoarivo Atsinanana). Recherches pour le développement. Série Sciences Biologiques No. 14. Centre d'Information et de Documentation Scientifique et Technique: Antananarivo, 1998.
35. Steinmetz, R.; Chutipong, W.; Seuaturien, N.; Chirngsaard, E.; Khaengkhetkarn, M. Population recovery patterns of Southeast Asian ungulates after poaching. Biol. Conserv. **2010**, *143*, 42–51.

Ecological Security: Land Use Pattern and Simulation Modeling

Xun Shi
Department of Geography, Dartmouth College, Hanover, New Hampshire, U.S.A.
Qingsheng Yang
Department of Resources and Environment, Guangdong University of Business Studies, Guangzhou, China

Abstract
While a consensus definition has yet to be reached, studies under the topic of *ecological security* have been active in the past two decades. This entry summarizes the diverse studies under this topic from perspectives of the *growth-limit* notion, national security concern, and geographic scale. At the city level, land use dynamics can be both the cause and consequence of a severe ecological security situation. Cellular automata (CA), integrated with geographic information systems (GIS), are an effective approach to: 1) predicting future land use dynamics of an area based on the historical trend; and 2) modeling the effect or consequence of certain planning measures or policies on urban land use. Therefore, CA-GIS models should be useful in decision support for planning and policy-making regarding urban ecological security. Case studies in Guangdong, one of the most economically active regions of China, are used to illustrate such a modeling process.

INTRODUCTION

The term "ecological security" appeared in the late 1980s.[1,2] There is no consensus yet on its definition,[3] but that has not prevented the research under this topic from growing into a fairly active area in the past two decades.[4–7] While the environmental, ecological, and resource problems addressed by the research are not new, its perspective and emphasis may be somewhat different from those of conventional environmental and ecological studies. The diverse work in this area can be summarized as in the following text.

Expanding the Growth-Limit Perspective

Represented by the book *The Limits to Growth*,[8] a classical conceptualization attributes contemporary environmental problems to the depletion of *resources*, including energy, food, and capability of dispersing waste, as a result of fast-growing population and industrialization. The growth-limit notion triggered fierce debate between the resource-pessimists and techno-optimists.[5] A few decades later, the crises caused purely by resource scarcity have not materialized, but humankind is not facing a less challenging environmental condition. While empirical observations indicate that for the foreseeable future resource scarcity might be a relatively minor source of human suffering, infectious disease, conflict among peoples, starvation, and various kinds of environmental disasters have become major causes of premature human deaths and disabilities, and meanwhile, the extinction of many flora and fauna species has become a serious problem. All these have raised the level of concern over future ecological security.[5] This general concern is presented by Pirages as four interrelated dynamic equilibriums:

1. Between human populations living at higher consumption levels and the ability of nature to provide resources and services
2. Between human populations and pathogenic microorganisms
3. Between human populations and those of other plant and animal species
4. Among human populations

Pirages considers that whenever any of these equilibriums is disrupted either by changes in human behavior or in nature, insecurity will increase.[5]

Raising to the Security Level

A highlight of the conception of ecological security is that it raises concern about the environmental/ecological issues to a qualitatively new level, the level of national security.[2] Conventionally, national security mainly refers to the protection of societies and states from predatory neighbors. However, ecosystemic challenges such as plagues, pestilence, pollution, blizzards, floods, and droughts, often aided and abetted by intemperate human behavior, over time, have been responsible for killing and injuring much larger numbers of human beings.[4] Thus, Westing defines a broader human security as being

Encyclopedia of Natural Resources DOI: 10.1081/E-ENRL-120048596

composed of two intertwined components: political security and environmental/ecological security, where political security includes military, economic, and social/humanitarian subcomponents, and ecological security has protection- and utilization-oriented subcomponents.[9] The protection requirement refers to safeguarding the quality of the human environment, and the utilization requirement means providing a sustaining basis for any exploitation (harvesting or use) of a renewable natural resource.

Geographic Scale

Ecological security is temporally dynamic and has spatial structure,[10] and is thus closely associated with the geographic scale. Typically, ecological security has been studied at the global, national, and regional/city levels. Globalization is making the boundaries formed by physical, political, and cultural barriers much more porous and less important, which seems to have brought about ecological risk and insecurity.[4] For example, the dramatically increased number of people and quantities of goods moving rapidly from place to place is facilitating the unintended and often destructive spread of plants and pests into new environments and ecosystems, which is disturbing the long-established equilibrium between people and pathogens, and is causing the rapid spread of new and resurgent diseases.[4] At the national and regional/city scales, over the background of resource constraints and climate change, national security, infrastructure "protection," and economic competitiveness are being overlaid with concerns around energy security, constraints on water resources, the growth of diseases, increased flood risks, and multiple aspects of demographic shifts. National and regional/city governments are seeking the ability to ensure *secure* access to the resources needed for the country's, region's, and city's replications.[6]

URBAN ECOLOGICAL SECURITY AND LAND USE PATTERN

Urban ecological security (UES) refers to the contemporary conditions within which cities must actively seek to reproduce their economic, social, and material fabric.[6] The term reflects the emergence of the situation that ecological security is increasingly becoming an issue particularly at an urban scale, which is the consequence of four interrelated sets of pressures:[6] First, cities are responsible for disproportionate levels of consumption of resources and greenhouse gas emissions. For example, despite accounting for around half of the world's population, cities are responsible for around 75% of energy consumption and 80% of greenhouse gas emissions. Second, consequently, cities are likely to be chief amongst the *victims* of resource constraints and climate change. For example, many coastal and river-side cities may suffer flooding and the health consequences of the urban heat island effect. Third, cities give the potential contexts for the response to issues of resource constraint and climate change. For example, cities may be ideal places to demonstrate and experiment with decentralized energy and water technologies, and new urban mobility and transportation systems based on, for example, hydrogen and biodiesel. Finally, in an era of intensified economic globalization and competition between places, urban coalitions should be able to anticipate, shape, and respond strategically to national priorities.

Urban land use and land cover changes (LUCC) can be both causes and results of ecological and environmental problems. Thus, strategic urban/regional plans, which are targeting and are to be reflected by LUCC, ought to support sustainable development and promote ecological security by addressing existing as well as newly introduced environmental problems.[10] Methods to assess urban planning and the resulting growth scenarios, as well as to forecast the resulting urban ecological security, are a critical requirement for the sustainable development of an area.[10]

Simulation and Modeling of Land Use Pattern for Assessing Urban Ecological Security

Cellular automata (CA), integrated with geographic information systems (GIS), are an effective technical approach to simulating complex urban land use.[11–17] For UES studies, the CA-GIS simulation can serve two purposes: 1) predicting an UES situation in the future based on the simulation of land use dynamics, and 2) experimenting with planning interferences that may affect UES by creating different "what-if" scenarios.

CA consist of a collection of discrete cells that represent spatial units, each in one of a finite number of states. The state of each cell evolves through a number of discrete time steps controlled by a set of transition rules. These rules define how a cell will evolve based on its own state and the states of its neighboring cells.[17] CA models can generate complex global patterns with simple local rules. While the rules are only applied at the neighborhood level, they may be able to represent the impact of factors at different (i.e., local, regional, and global) spatial scales.[13,14,18] The transition rules are key inputs in a CA model.

Gong et al. developed a CA model for predicting the ecological security of Guangzhou, a big city located in one of the most economically active regions of China.[10] The model couples CA with ecological security assessment to identify insecure cells within the region and simulate their future development. The *digital space* of the model is composed of square grids, each with a size of 100 × 100 m. The status of each cell was designated as 1 for *not secure* or 0 for *secure*. The cell values were determined according to a previously conducted ecological security assessment for Guangzhou for 1990–2005. During the modeling, the grids evolved as the simulation map generated from one

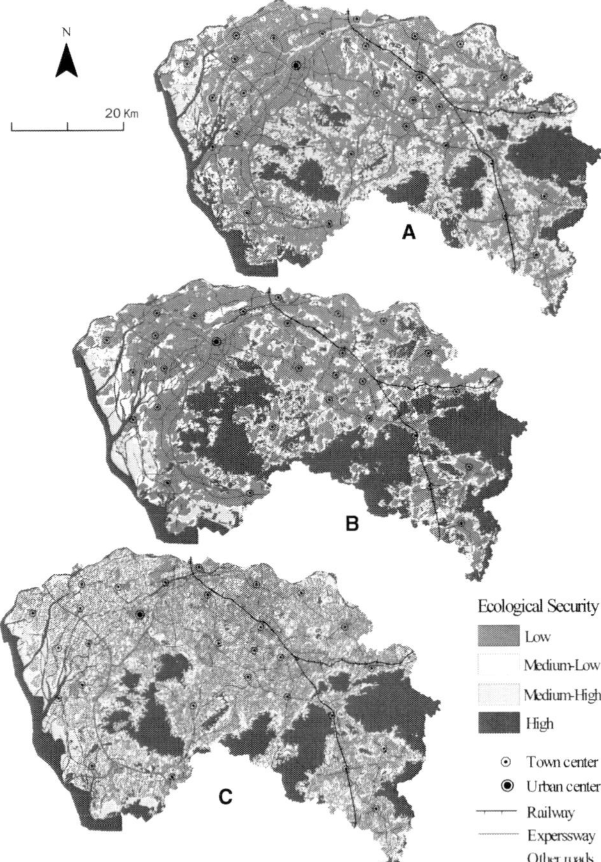

Fig. 1 GIS-CA-based simulation of urban ecological security under certain planning measures for Guangzhou City, China: (**A**) The result of ecological security assessment based on the actual land use situation in 2005; (**B**) The simulated ecological security for 2005 solely based on the historical land use trend (i.e., without any restrictions from planning); (**C**) The simulated ecological security for 2005 based on the historical trend and controlled by a local development threshold.

iteration sequentially being used as the input for the next iteration. The transition rules that determine if the status of a cell would change were represented by an integrated vector. These rules calculated the transition probability based on the cell's current status, the neighborhood effect, and the urban planning factor. Validation using historical data showed that the accuracy of the model could reach 72.1%. Using the situation of 2005 as the starting state, Gong et al. ran simulations with the model to predict ecological security for 2020. The predicted ecological security situation of Guangzhou shows that while certain measures in the governmental development planning may gentle the decreasing trend of UES of the city, such a trend will continue into 2020. A landscape pattern analysis suggested a more scattered and homogenous distribution of land use in Guangzhou's urban landscape, as well as considerable variation among different districts of the city. The modeled UES highlights the need to make ecological protection an integral part of urban planning.

The CA-GIS-based modeling can also be used to conduct experiments about effects of different planning measures on UES. This idea can be implemented by incorporating the planning measures into the transition rules of CA. For example, if a goal of the planning is to restrict excessive expansion of big cities, and meanwhile maintain the total magnitude of regional urban development, one can translate this policy into a rule that suppresses the transition probability of a non-urban cell if the number of urban cells within its neighborhood is above a threshold. The maps in Fig. 1 illustrate such a simulation.

SUMMARY

The concept of ecology security raises concern about environmental/ecological problems to the level of national security, and recognizes this security issue at different geographic scales, from global to regional and city. At the city level, land use dynamics can be both cause and consequence of a severe ecological security situation. CA integrated with GIS are an effective approach to: 1) predicting future land use dynamics of an area based on the historical trend; and 2) modeling the effect or consequence of certain planning measures or policies on urban land use. Therefore, CA-GIS models should be useful in decision support for planning and policy-making regarding urban ecological security.

REFERENCES

1. Myers, N. The environmental dimension to security issues. Environmentalist **1986**, *6* (4), 251–257.
2. Timoshenko, A.S. Ecological security: The international aspect. Pace Environ. Law Rev. **1989**, *7* (1), 151–160.
3. Zou, C.; Shen, W. Advances in ecological security. Rural Eco-Environ. **2003**, *19* (1), 56–59.
4. Pirages, D.C.; DeGeest, T.M. From international to global relations. In *Ecological Security: An Evolutionary Perspective on Globalization*; Rowman & Littlefield Publishers: Lanham, USA, 2004, 1–28.
5. Pirages, D. From limits to growth to ecological security. In *From Resource Scarcity to Ecological Security: Exploring New Limits to Growth*; Pirages, D., Cousins, K., Eds.; The MIT press: Cambridge, USA, 2005, 1–19.
6. Hodson, M.; Marvin, S. Urban ecological security: A new urban paradigm? Int. J. Urban Reg. Res. **2009**, *33* (1), 193–215.
7. Hodson, M.; Marvin, S. Urbanism in the anthropocene: Ecological urbanism or premium ecological enclaves? City **2010**, *14* (3), 298–313.
8. Meadows, D.H.; Meadows, D.L.; Randers, J.; Behrens, W. *The Limits to Growth*; University books: New York, USA, 1972.
9. Westing, A.H. The environmental component of comprehensive security. Bull. Peace Proposals **1989**, *20* (2), 129–134.
10. Gong, J.; Liu, Y.; Xia, B.; Zhao, G. Urban ecological security assessment and forecasting, based on a cellular automata model: A case study of Guangzhou, China. Ecol. Model. **2009**, *220* (24), 3612–3620.

11. Batty, M.; Xie, Y. From cells to cities. Environ. Plann. B Plann. Des. **1994**, *21* (7), 531–548.
12. Clarke, K.C.; Hoppen, S.; Gaydos, L. A self-modifying cellular automaton model of historical urbanization in the San Francisco Bay area. Environ. Plann. B Plann. Des. **1997**, *24* (2), 247–261.
13. Wu, F.; Webster, C.J. Simulation of land development through the integration of cellular automata and multicriteria evaluation. Environ. Plann. B Plann. Des. **1998**, *25* (1), 103–126.
14. Li, X.; Yeh, A.G. Neural-network-based cellular automata for simulating multiple land use changes using GIS. Int. J. Geogr. Inf. Sci. **2002**, *16* (4), 323–343.
15. Li, X.; Yeh, A.G. Data mining of cellular automata's transition rules. Int. J. Geogr. Inf. Sci. **2004**, *18* (8), 723–744.
16. Li, X.; Yeh, A.G. Modelling sustainable urban development by the integration of constrained cellular automata and GIS. Int. J. Geogr. Inf. Sci. **2000**, *14* (2), 131–152.
17. Yang, Q.; Li, X; Shi, X. Cellular automata for simulating land use changes based on support vector machines. Comput. Geosci. **2008**, *34* (6), 592–602.
18. Li, X.; Shi, X.; He, J.; Liu, X. Coupling simulation and optimization to solve planning problems in a fast developing area. Ann. Assoc. Am. Geogr. **2011**, *101* (5), 1032–1048.
19. Yu, P. *Multiple Criteria Decision Making: Concepts, Techniques and Extensions*; Plenum Press: New York, USA, 1985.

Ecology: Functions, Patterns, and Evolution

Fernando Valladares
Centro de Ciencias Medioambientales, Madrid, Spain

Biomes—Ecosystems

Abstract
Ecology is the study of living organisms and their environment in an attempt to explain and predict. The objectives of ecology in general and of functional ecology in particular are to develop predictive theories and to assemble the data to develop general models. Functional ecology is concerned with the links between structure and function, the existence of general patterns among species, and the evolutionary connections among these patterns.

INTRODUCTION

Ecology is the study of living organisms and their environment in an attempt to explain and predict. While natural history involves the accumulation of detailed data with emphasis on the autecology of each species, the objectives of ecology in general and of functional ecology in particular are to develop predictive theories and to assemble the data to develop general models. Functional ecology has three basic components: 1) constructing trait matrices through screening of various plant and animal species, 2) exploring empirical relationships among these traits, and 3) determining the relationships between traits and environments.[1] Studies in functional ecology encompass a wide range of approaches, from individuals to populations; from mechanistically detailed to deliberately simplified, black-box simulations; and from deductive to inductive.[2] Functional ecology is concerned with the links between structure and function, the existence of general patterns among species, and the evolutionary connections among these patterns. And functional ecology is, above all, timely and pertinent, because the environmental degradation associated with human development is rapidly destroying the very systems that ecologists seek to understand. If we are to anticipate the extent and repercussions of global change in natural habitats, we first need to understand how organisms and ecosystems function.

EXPLORING THE MECHANISMS: FROM PHYSIOLOGY TO ECOPHYSIOLOGY AND FUNCTIONAL ECOLOGY

The study of plant functions has largely followed a reductionistic approach aimed at explaining the functions in terms of the principles of physics and chemistry. However, plant physiology, which is focused on molecules, organelles, and cells, has not been able to provide reliable predictions of the responses of vegetation to changes in their environment, due to the multiple interactions involved in the responses of plants and the hierarchical nature of plant organization.[3] The realization of the fact that the responses of higher organizational levels are not predictable from the dynamics of those of smaller scales led to the advent of plant ecophysiology, which is focused on organs—a more relevant scale of organization to address questions regarding plant performance. Ecophysiologists have primarily focused on the structural and functional properties of leaves. Leaves display wide variation in morphology and physiology, including differences in specific mass, carbon and nitrogen investments, stomatal densities, optical properties, and hydraulic and photosynthetic characteristics. The emphasis of ecophysiological studies on leaves is due to the profound implications of the interactions between leaf structure and function for the performance of plants in natural habitats.[4,5] However, the traditional ecophysiological approach proved to be insufficient in predicting plant distribution and responses to changing environments, which led to the development of functional plant ecology.[6] Functional ecology is centered on whole plants as the unit of analysis, encompassing a range of scales of organization from organs to whole organism architecture, and is based on a much broader conception of plant functions than that formulated by the earlier practitioners of ecophysiology.[7]

THE SEARCH OF GENERAL PATTERNS: COMPARATIVE ECOLOGY

Since elucidation of the range of possible functional responses of plants is not possible with the use of model organisms, such as those typically used in plant physiology,

Encyclopedia of Natural Resources DOI: 10.1081/E-ENRL-120010498

functional ecology arises as an essentially comparative science. Ideally, functional ecology deals with traits measured on a large number of species in order to minimize the influence of the peculiarities of the autecology of each species. Two main approaches have been followed to find general ecological patterns in nature: 1) screening, that is, the design of bioassays for a trait or a set of traits measured simultaneously on a large number of species, as in the classic study by Grime and Hunt[8] of the relative growth rate of 132 species of British flora, and 2) empiricism, or the search for quantitative relationships between measurable dependent and independent variables (e.g., correlations among pairs of traits or traits and environments) producing quantitative models using traits and not species, as in the general revision of leaf traits by Reich et al.[4] Models using traits are more general than those based on species and can be more easily transferred to different floras.[1] The question arises as to which trait must be measured. A possible answer can be obtained by analyzing the basic functions that organisms perform (i.e., resource acquisition, the ability to tolerate environmental extremes, and the ability to compete with neighbors) and then either looking for traits which measure these functions or carrying out direct bioassays of them. The most common limitations of this kind of study are the difficulties in finding unambiguous linkages of a trait to a specific function, and the so-called phylogenetic constraints that are due to the fact that phylogenetic proximity among species can influence their functional similarities.

Individual plants demonstrate important degrees of phenotypic variation, which must be considered in comparative studies. No two individuals of the same species exhibit the same final shape or functional features, regardless of how similar the genotypes of two individuals may be.[9] Part of this variation is due to phenotypic plasticity, that is, the capacity of a given genotype to render different phenotypes under different environmental conditions, and part is due to other reasons (ontogenetic stage, developmental instability). Plants of the same chronological age can be ontogenetically different, so interpretation of differences in phenotypic traits will depend on whether comparisons are made as a function of age, size, or developmental stage.

EXPERIMENTAL ECOLOGY

Broad-scale comparisons must be driven by hypotheses and must be based on robust statistical designs. However, the finding of statistically significant patterns is no guarantee of underlying cause-and-effect relationships, which must be tested experimentally. Gradient analysis, where functional responses are examined along a clearly defined environmental gradient, is a powerful approach to exploring the relationships between plant function and environment, but it is prone to spurious relationships when there is a hidden factor covarying with the factor defining the gradient.[3] Inferences from gradient analysis and broad-scale comparisons are statistical in nature and must be confirmed experimentally.

Experiments are designed to test hypotheses, but sufficient knowledge must be available to specify more than a trivial hypothesis before thorough experimentation can be undertaken.[2] Unfortunately, this is not the case for many natural systems. Although experiments are a common practice in physiology and ecophysiology, where mechanisms can be explored and hypotheses tested under relatively well-controlled conditions, they are less common in functional ecology studies, especially when hypotheses must be tested in natural habitats. Experimental design should meet a number of standards that are not easy—and in certain cases not possible—to meet in field ecology. The controls should be randomly intermixed with the treatments, both in space and time, and both the control and the experiments must be replicated enough. Replicates are not easy to find in natural scenarios, and when they can be found they are frequently not truly independent of one another, leading to a weak experimental design due to pseudoreplication. In addition, multiple causality and indirect effects, which significantly complicate the interpretation of results, are commonplace in ecology. However, experimental ecology is burgeoning despite all this adversity, because experiments are irreplaceable elements in achieving relevant progress in our understanding of ecological processes, as has been revealed by a number of experimental manipulations of natural populations.[10]

EVOLUTIONARY ECOLOGY

Functional ecology eventually leads to evolutionary ecology. Trends in functional traits across species and mechanisms linking cause and effect contribute to our understanding of evolutionary processes, especially when they are considered on a time scale long enough to allow for changes in gene frequencies, the essence of evolution. Whereas functional ecology is interested in the immediate influence of environment on a given trait, evolutionary ecology is aimed at understanding why some individuals have left the most offspring in response to long-term consistent patterns of environmental conditions. Functional ecology, focused on an ecological time scale (*now* time), asks questions of "how?" and is concerned with the proximate factors influencing an event. Evolutionary ecology is focused on an evolutionary time scale (geological time), asking questions of "why?" and concerned with the ultimate factors influencing an event. Neither is more correct than the other, and they are not mutually exclusive because ecological events can always be profitably considered within an evolutionary framework, and vice versa.[10]

There are five agents of evolution: natural selection (differential reproductive success of individuals within a population); genetic drift (random sampling bias in small populations); gene flow (migration movements of individuals among and between populations with different gene frequencies); meiotic drive (segregation distortion of certain alleles that do not follow the Mendelian lottery of meiosis and recombination); and mutation. Of these agents, only natural selection is directed, resulting in conformity between organisms and their environments.[10] Darwin's theory of natural selection is a fundamental unifying theory, and the many studies that have been carried out over the last century in support of it demonstrate the power of the rigorous application of the genetic theory of natural selection to population biology.

REFERENCES

1. Keddy, P.A. A pragmatic approach to functional ecology. Funct. Ecol. **1992**, *6*, 621–626.
2. Mentis, M.T. Hypothetico-deductive and inductive approaches in ecology. Funct. Ecol. **1988**, *2*, 5–14.
3. Duarte, C. Methods in Comparative Functional Ecology. In *Handbook of Functional Plant Ecology*; Pugnaire, F.I., Valladares, F., Eds.; Marcel Dekker: New York, 1999; 1–8.
4. Reich, P.B.; Ellsworth, D.S.; Walters, M.B.; Vose, J.M.; Gresham, C.; Volin, J.C.; Bowman, W.D. Generality of leaf trait relationships: A test across six biomes. Ecology **1999**, *80*, 1955–1969.
5. Niinemets, U. Global-scale climatic controls of leaf dry mass per area, density, and thickness in trees and shrubs. Ecology **2001**, *82*, 453–469.
6. Pugnaire, F.I.; Valladares, F. *Handbook of Functional Plant Ecology*; Marcel Dekker: New York, 1999.
7. Valladares, F.; Pearcy, R.W. The functional ecology of shoot architecture in sun and shade plants of *Heteromeles arbutifolia* M. Roem., a Californian chaparral shrub. Oecologia **1998**, *114*, 1–10.
8. Grime, J.P.; Hunt, R. Relative growth rate: Its range and adaptive significance in a local flora. J. Ecol. **1975**, *63*, 393–422.
9. Coleman, J.S.; McConnaughay; Ackerley, D.D. Interpreting phenotypic variation in plants. Trends Ecol. Evol. **1994**, *9*, 187–191.
10. Pianka, E.R. *Evolutionary Ecology*; Benjamin-Cummings: San Francisco, 2000.

Ecosystem Services: Evaluation

James Boyd
Resources for the Future, Washington, District of Columbia, U.S.A.

Abstract

Evaluation of ecosystem services involves two types of linked analyses. The first is biophysical analysis—associated with ecology, hydrology, and the other natural sciences—that describes ecological conditions, and changes in those conditions, resulting from environmental management, stresses, and protection. The second is economic analysis to measure and communicate the social benefit or cost of those ecological changes. Spatial analysis is important to both types of evaluation, as are ecological outcome measures that facilitate the interpretation of ecological changes by communities, households, and other stakeholders.

INTRODUCTION

The term "ecosystem services" conveys the principle that natural systems provide socially and economically valuable goods and services deserving of protection, restoration, and enhancement. Ecosystem goods and services include the ecological features, qualities, or commodities we value, such as food, timber, clean drinking water, available water for irrigation, transportation, and industry, clean air, scenic beauty, and species important to us for recreational, ethical, or cultural reasons. Ecosystem goods and services also relate to nature's ability to protect us from harm. For example, wetlands and natural coastlines provide a service by protecting against flood damages. And trees and other plants provide a service by sequestering carbon and thereby reducing the costs of greenhouse gas emissions.

Despite the fact that we do not typically trade them in markets or see their value via prices, ecosystem goods and services are valuable inputs to most aspects of our economy and well-being. In fact, they are the foundation on which all economic activity and social welfare depend. Wise decisions about ecosystem goods and services thus require both biophysical and economic evaluation. Not only that, the biophysical and economic evaluation must be coordinated. The relationship between changes in natural systems and corresponding changes in human welfare is today a focus of both ecology and economics.[1–8]

Many government, nongovernmental organization (NGO), and private sector decisions affect the delivery of ecosystem goods and services. Carbon sequestration practices, water allocations, land uses, air and water quality regulations, and resource management choices all affect nature's delivery of ecosystem goods and services. When we protect or conserve natural resources, we are protecting and conserving ecosystem goods and services.

Evaluation of ecosystem goods and services is necessary if we are to protect the social wealth they provide. To counter the economic invisibility of ecosystem goods and services, natural and social scientists are working together to depict nature's role in our social and economic well-being. Evaluation requires extensive knowledge of biophysical systems, the ways in which human activity alters those systems, and our ability to influence them positively via restoration, management, and protection. It also requires knowledge of nature's role in our economic and social lives and the value of those goods and services.

BIOPHYSICAL EVALUATION

The evaluation of ecosystem services involves two linked missions, one biophysical and the other, economic and social. Biophysical evaluation is associated with the natural sciences including ecology, hydrology, biology, oceanography, and soil science. If we want clean air and water, healthy and abundant species populations, pollination, irrigation, and protection from floods and fires, how can we take action to preserve these things? The natural sciences describe how ecosystems function, the factors that threaten ecological functions, and our ability to restore, enhance, or protect functions in decline. This knowledge is necessary if we are to manage, protect, or enhance the goods and services ecosystems provide.

Any natural process, by definition, transforms a set of inputs into a different set of outputs, much like an industrial process transforms inputs like labor and capital into

Encyclopedia of Natural Resources DOI: 10.1081/E-ENRL-120047459

outputs like cars and loaves of bread. Hydrological processes transform rainfall into ground and surface water. Biological and chemical processes transform water of one quality into water of a different quality. Reproductive, forage, and migratory processes relate biotic and physical conditions to the abundance of species. Food webs convert one form of biomass into another. Wetland processes transform the scale, location, and speed of flood pulses. Sequestration processes affect the release and in some cases, transformation of chemical inputs into water or the atmosphere. Even a process as simple as "shading" relates tree canopies to the regulation of water temperature.

When ecologists or other natural scientists speak of ecological processes or functions, they are referring to the transformation of one set of biophysical conditions into another. Ecologists and economists refer to these processes as *biophysical production functions*.[9–11] Biophysical production functions provide the causal link between ecological stresses—from land development, climate change, and invasive species—and losses in ecosystem goods and services. They also provide the link between ecological protection or enhancement—from conservation, regulation, and restoration—and improvements in ecosystem services.

For example, consider a wetland ecosystem. A wetland ecosystem has a structure, size, and boundaries that contain a bundle of biophysical inputs including soils, water volumes, water quality, and plant and animal populations. The socially valuable outputs of this ecosystem include more abundant species, cleaner water, and lower flood risks. Biophysical production functions describe how changes in the wetland system's inputs (size, boundaries, soil type, water volume, and vegetation) affect its ability to support species, absorb floodwater, and remove contaminants (the system's goods and services). Knowledge of these production functions is necessary to answer questions like the following: How many wetland acres do we need to protect downstream communities from flood damage? Or, what kinds of wetlands do we need to prevent the extinction of some bird or amphibian species? Production functions provide the link between actions we can take as resource managers and conservationists and the socially desirable biophysical benefits those actions create.

Measurement of these production functions requires investment in evidence-based ecology that explicitly focuses on the relationship between observable stresses to ecological systems and outputs (goods and services) that are socially important. Like clinical medicine, ecosystem services scientists are developing ways to consistently relate stresses (e.g., impervious land cover) and interventions (e.g., stream bank restoration) to ecosystem service outcomes (e.g., water quality, fish abundance). An important feature of these activities is that they are inherently interdisciplinary, requiring collaboration among disciplines such as hydrology, water chemistry, population ecology, and ecotoxicology.

THE IMPORTANCE OF SPATIAL ANALYSIS

Biophysical evaluation of ecosystem services has another important characteristic: it almost always involves spatial analysis because both the production of biophysical functions and the social determinants of the service's benefits depend upon the landscape context in which those functions and services arise.[12,13] Numerous efforts are underway to map where ecosystem goods and services are provided and are most threatened.[14,15] From the biophysical perspective, geographic context matters for several broad reasons. First, ecological production can exhibit non-linearities in scale and configuration—for example, where a whole protected landscape produces more than the sum of the unconnected parts. Second, natural systems are often characterized by movement: air circulates, water runs downhill, species migrate, and seeds and pollen disperse. Moreover, the movement of one biophysical feature—say water—tends to trigger the movement of other things—like birds and fish.[16] In general, the ecosystem goods and services we care about often depend on biophysical conditions and processes that occur over large distances.

Put another way, if we want to enjoy ecosystem services in one location, we must usually protect, enhance, or manage ecosystems in other locations. Evaluation thus involves understanding the spatial phenomena and interdependencies. Examples of these spatial production relationships describe the dependence of the following: species on the configuration of lands needed for their reproduction, forage, and migration; surface and aquifer water volume and quality on land cover configurations and land uses; flood and fire protection services on land cover configurations; soil quality on climate variables and land uses; and air quality on pollutant emissions, atmospheric processes, and natural sequestration.

In particular, the pattern of conservation of lands on the landscape can significantly impact the value of both the ecological and economic attributes.[17] It should also be noted that temporal analysis of production functions is also often important. First, many ecosystem processes are time dependent (e.g., runoff from melting snowpacks in spring, the need for forage food when migratory species are present). Second, the value of ecosystem goods and services is often a function of when those services are provided (e.g., irrigation water during the growing season).

ECONOMIC EVALUATION

Ecosystem services evaluation also involves some form of social assessment to measure or communicate the importance of ecosystem services to households, communities, or other decision-makers. Often, social assessment of ecosystem services takes the form of economic, monetary evaluation of the services' value. It should be noted that there are also non-economic, non-monetary approaches to ecosystem services evaluation, a review of which is beyond the scope of this entry (but for examples, see Sherrouse et al.,[18] Blamey et al.,[19] Hagerhall,[20] Pereira et al.,[21] and Bender et al.[22]).

The economic value of natural resources and ecological systems goes well beyond the value of consuming natural resources such as water, lumber, and food. It also includes the value associated with non-consumptive natural resource uses (e.g., for aesthetic or recreational enjoyment), the intrinsic value of protecting nature for its own sake, and the bequest value of passing a healthy environment along to our descendants. In principle, *any* contribution of nature to our objective or subjective well-being can be expressed in economic terms.

Economic valuation is a way to depict something's importance or desirability. Economic values can be thought of as rankings, weights, or priorities. Values are detected or measured by examining people's choices. Whenever we choose one thing over another, as individuals do everyday, we are engaged in valuation. Choices are a particularly reliable form of evidence when it comes to detecting preferences and values. When we make choices, we reveal our preferences for one thing over another. Paying for something is a choice. When we pay for something, we are deliberately choosing that something over the amount of money paid. Assuming people are rational, they will only pay the price if the thing they are buying is worth at least that much to them. The higher the price paid, the higher the valuation we can infer.

Prices are desirable, not because they are the ideal measure of value, but because they are readily available. We can use market prices to value some ecosystem goods and services, but only those that are bought and sold as private goods. Examples of ecosystem goods that are bought and sold in private markets include timber, commercial fish harvests, and carbon sequestration credits (if a credit market exists). Often, however, ecosystem goods and services are public, non-market commodities for which there is no market price. Without market prices, economists must resort to so-called non-market valuation methods, including the methods described in the following text.

Hedonic valuation methods examine the prices people pay for things that have an environmental component. For example, when people purchase a home near an aesthetically pleasing ecosystem, home prices reflect that environmental amenity.[23] The price premium of living near the ocean, having a mountain view, or being in close proximity to urban parks can be measured via statistical analysis. Similarly, farm values are related to the availability of groundwater, precipitation, and soil quality. The premium due to those features can be estimated by controlling for other factors that affect farm value.

Evidence of conservation value can also be inferred from political choices such as prohibitions on drilling, development, and other land use changes associated with public lands or conservation referenda that approve local or state financing of land acquisitions.[24]

Travel cost methods examine the costs people are willing to bear to enjoy natural resources. When we spend time and money to enjoy nature, we are revealing something about its value. Again, if we are willing to pay the price (the cost), we must value the experience, enjoyment, or use of the resource more than the cost.[25] The travel cost method requires data and analysis linking the number of trips to a site to its quality, size, or location. Changes in these attributes can be valued if there is a perceptible change in the number, length, or cost of trips taken to the site.

Another technique is to examine costs avoided by the presence of an ecological feature or service. For example, if we lose wetlands and lose their water purification or flood damage reduction benefits, we may have to invest in water treatment facilities or levees and dams. If instead we protect the wetlands, we avoid the costs associated with built infrastructure alternatives. Similarly, private firms can conduct engineering and economic analyses that calculate the costs associated with, for example, the loss of surface waters for cooling, where the cost might be associated with new refrigeration technologies.

Another approach, called the stated preference method, is to present people with hypothetical scenarios that ask them to choose—in a survey format—between ecosystem goods or services and something with a clear dollar value, such as an increase in property tax. To pass academic muster, these studies are much more structured and carefully designed than simple opinion polling.[26] Stated preference methods are controversial because people's choices are undisciplined by the need to spend their own, real money, which in principle may lead them to overstate their willingness to pay. Care must also be given to clearly defining and isolating the goods or services in question and framing the choice problem in a way that does not bias people's responses. Nevertheless, stated methods are an improvement relative to evaluation techniques that ignore social preferences.[27]

Finally, mention should be made of benefit transfer methods. Benefit transfer methods take existing valuations derived from any of the aforementioned methods and transfer them to new landscape and resource contexts. Benefit transfer studies are desirable because they avoid the costs of conducting original valuation research. However, the transfer of valuations from one ecological and social context to another is dangerous because ecosystem values are highly dependent on location. Benefit transfer involves statistical methods designed to control for similarities and differences in spatial contexts and adjust the transferred valuation accordingly.

A wide range of ecosystem goods and services, in specific spatial and social contexts, has been given economic values by researchers. A review of existing valuation studies is beyond the scope of this entry (see Boyd and

Krupnik,[28] the Environmental Valuation Reference Inventory,[29] and Kumar[30] for a review and organized collection of studies, respectively). In general, the non-market valuation methods described earlier have a long history and are considered within economics to be valid, if imperfect, approaches to the missing prices problem associated with public environmental goods.[31]

WHAT AFFECTS THE VALUE OF ECOSYSTEM SERVICES?

From an economic perspective, we can make several broad statements about the value of ecosystem goods and services: the scarcer an ecological feature, the greater its value; the scarcer are substitutes for an ecological feature, the greater its value; the more abundant are complements to an ecological feature, the greater its value; the larger the population benefiting from an ecological feature, the greater its value; and the larger the economic value protected or enhanced by the feature, the greater its value. All of these factors relate to the geographic context in which the services are delivered and enjoyed.

As economic commodities, ecosystem goods and services resemble real estate, rather than cars or bottles of dish soap. The value of real estate is highly dependent on its location—the features of the surrounding neighborhood. This is because a given house or building cannot be easily transported to another neighborhood. In contrast, cars or soap can be easily moved around (shipped from one location to another); so, their value tends to be independent of their geographic location.

For example, the value of irrigation and drinking water quality depends on how many people depend on the water—a function of proximity to the water. Flood damage-avoidance services are more valuable, the larger the value of lives, homes, and businesses protected from flooding in a given watershed. Species important to recreation (for anglers, hunters, and birders) are more valuable when more people have proximity and access to recreational lands.

The scarcity of an ecosystem service is also usually a function of geography. Scarcity—and the availability of substitutes—matters to economic assessment because, all else equal, value increases with scarcity. For example, the value of irrigation water depends on the availability (and hence location) of alternative water sources. If wetlands are plentiful in an area, then a given wetland may be less valuable as a source of flood pulse attenuation than it might be in a region in which it is the only such resource. Finally, many ecosystem goods and services are valuable only if they are bundled with certain man-made assets. These assets are called "complements" because they complement the value of the ecosystem service. Recreational fishing and kayaking require docks or other forms of access. For example, a beautiful vista yields social value when people have access to it. Access may require infrastructure—roads, trails, parks, and housing, all of which are spatially configured.

COORDINATED BIOPHYSICAL AND SOCIAL EVALUATION

This entry has emphasized that ecosystem services evaluation involves both biophysical and economic analyses. An important corollary is that those two types of analyses need to be linked and coordinated, so that the delivery of ecosystem goods and services can be described in economic terms. Biophysical analysis describes nature's condition and how that condition changes in response to interventions or stresses. Social analysis describes the benefits or costs of changes in those biophysical conditions.

One way to facilitate this coordination is by having ecological and economic researchers codefine *ecological end points*.[32] Ecological end points are a distinct subset of the larger universe of biophysical outcome measures. The test of such measures is that they are meaningful and understandable to communities, businesses, households, planners, and other stakeholders. In other words, they are biophysical outcome measures that require little further biophysical translation to make clear their relevance to human welfare.

Many outcomes of interest to natural science do not have this property. Outcomes like biotic integrity indices, chemical water quality concentrations, hydrogeomorphic classifications, and biological productivity measures are of scientific interest, are related to ecosystem services measurement, and establish the scientific basis for accurately modeling ecosystem functions and services. But their effect on people's well-being is unclear.

Consider a commonly used measure such as an "aquatic macroinvertebrate community index."[33] This index is a useful ecological measure because it acts as a proxy for food availability for fish and birds, indicates level of chronic stress on aquatic organisms, and may be correlated with abundance of fish species or fish diversity. But this measure does not satisfy the conditions of an end point: it is technical and not directly meaningful to non-ecologists. Consequently, it is by itself almost impossible to evaluate socially or economically. To be clear, that does not mean the information conveyed by the measure is not useful or important. On the contrary, such information may be extremely relevant to our understanding of ecosystem functions. But alone, this kind of measure is difficult to interpret in social or economic terms.

Note, though, that these indicators are used because they are thought to be proxies for the goods and services we can evaluate socially (such as fish abundance or reduced risks of waterborne illness). What are needed are biophysical production functions that translate technical outcome measures into ecological end points amenable to social evaluation. Other examples of measures that require further biophysical translation to foster social evaluation include, but are not limited to, criteria such as carbon sequestration potential, dissolved oxygen, nutrient, and other chemical concentrations, bacterial and other pathogen concentrations, biotic tissue burdens and pathological measures, and biodiversity indices.

CONCLUSION

Ecosystem services evaluation links biophysical and social analyses to measure the economic benefits and costs of decisions that affect natural resources. Biophysical evaluations describe the ways in which environmental stresses (e.g., from land use or climate change) or protections (e.g., conservation or restoration) change ecosystems' ability to deliver beneficial outcomes such as cleaner air, water for irrigation, drinking, and commerce, more productive soils, and more abundant species. Economic evaluations measure the value of these ecological outcomes (or costs of their loss) in monetary or other social terms. Together, the two forms of analyses allow decision-makers to describe the economic consequences of ecological change. Because ecological losses often lead to economic costs, ecosystem service evaluation can help society perceive, and thus hopefully manage and avoid those costs.

REFERENCES

1. Daily, G. *Nature's Services: Societal Dependence on Natural Ecosystems*; Island Press: Washington, DC, 1997.
2. Wainger, L.; Dennis, K.; Salzman, J.; Boyd, J. Wetland value indicators for scoring mitigation trades. Stanford Environ. Law J. **2001**, *20* (2), 413–478.
3. Ricketts, T.H.; Daily, G.C.; Ehrlich, P.R.; Michener, C.D. Economic value of tropical forest to coffee production. Proc. Natl. Acad. Sci. USA **2004**, *101* (34), 12579–12582.
4. Polasky, S.; Nelson, E.; Lonsdorf, E.; Fackler, P.; Starfield, A. Conserving species in a working landscape: land use with biological and economic objectives. Ecol. Appl. **2005**, *15* (4), 1387–1401.
5. Boyd, J. The nonmarket benefits of nature: what should be counted in green GDP? Ecol. Econ. **2006**, *61* (4), 716–723.
6. Carpenter, S.R.; DeFries, R.; Dietz, T.; Mooney, H.A.; Polasky, S.; Reid, W.V.; Scholes, R.J. Millennium ecosystem assessment: research needs. Science **2006**, *314* (5797), 257–258.
7. Boyd, J.; Banzhaf, S. What are ecosystem services? Ecol. Econ. **2007**, *63* (2–3), 616–626.
8. Barbier, E.B.; Koch, E.W.; Silliman, B.R.; Hackery, S.D.; Wolanski, E.; Primavera, J.; Granek, E.F.; Polasky, S.; Aswani, S.; Cramer, L.A.; Stoms, D.M.; Kennedy, C.J.; Bael, D.; Kappel, C.V.; Perillo, G.M.; Reed, D.J.; Coastal ecosystem-based management with nonlinear ecological functions and values. Science **2008**, *319* (5861), 321–323.
9. U.S. EPA (Environmental Protection Agency) Science Advisory Board. Valuing the Protection of Ecological Systems and Services: A Report of the EPA Science Advisory Board. EPA-SAB-09-012. U.S. EPA: Washington, DC, 2009.
10. Daily, G., Matson, P. Ecosystem services: from theory to implementation. Proc. Natl. Acad. Sci. USA **2008**, *105* (28), 9455–9456.
11. Boyd, J. Counting Non Market, Ecological Public Goods: The Elements of a Welfare Significant Ecological Quantity Index. Discussion paper 07 42. Resources for the Future: Washington, DC, 2007.
12. Bockstael, N. Modeling economics and ecology: the importance of a spatial perspective. Am. J. Agric. Econ. **1996**, *78*, 1168.
13. Potschin, M.; Haines-Young, R. Ecosystem services: exploring a geographical perspective. Prog. Phys. Geography. **2011**, *35* (5), 579–598.
14. Raymond, C.M.; Bryan, B.A.; MacDonald, D.H.; Cast, A.; Strathearn, S.; Grandgirard, A.; Kalivas, T. Mapping community values for natural capital and ecosystem services. Ecol. Econ. **2009**, *68*, 1301–1315.
15. Egoh, B., Reyers, B.; Rouget, M.; Richardson, D.M.; LeMaitre, D.C.; van Jaarsveld, A.S.; Mapping ecosystem services for planning and management. Agriculture Ecosystems Environ. **2008**, *127*, 135–140.
16. Semmens, D.; Diffendorfer, J.E.; Lopez-Hoffman, L.; Shapiro, C.; Accounting for the ecosystem services of migratory species: quantifying migration support and spatial subsidies. Ecol. Econ. **2011**, *70*, 2236–2242.
17. Polasky, S.; Nelson, E.; Camm, J.; Csuti, B.; Fackler, P.; Lonsdorff, E.; Montgomery, C.; White, D.; Arthur, J.; Garber-Yonts, B.; Haight, R.; Kagan, J.; Starfield, A.; Tobalske, C.; Where to put things? spatial land management to sustain biodiversity and economic returns. Biol. Conserv. **2008**, *141*, 1505–1524.
18. Sherrouse, B.C.; Clement, J.M.; Semmens, D.J. A GIS application for assessing, mapping, and quantifying the social values of ecosystem services. Appl. Geography **2011**, *31* (2), 748–760.
19. Blamey, R.K.; James, R.F.; Smith, R.; Niemeyer, S. Citizens' Juries and Environmental Value Assessment. Research School of Social Sciences, Australian National University: Canberra, 2000.
20. Hagerhall, C.M. Consensus in landscape preference judgements. J. Environ. Psychol. **2001**, *21* (1), 83–92.
21. Pereira, E.; Queiroz, C.; Pereira, H.; Vicente, L. Ecosystem services and human well-being: a participatory study in a mountain community in Portugal. Ecol. Soc. **2005**, *10* (2), Art. 14.
22. Bender, O.; Boehmer, H.J.; Jens, D.; Schumacher, K.P. Using GIS to analyse long-term cultural landscape change in Southern Germany. Landscape Urban Plan. **2005**, *70* (1–2), 111–125.
23. Brent, M.; Polasky, S.; Adams, R. Valuing Urban Wetlands: A property price approach. Land Econ. **2000**, *76* (1), 100-113.
24. Banzhaf, H.; Wallace Oates, S.; Sanchirico, J. Success and design of local referenda for land conservation. J. Policy Anal. Manag. **2010**, *29*, 769–798.
25. McConnell, K. On-site time in the demand for recreation. Am. J. Agric. Econ.**1992**, *74*, 918.

26. Kopp, R.J.; Pommerehne, W.; Schwarz, N., Eds.; *Determining the Value of Non-Marketed Goods*. Kluwer, Boston, MA, 1997.
27. Carson, R.; Flores, N.; Meade, N. Contingent valuation: controversies and evidence. Environ. Resource Econ. **2001**, *19*, 173–210.
28. Boyd, J.; Alan, K. *The Definition and Choice of Environmental Commodities for Nonmarket Valuation*. Resources for the Future, Washington, D.C., USA, DP-09-35, 2009.
29. The Environmental Valuation Reference Inventory. http://www.evri.ca/Global/Splash.aspx.
30. Kumar, P. *The Economics of Ecosystems and Biodiverstiy: Ecological and Economic Foundations*. Earthscan, London: Washington, DC, 2010.
31. Freeman, A.M. *The Measurement of Environmental and Resource Values: Theory and Methods*. Resources for the Future: Washington, D.C., 1993.
32. Boyd, J.; Krupnick, A. *The Definition and Choice of Environmental Commodities for Nonmarket Valuation*. Resources for the Future Washington, D.C., USA, DP-09-35, 2009.
33. Plafkin, J.L.; Barbour, M.T.; Porter, K.D.; Gross, S.K.; Hughes, R.M. *Rapid Bioassessment Protocols for Use in Streams and Rivers: Benthic Macroinvertebrates and Fish*. U.S. Environmental Protection Agency, Office of Water Regulations and Standards, EPA 440-4-89-001: Washington, D.C., 1989.

Ecosystem Services: Land Systems Approach

Richard Aspinall
Honorary Research Fellow, James Hutton Institute, Aberdeen, U.K.

Abstract
Land systems are coupled human–environment systems that are the source of a wide variety of ecosystem services used by individuals and society. The dynamics of land systems are influenced by natural, technological, financial, and human capital and flows, and by land management and decision-making at scales ranging from local to global. Supporting, regulating, provisioning, and cultural ecosystem services are made available in land systems through the dynamics of human activity linked to environmental processes, and are recovered through a variety of land-management practices. The global scope of land change and pressures on land systems provide compelling reasons for using a land systems approach to move to understanding the roles of land management and other human activities in providing ecosystem services.

INTRODUCTION

Land systems are a category of coupled human–environment system in which the functions and purposes of land are understood and viewed as the results of human activities and environmental processes working together. Ecosystem services are the goods and services that societies recover from both natural and managed ecosystems. Many ecosystem goods and services are, therefore, a direct product of management and use of land systems. Most analyses of ecosystem services focus on natural processes and functions that are based in biodiversity, although a wide variety of biogeochemical, hydrological, and other processes, and the use of technologies, financial capital, and labor and energy, influence the functioning of ecosystems and the quantity and quality of ecosystem services recovered. Ecosystem services can be usefully considered as the product of land systems, since land systems reflect not only ecosystem processes and functions, but also human management and financial, social, and technological inputs that influence processes and functions, especially in terrestrial ecosystems.

LAND SYSTEM AS A COUPLED HUMAN–ENVIRONMENT SYSTEM

Land systems are a category of coupled human and environment system,[1,2] also known as a socio-ecological system, that incorporates both human and environment systems and their interaction. As such, they represent not only the structures and processes that link human and environment systems, but also the driving factors and processes producing change and dynamics[3] and that influence the multiple goods and services societies derive from natural resources. A land system approach necessarily embeds the different spatial, temporal, and organizational scales that the environment[4,5] and human[6] system processes to provide an understanding of local, regional, and global changes.[7,8] The environment system includes the Earth system, which is made up of biological, biogeochemical, hydrological, and other physical environment sub-systems. The human system includes economic, institutional, demographic, technological, and political factors and subsystems. Processes in the environment system are largely driven by biodiversity, energy, and material flows, while processes in the human system are driven by flows of finance, ideas, innovation, and human activity, including choices and decisions. Human and environment systems link most directly through the interaction of land cover and land use,[8,9] and through the influence and impacts of land management.[10] Land systems are studied with methods of interdisciplinary research including Earth observation and monitoring, survey, and modeling.[11]

A land systems approach emphasizes the system-level interactions, not only of factors that affect land but also of the different activities and processes that have an impact on food, water, and other ecosystem goods and services provided by land systems. Land use and land-produced ecosystem services are thus not solely the product of environmental systems; active land management is central to the delivery of ecosystem services.

ECOSYSTEM GOODS AND SERVICES

Ecosystem goods and services are defined as the benefits society receives from environmental systems.[12] Global-scale environmental changes have raised the profile and importance of ecosystem goods and services,[13] leading to reviews of the state of ecosystems and the goods and

Encyclopedia of Natural Resources DOI: 10.1081/E-ENRL-120048594

Table 1 Definitions and examples of the four main categories of ecosystem goods and services

Provisioning Services: The goods and other products obtained from ecosystems.
Examples include food, fiber, fuel, fresh water, genetic and ornamental resources, biochemicals, natural medicines, and pharmaceuticals. Since the environment also provides sources of energy, renewable and fossil sources of energy can also be considered among provisioning services.

Regulating Services: The goods, services, and benefits obtained from environmental regulation functions provided by ecosystem processes.
Examples include air quality, climate regulation, flood and hazard regulation, erosion regulation, regulation of diseases and pests, regulation of water quality (water purification), and waste treatment.

Supporting Services: The underpinning system functions necessary for the production of all other ecosystem goods and services. Supporting services refer to the products and results of biogeochemical, physical, chemical, and biological functions that operate in environmental systems.
Examples include soil formation, photosynthesis, primary production, nutrient cycling, and water cycling.

Cultural Services: The non-material benefits people obtain from ecosystems through education, exercise and recreation, spiritual enrichment, cognitive development, reflection, and aesthetic experience.
The UK National Ecosystem Assessment provides a rich suite of definitions of cultural services that expand the range of cultural services defined by the MEA.[33]

Source: Adapted from Millennium Ecosystem Assessment,[12] UK National Ecosystem Assessment,[14] EEA,[16] and Church et al.[33]

Biomes—Ecosystems

services at global[12] and national[14] scales. Terrestrial ecosystems are recognized as the primary source of goods and services that support human well-being and society.[12,15]

There are numerous taxonomies of ecosystem goods and services. The Millennium Ecosystem Assessment (MEA), for instance, has popularized their classification into four main types: supporting, regulating, provisioning, and cultural services (Table 1). This taxonomy has been continued in other classifications, including the Common International Classification of Ecosystem Services, which also explicitly recognizes a variety of services from the abiotic elements of the environment.[16]

Ecosystem assessment, a social process of evaluating ecosystem services and their value,[14] is generally based on a conceptual model that emphasizes a foundation in biodiversity and natural capital.[12,17–20] In this approach, human and technological capitals are considered exogenous factors. Because ecosystem goods and services are not, however, produced solely by biodiversity and natural capital, there is increasing recognition of the importance of evaluating ecosystem goods and services using a land systems approach that explicitly includes land management.[2,20]

LAND SYSTEMS CHANGE AND ECOSYSTEM SERVICES

Land use change and the consequences of land management activities are global in their scope and scale of impact.[7,21,22] About 140 million km^2 of the Earth, some 28% of its total surface area, is land.[12] Cropland and pastures/rangelands together occupy about 43 million km^2,[23] or about 40% of the ice-free land area.[24] Human land use and management practices and activities have increased the intensity of resource use and impact,[25] and dominate biogeochemical[26,27] and water[28,29] fluxes. Human use or approximately 24% (15.6 Pg C yr^{-1}) of the total potential net primary productivity in terrestrial systems.[30]

These demands on land systems, combined with global population growth and changing climate, lead to international and national concerns over food, water, and energy security. Although land use and land use change are phenomena with global scope and impact, land use and land management are subject to the influences of a wide variety of human and environmental processes operating at a range of scales.[31] These influences include global-scale political and economic forces. Internationally, the land systems science community and land managers recognize that global forces become the main determinants of land use change primarily through their ability to amplify or attenuate local factors[3,8] and via global teleconnections.[32] Local factors, including land management practices, are, therefore, of primary importance not only in driving land changes that are local, regional, and global in scope and impact, but also in resolving land use issues and the delivery of ecosystem services.

CONCLUSION

Land systems offer a basis for a more holistic understanding of the generation and recovery of ecosystem services by individuals and society. Land systems include both human and environment system influences on the dynamics that produce ecosystem services, linking natural, technological, financial, and human capital and flows, as well as land management and decision-making to ecosystem services.

REFERENCES

1. Foresight Land Use Futures Project. *Land Use Futures: Making the Most of Land in the 21st Century*, Final Project Report; London, 2010.
2. *Global Land Project*: *Science Plan and Implementation Strategy*, IGBP Report No. 53/IHDP Report No. 192005; IGBP Secretariat: Stockholm, 2005; p 64.
3. Lambin, E.F.; Turner, B.L.; Geist, H.J.; Agbola, S.B.; Angelsen, A.; Bruce, J.W.; Coomes, O.T.; Dirzo, R.; Fischer, G.; Folke, C.; George, P.S.; Homewood, K.; Imbernon, J.; Leemans, R.; Li, X.B.; Moran, E.F.; Mortimore, M.; Ramakrishnan, P.S.; Richards, J.F.; Skanes, H.; Steffen, W.; Stone, G.D.; Svedin, U.; Veldkamp, T.A.; Vogel, C.; Xu, J.C. The causes of land-use and land-cover change: Moving beyond the myths. Global Environ. Change-Hum. Policy Dimens. **2001**, *11* (4); 261–269.
4. Chave, J.; Levin, S. Scale and scaling in ecological and economic systems. Environ. Resour. Econ. **2003**, *26* (4), 527–557.
5. Levin, S.A. The problem of pattern and scale in ecology. Ecology **1992**, *73* (6), 1943–1967.
6. Gibson, C.C.; Ostrom, E.; Ahn, T.K. The concept of scale and the human dimensions of global change: A survey. Ecol. Econ. **2000**, *32* (2), 217–239.
7. Foley, J.A.; DeFries, R.; Asner, G.P.; Barford, C.; Bonan, G.; Carpenter, S.R.; Chapin, F.S.; Coe, M.T.; Daily, G.C.; Gibbs, H.K.; Helkowski, J.H.; Holloway, T.; Howard, E.A.; Kucharik, C.J.; Monfreda, C.; Patz, J.A.; Prentice, I.C.; Ramankutty, N.; Snyder, P.K. Global consequences of land use. Science **2005**, *309* (5734), 570–574.
8. Lambin, E.F.; Geist, H., Eds. Local processes and global impacts. In *Land-Use and Land-Cover Change*; Springer-Verlag: Berlin, 2006; 222.
9. Gutman, G.; Janetos, A.C.; Justice, C.O.; Moran, E.F.; Mustard, J.F.; Rindfuss, R.R.; Skole, D.; Turner, B.L.; Cochrane, M.A., Eds. *Land Change Science. Observing, Monitoring and Undertanding Trajectories of Change on the Earth's Surface*; Kluwer Academic Publishers: Dordrecht/Boston/London, 2004; 457.
10. Lambin, E.F.; Rounsevell, M.D.A.; Geist, H.J. Are agricultural land-use models able to predict changes in land-use intensity? Agric. Ecosyst. Environ. **2000**, *82* (1–3), 321–331.
11. Turner, B.L.; Lambin, E.F.; Reenberg, A. The emergence of land change science for global environmental change and sustainability. Proc. Natl. Acad. Sci. U.S.A. **2007**, *104* (52), 20666–20671.
12. Millennium Ecosystem Assessment. *Ecosystems and Human Well-being: Synthesis*; Island Press: Washington, DC, 2005; 137.
13. Daily, G.C.; Matson, P.A. Ecosystem services: From theory to implementation. Proc. Natl. Acad. Sci. U.S.A. **2008**, *105* (28), 9455–9456.
14. UK National Ecosystem Assessment. *The UK National Ecosystem Assessment Technical Report 2011*; UNEP-WCMC: Cambridge.
15. Costanza, R.; dArge, R.; de Groot, R.; Farber, S.; Grasso, M.; Hannon, B.; Limburg, K.; Naeem, S.; ONeill, R.V.; Paruelo, J.; Raskin, R.G.; Sutton, P.; van den Belt, M. The value of the world's ecosystem services and natural capital. Nature **1997**, *387* (6630), 253–260.
16. EEA. Proposal for a Common International Classification of Ecosystem Goods and Services (CICES) for Integrated Environmental and Economic Accounting. In Paper prepared for the EEA by Centre for Environmental Management; Background document, ESA/STAT/AC.217, UNCEEA/5/7/Bk, Fifth Meeting of the UN Committee of Experts on Environmental-Economic Accounting, NewYork, June 23–25, 2010; University of Nottingham: United Kingdom, 2010.
17. Mace, G.M.; Bateman, I.; Albon, S.; Balmford, A.; Brown, C.; Church, A.; Haines-Young, R.; Pretty, J.; Turner, K.; Vira, B.; Winn, J. Conceptual framework and methodology. In *The UK National Ecosystem Assessment Technical Report*; UK National Ecosystem Assessment, Ed.; UNEP-WCMC: Cambridge, 2011.
18. Pe'er, G.; McNeely, J.A.; Dieterich, M.; Jonsson, B.-G.; Selva, N.; Fitzgerald, J.M.; Nesshover, C. IPBES: Opportunities and challenges for SCB and other learned societies. Conserv. Biol. **2013**, *27* (1), 1–3.
19. Turnhout, E.; Bloomfield, R.; Hulme, M.; Vogel, J.; Wynne, B. Conservation policy: Listen to the voices of experience. Nature **2012**, *488* (7412), 454–455.
20. Carpenter, S.R.; Mooney, H.A.; Agard, J.; Capistrano, D.; DeFries, R.S.; Diaz, S.; Dietz, T.; Duraiappah, A.K.; Oteng-Yeboah, A.; Pereira, H.M.; Perrings, C.; Reid, W.V.; Sarukhan, J.; Scholes, R.J.; Whyte, A. Science for managing ecosystem services: Beyond the Millennium Ecosystem Assessment. Proc. Natl. Acad. Sci. U.S.A. **2009**, *106* (5), 1305–1312.
21. Rockström, J.; Steffen, W.; Noone, K.; Persson, A.; Chapin, F.S., III; Lambin, E.F.; Lenton, T.M.; Scheffer, M.; Folke, C.; Schellnhuber, H.J.; Nykvist, B.; de Wit, C.A.; Hughes, T.; van der Leeuw, T.; Rodhe, H.; Sorlin, S.; Snyder, P.K.; Costanza, R.; Svedin, U.; Falkenmark, M.; Karlberg, L.; Corell, R.W.; Fabry, V.J.; Hansen, J.; Walker, B.; Liverman, D.; Richardson, K.; Crutzen, P.; Foley, J.A. A safe operating space for humanity. Nature **2009**, *461* (7263), 472–475.
22. Vitousek, P.M.; Mooney, H.A.; Lubchenco, J.; Melillo, J.M. Human domination of Earth's ecosystems. Science **1997**, *277* (5325), 494–499.
23. Ramankutty, N.; Evan, A.T.; Monfreda, C.; Foley, J.A. Farming the planet: 1. Geographic distribution of global agricultural lands in the year 2000. Global Biogeochem. Cycles **2008**, *22*, GB1003, doi:10.1029/2007GB002952.
24. Ramankutty, N.; Foley, J.A. Estimating historical changes in global land cover: Croplands from 1700 to 1992. Global Biogeochem. Cycles **1999**, *13* (4), 997–1027.
25. Monfreda, C.; Ramankutty, N.; Foley, J.A. Farming the planet: 2. Geographic distribution of crop areas, yields, physiological types, and net primary production in the year 2000. Global Biogeochem. Cycles **2008**, *22*, GB1022, doi:10.1029/2007GB002947.
26. Matson, P.A.; Parton, W.J.; Power, A.G.; Swift, M.J. Agricultural intensification and ecosystem properties. Science **1997**, *277* (5325), 504–509.
27. Vitousek, P.M.; Aber, J.D.; Howarth, R.W.; Likens, G.E.; Matson, P.A.; Schindler, D.W.; Schlesinger, W.H.; Tilman, D. Human alteration of the global nitrogen cycle: Sources and consequences. Ecol. Appl. **1997**, *7* (3), 737–750.
28. Gleick, P.H. Water use. Annu. Rev. Environ. Resour. **2003**, *28*, 275–314.

29. Hoekstra, A.Y.; Mekonnen, M.M. The water footprint of humanity. Proc. Natl. Acad. Sci. U.S.A. **2012**, *109* (9), 3232–3237.
30. Haberl, H.; Erb, K.H.; Krausmann, F.; Gaube, V.; Bondeau, A.; Plutzar, C.; Gingrich, S.; Lucht, W.; Fischer-Kowalski, M. Quantifying and mapping the human appropriation of net primary production in earth's terrestrial ecosystems. Proc. Natl. Acad. Sci. U.S.A. **2007**, *104* (31), 12942–12945.
31. Lambin, E.F.; Geist, H.J.; Lepers, E. Dynamics of land-use and land-cover change in tropical regions. Annu. Rev. Environ. Resour. **2003**, *28*, 205–241.
32. Seto, K.C.; Reenberg, A.; Boone, C.G.; Fragkias, M.; Haase, D.; Langanke, T.; Marcotullio, P.; Munroe, D.K.; Olah, B.; Simon, D. Urban land teleconnections and sustainability. Proc. Natl. Acad. Sci. U.S.A. **2012**, *109* (20), 7687–7692.
33. Church, A.; Burgess, J.; Ravenscroft, N. Cultural services. In *The UK National Ecosystem Assessment Technical Report*; UK National Ecosystem Assessment, Ed.; UNEP-WCMC: Cambridge, 2011; 633–691.

Ecosystem Services: Pollinators and Pollination

Victoria A. Wojcik
Pollinator Partnership, San Francisco, California, U.S.A.

Biomes—Ecosystems

Abstract

An examination of the origins of products and of the interconnections of natural systems quickly reveals the support that pollinators provide to our natural landscapes and resources. Pollinators are keystone species across all terrestrial ecosystem types that help provide food, raw products, and further ecosystem services to humans and other living things. The roles that pollinators play in natural resource provisioning are complex, ranging from primary direct services (basic plant reproduction, fruit development) to indirect supportive services (e.g., soil stabilization, nutrient cycling, and air filtration). Valuing the total contributions that pollinators provide to ecosystem services is a complicated task that involves quantifying the economic benefits seen from goods and commodities and qualifying the value of biodiversity and ecosystem stability.

INTRODUCTION

Pollination is unique as an individual ecosystem service in and of itself, as well as a supportive factor to other ecosystem services. The sustainability of terrestrial ecosystems relies on the maintenance of plant reproductive cycles, evolutionary patterns, and ecosystem functions.[1] The vast majority of the botanical diversity that constitutes ecosystems, defines biomes, and structures faunal communities is dependent on animal vectors for successful reproduction.[2–4] This direct role in plant reproduction makes pollinators keystone species.

In addition to their direct role in plant reproduction and food production in both wild and managed landscapes, pollinators support the full scope of global ecosystem services. The supportive roles of pollinators in larger global systems include climate regulation, nutrient cycling, water quality management, and air filtration.

ECOSYSTEM SERVICES IN NATURAL RESOURCE SYSTEMS

Ecosystem services are those benefits provided by the biosphere that maintain, support, and enhance life on Earth. Ecosystem services are defined as the set of direct and indirect benefits that are derived from healthy, sustainable, and functioning ecosystems.[5,6] Efforts to create a tangible, manageable system for classification have sorted ecosystem services into groups of functions that interact with all aspects of life. *Supportive services* connect ecosystems and facilitate the movement of energy and matter. *Provisional services* are responsible for the bounty of food and resources provided to humans and animals. *Regulating services* provide resilience to the environment, regulating and buffering against extreme shifts and supporting the status quo. *Cultural services* support creativity and esthetic landscapes, provide the inspiration for arts, literature, belief systems, and deliver opportunities for recreation.[7]

The spatial extent of ecosystem services offered by the biosphere, and pollinators, can be further subdivided into their geographic scale of interaction with humans and other species.[8] Services can be *proximal* (local), derived and applied at a source, or *distal* (global), derived at a source and distributed broadly. Global services include the regulation of climate and the atmosphere, the transition of energy and materials from one site to another, and the general esthetic value of diversity to human culture. Local services provide environmental regulation at the local scale, mediating disturbance and local climate, filtering water, buffering pollution, regulating pests and disease, and providing habitat.

POLLINATION

Defining Pollination

What is pollination? By definition, and in the most robust sense, it is the transfer of a pollen grain, the mobile gamete, to the ovary that contains the immobile gamete. Plant species that reproduce by sexual means (either exclusively or in addition to vegetative growth) must have this movement of pollen facilitated.[1] Pollen is a male gamete that can be compared to sperm, but unlike sperm cannot move on its own. The pollen vector can be abiotic (water or wind) or biotic (an animal vector). Animal pollen vectors include bees, birds (hummingbirds, honey creepers), bats, butterflies, moths, beetles, flies, wasps, small scansorial mammals (honey possums), larger fruit-eating mammals (lemurs, perhaps even chimpanzees), and even geckos.[9]

Encyclopedia of Natural Resources DOI: 10.1081/E-ENRL-120049039

To be a pollinator an animal must somehow be able to transport pollen; this usually occurs through contamination when an animal visits a flower for a nectar meal.

Defining Pollination Services

An examination of the origins of products the interconnections of natural systems quickly reveals the support that pollinators provide to our natural landscapes and resources. Many of the direct and indirect services that humans and all other living beings benefit from are a result of intimate long-standing pollinator–plant relationships.

The Pollination of Ecosystem Services

Pollinators and pollination have primary or direct roles in ecosystem service provisioning and secondary or indirect roles in ecosystem service delivery. Tertiary, or associated, roles can also be defined in some systems. For example, pollinators can have a distant role in water purification in ecosystems where riparian vegetation or other ground cover (that depends on pollinators for reproduction) provides the main role in sediment retention. Table 1 lists various terrestrial ecosystem services, their reliance on pollinators, and their spatial extent. This list is comprehensive, but by no means exhaustive, as some services can be segregated into further subcategories and future study will likely reveal the connection of pollinators to other services. The aim of this table is to highlight this complex network of energy transfer and support.

Supportive services

Supportive ecosystem services provided or enhanced by pollinators primarily include the structuring of ecosystems that define local ecotypes, wildlife patterns, and resource bases. Between 67% and 96% of all terrestrial species of angiosperms require animal pollinators for reproduction.[10] Considering the act of pollination itself, even when vectored by wind or water, the entirety of sexually reproducing plants (angiosperms and gymnosperms) that are not capable of self fertilization are dependent on pollination. It follows that all consumers of natural resources are reliant on pollination.

Diversity and local endemism are natural resources that define biomes, local communities, and cultures. The plant communities that define our natural systems and structure our landscape are reliant on pollination. For example, the Sonoran Desert is filled with vast plains spotted with saguaro cactus (*Carnegiea gigantea*). The giant saguaro is

Table 1 Terrestrial ecosystem services and their connections to pollinators and pollination services

Service type	Service	Service scale	Pollinator role
Supportive services	Nutrient cycling	Local	Primary/direct
	Seed dispersal	Local/global	Tertiary/associated
	Genetic diversity	Local	Primary/direct
	Primary production	Local/global	Tertiary/associated
Provisioning services	Pollination[a] – plant reproduction	Local	Primary/direct
	Crop production[a] (food for agrarian human population)	Local	Primary/direct
	Food production[a] (food-gathering human populations, wildlife)	Local	Primary/direct
	Biomass[a]	Local	Primary/direct
	Pharmaceuticals	Local	Primary/direct
	Industrial products (hardwoods and plant fibers)	Local	Primary/direct
	Maintenance of genetic diversity	Local	Primary/direct
Regulating services	Carbon sequestration	Local/global	Secondary/indirect
	Climate regulation[b]	Local/global	Primary/direct and secondary/indirect
	Pest/disease control	Local	Tertiary/associated
	Air purification[b]	Local/global	Primary/direct and secondary/indirect
	Water purification	Local	Secondary/indirect
Cultural services	Recreational experiences	Local	Primary/direct and secondary/indirect
	Cultural inspiration	Local	Primary/direct and secondary/indirect

Source: From Ollerton et al. 2011.[10]

[a]Between 67% and 96% of all terrestrial plants are believed to require pollinators for reproduction, or see significant increases in yields/seed development, when pollinators are present.

[b]The spatial extent of pollinators' roles in some services can shift depending on the time of year and nearness to equatorial regions.

a keystone species that supports life in this arid land. But the saguaro is incapable of self-fertilization, dependent on bats, such as the endangered lesser long-nose bat, *Leptonycteris curasoae yerbabuenae* (Fig. 1), bees, and birds for pollination. Like the pollinator visitors of succulents in arid environments, pollinators across the globe help perpetuate local diversity by ensuring the reproduction of unique and keystone plants.

The cycling of nutrients is another primary supportive service. Some pollinator guilds (functional groups), such as ground nesting bees and wasps, assist in nutrient cycling and soil aeration by their digging actions. About 60% of bee species nest in the ground.[11,12] For example, large ground-nesting bees in the genus *Habropoda* (pictured emerging from a nest in Fig. 2) tunnel into hillsides and slopes, loosening soil. Pollinators also have a role in maintaining reproduction in plants that fix and cycle nitrogen as many legumes are bee-pollinated.[13]

Provisional services

The pollination of food crops is the primary and best known example of pollinator services. Two of every three bites of food consumed by humans are the direct result of pollinator activities.[9] Corn, wheat, and rice—the carbohydrate stables of most cultural diets—are wind pollinated, but the food products that provide the necessary and essential nutrients for the human diet require pollinator visitation for fruiting and reproduction.[14] The variety of agricultural species that are dependent on pollinators is high, nearly 1000 of the 1200 common crop varieties planted across the globe.[13,15] The roles of pollinators and pollination services in food production systems are often more directly quantifiable than those in resource-based systems. The global value of pollination to agriculture has been estimated at €153 billion ($210 billion).[16] The individual contribution of insect pollinators is $57 billion (€41 billion) to the agricultural economy of the United States.[17]

The influence of pollinators on the daily lives of humans extends to the products that we interact with. Many of the resources and substrates that are used to develop and advance societies are derived from the living components of ecosystems. Hardwood products, fibers, textiles, dyes, scents, plant-derived chemical, and pharmaceutical products are all dependent on reproductive lineages of flowering plants,[5,6] the majority of which are dependent on animal pollinators. Softwood timber species are not dependent on animal vectors for pollination, but require pollination for continued reproduction. Without pollination we would see a decline in woody species that provide lumber, fibers, and other resources that have helped humans industrialize.

Regulating services

Regulating services, like the maintenance of climatic patterns, carbon sequestrations, air and water purification, and resistance to disturbances, are all characteristics of biomes that are maintained by the actions of pollinators. The sequestration of carbon by terrestrial plants depends on the continued growth and reproduction of these species. Tropical forest systems play a key role in air filtration and carbon regulation, especially in winter periods when deciduous species in temperate regions are in a resting stage. These tropical forest communities are dominated by pollinator-dependent species, with an estimate of nearly 95% dependence.[10]

Regulating services that prevent or buffer against environmental stochasticity are supported by healthy pollinator

Fig. 1 The Lesser long-nose bat, *Leptonycteris curasoae yerbabuenae,* feeding at flowers of *Agave palmeri palmeri* blossom during the night in the Sonoran Desert. *L. curasoae yerbabuenae* is also an essential pollinator of saguaro cactus (*Carnegiea gigantea*) that defines the Sonoran ecosystem.
Source: Photo credit Steven Buchmann.

Fig. 2 A male *Habropoda pallida* emerging from a ground nest, which is a network of tunnels beneath the surface that house brood chambers. Ground nesting aids in the aeration of soil and nutrient cycling.
Source: Photo credit Steven Buchmann.

communities. The California chaparral ecosystem is one example. Plant species responsible for soil stabilization along hillsides require pollinators for reproduction.[18] A loss of pollinators, and the eventual senescence of these plant species can cause vulnerability to landslides during annual wet periods. Similarly, coastal mangrove communities that provide protection during storm events need pollinators for population maintenance.[19]

Cultural services

Scientist and philosopher E.O. Wilson has written extensively on the topic of biophilia, meaning the love of the living world.[20] Humans are enriched and simply happier when surrounded by an abundance of living things. Species richness and landscape diversity inspire creativity. Cultural, recreational, and spiritual enhancement is provided by ecosystems that depend on pollinators. Recreational landscapes frequented by hikers, wildlife viewers, and hunters have primary or secondary dependencies on pollinators. Every few years, massive wildflower blooms in Death Valley National Park (Californian, USA) draw tourists from around the world to see the phenomenon of life in an unforgiving landscape. These desert species bloom each year because pollinators maintain their population. Another example is the indirect role that pollinators play in supporting seed dispersal by supporting populations of songbirds. These seed-dispersing birds feed directly off of many insect pollinators (adults and larvae), especially during juvenile development; thus pollinators support the growth and development of bird populations that eventually play a role in seed dispersal. Similarly, common game species such as deer, wild turkey, pheasant, and quail rely on angiosperm-dominated landscapes for habitat and food; threatened and rare species have the same dependence.

While pollinators and pollination services are generally restricted to terrestrial ecosystems there are some examples of interactions with aquatic systems. For example, a salmon fishery is dependent on the quality of upland river systems for spawning; as mentioned above, pollinators have indirect roles in maintaining water quality by maintaining natural filtration systems. Additionally, many aquatic fly larvae (the larvae of some pollinators) are eaten by salmon fry—a case can be made for the role that pollinators have in supporting salmon fisheries. These examples, while relevant, are difficult to quantify, but certainly should be qualified when the total value of pollination to natural resource systems is considered.

MAINTAINING AND SUPPORTING POLLINATION SERVICES

When productivity and service provision are concerned, increased biodiversity corresponds to increased stability, resistance, longevity, and product derivation.[5,21–23] The same is true of pollinator-dependent ecosystem services. Ecosystems with compromised or reduced diversity are susceptible to regime shifts that can change their composition and evolutionary trajectory.[24] For example, disturbances of ecosystems by anthropogenic factors (habitat loss, pollution, climate change, invasive species introduction, etc.) that consequently remove large numbers of species can eliminate functional groups, even entire trophic levels. A loss of biodiversity and the resultant loss in ecosystem function can have substantial effects on the output of ecosystem services.[25,26] In resource-based economic systems, a loss of function translates into a loss of productivity and a further loss of livelihood.[27] Losing pollinators will first result in decreased food production;[28] subsequently, each ecosystem service provided or supported by pollinators will suffer.

CONCLUSION

Our natural resources are those environmental products or elements on which we rely for survival. They support and enhance our lives. As consumers of natural resource we are reliant on pollination and pollinators. Understanding the local and global, as well as the primary, secondary, and tertiary connections that natural systems have with pollinators enhances our ability to manage and maintain them.

Healthy and diverse pollinator communities are an important factor for the functioning and sustainability of ecosystems and for the future of our societies.

REFERENCES

1. Kevan, P. Pollination: a plinth, pedestal, and pillar for terrestrial productivity: the why, how, and where of pollination protection, conservation, and promotion. Bees and Crop Pollination-Crisis, Crossroads, and Conservation, 2001; 7–68.
2. Bierzychudek, P. Pollinator limitation of plant reproductive effort. Am. Nat. **1981**, *117*, 838–840.
3. Kevan, P.G.; Baker, H.G. Insects as flower visitors and pollinators. Ann. Rev. Entomol. **1983**, *28*, 407–453.
4. Knight, T.M.; Steets, J.A.; Varmosi, J.C.; Mazer, S.J.; Burd, M.; Campbell, D.R.; Dudash, M.O.; Johnston, R.J.; Mitchell, R.J., Ashman, T.L. Pollen limitation of plant reproduction: Pattern and process. Ann. Rev. Ecol. Evol. Syst. **2005**, *36*, 467–497.
5. Daily, G.C., Ed.; *Nature's Services: Societal Dependence On Natural Ecosystems*. Island Press: Washington DC, 1997.
6. Costanza, R.; d'Arge, R.; de Groot, R.; Faber, S.; Grasso, M.; Hannon, B.; Limburg, K.; Naeem, S.; O'Neill, R.V.; Paruelo, J.; Raskin, R.G.; Sutton, P.; van den Belt, M. The value of the world's ecosystem services and natural capital. Nature **1997**, *387*, 253–260.
7. Millennium Ecosystem Assessment (MEA). *Ecosystems and Human Well-Being: Synthesis;* Island Press: Washington, 2005; 155 pp.
8. Costanza, R. Ecosystem services: Multiple classification systems are needed. Biol. Conserv. **2008**, *141*, 350–352.

9. Buchmann, S.; Nabhan, GP. *The Forgotten Pollinators*; Island Press: New York, 1996.
10. Ollerton, J.; Winfree, R.; Tarrant, S. How many flowering plants are pollinated by animals? Oikos **2011**, *120*, 321–326.
11. Linsley, EG. The ecology of bees. Hilgardia **1958**, *27*, 543–599.
12. Michener, CD. *The Bees of the World*; Johns Hopkins University Press: Baltimore, 2000.
13. Free, JB. *Insect Pollination of Crops*; Academic Press: London, UK, 1993.
14. Kevan, P.; Wojcik, VA. Pollinator services. In: *Managing Biodiversity in Agricultural Ecosystems*; Jarvis, D.I., Padoch, C., Cooper, HD., Eds.; Columbia University Press: New York, 2007; 200–249 pp.
15. Klein, AM.; Vaissière, B.E.; Cane, J.H.; Steffan-Dewenter. I.; Cunnigham, S.A.; Kremen, C.; Tscharntke, T. Importance of pollinators in changing landscapes for world crops. Proc. R. Soc. **2007**, *274*, 303–313.
16. Gallai, N.; Salles, J.M.; Settele. J.; Vaissière, B.E. Economic valuation of the vulnerability of world agriculture confronted with pollinator decline. Ecol. Econ. **2009**, *68*, 810–821.
17. Losey, J.E.; Vaughan, M. The economic value of ecosystem services provided by insects. BioScience **2006**, *56* (4), 311–323.
18. Kremen, C.; Bugg, R.L.; Nicola, N.; Smith, S.A.; Thorp, R.W.;Williams, N.M. Native bees, native plants, and crop pollination in California. Fremontia **2002**, *3–4* (30), 41–49.
19. Tomlinson, T.B. *The Botany of Mangroves*. Cambridge University Press: New York, 1986.
20. Wilson, E.O. *Biophilia*. Harvard University Press: Cambridge 1986; 1–176 pp.
21. Ehrlich, R.; Ehrlich, A.H.; Holdren, J.P. *Ecoscience: Population, Resources, Environment;* Freeman, W.H.: San Francisco, 1977.
22. Aylward, B.A.; Barbier, E.B. Valuing environmental functions in developing countries. Biodivers. Conserv. **1992**, *1*, 34.
23. Costanza, R.; Daly, H.E. Natural capital and sustainable development. Conserv. Biol. **1992**, *6*, 37–46.
24. Folke, C.; Carpenter, S.; Walker, B.; Scheffer, M.; Elmqvist, T.; Gunderson, L.; Holling, C.S. Regime shifts, resilience, and biodiversity in ecosystem management. Annu. Rev. Ecol. Evol. Syst. **2004**, *35*, 557–581.
25. Perrings C. The economic value of biodiversity. In *The Economic Value of Biodiversity*; Heywood, V.H., Gardner, K., Eds.; The United Nations Environment Programme by the Press Syndicate of the University of Cambridge: Cambridge, UK, **1995**; 822–913 pp.
26. Perrings, C.; Walker, B. Biodiversity, resilience and the control of ecological-economic systems: the case of fire-driven rangelands. Ecol. Econ. **1997**, *22*, 73–83.
27. Kevan, PG.; Phillips, TP. The economic impacts of pollinator declines. Conserv. Ecol. **2001**, *5*, 8. http://www.consecol.org/vol5/iss1/art8.
28. Aizen, M.A.; Garibaldi, L.A.; Cunningham. S.A.; Klein, A.M. How much does agriculture depend on pollinators? Lessons from long-term trends in crop production. Ann. Bot. **2009**, *103*, 1579–1588.

BIBLIOGRAPHY

1. Eardley, C.; Roth, D.; Clarke, J.; Buchmann, S.; Gemmill, B.; Eds.; Pollinators and Pollination: A Resource Book for Policy and Practice; African Pollinator Initiative (API): Johannesburg. 2006; viii–77 pp.
2. National Research Council (NRC): Committee on the Status of Pollinators in North America NRC. Status of Pollinators in North America. The National Academies Press, Washington, D.C., 2007.

Ecosystems: Diversity

L. M. Chu
School of Life Sciences, Chinese University of Hong Kong, Hong Kong, China

Abstract
Ecological complexity and stability are maintained by a diversity of interacting species or populations that have profound effects on the structure and function of ecosystems. At the ecosystem level, biodiversity, productivity, and nutrient cycling are closely linked, and they change dynamically across time through natural succession to form more complex and stable ecosystems. A variety of ecosystems can exist in a given place, forming a mosaic of interconnected patterns. The summation of the earth's ecosystems forms the biosphere, the global geographical distribution of which is largely determined by climatic factors, mainly rainfall and temperature, resulting in a spectrum of plant types and productivity. Latitude and altitude are highly influential, affecting the formation of the dominant vegetation. Among the various ecosystems, tropical forests are the most productive and support the highest biodiversity. However, they are under great pressure from deforestation, which represents a major threat to biodiversity and ecosystems. Proper ecosystem management is essential to maintain our planet as a provider of different ecosystem services.

INTRODUCTION

As the basic functional unit of nature, an ecosystem is a community of living organisms (microorganisms, plants, and animals) that interact among themselves and with the abiotic components of their environment.[1,2] Cultural components may also be included as part of this community, as ecosystems are shaped by humans. Ecosystems are defined by the network of such interactions[3] and form the basic level at which the properties required to support life exist. Ecosystems have system boundaries but may differ in their spatial and temporal scale and components to form systems with unique compositions and properties. They are adaptive units as they are evolving together with the species within them. They have the attributes of structure, function, heterogeneity, and process dynamics. They range in size, productivity, and complexity, but usually encompass specific, limited spaces.[4] However, ecosystems have also been viewed as functional, not spatial, units, as defined from a process-function perspective.[5] Due to biological hierarchy and functional continuity, the boundaries of ecosystems can be chosen arbitrarily such that ecosystems exist at all scales of organization and complexity, and should include communities and landscapes.[6,7]

Ecosystems can be broadly divided into two major types—terrestrial and aquatic. Examples include forests, grasslands, ponds, and estuaries. Occasionally, they can be grossly categorized into lowland and upland ecosystems based on elevations. Ecosystems usually exist as spatially heterogenous land areas of interconnected systems where natural processes and human activities interact to produce ever-changing mosaics,[8] such as a pond in a forest with adjacent grassland. Although borders between ecosystems may be clear and well defined, such as a freshwater lake, the transition from one ecosystem to another can also be gradual. Compared with aquatic ecosystems, in particular marine ecosystems, terrestrial ecosystems are characterized by greater light availability, higher gaseous concentration, lower water content, and greater seasonal and diurnal temperature fluctuations.

BIODIVERSITY AT THE ECOSYSTEM LEVEL

Ecological diversity refers to the species richness in a biological community—it is the diversity of a site at the level of the ecosystem. Together with genetic and species diversity, it forms the third level of biodiversity. In general, the diversity and functional complexity of an ecosystem are governed by the population and diversity of the species that are present, the interactions between these species and their interactions with the environment, and the complexity of the physical environment.[9]

The diversity of interacting types at the species or population level maintains ecological complexity and stability. It is known that species diversity has pronounced effects on ecosystem structures and functions. However, ecosystems are dynamic in the sense that they experience periodic perturbations from external forces such as environmental stochasticity and catastrophic events, and internal processes such as nutrient cycling. They have the capacity of resistance and/or resilience to disturbances, which is a function of their ecological attributes. Seasonality is a driving force behind annual changes in community composition and is

Encyclopedia of Natural Resources DOI: 10.1081/E-ENRL-120047427

important in shaping the life history of individual species in terms of breeding behavior and habits. Plant and animal communities constantly and gradually alter and develop over time through a directional, non-seasonal process called succession, as a result of endogenic processes and exogenic influences, thus changing the diversity and structural complexity of the ecosystem. Succession drives natural ecosystem development and assists in ecosystem restoration, which can be manipulated with human interference. Primary succession occurs naturally on glacier moraines or volcanic ashes, which are changed by weathering, plant colonization, organic matter accumulation, and soil formation, which subsequently affect the community and ecosystem levels.

Ecosystem properties are influenced by the diversity of species and their functional characteristics (functional traits, functional types, and functional diversity). Resource partitioning and niche differentiation allow the coexistence of species competing for similar resources. This coexistence increases the combinations of species that are complementary, and hence increases the biodiversity, productivity, and nutrient cycling of the ecosystem. Nevertheless, biological richness may differ in ecosystems that have similar basic processes of matter and energy fluxes.

ECOSYSTEM TYPES

Ecological diversity should be distinguished from ecosystem diversity, which relates to the assortment of biological communities and their habitats, and the physical environment and ecological processes within an ecosystem and between different ecosystems. Ecosystem diversity can be defined as the variety and relative extent of ecosystem types, including their composition, structure, and processes. It refers to the spectrum of ecosystems—i.e., the diversity of ecosystems, sometimes called ecodiversity—present in the biosphere. Ecosystem diversity is high in Southeast Asia, Africa, and South America, which have a massive land area with variable climate, complex geomorphic types, a long shoreline, and an extensive river system.

Our biosphere is viewed as a giant system that contains a diversity of ecosystems. Ecosystems can be divided into two distinct groups—natural and man-made—that vary in different parts of the world. Terrestrial ecosystems can be classified on the basis of the physiognomy and physical environment of their vegetation, defined by the external appearance and the structure of the dominant vegetation, such as deciduous forests, coniferous forests, savannas, and deserts. More often, ecosystems are classified using multi-factor systems that include a combination of biotic and abiotic factors such as vegetation, climate, soil, geology, physiography, and hydrology. Ecosystems can also be categorized in terms of their forcing functions (external variables) that determine the ecosystem conditions (Table 1).[10]

There are many different ecosystems—ponds and coral reefs, deserts and grasslands, tundra and rainforests. The types of ecosystems found on Earth are largely determined by climatic factors, which are influenced by latitude and altitude and affect the formation of the dominant vegetation. Ecosystems can be classified by latitude, e.g., temperate forests and tropical forests, which causes variations in temperature and precipitation. These variations give rise to ecosystems with different magnitudes of productivity. There are spatial gradients in biodiversity and energy flux among ecosystems with different climates and geographical locations. Nonetheless, the Earth's ecosystems are linked by global atmospheric and water circulation to form the biosphere. The occurrence of an organism in a particular habitat is a reflection of its adaptation to environmental stresses or disturbances, regardless of whether or not its population distribution is restricted, though dispersal ability is as well important, e.g., high salinity in salt marshes or mangroves, heat and drought in hot deserts, physiological drought in montane ecosystems, shade in forest understory, and cold in polar ecosystems. Because stresses impose different effects on plants and animals in different ecosystems, living organisms that survive will have different adaptation strategies to cope with/exploit these stresses. These specialist organisms may pay an ecological price—e.g., extreme biomass turnover rate and shifted reproductive mode—for their adaptation to those high-stress ecosystems. An important outcome of this adaptation is that stressed ecosystems provide a refuge for these specialists, resulting in greater overall biodiversity than if all organisms were generalists.

Table 1 Ecosystem categories according to external forcing functions

Category	Forcing functions	Unique properties	Examples
1	Natural	Buffer capacity, biodiversity, ecosystem health	Polar terrestrial ecosystems, tundra, deserts, dunes, alpine forests, caves
2	Pollution	Buffer capacity, biodiversity, adaptation	Boreal forests, savannas, steppes, prairies, rainforests
3	Pollution and climate changes	Buffer capacity, adaptation, resilience	Estuaries, lakes, marshes, floodplains, swamps, mangroves, rivers
4	Anthropogenic	Management-oriented	Agroecosystems, plantations, urban ecosystems, landfills, quarries

Source: Adapted from Jorgensen.[10]

WORLD ECOSYSTEMS

The biosphere is occupied by biomes, which represent distinctive plant and animal communities at the continental level. The geographical pattern of biomes on a global scale is determined by primary productivity, which is governed mainly by the temperature and rainfall in the terrestrial environment. Biomes are regional-scale assemblages of ecosystems (continental-level ecosystems). The biome approach to ecosystem diversity is linked directly to the spatial pattern of ecosystems throughout the biosphere. The distribution of biomes is heavily influenced by climate—precipitation produces humid, semihumid, semiarid, and arid ecosystems and thus has an overriding effect. Where rainfall is plentiful, vegetation is abundant and tall trees dominate. With decreasing rainfall, the biomass of vegetation is reduced and becomes scattered and is dominated by shrubs and grasses. Temperature is another limiting factor, and its ecological effects are manifested at higher latitudes and altitudes. Nival, alpine, subalpine, and montane ecosystems vary by altitude, while polar, boreal, temperate, subtropical, and tropical ecosystems vary by latitude. Interestingly, there are two types of rainforests—tropical and temperate—and grasslands are found on all continents except Antarctica.

Forest Ecosystems

Forest ecosystems are characterized by their vertical structure, which usually consists of several layers: overstoreys of trees and understoreys of shrubs and herbs. The overstoreys can differentiate to form a non-uniform crown canopy. Forest ecosystems can be pure or mixed stands. Young ecosystems are usually open stands, while mature ones generally have closed crown canopies. Tree species assemblages vary between regions, with the crown canopy structure greatly affected by age, the prevailing climatic conditions, and ecosystem management. Changes in forest vegetation are strongly associated with climatic factors due to latitudinal and elevation gradients as well as edaphic factors such as geology and soil. Forest ecosystems provide various ecological services, although the focus is often on their value for timber production. Many forests disappear as a result of mismanagement, causing carbon dioxide release and exacerbating global warming. Despite global and local regulatory measures to halt the negative effects of forest degradation and deforestation, their efficacy is yet to be proven.

Tropical rainforests

Tropical rainforests are equatorial evergreen rainforests located in humid tropical areas in Central and South American, Southeast Asia, and West Africa. The annual rainfall in such areas exceeds 3000–4000 mm and is evenly distributed throughout the year. Humidity and temperature are relatively and constantly high. They have the greatest biodiversity, highest primary productivity, and most complex structure of all terrestrial ecosystems. They contribute 30% of the total net primary productivity on land.[11] The Amazonian tropical rainforests are megadiverse hotspots, with over 5000 vascular plant species per 10,000 km^2.[12] Tropical rainforests are the most complex terrestrial ecosystems, as they are species-rich and stratified vertically. Their canopies are thick and continuous, and competition for light is intense. There are typically three to four vertical layers—forest floor, understory, main canopy, and emergent. Most trees are about 30 m tall, with some reaching 60–80 m, and they can support massive epiphytic growths on their trunks; this stratification creates a variety of microhabitats in terms of light intensity, moisture content, and wind velocity. Plant species are adapted to deal with the overwhelming competition from tall trees for light, water, and nutrients. Soils are surprisingly thin and nutrient poor. Nutrient loss due to leaching by heavy rains is severe and decomposition is rapid, keeping the organic matter content of the soil low. However, these soils can sustain high biomass and biodiversity as rainforest plants are adapted to low levels of nutrients and can recycle nutrient efficiently due to the presence of mycorrhizae in the roots, which enhance nutrient absorption. Faunal diversity is enormous and includes insects, amphibians, reptiles, bats, and primates, most of which are arboreal and canopy-dwellers and are supported by the diverse food sources from trees and epiphytes.

Tropical seasonal forests

These are commonly called tropical dry forests and are located in regions further away from the equator, between 10° and 25° latitude north and south of the Amazon rainforests, northern Mexico, north and south of the central African rainforests, India, Indochina, northern Australia, and some tropical islands. They are tropical forests with an annual rainfall cycle of a 5–7-month wet season with pulses of torrential rains, and a 5–7-month dry season. Soil is nutrient-rich but highly susceptible to erosion. Rainfall seasonality is a dominant ecological force that drives biological activities, with tall deciduous trees forming a dense canopy during the wet season, and their height highly correlated with average rainfall.[13] Trees are shorter and shed their leaves during dry months. Seasonal migration is a phenomenon in animal populations, which move according to water availability in dry seasons. Many tropical dry forests are increasingly threatened by human activities: habitat fragmentation and fire are important in African forests, agricultural conversion is most notable in Eurasia, and climate change is more influential in the Americas. Globally, they are considered the most endangered tropical ecosystem and should be accorded higher conservation priority.[14]

Deciduous forests

Deciduous forest is a common ecosystem that used to cover most of the more humid areas at temperate latitudes. The primary productivity of temperate deciduous forests is about 40% of that of tropical rainforests, due to lower growth and activity during the cold winter months. They exhibit strong seasonality in terms of foliage cover and flowering, with distinct vegetative and dormant periods. Summer-green forests are found in regions with pronounced seasonal temperature changes, and contain tall broad-leafed trees that shed their leaves in winter. Their biodiversity is lower and their vegetation structure less complex. Conifers may comprise a considerable portion of deciduous forests as we move northward in latitude or upward in elevation. The original vegetation in this type of ecosystem has been severely altered by humans as a result of logging, livestock grazing, agriculture, and plantations, independent of successional changes.

Grasslands

Grasslands are ubiquitous ecosystems, occupied exclusively by low-growth vegetation that contributes to its open nature. Grasses and sedges are the dominant species, together with flowering herbs, including many kinds of composites and legumes. Grasslands are usually devoid of woody species. However, while temperate grasslands are usually treeless, tropical grasslands have scattered small trees. Temperate species are C3 plants while tropical species are C4 plants based on their photosynthetic pathways. Grasslands can be classified by the unidirectional approach of successional linear replacement into types such as pioneer and climax, or less commonly into increasers, decreasers, neutrals, invaders, and retreaters,[15] depending upon their responses to factors such as grazing, drought, and fire. Annuals or short-lived perennials are pioneer grassland species that readily invade and colonize exposed soils and thrive in full sunlight. Grasslands are found in areas where the moisture content of soil is too low for a closed forest canopy, and grass-dominated communities are maintained through vegetative growth cycles in a seasonal pattern. Grassland areas include prairie in North America, pampas in South America, steppe in central Asia, and savanna in tropical Africa, Australia, South America, and Asia. They are the grazing grounds for both wild and domestic herbivores. Their primary production supports diverse fauna. Plants respond morphologically, structurally, and chemically to herbivory, and herbivores evolve by developing different feeding strategies regarding food source, body size, and feeding behavior. Plant–herbivore interactions and the effects of grazing by large generalist herbivores are generally accounted for by predation, disturbances arising from the evolutionary history and the selective pressures of grazing,[16] and the feedback mechanism between plants and grazers that is manifested in the switching capabilities of plant species and competition modes, especially in grasslands with a long grazing history.[17] Grasses are highly adapted to repetitive grazing and fire, which stimulate shoot regeneration from leaf meristems near the soil surface or by tillering after defoliation. Grassland can be maintained by moderate grazing intensity or fire, which prevents fire-intolerant woody species from successional recovery, and by mowing and other management practices. The long-term sustainability of pastures is threatened by grassland degradation attributable to climate change, desiccation, and over-grazing.

Deserts

Dryland ecosystems constitute the largest biome on Earth, occupying about one-third of its surface. Major deserts are found on all continents except Europe and are geographically located at 25–40° North and South on continental inlands, though there are also coastal deserts in Central America and Southern Africa. They range from semi-arid rangelands to semi-deserts, deserts, and hyper-arid deserts, with decreasing rainfall intensity which restricts primary productivity (Table 2).[18] Deserts are characterized by bioclimatic aridity, defined by the ratio of precipitation to evapotranspiration, rainfall seasonality, erosion, and deposition. Climatologically, deserts are influenced by descending dry air currents that reduce rainfall formation. They are found in areas with annual precipitation of less than 250 mm. Rainfall is usually highly irregular, both spatially and temporally.

Deserts can be broadly divided into two groups based on their geological structure and elevational setting: a) shield and platform deserts found mainly in Africa, Arabia, Australia, and India; and b) basin and mountain deserts located in Asia and America. Geographically, they

Table 2 Arid lands categorized by bioclimatic aridity resulting from precipitation (P) and evapotranspiration (ET)

	Bioclimatic aridity	Rainfall	Vegetation
Hyper-arid	P/ET < 0.03	Low, irregular, aseasonal	Perennial; shrubs in riverbeds
Arid	0.03 < P/ET < 0.2	80–350 mm (annual)	Perennial; woody succulents; thorny or leafless
Semi-arid	0.2 < P/ET < 0.5	300–800 mm (summer) 200–500 mm (winter)	Shrubs; steppes, savannas, tropical scrub
Sub-humid	0.5 < P/ET < 0.75		Tropical savanna, chaparral, steppes

Source: From UNESCO.[18]

are categorized into four main types: subtropical, cool coastal, rain shadow, and interior continental.[19] Deserts may be devoid of vegetation when rainfall is totally absent. High temperature is usually associated with aridity, which further limits productivity. The composition and structure of the vegetation are largely determined by the landscape's ecology, lithology, and geomorphology,[20,21] which are a function of the topography, gradient, and aspect.

Water supply in deserts is erratic and sporadic, and is aggravated by the high evaporation rate. The resulting water shortage severely restricts life and drives adaptive evolution. Desert organisms are usually tolerant to drought and heat. Desert plants have a scattered distribution and are adapted to arid conditions by means of escape, evasion, resistance, and endurance to drought. Plants that become established usually develop water-conserving features to reduce moisture loss and deep roots to access water at greater soil depths. Only drought-resistant plants such as aloe and sagebrush and succulents such as cacti can survive. Desert shrubs and trees are morphologically adapted to enhance stem flow and root channelization.[22] Plants may also develop ecophysiological strategies to overcome water scarcity, which to a certain extent may generate more efficient photosynthesis. Desert animals survive by escape, burrowing, physiological resistance, and aestivation. They have evolved a range of structural, physiological, and behavioral adaptations in body temperature regulation to cope with the heat of the desert.

Water is the major limiting factor, as illustrated by the "pulse-reserve" life strategy,[23] which responds to abiotic conditions modified by biotic interactions. Many desert plants are ephemerals, which complete their life cycles by taking advantage of short wet periods. Some species may germinate, grow, flower, and set seeds after a single rainfall event. Competition and predation have major effects on water–nutrient availability, seed dispersal, and herbivory, which alter carbon allocation patterns and plant morphologies. Most organisms live at or near the threshold for surviving the desert's climatic extremes and resource availability. Competitive interactions determine the patch-mosaic pattern and dynamics, resulting in a patchy distribution of plants and animals.[24] The association between plants, though may be weak, provides subcanopy microsites for seed germination and plant establishment. Patches are attributed to differential germination and establishment on sites modified by the subcanopy microclimate and animal activities (animal dispersal). Nurse plant–commensal relationships facilitate a number of cacti and stem succulents, and some grasses. Processes such as herbivory affect vegetation patchiness as a result of the differential spatial pattern of plant material consumption and fecal matter accumulation. The spatial distribution of plant life forms is also governed by the elevation, soil texture and soil depth, and the availability of nutrients and water. Despite low primary productivity, biodiversity can be higher than expected. Arid land is expanding as a result of desertification caused by climate change and poor land management.

Wetlands

Wetland broadly refers to areas of marshland, fens, peatlands, or water, whether natural or artificial, permanent or temporary. Water may be static or flowing, fresh, brackish, or salt, including marine water areas in which the depth at low tide does not exceed 6 m.[25] Wetland ecosystems are a transition between terrestrial and aquatic systems, with hydric soil that results in prolonged anaerobic conditions. Wetland ecosystems can be classified into coastal and inland wetlands. Coastal wetlands include tidal salt marshes, tidal freshwater marshes, and mangrove wetlands, while inland wetlands consist of freshwater marshes, freshwater swamps, riparian forests, and peatlands. Freshwater wetlands can be further divided into herbaceous wetlands that include lacustrine (ponds and lakes), riverine, and palustrine (marshes and peatlands) wetlands, as well as woody wetlands, i.e., forested wetlands such as swamps. Vegetation is dominated by hydrophytes, which include submerged, floating, and emergent plants. Floating hydrophytes have their roots in water. Emergents include reed, cattail, and cypress; they have their rhizomes and roots in substrates, while their shoots protrude into the air.

Wetlands are highly productive and wetland plants are efficient in absorbing nutrients and retaining them. They have been described as the kidneys of the Earth because of their capacity to purify the water that flows through them.[26] Wetlands play significant roles in modifying the hydrology and reducing flood flows. They act like a sponge during periods of high water inflow, which has important implications for downstream ecosystems. Wetland vegetation retains sediments and reduces erosion. Wetlands are of international conservation value and are significant in carbon sequestration. They have high biodiversity and attract a great variety of fauna, including migratory birds that use them as fueling stations.

Mangrove ecosystems are forested wetlands found in estuaries and sheltered inlets exclusively in the intertidal zones of the tropical and subtropical coastline regions. True mangroves are taxonomically diverse and comprise about 60 species. They are woody species that, together with the mangrove associates, make up the mangals in tidal sediment habitats. Their mechanisms for salinity tolerance are exclusion, secretion, and/or accumulation of salts. To adapt to waterlogging and fluctuating salinity, mangroves have developed complex structural and physiological specializations such as aerial roots, pneumatophores, salt glands in their leaves, and viviparous floating propagules, allowing them to survive in brackish water. Their growth and productivity as well as litter dynamics are governed by hydroperiod and nutrient biogeochemistry, which produce different stress combinations associated with coastal wetlands.[27] Mangroves support a variety of food webs that involve a wide range of halophytes and create a habitat for a rich heterogeneous faunal community, and are used by a vast array of marine and terrestrial animals as feeding and breeding grounds. They have great economic and educational value and protect the shoreline by acting as a natural barrier against water erosion. They are under high deforestation pressure as a result of human impacts; one-fifth of the world's mangrove ecosystems have been lost since 1980.[28]

Polar Ecosystems

Polar ecosystems are characterized by extreme environments because of their high latitudes and low temperatures (below 5°C). Most are glaciated and permanently covered with a thick layer of ice—permafrost that reaches a thickness >600 m.[29] Geographically, the Arctic region in the Northern Hemisphere is primarily an ice-covered ocean surrounded by land, whereas the Antarctic in the Southern Hemisphere is a landmass surrounded by ocean. The Arctic and Antarctic regions differ in various ways regarding their geography, physical features, vegetation composition, and wildlife diversity. However, both regions have cold, dry, and windy climate, with short-growing seasons and prolonged darkness in the winter. They are sometimes referred to as polar deserts because of the lack of precipitation and low temperatures. Polar life survives through unique adaptations to withstand the cold. They contain 3–5% of global biodiversity and net primary productivity ranges from 1 to 1000 $g/m^2/yr$.[30] Polar ecosystems are typically treeless, and in warmer and wetter places with organic soil, tundra vegetation is found and is dominated by simple plant communities of algae, lichens, mosses, and liverworts. They contain large mammals such as bears, caribou, wolves, foxes, walruses, and seals, and a variety of birds including penguins and migratory species. Most of them are homeotherms that have adaptations to allow them to thrive in cold environments. Polar ecosystems are vulnerable to human disturbances such as global climate change, pollution, and resource extraction. Global warming has facilitated the invasion of exotic species to polar regions, which used to be physiological filters.[31,32]

Agroecosystems

Agroecosystems are managed ecosystems designed for the production of crops and livestock. Conventional agroecosystems have lower species diversity, a more homogenous ecological structure, and modified growth dynamics and agricultural input. To overcome the climatic and soil constraints, the productivity of agroecosystems can be enhanced through the use of machinery and human practices such as irrigation and the application of fertilizers, pesticides, herbicides, and fungicides. Arable fields are composed of crop species that are cultivated for biomass harvest and usually have a short life-cycle and small size; weeds, which are unwanted species that compete with the crops for sunlight, space, water, and nutrients; and pests such as insects that cause harm to crops in terms of health and productivity. The organic matter turnover and nutrient cycling of agricultural fields is manipulated. Biodiversity can be remarkably low in monocultures, in contrast with the general relationship between biodiversity, productivity, and ecological stability. Agricultural systems can be classified according to farming intensity, the degree of mechanization and the use of agrochemicals. Industrial farming is a conventional system, while integrated and organic farming use less or no pesticides and chemical fertilizers. The majority of agricultural land is formed by the conversion of forests. In addition to food production, agricultural land provides ecosystem services such as organic recycling, carbon sequestration, and landscape attributes.

Urban Ecosystems

Urban areas are essentially built environment covered by concrete structures and asphalt surfaces, which drastically change their metabolisms, material flows, and energy cycling.[33] With the concentration of populations, cities usually have high ecological loads and footprint with respect to food and water consumption, waste and wastewater disposal, as well as pollutant discharges. Urban ecosystems have a human dimension and are created as a result of urbanization, which is considered a major driving factor in the extinction of many fauna and flora. They have a unique community structure and ecosystem function due to their altered soil and water status, climate change, and local air pollution. Pollution problems are widespread particularly for those with heavy industrial and traffic emissions to result in ecosystem responses typically along a rural–urban gradient.[34,35] Urban growth and sprawl are converting natural habitats to isolated fragments with great reduction in biodiversity. Biogeochemically, urban ecosystems have modified soil carbon pools and fluxes;[36] physically, they have elevated temperatures and are urban heat islands that have ecological, energy, and climate implications. Urban landscapes and green areas are becoming important, and the more common use of alien species that fill up their specific urban niches have significantly modified urban floristic diversity and the ecological processes in such artificial ecosystems.[37,38]

Urban ecosystems range from roof gardens, parks, and greenbelts to urban forests and entire cities; they can be very diverse and include cemeteries, golf courses, and slopes. They sometimes encompass man-made ecosystems that are formed by the rehabilitation of degraded and damaged habitats such as mine sites, quarries, landfills, and coal ash disposal sites. However, these habitat patches represent a barrier to animal and plant movement unless they are connected by ecological corridors to maintain biological interactions. Cities will have distinctive species assemblages and support special fauna and flora that are highly adapted to the urbanized environmental conditions.[39,40] Wildlife–human conflicts do exist and wildlife management is an integral part of urban ecosystem management.[41] There are indeed variations in their management, which affect their role in city greening and biodiversity conservation. Overall, urban ecosystems benefit human populations through the provision of ecosystem services, among which the regulation of air quality, city climate, and water cycling are the most important.

ECOSYSTEM MANAGEMENT

Ecosystem diversity is the focal point of biodiversity, and ecosystem function is regarded as a core element in biodiversity assessment.[42] Likewise, the ecosystem approach to conservation and sustainable development is a strategy for the integrated management of water, land, atmosphere, and biological resources at the appropriate spatial and temporal scales in a fair and equitable manner. This requires the adaptive management of ecosystems. However, ecosystems of all kinds are being threatened worldwide. Many types of forests, grasslands, coastal and lowland areas, and wetlands have been under severe pressure from human destruction or degradation. These pressures affect biodiversity in the ecosystems and ecosystem properties which in turn depend on biodiversity and ecosystem functioning.

A coarse filter conservation approach has been proposed to maintain ecosystem diversity, which is the key to ecological sustainability. It is argued that preserving a good representation of all ecosystems helps to meet the habitat requirements of all native species in a locality, hence protecting biodiversity on both a local and regional scale.[43] Ecosystems include both physical and biological dimensions, and may be expanded to include economic, social, cultural, and ethical factors. With the challenges from a multitude of development threats, ecosystem management is vital for setting priorities and accommodating stressors, whether natural or anthropogenic, to sustain our planet as a provider of resources and other essential ecological services. Our knowledge of the relationship between biodiversity and ecosystem functioning will certainly benefit nature conservation and management as a whole.

REFERENCES

1. Tansley, A.G. The use and abuse of vegetational concepts and terms. Ecology **1935**, *16* (3), 284–307.
2. Likens, G.E. *The Ecosystem Approach: Its Use and Abuse*; Ecology Institute: Oldendorf/Luhe, Germany, 1992.
3. Schulze, E.D.; Beck, E.; Müller-Hohenstein, K. *Plant Ecology*; Springer: Berlin, 2005.
4. Chapin, F.S., III; Matson, P.A.; Mooney, H.A. *Principles of Terrestrial Ecosystem Ecology*; Springer: New York, 2002.
5. O'Neill, R.V.; DeAngelis, D.L.; Waide, J.B.; Allen, T.F.H. *A Hierarchical Concept of Ecosystem*; Princeton University Press: Princeton, 1986.
6. Haufler, J.B.; Crew, T.; Wilcove, D. Scale considerations for ecosystem management. In *Ecological Stewardship. A Common Reference for Ecosystem Management*; Szaro, R.C., Johnson, N.C., Sexton, W.T., Malk, A.J., Eds.; Elsevier: Oxford, 1999; Vol. 2, 331–342.
7. Lugo, A.E.; Baron, J.S.; Frost, T.P.; Cundy, T.W.; Dittberner, P. Ecosystem processes and functioning. In *Ecological Stewardship. A Common Reference for Ecosystem Management*; Szaro, R.C., Johnson, N.C., Sexton, W.T., Malk, A.J., Eds.; Elsevier, Oxford, 1999; Vol. 2, 219–254.
8. Forman, R.T.T. *Land Mosaics: The Ecology of Landscapes and Regions*; Cambridge University Press: Cambridge, 1995.
9. Smith, R.L. *Ecology and Field Biology*, 4th Ed.; Harper Collins: New York, 1990.
10. Jorgensen, S.E. Overview of ecosystem types, their forcing functions and most important properties. In *Ecosystem Ecology*; Jorgensen, S.E., Ed.; Elsevier: Amsterdam, 2009; 140–141.
11. Whittaker, R.H. *Communities and Ecosystems*, 2nd Ed.; The Macmillan Co.: Toronto, 1975.
12. Barthlott, W.; Hostert, A.; Kier, G.; Küper, W.; Kreft, H.; Mutke, J.; Rafiqpoor, D.; Sommer, J.H. Geographic patterns of vascular plant diversity at continental to global scales. Erdkunde **2007**, *61* (4), 305–315.
13. Murphy, P.G.; Lugo, A.E. Ecology of tropical dry forest. Annu. Rev. Ecol. Syst. **1986**, *17*, 67–88.
14. Miles, L.; Newton, A.C.; DeFries, R.S.; Ravilious, C.; May, I.; Blyth, S.; Kapos, V.; Gordon, J.E. A global overview of the conservation status of tropical dry forests. J. Biogeogr. **2006**, *33* (3), 491–505.
15. Vogl, R.J. The effects of fire on the vegetational composition of bracken-grasslands. Trans. Wis. Acad. Sci. Arts Lett. **1964**, *53*, 67–82.
16. Fox, J.F. Intermediate-disturbance hypothesis. Science **1979**, *204* (4399), 1344–1345.
17. Milchunas, D.G.; Sala, O.E.; Lauenroth, W.K. A generalized model of the effects of grazing by large herbivores on grassland community structure. Am. Nat. **1988**, *132* (1), 87–106.
18. United Nations Educational, Scientific and Cultural Organization (UNESCO). *World Map of Arid Regions*; UNESCO: Paris, 1977.
19. Cloudsley-Thompson, J.L. *Man and the Biology of Arid Zones*; Edward Arnold: London, 1977.
20. Schulz, E.; Whitney, J.W. Vegetation in north-central Saudi Arabia. J. Arid Environ. **1986**, *10* (3), 175–186.
21. Wondzell, S.M.; Cunningham, G.L.; Bachelet, D. Relationship between landforms, geomorphic processes and plant communities on a watershed in the northern Chihuahuan Desert. Landscape Ecol. **1996**, *11* (6), 351–362.
22. DeSoyza, A.G.; Whitford, W.G.; Martinez-Meza, E.; Van Zee, J.W. Variation in creosotebush (*Larrea tridentata*) canopy morphology in relation to habitat, soil fertility and associated annual plant communities. Am. Midl. Nat. **1997**, *137* (1), 13–26.
23. Noy Meir, I. Desert ecosystems: Environment and producers. Annu. Rev. Ecol. Syst. **1973**, *5*, 195–214.
24. Whitford, W. *Ecology of Desert Systems*; Academic Press: San Diego, 2002.
25. Ramsar Convention. Convention on Wetlands of International Importance especially as Waterfowl Habitat, http://www.ramsar.org/cda/en/ramsar-documents-texts-convention-on/main/ramsar/1-31-38%5E20671_4000_0, 1971 (accessed November 2012).
26. Mitsch, W.J.; Gosselink, J.G. *Wetlands*, 2nd Ed.; John Wiley & Sons: Hoboken, New Jersey, 2007.

27. Twilley, R.R.; Rivera-Monroy, V.H. Developing performance measures of mangrove wetlands using simulation models of hydrology, nutrient biogeochemistry, and community dynamics. J. Coast. Res. **2005**, SI40, 79–93.
28. Spalding, M.; Kainuma, M.; Collins, L. *World Atlas of Mangroves*; Routledge: Oxford, 2010.
29. Stonehouse, B. *Polar Biology*; Chapman and Hall: New York, 1998.
30. Callaghan, T.V. Polar terrestrial ecology. In *Ecosystem Ecology*; Jorgensen, S.E. Ed.; Elsevier: Amsterdam, 2009; 339–342.
31. Walther, G.-R.; Post, E.; Convey, P.; Menzel, A.; Parmesan, C.; Beebee, T.J.C.; Fromentin, J.-M.; Hoegh-Guldberg, O.; Bairlein, F. Ecological responses to recent climate change. Nature **2002**, *416* (6879), 389–395.
32. Robinson, S.A.; Wasley, J.; Tobin, A.K. Living on the edge – Plants and global change in continental and maritime Antarctica. Glob. Chang. Biol. **2003**, *9* (12), 1681–1717.
33. Kennedy, C.; Cuddihy, J.; Engel-Yan, J. The changing metabolism of cities. J. Ind. Ecol. **2008**, *11* (2), 43–59.
34. Gilbert, O.L. *The Ecology Of Urban Habitats*; Chapman and Hall: London, 1989.
35. McDonnell, M.J.; Pickett, S.T.A.; Groffman, P.; Bohlen, P.; Pouyat, R.V.; Zipperer, W.C.; Parmelee, R.W.; Carreiro, M.M.; Medley, K.Y. Ecosystem processes along an urban-to-rural gradient. Urban Ecosyst. **1997**, *1* (1), 21–36.
36. Pouyat, R.; Groffman, P.; Yesilonis, I.; Hernandex, L. Soil carbon pools and fluxes in urban ecosystems. Environ. Pollut. **2002**, *116* (Suppl. 1), S107–S118.
37. Kowarik, I. On the role of alien species in urban flora and vegetation. In *Plant Invasions: General Aspects and Special Problems*; Pysek, P., Prach, K., Rejmanek, M., Wade, M., Eds.; SPB Academic Publishing: Amsterdam, 1995; 85–103.
38. Savard, J.-P.L.; Clergeau, P.; Mennechez, G. Biodiversity concepts and urban ecosystems. Landscape Urban Plann. **2000**, *48* (3–4), 131–142.
39. Blair, R.B. Birds and butterflies along an urban gradient: Surrogate taxa for assessing biodiversity? Ecol. Appl. **1999**, *9* (1), 164–170.
40. McKinney, M.L. Effects of urbanization on species richness: A review of plants and animals. Urban Ecosyst. **2008**, *11* (2), 161–176.
41. Adams, C.E.; Lindsey, K.J.; Ash, S.J. *Urban Wildlife Management*; CRC Press: Boca Raton, 2006.
42. United Nations Environment Program (UNEP). *Global Biodiversity Assessment*; Cambridge University Press: Cambridge, 1996.
43. Hunter, M.L.; Jacobson, G.L.; Webb, T. Paleoecology and the coarse-filter approach to maintaining biological diversity. Conserv. Biol. **1988**, *2* (4), 375–385.

Ecosystems: Forest Nutrient Cycling

Neil W. Foster
Canadian Forest Service, Ontario, Sault Ste. Marie, Ontario, Canada
Jagtar S. Bhatti
Canadian Forest Service, Northern Forestry Centre, Edmonton, Alberta, Canada

Abstract
Forest trees make less demand on the soil for nutrients than annual crops because a large proportion of absorbed nutrients are returned annually to the soil in leaf and fine root litter and are reabsorbed after biological breakdown of litter materials. This entry provides an overview of that cycle.

INTRODUCTION

Nutrients are elements or compounds that are essential for the growth and survival of plants. Plants require large amounts of nutrients such as nitrogen (N), phosphorus (P), carbon (C), hydrogen (H), oxygen (O), potassium (K), calcium (Ca), and magnesium (Mg), but only small amounts of others such as boron (B), manganese (Mn), iron (Fe), copper (Cu), zinc (Zn) and chlorine (Cl) (micronutrients). Forest nutrient cycling is defined as the exchange of elements between the living and nonliving components of an ecosystem.[1] The processes of the forest nutrient cycle include: nutrient uptake and storage in vegetation perennial tissues, litter production, litter decomposition, nutrient transformations by soil fauna and flora, nutrient inputs from the atmosphere and the weathering of primary minerals, and nutrient export from the soil by leaching and gaseous transfers.

Each nutrient element is characterized by a unique biogeochemical cycle. Some of the key features of the major nutrients are shown in Table 1. Forest trees make less demand on the soil for nutrients than annual crops because a large proportion of absorbed nutrients are returned annually to the soil in leaf and fine root litter and are reabsorbed after biological breakdown of litter materials. Also, a large portion of nutrient requirement of trees are met through internal cycling as compared with agricultural crops.

Nutrient cycling in forest ecosystems is controlled primarily by three key factors: climate, site, abiotic properties (topography, parent material), and biotic communities. The role of each factor in ecosystem nutrient dynamics is discussed and illustrated with selected examples from boreal, temperate, and tropical zones. The importance of ecosystem disturbance to nutrient cycling is examined briefly, since some nutrients are added or lost from forest ecosystems through natural (e.g., fire, erosion, leaching) or human activity (harvesting, fertilization).

INFLUENCE OF CLIMATE

Large-scale patterns in terrestrial primary productivity have been explained by climatic variables. In above-ground vegetation, nutrient storage generally increases in the order: boreal < temperate < tropical forests (Table 2). In contrast, forest floor nutrient content and residence time increase from tropical to boreal forests, as a result of slower decomposition in the cold conditions of higher latitudes.

In subarctic woodland soils and Alaskan taiga forests, nutrient cycling rates are low because of extreme environmental conditions.[2] Arctic and subarctic forest ecosystems have lower rates of nutrient turnover and primary production because of low soil temperature, a short growing season, low net AET and the occurrence of permafrost. Low temperature reduces microbial activity, litter decomposition rates, and nutrient availability and increases C accumulation in soil.

In contrast with high latitudes, conditions in a tropical forest favor microbial activity throughout the year, which generally results in faster decomposition except in situations with periodic flooding, soil dessication, and low litter quality.[3] Rates of plant material decay are an order of magnitude higher in tropical soils than in subarctic woodland soils. The low storage of C and high amount of litter production in highly productive tropical forests contrasts with the high C storage and low litter production in boreal forests (Table 2).

INFLUENCE OF BIOTIC FACTORS

Nutrient cycles are modified substantially by tree species-specific controls over resource use efficiency (nutrient use per unit net primary production). Species vary widely in their inherent nutrient requirements and use.[4] These effects can be split into two categories: accumulation into living phytomass and production of various types of

Encyclopedia of Natural Resources DOI: 10.1081/E-ENRL-120001709

Table 1 Features of the major nutrient cycles

Element	Uptake by the trees	Major sources for tree uptake	Limiting situations
Carbon	Atmosphere	Atmosphere	Atmospheric concentration may limit growth
Oxygen	Atmosphere	Atmosphere	Waterlogged soils
Hydrogen	Atmosphere	Atmosphere	Extremely acidic and alkaline conditions
Nitrogen	Soluble NO_3 and NH_4; N_2 for nitrogen fixing species	Soil organic matter; atmospheric N_2 for nitrogen fixing species	Most temperate forests, many boreal forests and some tropical forests
Phosphorus	Soluble phosphorus	Soil organic matter; adsorbed phosphate and mineral phosphorus	Old soils high in iron and aluminum, common in subtropical and tropical environment
Potassium, calcium, magnesium	Soluble K^+/soluble Ca^{2+}/soluble Mg^{2+}	Soil organic matter; exchange complex and minerals	Miscellaneous situations and some old soils

Table 2 Nutrient distribution in different forest ecosystems

	Vegetation ($Mg\,ha^{-1}$)	Forest floor ($Mg\,ha^{-1}$)	Soil ($Mg\,ha^{-1}$)	Residence time (year)
Carbon				
Boreal coniferous	78–93	37–113	41–207	800
Temperate deciduous	103–367	42–105	185–223	200
Tropical rain forest	332–359	7–72	2–188	120
Nitrogen				
Boreal coniferous	0.3–0.5	0.6–1.1	0.7–2.87	200
Temperate deciduous	0.1–1.2	0.2–1.0	2.0–9.45	6
Tropical rain forest	1.0–4.0	0.03–0.05	5.0–19.2	0.6
Phosphorus				
Boreal coniferous	0.033–0.060	0.075–0.15	0.04–1.06	300
Temperate deciduous	0.06–0.08	0.20–0.10	0.91–1.68	6
Tropical rain forest	0.2–0.3	0.001–0.005	0.06–7.2	0.6
Potassium				
Boreal coniferous	0.15–0.35	0.3–0.75	0.07–0.8	100
Temperate deciduous	0.3–0.6	0.050–0.15	0.01–38	1
Tropical rain forest	2.0–3.5	0.020–0.040	0.05–7.1	0.2

nutrient-containing dead phytomass. Rapid accumulation of phytomass is associated with a net movement of nutrients from soil into vegetation. More than half of the annual nutrient uptake by a forest is typically returned to forest floor (litterfall) and soil (fine-root turnover). The subsequent recycling of these nutrients is a major source of available nutrients for forest vegetation. The mean annual litterfall from above-ground vegetation increases from boreal regions to the tropics following the gradient of productivity (Table 2).

Nutrient availability is strongly influenced by the quantity and quality of litter produced in a forest. A high proportion of the variation in foliar N concentrations at the continental scale has been explained by differences between forest types, which in turn has large impact on litter quality and the nutrient content of forest floors. In many temperate and boreal forest ecosystems, microbial requirement for N increases or decreases with labile supplies of soil C. Increased microbial demand for N may temporarily decrease the N availability to trees during the initial decomposition of forest residues with a wide C/N ratio. Microbes immobilize N from the surrounding soil, relatively rapid for readily decomposable organic matter (needle litter), and more slowly for recalcitrant material (branches, boles).

Rates of net N mineralization are higher and retention of foliar N is lower in temperate and tropical than in boreal forest soils. Nitrogen limitation of productivity, therefore, is weak in tropical forests and increases from temperate to boreal and tundra forest systems. Trees may obtain organic N and P from the soil via mycorrhizae or by relocation from older foliage prior to abscission, and thereby, partly reduce their dependence on soil as a source of inorganic nutrients. Increased understanding of the fundamental relationships between soil properties and plant nutrients requirements will most likely come from examination of plant–fauna–microbe interactions at root surfaces (rhizosphere), rather than in the bulk soil.

INFLUENCE OF ABIOTIC FACTORS

Forests have distinctive physiographic, floristic, and edaphic characteristics that vary predictably across the landscape within a climatically homogeneous region. Differences in the elemental content of parent material influence the tree species composition between and within a landscape unit. For example, wind deposited soils, which support hardwood or mixed wood forest, are likely to be fine textured with high nutrient supplying capacity. In contrast, outwash sands that often support pine forests are coarse textured and infertile.

Heterogeneity within the landscape results in sites differing in microclimatic conditions, and physical and chemical properties, which produces different geochemical reaction rates and pools of available nutrients in soil. Soil type and topographic–microclimate interactions are important feedbacks that influence biological processes, such as the rate of N mineralization in soil. Low P availability is a characteristic of geomorphically old, highly weathered tropical, subtropical, and warm temperate soils.[3] The type and age of parent material from which the soil is derived can influence the base status and nutrient levels in soil. Soils in glaciated regions are relatively young and rich in weatherable minerals. Mineral weathering is an important source of most nutrients for plant uptake, with the exception of N. Nutrient availability is regulated by the balance between weathering of soil minerals and precipitation, adsorbtion, and fixation reactions in soil.

Edaphic conditions can exert a strong influence on forest productivity and produce considerable variation in nutrient cycling processes. Soils with low N, P, or pH support trees with low litter quality (high in lignin and tannins that bind N) that decomposes slowly. Edaphic limitations on growth may be compensated for by an increase in rooting density and depth. Some late-succession or tolerant species have a shallower root distribution relative to intolerant pioneer species and are adapted to sites where nutrients and moisture are concentrated at the soil surface. In contrast, nutrient uptake from sub-soil horizons is more important in highly weathered warm temperate soils where nutrient depletion takes place deeper in the soil.

ROLE OF DISTURBANCE

Disturbances such as fire, harvesting, hurricanes, or pests affect nutrient cycling long after the event. In fire-dominated ecosystems, intensive wildfire results in a horizontal and vertical redistribution of ecosystem nutrients. Redistribution results from the combined effects of the following processes: 1) oxidation and volatilization of live and decomposing plant material; 2) convection of ash particles in fire generated winds; 3) water erosion of surface soils; and 4) percolation of solutes through and out of the soil. The relative importance of these processes varies with each nutrient and is modified by differences in fire intensity, soil characteristics, topography, and climatic patterns. Expressed as a percentage of the amount present in vegetation and litter before fire, the changes often follow the order:

$$N > K > Mg > Ca > P$$

Harvesting removes nutrients from the site and interrupts nutrient cycling temporarily. The recovery of the nutrient cycle from harvest disturbance is dependent partly on the rate of re-establishment of trees and competing vegetation. Re-vegetation may occur within months in the tropics, 2–5 years in temperate regions, and longer in boreal and tundra regions.[5] Recovery assumes that the soil's ability to supply nutrients to plant roots has not been altered by disturbance. If nutrients cannot be supplied by the soil at rates sufficient to at least maintain the rate of growth of the previous forest then fertilization may be necessary to maintain site productivity.

Nutrient cycling and the impacts of disturbance on nutrient cycling have been examined thoroughly in many representative world forests. The impact of natural disturbances and management practices on nutrient cycling processes are generally characterized of the stand or occasionally on a watershed basis. There is a growing demand from policy makers and forest managers for spatial estimates on nutrient cycling at local, regional, and national scales.

The availability of N, P, and K in soil largely determines the leaf area, photosynthetic rate, and net primary production of forest ecosystems. Forest management practices that produce physical and chemical changes in the soil that accentuate the cycle of nutrients between soil and trees, may increase forest productivity. Clear-cut harvesting and site preparation practices (mechanical disturbance, slash burning) remove nutrients from soil in tree components and by increased surface runoff, soil erosion, and off-site movement of nutrients in dissolved form or in sediment transport. In the tropics, potential negative impacts associated with complete forest removal and slash burning are greatest because a larger proportion of site nutrients are contained in the living biomass. Environmental impacts associated with clear-cutting and forest management in general, are confounded by climatic, topographic, soil, and vegetation diversity associated with the world's forests. Best forest management practices can be utilized to control negative impacts on nutrient cycling.

REFERENCES

1. Soil Science Society of America. *Glossary of Soil Science Terms*; Soil Science Society of America: Madison, WI, 1996; 134.

2. Van Cleve, K.; Chapin, F.S. III; Dyrness. C.T. III; Viereck, L.A. Element cycling in taiga forests: State factor control. Bioscience **1991**, *41* (2), 78–88.
3. Vitousek, P.M.; Stanford, R.L. Jr. Nutrient cycling in moist tropical forest. Ann. Rev. Ecol. Syst. **1986**, *17* (1), 137–167.
4. Cole, D.W.; Rapp, M. Elemental cycling in forest ecosystems. In *Dynamic Properties of Forest Ecosystems*; Reichle, E.D. Ed.; Cambridge University Press: London, 1981, 341–409.
5. Keenan, R.J.; Kimmins, J.P. The ecological effects of clear-cutting. Environ. Rev. **1993**, *1* (2), 121–144.

Ecosystems: Functions and Services

Robert Costanza
Institute for Sustainable Solutions, Portland State University, Portland, Oregon, U.S.A.

Abstract

Making decisions regarding trade-offs among ecological systems and services and other contributors to human well-being is a difficult and critical, but unavoidable process that requires valuation. This entry clarifies some of the definitional controversies surrounding the contributions to human well-being from functioning ecosystems, whether people are aware of those benefits or not. It also classifies and describes the areas of applicability of the various valuation methods that can be used in estimating the benefits of ecosystem services. Finally, it describes some recent case studies and the research agenda for valuation going forward in ecosystem services.

ECOSYSTEM SERVICES

"Ecosystem services" (ESs) are the ecological characteristics, functions, or processes that directly or indirectly contribute to human well-being, the benefits people derive from functioning ecosystems.[1,2] Ecosystem processes and functions may contribute to ESs but they are not synonymous. Ecosystem processes and functions describe biophysical relationships and exist regardless of whether humans benefit from or not.[3,4] ESs, on the other hand, only exist if they contribute to human well-being and cannot be defined independently.

The ecosystems that provide the services are sometimes referred to as "natural capital," using the general definition of capital as a stock that yields a flow of services over time.[5] In order for these benefits to be realized, natural capital (which does not require human activity to build or maintain) must be combined with other forms of capital that *do* require human agency to build and maintain. These include: (1) built or manufactured capital; (2) human capital; and (3) social or cultural capital.[6]

These four general types of capital are all required in complex combinations to produce any and all human benefits. ESs thus refer to the relative contribution of natural capital to the production of various human benefits, in combination with the three other forms of capital. These benefits can involve the use, non-use, option to use, or mere appreciation of the existence of natural capital.

The following categorization of ESs has been used by the Millennium Ecosystem Assessment:[2]

Provisioning services: ESs that combine with built, human, and social capital to produce food, timber, fiber, or other "provisioning" benefits. For example, fish delivered to people as food require fishing boats (built capital), fisherfolk (human capital), and fishing communities (social capital) to produce.

Regulating services: Services that regulate different aspects of the integrated system. These are services that combine with the other three capitals to produce flood control, storm protection, water regulation, human disease regulation, water purification, air quality maintenance, pollination, pest control, and climate control. For example, storm protection by coastal wetlands requires built infrastructure, people, and communities to be protected. These services are generally not marketed but have clear value to society.

Cultural services: ESs that combine with built, human, and social capital to produce recreation, esthetic, scientific, cultural identity, sense of place, or other "cultural" benefits. For example, to produce a recreational benefit requires a beautiful natural asset (a lake), in combination with built infrastructure (a road, trail, dock, etc.), human capital (people able to appreciate the lake experience), and social capital (family, friends, and institutions that make the lake accessible and safe). Even "existence" and other "non-use values" require people (human capital) and their cultures (social and built capital) to appreciate.

Supporting services: Services that maintain basic ecosystem processes and functions such as soil formation, primary productivity, biogeochemistry, and provisioning of habitat. These services affect human well-being *indirectly* by maintaining processes necessary for provisioning, regulating, and cultural services. They also refer to the ESs that have not yet, or may never be intentionally combined with built, human, and social capital to produce human benefits that support or underlie these benefits and may sometimes be used as proxies for benefits when the benefits cannot be easily measured directly. For example, net primary

Encyclopedia of Natural Resources DOI: 10.1081/E-ENRL-120047507

production (NPP) is an ecosystem function that supports carbon sequestration and removal from the atmosphere, which combines with built, human, and social capital to provide the benefit of climate regulation. Some would argue that these "supporting" services should rightly be defined as ecosystem "functions," since they may not yet have interacted with the other three forms of capital to create benefits. We agree with this in principle, but recognize that supporting services/functions may sometimes be used as proxies for services in the other categories.

This categorization suggests a very broad definition of services, limited only by the requirement of a contribution to human well-being. Even without any subsequent valuation, explicitly listing the services derived from an ecosystem can help ensure appropriate recognition of the full range of potential impacts of a given policy option. This can help make the analysis of ecological systems more transparent and can help inform decision makers of the relative merits of different options before them.

VALUATION

Many ESs are public goods. This means they are nonexcludable and multiple users can simultaneously benefit from using them. This creates circumstances where individual choices are not the most appropriate approach to valuation. Instead, some form of community or group choice process is needed. Furthermore, ESs (being public goods) are generally not traded in markets. We therefore need to develop other methods to assess their value.

There are a number of methods that can be used to estimate or measure benefits from ecosystems. Valuation can be expressed in multiple ways, including monetary units, physical units, or indices. Economists have developed a number of valuation methods that typically use metrics expressed in monetary units[7] while ecologists and others have developed measures or indices expressed in a variety of nonmonetary units such as biophysical trade-offs.[8]

There are two main methods for estimating monetary values: revealed and stated preferences. Both these methods typically involve the use of sophisticated statistical methods to tease out the values.[9] Revealed preference methods involve analyzing individuals' choices in real-world settings and inferring value from those observed choices. Examples of such methods include production-oriented valuation that focuses on changes in direct use values from products actually extracted from the environment (e.g., fish). This method may also be applicable to indirect use values, such as the erosion control benefits forests provide to agricultural production. Other revealed preference methods include hedonic pricing, which infers ecosystem service values from closely linked housing markets. For example, urban forest ecosystems and wetlands may improve water quality and that may be (partially) captured in property values.[10] The travel cost valuation method is used to value recreation ESs and estimate values based on the resources, money, and time visitors spend to visit recreation sites.

Stated preference methods rely on individuals' responses to hypothetical scenarios involving ESs and include contingent valuation and structured choice experiments. Contingent valuation utilizes a highly structured survey methodology that acquaints survey respondents with ecosystem improvements (e.g., better stream quality) and the ESs they will generate (e.g., increased salmon stocks). Respondents are then asked to value ecosystem improvements usually using a referendum method.[11]

Choice experiments are sometimes called conjoint analysis. This method presents respondents with combinations of ESs and monetary costs and asks for the most preferred combinations. Based on these choices, ecosystem service values are inferred.

A key challenge in any valuation is imperfect information. Individuals might, for example, place no value on an ecosystem service if they do not know the role that the service is playing in their well-being.[12] Here is an analogy. If a tree falls in the forest and there is no one around to hear it, does it still make a sound? Assume in this case that the "sound" is the ecosystem service. The answer to this old question obviously depends on how one defines "sound". If "sound" is defined as the perception of sound waves by people, then the answer is no. If "sound" is defined as the pattern of physical energy in the air, then the answer is yes. In this second case, choices in both revealed and stated preference models would not reflect the true benefit of the ecosystem service. Another key challenge is accurately measuring the functioning of the system to correctly quantify the amount of a given service derived from that system.[13,14]

But recognizing the importance of information does not obviate the limitations of human perception-centered valuation. As the tree analogy demonstrates, perceived value can be a quite limiting valuation criterion, because natural capital can provide positive contributions to human well-being that are either never (or only vaguely) perceived or may only manifest themselves at a future time. A broader notion of value allows a more comprehensive view of value and benefits, including, for example, valuation relative to alternative goals/ends, like fairness and sustainability, within the broader goal of human well-being.[15] Whether these values are perceived or not and how well or accurately they can be measured are separate (and important) questions.

CASE STUDIES

Early Valuation Syntheses

Scientists and economists have discussed the general concepts behind natural capital, ESs, and their value for decades, with some early work as far back as the 1920s.

However, the first explicit mention of the term "ESs" was in Ehrlich and Mooney.[16] More than 6000 papers have been published on the topic of ESs since then. The first mention of the term "natural capital" was by Costanza and Daly.[5]

One of the first studies to estimate the value of ESs globally was published in *Nature* entitled "The value of the world's ESs and natural capital."[1] This paper estimated the value of 17 ESs for 16 biomes to be in the range of US$16–54 trillion per year, with an average of US$33 trillion per year, a figure larger than annual GDP at the time. Some have argued that global society would not be able to pay more than their annual income for these services, so a value larger than global GDP does not make sense. However, not all benefits are picked up in GDP and many ESs are not marketed, so GDP does not represent a limit on real benefits.[17]

In this study, estimates of global ESs were derived from a synthesis of previous studies that utilized a wide variety of techniques like those mentioned above to value-specific ESs in specific biomes. This technique, called "benefit transfer," uses studies that have been done at other locations or in different contexts, but can be applied with some modification. See Costanza[18] for a collection of commentaries and critiques of the methodology. Such a methodology, although useful as an initial estimate, is just a first cut and much progress has been made since then.[19,20]

MAJOR WORLD REPORTS ON ECOSYSTEM SERVICES

More recently the concept of ESs gained attention with a broader academic audience and the public when the Millennium Ecosystem Assessment (MEA) was published.[2] The MEA was a 4-year, 1300 scientist study commissioned by the United Nations in 2005. The report analyzed the state of the world's ecosystems and provided recommendations for policymakers. It determined that human actions have depleted the world's natural capital to the point that the ability of a majority of the globe's ecosystems to sustain future generations can no longer be taken for granted.

In 2008, a second international study was published on the Economics of Ecosystems and Biodiversity (TEEB), hosted by United Nations Environment Programme. TEEB's primary purpose was to draw attention to the global economic benefits of biodiversity, to highlight the growing costs of biodiversity loss and ecosystem degradation, and to draw together expertise from the fields of science, economics, and policy to enable practical actions moving forward. The TEEB report was picked up extensively by the mass media, bringing ESs to a broad audience.

THE ECOSYSTEM SERVICES' PARTNERSHIP AND ONGOING WORK

With such high profile reports being published, ESs have entered not only the public media[21] but also into business. Dow Chemical established a $10 million collaboration with The Nature Conservatory to tally up the ecosystem costs and benefits of every business decision.[22] Such collaborations will provide a significant addition to ES valuation knowledge and techniques. However, there is significant research that is still required. Our scientific institutions can help lead this process through transdisciplinary graduate education, such as the ESs for the Urbanizing Regions program funded by the National Science Foundation's Integrative Graduate Research and Education Traineeship program.[23]

Hundreds of projects and groups are currently working toward better understanding, modeling, valuation, and management of ESs and natural capital. It would be impossible to list all of them here, but the new Ecosystem Services Partnership (ESP; http://www.es-partnership.org) is a global network that does just that and helps to coordinate the activities and build consensus.

The following lays out the research agenda as agreed to by a group of 30 participants at a meeting in Salzau, Germany, in June 2010, at the launch of the ESP.

Integrated Measurement, Modeling, Valuation, and Decision Science in Support of Ecosystem Services

The scientific community needs to continue to develop better methods to measure, monitor, map, model, and value ESs at multiple scales. Ideally, these efforts should take place using interdisciplinary teams and strategies and in close collaboration with ecosystem stakeholders. Moreover, this information must be provided to decision makers in an appropriate, transparent, and viable way, to clearly identify differences in outcomes among policy choices. At the same time, we cannot wait for high levels of certainty and precision to act when confronting significant irreversible and catastrophic consequences. We must synergistically continue to improve the measurements with evolving institutions and approaches that can effectively utilize these measurements.

Trade-offs

Ecological conflicts arise from two sources: (1) scarcity and restrictions in the amount of ESs that can be provided and (2) the distribution of the costs and benefits of the provisioning of the ES. ES science makes trade-offs explicit and, thus, facilitates management and planning discourse. It enables stakeholders to make sound value judgments. ES science thus generates relevant social–ecological knowledge for stakeholders and policy decision makers and sets of planning options that can help resolve sociopolitical conflicts.

Accounting and assessment

Accounting attempts to look at the flow of materials with relative objectivity, while assessment evaluates a system or process with a goal in mind and is more normative. Both are integrating frameworks with distinctive roles. Both ecosystem service accounting and assessment need to be developed and pursued using a broader socioecological lens. Within the broader lens we also need to balance expert and local knowledge across scales.

Modeling

We need modeling to synthesize and quantify our understanding of ESs and to understand dynamic, nonlinear, spatially explicit trade-offs as part of the larger socioecological systems. Stakeholders should be active collaborators in this model development process to assure relevancy. These models can incorporate and aid accounting and assessment exercises and link directly with the policy process at multiple time and space scales. In particular, modeling can quantify potential shifts in ESs under different environmental and socioeconomic scenarios.

Bundling

Most ESs are produced as joint products (or bundles) from intact ecosystems. The relative rates of production of each service vary from system to system, site to site, and time to time. We must consider the full range of services and the characteristics of their bundling in order to prevent creating dysfunctional incentives and to maximize the net benefits to society.[24,25] For example, focusing only on the carbon sequestration service of ecosystems may, in some instances, reduce the overall value of the full range of ESs.

Scaling

ESs are relevant over a broad range of scales in space, time, governance, and complexity, including the legacy of past behavior. We need measurement, models, accounts, assessments, and policy discussions that address these multiple scales, as well as interactions, feedbacks, and hierarchies among them.

Adaptive Management and New Institutions for Ecosystem Services

Given that pervasive uncertainty always exists in ecosystem service measurement, monitoring, modeling, valuation, and management, we should continuously gather and integrate appropriate information regarding ESs, with the goal of learning and adaptive improvement. To do this we should constantly evaluate the impacts of existing systems and design new systems with stakeholder participation as experiments from which we can more effectively quantify performance and learn ways to manage such complex systems.

Property Rights

Given the public goods nature of most ESs, we need institutions that can effectively deal with this characteristic using a sophisticated suite of property rights regimes. We need institutions that employ an appropriate combination of private, state, and common property rights systems to establish clear property rights over ecosystems without privatizing them. Systems of payment for ecosystem services and common asset trusts can be effective elements in these institutions.

Scale Matching

The spatial and temporal scale of the institutions to manage ESs must be matched with the scales of the services themselves. Mutually reinforcing institutions at local, regional, and global scales over short-, medium- and long-time scales will be required. Institutions should be designed to ensure the flow of information across scales, to take ownership regimes, cultures, and actors into account, and to fully internalize costs and benefits.

Distribution Issues

Systems should be designed to ensure inclusion of the poor, since they are generally more dependent on common property assets like ESs. Free riding, especially by wealthier segments of society, should be deterred and beneficiaries should pay for the services they receive from biodiverse and productive ecosystems.

Information Dissemination

One key limiting factor in sustaining natural capital is lack of knowledge of how ecosystems function and how they support human well-being. This can be overcome with targeted educational campaigns that are tailored to disseminate success and failures to both the general public and elected officials and through true collaboration among public, private, and government entities.

Participation

Relevant stakeholders (local, regional, national, and global) should be engaged in the formulation and implementation of management decisions. Full stakeholder awareness and participation not only improves ES analyses but contributes to credible accepted rules that identify and assign the corresponding responsibilities appropriately, and that can be effectively enforced.

Science–Policy Interface

ES concepts can be an effective link between science and policy by making the trade-offs more transparent.[4] An ES framework can, therefore, be a beneficial *addition* to policy-making institutions and frameworks and to integrating science and policy.

CONCLUSIONS

Natural capital and ESs are key concepts that are changing the way we view, value, and manage the natural environment. They are changing the framing of the issue away from "jobs vs. the environment" to a more balanced assessment of all the assets that contribute to human well-being. Significant transdisciplinary research has been done in recent years on ESs, but there is still much more to do and this will be an active and vibrant research area for the coming years, because better understanding of ESs is critical for creating a sustainable and desirable future. Placing credible values on the full suite of ESs is key to improving their sustainable management.

REFERENCES

1. Costanza, R.; d'Arge, R.; de Groot, R.; Farber, S.; Grasso, M.; Hannon, B.; Naeem, S.; Limburg, K.; Paruelo, J.; O'Neill, R.V.; Raskin, R.; Sutton, P.; van den Belt, M. The value of the world's ecosystem services and natural capital. Nature **1997a**, *387*, 253–260.
2. *Millennium Ecosystem Assessment*; United Nations, 2005.
3. Boyd, J.; Banzhaf, S. What are Ecosystem Services? Ecol. Econ. **2007**, *63*, 616–626.
4. Granek, E.F.; Polasky, S.; Kappel, C.V.; Stoms, D.M.; Reed, D.J.; Primavera, J.; Koch, E.W.; Kennedy, C.; Cramer, L.A.; Hacker, S.D.; Perillo, G.M.E.; Aswani, S.; Silliman, B.; Bael, D.; Muthiga, N.; Barbier, E.B.; Wolanski, E. Ecosystem services as a common language for coastal ecosystem-based management. Conserv. Biol. **2010**, *24*, 207–216.
5. Costanza, R.; Daly, H.E. Natural capital and sustainable development. Conserv. Biol. **1992**, *6*, 37–46.
6. Costanza, R.; Cumberland, J.C.C.; Daly, H.E.; Goodland, R.; Norgaard., R. *An Introduction to Ecological Economics*; St. Lucie Press: Boca Raton,1997b; 275.
7. Freeman, A.M. *The Measurement of Environmental and Resource Values: Theories and Methods* 2nd edition; RFF Press: Washington, DC, 2003.
8. Costanza, R. Value Theory and Energy. In *Encyclopedia of Energy*. C. Cleveland Ed.; Elsevier: Amsterdam, 2004, Vol. no. 6: 337–346.
9. Haab, T.; McConnell, K. *Valuing Environmental and Natural Resources: The Econometrics of Non-Market Valuation*; Edward Elgar Publishing Ltd.: Cheltenham, UK, 2002.
10. Phaneuf, D.J.; Smith, V.K.; Palmquist, R.B.; Pope, J.C. Integrating property value and local recreation models to value ecosystem services in urban watersheds. Land Econ. **2008**, *84*, 361–381.
11. Boardman, A.D.; Greenberg, A.; Vining; Weimer, D. *Cost-Benefit Analysis: Concepts and Practice*, 4th Edition; Prentice Hall, Inc: Upper Saddle River, NJ, 2006.
12. Norton, B.; Costanza, R.; Bishop, R. The evolution of preferences: why "sovereign" preferences may not lead to sustainable policies and what to do about it. Ecol. Eco. **1998**, *24*, 193–211.
13. Barbier, E.B.; Koch, E.W.; Silliman, B.R.; Hacker, S.D.; Wolanski, E.; Primavera, J.; Granek, E.F.; Polasky, S.; Aswani, S.; Cramer, L.A.; Stoms, D.M.; Kennedy, C.J.; Bael, D.; Kappel, C.V.; Perillo, G.M.E.; Reed, D.J. Coastal ecosystem-based management with non-linear ecological functions and values. Science **2008**, *319*, 321–323.
14. Koch, E.W.; Barbier, E.D.; Silliman, B.R.; Reed, D.J.; Perillo, G.M.E.; Hacker, S.D.; Granek, E.F.; Primavera, J.H.; Muthiga, N.; Polasky, S.; Halpern, B.S.; Kennedy, C.J.; Wolanski, E.; Kappel, C.V. Non-linearity in ecosystem services: temporal and spatial variability in coastal protection. Front. Ecol. Environ. **2009**, *7*, 29–37.
15. Costanza, R. Social goals and the valuation of ecosystem services. Ecosystems **2000**, *3*, 4–10.
16. Ehrlich, P.R.; Mooney, H. Extinction, substitution, and ecosystem services. BioScience **1983**, *33*, 248–254.
17. Costanza, R., d'Arge, R.; de Groot, R.; Farber, S.; Grasso, M.; Hannon, B.; Naeem, S.; Limburg, K.; Paruelo, J.; O'Neill, R.V.; Raskin, R.; Sutton, P.; van den Belt. M. The value of the world's ecosystem services: putting the issues in perspective. Ecol. Eco. **1998**, *25*, 67–72.
18. Costanza, R. The Value of Ecosystem Services. Special Issue Eco. Eco. **1998**, *25* (1), 139.
19. Boumans, R.; Costanza, R.; Farley, J.; Wilson, M.A.; Portela, R.; Rotmans, J.; Villa, F. Grasso, M. Modeling the Dynamics of the Integrated Earth System and the Value of Global Ecosystem Services Using the GUMBO Model. Ecol. Eco. **2002**, *41*, 529–560.
20. U.S. Environmental Protection Agency Science Advisory Board. Valuing the Protection of Ecological Systems and Services: A Report of the EPA Science Advisory Board. EPA-SAB-09-012. Washington, DC: EPA, 2009 http://yosemite.epa.gov/sab/sabproduct.nsf/WebBOARD/ValProtEcolSys&Serv?OpenDocument
21. Schwartz, J. D. Should We Put A Dollar Value On Nature?; Time Magazine. Time Inc.: http://www.time.com/time/business/article/0,8599,1970173,00.html, 2010.
22. Walsh, B. Paying for Nature. Time Magazine. Time Inc. http://www.time.com/time/magazine/article/0,9171,2048324,00.html, 2011.
23. Ecosystem Services for Urbanizing Regions (ESUR). Portland State University, http://www.pdx.edu/esur-igert, 2011.
24. Nelson, E.; Mendoza, G.; Regetz, J.; Polasky, S.; Tallis, H.; Cameron, D.R.; Chan, K.M.A.; Daily, G.; Goldstein, J.; Kareiva, P.; Lonsdorf, E.; Naidoo, R.; Ricketts, T.H.; Shaw, M.R. Modeling multiple ecosystem services, biodiversity conservation, commodity production, and tradeoffs at landscape scales. Front. Ecol. Environ. **2009**, *7*, 4–11.
25. Polasky, S.; Nelson, E.; Pennington, D.; Johnson, K. The impact of land-use change on ecosystem services, biodiversity and returns to landowners: a case study in the State of Minnesota. Env. Resour. Eco. **2011**, *48*, 219–242.

BIBLIOGRAPHY

1. Daily, G.; Polasky, S.; Goldstein, J.; Kareiva, P.M.; Mooney, H.A.; Pejchar, L.; Ricketts, T.H.; Salzman, J.; Shallenberger, R. Ecosystem services in decision-making: time to deliver. Front. Ecol. Environ. **2009**, *7*, 21–28.
2. de Groot, R.S.; Wilson, M.A.; Boumans, R. A typology for the classification, description and valuation of ecosystem functions, goods and services. Ecol. Eco. **2002**, *41*, 393–408.
3. Polasky, S.; Segerson, K. Integrating ecology and economics in the study of ecosystem services: some lessons learned. Annu. Rev. Resour. Econ. **2009**, *1*, 409–434.
4. Sukhdev, P.; Kumar, P. The economics of ecosystems & biodiversity (TEEB). http://www.teebweb.org/, 2008.

Ecosystems: Soil Animal Functioning

Alan J. Franzluebbers
Agricultural Research Service, U.S. Department of Agriculture (USDA-ARS), Watkinsville, Georgia, U.S.A.

Abstract
Soil animals (i.e., fauna) are represented by a diverse array of creatures living in or on soil for at least a part of their life cycle. Soil animals can be classified according to body size, where they inhabit the soil, and feeding activity. This entry discusses the effect of these animals on soil composition, condition, and ecosystem interactivity.

INTRODUCTION

Soil animals (i.e., fauna) are represented by a diverse array of creatures living in or on soil for at least a part of their life cycle. Many animals have influences on soil properties, but should not be considered soil dwellers since only a minor portion of their life cycle is spent in the soil (Fig. 1).

Based on body size, soil animals can be divided into three categories:

1. microfauna (<200 μm length, <100 μm width) including protozoa, rotifers, and nematodes
2. mesofauna (0.2–10 mm length, 0.1–2 mm width) including tartigrades, collembola, and mites
3. macrofauna (>10 mm length, >2 mm width) including millipedes, spiders, ants, beetles, and earthworms

Soil animals can also be classified according to where they inhabit the soil. The aquatic fauna (e.g., protozoa, rotifers, tartigrades, and some nematodes) live primarily in the water-filled pore spaces and surface water films covering soil particles. Earthworms are divided into species that occupy the surface litter of soil (epigeic), that are found in the upper soil layers (endogeic), or that burrow deep into the soil profile (anecic).

A further classification of five groups of soil animals is based on feeding activity, which can be useful in distinguishing how different groups affect soil ecosystem functions:

1. Carnivores feed on other animals. This group can be subdivided into: i) predators (e.g., centipedes, spiders, ground beetles, scorpions, ants, and some nematodes), who normally engulf and digest their smaller prey and ii) parasites (e.g., some flies, wasps, and nematodes), who feed on or within their typically larger host organism.
2. Phytophages feed on living plant materials, including those that feed on above-ground vegetation (e.g., snails and butterfly larvae), roots (e.g., some nematodes, fly larvae, beetle larvae, rootworms, and cicadas), and woody materials (e.g., some termites and beetle larvae).
3. Saprophages feed on dead and decaying organic material and include many of the earthworms, enchytraeids, millipedes, dung beetles, and collembola (or springtails). Saprophages are often referred to as scavengers, debris-feeders, or detritivores.
4. Microphytic feeders consume bacteria, fungi, algae, and lichens. Typical microphytic feeders include mites, collembola, ants, termites, nematodes, and protozoa.
5. Miscellaneous feeders are not restrictive in their diet and consume a range of the previously mentioned sources of food. This group includes certain species of nematodes, mites, collembola, and fly larvae.

The arrangement of these feeding groups can be visualized as a soil food web with multiple trophic levels, beginning with the autotrophic flora (Fig. 2). Trophic levels describe the order in the food chain. The first trophic level is composed of photosynthetic organisms, including plants, algae, and cyanobacteria, which fix CO_2 from the atmosphere into organic compounds. Organisms that consume the photosynthesizers are in the second trophic level, which includes bacteria, actinomycetes, fungi, root-feeding nematodes and insects, and plant pathogens and parasites. The third trophic level feeds on the second trophic level, including many of the dominant soil animals, including bacterial- and fungal-feeding arthropods, nematodes, and protozoa. The soil food web can be continued to include various vertebrates, including amphibians, reptiles, and mammals.

SPATIAL DISTRIBUTION OF SOIL ANIMALS

Soil animals are not uniformly distributed in soil. Unlike the soil microflora, which could be considered ubiquitous, the proliferation of soil animal communities is more

Encyclopedia of Natural Resources DOI: 10.1081/E-ENRL-120001731

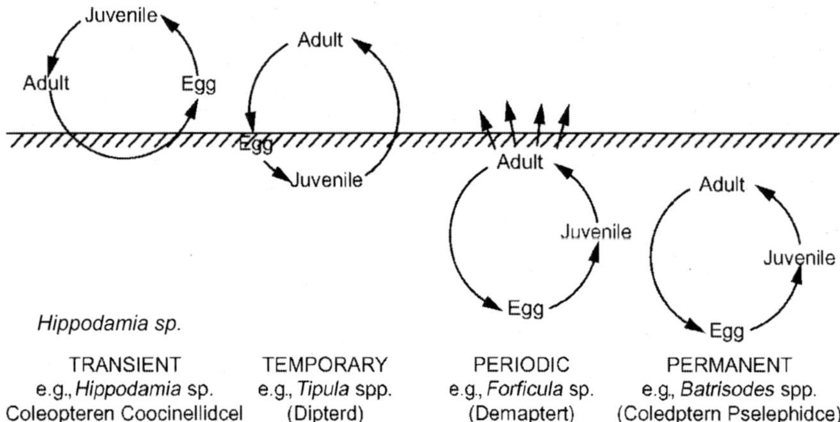

Fig. 1 Categories of soil animals defined according to degree of presence in soil, as illustrated by some insect groups.
Source: From Wallwork.[1]

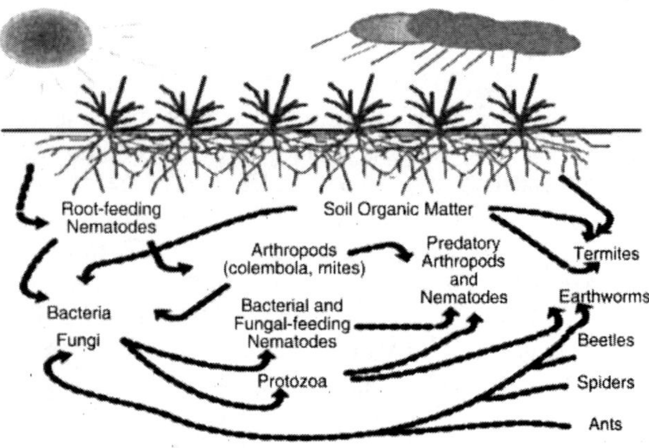

Fig. 2 Generalized diagram of a soil food web.

sensitive to environmental disturbances and ecological interactions. Gross climatic differences afford opportunities for unique assemblages of organisms. Even within a specific climatic region, large differences occur in the community of organisms present based upon type of vegetation, soil, availability of water, land use, and presence of xenobiotics. Within the confines of a seemingly uniform pedon, "hot spots" of soil organism activity can be isolated based on localized availability of resources and environmental conditions (Fig. 3).

INFLUENCE OF SOIL ANIMALS ON SOIL FUNCTIONS

Decomposition and Nutrient Cycling

Soil animals work directly and indirectly with the soil microflora (i.e., bacteria, actinomycetes, fungi, and algae) to decompose organic matter and mineralize nutrients.[3] The primary consumers of organic materials are the soil microflora. Soil animals, like many of the microflora, are heterotrophs and therefore consume organic materials to gain energy for growth and activity. Soil animals make important contributions to decomposition by

1. shredding organic materials, thereby exposing a greater surface area for enhancing the activities of other organisms, especially microorganisms;
2. consuming resistant plant materials that would decompose slowly otherwise, such as wood, roots, and dung, and transforming these materials into more decomposable constituents;
3. dispersing soil microorganisms (i.e., inoculation) within the soil profile by transporting them on their bodies and through their intestinal tracts;
4. creating burrows in soil to increase aeration, which stimulates microbial activity;
5. transporting organic materials from the soil surface to deeper in the soil profile, thereby improving environmental conditions for decomposition and increasing biological interactions deeper in the soil profile;
6. consuming bacteria and fungi, thereby releasing nutrients and stimulating the regeneration of microbial populations; and
7. providing unique food sources themselves for consumption by other soil fauna and microflora.

Water Cycling

Soil animals are active participants in the formation of soil structure, which is an important characteristic that influences water infiltration, soil water retention, and percolation.[4] The biochemical activity of soil organisms transforms organic materials into soil-stabilizing cementing agents, which bind the primary soil particles (i.e., sand, silt, and clay) into aggregates. In addition, the burrowing activity of soil animals creates larger pores alongside water-stable aggregates to increase total porosity of soil, which aids water flow without decreasing overall water retention capacity and improves the plant rooting environment.

Both aggregates and porosity are important components of soil structure. Poor soil structure due to disruption of aggregates, which fills pores with disaggregated primary particles and causes crusting of the soil surface, results in more rainfall that runs off land (i.e., less infiltration),

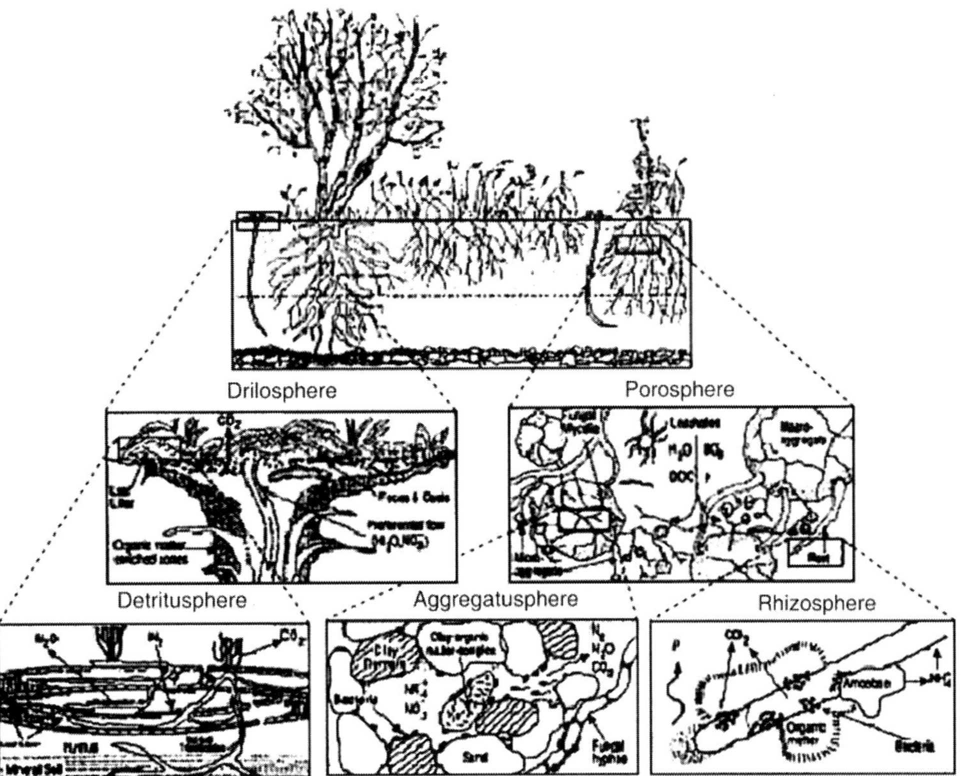

Fig. 3 Key locations of soil organism activity.
Source: From Beare et al.[2]

potentially carrying with it sediment, nutrients, and pesticides that can contaminate surface waters. Reduced infiltration with poor aggregation reduces available water for plant growth (i.e., reduces net primary productivity and the potential to fix atmospheric CO_2) and reduces percolation of water through the soil profile, essential for purification and recharge of groundwater.

Those animals that create burrows in soil also create conduits for water movement through the soil profile. These biopores can be important for improving water percolation and improving rooting below claypans and other restrictive soil layers.

Many different soil animals deposit fecal pellets, which become stable soil aggregates when the organic material is mixed with soil mineral particles. These aggregates are able to retain more water because of the high water-holding capacity of soil organic matter.

Pest Control

Intense competition among soil organisms keeps an ecosystem healthy by preventing one organism from becoming dominant. Potential plant pathogens, such as root-feeding nematodes, are often held below damaging levels because of consumption by predatory nematodes and arthropods. With a healthy food web rich in species diversity, the predatory activity of many arthropods can keep crop pests below economic thresholds.

Impact of Key Soil Animals

Earthworms

Earthworms are well-known soil animals inhabiting many environments, most prominently found in moist-temperate ecosystems. As earthworms ingest organic materials and mineral particles, they excrete waste as casts, which are a particular type of soil aggregate that is rich with organic matter and mineralizable nutrients. It is estimated that a healthy population of earthworms can consume and aerate a 15 cm surface of soil within one or two decades. Anecic or deep-burrowing species of earthworms can create relatively permanent vertical channels for improving root growth and water transport. Important attributes of earthworm activities are increased surface soil porosity, enhanced water infiltration and nutrient cycling, and distribution of organic matter within the soil profile to increase soil microbial activity.

Termites

Termites are important soil animals in grasslands and forests of tropical and subtropical regions. They often build mounds by excavating subsoil and depositing it above ground to build a city of activity with a complex social system. Termites are able to decompose cellulose in wood because they harbor various microorganisms (protozoa, bacteria, or fungi) to aid in decomposition. Better drainage and aeration of termite mounds may be beneficial to nearby plant growth in soils with a high water table. Stable macrochannels created by termites can improve water infiltration into soils that otherwise would form impermeable surface crusts.

Protozoa

Protozoa are single-celled animals that generally consume bacteria and soluble organic matter. Protozoa are more numerous in marine and freshwater environments, but do occur widely in water films of many soils.[5] Their principal soil function is predation on soil bacteria, which releases nutrients for potential plant uptake; increases decomposition and soil aggregation by stimulating their bacterial prey; and prevents some bacterial pathogens from developing on plant roots.

SOIL BIODIVERSITY

There has been a great deal discovered about soils and the organisms that live in them, yet it is estimated that <10% of the species of soil organisms have been identified globally. A rich diversity of genetic information resides in soil. An awaiting challenge is to discover the ecological consequences of soil biodiversity. A critical understanding of how organisms interact has begun, yet there will be much more to learn about how species diversity interacts with functional diversity. There may be soil functions provided by organisms that we are unaware of even today.

How biodiversity relates to ecosystem functioning is an intensive area of current research in both above- and below-ground ecology.[6,7] Recent experimental evidence suggests that loss of species richness due to perturbations may not always lead to loss of ecosystem functioning, especially in initially species-diverse functional groups.[8] Ecosystem functions can be performed by a number of different species within a trophic level, suggesting that functional redundancy is a mechanism to insure stability. However, loss of functional groups in trophic levels closest to the base of the detrital food web would be most detrimental to ecosystem stability. Future work on soil biodiversity should unravel the relative importance of species richness on the resistance and resilience of ecological processes under various short- and long-term stresses and environmental conditions.

REFERENCES

1. Wallwork, J.A. *Ecology of Soil Animals*; McGraw-Hill: London, 1970; 283 pp.
2. Beare, M.H.; Coleman, D.C.; Crossley, D.A., Jr.; Hendrix, P.F.; Odum, E.P. A hierarchical approach to evaluating the significance of soil biodiversity to biogeochemical cycling. Plant Soil **1995**, *170* (1), 5–22.
3. Coleman, D.C.; Crossley, D.A., Jr. *Fundamentals of Soil Ecology*; Academic Press: San Diego, CA, 1996; 205 pp.
4. Brady, N.C.; Weil, R.R. *The Nature and Properties of Soils*, 12th Ed.; Prentice Hall: Upper Saddle River, NJ, 1999; 881 pp.
5. Dabyshire, J.F., Ed.; *Soil Protozoa*; CAB International: Wallingford, Oxon, UK, 1994; 209 pp.
6. Tilmann, D.; Knops, J.; Wedin, D.; Reich, P.; Ritchie, M.; Siemann, E. The influence of functional diversity and composition on ecosystem processes. Science **1997**, *277*, 1300–1302.
7. Wardle, D.A. How soil food webs make plants grow. Trends Ecol. Evol. **1999**, *14*, 418–420.
8. Laakso, J.; Setälä , H. Sensitivity of primary production to changes in the architecture of belowground food webs. Oikos **1999**, *87*, 57–64.

Ecotone

Finn C. Pillsbury
Debra P. C. Peters
Jornada Experimental Range, Agricultural Research Service, U.S. Department of Agriculture (USDA-ARS), Las Cruces, New Mexico, U.S.A.

Abstract
Any two ecological states have an ecotone, or zone of transition, between them. Recent conceptual advances have demonstrated that rather than being an average or combination of two communities, ecotones occur at multiple scales of ecological organization, are dynamic in space and time, and can exhibit emergent properties. The original concept of the ecotone was related to a spatial scale that encompassed two distinct vegetation communities, but recent research has emphasized the hierarchy of scale within ecotones, from transitions between individual plants (or smaller) to transitions from one biome to another. There are three basic types of ecotone dynamics in space and time: directional, stationary, and shifting, depending on the relative strength of different abiotic and biotic factors in the system. Ecotones modulate flows of material, energy, and organisms across landscapes, which has ramifications for a host of ecological processes such as dispersal, predation, and reproduction. As a result, ecotones can exhibit behaviors, or emergent properties, that are not simply a combination or average of the ecological states between which they are situated.

INTRODUCTION

Any two adjacent ecological states, be they tundra and taiga, field and forest, or bunchgrass and sand dune, have an area of transition between them. This transition can be gradual and indistinct, like that between shortgrass and tallgrass prairie, or it can be abrupt, like the transition between a hayfield and a neighboring woodlot. These transitions are collectively known as ecotones. Coined by Clements,[1] the term's original conception was of a short transition between stable climax communities, although Weaver and Clements[2] concede that ecotones can be "broad and indefinite." The concept remained something of a mere curiosity until the middle of the 20th century, with the advent of quantitative methods to measure and evaluate transitions between adjacent vegetation associations.[3,4] The essential idea to emerge from this period was that natural systems that are heterogeneous in space can be characterized by a set of ecological gradients underlying those various spatial differences.[5] Adjacent ecological states and the ecotones between them can thus be characterized by a series of gradients in soil characteristics, rainfall, vegetation structure, or other environmental conditions that illustrate the magnitude and abruptness of the ecotone.

Understanding the nature of these gradients is important because ecotones are expected to be sensitive to climate change.[6–8] With changing temperature and precipitation patterns, species' ranges are shifting worldwide, and these shifts are, by definition, showing the greatest flux at range boundaries, which coincide with broad-scale ecotones among biomes.[8] Being able to understand ecotone dynamics will be a key component of conservation planning in the coming decades, as scientists and managers look for effective ways of incorporating climate change into natural resource conservation efforts.[6]

The concept of the ecotone falls into a broader class of "biotic transitions"[9] or "ecological boundaries"[10,11] and has been historically associated with the plant ecology literature because it is conceptually consistent with observed patterns of plant distributions across a landscape. Others have focused on "habitat edges," which are also definable transitions between ecological states.[12] The difference between the ecotone and edge concepts is in the conceptual frameworks for measuring and evaluating them. Although ecotones are generally defined by their structural properties, edges are usually defined in relation to how they modify the flow of materials and energy from one adjacent state to another. In many ways, it is a purely semantic distinction, but it is illustrative of the fact that ecotones and other types of transitions are, ultimately, defined by the questions and scales relevant to the organisms and processes of interest.

Until the late 20th century, ecotones were assumed to be an "average" of two neighboring vegetation patches, showing a gradient of conditions bounded by those two patch types. However, recent conceptual advances have demonstrated that rather than being simple transitions

Encyclopedia of Natural Resources DOI: 10.1081/E-ENRL-120047428

between adjacent communities, ecotones occur at multiple scales of ecological organization, are dynamic in space and time, and can exhibit emergent properties.[9–11]

ECOTONE HIERARCHIES

The original concept of the ecotone was related to a spatial scale that encompassed two distinct vegetation communities, but recent conceptual development has emphasized the hierarchy of scale within ecotones, from transitions between individual plants (or smaller) to transitions from one biome to another.[9,13] Though there is a continuum of spatial scales at which ecotones occur, we can condense them into a few basic types: plant scale, patch scale, and landscape scale.[9,13] Transitions at these different scales are influenced by distinct ecological patterns and processes and exhibit interactions with transitions at higher and lower levels of the ecotone hierarchy. Although landscape-scale ecotones are generally more gradual than those at plant and patch scales, the relative abruptness of the transition is directly related to the ecological processes and climatological and edaphic context in which that ecotone occurs. To illustrate the hierarchical nature of ecotones, we draw upon transitions within a landscape-scale ecotone between a black grama (*Bouteloua eriopoda*) grassland and a mesquite (*Prosopis glandulosa*) shrubland at the Jornada Experimental Range in southern New Mexico, United States of America.

Plant-Scale Ecotones

The basic unit of the ecotone hierarchy is the individual plant, because it is the nonrandom distribution of individuals, populations, and species that make up extant ecological patterns and from which landscape structure is characterized. In a given area, the location of an individual plant is constrained by an interaction between physiology, microsite characteristics, and competition with other plants. At this spatial scale, an ecotone can be characterized by a transition in microclimate, soil characteristics, or other fine-scale factors. For example, the plant-scale ecotone of a mesquite shrub at the Jornada can be characterized by a gradient in light transmission at a spatial scale of centimeters to a few meters (Fig. 1A). Light transmission is an abiotic factor that influences soil temperature, soil moisture, and photosynthesis by understory species, therefore playing an important role in soil community differences across the ecotone at the edge of an individual mesquite shrub.

Patch-Scale Ecotones

A landscape can be characterized as a mosaic of different patches of vegetation, where a patch is defined as an area that is compositionally distinct from adjacent areas. Patch-scale ecotones are the transitions between adjacent patches

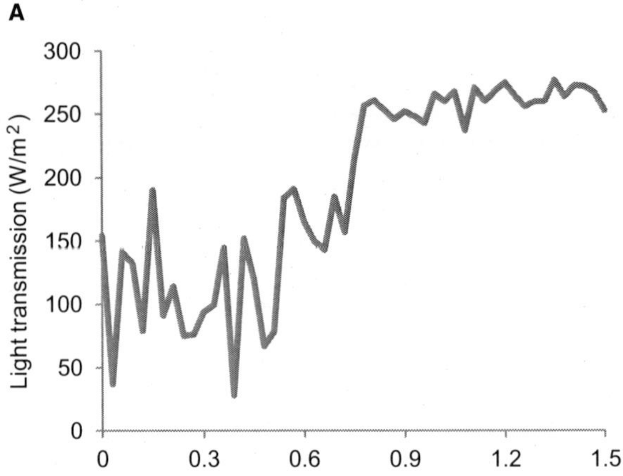

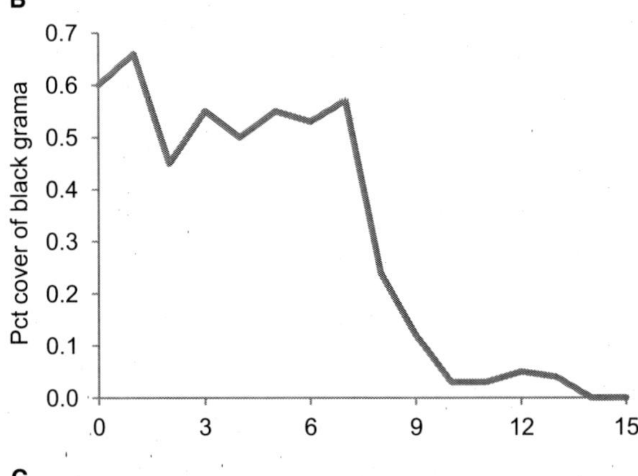

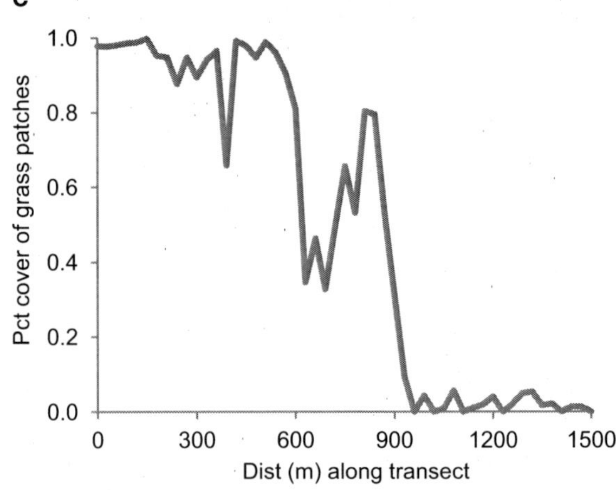

Fig. 1 Examples of different ecotone scales at the Jornada Experimental Range in southern New Mexico. The plant-scale ecotone (**A**) is a measure of light transmission along a 1.5 m transect from the middle of a mesquite (*Prosopis glandulosa*) dune to an adjacent bare gap, demonstrating a fine-scale gradient in an abiotic ecological parameter. The patch-scale ecotone (**B**) is a measure of proportional cover of black grama (*Bouteloua eriopoda*) in 1 m^2 quadrats along a 15 m transect between a patch of black grama and an adjacent bare gap, demonstrating a gradient of cover in a foundational species. The landscape-scale ecotone (**C**) is a measure of proportional cover of grass patches within 900 m^2 grid cells along a 1500 m transect, measured using a digitized satellite image, showing a gradient in the distribution of grass patches across a broad area.

and can be characterized by a gradient of population density of a dominant species. An example of a patch-scale ecotone is that between individual patches of black grama and adjacent bare patches at the Jornada. The patch-scale ecotone can be characterized by a gradient in the density of black grama plants across intermediate scales of tens of meters (Fig. 1B). Plant-scale ecotones are nested within this scale.

Landscape-Scale Ecotones

Vegetation patches are not distributed randomly in space; they aggregate based on climatic conditions, geomorphology, dispersal, competition, and other factors. A landscape-scale ecotone is one that can be characterized by a change in the distribution of patches of a given type across space. The dominant landscape-scale ecotone at the Jornada is between black grama grasslands and mesquite shrublands, which is characterized by a gradient in grass patch cover at scales of hundreds to thousands of meters (Fig. 1C). Plant- and patch-scale ecotones are nested within this scale.

SPATIAL AND TEMPORAL DYNAMICS OF ECOTONES

An ecotone has been described as a "zone of tension" between two adjacent ecological states.[14,15] This tension occurs when there are strong internal dynamics within those two states, produced by positive feedbacks that maintain one state over another. This causes some ecotones to be inherently unstable, because they operate outside the biotic feedbacks that maintain a stable ecological state. Other ecotones, however, do not display such instability; these spatial and temporal dynamics are a key component to understanding ecotone function and future behavior.

Peters et al.[9] identify three basic types of ecotone dynamics in space and time: directional, stationary, and shifting, depending on the relative strength of different abiotic and biotic factors in the system. Abiotic factors could be drivers like precipitation or wind erosion, inherent constraints such as geomorphology, or feedbacks such as soil accumulation or microclimate. Biotic factors in this context are feedbacks such as competition or recruitment.

Directional Transitions

Landscapes undergoing active invasion of one ecological state into another are examples of a directional transition, and this has several consequences for ecotone dynamics in space and time. In this scenario, abiotic constraints are generally weak (i.e., underlying environmental conditions are suitable to both patch types), and the ecotone shifts based on abiotic and biotic feedbacks within the invading patch type (Fig. 2A). For example, in the northern Chihuahuan Desert, mesquite (*Prosopis glandulosa*) dunes are actively invading historic black grama (*Bouteloua eriopoda*) grasslands. The ecotone between these two states is characterized by a mixture of small grass patches surrounded by bare ground and small shrubs, which facilitates wind erosion and further grass loss (an abiotic feedback). The asymmetry in the strength of these processes maintains a directional shift in landscape structure, exemplified by a directional shift in ecotone location through time.

A

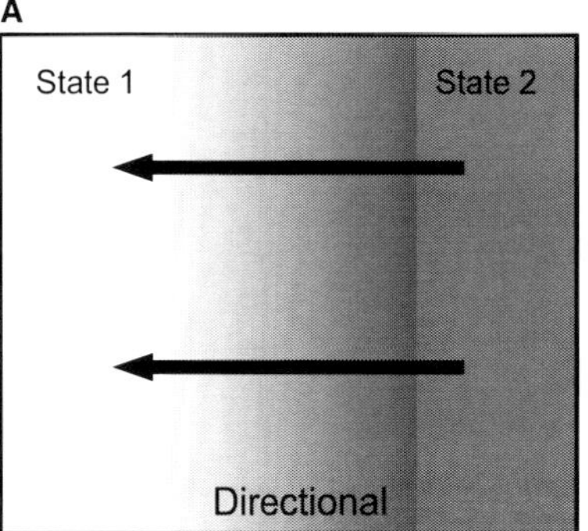

B

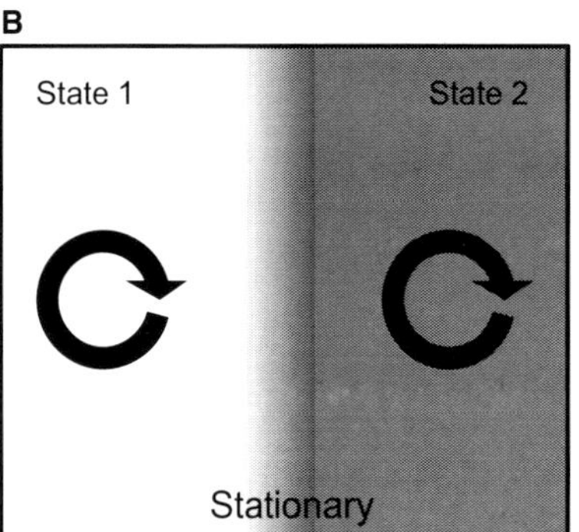

C

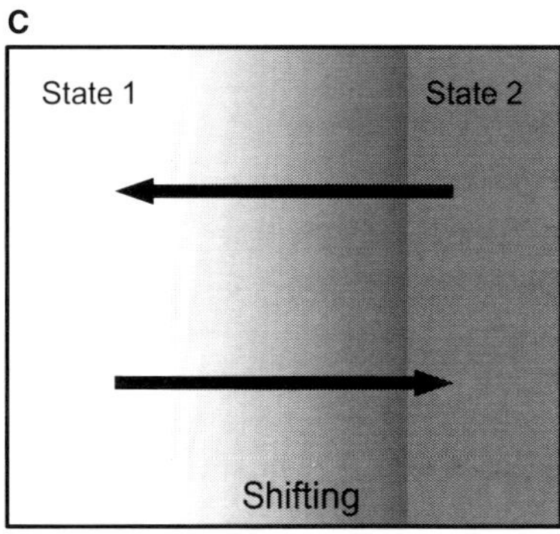

Fig. 2 Ecotone dynamics in time and space. Directional transitions exhibit asymmetry in the strength of drivers and feedbacks between two states. Stationary transitions occur when the internal dynamics of each state are dominated by feedbacks that reinforce the boundary between them. Shifting transitions occur when outside abiotic drivers overwhelm internal patch dynamics.

Stationary Transitions

Where two stable ecological states meet, there is often a stationary ecotone between them. In this situation, each state is maintained by strong abiotic and biotic feedbacks,

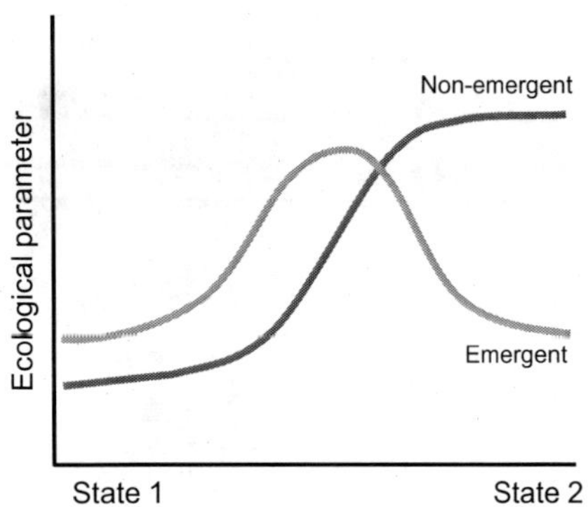

Fig. 3 Defining characteristic of emergent and non-emergent properties of ecotones.

as well as the presence of strong abiotic constraints limiting the distribution of one patch type or the other (Fig. 2B). These ecotones occur where there is an abrupt geomorphic transition, like between a valley floor and steep mountainside, or where human activities dominate. For example, the ecotone between a field and forest is stationary when the field is actively grazed by livestock. Dispersal of woody species from the forest edge is kept low by grazing or other human management activities.

Shifting Transitions

Some landscapes exhibit a "shifting mosaic" of patch types through time, when abiotic drivers such as precipitation and fire fluctuate through time (Fig. 2C). In this scenario, internal feedbacks maintaining one patch type or another are secondary to abiotic drivers. Ecotones between patches can be very abrupt, if spatially variable fire is the dominant driver, or much more indistinct, if multiyear fluctuations in precipitation promote one or another patch type. For example, prescribed fire can be used to maintain a shifting mosaic of different grassland states, wherein an abiotic driver (fire) overwhelms the internal dynamics of each grass patch and maintains very short ecotones between patches.[16]

EMERGENT PROPERTIES OF ECOTONES

Ecotones modulate flows of material, energy, and organisms across heterogeneous landscapes, which has ramifications for a host of ecological processes such as dispersal, predation, and reproduction.[10,12,17] As a result, ecotones can exhibit behaviors, or emergent properties, that are not simply a combination or average of the ecological states between which they are situated.[11,18] One common emergent property is an increase in population density of certain species of plants and animals, termed "edge species" because of their association with ecotones between ecological states.[12] Such species might require multiple complementary habitats, and traveling along ecotones offers an efficient means of acquiring resources.[19] Other species might exhibit a preference for ecotones if their forage or prey species are associated with ecotones, a phenomenon known as "resource mapping."[12] Increased abundance of edge species in ecotones may thus have a cascading effect on other ecological properties, making ecotones distinct in numerous ways from adjacent ecological states (Fig. 3).

CONCLUSIONS

Ecotones, the boundaries between adjacent ecological states, are a key component of all landscapes. Much more than being just gradual transitions between states, they are dynamic and exist at multiple scales of ecological organization. They can also exhibit emergent properties distinct from adjacent states, thus making them important landscape components in their own right. Understanding the abiotic and biotic processes associated with ecotones are especially important to natural resource conservation; under ongoing climate change, it is these biotic transitions that will exhibit the leading edge of broader landscape change. Understanding and predicting landscape change at these boundaries is a key component to understanding and predicting broader effects of climate change in an unpredictable future.

REFERENCES

1. Clements, F.E. *Research Methods in Ecology*; University Publishing Company: Lincoln, NE, 1905; 161–183 pp.
2. Weaver, J.E.; Clements, F.E. *Plant Ecology*; Second Edition. McGraw-Hill Book Company: New York, 1938; 601 pp.
3. Whittaker, R.H. Vegetation of the great smoky mountains. Ecol. Monogr. **1956**, *26*, 1–80.
4. Whittaker, R.H. Gradient analysis of vegetation. Biol. Rev. **1967**, *4*, 207–264.
5. Risser, P.G. The status of the science examining ecotones. BioScience. **1995**, *45*, 318–325.
6. McCarty, J.P. Ecological consequences of recent climate change. Conserv. Biol. **2001**, *15*, 320–331.
7. Parmesan, C; Yohe, G. A globally coherent fingerprint of climate change impacts across natural systems. Nature **2003**, *421*, 37–42.
8. Parmesan, C. Ecological and evolutionary responses to recent climate change. Annu. Rev. Ecol. Evol. Syst. **2006**, *37*, 637–669.
9. Peters, D.P.C.; Gosz, J.R.; Pockman, W.T.; Small, E.E.; Parmenter, R.R.; Collins, S.S.; Muldavin, E. Integrating patch and boundary dynamics to understand and predict biotic transitions at multiple scales. Landscape Ecol. **2006**, *21*, 19–33.
10. Cadenasso, M.L.; Pickett, S.T.A.; Weathers, K.C.; Jones, C.G. A framework for a theory of ecological boundaries. BioScience **2003**, *53*, 750–758.
11. Yarrow, M.M.; Marin, V.H. Toward conceptual cohesiveness: A historical analysis of the theory and utility of ecological boundaries and transition zones. Ecosystem. **2007**, *10*, 462–476.

12. Ries, L.; Fletcher, R.J.; Battin, J; Sisk, T.D. Ecological responses to habitat edges: mechanisms, models, and variability explained. Annu. Rev. Ecol. Evolu. Syst. **2004**, *35*, 491–522.
13. Gosz, J.R. Ecotone hierarchies. Ecol. Appl. **1993**, *3*, 369–376.
14. Clements, F.E. Peculiar zonal formations of the Great Plains. Am. Nat. **1897**, *31*, 968–970.
15. Lidicker, W.Z. Responses of mammals to habitat edges: An overview. Landscape Ecol. **1999**, *14*, 333–343.
16. Fuhlendorf, S.D.; Engle, D.M. Restoring heterogeneity on rangelands: ecosystem management based on evolutionary grazing patterns. BioScience **2001**, *51*, 625–632.
17. Wiens, J.A.; Crawford, C.S.; Gosz, J.R. Boundary dynamics: a conceptual framework for studying landscape ecosystems. Oikos **1985**, *45*, 421–427.
18. Strayer, D.L.; Power, M.E.; Fagan, W.F; Pickett, S.T.A; Belnap, J. A classification of ecological boundaries. BioScience **2003**, *53*, 723–729.
19. Dunning, J.B.; Danielson, B.J.; Pulliam, H.R. Ecological processes that affect populations in complex landscapes. Oikos **1992**, *65*, 169–175.

Edge Effects on Wildlife

Peter W. C. Paton
Department of Natural Resources Science, University of Rhode Island, Kingston, Rhode Island, U.S.A.

Abstract
Aldo Leopold, founder of the science of wildlife management, coined the term "edge effects" to refer to the biological phenomenon that the densities of many game species are higher near the juxtaposition of adjacent habitats. For decades, biologists created more edge habitat to enhance the populations of game species. However, edge habitat is not necessarily beneficial to all species, particularly for forest-interior specialists that are sensitive to habitat fragmentation. Most research investigating the effects of habitat fragmentation on avian nesting success has investigated this question by looking just at edges and adjacent habitats. However, studies investigating larger spatial scales have found that landscape context helps to determine whether edge effects will be documented. Edge effects in avian nesting success are more likely to occur in highly fragmented landscapes, whereas edge effects are less likely to be documented in unfragmented landscapes. The underlying mechanisms that determine edge effects are still unclear.

INTRODUCTION

Ecologists refer to the boundary between adjacent habitat types as the edge or ecotone. At the juxtaposition of adjacent habitat types, there is a tendency for increased species diversity and increased densities of certain game species. In 1933, Aldo Leopold[1] coined the phrase "*edge effect*" to refer to this relationship in his classic book on game management, "Carrying capacity in species of high type requirements and low radius varies directly with the interspersion of the types, which is proportional to the sum of the type peripheries. Such game is an *edge effect.*" Leopold, who was the founder of the science of wildlife management, developed the "law of interspersion," which hypothesizes that increases in the amount of edge habitat result in higher population densities of some game species.[1] There are a number of hypotheses to explain this increase in density of game species at edges including proximity to two different habitat types preferred for different uses (e.g., roosting, feeding, or nesting) microclimatic changes in light, temperature, and humidity; changes in plant composition due to increased competition from shade-intolerant species; increased colonization rates by invasive plant species; increased colonization by insects; and predation, parasitism, and competition by "weedy" birds and mammals.[2]

As one of the United States' first spatial ecologists, Leopold believed that the composition and interspersion of essential habitat types in relationship to the daily dispersal capabilities of target species were vital to long-term population persistence.[3] Leopold's ideas led many wildlife biologists to create edges to enhance habitat for game species, which in some instances meant fragmenting existing contiguous habitats.[4] In the 1970s and 1980s, wildlife biologists began to realize that Leopold's hypothesis pertained only to game species of low mobility and specific habitat requirements and challenged the idea that edges benefited most species of wildlife.[5,6] Yet, Leopold knew that the relationship between edge and density was complex and that the habitat requirements of forest-interior specialists were the inverse of game species that preferred early-succession habitats, such as old fields.[3]

In 1978, the "*ecological trap hypothesis*" postulated that avian nest predation rates were density dependent, and that the greater density of nests near edges leads to a concomitant increase in depredation rates of avian nests near edges.[7] Much of the interest in the negative influence of edge effects in the 1980s was due to population declines in neotropical migratory birds that nest in North America.[8] Nest success is vital to long-term avian population dynamics,[9] therefore determining which factors influence nest predation is essential for biologists hoping to successfully manage avian populations. Reviews of avian nest predation studies in the mid- to late-1990s found that most, but not all, studies found evidence of increased nest predation near edges.[10,11] However, evidence of an edge effect on nest predation rates is not restricted to birds, as painted turtles (*Chrysemys picta*) can have greater nest predation rates near water edges.[12]

Although there is strong evidence that avian (and other taxa) nest success can be lower for some species near edges, there appear to be many factors that affect nest

Encyclopedia of Natural Resources DOI: 10.1081/E-ENRL-120047452

success,[10] which can include nest type,[13] plot age,[14] predator densities,[15] plot size,[16] and among the most important factors–landscape context.[17] One of the major challenges for biologists hoping to understand the impact of edge effects on wildlife populations is disentangling the effects of habitat fragmentation on avian nesting success at multiple spatial scales, including edge, patch, and landscape.[17,18] There is considerable confusion or disagreement about whether or how different stand types or ages of forest (e.g., recently harvested patches surrounded by mature forest) constitute a "fragmentation" of that larger habitat patch or landscape. In forested landscapes, some biologists feel that forest harvesting does not necessarily "fragment" larger forest patches or forested landscapes, but rather represents a patch with a different disturbance history and resulting structure.

Most research that has examined the effects of habitat fragmentation on avian nesting success used either real or artificial nests at the edge scale, whereas fewer studies have investigated nest success at either patch or landscape scales.[18] Research in the mid-western United States found that avian nest predation and Brown-headed Cowbird (*Molothrus ater*) nest parasitism rates were much higher in fragmented landscapes than unfragmented landscapes, and nest predation rates were much higher in edge habitats than core areas far from edges.[17,19] More importantly, this research clearly showed that the evidence of edge effects on nest predation and cowbird parasitism rates differed among landscapes, with greater depredation rates near edges in highly and moderately fragmented landscapes, but not in unfragmented landscapes.[17,19] At a landscape scale, nest predation rates of many species of forest birds have a negative relationship with the amount of forest cover; that is as forest cover declines, nest predation rates tend to increase.[19] Therefore, it can be challenging to separate the effects of habitat loss, habitat fragmentation, and edge effects on nest predation rates.

Recently, there is considerable evidence that early succession patches within larger forests may be extensively used by forest-interior birds during the post-fledging period for juvenile and postnesting period by adults. Many researchers now believe that the value of these clearcut patches for increasing juvenile and adult survival rates may outweigh the possible negative "edge effects" on nest predation rates, especially in areas with high renesting rates and moderate-to-high production overall.[20]

CONCLUSION

Much is known about the potential impacts of edge effects on wildlife populations, with most research focusing on the potential negative impact of edges on avian nest success.[10,11] However, research on other taxa has shown that small mammal populations [e.g., red-backed voles (*Clethrionomys californicus*)],[21] and turtles[12] also can be sensitive to habitat edges. Research conducted to date suggests that avian nest predation rates tend to be significantly higher nearer edges, with the most conclusive studies showing that elevated depredation rates generally occur within 50 m of an edge.[10] However, edge effects are most likely to be detected in forests surrounded by farmland and less likely to be detected in landscapes that are mostly forested mosaics.[22] There is strong evidence that landscape context is critical when trying to understand the mechanisms affecting depredation rates.[17–19] In particular, edge effects on nest predation rates are much more likely to be detected in fragmented landscapes; however, the mechanisms leading to increased depredation rates near edges still remain uncertain.[23]

REFERENCES

1. Leopold, A. *Game management*; University of Wisconsin Press: Madison, WI, 1933; 135.
2. Alverson, W.S.; Waller, D.M.; Solheim, S.L. Forests too deer: Edge effects in Northern Wisconsin. Conserv. Biol. **1998**, *2* (4), 3458–358.
3. Silbernagel, J. Spatial theory in early conservation design: examples from Aldo Leopold's work. Landscape Ecol. **2003**, *18* (2), 635–646
4. Yoakum, J.; Dasmann, W.P.; Sanderson, H.R.; Nixon, C.M.; H. S. Crawford, H. S. Habitat improvement techniques. In *Wildlife Management Techniques Manual*. Schemnitz, S. D. Ed.; The Wildlife Society: Washington, D.C., 1980; 329–404.
5. Yahner, R.H. Changes in wildlife communities near edges. Conserv. Biol. **1988**, *2* (4), 333–339.
6. Reese, K.P.; Ratti, J.T. Edge effect: A concept under scrutiny. 53rd Transactions of the North American Wildlife Resources Conference, 1988; 127–136.
7. Gates, J. E.; Gysel, L.W. Avian nest dispersion and fledging success in field-forest ecotones. Ecology **1978**, *59* (5), 871–883.
8. Askins, R.A.; Lynch, J.F.; Greenburg, R. Population declines in migratory birds in eastern North American. Curr. Ornithol. **1990**, *7* (1), 1–57.
9. Lack, D.L. *The natural regulation of animal numbers*; Clarendon Press: Oxford, England, 1954; 343 p.
10. Paton, P.W.C. The effect of edge on avian nest success: How strong is the evidence? Conserv. Biol. **1994**, *8* (1), 17–26.
11. Hartley, M.J.; Hunter, M.L. Jr. A meta-analysis of forest cover, edge effects, and artificial nest predation rates. Conserv. Biol. **1998**, *12* (2), 465–469.
12. Kolbe, J.J.; Janzen, F.J. Spatial and temporal dynamics of turtle nest predation: edge effects. Oikos **2002**, *99* (3), 538–444.
13. Møller, A.P. Nest site selection across field-woodland ecotones: the effect of nest predation. Oikos **1989**, *56* (2), 240–246.
14. Yahner, R.H.; Wright, A.L. Depredation of artificial ground nests: effects of edge and plot age. J. Wildlife Manag. **1985**, *49* (3), 508–513.

15. Angelstam, P. Predation on ground-nesting birds' nests in relation to predator densities and habitat edge. Oikos **1986**, *47* (3), 365–373.
16. Wilcove, D.S. Nest predation in forest tracts and the decline of migratory songbirds. Ecology **1985**, *66* (4), 1211–1214.
17. Donovan, T.E.; Jones, P.W.; Annand, E.M.; Thompson, F.R. Variation in local-scale edge effects: mechanisms and landscape context. Ecology **1997**, *78* (7), 2064–2075.
18. Stephens, S.E.; Koons, D.N.; Rotella, J.J.; Willey, D.W. Effects of habitat fragmentation on avian nesting success: a review of the evidence at multiple spatial scales. Biol. Conserv. **2003**, *115* (1), 101–110.
19. Robinson, S.K.; Thompson, F.R.; Donovan, T.M.; Whitehead, D.R.; Faaborg, J. Regional forest fragmentation and the nesting success of migratory birds. Science **1995**, *267* (5206), 1987–1990.
20. Vitz, A.C., Rodewald, A.D. Can regenerating clearcuts benefit mature-forest songbirds? An examination of post-breeding ecology. Biol. Conserv. **2006**, *127* (4), 477–486.
21. Mills, L.S. Edge effects and isolation: red-backed voles on forest remnants. Conserv. Biol. **1995**, *9* (2), 395–403.
22. Andrén, H. Effects of landscape composition on predation rates at habitat edges. In *Mosaic landscapes and ecological processes*. Hansson, L.; Fahrig, L.; Merriam, G., Eds.; Chapman and Hall: London. 1995; 225–242.
23. Driscoll, M.J.L.; Donovan, T.M. Landscape context moderates edge effects: Nesting success of Wood Thrushes in Central New York. Conserv. Biol. **2004**, *18* (5), 1330–1338.

Environmental Goods and Services: Economic and Non-Economic Methods for Valuing

Daniel E. Campbell
Atlantic Ecology Division (AED), National Health and Environmental Effects Research Laboratory (NHEERL), U.S. Environmental Protection Agency (EPA), Narragansett, Rhode Island, U.S.A.

Abstract

One of the greatest problems that global society faces in the 21st century is to accurately determine the value of the work contributions that the environment makes to support society. This work can be valued by economic methods, both market and nonmarket, as well as by accounting methods that determine the inputs required for the production of a good or service. All commonly used economic valuation methods are subjective and attempt to capture the value that people assign to ecosystem products and services based on their preferences; whereas, emergy methods are objective and take an accounting perspective that determines value based on the sum of the energy, material, and information inputs required for the production of a good or service without double-counting. Non-economic value as measured by emergy exceeds the market value of the same product or service in a predictable manner with the difference between the two narrowing for more highly processed products and services. When nonmarket estimates of value are added to market value to get the total economic value of an ecosystem's contributions to society, in theory, the values obtained can be either more or less than those found from an emergy evaluation of the same system. Values of environmental goods and services determined through emergy evaluations can promote better decision-making by providing an objective check on the subjective values assigned by economic methods using human perceptions of value, which are invariably based on incomplete information.

INTRODUCTION

One of the most important problems facing global society in the 21st century is to accurately determine the value of the contributions that the environment makes to support society. These contributions can be valued by economic methods, both market and nonmarket, as well as by accounting methods that determine the inputs required for the production of a good or service. In the past, the determination of value has been seen as a uniquely economic problem, which has been quantified in neoclassical economics by willingness to pay or accept payment for an item (receiver value). However, at least from the development of Marx's labor theory of value during the 19th century, the idea of determining value from the required inputs to a production process (donor value) has been understood, if not widely accepted among economists. Of course, labor is not the only important input to economic production, and thus Marx's labor theory of value can be easily refuted based on practical arguments. However, in the latter half of the 20th century, H.T. Odum with his students and colleagues used the principles of Energy Systems Theory[1] to develop theory and methods to account for all the various inputs to any production process in terms of the past use of available energy (i.e., energy with the potential to do work) required for the production of any product or service. This donor-based method tracks the transformations of available energy that are required for the production of a product or service. Thus, any ecological, economic, or social production process, for which all the required inputs are known, can be quantified in this manner. Odum's complete donor-based accounting was performed using a new quantity, emergy,[2] expressed with a new unit, the solar emjoule (semj), where "em" is a prefix meaning "within" or "inside" and is sometimes interpreted as a mnemonic for "energy memory." The emergy measure accounts for the solar equivalent joules used in the past, both directly and indirectly, to create a product or service. In this way, the emergy of the inputs to a production process in the economy or in the environment can be summed to determine the total work required to have a product or service as part of the system. This total work requirement is then taken as an estimate of the value of the product or service.

ECONOMIC VALUATION

Figure 1 shows a diagram of economic (i.e., monetary) valuation methods expanded to include emergy evaluation as an example of a quantitative non-economic (i.e., non-monetary) method of valuation. The economic model for value is shown to the left of the dotted line. Note that total economic value (TEV) includes both real money flows, measured directly by markets and real or hypothetical

Encyclopedia of Natural Resources DOI: 10.1081/E-ENRL-120047465

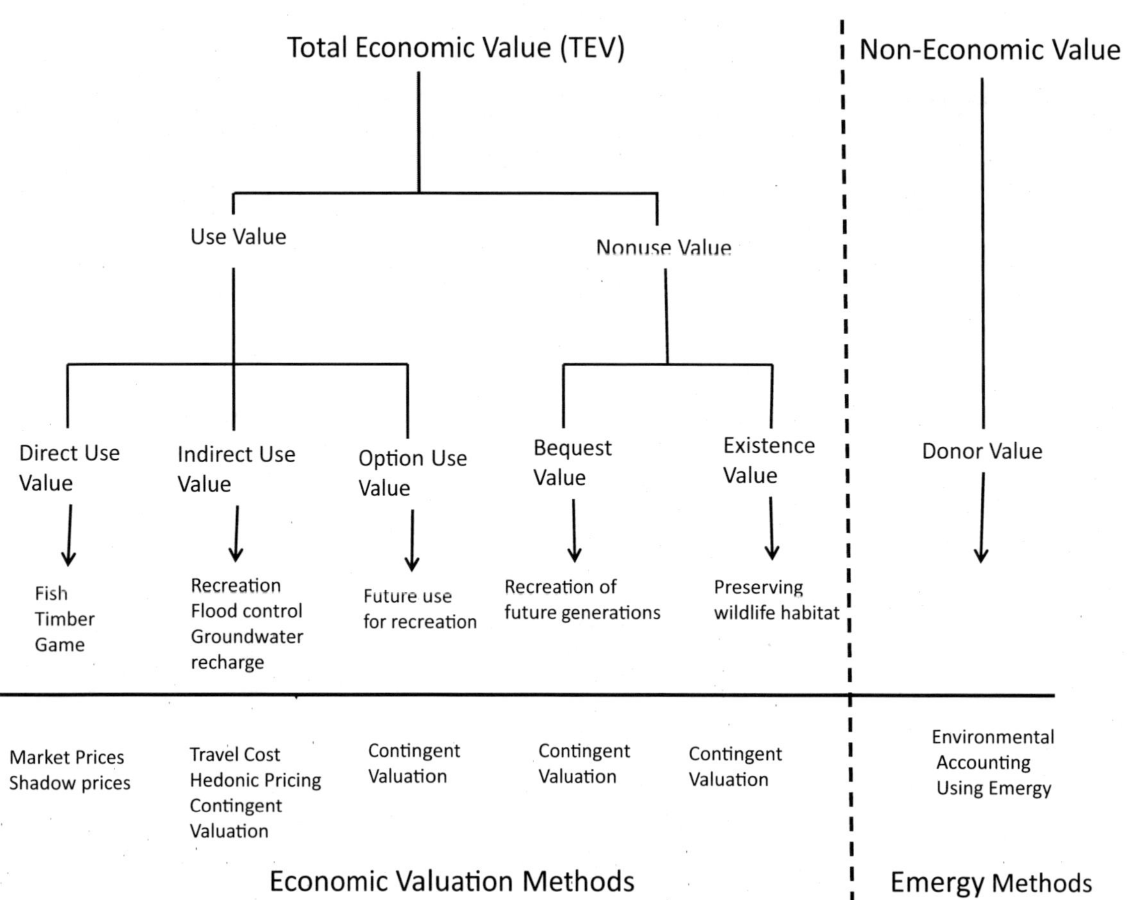

Fig. 1 The elements of total economic value are shown in relationship to one another and in comparison with the primary non-economic method of determining the total value of an item, Environmental Accounting Using Emergy.
Source: Adapted from Bateman and Langford.[3]

money flows determined by various nonmarket methods. Some nonmarket values are assigned from money flows that are indirectly revealed by markets, that is, hedonic values or travel costs. Other nonmarket methods depend on stated preferences, which are usually measured as hypothetical dollar flows determined by surveys, that is, willingness to pay or accept payment as inferred from the answers given to survey questions. The contingent valuation and contingent choice methods are the most commonly used stated preference methods of determining value, but the revealed preference methods are also commonly used nonmarket methods of valuing ecological goods and services. The economic value model is designed to elicit the total value, including both use and nonuse values, that people assign to a given product or service. The various aspects of TEV are additive, as long as double counting is avoided, and because economic value can be concerned with value received by people, who benefit from the goods or services, regardless of their location, it is not necessarily constrained by the geographic boundaries or area of influence of the system providing the products and services, whose value is being assessed. In practice, double counting within TEV is a difficult problem that is the subject of current research.

The methods of economic valuation have been applied to attempt to quantify the benefits that humans derive from ecosystems (i.e., ecosystem services) and published papers related to this topic have grown exponentially since the early 1980s.[4] Several classification schemes for "ecosystem services" have been proposed;[4,5–8] however, there is still some debate concerning the correctness and usefulness of these ideas.[9–11] While most valuations of "ecosystem services" have been approached using one or more of the economic methods mentioned above, their value can also be determined using donor-based methods.[12]

NON-ECONOMIC VALUATION

Donor-based quantitative methods are shown to the right of the dotted line in Fig. 1 Environmental accounting using emergy is the only example given here, although other biophysical accounting methods might be used, for example, energy, exergy, material, or information flow. However, emergy evaluation includes information on all other kinds of flows and thus it is the most complete method of assessing donor value; therefore, emergy will be used to represent all

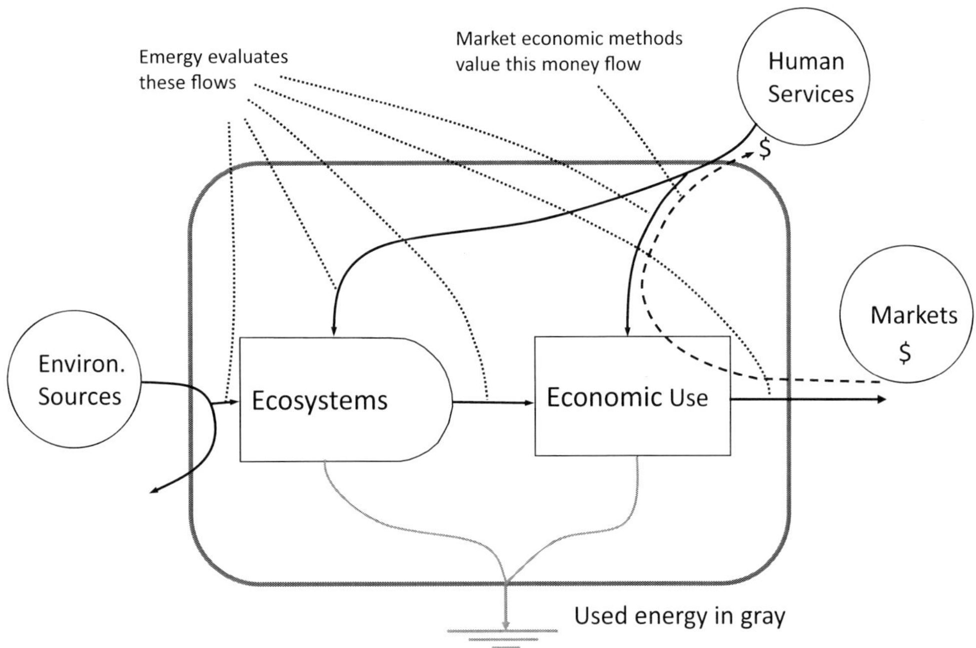

Fig. 2 Environmental system flows evaluated by emergy and economic market methods. The dashed line is money flow; the dotted lines indicate that information is being read.
Source: Modified from Odum and Odum.[13]

non-economic biophysical methods of determining value. Note that determining the value of an item using emergy relies on the creation and use of the item in a real system. Emergy like economics has nothing to say about the intrinsic value of a thing in itself, that is, an entity divorced from the system of which it is a part and within which it was created.

Odum and Odum[13] demonstrated that emergy evaluation was a more complete method of determining the value of products and services in an environmental system than was market valuation (Fig. 2). Note that emergy evaluations include the flows of energy, matter, and information from the entire system as well as the flow of products going to market. Markets only capture the money that flows as a countercurrent to the useful products of the environmental system, for example, the logs mentioned in the example below. Campbell and Brown[12] found that economic estimates of the market value of goods and services produced by the U.S. Forest Service (USFS) lands captured 46% of the value obtained by emergy evaluation of the same systems. When the emergy value of nonmarket services was included, only one-third of the value assessed using emergy methods was captured by the economic analyses; however, economic assessments of nonmarket values (e.g., clean air, clean water, gross primary production, etc.) were not available for the USFS system at that time.

MODERN ORIGIN OF THE TWO PERSPECTIVES ON VALUING THE WORK OF THE ENVIRONMENT

The origins of environmental economics can be traced to the 1950s when Resources for the Future was established in Washington DC.[14] It is interesting that the origin of the modern field of ecological economics can be traced to about this same time (the late 1950s) in the study that H.T. Odum was doing and sponsoring as Director, Port Aransas Laboratory, University of Texas.[15] Although these two different perspectives on valuing the environment, one rooted in economics and the other in ecology, began to develop at about the same time, there was little or no interaction between them until the ecological economic approach was organized formally around 1989[16] by an Odum student, Robert Costanza,[17] working with Herman Daly, the proponent of Steady-State Economics[18] and others. Costanza transferred many of Odum's synthetic ideas on general systems' theory and ecology[1] to the new field, but one of the key ideas of Energy Systems Theory (i.e., the maximum empower principle) and the emergy evaluation methods were left behind, perhaps because they were too radical for general acceptance at that time. To see some of what was left behind compare Farber et al.[19] with Campbell.[20]

THE RELATIONSHIP BETWEEN MONEY AND EMERGY FLOW

Both receiver and donor methods can be used to determine the value of the work of the environment to society, albeit from different perspectives. Since both approaches can be used to inform decision-making and each one uses a different unit for the evaluation (i.e., money flow vs. emergy flow), it is important to understand the relationship between the two methods and the values that they assign to the work of the environment. One method of comparison is to relate money flows in an economy to the emergy flows supporting that economy by converting emergy (semj) to emdollars (Em$). This is accomplished by dividing an emergy flow (semj/yr) by the Emergy to Money Ratio (EMR) of the economy for the year in which the values are being determined. The EMR for an economy is the ratio of the system's emergy flow to the system's money flow (e.g., semj/$ of U.S. GDP) for a given year. If all the emergy flows supporting an economy in a given year are converted to Em$, the dollar value of the GDP is redistributed or prorated according to the fraction of the total emergy flow represented by each individual flow in the system, regardless of whether that flow is from the environment or from the economy. So, the rainfall in Texas in a given year can be assigned an Em$ value, which represents the fraction of the total buying power in the state accounted for by the rain, even though there is no market where rain is bought and sold. Recently, Pulselli et al. suggested comparing the value of ecosystem services determined in monetary units per hectare per year to the renewable emergy flow (semj/ha/yr) needed to support that flow of services as a method for showing the relative efficiency with which ecosystem services are being produced in different systems and as a way to track trajectories of efficiency and evaluate changes in the state of a system, when these values are plotted over time.[21]

MARKET VALUE AND EMERGY VALUE OF ENVIRONMENTAL GOODS AND SERVICES

Emergy evaluations determine value based on a specific system's spatial boundaries and the values of products and services are constrained by the emergy of total resource flows that enter the system from outside and by the internal emergy flows derived from the natural capital storages within the system. Odum et al.[22] provided the first comparison of economic market value and emergy value of environmental resources by analyzing the water supply system of Texas (Em$ values were obtained by dividing measured emergy flows in rain, groundwater, etc. by the EMR of the United States in 1985, 2.2E+12 semj/$). They found that environmental products like the rain (Em$ 0.035 m^{-3}), rivers (Em$ 0.09 m^{-3}), and ground water (Em$ 0.25 m^{-3}) on the Texas plains represented quantifiable values using emergy, but water flows were not valued by markets until they were extracted for human use. The market value of irrigation water for Texas agriculture in the 1980s ($0.04 m^{-3}) was about one-eleventh of its value as indicated by emergy valuation (Em$ 0.44 m^{-3}). Campbell and Cai [23] recognized that the Em$ value of any item would always be greater than the market value, given that the market was not perturbed by psychological factors. Using data on U.S. forest resources from Tilley,[24] they showed a pattern in which the monetary value of a forest resource was less than the Em$ value of the same resource by the greatest amount at the point of extraction from nature. Money captured progressively more of the value of an item as measured by Em$ as the resource moved through the economic production chain. This relationship was observed because the work required for all products is either from the environment or the economy and only the work of people is tracked by money. When a product is first extracted from nature most of the work required for that product still comes from the environment. For example, the market value (monetary value) of logs harvested from a forest only reflects the work that people supplied to support the harvest; which, in general, is smaller than the environmental work of the rain, soil nutrients, etc. that was required for the growth of the tree over the number of years needed to reach a harvestable size. As the available energy in a raw product (e.g., pulp logs) is transformed as it passes through a chain of industrial and commercial operations (e.g., from wood pulp, to paper, and then to the pages of a book) more and more of the work required for the end product is supplied by people, who are paid for their work. Thus, there is a transition along a manufacturing chain, in which the emergy directly associated with the money paid to workers, investors, and suppliers becomes greater than the emergy of the raw materials (e.g., the logs). Emergy evaluation tracks the work processes associated with environmental and economic sources on an equal basis.

NONMARKET VALUE AND EMERGY VALUE OF ENVIRONMENTAL GOODS AND SERVICES

When nonmarket values are assessed so that economic value is measured as TEV rather than market value alone, it is possible for economic methods to assign values to the pathways formerly identified as outside of the market (Fig. 3). On one hand, since the components of TEV are additive and the boundaries of an economic evaluation are not restricted necessarily to those of the system producing the goods and services, in theory, the sum of market and nonmarket value could exceed the value of the system as determined by emergy evaluation. On the other hand, if information is only available on one or two aspects of TEV (e.g., the hedonic value of housing near a stream or the recreational benefit of fishing) that are then used to assess the value of ecosystem services delivered by a stream, the value ascribed to the ecosystem services could be considerably less than the value determined by emergy methods. Fig. 3 shows that double counting is a real problem for such comprehensive economic evaluations, as the products sold on the market generate a money flow that may well depend on all of the internal flows of the system. Emergy evaluations avoid double counting by applying the rules of emergy algebra to trace the inflows of emergy through the network[2] to the outputs.

SUMMARY

All commonly used methods of economic valuation are subjective, because they are designed to capture the value

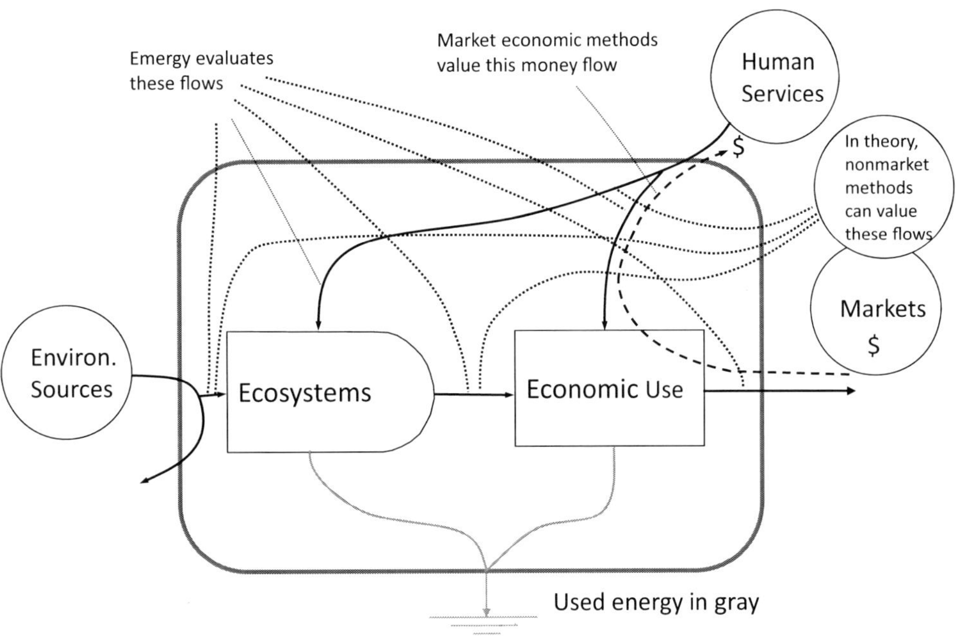

Fig. 3 The application of market and nonmarket economic methods to value ecosystem flows using the analogy of ecosystem services. The dashed line is money flow; the dotted lines indicate that information is being read.
Source: Modified from Odum and Odum.[13]

that people assign to environmental products and services based on their preferences. In contrast, emergy methods are objective, because they take an accounting perspective, which determines value based on the sum of the normalized energy, material and information inputs required for the production of a good or service. Furthermore, the principles of Energy Systems Theory indicate that maximizing the emergy flow in a system is nature's decision criterion.[20] In this regard, developing an understanding of the relationship between emergy value and economic value may not be simply a matter of academic interest, but may be critical to improve decision-making and to set reasonable public policy values for environmental products and services.

For example, in comparing the results of analyses using the two different methods, we might say that where emergy and economic methods agree on the value of an environmental good or service, decisions can be made with greater certainty. On the contrary, if they disagree, a cautionary flag should be raised and more research recommended before implementing policies. In this case, disagreement indicates that people's values are not consistent with the thermodynamic and biophysical conditions governing the system, and thus there will be both constraints on how long such an activity can continue and possible negative consequences for the choices made.

In summary, non-economic value as measured by emergy exceeds the market value of the same product or service in a predictable manner with the difference between the two narrowing for more highly processed products and services. When nonmarket value estimates are added to market values to get TEV of an ecosystem's contributions to society, the values obtained can be either more or less than those found from an emergy evaluation of the same system. Furthermore, both market and nonmarket economic methods are useful in identifying the different values that people assign to environmental goods and services. In addition, emergy evaluation provides an independent biophysically based assessment of the value of the same goods and services. Values of environmental goods and services determined through emergy evaluations can promote better decision-making by providing an objective check on the subjective values assigned by economic methods using human perceptions of value, which are invariably based on incomplete information. More research needs to be done to better understand the relationship between these two very different perspectives on determining the value of the environment to society and on their joint application in developing environmental policies that are conducive to both human and environmental well-being in the long run.

REFERENCES

1. Odum H.T. *Systems Ecology: An Introduction*; John Wiley & Sons: New York, NY 1983; 644 p.
2. Odum, H.T. *Environmental Accounting, Emergy and Environmental Decision Making*; John Wiley & Sons: New York, NY. 1996; 370 p.
3. Bateman, I.J.; Langford, I.H. Non-users' willingness to pay for a national park: An application and critique of the contingent valuation method. Reg. Stud. **1997**, *31* (6), 571–582.
4. Fisher, B.; Turner, R.K.; Moring, P. Defining and classifying ecosystem services for decision making. Ecol. Econ. **2009**, *68*, 643–653.

5. Costanza, R.; d'Arge, R.; de Groot, R.; Farber, S.; Grasso, M.; Limburg, K.; Naeem, S.; O'Neill, R.V.; Paruelo, J.; Raskin, R.G.; Sutton, P.; van den Belt, M. The value of the world's ecosystems services and natural capital. Nature **1997**, *387*, 253–260.
6. De Groot, R.S.; Wilson, M.A.; Boumans, R.M.J. A typology for the classification, description and valuation of ecosystem functions, goods and services. Ecol. Econ. **2002**, *41*, 393–408.
7. MA, Millennium Assessment, *Millennium Ecosystem Assessment*; Island Press: Washington DC, USA, 2005.
8. Wallace, K.J. Classification of ecosystem services: Problems and solutions. Biol. Conserv. **2007**, *139*, 235–246.
9. Costanza, R. Ecosystem services: Multiple classification systems are needed. Biol. Conserv. **2008**, *141*, 350–352.
10. Wallace, K.J. Ecosystem services: Multiple classifications or confusion? Biol. Conserv. **2008**, *141*, 353–354.
11. Peterson, M.J.; Hall, D.M.; Feldpausch-Parker, A.M.; Peterson, T.R. Obscuring ecosystem function with application of the ecosystem services concept. Conserv. Biol. **2009**, *24* (1), 113–119.
12. Campbell, E.T.; Brown, M.T. Environmental accounting of natural capital and ecosystem services for the U.S. National Forest System. Environ. Dev. Sustain. **2012**, DOI 10.1007/s10668-012-9348-6.
13. Odum, H.T.; Odum, E.P. The energetic basis for valuation of ecosystem services. Ecosystems **2000**, *3*, 21–23.
14. Pearce, D. An intellectual history of environmental economics. In Ann. Rev. Energy Environ. **2002**, 57–81 pp, DOI: 10.1146/annurev.energy.27.122001.083429.
15. Kangas, P. Ecological economics began on the Texas bays in the 1950s. Ecol. Model. **2004**, *178* (1,2), 179–181.
16. Ropke, I. An early history of modern ecological economics. Ecol. Econ. **2004**, *50*, 293–314.
17. Costanza, R. What is ecological economics? Ecol. Econ. **1989**, *1*, 1–7.
18. Daly, H.E. *Steady State Economics*; W.H. Freeman and Co.: San Francisco 1977.
19. Farber, S.C.; Costanza, R.; Wilson, M.A. Economic and ecological concepts for valuing ecosystem services. Ecol. Econ. **2002**, *41*, 375–392.
20. Campbell, D.E. Proposal for including what is valuable to ecosystems in environmental assessments. Environ. Sci. Technol. **2001**, *35*, 2867–2873.
21. Pulselli, F.M.; Coscieme, L.; Bastianoni, S. Ecosystem services as a counterpart of emergy flows to ecosystems. Ecol. Model. **2012**, *222*, 2924–2928.
22. Odum, H.T.; Odum, E.C.; Blissett, M. *Ecology and Economy: Emergy Analysis and Public Policy in Texas;* Lyndon B. Johnson School of Public Affairs, University of Texas at Austin: Policy Research Project Report 1987; 78, 178 p.
23. Campbell, D.E., Cai, T.T., Emergy and economic value, 2007, pp 24-1–24-16. In: Brown, M.T., Bardi, E., Campbell, D.E., Huang, S.L., Rydberg, T., Tilley, D.R., Ulgiati, S. Eds.; *Emergy Synthesis 4, Theory and Applications of the Emergy Methodology*. Proceedings of the 4th Biennial Emergy Research Conference, the Center for Environmental Policy, University of Florida, Gainesville, FL.
24. Tilley, D.R. *Emergy Basis of Forest Systems*; Doctoral Dissertation, University of Florida, UMI, Ann Arbor: MI, 1999; 296 pp.

Exotic and Invasive Species

Jan Pergl
Petr Pyšek
Department of Invasion Ecology, Institute of Botany, Academy of Sciences of the Czech Republic (CAS), Průhonice, Czech Republic

Abstract

Alien species are significant component of global change, and their occurrence is often associated with a negative perception of the impact on humans and biodiversity. Even though the numbers of introduced species across all taxonomic groups are high, counts of invasive species, those that are spreading and have impact, are much lower. Since the 1980s, invasions by exotic species have been considered across different groups of policymakers, researchers, and stakeholders as a significant problem for nature and economy. There is progress in understanding the basic questions of invasion ecology, management, and prevention issues. Unfortunately, they are rarely transformed into functional policy measures. In this entry, we review the current state-of-the-art understanding of biological invasions, impact assessment, and management of exotic species.

EXOTIC OR INVASIVE: WHO IS WHO?

What do ecologists mean when they describe a species as native or exotic? Is making this distinction important? The vast majority of invasion ecologists and environmental managers consider the status of a species (native or exotic) as the key attribute that we need to know. Although opinions may differ, most people agree that native (indigenous) species are those that originated in a given area. Conversely, exotic (alien; introduced; non-native; non-indigenous) species are present in a given area due to intentional or unintentional human involvement. The classification as native or exotic depends on the scale at which the assessment is made. For example, a species can be considered alien to the whole continent if it arrived from overseas or to its part if it is native in a different region on the same continent. Distinction between native and exotic is not, however, fully consistent in different parts of the world and changes according to dates of human intervention. In Europe, species are considered native if they were present before the Neolithic period (the onset of widespread crop growing), but in the Americas, the year 1492 (the discovery of America by Europeans) is generally considered the demarcation point. Invasion ecologists in Europe further distinguish exotic plants into archaeophytes, introduced since the beginning of Neolithic agriculture until the end of the middle Ages, and neophytes, the modern invaders introduced after the discovery of America in 1492. This division is supported by the fact that both groups have distinct habitat characteristics, a response to climate and invasion patterns; most serious invasive plants of today recruit from neophytes, whereas archaeophytes are often agricultural weeds. For pragmatic reasons, native species expanding their distribution ranges and spreading into new habitats are not called invasive (this term is reserved for species introduced by humans) but instead are described as expansive. The proper terms referring to their behavior are migration, spread, range expansion, or range extension.

There are two approaches to defining invasive species. The first one, more scientifically rigorous, breaks the process of a biological invasion into a series of consecutive stages (the so-called introduction–naturalization–invasion continuum), defined on the basis of barriers a species needs to negotiate or overcome to reach the next stage. The process begins with transport and introduction, followed by the establishment of self-reproducing populations, and in the final stage, the species spreads through the landscape and reaches a high abundance in invaded communities. Within this framework, an invasive species is defined based on ecological criteria and species population dynamics, but the definition does not distinguish the type of habitat invaded (natural vs. human made). The second perspective, more pragmatic, considers biodiversity [e.g., Convention on Biological Diversity, Article 8 (h)] and labels invasive only those exotics that threaten native ecosystems, habitats, species, or humans and their activities. This definition, however, relies on detection of a negative impact by the invading exotic species which is problematic to define and difficult to measure (see section Impact of exotic species). In this entry, we follow the ecological definition.[1–3]

Not all introduced species are successful; only a fraction of each species reaches the next stage of invasion as they overcome barriers or filters (Fig. 1). This is reflected in a

Encyclopedia of Natural Resources DOI: 10.1081/E-ENRL-120047429

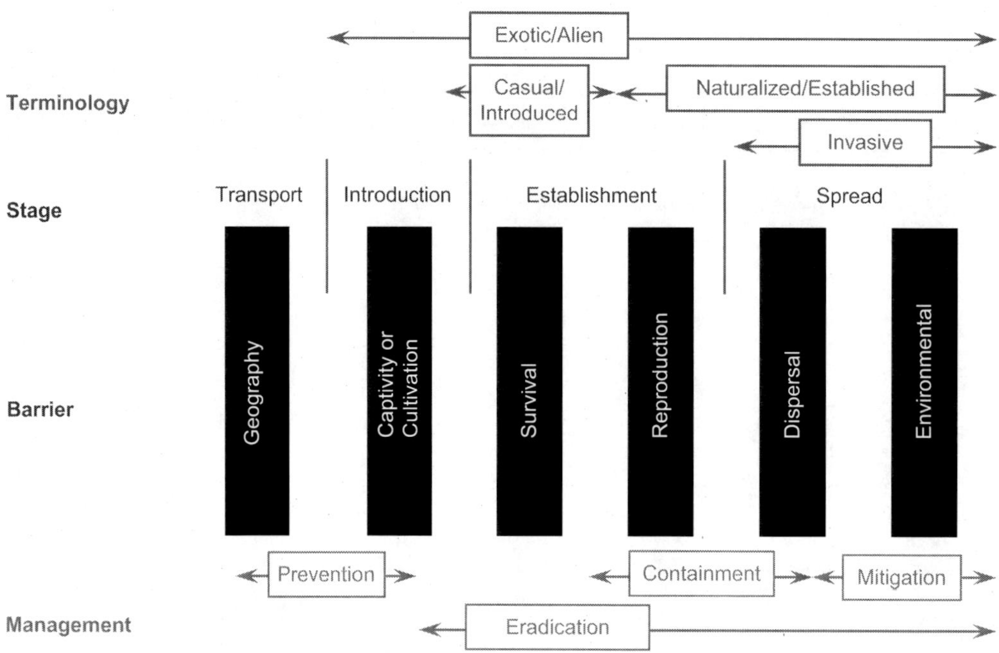

Fig. 1 A scheme showing individual geographic and environmental barriers with the corresponding stages a species passes along the introduction–naturalization–invasion continuum, as its status changes from casual to naturalized to invasive. Management intervention possibilities along the stages of the process are indicated.
Source: Redrawn from Blackburn et al.[1]

general rule of thumb called the "Tens Rule," which posits that approximately 10% of introduced species are able to escape from cultivation or captivity, 10% of the escapees can establish, and only 10% of these established species exert economic impact (termed "weeds" or "pests").[4] Exotic species that rely on repeated introductions for their persistence, flourish and reproduce occasionally outside cultivation but eventually die out because they do not form self-replacing populations and are called casuals (subspontaneous, occasional escapes, ephemeral taxa). Those that overcome abiotic and biotic barriers to survival and regular reproduction are called naturalized (established). Invasive species are a subset of naturalized species that are able to spread far from their place of introduction and often become abundant in invaded communities.

In addition, the anthropocentric view generates more terms like pests, weeds, harmful species, problem species, or noxious species for those that are not necessarily alien but grow in sites where they are not wanted and have detectable economic and/or environmental impact.[1–3]

WHAT MAKES A SPECIES INVASIVE?

There are three classical questions in invasion biology: Which traits make the species invasive? What are the attributes of a community that make it more or less prone to invasions? And what is the impact of an alien species and how can it be diminished? A number of hypotheses and frameworks have emerged in the last two decades that attempt to identify why some species become successful and others do not.[5,6] The most sophisticated hypotheses reflect the fact that establishment and invasion are complex processes that involve many factors (Fig. 2). Some of the concepts relate to species invasiveness, whereas others address habitat invasibility. The most popular of the former is the Enemy Release Hypothesis, which suggests that upon introduction, exotic species are freed from their natural enemies with common evolutionary history, such as parasites, pathogens, herbivores, or predators that exert pressure and thus regulate their populations in the native range.[7] This provides them with a competitive advantage in the new range. It has also been suggested and widely tested that populations respond to the lack of enemies in the process of postinvasion evolution, by allocating resources no longer needed for defense to growth and reproduction, thus increasing fitness [formalized in the Evolution of Increased Competitive Ability (EICA) Hypothesis].[8] The phenomenon of enemy release is a basis for classical biocontrol where natural enemies from the native range are deliberately introduced to the new range to control invading populations.[9,10]

The Fluctuating Resources Hypothesis suggests that invasion of a new species is only possible when resources are available. The level of resources in an ecosystem is determined by availability resulting from disturbances and uptake from resident species. Therefore, high levels of disturbance combined with high levels of ecosystem productivity may lead to successful invasion or even expansion by

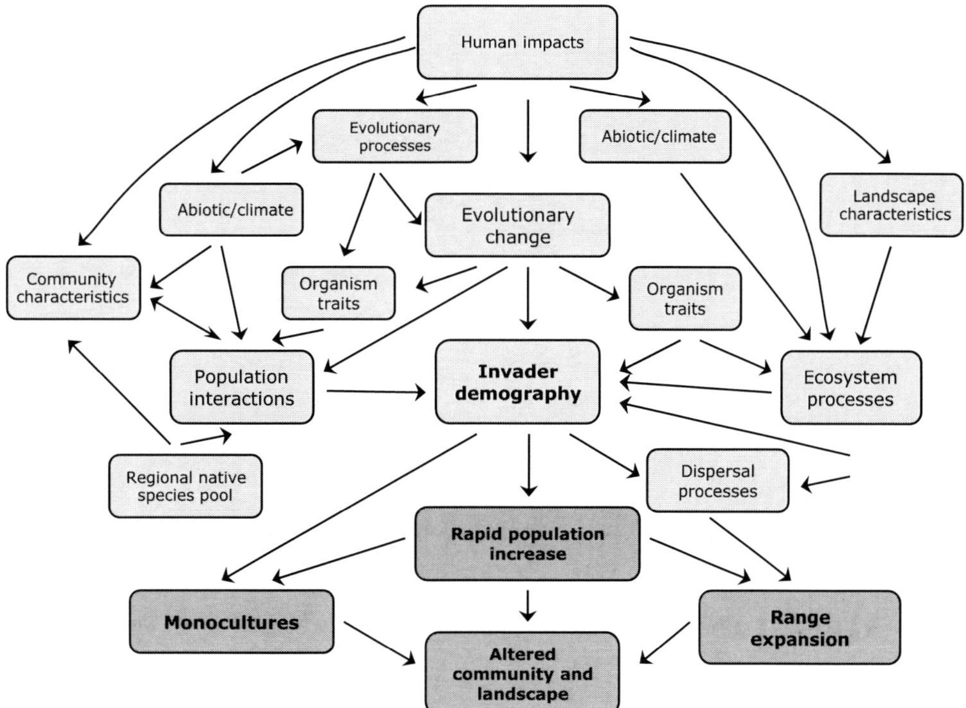

Fig. 2 A conceptual synthetic invasion framework capturing fundamental ecological and evolutionary processes that interact to determine the outcome of invasion. The demography of the invader is central to the concept; in successful invaders, it results in population increase and range expansion and affects community and landscape characteristics.
Source: From Gurevitch et al.[6]

Ecotone—Fragmentation

native species, especially in the early stages of succession. By relating the availability of resources to the ability of a species to capitalize on free resources, the hypothesis introduces a link between the invasibility of a community and species traits.[11]

Since Darwin, it has been proposed that species-rich communities do not harbor as many exotic species as species-poor communities. This relationship between species diversity and invasibility has been thoroughly investigated within the framework of the Biotic Resistance Hypothesis.[12] It is based on a notion that high diversity of native species results in lower niche opportunities for invaders, making the invasion less likely. However, it is now understood that the relationship between native and exotic species richness is scale dependent. Competition and niche availability results in negative relationship at the community scale, but on larger scales, there is a positive relationship because both groups of species respond to major environmental factors affecting species richness in the same way (Fig. 3).[13,14]

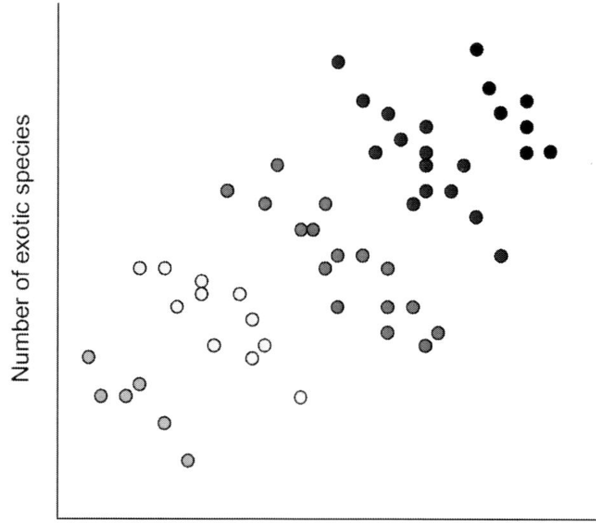

Fig. 3 Alien–native species richness relationship. Within clusters with the same environmental conditions, the correlation is negative but in general, the relationship is positive across the clusters.
Source: From Shea & Chesson.[14]

The search for biological and ecological traits that distinguish successful exotic species from unsuccessful ones, or exotic from native species, has been an important line of research since the onset of invasion ecology in the 1980s and was therefore transformed into modern and sophisticated pest- or weed-risk assessment schemes (PRA, WRA).[15–17] It is now widely acknowledged that species traits related to reproduction, growth, and dispersal affect the outcome of the invasion. Recent analyses of large data sets point out that invasive and non-invasive plant species differ in traits such as those related to physiology, leaf-area allocation, shoot allocation, growth rate, size, and fitness.[18–20]

However, the role of species traits is context dependent. Different traits are important at different stages of the process, and their effects interact with many other factors related to invasion dynamics such as initial population size, residence time (how long ago the species was introduced), propagule pressure (the intensity with which the species is introduced to the region, e.g., the number of introduction events, extent of planting), climate match between source and target area, availability of habitats determining the spatial distribution, or the vectors and pathways of introduction that may differ in their efficiency. It is a complex interaction between these factors that codetermines whether a species becomes invasive. For example, it is now widely acknowledged that the effect of residence time needs to be controlled for in analyses because the longer the alien species is present in a territory, the greater their chance to pass successfully through the stage of casual occurrence and become naturalized, more widely distributed, and invade over a larger range.[21–24]

IMPACT OF EXOTIC SPECIES

As predicted by the Tens Rule, the majority of exotic species have little or no measurable impact. In Europe, about 5% of naturalized (established) exotic species are documented to have impact (Table 1)[25] and it is estimated that 20–30% of the species introduced into the United States, United Kingdom, Australia, India, South Africa, and Brazil are pests and cause major environmental problems. Similarly, only relatively few exotic species become serious pests with significant damage to natural and managed ecosystems or cause health problems to humans.[26] However, reliable information on how many invasive species exert significant impact is still lacking for most regions of the world due to the difficulties linked to defining what an impact actually is and how to detect it. This is because there is no universal measure of impact that is context dependent and multifaceted.[27] For example, the relatively small impact of the tall and fast-growing Asian plant *Impatiens glandulifera* invading in Europe can be attributed to the fact that the riparian communities colonized are dominated by native tall nitrophilous species, and therefore the impact of the invader on species diversity is not much different from the competitive effect of these native dominants on resident species. However, by affecting the behavior of insect pollinators, this species decreases the fitness of co-occurring native plants.[28,29]

The impacts of exotic species range from those on the diversity of native communities to the survival of rare species or ecosystem functioning to human health, predation by introduced pests, habitat transformation, or fire regime (Fig. 4).[2,19,25,27] Some of the impacts are possible to quantify easily irrespective of whether they affect native or exotic species or are directly affecting the economic sphere. Although the ecological and economic impacts of alien species are usually studied separately, they are closely correlated. Because of the lack of existing data on impacts on other services, current economic valuations focus primarily on provisioning ecosystem services.[30] Overall, the impact of exotic species transformed into economic terms has been estimated to reach 5% of the gross domestic product (GDP),[26] and estimates of costs of invasions by exotic species also exist for world regions. For Europe, they exceed €12.7 billion annually,[31] and for the United States, the figure of $30 billion per year was reported, with most of the costs associated with agriculture.[26] It should be noted that introduced species with the highest documented negative impact and resulting costs are human pathogens, including the bubonic plague or HIV causing AIDS.[30]

PREVENTION, CONTROL, MANAGEMENT, AND RISK ASSESSMENT

Pathways, the different ways by which alien species become introduced, not only affect the probability of successful species invasion but also represent a promising tool to prevent invasions by pathway management. Some species are associated with a unique pathway, whereas others are introduced by multiple means.[32,33] The knowledge of invasion pathways is therefore crucial for developing preventive

Table 1 Total numbers and percentages of alien species known to have an ecological or economic impact for different taxonomic groups in Europe. Numbers in brackets show percentages from recorded species in the DAISIE database (http://www.europe-aliens.org)

		Species with	
Taxonomic group	**No. of exotic species**	**Ecological impact**	**Economic impact**
Terrestrial plants	5789	326 (5.6)	315 (5.4)
Terrestrial invertebrates	2481	342 (13.8)	601 (24.2)
Terrestrial vertebrates	358	109 (30.4)	138 (38.5)
Freshwater flora and fauna	481	145 (30.1)	117 (24.3)
Marine flora and fauna	1071	172 (16.1)	176 (16.4)

Source: From Vilà et al.[25]

Abbreviation: DAISIE, Delivering Alien Invasive Species In Europe.

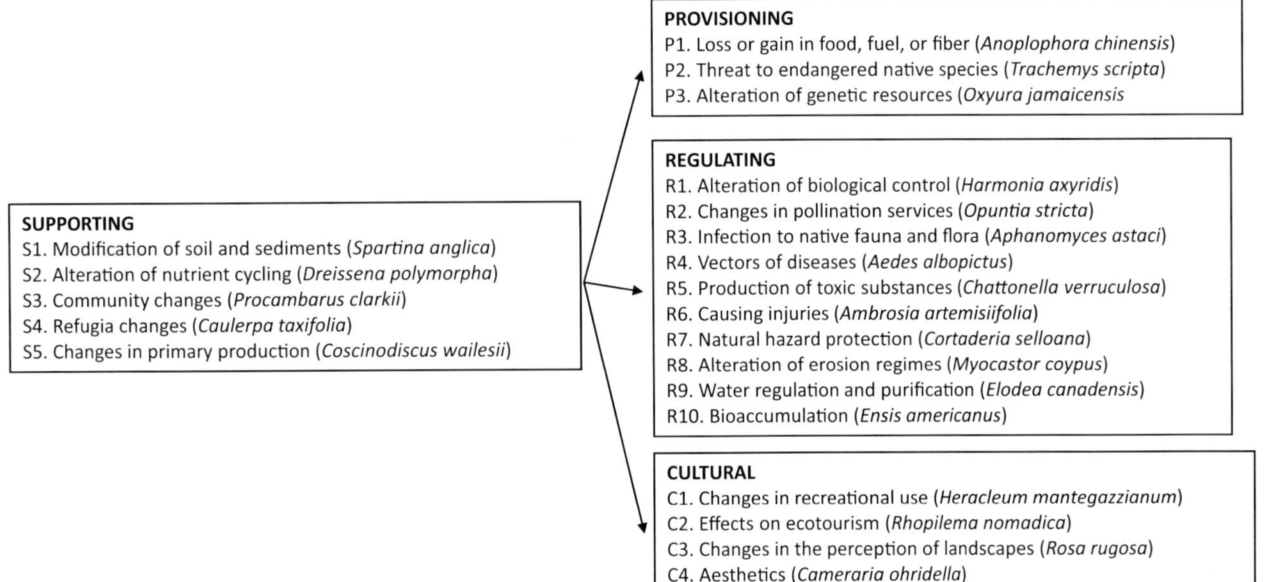

Fig. 4 Impact of exotic species on ecosystem services classified into four categories. Individual types of impact are followed by corresponding example species.
Source: From Vilà et al.[25]

methods such as screening systems, interception programs, early warning strategies, and import regulations.[34–36] Alien species may enter the new region as a direct or indirect result of human activities that can be classified under three broad mechanisms: the importation of a commodity, the arrival of a transport vector, and/or natural spread from a neighboring region where the species is alien. The three mechanisms have been incorporated into a framework consisting of six principal pathways that reflect the degree of human involvement: exotic species are introduced by release, escape, contaminant, stowaway, corridor, and unaided (Fig. 5). Such a framework is applicable to a wide range of taxa and invaded environments and can be used as a basis for developing legislation and codes of practice for different sectors responsible for alien introductions.[35]

Prevention is the cheapest and most preferable method of defense against alien species. Although only a few exotic species become environmental or economic pests, it would be plausible to screen all potential invaders at national borders or before approving import, should those imposing a high risk of invasion be identified. However, the huge numbers of species being introduced for ornamental purposes by horticultural trade makes full screening on a regular basis impossible. Similarly, the magnitude of transported material over borders greatly reduces the efficiency of preventive screening controls. Nevertheless, a range of methods are generally used in assessing the invasion risk; the most common are based on traits of exotics that proved to successfully invade elsewhere, taking also the climate and habitat matching between the native and invaded distribution ranges into account. Risk assessments (RAs) aim at determining the likelihood of risks associated with the introduction of a species to a new region as well as the probability of it becoming naturalized and exerting significant impacts. In the last several years, the precision of RA schemes have improved due to accumulation of data in large species databases so that they are used in some regions in preborder screening protocols or become a part of legal instruments and policies.[1,10,17,37,38]

When dealing with a suite of exotic species that have successfully established in a region, it is important to realize that it is impossible and implausible to attempt at managing all exotics within the given area. This points to the necessity of differentiating between exotic species in general and the subset of those that are invasive and exert impact; most invasion ecologists are aware that only the latter need to be targeted by management measures and that there is no reason for xenophobia against all exotics or a fundamentalistic fight against all species of a foreign origin. This does not mean that the precautionary principle should be discarded, especially when assessing new introductions. Here again, RA represents a cost-effective approach toward deciding which species, where and how should be eradicated or mitigated. In the first place, rapid response to emerging invasions is crucial for eradication success; small and isolated populations are easier to manage.[39,40] Full eradication, i.e., the extirpation of an entire population of a species including its dormant stages such as soil seed banks of invasive plants, is usually not feasible; therefore, partial mitigation or eradication from a small defined area (e.g., a nature reserve) is often the target. If eradication is not possible, containment should be considered as another management option to minimize the impact of the target alien species.

Several methods of managing invasive exotic species are in use; mechanical (e.g., in plants: mowing, cutting, pasturing; animals: hunting, physical removing of alien pests) and chemical (plants: usage of herbicides; animals: poisoning or

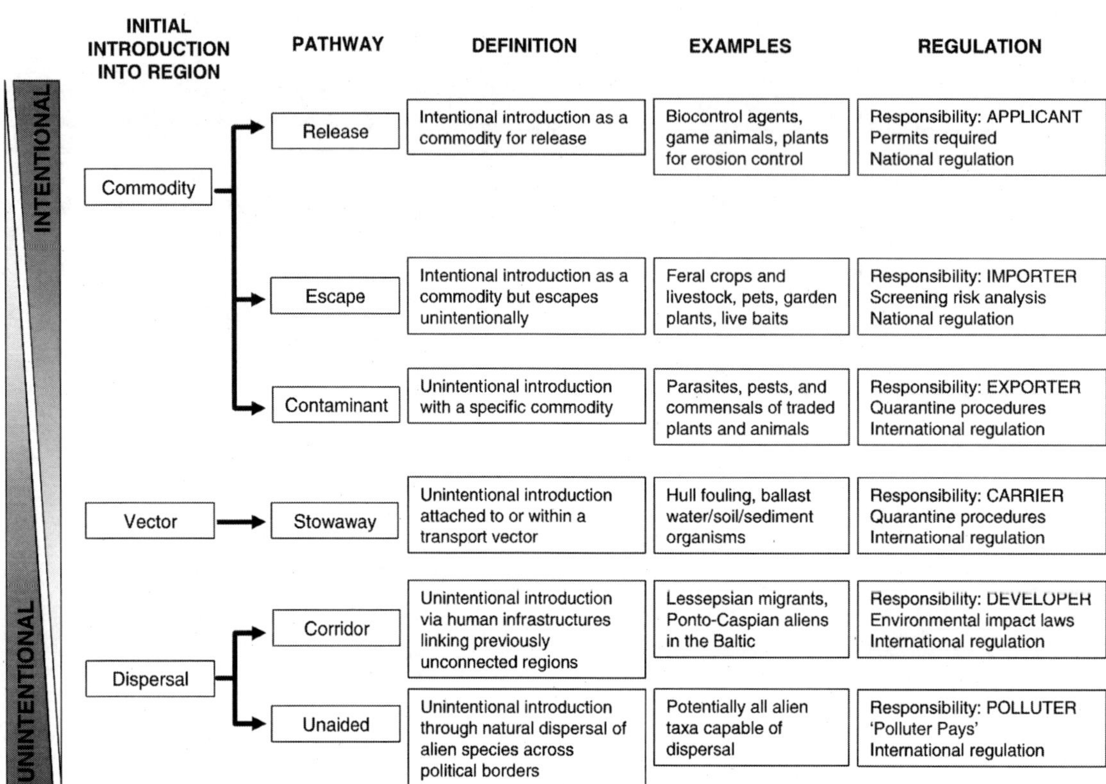

Fig. 5 A framework categorizing existing pathways of introduction of alien species with indication of the mechanisms of introduction, brief description of the pathway and possible ways of regulation.
Source: From Hulme et al.[35]

usage of protective coat) control or their combination, and biocontrol. Biocontrol is a deliberate import and release of potential natural enemies of the targeted pest or weed, termed biocontrol agents. Many exotic species lack their natural enemies in a new area, especially introduced crop or ornamental plant species are imported free of insects, mites, pathogens, and diseases. Although biocontrol has been widely used in greenhouses around the world, and there are numerous success stories in the wild, too, it is still regarded as controversial in open environments, especially in some continents. The reason for this skepticism is that an additional exotic species introduced into a system may itself become invasive, attack non-target native species, and exert various side effects. Examples of animals getting out of control and becoming invasive themselves include the cane toad (*Rhinella marina*), which was introduced to Australia to control sugar cane pests, or the harlequin ladybug (*Harmonia axyridis*), which was introduced to North America and Europe for the control of aphids and coccids. In the field of plant biocontrol, there is a well-known case of head weevil (*Rhinocyllus conicus*) that attacked the native *Cirsium* species in the United States or that of the cactus moth (*Cactoblastis cactorum*) impacting native *Opuntia* cacti in Florida and Mexico.[41,42] On the other hand, compared to the classical mechanical or chemical methods, biocontrol provides a sustainable and long-term management tool because once the agent is released, it reproduces itself in the ecosystem and regulates the pest. Even if the biocontrol does not lead to complete eradication of the targeted pest or weed, it may lower its abundance to an acceptable level.[43] It has to be borne in mind, however, that eradication or containment of the invading species creates an empty niche in the invaded community and that restoration of the managed ecosystem needs to follow eradication to prevent reinvasion and secondary invasions of other species. The restoration aims at reintroduction of native species through improving habitat conditions, supporting the survival and establishment of introduced native species, or directing succession toward desired target communities.[44]

EXOTIC SPECIES LISTS AND DATA SOURCES

There are two major kinds of lists of alien species: more or less complete inventories for defined regions, and shorter and focused species lists of important invaders, depending on expert opinion and management viewpoint. The former are very useful for early warning systems, prioritization of management, or developing the decision schemes for PRA or WRA. The latter lists of the "worst" invasive species that pose a major threat to native diversity or interfere with human interests include global accounts such as the IUCN (International Union for Conservation of Nature) list of the world's 100 worst invasive alien species[45] or continental datasets such as the DAISIE (Delivering Alien Invasive Species In Europe) list of "one hundred of the most invasive alien species in Europe" for Europe, or the SEBI (Streamlining European Biodiversity Indicators) list of 166 "worst invasive alien species threatening biodiversity in Europe."[46,47] The aim of such lists is to highlight the problems associated with invasive species and illustrate various pathways of introduction, taxon groups, invaded habitats, and impacts. Besides complete lists of alien species, the so-called "black" or "gray lists" are being compiled, including invasive species that are banned to be sold, planted, or used in forestry or agriculture. Such lists of noxious weeds have a long history of use in the United States and Australia at both state and federal levels.[48,49]

Several milestones have led to the establishment of invasion biology itself and have initiated and stimulated the development of the field. Although Elton's 1958 book *The Ecology of Invasions by Animals and Plants* is generally thought to have sparked the interest in biological invasions, the real stimulus was the SCOPE (Scientific Committee on Problems of the Environment) program addressing invasive species in the 1980s that generated synthetic publication from many regions of the world.[50] Currently, there are many books dealing with biological invasions ranging from popular science to textbooks. Several scientific journals specialize in biological invasions and their consequences (e.g., *Biological Invasions*, *Diversity and Distributions*, *NeoBiota*, *International Journal of Plant Management*, *Weed Management*, *BioControl*). Many other journals also publish articles on biological invasions and the total number of papers published on this topic each year is estimated at several thousands, and still increasing.

CONCLUSIONS

Considering the fact that both intentional as well as unintentional introductions are important for all taxonomic groups, we can expect that based on current levels of international trade, human travel, and quarantine procedures, the trend of increasing alien introductions will continue at least at current levels. Even if new introductions cease, we need to be aware that some of the already-present not-yet-established casual exotics will become naturalized and some of the currently naturalized species will became invasive pests and weeds.[51] This reflects the phenomenon of lag phases typically associated with invasion dynamics, and that of the so-called invasion debt, reflecting that exotic species respond to human activities with a delay of several decades.[52,53] Other important factors that are likely to influence the trends of species introductions in the future and their ability to naturalize are climate and land-use changes together with agronomy and forestry, which are responsible for the introduction of new drought-resistant crops, as well as increasing stress on renewable resources of energy (new fast-growing species).[54] Actual risks that have to be considered with respect to newly introduced species and nature protection are the increased use of genetically modified organisms (GMOs), which creates the possibility of hybridization with their wild relatives or transfer of herbicide resistance to weeds. Therefore, with respect to the distribution and impact of already naturalized species, it is crucial to prioritize management, enhance the role of early warning systems and rapid response, and be more careful in the sense of precautionary measures when intentionally introducing species.[10] A key challenge is to improve the cooperation between different bodies responsible for introduction and management of alien species (agriculture, forestry, and fisheries) and those whose role is to protect nature both at the national and international levels.[51]

ACKNOWLEDGMENTS

We thank two anonymous reviewers for valuable comments and Laura Meyerson for editing our English. Work on this contribution was funded by the long-term research development project no. RVO 6798593, Praemium Academiae award from the Academy of Sciences of the Czech Republic and project P504/11/1028.

REFERENCES

1. Blackburn, T.M.; Pyšek, P.; Bacher, S.; Carlton, J.T.; Duncan, R.P.; Jarošík, V.; Wilson, J.R.U.; Richardson, D.M. A proposed unified framework for biological invasions. Trends Ecol. Evol. **2011**, *26*, 333–339.
2. Lockwood, J.L.; Hoopes, M.F.; Marchetti, M.P. *Invasion Ecology*; Blackwell Publication: 2007; 466 pp.
3. Richardson, D.M.; Pyšek, P.; Rejmánek, M.; Barbour, M.G.; Panetta F.D.; West, C.J. Naturalization and invasion of alien plants: concepts and definitions. Divers. Distrib. **2000**, *6*, 93–107.
4. Williamson, M.; Fitter, A. The varying success of invaders. Ecology **1996**, *77*, 1661–1666.
5. Catford, J.A.; Jansson, R.; Nilsson, C. Reducing redundancy in invasion ecology by integrating hypotheses into a single theoretical framework. Divers. Distrib. **2009**, *15*, 22–40.
6. Gurevitch, J.; Fox, G.A.; Wardle, G.M.; Inderjit; Taub, D. Emergent insights from the synthesis of conceptual frameworks for biological invasions. Ecol. Let. **2011**, *14*, 407–418.
7. Blossey, B.; Nötzold, R. Evolution of increased competitive ability in invasive nonindigenous plants: a hypothesis. J. Ecol. **1995**, *83*, 887–889.
8. Willis, A.J.; Memmott, J.; Forrester, R.I. Is there evidence for the post-invasion evolution of increased size among invasive plant species? Ecol. Let. **2000**, *3*, 275–283.
9. McFadyen, R.E.C. Biological control of weeds. Ann. Rev. Entomol. **1998**, *43*, 369–393.

10. Pyšek, P.; Richardson, D.M. Invasive species, environmental change and management, and health. Ann. Rev. Envir. Res. **2010**, *35*, 25–55.
11. Davis, M.A.; Grime, J.P.; Thompson, K. Fluctuating resources in plant communities: a general theory of invasibility. J. Ecol. **2000**, *88*, 528–536.
12. Elton, C.S. *The Ecology of Invasions by Animals and Plants*. (reprinted 2000, University of Chicago Press): London, 1958.
13. Herben, T. Species pool size and invasibility of island communities: a null model of sampling effects. Ecol. Let. **2005**, *8*, 909–917.
14. Shea, K.; Chesson, P. Community ecology theory as a framework for biological invasions. Tr. Ecol. Evol. **2002**, *17*, 170–176.
15. Gordon, D.R.; Riddle, B.; Pheloung, P., Ansari, S.; Buddenhagen, C.; Chimera, C.; Daehler, C.C.; Dawson, W.; Denslow, J.S.; Jaqualine, T.N.; LaRosa, A.; Nishida, T.; Onderdonk, D.A.; Panetta, F.D.; Pyšek, P.; Randall, R.P.; Richardson, D.M.; Virtuc, J.G.; Williams, P.A. Guidance for addressing the Australian Weed Risk Assessment questions. Plant Prot. Q. **2010**, *25*, 56–74.
16. Kolar, C.S.; Lodge, D.M. Ecological predictions and risk assessment for alien fishes in North America. Science **2002**, *298*, 1233–1236.
17. Pheloung, P.C.; Williams, P.A.; Halloy, S.R. A weed risk assessment model for use as a biosecurity tool evaluating plant introductions. J. Environ. Manag. **1999**, *57*, 239–251.
18. Grotkopp, E.; Rejmánek, M.; Rost, T.L. Towards a causal explanation of plant invasiveness: seedling growth and life-history strategies of 29 pine (*Pinus*) species. Am. Nat. **2002**, *159*, 396–419.
19. Pyšek, P.; Richardson, D.M. Traits associated with invasiveness in alien plants: where do we stand? In *Biological Invasions*; Nentwig, W., Ed.; Springer-Verlag: Berlin, Heidelberg, 2007; 97–125.
20. van Kleunen, M.; Weber, E.; Fischer M. A meta-analysis of trait differences between invasive and non-invasive plant species. Ecol. Let. **2010**, *13*, 235–245.
21. Phillips, M.M.; Murray, B.R.; Leishman, M.R.; Ingram, R. The naturalization to invasion transition: Are there introduction-history correlates of invasiveness in exotic plants of Australia? Austral Ecol. **2010**, *35*, 695–703.
22. Pyšek, P.; Jarošík, V. Residence time determines the distribution of alien plants. In *Invasive Plants: Ecological and Agricultural Aspects*; Inderjit, Ed.; Birkhäuser Verlag-AG: Basel, 2005; 77–96.
23. Pyšek, P.; Křivánek, M.; Jarošík, V. Planting intensity, residence time, and species traits determine invasion success of alien woody species. Ecology **2009**, *90*, 2734 –2744.
24. Williamson, M.; Dehnen-Schmutz, K.; Kühn I.; Hill, M.; Klotz, S.; Milbau, A.; Stout, J.; Pyšek, P. The distribution of range sizes of native and alien plants in four European countries and the effects of residence time. Divers. Distrib. **2009**, *15*, 158–166.
25. Vilà, M.; Basnou, C.; Pyšek, P.; Josefsson, M.; Genovesi, P.; Gollasch, S.; Nentwig, W.; Olenin, S.; Roques, A.; Roy, D.; Hulme, P.E.; DAISIE partners How well do we understand the impacts of alien species on ecological services? A pan-European cross-taxa assessment. Front. Ecol. Envir. **2010**, *8*, 135–144.
26. Pimentel, D.; McNair, S.; Janecka, J.; Wightman, J.; Simmonds, C.; O'Connell, C.; Wong, E.; Russel, L.; Zern, J.; Aquino, T.; Tsomondo, T. Economic and environmental threats of alien plant, animal, and microbe invasions. Agric. Ecosyst. Envir. **2001**, *84*, 1–20.
27. Levine, J.M.; Vilà, M.; D'Antonio, C.M.; Dukes, J.S.; Grigulis, K.; Lavorel, S. Mechanisms underlying the impacts of exotic plant invasions. Proc. R. Soc. London, B **2003**, *270*, 775–781.
28. Chittka, L.; Schürkens, S. Successful invasion of a floral market. Nature (London) **2001**, *411*, 653.
29. Hejda, M.; Pyšek, P.; Jarošík V. Impact of invasive plants on the species richness, diversity and composition of invaded communities. J. Ecol. **2009**, *97*, 393–403.
30. Perrings, C.; Mooney, H.; Williamson, M. Bioinvasions & globalization. *Ecology, Economics, Management, and Policy;* Oxford University Press: 2010; 247–258.
31. Kettunen, M.; Genovesi, P.; Gollasch, S.; Pagad, S.; Starfinger, U.; ten Brink, P.; Shine, C. Technical support to EU strategy on invasive species (IAS)—Assessment of the impacts of IAS in Europe and the EU (final module report for the European Commission). Institute for European Environmental Policy (IEEP), Brussels, Belgium; 2008.
32. Nentwig, W. Pathways in animal invasions. In *Biological Invasions, Ecological Studies*, Vol. 193; Nentwig, W., Ed.; Springer-Verlag: Berlin Heidelberg, 2007; 11–27.
33. Gollasch, S. Is ballast water a major dispersal mechanism for marine organisms? In *Biological Invasions*; Nentwig, W., Ed.; Springer-Verlag: Berlin, Heidelberg, 2007; 48–57.
34. Carlton, J.T.; Ruiz, G.M. Vector science and integrated vector management in bioinvasion ecology: conceptual frameworks. In *Invasive Alien Species*; Mooney, H.A.; Mack, R.N.; McNeely, J.A.; Neville, L.E.; Schei, P.J.; Waage, J.K., Eds.; Island Press: Washington, DC, 2005; 36–58.
35. Hulme, P.E.; Bacher, S.; Kenis, M.; Klotz, S.; Kühn, I.; Minchin, D.; Nentwig, W.; Olenin, S.; Panov, V.; Pergl, J.; Pyšek, P.; Roque, A.; Sol, D.; Solarz, W.; Vilà, M. Grasping at the routes of biological invasions: a framework for integrating pathways into policy. J. Appl. Ecol. **2008**, *45*, 403–414.
36. Lockwood, J.L.; Cassey, P.; Blackburn, T.M. The more you introduce the more you get: the role of colonization pressure and propagule pressure in invasion ecology. Divers. Distrib. **2009**, *15*, 904–910.
37. Touza, J.; Dehnen-Schmutz, K.; Jones, G. Economic analysis of invasive species policies. In *Biological Invasions*; Nentwig, W., Ed.; Springer-Verlag: Berlin, Heidelberg, 2007; 353–366.
38. Hallman, G.J. Phytosanitary measures to prevent the introduction of invasive species. In *Biological Invasions*; Nentwig, W., Ed.; Springer-Verlag: Berlin, Heidelberg, 2007; 367–384.
39. Pluess, T.; Cannon, R.; Jarošík, V.; Pergl, J.; Pyšek, P.; Bacher, S. When are eradication campaigns successful? A test of common assumptions. Biol. Invas. **2012**, *14*, 1365–1378.
40. Veitch, C.R.; Clout, M. *Turning the tide: the eradication of invasive species*. IUCN SSC Invasive Species Specialist Group, Gland and Cambridge, 2002.
41. Frank, J.H.; Van Driesche, R.G.; Pitcairn, M.J. Biological control of animals. In *Encyclopedia of Biological Invasions*;

Simberloff, D.; Rejmánek, M., Eds.; University of California Press: 2011; 58–63.
42. Hoddle, M.S.; McCoy, E.D. Biological control of plants. In *Encyclopedia of Biological Invasions;* Simberloff, D.; Rejmánek, M., Eds.; University of California Press: 2011; 63–70.
43. Knochel, D.G.; Seastedt, T.R. Sustainable control of spotted knapweed. In *Management of Invasive Weeds*; Inderjit, Ed.; Springer: 2009; 211–225.
44. Hobbs, R.J.; Richardson, D.M. Invasion ecology and restoration ecology: parallel evolution in two fields of endeavour. In *Fifty Years of Invasion Ecology: The Legacy of Charles Elton*; Richardson, D.M., Ed.; Blackwell: 2011; 61–69.
45. Lowe, S.; Browne. M.; Boudjelas, S.; De Poorter, M. 100 of the World´s worst invasive alien species. A selection from the Global Invasive Species Database. ISSG, 2000.
46. EEA. SEBI Assessing biodiversity in Europe: the 2010 report. EEA, Copenhagen, 2010.
47. DAISIE. *Handbook of Alien Species*; Springer; Europe, 2009.
48. http://plants.usda.gov/java/noxiousDriver (accessed March 2012)
49. http://www.weeds.org.au/noxious.htm (accessed March 2012)
50. Simberloff, D. Charles Elton: neither founder nor siren, but prophet. In *Fifty Years of Invasion Ecology: The Legacy of Charles Elton*; Richardson, D.M., Ed.; Blackwell, 2011; 11–24.
51. Hulme, P.; Pyšek, P.; Nentwig, W.; Vilà, M. Will threat of biological invasions unite the European Union? Science **2009**, *324*, 40–41.
52. Aikio, S.; Duncan, R.P.; Hulme, P.E. Lag-phases in alien plant invasions: Separating the facts from the artefacts. Oikos **2010**, *119*, 370–378.
53. Essl, F.; Dullinger, S.; Rabitsch, W.; Hulme, P.E.; Hülber, K.; Jarošík, V.; Kleinbauer, I.; Krausmann, F.; Kühn, I.; Nentwig, W.; Vilà, M.; Genovesi, P.; Gherardi, F.; Desprez-Lousteau, M.-L.; Roques, A.; Pyšek, P. Socioeconomic legacy yields an invasion debt. Proc. Natl. Acad. Sci. USA **2011**, *108*, 203–207.

Fires: Wildland

Jian Yang
State Key Laboratory of Forest and Soil Ecology, Institute of Applied Ecology, Chinese Academy of Sciences, Shenyang, China

Abstract
Wildland fire is any fire burning vegetation (e.g., forests, grasslands) that occurs in wildland areas. Earth has had wildland fires for more than 400 million years, with about 2% of terrestrial land being burned annually in the modern period. Although wildland fires have been viewed as a damaging agent to nature and social systems, recent advances in fire ecology have raised awareness that wildland fire is an important natural disturbance that plays an important role in shaping global biome distribution, evolution of species, plant traits, and biodiversity. This entry discusses causes, characteristics, consequences, and management of wildland fire. Special focus areas include dominant controls of wildland fire (i.e., fire triangles) at multiple spatiotemporal scales, parameters to describe characteristics of a single fire event (i.e., fire behavior) and recurring fire events in a landscape over an extended period of time (i.e., fire regime), the ecological effects of wildland fire, and the history of wildland fire policy.

INTRODUCTION

Wildland fire is any fire burning vegetation that occurs in wildland areas.[1] It is a worldwide phenomenon since the appearance of terrestrial plants about 400 million years ago.[2] Wildland fires co-evolved with vegetation and has become an integral component of the Earth systems,[3] helping to shape global biome distribution[4] and plant traits.[5] It also influences climate systems via feedback to land–atmosphere energy exchange.[6] Wildland fires are widely distributed in many terrestrial ecosystems (Fig. 1), particularly in regions where the climate is sufficiently moist to allow for growth of vegetation but also characterized by extended dry, hot periods (e.g., Mediterranean and tropical ecosystems).[7] It has been estimated that about 2% of terrestrial land is burned annually in the modern period.[8] Due to climate warming and fuel buildup resulting from fire exclusion, many places in the world have experienced more and larger wildland fires over the past few decades.[9] Consequently, wildland fire research has drawn considerable attention in recent years.

Wildland fires are regulated by a wide range of controls that vary greatly across the gradients of spatial and temporal scales. Fire triangles are often used to depict dominant controls at different scales. The characteristics of a single fire event are described by fire behavior parameters, while the characteristics of recurring fire events in a large landscape over an extended period of time are described by fire regime parameters. Wildland fire is an important natural disturbance, which greatly affects ecosystem productivity, biodiversity, forest landscape dynamics, and climate systems.[10] Humans have traditionally thought that fire was a damaging agent to nature, human lives, and economy. Recent advances in fire ecology have raised public awareness that fire is also an integral component of the Earth systems and has helped foster a shift in fire management policy from fire exclusion to prescribed burning, mechanical fuel treatment, and increased utilization of the existing Wildland Fire Use Policy.[11] This entry discusses causes (and controls), characteristics, consequences, and management of wildland fire.

CAUSES AND CONTROLS

Wildland fire occurs as a function of a suitable combination of biomass resources (fuel quantity and fuel quality such as fuel moisture and fuel continuity), conditions conducive to combustion and spread (e.g., fire weather and climate), and ignition agents.[12] Causes of wildland fire starts can be either natural (e.g., lightning, volcanic eruptions, sparks from rockfalls, and spontaneous combustion) or anthropogenic (e.g., arson, discarded cigarettes, sparks from equipment and power line arcs, out-of-control prescribed burns).[13] There are usually more human-caused than naturally caused wildland fires in many regions during the modern periods. For example, lightning-caused fires contributed only 35% of the fire occurrences in the boreal forest of Canada[14] and less than 1% of the fires in the Central Hardwood Forest of Missouri, United States, reported in the late 20th century.[15]

Although ignition is a necessary precursor to wildland fire, wildland fire is not a necessary consequence of an ignition source. The factors that regulate wildland fire activity are more complex; subsequent fire growth is mainly controlled by weather conditions, topography, fuel

Encyclopedia of Natural Resources DOI: 10.1081/E-ENRL-120049031

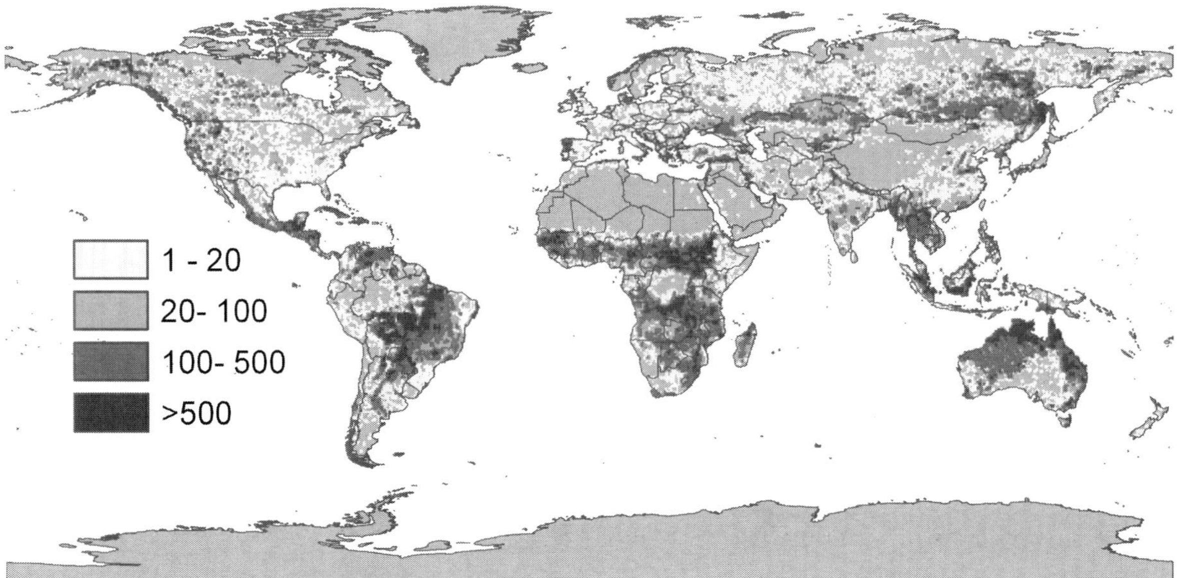

Fig. 1 Spatial distribution of fire occurrence density (the number of active fire counts detected from the satellite using ATSR instrument between 1996 and 2006 within a 0.5 × 0.5 degree grid cell).
Source: Data adapted from http://due.esrin.esa.int/wfa/.

amount, and fuel continuity. Dominant controls of wildland fire process vary at different spatial and temporal scales, often conceptualized as *fire triangles* (Fig. 2). For example, at the finest microsite scale, the three dominant controls that determine a combustion process are fuel, heat, and oxygen. For a single wildland fire event, the three legs of the fire triangle are fuel, weather, and topography. At the landscape scale, the three dominant factors are vegetation, climate, and ignition.[16,17]

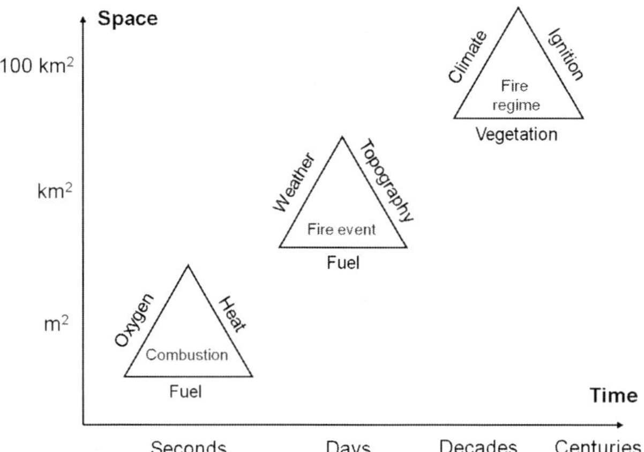

Fig. 2 Dominant controls of wildland fire at multiple temporal and spatial scales conceptualized as fire triangles.
Source: Modified from Moritz et al.,[16] Whitlock et al.[17]

CHARACTERISTICS OF WILDLAND FIRE

Characteristics of a single wildland fire event are often described by fire behavior parameters such as duration, flame length and intensity (i.e., rate of released energy), rate of spread, and fire type (e.g., ground fire, surface fire, crown fire). Ground fire consumes the organic material beneath the surface litter ground, while surface fire burns loose debris (e.g., dead branches, leaves, and low vegetation) on the surface, and crown fire advances from top to top of trees or shrubs independent of a surface fire (http://www.nwcg.gov/pms/pubs/glossary/index.htm). In contrast, a *fire regime* is used to describe characteristics of cumulative wildland fire events over a broad spatiotemporal domain. Parameters to describe a fire regime include, but are not limited to, frequency, size, seasonality, and severity. *Fire frequency* is the number of fires per unit time in a specific point. *Fire return interval* is the average number of years between fires at a given location. *Fire rotation*, called by some authors the *fire cycle*, is the number of years necessary for the cumulative burned area to equal the entire area of interest. *Fire severity* is defined broadly as the immediate effects of fire on various components of an ecosystem (e.g., vegetation and soil), which is open to individual interpretation. Empirical studies have often defined fire severity operationally as the loss of or change in organic matter aboveground and belowground, although the precise metric varies with management needs.[18] Fire severity differs from *fire intensity* in that the latter term describes only the energy released during various phases of a fire. Fire severity describes in large part the mortality of dominant vegetation and changes in the aboveground structure of the plant community. Severe fires in forests are often called "*stand-replacement*" fires (Fig. 3).

Fire regimes often vary significantly in different ecosystems and climatic conditions. Wildland fire interacts with ecosystem structures to create fire regimes. For example, low- to mid-elevation ponderosa pine forests in the western United States that were formerly open woodlands with

Fig. 3 Stand-replacement wildfire in a Chinese boreal forest.

abundant, contiguous fine fuels in the understory historically experienced low- and mixed-severity fire regimes with average fire return intervals of 5–30 years.[19] High-elevation lodgepole pine, Douglas fir, and subalpine fir forests may exhibit dense stand structures and historically have experienced mixed and stand-replacement fire regimes. Fire regime can be relatively constant if climate and vegetation are in an equilibrium state. However, due to anthropogenic activities and climate changes, fire regimes are being modified rapidly in many ecosystems.[10]

ECOLOGICAL EFFECTS OF WILDLAND FIRE

Wildland fires can exert strong effects on the abiotic and biotic components of an ecosystem, particularly the soil and vegetation. Many physical, chemical, mineralogical, and biological soil properties can be affected by wildland fires.[20] Extended residence time of burning, which is often a result of high levels of coarse fuels, may raise soil temperature substantially. After a burning, wildland fire can still lead to increased diurnal soil surface temperature by removing the shading effect of vegetation and insulating effect of litter, as well as by reducing albedo of the soil surface. It may also lead to greater cooling at night through the loss of radiative heat. Soil moisture may be increased in a burned environment where aboveground vegetation is largely consumed due to reduced rain interception and transpiration by plants. However, exposure to sunlight, wind and evaporation could cause soil to dry as well. Wildland fire can increase soil bulk density, reduce water holding capacity, and create a discreet or continuous water-repellent crust at the soil surface, all of which will lead to surface run-off and soil erosion.[1]

High-severity wildland fires generally cause significant removal of soil organic matter and considerable loss of nutrients through volatilization, ash entrapment in smoke columns, leaching, and erosion. Soil pH is often increased after fire due to acid combustion. Wildland fires can also cause a marked alteration of both quantity and composition of microbial and soil-dwelling invertebrate communities. However, the prefire level of most soil properties can be recovered and even enhanced within a relatively short time after the fire if plants succeed in promptly recolonizing the burnt area.[20]

Wildland fire is a global "herbivore" in that it feeds on vegetation (as do herbivores) and converts them to organic and mineral products.[4] As a consumer that has broad dietary preferences, wildland fire interacts with resources (i.e., climate and soil) to determine plant distribution and abundance in flammable terrestrial ecosystems. Many plants exhibit functional traits (e.g., flammability and persistence) that are adaptive to a particular fire regime. For example, Ponderosa pine forests in the southwest United States subject to frequent low-severity surface fires have thick bark and insulated buds, which function to protect living tissues from heat damage; they also self-prune their lower dead branches, which ensure a gap in fuels between the dead surface litter and live canopy to discourage torching. In contrast, lodgepole pine and jack pine forests subject to a crown fire regime have a thin bark, retention of lower dead branches, and serotinous cones that synchronously release seeds following high-severity stand replacing fires. These traits are adaptive in fire-prone environments and are the key to providing resilience to specific fire regimes.[5] The continued existence of particular species and communities often requires wildland fires, but too few or too many fires may both drive the landscape characteristics outside their historical range and variability[21] and/or lead to degradation of natural communities.[22]

WILDLAND FIRE POLICY

Humans and their ancestors have exerted myriad influences on wildland fire. Modern humans can increase or decrease levels of wildland fire activity by forest fire policy making. In the United States, federal forest-fire management began in the late 19th century and focused on suppressing all fires on national forests with the goal of protecting timber resources and rural communities. Decades of fire exclusion have produced uncharacteristically dense forests in many areas. Some forests, which previously burned frequently in low-severity surface fires, are now experiencing fires of much higher intensity than historical levels. The prevailing policy began to be questioned in the 1940s, when many foresters recognized the value of light controlled burning to clear out understory vegetation and reduce fire hazards. U.S. fire policy has recognized fire as a critical ecosystem process that must be reintroduced to restore forested ecosystems, particularly in regions that historically experienced more frequent low-intensity fires.[23] Improved utilization of the existing Wildland Fire Use Policy can provide for careful and gradual reintroduction of fire into landscapes, with special consideration to diversity of fire regimes and necessary spatial scale and arrangement of fuel treatment.[11,24]

CONCLUSIONS

Wildland fires have been a great natural force in shaping global biome distribution, plant traits, biodiversity, and ecosystem functions. Humans have greatly affected natural fire disturbances both directly by igniting and suppressing fires, and indirectly by altering vegetation and climate. Fire behavior is a major aspect of a single wildland fire event, while fire regime describes characteristics of recurring wildland fire events over a longer period. Many plant species are adapted to specific fire regimes. However, modern fire regimes are currently undergoing rapid change due to climate warming and anthropogenic influences. Changing fire regimes may in turn lead to degradation of ecosystem properties and services, novel trajectories in landscape dynamics, and feedbacks to global drivers.

ACKNOWLEDGMENT

Funding support for this work is provided by the National Natural Science Foundation of China (project no. 41222004, 41071121, and 31270511) and the Hundred Talent Program of the Chinese Academy of Sciences.

REFERENCES

1. Whelan, R.J. *The Ecology of Fire*; Cambridge University Press: 1995.
2. Scott, A.C. The Pre-Quaternary history of fire. Palaeogeogr. Palaeoclimatol. Palaeoecol. **2000**, *164* (1), 281–329.
3. Bowman, D.M.J.S.; Balch, J.K.; Artaxo, P.; Bond, W.J.; Carlson, J.M.; Cochrane, M.A.; D'Antonio, C.M.; DeFries, R.S.; Doyle, J.C.; Harrison, S.P. Fire in the Earth system. Science **2009**, *324* (5926), 481–484.
4. Bond, W.J.; Keeley, J.E. Fire as a global "herbivore": the ecology and evolution of flammable ecosystems. Trends Ecol. Evol. **2005**, *20* (7), 387–394.
5. Keeley, J.E.; Pausas, J.G.; Rundel, P.W.; Bond, W.J.; Bradstock, R.A. Fire as an evolutionary pressure shaping plant traits. Trends Plant Sci. **2011**, *16* (8), 406–411.
6. Running, S.W. Ecosystem disturbance, carbon, and climate. Science **2008**, *321* (5889), 652–653.
7. Pyne, S.J. Fire in America. *A Cultural History of Wildland and Rural Fire*; Princeton University Press; 1982.
8. Giglio, L.; van der Werf, G.R.; Randerson, J.T.; Collatz, G.J.; Kasibhatla, P. Global estimation of burned area using MODIS active fire observations. Atmos. Chem. Phys. Discuss. **2005**, *5* (6), 11091–11141.
9. Westerling, A.L.; Hidalgo, H.G.; Cayan, D.R.; Swetnam, T.W. Warming and earlier spring increase western U.S. forest wildfire activity. Science **2006**, *313* (5789), 940–943.
10. Turner, M.G. Disturbance and landscape dynamics in a changing World Ecol. **2010**, *91* (10), 2833–2849.
11. Stephens, S.L.; Ruth, L.W. Federal forest-fire policy in the United States. Ecol. Appl. **2005**, *15* (2), 532–542.
12. Krawchuk, M.A.; Moritz, M.A. Constraints on global fire activity vary across a resource gradient. Ecology **2011**, *92* (1), 121–132.
13. NWCG [National Wildfire Coordinating Group]. Wildfire prevention strategies; National Interagency Fire Center: Boise, Idaho, USA, 1998.
14. Weber, M.G.; Stocks, B.J. Forest fires and sustainability in the boreal forests of Canada. Ambio **1998**, *27* (7), 545–550.
15. Yang, J.; He, H.S.; Shifley, S.R.; Gustafson, E.J. Spatial patterns of modern period human-caused fire occurrence in the Missouri Ozark Highlands. Forest Sci. **2007**, *53* (1), 1–15.
16. Moritz, M.A.; Morais, M.E.; Summerell, L.A.; Carlson, J.M.; Doyle, J. Wildfires, complexity, and highly optimized tolerance. Proc. Nat. Acad. Sci. U.S.A. **2005**, *102* (50), 17912–17917.
17. Whitlock, C.; Higuera, P.E.; McWethy, D.B.; Briles, C.E. Paleoecological perspectives on fire ecology: revisiting the fire-regime concept. Open Ecol. J. **2010**, *3*, 6–23.
18. Keeley, J.E. Fire intensity, fire severity and burn severity: A brief review and suggested usage. Int. J. Wildland Fire **2009**, *18* (1), 116–126.
19. Schoennagel, T.; Veblen, T.T.; Romme, W.H. The interaction of fire, fuels, and climate across Rocky Mountain forests. BioScience **2004**, *54* (7), 661–676.
20. Certini, G. Effects of fire on properties of forest soils: A review. Oecologia **2005**, *143* (1), 1–10.
21. Keane, R.E.; Hessburg, P.F.; Landres, P.B.; Swanson, F.J. The use of historical range and variability (HRV) in landscape management. Forest Ecol. Manag. **2009**, *258* (7), 1025–1037.
22. Hobbs, R.J.; Huenneke, L.F. Disturbance, diversity, and invasion: implications for conservation. Conserv. Biol. **1992**, *6* (3), 324–337.
23. NWCG [National Wildfire Coordinating Group]. *Review and Update of The 1995 Federal Wildland Fire Management Policy*; National Interagency Fire Center: Boise, Idaho, USA, 2001.
24. Finney, M.A. Design of regular landscape fuel treatment patterns for modifying fire growth and behavior. Forest Sci. **2001**, *47* (2), 219–228.

Forests: Temperate Evergreen and Deciduous

Lindsay M. Dreiss
John C. Volin
Department of Natural Resources and the Environment, University of Connecticut, Storrs, Connecticut, U.S.A.

Abstract

Temperate forests represent one of the major biomes on Earth. They are most common in eastern North America, western and central Europe, and northeastern Asia, where the climate is defined by warm summers, cold winters, and intermediate levels of precipitation. To a lesser extent, they are also present in this same climate in Australia, New Zealand, South America, and South Africa. Temperate forests are dominated by either deciduous or evergreen canopies. In temperate deciduous forests, more commonly found in the Northern Hemisphere, plants drop their leaves in autumn, allowing for high seasonal variation in light availability to the understory. By contrast, in temperate evergreen forests, which are more commonly found in the Southern Hemisphere, plants keep their leaves year round. The temperate forest biome is rich in geologic and anthropocentric history, with much of the land shaped by glacial and human activity. Located in some of the most industrialized places in the world, temperate forests have been instrumental to socioeconomic development in these regions. As a consequence, these forests have been heavily impacted by expanding human populations, which remain the primary continuous threat to the structure and function of these forests.

INTRODUCTION

Temperate forests represent one of the major biomes on Earth, covering ~14% of Earth's terrestrial land surface[1] (Fig. 1). Along with boreal and subtropical/tropical wet and dry forests, temperate forests are one of the Earth's dominant forest types. In comparison to other types, temperate forests are intermediate in latitude, temperature, and precipitation. By contrast, they generally exhibit much stronger seasonality, favoring those plant and animal species that were able to adapt to short-term climatic variation. Environmental stresses from seasonal change include extreme temperatures and variable access to moisture (in the form of rain and/or snow), light, and nutrients (due to the deciduous nature of some canopies). As a result, significant changes in microclimates and growing conditions throughout the year are common, creating adaptive opportunities and resulting in unique structure and composition of ecosystems.[2–4]

Temperate forests are located in some of the most heavily populated and developed regions on Earth, including much of eastern North America, western and central Europe, northeastern Asia and, to a lesser extent, Australia, New Zealand, South America, and South Africa[5–7] (Fig. 1). Temperate forests are an established part of the history and culture of these regions and continue to play an important role in providing ecosystem services to human populations, including timber products, carbon storage, clean drinking water, erosion prevention, recreation, tourism, aesthetics, property value security, and others. On the other hand, the unsustainable use of such services has led to negative consequences for temperate forest ecosystems worldwide.

In this entry, we provide an overview of temperate forests from both an ecological and societal perspective. We describe the unique characteristics of temperate deciduous and evergreen forests, the major anthropogenic agents that have impacted them in the past and some of the emerging threats they face today.

CLIMATE

Climate is a fundamental factor in defining the temperate forest biome. Generally, a forest is considered temperate if it is located in a region with hot summers and cold winters. The annual range in temperature can be ±30°C, and the mean annual temperature is between 3 and 18°C, with mean midwinter temperature below 8°C and mean midsummer temperature about 18°C.[8] In the Northern Hemisphere, the length of the frost-free period ranges from 120 to over 250 days, with growing season and temperature increasing closer to the equator. As a result of less land mass in the Southern Hemisphere, the climate is milder than at similar latitudes in the Northern Hemisphere. Temperatures in the Southern Hemisphere can be cold, but typically far less than 1% of the hours of the year are

Encyclopedia of Natural Resources DOI: 10.1081/E-ENRL-120047447

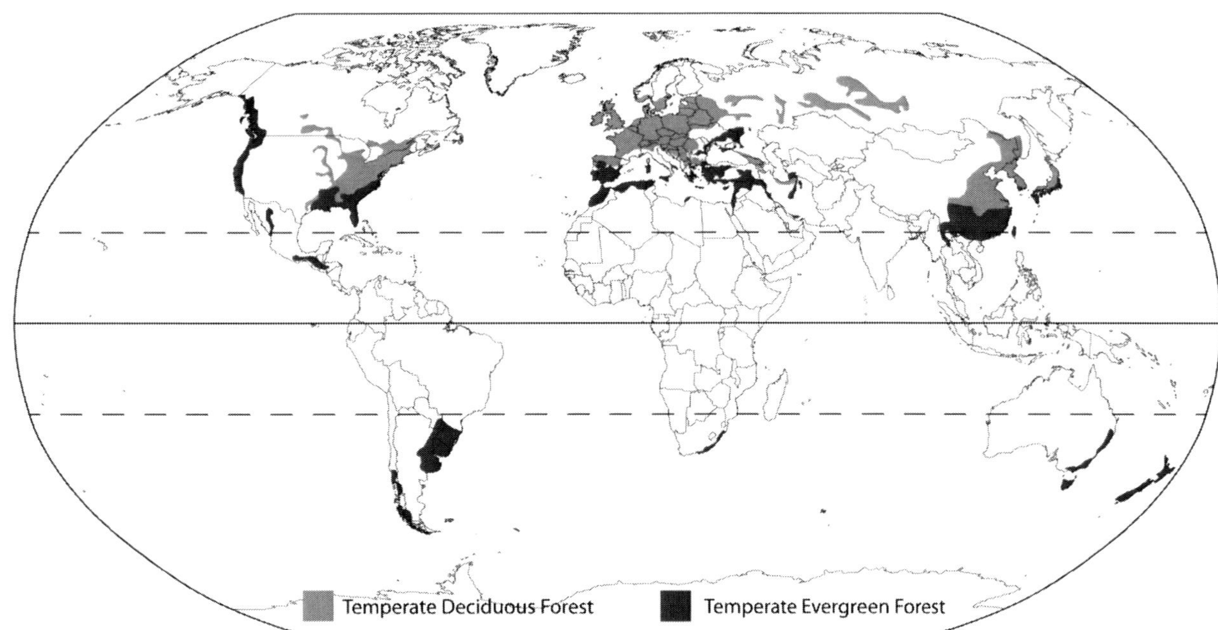

Fig. 1 Global temperate deciduous and evergreen forest biomes.

subject to frost, while in the Northern Hemisphere subfreezing temperatures are typically reached many days during the winter.[6]

Temperate forests receive even precipitation year round averaging 750–1500 mm/yr. Snowfall can range from nonexistent in southern regions to extremely heavy in northern regions. In the Northern Hemisphere, temperate forests are geographically located between the boreal forest/tundra and the tropical/subtropical forests while in the Southern Hemisphere they are found south of the tropical/subtropical forests up to tree line in some areas (Fig. 1). Correspondingly, the climate, productivity and species diversity of temperate forests are intermediate in relation to other biomes.[9]

The seasonal climatic regime in the temperate forest regions has changed very little in the past 1000–2000 years. This is reflected in the structure and ecosystem dynamics of the forests.[10] Winter deciduous trees or coniferous evergreen species that are physiologically dormant during winter months dominate the majority of the forests. Genera differ between Northern and Southern hemispheres. Examples include *Acer*, *Fagus*, *Tilia*, *Quercus*, *Carya*, *Populus*, *Ulnus*, *Betula*, *Fraxinus*, *Magnolia*, *Cornus*, *Robinia*, and *Juglans* in the Northern Hemisphere and *Eucalyptus*, *Acacia*, *Quercus*, and *Nothofagus* in the Southern Hemisphere.

At higher latitudes and altitudes, where temperatures range from –30 to 20°C annually, temperate forests become more dominated by coniferous evergreen tree species. These regions are generally drier, with annual precipitation from 300 to 900 mm. Seasons are defined by cold, long snowy winters and warm, humid summers with at least four to six frost-free months.[11] Evergreen temperate forests may also occur in regions with a stronger maritime climatic influence. Precipitation can be very high (typically 700–1000 mm) and snow is uncommon.[12,13] Summer fog is often an important contributor to moisture uptake, reducing stresses from evapo-transpiration. In addition, the growing season is much longer and sometimes continuous in forests closer to the equator. In the midst of global climate change, however, a shift is occurring in the location of ecotone and temperate forest species.

Climate-linked range shifts have been observed in the northern hardwood–boreal forest ecotone where there has been a decrease in boreal and an increase in northern hardwood basal area.[14] The velocity of temperature change is projected to be much faster in temperate broadleaf and mixed forests than in temperate coniferous forests.[15] Temperature being considered the primary control on treeline formation and maintenance, global warming is also facilitating treeline advances to greater altitudes and latitudes.[16,17] The synergistic effects of warming temperatures and intensifying droughts have also triggered concerns. The sensitivity of tree mortality and range shifts can result from extreme water stress and lead to other ecological consequences including changes in carbon stores and dynamics, changes in microclimate, and changes in future production of important habitat and food sources.[18,19]

LAND FORMATION AND HISTORY

On the geological time scale, climate has shaped the land on which temperate forests now grow. Between 110,000 and 10,000 years ago, during the Late Pleistocene age, glacial ice sheets left large parts of the Eurasian and North American continents as treeless, frozen tundra.[20,21] Below the tundra stretched the temperate forests, and beyond these, temperate, grassy plains. As the glaciers retreated, they left behind glacial till and lake-bottom sediments. As the zones of tundra followed the shrinking glaciers, the boreal and temperate forests encroached on the tundra belt until the matrix of ecoregions looked much like it does today. Since the end of the last ice age, human activity has become a primary factor in shaping temperate forests.

The three largest temperate forest regions, eastern North America, western and central Europe, and northeastern Asia, have played an important role in the development of human society. In North America, temperate forests were heavily impacted by the spread of European settlements in the 1700s. Much of the temperate forest was cleared, logged, and/or burned to make way for agricultural needs and human settlements.[22,23] In the 18th century, the temperate forests of Europe were the birthplace of new lines of scientific study and management, becoming what is now modern forestry.[24,25] These practices were applied in Europe and North America, where temperate forests provided the wood for fuel, building materials, and raw material in industrial processing during the Industrial Revolution of the 19th and 20th centuries.[26,27] With the spread and development of forestry practices, large amounts of temperate forest were logged for the production of timber, pulp, paper goods, and other forest products. During the 19th and into the 20th century, much of the agricultural land in the temperate forest range in North America was abandoned and the temperate forest grew back, peaking in its coverage in the mid-twentieth century.[28] Today it is once again under threat from continuous loss and fragmentation primarily through human developments.

In Europe, many native temperate deciduous forests have been extirpated and replaced with planted monocultures, many of which are not representative of the native vegetation.[29] Most of the remaining European temperate forests display a simplified structure and composition resulting from centuries of tree harvesting and forest grazing. In other places, land use change and conversion still continue to affect forest ecosystems. For example, the temperate forests of Asia, which cover parts of China, Japan, and Korea, were largely unaffected by glaciation through the Pleistocene.[30,31] Because of this, Asian temperate forests have higher levels of biodiversity than their counterparts in Europe or North America and are home to species with ancestries that extend much further back in geologic history. However, more recent heavy logging and fires have reduced forested land area and caused species such as the red-crowned crane (*Grus japonensis*) and the red panda (*Ailurus fulgens*) to be listed as endangered.[32] Collectively, temperate forests across the globe may have experienced more severe impacts over broader areas than any other large forest biome.[33]

STRUCTURE AND COMPOSITION

In contrast to grasslands or desert ecosystems, temperate forests have a more complex structure composed of different layers of vegetation. Temperate forest layers typically include a layer of mosses and lichens on the forest floor (and often on tree trunks and limbs), herbaceous and shrub layers, a subcanopy of trees, and a taller dominant canopy tree layer.[34] The complexity of the subcanopy and canopy layers can vary and often depend on the light requirements of the dominant tree species. Trees are the dominant life form of these ecosystems and can be either deciduous or evergreen.

Temperate forests undergo different stages in ecosystem succession, allowing for the coexistence of plant species with different life history strategies. These adaptations are often linked to specific growing conditions associated with different stages of forest growth. As a result, gradual transitions in plant community composition and forest structure change over time.[35,36] A classic example of succession in northeastern temperate forests in North America is when pioneer tree/shrub species such as birch (*Betula* spp.) and poplar (*Populus* spp.), which are fast growing and require high light availability, are often the first to establish following a disturbance. As trees grow and the canopy begins to close, conditions become more favorable for more long-lived, shade-tolerant species such as maple (*Acer* spp.), oak (*Quercus* spp.), beech (*Fagus* spp.), and hemlock (*Tsuga* spp.). These later-successional trees become dominant or codominant and eventually out-compete early successional species, reaching heights of 35–40 m.[37] This process takes place over hundreds of years and can be reset at any time by a disturbance of natural or human origin. Indeed, within a relatively large and healthy naturally occurring forest area, all of these successional stages are present at once.

Disturbance regimes, which refer to the type, intensity, and frequency of disturbances in a particular region, have a substantial impact on the composition and age structure of vegetation.[38–40] In temperate forests, eolic events such as hurricanes, tornadoes, and thunderstorm downbursts are a major contributor to forest dynamics. Fire was once also an important natural regulator, but with land use changes and suppression of fire by humans, many temperate forests have become increasingly dominated by late successional species.[41] Other natural disturbances include pest/disease outbreaks, flood or drought, and snow and landslides.

Temperate Deciduous Forest

The temperate deciduous forest canopy is composed mostly of broadleaved angiosperm tree species with the inclusion of some coniferous tree species. Temperate deciduous broadleaved forests are often defined by the associations of species. Association names are most useful for distinguishing broad differences in forest type and are often representative of variation in soils, topography, and climate. For example, in eastern North America, some common associations are

Oak–Hickory and Beech–Maple.[42] The temperate deciduous broadleaved forest that extends across East Asia (from 30° to 60°N) displays the greatest diversity of vegetation. The greater diversity in Asian temperate deciduous forests is likely the result of being less severely glaciated during the last ice age. This was largely a result of the more connected landmass, which allowed temperate forests to retreat from the advancing glaciers. For instance, in North America, the Florida peninsula and Gulf of Mexico coastal region was the southern limit for temperate forest refuge, while in Europe the east to west mountain range of the Alps limited southern migration. Other locations of deciduous broadleaved forests include areas around the eastern Black Sea, mountainous regions in Iran and Caspian Sea, and a narrow strip of South America including southern Chile and Argentina.[34,43–45]

The composition of deciduous temperate forest depends on latitude. In North America, birches (*Betula* spp.), aspen (*Populus* spp.), and maples (*Acer* spp.) are common, along with needle-leaved conifers such as pines (*Pinus* spp.). Coniferous species tend to decline in abundance further south, but they remain in localized areas that exhibit drier, more nutrient-poor conditions and higher altitudes. In Europe, forests at higher elevations are composed mainly of conifers (*Picea* spp. and *Abies* spp.) and *Fagus* spp., as well as larch (*Larix decidua*).

In central temperate deciduous forests, broadleaved deciduous species in Europe include oaks (*Quercus* spp.), European beech (*Fagus sylvatica*), and hornbeam (*Carpinus betulus*) [with *Fraxinus* spp. and *Castanea* spp. often just as common as *Carpinus*]. Common trees in North America include sugar maple (*Acer saccharum*), American beech (*Fagus grandifolia*), basswoods (*Tilia* spp.), oaks, and hickories (*Carya* spp.). In Asia, Japanese beech (*Fagus crenata*), oaks, maples, ashes (*Fraxinus* spp.) and basswoods are most common. At lower latitudes, broadleaved deciduous species are joined by broadleaved evergreen angiosperm trees, such as oaks, beeches (*Northofagus* spp.), eucalyptus (*Eucalyptus* spp.), and pines (*Podocarpus* spp.).

Plant life

Deciduous leaves are the most distinctive feature of temperate deciduous broadleaved forests. Autumn changes in leaf color and senescence have become very important to these regions both economically and ecologically. "Leaf-peeping" tourists travel to deciduous forest destinations to view displays of reds, oranges, and yellows and invest in local markets.[46,47] The dropping of leaves in the fall is also a phenological event or recurring biological cycle that changes the seasonal microhabitat, creating unique understory conditions for other plants and animals.

The onset of plant phenophases such as spring leaf flush and autumn leaf senescence are controlled by two major environmental signals: photoperiod (relative length of days/nights) and total degree days (relative measure of heating/cooling). However, with the onset of global climate change, recent studies recognize that temperature plays a more important role in many phenological events.[48,49] Plant hormonal signals to leaves cause chlorophyll, the compound that absorbs light for photosynthesis and makes the leaves look green, to degrade. Some nutrients from the leaves are partially recovered and moved (i.e., retranslocated) to stem and trunk tissues before the leaves fall off.[50–53] Buds are formed and go into dormancy until the following spring, when the accumulation of degree days over a temperature threshold triggers leaf growth. Canopy tree species differ in their timing of bud break and leaf expansion, with the development of the forest canopy taking place over a period of a month or more.[54] Once intact canopies are in full leaf flush and the growing season is underway, typically only 1–5% of sunlight reaches the forest floor.[55–57] Therefore, the portion of the year when canopy trees are leafless creates a unique opportunity of increased light availability for understory vegetation.

With the beginning of the warm season, herbaceous plants, shrubs and small trees begin growth before the canopy does. Because maximum incident solar radiation flux and zenith angle of the sun are larger in the spring months, there is greater transmission of light and a greater likelihood for understory plants to reach photosynthetic capacity before canopy leaf flushing.[58,59] In an evolutionary adaptation designed to maximize the amount of light received, some plants, known as spring ephemerals, complete most of their annual growth cycle in the few weeks between snowmelt and the closure of the tree canopy. Other more shade-tolerant species begin or continue to grow after canopy closure.[60]

Animal life

Many animals in temperate deciduous forests are mast-eaters (nuts and acorns) or omnivores. Herbivores (e.g., deer, hare, and rodents) perform important tasks in the ecosystem, such as aiding in plant seed dispersal, but may cause stress when their populations are overly abundant. Small rodents, such as squirrels and chipmunks (*Tamias* spp.), and some birds, such as blue jays (*Cyanocitta cristata*), cache large seeds/nuts in the autumn for sustenance in the winter.[61] Seeds that animals fail to recover may grow in their buried location. Omnivores include raccoon, opossum, skunk, fox, and black bear in North America and badger, mink, and marten in Europe. Asian forest fauna also include primates such as macaques. In temperate forests around the world, humans have drastically reduced the populations of many large carnivores such as wolves in North America and Europe and tigers in Asia, and now smaller carnivores such as coyote (*Canis latrans*) and lynx (*Lynx* spp.) fulfill part of this niche.[62,63]

Some animals have adapted unique behaviors to better suit them for the varied seasons. For example, some mammals store fat and hibernate or go into torpor during the cold months. Some birds take shelter in the trees by becoming cavity-nesters, while others are migratory and avoid the cold winters by flying to more southerly climes.

Soils and nutrients

Temperate deciduous forest soils show variation with latitude and altitude, as well as with moisture. At the northern or higher extent of the range, where there is a greater presence of needle-leaf evergreen tree species, soils tend to be relatively acidic and nutrient poor. By contrast, soils are strongly leached and less productive in the southern or lower ranges. Temperate deciduous forest soils, however, tend to be deep and fertile and do not freeze year-round. Brown forest soils (alfisols, in the American soil taxonomy) develop under the nutrient-demanding broadleaf trees whose leaves bind the major nutrient bases.[64,65] This causes leaf litter to be less acidic and provide a rich humus layer. Along with the favorable summer growing season, this is one of the reasons why this biome has historically been popular for agricultural use.

Temperate Evergreen Forest

Temperate broadleaved evergreen forests occur at the warm, moist end of the latitudinal gradient of temperate forests and are characterized by long, hot, humid growing seasons, and relatively milder winters. Precipitation is high (1000–1750 mm) and the forest is dominated by broadleaved evergreen angiosperms. These forests occur in eastern Asia, coastal regions of New Zealand and Australia, western coastal regions of North America and parts of Chile, South Africa, and warmer European regions.[34, 43–45] There are several forest variants that occur in regions of maritime and dry-season climates. Maritime broadleaved forests are dominated by tall, broadleaved evergreen angiosperms and narrow- to broadleaved conifers and exhibit very high precipitation (2000–3000 mm). Maritime needle-leaved forests occur only in a narrow band along the coastline of North America from California to Alaska and throughout coastal southern Europe and are known for needle- and scale-leaved evergreen conifers such as giant sequoias in North America or angiosperms such as evergreen oaks in Europe. Dry-season broad-sclerophyll forests occur in areas with periodic drought, exhibiting more of a Mediterranean climate, with hot dry summer and cool wet winters.

Needle-leaved conifers can be interspersed within deciduous canopies or occur in patches. They tend to grow at higher altitudes near treeline or in drier regions. Conditions can be highly seasonal, with severe winters and warm, humid summers. Conifer tree species include fir (*Abies* spp.), spruce (*Picea* spp.), pine (*Pinus* spp.), cedar (*Thuja* spp.), hemlock (*Tsuga* spp.), and larch (*Larix decidua*) among others.

Plant life

While broadleaved trees in the Northern Hemisphere are exposed to variable environmental stress, broadleaved trees in the Southern Hemisphere are under weaker seasonal forces. Because of milder climates and less extreme variability in the Southern Hemisphere, it is more advantageous for trees to expose leaves all year round in order to maximize photosynthesis and carbon gain.[66,67] Temperate broadleaved evergreen tree leaves are generally thick with smooth margins as opposed to deciduous leaves, which by comparison are generally thin and lobed. These differences are in part due to higher levels of rainfall and humidity. Southern Hemisphere species maintain a waxy outer layer on leaves to repel water. In these mild temperate evergreen forests, species compositions are mainly from the families of Lauraceae, Fagaceae, Theaceae, Magnoliaceae, and Hamamelidaceae, all of which have a relatively low resistance to cold and freezing temperatures.[68,69] In the Southern Hemisphere, the very dry-adapted forests of Australia are also found. Fire disturbance in this dry region has allowed for the domination of *Eucalyptus* spp. which have thick bark for fire-resistance.[70,71]

Also relatively common in many temperate evergreen forests are narrow-leaved evergreen trees. These species produce their seeds in compact structures called cones, which allow for greater protection of reproductive material from predation as well as some unique adaptations to environmental conditions.[72] Some evergreen conifers have flat, triangular scale-like leaves, such as found in the Cupressaceae and some Podocarpaceae, while many evergreen trees also produce needle-like leaves, which increase the leaf area for photosynthesis and are thicker and smaller to avoid desiccation.[73,74] Rates of area- or massed-based leaf photosynthesis in temperate needle-liked evergreen trees are typically much lower than for leaves of temperate deciduous tree species, except when amortized over the life of the leaf.[75] Temperate deciduous trees only keep their leaves for one growing season, while temperate evergreen forests keep their leaves for much longer.

The moist temperate coniferous forests dominated by tree species such as giant sequoias (*Sequoiadendron giganteum*), coastal redwoods (*Sequoia sempervirens*), Douglas firs (*Pseudotsuga menziesii*), and kauris (*Agathis australis*) sustain the highest levels of biomass in any terrestrial ecosystem.[76] These forests are rare and known for a rich variety of plant and animal species. Typically, two or more tree layers of large broadleaved, dome-shaped evergreen angiosperms are present. These forests, which grow in moist regions, are also known for higher abundance and diversity of epiphytes and vines.

In maritime climates, fog can provide a significant amount of the annual water available to vegetation.[77] Forming over adjacent oceans and water bodies, the moist air travels with both humidity and nutrients that can be intercepted by plants, reducing water stress caused by transpiration. Some studies suggest that some leaf shapes have evolved for more efficient collection of fog drip. It has also been hypothesized that fog in dry summer months is essential in supporting the sustained growth of large trees such as the redwoods of California.[78,79]

Temperate coniferous forests at higher altitudes often have narrow tree morphologies which allow heavy snow to slough off branches before breaking them.[72,73] These forests are dense, with trees growing close together to minimize wind damage; wind minimization is important because conifers often have very shallow root systems due to higher allocation of nutrients in the top layers of soil.[80,81] Evergreen trees in higher, drier areas allocate carbon to thick bark for greater protection from low-heat summer fires. Some conifers have adapted to take advantage of fire events, producing serotinous cones in which seeds are maintained in a dormant state until a fire occurs, opening the cones and releasing the seeds for germination.[82,83]

Because the canopy of these temperate forests is evergreen, the vegetation in the understory experiences low light availability. Understory plants have adapted to maximize their light/energy capture as well as their photosynthetic capacity under such conditions. Shaded plants tend to allocate more growth to leaves, displaying larger, thinner leaves which take up more horizontal area for light capture.[75,84] Physiologically, shade leaves saturate at lower light levels and have lower light compensation points, which allows them to maintain positive rates of net photosynthesis at low light levels. Understory plants also adapt to respond quickly to environmental changes in order to make use of sunflecks or brief, unpredictable periods of high sunlight usually lasting only seconds.[85–87] Instantaneous adjustments in photosynthetic rates allow plants to be efficient with the limited light they receive to put toward carbon gain.

Animal life

The fauna of the temperate evergreen forest is similar to that of the deciduous forest ecosystems, with many smaller mammals, large ungulates and a few larger omnivorous mammals contributing to biological interactions. Many of these species, mostly birds and small rodents, have adapted skills to compensate for the scarcity of food in these nutrient poor systems. The red squirrel (*Sciurus vulgaris*) removes cone scales to reach the nutrient rich seeds of the coniferous trees.[88,89] They often hoard these seeds, many of which they do not recover, resulting in seed dispersal. In dry forests at higher latitudes, large ungulates such as moose (*Alces alces*) browse nutritious spring vegetation and wetland foliage in warmer seasons and on young tree shoots when the ground is frozen and covered with snow.[90,91] Some fauna in the coniferous evergreen forests located at higher latitudes have also adapted to the cold, snowy winters. For example, bears (*Ursus* spp.) are able to go into long periods of dormancy, maintaining both body temperature and metabolism without eating, drinking, defecating, or urinating.[92] In austral regions, temperate evergreen forests are also home to many endemic species such as Leadbeater's possum (*Gymnobelideus leadbeaten*), Parma wallaby (*Marcopus parma*) and Albert's lyrebird (*Menura alberti*).

Soil and nutrients

Soils of temperate evergreen forests are well leached and low in productivity. Ultisols represent older, unglaciated soils that have been weathered to a much greater degree than those of temperate deciduous forests. Ultisols are generally less fertile than most temperate deciduous forest soils and were further degraded under plantation and subsistence agriculture in both the colonial and postcolonial periods.[65,93]

THREATS

Threats to the temperate forest biome stem from direct or indirect human activity. Historically, farming has been one of the main reasons for deforestation in temperate biomes. Today, an increasing number of roads, residential and other developments have left forests highly fragmented, raising concerns over the healthy functioning of these ecosystems.[94] Many of the plant species living in temperate forests are well-adapted and function optimally in soils with fairly low nutrient availability. However, more impervious surfaces as well as modern agricultural and industrial practices can alter nutrient availability and contribute to a decline in forest health. Atmospheric deposition of nitrogen, acid deposition and fertilizer applications are changing the rate of input of nutrients along with the growth and dynamics of temperate forest communities.[95–97] This also leads to alterations in soil properties, nutrient cycling and microbial community composition as well as nutrient imbalances in plants and increases in forest mortality. These nutrient imbalances in plants make them more vulnerable to other outside stressors such as diseases and introduced invasive species.[98–100]

Human activities are also altering global climate, a foundation for the characterization and distribution of temperate forests. For deciduous canopies, observational evidence has linked global climate change to differences in the timing of spring leaf growth and autumn leaf drop. Warm springs cause leaves to grow earlier, sometimes by up to one month (http://www.usanpn.org), lengthening the duration of the growing season in some temperate forests.[101–103] Climate variation may also result in a shift in flowering periods, potentially causing disruption in timing with pollinators/dispersers. In Japan, cherry blossom (*Prunus serrulata*) trees are blooming earlier than they did historically, causing concerns regarding the traditional festivals and tourism that surround the annual event.[104] Evidence of warming temperatures also suggests a shift in species distributions toward the poles, which could result in the extinction of some plant and animal species.[105,106]

Rapid global climate change creates conditions less suitable for many species adapted to their particular environment, increasing the amount of stress put on individuals as well as their vulnerability to certain pathogens or pests.[107] Climate also has a strong influence on the success of certain fungi, bacteria, and viruses, affecting infection severity and spread. For example, in the northwestern United States, heavy rains and wet weather during warm periods are optimal conditions for sudden oak death infection.[108] In many cases, humans accidentally introduce these diseases. Chestnut blight (*Cryphonectria parasitica*), a fungus introduced to America from Europe in 1904, is one of the best-known examples. In four decades, the chestnut blight largely eliminated American chestnut (*Castanea dentata*) from the continent, one of the most important and widespread tree species across the eastern temperate forest of North America.[109,110]

Other introductions to temperate forests include nonnative plant invaders that have, in some cases, taken over large portions of forested land, changing species, and nutrient compositions as well as the identity of the landscape.[111–113] In some cases, invasive species have the potential to reduce biodiversity, extirpate native species and negatively impact ecosystem structure and services, thereby negatively affecting the economy and human health and well-being. Species such as garlic mustard (*Alliaria petiolata*), native to Europe and introduced to forests of North America, exude chemicals that may disturb natural plant–plant and plant–soil processes and interactions.[114] Many invasives may be unpalatable to native fauna such as Japanese knotweed (*Fallopia japonica*)[115] which is nonnative to forests of Europe and North America or even poisonous like St. John's Wort (*Hypericum perforatum*) in Australia.[116] Negative effects of invasive plants on human populations include economic losses[117] and even disease as in the case of Japanese barberry (*Berberis thunbergii*), a shrub introduced to North America that creates favorable habitat for vectors of Lyme disease (*Borrelia burgdorferi*).[118]

CONCLUSION

Temperate forests are ecologically distinct biomes and provide important regional and global ecosystem services. The species in these ecosystems have evolved unique physiological traits to function under a variety of seasonal stresses. Being closely related to variation in environmental conditions, temperate forests may act as an ecological indicator of global climate change, changes in nutrient cycling and other environmental alterations brought on by human activities. Temperate forests have been and are of important use to industry and tourism. As they continue to play an important role in regional cultures and economies, sustainable use of these resources and proper protection and management of temperate forests will be important in ensuring the health and productivity of the biome.

REFERENCES

1. Thomas, S.C.; MacLellan, J. Boreal and Temperate Forests, in Forests and Forest Plants. In *Encyclopedia of Life Support Systems (EOLSS)*; UNESCO, Eolss Publishers: Oxford, UK, 2004.
2. Chabot, B.F.; Mooney, H.A., Eds.; *Physiological Ecology of North American Plant Communities*; Chapman and Hall: New York, NY, 1985.
3. Small, C.J.; McCarthy, B.C. Spatial and temporal variation in the response of understory vegetation to disturbance in a central Appalachian oak forest. J. Torrey Bot. Soc. **2002**, *129* (2), 136–153.
4. Gendron, F.C.; Comeau, P.G. Temporal variation in the understory photosynthetic photon flux density of a deciduous stand: the effects of canopy development, solar elevation and sky conditions. Agric. Forest Meteorol. **2001**, *106*, 23–40.
5. Woodward, F.I. *Climate and Plant Distribution*; Cambridge University Press: New York, NY, 1987.
6. Axelrod, D.I. Origin of deciduous and evergreen habits in temperate forests. Evolution **1966**, *20*, 1–15.
7. Food and Agriculture Organization of the United Nations. State of the World's Forests, 2009.
8. Vankat, J.L. Boreal and Temperate Forests. In *eLS*; John Wiley & Sons Ltd: Chichester, England, 2002.
9. Reich, P.B.; Bolstad, P. Temperate Forests: Evergreen and Deciduous. In *Terrestrial Global Productivity*; Mooney, H., Roy, J., Saugier, B., Eds.; Academic Press: San Diego, CA, 2001.
10. Graumlich, L.J. A 1000-year record of temperature and precipitation in the Sierra Nevada. Quarter. Res. **1993**, *39*, 249.
11. Elsevier. Coniferous Forests. In *Ecosystems of the World*, Vol. 6.; Elsevier: Amsterdam, the Netherlands, 2005.
12. Ovington, J.D. *Temperate Broad-Leaved Evergreen Forests*; Elsevier: Amsterdam, the Netherlands, 1983.
13. Ohte, N; Tokuchi, N. Hydrology and biogeochemistry of temperate forests. In *Forest Hydrology and Biogeochemistry: Synthesis of Past Research and Future Directions*; Levia, D.F; Carlyle, D; Tanaka, T., Eds.; Springer Science+Business Media: New York, NY, 2011.
14. Beckage, B.; Osborne, B.; Gavin, D.G.; Pucko, C.; Siccama, T.; Perkns, T. A rapid upward shift of a forest ecotone during 40 years of warming in the Green Mountains of Vermont. Proc. Nat. Acad. Sci. **2008**, *105* (11), 4197–4202.
15. Loarie, S.R.; Duffy, P.B.; Hamilton, H.; Asner, G.P.; Field, C.B.; Ackerly, D.D. The velocity of climate change. Nature **2009**, *462* (24), 1052–1055.
16. Holtmeier, F.K.; Broll, G. Sensitivity and response of northern hemisphere altitudinal and polar treelines to environmental change at landscape and local scales. Global Ecol. Biogeogr. **2005**, *14*, 395–410.
17. Harsch, M.A.; Hulme, P.E.; McGlone, M.S.; Duncan, R.P. Are treelines advancing? A global meta-analysis of treeline response to climate warming. Ecol. Lett. **2009**, *12*, 1040–1049.

18. Breshears, D.D.; Cobb, N.S.; Rich, P.M.; Price, K.P.; Allen, C.D.; Balice, R.G.; Romme, W.H.; Kastens, J.H.; Floyd, M.L.; Belnap, J.; Anderson, J.J.; Myers, O.B.; Meyer, C.W. Regional vegetation die-off in response to global-change-type drought. Proc. Nat. Acad. Sci. **2005**, *102* (42), 15144–15148.
19. Adams, H.D.; Guardiola-Claramonte, M.; Barron-Gafford, G.A.; Villegas, J.C.; Breshears, D.D.; Zou, C.B.; Troch, P.A.; Huxman, T.E. Temperature sensitivity of drought-induced tree mortality portends increased regional die-off under global-change-type drought. Proc. Nat. Acad. Sci. **2009**, *106* (17), 7063–7066.
20. Denton, G.H.; Hughes, T.J. The Arctic ice sheet: an outrageous hypothesis. In *The Last Great Ice Sheets*; Denton, G.H., Hughes, T.J., Eds.; John Wiley: New York, NY, 1981.
21. Cox, C.B.; Moore, P.D. *Biogeography: An Ecological and Evolutionary Approach*; John Wiley: New York, NY, 2010.
22. Foster, D.R. Land Use History (1730–1990) and Vegetation Dynamics in Central New England. J. Ecol. **1995**, *76*, 135–151.
23. Foster, D.R.; Motzkin, G.; Slate, B. Land-use history as long-term broad-scale disturbance: regional forest dynamics in central New England. Ecosystems **1998**, *1L*, 96–119.
24. Ferrow, B.E. *A Brief History of Forestry in Europe, the United States and Other Countries*; University Press: Cambridge, MA, 1991.
25. Volin, V.C.; Buongiorno, J. Effects of alternative management regimes on forest stand structure, species composition, and income: a model for the Italian Dolomites. Forest Ecol. Manag. **1996**, *87* (1–3), 107–125.
26. Perlin, J.A. *Forest Journey: The Role of Wood in the Development of Civilization*; Harvard University Press: London, UK, 1991.
27. Lehtinen, A.A.; Donnor-Amnell, J.; Saether, B., Eds. *Politics of Forests: Northern Forest-Industrial Regimes in the Age of Globalization*; Ashgate Publishing: Great Britain, 2004.
28. USDA Forest Service. *U.S. Forest Facts and Historical Trends*; FS-696. USDA Forest Service: Washington, DC, 2001.
29. Zerbe, S. Restoration of natural broad-leaved woodland in Central Europe on sites with coniferous forest plantations. Forest Ecol. Manag. **2001**, *167* (1–3), 27–42.
30. Grichuk, V.P. Late Pleistocene vegetation history. In *Late Quaternary Environments of Soviet Union*; Vwlichko, A.A., Eds.; University of Minnesota Press: Minneapolis, MN, 1984.
31. Ehlers, J; Gibbard, P.L. The extent and chronology of Cenozoic global glaciations. Quarter. Int. 2007, 164–165, 6–20.
32. http://www.iucnredlist.org (accessed March 2012).
33. Williams, M. *Americans and Their Forests*; Cambridge University Press: Cambridge, MA, 1989.
34. White, M.A. Deciduous Forests. Encyclopedia.com 2001 (accessed January 2012).
35. Thompson, J.N. Treefalls and colonization patterns of temperate forest herbs. Am. Midland Nat. **1980**, *104*, 176–184.
36. Rees, M.; Condit, R.; Crawley, M.; Pacala, S.; Tilman, D. Long-term studies of vegetation dynamics. Science **2001**, *293* (5530), 650–655.
37. Allaby, M. *Temperate Forests. Biomes of the Earth Series*; Chelsea House: New York, NY, 2006.
38. Oliver C.D.; Larson, B.C. Forest Stand Dynamics. In *Biological Resource Management Series*; McGraw Hill: New York, NY, 1990.
39. Schelhaas, M.; Nabuurs, G.; Schuck, A. Natural disturbances in the European forests in the 19th and 20th centuries. Global Change Biol. **2003**, *9* (11), 1620–1633.
40. Lorimer, C.G.; White, A.S. Scale and frequency of natural disturbances in the northeastern U.S. implications for early successional forest habitats and regional age distributions. Forest Ecol. Manag. **2003**, *185*, 41–64.
41. Aber, J.D.; Nadelhoffer, K.J.; Steudler, P.; Melillo, J.M. Nitrogen Saturation in Northern Forest Ecosystems. BioScience **1989**, *39*, 378–386.
42. Braun, E.L. *Deciduous Forests of Eastern North America*; Hafner Publishing Co.: New York, 1972.
43. Röhig, E.; Ulrich, B. Temperate Deciduous Forests. In *Ecosystems of the World*; Röhig, E.; Ulrich, B. Eds.; Elsevier: Amsterdam, the Netherlands, 1991.
44. Archibold, O.W. *Ecology of World Vegetation;* Chapman and Hall: London, UK, 1995.
45. Sayre, A.P. *Temperate Deciduous Forest.* Twenty-First Century Books: New York, NY, 1994.
46. Beckley, T.M. The nestedness of forest dependence: a conceptual framework and empirical exploration. Soc. Nat. Resour.: An Int. J. **1998**, *11* (2), 101–120.
47. Norris, G.A. *The Economical Impact of Climate Change on the New England Region.* New England Regional Assessment 2008.
48. Mahall, B.E.; Bormann, F.H. A quanititative description of the vegetative phenology of herbs in a northern hardwood forest. Bot. Gazette **1978**, *139*, 467–481.
49. Korner, C.; Basler, D. Phenology under global warming. Science **2010**, *327* (5972), 1461–1462.
50. Havrenek, W.M.; Tranquillini, W. Physiological processes during winter dormancy and their ecological significant. In *Ecophysiology of Coniferous Forests*; Smight, W.K., Hinckley T.M. Eds.; Springer-Verlag: Berlin, 1995.
51. Hutchison, B.A.; Matt, D.R. The distribution of solar radiation within a deciduous forest. Ecol. Monogr. **1977**, *47*, 185–207.
52. Reifsnyder, W.E.; Furnival, C.M.; Horowitz, J.L. Spatial and temporal distribution of solar radiation beneath forest canopies. Agric. Meteorol. **1971**, *9*, 21–38.
53. Hoch, W.A.; Singsaas, E.L.; McCown, B.H. Resorption protections. Anthocyanins facilitate nutrient recovery in autumn by shielding leaves from potentially damaging light levels. Plant Physiol. **2003**, *113* (3), 1296–1305.
54. Dreiss, L.M.; Volin, J.C. Influence of leaf phenology and site nitrogen on invasive species establishment in temperate deciduous forest understories. Forest Ecol. Manag. **2013**, *296*, 1–8.
55. Gill, D.S.; Amthor, J.S.; Bormann, F.H. Leaf phenology, photosynthesis and the persistence of saplings and shrubs in a mature northern hardwood forest. Tree Physiol. **1998**, *18*, 281–289.
56. Sefcik, L.T.; Zak, D.R.; Ellsworth, D.S. Photosynthetic responses to understory shade and elevated carbon dioxide concentration in four northern hardwood tree species. Tree Physiol. **2006**, *26*, 1589–1599.

57. Hattenschwiler, S.; Korner, C. Tree seedling responses to in situ CO_2-enrichment differ among species and depend on understorey light availability. Global Change Biol. **2000**, *6*, 213–226.
58. Richardson, A.D.; O'Keefe, J. Phenological differences between understory and overstory: a case study using the long-term Harvard Forest Records Book. In *Phenology of Ecosystem Processes: Applications in Global Change Research*; Noormets, A., Eds.; Springer: New York, NY, 2009.
59. Augspurger, C.K. Early spring leaf out enhances growth and survival of saplings in a temperate deciduous forest. Oecologia **2008**, *156*, 281–286.
60. Kudo, G.; Ida, T.Y.; Tani, T. Linkages between phenology, pollination, photosynthesis and reproduction in deciduous forest understory plants. Ecology **2008**, *89*, 321–331.
61. Li, H.; Zhang, Z. Effects of mast seeding and rodent abundance on seed predation and dispersal by rodents in *Prunus armeniaca* (Rosaceae). Forest Ecol. Manag. **2008**, *242*, 511–517.
62. Nowak, R.M. Evolution and taxonomy of coyotes and related *Canis*. In *Coyotes: Biology, Behavior, and Management*; Bekoff, M., Ed.; Academic Press: New York, 1978.
63. Thurber, J.M.; Peterson, R.O. Changes in body size associated with range expansion in the coyote (*Canis latrans*). J. Mammal. **1991**, *72* (4), 750–755.
64. Ulrichm B. Stability, elasticity and resilience of terrestrial ecosystems with respect to matter balance. In *Potentials and Limitations of Ecosystem Analysis*; Schulze, E.D., Zwölfer, H. Eds.; Springer Verlag: Berlin, 1987.
65. Hendrick, R.L. Forest Types and Classification. In *The Forests Handbook, Volume 1: An Overview of Forest Science*; Evans, J., Ed.; Blackwell Science Ltd: Oxford, UK, 2008.
66. Chabot, B.F.; Hicks, D.J. The ecology of leaf life spans. Annu. Rev. Ecol. Syst. **1982**, *13*, 229–259.
67. Hallik, L.; Niinements, U.; Wright, I.J. Are species shade and drought tolerance reflected in leaf-level structural and functional differentiation in Northern Hemisphere temperate woody flora? New Phytol. **2009**, *184*, 257–274.
68. Lieth, H.; Box, E.O.; Peet, R.K.; Masuzawa, T.; Yamada, I.; Fujiwara, K.; Maycock, P.F. *Handbook of Vegetation Science*; Kluwer Academic Publishers: the Netherlands, 1995.
69. Bannister, P. Godley review: a touch of frost? Cold hardiness of plants in the southern hemisphere. New Zealand J. Bot. **2007**, *45* (1), 1–33.
70. Beard, J.S. Temperate forests of the Southern Hemisphere. Vegio **1990**, *89* (1), 7–10.
71. Fire in ecosystems of south-west Western Australia: impacts and management. Symposium proceedings (Vol. 1): 225–268, Perth, Australia, 2002.
72. Leslie, A.B. Predation and protection in the macroevolutionary history of conifer cones. Proc. R. Soc. B **2011**, *278*, 3003–3008.
73. Sprugel, D.G. The relationship of evergreenness, crown architecture, and leaf size. Am. Nat. **1989**, *133* (4), 465–479.
74. Brodribb, T.J.; Field, T.S.; Sack, L. Viewing leaf structure and evolution from a hydraulic perspective. Func. Plant Biol. **2010**, *37*, 499–498.
75. Busing, R.T.; Fujimori, T. Biomass, productions and woody detritus in an old coast redwood (*Sequoia sempervirens*) forest. Plant Ecol. **2005**, *177*, 177–188.
76. Ewing, H.A.; Weathers, K.C.; Templer, P.H.; Dawson, T.E.; Firestone, M.K.; Elliott, A.M.; Boukili, V.K. Fog water and ecosystem function: heterogeneity in a California redwood forest. Ecosystems **2009**, *12*, 417–433.
77. Canadell, J.; Jackson, R.B.; Ehleringer, J.R.; Mooney, H.A.; Sala, O.E.; Schulze, E.D. Maximum rooting depth of vegetation types at the global scale. Oecologia **1996**, *108* (4), 583–595.
78. Burgess, S.S.O.; Dawson, T.E. The contribution of fog to the water relations of *Sequoia sempervirens* (D. Don): foliar uptake and prevention of dehydration. Plant, Cell Environ. **2004**, *27* (8), 1023–1034.
79. Simonin, K.A.; Santiago, L.S.; Dawson, T.E. Fog interception by *Sequoia sempervirens* (D. Don) crowns decouples physiology from soil water deficit. Plant, Cell Environ. **2009**, *32*, 882–892.
80. Enright N.J.; Marsula R.; Lamont B.B.; Wissel C. The ecological significance of canopy seed storage in fire-prone environments: a model for non-sprouting shrubs. J. Ecol. **1998**, *86*, 946–959.
81. Echenwalder, J.E. *Conifers of the World*; Timber Press, Inc.: Portland, OR, 2009.
82. Givnish, T.J. Adaptation to sun and shade: a whole plant perspective. Aust. J. Plant Physiol. **1988**, *15*, 63–92.
83. Midgley, J.; Bond, W. Pushing back in time: the role of fire in plant evolution. New Phytol. **2011**, *191* (1), 5–7.
84. Boardman, N.K. Comparative photosynthesis of sun and shade plants. Annu. Rev. Plant Physiol. **1977**, 28, 355–377.
85. Chazdon, R.L.; Pearcy, R.W. The importance of sunflecks for understory plants. BioScience **1991**, *41* (11), 760–766.
86. Brantley, S.T.; Young, D.R. Contribution of sunflecks is minimal in expanding shrub thickets compared to temperate forest. Ecology **2009**, *90*, 1021–1029.
87. Lertzman, K.P.; Sutherland, G.D.; Inselberg, A.; Saunders, S.C. Canopy gaps and the landscape mosaic in a coastal temperate rain forest. Ecology **1996**, *77*, 1254–1270.
88. Miyaki, M. Seed dispersal of the Korean pine, *Pinus koraiensis*, by the red squirrel, *Sciurus rulgaris*. Ecol. Res. **1987**, *2* (2), 147–157.
89. Wauters, L.A.; Gurnell, J.; Martinoli, A.; Tosi, G. Does interspecific competition with introduced grey squirrels affect foraging and food choice of Eurasian red squirrels? Animal Behav. **2001**, *61* (6), 1079–1091.
90. Hjeljord, O.; Sundstol, F.; Haagenrud, H. The nutritional value of browse to moose. J. Wildlife Manag. **1982**, *46* (2), 333–343.
91. Wam, H.K.; Hjeljord, O. Moose summer and winter diets along a large scale gradient of forage availability in southern Norway. Eur. J. Wildlife Res. **2010**, *56*, 745–755.
92. Hellgren, E.C. Physiology of hibernation in bears. Int. Assoc. Bear Res. Manag. **1995**, *10*, 467–477.
93. Mbagwu, J.S.C.; Lal, R.; Scott, T.W. Effects of desurfacing alfisols and ultisols in southern Nigeria: I. Crop Performance. Soil Sci. Soc. Am. J. **1984**, *48*, 828–833.
94. Foley, J.A.; DeFries, R.; Asner, G.P.; Barford, C.; Bonan, G.; Carpenter, S.R.; Chapin, F.S.; Coe, M.T.; Daily, G.C.; Gibbs, H.K.; Hellowski, J.H.; Holloway, T.; Howard, E.A.; Kucharik, C.A.; Monfreda,C.; Patz, J.A.; Prentice,

I.C.; Ramankutty, N.; Snyder, P.K. Global consequences of land use. Science **2005**, *309* (5734), 570–574.
95. Vitousek, P.M.; Aber, J.D.; Howarth, R.W.; Likens, G.E.; Matson, P.A.; Schindler, D.W.; Schlesinger, W.H.; Tilman, D.G Human alteration of the global nitrogen cycle: sources and consequences. Ecol. Appl. **1997**, *7* (3), 737–750.
96. Gundersen, P.; Schmidt, I.K.; Raulund-Rasmussen, K. Leaching of nitrate from temperate forests: effects of air pollution and forest management. Environ. Rev. **2006**, *14* (1), 1–57.
97. Aber, J.D.; Goodale, C.L.; Ollinger, S.V.; Smith, M.-L.; Magill, A.H.; Martin, M.E.; Hallett, R.A.; Stoddard, J.L. Is nitrogen deposition altering the nitrogen status of Northeastern forests? BioScience **2003**, *53* (4), 375–389.
98. Waring, R.H. Characteristics of trees predisposed to die. BioScience **1987**, *37* (8), 569–574.
99. Bytnerowicz, A.; Omasa, K.; Paoletti, E. Integrated effects of air pollution and climate change on forests: a northern hemisphere perspective. Environ. Poll. **2007**, *14*, 438–445.
100. Turner, J.; Lambert, M. Soil and nutrient processes related to eucalypt forest dieback. Aust. Forest. **2005**, *68* (4), 251–256.
101. Menzel, A.; Fabian, P. Growing season extended in Europe. *Nature* **1999**, *397*, 659.
102. Cleland, E.E.; Chuine, I.; Menzel, A.; Mooney, H.A.; Schwartz, M.D. Shifting plant phenology in response to global change. Trends Ecol. Evol. **2007**, *22* (7), 357–365.
103. Badeck, F-W.; Bondeau, A.; Bottcher, K.; Doktor, D.; Lucht, W.; Schaber, J.; Sitch, S. Responses of spring phenology to climate change. New Phytol. **2004**, *162* (2), 295–309.
104. Aono, Y.; Kazui, K. Phenological data series of cherry tree flowering in Kyoto, Japan, and its application to reconstructions of springtime temperatures since the 9th century. Int. J. Climatol. **2008**, *28* (7), 905–914.
105. Malcolm, J.R.; Markham, A. *Global Warming and Terrestrial Biodiversity Decline*. World Wildlife Fund Report, WWF U.S.: Washington, DC, 2000.
106. Parmesan, C. Ecological and evolutionary responses to recent climate change. Annu. Rev. Ecol. Evol. Syst. **2006**, *37*, 637–669.
107. Ayres, M.P.; Lombardero, M.J. Assessing the consequences of global change for forest disturbance from herbivores and pathogens. Sci. Total Environ. **2000**, *262*, 263–286.
108. Kelly, M.; Meentemeyer, R.K. Landscape dynamics of the spread of Sudden Oak Death. Photogrammatic Eng. Remote Sens. **2002**, *68* (10), 1001–1009.
109. Anagnostakis, S.L. Chestnut blight: the classical problem of an introduced pathogen. Mycologia **1987**, *79* (1), 23–37.
110. Paillet, F.L. Chestnut: history and ecology of a transformed species. J. Biogeogr. **2002**, *29*, 1517–1530.
111. Walker L.R.; Smith S.D. Impacts of invasive plants on community and ecosystem properties. In *Assessment and Management of Plant Invasions*; Luken, J.O., Thieret, J.W., Eds.; Springer: New York, NY, 1997.
112. Kohli, R.K.; Jose, S.; Singh, H.P.; Batish, D.R. *Invasive Plants and Forest Ecosystems*.; CRC Press: Boca Raton, FL USA, 2009.
113. Daehler, C.C. Performance comparisons of co-occuring native and alien invasive plants: implications for conservation and restorations. Annu. Rev. Ecol. Evol. Syst. **2003**, *34*, 183–211.
114. Stinson, K.A.; Campbell, S.A.; Powell, J.R.; Wolfe, B.E., Callaway, R.M.; Thelen, G.C.; Hallet, S.G.; Prati, D.; Klironomos, J.N. Invasive plant suppresses the growth of native tree seedlings by disrupting belowground mutualisms. PLoS Biol. **2006**, *4* (5), e140.
115. Krebs, C.; Gerber, E.; Mathies, D.; Schaffner, U. Herbivore resistance of invasive species and their hybrids. Oecologia **2011**, *167* (4), 1041–1052.
116. Southwell, I.A.; Bourke, C.A. Seasonal variation in hypericin content of *Hypericum perforatum* (St. John's Wort). Phytochemistry **2001**, *56* (5), 437–441.
117. Pimentel, D.; Lach, L.; Zuniga, R.; Morrison, D. Environmental and economic costs associated with non-indigenous species in the United States. BioScience **2000**, *50* (1), 53–65.
118. Williams, S.C.; Ward, J.S.; Worthley, T.E.; Stafford, K.C. Managing Japanese barberry (Ranunculales: Berberidacea) infestations reduces blacklegged tick (Acari: Ixodidae) abundance and infection prevalence with *Borrelia burgdorferi* (Spirochaetales: Spirochaetaceae). Environ. Entomol. **2009**, *38* (4), 977–984.

Forests: Tropical Rain

Richard T. Corlett
Xishuangbanna Tropical Botanical Garden, Chinese Academy of Sciences, Yunnan, China

Abstract
Tropical rain forests are the tall, dense, evergreen forests that grow in parts of the tropics that are continuously hot and wet. The five major regions with tropical rain forests (South and Central America, Africa, Asia, Madagascar, and New Guinea) have been isolated from each other for a long time and support distinctly different plant and animal communities. Tropical rain forests support more plant and animal species and contain more carbon than any other ecosystem. They are also a major source of timber and other economic products and provide clean water and other services for local people. They are currently threatened by clearance for agriculture as well as illegal logging and hunting. Well-managed protected areas are the best way of ensuring that intact tropical rain forests survive. Financial rewards for countries that reduce their greenhouse emissions from deforestation and forest degradation could help fund the protection and restoration of tropical rain forests.

INTRODUCTION

Tropical rain forests cover only 6% of the Earth's total land area but have attracted more public attention than any other terrestrial ecosystem. The key roles these forests play in the leading environmental concerns of the 21st century—biodiversity loss and climate change—will ensure that this attention will continue. The public image of tropical rain forests is oversimplified, however, and a major aim of this entry is to introduce the diversity of rain forests that exists today. This is followed by a consideration of their importance, the threats to their survival, and the options for their conservation.

WHAT IS (AND WHAT IS NOT) A TROPICAL RAIN FOREST?

Tropical rain forests are the tall, dense, evergreen forests that form the natural vegetation of those areas of the lowland tropics that are hot and wet all (or almost all) of the time.[1] Tropical rain forests extend into areas with a short (1–3 month) dry season, particularly on deep, water-retaining soils or where the dry season is moderated by fog or lower temperatures, but with a progressively longer dry season, the proportion of deciduous trees increases and rain forests are eventually replaced by tropical dry forests (tropical deciduous forests) of various types. The forest also changes with an increase in altitude on high mountains and, at around 1000 meters above sea level, tropical lowland rain forest is replaced by tropical montane forest, which is shorter and generally less species rich. An alternative term, tropical moist forest, has often been used in a broader sense to include semi-evergreen and montane forests as well as "true" tropical rain forests, but neither term is used consistently.

WHERE DO TROPICAL RAIN FORESTS OCCUR?

On a more symmetrical planet, tropical rain forests would form a broad band 5–10° north and south of the equator, interrupted only by the oceans. On our complex Earth, however, interactions between wind direction and mountain ranges, along with a variety of other factors, cause gaps in this band, most notably in East Africa, as well as places where the tropical rain forest extends north or south of the expected latitude (Fig. 1).[1] More recently, huge areas of tropical rain forest have been converted to crops and other land uses, creating further fragmentation.

Tropical rain forests occur in five major regions and as isolated areas on many small oceanic islands (Fig. 1).[1] Exchanges of species between the five regions have been difficult for tens of millions of years, so they have very few species in common and many genera and even families are confined to only one or two regions. In general, the differences between rain forest regions are greater for animals than plants, reflecting both the greater age of plant lineages and the greater dispersal ability of many plants. Asian and African tropical rain forests share most of their plant and animal families, as well as numerous genera, because they have been connected by land since Africa collided with Eurasia 20 million years ago, but they share few species, because the land link has been too dry for forest for most of

Encyclopedia of Natural Resources DOI: 10.1081/E-ENRL-120047448

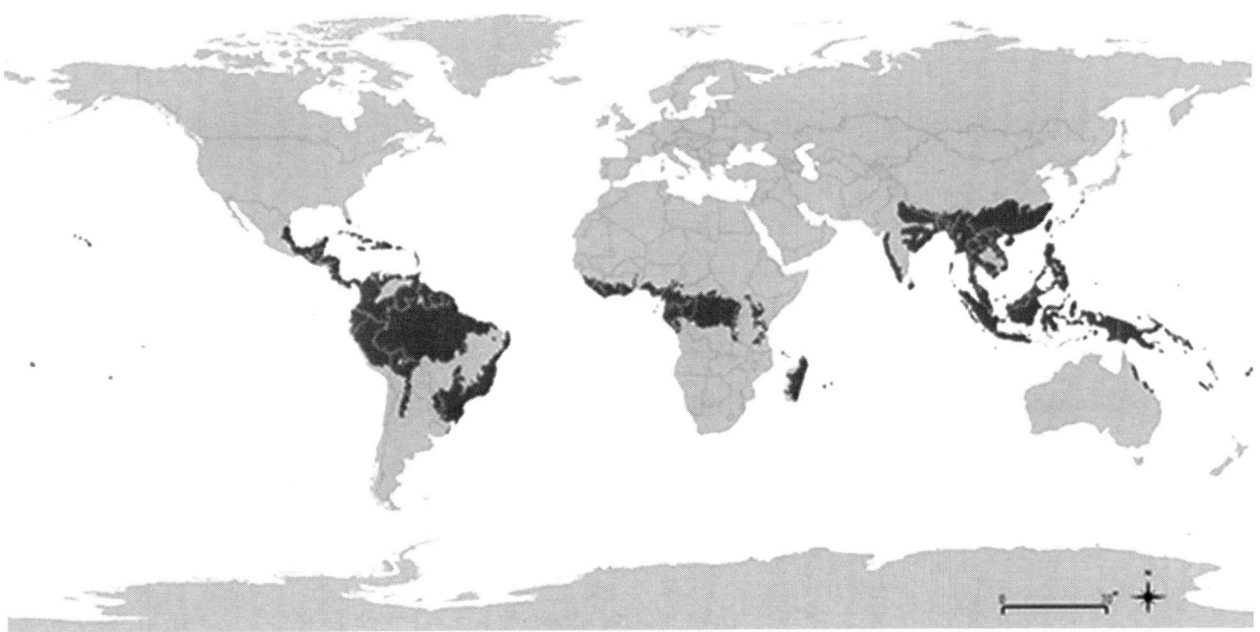

Fig. 1 Map showing the global extent of tropical rain forests.
Source: Adapted from Wikimedia Commons.

that period. Asia and New Guinea have never had a land connection but have been close enough for at least 15 million years to exchange species able to cross relatively short water gaps. South America was an island until three million years ago, when it became connected to the north through the Isthmus of Panama, resulting in a very recent influx of northern animals: the Great American Biotic Interchange.[2] Madagascar, in contrast, has been an island for 90 million years and was reached only by species able to make substantial sea crossings. Oceanic islands, by definition, have never been connected to other land masses and have acquired their entire flora and fauna across the water by dispersal.

South and Central America

The tropical rain forests of South and Central America—the Neotropics—are the most distinctive, reflecting their long period of isolation.[1] Moreover, for many groups of organisms, including trees, birds, bats, and butterflies, the richest Neotropical sites support more species than the richest sites in the other regions. There are three main blocks of rain forest in the Neotropics. The largest block—indeed, the largest block of tropical rainforest in the world—is centered on the Amazon basin and stretches from the foothills of the Andes to the Atlantic Ocean. Most of this block is in Brazil, but there are also large areas in Bolivia, Peru, Ecuador, Colombia, Venezuela, Guyana, Surinam, and French Guiana. A second block extended as a narrow band along the Atlantic coast of Brazil and was separated from the Amazon block by a broad swath of drier vegetation types. Most of this block has now been cleared, leaving only small, isolated fragments. The third block extended from the Pacific coasts of Ecuador and Columbia through Central America to southern Mexico and has also been badly fragmented. Characteristic organisms of Neotropical rain forests include the following: the epiphyte family Bromeliaceae, the New World Monkeys (Platyrrhini), the New World fruit bats (Phyllostomidae), the sloths (Pilosa), the coatis and other members of the raccoon family (Procyonidae), the agoutis and other caviomorph rodents (Dasyproctidae, Agoutidae), the antbirds and other suboscine passerines (Tyranni), the hummingbirds (Trochilidae), and the leafcutter ants (Attini). Neotropical rain forests are also characterized by the absence of any really large mammals and of such characteristic Old World groups as the apes, hornbills, and honeybees (although these were recently introduced).

Africa

The second largest block of tropical rain forest is in Central Africa, centered on the Congo River basin and extending over much of the Democratic Republic of the Congo, Republic of the Congo, Gabon, and Cameroon. A separate block, now heavily fragmented, extended along the coast of West Africa, from Ghana, through the Ivory Coast and Liberia to eastern Sierra Leone. There are also small patches of tropical rain forest in East Africa that have been isolated from the larger blocks for millions of years.

The tropical rain forests of Africa share many families and genera with Asia but also support an endemic family of fruit-eating birds, the turacos (Musophagidae).[1] Diversity is relatively low in many groups of organisms, particularly plants, as a result of the gradual drying of the climate over the last 30 million years. The termites, however, are exceptionally diverse, and these forests are notable for the abundance and diversity of large, ground-living herbivorous mammals, including the lowland gorilla, mandrill, African elephant, okapi, and many species in the cattle family, including the bongo and forest buffalo.

Asia

The tropical rain forests of Asia consist of a relatively uniform block covering much of mainland South East Asia and the large islands of Borneo, Sumatra, and Java, and more or less distinctive outlying areas in the Philippines, Sulawesi, the Western Ghats of India, and the wet zone of Sri Lanka. Indonesia and Malaysia have the most tropical rain forests, but most countries in the Asian tropics have some. Most families in Asian rain forests are shared with Africa, but several mammal families are unique to Asia: the gibbons (Hylobatidae), tarsiers (Tarsiidae), tree shrews (Tupaiidae), and colugos (Cynocephalidae).[1] These are also the only rain forests with rhinoceroses, although both species are on the verge of extinction, and the only lowland rain forests with bears. There are also three small families of birds that are unique to the region. Botanically, the most distinctive feature is the dominance of the upper layers of the forests by the tree family Dipterocarpaceae. Dipterocarps dominate both the biomass and the ecology of these forests in a way that has no parallel in the other rain forest regions. The most dramatic manifestation of this is the tendency of the majority of dipterocarp species over large areas to flower and fruit together at irregular, multiyear intervals.[1] Many non-dipterocarps join in these mass-flowering and fruiting events, but dipterocarps usually dominate.

New Guinea and Australia

The island of New Guinea supports the fourth largest block of tropical rain forest, divided politically between the Indonesian provinces of Papua and West Papua and the independent country of Papua New Guinea. There are also several, much smaller, areas of tropical rain forest in the northeast of neighboring Australia. Although the rain forests of Australia and New Guinea have many similarities, there are also differences that reflect their very different histories.[1] Rain forest covered much of northern Australia until around 15 million years ago, since then it has been progressively restricted by the drying of the continent while most of the rain forest in New Guinea is on land that was uplifted above sea level only 10–15 million years ago.[1] As a result, while the remnants of Australian rain forests have a very distinctive flora, including two woody climbers of an ancient family of flowering plants that is found nowhere else (Austrobaileyaceae), the younger rain forests in New Guinea are dominated by rain forest plants of Asian origin. The faunas are more similar, however, and have fewer elements of Asian origin. Rats and bats are the only native placental mammals and there are no primates, deer, squirrels, or placental carnivores. Marsupials, including possums, wallabies, and tree kangaroos, occupy most mammalian niches, but there are no large carnivores. The distinctive bird fauna includes the flightless, fruit-eating cassowaries (Casuariidae), which are the largest of all rain forest birds, the birds of paradise (Paradisaeidae), the honeyeaters (Meliphagidae), and a number of smaller families that are found nowhere else. There are also representatives of many birds families found in Asia and Africa, but no bulbuls, barbets, woodpeckers, or babblers.

Madagascar

Until recently, there was a band of tropical rain forest along the east coast of this large and mostly dry island, 425 km (266 miles) from Africa. Most plant families and genera are shared with African rain forests, but the fauna is far more distinctive, with many groups showing evidence of a single colonization event followed by adaptive radiation into the many empty niches on the island.[1] The best known example is the lemurs (Lemuriformes), whose ancestors arrived from Africa around 65 million years ago.[3] Apes, monkeys, and squirrels never reached Madagascar, so the lemurs have radiated into a great diversity of forms, many of which became extinct soon after people first arrived on the island 2000 years ago. Similar radiations following single colonization events account for all the rodents, insectivores, and carnivores. There were also some spectacular bird radiations, with one group, the vangas, which arrived around 25 million years ago, now occupying foraging niches that are filled by a variety of different bird families elsewhere.[4] Frogs and reptiles show similar patterns, as do some groups of insects.

Island Rain Forests

Although the rain forests on each island usually support only a fraction of the species diversity found in the major rain forest regions, most species are confined to one island, so that the island rain forests together support a substantial fraction of the total global rain forest diversity.[1] The larger Caribbean islands were colonized by bats, rodents, and primates, like Madagascar, as well as the Neotropical sloths, although all the larger mammals are now extinct. They also had spectacular radiations in the *Anolis* lizards,[5] the *Sphaerodactylus* geckos, and the *Eleutherodactylus* frogs. More distant islands, such as Fiji, were reached by frogs, snakes, and lizards, but bats are the only mammals, and many bird families are missing. The most remote islands with tropical rain forest are in the Hawaiian archipelago, 3765 km (2340 miles) from the nearest major land mass.[1] Terrestrial mammals are represented by a single species of insectivorous bat, the birds by only 13 families, and there were no ants, termites, mosquitoes, cockroaches, or swallowtail butterflies until people introduced them. The birds and many insect groups are dominated by a few, highly diverse, evolutionary radiations originating from single colonists. Well-studied examples include the Hawaiian honeycreepers (Drepanidinae)[6] and the Hawaiian fruit flies (*Drosophila*).[7] Many plants disperse better than most rain forest animals, so floral diversity declines more slowly with distance from the mainland, but the Hawaiian rain forests lack many widespread plant groups, while one-eighth of the native flora (126 species) is made up of the descendants of a single ancestral species of Campanulaceae (the bell flower family) that arrived in the archipelago 13 million years ago.[8]

THE IMPORTANCE OF TROPICAL RAIN FORESTS

Intact tropical rain forests have multiple values, with their relative importance depending on your particular perspective. Most values can be included under one of four general headings: biodiversity, carbon, products, and services. Tropical rain forests support more plant and animal species than any other vegetation type on Earth. An area of 500 hectares (1250 acres) of rain forest at Garza Cocha in Ecuador supports more butterfly species (676) than the whole of North America[9] while a mere 52 hectares (128 acres) at Lambir in Malaysian Borneo contains as many tree species (1175) as the entire temperate forest of the northern hemisphere (Asia, Europe, and North America; 1166 species).[10] Similar patterns occur in a range of other organisms, giving rise to suggestions that tropical rain forests, which cover about 6% of Earth's land surface, support 50% or more of all non-marine species on Earth. Although widely repeated, this number is little more than a guess, and we do not have a complete species list for any area of tropical rain forest. Many commercial crop species (fruits, nuts, timber, spices, etc.) have close relatives in the rain forest that are potentially critical resources for breeding.

Concern for global warming is responsible for another way of looking at tropical rain forest: as a store of carbon compounds that are easily converted to and from the major greenhouse gas, carbon dioxide.[11] All organisms contain carbon, as do soils, but in tropical rain forests, most is in the large trees. Plants remove carbon dioxide from the atmosphere when they photosynthesize in sunlight, and return it through respiration and when a part or all of the plant dies. An undisturbed rain forest is a massive carbon store, more or less in equilibrium, with plant photosynthesis balanced by the respiration of plants, animals, and microbes. Disturbances, such as logging, release part of this store into the atmosphere. Conversion to another land use, such as a soybean field, releases an amount of carbon (as carbon dioxide) equal to the difference between the amount stored in the forest, including the soil, and the amount in the crop. Recent estimates suggest that the degradation and clearance of tropical forests is a source of carbon dioxide second in magnitude only to fossil fuels, but that these very large emissions are more or less offset by an equally large uptake by the surviving forests.[11] Most of this uptake is in forests regrowing after logging or clearance, which make up about a third of all tropical forests, but even intact forests appear to be currently taking up more than they release.

The most valuable product from tropical rain forests in international trade is timber, but a huge variety of non-timber products (edible plants, resins, fibers, honey, medicines, meat, etc.) are also harvested and may be locally more valuable as sources of both income and employment.[1] Apart from carbon fixation, considered separately in the preceding text, the most valuable service performed by tropical rain forests is usually the maintenance of a reliable supply of clean water. Although this service is generally viewed as "free," the benefits of maintaining forest cover for this purpose are easily understood, and payments from large landowners and urban areas are a potential source of revenue for protecting the forest. Other services, such as the provision of wild pollinators for agriculture crops, are more local and the benefits less visible, although not necessarily less important.

THREATS TO TROPICAL RAIN FORESTS

The fact that most tropical rain forest species cannot live outside the forest means that conversion to other land uses is the greatest threat. The world is currently losing about 0.5% of its tropical rain forests every year, with rates of loss varying hugely from place to place.[1] The highest rates are currently in Indonesia and the south and southeast of the Brazilian Amazon while most of Central Africa still has relatively low rates. Until fairly recently, most clearance was by poor farmers, but increasingly the lowland tropics are becoming dominated by the plantations and ranches of big landowners. This is particularly true in Indonesia, where a single crop, oil palm, has replaced vast areas of lowland rain forest in Sumatra and Borneo, and in parts of the Amazon region, where cattle ranching and soybeans are the major threats. Deforestation leads to biodiversity losses long before all the forest has gone because the forest fragments that persist in many human-dominated landscapes are not large enough to support all the species present in continuous forest.

Much of the remaining tropical rain forest has been exploited in other ways through logging (i.e., timber harvest), hunting, and/or the harvesting of a wide range of non-timber forest products. Logging has the most visible impact, because the biggest trees are removed, but studies suggest that the impacts on biodiversity are relatively small, compared with clearance, and largely reversible if the logged forest is protected from additional impacts such as fire. The impact of overhunting is less visible because most rain forest animals are hard to see anyway but can be more pervasive because so many processes within the forest, such as seed dispersal, depend on animals.[12] Dramatic evidence for this indirect impact comes from Sarawak, Malaysia, where hunting has led to a decline in tree diversity and other changes in the forest[13] and the Brazilian Atlantic forest where the loss of large fruit eaters has resulted in a reduction in fruit sizes.[14] Hunting may also lead to local, regional, and, eventually, global extinction of favored animal species, such as the critically endangered forest rhinoceroses in tropical Asia. Other widespread human impacts on tropical rain forests include fires (usually only after fragmentation or logging have opened up the canopy), invasive species (particularly on oceanic islands), and, increasingly, air pollution. Climate change is an additional threat over the next few decades. By 2100, the regions occupied by tropical rain forests today will be 2–5°C warmer, creating climates that do not exist today anywhere on Earth.[15] Changes in rainfall are also expected but more difficult to predict. The potential biological impacts of the expected changes in temperature and rainfall are currently poorly understood but could be severe.[16]

CONSERVATION OF TROPICAL RAIN FORESTS

Some rain forest areas are still protected by their remoteness from human populations, but these areas are shrinking, and eventually the only unexploited forests that remain will be in protected areas (national parks, nature reserves, etc.). Existing protected areas vary hugely in their size and in the aims and effectiveness of their management, but there is ample evidence that well-managed protected areas can work.[1] In much of the tropics, particularly in Asia, there are no large, unexploited areas of tropical rain forest left for protection, and new protected areas will need to incorporate logged or hunted forests that have the potential for recovery. Many of the values of tropical rain forests can also persist in areas that are well managed for sustainable timber production and, potentially, for other forms of resource exploitation. Most logging and hunting in tropical rain forests are currently illegal and thus unregulated, so the control of these activities is another key to conservation both in and outside the protected areas. This requires effective enforcement, both on the ground and along the supply chain that links producers through local and regional markets to the consumers, often in other countries. Unfortunately, successful conservation management requires higher standards of governance than currently exist in many tropical rain forest countries, where corrupt practices often limit the ability of governments to convert good intentions into effective action.[17]

Tropical rain forests are mostly in developing countries with many priorities other than biodiversity protection, so funding rain forest conservation is a major problem. The massive amount of carbon stored in intact rain forests is the basis for plans to financially reward countries that reduce their carbon emissions by reducing deforestation and forest degradation. These policies are collectively known by the acronym REDD+, standing for Reduced Emissions from Deforestation and Degradation.[18] REDD+ is a complex and controversial issue, but the potentially large amounts of money involved make it an attractive option for tropical forest conservation although conserving carbon and biodiversity are not necessarily the same thing. Where most rain forest has already been lost, restoring forest on marginal land of little agricultural value has both carbon and biodiversity benefits. A variety of reforestation techniques has been developed, ranging from promoting natural forest succession to actively planting the site with nursery-raised saplings.[19] So far, the areas restored have been tiny in comparison with the total area of rain forest lost each year, but these projects could be scaled up if funding becomes available from REDD+ or other sources.

CONCLUSION

The protection and sustainable use of the remaining tropical rain forests are critical for both biodiversity conservation and reducing greenhouse gas emissions and can also contribute to the livelihoods of tropical people. The diversity of tropical rain forests will require diverse solutions, but well-managed protected areas and the control of illegal hunting and logging are likely to be effective everywhere. International concerns for the carbon dioxide emitted by the clearance and degradation of tropical rain forests may provide the funding to make this possible.

REFERENCES

1. Corlett, R.T.; Primack, R.B. *Tropical Rainforests: An Ecological and Biogeographical Comparison*, 2nd Ed.; Wiley-Blackwell: Oxford, UK, 2001.
2. Cody, S.; Richardson, J.E.; Valenti, R.; Ellis, C.J.; Pennington, T.R. The great American biotic interchange revisited. Ecography **2010**, *33* (2), 326–332.
3. Horvath, J.E.; Weisrock, D.W.; Embry, S.L.; Fiorentino, I., Balhoff, J.P.; Kappeler, P.; Wray, G.A.; Willard, H.F.; Yoder, A.D. Development and application of a phylogenomic toolkit: Resolving the evolutionary history of Madagascar's lemurs. Genome. Res. **2008**, *18* (3), 489–499.
4. Jønsson, K.A.; Fabre, P.H.; Fritz, S.A.; Etienne, R.S.; Ricklefs, R.E.; Jørgensen, T.B.; Fjeldså, J.; Rahbek, C.; Ericson, P.G.P.; Woog, F.; Pasquet, E.; Irestedt, M. Ecological and evolutionary determinants for the adaptive radiation of the Madagascan vangas. Proc. Nat. Acad. Sci. **2012**, *109* (17), 6620–6625.

5. Losos, J.B. *Lizards in an Evolutionary Tree: Ecology and Adaptive Radiation of Anoles*; University of California Press: Berkeley, 2009.
6. Lerner, H.R.L.; Meyer, M.; James, H.F.; Hofreiter, M.; Fleischer, R.C. Multilocus resolution of phylogeny and timescale in the extant adaptive radiation of Hawaiian honeycreepers. Curr. Biol. **2011**, *21* (21), 1838–1844.
7. O'Grady, P.M.; Lapoint, R.T.; Bonacum, J.; Lasola, J.; Owen, E.; Wu, Y.; DeSalle, R. Phylogenetic and ecological relationships of the Hawaiian *Drosophila* inferred by mitochondrial DNA analysis. Mol. Phylogen. Evol. **2011**, *58* (2), 244–256.
8. Givnish, T.J.; Millam, K.C.; Mast, A.R. Origin, adaptive radiation and diversification of the Hawaiian lobeliads (Asterales: Campanulaceae). Proc. Roy. Soc. B **2009**, *276* (1656), 407–416.
9. DeVries, P.J. Butterflies. In *Encyclopedia of Biodiversity*; Levin, S.A., Ed.; Academic Press: San Diego, 2001; 559–573.
10. Wright, S.J. Plant diversity in tropical forests: A review of mechanisms of species coexistence. Oecologia **2002**, *130*, 1–14.
11. Pan, Y.; Birdsey, R.A.; Fang, J.; Houghton, R.; Kauppi, P.E.; Kurz, W.A.; Phillips, O.L.; Shvidenko, E.; Lewis, S.L.; Canadell, J.G.; Ciais, P.; Jackson, R.B.; Pacala, S.W.; McGuire, A.D.; Piao, S.; Rautiainen, A.; Sitch, S.; Hayes, D. A large and persistent carbon sink in the world's forests. Science **2011**, *333* (6045), 988–993.
12. Corlett, R.T. The importance of animals in the forest. In *Managing the Future of Southeast Asia's Valuable Tropical Rainforests: A Practitioner's Guide to Forest Genetics*; Wickneswari, R., Cannon, C., Eds.; Springer: Dordrecht, Germany, 2011; 83–92.
13. Harrison, R.D.; Tan, S.; Plotkin, J.B.; Slik, F.; Detto, M.; Brenes, T.; Itoh, A.; Davies, S.J. Consequences of defaunation for a tropical tree community. Ecol. Lett. **2013**, *16* (5), 687–694.
14. Galetti, M.; Guevara, R.; Côrtes, M.C.; Fadini, R.; Von Matter, S.; Leite, A.B.; Labecca, F.; Ribeiro, T.; Carvalho, C.S.; Collevatti, R.G.; Pires, M.M; Guimarães, P.R., Jr.; Brancalion, P.H.; Ribeiro, M.C.; Jordano, P. Functional extinction of birds drives rapid evolutionary changes in seed size. Science **2013**, *340* (6136), 1086–1090.
15. Corlett, R.T. Climate change in the tropics: The end of the world as we know it? Biol. Conserv. **2012**, *151*, 22–25.
16. Corlett, R.T. Impacts of warming on tropical lowland rainforests. Trends Ecol. Evol. **2011**, *26* (11), 606–613.
17. Peh, K.S.H. Governance and conservation in the tropical developing world. In *Conservation Biology: Voices from the Tropics*; Sodhi, N. S., Gibon, L., Raven, P.H., Eds.; Wiley: Oxford, UK, 2013; 216–225.
18. Phelps, J.; Webb, E.L.; Adams, W.A. Biodiversity co-benefits in forest carbon emissions reduction policy. Nat. Clim. Change **2012**, *2*, 497–503.
19. Lamb, D. *Regreening the Bare Hills: Tropical Forest Restoration in the Asia-Pacific Region*; Springer: Heidelberg, Germany, 2011.

Fragmentation and Isolation

Anne Kuhn
Atlantic Ecology Division, U.S. Environmental Protection Agency (EPA), Narragansett, Rhode Island, U.S.A.

Abstract

Landscapes are being modified by humans at ever-increasing rates worldwide. This landscape level modification has resulted in changes in ecological patterns and processes, including species distributions. The rate at which humans are altering both terrestrial and aquatic habitats far exceeds the capacity for most species to respond to the increasing loss of habitat and isolated remains of these habitat patches. Fragmentation and isolation of habitat are occurring alongside other environmental stressors and may be interacting synergistically rather than independently. To have any real or sustainable impact, the interacting effects and responses to anthropogenically altered landscapes need to be addressed at an ecosystem level. Understanding and mitigating the causal effects of habitat fragmentation and isolation will require integrated multi-scale landscape level approaches that consider ecological processes occurring within these landscapes, and the synergistic interactions between fragmentation, isolation, and other drivers of environmental change.

INTRODUCTION

Habitat fragmentation is the landscape-level process by which large contiguous native habitat is segmented or "fractured" into smaller and more isolated remnants.[1–4] This fracturing of native habitat can be caused by natural forces such as fire, hydrologic and geological processes, or as a result of direct and indirect anthropogenic activities including land transformation. An ecological consequence of human-induced fragmentation is that the remaining smaller habitat fragments or patches become isolated in space and time from the original larger intact habitat, disrupting species interactions and distribution patterns. This ecological or "effective isolation" is species dependent with a range of responses linked closely to the dispersal traits of a species.[5,6] Habitat loss per se is another ecological consequence of fragmentation and is an interrelated process associated with fragmentation and isolation. The five primary aspects of fragmentation are 1) reduction of area or habitat loss; 2) increase in boundary or edge habitat; 3) increased isolation of fragmented habitat patches; 4) altered patch shape; and 5) altered matrix structure. The relative intensity of each of these factors is mediated by the shape and spatial arrangement of the remnant habitats and the structure and quality of the surrounding area or matrix habitat.[6] Matrix habitat is described as the most predominant, most connected, or most influential landscape element of an area.[7]

The fragmentation and isolation of natural habitat have been identified as two of the leading threats to biodiversity and have contributed to population decline in many species. Fragmentation along with habitat loss has also led to significant alterations in species interaction and composition within communities, disrupting ecological flows and ecosystem functioning. Habitat fragmentation is not an isolated phenomenon and co-occurs and interacts synergistically with many other anthropogenic threats to biodiversity and species persistence. Some of these threats may be non-mechanical disturbances such as chemicals, pesticides, herbicides, hydrological changes, livestock grazing, emerging infectious diseases, and invasive species all of which can further disrupt and degrade already fragmented habitats.[8]

The global scientific community, largely ecologists and especially conservation biologists and landscape ecologists, have been studying and measuring the effects of human-induced habitat fragmentation and isolation since the 1960s. This focus on the effects of fragmentation and isolated habitats coincided with the "environmental awakening" during the 1960s and awareness that human activities were inducing ecological deterioration of the global environment.[9] Ecologists began to apply one of the most important and influential contemporary ecological concepts, Island Biogeography Theory (IBT)[10] (described in the next section) to fragmented habitats, using an explicit analogy between habitat fragments and islands, first noted by Preston (1960).[11] From this foundational theory, fragmentation research evolved and diversified into many interrelated subdisciplines ranging from landscape ecology and conservation reserve design to metapopulation dynamics, and from conservation genetics to population viability analysis (PVA).[8]

Encyclopedia of Natural Resources DOI: 10.1081/E-ENRL-120047453

This entry provides a broad overview of the theoretical and conceptual foundation and progression of the science involved in measuring the effects of habitat fragmentation and isolation, as well as the major landscape issues confronting researchers, practitioners, and conservationists. Future research directions along with advanced innovative methods currently being used to measure the effects of human-induced fragmentation and isolation are presented with the aim of providing a more pragmatic understanding of the ecological consequences of fragmented terrestrial and aquatic ecosystems.

THEORETICAL FOUNDATION FOR FRAGMENTATION AND ISOLATION

Island Biogeography Theory

IBT[10] provides the theoretical foundation for understanding habitat fragmentation. The conceptual basis of IBT describes the relationship between species abundance and the size and isolation of available habitat or island.[12] IBT predicts an expected number of species when the opposing forces of immigration and local extinction achieve a dynamic balance or equilibrium as a function of the island's size and degree of isolation. The theory predicts that islands located farther from the mainland (more isolated) will have lower immigration rates than those close to the mainland, and dynamic equilibrium of species will occur with fewer species on distant, more isolated islands. Islands located closer to the mainland will have higher immigration rates and are predicted to support more species. IBT also logically postulates that larger islands will have more species than smaller islands (which could also be attributed to a greater variety of habits on larger islands).

In the late 1960s, IBT captured the imagination of ecologists and population biologists who began applying and testing the general concepts with "habitat islands" or terrestrial patches of habitat within landscapes surrounded by an inhospitable or resistant matrix. These fragmented and isolated habitat patches could be metaphorically considered islands. One of the most important aspects of IBT was the notion that species composition could and would fluctuate over time, but overall the number of species or species richness would remain relatively stable.[12] This dynamic equilibrium concept has since been applied to a range of environments from "sky islands" or isolated mountain tops (Fig. 1) to fragmented forest patches, to freshwater lakes and aquatic streams disconnected by dams, to conservation reserves to molecular levels, such as genetically isolated populations. Another overarching concept of IBT that has proved to be most pragmatic is the relationship between habitat area and species persistence. This one crucial concept spawned decades of empirical studies measuring the effects of habitat size on a wide range of plant, terrestrial, and aquatic species and their

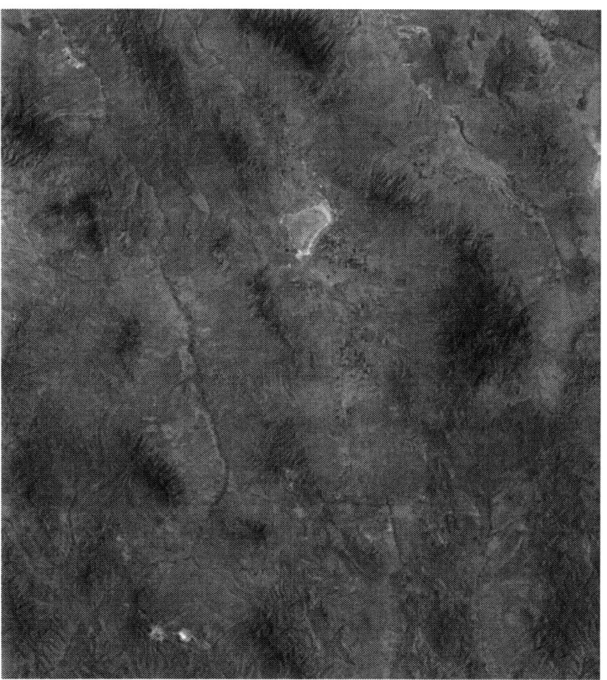

Fig. 1 Madrean Sky Islands: These mountain "islands," forested ranges separated by vast expanses of desert and grassland plains, are among the most diverse ecosystems in the world because of their great topographic complexity and unique location at the meeting point of several major desert and forest biological provinces, southwestern United States and northwestern Mexico.

ability to sustain viable population sizes. The conservation community adopted the basic principles of IBT and applied the theory to the design of protected areas and reserves, advancing the notion that single large reserves were better for preserving population-level persistence than several small reserves of comparable area. This idea became known as single large or several small reserves ("SLOSS") and ignited more research and an enduring debate within the conservation design community.[13–17]

Transcending IBT

By the mid-1980s the habitat fragmentation theoretical foundation shifted as some of the basic tenets of IBT were challenged. One of the main criticisms of IBT was with the "closed community" concept where community equilibrium is controlled by inter-species competition alone.[18] Evidence from empirical field studies suggested that species composition and diversity of local assemblages were regulated by local or regional processes. These findings led to a broadened perspective of species distributions and a focus on issues of scale specifically: design of sampling relative to ecological processes and the hierarchical organization of ecological processes in space and time.[9] These issues of scale were being directly addressed in the emerging research field of "Landscape Ecology." From a landscape perspective, new hypotheses were suggesting that non-equilibrium processes were determining landscape dynamics and that fragmentation of landscapes must be assessed at multiple scales and be defined from an organismal perspective.[19] A landscape perspective also shifted the focus from species equilibrium to population level studies at landscape scales in heterogeneous environments or metapopulation dynamics [20,21] (Fig. 2).

Field-based evidence also suggested that using the IBT to equate habitat fragments with islands was problematic. Field studies demonstrated that "matrix matters" and that not all fragmented patches were surrounded by inhospitable and resistant matrix and sharp boundaries, and that connectivity among fragmented habitats and the ability to traverse the surrounding matrix was determined by species-specific dispersal traits, habitat requirements, and general adaptability (e.g., habitat generalist vs. specialist). This species-specific "effective isolation" has important conservation implications relative to the genetic and demographic risk of isolation.[5] Another observed effect of fragmentation that was not considered by IBT was the "edge effect," a non-discrete boundary line around a patch, which can have negative or positive effects and is species and context dependent. Despite the discrepancies found between IBT and empirical studies, IBT has had an enormous impact on the study of human-altered landscapes and still provides a heuristic and conceptual, albeit a simplified framework for understanding habitat fragmentation.

Fragmentation research today has advanced far beyond the original intended scope of IBT [8] and approaches habitat fragmentation as a landscape-level phenomenon incorporating more complex processes to elucidate the direct and indirect causal relationships of species responses to landscape change.[22] Currently, a conceptual framework of "interdependence" in both landscape effects on species and species responses to landscape change is being proposed as an integrative perspective to advance the mechanistic understanding of fragmentation and isolation processes.[23] This conceptual framework incorporates the "integrated community" concept [24] recognizing a range of interdependence in natural communities' responses to habitat fragmentation from highly individualistic to highly interdependent. This context-dependence approach recognizes that the patterns, processes, and consequences of habitat fragmentation and isolation are interrelated. Using this interdependence framework as a basis, Didham et al. (2012)[23] developed a hierarchical conceptual model of potential direct and indirect causal paths by which the amount and spatial arrangement of habitat can affect a measured response variable (Fig. 3). These most recent advances in fragmentation research acknowledge context dependence in ecosystem responses at multiple spatial and temporal scales, along with trait-dependent species responses and synergistic interactions between fragmentation and other components of global environmental change.[22]

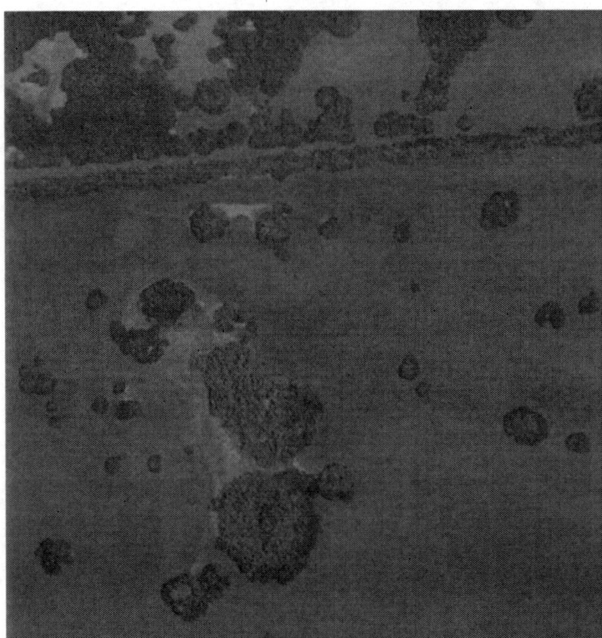

Fig. 2 Example of metapopulation patch dynamics analogy using mangrove islands in Everglades National Park, Florida, USA. These island patches illustrate how focal populations occur within a network of local populations and may interact via migration within a landscape or region.

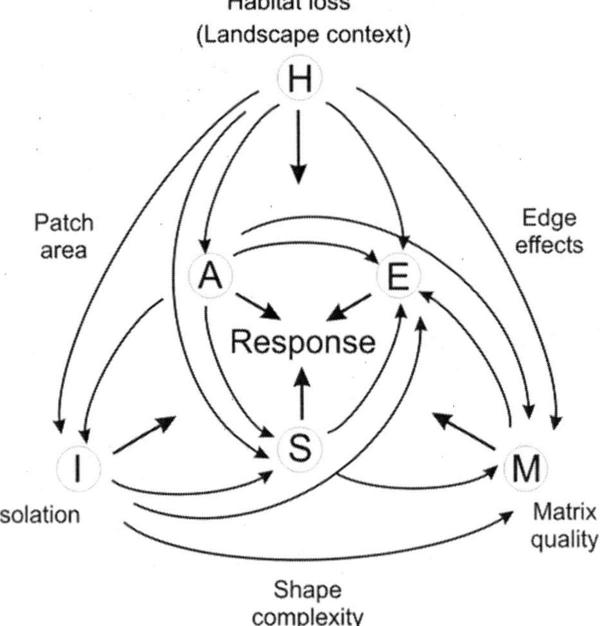

Fig. 3 A hierarchical conceptual model of potential direct and indirect causal paths by which the amount and spatial arrangement of habitat can affect a measured response variable (termed "Response").

Source: © Oikos 2012 as included in Rethinking the conceptual foundations of habitat fragmentation research, R. K. Didham; V. Kapos; R. M. Ewers, Vol. 121: 161–170.

MEASURING THE EFFECTS OF FRAGMENTATION AND ISOLATION

Scale of Assessment

The importance of scale is widely recognized and considered throughout the broad field of ecology and is of specific relevance to the field of Landscape Ecology. Landscape Ecology specifically addresses broad and fine-scale spatial patterns and ecological processes and the hierarchical nature of these ecological processes.[25–28] Fragmentation and the isolation of remnant habitat can occur on different spatial scales and the effects may occur at different temporal scales (i.e., time lags between fragmentation and effect). The choice of scale assessment can affect the analysis of fragmentation and there is no single correct scale for analysis; the relevant scale being different for each species. Many forest fragmentation studies have been conducted to specifically investigate the effect of scale on the analysis of fragmentation.[29–33] It is widely accepted that any analysis of fragmentation and isolation should occur at multiple scales so the effect of scale on the response (e.g., species abundance, biodiversity, community indices) can be assessed at scales ranging from small (fine grain resolution) to large (landscape level extents). The temporal scale should also be considered by using long-term studies or meta-analyses that incorporate population-level responses over time to account for time lags between the fragmentation of habitat and the manifestation of species responses.

RESPONSE MEASURES TO FRAGMENTATION AND ISOLATION

There are a multitude of response measures for assessing the effects of fragmentation and isolation ranging from molecular-level genetics to species' population-level effects to community dynamics to measures of biodiversity and landscape patterns and processes. Habitat fragmentation in both aquatic and terrestrial ecosystems has led to the genetic isolation of populations in numerous species, and can be assessed using measures of population genetic differentiation, dispersal and gene flow, genetic diversity, and effective population size. At higher levels of biological organization, population level measures, such as population density, dispersal, recolonization, abundance, fitness and productivity, extinction proneness, PVA, metapopulation dynamics, and patch model occupancy are used to predict a species' probability of extinction relative to the spatial arrangement across a network of habitat patches.[34–37] One conclusion from metapopulation modeling is that the probability of population extinction depends not only on habitat area or quality, but on the degree of isolation and spatial location within the metapopulation network.[38] The next response level of increasing complexity is the measure of species assemblages or community-level responses, including species richness and interactions. These community-level measures attempt to identify the general species distribution patterns from focal indicator and/or umbrella species,[39,36,40] seeking to strike a balance between detecting generalities that apply across landscapes and species with accurately predicting the effects of fragmentation for a single species. The highest level of response measure for evaluating the effects of fragmentation is the landscape level, where the altered ecological processes (e.g., flow of materials, magnitude, and timing of disturbance) are examined alongside altered patterns of habitat (e.g., forests, vegetative cover, and stream connectivity) to better understand how ecological processes and patterns emerge so that extrapolations from one landscape may be applied to another landscape. Landscape indices are used to characterize spatial patterns of habitat created by landscape alteration and typically include habitat patch metrics that describe the composition (patch habitat type), configuration (spatial arrangement of patches), and connectivity of habitat (from either a human or species-specific perspective).[41,36]

Habitat Loss Effects

Fragmentation involves two interrelated processes: habitat loss and isolation. As fragmentation occurs the total amount of habitat in the landscape is reduced and the remaining habitat is chopped up into fragments of varying sizes and degrees of isolation.[8] Empirical studies have demonstrated that it is difficult to separate these two interrelated processes as they generally co-vary.[42] Fahrig (2003)[42] concludes that habitat loss has a much greater negative effect on biodiversity than the breaking apart of habitat or fragmentation per se. This finding is only elucidated when empirical studies have been carefully designed to separate the effects of habitat loss and fragmentation. Because it has been challenging to separate the confounding effects of habitat loss and isolation from fragmentation, researchers have tried to experimentally control for habitat amount while varying fragmentation. Another research approach has been to compare many different landscapes extracting indices of fragmentation that are independent of the amount of habitat in each landscape.[43,44] A caveat to consider when measuring habitat loss is that the definition of habitat is species specific and therefore habitat loss is a species-specific phenomenon. Additionally, habitat loss often co-occurs with habitat degradation, isolation, and changes in species interactions so it becomes quite challenging to measure the effects of habitat loss alone. Another important consideration is that habitat loss is a highly non-random process,[45–47] and the habitat remnants are often a biased subset of the original landscape.[8] Given all of these caveats and confounding complications it appears that the most pragmatic approach to studying these effects is to use a conceptual framework that incorporates the interdependence in both effects and responses to fragmentation.[23]

Edge Effects

Human modified habitat often results in an increase in perimeter-to-area ratios of the remaining patches creating boundaries or marginal zones of altered microclimatic and ecological conditions (Fig. 4). Edges can also occur naturally at the interface of two ecological communities or ecotones. Edge effects can be classified by their impacts in two broad categories: abiotic and biotic. Abiotic edge effects can modify microclimatic conditions such as light, air temperature and humidity, soil moisture, wind speed, and altered fire regimes. Other abiotic edge alterations could increase habitat fragment exposure to chemicals such as fertilizers, herbicides, road salt, and acid rain.[48] Biotic edge effects can be manifested in a number of ways, for example, increases in predators, invasive species, diseases, altered animal and plant dispersal, species interactions, reduced reproductive success, and habitat quality for individual species.[49–51] Also, not all species respond negatively to edges and some species' demographic rates are enhanced by edge habitat. Edge effects are species dependent and are directly influenced by the surrounding matrix.[33]

Isolation and Measures of Connectivity

An obvious spatial consequence of habitat fragmentation is that remnant patches may become isolated in space and time from other patches of suitable habitat (Fig. 5). There is a range of species-specific responses to isolation with some species having a naturally patchy distribution across the landscape (e.g., mountain top species, freshwater ponds), whereas other species have a more continuous distribution across the landscape and may be more affected by habitat isolation and loss of connectivity. "Effective isolation" is also scale and matrix dependent with a range of species tolerances for traversing the surrounding matrix that may allow dispersal to other remnant habitat patches.[5] Landscape connectivity can be defined as the degree to which a landscape facilitates or impedes movement among resources patches.[52] The two types of connectivity that are generally considered and measured by ecologists are structural and functional.[53,54] Structural connectivity assesses the spatial arrangement of habitat types and ecological systems in a landscape without considering the movement of organisms through the landscape.[55] Alternatively, functional connectivity incorporates the likelihood of species movement and behavioral responses as well as ecological processes within the landscape. For example, corridors or narrow strips of lands that differ from the surrounding matrix can be used to connect patches of suitable habitat, facilitating movement of organisms and ecological flows, increasing patch connectivity. There have been many recent landscape-level studies and technical advancements for measuring and exploring the importance of maintaining and restoring functional connectivity within landscapes, including aquatic and terrestrial ecosystems.[56–63]

Fig. 4 Example of anthropogenically induced edge effect.

Fig. 5 Example of intact naturally connected landscape and human-induced highly fragmented and isolated landscape.

Matrix Effects

Forman (1995) [7] described the matrix as the largest background patch in a landscape, which exerts a dominant role on ecological processes. An increasing number of studies highlight the importance of matrix quality in modified landscapes as a major influence on fragment connectivity and as an influential force for mitigating biotic and abiotic edge effects.[64,5,65,66,54] Laurance (2004)[67] and others have found that a species tolerance or ability to exploit the surrounding matrix is positively associated with its tolerance of edge habitats, highlighting again a species-dependent response phenomenon. Ricketts (2001)[5] suggests that while many efforts have focused on reconnecting fragmented habitats using wildlife corridors and stepping stones (relatively small patches of native habitat facilitating connectivity across a fragmented landscape) (Fig. 6), it may be more feasible to reduce effective isolation of fragments by altering management practices in the surrounding matrix. In the longest-running field experimental study of habitat fragmentation in Central Amazonia, Laurance et al. (2011)[33] found that the matrix in the study area had changed markedly and in turn has strongly influenced fragment dynamics and faunal persistence.

CONCLUSIONS

Habitat fragmentation and the resulting isolation of habitat remnants affects far more than biodiversity and interactions among species. As human-induced fragmentation increases many ecological processes are being altered within fragmented landscapes, including gene flow, hydrological flows, species dispersal patterns, forest biomass, and carbon and nitrogen storage. It is clear that to enhance the understanding of the effects of habitat fragmentation and isolation, researchers need to deploy an "interdependence" perspective for the multiple interacting drivers of habitat fragmentation, within a community and landscape context over larger spatial and temporal scales. Land managers will need to educate and elicit support from the general public to reduce the effects of fragmentation on privately owned land, as much of the remaining protected areas of natural habitat are surrounded by privately held land. Conservation conducted in a broader landscape context that includes incentives for private land owners has the potential to enhance the management objectives of protected areas and reduce the effects of fragmentation.[68] To mitigate the effects of anthropogenic fragmentation, ecologists and conservationists will need to expand beyond traditional reserve design principles (conserving large natural areas connected to other protected areas by corridors) and incorporate restorative practices, such as enhancing and managing the matrix surrounding habitat fragments.

Fig. 6 Highly fragmented landscape with marginal connectivity provided by remaining wetland fragments, surrounded by dense urban and recreational land uses.

ACKNOWLEDGMENTS

The author wishes to acknowledge the far too many to list landscape ecologists, conservationists, field biologists, land practitioners, resource managers, bioreserve specialists who have contributed to the science and understanding of the effects of habitat fragmentation and isolation.

REFERENCES

1. Wilcove, D. S.; McLellan, C.H.; Dobson, A.P. Habitat fragmentation in the temperate zone. In *Conservation Biology. The Science of Scarcity and Diversity*; Soulé, M. E. Ed.; Sinauer: Sunderland, 1986; 237–256.
2. Ranta, P.; Blom, T.; Niemela, J.; Joensuu, E.; Siitonen, M. The fragmented Atlantic rain forest of Brazil: size, shape and distribution of forest fragments. Biodivers. Conserv. **1998**, *7*, 385–403.
3. McGarigal, K.; McComb, W.C. Forest fragmentation effects on breeding birds in the Oregon Coast Range. In *Forest Fragmentation: Wildlife and Management Implications*. J.A. Rochelle, L.A. Lehman and J. Wisniewski, Eds.; Koninklijke Brill NV: Leiden, the Netherlands, 1999; 223–246.
4. Fahrig, L. Effects of habitat fragmentation on biodiversity. Annu. Rev. Ecol., Evol. Syst. **2003**, *34*, 487–515.
5. Ricketts, T. H. The matrix matters: effective isolation infragmented landscapes. Am. Nat. **2001**, *158*, 87–99.
6. Ewers, R. M.; R. K. Didham. Confounding factors in the detection of species responses to habitat fragmentation. Biol. Rev. **2006**, *81*, 117–142 pp.
7. Forman, R. T. T. *Land Mosaics: The Ecology of Landscapes and Regions*; Cambridge University Press: New York, USA, 1995; 632 p.
8. Laurance, W. F. Theory meets reality: how habitat fragmentation research has transcended island biogeographic theory. Biol. Conserv. **2008**, *141*, 1731–1744.
9. Haila, Y. A conceptual genealogy of fragmentation research: from island biogeography to landscape ecology. Ecol. Appl. **2002**, *12*(2), 321–334.
10. MacArthur, R.H.; Wilson, E. O. *The Theory of Island Biogeography*; Princeton University Press: Princeton, USA, 1967.
11. Preston, F. W. Time and space and the variation of species. Ecology **1960**, *41*, 611–627.

12. Powledge, F. Island biogeography's lasting impact. BioScience **2003**, *53*, 1032–1038.
13. Simberloff, D.S.; Abele, L. G. Island biogeography theory and conservation practice. Science, **1976**, *191*, 285–286.
14. Diamond, J.M. The island dilemma: lessons of modern biogeographic studies for the design of natural reserves. Biol. Conserv. **1975**, *7*, 129–146.
15. Diamond, J.M. Island biogeography and conservation: strategy and limitations. Science, **1976**, *193*, 1027–1029.
16. Saunders, D.A.; Hobbs, R.J.; Margules, C.R. Biological consequences of ecosystem fragmentation: a review. Conserv. Biol. **1991**, *5*, 18–32.
17. Ovaskainen, O. Long-term persistence of species and the SLOSS problem. J. Theor. Biol. **2002**, *218*, 419–433.
18. Wiens, J. Habitat fragmentation: island vs. landscape perspectives on bird conservation. Ibis **1994**, *137*, S97–S104.
19. Haila, Y. Implications of landscape heterogeneity for bird conservation. Int. Ornithol. Congr. **1991**, *20*, 2286–2291.
20. Lefkovitch, L.P.; Fahrig, L. Spatial characteristics of habitat patches and population survival. Ecol. Model. **1985**, *30*, 297–308.
21. Hanski, I. A.; M. E. Gilpin, Eds. *Metapopulation Biology. Ecology, Genetics, and Evolution*; Academic Press: New York, USA, 1997.
22. Didham, R. K. *Ecological Consequences of Habitat Fragmentation. Encyclopedia of Life Sciences*; Wiley: New York, USA, 2010.
23. Didham, R.K.; Kapos, V. and Ewers, R. M. Rethinking the conceptual foundations of habitat fragmentation research. Oikos **2012**, *121*, 161–170.
24. Lortie, C. J.; Brooker, R.W.; Choler, P.; Kikvidze, Z.; Michalet, R.; Pugnaire, F.I.; Callaway, R.M. Rethinking plant community theory. Oikos **2004**, *107*, 433–438.
25. Turner, M.G. Landscape ecology: the effect of pattern on process. Annu. Rev. Ecol. Syst. **1989**, *20*, 171–197.
26. Wiens, J.A. Spatial scaling in ecology. Funct. Ecol. **1989**, *3*, 385–397.
27. Levin, S.A. The problem of pattern and scale in ecology. Ecology **1992**, *73*(6), 1943–1967.
28. O'Neill, R.V.; Hunsaker, C.T.; Timmins, S.P.; Jackson, B.L.; Riitters, K.L.; Wickham, J.D. Scale problems in reporting landscape pattern at the regional scale. Landscape Ecol. **1996**, *11*(3), 169–180.
29. Campbell, D. J. Scale and patterns of community structure in Amazonian forests. In *Large-Scale Ecology and Conservation Biology*, P. J. Edwards, R. M. May, and N. R. Webb, Eds.; Blackwell: Oxford, 1994; 1-17.
30. Keitt, T. H.; D. L. Urban; B. T. Milne. Detecting critical scales in fragmented landscapes. Conserv. Ecol. **1997**, *1*, 4.
31. Riitters, K.; Wickham, J.; O'Neill, R.; Jones, B.; Smith, E. Global-scale patterns of forest fragmentation. Conserv. Ecol. **2000**, *4*, 3.
32. Radford, J.Q.; Bennett, A.F.; Cheers, G.J. Landscape-level thresholds of habitat cover for woodland-dependent birds. Biol. Conserv. **2005**, *124*, 317–337.
33. Laurance, W.; Camargo, J.L.C; Luizão, R.C.C. et al. The fate of Amazonian forest fragments: A 32-year investigation. Biol. Conserv. **2011**, *144*, 56–67.
34. Laurance, W.F.; Camargo, J.L.C.; Luizãoa, R.C.C.; Laurence, S.G.; Pimm, S.L.; Bruna, E.M.; Stouffer, P.C.; Williamson, G.B.; Benítez-Malvido, J.; Vasconcelos, H.L.; Houtan, K.S.V.; Zartman, C.E.; Boyle, S.A.; Didham, R.K.; Andrade, A.; Lovejoy, T.E. Ecological correlates of extinction processes in Australian tropical rainforest mammals. Conserv. Biol. **1991**, *5*, 79–89.
35. Simberloff, D.; Farr, J. A.; Cox, J.; Mehlman, D.W. Movement corridors: conservation bargains or poor investments? Conserv. Biol. **1992**, *6*, 493–504.
36. Lindenmayer, D.; Fischer, J. *Habitat Fragmentation and Landscape Change: An Ecological and Conservation Synthesis*; Island Press: Washington, D.C., USA, 2006.
37. Burns, C.E.; Grear, J. Effects of habitat loss on populations of white-footed mice: testing matrix model predictions with landscape-scale perturbation experiments. Landscape Ecol. **2008**, *23*, 817–831.
38. Ovaskainen, O.; Hanski, I. Metapopulation dynamics in highly fragmented landscapes. In *Ecology, Genetics, and Evolution of Metapopulations*; I. Hanski and O. E. Gaggiotti, Eds.; Elsevier Academic Press: San Diego, USA, 2004; 73–103.
39. Lambeck, R. Focal species: a multi-species umbrella for nature conservation. Conserv. Biol. **1997**, *11*(4), 849–856.
40. Watts, K.; Eycott, A.E.; Handley, P.; Ray, D.; Humphrey, J.W.; Quine, C.P. Targeting and evaluating biodiversity conservation action within fragmented landscapes: an approach based on generic focal species and least-cost networks. Landscape Ecol. **2010**, *25*, 1305–1318.
41. McGarigal, K.; Cushman, S.A.; Neel, M.C.; Ene, E. FRAGSTATS 3.0: Spatial Pattern Analysis Program for Categorical Maps; University of Massachusetts: Amherst, MA, 2002.
42. Fahrig, L. Effects of habitat fragmentation on biodiversity. Annu. Rev. Ecol. Syst. **2003**, *34*, 487–515.
43. McGarigal, K.; McComb, W.C. Relationships between landscape structure and breeding birds in the Oregon Coast Range. Ecol. Monogr. **1995**, *65*, 235–260.
44. Villard, M.A.; Trzcinski, M.; Merriam, G. Fragmentation effects on forest birds: relative influence of woodland cover and configuration on landscape occupancy. Conserv. Biol. **1999**, *13*, 774–783.
45. Laurance, W.F.; Cochrane, M.A. Synergistic effects in fragmented landscapes. Conserv. Biol., **2001**, *15*, 1488–1489.
46. Seabloom, E.W.; Dobson, A.P.; Stoms, D.M. Extinction rates under nonrandom patterns of habitat loss. Proc. Natl. Acad. Sci. USA **2002**, *99*, 1129–11234.
47. August, P.; Iverson, L.; Nugranad, J. Human conversion of terrestrial habitats. In *Applying Landscape Ecology in Biological Conservation*; Gutzwiller, K. J. Ed.; Springer-Verlag: New York, USA, 2002; 198–224.
48. Soulé, M.E.; Simberloff, D.S. What do genetics and ecology tell us about the design of nature reserves? Biol. Conserv. **1986**, *35*, 19–40.
49. Paton, P.W.C. The effect of edge on avian nest success: how strong is the evidence? Conserv. Biol. **1994**, *8*, 17–26.
50. Van Horn, M. A.; Gentry, R. M.; Faaborg, J. Patterns of Ovenbird (*Seiurus aurocapillus*) pairing success in Missouri forest tracts. Auk **1995**, *112*, 98–106.
51. Meentemeyer, R. K.; Haas, S.E.; Vaclavık, T. Landscape Epidemiology of emerging infectious diseases in natural and human-altered ecosystems. Annu. Rev. Phytopathol. **2012**, *50*, 19.1–19.24

52. Taylor, P. D.; Fahrig, L.; Henein, K.; Merriam, G. Connectivity is a vital element of landscape structure. Oikos **1993**, *68*, 571–573.
53. Crooks, K. R.; M. A. Sanjayan, Eds.; *Connectivity Conservation*. Cambridge University Press; New York, USA, 2006.
54. Watts K.; Handley P. Developing a functional connectivity indicator to detect changes in fragmented landscapes. Ecol. Indicators **2010**, *10*, 552–557.
55. Theobald, D. M.; Crooks, K. R.; Norman, J.B. Assessing effects of land use on landscape connectivity: loss and fragmentation of western U.S. forests. Ecol. Appl. **2011**, *21*(7), 2445–2458.
56. Urban, D. L.; Keitt, T. H. Landscape connectedness: A graph theoretic perspective. Ecology **2001**, *82*, 1205–1218.
57. Theobald, D. M. Exploring the functional connectivity oflandscapes using landscape networks. In *Connectivity Conservation: Maintaining Connections for Nature*; Crooks, K. R. and Sanjayan, M. A., Eds.; Cambridge University Press: Cambridge, UK, 2006; 416–443.
58. McRae, B. H.; Beier, P. Circuit theory predicts gene flow in plant and animal populations. PNAS, **2007**, *104*(50), 19885–19890.
59. Sterling, K.A.; Reed, D. H.; Noonan, B.P.; Warren Jr., M.L. Genetic effects of habitat fragmentation and population isolation on *Etheostoma raneyi* (Percidae). Conserv. Genet. **2012**, *13*, 859–872.
60. Hitt, N. P.; Angermeier, P. L. River-stream connectivity affects fish bioassessment performance. Environ. Manag., *42*, 132–150.
61. Cote, D.; Kehler, D. G.; Bourne, C.; Wiersma, Y.F. A new measure of longitudinal connectivity for stream networks. Landscape Ecol. **2009**, *24*, 101–113.
62. Pinto, N.; Keitt, T. H. Beyond the least-cost path: evaluating corridor redundancy using a graph-theoreticapproach. Landscape Ecol. **2009**, *24*, 253–266.
63. Pinto, N.; Keitt, T.H.; Wainright, M. LORACS: JAVA software for modeling landscape connectivity and matrix permeability. Ecography **2012**, *35*, 001–005.
64. Didham, R.K.; Lawton, J.H. Edge structure determines the magnitude of changes in microclimate and vegetation-structure in tropical forest fragments. Biotropica **1999**, *31*, 17–30.
65. Cook, W. M.; Lane, K. T.; Foster, B. L.; Holt, R. D. Island theory, matrix effects and species richness patterns inhabitat fragments. Ecol. Lett. **2002**, *5*, 619–623.
66. Perfecto, I.; Vandermeer, J. Quality of agroecological matrix in a tropical montane landscape: ants in coffee plantations in southern Mexico. Conserv. Biol. **2002**, *16*, 174–182.
67. Laurance, W.F. Forest–climate interactions in fragmented tropical landscapes. Phil. Trans. Roy. Soc. B, **2004**, *359*, 345–352.
68. Goetz, S.J.; Jantz, P.; Jantz, C.A. Connectivity of core habitat in the Northeastern United States: parks and protected areas in a landscape context. Remote Sensing Environ. **2009**, *113*, 1421–1429.

BIBLIOGRAPHY

1. Gutzwiller, K. *Applying Landscape Ecology in Biological Conservation*; Springer-Verlag: New York, USA, 2002.
2. Hanski, I. *Metapopulation Ecology*; Oxford University Press: New York, USA, 1999.
3. Lindenmayer, D.B.; Fischer, J. *Habitat Fragmentation and Landscape Change: An Ecological and Conservation Synthesis*; Island Press: Washington, D.C., USA, 2006.
4. Losos, J.B.; Ricklefs, R.E. *The Theory of Island Biogeography Revisited*; Princeton University Press: Princeton, New Jersey, USA, 2010.
5. MacArthur, R.H.; Wilson, E.O. *The Theory of Island Biogeography*; Princeton University Press: Princeton, New Jersey, USA, 1967.
6. Quammen, D. *Song of the Dodo: Island Biogeography in an Age of Extinctions*; Scibner: New York, USA, 1996.
7. Tilman, D.; Kareiva, P. *Spatial Ecology: The Role of Space in Population Dynamics and Interspecific Interactions*; Princeton University Press: Princeton, New Jersey, USA, 1997.
8. Wiens, J.A.; Moss, M.R.; Turner, M.G.; Mladenoff, D.J. *Foundation Papers in Landscape Ecology*; Columbia University Press: New York, USA, 2007.

Genetic Diversity in Natural Resources Management

Thomas Joseph McGreevy, Jr.
Biology Department, Boston University, Boston, Massachusetts, U.S.A.

Jeffrey A. Markert
Biology Department, Providence College, Providence, Rhode Island, U.S.A.

Abstract

Genetic diversity is the raw material used by natural selection to allow a population to adapt to a changing environment. It is an integral component of biodiversity that must be conserved. Molecular genetic techniques have been used to quantify genetic diversity and inform management decisions for wild and captive populations since the 1980s, but the full utility of these tools is yet to be realized. One major limitation has been the communication disconnect between research scientists and resource managers in some jurisdictions. The two main goals of this entry are 1) to help bridge this gap by providing an overview of the type of information that can be gained with molecular genetic techniques to inform a wide range of management decisions and 2) to discuss future directions that will reduce the cost and expand the utility of molecular genetic techniques in the management of natural resources.

INTRODUCTION

What Is Genetic Diversity?

Genetic diversity is the amount of variation in the sequence of four nucleotides (adenine, thymine, cytosine, and guanine) that comprise deoxyribonucleic acid (DNA) within individuals, populations, or species. The particular combination of nucleotides within an individual is called a genotype.[1] From an information theoretic approach, population genetic diversity represents the amount of information stored in DNA within a population above and beyond the information common to all individuals.[2,3] Thus, genetic diversity represents the maximum amount of information available to code for distinct physical characteristics (phenotypes) within a group. Less information means fewer distinct phenotypes, decreasing the likelihood that some individuals in a population have characteristics suited to the next environmental challenge. These challenges may include things like new diseases, introduced predators, or altered physical environments. In diploid organisms, inbreeding depression-like effects also may reduce individual fitness when both copies of a segment of DNA within an individual are identical. The probability of this occurring is higher in populations with lower genetic diversity.

From a wildlife management perspective, we would expect that groups with lower levels of genetic diversity are at greater risk for extinction. Experimental studies support this expectation. For example, when genetic diversity is experimentally manipulated in isolated plant populations, those with lower genetic diversity are dramatically less fit.[4] Similarly, work using short-lived crustaceans as a model shows that even minor losses of genetic diversity greatly increase extinction risk when low-diversity populations are challenged with suboptimal environments.[5]

The importance of conserving genetic diversity is recognized by the International Union for Conservation of Nature (IUCN) as a major component of biodiversity.[1] In the United States, genetic diversity has been recognized as an important resource that should be conserved since the drafting of the Endangered Species Act (ESA) of 1973. Further rationale for why genetic diversity matters was explained by the House of Representatives in House Resolution 37, the precursor to the ESA of 1973:[6]

> "From the most narrow possible point of view, it is in the best interests of mankind to minimize the losses of genetic variations. The reason is simple: they are potential resources. They are keys to puzzles which we cannot yet solve, and may provide answers to questions which we have not yet learned to ask."

There are still important puzzles to solve and unasked questions in conservation genetics. In the past 40 years, researchers have developed and refined a strong set of genetic tools that address key aspects of conservation biology. These techniques have been used to inform management decisions for wild and captive populations since the 1980s, but the full utility of these tools is yet to be realized.[7] One major

Encyclopedia of Natural Resources DOI: 10.1081/E-ENRL-120047430

limitation has been the communication disconnect between research scientists and resource managers in some jurisdictions. The first goal of this entry is to help bridge this gap by providing an overview of the type of information that can be gained with molecular genetic techniques to inform a wide range of management decisions. The second goal is to discuss future directions that will reduce the cost and expand the utility of molecular genetic techniques in the management of natural resources.

HOW CAN GENETIC TECHNIQUES INFORM MANAGEMENT DECISIONS?

The development of molecular genetic tools to assist management decisions has increased with the emergence of the field of conservation genetics in the early 1990s. Conservation geneticists are concerned with measuring genetic diversity within groups and determining how it is partitioned among groups. These techniques are used to address a number of important wildlife management issues, including the following:

Demographics and Extinction Risk

Molecular markers can be used to detect the presence of a species, estimate census and effective population sizes, determine sex ratios, and measure other key demographic parameters. Detecting the presence of a species is important to determine their geographic distribution and potential range decline. The concept of an effective population size was introduced by Wright[8] and in general terms is the number of individuals in an ideal population that would lose genetic diversity through sampling losses at the same rate as the observed population. It is generally smaller than the census size because not all individuals in a population contribute the same number of offspring to the next generation. It is critical to understand how quickly genetic diversity will be lost in a population, which in part is dictated by their effective population size.[1] Two important demographic parameters for managing a population are census size and density. Molecular genetic techniques can be used to identify individuals by determining their unique genotype and quantify a population size using traditional mark recapture methods, which also can be used to estimate the density of individuals per area. This technique can be used with non-invasively collected samples and replace the physical marking of individuals using unique markers, such as ear tags. Sex chromosome-specific markers can be used to determine sex ratios and model population dynamics.[7]

The demographic history of a population can impact its population dynamics and adaptive potential. When the size of a population is severely reduced, it is said to have undergone a bottleneck.[1] Similarly, founding events involve the establishment of a new population from a small number of individuals. Both bottlenecks and founding events result in populations with low genetic diversity and can impact their evolutionary trajectory because genetic information is lost randomly.[9] Lost genetic diversity in a population also can lead to inbreeding depression, which occurs when related or genetically similar animals mate and produce offspring. Inbreeding depression can manifest as reduced fitness of quantitative traits such as number of offspring produced or offspring survival rate.[10] All these forces can result in reduced adaptive potential and increased risk of extinction.[1]

Two well-known examples of the negative effects of inbreeding depression that resulted from a bottleneck occurred in the African cheetah (*Acinonyx jubatus*)[11] and the Florida panther (*Puma concolor coryi*).[12] The genetic diversity of African cheetahs was reduced during the Pleistocene because their population size was dramatically reduced over several generations. Their extremely low levels of genetic diversity are manifested in extant populations by their high infant mortality rates and sperm deformities. Due to the genetic similarity among individuals, cheetahs are unable to reject skin grafts from nominally unrelated individuals.[11]

The Florida panther endured two bottlenecks, one during the same time as the African cheetah and a second bottleneck during the 19th and early 20th century as a result of human activities.[11] By the early 1990s, the Florida panther population had declined to an estimated 20 to 25 individuals and had very low levels of genetic diversity. The animals showed many signs of inbreeding depression, including undescended testicles and low sperm quality, all of which suggested that extinction was very likely. Because genetic diversity takes a long time to increase by mutation, the only source of additional alleles (variable forms of a gene) was from a sister subspecies from Texas (*P. c. stanleyana*). Females from this sister group were transported to Florida where they bred with Florida panthers with positive initial results. Genetic diversity and apparent fitness both increased; however, the long-term success of this genetic rescue remains to be demonstrated. Further, because the rescue was effected by introducing alien genes, and the new genomes may swamp those of the original subspecies, philosophical questions remain about precisely what has been conserved.[13]

Resolving Taxonomic Status and Determining Relevant Conservation Units

One of the top priorities of a conservation program should be to resolve any existing taxonomic uncertainties. Funding resources and personnel time that can be devoted to a given project are limited, and it is vital to ensure that the species status of a focus organism is well established. If a focal organism is really the same species as another organism that is not of conservation concern, then precious funds and resources will be wasted. This is fundamentally a question of understanding how genetic diversity is distributed among groups. The expectation is that distinct species will have higher levels of genetic diversity between the taxa than within each species because of the homogenizing effect of gene flow.

One example is the dusky seaside sparrow (*Ammodramus maritimus nigrescens*), which was initially one of nine recognized subspecies.[14] Genetic analyses, however, only supported the designation of two subspecies.[15,16] Conversely, if a species is not recognized as a separate entity, then a species could be lost or identified at a point where the number of organisms remaining precludes conserving the species. This was the case with the New Zealand tuatara, a lizard that was once widely distributed throughout New Zealand but became extinct on the mainland before the arrival of European settlers. Initially, it was managed as a single species. More recently, genetic and morphological analyses identified two species, *Sphenodon punctatus* and *S. guntheri*,[16,17] that are now managed as separate taxa because they do not currently exchange genetic material.

The identification of conservation units is used to determine the fundamental units that will be the focus of management efforts. Relevant conservation units include consideration of species boundaries as described in the preceding text, but it goes further by identifying subgroups that may be on distinct evolutionary trajectories from other populations within the species. One example is the mountain yellow-legged frog (*Rana muscosa*) that has undergone a rapid decline in recent decades due to predation by an introduced species, the virulent chytrid fungus (*Batrachochytrium dendrobatidis*), and habitat destruction related to wildfires. Currently, these frogs exist as isolated mountain populations. When once-widespread species are reduced to a collection of isolated populations, it is useful to understand whether these populations have always been isolated (and are therefore potentially adapted to local conditions) or whether they are merely relics of a previously widespread but uniform population. Genetic analyses determined that existing populations have long been isolated from each other, suggesting that they should be managed as separate units,[18] and this will need to be considered when frogs from captive breeding programs are reintroduced to the wild.

Determining Population Structure

An important part of identifying conservation units is testing for population structure, which involves the level of dispersal between geographically distinct areas. When widely distributed populations consist of several distinct genetic subunits, population structure is said to exist. It is also important to understand metapopulation dynamics (periodic local extinction and recolonization events) when they are present. The traditional approach is to group individuals into researcher-determined geographic units and measure the distribution of genetic diversity within and between these groupings.[19] Assignment test and clustering programs like STRUCTURE[20] are being used to determine whether cryptic population structure exists without any *a priori* groupings of individuals.

The presence of distinct population units has important consequences for wildlife management plans because a widely distributed species with distinct and isolated population segments might have little potential to recolonize habitat after a local population crash. Sometimes the presence of structured populations is indicated by obvious biogeographic barriers or the presence of distinct phenotypes. However, population structure often exists without these telltale signs. Cryptic population structure is rather common in nature, even in large, mobile animals such as the Eurasian lynx (*Lynx lynx*) in Scandinavia where genetic markers indicate that the species is divided into three distinct genetic units.[21]

Captive Breeding Programs

Molecular techniques can be used to inform management decisions during all phases of a captive breeding program. Captive-bred populations ideally should be representative of the genetic diversity found in wild populations. Genetic techniques can be used to define population structure in wild populations to help identify appropriate source populations, confirm the correct species identification for animals included in the captive breeding program, assist initial pairing decisions to ensure that closely related individuals are not mated, help determine appropriate release sites, and monitor the success of reintroduced animals by monitoring their population size and reproductive contributions. Genetic techniques also can be used several years after a captive breeding program is established to reconstruct pedigrees for organisms that live in groups[22] or determine the amount of genetic diversity that a captive population has retained from wild population.[23,24]

A combination of genetic and morphological analyses were used to discover that the Association of Zoos and Aquarium's Asiatic lion (*Panthera leo persica*) captive breeding program included African lion (*P. leo leo*) founders.[25] Asiatic lions have low levels of genetic diversity, so the interbreeding of a closely related subspecies increased their genetic diversity. However, the crossing of two subspecies could reduce adaptations that may have developed in each taxon. The previous example of interbreeding Texas puma with Florida panther may have reduced some of their adaptations to their environment—but if nothing had been done, extinction was the likely outcome.

Managing Invasive Species

Invasive species can have a tremendous negative impact on biodiversity.[1] Molecular genetic techniques can be used to identify the source population of an invasive species, monitor potential hybridization with closely related taxa, determine the geographic distribution of an invasive species, and monitor their population dynamics. Genetic techniques also can be used to predict whether an invasive species will have a good chance at becoming established and resisting control agents. Sometimes, invasive species have more genetic diversity than the suspected source populaiton, suggesting that there may have been multiple source populations. The higher amount of genetic diversity created by interbreeding multiple source populations in turn could increase the adaptive potential of the invasive species to resist strong selective pressures, such as herbicides and pesticides.[26]

Wildlife Forensics

The field of wildlife forensics has used molecular genetic techniques to identify the species of origin of unknown samples, identify the gender of an individual, and determine the geographic location from which an organism was obtained.[26] These techniques have been applied to a wide range of animals that include cetaceans,[27] sharks,[28] elephants,[29] and sturgeon.[30] DNA sequence information can be used like grocery store barcodes to match an unknown sample to a database of known samples to determine the species identity. If the population structure of an organism is known, then the most likely geographic source of the organism can be determined by identifying the source population with the most similar genotype to the organism's genotype of interest, and this information could be used to determine if an animal was taken from a protected population.

Crop and Livestock Management

Genetic diversity is a key aspect of food security. The industrial-scale food production necessary to feed the increasing human population requires a high level of standardization and automation, leading to a focus on a relatively few varieties of crops and animals.[31] While standardization and mass production are important in feeding the world's population, this strategy is not without risk. When a single variety dominates the food production system in a region, there is a risk of total crop failure when something goes wrong.[32] In low-diversity varieties, all individuals are likely equally vulnerable to diseases or environmental challenges.

A classic manifestation of this risk is the southern U.S. corn blight that occurred in 1970. Commercial corn seeds are produced by crossing different inbred lines of corn. One of these lines was used in seed corn that was widely distributed in the 1960s, which initially contributed to robust and healthy corn crops. However, parasites with short generation times can often out-evolve their relatively long-lived hosts. One such parasite is the fungus *Helmithosporium maydis*, which was a minor agricultural pest until 1970. It is believed that natural mutation created a more virulent form of the fungus which, when combined with the warm, wet summer weather, reduced U.S. corn yields by perhaps 15% that year.[33]

Strategies to mitigate these effects in the long term include seed banks and the preservation of heirloom varieties and landraces.[34] Genetic and genomic technologies are useful in mitigating these risks. For example, molecular markers have been used to determine that Iberian landraces of pigs (*Sus scrofa domesticus*) represent important reservoirs of genetic diversity.[35]

Genetic Ecotoxicology

The field of genetic ecotoxicology studies the impact of contaminants on an organism's DNA.[36] Pollutants can increase the mutation rate of DNA and increase the amount of genetic diversity in a population. However, most mutations that affect fitness will be harmful.[26] The genetic diversity of populations from control and contaminated sites can be compared to quantify the impact of a particular contaminant.[37,38] Ellegren et al.[39] compared the occurrence of a mutation that causes partial albinism in barn swallows (*Hirundo rustica*) from a site near the nuclear accident at Chernobyl to barn swallows from uncontaminated sites. They found a higher incidence of partial albinism in barn swallows that bred near Chernobyl compared to their control sites. Quantifying the genetic diversity of a population that was exposed to a pollutant can be used to predict the potential negative impacts of the pollutant on the viability of the population.

HOW IS GENETIC DIVERSITY MEASURED AND INTERPRETED?

DeYoung and Honeycutt[38] provided an overview of the different types of molecular genetic tools or molecular markers that can be used to generate genetic data. These molecular markers include 1) expressed proteins; 2) Restriction Fragment Length Polymorphisms (RFLPs); 3) Random Amplified Polymorphic DNA (RAPDs); 4) Amplified Fragment Length Polymorphisms (AFLPs[40]); 5) nucleotide sequence data from nuclear, mitochondrial, or chloroplast DNA; 6) Simple Sequence Repeats (SSRs); and 7) Single-Nucleotide Polymorphisms (SNPs).

The first three methods are now obsolete.[41] While they yielded important data in early studies, they are expensive per individual, are difficult to automate, and require considerable skill to generate reproducible results. AFLPs, SSRs, and sequencing are all still widely used in conservation genetics. The AFLP technique uses restriction enzymes to break the genome into small fragments and a modified version of the polymerase chain reaction (PCR) to visualize a subset of these fragments. As with RFLPs, pairs of homologous restriction sites will generate identically sized fragments. The PCR primers used contain fluorescent labels that allow fragment sizes to be measured using high-resolution DNA capillary electrophoresis. A high degree of automation increases the objectivity of the analysis.

Similarly, genotypes determined using microsatellite loci (also known as SSR loci, Single Tandem Repeat loci, Variable Number of Tandem Repeat loci, and other names) are quite reproducible and easy to automate. SSR loci are repeated nucleotide motifs, typically one to six nucleotides that are repeated dozens of times. In the nuclear genome, they are usually about 75 to 300 base pairs long. These short motifs are repeated, and mutations occur frequently, manifesting themselves as different numbers of motif repeats that are called alleles.[26] Microsatellite markers quickly gained in popularity in the mid-1990s, and they are one of the most widely used markers in conservation genetics.[41] A key advantage of this technique is that when combined with high-resolution electrophoresis, the data are highly reproducible and the scoring of genotypes is quite objective. Traditionally, the cost of developing a set of microsatellite loci and testing their reliability for a new species has been quite high; however, next-generation DNA sequencing methods (also known as massively parallel or deep sequencing methods) have made locus discovery more affordable.[42]

Direct sequencing is perhaps the most objective and reproducible method for measuring genetic diversity. However, the level of diversity in a short region is generally low; so long stretches of DNA must be sequenced to provide useful information. The cost of determining sequences has historically been high relative to the other markers. One popular strategy is to sequence a small number of individuals for a large number of base pairs to identify positions where more than one nucleotide is variable within a population. These SNPs are single base pair differences either between homologous chromosomes within an individual or between individuals. Once these positions are identified, PCR-based techniques may be used to assay these SNPs without having to resequence long stretches of non-variable DNA. The identification of SNPs using PCR or other assay techniques is labor intensive. However, the reduced cost of next-generation DNA sequencing methods suggests in the future that broad and representative sections of the genomes of many individuals may be sequenced at a reasonable cost, thus collapsing the processes of locus discovery and population assays.[43] In the near future, SNP markers will likely replace microsatellite markers as the most common type of markers used in conservation genetic studies, and these will be derived from next-generation DNA sequences.

Next-generation DNA sequencing is only the latest innovation to reduce the cost of generating sequencing data, which has been steadily falling since the development of Sanger sequencing in the 1970s.[44] New sequencing technologies in the field of human genomics are the main drivers for the reduction in cost for generating enormous amounts of nucleotide sequence data.[45] Sequence information from partial or entire genomes can be generated for multiple individuals. Miller et al.[46] developed a restricted site-associated DNA (RAD) marker method that uses restriction enzymes to digest genomic DNA and determine the presence or absence of a restriction enzyme's recognition site across an entire genome. This method was substantially improved by Baird et al.[47] by adding adapters to the fragmented pieces of DNA and using next-generation DNA sequencing to determine the nucleotides in each fragmented piece of DNA. This RAD-sequenced (RAD-seq) method can generate tens of thousands of SNP markers. An initial limitation of next-generation methods has been the requirement of substantial genomic resources, such as a genome map, to fully use these methods. Peterson et al.[48] have greatly improved the RAD-seq method by developing a cost-effective double-digest RAD-seq method for identifying tens to hundreds of thousands of SNP markers in non-model organisms that have no previous genomic resources available. We expect that going forward, these partial genome sequences or RAD-seq markers will be widely used to directly assay both SNP and microsatellite variation in population studies.

CONSERVATION GENOMICS

Genomics is a field concerned with the study of an organism's whole genome (or at least substantial portions of it). The major components of genomics are data acquisition, bioinformatics, and archiving data in databases.[49] The field of genetics in general is rapidly expanding with the advances in technology. As a consequnce, conservation genetics is transitioning into the field of conservation genomics.[41,50–52] Advances in human genomics are greatly reducing the cost of generating genomic data. A challenge grant was initiated in 2003 by the J. Craig Venter Science Foundation to generate an entire human genome for $1,000.[53] Once this goal is achieved, the whole genome analysis of wild populations (i.e., population genomics) will become routine. This transition represents a shift from analyzing a limited number of genes to a more comprehensive survey of the entire genome of an organism. The conceptual difference between a conservation genetic and genomic approach is depicted by Ouborg et al.[41] One main difference is that a genomic approach can quantify the impact of selection and include gene expression. One of the major contributors to the founding of the field of conservation genetics and its transformation into conservation genomics is Dr. Stephen J. O'Brien, former Chief of the Genomic Diversity Laboratory at the National Cancer Institute. O'Brien was directly involved with applying the latest developments in human genomics to endangered species[25] and began an internationally recognized annual workshop focused on conservation genetics in 1996. In 2010, the American Genetics Association held a conference on integrating genomics into the field of conservation genetics.[54]

Conservation genetics is mainly interested in characterizing levels of quantitative variation or how individuals differ in traits that are important for reproductive fitness.

A major assumption of conservation genetics is that neutral genetic variation correlates with levels of quantitative genetic variation, and if neutral genetic variation is preserved, then an organism's adaptive potential is retained. However, the correlation between neutral and quantitative genetic variation is low.[55] The development of genomic tools has exponentially increased the number of genetic markers that can be produced from typically dozens to more than tens of thousands and has catapulted the field of conservation genetics beyond just neutral markers to direct measures of quantitative genetic variation. The enormous increase in the number of markers will greatly improve the conclusions made from genetic analyses by providing more comprehensive measures of genetic diversity across the entire genome of a given species.[52] This also has allowed for the direct measurement of detrimental and adaptive genetic variation[51] and the development of adaptive potential metrics.[56]

The transition of conservation genetics to conservation genomics represents a major shift in the types of questions that can be addressed and how the questions can be addressed. Conservation genetic studies have mainly been limited to correlative studies. The development of genomic tools has allowed for a shift to a causative approach that can directly investigate evolutionary processes.[41] Genomic tools can be used to investigate gene expression using transcriptomics,[57] rather than just measuring nucleotide variation. Transcriptomics is a field of study that investigates genes that are actually expressed in an organism. Ouborg et al.[41] identified four main questions that can be addressed using genomic tools that involve: 1) measuring the impact of small population size on levels of adaptive variation; 2) identifying the underlying mechanisms that connect adaptive genetic variation and fitness; 3) determining how genes and the environment interact; and 4) estimating the impact of inbreeding and genetic drift on gene expression. Genomic tools will greatly improve the quantity and quality of information gained from genetic analyses; however, genomic tools will not be a panacea. Identifying the genetic mechanisms of adaptation is a difficult endeavor and will require a multidisciplinary approach.[50]

Genomic techniques also have so far been restricted to the analysis of relatively high-quality DNA, which will limit their application in cases where DNA is obtained from non-invasively collected hair and scat samples or specimens in museum collections. However, recently developed whole-genome amplification techniques are being used to analyze low quantities of extracted DNA in the field of human forensics. Giardina et al.[58] tested the efficacy of using whole-genome amplification on low quantities of extracted DNA to produce accurate SNP marker genotypes.

The use of genomic tools to inform management decisions is beginning to be used in fishery management.[59–61] Neutral genetic markers have been the main tool used to identify conservation units; however, with the advent of genomic tools, adaptive genetic variation can now be included in these analyses.[62] For example, Tymchuk et al.[60] used microsatellite markers and gene expression assays (i.e., transcriptomics) to identify potential adaptive differences in Atlantic salmon (*Salmo salar*) from different rivers in the Bay of Fundy. Hansen[60] advocates using Tymchuk et al.'s[59] approach to identify conservation units. Rusello et al.[61] used a genome scan of over 10,000 nuclear markers to demonstrate the efficacy of using adaptive loci (i.e., outlier loci) to identify ecologically divergent populations of Okanagan Lake kokanee, landlocked sockeye salmon (*Oncorhynchus nerka*). These researchers have identified eight potentially adaptive loci that identify the two kokanee ecotypes more reliably than 44 neutral markers. Miller et al.[46] used 4590 SNP markers to quantify potentially adaptive differences in rainbow trout (*O. mykiss*) and recommend the use of adaptive genetic variation to identify conservation units.

SPATIAL ANALYSES OF GENETIC DIVERSITY

The management of wild populations is inherently a spatial endeavor that is directly connected to an organism's environment. The spatial analysis of genetic diversity has increased with the development of the field of landscape genetics. Landscape genetics is a nascent field that combines two other fields, population genetics and landscape ecology, to study the processes that influence the spatial patterns of an organism's genetic diversity in the environment.[63] Landscape genetic studies can be used to detect discontinuities in an organism's spatial pattern of genetic diversity and correlate these gene flow discontinuities to environmental variables.[64] The spatial analyses of genetic diversity patterns also can be used to identify areas that should be designated as reserves. Vandergast et al.[65] spatially analyzed the pattern of genetic diversity of multiple taxa to identify areas that harbor a large amount of genetic discontinuities relative to other areas. These "hotspots" of evolutionary potential can then be the focus of management efforts to conserve biodiversity.

Understanding and mitigating the effects of habitat fragmentation are among the chief concerns of applied conservation genetics. Humans continue to impact the environment by increasing the amount of fragmentation. A landscape genetics approach can be used to quantify the impact of fragmentation on the movement of an organism's genes. One example is the Jerusalem cricket species *Stenopelmatus* n. sp. *"santa monica"* described by Vandergast et al.[66] from Los Angeles and Ventura Counties in southern California. These large, wingless insects are thought to be poor dispersers, and current taxonomy suggests that the family Stenopelmatidae is particularly species rich. Most human development in southern California has occurred relatively recently, and the origin dates of major settlements and highways are well documented. Using both mitochondrial DNA sequences and nuclear sequences between microsatellite loci as genetic markers, the authors identified highway presence and age as a critical contributor to levels of population isolation above and beyond that predicted based on geographic distance alone. It is intuitive to expect that small, earthbound creatures will find it difficult to cross a multilane freeway, but why should the age of the freeways be a factor? The answer is genetic drift. When populations are small, random factors cause allele and haplotype frequencies to vary. When a patch of habitat is cut in half by a barrier, each new subpopulation is more susceptible to the effects of genetic drift than the previous, larger population. Because the new parts of a bisected population are likely to drift in different directions, the genetic distance between them will increase over time. Newly isolated subpopulations, however, will be similar until the effects of genetic drift have manifested themselves, and so populations on either side of a new highway will be similar at first but then diverge over time. The same is true for lost genetic diversity, as new population fragments may be highly diverse but eventually lose allelic diversity over time through drift and selection. Another key aspect of genetic diversity and divergence is that they may reflect divergent selection between distinct types of habitat. When local adaptation occurs, it may be important to preserve it.

The field of landscape genetics is at the cusp of rapidly transforming into the field of landscape genomics, similar to the transformation that is occurring in the field of conservation genetics to conservation genomics. The advantages of this transformation are similar to the ones discussed for the transformation from conservation genetics to conservation genomics. Landscape genetic studies have been focused on the patterns of a limited number of neutral molecular markers. The expansion to landscape genomics will allow researchers to analyze thousands of neutral markers and include adaptive loci.[67] Joost et al.[68] developed a method for identifying adaptive loci in a wild population by using a landscape genomics approach. Landscape genomic studies have the potential to identify historic processes that have influenced the currently observed patterns of adaptive genetic variation in wild populations and predict how future processes, such as climate change, will influence a population's viability.

Examples of applied landscape genomic studies are limited. Several studies have described their approach as a landscape genomics approach, but they have only used a limited number of AFLPs,[69] microsatellites,[70] or SNPs[71] and should still be described as landscape genetic studies. The best example of a landscape genomics study is by Poncet et al.[72] This group identified potentially adaptive loci in the alpine plant *Arabis alpina* by correlating their AFLP genotypes to environmental variables. This allows management efforts to focus on preserving *A. alpina*-adaptive genetic variation. This study included 825 AFLP markers, a significant improvement upon the typical number or AFLP markers; however, using this number of markers in such a study will soon become outdated with the near eventuality that the analysis of tens to hundreds of thousands of markers in a study will be affordable.

CONCLUSION

Conservation genetics is a young field;[52] however, genetic techniques already have been used to inform a wide range of management decisions. For the field to reach its full potential, conservation geneticists need to continue to educate resource managers about the important contributions that genetic analyses can make to improve the efficacy of management decisions. Collaborations should be established at the initial stage in a project between research scientists and resource managers to bridge the communication barrier and make full use of the genetic toolbox.

The quality and quantity of genetic information are expanding exponentially with the transition of conservation genetics into conservation genomics. The increased amount of information will significantly increase the accuracy and precision of conclusions made using genomic tools. The transition to a genomics approach also will expand the questions that can be addressed and fundamentally change the approach that is used to answer questions. Genomic tools will allow the direct quantification of a population's adaptive potential and transcend the field beyond a correlative field to one that uses a causative approach to discover ultimate rather than proximate causes of natural processes.

ACKNOWLEDGMENTS

We thank Matthew DeBlois, Kelsey Garlick, and the reviewers for their comments. McGreevy is supported by the National Science Foundation under Award Number 1003226. Any opinions, findings, and conclusions or recommendations expressed in this material are those of the authors and do not necessarily reflect the views of the National Science Foundation.

REFERENCES

1. Frankham, R.; Ballou, J.D.; Briscoe, D.A. *Introduction to Conservation Genetics,* 2nd Ed.; Cambridge University Press: New York, 2010.
2. Lewontin, R.C. The apportionment of human diversity. Evol. Biol. **1972**, *6* (381), 391–398.
3. Sherwin, W.B.; Jabot, F.; Rush, R.; Rossetto, M. Measurement of biological information with applications from genes to landscapes. Mol. Ecol. **2006**, *15* (10), 2857–2869.
4. Newman, D.; Tallmon, D.A. Experimental evidence for beneficial fitness effects of gene flow in recently isolated populations. Conserv. Biol. **2001**, *15*, 1054–1063.
5. Markert, J.A.; Champlin, D.M.; Gutjahr-Gobell, R.; Grear, J.S.; Kuhn, A.M., McGreevy, T.J.; Roth, A.; Bagley, M.J.; Nacci, D.E. Population genetic diversity and fitness in multiple environments. BMC Evol. Biol. **2010**, *10* (205), http://www.biomedcentral.com/1471-2148/1410/1205

6. Waples, R.S. *Definition of "species" under the endangered species act: Application to pacific salmon. NOAA Technical Memorandum NMFS F/NWC-194.* 1991.
7. Schwartz, M.K.; Luikart, G.; Waples, R.S. Genetic monitoring as a promising tool for conservation and management. Trends Ecol. Evol. **2007**, *22* (1), 25–33.
8. Wright, S. Evolution in Mendelian populations. Genetics. **1931**, *16* (2), 97–159.
9. Kolbe, J.J.; Leal, M.; Schoener, T.W.; Spiller, D.A.; Losos, J.B. Founder effects persist despite adaptive differentiation: a field experiment with lizards. Science **2012**, *335* (6072), 1086–1089.
10. Ralls, K.; Ballou, J.D.; Templeton, A. Estimates of lethal equivalents and the cost of inbreeding in mammals. Conserv. Biol. 1988, *2* (2), 185–193.
11. O'Brien, S.J.; Johnson, W.E. Big cat genomics. In *Annual Review of Genomics and Human Genetics*; Chakravarti, A., Green, E., Eds.; Annual Review, Inc.: Palo Alto, California, 2005; pp. 407–429.
12. Johnson, W.E.; Onorato, D.P.; Roelke, M.E.; Land, E.D.; Cunningham, M.; Belden, R.C.; McBride, R.; Jansen, D.; Lotz, M.; Shindle, D.; Howard, J.; Wildt, D.E.; Penfold, L.M.; Hostetler, J.A.; Oli, M.K.; O'Brien, S.J. Genetic restoration of the Florida panther. Science **2010**, *329* (5999), 1641–1645.
13. Hedrick, P. Genetic future for Florida panthers. Science **2010**, *330* (6012), 1744; author reply 1744.
14. Avise, J.C. Toward a regional conservation genetics perspective: phylogeography of faunas in the southeastern United States, In *Conservation Genetics Case Histories from Nature*; Avise, J., Hamrick, J., Eds.; Chapman and Hall: New York, New York, 1996; pp. 431–470.
15. Avise, J.C.; Nelson, W.S. Molecular genetic relationships of the extinct dusky seaside sparrow. Science **1989**, *243* (4891), 646–648.
16. Mills, L.S. *Conservation of Wildlife Populations*; Blackwell Publishing: Malden, Massachusetts, 2007.
17. Daugherty, C.H.; Cree, A.; Hay, J.M.; Thompson, M.B. Neglected taxonomy and the continuing extinctions of tuatara (*Sphenodon*). Nature **1990**, *347* (6289), 177–179.
18. Schoville, S.D.; Tustall, T.S.; Vredenburg, V.T.; Backin, A.R.; Gallegos, E.; Wood, D.A.; Fiser, R.N. Conservation genetics of evolutionary lineages of the endangered mountain yellow-legged frog, *Rana muscosa* (Amphibia: Ranidae) in southern California. Biol. Conserv. **2011**, *144* (7), 2031–2040.
19. Wright, S. *Evolution and the Genetics of Populations, Volume 4: Variability within and among Natural Populations*; University of Chicago Press: Chicago, 1978.
20. Pritchard, J.K.; Stephens, M.; Donnelly, P. Inference of population structure using multilocus genotype data. Genetics **2000**, *155* (2), 945–959.
21. Rueness, E.K.; Jorde, P.E.; Hellborg, L.; Stenseth, N.C.; Ellegren, H.; Jakobsen, K.S. Cryptic population structure in a large, mobile mammalian predator: The Scandinavian lynx. Mol. Ecol. **2003**, *12* (10): 2623–2633.
22. Ivy, J.A.; Lacy, R.C. Using molecular methods to improve the genetic management of captive breeding programs for threatened species, In *Molecular Approaches in Natural Resource Conservation and Management*; DeWoody, J.A., Bickham, J.W., Michler, C.H., Eds.; Cambridge University Press: New York, 2010; pp. 267–295.
23. McGreevy, T.J.; Dabek, L.; Gomez-Chiarri, M.; Husband, T.P. Genetic diversity in captive and wild Matschie's tree kangaroo (*Dendrolagus matschiei*) from Huon Peninsula, Papua New Guinea, based on mtDNA control region sequences. Zoo Biol. **2009**, *28* (3), 183–196.
24. McGreevy, T.J.; Dabek, L.; Husband, T.P. Genetic evaluation of the association of zoos and aquariums Matschie's tree kangaroo (*Dendrolagus matschiei*) captive breeding program. Zoo Biol. **2011**, *30* (6), 636–646.
25. O'Brien, S.J.; Wienberg, J.; Lyons, L.A. Comparative genomics: Lessons from cats. Trends Genet. **1997**, *13* (10), 393–399.
26. Allendorf, F.W.; Luikart, G. *Conservation and the Genetics of Populations*; Blackwell Publishing: Malden, Massachusetts, 2007.
27. Baker, C.S.; Cipriano, F.; Palumbi, S.R. Molecular genetic identification of whale and dolphin products from commercial markets in Korea and Japan. Mol. Ecol. **1996**, *5* (5), 671–685.
28. Magnussen, J.E.; Pikitch, E.K.; Clarke, S.C.; Nicholson, C.; Hoelzel, A.R.; Shivji, M.S. Genetic tracking of basking shark products in international trade. Anim. Conserv. **2007**, *10* (2), 199–207.
29. Wasser, S.K.; Joseph Clark, W.; Drori, O.; Stephen Kisamo, E.; Mailand, C.; Mutayoba, B.; Stephens, M. Combating the illegal trade in African elephant ivory with DNA forensics. Conserv. Biol. **2008**, *22* (4), 1065–1071.
30. Birstein, V.J.; Desalle, R.; Doukakis, P.; Hanner, R.; Ruban, G.I.; Wong, E. Testing taxonomic boundaries and the limit of DNA barcoding in the Siberian sturgeon, *Acipenser baerii*. Mitochondrial DNA. **2009**, *20* (5–6), 110–118.
31. Groeneveld, L.F.; Lensstra, J.A.; Eding, H.; Toro, M.A.; Scherf, B.; Pilling, D.; Negrini, R.; Finlay, E.K.; Jianlin, H.; Groeneveld, E.; Weigend, S.; Consortium, G. Genetic diversity in farm animals - A review. Anim. Genet. **2010**, *41* (Suppl. 1), 6–31.
32. Esquinas-Alcazar, J. Science and society: Protecting crop genetic diversity for food security: Political, ethical and technical challenges. Nat. Rev. Genet. **2005**, *6* (12), 946–953.
33. NRC. *Genetic Vulnerability of Major Crops*, N.R.C.N.A.O. Sciences; Ed.; Washington, D.C, 1972.
34. Food and Agriculture Organization of the United Nations. Report of the commission on genetic resources for food and agriculture. CGRFA-12/09/Report. 2009, ftp://ftp.fao.org/docrep/fao/meeting/017/k6536e.pdf.
35. Vicente, A.A.; Carolino, M.I.; Sousa, M.C.; Ginja, C.; Silva, F.S.; Martinez, A.M.; Vega-Pla, J.L.; Carolino, N.; Gama, L.T. Genetic diversity in native and commercial breeds of pigs in Portugal assessed by microsatellites. J. Anim. Sci. **2008**, *86* (10), 2496–2507.
36. Shugart, L.R.; Theodorakis, C.W.; Bickham, J.W. Evolutionary toxicology, In *Molecular Approaches in Natural Resource Conservation and Management*, DeWoody, J.A., Bickham, J.W., Michler, C.H., Eds.; Cambridge University Press: New York, 2010; pp. 320–362.
37. Bickham, J.W.; Sandhu, S.; Hebert, P.D.; Chikhi, L.; Athwal, R. Effects of chemical contaminants on genetic diversity in natural populations: implications for biomonitoring and ecotoxicology. Mutat. Res. **2000**, *463* (1), 33–51.
38. DeYoung, R.W.; Honeycutt, R.L. The molecular toolbox: genetic techniques in wildlife ecology and management. J. Wildlife Manag. **2005**, *69* (4), 1362–1384.

39. Ellegren, H.; Lindgren, G.; Primmer, C.R.; Moller, A.P. Fitness loss and germline mutations in barn swallows breeding in Chernobyl. Nature **1997**, *389* (6651), 593–596.
40. Vos, P.; Hogers, R.; Bleeker, M.; Reijans, M.; van de Lee, T.; Hornes, M.; Frijters, A.; Pot, J.; Peleman, J.; Kuiper, M.; Zabeau, M. AFLP: A new technique for DNA fingerprinting. Nucleic Acids Res. **1995**, *23* (21), 4407–4414.
41. Ouborg, N.J.; Pertoldi, C.; Loeschcke, V.; Bijlsma, R.K.; Hedrick, P.W. Conservation genetics in transition to conservation genomics. Trends Genet. **2010**, *26* (4), 177–187.
42. Davey, J.W.; Hohenlohe, P.A.; Etter, P.D.; Boone, J.Q.; Catchen, J.M.; Blaxter, M.L. Genome-wide genetic marker discovery and genotyping using next-generation sequencing. Nat. Rev. Genet. **2011**, *12* (7), 499–510.
43. Glenn, T.C. Field guide to next-generation DNA sequencers. Mol. Ecol. Resour. **2011**, *11* (5), 759–769.
44. Sanger, F.; Coulson, A.R. A rapid method for determining sequences in DNA by primed synthesis with DNA polymerase. J. Mol. Biol. **1975**, *94* (3), 441–446.
45. Pareek, C.S.; Smoczynski, R.; Tretyn, A. Sequencing technologies and genome sequencing. J. Appl. Genet. **2011**, *52* (4), 413–435.
46. Miller, M.R.; Brunelli, J.P.; Wheeler, P.A.; Liu, S.; Rexroad, C.E., 3rd; Palti, Y.; Doe, C.Q.; Thorgaard, G.H. A conserved haplotype controls parallel adaptation in geographically distant salmonid populations. Mol. Ecol. **2012**, *21* (2), 237–249.
47. Baird, N.A.; Etter, P.D.; Atwood, T.S.; Currey, M.C.; Shiver, A.L.; Lewis, Z.A.; Selker, E.U.; Cresko, W.A.; Johnson, E.A. Rapid SNP discovery and genetic mapping using sequenced RAD markers. PLoS One **2008**, *3* (10), e3376.
48. Peterson, B.K.; Weber, J.N.; Kay, E.H.; Fisher, H.S.; Hoekstra, H.E., Double digest RADseq: an inexpensive method for de novo SNP discovery and genotyping in model and non-model species. PLoS One **2012**, *7* (5), e37135.
49. Amato, G.; DeSalle, R. *Conservomics? The Role of Genomics in Conservation Biology. Conservation Genetics in the Age of Genomics, New Directions in Biodiversity Conservation*; Columbia University Press: New York, 2009; 169–178.
50. Allendorf, F.W.; Hohenlohe, P.A.; Luikart, G. Genomics and the future of conservation genetics. Nat. Rev. Genet. **2010**, *11* (10), 697–709.
51. Kohn, M.H.; Murphy, W.J.; Ostrander, E.A.; Wayne, R.K. Genomics and conservation genetics. Trends Ecol. Evol. **2006**, *21* (11), 629–637.
52. Primmer, C.R. From conservation genetics to conservation genomics. In *The Year in Ecology and Conservation Biology, Annals of the New York Academies of Sciences*, Ostfeld, R.S., Schlesinger, W.H. Eds.; Wiley-Blackwell: Hoboken, New Jersey, 2009; pp. 357–368.
53. Service, R.F., Gene sequencing. The race for the $1000 genome. Science **2006**, *311* (5767), 1544–1546.
54. AGA, American Genetics Association Annual Symposium, 2010.
55. Reed, D.H.; Frankham, R. How closely correlated are molecular and quantitative measures of genetic variation? A meta-analysis. Evolution **2001**, *55* (6), 1095–1103.
56. Bonin, A.; Nicole, F.; Pompanon, F.; Miaud, C.; Taberlet, P. Population adaptive index: a new method to help measure intraspecific genetic diversity and prioritize populations for conservation. Conserv. Biol. **2007**, *21* (3), 697–708.
57. Anonymous. Proteomics, transcriptomics: What's in a name? Nature **1999**, *402* (6763), 715.
58. Giardina, E.; Pietrangeli, I.; Martone, C.; Zampatti, S.; Marsala, P.; Gabriele, L.; Ricci, O.; Solla, G.; Asili, P.; Arcudi, G.; Spinella, A.; Novelli, G. Whole genome amplification and real-time PCR in forensic casework. BMC Genomics **2009**, *10*, 159.
59. Tymchuk, W.; O'Reilly, P.; Bittman, J.; Macdonald, D.; Schulte, P. Conservation genomics of Atlantic salmon: variation in gene expression between and within regions of the Bay of Fundy. Mol. Ecol. **2010**, *19* (9), 1842–1859.
60. Hansen, M.M. Expression of interest: transcriptomics and the designation of conservation units. Mol. Ecol. **2010**, *19* (9), 1757–1759.
61. Rusello, M.A.; Kirk, S.L.; Frazer, K.K.; Askey, P.J. Detection of outlier loci and their utility for fisheries management. Evol. Appl. **2011**, *5* (1), 39–52.
62. Nosil, P.; Funk, D.J.; Ortiz-Barrientos, D. Divergent selection and heterogeneous genomic divergence. Mol. Ecol. **2009**, *18* (3), 375–402.
63. Manel, S.; Schwartz, M.K.; Luikart, G.; Taberlet, P. Landscape genetics: combining landscape ecology and populaiton genetics. Trends Ecol. Evol. **2003**, *18* (4), 189–197.
64. McKelvey, K.S.; Cushman, S.A.; Schwartz, M.K. Landscape genetics. In *Spatial Complexity, Informatics, and Wildlife Conservation*, Cushman, S.A., Huettmann, F., Eds.; Springer: New York, 2010; pp. 313–328.
65. Vandergast, A.G.; Bohonak, A.J.; Hathaway, S.A.; Boys, J.; Fisher, R.N. Are hotspots of evolutionary potential adequately protected in southern California? Biol. Conserv. **2008**; *141* (6), 1648–1664.
66. Vandergast, A.G.; Lewallen, E.A.; Deas, J.; Bohanak, A.J.; Weissman, D.B.; Fisher, R.N. Loss of genetic connectivity and diversity in urban microreserves in a southern California endemic Jerusalem cricket (Orthoptera: Senopelmatidae: *Stenopelmatus* n. sp. "santa monica"). J. Insect Conserv. **2009**, *13* (3), 329–345.
67. Schwartz, M.K.; Luikart, G.; McKelvey, K.S.; Cushman, S.A. Landscape genomics: A brief perspective. In *Spatial Complexity, Informatics, and Wildlife Conservation*, Cushman, S.A., Huettmann, F., Eds.; Springer: New York, 2010; pp. 165–174.
68. Joost, S.; Kalbermatten, M.; Bonin, A. Spatial analysis method (sam): a software tool combining molecular and environmental data to identify candidate loci for selection. Mol. Ecol. Resour. **2008**, *8* (5), 957–960.
69. Parisod, C.P.; Christin, P-A. Genome-wide association to fine-scale ecological heterogeneity within a continuous population of *Biscutella laevigata* (Brassicaceae). New Phytol. **2008**, *178* (2), 436–447.
70. Meier, K.; Hansen, M.M.; Bekkevold, D.; Skaala, O.; Mensberg, K-LD. An assessment of the spatial scale of local adaptation in brown trout (*Salmo trutta* L.): Footprints of selection at microsatellite DNA loci. Heredity **2011**, *106* (3), 488–499.
71. Pariset, L.; Joost, S.; Marsan, P.A.; Valentini, A.; Consortium, E. Landscape genomics and biased Fst approaches reveal single nucleotide polymorphisms under selection in goat breeds of North-East Mediterranean. BMC Genet. **2009**, *10* (7), 1–8.
72. Poncet, B.N.; Herrmann, D.; Gugerli, F.; Taberlet, P.; Holderegger, R.; Gielly, L.; Rioux, D.; Thuiller, W.; Aubert, S.; Manel, S. Tracking genes of ecological relevance using a genome scan in two independent regional population samples of *Arabis alpina*. Mol. Ecol. **2010**, *19* (14), 2896–2907.

BIBLIOGRAPHY

1. Bertorelle, G.; Bruford, M.W.; Hauffe, H.C.; Rizzoli, A.; Vernesi, C. *Population Genetics for Animal Conservation, Conservation Biology 17*; Cambridge University Press: New York, 2009.
2. Cassidy, B.G.; Gonzales, R.A. DNA testing in animal forensics. J. Wildl. Manage. **2005**, *69* (4), 1454–1462.
3. Leberg, P. Genetic approaches for estimating the effective size of populations. J. Widl. Manage. **2005**, *69* (4), 1385–1399.
4. Scribner, K.T.; Blanchong, J.A.; Bruggeman, D.J.; Epperson, B.K.; Lee, C-Y.; Pan, Y-W.; Shorey, R.I.; Prince, H.H.; Winterstein, S.R.; Luukkonen, D.R.; Geographical genetics: conceptual foundations and empirical application of spatial genetic data in wildlife management. J. Wildl. Manage. **2005**, *69* (4), 1434–1453.
5. Waits, L.P.; Paetkau, D. Noninvasive genetic sampling tools for wildlife biologists: a review of applications and recommendations for accurate data collection. J. Wildl. Manage. **2005**, *69* (4), 1419–1433.
6. Wilson, E.O.; Bossert, W.H. *A Primer of Population Biology*; Sinauer Associates, Inc.: Sunderland, Massachusetts, 1971.

Genetic Resources Conservation: Ex Situ

Mary T. Burke
UC Davis Arboretum, University of California, Davis, California, U.S.A.

Abstract

Botanical gardens, arboreta, and herbaria hold nearly 50 million documented ex situ plant specimens in scientific collections. As educators, scientists, and conservationists, the staff of these institutions are in a special position to reach the public with ideas, sound information, and inspiration. In addition to their considerable scientific merit, these collections help educate a community of people worldwide and engage with them to preserve and restore the damaged ecosystems of the world.

INTRODUCTION

Botanical gardens and arboreta are collections of living plants used for research, study, and education. Arboreta traditionally focus on collections of large trees and shrubs (woody plants) while botanical gardens include research and display collections of both herbaceous (bulbs, biennials, perennials) and woody (trees and shrubs) plants. In actual practice, the boundaries between these two collection types have become less clear during the past century; for instance, many arboreta have expanded their collections to include herbaceous plants but retain their historical names. Both botanical gardens and arboreta are distinguished from public parks and other recreational landscapes by collection policies that emphasize wild-collected plants, by collections organized for scientific and educational purposes, and by extensive documentation; that is, a plant record system that includes information on provenance, nomenclature, and other taxonomic and cultural information of interest to researchers. A herbarium (herbaria, pl.) is a collection of dried, pressed, or preserved plant specimens with associated relevant collection information. Many herbaria are associated with botanical gardens and arboreta; others are associated with universities and other plant science research facilities. Some botanical gardens also maintain or are closely associated with seed banks, facilities where seeds are stored under cold and dry conditions in order to preserve the seed viability for future use.[1] As extinction rates for plant species have increased, these institutions have begun to play important roles in ex situ conservation, or conservation outside the native habitat.

CENTERS FOR EDUCATION

Botanical gardens and arboreta are "living museums"; in addition to their scientific mission, these plant collections are important and popular centers of informal science education. In the last 20 years, botanical gardens and arboreta have begun to switch their focus away from simply displaying and identifying plants, and more toward conveying critical scientific ideas and significant issues. Exhibits and labels are being redesigned to teach not only about plants as beautiful or interesting organisms, but about the role plants play in the real world: their ecosystem functions, the dependence of all life upon them, information about invasive species and their threats, the need to survey and understand unknown areas, and how people can manage native plant ecosystems. Effective conservation exhibits give visitors more of the fascinating and complex story of which plants are a critical part, rather than just a little information about a particular plant.

Globally, botanical gardens and arboreta are visited by more than 150 million people a year.[2] Recognizing the opportunity this represents, botanical gardens have assumed a new responsibility to educate people in a way that will help local citizens go on to influence policy makers and create a locally active group of people that have global concerns.

Examples of gardens and displays organized around conservation themes include the two-acre New England Garden of Rare and Endangered Plants at the New England Wild Flower Society's Garden in the Woods; the large display of rare and endangered California native plants, as well as 210 rare taxa from around the world, at the University of California Botanical Garden at UC Berkeley; and the exhibits highlighting the impact of invasive plants in Australia at the Botanic Gardens of Adelaide. Even the work of tissue culture as a method for preserving genetic resources has been showcased as a public exhibit at the Center for Research of Endangered Wildlife, the research program of the Cincinnati Zoo and Botanical Gardens. Here, tour groups can walk through the scientific facility and look through glass-lined walls at tissue culture incubators, a working greenhouse, and into a laboratory with liquid nitrogen storage tanks—The Frozen Garden. These

Encyclopedia of Natural Resources DOI: 10.1081/E-ENRL-120020305

exhibits introduce the general public to some aspects of science and technology currently focused on the preservation of rare and endangered species.[3]

CENTERS FOR RESEARCH AND DISCOVERY

As scientific institutions, botanical gardens and arboreta serve another important conservation role by conducting research or plant surveys on critical habitats, endangered plants, and invasive species. Collaborating with universities and other centers for plant studies, botanical gardens and arboreta are a place where people come together to study species of special concern. In many countries, public gardens are among the leading, and sometimes the only, institutions capable of undertaking the extensive work needed in plant research and conservation.[2]

There are about 2200 botanical gardens and arboreta in the world; it is estimated that about one-fourth of the world's flowering plants and ferns are included within their collections.[4] The extensive ex situ taxonomic and geographic collections that botanical gardens have amassed also play a critical role in botanical research, as they gather together in one place a wide variety of plants that would be difficult to study in the wild. These documented collections provide scientists with accurately identified plants of known provenance. Information about the locality in which the plant was originally collected is maintained by the scientific staff to enhance the value of the collection for researchers. The genetic resource collections in botanical gardens are often supplemented at major plant research centers by ancillary collections, including extensive herbarium collections of preserved plant specimens and seed and tissue banks. Herbaria deserve special mention as repositories of plant genetic resources: these museum specimens serve as a permanent global reference on the plant diversity of the world.

Efforts are currently underway to consolidate the inventories of ex situ plant genetic resource collections around the world, held in seed banks, field collections, herbaria, and other collections. These efforts have been hampered to some extent by not simply the scale of the undertaking, but the wide variety of ways in which collection managers have independently tracked information about their scientific holdings. Protocols for information management in natural history collections are gradually evolving in the scientific community (Darwin protocol,[5] Dublin metadata core[6]). Currently, some botanical gardens and arboreta that use a common database standard (BG-Base) list their holdings on the web in a common format that can be searched in all the consolidated inventories through a single search query.

Traditionally, botanical gardens, arboreta, and herbaria have provided open access to their collections and holdings for research or conservation purposes. In the spirit of willing compliance with the Convention on Biologic Diversity (1993) on the issue of ownership of genetic resources,[7] many gardens have begun to implement policies and procedures to address obligations regarding access to their collections and sharing the benefits of research based on them with the nations of origin. An estimated 90% of all living plant collections in botanical gardens predate the Convention on Biological Diversity (CBC). Although, technically, such collections are not covered by its provisions,[2] most scientific collections attempt to meet the CBC's goals and comply with the spirit of the convention.

CENTERS FOR PLANT PROTECTION

In addition to education and research, botanical gardens have another critical role to play in plant conservation: they often serve as the sole source of horticultural expertise for specialty plant groups, particularly for wild plants that are rarely used for display or ornamental purposes.[2] Highly accomplished nursery staff at botanical gardens often are experienced in the propagation of unusual plants. The value of this highly specialized skill to global plant conservation cannot be overemphasized, for knowing how to grow a plant may be a key to its survival or its return from the brink of extinction. In Hawaii, tissue culture and micropropagation have also been critical tools in the attempt to reestablish plants from in situ populations with less than 10 or 20 individuals.

Working within national or local consortiums or with state or federal agencies, botanical gardens may grow large collections of endangered plants, sometimes receiving them in the wake of new development or urbanization of former wildlands, holding them safely in cultivation or in seed banks, and occasionally reintroducing them back into the wild as part of species-recovery programs. Other endangered plants are featured in display gardens as part of the educational program of the garden. These displays vividly illustrate the disappearing flora to visitors and educate them about their own disappearing local ecosystems. For example, the Coco De Mer Gardens Reserve on Praslin Island and the National Botanic Garden on Mahe in the Seychelles both feature the unusual rare and endemic plants native to these islands.

Many botanic gardens are also involved in in situ conservation efforts, as they manage natural reserves or work with associated scientists and activists to study, monitor, and conserve plants in the wild. Restoration ecology and its practical applications is an important area of study and application for many botanical gardens in the United States.

The Royal Botanical Garden in Hamilton, Ontario, Canada, has been working for 10 years on Project Paradise, the largest habitat restoration project in North America, to restore the wetlands and fisheries included within the 2200 acres of natural land the garden manages on the shores of Lake Ontario.[8] Staff of botanical gardens often work as champions of natural reserve systems that protect the flora of their region. Examples include the work of Dr. Charles Lamoureux, Director of the University of Hawaii Lyon

Arboretum, in establishing the 19 reserves, encompassing more than 109,000 acres, of the State of Hawaii Natural Reserves System, and the work of Dr Mildred Mathias, Director of the UC Irvine Botanical Garden (now named in her honor), in establishing the University of California Natural Reserve System that protects 130,000 acres.

A NEW ERA OF INTERNATIONAL AND NATIONAL COLLABORATION

During the last 20 years, botanical gardens, arboreta, and public gardens worldwide have begun to link together in a highly focused effort to safeguard the world's genetic resources. Clearly, in an era of declining habitats and disappearing plant species, intensive and deliberate collaboration on an international scale is needed to identify plants most in need of protection in ex situ collections. In 1987, Botanical Gardens Conservation International (BGCI) was founded as a network of cooperating botanical gardens dedicated to effective plant conservation.

At present, with over 450 member gardens in 100 countries, BGCI works to establish clear and effective plant conservation goals for a worldwide community of scientists and professions and to assist gardens in working collaboratively to meet these goals. In its first 15 years, BCGI has supported a wide range of activities for the international community: it provides technical guidance and support for botanical gardens engaged in plant conservation efforts; assists in the development and implementation of the worldwide Botanical Gardens Conservation Strategy for plant conservation (the "International Agenda for Botanic Gardens in Conservation," June 2000); helps to create and strengthen national and regional networks of gardens in many parts of the world; and organizes major meetings, workshops, and training courses. Recognizing the primary role public gardens play in conservation education, BCGI also publishes a newsletter, *Roots*, to share experience and information about plant conservation specifically written for educators.

Just as importantly, BGCI has a support group, the Plant Charter Group, to reach out to the business community to educate them about the importance of plants and help develop partnerships for plant conservation projects. These efforts have led to establishment of a unique $50 million "eco-partnership" between HSBC, an investment bank, and BCGI, Earthwatch, and the World Wildlife Federation to assist botanical gardens internationally in conserving and managing plant genetic resources.[9]

In addition to working tirelessly to increase communication and collaboration between the botanical gardens and arboreta internationally, BGCI also conducts inventories and surveys on issues in plant conservation, supplying much needed data for good decision making. A critical first step was the 2001 survey of botanical gardens of the world and their ex situ plant collections. This survey documented the status of 2178 botanical gardens in 153 countries. These research and teaching gardens include approximately 6.13 million accessions in their living collections and another 42 million herbarium specimens in botanical garden herbaria. The majority of these botanical gardens are in the developed countries in Europe and North America (850), while another 200 are found in East and Southeast Asia. There are relatively few botanical gardens or centers of plant conser vation in North and Southern Africa, the Caribbean islands, South West Asia, and the Middle East. In some tropical countries, new gardens have been created in conjunction with national parks and are designed to play roles in conservation, sustainable development, and public education.[4]

Similar collaborative efforts are underway in the United States. The American Association of Botanical Gardens and Arboreta has an active Plant Conservation Committee and regularly publishes articles on the plant conservation activities of its member gardens in its journal, *The Public Garden*. The North American Plant Collections Consortium is a voluntary association of botanical gardens and arboreta in North America that focuses each participating garden's collection development in way that will ensure effective conservation of genetic resources.[10] National collections of important genera are protected in a coordinated and complementary network of public gardens across North America, where each garden is encouraged to enrich its collection with a targeted list of plants that have been identified as top priority for conservation and are appropriate to that garden's climate and site. The Center for Plant Conservation conducts similar collaborative efforts, enjoining its participant gardens to step forward and assume coordinated ex situ protection of the rare, endangered, and threatened plants in its local flora.

CONCLUSIONS

Botanical gardens, arboreta, and herbaria hold nearly 50 million documented ex situ plant specimens in scientific collections. As educators, scientists, and conservationists, the staff of these institutions are in a special position to reach the public with ideas, sound information, and inspiration. In addition to their considerable scientific merit, these collections help educate a community of people worldwide and engage with them to preserve and restore the damaged ecosystems of the world.

REFERENCES

1. Schoen, D.J.; Brown, A.H.D. The conservation of wild plant species in seed banks. BioScience **2001**, *51*, 960–966.
2. *Introduction: Botanic Gardens & Conservation*. BGCI Online. Botanic Gardens Conservation International website, 2003; http://raq43.nildram.co.uk/bgci/action/viewDocument?id¼43&location¼searchresults#Introduction (accessed August 2003).

3. Marinelli, J. Bringing plant conservation to life. Public Garden **2001**, *16* (1), 8–11.
4. Wyse Jackson, P.S. *An International Review of the Ex Situ Plant Collections of the Botanic Gardens of the World*, Botanic Gardens Conservation International, 2001; http://www.biodiv.org/programmes/socio-eco/benefit/bot-gards.asp (accessed August 2003).
5. Darwin Core (minimum set of standards for search and retrieval of natural history collections and observation databases), 2001; http://tsadev.speciesanalyst.net/DarwinCore/darwin_core.asp (accessed August 2003).
6. Dublin Core Metadata Initiative, 2003; http:// www.dublincore.org/index.shtml (accessed August 2003).
7. Convention on Biological Diversity, United Nations Environment Programme; *Global Strategy for Plant Conservation*, 1993; http://www.biodiv.org/convention/articles.asp (accessed August 2003).
8. Stewart, S. Project paradise: Restoring biodiversity to Lake-Ontario. Public Garden **2001**, *16*(1), 14–17.
9. Bond, J. Investing in Nature, website home page, 2003; http://www.investinginnature.com/ (accessed August 2003).
10. North American Plant Collections Consortium, website home page, 2003; http://www. aabga.org/napcc/ (accessed August 2003).

Genetic Resources Conservation: In Situ

V. Arivudai Nambi
L. R. Gopinath
M.S. Swaminathan Research Foundation, Chennai, India

Abstract

Natural systems are complex and multifunctional, providing differential services to different sections of the populations. While financial capital is crucial to the continued management of plant and animal genetic resources, social capital involving multiple stakeholders to manage natural capital at the landscape level is essential for sustainable natural resource management.

INTRODUCTION

The concept of "reserve" with regard to natural resources has been to set aside areas for exclusive current use as well as unperceived future uses. Precolonial states had several reserves where royalty protected valuable species like elephants, deer, teak, or sandalwood. Systematic and organized conservation efforts have largely been undertaken only in the last two centuries or so. Such conservation efforts closely followed conquest, expanding trade, commerce, and capital accumulation by colonial powers. Colonial expansion clubbed with the Industrial Revolution opened up the possibility of large-scale movement of raw material and goods from one region to another. What was considered an inexhaustible resource at one point in time turned out to be exhaustible, leading to threats of scarcity. Such a threat led to the emergence of systematic and organized efforts at conservation of valuable resources.[1,2]

Conservation efforts can broadly be classified into the following: managed in situ conservation was pursued through National Parks, Protected Areas, Biosphere Reserves, and World Heritage Sites while ex situ conservation was through botanical gardens and gene banks. The above are widely recognized efforts; conservation in the public domain, community conservation, or in situ on-farm conservation by rural and tribal women and men remain largely unrecognized (Fig. 1).[3,4] These communities continue to possess multifaceted traditional knowledge related to biodiversity, such as medicinal properties or food value of plants and animals.[3]

CONSERVATION STRATEGIES IN AGRICULTURE AND FORESTRY SECTORS

The last century saw worldwide conservation efforts that focused on a package of physical and biological entities such as soil, water, and individual species within ecosystems and excluded humans. The present century poses human societies with a formidable challenge: sustainable management of physical and biological resources where agriculture and forestry sectors move in opposite directions with regard to conservation. The agriculture sector has moved from the level of species to varieties and gene, while the forestry sector has moved from the level of species to ecosystem and landscape. A balance of competing interests of individuals and society in the multiple uses of natural resources is required.

Community conservation encompasses not only genetic resources such as landraces, folk varieties, cultivars, and breeds, but also ecosystems and landscapes in which sacred species, groves, and landscapes are embodied. There is a growing realization about the biodiversity–cultural diversity link and its importance. Changes in the social values with regard to traditional lifestyles, emerging new economies, livelihoods, and lifestyles have weakened community conservation to a significant extent. There is a pressing need to revitalize on-farm in situ conservation, as it is the key link between traditional knowledge, livelihoods, and lifestyles. Human beings always remained as managers of natural resources, through which they gained knowledge on animals, plants, climate, soil, and water and evolved various techniques and technologies for survival. This process continues to the present time. Communities living in hilly and inaccessible forested regions continue to hold traditional knowledge as well as generate new knowledge for the management of natural resources. It was through centuries of conservation and use by communities across continents that facilitated transfer of plants, such as rubber, tea, coffee, cotton, and groundnut to new locations for human use. The noted Russian scientist N.I. Vavilov undertook theoretical and applied work on economically important plants, to identify center of origin climatic analogies and wild relatives.

Encyclopedia of Natural Resources DOI: 10.1081/E-ENRL-120020309

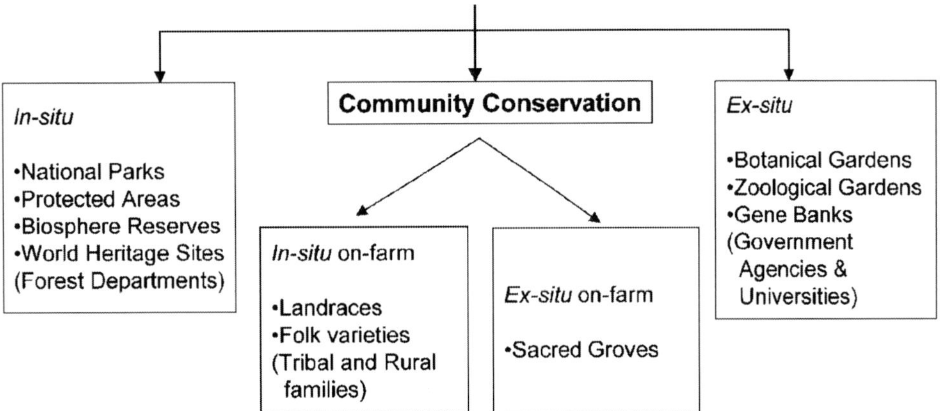

Fig. 1 The integrated gene management system.
Source: Adapted from Swaminathan.[3]

Since the Second World War, community-conserved genetic material was the feedstock for the crop varieties of the Green Revolution. The Green Revolution and the global environmental movement since 1960 together shaped conservation strategies for the domesticated and the wild species. With the passage of time, a number of global agreements were signed on realization that community-conserved genetic wealth (domesticated and wild) and its associated traditional knowledge would leave more options for the morrow.

ECOLOGY IN COMMUNITY CONSERVATION

Species coexist under different relationships like competition, predation, parasitism, symbiosis, commensalism, and mutualism. Although humans are a part of this relationship, they have been able to control nature because of their intelligence. In the light of the above, community-conserved biological diversity has three important functions (Fig. 2): 1) ecological functions that maintain soil fertility and conserve water, leading to synchronized utilization of natural resources; 2) economic functions include food, fuel, fodder, timber, fiber, and medicine that contribute to local incomes both on-farm and off-farm; and 3) socio-functions that lead to diversity in resource use, tenures (private or public property), and social customs. Such relationships have a major impact on the demography of the region, political system, and social customs that are reflected in cultural diversity. Long-term sustainability of a species or diversity of species in a system depends upon the level of synergy among these three functions.[5]

However, the economic function of a system assumes a vital role, since transactions among various social groups are determined by economics. The synergy of the above system is lost when there is a reduction in the economic function. A reduction in the local economy leads to direct negative environmental impacts that further reduce local biodiversity. It is within this context that institutions can contribute to biodiversity conservation.

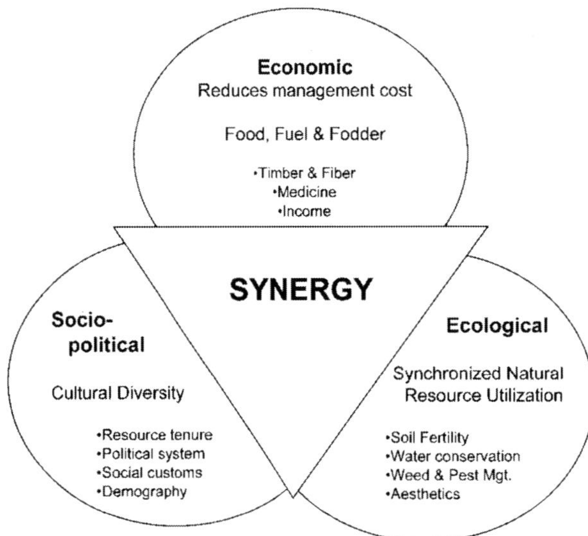

Fig. 2 Functional role of biodiversity of a system.

NATURAL RESOURCE MANAGEMENT: THE ROLE OF COMMUNITIES

Although initial conservation at the species level largely excluded human activity, the establishment of national parks (Fig. 3) entailed some amount of inclusion of the human element. The evolution of the concept of the biosphere at the level of the landscape considers humans and their activities as part of nature. Such conservation efforts at the landscape level undertaken in the recent past have proved to be far more effective and sustainable.[6]

In general, biosphere reserves act as havens for endangered animals and plants with a core zone in the center surrounded by buffer and transition zones. Traditional communities live in the core zone, and the buffer zone depends

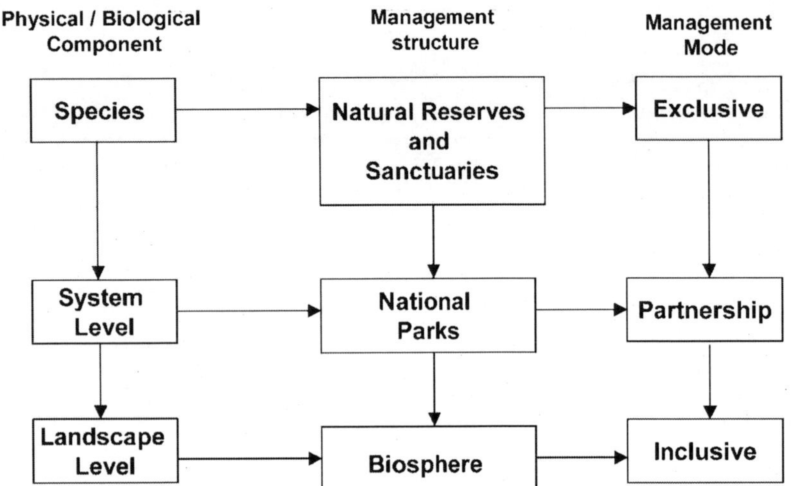

Fig. 3 Management strategies in natural resources: historical perspective.

on forest resources such as wild food, minor forest products, and traditional agriculture. The buffer zone and the core zones are restricted in access to nonlocal people. The buffer zone withstands changes from inner core and the outer transition zones. The transition zone is sandwiched between the human modified modern industrial world and the buffer zone. It is this zone that assumes importance with regard to community conservation of crop genetic resources.

AN INTEGRATED APPROACH TO CONSERVATION

The biosphere concept considers humans as a natural component of the landscape, but there is no legal basis for its implementation. As a result, biosphere reserves share the rules and regulations of national parks, reserves, and sanctuaries as their management strategy. However, such overlap of different management strategies in the same region may lead to conflicts. Therefore, the concept of biosphere reserve needs a more synchronized pathway for operationalization.

A multistakeholder approach trusteeship model and participatory conservation system (PCS) evolved by the M.S. Swaminathan Research Foundation (MSSRF) is an attempt to balance several issues addressed above through a process of share and care. This model is being implemented in the Gulf of Mannar Biosphere Reserve, a coastal aquatic system in southern India. For example, in the Gulf of Mannar Biosphere Trust in the Indian state of Tamil Nadu, MSSRF partners with the local communities; fisheries, forestry, and agricultural departments; and research institutes such as the Central Marine Fisheries Research Institute to enhance sustainable harvest of natural resources. The activities in the area concentrate on providing land- and water-based income generation activities to local communities to reduce overexploitation of the bioresources of the Gulf of Mannar. Charcoal production, dairy farming, and renovation of freshwater tanks are the land-based activities. Community-based agar production and pickled fish unit, creation of artificial reefs, and pearl culture are some of the marine-based livelihoods.

An integrated conservation strategy that combines in situ and ex situ conservation by blending traditional knowledge with modern conservation science should be developed in centers of high natural biological and agrobiodiversity. In situ conservation cannot be pursued without the component of traditional knowledge of communities associated with genetic resources. Moreover, given the present genetic erosion due to industrial agriculture, it makes practical sense to create agrobiodiversity sanctuaries in areas that are rich in agrobiodiversity and have a living traditional knowledge. In addition, eco-agriculture strategies[7] may be required to integrate at landscape-level agricultural and forestry concerns.

CONCLUSIONS

Natural systems are complex and multifunctional, providing differential services to different sections of the populations. While financial capital is crucial to the continued management of plant and animal genetic resources, social capital involving multiple stakeholders to manage natural capital at the landscape level is essential for sustainable natural resource management. This will be possible through a three-pronged strategy for the conservation and use of genetic resources: 1) regulations on conservation, enhancement, sustainable use, and equitable sharing of benefits; 2) social mobilization and community participation in conservation, enhancement, use, and sharing of benefits; and 3) public education on the importance of conservation, enhancement, sustainable use, and equitable sharing of benefits.

ACKNOWLEDGMENTS

The authors wish to express their profound gratitude to Prof. M.S. Swaminathan, Chairman, who gave us an opportunity to write this entry and offered his valuable comments and suggestions on the it and to Prof. P.C. Kesavan, Executive Director, M.S. Swaminathan Research Foundation, Chennai, for his constant encouragement and support and critical comments on an earlier version.

REFERENCES

1. Rangarajan, M. *Fencing the Forest: Conservation and Ecological Change in India's Central Province 1866–1914*; Oxford University Press: New Delhi, 1996; 245.
2. Poffenberger, M. *Keepers of the Forest: Land Management Alternatives in Southeast Asia*; Kumarian Press: Connecticut, 1990; 292.
3. Swaminathan, M.S. Government-Industry-Civil Society: partnerships in integrated gene management. Ambio **2000**, *29*(2), 115–121.
4. Qualsat, C.O.; Damania, A.B.; Zanatta, A.C.A.; Brush, S.B. Locally-based crop plant conservation. In *Plant Conservation: The In Situ approach*; Maxted, N., Ford-Lloyd, B.V., Hawkes, J.G., Eds.; Chapman and Hall: London, 1997; 160–175.
5. Gopinath, L.R. *SHGs: Effective Pathway to Biodiversity Conservation*; M.S. Swaminathan Research Foundation: Chennai, 2002 (Unpublished report).
6. Ramakrishnan, P.S.; Rai, R.K.; Katwal, R.P.S.; Mehmdiratta, S. *Traditional Ecological Knowledge for Managing Biosphere Reserves in South and Central Asia*; Oxford & IBH Publishers: New Delhi, 2002.
7. McNeely, J.; Scherr, J.S. *Ecoagriculture: Strategies to Feed the World and Save Wild Biodiversity*; Island press: Washington, 2002; Vol. 323.

Genetic Resources: Farmer Conservation and Crop Management

Daniela Soleri
Geography Department, University of California, Santa Barbara, California, U.S.A.
David A. Cleveland
Environmental Studies Program, University of California, Santa Barbara, California, U.S.A.

Abstract
Crop genetic variation (V_G) measures the number of alleles, differences between them, and their arrangement in plants and populations. Farmers and the biophysical environment select plants within populations and farmers choose between populations. Together, phenotypic selection and choice determine the extent of population change between generations, and evolution over generations. With in situ conservation on farm there is evolution in response to local selection pressures, often maintaining a high level of V_G. In contrast, ex situ conservation in gene banks conserves the V_G present at a given time and place. Sometimes farmers carry out selection or choice intentionally to change or conserve V_G. Yet much of farmer practice is for production and consumption goals, affecting crop evolution unintentionally if at all. The need to understand farmer selection and conservation is increasing with the loss of genetic resources, spread of transgenic varieties with limited V_G, development of a global intellectual property rights in crop genetic resources, global climate change, and efforts to make formal plant breeding relevant to traditional farmers.

INTRODUCTION

Food production is essential to support human society yet agriculture is one of the largest contributors to global environmental destruction through loss of habitat and diversity, and greenhouse gas emissions that are driving climate change.[1] Identifying options for higher but more sustainable production requires consideration of diverse strategies, including understanding farmer management and conservation of crop varieties that have been the basis of our food system for nearly all of settled human history. About 2 billion people live on 500 million small-scale farms (under 2 ha) globally,[2] most of these in traditionally based agricultural systems (TBAS), and the number will grow dramatically with population growth in the coming decades. Many farmers in TBAS save their own seed to grow at least a portion, and for some most, of the food they eat, conserving valuable genetic resources in the process. Plant breeders working with TBAS farmers consider diversity at many spatial levels of the agrifood system a key to alternatives such as organic and low-input agriculture.[3]

The beginning of agriculture with the Neolithic revolution initiated a dramatic reduction in the diversity of species humans used for food. After domestication, crop species were often transported widely, and many genetically distinct farmers' varieties (FVs, crop varieties traditionally maintained and grown by farmers) developed in specific locations, greatly increasing intraspecific diversity.[4] As Simmonds stated, "Probably, the total genetic change achieved by farmers over the millennia was far greater than that achieved by the last hundred or two years of more systematic science-based effort,"[5] an insight verified by a genome-wide review of maize wild relatives, FVs, and modern varieties (MVs) created by professional plant breeders.[6] FVs continue to be grown today by many small-scale farmers in TBAS, providing for both local consumption and the conservation of genetic diversity for global society.[7]

Crop genetic variation (V_G) is a measure of the number of alleles and degree of difference between them, and their arrangement in plants and populations. For our purposes, a cumulative change in crop population V_G over generations is called microevolution (E_V). Farmers and the biophysical environment select plants within populations based on their phenotypic variation (V_P). Farmers also choose between populations or varieties. This phenotypic selection and choice together determine the degree to which varieties change between generations, evolve over generations, or stay the same. With in situ conservation in farmers' fields specific alleles and genetic structures contributing to V_G may evolve in response to changing local selection pressures, while still maintaining a high level of V_G.[8] In contrast, ex situ conservation in gene banks is more narrowly defined as conserving the specific alleles and structures of V_G present at a given location

Encyclopedia of Natural Resources DOI: 10.1081/E-ENRL-120049217

and moment in time. Thus, different forms of conservation include different amounts and forms of change.

Sometimes, farmers carry out selection or choice intentionally to change or conserve V_G. However, much of farmer practice is intended to further production and consumption goals, affecting crop evolution unintentionally if at all. Thus, in order to understand farmer selection and conservation, it is important to understand the relationship between production, consumption, selection, and conservation in TBAS.[4] This in turn involves understanding the relationship between farmer knowledge and practice in terms of the basic genetics of crop populations and their interactions with growing environments (genetic variation, environmental variation, variation due to genotype-by-environment interaction [$V_{G \times E}$], and response to selection[R])[7,9] (Table 1).

FARMERS AND FVS IN TRADITIONALLY BASED AGRICULTURE SYSTEMS

TBAS are characterized by integration within the household or community of production, consumption, selection, and conservation, whereas in industrial agriculture these functions are spatially and structurally separated. Farm households in TBAS typically rely on their own food production for a significant proportion of their consumption; this production is essential for feeding the population in TBAS now and in the future, even with production increases in industrial agriculture.[10]

TBAS are also characterized by marginal growing environments (relatively high stress, high temporal and spatial variability, and low external inputs), and by the continued use of FVs, even when MVs are available.[11] FVs include landraces, traditional varieties selected by farmers, MVs adapted to farmers' environments by farmer and natural selection, and progeny from crosses between landraces and MVs (sometimes referred to as "creolized" or "degenerated" MVs).

TBAS farmers value FVs for agronomic traits, such as drought resistance, pest resistance, and photoperiod sensitivity. Because farmers grow some or most of the food they eat, storage and culinary criteria are frequently important; for example, families who make the traditional maize beverage *tejate* maintain more varieties of maize than families who do not, using them in preparation of that drink.[12]

The V_G of farmer-managed FVs is often much higher than that of MVs, and is presumed to support broad resistance to multiple biotic and abiotic stresses.[13,14] This makes FVs valuable not only for farmers, because they decrease the production risks in marginal environments especially with climate change,[15] but also for plant breeders and conservationists as the basis for future production in industrial agriculture.[8,16]

FARMER CHOICE: GENETIC VARIATION, CLASSIFICATION, GENOTYPE-BY-ENVIRONMENT INTERACTION, AND RISK

Farmers classify and value traits in their crops, and this can vary between women and men,[7] and between households in a community.[17–19] This variation affects their definition of varieties and populations, and thus the degree of intraspecific V_P (and V_G) they are willing to accept, as a result for example of intravarietal gene flow. These definitions in turn affect farmers' choice, such as which crops, varieties, and populations to adopt or abandon and thus the total V_G they manage, and the number of populations from which they select plants. Experimental evidence indicates that farmers can choose among large numbers of genotypes. In Syria, farmers were able to effectively identify high yielding barley populations from among 208 entries, including 100 segregating populations.[20]

Farmers' choice of varieties and populations without discriminating between individual propagules when adopting or abandoning them from their repertoires, saving seed for planting, and in seed procurement, does not change the genetic makeup of those units directly, and there is no evidence that farmers expect to change them. However, genetic structure may be altered due to sampling error, if the number of seeds required to plant an area is small, and many of these may be half sibs in a crop like maize, with <143 ears ha^{-1} in the case of Oaxaca, Mexico and many farmers planting much smaller areas.[21]

FV crop mating systems in combination with farmers' propagation methods are important determinants of inter- and intraspecific V_G. These also affect differences in phenotypic consistency over generations and therefore farmers' perception and management.[22] Apart from low-frequency somatic mutations, V_G in asexually propagated outcrossing crops, such as cassava, is unchanged between generations, and discrete, fixed types (clones) or groups of types are maintained as distinct varieties[23,24] that may be either homo- or heterogeneous. Intrapopulation V_G increases and genetic structures become more variable and dynamic with the intentional inclusion by farmers of sexually propagated individuals into clonal populations based on morphological similarity or heterosis.[24]

The same increase in dynamism occurs with increasing rates of outcrossing in sexually propagated crops, because variation can be continuous within a population. Moreover, segregation, crossing-over, recombination, and other events during meiosis and fertilization, result in much change in V_G between generations. In highly allogamous crops, such as maize, heterozygosity can be high, making it difficult to discern discrete segregation classes, particularly in the presence of environmental variation and retaining distinguishing varietal characteristics requires maintenance selection[25] (see below). Highly autogamous crops such as rice are predominantly homozygous, making exploitation of V_G and retention of varietal distinctions easier, even if varieties are composed of multiple, distinct lines.

Table 1 Farmer selection and choice and the change and conservation of crop varieties

Farmer knowledge (including values) on which practice may be based	Farmer practice	Potential effect of farmer practice on selection and conservation of populations/varieties	Example
Indirect selection/conservation by farmer-managed growing and storage environment			
Understanding of G × E	Allocation of varieties to spatial, temporal, and management environments	Selection pressures in environments result in maintenance of existing, or development of new populations/varieties, including evolution of wide or narrow adaptation	*Spatial*: varieties specified for different soil or moisture types; rice, Nepal; pearl millet, India *Temporal*: varieties with different cycle lengths, maize, Mexico
	Management of growing environments	Changing selection pressures	Changes in fertilizer application, maize, Mexico
Risk, values, G × E	Choice of environments for testing new populations/ varieties	↑ or ↓ V_G	High stress, rice, Nepal; optimal conditions, barley, Syria
Escape from economic or political pressure; desire for different ways of life	Abandonment of fields or farms, reduced field size	↑ V_G within due to reduced area for planting, ↓ effective population size, genetic drift	Pooling of subvarieties, maize, Hopi and Zuni Reduction in area, potatoes, Peru; maize, Mexico
Direct selection/conservation, intentional re. population change			
Discount rate (values re. future), altruism (values re. community)	Conservation of varieties for the future, for other farmers	Intraspecific V_G	Rice, Thailand; maize, Hopi
Interest and expertise in experimentation	Deliberate crossing	↑V_G	Maize-teosinte, Mexico; MV-FV pearl millet, India; MV-FV and FV-FV, maize, Mexico
Understanding of h^2	Selection of individuals (plants, propagules) from within parent population	↑ or ↓ V_G via R	Among seedlings, cassava, Guyana; among panicles, pearl millet, India
Direct, selection/conservation, unintentional re. population change, but intentional re. other goals, as result of production/ consumption practices			
Attitudes towards risk re. yield stability	Adoption and abandonment of FVs, MVs	↑ or ↓ intraspecific diversity	Maize, Hopi; rice, Nepal
	Adoption and abandonment of lines in multiline varieties of self-pollinated crops; seed lots in cross-pollinating crops	↑ or ↓ intravarietal diversity	Common bean, East Africa; maize, Mexico
Agronomic, storage, culinary, esthetic and ritual criteria, implicit and explicit	Selection or choice based on production/consumption criteria	↑ or ↓ intra- and intervarietal diversity	Storage and culinary criteria: maize, Mexico; and ritual criteria, rice, Nepal
Choice criteria	Acquisition of seed, seed lots	Gene flow via seed then pollen flow, hybridization, recombination within varieties	Cycle length, maize, Mexico; cuttings and seedlings, cassava, Guyana

Abbreviations: FV, farmer developed crop variety; G × E, genotype-by-environment interaction; h^2, heritability in the narrow sense; MV, modern crop variety, product of formal breeding system; R, response selection; V_G, genetic variation; ↑, increase; ↓, decrease.

Farmers' choices depend in part on the range of spatial, temporal, and management environments present, the V_G available to them, and the extent to which genotypes are widely versus narrowly adapted. In turn, environmental variation (V_E) in these growing environments interacts with V_G ($V_{G \times E}$) to produce variation in yield of grain, straw, roots, tubers, leaves, and other characteristics over space and time. As a result, farmers may have different choice criteria for different environments, as in Rajasthan, India, where pearl millet farmers realize there is a trade-off between panicle size and tillering ability. So, farmers in a less stressful environment prefer varieties producing larger panicles, whereas those in a more stressful environment prefer varieties with high tillering.[26]

Patterns of variation in yield affect farmers' choice of crop variety via their attitude toward risk. In response to scenarios depicting varietal $V_{G \times E}$ and temporal variation, farmers from more marginal growing environments were more risk averse compared to those from more favorable environments. The former preferred a crop variety with low but stable yields across environments while the latter chose a variety highly responsive to favorable conditions but with poor performance under less favorable conditions.[27] Sorghum farmers in Mali tend to choose varieties to optimize outputs in the face of variation in rainfall, level of striga infestation, and availability of labor and other production resources, especially cultivator and seeder plows.[28] As a result, they choose combinations of long- and short-cycle sorghum varieties to optimize yield, yield stability, and post-harvest traits like taste. For example, when rains are better, farmers choose a greater number of long-cycle varieties.

FARMER SELECTION: HERITABILITY, PHENOTYPIC SELECTION DIFFERENTIAL, AND RESPONSE

Phenotypic selection, operating on V_P, is identification of the individual plants within a population that will contribute genetic material to the next generation. Phenotypic selection of FVs in TBAS can be classified according to the agent of selection (natural environment, farmer-managed environment, or farmer), and according to farmers' goals for selection (Fig. 1). Farmer selection can also be classified according to the outcome (Fig. 2). Geneticists and plant breeders tend to think of phenotypic selection as seeking to produce genetic change, but farmers often do not.[29] Whether or not farmer selection does change the genetic makeup of the population (i.e., effects genetic response between generations [R], or cumulative multigenerational microevolution [E_V]) depends on heritability (h^2), or the proportion of phenotypic variation that is genetic and can be inherited; and the selection differential (S), or the difference in mean between parental population and sample selected from it: $R = h^2S$.

$S > 0, R \approx 0$. Heritability is often understood by farmers who distinguish between high and low heritability traits, consciously selecting the former, while often considering it not worthwhile or even possible to select the latter, especially in cross-pollinating crops.[27] When farmers' selection criteria center on relatively low heritability traits such as large ear and seed size in maize, they may achieve high S, and little or no R. However, they persist with that selection because their goal is high-quality seed for planting.[25,30] A study across four sites, each with different crops, found that often a majority of farmers at a site did not see their seed selection as a process of cumulative, directional change.[27] However, intentional phenotypic selection for goals other than genetic response was practiced by nearly all farmers in that study. Selection exercises with maize in Oaxaca, Mexico[30] found farmers' selections to be significantly different from that of the original population in their selection criteria, resulting in high S values. However, R values were zero for these as well as other morphophenological traits. The reasons documented to date are for seed quality (germination and early vigor) and purity, and because of "custom," that is, not to change or improve a variety. To understand this from the farmers' perspective, it is necessary to take into account the multiple functions of crop populations in TBAS (production of food and seed, consumption, conservation, improvement).

$R > 0, E_V \approx 0$. Farmers also select seed to maintain defining, desirable, heritable varietal traits that change as a result of gene flow and indirect selection by environmental factors in fields and storage containers, especially in allogamous species. When successful, this results in R and, over time, prevents unwanted E_V. Selection exercises with maize in Jalisco, Mexico, found that farmers' selection served to diminish the impact of gene *flow* and maintain varieties' morphological characteristics, but not to change the population being selected on. Indian farmers were able to maintain the distinct ideotypes of introduced FVs of their allogamous crop pearl millet via intentional selection of panicles for their unique phenotypes.[31]

$E_V > 0$. In seeking cumulative genetic response or E_V, farmers may practice intentional selection either to create new varieties, best documented in vegetatively propagated and self-pollinating crops,[32] or for varietal improvement, although much evidence for this is anecdotal. Most often this is selection for heritable, qualitative traits; for example, farmers in central Mexico have selected for and maintained a new landrace, based on seed and ear morphology, among segregating populations resulting from the hybridization of two existing landraces.[33]

CONCLUSIONS

Selection and conservation in TBAS contrast substantially with industrial agricultural systems. Therefore, understanding farmers' practices, and the knowledge and goals underlying them, is critical for supporting food production, food consumption, crop improvement, and crop genetic

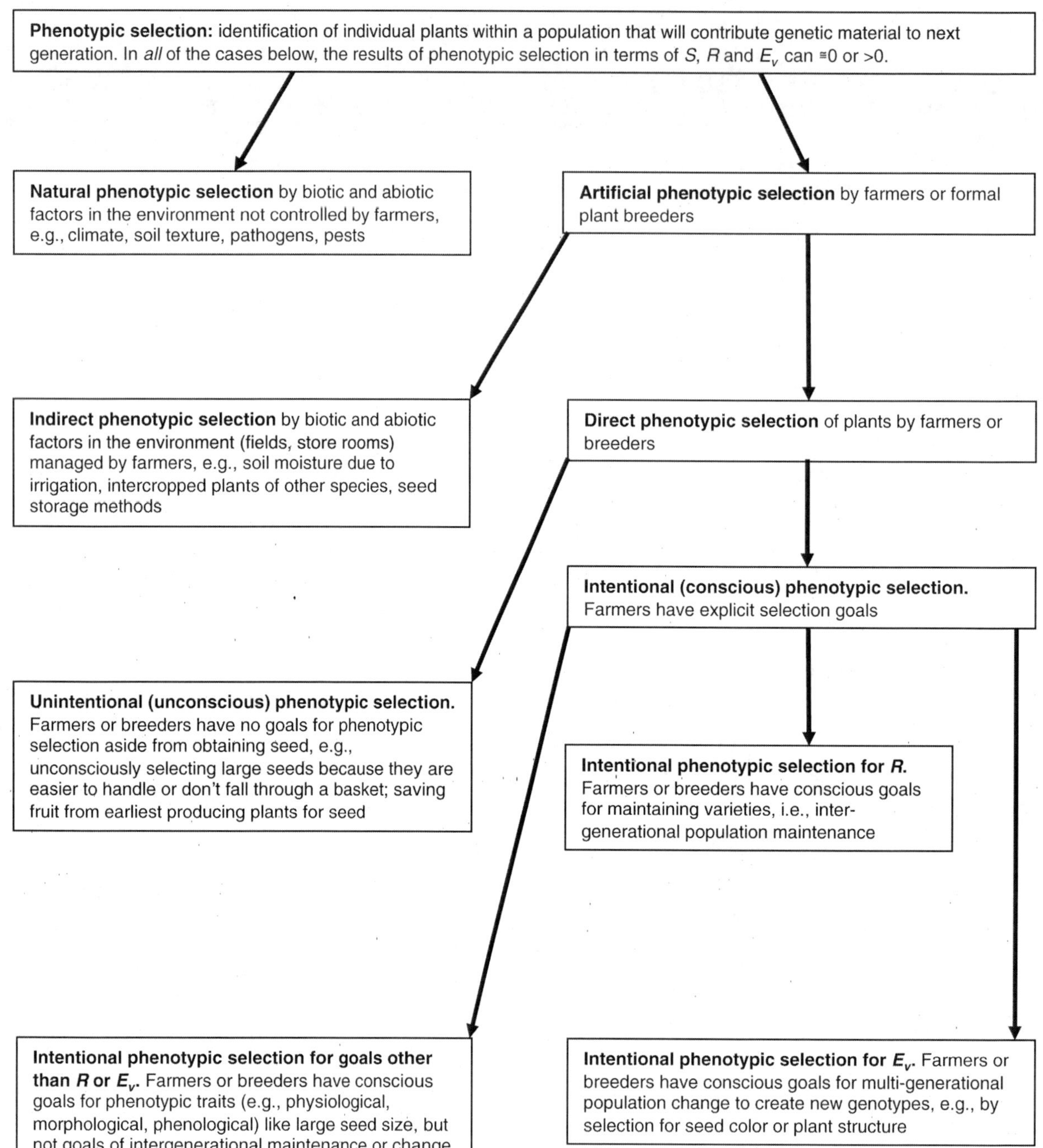

Fig. 1 Phenotypic selection classified according to the agent of selection, and intention of the farmer or plant breeder as agent. See text for definition of abbreviations. © D. Soleri & D.A. Cleveland, 2013.

resource conservation for farm communities in TBAS, and for long-term global food security. The urgency of understanding farmer selection and conservation will increase in the future with the ongoing loss of genetic resources, the rapid spread of transgenic crop varieties with limited genetic diversity, the development of a global system of intellectual property rights in crop genetic resources, and the movement to make formal plant breeding more relevant to farmers in TBAS through plant breeding and conservation based on direct farmer and scientist collaboration.

At present, attention and investment in transgenic genetic engineering dominate crop improvement globally. Yet, schemes that are to some extent modeled on and make use of farmer management and the V_G of their FVs show good potential for increasing yields, conserving genetic resources, and supporting adaptation to growing environments that are changing at an accelerating pace.[34] Farmer management and conservation of crop varieties developed in situ is a form of precision agriculture that, when combined with formal scientific methods and research support, may be the strategy most

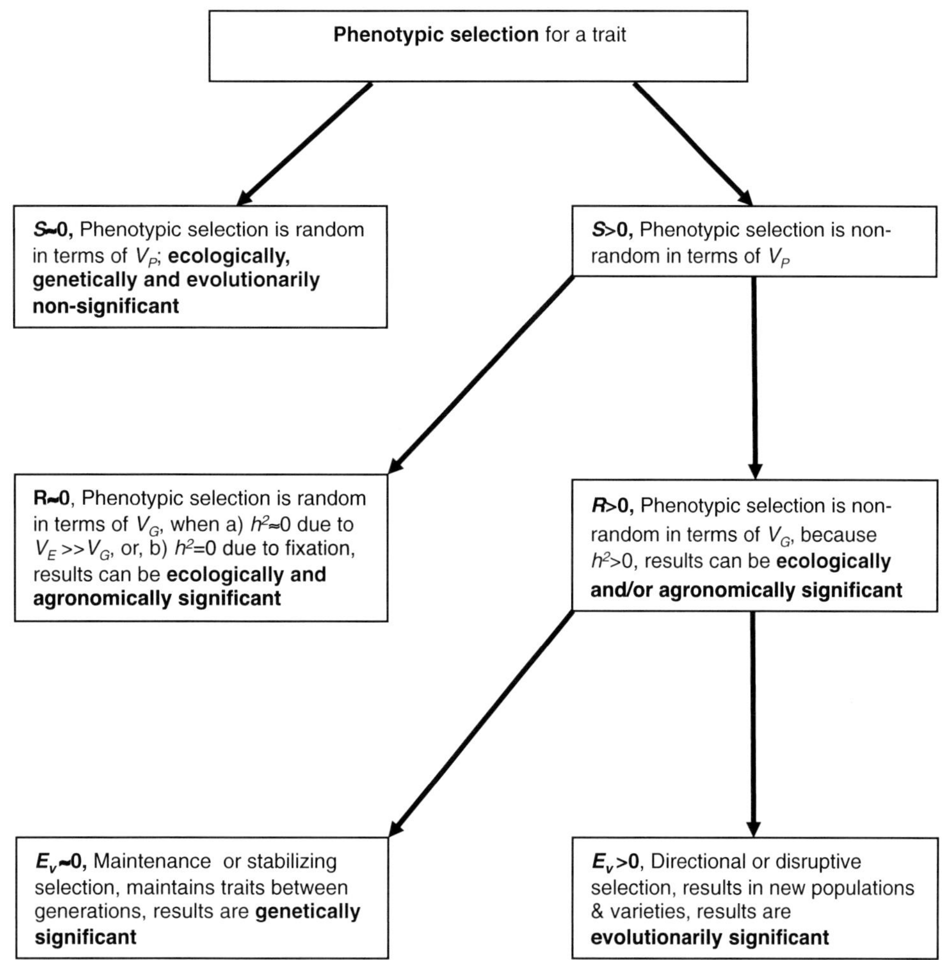

Fig. 2 Phenotypic selection classified according to outcome of selection. See key to symbols below. © D.A. Cleveland & D. Soleri, 2013. Abbreviations: FV, farmer developed crop variety; h^2, heritability in the narrow sense; MV, modern crop variety, product of formal breeding system; R, response to selection; V_E, environmental variation; V_G, genetic variation; $V_{G\times E}$, genotype-by-environment interaction variation; V_P, phenotypic variation; ↑, increase; ↓, decrease.

likely to address multiple criteria of environmental, economic, and social sustainability in the global food system.[35]

ACKNOWLEDGMENTS

This entry is an update of an entry in the *Encyclopedia of Plant and Crop Science*.[36] The research on which some of this entry is based was supported in part by NSF grant nos. SES-9977996 and DEB-0409984, and a grant from the Wallace Genetic Foundation.

REFERENCES

1. Power, A.G. Ecosystem services and agriculture: Tradeoffs and synergies. Philos. Trans. R. Soc. B-Biol. Sci. **2010**, *365* (1554), 2959–2971.
2. IFAD. Conference on New Directions in *Smallholder Agriculture*, Proceedings of the Conference, 24–25 January 2011; Rome, IFAD HQ. In IFAD: Rome, 2011.
3. SOLIBAM Newsletter 1. http://www.solibam.eu/modules/addresses/viewcat.php?cid=1 (2011 November 11).
4. Harlan, J.R. *Crops and Man*. Second ed.; American Society of Agronomy, Inc. and Crop Science Society of America, Inc.: Madison, Wisconsin, 1992; p 284.
5. Simmonds, N.W.; Smartt, J. *Principles of Crop Improvement*. Second ed.; Blackwell Science Ltd.: Oxford, UK, 1999; p xii, 412.
6. Hufford, M.B.; Xu, X.; van Heerwaarden, J.; Pyhajarvi, T.; Chia, J.-M.; Cartwright, R.A.; Elshire, R.J.; Glaubitz, J.C.; Guill, K.E.; Kaeppler, S.M.; Lai, J.; Morrell, P.L.; Shannon, L.M.; Song, C.; Springer, N.M.; Swanson-Wagner, R.A.; Tiffin, P.; Wang, J.; Zhang, G.; Doebley, J.; McMullen, M.D.; Ware, D.; Buckler, E.S.; Yang, S.; Ross-Ibarra, J. Comparative population genomics of maize domestication and improvement. Nat Genet. **2012**, *44* (7), 808–811.
7. Smale, M. Economics perspectives on collaborative plant breeding for conservation of genetic diversity on farm. In *Farmers, Scientists and Plant Breeding: Integrating Knowledge and Practice*, Cleveland, D.A., Soleri, D., Eds. CAB International: Oxon, UK, 2002; pp 83–105.

8. Brown, A.H.D. The genetic structure of crop landraces and the challenge to conserve them *in situ* on farms. In *Genes in the Field: On-Farm Conservation of Crop Diversity*, Brush, S.B., Ed. Lewis Publishers; IPGRI; IDRC: Boca Raton, Florida; Rome; Ottawa, 1999; pp 29–48.
9. Soleri, D.; Cleveland, D.A. Farmers' genetic perceptions regarding their crop populations: an example with maize in the Central Valleys of Oaxaca, Mexico. Econ. Bot. **2001**, *55*, 106–128.
10. Heisey, P.W.; Edmeades, G.O. Part 1. Maize production in drought-stressed environments: technical options and research resource allocation. In *World Maize Facts and Trends 1997/98*, CIMMYT, Ed. CIMMYT: Mexico, D.F., 1999; pp 1–36.
11. Brush, S.B.; Taylor, J.E.; Bellon, M.R. Technology adoption and biological diversity in Andean potato agriculture. J. Dev. Econ. **1992**, *39*, 365–387.
12. Soleri, D.; Cleveland, D.A.; Aragón Cuevas, F. Food globalization and local diversity: The case of tejate, a traditional maize and cacao beverage from Oaxaca, Mexico. Curr. Anthropol. **2008**, *49*, 281–290, http://www.journals.uchicago.edu.proxy.library.ucsb.edu:2048/doi/full/10.1086/527562?cookieSet=1#apb.
13. Ceccarelli, S. Landraces: Importance and Use in Breeding and Environmentally Friendly Agronomic Systems. *Agrobiodiversity Conservation: Securing the Diversity of Crop Wild Relatives and Landraces* 2012; 103.
14. Cooper, H.D.; Spillane, C.; Hodgkin, T. Broadening the genetic base of crops: an overview. In *Broadening the Genetic Base of Crop Production*, Cooper, H.D., Spillane, C., Hodgkin, T., Eds. CABI: Wallingford, Oxon, UK, 2001; pp 1–23.
15. Ceccarelli, S.; Grando, S.; Maatougui, M.; Michael, M.; Slash, M.; Haghparast, R.; Rahmanian, M.; Taheri, A.; Al-Yassin, A.; Benbelkacem, A.; Labdi, M.; Mimoun, H.; Nachit, M. Plant breeding and climate changes. J. Agri. Sci. **2010**, *148*, 627–637.
16. Newton, A.C.; Akar, T.; Baresel, J.P.; Bebeli, P.J.; Bettencourt, E.; Bladenopoulos, K.V.; Czembor, J.H.; Fasoula, D.A.; Katsiotis, A.; Koutis, K.; Koutsika-Sotiriou, M.; Kovacs, G.; Larsson, H.; de Carvalho, M.; Rubiales, D.; Russell, J.; Dos Santos, T. M.M.; Patto, M.C.V. Cereal landraces for sustainable agriculture. A review. Agron. Sustain. Dev. **2010**, *30* (2), 237–269.
17. Barry, M.B.; Pham, J.L.; Courtois, B.; Billot, C.; Ahmadi, N. Rice genetic diversity at farm and village levels and genetic structure of local varieties reveal need for in situ conservation. Genet Resour Crop Evol. **2007**, *54*, 1675–1690.
18. Delêtre, M.; McKey, D.B.; Hodkinson, T.R. Marriage exchanges, seed exchanges, and the dynamics of manioc diversity. Proc. Nat. Acad. Sci. **2011**, *108* (45), 18249–18254.
19. Worthington, M.; Soleri, D.; Aragón-Cuevas, F.; Gepts, P. Genetic composition and spatial distribution of farmer-managed Phaseolus bean plantings: An example from a village in Oaxaca, Mexico. Crop Sci. **2012**, *52*, 1721–1735.
20. Ceccarelli, S.; Grando, S.; Tutwiler, R.; Bahar, J.; Martini, A.M.; Salahieh, H.; Goodchild, A.; Michael, M. A methodological study on participatory barley breeding I. Selection phase. Euphytica **2000**, *111*, 91–104.
21. Cleveland, D.A.; Soleri, D.; Aragón Cuevas, F.; Crossa, J.; Gepts, P. Detecting (trans)gene flow to landraces in centers of crop origin: lessons from the case of maize in Mexico. Env. Biosafety Res. **2005**, *4*, 197–208.
22. Cleveland, D.A.; Soleri, D.; Smith, S.E. A biological framework for understanding farmers' plant breeding. Econ. Bot. **2000**, *54*, 377–394.
23. Gibson, R.W. A review of perceptual distinctiveness in landraces including an analysis of how its roles have been overlooked in plant breeding for low-input farming systems. Econ. Bot. **2009**, *63* (3), 242–255.
24. McKey, D.; Elias, M.; Pujol, B.; Duputie, A. The evolutionary ecology of clonally propagated domesticated plants. New Phytol. **2010**, *186* (2), 318–332.
25. Louette, D.; Smale, M. Farmers' seed selection practices and maize variety characteristics in a traditional Mexican community. Euphytica **2000**, *113*, 25–41.
26. Weltzien R.E.; Whitaker, M.L.; Rattunde, H.F.W.; Dhamotharan, M.; Anders, M.M. Participatory approaches in pearl millet breeding. In *Seeds of Choice*, Witcombe, J., Virk, D., Farrington, J., Eds.; Intermediate Technology Publications: London, 1998; pp 143–170.
27. Soleri, D.; Cleveland, D.A.; Smith, S.E.; Ceccarelli, S.; Grando, S.; Rana, R.B.; Rijal, D.; Ríos Labrada, H. Understanding farmers' knowledge as the basis for collaboration with plant breeders: methodological development and examples from ongoing research in Mexico, Syria, Cuba, and Nepal. In *Farmers, Scientists and Plant Breeding: Integrating Knowledge and Practice*, Cleveland, D.A., Soleri, D., Eds.; CAB International: Wallingford, Oxon, UK, 2002; pp 19–60.
28. Lacy, S.; Cleveland, D.A.; Soleri, D. Farmer choice of sorghum varieties in southern Mali. Human Ecol. **2006**, *34*, 331–353.
29. Cleveland, D.A.; Soleri, D. Extending Darwin's analogy: Bridging differences in concepts of selection between farmers, biologists, and plant breeders. Econ. Bot. **2007**, *61*, 121–136.
30. Soleri, D.; Smith, S.E.; Cleveland, D.A. Evaluating the potential for farmer and plant breeder collaboration: A case study of farmer maize selection in Oaxaca, Mexico. Euphytica **2000**, *116*, 41–57.
31. vom Brocke, K.; Weltzien, E.; Christinck, A.; Presterl, T.; Geiger, H.H. Farmers' seed systems and management practices determine pearl millet genetic diversity in semiarid regions of India. Crop Sci. **2003**, *43*, 1680–1689.
32. Boster, J.S. Selection for perceptual distinctiveness: Evidence from Aguaruna cultivars of Manihot esculenta. Econ. Bot. **1985**, *39*, 310–325.
33. Perales, H.; Brush, S.B.; Qualset, C.O. Dynamic management of maize landraces in Central Mexico. Econ. Bot. **2003**, *57* (1), 21–34.
34. IAASTD *Synthesis Report with Executive Summary: A Synthesis of The Global and Sub-Global Iaastd Reports*; Washington, DC, 2009; p 97.
35. De Schutter, O. *Agroecology and the Right to Food. Report Submitted by the Special Rapporteur on the Right to Food*; Human Rights Council, United Nations General Assembly: 2010.
36. Soleri, D.; Cleveland, D.A. Farmer selection and conservation of crop varieties. In *Encyclopedia of Plant & Crop Science*; Goodman, R.M., Ed.; Marcel Dekker: New York, 2004; pp. 433–438.

Genetic Resources: Seeds Conservation

Florent Engelmann
Institute of Research for Development (IRD), Montpellier, France

Abstract
Storage of orthodox seeds is the most widely practiced method of ex situ conservation of plant genetic resources as 90% of the 7.4 million accessions stored in genebanks are maintained as seed. In contrast to orthodox seeds, a considerable number of species produce nonorthodox seeds, which are unable to withstand much desiccation and are often sensitive to chilling. Cryopreservation (liquid nitrogen, −196°C) is the only safe and cost-effective option for long-term conservation of genetic resources of these species. However, cryopreservation research is still at a preliminary stage for nonorthodox species. Research priorities for orthodox seed species include improving the longevity of seeds under standard genebank storage conditions, determining critical seed moisture content, developing low-input storage techniques, and improving and monitoring viability. In case of nonorthodox seed species, research should focus on understanding seed recalcitrance and developing improved conservation techniques. Improved germplasm management procedures should also be established.

INTRODUCTION

In the 1950s and 1960s, major advances in plant breeding brought about the "green revolution," which resulted in wide-scale adoption of high-yielding varieties and genetically uniform cultivars of staple crops, particularly wheat and rice. Consequently, global concern about the loss of genetic diversity in these crops increased, as farmers abandoned their locally adapted landraces and traditional varieties, replacing them with improved, yet genetically uniform modern ones. The International Agricultural Research Centers (IARCs) of the Consultative Group on International Agricultural Research (CGIAR) started to assemble germplasm collections of the major crop species within their respective mandates. The International Board for Plant Genetic Resources (IBPGRs) was established in 1974 in this context to coordinate the global effort to systematically collect and conserve the world's threatened plant genetic diversity. Today, as a result of this effort, over 1750 genebanks and germplasm collections exist around the world, maintaining ~7,400,000 accessions, largely of major food crops including cereals and some legumes, i.e., species that can be conserved easily as seed.[1]

This entry reviews the current storage technologies and management procedures developed for seeds, describes the problems and achievements with seed storage, and identifies priorities for improving the efficiency of seed conservation.

CONSERVATION OF ORTHODOX SEEDS

Many of the world's major food plants produce so-called orthodox seeds, which tolerate extensive desiccation and can be stored dry at low temperature. Storage of orthodox seeds is the most widely practiced method of ex situ conservation of plant genetic resources,[1] as 90% of the accessions stored in genebanks are maintained as seed. Following drying to low moisture content (3–7% fresh weight basis, depending on the species), such seeds can be conserved in hermetically sealed containers at low temperature, preferably at −18°C or cooler, for several decades.[1] All relevant techniques are well established, and practical documents covering the main aspects of seed conservation, are available, including design of seed storage facilities for genetic conservation, principles of seed testing for monitoring viability of seed accessions maintained in genebanks, methods for removing dormancy and germinating seeds, and suitable methods for processing and handling seeds in genebanks.[2]

In addition to being the most convenient material for genetic resource conservation, seeds are also a convenient form for distributing germplasm to farmers, breeders, scientists, and other users. Moreover, since seeds are less likely to carry diseases than other plant material, their use for exchange of plant germplasm can facilitate quarantine procedures.

CONSERVATION OF NONORTHODOX SEEDS

In contrast to orthodox seeds, a considerable number of species, predominantly from tropical or subtropical origin, such as coconut, cacao, and many forest and fruit tree species, produce so-called nonorthodox seeds, which are unable to withstand much desiccation and are often sensitive to chilling. Nonorthodox seeds have been further

Encyclopedia of Natural Resources DOI: 10.1081/E-ENRL-120049218

subdivided in recalcitrant and intermediate seeds based on their desiccation sensitivity, which is high for the former and lower for the latter group. Nonorthodox seeds cannot be maintained under the storage conditions described above, i.e., low moisture content and temperature, and have to be kept in moist, relatively warm conditions to maintain viability. Even when stored in optimal conditions, their lifespan is limited to weeks, occasionally months. Of over 7000 species for which published information on seed storage behavior exists,[3] ~10% are recorded as nonorthodox or possibly nonorthodox.

Genetic resources of nonorthodox species are traditionally conserved as whole plants in field collections. This mode of conservation is faced with various problems and limitations, and cryopreservation (liquid nitrogen (LN), −196°C) currently offers the only safe and cost-effective option for the long-term conservation of genetic resources of problem species. However, cryopreservation research is still at a preliminary stage for nonorthodox species.[4]

MANAGEMENT OF SEED COLLECTIONS

Only a limited number of genebanks operate at very high standards, whereas many others face difficulties owing to inadequate infrastructures, lack of adequate seed processing and storage equipment, unreliable electricity supply, funding and staffing constraints, and inadequate management practices. Therefore, seeds are often stored under suboptimal conditions and require more frequent regeneration, thus bearing additional costs on often already insufficient operating budgets of genebanks. Great difficulties are faced, in particular, by many countries with regeneration of seed collections.[1] Seed storage technologies are relatively easy to apply. The problems relate more to resource constraints that impact the performance of essential operations. Efficient and cost-effective genebank management procedures have now become key elements for long-term ex situ conservation of plant genetic resources.

Other important aspects of genebank operations concern germplasm characterization and documentation. The extent to which germplasm collections are characterized varies widely between genebanks and species[1] but is far from complete in many instances. Concerning documentation, the situation is highly contrasted.[1] Some, mainly developed, countries have fully computerized documentation systems and relatively complete accession data, while many others lack information on the accessions in their collections, including the so-called passport data.

MAIN CHALLENGES AND PRIORITIES FOR IMPROVING SEED CONSERVATION

Various priority areas have been identified for orthodox and nonorthodox seed conservation research and for improving seed genebank management procedures.

Research Priorities for Orthodox Seed Species

Longevity of seeds under standard genebank storage conditions

One priority research topic is the longevity of seeds under standard genebank storage conditions. Indeed, over the past 30 years, relatively widespread evidence has emerged of less than expected longevity at conventional seed bank temperatures.[5] Across ~200 species, species from drier (total rainfall) and warmer temperature (mean annual) locations tended to have greater seed P_{50} (time taken in storage for viability to decline to 50%) under accelerated aging conditions than species from cool, wet conditions.[6] Moreover, species P_{50} values were correlated with the proportion of accessions (not necessarily the same species) in that family that lost a significant amount of viability after 20 years under conditions for long-term seed storage, that is, seeds pre-equilibrated with 15% relative humidity air and then stored at –20°C.[7] Such relative underperformance at –20°C was apparent in 26% of the accessions.[6] Similarly, it has been estimated that half-lives for the seeds of 276 species held for an average of 38 years under cool (–5°C) and cold (25 years at −18°C) temperature was >100 years only for 61 (22%) of the species.[8] Nonetheless, cryogenic storage did prolong the shelf life of lettuce (*Lactuca*) seeds with projected half-lives in the vapor and liquid phases of LN of 500 and 3,400 years, respectively,[9] up to 20 times greater than that predicted for that species in a conventional seed bank at –20°C.[10,11] Although 25 species (from 19 genera) of Cruciferae had high germination (often >90%) after ~40 years storage at –5 to −10°C,[12] it is hard not to conclude that, as an extra insurance policy for conservation, cryopreservation should be considered appropriate for all orthodox seeds, and one subsample of any accession systematically stored in LN, in addition to the samples stored under classical genebank conditions.[13] As highlighted by,[14] such loss of viability during seed storage under standard conditions reflects molecular mobility within the system, i.e., relaxation of glassy matrixes. Stability of biological glasses is currently not fully understood and research in the thermodynamic principles, which contribute to temperature dependency of glassy relaxation as the context for understanding potential changes in viability of cryogenically stored germplasm is urgently needed.

Determining critical seed moisture content

The preferred conditions recommended for long-term seed storage are 3–7% moisture content, depending on the species, at −18°C or lower.[7] However, it has been shown that drying seeds beyond a critical moisture content provides no additional benefit to longevity and may even accelerate seed aging rates, and that interactions exist between the critical relative humidity and storage temperature. Research should be pursued on this topic of high importance.

Developing low-input storage techniques

Various research projects have focused on the development of the ultra-dry seed technology, which allows storing seeds desiccated to very low moisture contents at room temperature, thereby suppressing the need for refrigeration equipment. Although drying seed to very low moisture prior to storage seems to have fewer advantages than was initially expected, ultra-dry storage is still considered to be a useful, practical, low-cost technique in circumstances where no adequate refrigeration can be provided.[15] Research on various aspects of the ultra-dry seed storage technology and on its applicability to a broader number of species should therefore be continued.

Improving and monitoring viability

The viability of conserved accessions depends on their initial quality and how they have been processed for storage, as well as on the actual storage conditions. There is evidence that a very small decrease in initial seed viability can result in substantial reduction in storage life. This needs further investigation, and research should be performed notably on the effect of germplasm handling in the field during regeneration and during subsequent processing stages prior to its arrival at the genebank, as well as on growing conditions, disease status, and time of harvest of the plants.

Research Priorities for Nonorthodox Seed Species

Understanding seed recalcitrance

A number of physical and metabolic processes or mechanisms have been suggested to confer, or contribute to, desiccation tolerance.[16] Different processes may confer protection against the consequences of loss of water at different hydration levels, and the absence, or ineffective expression, of one or more of these could determine the relative degree of desiccation sensitivity of seeds of individual species. Additional research is needed to improve our understanding of the mechanisms involved in seed recalcitrance.

Developing improved conservation techniques

Various technical options exist for improving storage of nonorthodox species. Especially with species for which no or only little information is available, it is advisable, before undertaking any "high-tech" research, to examine the development pattern of seeds and to run preliminary experiments to determine their desiccation sensitivity as well as to define germination and storage conditions. IPGRI, in collaboration with numerous institutions worldwide, has developed a protocol for screening tropical forest tree seeds for their desiccation sensitivity and storage behavior,[17] which might be applicable to seeds of other species after required modification and adaptation. For long-term storage of nonorthodox species, cryopreservation represents the only option. Numerous technical approaches exist, including freezing of seeds, embryonic axes, shoot apices sampled from embryos, adventitious buds, or somatic embryos, depending on the sensitivity of the species studied.[4]

Germplasm Management Procedures

The objective of any genebank management procedure is to maintain genetic integrity and viability of accessions during conservation and to ensure their accessibility for use in adequate quantity and quality at the lowest possible cost. As mentioned earlier, many genebanks face financial constraints that hamper their efficient operation. It is therefore very important to improve genebank management procedures to make them more efficient and cost-effective. In this aim, the use of molecular markers should be increased in order to improve characterization and evaluation, improved seed regeneration and accession management procedures should be established, the use of collections should be enhanced through the development of core collections, germplasm health aspects including new biotechnological tools for detection, indexing, and eradication of pathogens should be better integrated in routine genebank operations, and improved documentation tools should be developed. Special attention should be given to developing appropriate techniques for genebanks in developing countries, where specialized equipment is frequently lacking and resources are usually limited.

CONCLUSION

The improvements in the seed storage techniques and genebank management procedures resulting from the research performed on the priority areas identified earlier will further increase the key role of seeds in the ex situ conservation of genetic resources of many species. However, it is now well recognized that an appropriate conservation strategy for a particular plant genepool requires a holistic approach, combining the different ex situ and in situ conservation techniques available in a complementary manner.[18] Selection of the appropriate methods should be based on a range of criteria, including the biological nature of the species in question, practicality and feasibility of the particular methods chosen (which depends on the availability of the necessary infrastructures), their efficiency, and the cost-effectiveness and security afforded by their application. An important area in this is the linkage between in situ and ex situ components of the strategy, especially with respect to the dynamic nature of the former and the static, but potentially more secure, approach of the latter.

REFERENCES

1. FAO, The Second Report on the State of the World's Plant Genetic Resources for Food and Agriculture; Food and Agriculture Organization of the United Nations: Rome, 2010.
2. FAO/IPGRI Genebank Standards; Food and Agriculture Organization of the United Nations: Rome and International Plant Genetic Resources Institute: Rome, 1994.
3. Hong, T.D.; Linington, S.; Ellis, R.H. Seed Storage Behaviour: A Compendium; Handbooks for Genebanks; International Plant Genetic Resources Institute: Rome, 1996; Vol. 4.
4. Engelmann, F.; Takagi, H. Cryopreservation of Tropical Plant Germplasm—Current Research Progress and Applications; Japan International Center for Agricultural Sciences: Tsukuba, Japan and International Plant Genetic Resources Institute: Rome, 2000.
5. Li, D.Z.; Pritchard, H.W. The science and economics of ex situ plant conservation. Trends Plant Sci. **2009**, *14*, 614–621.
6. Probert, R.J.; Daws, M.I.; Hay F. Ecological correlates of ex situ seed longevity: a comparative study on 195 species. Ann. Bot. **2009**, *104*, 57–69.
7. http://www.ipgri.cgiar.org/system/page.asp?theme=7 (accessed March 2003).
8. Walters, C.; Wheeler, L.J.; Grotenhuis, J.M. Longevity of seeds stored in a genebank: species characteristics. Seed Sci. Res. **2005**, *15*, 1–20.
9. Walters, C.; Wheeler, L.J.; Stanwood, P.C. Longevity of cryogenically stored seeds. Cryobiology **2004**, *48*, 229–244.
10. Roberts, E.H.; Ellis, R.H. Water and seed survival. Ann. Bot. **1989**, *63*, 39–52.
11. Dickie, J.B.; Ellis, R.H.; Kraak, H.L.; Ryder, K.; Tompsett, P.B. Temperature and seed storage longevity. Ann. Bot. **1990**, *65*, 197–204.
12. Perez-Garcia, F.; Gonzalez-Benito, M.E.; Gomez-Campo, C. High viability recorded in ultra-dry seeds of 37 species of Brassicaceae after almost 40 years of storage. Seed Sci. Technol. **2007**, *35*, 143–155.
13. Pritchard, H.W.; Ashmore, S.; Berjak, P.; Engelmann, F.; González-Benito, M.E.; Li D.Z.; Nadarajan, J.; Panis, B.; Pence, V.; Walters, C. Storage stability and the biophysics of preservation. In: Proc. Plant conservation for the next decade: a celebration of Kew's 250th anniversary; Royal Botanic Garden Kew: London, UK, 2009; 12–16.
14. Walters, C.; Volk, G.M.; Stanwood, P.C.; Towill, L.E.; Koster, K.L.; Forsline, P.L. Long-term survival of cryopreserved germplasm: contributing factors and assessments from thirty-year-old experiments. Acta Hort. **2011**, *908*, 113–120.
15. Walters, C. Ultra-dry seed storage. Seed Sci. Res. **1998**, *8*, Suppl. No 1.
16. Pammenter, N.W.; Berjak, P. A review of recalcitrant seed physiology in relation to desiccation-tolerance mechanisms. Seed Sci. Res. **1999**, *9*, 13–37.
17. IPGRI/DFSC Desiccation and Storage Protocol. The Project on Handling and Storage of Recalcitrant and Intermediate Tropical Forest Tree Seeds, Newsletter, 1999; 23–39.
18. Maxted, N.; Ford-Lloyd, B.V.; Hawkes, J.G. Complementary Conservation Strategies. In *Plant Genetic Resources Conservation*; Maxted, N.; Ford-Lloyd, B.V.; Hawkes, J. G., Eds.; Chapman & Hall: London, 1997; 15–39.

Herbicide-Resistant Crops: Impact

Micheal D. K. Owen
Department of Agronomy, Iowa State University, Ames, Iowa, U.S.A.

Abstract

The development of herbicide-resistant (HR) crops, specifically as a result of genetic engineering and transgene movement and subsequent global adoption is unprecedented in agriculture. The adoption of these technologies has increased 94-fold since introduction in 1996. HR crops represent the most widely utilized of the transgene technologies. Generally, the impact of the transgene HR crops has been reported as favorable on agriculture; however, there has been no consensus, and the utilization of the technologies continues to be controversial despite a 16-year history. Benefits of the HR crop technologies include, but are not limited to, simple, convenient, and inexpensive weed control based on an environmentally friendly herbicide and stewardship of agriculture by better management of soil erosion and improved water quality. Risks of the HR crops include, but are not limited to, the lack of the development of new herbicides and conventional crop cultivars and the focus on one herbicide for weed control generally without the inclusion of more diverse tactics, which has resulted in weed populations shifts and evolved HR weed biotypes.

INTRODUCTION

The most important constraint on efficient production of food, fiber, and biofuel is the ability to manage weeds effectively. Weeds represent the most widespread and consistent pest to economic crop production. As such, weeding crops has required more global expenditure of energy and time than any other agronomic task. Historically, mankind has managed weeds by hand. Mechanical weed management began a new era in crop production efficiency. In the mid-20th century, the control of weeds with herbicides dramatically improved weed management and thus crop production efficiency. The next technological step in weed management was the development of herbicide-resistant (HR) crops, which were adopted at an unprecedented rate compared to any previous agricultural technology. HR crops, specifically those with resistance(s) attributable to genetic engineering (GE) have become a dominant feature in global agriculture since the mid-1990s, and their impact on mankind has been important.

TRANSGENES AND CROPS

The ability to transform plants by the introduction of selected genetic characteristics has dramatically changed the ability of plant breeders to improve agronomic crops.[1] The first GE crops contained transgenes that conferred resistance to herbicides, diseases, or insects.[2] Commercial utilization of GE crops has increased to over 160 million hectares worldwide of which 126 million hectares are GE crops with HR traits.[3,4] Although most of the GE HR crops are utilized in the industrialized countries, a total of 29 countries worldwide have planted GE crops.[4] Proponents of biotechnology suggest that increases in yields and food quality attributable to GE crops are critically important to developing countries.[3,5–9]

Although GE crops have been positioned by some to be environmentally benign and more beneficial than conventional crops, there continues to be concerns and debate that there may be undesirable characteristics and resultant negative impacts with the commercial-scale production of GE crops that has dominated agriculture in the last 16 years.[2] Specifically, the concerns reflect increased pesticide use, movement of transgenes to compatible weedy species, evolved resistance(s) in pest complexes, transgene contamination of food, concerns about soil and plant health attributable to pesticide use, and other perceived risks.[10–15] The concerns about increased pesticide use must be considered from the perspective of what is the appropriate comparison. In the case of HR crops, specifically those that include a transgene that provides resistance to glyphosate compared to previously used herbicides were used at a fraction of the active ingredient that glyphosate is used; so it is intuitively obvious that the herbicide amount used per treated field increased. However, the number of hectares treated with herbicides did not change.[3,15–18] Consider that inclusion of transgenes that code of the *Bacillus thuringiensis* (Bt) protein(s) has resulted in reduced use of insecticides. However, the recent evolution of resistance to Bt in specific insects pests suggests that there are issues with the technology.[14]

Transgenes that accrue glyphosate resistance in crops allow growers to use glyphosate more effectively and efficiently for weed management. Glyphosate is suggested to

Encyclopedia of Natural Resources DOI: 10.1081/E-ENRL-120049211

be an environmentally safer herbicide. However, there are concerns that the inclusion of transgenes that code for proteins may cause a hypoallergenic reaction in some people and other issues reported to be of great import.[12,19,20] Interestingly, there has not been a societal consensus that reflects the attitudes toward GE crops despite the unprecedented adoption of the technology by industrialized agriculture.[3,11,18,21,22]

As previously stated, one of the most important and widely successful concepts resulting from transgene technology is the introduction of herbicide resistance to crops. However, HR crops have also been developed with traditional breeding techniques. HR maize (*Zea mays*), soybean (*Glycine max*), canola (*Brassica napus*), rice (*Oryza sativa*), cotton (*Gossypium hirsutum*), and sugar beets (*Beta vulgaris*) are widely planted in North America, whereas HR wheat (*Triticum aestivum*) and sunflowers (*Helianthus annuus*) are commercially available albeit not widely adopted.[3] A primary benefit attributed to some HR crops is the ability to use environmentally benign herbicides such as glyphosate.

Crops with HR cultivars resistant to glyphosate include maize, soybean, cotton, canola, sugar beet, and alfalfa. Given the public concerns about herbicide use, the ability to use a herbicide with a favorable environmental profile is an important consideration supporting the adoption of HR crops.[3,12,17,21,23,24] Other pecuniary and non-pecuniary benefits supporting the adoption of HR crops include GE-conferred herbicide selectivity where no selective herbicides were previously available, choices of cheaper herbicides, ease or simplicity of weed management or both, increased adoption of tillage practices that minimize soil erosion and improve soil quality, and the elimination of herbicide injury to crops.[3,24–26]

Herbicides are the primary, if not sole tactic used to control weeds in most crops including HR crops. Farmers in the United States applied herbicides to almost 9 million crop hectares at a cost of $6.6 billion in 2003.[27] Herbicides are used annually on more than 90% of the row crop acres, and it has been suggested that without herbicides, crop production would decline by more than 20% and accounts for the efforts of 70 million laborers.[16] The adoption of HR crops has increased the emphasis on herbicidal weed control and minimized the use of alternative weed management tactics.[15,17,18] The unprecedented adoption of GE HR crops has changed herbicide use practices and the number of herbicides used, and resulted in a limited number of herbicides (e.g., glyphosate) being applied recurrently to a majority of the row crop acres.[28,29] This entry will provide an overview of the implications of GE HR crops on agriculture and describe in a broad sense the impact of these technologies on current agroecosystems and the socioeconomic considerations of an agriculture that is based on GE HR crop technologies.

WEED MANAGEMENT WITH HR CROPS

Fundamentally, weed management with HR crops is no different than using other selective herbicide technologies—a herbicide is applied to the crop, and the weeds are selectively controlled. The difference is how selectivity between crops and weeds is achieved.[30] For most herbicides, selectivity between crops and weeds is the result of differential abilities to metabolize the herbicide to non-active products. Crops are able to more effectively and efficiently metabolize the herbicide than the weeds.

Differential metabolism of herbicides is influenced by a number of factors including application timing relative to crop and weed stage of development and the overall growth conditions of the plants. The insertion of transgenes, or the selection for critical plant enzymes through traditional breeding programs, greatly elevates the differential response of the HR crop and weeds compared to conventional crops. The mechanisms by which the HR crop is protected from the herbicide is either by an alteration of the target enzyme (i.e., glyphosate resistance in soybean) or by the addition of genetic code that allows the crop plant to metabolize the herbicide (i.e., glufosinate in maize).[3,17,31,32]

As the utilization of HR crops does nothing to change the emphasis on herbicides for weed control and in fact has lessened the use of alternative tactics, public concerns for the negative impacts of herbicides on health and the environment cannot be discounted.[18,33] Furthermore, whereas the use of herbicides and HR crops is described to be generally beneficial to the environment, provides lower cost of weed management, and allows the farmer to use simpler weed management tactics, there are consequences from using HR crops and herbicides on weed management and weed populations.[15]

These consequences include weed population shifts that favor weed species not effectively managed by the target herbicide, the evolution of weed biotypes resistant to the target herbicide, the hybridization of the HR crops with closely related weed species and the potential to confer resistance to the resultant progeny, the likelihood that volunteer crops will become HR weeds in rotational crops, and the potential for the HR crops to contaminate non-HR and organically produced crops because of the inability to segregate grain or via pollen movement.[13,34–42] Weeds with evolved resistance to glyphosate have increased significantly since the introduction of HR crops with GE resistance to glyphosate and now represent a major problem for HR crop production in specific crop production systems[18,43–47] In addition, given that multiple applications of the target herbicides (e.g., glyphosate) are typically required to provide acceptable weed control, there is a greater opportunity for off-target movement of the herbicide.[37]

Generally, the adoption of HR crops and subsequent weed control has been an unprecedented major commercial success.[31] The ability to use registered herbicides that previously did not provide sufficient selectivity between crops and weeds provides an excellent potential to improve the profitability of agriculture and increase weed management options for farmers.[3,17] Furthermore, many crops do not represent a large-enough market opportunity for the agrochemical industry either because of the limited hectares grown or an unacceptable risk of crop injury relative to the profit potential from selling an existing herbicide. Thus, weed control options are limited in these crops.

Developing HR varieties of these crops will improve weed management options and thus increase yields, grain quality, and profitability.[39] Developing HR cultivars in "minor" crops could represent an important opportunity to growers who farm smaller areas and grow minor crops to achieve more efficient and less expensive weed management.

IMPACT OF HR CROPS ON CROP PRODUCTION AND ECONOMICS

The adoption of HR crops has largely been driven by the perceived production benefits ascribed to the HR traits. For example, in Western Canada, an estimated 90% of the hectares seeded to canola are HR varieties.[48] The overall adoption of GE crops has increased 94-fold since the commercial introduction of the first GE crop in 1996 and represented an accumulated hectarage of 1.25 billion hectares.[4]

However, most of the HR crop adoption reflects the unprecedented acceptance of HR soybean. An estimated 73.8 million hectares of HR soybeans were seeded in 2010 and represented 50% of the GE crops grown worldwide.[4] HR alfalfa, maize, canola, cotton, sugar beet, and soybean accounted for 83% or 122 million hectares of the total 148 million hectares planted to GE crops in 2010.[4] Other HR crops that are approved for production include carnation (*Dianthus caryophyllus*), chicory (*Cichorum intybus*), linseed (*Linum usitatissimum*), and tobacco (*Nicotiana tabacum*).[49] As many as 14 other crops have transgenic HR cultivars currently under development worldwide.[2]

In total, the inclusion of HR crops represents a major change in crop production systems and specifically in weed management tactics, and has required major effort by regulative groups in the federal government to insure that food safety has not changed because of the inclusion of transgenes.[40] From a regulatory perspective, HR and non-HR crops are identical. Importantly, the Environmental Protection Agency has determined that the major HR transgenes are safe, and food developed from HR crops was determined to be no different than food derived from non-HR crops. Canadian regulatory authorities determined that the glyphosate-resistant canola was the same as other cultivars after conducting an extensive series of compositional, processing, and feeding studies.[50] There can be, however, pleiotropic consequences to the addition of the transgene in HR crops.[51]

The economic impact of HR crops on the global agricultural economy is clearly positive.[4] Farmers and manufacturers of the HR crops report that there are clear and consistent savings in weed management costs compared with weed management not based on HR technology. The 2011 global market value of GE crops was US$13.3 billion, of which 77% was in industrial countries and 23% in developing countries.[4] However, movement of the HR pollen into non-HR crops can result in contamination of the grain.[34,35,52] In addition, the grain-handling industry is not currently able to segregate HR grain from non-HR grain, further increasing the risk of contamination and subsequent economic penalty to the producers.

CONCLUSIONS

There can be no question that the development of HR technologies has been one of the most important and yet contentious changes in global agriculture. Whereas the adoption of HR technology by farmers has been unprecedented worldwide, acceptance of the technology by consumers is less clear.[53–55] In fact, the issues about the benefits and risks of HR crops are similar to those voiced by consumers about the adoption of pesticide technologies.[56] On the other hand, the introduction of HR traits by transgene technology could not be readily or efficiently accomplished by traditional crop-breeding tactics, with the possible exception of resistance to imidazolinone herbicides in several crops.

Regardless, the resultant HR traits are portrayed as environmentally and economically beneficial;[24,28,57] a major component of the debate is that the HR trait is based on herbicidal weed management.

The historic baggage of the herbicide-use debate has not placed the HR crops in a positive position with many consumers, particularly in the European Union.[20,22] However, it is clear that farmers have decided that HR crops represent an important benefit to agriculture.[25]

It is interesting to note that the technology that allowed the development of HR crops is predicted to provide new weed management tools not based on herbicides.[30,58] Three novel approaches to utilizing transgene technology in weed management were reported by Gressel.[59] These were the alteration of biocontrol agents to improve their effectiveness, the use of transgenes to develop more competitive crop either by improving the growth habit or allowing the production of natural allelochemicals, or the alteration of cover crops to make them more applicable in weed management.

Thus, whereas it is unclear what the future impact of HR crops will be on world food production, it is clear that transgene technology will continue to play an important role.

REFERENCES

1. Noteborn, H.P.J.M.; A.A.C.M. Peijnenburg, and R. Zeleny, Strategies for analysing unintended effects in transgenic food crops. In *Genetically Modified Crops: Assessing Safety;* Atherton, K.T., Ed.; Taylor and Francis: London, 2002; 74–93.
2. Snow, A.A.; Palma, P.M. Commercialization of transgenic plants: potential ecological risks. BioScience **1997**, *47* (2), 86–96.
3. Green, J.M. The benefits of herbicide-resistant crops. Pest Manag. Sci. **2012**, *68* (10), 1323–1331.
4. James, C. *Global Status of Commercialized Biotech/GM crops:* 2011; ISAAA Brief No. 43: Ithaca, NY, 2011.
5. Chrispeels, M.J. Biotechnology and the poor. Plant Physiol. **2000**, *124* (1), 3–6.
6. Evenson, R.E.; Gollin, D. Assessing the impact of the green revolution, 1960 to 2000. Science **2003**, *300*, 758–762.
7. Gressel, J. Biotech and gender issues in the developing world. Nat. Biotechnol.**2009**, *27*, 1085–1086.
8. Gressel, J. Global advances in weed management. J. Agrl-cultural Sci. **2011**, *149*, 47–53.
9. Qaim, M.; Zilberman, D. Yield effects of genetically modified crops in developing countries. Science **2003**, *299*, 900–902.
10. Benbrook, C.M. Impacts of genetically engineered crops on pesticide use in the U.S.—the first sixteen years. Environ. Sci. Eur. **2012**, *24*, 24.
11. Bonny, S. Herbicide-tolerant transgenic soybean over 15 years of cultivation: pesticide use, weed resistance and some economic issues. The case of the USA. Sustainability **2011**, *3* (9), 1302–1322.
12. Duke, S.O.; Lydon, J.; Koskinen, W.C.; Moorman, T.B.; Chaney, R.L.; Hammerschmidt, R. Glyphosate effects on plant mineral nutrition, crop rhizosphere microbiota, and plant disease in glyphosate-resistant crops. J. Agricultural Food Chem. **2012**, *60*, 10375–10397.
13. Ellstrand, N.C. Evaluating the risks of transgene flow from crops to wild species. In *Gene Flow among Maize Landraces, Improved Maize Varieties, and Teosinte: Implications for Transgenic Maize*; CIMMYT: Mexico City, 1997.
14. Gassman, A.J.; Petzold-Maxwell, J.L.; Keweshan, R.S.; Dunbar, M.W. Field-evolved resistance to Bt maize by western corn rootworm. Plos ONE **2011**, *6* (7), e22629.
15. Owen, M.D.K.; Young, B.G.; Shaw, D.R.; Wilson, R.G.; Jordan, D.L.; Dixon, P.M.; Weller, S.C.; Benchmark study on glyphosate-resistant cropping systems in the USA. II. Perspective. Pest Manag. Sci. **2011**, *67* (7), 747–757.
16. Gianessi, L.P.; Reigner, N.P. The value of herbicides in U.S. crop production. Weed Technol. **2007**, *21*, 559–566.
17. Green, J.M.; Owen, M.D.K. Herbicide-resistant crops: Utilities and limitations for herbicide-resistant weed management. J. Agricultural Food Chem. **2011**, *59*, 5819–5829.
18. Owen, M.D.K. Weed resistance development and management in herbicide-tolerant crops: experiences from the USA. J. Consum. Prot. Food Safety **2011**, *6* (1), 85–89.
19. Johal, G.S.; Huber, D.M. Glyphosate effects on diseases of plants. Eur. J. Agronomy. **2009**, *31*, 144–152.
20. Madsen, K.H.; Sandoe, P. Ethical reflections on herbicide-resistant crops. Pest Manag. Sci. **2005**, *61*, 318–325.
21. Duke, S.O. Comparing conventional and biotechnology-based pest management. J. Agric. Food Chem. **2011**, *59*, 5793–5798.
22. Ehlers, U. Interplay between GMO regulations and pesticide regulation in the EU. J. Consum. Prot. Food Safety. **2011**, *4*, 4.
23. Kudsk, P.; Streibig, J.C. Herbicides - a two-edged sword. Weed Res. **2003**, *43* (2), 90–102.
24. Cerdeira, A.L.,Duke, S.O., The current status and environmental impacts of glyphosate-resistant crops; a review. J. Environ. Qual. **2006**, *35*, 1633–1658.
25. Ervin, D.E.; Carriere, Y.; Cox, W.J.; Fernandez-Cornejo, J.; Jussaume R.A., Jr.; Marra, M.C.; Owen, M.D.K.; Raven, P.H.; Wolfenbarger, L.L.; Zilberman, D. *The Impact of Genetically Engineered Crops on Farm Sustainability in the United States, in National Research Council Report*; Grossblatt, N., Ed.; National Research Council: Washington, D.C., 2010; 250.
26. Frisvold, G.B.; Hurley, T.M.; Mitchell, P.D. Overview: Herbicide resistant crop - diffusion, benefits, pricing, and resistance management. AgBioForum **2009**, *12* (3, 4), 244–248.
27. Gianessi, L.P.; Sankula, S. Executive Summary: The value of herbicides in U.S. crop production. [computer website] 2003. http://www.ncfap.org (cited 2003 10/6/03).
28. Gianessi, L.P. Economic impacts of glyphosate-resistant crops. Pest Manag. Sci. **2008**, *64* (4), 346–352.
29. Young, B.G. Changes in herbicide use patterns and production practices resulting from glyphosate-resistant crops. Weed Technol. **2006**, *20*, 301–307.
30. Duke, S.O.; Weeding with transgenes. Trends Biotechnol. **2003**, *21* (5), 192–195.
31. Duke, S.O.; Powles, S.B. Glyphosate: a once-in-a-century herbicide. Pest Manag. Sci. **2008**, *64* (4), 319–325.
32. Green, J.M.; Hale, T.; Pagano, M.A.; Andreassi II, J.L.; Gutteridge, S.A. Response of 98140 corn with gat4621 and hra transgenes to glyphosate and ALS-inhibiting herbicides. Weed Sci. **2009**, *57*, 142–148.
33. Phipps, R.H.; Park, J.R. Environmental benefits of genetically modified crops: Global and European perspectives on their ability to reduce pesticide use. J. Anim. Feed Sci. **2002**, *11*, 1–18.
34. Legere, A. Risks and consequences of gene flow from herbicide-resistant crops: canola (*Brassica napus* L.) as a case study. Pest Manag. Sci. **2005**, *61*, 292–300.
35. Mallory-Smith, C.; Zapiola, M. Gene flow from glyphosate-resistant crops. Pest Manag. Sci. **2008**, *64* (4), 428–440.
36. Mallory-Smith, C.A.; Olguin, E.S. Gene flow from herbicide-resistant crops: It's not just for transgenes. J. Agricultural Food Chem. **2010**, *59*, 5813–5818.
37. Owen, M.D.K. Current use of transgenic herbicide-resistant soybean and corn in the USA. Crop Prot. **2000**, *19*, 765–771.
38. Owen, M.D.K. Weed species shifts in glyphosate-resistant crops. Pest Manag. Sci. **2008**, *64*, 377–387.
39. Orson, J. *Gene Stacking in Herbicide Tolerant Oilseed Tape: Lessons from the North American Experience*; English Nature: Peterbrough, UK, 2002; 17.
40. Goldman, K.A. Bioengineered foods - safety and labeling. Science **2000**, *290* (5491), 457–459.

41. Hodgson, E. Genetically modified plants and human health risks: can additional research reduce uncertainties and increase public confidence? Toxicol. Sci. **2001**, *63*, 153–156.
42. Kimber, I.; Dearman, R.J. Approaches to assessment of the allergenic potential of novel proteins in food from genetically modified crops. Toxicol. Sci. **2002**, *68*, 4–8.
43. Culpepper, A.S. Glyphosate-induced weed shifts. Weed Technol. **2006**, *20*, 277–281.
44. Culpepper, A.S.; Webster, T.M.; Sosnoskie, L.M.; York, A.C. Glyphosate-resistant Palmer amaranth in the United States. In *Glyphosate Resistance in Crops and Weeds*; Nadula, V.K., Ed.; John Wiley & Sons: New York, NY, 2010; 195–212.
45. Heap, I. The international survey of herbicide resistant weeds. 2012. http://www.weedscience.com.
46. Powles, S.B. Evolution in action: glyphosate-resistant weeds threaten world crops. Outlooks Pest Manag. **2008**, *12* (December), 256–259.
47. Powles, S.B. Evolved glyphosate-resistant weeds around the world: lessons to be learnt. Pest Manag. Sci. **2008**, *64* (4), 360–365.
48. Zhu, B.; Ma, B.-l. Genetically-modified crop production in Canada: agronomic, ecological and environmental considerations. Am. J. Plant Sci. Biotechnol. **2011**, *5*(1), 90–97.
49. Nap, J.-P.; Metz, L.J.; Escaler, M.; Conner, A.J. The release of genetically modified crops into the environment. Part I. Overview of current status and regulations. Plant J. **2003**, *33*, 1–18.
50. Taylor, M.; Stanisiewski, E.; Riordan, S.; Nemeth, M.; George, B.; Hartnell, G. Comparison of broiler performance when fed diets containing roundup ready-event RT73-, non-transgenic control or commercial canola meal. Poult. Sci. **2004**, *83*, 456–461.
51. Pline, W.A.; Wu, J.; Hatzios, K.K. Effects of temperature and chemical additives on the response of transgenic herbicide-resistant soybeans to glufosinate and glyphosate applications. Pest Biochem. Physiol. **1999**, *65* (2), 119–131.
52. Snow, A.A.; Andow, D.A.; Gepts, P.; Hallerman, E.M.; Power, A.; Tiedje, J.M.; Wolfenbarger, L.L.; Genetically engineered organisms and the environment: current status and recommendations. Ecol. Appl. **2005**, *15* (2), 377–404.
53. Jayaraman, K.; Jia, H. GM phobia spreads to South Asia. Nat. Biotechnol. **2012**, *30*, 1017–1019.
54. Meldolesi, A. Media leaps on French study claiming GM maize carcinogenicity. Nat. Biotechnol. **2012**, *30*, 1018.
55. Nording, L. Opposition thaws for GM crops in Africa. Nat. Biotechnol. **2012**, *30*, 1019.
56. William, R.D.; Ogg, A.; Baab, C.; My view. Weed Sci. **2001**, *49*, 149.
57. Strauss, S.H. Genomics, genetic engineering, and domestication of crops. Science **2003**, *300* (5616), 61–62.
58. Gressel, J.; Levy, A.A. Stress, mutators, mutations and stress resistance. In *Abiotic Stress Adaptation in Plants: Physiological, Molecular and Genomic Foundation*; Pareek, A., Sopory, S.K., Bohnert, H.J., Govindjee, Eds.; Springer Science: Dordrecht, 2009; 471–483.
59. Gressel, J. *Molecular Biology of Weed Control. Frontiers in Life Science*; Taylor and Francis: London, 2002; Vol. 1, 504.

Herbicide-Resistant Weeds

Carol Mallory-Smith
Oregon State University, Corvallis, Oregon, U.S.A.

Abstract
The introduction of synthetic herbicides revolutionized weed control. However, the repeated use of herbicides led to the selection of herbicide-resistant weeds. The selection of resistant weeds reduces control options and may increase control costs. In many cases, the most effective and economical herbicide is lost because of resistance. Herbicide-resistant weeds are one of the major challenges in weed management. Although resistance presents a challenge, herbicide-resistant weeds can be controlled with herbicides with different sites of action or other methods of weed control.

BACKGROUND INFORMATION

Herbicide resistance is the inherited ability of a biotype to survive and reproduce following exposure to a dose of a herbicide that is normally lethal to the wild type.[1] Herbicide resistance is an evolved response to selection pressure by a herbicide.

Herbicide-resistant biotypes may be present in a weed population in very small numbers. The repeated use of one herbicide or herbicides with the same site of action allows these resistant plants to survive and reproduce. The number of resistant plants increases until the herbicide is no longer effective. There is no evidence that the herbicide causes the mutations that lead to resistance.

Cross-resistance is the expression of one mechanism that provides plants with the ability to withstand herbicides from different chemical classes.[2] For example, a single point mutation in the enzyme acetolactate synthase (ALS) may provide resistance to five different chemical classes including the widely used sulfonylurea and imidazolinone herbicides.[3] However, cross-resistance at the whole-plant level is difficult to predict because a different point mutation in the ALS enzyme may provide resistance to one chemical class and not others. Cross-resistance also can result from increased metabolic activity that leads to detoxification of herbicides from different chemical classes.

Multiple-resistance is the expression of more than one mechanism that provides plants with the ability to withstand herbicides from different chemical classes.[2] Weed populations may have simultaneous resistance to many herbicides. For example, a common waterhemp (*Amaranthus rudis*) population in Illinois is resistant to triazine and ALS-inhibiting herbicide classes. However, resistance to these two different classes of herbicides is endowed by two different mechanisms within the same plant.[4] The weed species has two target site mutations, one for each herbicide class. An annual ryegrass (*Lolium rigidum* Gaud.) population in Australia is resistant to at least nine different herbicide classes.[5] In this case, herbicide options may become very limited. As with cross-resistance, multiple-resistance is difficult to predict; therefore, management of weeds with these types of resistance is complicated.

HISTORY OF HERBICIDE RESISTANCE

It was not long after the commercial use of herbicides that Abel[6] and Harper[7] discussed the potential for weeds to evolve resistance. However, the first well-documented example for the selection of a herbicide-resistant weed was triazine-resistant common groundsel (*Senecio vulgaris* L.), which was identified in 1968 in Washington State.[8] The resistant biotype was found in a nursery that had been treated once or twice annually for 10 years with triazine herbicides. There were earlier reports of differential responses within weed species, but these variable responses to herbicides were not necessarily attributed to resistance.[8]

To date, 281 herbicide-resistant biotypes from 168 species have been identified.[9] The reason that the biotype number and the species number are different is that the same species has been identified with resistance to different herbicides in different locations. Of the 168 species, 100 are dicots and 68 monocots. Resistance has occurred to most herbicide chemical families.

MECHANISMS RESPONSIBLE FOR RESISTANCE

Several mechanisms theoretically could be responsible for herbicide resistance. Those mechanisms include reduced herbicide uptake, reduced herbicide translocation, herbicide sequestration, herbicide target-site mutation, and herbicide

Encyclopedia of Natural Resources DOI: 10.1081/E-ENRL-120020280

detoxification. In the cases where the resistance mechanism has been determined, the mechanism responsible in most instances has been either target-site mutations or detoxification by metabolism.[10]

Target-site mutations have been identified in weeds resistant to herbicides that inhibit photosynthesis, microtubule assembly, or amino acid production. Most often there is a point mutation, a single nucleotide change, which results in an amino acid change and is responsible for the resistance.[10] The shape of the herbicide binding site is modified and the herbicide can no longer bind.

Metabolism-based resistance does not involve the binding site of the herbicide, but instead the herbicide is broken down by biochemical processes that make it less toxic to the plant. Several groups of enzymes are involved in the process. Enzymes that are thought to be most important in herbicide metabolism are glutathione *S*-transferases and cytochrome P450 monooxygenases.[11,12]

FACTORS THAT INFLUENCE THE SELECTION OF RESISTANT BIOTYPES

Resistance usually occurs when a herbicide has been used repeatedly, either year after year or multiple times during a year, and the herbicide is highly effective, killing more than 90% of the treated weeds. The more effective a herbicide, the higher the selection pressure for resistance. Therefore, all herbicides do not exert the same selection pressure.

If a herbicide has only one site of action, it is easier to select a resistant biotype because only one mutation is needed. A herbicide that has soil residual activity will provide more selection pressure because the herbicide remains in the environment and any new seedlings will be exposed to the herbicide. If a herbicide has a very short residual, repeated applications during one growing season can have the same effect.

Herbicide factors influence the selection of herbicide-resistant weeds, but agronomic factors can also be important. Many resistant weed species have been selected in monoculture production systems. These systems result in the repeated use of the same herbicide. The increased reliance on herbicides for weed control along with a concurrent decrease in other weed management tactics further increases selection of resistant biotypes. The introduction of herbicide-resistant crops also increased the use of a single herbicide with decreased alternative controls.

Some weed species seem to be more prone to herbicide resistance than others. Some species have increased mutation rates or increased genetic variability. Increased variability is usually found in cross-pollinating species. Selection of resistant biotypes varies depending upon how likely it is for the resistance mutation to be lethal.

Weeds that produce more than one generation per year may be exposed to herbicides with the same site of action more often than those that produce only one generation per year. Because the selection of a resistant biotype is dependent on the number of individuals that are exposed to the herbicide, those weed species that produce more seeds may also produce a resistant individual more quickly.

Breeding systems and inheritance of the trait will influence how fast resistance spreads once it occurs. If the trait is controlled by one recessive gene, the heterozygote and the homozygote dominant plants will be susceptible, so only ¼ of the population will survive herbicide treatment. If the trait is dominant, the heterozygote and the homozygote dominant plants will be resistant, and the population will build quickly because ¾ of the plants will survive herbicide treatment. The trait will be readily moved within and between populations if the weed species outcrosses. In a selfing population, the trait will have reduced movement with pollen. Maternal inheritance will prevent herbicide resistance from moving with the pollen. An example of maternal inheritance is triazine resistance.[13]

Fitness is the reproductive ability of an individual, and competitive ability is the capacity of a plant to acquire resources. Initially, researchers assumed that herbicide-resistant weed species would have reduced fitness and competitive ability. Indeed, triazine-resistant weed species do have reduced growth and competitive ability.[14–16] However, weeds resistant to ALS-inhibiting herbicides have not consistently had reduced fitness or competitive ability.[17,18]

Refuges are a common tactic for the management of insect and pathogen resistance. Susceptible populations are maintained in surrounding areas and can be used to swamp resistance alleles. Generally, pollen and seed are not sufficiently mobile to swamp resistant weed populations. Migration is probably more important in the movement of resistance to a susceptible population than vice versa. This is particularly true with the tumbleweeds. Long-distance movement of a herbicide resistance gene out of an area is most likely to occur through seed movement.

PREVENTION AND MANAGEMENT OF HERBICIDE-RESISTANT WEEDS

Any management strategy that reduces the selection pressure from a herbicide will reduce the selection for a herbicide-resistant weed in the system. Recommendations for the prevention or management of herbicide-resistant weeds are often the same.[1] The recommendations from the herbicide industry and university personnel include many common factors. Common recommendations are to rotate herbicides with different sites of action to reduce selection pressure, to use short-residual herbicides so that selection pressure during a cropping season is reduced, to use multiple weed-control methods in conjunction with herbicides, and to plant certified crop seed so that herbicide-resistant weed seeds are not introduced into a field. Growers need to keep accurate records of herbicides that have been used on a field so that they can adopt a weed management plan that reduces the selection of herbicide-resistant weeds. The integration of these recommendations will reduce selection pressure for herbicide-resistant weeds.

CONCLUSION

Herbicide-resistant weeds will continue to be an issue for weed management as long as herbicides are used for weed control. Herbicide-resistant weeds can be managed, and no herbicides have been removed from the marketplace because of resistance. When growers use multiple weed-management techniques, the selection of herbicide-resistant weeds is reduced. An integrated approach using crop rotation, herbicides with different sites of action, and alternative weed control such as physical and mechanical weed control is useful in both the prevention of herbicide-resistant weeds and the management of herbicide-resistant weeds if they do occur.

REFERENCES

1. Retzinger, E.J.; Mallory-Smith, C. Classification of herbicides by site of action for weed resistance management strategies. Weed Sci. **1997**, *11* (2), 384–393.
2. Hall, L.M.; Holtum, J.A.M.; Powles, S.B. Mechanisms responsible for cross resistance and multiple resistance. In *Herbicide Resistance in Plants*; CRC Press: Boca Raton, FL, 1994; 243–261.
3. Tranel, P.J.; Wright, T.R. Resistance of weeds to ALS-inhibiting herbicides: What have we learned? Weed Sci. **2002**, *50* (6), 700–712.
4. Foes, M.J.; Liu, L. A biotype of common waterhemp (*Amaranthus rudis*) resistant to triazine and ALS herbicides. Weed Sci. **1998**, *46* (5), 514–520.
5. Burnet, M.W.M.; Hart, Q.; Holtum, J.A.M.; Powles, S.B. Resistance to nine herbicide classes in a population of rigid ryegrass (*Lolium rigidum*). Weed Sci. **1994**, *42* (3), 369–377.
6. Abel, A.L. The rotation of weedkillers. Proc. Brit. Weed Control Conf. **1954**, *2*, 249–255.
7. Harper, J.C. The evolution of weeds in relation to resistance to herbicides. Proc. Brit. Weed Control Conf. **1956**, *3*, 179–188.
8. Ryan, G.F. Resistance of common groundsel to simazine and atrazine. Weed Sci. **1970**, *18* (5), 614–616.
9. Heap, I. The International Survey of Herbicide Resistant Weeds, http://www.weedscience.com (accessed August 2003).
10. Preston, C.; Mallory-Smith, C.A. Biochemical mechanisms, inheritance, and molecular genetics of herbicide resistance in weeds. In *Herbicide Resistant Weed Management in World Grain Crops*; CRC Press: Boco Raton, FL, 2001; 23–60.
11. Barrett, M. The role of cytochrome P450 enzymes in herbicide metabolism. In *Herbicides and Their Mechanisms of Action*; CRC Press: Boca Raton, FL, 2000; 25–37.
12. Edwards, R.; Dixon, D.P. The role of glutathione in herbicide metabolism. In *Herbicides and Their Mechanisms of Action*; CRC Press: Boca Raton, FL, 2000; 38–71.
13. Souza-Machado, V.; Bandeen, J.D.; Stephenson, G.R.; Lavigne, P. Uniparental inheritance of chloroplast atrazine tolerance in *Brassica campestris*. Can. J. Plant Sci. **1978**, *58* (4), 977–981.
14. Conard, S.G.; Radosevich, S.R. Ecological fitness of *Senecio vulgaris* and *Amaranthus retroflexus* biotypes susceptible and resistant to atrazine. J. Appl. Ecol. **1979**, *16* (1), 171–177.
15. Marriage, P.B.; Warwick, S.I. Differential growth and response to atrazine between and within susceptible and resistant biotypes of *Chenopodium album*. L. Weed Res. **1980**, *20* (5), 9–15.
16. Williams, M.M.I.I.; Jordan, N. The fitness cost of triazine resistance in jimsonweed (*Datura stramonium* L.). Am. Midl. Nat. **1995**, *133* (1), 131–137.
17. Alcocer-Rutherling, M.; Thill, D.C.; Shafii, B. Differential competitiveness of sulfonylurea resistant and susceptible prickly lettuce (*Lactuca serriola*). Weed Tech. **1992**, *6* (2), 303–309.
18. Thompson, C.R.; Thill, D.C.; Shafii, B. Growth and competitiveness of sulfonylurea-resistant and -susceptible kochia (*Kochia scoparia*). Weed Sci. **1994**, *42* (2), 172–179.

Herbicides in the Environment

Kim A. Anderson
Environmental and Molecular Toxicology, Oregon State University, Corvallis, Oregon, U.S.A.
Jennifer L. Schaeffer
Food Safety and Environmental Stewardship Program, Oregon State University, Corvallis, Oregon, U.S.A.

Abstract
Understanding (and predicting) the environmental fate of herbicides is critically important. The environmental fate of herbicides depends on the chemical transformations, degradation, and transport in each environmental compartment.

Genetic—Keystone

INTRODUCTION

Herbicides are an integral part of our society; they are used by the general public, governments, institutions, foresters, and farmers. The benefits of herbicides need to be delivered without posing unacceptable risk to non-target sites. Therefore, understanding (and predicting) the environmental fate of herbicides is critically important. Environmental fate is determined by the individual fate processes of transformation and transport. Chemical and physical processes are primary determinants of transformation and transport of herbicides applied to soils and plants. Transformation determines what herbicides are degraded to in the environment and how quickly, while transport determines where herbicides move in the environment and how quickly. Jointly, these processes affect how much of a pesticide and its metabolites (degradation products) are present in the environment, where, and for how long. Herbicide fate also varies in response to changes in environmental conditions and application management practices (e.g., spray drift, volatilization), so it is important to understand these variables as well.

Major environmental compartments for herbicides can be considered as surface waters, the subsurface (soil and groundwater), and the atmosphere. Each medium has its own unique characteristics; however, there are many similarities when considering herbicide movement. Herbicides are rarely restricted to only one medium; therefore, chemical exchange among the compartments must be considered.

Wind erosion, volatilization, photo-degradation, runoff, plant uptake, sorption to soil, microbial or chemical degradation and leaching are potential pathways for loss from an application site. Chemical and microbial degradation are critically important factors in the fate of herbicides. Chemical conditions in soil are important secondary determinants of herbicide transport and fate. The importance of interactions between herbicides and solid phases of soils, soil water, and air within and above soil depends on a variety of chemical factors. Adsorption of herbicides from soil water to soil particle is one of the most important chemical determinants that limit mobility in soils. Environmental fate of herbicides depends on the chemical transformations, degradation, and transport in each environmental compartment.

TRANSFORMATION

Transformations determine how long a herbicide will stay in the environment. Molecular interactions of herbicides are based in part on the herbicide's chemical nature and are predicated on the physical-chemical properties and reactivities of the herbicide. Several generalized exchange processes between compartments are shown in Fig. 1. In order to understand (and therefore predict) environmental fate, the physical-chemical properties of herbicides must be known. Physical-chemical properties such as molecular formula, molecular weight, boiling point, melting point, decomposition point, water solubility, organic solubility, vapor pressure (V_p), Henry's law constant (K_H), octanol/water partitioning (K_{ow}), acidity constant (pKa), and soil sorption (K_d, K_{oc}) are important in predicting transformations. How some of these chemical-physical properties affect herbicide fate is briefly discussed (Fig. 2). Table 1,[1,4] demonstrates the wide variation of chemical-physical properties of a selected group of herbicides. Water solubility, the solubility of a herbicide in water at a specific temperature, determines the affinity for aqueous media and affects movement between air, soil, and water compartments. Vapor pressure, the pressure of the vapor of a herbicide at equilibrium with its pure condensed phase, measures a herbicide's tendency to transfer to and

Encyclopedia of Natural Resources DOI: 10.1081/E-ENRL-120020276

from gaseous environmental phases. Vapor pressure is critical for predicting either the equilibrium distribution or the rate of exchange to and from natural waters. Henry's law constant (K_H), the air–water distribution ratio for neutral compounds in dilute solutions in pure water, determines how a chemical will distribute between the gas and aqueous phase at equilibrium. Henry's law constant, therefore, only approximates the air–water partition in natural waters. Octanol–water partition coefficient (K_{ow}), the partition of organic compounds between water and octanol, is used to estimate the equilibrium partitioning of nonpolar organic compounds between water and organisms. The octanol–water partition coefficient is also proportional to partitioning of organic compounds (herbicides) into soil humus and other naturally occurring organic phases.

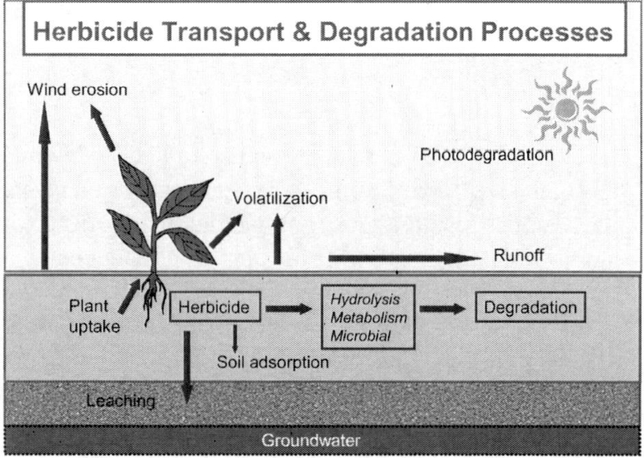

Fig. 1 Illustrated are pathways for loss of herbicide from a herbicide application site. Each pathway (arrow) illustrates a particular loss process occurring at some rate constant (k). The importance of each pathway and magnitude of each rate constant will vary substantially between herbicides and environmental conditions.

In the environment, many herbicides are present in a charged state (not neutral). Charged species have different properties and reactivities as compared to their neutral counterparts. Therefore, the extent to which a compound forms ions in environmental ecosystems is important. The pKa is a measure of the strength of an acid relative to water. Strong organic acids (pKa$\cong$0–3) in ambient natural waters (pH 4–9) will be present predominantly as anions. Conversely, very weak acids (pKa $\geq$12) in ambient natural waters will be present in their associated form (neutral). In an analogous fashion, strong bases (pKa $\geq$11) will be present as ions.[5] Examples of weak acids are 2,4-D and triclopyr, and examples of weak bases are atrazine, dicamba, and simazine (Table 1).

Sorption of a herbicide is when the herbicide binds to the soil or sediment particles (K_d). Soil sorption is an important process since it can dramatically affect herbicide fate. Figure 2 illustrates the relationship between herbicides sorbed to soil particles versus those dissolved in soil water. Soil sorption is dependent on soil type and influenced heavily by the organic matter content. Typically, soils high in clay and organic matter have a higher sorption capacity. To account for organic matter content, sorption is sometimes normalized for % organic matter, referred to as K_{oc}. K_d, K_H, and V_p describe the potential for exchange of herbicide compounds among soil, water, and air over short distances by diffusion. Transport over longer distances involves mass transfer (advection-dispersion).

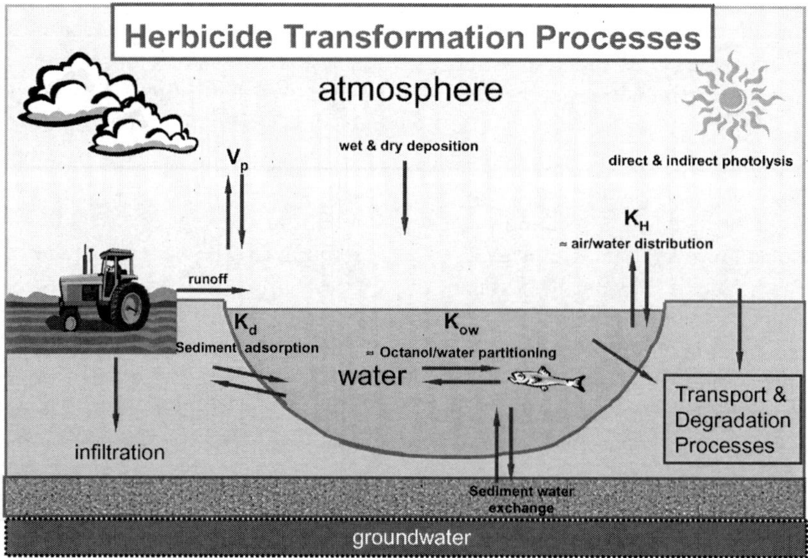

Fig. 2 Illustrated are pathways for transformation processes for loss of herbicide from a herbicide application site. Each pathway (arrow) illustrates a particular loss process occurring at some rate constant (k). The importance of each pathway and magnitude of each rate constant will vary substantially between herbicides and environmental conditions.

Table 1 Physiochemical Properties of Selected Herbicidesa

Chemical name	CAS number	Chemical class	Solubility (mg/L)	Vapor pressure (mm Hg)	K_d soil sorption K_{oc} normalized for organic intent	Log K_{ow} octanol/water partition constant	K_H Henry's constant (atm-m^3/mole)	pKa acidity constant
2,4-D	94-75-7	Chlorphenoxy acid	677	8.25×10^{-5} at 20°C	0.08–0.94	2.81	3.54×10^{-8}	2.73
Atrazine	1912-24-9	Triazine	34.7	2.89×10^{-7}	K_{oc} 100	2.61	2.36×10^{-9}	1.7
Bromoxynil	1689-84-5	Hydroxybenzonitrile	130	4.8×10^{-6}	K_{oc} 2.48	2	1.4×10^{-6}	4.06
Carbaryl	63-25-2	Carbamate	110 at 22°C	1.36×10^{-6}	2.45–4.69	2.36	3.27×10^{-9}	NA
Chlorsulfuron	64902-72-3	Sulfonylurea	2.8	2.3×10^{-11}	K_{oc} 1.02	2	3.9×10^{-15}	3.6
Dicamba	1918-00-9	Benzoic acid	8310	3.38×10^{-5}	0.07–0.53	2.21	2.18×10^{-9}	1.97
Diuron	330-54-1	Substituted urea	42	6.9×10^{-8}	2.9–14	2.68	5.04×10^{-10}	−1
Metolachlor	051218-45-2	Chloroacetanilide	530 at 20°C	3.14×10^{-5}	1.5–10	3.13	9×10^{-9} at 20°C	NA
Metsulfuron	74223-64-6	Sulfonylurea	9500	2.5×10^{-12}	0.36–1.40	2.2	1.32×10^{-16}	3.64
Pendimethenalin	40487-42-1	2,4-Dinitroanaline	0.3 at 20°C	3×10^{-5}	30–380	5.18	8.56×10^{-7}	NA
Pentachlorophenol	87-86-5	Chlorinated phenol	14	1.1×10^{-4}	K_{oc} 262–38905	5.12	2.45×10^{-8} at 22°C	4.7
Prometon	1610-18-0	Triazine	750	2.3×10^{-6} at 20°C	0.373–2.61	2.99	9.09×10^{-10} at 20°C	4.3
Simazine	122-34-9	Triazine	6.2 at 22°C	2.21×10^{-8}	0.48–4.31	2.18	9.42×10^{-10}	1.62
Thifensulfuron	79277-27-3	Sulfonylurea	230	1.28×10^{-10}	0.08–1.38	1.56	4.08×10^{-14}	4
Triasulfuron	82097-50-5	Sulfonylurea	815	1.5×10^{-8}	K_{oc} 73.4–190.6	0.58	9.9×10^{-7} at 20°C	4.64
Tribenuron	101200-48-0	Sulfonylurea	2040	3.97×10^{-10}	0.19–2.0	1.17	1.01×10^{-8}	5
Trifluralin	1582-09-8	Dinitroanaline	0.184	4.58×10^{-5}	18.6–54.8	5.34	1.03×10^{-4} at 20°C	NA

[a]Data at 25°C unless otherwise noted.

TRANSPORT

Transport of herbicides from runoff from soil is an important route of entry into surface waters. Herbicide runoff can be classified into two categories: herbicides that are either dissolved or suspended in runoff waters. Dissolved herbicides are characterized by low adsorption and high water solubility, while suspended herbicides are characterized by high soil sorption. Leaching depends strongly on local environmental conditions, such as percolation rates through local soils. Adsorption decreases herbicide leaching mobility by reducing the amount of herbicide available to the percolating soil water. However, herbicide runoff and leaching also are controlled by the amount of herbicide degradation.

DEGRADATION

Herbicide degradation ultimately ends with the formation of simple stable compounds, such as carbon dioxide; however, there are intermediates of varying stability on the way to complete mineralization (e.g., H_2O and CO_2). The rate of degradation of a particular herbicide can vary widely. The chemical nature of the herbicide is important, as discussed previously, but the degradation rate also depends on the availability of other reactants, as well as environmental factors. There are three major degradation pathways for herbicides: photo-degradation, chemical degradation, and microbial degradation.

Photo-degradation is the breakdown of a herbicide by sunlight at the plant, soil, or water surface. Direct photolysis occurs when the herbicide absorbs light (some portion of the solar spectrum), and this leads to dissociation of some kind (e.g., bonds break). Indirect photolysis occurs when a sensitizer molecule is radiatively excited and is sufficiently long-lived to transfer energy, such as an electron, a hydrogen atom, or a proton to another "receptor" molecule. The receptor molecule (herbicide), without directly absorbing radiation, can be activated via the "sensitizer" molecule to undergo dissociation or other kinds of chemical reactions leading to photo-degradation.

Chemical degradation is the breakdown of herbicides by processes not involving living organisms (abiotic). Hydrolysis may be one of the more important mechanisms of degradation for herbicides. Hydrolysis is the reaction of a herbicide where water interacts with the herbicide, replacing a portion of the molecule with OH. The following functional groups and chemical classes are known to be susceptible to hydrolysis: ethers, amides, phenylurea compounds, nitrile, carbamates, thiocarbamates, and triazines. Hydrolysis is influenced by environmental conditions including pH, water hardness, dissolved organic matter, dissolved metals, and temperature. Half life ($t_{1/2}$) is the amount of time it takes the parent compound (herbicide) to decay to half its original concentration. The half life for chemical degradation via the hydrolysis pathway is strongly dependent on environmental conditions. For example, the $t_{1/2}$ for 2,4-D at pH 6 is four years, while the $t_{1/2}$ at pH 9 is 37 hours.[6]

Microbial degradation is the breakdown of compounds (herbicides) by microorganisms. Bacteria and fungi are the primary microorganisms responsible for biotransformations (biotic reactions). Rates of microbial degradation are largely determined by environmental conditions such as temperature, pH, reduction-oxidation conditions, moisture, oxygen, organic matter, and food. Complete biodegradation, mineralization, yields carbon dioxide, water, and minerals.

CONCLUSION

The perfect herbicide would be one that controls weeds as necessary and then quickly breaks down and never moves off-site. The development of new herbicides over the last decade has focused in part on minimizing unacceptable risk to non-target sites. This has resulted in the development of more biologically active herbicides that greatly reduce application rates. Other research developments include minimizing the leaching of herbicides. When herbicides are applied in the environment, many transport and transformation processes are involved in their dissipation. Predicting herbicide environmental fate and behavior is complicated and is determined by the chemical processes for each environmental compartment.

In addition to understanding the fate of herbicides in the environment, emerging research also endeavors to look at the risk to non-target organisms. The risk of herbicides to organisms is relative to the bioavailability of the herbicide. Bioavailability may be approximately inversely proportional to K_d. Herbicides with larger K_ds tend to have lower bioavailability. Along with the determination of transport and transformation chemical processes, understanding the complete risk to non-target sites, therefore, should include the determination of the bioavailability of the herbicide.

ACKNOWLEDGMENTS

The authors thank Dr. Jeffery Jenkins, Oregon State University, for helpful discussions.

REFERENCES

1. University of Guelph Ontario Agricultural College. http://www.uoguelph.ca/OAC/env/bio/PROFILES.pdf, (accessed on March 2003).
2. The Extension Toxicology Network. http://ace.orst.edu/info/extoxnet/pips/ghindex.html, (accessed March 2003).

3. United States Department of Agriculture, Agricultural Research Services. http://wizard.arsusda.gov/acsl/ppdb3.html (accessed March 2003).
4. Environmental Science Syracuse Research Corporation. http://esc.syrres.com/interkow/physdemo.htm (accessed March 2003).
5. Montgomery, J.H. *Agrochemicals Desk Reference: Environmental Data*; Lewis Publishers: Ann Arbor, 1993.
6. Kamrin, M.A. *Pesticide Profiles: Toxicity, Environmental Impact, and Fate*; CRC Press LCC: New York, 1997.

Insects: Economic Impact

David Pimentel
Department of Entomology, Cornell University, Ithaca, New York, U.S.A.

Abstract

Insects, weeds, and plant pathogens destroy ~40% of all food in the world despite the use of ~3 billion kilograms of pesticides each year and cost ~ $32 billion per year. Most or 68% of the pesticides applied in the United States are herbicides.

Genetic—Keystone

INTRODUCTION

Insects, plant pathogens, and weeds are major pests of crops in the United States and throughout the world. Approximately 70,000 species of pests attack crops, with about 10,000 species being insect pests worldwide. This entry focuses on the economic consequences for agriculture, but it is important to establish that there are serious consequences for humans and other animals in the food chain from insects—topics that cannot be explored in detail in this entry.

SCOPE OF PROBLEM

There are ~2500 insect and mite species that attack U.S. crops. About 1500 of these species are native insect/mite species that moved from feeding on native vegetation to feeding on our introduced crops. Thus, an estimated 60% of the U.S. insect pests are native species and ~40% are introduced invader species.[1,2] Both groups of pests contribute almost equally to the current annual 13% crop losses to insect pests, despite all the insecticides that are applied plus other types of controls practiced in the United States. Most or ~99% of U.S. crops are introduced species; therefore, many U.S. native insect species find these introduced crops an attractive food source.[3] For example, the Colorado potato beetle did not feed originally on potatoes but fed on a native weed species. Although the center of origin of the potato is the Andean regions of Bolivia and Peru, the plant was introduced into the United States from Ireland, where it has been cultivated since the mid-17th century.[3] Today, the Colorado potato beetle is the most serious pest of potatoes in the United States and elsewhere in the world where both the potato and potato beetle have been introduced.

An estimated 3 billion kg of pesticides are applied annually in the world in an attempt to control world pests.[4] Approximately 40% of the pesticides are insecticides, 40% herbicides, and 20% fungicides. About 1 billion kg of pesticides are applied within the United States; however, only 20% are insecticides.[4] Most or 68% of the pesticides applied in the United States are herbicides.

Despite the application of 3 billion kg of pesticides in the world, >40% of all crop production is lost to the pest complex, with crop losses estimated at 15% for insect pests, 13% weeds, and 12% plant pathogens.[4] In the United States, total crop losses are estimated at 37% despite the use of pesticides and all other controls, with crop losses estimated at 13% due to insect pests, 12% weeds, and 12% plant pathogens.

The 3 billion kg of pesticides applied annually in the world costs an estimated $32 billion per year or slightly >$10 per kg. With ~1.5 billion ha of cropland in the world, it follows that ~2 kg of pesticides is applied per ha, or $20 per ha is invested in pesticide control.

INSECT PESTS

Note that despite a tenfold increase in total quantity (weight) of insecticide used in the United States since about 1945 when synthetic insecticides were first used, crop losses to insect pests have nearly doubled from ~7% in 1945 to 13% today.[1] Actually, the situation is a great deal more serious because during the past 40 years, the toxicity of pesticides has increased 100–200 fold. For example, in 1945 many of the insecticides were applied at ~1 to 2 kg per ha; however, today many insecticides are applied at dosages of 10–20 grams per ha.[1]

The increase in crop losses to insect pests from 7% in 1945 to 13% today, despite a 100–200-fold increase in the toxicity of insecticides applied per ha, is due to changes in agricultural technologies over the past 50 years.[2] These

Encyclopedia of Natural Resources DOI: 10.1081/E-ENRL-120049215

changes in agricultural practices include the following: 1) the planting of some crop varieties that are more susceptible to insect pests than those planted previously; 2) the destruction of natural enemies of some pests by insecticides (e.g., destruction of cotton bollworm and budworm natural enemies), thereby creating the need for additional insecticide applications; 3) insecticide resistance developing in insect pests, thus requiring additional applications of more toxic insecticides; 4) reduction in crop rotations, which caused further increases in insect pest populations (e.g., corn rootworm complex); 5) lowering of the Food and Drug Administration (FDA) tolerances for insects and insect parts in foods, and enforcement of more stringent "cosmetic standards" for fruits and vegetables by processors and retailers; 6) increased use of aircraft application of insecticides for insect control, with significantly less insecticide reaching the target crop; 7) reduced field sanitation and more crop residues for the harboring of insect pests; 8) reduced tillage that leaves more insect-infested crop residues on the surface of the land; 9) culturing some crops in climatic regions where insect pests are more severe; and 10) the application of some herbicides that improve the nutritional makeup of the crop for insect invasions.[2]

The 13% of potential U.S. crop production represents a crop value lost to insect pests of ~$35 billion per year. In addition, nearly $2 billion in insecticides and miticides are applied each year for control. Ignoring other control costs, combined crop losses and pesticides thus total ~$37 billion per year.[3]

Total crop losses to insects and mites worldwide are estimated to be ~$400 billion per year. Given that >4.5 billion of the 7 billion people on earth are malnourished, this loss of food to insect and mite pests each year, despite the use of 3 billion kg of pesticides, is an enormous loss to society.

INSECTICIDE AND MITICIDE PEST CONTROLS

In general, insecticide and miticide control of insects and mites return ~$4 for every $1 invested in chemical control.[1] This is an excellent return, but not as high as some of the nonchemical controls. For example, biological pest control has reported earnings of $100–$800 per $1 invested in pest control. It must be recognized that the development of biological controls, although highly desirable, are not easy to develop and implement.

Both in crops and nature, host plant resistance and natural enemies (parasites and predators) play an important role in pest control. In nature, seldom do insect pests and plant pathogens remove >10% of the resources from the host plant. Host plant resistance, consisting of toxic chemicals, hairiness, hardness, and combinations of these, prevents insects from feeding intensely on host plants.[5] Some of the chemicals involved in plants resisting insect attack include cyanide, alkaloids, tannins, and others. Predators and parasites that attack insects play an equally important role in controlling insect attackers on plants in nature and in agroecosystems.

INSECT TRANSMISSION OF PLANT PATHOGENS

Insects with sucking mouthparts, such as aphids and plant bugs, play a major role in the transmission of plant pathogens from plant to plant. It is estimated that ~25% of the plant pathogens are transmitted by insects. The most common pathogens transmitted are viruses. These pathogens include lettuce yellows and pea mosaic virus. Fungal pathogens are also transmitted by insects. For example, Dutch elm disease is transmitted by two bark beetle species that live under the bark of elm trees. In infected trees, the bark beetles become covered with fungal spores. When the beetles disperse and feed on uninfected elm trees, they leave behind fungal spores that in turn infect the healthy elm trees.

ENVIRONMENTAL AND PUBLIC HEALTH IMPACTS OF INSECTS

Some insect species have become environmental and public health pests. For example, the imported red fire ant kills poultry chicks, lizards, snakes, and ground-nesting birds. Investigations suggest that the fire ant has caused a 34% decline in swallow nesting success as well as a decline in the northern bobwhite quail populations in the United States. The ant has been reported to kill infirm people and people who are highly sensitive to ant sting. The estimated damages to wildlife, livestock, and public health in the United States is >$1 billion per year, with these losses occurring primarily in southern United States.[3]

In another example, the Formosan termite that was introduced into the United States has been reported to cause >$1 billion per year in property damage, repairs, and controls. As it spreads further in the nation, the damages will increase.[3]

CONCLUSION

Insect and mite pests in the United States and throughout the world are causing significant crop, public health, and environmental damages. Just for crop losses in the United States, it is estimated that insect and mite species are causing $37 billion per year, if control costs are included. Worldwide crop losses to insects and mites are estimated to be $400 billion per year. The public health and environmental damages in the United States and throughout the world are estimated to be valued at several hundred billion dollars per year.

REFERENCES

1. Pimentel, D. *Handbook on Pest Management in Agriculture*; Three Volumes; CRC Press: Boca Raton, FL, 1991; 784 pp.
2. Pimentel, D.; Lehman, H. *The Pesticide Question: Environment, Economics and Ethics*; Chapman and Hall: New York, 1993; 441 pp.
3. Pimentel, D.; Lach, L.; Zuniga, R.; Morrison, D. Environmental and economic costs of non-indigenous species in the United States. BioScience **2000**, *50*(1) 53–65.
4. Pimentel, D. *Techniques for Reducing Pesticides: Environmental and Economic Benefits*; John Wiley & Sons: Chichester, UK, 1997; 456 pp.
5. Pimentel, D. Herbivore population feeding pressure on plant host: Feedback evolution and host conservation. Oikos **1988**, *53*, 289–302.

Insects: Flower and Fruit Feeding

Antônio R. Panizzi
Embrapa Trigo, Passo Fundo, Brazil

Abstract
Hundreds of different species of insects feed on flowers and fruits. The impact of such insects can be detrimental, beneficial, or mutualistic. Mutualistic interactions between flowers and honey bees, for example, facilitate pollination in flowers and provide pollen and nectar for developing bees in hives.

INTRODUCTION

Flowers are frequently visited by insects searching for food, such as pollen and nectar.[1] The impact of such insects can be detrimental, beneficial, or mutualistic. Some insects will feed on flowers and damage them. Fruits are damaged by a great variety of insects, which look for seeds that are very rich in nitrogen.[2] Others will feed on the fruit itself.[3,4]

Every year, millions of dollars are spent to control the insect pests of flowers, and, in particular, those feeding on fruits and seeds. However, insects visiting flowers are also beneficial, by dispersing seeds and pollinating several major crops and vegetables.[5] Mutualistic interactions occur between flowers and honey bees, *Apis mellifera* L. Whereas bees facilitate pollination, flowers provide pollen and nectar for developing bees in hives. This mutualistic interaction led to domestication of the honey bee.

DETRIMENTAL INSECTS

Hundreds of different species of insects feed on flowers and fruits. These include larvae and adults of beetles (Coleoptera), butterflies and moths (Lepidoptera), flies (Diptera), sawflies, wasps, ants, and bees (Hymenoptera), and the "true bugs" (Heteroptera) (Fig. 1). In addition, scale insects (Homoptera) feed on the surface of fruits, and the so-called thrips (Thysanoptera) damage flowers.[3,4,6,9]

Mouthparts

Insects feeding on flowers, fruits, and seeds have different types of mouthparts that allow them to obtain different types of food. The most common one is the chewing mouthparts, when insects use their mandibles to cut the plant tissue to penetrate the hard surface of fruits and seeds. Chewing mouthparts also allow some beetles to feed on flowers, by cutting the petals.

A second type is the sucking mouthparts where mouthparts are modified into a long sharp tube with two internal openings, one used to pump in saliva into the plant tissue, and the other is used to suck up the fluids. Sucking insects will feed in the following ways: stylet-sheath feeding, lacerate-and-flush feeding, macerate-and-flush feeding, and osmotic pump feeding.[7] In the stylet-sheath feeding the bugs insert their stylets in the tissue, mostly in the phloem, destroying few cells, and a stylet sheath is produced, which remains in plant tissues. In the lacerate-and-flush feeding, the bugs move their stylets vigorously back and forth, and several cells are lacerated. In the macerate-and-flush feeding type the cells are macerated by the action of salivary enzymes. In the last two cases, the cell contents are injected with saliva, resulting in several cells damaged. Finally, the osmotic pump feeding occurs through the secretion of salivary enzymes injected in the plant tissue to increase osmotic concentration of intercellular fluids, which are then sucked, leaving empty cells around the stylets.

A third type of mouthparts is the rasping-sucking. In this case the insects scratch the plant tissue and suck up the emerging plant fluid. This type of mouthpart is less common and is typical of the small insects called "thrips."

Parts Damaged

Flowers, fruits, and seeds are damaged in many different ways. For instance, flowers may be damaged by chewing insects that either destroy the petals or may feed on the pollen. Many insects feed on the nectar, but in this case they do not damage the flowers.

Fruits are damaged in many ways, such as on the surface, by scale insects (Homoptera) or chewing insects, or to the inner parts, in particular by boring pests.[3,4]

Seeds on the plant and in storage are a preferred feeding site, because of their great nutritional value.[2] Seed damage is variable, depending on the type of feeding. For instance, sucking insects feed on the cotyledons, and if the feeding

Encyclopedia of Natural Resources DOI: 10.1081/E-ENRL-120010485

Fig. 1 The southern green stink bug, *Nezara viridula* (L.) (Heteroptera). It feeds on seeds and on fruits of a wide variety of crops, vegetables, and fruit trees worldwide.

punctures reach the radicule or the hypocotyl it can prevent the seed germination.[2,10] Chewing insects will consume the seed/fruit tissues either by feeding on the outer surface or by boring into the seeds and fruits.

Types of Damage

There are several types of damage caused by insects feeding on flowers, fruits, and seeds. Larvae of chewing bruchids bore into the seeds and the adults emerge through round holes cut in the seed. The "sap beetles" cause damage by feeding and are also known to transmit fruit-degrading microorganisms, such as brown rot *Monolinia fruticola*, in stored fruits.[3,4]

Sucking insects feed by inserting their stylets into the food source to suck up the nutrients. By doing this, they cause injury to plant tissues, resulting in plant wilt and, in many cases, abortion of fruits and seeds. The damage to plant tissues, including seeds and fruits, results from the frequency of stylet penetration and feeding duration, associated with salivary secretions that can be toxic and cause tissue necrosis.[2,6,7,10]

During the feeding process, sucking insects also may transmit plant pathogens, which increase their damage potential. For instance the "cotton stainers" transmit a series of fungi and bacteria, the most important one called the internal boll disease is caused by *Nematospora gossypii*. The feces of adults are yellowish and stain the cotton lint, which is another type of damage they cause. The direct damage varies with the age of the boll, with young bolls having the seeds destroyed and drying up as a result of the feeding activity.[7]

BENEFICIAL INSECTS

Beneficial insects involved with flowers and fruits include mostly species associated with pollination (see "Mutualistic Associations" section) and production of honey. These insects in general belong to the Order Hymenoptera.

This order includes all different species of bees, such as the honey bee (*A. mellifera* L.), the stingless bees (*Melipona* and *Trigona*), the bumble bees (*Bombus*), the carpenter bees (*Xylocopa*), the mining and cuckoo bees (Anthophoridae), the leaf cutting bees (Megachilidae), and others, which are frequent visitors of flowers.[1,5] Among the hymenopterans there are the so-called pollen wasps (masarine wasps) which are the only wasps that provision their nest cells with pollen and nectar, as bees do.[11]

Honey is perhaps the greatest benefit produced by insects as they visit the reproductive structure of plants. Honey is used as food by bees, and it is explored by man as one of the most popular and beneficial food.

Another benefit that insects associated with flowers and fruits may do to plants is by disperse their seeds. Small seeds might get attached to the insects' body and be taken elsewhere, helping the dispersion of the plant species.

MUTUALISTIC ASSOCIATIONS

Insects may have a mutualistic relationship with plants, where both insects and plants have advantages of the association.[1,5] Plants are the source of food for insects, and in return they pollinate the flowers. Pollination is the transfer of the male parts (anthers) of a flower to the female part (stigma) of the same or different flowers. More seeds develop when large numbers of pollen grains are transferred.

Insects, in general, play an important role in pollination (entomophily), as well as wind (anemophily). Beyond the hymenopterans (bees, wasps and ants), dipterans (flies), lepidopterans (butterflies and moths), coleopterans (beetles), thysanopterans (thrips), and other minor orders are also engaged in pollination.[5] Other pollination agents include birds and bats.

About 130 agricultural plants are pollinated by bees in the United States. The annual value of honey bee pollination to U.S. agriculture has been estimated at over $9 billion. The honey bee pollination in Canada is estimated at Can$443 million, and over 47,000 colony rentals are taken every year. In the United Kingdom at least 39 crops grown for fruit and seed are insect pollinated, by honey bees and bumble bees. The European Union has an estimated annual value of 5 billion euros, with 4.3 billion attributable to honey bees.[5]

As flowers are visited by insects and get pollinated, flowers provide proteinaceous pollen and sugary nectar, which are used by insects as food source to larvae and adults.

CONCLUSION

Insects feeding on flowers/fruits/seeds can cause severe damage to many plants of economic importance worldwide.

Despite the many efforts to control these insects, they still remain a challenge to economic entomologists. Because of their feeding habits, i.e., feeding on the reproductive structures of plants, their economic importance is much greater than that of insect pests that feed on other plant parts, such as leaves, from whose damage some plants can compensate for at least partially. Moreover, for some pests such as the heteropterans that suck plant juices, the damage is difficult to detect early on compared to damage by leaf chewing insects. Therefore, sometimes, damage by sucking insects is noticed only at harvest, when control measures are too late to avoid economic losses.

In conclusion, while insects feeding on reproductive structures of plants cause enormous damage, they can also be of enormous benefit to humankind. Honey production is a source of a healthy and nutritional food for humans. Pollination caused by insects feeding on flowers is responsible for the development of fruit/seeds of hundreds of plant species, several of them of highly economic importance. Moreover, without pollination, many of the plant species we know today would have disappeared a long time ago.

REFERENCES

1. Barth, F.G. *Insect and Flowers. The Biology of a Partnership*; Princeton University Press: Princeton, NJ, USA, 1996.
2. Slansky, F., Jr.; Panizzi, A.R. Nutritional Ecology of Seedsucking Insects. In *Nutritional Ecology of Insects, Mites, Spiders and Related Invertebrates*; Slansky, F., Jr, Rodriguez, J.G., Eds.; Wiley: New York, 1987; 283–320.
3. Borror, D.J.; Triplehorn, C.A.; Johnson, N.F. *An Introduction to the Study of Insects*, 6th Ed.; Saunders College Publishing: Philadelphia, USA, 1989.
4. Hill, D.S. *The Economic Importance of Insects*; Chapman & Hall: London, UK, 1997.
5. Delaplane, K.S.; Mayer, D.F. *Crop Pollination by Bees*; CABI Publishing: New York, USA, 2000.
6. McPherson, J.E.; McPherson, R.M. *Stink Bugs of Economic Importance in America North of Mexico*; CRC Press: Boca Raton, Florida, 2000.
7. *Heteroptera of Economic Importance*; Schaefer, C.W., Panizzi, A.R., Eds.; CRC Press: Boca Raton, Florida, USA, 2000.
8. Schuh, R.T.; Slater, J.A. *True Bugs of the World (Hemiptera: Heteroptera). Classification and Natural History*; Cornell University Press: Ithaca, New York, USA, 1995.
9. White, I.M.; Elson-Harris, M.M. *Fruit Flies of Economic Significance: Their Identification and Bionomics*; CAB International: Oxon, UK, 1992.
10. Panizzi, A.R. Wild hosts of pentatomids: Ecological significance and role in their pest status on crops. Annu. Rev. Entomol. **1997**, *42*, 99–122.
11. Gess, S.K. *The Pollen Wasps. Ecology and Natural History of the Masarinae*; Harvard University Press: Cambridge, Massachusetts, USA, 1996.

Integrated Pest Management

Marcos Kogan
Integrated Plant Protection Center, Oregon State University, Corvallis, Oregon, U.S.A.

Abstract

The sustainable production of food and fiber crops requires knowledge about the economic impact of pests and the most efficient and ecologically sound means for the control of those pests. Integrated Pest Management (IPM) emerged in late 1960s as a system to support decisions in the selection and use of pest control tactics singly or in various combinations to optimize economic benefits to growers while protecting the integrity of the environment. Development of IPM systems depends on the best available ecological information on the crop production system and the complex of pests that impacts the crop. Favorable cost/benefit analyses are essential to lead producers to adopt IPM.

INTRODUCTION

In the context of integrated pest management (IPM), a pest is any living organism whose life system conflicts with human interests, economy, health, or comfort. Pests belong to three main groups: 1) invertebrate and vertebrate animals; 2) disease-causing microbial pathogens; and 3) weeds or undesirable plants growing in desirable sites.

The concept of pest is entirely anthropocentric. There are no pests in nature, in the absence of humans. An organism becomes a pest only if it causes injury to crop plants in the field or to their products in storage or, outside an agricultural setting, if it affects structures built to serve human needs. This entry focuses on the impact of pests on plants.

BACKGROUND

Since the beginnings of agriculture some 10,000 years ago, humans competed with other organisms for the same food resources. Primitive tools to fight pests were probably limited to handpicking and methods such as flooding or fire, although at times of despair, agriculturists resorted to magic incantations or religious proclamations.[1] But as early as 2500 BC, Sumerians used sulfur to control insects and by 1000 BC, Egyptians and Chinese were using plant-derived insecticides to protect stored grain.[2] Real progress in the fight against agricultural pests occurred with advances in biological sciences, resulting in a better understanding of arthropod pests and the relationships between microbial pathogens and plant diseases. Until the middle of the 20th century, insect pest control ingeniously combined knowledge of pest biology and cultural and biological control methods. Trichlorodiphenyltrichloroethane (DDT) in the early 1940s ushered in an era of organosynthetic insecticides. Initial results were spectacular, and many entomologists foresaw the day when all insect pests would be eradicated. This optimistic view was premature, as insects quickly evolved mechanisms to resist the new chemicals, requiring more frequent applications, higher dosages, or the replacement of old insecticides with more powerful ones. This cycle has become known as the insecticide treadmill.[1] At the end of the 1950s, scientists around the world became aware of the risks associated with the application of powerful insecticides as resistance rendered one chemical after the other useless and as beneficial organisms were indiscriminately killed along with target pest species. Such killing of insect parasitoids and predators caused the resurgence of pests at even higher levels. In addition, populations of other plant-eating species, usually balanced by the pests' natural enemies, exploded in outbreak proportions, causing upsurges of secondary pests and pest replacements. As a result of mounting concern about resistance, resurgence, and upsurges of secondary pests, the concept of integration emerged—an attempt to reconcile pesticides with the preservation of natural enemies.[3]

INTEGRATION

The next phase in the evolution of IPM was the expansion of the term "integrated" to include all available methods (biological, cultural, mechanical, physical, legislative, as well as chemical) and to span all pest categories of importance in the cropping system (invertebrates, vertebrates, microbial pathogens, and weeds). Because of the multiple meanings, integration in IPM was conceived at three levels of increasing complexity. Level I integration refers to multiple control tactics against single pests or pest complexes; level II is integration of all pest categories and the methods for their control; level III integration applies the steps in level II to the entire cropping system. At level III, the principles of IPM and sustainable agriculture converge.[4]

Encyclopedia of Natural Resources DOI: 10.1081/E-ENRL-120049210

THE ECOLOGICAL BASES OF IPM

Ecology offered the tools to describe how populations expand and contract under natural conditions; how biotic communities are organized and how their component organisms interact; how abiotic (nonliving) factors (climate, soil, topography, hydrology) shape biotic communities; how biodiversity stabilizes ecological systems; and, perhaps most significantly, how to understand the effects of disturbances in the dynamics of biotic communities.[5,6] Thus, IPM relied upon a solid ecological foundation.[4,5,7]

THE TOOLS OF PEST MANAGEMENT

Management of pests in an IPM system is accomplished through selection of control tactics singly or carefully integrated into a management system. These tactics usually fall into the following main classes: 1) chemical; 2) biological; 3) cultural/mechanical; 4) physical control methods (these four methods apply to all pests); 5) host plant resistance (applicable to invertebrate and microbial pests); 6) behavioral; and 7) sterile insect technique, a genetic control method (relevant only to insects).

Chemical Control

Chemical pesticides are used to kill or interfere with the development of pest organisms. Pesticides are called insecticides, acaricides, nematicides, fungicides, or herbicides if the target pests are insects, mites, nematodes, fungi, or weeds, respectively. Pesticides are useful tools in IPM if used selectively and if the pest population threatens to reach the economic injury level (described later). Much care is necessary to handle pesticides and avoid their potentially negative effect on the environment.[8]

Biological Control

Biological control refers to the action of parasitoids, predators, and pathogens in maintaining another organism's population density at a lower average than would occur in their absence.[9] In IPM practice, biological control includes intentional introduction of natural enemies to reduce pest population levels (classical biological control), the mass release of biocontrol organisms (inundative biocontrol), or the timely inoculation of natural enemies (augmentative biocontrol).[10]

Cultural/Mechanical Control

Cultural practices become cultural controls when adopted for intentional effect on pests. Thus, tillage and other operations for soil preparation have a direct impact on weeds and on soilborne insect pests. The manipulation of the crop environment to favor natural enemies is called habitat management and is of growing interest as a form of conservation biological control.[10,11]

Physical Control

Fire, water, electricity, and radiation are some of the main physical forces used in IPM. Flooding in paddy rice cultivation is a major adjuvant method in many parts of the world.

Host Plant Resistance

This means of control is based on breeding crop plants to incorporate genes whose products cause plants to be unsuitable for insects feeding on them (antibiosis) or impair insects' feeding behavior (antixenosis).[12] Resistance is the only effective method for control of plant viral diseases. Techniques of genetic engineering have opened new opportunities to expand plant resistance in IPM, but the approach is not without risks and should be adopted cautiously, within a strict IPM framework.[13]

Behavioral Control

Behavioral control uses chemicals (allelochemicals) to interfere with normal patterns of mainly sexual (mating) and feeding behaviors of arthropods. Sex pheromones are used for mating disruption.[14,15] Feeding excitants or deterrents are used to disrupt normal feeding behavior or to attract and kill insects.[16]

Sterile Insect Technique

This genetic control method disrupts normal progeny production of the target species by the mass release of sterile insects.[17]

THE SCALE OF IPM SYSTEMS

Most IPM programs focus on single fields because of local variability in physical, crop, and pest conditions. Some pests, however, are highly mobile and require a regional approach for their control. Furthermore, certain control tactics are effective only if deployed over large areas. Mating disruption for control of the codling moth in apple and pear orchards in the western United States used an area-wide approach that required a minimum operational unit of about 160 hectares. Advancement of IPM to higher levels of integration will require planning and implementation at the landscape or even ecoregional levels. Advanced technologies of geographic information systems (GISs) and remote sensing are essential for development of such programs. Such planning is still at its infancy in IPM.

DECISION SUPPORT SYSTEMS FOR IPM IMPLEMENTATION

IPM uses objective criteria for making decisions about the need for a control action and selection of the most appropriate tactics. In arthropod pest control, the criteria are based on assessments of the extant pest population in the field, its damage potential (if not controlled), and the relative costs of treatment and crop value. The main parameters are the economic injury level (EIL) and economic threshold (ET). ET is the population level that will trigger a control action. Figure 1 describes these parameters. To use ETs, it is necessary to monitor pest populations in the field using well-defined sampling methods, which vary with the pest species. Although useful, the concept of EIL is less effective as a decision-making support tool for weed management and is partially applicable for microbial disease management.[18]

IPM EDUCATION AND EXTENSION

Since the establishment of the Land Grant Universities (1890), teaching of entomology and plant pathology was gradually introduced in the curricula. Courses on insect science that included the control of agricultural pests were variously designated applied-, economic- or agricultural entomology. As faculties expanded in plant protection departments, specialized courses were introduced in chemical pesticides, biological control, host plant resistance, and others. With the introduction of IPM in the late 1960s, most entomology departments started offering basic IPM courses at the undergraduate level and advanced IPM for graduate students. One of the first entomology textbooks aimed at teaching IPM was by Metcalf and Luckmann.[19] Other texts followed. Pedigo's[20] *Entomology and Pest Management*, first published in 1989, became the standard for introductory courses in insect IPM in the United States, and Dent's[21] *Integrated Pest Management* probably had the same role in the United Kingdom. Although IPM continues to be taught mostly along disciplinary boundaries, a few universities offer multidisciplinary courses combining entomology, plant pathology, and weed science within integrated curricula. The book by Norris, Caswell-Chan, and Kogan[2] was the first to offer an interdisciplinary text for IPM teaching.

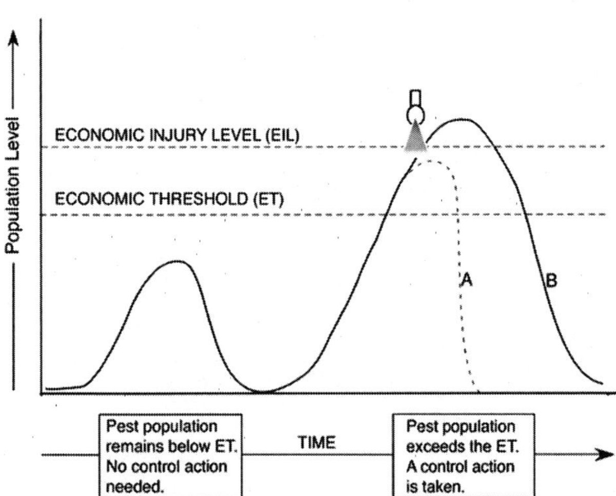

Fig. 1 Fundamental decision-making concepts for insect pest management: economic threshold (ET) and economic injury level (EIL). Illustrated is an insect pest population with two peaks in one year. The first peak remains below ET and no treatment is needed; the second peak exceeds ET. If a treatment is applied, the population crashes (curve A); if a treatment is not applied, the population exceeds EIL and an economic crop loss occurs (curve B).

It is the role of agricultural extension specialists to bring to growers the most up-to-date information on IPM technological advances and promote adoption of effective IPM systems. Extension specialists use demonstration plots, field days, intensive short courses, special publications, and increasingly, the power of the Internet, to expose growers to the latest in IPM. Formerly printed IPM information is now available online, which allows for more timely updates and ease of access to the information. An example is the Pacific Northwest IPM Handbooks that had been traditionally issued once a year in three volumes for insect pests, plant diseases, and weeds. These are now entirely published online.[22] One critical area for IPM implementation involves decision support systems. Online degree day-driven pest development models are available to help growers in the optimization of sampling and timing of control actions.[23] In developing regions of the world where computers are still rare in rural areas, extension has been highly effective with a model developed by specialists of the Food and Agriculture Organization (FAO) in collaboration with international research centers. The model known as farmer field schools is a participatory approach in which extension specialists and farmers meet in the field to identify pest problems and discuss possible solutions. First tested in Southeast Asia,[24] the approach has been successful in Africa also.[25] Maintaining the flow of information from research laboratories and experimental fields into the classroom and the farm is essential to expand IPM adoption.

CONCLUSION

It is difficult to assess the global impact of IPM. It is safe, however, to state that the concept drastically changed the approach to pest control among progressive growers, the chemical industry, research establishments, and policymakers. One of the early economic analyses of the impact of IPM was done in Brazil on soybean production costs. The study suggested that technologies available in 1980 (date of the study), if adopted throughout the entire production area (8.5 million ha), would have resulted in a saving of U.S. $216 million annually mainly in the cost of insecticides and application costs. A study on the environmental benefits of changes in pesticide use conducted in Ontario, Canada, estimated benefits to amount to $711 million.[26] Farmers in most agricultural production regions of the world have benefited from the adoption of IPM. Although IPM is still in its infancy, major advances are to be expected as modern technologies are incorporated into the management of all pests.[27]

REFERENCES

1. Van den Bosch, R. *The Pesticide Conspiracy*; Doubleday: Garden City, NY, 1978; Vol. 8, 226 pp.
2. Norris, R.F.; Caswell-Chen, E.P.; Kogan, M. *Concepts in Integrated Pest Management*; Prentice Hall: Upper Saddle River, NJ, 2003; 586.
3. Kogan, M. Integrated pest management: Historical perspectives and contemporary developments. Annu. Rev. Entomol. **1998**, *43*, 243–277.
4. Kogan, M. Integrated pest management theory and practice. Entomol. Exp. Appl. **1988**, *49*, 59–70.
5. Levins, R.L.; Wilson, M. Ecological theory and pest management. Annu. Rev. Entomol. **1980**, *25*, 287–308.
6. Kogan, M. Ecological theory and integrated pest management practice. In *Environmental Science and Technology*; Wiley: New York, 1986; Vol. 84, 362 pp.
7. Huffaker, C.B.; Gutierrez, A.P. *Ecological Entomology*; Wiley: New York, 1999; Vol. 19, 756 pp.
8. Wheeler, W.B. *Pesticides in Agriculture and the Environment*; Marcel Dekker: New York, 2002; Vol. 10, 330 pp.
9. Bellows, T.S.; Fisher, T.W. *Handbook of Biological Control: Principles and Aplications of Biological Control*; Academic Press: San Diego, 1999; Vol. 23, 1046 pp.
10. Gurr, G.M.; Wratten, S.D. Integrated biological control: A proposal for enhancing success in biological control. Int. J. Pest Manag. **1999**, *45*, 81–84.
11. Landis, D.A.; Wratten, S.D.; Gurr, G.M. Habitat management to conserve natural enemies of arthropod pests in agriculture. Annu. Rev. Entomol. **2000**, *45*, 175–201.
12. Smith, C.M. *Plant Resistance to Insects: A Fundamental Approach*; Wiley: New York, 1989; Vol. 8, 286 pp.
13. Atherton, K.T. *Genetically Modified Crops: Assessing Safety*; Taylor & Francis: London, UK, 2002; 256.
14. Carde, R.T.; Bell, W.J. *Chemical Ecology of Insects 2*; Chapman & Hall: New York, 1995; Vol. 8, 433 pp.
15. Carde, R.T.; Minks, A.K. *Insect Pheromone Research: New Directions*; Chapman & Hall: New York, 1997; Vol. 22, 684 pp.
16. Metcalf, R.L.; Deem-Dickson, L.; Lampman, R.L. Attracticides for the Control of Diabroticite Rootworms. In *Pest Management: Biologically Based Technology*; Vaughn, J.L., Ed.; American Chemical Society: Beltsville, MD, 1993; 258–264.
17. Calkins, C.O.; Klassen, W.; Liedo, P. *Fruit Flies and the Sterile Insect Technique*; CRC Press: Boca Raton, FL, 1994; 258.
18. Higley, L.G.; Pedigo, L.P.; Eds. *Economic Thresholds for Integrated Pest Management*; The University of Nebraska Press: Lincoln, NE, 1996; 327.
19. Metcalf, R.L.; Luckmann, W.H. *Introduction to Insect Pest Management*, 1st Ed.; Wiley: New York, 1975; 587 p.
20. Pedigo, L.P. *Entomology and Pest Management*, 1st Ed.; Macmillan: New York, 1989.
21. Dent, D.; Ed. *Integrated Pest Management*. Springer: London, 1995; 372 p.
22. Integrated Plant Protection Center. *On-line IPM Handbooks*; Corvallis, Oregon, 2012. http://ipmnet.org/IPM_Handbooks.htm (Accessed December 2013).
23. Coop, L. IPM pest and plant disease models and forecasting for agricultural pest management, and plant biosecurity decision support in the U.S. In *Pest and Crop Models*. pnwpest.org/wea (Accessed December 2013).
24. Pontius, J.C.; Dilts, R.; Bartlett, A. *From Farmer Field School to Community IPM: Ten Years of IPM Training in Asia*. RAP/2002/15; FAO Regional Office for Asia and the Pacific: Bangkok, 2002; 106 p.
25. SUSTAINET EA. Technical Manual for farmers and Field Extension Service Providers: Farmer Field School Approach. Sustainable Agriculture Information Initiative, Nairobi. 2010. http://www.fao.org/ag/ca/CA-Publications/Farmer_Field_School_Approach.pdf (accessed December 2013).
26. Brethour, C.; Weersink, A. An economic evaluation of the environmental benefits from pesticide reduction. Ag. Econ. **2001**, *25*, 219–226.
27. Kogan, M.; Jepson, P. *Perspectives in Ecological Theory and Integrated Pest Management*; Cambridge Univ. Press: Cambridge, UK, 2007; 570 pp.

Keystone and Indicator Species

Barry R. Noon
Department of Fish, Wildlife, and Conservation Biology, Colorado State University, Fort Collins, Colorado, U.S.A.

Abstract

Evaluations of the responses of ecological systems to management actions and environmental disturbance are generally conducted at broad spatial scales by assessing changes in the distribution and abundance of dominant land-cover types. However, these measures often fail to provide insights into how individual populations of plant and animals species respond to environmental change. In addition to broad-scale assessments, direct measures of species' responses to environmental stressors and landscape change are essential to assess the state of ecological systems. Unfortunately, assessing all plant and animal species of interest within an ecosystem is logistically impossible. As a result, a surrogate-based approach where only a small number of species are directly measured becomes a pragmatic necessity. This approach requires a logical, step-down process to select a small number of species whose measurement will provide insights to the majority of the unmeasured species and allow inference to changes in ecological processes. No unambiguous body of ecological theory or empiricism exists to precisely guide the selection of which species to measure. But, assessments of some types of species, such as keystone and indicator species, have a history of providing information beyond the individual species' measurements and allowing inference to the status and trend of the ecological systems to which they belong. For a comprehensive ecological assessment, the key property of the surrogate species set is that their information content, to the extent possible, be complementary and comprehensive.

INTRODUCTION

A surrogate-based approach to assessing the effects of land use and land change on biological diversity is a pragmatic necessity. It is impossible to manage or monitor changes directly in the state of populations of all the species that we are legally obligated to conserve or that are important to us for economic, aesthetic, or ethical reasons. Current approaches to sustain plant and animal diversity at broad spatial scales combine coarse-filter and fine-filter strategies.[1] The coarse-filter approach, operative at a landscape scale, proposes to sustain biological diversity indirectly by maintaining the diversity of habitats (i.e., vegetation communities and their successional stages) expected under the historic range of variability for that landscape. In contrast, the fine-filter approach is focused at the species level and is intended to address the needs of species not adequately protected by the coarse-filter strategy.[2] Indicator and keystone species concepts are examples of fine-filter strategies.

The great diversity of plant and animal species defies any simple prioritization. For managers and conservation biologists charged with sustaining biological diversity in human-dominated landscapes, some simplifying characterizations are needed. One uncomplicated classification is to characterize species by 1) their utilitarian value, 2) their information content—what do changes in the status and trends of their population tell us about the state of the ecosystems to which they belong, or 3) their functional role in ecological systems—how, and to what extent, does a given species contribute to key processes. Those with obvious utilitarian value would include species with direct economic benefits such as many commercial marine fish species and species hunted for subsistence consumption or recreation. Species that provide information beyond their own measurement may act as sentinels of ecosystem change and thus provide early warning of impending adverse system effects prior to unacceptable or irreversible state changes. Species with known and important functional roles would include, for example, dominant plant species with high levels of primary productivity, consumer species which provide top-down regulation of food webs, and species that alter the structure and composition of ecosystems and thereby promote high levels of species diversity.

The goal of this entry is to discuss the characteristics of two categories of species mentioned in the preceding text. First, those species that provide significant insights into the state of the ecological systems in which they reside, often indicating impending state changes and generically referred to as indicator species. Second, those species with highly significant functional roles in the structuring and stability of ecological systems, referred to as keystone species.[3]

INDICATOR SPECIES

I view the indicator species concept broadly—that is, they are species whose status and trends provide insights into environmental conditions that extend beyond their own

Encyclopedia of Natural Resources DOI: 10.1081/E-ENRL-120047434

measurement.[3,4] Indicator species can be used as surrogates for ecosystem processes or conditions that are too difficult or costly to measure directly. The primary role of ecological indicators, in general, and indicator species in particular, is to measure the response of the ecosystem to anthropogenic disturbance.[5] Selection of indicator species is a critical aspect of most environmental monitoring programs. For example, indicator species with known stress–response relationships are often chosen as early warning sentinels of impending stressor effects so as to avoid undesirable or irreversible environmental outcomes.

There is no topology that clearly discriminates among the various types of indicator species identified by environmental scientists—for example, keystone, umbrella, focal, ecological engineers, and flagship species all have attributes that both overlap with one another[6] and can possibly qualify them as ecological indicators. Selection of indicator species is difficult and is often based more on their ease of measurement or historic precedent than their ecological importance or responsiveness to management.[7] Fortunately, there is now an extensive literature on the process of indicator species selection applicable to a wide range of resource management issues.[4,7–10]

The concept of indicator species was originally associated with the assumption that the status and trends of the indicator reflected the dynamics of species with similar ecological requirements.[11] This assumption has been strongly criticized,[6,11–14] and current understanding is that species surrogacy can only be assumed in limited cases where a set of species share a common set of environmental drivers.[14] Though surrogacy for unmeasured species within a study or management area is no longer a tenable assumption for indicator species, they remain useful to the extent that they reflect the biotic or abiotic state of the environment, reveal the impacts of environmental change, or provide indirect insights into the diversity of other species in the area.[5,15] For many natural resource agencies, the judicious selection of indicator species is an essential component of their environmental monitoring programs to assess the effects of management.[1]

Table 1 Desirable properties of indicator species

- Species dynamics parallel those of the ecosystem to which they belong
- Species show a short-term but persistent response to change in ecosystem state
- Species status and trends can be accurately and precisely estimated
- Significant changes in the state of the indicator species can readily be distinguished from change induced by background variation in ecosystem state
- Are measurable in a cost-effective manner
- Are relevant to important societal values
- Are understandable by multiple audiences (scientists, policy-makers, managers, and the public)

Not all candidate indicator species are equally informative—one of the key challenges to environmental assessment and monitoring is to select for measurement species that will provide insights into ecological integrity at the scale of the ecosystem. An ecosystem with integrity is defined as one that has the capacity to support and maintain an integrated, adaptive community of organisms having a full range of components (genes, species, communities) and processes (birth, survival, movement, biotic interactions, energy dynamics) expected from the natural habitat of the region.[16,17] Table 1 lists a set of indicator species properties that reflect changes in a system's ecological integrity and make them useful to policy makers.[3]

In diverse ecosystems, multiple indicator species will often be selected for measurement.[18] Collectively, the set of species selected for monitoring should be complementary and comprehensive.[4] Indicator species are complementary if their information content is largely independent of other species in the set. The set is comprehensive if they collectively have ecologies and life history attributes that span the biological diversity of the ecosystem and encompass the system's temporal and spatial domain. These properties are difficult to achieve in a minimal indicator set but are useful goals to strive for. Given the complexity and dimensionality of an ecosystem, a surrogate-based approach to ecosystem assessment based on a small number of indicator species is often a pragmatic requirement.[6,19]

Selection of the indicator species set is the first step. Use of an indicator species in a monitoring program, for example, requires that a specific variable reflecting the state of the indicator species be identified and measured. That is, will the state of the indicator species be assessed by its distribution, abundance, density, birth rate, or some other demographic attribute? Interpreting the state of an indicator species so as to make inferences about the state of the environment in which it resides also requires a statistically defensible survey design.[3,20–22]

KEYSTONE SPECIES

It is obvious that abundant species play major ecological roles in controlling the rates of ecosystem processes simply as a result of their numerical or material dominance. Keystone species differ from dominant species in that their effects are much greater than would be predicted from their abundance.[23] Power et al.[23] measure a species' impact by its community importance (CI)—that is, the observed change in a community or ecosystem trait (e.g., primary productivity, species richness) per unit change in the abundance (or biomass) of the putative keystone species. CI can be expressed mathematically as

$$CI = \left[\frac{d(trait)}{dp}\right]\left[\frac{1}{trait}\right]$$

where p is the proportional abundance (or biomass) of the keystone species experiencing a change.

A set of experiments conducted by Robert Paine,[24] who first proposed the keystone species concept, illustrates this mathematical model. Working in the intertidal zone of Washington state, he found a sea star to have a keystone effect on macroinvertebrate species diversity mediated by its predation on a dominant mussel consumer species. When he experimentally deleted the sea star (lowering p), mussel abundance greatly increased and macroinvertebrate diversity (the ecosystem trait) substantially decreased.

Operationally, it is difficult to estimate whether a species is a keystone. However, management "experiments" often inadvertently remove or greatly reduce the abundance of an assumed keystone species. A measure of a species' "keystoneness" can be computed as the change in the ecosystem trait before and after its elimination,[18]

$$CI = \left[\frac{(t_n - t_p)}{t_n}\right]\left[\frac{1}{p_i}\right]$$

where t_n and t_p are quantitative measures of an ecosystem trait (e.g., species diversity) before and after deletion of the keystone species, respectively, and p_i is the relative abundance of the keystone before deletion. Hurlbert,[25] also proposed a heuristic model for estimating the relative importance of species i (I_i) in a community of S species as,

$$I_i = \sum_{j=1}^{S} \left| A_{j,t=1} - A_{j,t=0} \right|$$

where, A_j is the abundance of species j (or some other trait) before ($t = 0$) and after ($t = 1$) removal of the ith species.

The keystone concept has been heavily criticized, initially by Hurlbert and more recently by Caro.[6] Hurlbert's criticism focuses on the fact that both the I and CI metrics are not measurable for even a single species, let alone an entire community of species. Thus, these metrics have theoretical but not practical value. Hurlbert concludes that the keystone species notion lacks a clear definition and has never been convincingly demonstrated. Caro's criticisms focuses on the context dependence of the keystone concept—under one set of conditions, a species may play a keystone role but under another, it has no substantial effect on any ecosystem trait.

Recently, the keystone concept has been recast as a search for strongly or weakly interacting species measured by the strength of their trophic links within food webs.[26–28] In deterministic analyses of food web dynamics, importance indices[25] indicate that either top predators or producers are of primary importance.[27] However, when the variance of food web responses to species deletions are evaluated in stochastic simulations, strongly interacting species can appear at multiple trophic levels including the middle of food webs.[28]

The loss of strongly interacting species from food webs has been used to explain the emergence of trophic cascades, broadly defined as the propagation of impacts by apex consumers on their prey down through food webs.[29,30] Even in the context of the practical imitations of the keystone concept, recent reports on the adverse effects of the loss of top-down consumers from freshwater, marine, and terrestrial ecosystems, and subsequent trophic cascades, supports the idea that species are often unequal in their functional importance to ecosystem processes and resilience.[29] Recent research on the unequal role of species in food web dynamics may have revitalized the keystone species concept.

EXPECTED VALUES AND BENCHMARKS FOR ECOLOGICAL INDICATORS

An essential component of an environmental monitoring program is the expected values (e.g., baselines) or temporal trends of the ecological indicators. Only by comparing observed with expected values or trends can a determination be made about the current state of the ecosystem or the effectiveness of management practices. The close approach to, or the passing of, an expected value is the threshold point that triggers a change in management practices. In the ecological literature, these expected values have been referred to as reference conditions—"the condition in the absence of human disturbance which is used to describe the standard, or benchmark, against which the current condition is compared."[30] It is often difficult, however, to standardize reference conditions across indicator species. Stoddard et al.[31] have categorized ecosystem reference conditions usefully along a continuum from minimally disturbed to historical condition to least disturbed to best attainable. In terms of indicator or keystone species, these concepts translate into expected demographic attributes including birth and survival rates, population sizes, and geographic distributions.

Estimating reference conditions for indicator species is difficult and imprecise for several reasons: (1) our limited understanding of the demographic patterns expected in pristine, undisturbed ecosystems to provide insights into benchmark conditions; (2) an incomplete understanding of the relationship between the state of an indicator species and the desired ecosystem state(s); (3) inadequate knowledge of the expected variability, over time and space, of the indicator species' populations; (4) the non-linear relationships between indicator values and ecosystem processes, including the existence of sharp threshold regions;[32] and (5) the fact that indicator benchmarks may be best represented by probability distributions rather than single target values.[3]

Expected demographic values and thresholds for indicator species implicitly assume that an ecosystem will evolve to (or was historically at) a dynamic steady state with regard to its species composition, disturbance regimes, and climatic processes. This assumption is often false. The dynamic nature of ecosystems argues for specifying a distribution of indicator values (e.g., population sizes, birth rates, survival rates, etc.) rather than an expected value at a single point in time or space. That is, the current state of an ecosystem is best described by a probability distribution rather than a specific target value. In practice, this makes the management decision process less certain, but the probability concept is a more realistic characterization of nature.

Determining threshold values and reference distributions requires the selection of both a spatial and temporal scale to assess the status and trends of the indicator species. If the spatial scale is a point in space, an indicator threshold may be specified as a single value (e.g., a population density >75 individuals/ha). In practice, it will be uncommon for the state of an indicator species, and thus the integrity of the ecosystem to which it belongs, to be characterized and interpreted in terms of a single target value. More commonly, the spatial scale of interest would be large landscapes, and indicator values would be estimated at multiple sample locations. Replicate estimates of population size, for example, would provide a distribution of indicator values. This distribution is critically dependent on the spatial scale and temporal window over which the distribution is estimated.

Once measured, the observed distribution of indicator values would be compared to the expected (benchmark) distribution to detect both the magnitude and pattern of deviation from desired conditions. The concept of a spatial distribution of indicator values as the appropriate evaluative statistic is critical to the monitoring and assessment of ecological systems.[3,31]

Despite the importance of establishing benchmark distributions, specifying reference conditions is subject to some degree of arbitrariness. For example, there is uncertainty about how benchmark conditions are to be estimated.[31] There is no clear guidance on how far back in time, or to what spatial extent, one should go to find an appropriate point of reference or what constitutes pristine conditions. There may be no current analogs for past conditions. In addition, it is clear that the concept of benchmarks for ecological indicators can be reconciled with the dynamic nature of ecosystems only in terms of frequency distributions, especially when viewed over long temporal or broad spatial scales.[3] Thus, all decisions about the proximity of the status or trend of an indicator or keystone species to its expected value (e.g., abundance, density) must be made in probabilistic terms.

CONCLUSIONS

Assessing the status and trends of natural resources requires monitoring the components of ecosystems that give rise to these resources. The key reason to assess ecological systems at the species level is that species are the fundamental agents of production and transfer of matter and energy in ecosystems. However, rarely can all the components of biological diversity be monitored—some prioritization of species must occur.[1,33] The challenge is to strategically select a small subset of species to monitor, whose change in state provides insights into a broader set of species or ecological processes.[1]

Because the dynamics of ecosystems are often driven by a small number of species that have uneven effects on ecosystem processes, a surrogate-based approach is justified.[26–29]

Knowledge of changes in the status and trends of biological diversity is essential to effective ecosystem management. Many empirical and theoretical studies have demonstrated that more diverse plant and animal communities support less variable ecosystem outputs and provide more ecosystem services.[34] Ecosystem resilience to contemporary environmental stressors and future climate change is strongly related to native species diversity.[34–36] Finally, species are useful early-warning indicators of impending environmental change, because they integrate the effects of multiple stressors across broad temporal and spatial scales. Multiple categories of species provide useful insights into ecosystem integrity and the ability of the ecosystem to sustainably provide natural resources that people depend on.[6] Indicator and keystone are two logical species categories whose change in state can be directly linked to management impacts, environmental stressors, and ecological consequences.

REFERENCES

1. Noon, B.R.; McKelvey, K.S.; Dickson, B.G. Multispecies conservation planning on U.S. federal lands. In *Models for Planning Wildlife Conservation in Large Landscapes*; Millspaugh, J.J., Thompson, F.R. III, Eds.; Academic Press: New York, 2009; 51–84.
2. Haufler, J.B.; Meh, C.A.; Roloff, G.J. Using a coarse-filter approach with species assessment for ecosystem management. Wildl. Soc. Bull. **1996**, *24*, 200–208.
3. Noon, B.R. Conceptual issues in the monitoring of ecological resources. In *Monitoring Ecosystems: Interdisciplinary Approaches for Evaluating Ecoregional Initiatives*; Busch, D.E., Trexler, J.C., Eds.; Island Press: Washington, DC, 2003; 27–71.
4. Noon, B.R.; McKelvey, K.S. *The Process of Indicator Selection. USDA Forest Service Proceedings RMRS-P-42CD*; U.S. Department of Agriculture Forest Service: Washington, DC, 2006; 944–951.
5. Niemi, G.J.; McDonald, M.E. Application of ecological indicators. Annu. Rev. Ecol. Syst. **2004**, *35*, 89–111.
6. Caro, T.M. *Conservation by Proxy: Indicator, Umbrella, Keystone, Flagship, and Other Surrogate Species*; Island Press: Washington, DC, 2010.
7. Tulloch, T.; Possingham, H.P.; Wilson. K. Wise selection of an indicator for monitoring the success of management actions. Biol. Conserv. **2011**, *144*, 141–154.
8. Carignan, V.; Villard, M.A. Selecting indicator species to monitor ecological integrity: A review. Environ. Manag. Assess. **2002**, *78*, 45–61.

9. Niemeijer, D.; DeGroot, R.S. A conceptual framework for selecting environmental indicator sets. Ecol. Indicators **2008**, *8*, 14–25.
10. Rochet, M-J.; Rice, J.C. Do explicit criteria help in selecting indicators for ecosystem-based fisheries management? ICES J. Mar. Sci. **2005**, *62*, 528–539.
11. Landres, P.B.; Verner, J.; Thomas, J.W. Ecological use of vertebrate indicators: A critique. Conserv. Biol. **1988**, *2*, 316–328.
12. Landres, P.B. Ecological indicators: Panacea or liability? In *Ecological Indicators*; Hyatt, D.E., McDonald, V.J., Eds.; Elsevier Applied Science: New York, 1992; Vol. 2, 1295–1318.
13. Rolstad, J.; Gjerde, I.; Gunderson, V.S.; Sætersdal, M. Use of indicator species to assess forest continuity: A critique. Conserv. Biol. **2002**, *18*, 76–85.
14. Cushman, S.A.; McKelvey, K.S.; Noon, B.R.; McGarigal, K. Use of the abundance of one species as a surrogate for the abundance of others. Conserv. Biol. **2010**, *24*, 830–840.
15. Lawton, J.H.; Gaston, K.J. Indicator species; In *Encyclopedia of Biodiversity*; Levin, S., Ed.; Academic Press: California, 2001; 437–450.
16. Karr, J.R. Assessment of biological integrity using fish communities. Fisheries **1981**, *6*, 21–27.
17. Karr, J.R. Ecological integrity and ecological health are not the same. In *Engineering Within Ecological Constraints*; Schultz, P.C., Ed.; National Academy Press: Washington, DC, 1996; 97–109.
18. Hierl, L.A.; Franklin, J.; Deutschman, D.H.; Regan, H.M.; Johnson, B.S. Assessing and prioritizing ecological communities for monitoring regional habitat conservation plans. Environ. Manag. **2008**, *42*, 165–179.
19. Wiens, J.A.; Hayward, G.D.; Holthausen, R.S.; Wisdom, M.J. Using surrogate species and groups for conservation and management. Bioscience **2008**, *58*, 241–252.
20. Osenberg, C.W.; Schmitt, R.J.; Holbrook, S.J.; Abu-Saba, K.E.; Flegal, A.R. Detection of environmental impacts: Natural variability, effect size, and power analysis. Ecol. Appl. **1994**, *4*, 16–30.
21. Thompson, W.L.; White, G.W.; Gowan, C. *Monitoring Vertebrate Populations*; Academic Press: New York, 1998.
22. Noon, B.R., Bailey, L.L.; Sisk, T.D.; McKelvey, K.S. Efficient species-level monitoring at the landscape scale. Conserv. Biol. **2012**, *26*, 432–441.
23. Power, M.E.; Tilman, D.; Estes, J.A; Menge, B.A.; Bond, W.J.; Mills, L.S.; Daily, G.; Castilla, J.C.; Lubchenco, J.; Paine, R.T. Challenges in the quest for keystones. Bioscience **1996**, *46*, 609–620.
24. Paine, R.T. Food web complexity and species diversity. Am. Nat. **1966**, *100*, 65–75.
25. Hurlbert, S.H. Functional importance vs keystoneness: Reformulating some questions in theoretical biocenology. Aust. J. Ecol. **1997**, *22*, 369–382.
26. Soule, M.E.; Estes, J.A.; Miller, B.; Honnald, D.L. Strongly interacting species: Conservation policy, management, and ethics. Bioscience **2005**, *55*, 168–176.
27. Libralato, S.; Christense, V.; Pauly, D. A method for indentifying keystone species in food web models. Ecol. Model. **2006**, *195*,153–171.
28. Livi, C.M.; Jordan, F.; Lecca, P.; Okay, T.A. Identifying key species in ecosystems with stochastic sensitivity analysis. Ecol. Model. **2011**, *222*, 2542–2551.
29. Estes, J.A.; Terborgh, J.; Brushares, J.S.; Power, M.E.; Berger, J.; Bond, W.J.; Carpenter, S.R.; Essington, T.E.; Holt, R.D.; Jackson, J.B.C.; Marquis, R.J.; Oksanen, L.; Oksanen, T.; Paine, R.T.; Pikitch, E.K.; Ripple, W.J.; Sandin, S.A.; Scheffer, M.; Schoener, T.W.; Shurin, J.B.; Sinclair, A.R.E.; Soulé, M.E.; Virtanen, R.; Wardle, D.A. Trophic down-grading of planet Earth. Science **2011**, *333*, 301–306.
30. Sanchez-Montoya, M.M.; Vidal-Abarca, M.R.; Punti, T.; Poquet, J.M.; Prat, N.; Rieradevall, M.; Alba-Tercedor, J.; Zamora-Muñoz, C.; Toro, M.; Robles, S.; Álvarez, M.; Suárez, M.L. Defining criteria to select reference sites in Mediterranean streams. Hydrobiologica **2009**, *619*, 39–54.
31. Stoddard, J.; Larsen, J; Hawkins, D.; Johnson, R.K.; Norris, R.H. Setting expectations for the ecological condition of streams: the concept of reference condition. Ecol. Appl. **2006**, *16*, 267–1276.
32. Groffman, P.M.; Baron, J.S.; Blett, T.; Gold, A.J.; Goodman, I.; Gunderson, L.H.; Levinson, B.M.; Palmer, M.A.; Paerl, H.W.; Peterson, G.D.; Poff, N.L.; Rejeski, D.W.; Reynolds, J.F.; Turner, M.G.; Weathers, K.C.; Weins, J. Ecological thresholds: the key to successful environmental management, or an important concept with no practical application? Ecosystems **2006**, *9*, 1–13.
33. Possingham, H.P.; Andelman, S.J.; Noon, B.R.; Trombulak, S.; Pulliam, H.R. Making smart conservation decisions. In: *Conservation Biology: Research Priorities for the Next Decade*; M.E. Soule, G.H. Orians, Eds.; Island Press: Washington, D.C., 2001; 225–244.
34. Naeem, S., Bunker, D.E.; Hector, A.; Loreau, M.; Perrings, C. *Biodiversity, Ecosystem Functioning, and Human Wellbeing: An Ecological and Economic Perspective*; Oxford University Press: New York, 2009.
35. Cottingham, K.L.; Brown, B.L.; Lennon, J.T. Biodiversity may regulate the temporal variability of ecological systems. Ecol. Lett. **2001**, *4*, 72–85.
36. Hooper, D.U.; Chapin, F.S.; Ewel, J.J.; Hector, A.; Inchausti, P.; Lavore, S.; Lawton, J.H.; Lodge, D.M.; Loreau, M.; Naeem, S.; Schmid, B.; Setälä, H.; Symstad, A.J.; Vandermeer, J.; Wardle, D.A. Effects of biodiversity on ecosystem functioning: A consensus of current knowledge. Ecol. Monogr. **2005**, *75*, 3–35.

Land Capability Analysis

Michael J. Singer
Department of Land, Air, and Water Resources, University of California, Davis, California, U.S.A.

Abstract
The amount of land available for production of food, fiber, and other necessary products for human sustainability is limited. Land capability systems are designed to help determine appropriate uses and management of available land. Land rating systems for agriculture include those that evaluate the potential for bringing new lands into production and others that evaluate the production potential and conservation needs for those already in production. Additional land capability systems rate lands for nonagricultural uses.

INTRODUCTION

All soils are not the same and they are not of the same capability for every use. Because the Earth's human population exceeds seven billion on its way to nine or ten billion by the mid-21st century, the proper use of land becomes a critical component in human sustainability. Land capability implies that the choice of land for a particular use contributes to the success or failure of that use. It further implies that the choice of land for a particular use will determine the potential impact of that use on surrounding resources such as air and water. To make the best use of land and to minimize the potential for negative impacts on surrounding lands, land capability analysis is needed. The assessment of land performance for specific purposes is land evaluation.[1] A system that organizes soil and landscape properties into a form that helps differentiate among useful and less useful soils for a purpose is land capability classification. Land capability is a broader concept than soil quality, which has been defined as the degree of fitness of a soil for a specific use.[2] Bouma[3] points out that land capability or land potential needs to be evaluated via various scales and gives the example of precision agriculture, which requires land capability analysis in more detail than that required to determine if an investment should be made to initiate agriculture.

Land capability or suitability classification systems have been designed to rate land and soil characteristics for specific uses (Table 1). Huddleston[4] has reviewed many of these systems. They may also rate land qualities. Land qualities have been defined by the U.N. Food and Agricultural Organization[1] as "attributes of land that act in a distinct manner in their influence on the function of land for a specific kind of use." An example of a land quality is the plant available water stored in soil, and an example of a soil characteristic is the clay content that contributes to the plant available water holding capacity.

It is generally recognized that a single soil characteristic is of limited use in evaluating differences among soils,[5] and that use of more than one quantitative variable requires a system for combining the measurements into a useful index.[6] Gersmehl and Brown[7] advocate regionally targeted systems.

KINDS OF SYSTEMS

Land rating systems for agriculture include those that are used to evaluate the potential for agricultural development of new areas, and others that evaluate the potential for agriculture in already developed areas, for example, land capability for grazing in Australia[8] and fertility capability classification for the tropics.[9] Many other land capability assessments exist to help planners rate suitability of agricultural lands for nonagricultural uses; for example, see Steiner et al.[10] Some examples of agricultural land capability systems are described in this entry. All systems have, in common, a set of assumptions on which the analysis is based and each system answers the question "capability for what use."

Development Potential

Examples of systems designed to determine the potential for agricultural development include the FAO framework for land evaluation and the U.S. Bureau of Reclamation (USBR) irrigation suitability classification. The FAO framework combines soil and land properties with a climatic resources inventory to develop an agroecological land suitability assessment.

The USBR capability classification was frequently used to evaluate land's potential for irrigation in the Western U.S. during the period of rapid expansion of water delivery systems.[11–14] It combines social and economic evaluations

Encyclopedia of Natural Resources DOI: 10.1081/E-ENRL-120049142

Table 1 Examples of land capability systems

System	Purpose	Property	References
FAO framework	Development potential	Used for large-scale development of agriculture	[1,13]
USBR irrigation suitability	Potential for irrigation development	Used for determining potential to repay costs of developing irrigation	[11,12]
USDA land capability classification	Land capability for agriculture	Uses 13 soil, climate, and landscape properties to determine agricultural capability	[4,15]
SIR	Land capability for agriculture	Uses nine soil and management factors to determine agricultural capability	[20–22]
Soil potential	Soil suitability for specific uses	Uses a cost index to rate land for any potential use	[19]
Soil quality	Determining status of soil profile	Used for determining the status of selected soil properties	[6,19,24]

with soil and other ecological variables to determine whether the land has the productive capacity, once irrigated, to repay the investment necessary to bring water to an area. It recognizes the unique importance of irrigation to agriculture and the special qualities of soils that make them irrigable.

Land Capability

The USDA Land Capability Classification (LCC) is narrower in scope than either the FAO or USBR capability rating systems. The purpose of the LCC is to place arable soils into groups based on their ability to sustain common cultivated crops that do not require specialized site conditioning or treatment.[15] Nonarable soils, unsuitable for long-term, sustained cultivation, are grouped according to their ability to support permanent vegetation, and according to the risk of soil damage if mismanaged.

Several studies have shown that lands of higher LCC have higher productivity with lower production costs than lands of lower LCC.[16–18] In a study of 744 alfalfa-, corn-, cotton-, sugar beet- and wheat-growing fields in the San Joaquin Valley of California, those with LCC ratings between 1 and 3 had significantly lower input/output ratios than fields with ratings between 3.01 and 6.[18] The input/output ratio is a measure of the cost of producing a unit of output and is a better measure of land capability than output (yield) alone. This suggests that the LCC system provides an economically meaningful assessment of agricultural soil capability.

Quantitative systems result in a numerical index, typically with the highest number being assigned to the land or soil with the highest capability for the selected use. The final index may be additive, multiplicative, or more complex functions of many land or soil attributes. Quantitative systems have two important advantages over nonquantitative systems: 1) they are easier to use with GIS and other automated data retrieval and display systems and 2) they typically provide a continuous scale of assessment.[19] No single national system is presently in use, but several state or regional systems exist.

One example of a quantitative system is the Storie Index Rating (SIR). Storie[20] determined that land productivity is dependent on 32 soil, climate, and vegetative properties. He combined only nine of these properties into the SIR, to keep the system from becoming unwieldy. The nine factors are soil morphology (A), surface texture (B), slope (C), and management factors drainage class (X_1), sodicity (X_2), acidity (X_3), erosion (X_4), microrelief (X_5), and fertility (X_6). Each factor is rated from 1 to 100%. These are converted to their decimal value and multiplied together to yield a single rating for a soil map unit:[20–22]

$$\mathrm{SIR} = \left(A \times B \times C \times \prod_{i=1}^{6} X_i \right) \times 100$$

An area-weighted SIR can be calculated by multiplying the SIR for each soil map unit within a parcel by the area of the soil unit within the parcel, followed by summing the weighted values and dividing by the total area:

$$\text{Area} - \text{weighted SIR} = \frac{1}{\text{totalarea}} \times \sum_{i=1}^{n} \text{SIR soil}_i \times \text{area soil}_i$$

Values for each factor were derived from Storie's experience mapping and evaluating soils in California, and in soil productivity studies in cooperation with California Agricultural Experiment Station cost-efficiency projects relating to orchard crops, grapes, and cotton. Soils that were deep, which had no restricting subsoil horizons, and held water well had the greatest potential for the widest range of crops. More recently, much of what was nonquantitative scoring has been modernized using computer-based decision support systems.[23]

Reganold and Singer[18] found that area-weighted average SIR values between 60 and 100 for 744 fields in the San Joaquin Valley had lower but statistically insignificant input/output ratios than fields with indices <60. The lack of statistical significance is scientifically meaningful, but may not be interpreted the same by planners or farmers for whom small differences in profit can be significant.

Designers of these systems have selected thresholds or critical limits that separate one land capability class from another. Frequently, little justification has been given for the critical limit selected. The issue of critical limits is a difficult one in soils because of the range of potential uses and the interactions among variables.[24]

Soil Potential

Soil potential is an interpretive system that is used for both agricultural and nonagricultural capability assessments. It is an example of a land capability system that is both quantitative and highly specific. Soil potential is defined as the usefulness of a site for a specific purpose using available technology at an indexed cost. Experts in the location where the system will be used determine the soil properties that most influence the successful use of a soil for a particular purpose. The well-suited soils for the use are given a rating of 100. Soil limitations that can be removed or reduced are identified and continuing limitations that occur even after limitations have been addressed are also identified. The cost of available practices and technologies used to remove limitations and the cost of continuing limitations are indexed, and these index values are subtracted from 100 to yield a soil potential rating for each soil.

$$SPI = PI - (CI + CL)$$

The best soils in an area are those that have the lowest indexed cost for operating or maintaining use, while ensuring the lowest possible environment.

CONCLUSIONS

All soils are not the same and are not equally capable of sustained use. Various systems have been created that rate soil properties for these uses. Examples include FAO land suitability, USBR irrigation suitability, USDA LCC, SIR, and soil potential. Each has strengths and weaknesses that the user must recognize before using the system.

REFERENCES

1. *A Framework for Land Evaluation; Soils Bulletin*; Food and Agricultural Organization: FAO: Rome, Italy, 1976; 32.
2. Gregorich, E.G.; Carter, M.R.; Angers, D.A.; Monreal, C.M.; Ellert, B.H. Towards a minimum data set to assess soil organic matter quality in agricultural soils. Can. J. Soil Sci. **1994**, *74*, 367–386.
3. Bouma, J. Land evaluation for landscape units. In *Hand-Book of Soil Science*; Sumner, M.E., Ed.; CRC Press: Boca Raton, FL, 2000; E-393–E-412.
4. Huddleston, J.H. Development and use of soil productivity ratings in the United States. Geoderma **1984**, *32*, 297–317.
5. Reganold, J.P.; Palmer, A.S. Significance of gravimetric versus volumetric measurements of soil quality under biodynamic, conventional, and continuous grass management. J. Soil Water Conserv. **1995**, *50*, 298–305.
6. Halvorson, J.J.; Smith, J.L.; Papendick, R.I. Integration of multiple soil parameters to evaluate soil quality: a field example. Biol. Fert. Soils **1996**, *21*, 207–214.
7. Gersmehl, P.J.; Brown, D.A. Geographic differences in the validity of a linear scale of innate soil productivity. J. Soil Water Conserv. **1990**, *45*, 379–382.
8. Squire, V.R.; Bennett, F. Land capability assessment and estimation of pastoral potential in semiarid rangeland in South Australia. Arid Land Res. Manage. **2004**, *18*, 25–39.
9. Sanchez, P.A.; Palm, C.A.; Buol, S.W. Fertility capability soil classification: a tool to help assess soil quality in the tropics. Geoderma **2003**, *114*, 157–185.
10. Steiner, F.; McSherry, L.; Cohen J. Suitability analysis for the upper Gila River watershed. Landscape Urban Plan. **2000**, *50*, 199–214.
11. USBR. Land Classification Handbook; Bur. Recl. Pub. V, Part 2; USDI: Washington, DC, 1953.
12. Maletic, J.T.; Hutchings, T.B. Selection and classification of irrigable lands. In *Irrigation of Agricultural Lands*; Hagen, R.M., Ed.; Soil Science Society of America: Madison, Wisconsin, 1967; 125–173.
13. Food and agricultural organization. *Land Evaluation Criteria for Irrigation*. Report of an Expert Consultation; World Soil Resourc. Rep. 50; FAO: Rome, Italy, 1979.
14. McRae, S.G.; Burnham, C.P. *Land Evaluation*; Clarendon Press: Oxford, UK, 1981.
15. Klingebiel, A.A.; Montgomery, P.H. *Land-Capability Classification; Agriculture Handbook No. 210*; Soil Conservation Service USDA: Washington, DC, 1973.
16. Patterson, G.T.; MacIntosh, E.E. Relationship between soil capability class and economic returns from grain corn production in southwestern Ontario. Can. J. Soil Sci. **1976**, *56*, 167–174.
17. Van Vliet, L.J.; Mackintosh, E.E.; Hoffman, D.W. Effects of land capability on apple production in southern Ontario. Can. J. Soil Sci. **1979**, *59*, 163–175.
18. Reganold, J.P.; Singer, M.J. Comparison of farm production input/output ratios of two land classification-systems. J. Soil Water Conserv. **1984**, *39*, 47–53.
19. Ewing, S,A.; Singer, M.J. Soil quality. In *Handbook of Soil Science*; Huang, P.M.; Li, Y.; Sumner, M.E., Eds.; CRC Press: Boca Raton, Florida, 2012; 26-1–26-28.

20. Storie, R.E. *An Index for Rating the Agricultural Value of Soils*; California Agr. Exp. Sta. Bull. 556; Berkeley, CA, 1932.
21. Wier, W.W.; Storie, R.E. *A Rating of California Soils*; Bull. 599; California Agr. Exp. Sta.: Berkeley, CA, 1936.
22. Storie, R.E. *Handbook of Soil Evaluation*; Associated Students Store: U. Cal. Berkeley, CA, 1964.
23. O'Geen, A.T.; Southard, S.B.; Southard, R.J. *A Revised Storie Index for Use with Digital Soils Information*; ANR Publication: 8355; 2008.
24. Arshad, M.A.; Coen, G.M. Characterization of soil quality: physical and chemical criteria. Am. J. Altern. Agric. **1992**, *7*, 25–31.

Land Capability Classification

Thomas E. Fenton
Soil Morphology and Genesis, Agronomy Department, Iowa State University, Ames, Iowa, U.S.A.

Abstract

Land classification may be defined as a classification of specific bodies of land according to their characteristics or to their capabilities for use. A natural land classification system is one in which natural land types are placed in categories according to their inherent characteristics. Some sort of land classification system was used in the transfer of public lands to private owners and users as the United States settled. However, a scientific approach was not possible until soil surveys, topographic maps, economic analyses of production, and other activities became available in the early 1900s. One of the most commonly used land classification systems for land management is the land capability classification. Hockensmith defined this classification as a systematic arrangement of different kinds of land according to those properties that determine the ability of the land to produce permanently. There are three major groupings: 1) capability classes; 2) capability subclasses; and 3) capability units.

INTRODUCTION

A common definition of land is the surface of the Earth and all its natural resources. This is interpreted to include the atmosphere, the soil, and underlying geology, hydrology, and plants on the Earth's surface.[1] In the *1938 Yearbook of Agriculture*, land is defined as the total natural and cultural environment within which production must take place. Its attributes include climate, surface configuration, soil, water supply, subsurface conditions, etc., together with its location with respect to centers of commerce and population. It should not be used as synonymous with soil or in the sense of the Earth's surface only.[2] Other definitions included the results of the past and present human activities as well as the animals within this area when they exert a significant influence on the present and future uses of land by man. Thus, the concept of land is much broader than soil. However, because soils are a major component of terrestrial ecosystems,[3] they integrate and reflect internal and external environmental factors. They have been considered an important factor—if not the major factor—in many land classification systems.

CLASSIFICATIONS

Land classification may be defined as a classification of specific bodies of land according to their characteristics or to their capabilities for use. A natural land classification system is one in which natural land types are placed in categories according to their inherent characteristics. A land classification according to its capabilities for use may be defined as one in which the bodies of land are classified on the basis of physical characteristics with or without economic considerations according to their capabilities for man's use.[2]

Olson[4] defined land classification as the assignment of classes, categories, or values to the areas of the Earth's surface (generally excluding water surfaces) for immediate or future practical use. The project, product, or proposal resulting from this activity may be also generally referred to as land classification. Any land classification involves two parts or phases: resource inventory and analysis and categorization. The inventory consists of gathering data and delineating land characteristics on maps. The analysis and categorization put the basic data into a form that can be used generally for a specific use. Thus, there are potentially many land classification systems that could be developed depending on the objectives of the system.

HISTORICAL PERSPECTIVE

Some sort of land classification system was used in the transfer of public lands to private owners and users as the United States settled. However, a scientific approach was not possible until soil surveys, topographic maps, economic analyses of production, and other activities became available in the early 1900s. These sources provided essential data for a scientific approach to land classification. This approach gained added momentum in the mid-1930s due to the need for land classification for many new government programs.

A report submitted to the President of the United States by the National Resources Board dated December 31, 1934,[5] included a section that dealt with the physical classification of the productivity of the land. The total land area of the United States was rated and divided into five grades

Encyclopedia of Natural Resources DOI: 10.1081/E-ENRL-120049143

based on the factors thought to affect productivity—soil type, topography, rainfall, and temperature. Norton[6,7] defined five classes of land in the regions of arable soils according to their use capability. Two other classes were defined for land that should not be cultivated. A National Conference on Land Classification (NCLC) was held in the campus of the University of Missouri in October 1940. The proceedings of this conference were published in the Bulletin 421 of the Missouri Agricultural Experiment Station.[8] At this conference, it was recognized that land classification in the United States had begun many years prior to 1940. Simonson[9] summarized the purpose of the NCLC meeting as follows: to encourage and provide opportunity for the discussion of land classification in all of its different aspects.

In the years following the conference, there were many entries published that examined questions related to land classification. General principles associated with technical grouping[10] or solution of soil management problems[11] using natural soil classification systems were discussed in detail. Hockensmith and Steele[12] presented a land capability classification developed for conservation and development of land and for land-use adjustments. The system used eight land capability classes with the first four suited for cultivation and the remaining four not suited for cultivation. Within a capability class, the subclasses were determined by the kind of limitation. Within each subclass, the land that required the same kind of management and the same kind of conservation treatment was called a land capability unit. The highest level of abstraction, the capability class, provided a quick, easy understanding of the general suitability of the land for cultivation. Hockensmith[13] discussed the use of soil survey information in farm planning. A land capability map was an essential part of the farm plan. This map was an interpretative map of the soil map together with other pertinent facts. It was used to help the farmers understand the limitations of their soils and to aid in selection of those practices that would preserve their soils and keep them productive. Klingebiel[14] defined capability classification as the grouping of individual kinds of soils, called soil mapping unit, into groups of similar soils, called capability units, within a framework of eight general capability classes that were divided into subclasses representing four kinds of conservation problems. This land classification system has been used as a basis for all the conservation farm plans developed by the Soil Conservation Service (now Natural Resources Conservation Service) in the United States since the 1950s and continues to be used today. Because of its widespread use, this system is discussed in more detail in the following section.

LAND CAPABILITY CLASSIFICATION

One of the most commonly used land classification systems for land management is the land capability classification. Hockensmith[15] defined this classification as a systematic arrangement of different kinds of land according to those properties that determine the ability of the land to produce permanently. Classification was based on the detailed soil survey (scale of 1:15,840 or 1:20,000). The system was described in detail in the Agriculture Handbook No. 210[16] and continues to be in use, with no significant changes except that presently Arabic numerals are used to identify the capability classes rather than Roman numerals. There are a number of land capability classifications used throughout the world but all these methods are patterned after the U.S. system.[17]

The capability grouping of soils is designed to

1. Help landowners and others use and interpret the soil maps
2. Introduce users to the detail of the soil map itself
3. To make possible broad generalizations based on soil potentialities, limitations in use, and management problems.

There are three major groupings: 1) capability classes; 2) capability subclasses; and 3) capability units. All map unit components, including miscellaneous areas, are assigned a capability class and subclass.[18]

Capability Classes

The broadest unit is the capability class and there are eight classes. The probability of soil damage or limitations in use become progressively greater from Class I to VIII. There are no additional subdivisions of soils in Class I because by definition the soils in this category have no limitations in use. The proper grouping of the soils assumes that the recommended use will not deteriorate the soil over time.

- Class I soils have few limitations that restrict their use.
- Class II soils have some limitations that reduce the choice of plants or require moderate conservation practices.
- Class III soils have severe limitations that reduce the choice of plants or require special conservation practices or both.
- Class IV has very severe limitations that restrict the choice of plants, require very careful management, or both.
- Class V soils have little or no erosion hazard but has other limitations that are impractical to remove and limit their use largely to pasture, range, woodland, or wildlife food and cover.
- Class VI soils have severe limitations that make them generally unsuited to cultivation and limit their use largely to pasture or range, woodland, or wildlife food and cover.
- Class VII soils have very severe limitation that make them unsuited to cultivation and that restrict their use largely to grazing, woodland, or wildlife.
- Class VIII soils and landforms limitations that preclude their use for commercial plant production and restrict their use to recreation, wildlife, water supply, or to aesthetic purposes.

Capability Subclasses

Subclasses are groups of capability units within classes with the same kind of major limitations for agricultural use. The capability subclass is a grouping of capability units having similar kinds of limitations and hazards. Four general kinds of limitations or hazards are recognized:

1. Erosion hazard
2. Wetness
3. Rooting-zone limitations
4. Climate

The limitations recognized and the symbols used to identify them are risk of erosion designated by the symbol e; wetness, drainage, or overflow, w; rooting-zone limitations, s; and climatic limitations, c. There is a priority of use among the subclasses if the limitations are approximately of the same degree. The order of use is e, w, s, and c. For example, if a group of soils has both erosion and excess water hazard, e takes precedence over w. However, for some uses, it may be desirable to show two kinds of limitations. This combination is rarely used but when used, the higher priority one is shown first, i.e., IIew.

Capability Units

Capability units provide more specific and detailed information than class or subclass. They consist of soils that are nearly alike in suitability for plant growth and responses to the same kinds of soil management. However, they may have characteristics that place them in different soil series or soil map units. Soil map units in any capability unit are adapted to the same kinds of common cultivated and pasture plants and require similar alternative systems of management for these crops. Another assumption is that the longtime estimated yields of adapted crops for individual soil map units within the unit under comparable management do not vary more than about 25%.

CONCLUSIONS

The grouping of soil map units as used in the land capability classification is a technical classification. The basis for this classification is the natural soil classification system in which the soils and soil map units are defined based on their characteristics and interpretations made based on those properties that affect use and management. For conservation planning, environmental quality, and generation of interpretive maps, it is important to know the kind of soil, its location on the landscape, its extent, and its suitability for various uses.

REFERENCES

1. Brinkman, R.; Smyth, A.J. *Land Evaluation for Rural Purposes*; Publication 17; International Institute for Land Reclamation and Improvement: Wageningen, the Netherlands, 1973; 116.
2. United States Department of Agriculture. *Soils and Men: Yearbook of Agriculture*; U.S. Govt. Printing Office: Washington, DC, 1938; 1232.
3. Jenny, H. *The Soil Resource: Ecological Studies*; Springer: New York, 1980; Vol. 37, 377.
4. Olson, G.W. *Land Classification*; Cornell University Agricultural Experimental Station Agronomy 4: Ithaca, New York, 1970.
5. National Resources Board Report. Part II. *Report of the Land Planning Committee, Section II*; U.S. Govt. Printing Office: Washington, DC, 1934; 108–152.
6. Norton, E.A. Classes of land according to use capability. Soil Sci. Soc. Am. Proc. **1939**, *4*, 378–381.
7. Norton, E.A. *Soil Conservation Survey Handbook*; Misc. Publication 352; U.S. Department Agricultural Soil Conservation Service: Washington, DC, 1939; 40.
8. Miller, M.F. *The Classification of Land*, Proceeding of First International Conference on Land Classification, Bulletin 421; Missouri Agricultural Experimental Station; 1940, 334.
9. Simonson, R.W. The National Conference on Land Classification. Soil Sci. Soc. Am. Proc. **1940**, *5*, 324–326.
10. Orvedal, A.C.; Edwards, M.J. General principles of technical grouping of soils. Soil Sci. Soc. Am. Proc. **1941**, *6*, 386–391.
11. Simonson, R.W.; Englehorn, A.J. Interpretation and use of soil classification in the solution of soil management problems. Soil Sci. Soc. Am. Proc. **1942**, *7*, 419–426.
12. Hockensmith, R.D.; Steele, J.G. Recent trends in the use of the land-capability classification. Soil Sci. Soc. Am. Proc. **1949**, *14*, 383–387.
13. Hockensmith, R.D. Using soil survey information for farm planning. Soil Sci. Soc. Am. Proc. **1953**, *18*, 285–287.
14. Klingebiel, A.A. Soil survey interpretation-capability groupings. Soil Sci. Soc. Am. Proc. **1958**, *23*, 160–163.
15. Hockensmith, R.D. *Classification of Land According to its Capability as a Basis for a Soil Conservation Program*, Reprinted from Proceedings of the Inter-American Conference on Conservation of Renewable Natural Resources, Denver, CO, Sep 7–20, 1948.
16. Klingebiel, A.A.; Montgomery, P.H. Land capability classification. In *Agricultural Handbook 210*; U.S. Department of Agricultural Soil Conservation Service: Washington, DC, 1961; 21.
17. Hudson, N. Land use and soil conservation. In *Soil Conservation*, 3rd Ed.; Iowa State University Press: Ames, IA, 1995; 391.
18. U.S. Department of Agriculture, Natural Resources Conservation Service. *National Soil Survey Handbook, title 430-VI.* Available online at http://soils.usda.gov/technical/handbook/ (accessed October 2012).

Land Change and Water Resource Vulnerability

Gargi Chaudhuri
Department of Geography and Earth Science, University of Wisconsin, La Crosse, Wisconsin, U.S.A.

Abstract
One of the biggest transformations caused by anthropogenic activities is alteration of the land surface, essentially, for food and shelter. Land change has evidently affected other natural resources on the surface of the earth. Water resources are one of the most important natural resource whose increasingly vulnerability is partly attributed to the land use change. This entry discusses the general relationship between land use and land cover change and increasing vulnerability of water resources and highlights the present status with special focus on agricultural and urban land.

INTRODUCTION

Since the beginning of human civilization, mankind has been dependent on land and water for sustenance. Over the years, human activities have modified the natural environment to the extent that at present environmental sustainability is at danger. The term "land cover" refers to the biophysical cover of the land, and "land use" refers to the socio-economic use of land. Land use and land cover change (LULCC), in general means conversion of land to feed, provide shelter and other essentials to the growing human population.[1,2] Among different natural resources that are threatened by the direct and indirect impacts of LULCC, the vulnerability of fresh water resources is critical.[3] Land-change science has contributed significantly to the understanding of land use dynamics, yet the human use of land continues to be at the center of the most complicated and pressing problems faced by policy makers around the world today.[4] The principal effects of land use change include consumption of freshwater, alteration in the amount of evaporation, groundwater infiltration, surface runoff, increasing sedimentation, and changes in the streamflow regime, which result in increased risk of flood or drought.[5] In a nutshell, land use change affects both the quantity and the quality of freshwater.

Water quantity means the presence of adequate supply of water to sustain human and natural systems and is measured in terms of total water withdrawal and consumption from both surface water and groundwater sources.[6,7] On the other hand, water quality refers to the suitability of the water supplied for its intended use.[7] Both water quality and amount of water use can be heavily affected by the types of land cover and land use activity and its geographical region, and are important elements to consider for any watershed management program.[6] Mustard and Fisher[7] defined water vulnerability as "human supply and demand problem, where climate and the landscape establish the supply and human systems make the demands on the supply." Thus, decrease in quantity and quality of water resource due to the modification of the landscape and increase in demand due to population growth lead to increase in its vulnerability.

The relationship between land use and hydrology is well known but is complex and nonlinear.[5] According to United Nation Environment Program (UNEP),[8] of the total 35 million km^3 of freshwater resources, the total usable freshwater supply for ecosystems and humans is about 200,000 km^3 or less than 1% of all freshwater resources. In the twentieth century, land use and land cover have changed rapidly both in terms of extent and intensity of use,[9] which have made its impact on water resources more intense. Global reports suggest that in the last 50 years, the world's agricultural production has increased between 2.5–3 times while the cultivated area has increased only by 12%. Thus, more than 40% of the increased food production came from irrigated areas, which have doubled in area.[10] Forest cover lost 135 million hectares between 1990 and 2010,[11] and at present only covers 31% of the land. In 30 years' time period (1970–2000), the world's urban area increased by 58,000 km^2,[12] where 52.1% of the global population lives at present.[13]

According to the World Water Development Report 4 (WWDR4),[14] the world population is predicted to grow from 6.9 billion in 2010 to 8.3 billion in 2030, resulting in an expected increase of 50% in food demand.[15] Thus, the major drivers of increased water demand are change in food, energy (biofuels), industry, and human settlement. In terms of present water use, agriculture accounts for 70% (approx.), industry accounts for 20% (approx.), and municipal water demand accounts for the rest 10% of the global freshwater withdrawals.[14] Of the total groundwater withdrawal (1000 km^3/year), irrigation accounts for 67% (approx.), domestic purposes for 22%, and industrial purposes account for 11%.[14,16] Thus, with the projected increase in food demand, irrigation will continue to play

Encyclopedia of Natural Resources DOI: 10.1081/E-ENRL-120051107

an important role in food production and is expected to result in 11% increase in irrigation water consumption from 2008 to 2050 in water scarce regions.[10] Among all the continents, Asia accounts for the largest freshwater withdrawal (includes agricultural, municipal, and industrial water withdrawal) (Fig. 1) but in terms of per person usage per year, North America is the biggest consumer (Fig. 2). Among different sectors (Fig. 3), agricultural activity accounts for the biggest water withdrawal in most of the world except Europe, where industrial activities withdraw more water than any other sector, and North America, where agricultural and industrial activity account for almost similar levels of water withdrawal. The impacts of agricultural and urban land use/land cover have been discussed further.

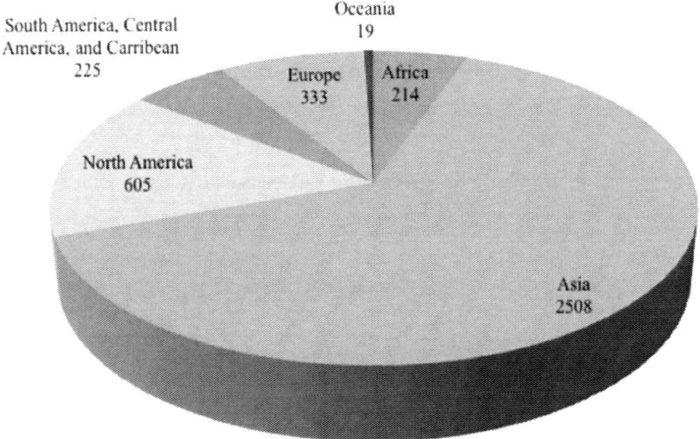

Fig. 1 Total freshwater withdrawal (km^3/year).
Source: From FAO.[17]

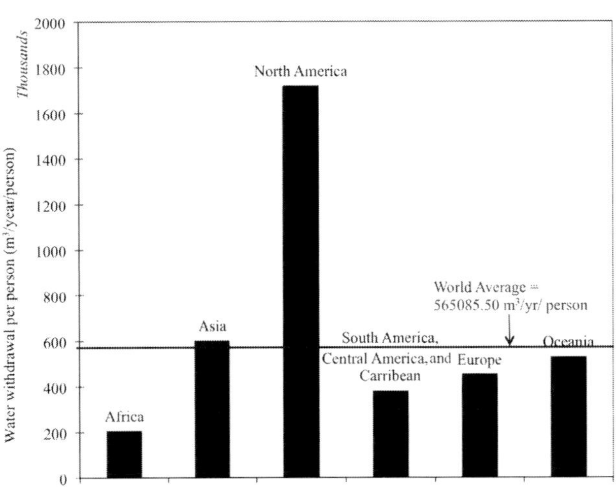

Fig. 2 Freshwater withdrawal (in m^3) per person per year.
Source: From FAO.[17]

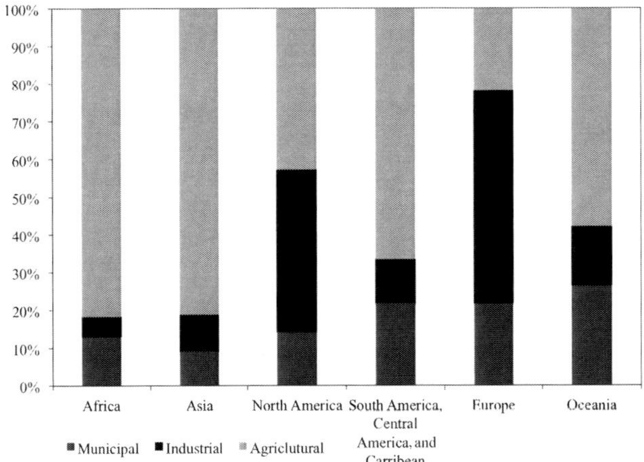

Fig. 3 Proportion of water withdrawal by each sector in each continent.
Source: From FAO.[17]

AGRICULTURE

Agriculture (croplands and pasturelands) has transformed the surface of the Earth at the expense of forests and grasslands.[5,18] Agricultural intensification and expansion have both contributed to the increasing vulnerability of the water resources, both in terms of quality and quantity. Worldwide, 60% of the crop production comes from agriculture that occupies 80% of the cultivated land, and the remaining 20% of the irrigated land accounts for 40% of the global crop production.[14] WWDR4[14] also projects an average increase of 0.6% a year in irrigated land until 2030 that will result in 36% more food being produced with 13% more water. Irrigation and intensive use of agricultural land may be an answer to the problem of global food security for the growing population, but it means increased risk on already stressed water resources. In addition to the food crops, the biofuel mandates in the developed countries of the world in 2007,[19,20] which encourage the increasing use of biofuels, are projected to increase the demand for freshwater, too.[14] The International Energy Agency (IEA) indicates that if only 5% of road transport is powered by biofuel, it will amount to 20% of the water used for agriculture globally.[14,21]

Generally, natural forested areas are being replaced by agriculture, thus the impact of this change varies depending on the type of fauna being replaced, the nature of agricultural crop, the type of agricultural practice, and the climatic condition of the geographical region.[18] For example, Muñoz-Arriola et al.[22] showed that in the Yaqui River Basin, an important wheat-producing region in Mexico, the streamflow is more sensitive to the climatic variability rather than land use change, to the extent that the streamflow is higher by 76% in winter and 16% in summer during El Nino years than La Nina years.[22] In terms of water quantity, conversion to agricultural land increases water consumption and the amount of water consumption varies highly between rain-fed and irrigated agricultural systems. Agriculture also results in a decreased amount of evapotranspiration from the land (Gordon et al.[23] showed that deforestation resulted in 4% increase in evapotranspiration globally), increase in groundwater recharge, and increase in streamflow.[7,18] In terms of water quality, an increased intensity of agricultural land use has led to the contamination of groundwater and run-off water with nitrate and other toxic agrochemicals and increased salinization.[7,18,24–27] For example, conversion of the semi-arid naturally forested land in Australia increased the groundwater recharge, but increased water levels in the shallow water tables also led to soil salinization and water logging.[28] Another example of negative impact of irrigation is the fate of the Aral Sea, the size of the water body reduced from 68,000 km^2 in 1960 to 14,280 km^2 in 2010[29] due to irrigation of the neighboring arid plains of Kazakhstan, Uzbekistan, and Turkmenistan for cotton plantation (Fig. 4). An enormous reduction of water level in the lake, increased salinization, and a degradation of the water quality due to agricultural chemicals affected the economy of the whole region. Agriculture is the predominant user of freshwater worldwide and the spatial scale of its impact on water resources varies from local catchment scale to regional basin scale.

URBAN

From the land use and land cover classification perspective, urban areas consist of residential, commercial, industrial, and mining areas.[30] Increase in urbanization involves the conversion of land surface into impervious surface, and high intensity of water use due to dense urban population. In terms of water quantity, an impervious cover leads to the direct modification and destruction of water bodies, increased storm flows, decreased base flows during dry periods, and a decrease in infiltration and increase in run-off, thus obstructing groundwater recharge.[6,31] Decreased infiltration and increased peak flows of storm water runoff also increase the magnitude of urban floods.[32] In terms of water quality, urbanization affects the water quality of both surface and groundwater from various point sources and non-point sources of pollution such as industrial and municipal waste discharges, leaky underground storage facilities, and spills of organic or inorganic contaminants, solvents, and fuels.[7] In a meta-analysis of global urban land expansion, Seto et al.[12] reported that global urban land has increased from 1.56% in 1970 to 3.89% in 2000, with the largest change in the total extent of urban land in North America and higher rates of urbanization in India, China, and Africa. The study also reports that in high-income countries, Gross Domestic Product (GDP) drives urbanization, whereas in countries like India and China, population growth is the main driver of change.[12] Figure 4, produced by the same study, shows that urban land expansion rates are higher than or equal to urban population growth rates across all regions in three decades (1970–2000).

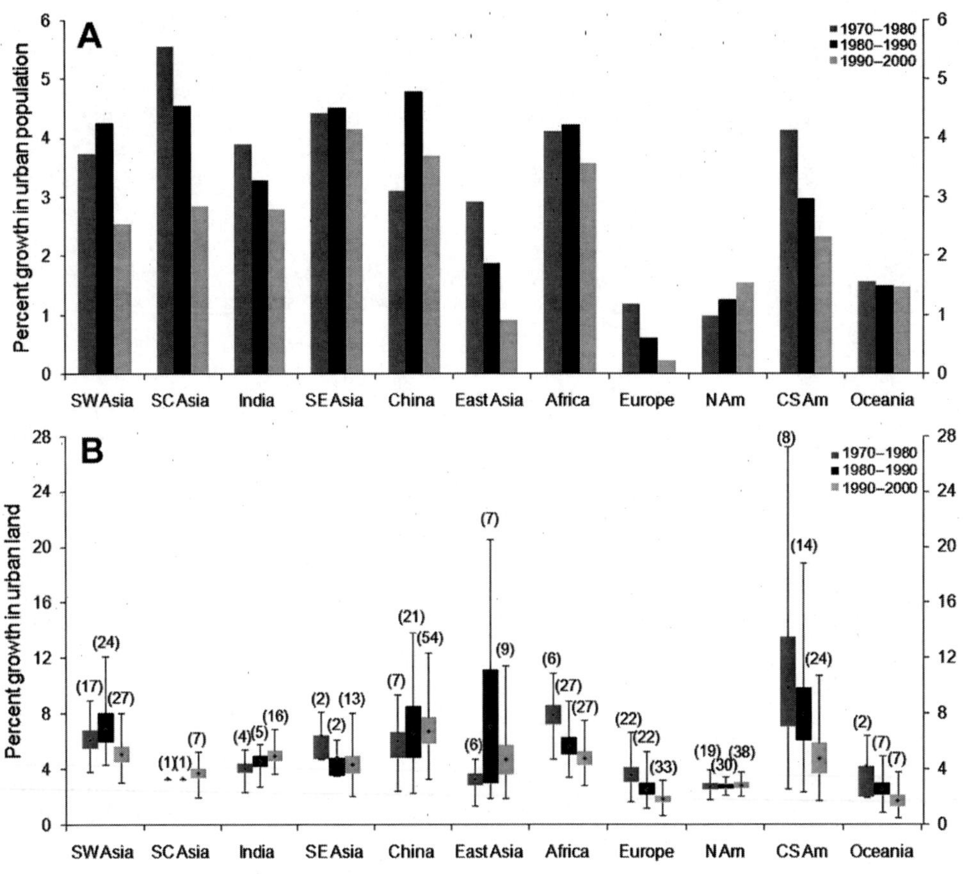

Fig. 4 Comparison of two different urban population growths (**A**) and urban land expansion by region (**B**) and by decade.
Source: Adapted from Seto et al.[12]

In 2011, the world urban population is recorded to be 3.6 billion and it is expected to increase by 72% by 2050, i.e., around 6.3 billion.[13] Growing urban population, especially in developing regions, has two broad effects on water resources: first a conversion of prime agricultural land or forested area into an impervious surface, thus affecting the water cycle, and second a better socio-economic status of urban population results in changing diets, which involves an increased consumption of protein. For example, in China during the 1980s, urban populations consumed 60% more pork in their diets than rural populations[33] and in India where a large number of urban population are vegetarian, a higher consumption of milk is reported.[34,35] Studies showed that, in industrialized countries, for an average person 3400 kcal/day[10] of meat diet uses 3600 liters of water per day compared to 3400 kcal/day of vegetarian diet (includes dairy) that uses 2300 liters of water per day.[36–38] Additionally, meat diets also entail the production of animal fodder in the form of crops, and during the period 2001–2007, on an average 37% of the cereals produced in the world were reportedly used for animal feed.[10,38] Thus, urbanization poses a dilemma for freshwater availability; on the one hand, an impervious surface covers the land thus hindering the natural water cycle and freshwater supply, and on the other hand it creates more demand for water due to the high density of human population and changing lifestyle.

CONCLUSION

Empirical studies on the impact assessment of LULCC on water resources have greatly enriched the body of literature by identifying and delineating regional variations, causes, and consequences of land use change on water resources, and temporal dynamics of the relationship between the two. In terms of spatial scale of effect, in humid and temperate regions with rain-fed agricultural system, the LULCC effect is limited to the drainage basin, whereas in arid and semi-arid regions, which are mostly dependent on the irrigated agricultural system, the impacts are more spatially extensive.[7] The temporal scales of the impact of LULCC on water resources are variable, too. At catchment scale, changes occur at an irregular time scale, whereas at plot scale for crop plantation and deforestation, changes are quite regular.[39–43] Moreover, there exists a time lag between LULCC and its effect on water resources that varies depending on the ecosystem, spatial extent, and climatic condition.[18] There are a number of water vulnerability indices[44] that show a global variation of water resource vulnerability measured on the basis of relative demand and supply of water. Figure 5, a water risk map produced by World Resource Institute, shows an overall global water risk,[45] and is calculated and mapped using indicators such as water quantity, water variability, water quality, public awareness of water issues, access to water, and ecosystem vulnerability. The regions that are mostly at "high to extremely high risk" class, such as western North America, northern part of Africa bordering the Saharas, Arabian Peninsula, Eurasian borders, northern China, and the Indian subcontinent, are areas of arid and semi-arid regions with high human demands and a high density of human population, respectively.

It is difficult to isolate the impacts of LULCC on water resources amidst influences of other biophysical forces, socio-political factors, and climatic variability, but the drivers of change are mostly region specific.[7,43] At present,

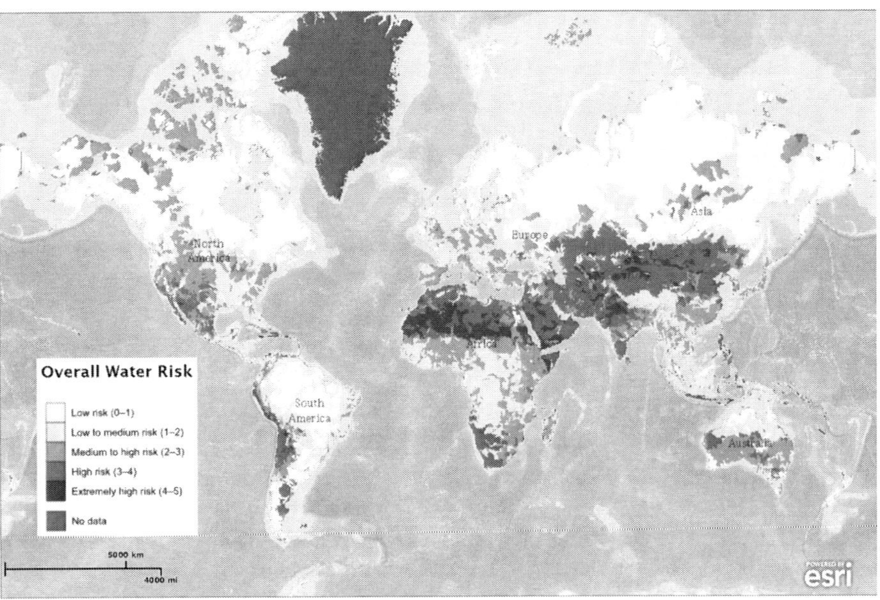

Fig. 5 Overall water risk map.
Source: Data from World Resource Institute; Mapping: Gassert et al.,[45] http://aqueduct.wri.org/atlas (accessed February 9, 2013).

the densely inhabited parts of the world are already heavily water stressed and these regions are also expected to experience a high population growth in the future. To support such a growing population, both land use activities and water resources are immensely important for human sustenance, thus an integrated approach to land use and water management is important for a sustainable future. Recently, there have been an increasing number of studies where scientists are using integrated assessment models to improve the understanding of the feedback mechanism between land use changes and water resources under multiple scenarios of climate change.[46–50] These integrative approaches explore linkages between the different components of earth systems and human systems to capture its dynamic relationship and assess the need for social, economic, and environmental coherence in the management of natural resources. Knowledge gained from such applications not only aids to better address policy issues, but also helps to develop a scientifically sound decision support system.

REFERENCES

1. DeFries, R.; Eshleman, K.N. Land-use change and hydrologic processes: A major focus for the future. Hydrol. Process. **2004**, *18* (11), 2183–2186.
2. Ellis, E. Land-Use and Land-Cover Change, Retrieved on 5/2/2013 from http://www.eoearth.org/view/article/51cbee4f7896bb431f696e92
3. Foley, J.A.; DeFries, R.; Asner, G.P.; Barford, C.; Bonan, G.; Carpenter, S.R.; Chapin, F.S.; Coe, M.T.; Daily, G.C.; Gibbs, H.K.; Helkowski, J.H.; Holloway, T.; Howard, E.A.; Kucharik, C.J.; Monfreda, C.; Patz, J.A.; Prentice, I.C.; Ramankutty, N.; Snyder, P.K. Global consequences of land use. Science **2005**, *309* (5734), 570–574.
4. Reid, R.S.; Tomich, T.P.; Xu, J.; Geist, H.; Mather, A.; DeFries, R.S.; Liu, J.; Alves, D.; Agbola, B.; Lambin, E.F.; Chabbra, A.; Veldkamp, T.; Kok, K.; van Noordwijk, M.; Thomas, D.; Palm, C.; Verburg, P.H. *Linking Land-Change Science and Policy: Current Lessons and Future Integration, Land-Use and Land-Cover Change: Local Processes and Global Impacts*; Springer-Verlag: Berlin, 2006; 157–171.
5. Bullard, W.E. Effects of land use on water resources. J. Water Pollut. Control Fed. **1966**, *38* (4), 645–659.
6. Lusch, D.P.; Wolfson, L.G. The Introductory Land and Water Learning Module (ILM) was developed by the Institute of Water Research, Michigan State University (MSU) with funding from MSU's All-University Outreach Grant and the U.S. Geological Survey. 1997 http://www.iwr.msu.edu/edmodule/ilmack.htm
7. Mustard, J.F.; Fisher, T.R. Land use and hydrology. In *Land Change Science: Observing, Monitoring and Understanding Trajectories of Change on the Earth's Surface*; Gutman, G., Janetos, A.C., Justice, C.O., Moran, E.F., Mustard, J.F., Rindfuss, R.R., Skole, D., Turner, B.L., II, Cochrane, M.A., Eds.; Kluwer Academic Publishers: Dordrecht, 2004; 257–276.
8. UNEP. *Vital Water Graphics - An Overview of the State of the World's Fresh and Marine Waters*, 2nd Ed.; UNEP: Nairobi, Kenya. ISBN: 92-807-2236-0.
9. Ramankutty, N.L.; Graumlich, F.; Achard, D.; Alves, A.; Chhabra, R.; DeFries, J.; Foley, H.; Geist, R.; Houghton, K.; Klein Goldewijk, E.; Lambin, A.; Millington, K.; Rasmussen, R.R.; Turner II, B.L. Global land cover change: recent progress, remaining challenges, In *Land Use and Land Cover Change: Local Processes, Global Impacts*; Lambin, E., Geist, H., Eds.; Springer Verlag: New York, 2006; 9–39.
10. Food and Agriculture Organization (FAO). *The State of the World's Land and Water Resources for Food and Agriculture: Managing Systems at Risk*; Earthscan: London, 2011a.
11. Food and Agriculture Organization (FAO). *Key Findings – Global Forest Resources Assessment*; Food and Agriculture Organization of the United Nations: Rome, Italy, 2010.
12. Seto, K.C.; Fragkias, M.; Güneralp, B.; Reilly, M.K. A meta-analysis of global urban land expansion. PLoS ONE **2011**, *6* (8), e23777.
13. World Urbanization Prospects (WUP): The 2011 Revision Highlights, United Nations Department of Economic and Social Affairs (UNDESA), Population Division special updated estimates of urban population as of October 2011. United Nation Publication ESA/P/WP/224, March 2012.
14. UN World Water Development Report (WWDR4), 4th Ed., http://www.unesco.org/new/en/natural-sciences/environment/water/wwap/wwdr/wwdr4-2012/, 2012. United Nation Publication.
15. Bruinsma, J. The resource outlook to 2050: By how much do land, water, and crop yields need to increase by 2050? FAO Expert Meeting on *How to feed the world in 2050*; Rome: FAO, June 2009; p. 24–26.
16. Food and Agriculture Organization of the United Nations (FAO). Food balance sheets. FAOSTAT, FAO, Rome, Italy, 2011b, http://faostat.fao.org (accessed January 2012).
17. FAO, AQUASTAT. Water Use, http://www.fao.org/nr/water/aquastat/water_use/index6.stm, 2010 (accessed August 2013).
18. Scanlon, B.R.; Jolly, I.; Sophocleous, M.; Zhang, L. Global impacts of conversions from natural to agricultural ecosystems on water resources: Quantity versus quality. Water Resour. Res. **2007**, *43* (3), W03437.
19. Energy Independence and Security Act (EISA) of 2007: Summary of Provisions, http://www.eia.gov/oiaf/aeo/otheranalysis/aeo_2008analysispapers/eisa.html (accessed August 2013).
20. CRS Report for Congress, Energy Independence and Security Act of 2007: A Summary of Major Provisions, December 21, 2007.
21. Comprehensive Assessment of Water Management in Agriculture. *Water for Food, Water for Life: A Comprehensive Assessment of Water Management in Agriculture*. Earthscan: London; International Water Management Institute: Colombo, 2007.
22. Muñoz-Arriola, F.; Avissar, R.; Zhu, C.; Lettenmaier, D.P. Sensitivity of the water resources of Rio Yaqui Basin, Mexico, to agriculture extensification under multiscale climate conditions. Water Resour. Res. **2009**, *45*, W00A20, DOI:10.1029/2007WR006783.
23. Gordon, L.J.; Steffen, W.; Jonsson, B.F.; Folke, C.; Falkenmark, M.; Johannessen, A. Human modification of global water vapor flows from land surface. Proc. Natl. Acad. Sci. USA **2005**, *102* (21), 7612–7617.
24. Scanlon, B.R.; Reedy, R.C.; Stonestrom, D.A.; Prudic, D.E.; Dennehy, K.F. Impact of land use and land cover change on

groundwater recharge and quality in the southwestern USA. Global Change Biol. **2005**, *11* (10), 1577–1593.
25. Schlesinger, W.H.; Reckhow, K.H.; Bernhardt, E.S. Global change: the nitrogen cycle and rivers. Water Resour. Res. **2006**, *42* (3), W03S06.
26. Schlesinger, W.H. On the fate of anthropogenic nitrogen. Proc. Natl. Acad. Sci. USA **2009**, *106* (1), 203–208.
27. Stonestrom, D.A.; Scanlon, B.R.; Zhang, L. Introduction to special section on impacts of land use change on water resources. Water Resour. Res. **2009**, *45*, W00A00, DOI:10.1029/2009WR007937.
28. Allison, G.B.; Cook, P.G.; Barnett, S.R.; Walker, G.R.; Jolly, I.D.; Hughes, M.W. Land clearance and river salinisation in the western Murray Basin, Australia. J. Hydrol. **1990**, *119*, 1–20, DOI:10.1016/0022-1694(90)90030-2.
29. Gaybullaev, B.; Chen, S.-C.; Gaybullaev, D. Changes in water volume of the Aral Sea after 1960. Appl. Water Sci. **2012**, *2* (4), 285–291.
30. Anderson, J.R.; Hardy, E.E.; Roach, J.T.; Witmer, R.J. A Land Use and Land Cover classification system for use with remote sensor data, Geological Survey Professional Paper 964, USGS: Washington, DC, 1976.
31. Dreher, D.W.; Price, T.H. *Best Management Practice Guidebook for Urban Development*; Northeastern Illinois Planning Commission: Chicago, Illinois, 1992.
32. *The Hydrologic Cycle*: U.S. Geological Survey, U.S. Department of the Interior, U.S. Government Printing Office: 1984-421-618/109.
33. Taylor, J.R.; Hardee, K.A. *Consumer Demand in China: A Statistical Factbook*; Westview Press: Boulder, Colorado and London, 1986.
34. Parikh, J.; Gokam, S. Consumption Patterns: The Driving Force of Environmental Stress, Indira Gandhi Institute of Development Research, presented at the United Nations Conference on Environment and Development, August 1991.
35. Torrey, B.B. *Urbanization: An Environmental Force to Be Reckoned with*; Population Reference Bureau, 2004. http://www.prb.org/Publications/Articles/2004/Urbanization-AnEnvironmentalForcetoBeReckonedWith.aspx
36. Falkenmark, M.; Rockström, J. *Balancing Water for Humans and Nature: The New Approach in Ecohydrology*; Earthscan: London, UK, 2004.
37. Hoekstra, A.Y.; Chapagain, A.K. *Globalization of Water: Sharing the Planet's Freshwater Resources*. Blackwell Publishing: Oxford, UK, 2008.
38. Hoekstra, A.Y. The hidden water resource use behind meat and dairy. Anim. Front. **2012**, *2* (2), 3–8. http://dx.doi.org/10.2527/af.2012-0038.
39. Versfeld, D.B. Overland flow on small plots at the Jonkershoek Forestry Research Station. S. Afr. Forest. J. **1981**, *119* (1), 35–40.
40. Calder, I.R. Water use by forests at the plot and catchment scale. Commonwealth Forest. Rev. **1996**, *75* (1), 19–30.
41. Calder, I.R. *Water-Resource and Land-Use Issues. SWIM Paper 3*; International Water Management Institute: Colombo, Sri Lanka, 1998.
42. Archer, D.R. Scale effects on the hydrological impact of upland afforestation and drainage using indices of flow variability: The River Irthing, England. Hydrol. Earth Syst. Sci. **2003**, *7* (3), 325–338.
43. Chhabra, A.; Geist, H.; Houghton, R.A.; Haberl, H.; Braimoh, A.K.; Vlek, P.L.G.; Patz, J.; Xu, J.; Ramankutty, N.; Coomes, O.; Lambin, E.F. Multiple impacts of land-use/cover change. In *Land-Use and Land-Cover Change: Local Processes and Global Impacts*; Lambin, E.F., Geist, H.J., Eds.; Springer-Verlag: Berlin, Heidelberg, 2006.
44. Plummer, R.; Loë, R.D.; Armitage, D. A systematic review of water vulnerability assessment tools. Water Resour. Manag. **2012**, *26* (15), 4327–4346.
45. Gassert, F.; Luck, M.; Landis, M.; Reig, P.; Shiao, T. *Aqueduct Global Maps 2.0. Working Paper*; World Resources Institute: Washington, DC, 2013. Available online at http://wri.org/publication/aqueduct-metadata-global.
46. Van De Giesen, N.; Kunstmann, H.; Jung, G.; Liebe, J.; Andreini, M.; Vlek, P.L.G. The GLOWA Volta project: Integrated assessment of feedback mechanisms between climate, land use, and hydrology. In *Climatic Change: Implications for the Hydrological Cycle and for Water Management*; Beniston, M., Ed.; Springer: New York, 2002; 151–170.
47. Bouwman, L.; Kram, T.; Klein-Goldewijk, K. *Integrated Modelling of Global Environmental Change. An Overview of IMAGE 2.4*; Netherlands Environmental Assessment Agency: Bilthoven, 2006.
48. Harrington, R.; Carroll, P.; Cook, S.; Harrington, C.; Scholz, M.; McInnes, R.J. Integrated constructed wetlands: water management as a land-use issue, implementing the 'Ecosystem Approach.' Water Sci. Technol. **2011**, *63* (12), 2929–2937.
49. Van Vuuren, D.P.; Bayer, L.B.; Chuwah, C.; Ganzeveld, L.; Hazeleger, W.; van den Hurk, B.; van Noije, T.; O'Neill, B.; Strengers, B.J. A comprehensive view on climate change: Coupling of earth system and integrated assessment models. Environ. Res. Lett. **2012**, *7*, 024012.
50. Antonellini, M.; Dentinho, T.; Khattabi, A.; Masson, E.; Mollema, P.N.; Silva, V.; Silveira, P. An integrated methodology to assess future water resources under land use and climate change: An application to the Tahadart drainage basin (Morocco). Environ. Earth Sci. **2013**, *61* (2), 241–252, doi: 10.1007/s12665-013-2587-5.

Land Plants: Origin and Evolution

Elizabeth R. Waters
Department of Biology, San Diego State University, San Diego, California, U.S.A.

Abstract
The origin of land plants is one of the most important events in biotic history. It is now well established that the ancestor of land plants was a fresh-water green alga (a charophyte alga). Based on fossil evidence, the land plants or embryophytes diverged from charophytes at least 475 million years ago (MYA). Once plants became established on land, they altered the terrestrial environment, and in doing so generated new niches that plants and other organisms could then inhabit. Plants faced numerous challenges on land, and they now possess many adaptations for terrestrial life. All land plants retain and nourish an embryo, have a complex body plan including an alternation of generations between sporophytes and gametophytes, and possess multicellular sporangia, antheridia, and archegonia.

INTRODUCTION

The origin of land plants, an event that occurred at least 475 million years ago (MYA), is one of the most important events in biotic history. Once established on land, plants (embryophytes) transformed the terrestrial environment, and by doing so created numerous new niches for plants and other groups of organisms to exploit. It is now known that plants evolved once from fresh-water algae. The first land plants faced numerous challenges that included but were not limited to desiccation, temperature stress, and increased exposure to UV radiation. These stresses or challenges were the selective forces that led to a number of innovations or adaptations.

ADAPTATIONS OF LAND PLANTS

The land plant radiation is marked by a long list of morphological and molecular innovations and adaptations.[1,2] The morphological adaptations include a waxy cuticle and desiccation-resistant haploid spores. One of the major innovations in land plants is a complex body plan including an alternation of generations between sprophytes and gametophytes and the presence of multicellular sporangia, antheridia, and archegonia.[2,3] In recent years, considerable effort has been made to understand the molecular and biochemical adaptations associated with land plants. It is clear that there are many of these adaptations, including an increased complexity in phenolics and other secondary plant compounds.[2] Analysis of the complete genome of the moss *Physcomitrella patens* has revealed that the complexity of gene families associated with desiccation and other stress responses, photoreceptors, and transport proteins occurred early in land plant evolution.[4]

WHEN DID LAND PLANTS EVOLVE?

There is considerable fossil evidence that land plants originated in the Ordovician at least 475 MYA.[1,3,5] Wellman et al.[6] reported the presence of spore-containing plant fragments in Ordovician rocks from Oman. This finding is important because for the first-time spores (microfossils) that suggested an Ordovician origin of plants were associated with macroscopic plant fossils. In fact, the spore wall ultrastructure of these new fossils shows some similarity to extant liverworts. These data provide conclusive evidence that land plants existed at least 475 MYA and suggest that liverworts may be the most basal (earliest diverging) group of land plants (for more details, see further).

The importance of the origin of land plants in evolutionary history and the scarcity of early land plant fossils have motivated some scientists to estimate the time of origin of land plants using a molecular clock analysis of DNA and amino acids sequences. This work, however, has not been without controversy. In 2001, Heckman et al. proposed a date of at least 700 MYA for the origin of land plants.[7] This date has not been confirmed by either other molecular clock analyses or by fossil evidence. In contrast, using another set of molecular data, Sanderson[8] estimated the origin of land plants to be between 483 and 425 MYA. These dates are similar to that from the fossil evidence. The disparity between these two molecular age estimates (700 vs. ~475 MYA) is most likely due to methodological differences in these two studies. Heckman et al.[7] used

Encyclopedia of Natural Resources DOI: 10.1081/E-ENRL-120049195

proteins that are encoded in the nucleus and calibrated their tree with divergence dates between animal lineages. Sanderson[8] used plastid-coding genes, utilized methods that did not rely on a clock, and used a calibration point within the plant lineage.

In short, determining the exact time of past evolutionary events is always problematic, and the origin of land plants is particularly problematic due to the scarcity of molecular sequence data from a number of different algal and bryophyte lineages and the lack of many early plant megafossils. However, at this time the best-supported date for the origin of land plants is at least 475 MYA.

WHICH ALGAE ARE THE SISTER GROUP TO LAND PLANTS AND HOW ARE THE EARLIEST LAND PLANTS RELATED TO THE REST OF THE EMBRYOPHYTES?

In the past ten years, numerous phylogenetic analyses of the embryophytes have been published. In fact, there are so many analyses that they cannot all be cited here. Within this literature, there are many controversies and many areas where there is general agreement. First, there is considerable morphological and molecular evidence that land plants or embryophytes (Embryobionta), consisting of the liverworts, mosses, and hornworts ("Bryophytes") and the tracheophytes (vascular plants), are monophyletic.[1,9] There is also general agreement that the closest extant algal relatives to land plants are the charophyte algae and that the charophyte algae with the embryophytes form the streptophytes (Streptobionta), which are a monophyletic group.[10–13] In summary, this wealth of data indicates that land plants evolved once from a fresh-water charophyte alga (Fig. 1).

Due to the evolutionary relationship of the charophyte algae to the land plants, our understanding of which groups comprise the charophyte algae and how they are related to each other is highly relevant to our understanding of the origin of land plants. Recent progress has been made in some parts of the charophycean phylogenetic tree; however, there is still much that is currently equivocal. The charophyte algae sensu Mattox and Stewart[14] have traditionally contained five orders of freshwater algae: Charales, Coleochaetales, Zygnematales, Klebsormidiales, and Chlorokybales.[15] However, it is now clear that the Mesostigmatales also belong within the charophyte algae.[10] There is also considerable evidence that *Mesostigma* and *Chlorokybus* represent the earliest diverging lineages within the Charophytes[10,11,13] (Fig. 2). One active area of research is the determination of which charophyte lineage might be the sister group to land plants. Morphological analysis has suggested a close relationship between the Coleochaetales, the Charales, and land plants.[1,15] However, analysis of molecular data has been somewhat contradictory.[16–19] Some molecular phylogenies also support *Coleochatate* as the sister group to embryophytes,[17] while others support either the Zygnematales alone[16,18] or the Zygnematales plus

Fig. 1 Representatives of important algal and plant lineages: (**A**) charophytes alga Coleochaete, a member of the Coleochaetales; (**B**) charophytes alga Chara, a member of the Charales; and (**C**) Anthoceros, a hornwort; (**D**) Marchantia, a liverwort.

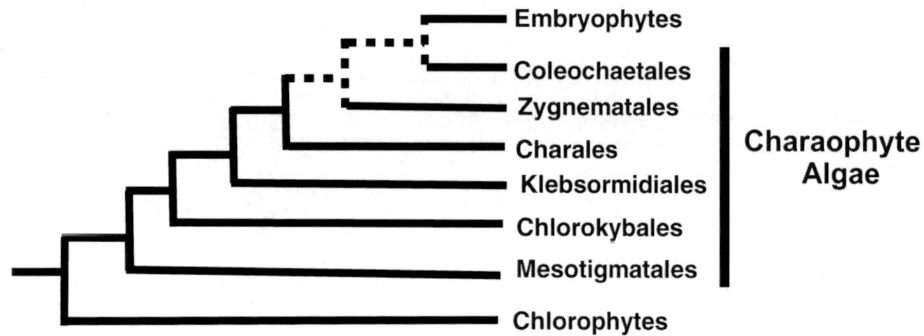

Fig. 2 Relationships of the charophyte algal lineages and land plants based on data and analysis found in (Sanderson,[8] Leliaert, et al.,[11] and Graham, et al.[12]). Phylogenetic relationships that are not strongly supported in all or most analyses are depicted by a dotted line.

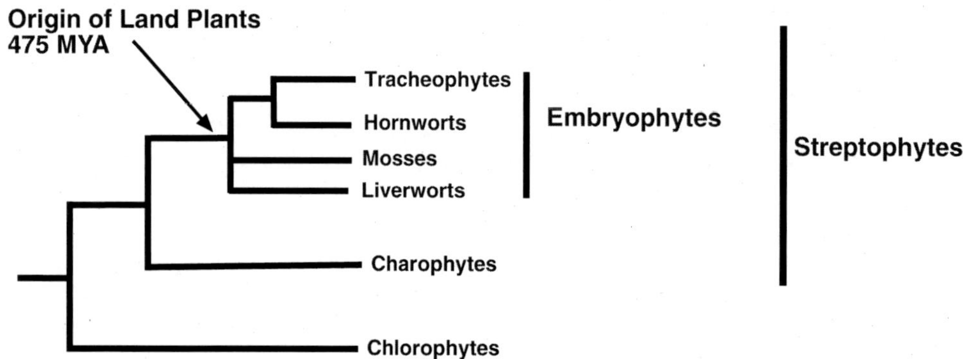

Fig. 3 Relationships of the Bryophyte (Liverwort, Hornwort and Moss) lineages to charophyte green algae and the tracheophytes based on data and analysis found in (Heckman, et al.,[7] Becker & Marin,[13] and Mattox & Stewart[14]).

Coleochaetate[19] as sister group to the embryophytes. All of these studies have included a large number of species and a wide range of DNA sequences. It is evident that the determination of which group(s) of charophytes is (are) the sister group to land plants is a very difficult problem. These evolutionary events occurred almost 500 MYA and the branching patterns among these groups may be short and difficulty to determine based on data from extant taxa. It is hoped that complete genome sequences of key taxa will soon become available and that these data will help shed light on the identity of the charophyte algae group that gave rise to land plants.

EVOLUTIONARY RELATIONSHIPS AMONG THE "BRYOPHYTES"

Determining the relationships among the extant land plant lineages and the identification of the lineage that branched off first in the evolution of land plants is crucial to our understanding of the transition from water to land. Thus, knowledge of the relationships among the basal land plant groups, the liverworts, hornworts, and mosses and the relationships of these groups to both the charophyte algae and the tracheophytes (vascular plants) is of great importance. There is clear agreement that all of the bryophytes are members of the embryophyte clade and that they are basal to the tracheophytes (vascular plants). However, it is also clear that the "Bryophytes" are a grade and are not a monophyletic group (Fig. 3).

There is now considerable evidence that the liverworts are the earliest lineages of extant land plants.[9,20–22] It is also now clear that the hornworts, and not the mosses as had been previously reported, are the sister group to the tracheophytes. Interestingly, the hornworts are the least diverse of the three nonvascular or bryophyte groups with just 14 genera and 200–240 species.[9] Phylogenies now place the mosses basal to both the hornworts and the tracheophytes (Fig. 3).

CONCLUSIONS

In summary, the embryophytes (Embryobionta) or land plants are a monophyletic group composed of the liverworts, mosses, hornworts, and vascular plants (tracheophytes). The origin of the land plants occurred at least 475 MYA, and their ancestor was a charophycean freshwater green alga, either a member of the Coleochaetales or the Zygnematales. The early evolution of the land plants is marked by a large number of important adaptations, including a desiccation-resistant cuticle, a complex body plan, the alternation of generations between sporophyte and gametophyte phases, the diversity of secondary plant compounds, and the increasing complexity of many gene and protein families.[4]

REFERENCES

1. Kenrick, P.; Crane, P.R. *The Origin and Early Diversification of Land Plants: A Cladistic Study*; Smithsonian Institution Press: Washington, DC, 1997; 441 pp.
2. Waters, E.R. Molecular adaptation and the origin of land plants. Mol. Phylogenet. Evol. **2003**, *29*, 456–463.
3. Graham, L.E.; Cook, M.E.; Busse, J.S. The origin of plants: Body plan changes contributing to a major evolutionary radiation. Proc. Natl. Acad. Sci. **2000**, *97*, 4535–4540.
4. Rensing, S.A.; Lang, D.; Zimmer, A.D.; Terry, A.; Salamov, A.; Shapiro, H.; Nishiyama, T.; Perroud, P.F.; Lindquist, E.A.; Kamisugi, Y.; Tanahashi, T.; Sakakibara, K.; Fujita, T.; Oishi, K.; Shin, I.T.; Kuroki, Y.; Toyoda, A.; Suzuki, Y.; Hashimoto, S.; Yamaguchi, K.; Sugano, S.; Kohara, Y.; Fujiyama, A.; Anterola, A.; Aoki, S.; Ashton, N.; Barbazuk, W.B.; Barker, E.; Bennetzen, J.L.; Blankenship, R.; Cho, S.H.; Dutcher, S.K.; Estelle, M.; Fawcett, J.A.; Gundlach, H.; Hanada, K.; Heyl, A.; Hicks, K.A.; Hughes, J.; Lohr, M.; Mayer, K.; Melkozernov, A.; Murata, T.; Nelson, D.R.; Pils, B.; Prigge, M.; Reiss, B.; Renner, T.; Rombauts, S.; Rushton, P.J.; Sanderfoot, A.; Schween, G.; Shiu, S.H.; Stueber, K.; Theodoulou, F.L.; Tu, H.; Van de Peer, Y.; Verrier, P.J.; Waters, E.; Wood, A.; Yang, L.; Cove, D.; Cuming, A.C.; Hasebe, M.; Lucas, S.; Mishler, B.D.; Reski, R.; Grigoriev, I.V.; Quatrano, R.S.; Boore, J.L. The Physcomitrella genome reveals evolutionary insights into the conquest of land by plants. Science **2008**, *319*, 64–69.
5. Gensel, P.G. The earliest land plants. Annu. Rev. Ecol. Evol. Syst. **2008**, *39*, 459–477.
6. Wellman, C.H.; Osterloff, P.L.; and Mohiuddin, U. Fragments of the earliest land plants. Nature **2003**, *425*, 282–285.
7. Heckman, D.S.; Geiser, D.M.; Eidell, B.R.; Stauffer, R.L.; Kardos, N.L.; Hedges, S.B. Molecular evidence for the early colonization of land by fungi and plants. Science **2001**, *293*, 1129–1133.
8. Sanderson, M.J.; Molecular data from 27 proteins do not support a precambrian origin of land plants. Amer. J. Bot. **2003**, *90*, 954–956.
9. Shaw, J.A.; Szoneyi, P.; Shaw, B. Bryophyte diversity and evolution: Windows into the early evolution of land plants. Am. J. Bot. **2010**, *98*, 352–369.
10. Lewis, L.A.; McCourt, R.M. Green algae and the origin of land plants. Am. J. Bot. **2004**, *91*, 15351556.
11. Leliaert, F.; Smith, D.R.; Moreau, H.; Herron, M.D.; Verbruggen, H.; Delwiche, C.F.; De Clerck, O. Phylogeny and molecular evolution of the green algae. Crit. Rev. Plant Sci. **2012**, *31*, 1–46.
12. Graham, J.; Wilcox, L.; Graham, L. *Algae*; 2nd ed.; Pearson: New York, 2008; 720.
13. Becker, B.; Marin, B. Streptophyte algae and the origin of embryophytes. Annu. Bot. **2009**, *103*, 999–1004.
14. Mattox, K.R.; Stewart, K.D. Classification of the green algae: A concept based on comparative cytology, In *Systematics of the Green Algae*, Irvine, G. E. G. and John, D.M. Eds.; Academic Press: London, 1984; 29–72 p.
15. Graham, L.E. *Origin of Land Plants*; John Wiley and Sons Inc.: New York, 1993; 287.
16. Chang, Y.; Graham S.W. Inferring the higher-order phylogeny of mosses (Bryophyta) and relatives using a large, multigene plastid data set. Am. J. Bot. **2011**, *98*, 839–849.
17. Finet, C.; Timme, R.E.; Delwiche, C.F.; Marletaz, F. Multigene phylogeny of the green lineage reveals the origin and diversification of land plants. Curr. Biol. **2010**, *20*, 2217–2222.
18. Timme, R.E., Bachvaroff, T.R.; Delwiche, C.F. Broad phylogenomic sampling and the sister lineage of land plants. PLoS One, **2012**, *7*, e29696.
19. Wodniok, S.; Brinkmann, H.; Glockner, G.; Heidel, A.J.; Philippe, H.; Melkonian, M.; Becker, B. Origin of land plants: do conjugating green algae hold the key? BMC Evol. Biol. **2011**, *104*, 11.
20. Qiu, Y.-L., Phylogeny and evolution of charophytic algae and land plants. J. Syst. Evol. **2008**, *46*, 287–306.
21. Qiu,Y-L.; Cho, Y.; Cox, J.C.; Palmer, J.D. The gain of three mitochondrial introns identifies liverworts as the earliest land plants. Nature **1998**, *394*, 671–674.
22. Qiu, Y-L; Li, L.B.; Wang, B.; Chen, Z.D.; Knoop, V.; Groth-Malonek, M.; Dombrovska, O. The deepest divergences in land plants inferred from phylogenomic evidence. PNAS **2006**, *103*, 15511–15516.

Land Surface Temperature: Remote Sensing

Yunyue Yu
Center for Satellite Applications and Research, National Environmental Satellite Data Information Service, National Oceanic and Atmospheric Administration (NOAA), College Park, Maryland, U.S.A.

Jeffrey P. Privette
National Climate Data Center, National Environmental Satellite Data Information Service, National Oceanic and Atmospheric Administration (NOAA), Asheville, North Carolina, U.S.A.

Yuling Liu
Cooperative Institute for Climate and Satellites, University of Maryland, College Park, Maryland, U.S.A.

Donglian Sun
Department of Geography and Geoinformation Science, George Mason University, Fairfax, Virginia, U.S.A.

Abstract

Land surface temperature (LST) is a key indicator of the Earth's surface energy and is used in a range of hydrological, meteorological, and climatological applications. Satellite remote sensed LST products, from geostationary-orbiting and polar-orbiting platforms, are the only resources providing sustainable regional and global coverage. Among LST measurements, those LSTs remote sensed from geostationary satellites can be up to every 5 minutes in temporal resolution and a few kilometers in spatial resolution; while that can be twice a day in temporal resolution and less than a kilometer in spatial resolution from the polar-orbiting satellites. This entry introduces techniques and products of the satellite remote sensed LST data.

INTRODUCTION

Most of the solar energy incident on the Earth is absorbed at the surface; this absorbed energy is used up in the evaporation of surface water or heating of the air near the surface and finally drives our weather and climate. Land Surface Temperature (LST) is a key indicator of the fraction of the absorbed energy. It is fundamental for estimating the net radiation budget at the Earth's surface in climate research as well as for monitoring the state of crops and vegetation in environmental studies. Long-term trends of LST are indicative of the rate of greenhouse warming. Therefore, LST data are widely utilized in a range of hydrological, meteorological, environmental management and climatological applications. Some of the most important applications of remote sensed LST data are to be assimilated into weather and climate models for the weather and climate predictions, and to be input into mesoscale atmospheric and land surface models to estimate sensible heat flux and latent heat flux. Commercial applications include the use of LST to evaluate water requirements for crops in summer and estimate where and when a damaging frost may occur in winter. High LST is a warning sign for possible forest and grass fires, as well as an indicator of possible drought. In 2011, World Meteorological Organization discussed to include LST as one of essential climate variables in Global Climate Observation System.[1]

Traditionally, LST is measured (or can be estimated from the measurements) from weather stations, which are usually located in relatively densely populated regions and are sparse and unevenly distributed; only by remote sensing of LST from satellites are feasible on a regional or global scale. Satellite LSTs have been routinely produced from geostationary and polar-orbiting satellites. As of 2012, more than 30 years of global satellite LST data had been accumulated. Geostationary satellites provide high temporal resolution (up to 15 minutes) LST data that are ideal for diurnal variation study, while polar-orbiting satellites produce high spatial resolution (up to 100 m) LST data that are valued for daily variation monitoring and climatology study.

PHYSICS OF LST REMOTE SENSING

Major techniques of LST remote sensing were inherited from sea surface temperature (SST) remote sensing. Compared to the SST remote sensing, however, LST remote sensing is more difficult in terms of derivation and validation, because of heterogeneity and anisotropy of land surface.

LST remote sensing is based on physics of radiative transfer process from Earth surface to satellite sensor. Infrared channels or microwave channels are used for the LST sensing. Advantage of using microwave channels is

Encyclopedia of Natural Resources DOI: 10.1081/E-ENRL-120049403

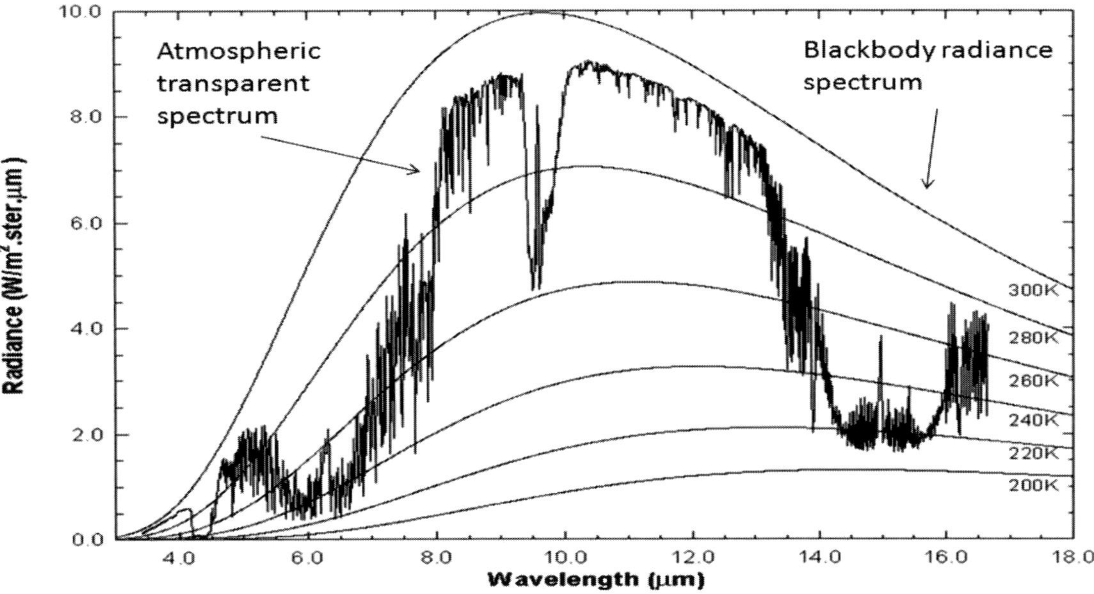

Fig. 1 Infrared spectral radiance distribution of black body at different temperatures (smooth curves) and spectral transmittance from Earth surface to top of atmosphere.

for its all-weather feature since the microwave band is atmospheric transparency. However, spatial resolution of the microwave LST remote sensing is very poor (about several tens of kilometers), and quality of the sensed LST is not so good because of its relatively low signal-to-noise ratio. Routinely, satellite LSTs are sensed from infrared channels within 3–12 μm spectrum where emission energy from Earth surface reaches the maximum. In particular, the best satellite LST measurements are performed in thermal infrared channels around 10–12 μm spectrum where atmospheric transmittance reaches the maximum as well (see Fig. 1). A disadvantage of the infrared LST remote sensing is that it must be under cloud-free sky condition as cloud is not transparent to infrared signal.

In infrared spectrum, satellite received radiance is mostly emission energy from Earth surface that is described by Planck's law:

$$B(\lambda,T) = \frac{2hc^2}{\lambda^5}\frac{1}{\frac{hc}{e\lambda\kappa T - 1}}$$

where B is the radiance at wavelength λ, T is the absolute temperature, k is the Boltzmann constant, h is the Planck constant, and c is the speed of light. Satellite received radiance can be explained in the following radiative transfer equation:

$$B(\lambda,T) = \varepsilon(\lambda)\tau(\lambda)B(\lambda,T_s) + I_{\text{atm}}(\lambda)^{\uparrow} + I_{\text{atm}}(\lambda)^{\downarrow}$$

where $B(\lambda,T_b)$ is the satellite sensor received radiance at wavelength λ, T_b, which is the so-called "brightness temperature" the satellite sensed, $\varepsilon(\lambda)$ and $\tau(\lambda)$ are the surface emissivity and the atmospheric transmittance at the wavelength, T_s is the surface temperature to retrieve; $I_{\text{atm}}(\lambda)^{\uparrow}$, and $I_{\text{atm}}(\lambda)^{\downarrow}$ are radiances of atmospheric emission and surface-reflected solar emission, respectively. Figure 2 illustrates the major components considered in the LST remote sensing.

All the LST remote sensing process is about how to derive surface temperature from the sensed brightness temperature. Note that satellite LST retrieval from infrared sensing is so-called "skin temperature" representing a few millimeter depth of land surface, as the infrared emission is blocked beyond the skin. Figure 3 shows a sample LST map derived from U.S. Earth Observation System (EOS) Terra satellite, processed at NASA Goddard Space Flight Center.

TECHNICAL APPROACHES

Difficulties of the LST retrieval are mostly due to nonlinearity of the Planck function, coupling of the emissivity and temperature in the radiative transfer equation, and correction of the atmospheric radiation and absorption. Linear relationship can be established between the remote sensed brightness temperature (T_b) and surface temperature (T_S) using Taylor expansion approach with the Planck function. This is a traditional regression technique of the LST remote sensing, in which coefficients of the linear relationship are determined from a training dataset, which can be generated from a radiative transfer model or from real measurements. Regression LST algorithms have been widely adopted in most satellite missions for LST production because of their simplicity and robustness.

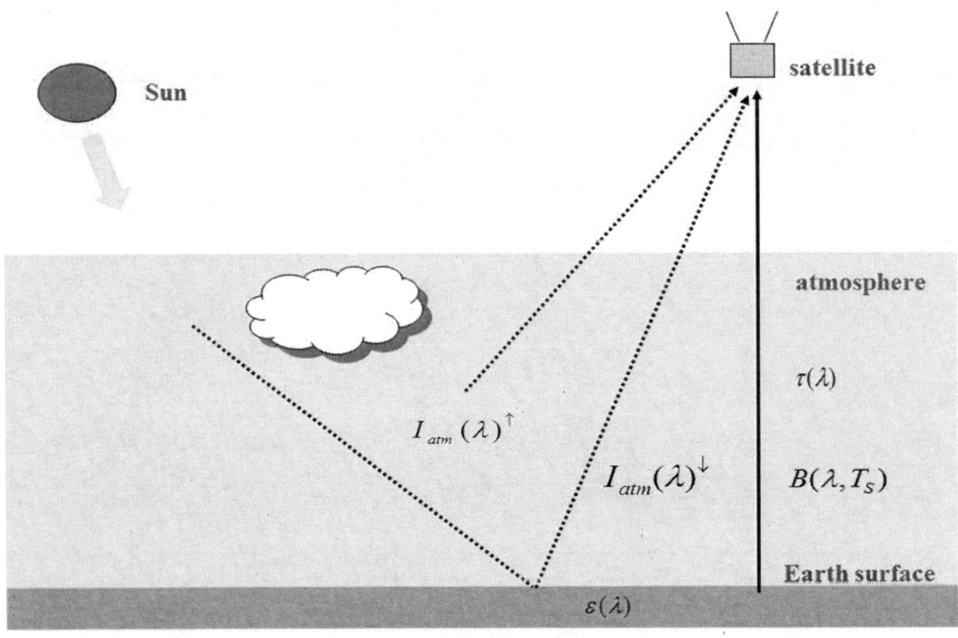

Fig. 2 A sketch of infrared radiances satellite sensor received: surface emitted radiance $B(\lambda,T_s)$, reflected atmospheric radiance $I_{atm}(\lambda)$, direct atmospheric radiance $I_{atm}(\lambda)$, surface emissivity $\varepsilon(\lambda)$, and atmospheric transmittance $\tau(\lambda)$.

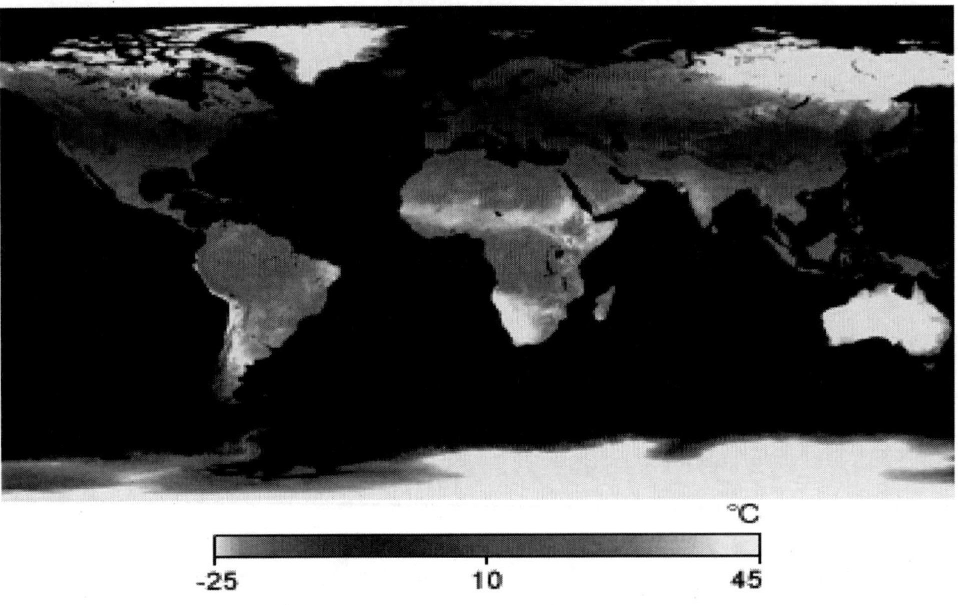

Fig. 3 Average global LST map obtained from the NASA Terra satellite MODIS (Moderate Resolution Imaging Spectroradiometer) sensor.
Source: From http://modis.gsfc.nasa.gov/gallery/individual.php?db_date=2009-04-30

The single-channel regression LST algorithm was developed in early 1980s,[2,3] which is a simple linear relationship between the satellite sensed brightness temperature at the thermal infrared channel and LST. Later, multichannel LST regression algorithms were developed, which in addition apply the shortwave infrared channels, in spectrum around 4 μm, for better atmospheric correction.[4,5] However, because of lower radiation energy compared to the thermal infrared spectrum (see Fig. 1), sensor noise in the shortwave infrared channels is a concern. A split-window sensor is specifically designed for solving the problem, in which the thermal infrared channel at 10–12 μm is split into two channels centered at around 10.5 and 11.5 μm, respectively. Different sensed brightness temperatures, but similar atmospheric transmittances, in the two adjacent channels provide an excellent measure of the atmospheric correction. Therefore, the split-window regression approach is the most popular technique for LST remote sensing,[6–8] as well as for SST remote sensing. The single-channel and multichannel regression LST algorithms have been successfully applied to a variety of satellite missions including the U.S. National Oceanic and Atmospheric Administration (NOAA) Polar-orbiting Operational Environmental Satellite (POES) series, Geostationary Operational Environmental Satellite (GOES) series, Feng-Yun meteorological satellite series of China, and European Organization for the Exploitation of Meteorological Satellite (EUMETSAT) series for operational LST productions.

Another approach of the atmospheric correction is to measure the LST from two different viewing angles. The difference of the satellite sensed brightness temperatures of the two angles, representing the atmospheric radiation/absorption difference along the two paths, provides excellent correction of the atmospheric impact.[9] The approach may be combined with the multichannel regression approach for optimal atmospheric correction. The multi-angle LST algorithm is applied to European Space Agency satellites, which are onboard a specifically designed sensor called Along Track Scanner Radiometer (ATSR) and later Advanced ATSR.[10,11]

Compared to the relative homogeneous ocean surface, emissivity of land surface may be significantly different from one surface species type to another. Coefficients of the regression LST algorithms can be emissivity dependent (such as the LST map shown in Fig. 3, derived from the MODIS sensor), or surface type dependent (such as the LST map shown in Fig. 4, derived from the Suomi-NPP VIIRS sensor), for correcting the emissivity variation impact.

Alternatively, atmospheric correction can be numerically computed using radiative transfer model. It requires a high performance radiative transfer model and prior accurate atmospheric profile information. In practice, such a radiative transfer model approach is applied in an emissivity–temperature separation technique,[12] in which the satellite-sensed brightness temperatures in a set of infrared channels are collected at two different times so that the number of spectral radiative transfer equations is doubled. Considering that LSTs vary significantly while land surface emissivity remains the same at those two different times, LSTs and land surface emissivities can be derived simultaneously by solving this well-posed problem. The approach is adopted in the U.S. EOS mission for the operational LST production.

VALIDATION AND EVALUATION OF REMOTE SENSED LST

Remote sensed LST needs to be validated before its applications. A four-stage validation process is defined by a land product validation subgroup of the Committee on Earth Observation Satellites working group on calibration and validation.[13] Traditionally, LST validation is performed by comparing the in situ LST measurements from ground stations or specific field campaigns. There are three major difficulties in the LST validation. First, within a satellite-sensed pixel, which is around a kilometer size in the infrared measurement, LST can vary significantly because of surface heterogeneity. In situ LST measurement must be evaluated or calibrated for its representativeness before it can be used for the satellite LST validation. Second, LST is

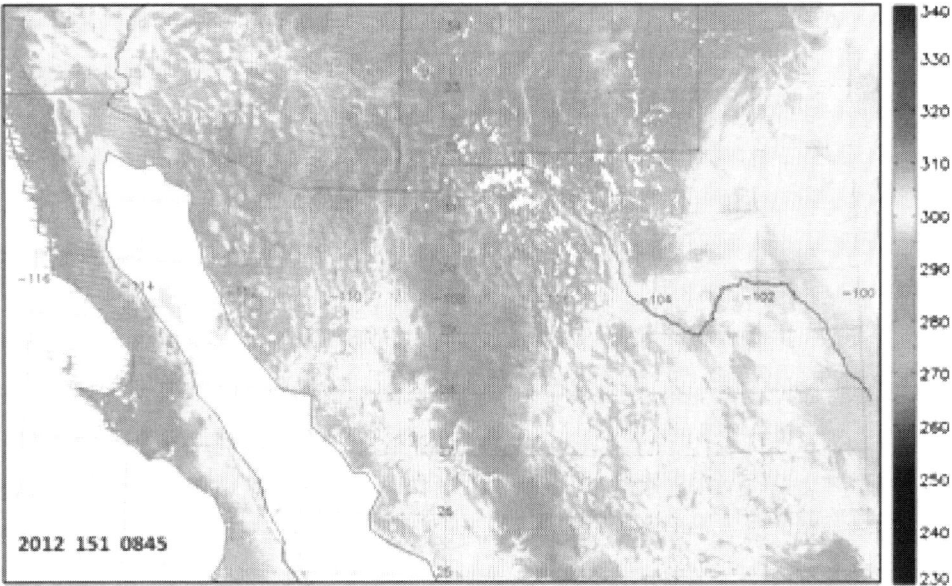

Fig. 4 Land surface temperature map of the west coast of United States.
Source: Obtained from Suomi-NPP satellite, VIIRS (visible and infrared imager radiometer suite) sensor at 8:45 am, May 30, 2012.

coupled with surface emissivity in the satellite sensed signal. Direct or indirect prior knowledge of the surface emissivity is required, which is not easy to obtain. Third, LST may have rapid temporal variation so that in situ LST measurements must be well matched to the satellite measurements. A comprehensive in situ validation of LST is impossible if those problems cannot be solved.

The radiative transfer model can be used for satellite LST validation.[14] Assuming that high-quality atmospheric profile and surface emissivity property are available, the satellite-sensed brightness temperature can be calculated by setting a certain LST as the model input. Validating the satellite LST is to compare it to the set LST. It is believed that an error of the radiative transfer model in such process is ignorable.

Instead, satellite LST is mostly evaluated rather than validated using limited high-quality in situ data.[15] Also, LST data from a new satellite are usually evaluated using existing satellite LST products;[16] it is considered as well evaluated if the new LST data are close enough to a LST product that is well validated or evaluated. A successful application is also a good evaluation of the satellite LST.

CONCLUSION

LST is listed as a baseline climate data record. For example, the U.S. National Climate Data Center uses global and hemispheric LST data as important indicators in its monthly report of global climate highlight.[17] Satellite remote sensed LST data are the only resource for this purpose. Infrared imaging LST measurement is the most popular manner for its accuracy and high spatial resolution. The accuracy requirement of the remote sensed LST data for many applications is about 1°C. Such a requirement is feasible through techniques (i.e., sensors and algorithms) developed in the past decades. However, validation of the satellite LST products is rather hard, mainly due to the variety, heterogeneity and emission isotropy of land surface. Satellite remote sensed LST products are widely applied in climatology, terrestrial and environment studies. The success of these applications lies in the best validation of the remote sensed LSTs.

REFERENCES

1. Designation of Essential Climate Variables, GCOS SC-XIX, Doc. 10.1, GCOS Steering Committee Nineteenth Session, 2011.
2. Price, J.C. Estimation of surface temperatures from satellite thermal infrared data - a simple formulation for the atmospheric effect. Remote Sens. Environ. **1983**, *13*, 353–361.
3. Susskind, J.; Rosenfield, J.; Reuter, D.; Chahine, M.T. Remote sensing of weather and climate parameters from HIRS2/Msu on TIROS-N. J. Geophys. Res. **1984**, *89*, 4677–4697.
4. McClain, E.P.; Pichel, W.G.; Walton, C.C. Comparative performance of AVHRR-based multichannel sea surface temperatures. J. Geophys. Res. **1985**, *90* (11), 587–11, 601.
5. Ulivieri, C.; Cannizzaro, G. Land surface temperature retrievals from satellite measurements. Acta Astronaut. **1985**, *12* (12), pp. 985–997.
6. McMillin, L.M. Estimation of sea surface temperature from two infrared window measurements with different absorption. J. Geophys. Res. **1975**, *80*, 5113–5117.
7. Price, J.C. Land surface temperature measurements from the split window channels of the NOAA-7/AVHRR. J. Geophys. Res. **1984**, *89*, 7231–7237.
8. Wan, Z.; Dozier, J. A Generalized split-window algorithm for retrieving land-surface temperature from space, IEEE Trans. Geosci. Remote Sens. **1996**, *34*, 892–905.
9. Chcdin, A.; Scott, N.; Berroir, A. A single-channel double viewing method for SST determination from coincident Meteosat and TIROS-N measurements. J. Appl. Meteorol. **1982**, *21*, 613–618.
10. Prata, A.J. Land surface temperatures derived from the AVHRR and the ATSR. 1, Theory J. Geophys. Res. **1993**, *98* (D9), 16,689–16,702.
11. Sobrino, J.A.; Li, Z.-L.; Stoll, M.P.; Becker, F. Multi-channel and multi-angle algorithms for estimating sea and land surface temperature with ATSR data. Int. J. Remote Sens. **1996**, *17*, 2089–2114.
12. Wan, Z.; Li, Z.-L. A physics-based algorithm for land-surface emissivity and temperature from EOS/MODIS data. IEEE Trans. Geosci. Remote Sens. **1997**, *35*, 980–996.
13. CEOS Working Group on Calibration and Validation, Land Product Validation Subgroup Website: http://lpvs.gsfc.nasa.gov/
14. Coll, C.; Wan, Z.; Galve, J.M. Temperature-based and radiance-based validations of the V5 MODIS land surface temperature product. J. Geophys. Res. **2009**, D20102, doi: 10.1029/2009JD012038.
15. Yu, Y.; Tarpley, D.; Privette, J.L.; Flynn, L.E.; Xu, H.; Chen, M.; Vinnikov, K.Y.; Sun, D.L.; Tian, Y.H. Validation of goes-r satellite land surface temperature algorithm using surfrad ground measurements and statistical estimates of error properties. IEEE Trans. Geosci. Remote Sens. **2012**, *50* (3), 704–713.
16. Yu, Y.; Privette, J.L.; Pinheiro, A.C. Analysis of the NPOESS VIIRS land surface temperature algorithm using MODIS data. IEEE Trans. Geosci. Remote Sens. **2005**, *43* (10), 2340–2350.
17. NOAA National Climate Data Center, State of Climate Global Analysis, http://www.ncdc.noaa.gov/sotc/global

Landscape Connectivity and Ecological Effects

Marinés de la Peña-Domene
Department of Biological Sciences, University of Illinois at Chicago, Chicago, Illinois, U.S.A.
Emily S. Minor
Department of Biological Sciences and Institute for Environmental Science and Policy, University of Illinois at Chicago, Chicago, Illinois, U.S.A.

Abstract
Landscape connectivity is the degree to which the landscape facilitates or impedes movement among resource patches. This movement is crucial for a number of different ecological processes including migration, dispersal, and colonization of locally extinct habitat patches. Hence, landscape connectivity has been of great interest to ecologists for at least two decades. Landscape connectivity is studied at the level of individual habitat patches as well as at the much larger landscape scale; it is modeled and empirically observed. Many different metrics have been developed to study the effect of landscape connectivity on a variety of organisms. Here, we give a brief overview of the different ways landscape connectivity has been defined and measured and highlight some important findings about the impact of connectivity on ecological and evolutionary processes, conservation, and natural resource management.

INTRODUCTION

Landscape connectivity is broadly defined as the "degree to which the landscape facilitates or impedes movement among resource patches."[1] Movement across the landscape is crucial for a number of different ecological processes including migration,[2] dispersal,[3,4] and colonization of locally extinct habitat patches.[5,6] These movements have serious implications for gene flow and long-term evolutionary processes[7–9] as well as biological conservation.[10,11] For that reason, landscape connectivity has been of great interest to ecologists for at least two decades.[1,12,13] However, it has also been the source of confusion because, as Calabrese and Fagan[14] wrote, "connectivity comes in multiple flavors." Here, we give a brief overview of the different ways landscape connectivity has been defined and measured and highlight some important findings about the impact of connectivity on ecological processes, conservation, and natural resource management.

DEFINITIONS

Structural versus Functional Connectivity

Structural connectivity is based on the amount and physical configuration of landscape elements, including habitat patches, corridors, and stepping stones (Table 1, Fig. 1). These elements are typically thought of as being embedded in a matrix of unsuitable habitat, through which movement may be limited or restricted.[15] Some studies have suggested that corridors or stepping stones can be established to increase connectivity between habitat patches.[3,16–20] However, these landscape elements may facilitate movement of some organisms more than others or more movement in some landscapes than others.[17,21,22] *Functional connectivity* incorporates an organism's behavior or response to elements such as corridors or stepping stones. Functional connectivity may also depend on a species' response to the matrix between landscape elements.[23,24] For example, corridors may be particularly important for species with a low capacity to move through the matrix.[25] Therefore, functional connectivity is species specific, and a single landscape can have different levels of connectivity depending on the species or process in question. Habitat that is functionally connected may not be structurally connected, and vice versa.[26]

Patch- versus Landscape-Level Connectivity

Connectivity can be measured at many different scales. On one hand, researchers or managers may want specific information about the connectivity of a single patch to the rest of the landscape. For example, studies of American pika (*Ochotona princeps*) have used incidence function models (IFMs)[27] to predict patch occupancy and population turnover in a large network of habitat patches.[28] On the other hand, researchers or managers may have questions about the connectivity of the entire landscape, perhaps to compare one landscape to another, or perhaps to identify whether

Encyclopedia of Natural Resources DOI: 10.1081/E-ENRL-120047451

Table 1 Landscape elements that affect landscape connectivity

Element	Definition
Habitat patch	A contiguous area of relatively homogeneous land cover that is suitable for the species of interest.
Corridor	Continuous strips of habitat that structurally connect two or more patches.
Stepping stone	Ecologically suitable patch (typically small in size) that may act to connect larger patches of habitat. Stepping stones may function at two different time scales. First, at the shortest time scale, they may provide a temporary stopping location for an organism moving through the landscape. Second, at the longer time scale, they may provide a breeding location so that an organism's offspring can move to more distant patches.
Matrix	The intervening area between habitat patches in a landscape mosaic, usually characterized by being the most extensive cover, having high spatial continuity, or having a major influence on the landscape dynamic.

Source: Modified from Forman.[15]

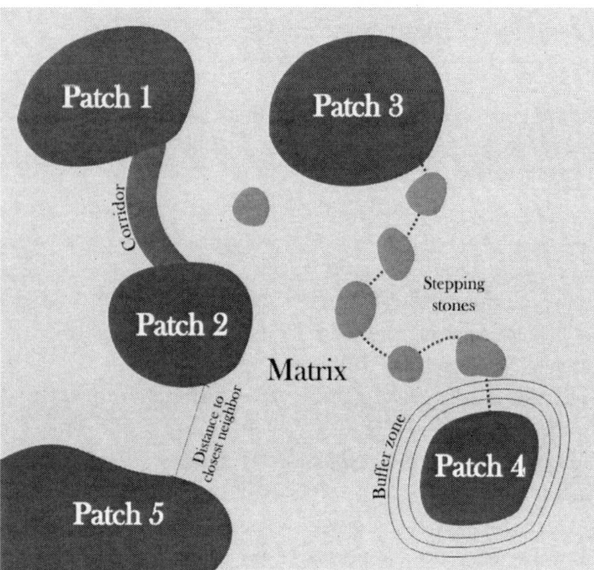

Fig. 1 Landscape structure and elements.

connectivity is sufficient for a species of concern. For instance, a study of protected area networks in the United States found that certain ecoregions had greater landscape-level connectivity than others and that protected area networks were consistently more connected for large mammals than for small mammals.[29] Different metrics are used for patch- and landscape-level connectivity, and we describe some of these in the following section.

MEASURING LANDSCAPE CONNECTIVITY

Dozens of different metrics exist for measuring landscape connectivity, and a number of reviews have been published on the topic.[26,30–35] Metrics may describe structural or functional connectivity, at the patch or the landscape level, or a combination of both (summarized in Table 2). One common approach that combines both patch- and landscape-level analyses involves patch-removal "experiments," in which researchers calculate a landscape-level connectivity metric, simulate the removal of a particular patch, and then recalculate the landscape-level metric once that patch has been removed. The difference between the first connectivity estimation and the second can be used to rank individual patches in terms of their importance for connectivity across the entire landscape.[36,37]

In general, structural connectivity measures may be calculated simply by examining a map. One of the simplest measures of patch-level structural connectivity is nearest-neighbor distance, which is the Euclidean distance to the nearest patch.[2] This measure was shown to be a useful predictor of the incidence of a frugivorous beetle in forest fragments.[38] This simple measure can be infused with more ecological information by measuring distance to the nearest occupied patch rather than distance to any patch,[2] but the more comprehensive measure is not always an improvement on the simpler one.[38] Another common way to measure patch-level structural connectivity is to examine the amount of habitat in a buffer around the focal patch. This measure was found to be much better than nearest-neighbor distance for predicting colonization events in two butterfly species but was very sensitive to buffer size.[32] At the landscape level, structural connectivity can be measured, for example, based on percolation or lacunarity. In these approaches, the landscape is considered as a two-dimensional grid where each cell in the grid is classified as either habitat or non-habitat. Landscape connectivity is then measured as the physical connection of habitat, in the case of percolation,[39] or as the variability and size of inter-patch distances, in the case of lacunarity.[40] The lacunarity index was shown to be a good predictor of dispersal success for simulated organisms in fractal landscapes.[41]

Because functional connectivity is specific to the species or process of interest, it is necessary to know something about species' movement behavior to measure functional connectivity. Species may perceive and respond to landscape pattern differently according to their dispersal characteristics, their preferred habitat type, or other life history traits.[26] For example, seed-dispersing birds differ in their presence in remnant trees within the matrix based on their frugivory levels. Fruiting trees could represent stepping stones across the matrix for birds with a completely fruit-based

Table 2 A subset of metrics used to measure landscape connectivity

Metric	Scale	Connectivity type	References
Nearest-neighbor distance	Patch	Structural	2, 38
Buffer area	Patch	Structural	32
Incidence function models	Patch	Structural	27, 28
Lacunarity	Landscape	Structural	39, 40, 41
Network centrality metrics	Patch	Potential	59
Cohesion index	Landscape	Potential	44
Least-cost paths	Dispersal route	Potential	36
Multiple shortest paths	Dispersal route	Potential	67
Isolation by resistance	Dispersal route	Potential	65
Probability of connectivity (PC)	Landscape	Potential	63, 64
Cell or patch immigration	Patch or landscape	Actual	48
Animal homing time	Landscape	Actual	53

diet, but may not for birds with mixed diets.[42] Functional connectivity measures are usually considered to be superior to structural connectivity measures.

Depending on how it is measured, functional connectivity can be divided into *potential connectivity*, where information about the movement ability of the organisms is limited, and *actual connectivity*, where there is detailed movement data for the organism of interest.[14] Potential connectivity measures may be based on attributes such as body size of the animal or dispersal mechanism of the plant. At the patch level, measuring potential connectivity may be as simple as using an ecologically meaningful distance when counting nearest neighbors.[43] At the landscape level, potential connectivity can be measured using the cohesion index, for example. This index integrates habitat quality, amount and configuration of habitat, and permeability of the landscape matrix to indicate species persistence.[44] Potential connectivity of both patches and landscapes can be calculated based on resistance or cost surfaces, which represent the willingness of a focal organism to cross the environment between habitat patches.[45] This approach emphasizes the importance of the matrix to connectivity and is based on potential "costs" of movement (usually in terms of energetic expenditures or mortality risks) through different regions of the landscape. Cost surfaces have been used to identify potential conservation corridors for jaguars[46] and were shown to help predict patch occupancy for prairie dogs in Colorado, United States of America.[47] Dispersal routes connecting habitat patches can be identified using least-cost path tools. This estimates the route with the least resistance between two points; however, by identifying a single route, alternative paths with comparable resistance costs may be ignored. This approach has been extended to include multiple routes, with methods such as circuit theory or Multiple shortest paths (MSPs), which are described in greater detail at the end of this section.

Actual connectivity measures are based on empirical, often spatially explicit, information about the movement of a particular organism. At the patch level, actual connectivity can be measured as the number of immigrants into a patch.[48] At the landscape level, actual connectivity may be measured as the number of patches visited by an organism or movement rates across the landscape.[49,50] However, these approaches can present a paradox in that animals may move more frequently between patches in lower-quality habitat, counterintuitively resulting in higher connectivity indices in these less-desirable environments.[51] Another approach to measuring actual connectivity draws heavily from behavioral ecology. For instance, Bélisle[51] suggests several different experimental methods for determining the motivation underlying movement of individuals through the landscape, including translocations, playback experiments, and measuring giving-up densities. Playback and homing experiments have been used to study the effect of roads[52] and other barriers[53] on animal movement. More recently, playback techniques have been used to parameterize graph theory models (described below) to explain the occurrence pattern of an Atlantic rainforest bird.[54]

Functional connectivity can also be inferred from genetic information. Dispersal influences gene flow between subpopulations,[55,56] which results in genetic differences among organisms occupying different parts of the landscape. Landscape genetics is a rapidly growing field and will likely continue to make large contributions to measuring and understanding the consequences of landscape connectivity.[57]

Graph theory, also called network analysis, is a flexible method for measuring landscape connectivity that has gained traction over the last decade. A graph is a set of nodes connected by links, where a link between nodes indicates a connection between them. In landscape ecology, the nodes typically represent habitat patches, and links indicate dispersal between patches.[58] Commonly used metrics from graph theory include various centrality measures, which determine the importance of individual nodes in the graph (i.e., patch-level connectivity). Some examples include degree centrality, which measures the number of links of a given node (akin to the number of neighboring patches), and betweenness centrality, which measures the number of shortest paths that pass through a given patch.[59] Graph theory can measure both potential and actual connectivity, at either the patch or the landscape level or a combination of both. To date, most applications of graph theory define links based on either an ecologically relevant measure of Euclidean distance[60,61] or a distance that incorporates cost or resistance to movement.[36,62]

Graph-based metrics have also been developed around the concept of measuring habitat availability or reachability at a landscape level. In this approach, movement between and within patches is combined in a single metric that describes the ability of species to reach resources across the landscape whether those resources come from the same patch (intrapatch connectivity), from connected neighboring patches (interpatch connectivity), or from a combination of both. Habitat availability metrics combine topological features with ecological characteristics of landscape elements, which has helped to place connectivity considerations in a broader and more informative context for conservation management alternatives.[63,64]

Some graph theory-based methods explicitly model multiple paths between two points of interest, extending the least-cost path approach described earlier. Circuit theory, which comes from the field of electrical engineering,[65] can be used to model the movement of individuals across a landscape based on the idea of isolation by resistance[65] and incorporates the effect of the matrix on movement across the landscape. In circuit theory, landscape "circuits" are a kind of graph with links defined in terms of resistances between nodes. Circuit theory techniques can be combined with genetic information, for example, to examine the influence of landscape composition and configuration on gene flow.[66] Additional graph theory methods that account for multiple paths across the landscape include calculating conditional minimum transit costs (CMTCs) and MSPs. Both methods have been used to enable visualization of multiple dispersal routes that, together, are assumed to form a corridor.[67]

RELEVANCE OF LANDSCAPE CONNECTIVITY TO CONSERVATION, RESTORATION, AND EXOTIC SPECIES MANAGEMENT

Habitat fragmentation and loss put many species at risk of local or regional extinction,[68] and persistence of many plant and animal populations depends on their ability to recolonize distant habitat patches.[6] One consequence of habitat fragmentation is that isolated populations tend to lose fitness though inbreeding depression and a loss of genetic diversity.[66] This reduces the ability of populations to adapt to environmental changes and could result in an increased risk of local extinction.[69–72] Habitat fragmentation may also prevent species from shifting their range in the case of climate change.[71] Recent studies have focused on how climate change will affect dispersal of individuals through the landscape and how populations will shift their distributions. For example, based on their movement capacities, it has been estimated that populations of many mammal species will be vulnerable following climate change.[73]

Many authors have suggested that increasing landscape connectivity is one of the best options for conservation in the face of habitat loss and climate change (reviewed in[74]). Corridors and stepping stones have been suggested as one way to increase connectivity[20, 75] and have been shown to direct the movement of a number of different species.[17,76,77] Designing strategic networks of patches and corridors that allow for dispersal between environmentally similar habitats[78] or between different climatic areas based on expected changes in climates[79] may help to counterbalance the effects of climate change on natural populations. Where habitat fragmentation is prevalent, restoring functional connectivity in the landscape can broaden species distributions, rescue genetically isolated populations, and assist in the conservation of animal and plant species.[17,19,20,80–82] Connectivity measures based on graph theory in particular have been used to assist with conservation planning for many species, including the European bison[83] and the gray wolf,[84] and have also been applied to freshwater[85] and marine[86] environments.

One concern about increasing landscape connectivity is the potential for also increasing the risk of invasion from exotic species and pathogens.[87,88] However, the degree to which landscape configuration constrains the spread of an exotic species may depend on dispersal characteristics of the focal species. For example, species considered to be "invasive" and species with frequent long-distance dispersal events are likely to spread across a landscape regardless of the configuration of landscape elements.[89,90] Fortunately, the methods described earlier may be useful for predicting and managing the spread of exotic species. For example, Minor and Gardner[89] used graph theory to identify critical points on the landscape where management could help contain the spread of invasive plants. Similarly, Wang et al.[91] identified particular types and spatial arrangements of land cover that were conducive to the spread of the invasive rice water weevil (*Lissorhoptrus oryzophilus*) in eastern China.

CONCLUSION

There is a large and expanding body of literature on the topic of landscape connectivity. Connectivity is known to be important for a number of ecological processes and thus for long-term biological conservation. There are currently dozens of methods for measuring landscape connectivity and new methods are proposed on a regular basis. However, because movement is difficult to observe, and large-scale experiments are expensive and logistically challenging, the field has lagged behind on empirically testing the effect of landscape configuration on the movement of plants and animals. Therefore, much remains to be learned about how organisms move around the landscape, how these movements influence population processes and gene flow, and how we can improve connectivity for species of conservation concern while minimizing the movement of exotic species. Future research can help us to find the balance between "desirable" movement, like gene flow or seed and pollen dispersal of native species, with "undesirable" movement of invasive species and pathogens. In both cases, understanding how landscape connectivity influences population dynamics will allow us to identify better conservation strategies and management plans.

REFERENCES

1. Taylor, P.D.; Fahrig, L.; Henein, K.; Merriam, G. Connectivity is a vital element of landscape structure. Oikos, **1993**, *68* (3), 571–573.
2. Matter, S.F.; Roslin, T.; Roland, J. Predicting immigration of two species in contrasting landscapes: effects of scale, patch size and isolation. Oikos, **2005**, *111* (2), 359–367.
3. Levey, D.J.; Bolker, B.M.; Tewksbury, J.J.; Sargent, S; Haddad, N.M. Effects of landscape corridors on seed dispersal by birds. Science **2005**, *309* (5731), 146–148.
4. Matlack, G.R.; Leu, N.A. Persistence of dispersal-limited species in structured dynamic landscapes. Ecosystems, **2007**, *10* (8), 1287–1298.
5. Alexander, H.M.; Foster, B.L.; Ballantyne, F.; Collins, C.D.; Antonovics, Jet al., Metapopulations and metacommunities: combining spatial and temporal perspectives in plant ecology. J. Ecol. **2012**, *100* (1), 88–103.
6. Gustafson, E.J.; Gardner, R.H. The effect of landscape heterogeneity on the probability of patch colonization. Ecology, **1996**, *77* (1), 94–107.
7. Amos, J.N.; Bennett, A.F.; Nally, R.M.; Newell, G. Pavlova, A. Radford, J.Q.; Thomson, J.R.; White, M; Sunnucks, P. Predicting landscape-genetic consequences of habitat loss, fragmentation and mobility for multiple species of woodland birds. Plos One **2012**, *7* (2), e30888.
8. Björklund, M.; Bergek, S. Ranta, E. Kaitala, V. The effect of local population dynamics on patterns of isolation by distance. Ecol. Info. **2010**, *5* (3), 167–172.
9. Hardesty, B.D.; Hubbell, S.P.; Bermingham, E. Genetic evidence of frequent long-distance recruitment in a vertebrate-dispersed tree. Ecol. Lett. **2006**, *9* (5), 516–525.
10. Berger, J. The last mile: How to sustain long-distance migration in mammals. Conserv. Biol. **2004**, *18* (2), 320–331.
11. Fahrig, L.; Merriam, G. Conservation of fragmented populations. Conserv. Biol. **1994**, *8* (1), 50–59.
12. Henein, K.; Merriam, G. The elements of connectivity where corridor quality is variable. Landscape Ecol. **1990**, *4* (2), 157–170.
13. Fahrig, L.; Merriam, G. Habitat patch connectivity and population survival. Ecology **1985**, *66* (6), 1762–1768.
14. Calabrese, J.M.; Fagan, W.F. A comparison-shopper's guide to connectivity metrics. Front. Ecol. Environ. **2004**, *2* (10), 529–536.
15. Forman, R.T.T. *Land Mosaics. The Ecology of Landscapes and Regions;* Cambridge University Press: Cambridge; 1995.
16. Loehle, C. Effect of ephemeral stepping stones on metapopulations on fragmented landscapes. Ecol. Complexity **2007**, *4* (1–2), 42–47.
17. Baum, K.A.; Haynes, K.J.; Dillemuth, F.P.; Cronin, J.T. The matrix enhances the effectiveness of corridors and stepping stones. Ecology, **2004**, *85* (10), 2671–2676.
18. Pardini, R.; de Souza, S.M.; Braga-Neto, R.; Metzger, J.P. The role of forest structure, fragment size and corridors in maintaining small mammal abundance and diversity in an Atlantic forest landscape. Biol. Conserv. **2005**, *124* (2), 253–266.
19. Tewksbury, J.J.; Levey, D.J.; Haddad, N.M.; Sargent, S.; Orrock, J.L.; Weldon, A. Danielson, B.J.; Brinkerhoff, J.; Damschen, E.I.; Townsend, P. Corridors affect plants, animals, and their interactions in fragmented landscapes. Proc. Nat. Acad. Sci. **2002**, *99* (20), 12923–12926.
20. Lander, T.A.; Boshier, D.H.; Harris, S.A. Fragmented but not isolated: Contribution of single trees, small patches and long-distance pollen flow to genetic connectivity for Gomortega keule, an endangered Chilean tree. Biol. Conserv. **2010**, *143* (11), 2583–2590.
21. Uezu, A.; Metzger, J.P.; Vielliard, J.M.E. Effects of structural and functional connectivity and patch size on the abundance of seven Atlantic forest bird species. Biol. Conserv. **2005**, *123* (4), 507–519.
22. Haddad, N.M.; Bowne, D.R.; Cunningham, A.; Danielson, B.J.; Levey, D.J.; Sargent, S. Spira, T. Corridor use by diverse taxa. Ecology **2003**, *84* (3), 609–615.
23. Ricketts, T.H. The matrix matters: Effective isolation in fragmented landscapes. Am. Nat. **2001**, *158* (1), 87–99.
24. Prevedello, J.A.; Vieira, M.V. Does the type of matrix matter? A quantitative review of the evidence. Biodivers. Conserv. **2010**, *19* (5), 1205–1223.
25. Martensen, A.C.; Pimentel, R.G.; Metzger, J.P. Relative effects of fragment size and connectivity on bird community in the Atlantic Rain Forest: Implications for conservation. Biol. Conserv. **2008**, *141* (9), 2184–2192.
26. Tischendorf, L.; Fahrig, L. On the usage and measurement of landscape connectivity. Oikos, **2000**, *90* (1), 7–19.
27. Hanski, I. A practical model of metapopulation dynamics. J. Anim. Ecol. **1994**, *63* (1), 151–162.
28. Moilanen, A.; Smith, A.T.; Hanski, I. Long-term dynamics in a metapopulation of the American pika. Am. Nat. **1998**, *152* (4), 530–542.
29. Minor, E.S.; Lookingbill, T.R. A multiscale network analysis of protected-area connectivity for mammals in the United States. Conserv. Biol. **2010**, *24* (6), 1549–1558.
30. Kindlmann, P.; Burel, F. Connectivity measures: a review. Landscape Ecol. **2008**, *23* (8), 879–890.

31. Prugh, L.R. An evaluation of patch connectivity measures. Ecol. Appl. **2009**, *19* (5), 1300–1310.
32. Moilanen, A.; Nieminen, M. Simple connectivity measures in spatial ecology. Ecology **2002**, *83* (4), 1131–1145.
33. Kool, J.T.; Moilanen, A.; Treml, E.A. Population connectivity: recent advances and new perspectives. Landscape Ecol. **2013**, *28* (2), 165–185.
34. Galpern, P.; Manseau M.; Fall, A. Patch-based graphs of landscape connectivity: A guide to construction, analysis and application for conservation. Biol. Conserv. **2011**, *144* (1), 44–55.
35. Pascual-Hortal, L.; Saura, S. Comparison and development of new graph-based landscape connectivity indices: towards the priorization of habitat patches and corridors for conservation. Landscape Ecol. **2006**, *21* (7), 959–967.
36. Bunn, A.G.; Urban D.L.; Keitt, T.H. Landscape connectivity: A conservation application of graph theory. J. Environ. Manag. **2000**, *59* (4), 265–278.
37. Bodin, O.; Saura, S. Ranking individual habitat patches as connectivity providers: Integrating network analysis and patch removal experiments. Ecol. Model. **2010**, *221* (19), 2393–2405.
38. Kehler, D.; Bondrup-Nielsen, S. Effects of isolation on the occurrence of a fungivorous forest beetle, *Bolitotherus cornutus*, at different spatial scales in fragmented and continuous forests. Oikos **1999**, *84* (1), 35–43.
39. Gardner, R.H.; Milne, B.T.; Turner, M.G.; O'Neill, R.V. Neutral models for the analysis of broad-scale landscape pattern. Landscape Ecol. **1987**, *1* (1), 19–28.
40. Plotnick, R.E.; Gardner, R.H.; Oneill, R.V. Lacunarity indexes as measures of landscape texture. Landscape Ecol. **1993**, *8* (3), 201–211.
41. With, K.A.; King, A.W. Dispersal success on fractal landscapes: a consequence of lacunarity thresholds. Landscape Ecol. **1999**, *14* (1), 73–82.
42. Lasky, J.R.; Keitt, T.H. The effect of spatial structure of pasture tree cover on avian frugivores in Eastern Amazonia. Biotrop. **2012**, *44* (4), 489–497.
43. Hanski, I.; Alho, J.; Moilanen, A. Estimating the parameters of survival and migration of individuals in metapopulations. Ecol. **2000**, *81* (1), 239–251.
44. Opdam, P.; Verboom, J.; Pouwels, R. Landscape cohesion: an index for the conservation potential of landscapes for biodiversity. Landscape Ecol. **2003**, *18* (2), 113–126.
45. Zeller, K.; McGarigal, K.; Whiteley, A. Estimating landscape resistance to movement: a review. Landscape Ecol. **2012**, *27* (6), 777–797.
46. Rabinowitz, A.; Zeller, K.A. A range-wide model of landscape connectivity and conservation for the jaguar, Panthera onca. Biol. Conserv. **2010**, *143* (4), 939–945.
47. Magle, S.; Theobald, D.; Crooks, K. A comparison of metrics predicting landscape connectivity for a highly interactive species along an urban gradient in Colorado, USA. Landscape Ecol. **2009**, *24* (2), 267–280.
48. Tischendorf, L.; Fahrig, L. How should we measure landscape connectivity? Landscape Ecol. **2000**, *15* (7), 633–641.
49. Jonsen, I.D.; Taylor, P.D. Fine-scale movement behaviors of calopterygid damselflies are influenced by landscape structure: an experimental manipulation. Oikos **2000**, *88* (3), 553–562.
50. Eycott, A.; Stewart, G.B.; Buyung-Ali, L.M.; Bowler, D.E.; Watts, K.; Pullin, A.S. A meta-analysis on the impact of different matrix structures on species movement rates. Landscape Ecol. **2012**, *27* (9), 1–16.
51. Bélisle, M. Measuring landscape connectivity: The challenge of behavioral landscape ecology. Ecology **2005**, *86* (8), 1988–1995.
52. Develey, P.F.; Stouffer, P.C. Effects of roads on movements by understory birds in mixed-species flocks in central Amazonian Brazil. Conserv. Biol. **2001**, *15* (5), 1416–1422.
53. Bélisle, M.; St. Clair, C.C. Cumulative effects of barriers on the movements of forest birds. Conserv. Ecol. **2002**, *5* (2), 9.
54. Awade, M.; Boscolo, D.; Metzger, J.P. Using binary and probabilistic habitat availability indices derived from graph theory to model bird occurrence in fragmented forests. Landscape Ecol. **2012**, *27* (2), 185–198.
55. Ibrahim, K.M.; Nichols, R.A.; Hewitt, G.M. Spatial patterns of genetic variation generated by different forms of dispersal during range expansion. Heredity **1996**, *77*, 282–291.
56. Sork, V.L.; Smouse, P.E. Genetic analysis of landscape connectivity in tree populations. Landscape Ecol. **2006**, *21* (6), 821–836.
57. Storfer, A.; Murphy, M.A.; Spear, S.F.; Holderegger, R.; Waits, L.P. Landscape genetics: where are we now? Molecular Ecol. **2010**, *19* (17), 3496–3514.
58. Urban, D.L.; Minor, E.S.; Treml, E.A.; Schick, R.S. Graph models of habitat mosaics. Ecol. Lett. **2009**, *12* (3), 260–273.
59. Estrada, E.; Bodin, O. Using network centrality measures to manage landscape connectivity. Ecol. Appl. **2008**, *18* (7), 1810–1825.
60. Minor, E.S.; Urban, D.L. Graph theory as a proxy for spatially explicit population models in conservation planning. Ecol. Appl. **2007**, *17* (6), 1771–1782.
61. Pascual-Hortal, L.; Saura, S. Integrating landscape connectivity in broad-scale forest planning through a new graph-based habitat availability methodology: application to capercaillie (*Tetrao urogallus*) in Catalonia (NE Spain). Eur. J. Forest Res. **2008**, *127* (1), 23–31.
62. Decout, S.; Manel, S.; Miaud, C.; Luque, S. Integrative approach for landscape-based graph connectivity analysis: a case study with the common frog (*Rana temporaria*) in human-dominated landscapes. Landscape Ecol. **2012**, *27* (2), 267–279.
63. Saura, S.; Rubio, L. A common currency for the different ways in which patches and links can contribute to habitat availability and connectivity in the landscape. Ecogr. **2010**, *33* (3), 523–537.
64. Luque, S.; Saura, S.; Fortin, M.-J. Landscape connectivity analysis for conservation: insights from combining new methods with ecological and genetic data PREFACE. Landscape Ecol. **2012**, *27* (2), 153–157.
65. McRae, B.H.; Dickson, B.G.; Keitt, T.H.; Shah, V.B. Using circuit theory to model connectivity in ecology, evolution, and conservation. Ecology **2008**, *89* (10), 2712–2724.
66. Manel, S.; Schwartz, M.K.; Luikart, G.; Taberlet, P. Landscape genetics: combining landscape ecology and population genetics. Trends Ecol. Amp. Evo. **2003**, *18* (4), 189–197.

67. Pinto, N.; Keitt, T.H. Beyond the least-cost path: evaluating corridor redundancy using a graph-theoretic approach. Landscape Ecol. **2009**, *24* (2), 253–266.
68. Fischer, J.; Lindenmayer, D.B. Landscape modification and habitat fragmentation: a synthesis. Global Ecol. Biogeogr. **2007**, *16* (3), 265–280.
69. Frankham, R.; Ralls, D.K. Conservation biology: inbreeding leads to extinction. Nature **1998**, *392* (6675), 441–442.
70. Frankham, R.; Genetics and extinction. Biol. Conserv. **2005**, *126* (2), 131–140.
71. Davis, M.B.; Shaw, R.G. Range shifts and adaptive responses to quaternary climate change. Science **2001**, *292* (5517), 673–679.
72. Opdam, P.; Wascher, D. Climate change meets habitat fragmentation: linking landscape and biogeographical scale levels in research and conservation. Biol. Conserv. **2004**, *117* (3), 285–297.
73. Schloss, C.A.; Nunez, T.A.; Lawler, J.J. Dispersal will limit ability of mammals to track climate change in the Western Hemisphere. Proc. Nat. Acad. Sci. Am. **2012**, *109* (22), 8606–8611.
74. Heller, N.E.; Zavaleta, E.S. Biodiversity management in the face of climate change: A review of 22 years of recommendations. Biol. Conserv. **2009**, *142* (1), 14–32.
75. Beier, P.; Noss, R.F. Do habitat corridors provide connectivity? Conserv. Biol. **1998**, *12* (6), 1241–1252.
76. Brudvig, L.A.; Damschen, E.I.; Tweksbury, J.J.; Haddad, N.M.; Levey, D.J. Landscape connectivity promotes plant biodiversity spillover into non-target habitats. Proc. Nat. Acad. Sci. Am. **2009**, *106* (23), 9328–9332.
77. Herrera, J.M.; Garcia, D. The role of remnant trees in seed dispersal through the matrix: Being alone is not always so sad. Biol. Conserv. **2009**, *142* (1), 149–158.
78. Alagador, D.; Trivino, M.; Cerdeira, J.O.; Brás, R.; Cabeza, M.; Araújo, M.B. Linking like with like: optimising connectivity between environmentally-similar habitats. Landscape Ecol. **2012,** *27* (2), 291–301.
79. Nunez, T.A.; Lawler, J.J.; McRae, B.H.; Pierce, D.J.; Krosby, M.B.; Kavanagh, D.M.; Singleton, P.H.; Tewksbury, J.J. Connectivity planning to address climate change. Conserv. Biol. **2013**, *27* (2), 407–416.
80. Powell, G.V.N.; Bjork, R. Implications of intratropical migration on reserve design: a case study using *Pharomachrus mocinno*. Conserv. Biol. **1995**, *9* (2), 354–362.
81. Vogt, P.; Ferrri, J.P.; Lookingbill, T.R.; Gardner, R.H.; Riitters, K.H.; Ostapowicz, K. Mapping functional connectivity. Ecol. Indicators **2009**, *9* (1), 64–71.
82. Martínez-Garza, C.; Howe, H.F. Restoring tropical diversity: beating the time tax on species loss. J. Appl. Ecol. **2003**, *40*, 423–429.
83. Ziółkowska, E.; Ostapowicz, K.; Kuemmerle, T.; Perzanowski, K.; Radeloff, V.C.; Kozak, J. Potential habitat connectivity of European bison (*Bison bonasus*) in the Carpathians. Biol. Conserv. **2012**, *146* (1), 188–196.
84. Carroll, C.; McRae, B.H.; Brookes, A. Use of linkage mapping and centrality analysis across habitat gradients to conserve connectivity of gray wolf populations in Western North America. Conserv. Biol. **2012**, *26* (1), 78–87.
85. Erös, T.; Olden, J.D.; Schick, R.S.; Schmera, D.; Fortin, M.-J. Characterizing connectivity relationships in freshwaters using patch-based graphs. Landscape Ecol. **2012**, *27* (2), 303–317.
86. Treml, E.A.; Halpin, P.N.; Urban, D.L.; Pratson, L.F. Modeling population connectivity by ocean currents, a graph-theoretic approach for marine conservation. Landscape Ecol. **2008**, *23*, 19–36.
87. Jules, E.S.; Kauffman, M.J.; Ritts, W.D.; Carroll, A.L. Spread of an invasive pathogen over a variable landscape: A nonnative root rot on Port Orford cedar. Ecology **2002**, *83* (11), 3167–3181.
88. Hess, G.R. Conservation corridors and contagious-disease-a cautionary note. Conserv. Biol. **1994**, *8* (1), 256–262.
89. Minor, E.S.; Gardner, R.H. Landscape connectivity and seed dispersal characteristics inform the best management strategy for exotic plants. Ecol. Appl. **2011**, *21* (3), 739–749.
90. Minor, E.S.; Tessel, S.M.; Engelhardt, K.A.M.; Lookingbill, T.R. The role of landscape connectivity in assembling exotic plant communities: a network analysis. Ecology **2009**, *90* (7), 1802–1809.
91. Wang, Z.; Wu, J.; Shang, H.; Cheng, J. Landscape connectivity shapes the spread pattern of the rice water weevil: a case study from Zhejiang, China. Environ. Manage. **2011**, *47* (2), 254–262.

Landscape Dynamics, Disturbance, and Succession

Hong S. He
School of Natural Resources, University of Missouri, Columbia, Missouri, U.S.A.

Abstract
A landscape is a place where natural and human forces interact. Both forces lead to landscape change over large spatial extents and long temporal spans. These forces are usually referred as landscape disturbances. This entry discusses both natural and anthropogenic disturbances and their roles in affecting landscape dynamics. The disturbances along with other spatial processes, such as seed dispersal and exotic species invasion, are termed landscape processes. Landscape processes are spatially continuous processes directly related to landscape position, spatial heterogeneity, and patch geometrics and adjacencies. They operate across a range of spatial extents (10^3–10^6 ha) and temporal spans (10^1–10^3 year) and are often the main forces shaping the landscapes and interact with landscape succession. Because the interaction of spatial and temporal factors across the landscape can be so complicated that it is beyond human comprehension, computer simulation modeling becomes a useful tool for understanding future trajectories of landscape dynamics. The lack of management experience at the landscape scales and the limited feasibility of conducting landscape-scale experiments have resulted in increasing the use of scenario modeling to analyze the effects of different management actions on focal forests and wildlife species.

INTRODUCTION

Traditionally, a landscape is considered to be an expanse of scenery extending as far as the eye can see.[1] It usually implies an area so large that a person cannot readily traverse it on foot. With the advancement of technologies such as remote sensing and geographical information systems, the capacity to acquire, process, and analyze spatial information has been greatly expanded. Consequently, landscapes analyzed in contemporary scientific literature can cover millions of hectares, expanding the geographic extent of what is considered a landscape.[2]

Within a landscape, there may be numerous components such as forest, grassland, water bodies, houses, and roads. Or there may be one primary component divided into subclasses, such as forest divided into subclasses of young, mature, or old-growth forest. How to classify a landscape depends on the purpose of the study. For example, for an urban sprawl study, a landscape may be classified into different housing density classes;[3] for a wildlife habitat study, a landscape may be classified into habitat, hospitable matrix, and inhospitable matrix;[4] for a study of land cover change, a landscape may be classified as forest, pasture, cropland, and other.[5]

A landscape is a place where natural and human forces interact. Both forces lead to landscape changes (succession) over large spatial extents and long temporal spans. These forces are usually referred to as landscape disturbances and investigated as landscape processes by scientific communities.[6] In this entry, I discuss landscape succession, disturbances, landscape processes, and landscape modeling.

LANDSCAPE SUCCESSION

Succession is a series change from one stage to the next. As landscape components change over time landscape conditions change, which is often referred to as *landscape succession*. Landscape succession usually occurs slowly over decades or centuries and is driven by natural or anthropogenic forces or both. For example, in the Changbai Mountains of Northeast China, early successional aspen and birch dominate the open lands resulted from volcanic activities, after about 40–50 years mid-successional tree species such as maple, oak, and basswood replace aspen and birch to become dominant. In about another 40–50 years the maple, oak, and basswood will be outcompeted by late successional Korean pine, which will reach mixed Korean pine hardwood climax status and will maintain the status for the next 300 years.[7] In this example, young forests grow to become mature forests, mature forests grow to become old-growth forests, and old-growth forests gradually regenerate to start a new forest succession cycle. The whole cycle takes hundreds of years and the main drivers are natural forces.

Increase in human population and the associated need for natural resources drive expansion, creation, and re-creation

Encyclopedia of Natural Resources DOI: 10.1081/E-ENRL-120047454

of human communities worldwide, and infrastructure developments also lead to more impervious surfaces. Through mapping historical impervious surface, researchers can reveal urban development for a region or state. For example, an urban sprawl study in the state of Missouri found that from 1980 to 2000, 129,853.2 ha of land were converted to impervious surface.[8] Although sprawl was very prominent on the urban fringe during 1980s in major metropolitan areas, the trend shifted to the rural landscapes in the 1990s and 2000s. Sprawl in rural areas (also called rural sprawl) may have greater impact on ecosystems due to the low density of development and larger affected areas. Increase in population and the expansion of human society can eventually cause a natural landscape to be transformed into an urban landscape.

Landscape changes are measured from two perspectives: landscape composition and landscape configuration. The former quantifies the proportions of various components (classes), the evenness, dominance, and diversity of the study landscape. The latter characterizes spatial arrangement of various landscape components such as degree of fragmentation, aggregation, association, and connectivity. The common indices for measuring landscape composition include fraction or proportion, relative richness, Shannon's Diversity index, and Shannon's Evenness index. The common indices for measuring landscape configuration include contagion, aggregation index, connectivity, and proximity index. Most landscape indices can be found in FRAGSTATS, a software package developed by McGarigal et al.[9]

LANDSCAPE DISTURBANCE

Disturbances include environmental fluctuations and destructive events in an otherwise relatively stable system.[10] It is often characterized by a set of parameters, including distribution (spatial, geographic, topographic, environmental, and community gradients), frequency (mean number of events per time period), return interval (mean time needed to disturb an area equivalent to the study area), mean size (average area per event), intensity (physical force of the event per time such as rate of spread), and severity (impact on the organism, community, ecosystem, or landscape). Disturbance can have endogenous and exogenous causes, whereas in the former, change is driven by the biological properties of the system (e.g., insect susceptibility due to aging), and in the latter an outside driving force (e.g., extreme weather) is present. Disturbance can be either anthropogenic or natural, whereas the former is always exogenous and the latter is mostly endogenous.

Anthropogenic disturbances include forest harvesting, grazing, agriculture, or residential development. Their effects are gradual and cumulative, and are often profound on natural landscapes. Anthropogenic disturbances are found to be the main causes of landscape fragmentation, loss of species diversity, altered hydrological functions, or other types of natural resource degradation.

Natural disturbances include extreme weather, landslides, wildfires, insects, and diseases, which are forces that may cause abrupt changes in natural landscapes. They can change the path of landscape succession in many ways. For example, fire disturbance is an integral part of many forest ecosystems. In boreal forests (e.g., larch, spruce, fir forests), high-intensity crown fire can kill most of the trees where they occur and reset the successional stage to that of a newly regenerating forest.[11] In central hardwood forests (e.g., oak-pine forests), low-intensity ground fires can reduce understory density, increase herbaceous species diversity beneath a mature forest overstory, kill pathogens, and ultimately maintain a healthy condition for forest to grow and regenerate.[12] Many studies report that natural disturbances are crucial in maintaining landscape heterogeneity, which in turn determines biodiversity.

LANDSCAPE PROCESSES

Biological organizations occur in a hierarchical manner across a wide spectrum of spatiotemporal scales. At a given level of resolution, a biological system is composed of interacting components (i.e., lower-level entities) and is itself a component of a larger system (i.e., higher level entity).[13] An equally wide range of ecological processes is associated with ecological components at each hierarchical level. For example, stomata conductance is an ecophysiological process at individual leaf level; growth and aging are at an individual organism level; competition is at the population level; inter-species competition is at the community level; water and nutrient cycling is at the ecosystem level; dispersal and fragmentation are at the landscape level; and adaptation, speciation, and extinction are at the biome level.[13] Processes at one scale often interact with processes at lower and higher scales.[14] An increasing level of organization leads to the increase in magnitude of both spatial and temporal scales. For example, at a single leaf scale, photosynthesis occurs on the order of minutes in a few square centimeters space, at a biome scale, species range distribution changes in response to climatic change, and species migration and extinction occur at a hundreds to thousands of years timeframe over a large region.[13]

In fact, any spatially continuous processes, including the above-mentioned natural and anthropogenic disturbances that are directly related to landscape position, spatial heterogeneity, and patch geometrics and adjacencies are landscape processes.[15] These processes operate across a range of spatial extents (10^3–10^6 ha) and temporal spans (10^1–10^3 year) and are often the main forces shaping the landscapes we have today.

LANDSCAPE MODELING

The challenges in predicting landscape change come from two fundamental aspects: the relevance of both long temporal and broad spatial dimensions of landscape processes. Temporally, a landscape may take hundreds of years to undergo significant successional change. Processes that operate on such long time spans may go undetected by many field experiments, which are often based on relatively short observation periods that may not capture the full range of the events. Spatially, landscape change can be strongly affected for centuries by environment heterogeneity; the current spatial pattern resulted from past human land use and natural disturbances. The interaction of spatial and temporal factors across the landscape can be so complicated that it is beyond human comprehension. Thus, computer simulation modeling becomes a useful tool for understanding these large (10^4–10^7 ha), long-term, complex systems. With modeling techniques, it is possible to describe the modeled components and relationships mathematically and logically and deduce results that cannot otherwise be investigated.[16,17]

A specific definition of a landscape model is one that simulates spatiotemporal characteristics of at least one recurrent landscape process in a spatially interactive manner.[15] The term spatially interactive means that a simulated entity (e.g., a pixel or polygon) is a function of neighboring, or spatially related, entities. A landscape model under the specific definition has the following characteristics: 1) it is a simulation model, 2) it simulates one or more landscape processes repeatedly, and (c) it operates at a large spatial and temporal extent that is adequate to simulate the landscape process.

Landscape models applications generally fall in the three categories: (1) spatiotemporal patterns of landscape processes, (2) sensitivities of model object to input parameters, and (3) comparisons of model simulation scenarios.[15]

The direct outputs of landscape models are the spatiotemporal patterns of the model objects because the modeled spatial processes are stochastic and complex, and understanding their manifestations over space and time is necessary. Thus, the most effective method is to simulate spatiotemporal patterns of the model objects using built-in model relationships and parameters related to the model objects.

The lack of management experience at the landscape scales and the limited feasibility of conducting landscape-scale experiments have resulted in increasing use of scenario modeling to analyze the effects of different management actions on focal forests and wildlife species. Model scenarios are created by altering input parameters to reflect changes in climate, disturbance, fuel and harvest alternatives. The built-in model relationships remain unchanged. Comparing results from different model scenarios provides relative measurements regarding the direction and magnitude of changes within the simulated landscape.

CONCLUSIONS

Landscape will continue to be a unit in nature where natural and anthropogenic disturbances operate and interact. Such interactions may alter natural landscape succession, lead to landscape fragmentation, result in loss of species diversity, reduce hydrological functions, and cause other types of natural resource degradation. Because of the large spatiotemporal dimensions and numerous variables and processes involved in landscape succession, landscape modeling becomes a necessary tool in defining modeled components and relationships mathematically and logically and deducing results that cannot otherwise be derived. Landscape models will be increasingly used in predicting the future landscape change under various climate change and management scenarios.

ACKNOWLEDGMENT

I thank Stephen Shifley for the helpful comments, which improved the clarity of this entry. Funding support of this work is come Missouri GIS Mission Enhancement Program and Chinese Academy of Sciences.

REFERENCES

1. Troll, C. *The Geographic Landscape and Its Investigation.* Translated by Conny Davidsen, Studium Generale, **1950**, *3*, no. 4/5, 163–181.
2. Riitters, K.H.; O'Neill, R.V.; Hunsaker, C.T.; Wickham, J.D.; Yankee, D.H.; Timmins, S.P.; Jones, K.B.; Jackson, B.L. A factor analysis of landscape pattern and structure metrics. Landscape Ecol. **1995**, *10* (1), 23–39.
3. Radeloff, V.C.; Hammer, R.B.; Stewart, S.I. Rural and suburban sprawl in the U.S. Midwest from 1940 to 2000 and its relation to forest fragmentation. Conserv. Biol. **2005**, *19* (3), 793–805.
4. Tischendorf, L. Can landscape indices predict ecological processes consistently? Landscape Ecol. **2001**, *16* (3), 235–254.
5. Lambin, E.F.; Turner, B.L.; Geist, H.J.; Agbola, S.B.; Angelsen, A.; Bruce, J.W.; Coomes, O.T.; Dirzo, R.; Fischer, G.; Folke, C.; George, P.S.; Homewood, K.; Imbernon, J.; Leemans, R.; Li, X.; Moran, E.F.; Mortimore, M.; Ramakrishnan, P.S.; Richards, J.F.; Skånes, H.; Steffen, W.; Stone, G.D.; Svedin, U.; Veldkamp, T.A.; Vogel, C.; Xu, J. The causes of land-use and land-cover change: Moving beyond the myths. Global Environ. Change **2001**, *11* (4), 261–269.
6. Turner, M.G. Landscape ecology: the effect of pattern on process. Annu. Rev. Ecol. syst. **1989**, *20*, 171–197.
7. Wang, Z.; Xu, Z.; Tan, Z.; Dai, H.; Li, X. The main forest types and their features of community structure in Northern slope of Changbai Mountain. Forest Ecol. Res. **1980**, *1*, 25–42.
8. Zhou, B.; He, H.S.; Nigh, T. A.; Schulz, J.H. Mapping and analyzing change of impervious surface for two decades using multi-temporal Landsat imagery in Missouri. Int. J. Appl. Earth Observ. Geoinf. **2012**, *18*, 195–206.
9. McGarigal, K.; Cushman, S.A.; Ene, E. FRAGSTATS v4: Spatial Pattern Analysis Program for Categorical and Continuous Maps. Computer software program produced by the authors at the University of Massachusetts, Amherst, 2012

Land-Use and Land-Cover Change (LULCC)

Burak Güneralp
Department of Geography, Texas A&M University, College Station, Texas, U.S.A.

Abstract
Land-use and land-cover change (LULCC) is an integral component of global environmental change. Land cover refers to the physical characteristics of Earth's surface. Land use refers to how the land is utilized by humans for various purposes. There is a large body of both theoretically and empirically grounded work on LULCC that falls under the umbrella of the relatively new discipline of land-change science. The biophysical processes and human activities interact in changing land use and land cover. The critical aspects underlying much of these recent debates on LULCC are further separation of the landscapes of consumption and those of production, and the ultimate scarcity of land. Although there is no overarching theory of land change, there have been many theories put forward to explain land-change dynamics. Land-change models, informed by the theoretical body on land-change processes, are essential in evaluating various scenarios of societal and biophysical changes in terms of land-change outcomes. The greatest challenge in the coming decades will be the formulation of an overarching theory of land change.

INTRODUCTION

Land-use and land-cover change (LULCC), or more succinctly land change, is an integral component of global environmental change.[1] Land cover refers to the physical characteristics of Earth's surface such as water, grass, trees, or concrete. Land use, on the other hand, refers to how the land is utilized by humans for various social and economic purposes. The same land cover may be used for different purposes. For example, a forest may be used for timber extraction or set aside as a protected area for conservation of wildlife. Land-use change refers to a change in the management of land by humans, which may or may not lead to land cover-change. LULCC is important for biogeochemical cycles such as nutrient cycling,[2,3] for biodiversity through its impacts on habitats,[4,5] and for climate by changing sources and sinks of greenhouse gases and land-surface properties.[6,7] LULCC also has important implications for the provision of ecosystem services on which both rural and urban societies depend.[8]

There is a large body of both theoretically and empirically grounded work on LULCC. These studies fall under the umbrella of the relatively new but growing discipline of land-change science. Several international science efforts such as the Global Land Project,[9] a joint research project of the International Geosphere-Biosphere Programme (http://www.igbp.net) and the International Human Dimensions Programme (http://www.ihdp.org), are devoted to issues revolving around LULCC in recognition of the pivotal role it plays in the functioning of the Earth system. These issues are treated at length in the many books and reports published specifically on the subject of LULCC.[10–20]

CAUSES AND CONSEQUENCES

The processes that lead to land change are numerous. These processes in many cases interact and collectively lead to larger impacts on the land cover than they would individually.[21,22] Although biophysical processes and predominant climatic conditions broadly determine the land cover in a particular location, human activities play a key role in creating the observed land-use and land-cover patterns (Fig. 1) and may even create land uses that could not be supported by the climatic conditions at that locale (e.g., irrigated agriculture in extremely arid environments).

Over geological time scales, atmospheric, hydrologic, and tectonic processes continuously alter the land cover across the surface of Earth. Starting with the Paleolithic Age, and especially with the use of fire, humans have had an increasing role in land change on the face of Earth.[23] In particular, the changes in land use and land cover due to human activities have dramatically accelerated with the onset of agrarianization of human societies about 12,000 years ago. Another accelerated phase of LULCC with much larger geographical spread was initiated about 300 years ago with the onset of the Industrial Revolution; then yet again, another surge after the 1960s and the Green Revolution.

Encyclopedia of Natural Resources DOI: 10.1081/E-ENRL-120047478

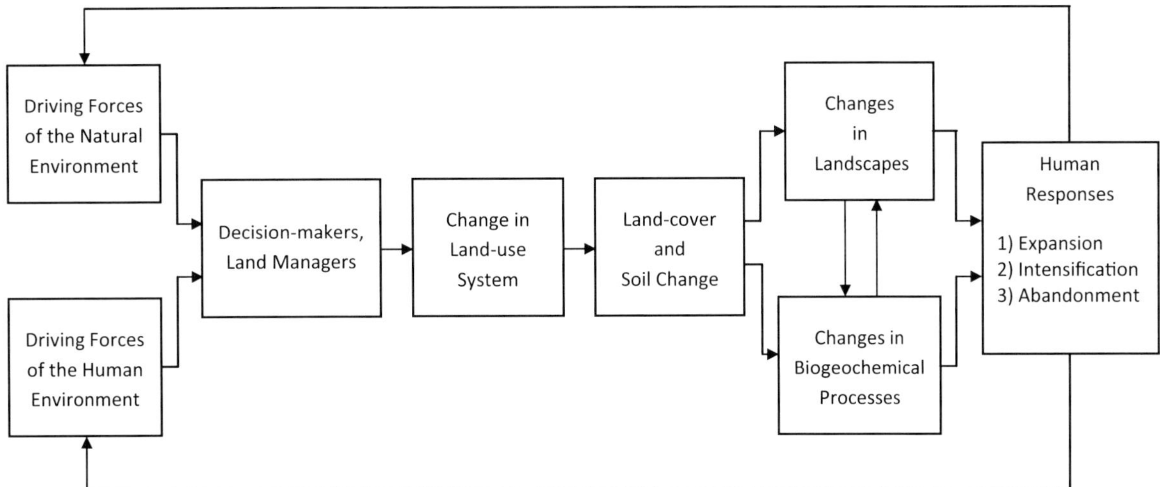

Fig. 1 Conceptual framework showing the interaction of biophysical and human-related factors creating land-use and land-cover change. © The University of Arizona Press.
Source: Adapted from Redman.[23]

Whenever humans are involved, land change may become a highly political activity[24] involving—across various spatial and temporal scales—institutional as well as social and economic dynamics in interaction with biophysical dynamics. Within these webs of interactions, one way to explain land change is to refer to proximate and underlying causes. Proximate (or direct) causes are those human activities and immediate actions that originate from the intended manipulation of land cover, whereas underlying (or indirect) causes are fundamental processes that compel the more proximate processes.[25,26] Thus, proximate causes are generally limited to a small set of activities such as agricultural expansion or infrastructure development, whereas underlying causes can consist of, depending on the context of land change, many factors—biophysical, demographic, economic, technological, institutional, and cultural—acting across multiple spatial and temporal scales. These factors can play varying and often interacting roles in creating the observed land-use and land-cover patterns on the ground[27] that may be either in the form of changing one land cover to another or changing the biophysical conditions of the existing land cover (i.e., land-use modification).[28] Modification of existing land uses leads to subtle changes in the character of the land but not necessarily a change in the land-use classification (e.g., agricultural intensification).

There are relatively few global-scale studies on long-term historical changes in land cover that are due to human activities. These studies harbor significant uncertainties in their estimates; nevertheless, they offer a comprehensive perspective.[29] According to one estimate, 75% of the ice-free surface of Earth has undergone noticeable land change due to human activities.[28] Another suggests that between A.D. 1700 and 2000, 42–68% of the global land surface was impacted by various land-use activities.[30] Up to half of Earth's net primary productivity is estimated, albeit with large error bounds, to be appropriated every year by humans as a result of these land-use activities.[31] A study that reports historical changes in both natural vegetation and agricultural land cover estimates that nearly 5 million km^2 of natural vegetation were converted to agriculture (both cropland and pasture) between A.D. 800 and 1700 (Table 1).[32] In comparison, the agricultural lands are estimated to have increased more than 40 million km^2 between A.D. 1700 and 1992.[32,33]

Deforestation has historically been a prominent feature of human-induced land-cover change.[34] By the time of European colonization, in much of Eurasia, extensive forested areas had already been cut down either to meet timber and fuel wood demand or to open space for croplands and pastures. Evidence suggests that tropical forests in Central and South America had witnessed periodic and localized deforestation with the rise and fall of many civilizations.[23,35,36] However, across much of the world, deforestation rates increased with the onset of colonization and especially after the onset of the Industrial Revolution. Much of this deforestation had its ultimate cause in the demand originating from Europe and also from North America for forest products as well as for agro-industrial produce such as coffee, bananas, and rubber. More recently, palm plantations, ranching, and soybean production have appeared as the more dominant proximate causes of ongoing deforestation due to increased demand not only from Europe and North America but also from rapidly developing countries such as India and China.[37] From the late 19th to the early 20th century, there has been a trend of reforestation in many of the previously deforested lands of Europe and New England,[38] which is facilitated by the timber and food imports. There are also extensive afforestation efforts in previously deforested lands across the world with huge efforts expended to this end in China. These reforestation and afforestation dynamics are linked in complicated ways

Table 1 A recent estimate of global extent of land-cover types over time[a]

		Year				
Land cover	**Potential**	**800**	**1100**	**1400**	**1700**	**1992**
Tropical forest	22.44	22.03	21.87	21.70	21.22	16.30
Temperate broadleaf forest	10.48	10.15	9.92	9.87	9.26	5.95
Temperate needleleaf forest	15.75	15.61	15.51	15.47	15.17	12.36
Total forest	*48.68*	*47.78*	*47.31*	*47.05*	*45.65*	*34.60*
Grass and shrubs	46.79	44.90	44.22	43.84	42.14	13.63
Tundra	4.08	4.07	4.06	4.06	4.05	2.94
Total natural vegetation	*99.55*	*96.75*	*95.60*	*94.95*	*91.84*	*51.16*
Crop	—	1.36	1.97	2.33	4.01	18.76
Pasture	—	1.44	1.98	2.27	3.70	29.63
Total agriculture	—	*2.80*	*3.95*	*4.60*	*7.71*	*48.39*

[a]Units are in 10^6 km^2.
Source: Adapted from Pongratz et al.[31]

to the ongoing deforestation in much of the tropical belt as well as boreal forests in Asia.[39,40] However, these afforestation efforts do not always take place in the same geographical locations where deforestation had occurred. Also, the newly forested landscapes are not necessarily composed of similar species composition and/or richness as the pristine forests they replace.[41–43]

Most of Earth's original grassland and savanna cover have been converted to cropland or grazing land since humans took up agriculture about 10,000 years ago. Broadly following the accelerated increase in human populations, croplands increased dramatically in every continent except Antarctica. Especially during the past 300 years, large areas of grasslands have been converted to croplands in North America and Eurasia.[44] By one estimate, half of the total land converted to croplands since A.D. 1700 has been savanna-like vegetation.[45] Grasslands of South America are undergoing a similar transformation fueled by the growing demand for soybeans from China and the European Union.[37] Intensification of agriculture on croplands can partly meet the increased demand for food, but this needs to be done with utmost care to biodiversity and ecosystem services.[46–49] Although some of these croplands such as the rice paddies in China have generated high yields for millennia, intensive cultivation practices and heavy grazing may lead to severe land degradation; indeed, such practices may have contributed to the shift of large expanses to semiarid and even arid conditions across much of the Middle East, Central Asia, and Africa that were already predisposed due to the particular geophysical characteristics and larger-scale climatic changes in those regions.[50,51] About 20% of global drylands, which constitutes about 41% of global ice-free land area, is undergoing degradation.[52] However, as is typical in studies on land-use and land-cover dynamics, it is challenging to attribute causality to individual factors. For example, it took the innovative use of remotely sensed data to show that the dreaded expansion of the Sahara as widely believed until the 1980s is actually part of the multiyear cycle of expansion and contraction of the desert.[53,54]

Urban areas cover less than 3% of the ice-free land surface and yet are home to more than half the total human population. In addition to the direct impacts of urbanization such as expansion of urban land,[55,56] which is predominantly composed of impervious surfaces across the landscape, urbanization in a particular location may also influence land change elsewhere indirectly through demands for natural resources such as sand, gravel, and timber for both construction as well as demand for food.[37,57–59] Although urbanization leads to depopulation of extensive rural areas as more people move to urban areas that are relatively more densely populated[60] urbanization cannot by itself save land for nature because of the increased demands for natural resources that are needed to support urban activities[57,59] and also because subsequent suburbanization may again reverse the trend.[61]

Recent debates on LULCC are dominated by the expansion of croplands due to increased demand for food—driven primarily by dietary changes of more affluent and more urban populations but also by the sheer increase in the world populace[58] as well as the potential increase in demand for biofuels,[62] large-scale land leases in less developed countries by their more developed counterparts,[38] future changes in the extent and species composition of forests,[63,64] and impacts on coastal settlements of sea-level rise.[65,66] The critical aspects underlying much of these debates are the unprecedented urbanization, increasing occurrence of indirect land-use change symptomized by the further separation of the landscapes of consumption and those of production, and the ultimate scarcity of land prompting more efficient uses of this resource—all of which are inter-related in complicated, even complex ways.[38,59,67]

THEORETICAL FOUNDATIONS

Although there is no overarching theory of land change, there have been many theories put forward to explain land-change dynamics.[68,69] While acknowledging the complicated nature of interactions among various processes that span multiple spatial and temporal scales to create the observed patterns of land use and land cover, each of these theories tends to highlight a particular subset of these processes and their interactions in bringing an explanation for the LULCC phenomenon in different places.[1] Thus, theories of land change collectively take a broad range of phenomena as their foci including, but not limited to, complexity and resilience concepts,[70] behavioral approaches based on economic theoretical frameworks,[71] globalization,[72] and institutions.[73]

Theorizing efforts can also be grouped under those focusing on particular types of land changes such as tropical deforestation, forest transition, desertification, agricultural change, and urban land-use change.[74–78] Yet, any empirical work, usually a case study, on land change is typically guided by more than one theory,[79] thus drawing upon the explanatory power of multiple perspectives.[80] The specific theories informing a particular case study are mainly determined by the disciplinary backgrounds of the researchers involved.[68] There are also differences in the theories of land change in the developed vs. developing world reflecting the differences between the two in terms of both the processes in question and the land-change outcomes.[24,81] None of these theories claim to have an explanatory power across the full spectrum of land-change dynamics. However, all bring something to the table that not only illuminates the particular conditions leading to observed land-change dynamics but also informs to an extent the land-change processes that are at play in other situations and places.

An overarching theory of land change will most likely bring together many aspects of these theories.[68] These include interactions within and between the biophysical and the human-related processes across spatial and temporal scales. Thus, it should reconcile the connections of local and regional with the broader world in which they are situated. Such a grand theory needs to be dynamic, meaning that it should explain land changes that occurred in the past as well as those that are being experienced in the present. The dynamic perspective will also have the desirable feature of allowing for anticipating potential land-change trajectories in the future in particular places under various scenarios pertaining to broader-scale changes.

MONITORING AND MODELING

Satellite-based observations have transformed monitoring land cover as they provide a consistent way to observe the change in land cover both across space and over time. Among the numerous space-based land-change monitoring programs, Landsat by the National Aeronautics and Space Administration (NASA) of the United States, in operation since 1972, is the longest running. Among other satellites whose images are widely utilized to monitor land-change are the SPOT by the French corporation Spot Image and IRS by the Indian Space Research Organisation (ISRO).[82] In addition to the medium-resolution sensors on these satellites, there are sensors with lower spatial resolution—but with higher temporal resolution—such as MODIS that are used for large regional and global monitoring of land-cover change. There are also several commercial satellites that provide very high-resolution imagery such as IKONOS and Quickbird. These high-resolution images allow for very detailed land-cover classification; however, being very expensive, they are generally used for calibration and validation of land-cover maps derived from Landsat or other moderate-resolution sensors.[83,84]

However, satellite-based monitoring of land change faces a number of limitations; foremost among these are, depending on the application, challenges in terms of the spatial, temporal, or radiometric resolution, inability to differentiate different land uses, and its availability only for the past few decades.[82] Because of these limitations, ancillary information such as aerial photography, survey maps, household surveys, census data, and official statistics are still frequently used along with or instead of satellite-based data.[85,71] Novel methods such as those for fusion of information from sensors with different spatial, temporal, and radiometric resolutions are also explored to overcome some of these limitations.[86,87]

Land-change models help organize the present knowledge about observed land-change dynamics in particular places, and they are essential in evaluating various scenarios regarding demographic, economic, institutional, and biophysical changes in terms of land-change outcomes in the future. As such, these models both are informed by and potentially inform the body of theory on land change.[71,88] In line with the multitude of interacting processes that are involved in land change in different parts of the world, there is a wide range of modeling approaches to capture the different aspects of the dynamics of land use and land cover depending on the research questions. The particular methods employed reflect their theoretical underpinnings, which span from economic theoretical frameworks to the complexity theory.[70,71] Methodologically, these approaches span a wide spectrum from econometric to agent-based to cellular automata (CA) models[89,90] and can broadly be divided into three classes in terms of how they deal with interacting processes across spatial scales: i) *Top-down* methods emphasize the influence of broader-scale processes such as regional economic or demographic changes on land-use patterns. There are many CA models[91,92] as examples of this type of methods; ii) *bottom-up* methods emphasize the emergent properties of the land-change system from the interaction of many individual entities such as smallholder farmers. Agent-based modeling[93,94] and spatially explicit economic models of land change[71] are examples of such methods; and iii) *hybrid* methods tend to draw from existing top-down and bottom-up approaches using them in complementary ways to reflect the bidirectional influences of broader-scale processes and local-level interactions on each other.[95,96]

In addition to numerous land-change modeling studies focusing on local or regional dynamics of land change, there are global-scale land-cover change modeling efforts going back to the early 1990s.[97–101] A new generation of these models that incorporate both past changes going back to year 1500[33] and scenario projections out to year 2100 have recently been developed in preparation for the Fifth Assessment Report of the Intergovernmental Panel on Climate Change (IPCC).[102] Another recent effort focused specifically on future urban-land expansion[103] and its likely impacts on biodiversity.[104]

There are usually significant differences among the modeling applications that make comparisons of findings and sharing insights difficult, if not impossible. These differences stem from many reasons such as the availability of data, the inclination of modelers toward a particular modeling methodology, type of land change (e.g., urban growth, deforestation) under question, and case-study locations (e.g., tropical deforestation in Amazonia vs. Southeast Asia). One of the pressing challenges for land-change modelers is then to increase compatibility among the available land-change modeling approaches including validation protocols[105] but also improving integrated models that simulate the interactions among social, economic, institutional, and biophysical components of a land-change system.[1] Such improvements can also help increase the utility of these models as practical tools that inform land-use policy and planning.[88]

CONCLUSION

Improving our understanding of land change as an integral component of the Earth system's functioning is a critical initial step in addressing the interactions between people and their environments.[9] As we move well into the 21st century, human activities continue to change the surface of Earth directly and indirectly as they have been doing for at least the last 12,000 years.[33] The big difference is in the realization that human activities over the past 200 years or so dramatically intensified and reached such a scale to affect the functioning of the whole Earth system.[106] Perhaps because of the recency of this collective realization, humans are yet to come to grips with all the various kinds of linkages that enmesh our societies with the larger Earth system. Likewise, the diversity of factors involved in land dynamics and linkages among these across multiple spatial and temporal scales present immense challenges to the formulation of an overarching theory of land change. Formulation of such a grand theory hinges on the evolution of land-change science toward a more integrative endeavor that spans across the natural and the social sciences. Therefore, the challenge in the coming decades will be moving toward the formulation of such a theory that will surely evolve together with the constantly changing relation of humans with the land.

REFERENCES

1. Turner, B.L., II; Lambin, E.F.; Reenberg, A. Land Change Science Special Feature: The emergence of land change science for global environmental change and sustainability. Proc. Natl. Acad. Sci. USA. **2007**, *104* (52), 20666–20671.
2. Fraterrigo, J.M.; Turner, M.G.; Pearson, S.M.; Dixon, P. Effects of past land use on spatial heterogeneity of soil nutrients in southern Appalachian forests. Ecol. Monogr. **2005**, *75* (2), 215–230.
3. Verchot, L.V.; Davidson, E.A.; Cattânio, J.H.; Ackerman, I.L.; Erickson, H.E.; Keller, M. Land use change and biogeochemical controls of nitrogen oxide emissions from soils in eastern Amazonia. Global Biogeochem. Cycles **1999**, *13* (1), 31–46.
4. Dupouey, J.L.; Dambrine, E.; Laffite, J.D.; Moares, C. Irreversible impact of past land use on forest soils and biodiversity. Ecology **2002**, *83* (11), 2978–2984.
5. Harding, J.S.; Benfield, E.F.; Bolstad, P.V.; Helfman, G.S.; Jones III, E.B.D. Stream biodiversity: The ghost of land use past. Proc. Natl. Acad. Sci. USA. **1998**, *95* (25), 14843–14847.
6. Pielke Sr., R.A.; Marland, G.; Betts, R.A.; Chase, T.N.; Eastman, J.L.; Niles, J.O.; Niyogi, D.D.S.; Running, S.W. The influence of land-use change and landscape dynamics on the climate system: Relevance to climate-change policy beyond the radiative effect of greenhouse gases. Philos. Trans. R. Soc. A **2002**, *360* (1797), 1705–1719.
7. Houghton, R.A. Revised estimates of the annual net flux of carbon to the atmosphere from changes in land use and land management 1850-2000. Tellus Ser. B **2003**, *55* (2), 378–390.
8. Naidoo, R.; Balmford, A.; Costanza, R.; Fisher, B.; Green, R.E.; Lehner, B.; Malcolm, T.R.; Ricketts, T.H. Global mapping of ecosystem services and conservation priorities. Proc. Natl. Acad. Sci. USA. **2008**, *105* (28), 9495–9500.
9. Global Land Project. *GLP Science Plan and Implementation Strategy*; IGBP Secretariat: Stockholm, 2005, 64.
10. Goudsblom, J.; de Vries, B. *Mappae Mundi: Humans and Their Habitats in a Long-Term Socio-Ecological Perspectives : Myths, Maps and Models*; Amsterdam University Press: Amsterdam, 2003.
11. Aspinall, R.J.; Hill, M.J., Eds.; *Land Use Change: Science, Policy, and Management*; CRC Press: Boca Raton, 2007; 185.

12. DeFries, R.S.; Asner, G.P.; Houghton, R.A., Eds.; *Ecosystems and land use change*; American Geophysical Union: Washington, D.C., 2004; 344.
13. Entwisle, B.; Stern, P.C., Eds.; *Population, Land Use, and Environment: Research Directions*; National Academies Press: Washington, D.C., 2005; 321.
14. Gutman, G.; Janetos, A.C.; Justice, C.O.; Moran, E.F.; Mustard, J.F.; Rindfuss, R.R.; Skole, D.; Turner, B.L.; Cochrane, M.A., Eds.; *Land Change Science: Observing, Monitoring and Understanding Trajectories of Change on the Earth's Surface*; Kluwer Academic Publishers: Dordrecht, 2004; 459.
15. Geist, H., Ed.; *Our Earth's Changing Land: An Encyclopedia of Land-Use and Land-Cover Change*; Greenwood Press: Westport, CT, 2006; 715.
16. Koomen, E.; Stillwell, J.; Bakema, A.; Scholten, H.J., Eds.; *Modelling Land-Use Change: Progress and Applications;* Springer: Dordrecht, 2007; 392.
17. Lambin, E.F.; Geist, H., Eds.; *Land-Use and Land-Cover Change: Local Processes and Global Impacts*; Springer: Berlin, 2006; 222.
18. Meyer, W.B.; Turner, B.L., Eds.; *Changes in Land Use and Land Cover: A Global Perspective*; Cambridge University Press: Cambridge, U.K., 1994; 537.
19. Walsh, S.J.; Crews-Meyer, K.A. *Linking People, Place, and Policy: A GIScience Approach*; Kluwer Academic Publishers: Dordrecht, 2002; 348.
20. Moran, E.F.; Ostrom, E., Eds.; *Seeing The Forest and The Trees: Human-Environment Interactions in Forest Ecosystems*; MIT Press: Cambridge, MA, 2005; 442.
21. Lambin, E.F.; Turner, B.L.; Geist, H.J.; Agbola, S.B.; Angelsen, A.; Bruce, J.W.; Coomes, O.T.; Dirzo, R.; Fischer, G.; Folke, C.; George, P.S.; Homewood, K.; Imbernon, J.; Leemans, R.; Li, X.; Moran, E.F.; Mortimore, M.; Ramakrishnan, P.S.; Richards, J.F.; Skanes, H.; Steffen, W.; Stone, G.D.; Svedin, U.; Veldkamp, T.A.; Vogel, C.; Xu, J. The causes of land-use and land-cover change: Moving beyond the myths. Global Environ. Change 2001; *11* (4), 261–269.
22. Vitousek, P.M.; Mooney, H.A.; Lubchenco, J.; Melillo, J.M. Human domination of Earth's ecosystems. Science 1997; *277* (5325), 494–499.
23. Redman, C.L. *Human Impact on Ancient Environments*; University of Arizona Press: Tucson, 1999; 288.
24. Turner, B.L.; Robbins, P. Land-change science and political ecology: Similarities, differences, and implications for sustainability science. Ann. Rev. Environ. Resources **2008**, *33*, 295–316.
25. Geist, H.J.; Lambin, E.F. *What Drives Tropical Deforestation?: A Meta-Analysis of Proximate and Underlying Causes of Deforestation Based on Subnational Case Study Evidence, in LUCC Report Series*; LUCC International Project Office, University of Louvain: Louvain-la-Neuve, Belgium, 2001; 116.
26. Ojima, D.S.; Galvin, K.A.; Turner B.L., II. The global impact of land-use change. Bioscience 1994; *44* (5), 300–304.
27. Geist, H.; McConnell, W. Causes and trajectories of land-use/cover change. In *Land-Use and Land-Cover Change: Local Processes and Global Impacts*; Lambin, E.F., Geist, H., Eds.; Springer: Berlin, 2006; 41–70.
28. Ellis, E.C.; Ramankutty, N. Putting people in the map: anthropogenic biomes of the world. Front. Ecol. Environ. **2008**, *6* (8), 439–447.
29. Ellis, E.C.; Kaplan, J.O.; Fuller, D.Q.; Vavrus, S.; Goldewijk, K.K.; Verburg, P.H. Used planet: A global history. Proc. Natl. Acad. Sci. USA. **2013**, *110* (20), 7978–7985.
30. Hurtt, G.C.; Frolking, S.; Fearon, M.G.; Moore, B.; Shevliakova, E.; Malyshev, S.; Pacala, S.W.; Houghton, R.A. The underpinnings of land-use history: Three centuries of global gridded land-use transitions, wood-harvest activity, and resulting secondary lands. Global Change Biol. **2006**, *12* (7), 1208–1229.
31. Haberl, H.; Erb, K.H.; Krausmann, F.; Gaube, V.; Bondeau, A.; Plutzar, C.; Gingrich, S.; Lucht, W.; Fischer-Kowalski, M. Quantifying and mapping the human appropriation of net primary production in earth's terrestrial ecosystems. Proc. Natl. Acad. Sci. USA. **2007**, *104* (31), 12942–12947.
32. Pongratz, J.; Reick, C.; Raddatz, T.; Claussen, M. A reconstruction of global agricultural areas and land cover for the last millennium. Global Biogeochemical Cycles **2008**, *22* (3): GB3018.
33. Goldewijk, K.K.; Beusen, A.; van Drecht, G.; de Vos, M. The HYDE 3.1 spatially explicit database of human-induced global land-use change over the past 12,000 years. Global Ecol. Biogeography **2011**, *20* (1), 73–86.
34. Williams, M. Forests. In *The Earth as Transformed by Human Action: Global and Regional Changes in the Biosphere over the Past 300 Years*; Turner, B.L., Clark, W.C., Kates, R.W., Richards, J.F., Mathews, J.T., Meyer, W.B., Eds.; Cambridge University Press with Clark University: New York, 1990, 179–201.
35. Heckenberger, M.J.; Russell, J.C.; Fausto, C.; Toney, J.R.; Schmidt, M.J.; Pereira, E.; Franchetto, B.; Kuikuro, A. Pre-Columbian urbanism, anthropogenic landscapes, and the future of the Amazon. Science **2008**, *321* (5893), 1214–1217.
36. McNeil, C.L.; Burney, D.A.; Burney, L.P. Evidence disputing deforestation as the cause for the collapse of the ancient Maya polity of Copan, Honduras. Proc. Natl. Acad. Sci. USA. **2010**, *107* (3), 1017–1022.
37. Grau, H.R.; Aide, M. Globalization and land-use transitions in Latin America. Ecol. Soc. **2008**, *13* (2), 16.
38. Lambin, E.F.; Meyfroidt, P. Global land use change, economic globalization, and the looming land scarcity. Proc. Natl. Acad. Sci. USA. **2011**, *108* (9), 3465–3472.
39. Meyfroidt, P.; Lambin, E.F. Forest transition in Vietnam and displacement of deforestation abroad. Proc. Natl. Acad. Sci. USA. **2009**, *106* (38), 16139–16144.
40. Mayer, A.L.; Kauppi, P.E.; Angelstam, P.K.; Zhang, Y.; Tikka, P.M. Importing timber, exporting ecological impact. Science **2005**, *308* (5720), 359–360.
41. Farley, K.A. Grasslands to tree plantations: Forest transition in the Andes of Ecuador. Ann. Assoc. Am. Geographers **2007**, *97* (4), 755–771.
42. Barlow, J.; Overal, W.L.; Araujo, I.S.; Gardner, T.A.; Peres, C.A. The value of primary, secondary and plantation forests for fruit-feeding butterflies in the Brazilian Amazon. J. Appl. Ecol. **2007**, *44* (5), 1001–1012.
43. Wilkie, D.S.; Bennett, E.L.; Peres, C.A.; Cunningham, A.A. The empty forest revisited. Ann. NY Acad. Sci. **2011**, *1223*, 120–128.
44. Ramankutty, N.; Foley, J.A. Estimating historical changes in global land cover: Croplands from 1700 to 1992. Global Biogeochemical Cycles **1999**, *13* (4), 997–1027.

45. Goldewijk, K.K. Estimating global land use change over the past 300 years: The HYDE Database. Global Biogeochemical Cycles **2001**, *15* (2), 417–433.
46. Godfray, H.C.J. Food and biodiversity. Science **2011**, *333* (6047), 1231–1232.
47. Tscharntke, T.; Klein, A.M.; Kruess, A.; Steffan-Dewenter, I.; Thies, C. Landscape perspectives on agricultural intensification and biodiversity - ecosystem service management. Ecol. Lett. **2005**, *8* (8), 857–874.
48. Zimmerer, K.S. The compatibility of agricultural intensification in a global hotspot of smallholder agrobiodiversity (Bolivia). Proc. Natl. Acad. Sci. USA. **2013**, *110* (8), 2769–2774.
49. Phelps, J.; Carrasco, L.R.; Webb, E.L.; Koh, L.P.; Pascual, U. Agricultural intensification escalates future conservation costs. Proc. Natl. Acad. Sci. USA. **2013**, *110* (19), 7601–7606.
50. Geist, H.J.; Lambin, E.F. Dynamic causal patterns of desertification. Bioscience **2004**, *54* (9), 817–829.
51. Issar, A.S. Climatic change and the history of the Middle East. Am. Sci. **1995**, *83* (4), 350–355.
52. Prince, S.D. Spatial and temporal scales for detection of desertification. In *Global Desertification: Do Humand Cause Deserts?*; Reynolds, J.F., Stafford Smith, D.M., Eds; Dahlem University Press: Berlin, 2002; 23–40.
53. Prince, S.D.; Wessels, K.J.; Tucker, C.J.; Nicholson, S.E. Desertification in the Sahel: A reinterpretation of a reinterpretation. Global Change Biol. **2007**, *13* (7), 1308–1313.
54. Tucker, C.J.; Dregne, H.E.; Newcomb, W.W. Expansion and contraction of the Sahara desert from 1980 to 1990. Science **1991**, *253* (5017), 299–301.
55. Seto, K.C.; Fragkias, M.; Güneralp, B.; Reilly, M.K. A meta-analysis of global urban land expansion. PLoS ONE **2011**, *6* (8), e23777.
56. Angel, S.; Parent, J.; Civco, D.L.; Blei, A.; Potere, D. The dimensions of global urban expansion: Estimates and projections for all countries, 2000–2050. Prog. Plann. **2011**, *75* (2), 53–107.
57. Fry, M. From crops to concrete: Urbanization, deagriculturalization, and construction material mining in central Mexico. Ann. Assoc. Am. Geographers **2011**, *101* (6), 1285–1306.
58. DeFries, R.S.; Rudel, T.; Uriarte, M.; Hansen, M. Deforestation driven by urban population growth and agricultural trade in the twenty-first century. Nat. Geosci. **2010**, *3* (3), 178–181.
59. Seto, K.C.; Reenberg, A.; Boone, C.G.; Fragkias, M.; Haase, D.; Langanke, T.; Marcotullio, P.; Munroe, D.K.; Olah, B.; Simon, D. Urban land teleconnections and sustainability. Proc. Natl. Acad. Sci. USA. **2012**, *109* (20), 7687–7692.
60. Díaz, G.I.; Nahuelhual, L.; Echeverría, C.; Marín, S. Drivers of land abandonment in Southern Chile and implications for landscape planning. Landscape Urban Plann. **2011**, *99* (3–4), 207–217.
61. Yackulic, C.B.; Fagan, M.; Jain, M.; Jina, A.; Lim, Y.; Marlier, M.; Muscarella, R.; Adame, P.; DeFries, R.; Uriarte, M. Biophysical and socioeconomic factors associated with forest transitions at multiple spatial and temporal scales. Ecol. Soc. **2011**, *16* (3), 23.
62. Wicke, B.; Verweij, P.; Van Meijl, H.; Van Vuuren, D.P.; Faaij, A.P.C. Indirect land use change: Review of existing models and strategies for mitigation. Biofuels **2012**, *3* (1), 87–100.
63. Wright, S.J. The future of tropical forests. Ann. N Y Acad. Sci. **2010**, *1195*, 1–27.
64. Meyfroidt, P.; Lambin, E.F. Global forest transition: Prospects for an end to deforestation. Ann. Rev. Environ. Resources **2011**, *36*, 343–371.
65. Syvitski, J.P.M.; Kettner, A.J.; Overeem, I.; Hutton, E.W.H.; Hannon, M.T.; Brakenridge, G.R.; Day, J.; Vörösmarty, C.; Saito, Y.; Giosan, L.; Nicholls, R.J. Sinking deltas due to human activities. Nat. Geosci. **2009**, *2* (10), 681–686.
66. Nicholls, R.J.; Cazenave, A. Sea-level rise and its impact on coastal zones. Science **2010**, *328* (5985), 1517–1520.
67. Foley, J.A.; DeFries, R.; Asner, G.P.; Barford, C.; Bonan, G.; Carpenter, S.R.; Chapin, F.S.; Coe, M.T.; Daily, G.C.; Gibbs, H.K.; Helkowski, J.H.; Holloway, T.; Howard, E.A.; Kucharik, C.J.; Monfreda, C.; Patz, J.A.; Prentice, I.C.; Ramankutty, N.; Snyder, P.K. Global consequences of land use. Science **2005**, *309* (5734), 570–574.
68. Lambin, E.F.; Geist, H.; Rindfuss, R.R. Introduction: Local processes with global impacts. In *Land-Use and Land-Cover Change: Local Processes and Global Impacts*; Lambin, E.F., Geist, H., Eds.; Springer: Berlin, 2006; 1–8.
69. Aspinall, R.J. Basic and applied land use science. In *Land Use Change: Science, Policy, and Management*; Aspinall, R.J., Hill, M.J., Eds.; CRC Press: Boca Raton, 2007; 3–16.
70. Manson, S.M. Challenges in evaluating models of geographic complexity. Environ. Plann. B **2007**, *34* (2), 245–260.
71. Irwin, E.G.; Geoghegan, J. Theory, data, methods: Developing spatially explicit economic models of land use change. Agriculture Ecosystems Environ. **2001**, *85* (1–3), 7–23.
72. Hecht, S. The new rurality: Globalization, peasants and the paradoxes of landscapes. Land Use Policy **2010**, *27* (2), 161–169.
73. Young, O.R.; Lambin, E.F.; Alcock, F.; Haberl, H.; Karlsson, S.I.; McConnell, W.J.; Myint, T.; Pahl-Wostl, C.; Polsky, C.; Ramakrishnan, P.S.; Schroeder, H.; Scouvart, M.; Verburg, P.H. A portfolio approach to analyzing complex human-environment interactions: Institutions and land change. Ecol. Soc. **2006**, *11* (2).
74. Barbier, E.B.; Burgess, J.C.; Grainger, A. The forest transition: Towards a more comprehensive theoretical framework. Land Use Policy **2010**, *27* (2), 98–107.
75. Geist, H. *The Causes and Progression of Desertification*; Ashgate Publishing: Aldershot, 2005, 258.
76. Webster, C.J.; Wu, F. Regulation, land-use mix, and urban performance. Part 1: Theory. Environ. Plann. A **1999**, *31* (8), 1433–1442.
77. Walker, R. Theorizing land-cover and land-use change: The case of tropical deforestation. Int. Reg. Sci. Rev. **2004**, *27* (3), 247–270.
78. Keys, E.; McConnell, W.J. Global change and the intensification of agriculture in the tropics. Global Environ. Change **2005**, *15* (4), 320–337.
79. VanWey, L.; Ostrom, E.; Meretsky, V. Theories underlying the study of human-environment interactions. In *Seeing The Forest and The Trees: Human-Environment Interactions in Forest Ecosystems*; Moran, E. and Ostrom, E., Eds.; MIT Press: Cambridge, MA, 2005; 23–56.
80. Chowdhury, R.R.; Turner II, B.L. Reconciling agency and structure in empirical analysis: Smallholder land use in the Southern Yucatán, Mexico. Ann. Assoc. Am. Geographers **2006**, *96* (2), 302–322.

81. Lambin, E.F.; Meyfroidt, P. Land use transitions: Socio-ecological feedback versus socio-economic change. Land Use Policy **2010**, *27* (2), 108–118.
82. Ramankutty, N.; Graumlich, L.; Achard, F.; Alves, D.; Chhabra, A.; DeFries, R.S.; Foley, J.A.; Geist, H.; Houghton, R.A.; Goldewijk, K.K.; Lambin, E.F.; Millington, A.; Rasmussen, K.; Reid, R.S.; Turner, B.L. Global land-cover change: Recent progress, remaining challenges land-use and land-cover change. In *Land-Use and Land-Cover Change: Local Processes and Global Impacts*; Lambin, E.F., Geist, H., Eds.; Springer: Berlin, 2006; 9–39.
83. Goward, S.N.; Davis, P.E.; Fleming, D.; Miller, L.; Townshend, J.R. Empirical comparison of Landsat 7 and IKONOS multispectral measurements for selected Earth Observation System (EOS) validation sites. Remote Sensing Environ. **2003**, *88* (1–2), 80–99.
84. Steven, M.D.; Malthus, T.J.; Baret, F.; Xu, H.; Chopping, M.J. Intercalibration of vegetation indices from different sensor systems. Remote Sensing Environ. **2003**, *88* (4), 412–422.
85. Flint, E.P.; Richards, J.F. Historical-analysis of changes in land-use and carbon stock of vegetation in South and Southeast Asia. Can. J. Forest Res. **1991**, *21* (1), 91–110.
86. Gray, J.; Song, C. Mapping leaf area index using spatial, spectral, and temporal information from multiple sensors. Remote Sensing Environ. **2012**, *119*, 173–183.
87. Hilker, T.; Wulder, M.A.; Coops, N.C.; Linke, J.; McDermid, G.; Masek, J.G.; Gao, F.; White, J.C. A new data fusion model for high spatial- and temporal-resolution mapping of forest disturbance based on Landsat and MODIS. Remote Sensing Environ. **2009**, *113* (8), 1613–1627.
88. Verburg, P.H.; Schot, P.P.; Dijst, M.J.; Veldkamp, A. Land use change modelling: Current practice and research priorities. GeoJournal **2004**, *61*, 309–324.
89. Agarwal, C.; Green, G.; Grove, J.; Evans, T.; Schweik, C. *A Review and Assessment of Land-Use Change Models: Dynamics of Space, Time, and Human Choice, in Gen Tech Rep NE-2972002*; U.S. Dept Agric, Forest Service, Northeastern Research Station: Burlington, VT, 2002; 61.
90. Verburg, P.H.; Kok, K.; Pontius Jr, R.G.; Veldkamp, A. Modeling land-use and land-cover change. In *Land-Use and Land-Cover Change: Local Processes and Global Impacts*; Lambin, E.F., Geist, H., Eds.; Springer: Germany, **2006**, 117–135.
91. Veldkamp, A.; Fresco, L.O. CLUE: A conceptual model to study the conversion of land use and its effects. Ecol. Model. **1996**, *85* (2–3), 253–270.
92. Pontius, R.G.; Cornell, J.D.; Hall, C.A.S. Modeling the spatial pattern of land-use change with GEOMOD2: application and validation for Costa Rica. Agriculture Ecosystems Environ. **2001**, *85* (1–3), 191–203.
93. Parker, D.C.; Manson, S.M.; Janssen, M.A.; Hoffmann, M.J.; Deadman, P. Multi-agent systems for the simulation of land-use and land-cover change: A review. Ann. Assoc. Am. Geographers **2003**, *93* (2), 314–337.
94. Brown, D.G.; Robinson, D.T.; An, L.; Nassauer, J.I.; Zellner, M.; Rand, W.; Riolo, R.; Page, S.E.; Low, B.; Wang, Z. Exurbia from the bottom-up: Confronting empirical challenges to characterizing a complex system. Geoforum **2008**, *39* (2), 805–818.
95. Verburg, P.H.; Overmars, K.P. Combining top-down and bottom-up dynamics in land use modeling: Exploring the future of abandoned farmlands in Europe with the Dyna-CLUE model. Landscape Ecol. **2009**, *24* (9), 1167–1181.
96. Veldkamp, A.; Fresco, L.O. CLUE-CR: An integrated multi-scale model to simulate land use change scenarios in Costa Rica. Ecol. Model. **1996**, *91* (1–3), 231–248.
97. Heistermann, M.; Müller, C.; Ronneberger, K. Land in sight? Achievements, deficits and potentials of continental to global scale land-use modeling. Agriculture Ecosystems Environ. **2006**, *114* (2–4), 141–158.
98. Alcamo, J.; Kreileman, G.J.J.; Bollen, J.C.; Van Den Born, G.J.; Gerlagh, R.; Krol, M.S.; Toet, A.M.C.; De Vries, H.J.M. Baseline scenarios of global environmental change. Global Environ. Change **1996**, *6* (4), 261–303.
99. Leemans, R.; Van Amstel, A.; Battjes, C.; Kreileman, E.; Toet, S. The land cover and carbon cycle consequences of large-scale utilizations of biomass as an energy source. Global Environ. Change **1996**, *6* (4), 335–357.
100. Woodward, F.I.; Smith, T.M.; Emanuel, W.R. A global land primary productivity and phytogeography model. Global Biogeochem. Cycles **1995**, *9* (4), 471–490.
101. Schaldach, R.; Alcamo, J.; Koch, J.; Kölking, C.; Lapola, D.M.; Schüngel, J.; Priess, J.A. An integrated approach to modelling land-use change on continental and global scales. Environ. Model. Software **2011**, *26* (8), 1041–1051.
102. Hurtt, G.C.; Chini, L.P.; Frolking, S.; Betts, R.A.; Feddema, J.; Fischer, G.; Fisk, J.P.; Hibbard, K.; Houghton, R.A.; Janetos, A.; Jones, C.D.; Kindermann, G.; Kinoshita, T.; Goldewijk, K.K.; Riahi, K.; Shevliakova, E.; Smith, S.; Stehfest, E.; Thomson, A.; Thornton, P.; van Vuuren, D.P.; Wang, Y.P. Harmonization of land-use scenarios for the period 1500–2100: 600 years of global gridded annual land-use transitions, wood harvest, and resulting secondary lands. Climatic Change **2011**, *109* (1), 117–161.
103. Seto, K.C.; Güneralp, B.; Hutyra, L.R. Global forecasts of urban expansion to 2030 and direct impacts on biodiversity and carbon pools. Proc. Natl. Acad. Sci. USA. **2012**, *109* (40), 16083–16088.
104. Güneralp, B.; Seto, K.C. Futures of global urban expansion: uncertainties and implications for biodiversity conservation. Environ. Res. Lett. **2013**, *8*, 014025.
105. Pontius, R.; Boersma, W.; Castella, J.-C.; Clarke, K.; de Nijs, T.; Dietzel, C.; Duan, Z.; Fotsing, E.; Goldstein, N.; Kok, K.; Koomen, E.; Lippitt, C.; McConnell, W.; Mohd Sood, A.; Pijanowski, B.; Pithadia, S.; Sweeney, S.; Trung, T.; Veldkamp, A.; Verburg, P. Comparing the input, output, and validation maps for several models of land change. Ann. Regional Sci. **2008**, *42* (1), 11–37.
106. IPCC. Climate change 2007. In *Synthesis Report Contribution of Working Groups I, II and III to the Fourth Assessment Report of the Intergovernmental Panel on Climate Change*; Core Writing Team, Pachauri, R.K., Reisinger, A., Eds.; IPCC: Geneva, Switzerland, 2007; 104.

BIBLIOGRAPHY

1. Baptista, S.R. Metropolitan land-change science: A framework for research on tropical and subtropical forest recovery in city-regions. Land Use Policy **2010**, *27* (2), 139–147.
2. Barnosky, A.D.; Koch, P.L.; Feranec, R.S.; Wing, S.L.; Shabel, A.B. Assessing the causes of late pleistocene extinctions on the continents. Science **2004**, *306* (5693), 70–75.
3. Blaikie, P. *The Political Economy of Soil Erosion in Developing Countries*; Longman Group Limited, Essex, 1985; 188.
4. Briassoulis, H. Land-use policy and planning, theorizing, and modeling: lost in translation, found in complexity? Environ. Plann. B **2008**, *35*, 16–33.
5. Brum, F.T.; Gonçalves, L.O; Cappelatti, L.; Carlucci, M.B.; Debastiani, V.J.; Salengue, E.V.; dos Santos Seger, G.D.; Both, C.; Bernardo-Silva, J.S.; Loyola, R.D.; da Silva Duarte, L. Land use explains the distribution of threatened new world amphibians better than climate. *PLoS ONE* **2013**, *8* (4), e60742.
6. Caldas, M.; Walker, R.; Arima, E.; Perz, S.; Aldrich, S.; Simmons, C. Theorizing land cover and land use change: The peasant economy of Amazonian deforestation. Ann. Assoc. Am. Geographers **2007**, *97* (1), 86–110.
7. Clarke, K.; Gazulis, N.; Dietzel, C.; Goldstein, N. A decade of SLEUTHing: Lessons learned from applications of a cellular automaton land use change model. In *Classics from IJGIS. Twenty Years of the International Journal of Geographical Information Systems and Science*; Fisher, P., Ed.; Taylor and Francis, CRC: Boca Raton, 2007; 413–425.
8. De Campos, C.P.; Muylaert, M.S; Rosa, L.P. Historical CO_2 emission and concentrations due to land use change of croplands and pastures by country. Sci. Total Environ. **2005**, *346* (1–3), 149–155.
9. Fischer, J.; Batáry, P.; Bawa, K.S.; Brussaard, L.; Chappell, M.J.; Clough, Y.; Daily, G.C.; Dorrough, J.; Hartel, T.; Jackson, L.E.; Klein, A.M.; Kremen, C.; Kuemmerle, T.; Lindenmayer, D.B.; Mooney, H.A.; Perfecto, I.; Philpott, S.M.; Tscharntke, T.; Vandermeer, J.; Wanger, T.C.; Von Wehrden, H. Conservation: Limits of land sparing. Science **2011**, *334* (6056): 593.
10. Friis, C.; Reenberg, A. *Land Grab in Africa: Emerging Land System Drivers in a Teleconnected World*; Global Land Project (GLP)-IPO: Copenhagen, 2010; 42.
11. Geoghegan, J.; Pritchard, L.J.; Ogneva-Himmelberger, Y.; Chowdury, R.; Sanderson, S.; Turner, B.L.I. 'Socializing the pixel' and 'pixelizing the social' in land-use/cover change. In *People and Pixels*; Liverman, E.F.M.D., Rindfuss, R.R., Stern, P.C. National Research Council: Washington, DC, 1998; 51–69.
12. Goldewijk, K.K.; Beusen, A.; Janssen, P. Long-term dynamic modeling of global population and built-up area in a spatially explicit way: HYDE 3.1. Holocene **2010**, *20* (4), 565–573.
13. Goldewijk, K.K.; Ramankutty, N. Land cover change over the last three centuries due to human activities: The availability of new global data sets. *GeoJournal* **2004**, *61* (4), 335–344.
14. Gorissen, L.; Buytaert, V.; Cuypers, D.; Dauwe, T.; Pelkmans, L. Why the debate about land use change should not only focus on biofuels. Environ. Sci. Technol. **2010**, *44* (11), 4046–4049.
15. Güneralp, B.; Reilly, M.K.; Seto, K.C. Capturing multiscalar feedbacks in urban land change: a coupled system dynamics spatial logistic approach. Environ. Plann. B **2012**, *39* (5), 858–879.
16. Gutman, G.; Reissell, A., Eds.; *Eurasian Arctic Land Cover and Land Use in a Changing Climate*; Springer: Dordrecht, 2011; 306.
17. Heilig, G. Neglected dimensions of global land-use change: reflections and data. *Population and* Dev. Rev. **1994**, *20* (4), 831–859.
18. Houghton, R.A. The worldwide extent of land-use change. Bioscience **1994**, *44* (5), 305–313.
19. Irwin, E. G. New directions for urban economic models of land use change: Incorporating spatial dynamics and heterogeneity. J. Reg. Sci. **2010**, *50* (1), 65–91.
20. Kaimowitz, D.; Angelsen, A. *Economic Models of Tropical Deforestation: A Review*; Center International Forestry Research (CIFOR): Bogor, **1998**, 139.
21. Kalnay, E.; Cai, M. Impact of urbanization and land-use change on climate. Nature **2003**, *423* (6939), 528–531.
22. Klosterman, R.E. Simple and complex models. Environ. Plann. B **2012**, *39* (1), 1–6.
23. Krausmann, F.; Erb, K.-H.; Gingrich, S.; Haberl, H.; Bondeau, A.; Gaube, V.; Lauk, C.; Plutzar, C.; Searchinger, T.D. Global human appropriation of net primary production doubled in the 20th century. Proc. Natl. Acad. Sci. USA. **2013**, *110* (25), 10324–10329
24. Matthews, E. Global vegetation and land use: new high-resolution data bases for climate studies. J. Climate Appl. Meteorol. **1983**, *22* (3), 474–487.
25. Metzger, M.J.; Rounsevell, M.D.A.; Acosta-Michlik, L.; Leemans, R.; Schröter, D. The vulnerability of ecosystem services to land use change. Agriculture Ecosystems Environ. **2006**, *114* (1), 69–85.
26. Meyfroidt, P.; Rudel, T.K. Lambin, E.F. Forest transitions, trade, and the global displacement of land use. Proc. Natl. Acad. Sci. USA. **2010**, *107* (49), 20917–20922.
27. Nolte, C.; Agrawal, A.; Silvius, K.M.; Soares-Filho, B.S. Governance regime and location influence avoided deforestation success of protected areas in the Brazilian Amazon. Proc. Natl. Acad. Sci. USA. **2013**, *110* (13), 4956–4961.
28. Ostrom, E.; Nagendra, H. Insights on linking forests, trees, and people from the air, on the ground, and in the laboratory. Proc. Natl. Acad. Sci. USA. **2006**, *103* (51), 19224–19231.
29. Oswalt, P.; Rieniets, T. *Atlas of Shrinking Cities*; Hatje Cantz Publishers: Ostfildern, 2006; 160.
30. Petrosillo, I.; Semeraro, T.; Zaccarelli, N.; Aretano, R.; Zurlini, G. The possible combined effects of land-use changes and climate conditions on the spatial-temporal patterns of primary production in a natural protected area. Ecol. Indicators **2013**, *29*, 367–375.
31. Potere, D.; Schneider, A. A critical look at representations of urban areas in global maps. GeoJournal **2007**, *69*, 55–80.

Available at the following web site: http://www.umass.edu/landeco/research/fragstats/fragstats.html
10. Pickett, S.T.A., White, P.S. Eds.; *The Ecology of Natural Disturbance and Patch Dynamics*; Academic Press: New York, USA 1985.
11. Hunter Jr., M.L. Natural fire regimes as spatial models for managing boreal forests. Biol. Conserv. **1993**, *65* (2), 115–120.
12. Lorimer, C.G. Historical and ecological roles of disturbance in Eastern North American forests: 9,000 years of change. Wildlife Soc. Bull. **2001**, *29* (2), 425–439.
13. Delcourt, H.R.; Delcourt, P.A. Quaternary landscape ecology: Relevant scales in space and time. Landscape Ecol. **1988**, *2* (1), 23–44.
14. Currie, W. S. Units of nature or processes across scales? The ecosystem concept at age 75. New Phytol. **2011**, *190,* 21–34.
15. He, H.S. Forest landscape models: Definitions, characterization, and classification. Forest Ecol. Manag. **2008**, *254* (3), 445–453.
16. Mladenoff, D.J.; Baker, W.L. *Advances in Spatial Modeling of Forest Landscape Change: Approaches and Applications;* Cambridge University Press: Cambridge, UK 1999; 163–185.
17. Perry, G.L.W.; Enright, N.J. Spatial modelling of vegetation change indynamic landscapes: A review of methods and applications. Progr Phys Geogr. **2006**, *30*, 47–72.

32. Ramankutty, N.; Foley, J.A. Characterizing patterns of global land use: An analysis of global croplands data. Global Biogeochemical Cycles **1998**, *12* (4), 667–685.
33. Riebsame, W.E.; Parton, W.J.; Galvin, K.A.; Burke, I.C.; Bohren, L.; Young, R.; Knop, E. Integrated modeling of land use and cover change. Bioscience **1994**, *44* (5),350–356.
34. Rindfuss, R.R.; Walsh, S.J.; Turner, B.L.; Fox, J. Mishra, V. Developing a science of land change: Challenges and methodological issues. Proc. Natl. Acad. Sci. USA. **2004**, *101* (39), 13976–13981.
35. Robinson, D.T.; Brown, D.G.; Currie, W.S Modelling carbon storage in highly fragmented and human-dominated landscapes: Linking land-cover patterns and ecosystem models. Ecol. Model. **2009**, *220* (9–10), 1325–1338.
36. Rulli, M.C.; Saviori, A.; D'Odorico, P. Global land and water grabbing. Proc. Natl. Acad. Sci. USA. **2013**, *110* (3), 892–897.
37. Schneider, A.; Woodcock, C.E. Compact, dispersed, fragmented, extensive? A comparison of urban growth in twenty-five global cities using remotely sensed data, pattern metrics and census information. *Urban Studies* **2008**, *45* (3), 659–692.
38. Seto, K.C.; de Groot, R.; Bringezu, S.; Erb, K.H.; Graedel, T.E; Ramankutty, N.; Reenberg, A.; Schmitz, O.; Skole, D. Stocks, flows, and prospects of land. In *Linkages of Sustainability*; Graedel, T., van der Voet, E., Eds.; MIT Press: Cambridge, MA, 2009, 71–96.
39. Seto, K.C.; Sánchez-Rodríguez, R.; Fragkias, M. The new geography of contemporary urbanization and the environment. Ann. Rev. Environ. Resources **2010**, *35* (1),167–194.
40. Stafford Smith, D.M.; McKeon, G.M.; Watson, I.W.; Henry, B.K.; Stone, G.S.; Hall, W.B.; Howden, S.M. Learning from episodes of degradation and recovery in variable Australian rangelands. Proc. Natl. Acad. Sci. USA. **2007**, *104* (52), 20690–20695.
41. Turner, B.L.; Clark, W.C.; Kates, R.W.; Richards, J.F.; Mathews, J.T.; Meyer, W.B., Eds.; *The Earth as Transformed by Human Action: Global and Regional Changes in the Biosphere over the Past 300 Years;* Cambridge University Press with Clark University: New York, 1990; 713.

Leaves: Elevated CO_2 Levels

G. Brett Runion
National Soil Dynamics Laboratory, Agricultural Research Service, U.S. Department of Agriculture (USDA-ARS), Auburn, Alabama, U.S.A.

Seth G. Pritchard
Department of Biology, College of Charleston, Charleston, South Carolina, U.S.A.

Stephen A. Prior
National Soil Dynamics Laboratory, Agricultural Research Service, U.S. Department of Agriculture (USDA-ARS), Auburn, Alabama, U.S.A.

Abstract

Burning fossil fuels and land use change have led to a rise in the concentration of carbon dioxide (CO_2) in the atmosphere since the onset of the Industrial Revolution. Carbon dioxide from the atmosphere enters leaves where it is concentrated and transformed into useful organic carbon compounds (food and fiber). In light of the fact that only plant leaves serve this critical function, we provide a brief summary of the effects of rising CO_2 on plant leaves at scales from the molecular to the whole plant community. Within leaves of C_3 plants, photosynthesis is stimulated by elevated CO_2 which generally results in increased biomass. Elevated CO_2 also increases water-use efficiency in both C_3 and C_4 plants by inducing partial closure of stomates which decreases transpiration. Growth in elevated CO_2 can alter leaf morphology, physiology, and chemistry which can affect interactions with pests and diseases. While increases in leaf-level photosynthesis usually translate to increased growth, decreased leaf-level transpiration does not necessarily result in lower whole plant water use; larger plants from exposure to elevated CO_2 can use more water on a whole plant level. Effects of elevated CO_2 scale up to the ecosystem level and gross primary productivity is usually increased. Reduced transpiration under high CO_2 can lead to higher leaf or canopy temperatures. Because not all plants respond in the same manner, growth in elevated CO_2 will benefit some plants to the detriment of others and plant community structure and function can be altered.

INTRODUCTION

The concentration of carbon dioxide (CO_2) in the atmosphere is increasing primarily because of fossil fuel use and land use change and is expected to double preindustrial levels during this century.[1] Leaves are the sites for the transfer of carbon from atmosphere to biosphere; thus, this increase has significant implications for plants. Carbon dioxide, along with sunlight and water, drive photosynthesis and it is the capture of carbon by leaves that enables plants to grow. This carbon capture by leaves forms the basis of agriculture and forestry and provides humans and other organisms with the food, clothing, and shelter they need to survive. This entry will provide a brief summary of the effects of the increasing atmospheric CO_2 on plant leaves at various scales from molecular to plant community.

MOLECULAR AND CELLULAR SCALE

Photosynthesis is stimulated by elevated atmospheric CO_2 in nearly all C_3 plant species, typically on the order of 20–60% for plants grown with 700 ppm compared to those grown at 365 ppm.[2] The increase in photosynthesis is commonly linked to 1) faster C fixation by the ribulose 1,5-bisphosphate carboyxlyase/oxygenase (rubisco) enzyme; 2) suppression of photorespiration caused by an increase in CO_2:O_2 within the chloroplast stroma; 3) decreased oxidative damage to the photosynthetic apparatus; and 4) amelioration of water stress by enhanced water-use efficiency (WUE; the ratio of carbon fixed to water transpired).

Predicting the photosynthetic response of C_4 species (such as maize, sugarcane, sorghum, and many other grasses) to elevated CO_2 is somewhat more difficult. Because C_4 plants possess a CO_2-concentrating mechanism, they are far less sensitive to changes in atmospheric CO_2 concentration compared to C_3 plants. Stimulation of photosynthesis in C_4 plants, therefore, is only expected to occur during periods of water stress.[3] When exposed to drought, both C_3 and C_4 plants benefit from exposure to elevated CO_2 because they are able to maintain sufficient leaf CO_2 concentrations to maintain photosynthesis while keeping stomatal conductance values low to preserve water (i.e., WUE is increased).

Encyclopedia of Natural Resources DOI: 10.1081/E-ENRL-120049205

Increased rates of carbon assimilation by leaves are accompanied by many biochemical and molecular adjustments.[4] For example, the stimulatory effect of elevated CO_2 on photosynthesis is often downregulated to some extent as a result of a reduction in rubisco content and maximum carboxylation capacity.[5] Alterations in expression patterns of genes coding for proteins involved in growth and development, the Calvin cycle, photorespiration, photosystem I and II subunits, carbon metabolism and utilization, respiration, aquaporins, antioxidants, and metabolism of secondary metabolites important for defense have all been reported. Whether specific leaf genes are upregulated or downregulated can vary depending on species and environmental conditions. Clearly, more work is needed at the transcriptomic, proteomic, and metabolomic levels to understand how leaves and whole plants will respond to global climate changes. Not only does research at this level of biological organization have the potential to explain higher level plant and canopy responses to elevated CO_2, these types of data may help guide researchers to select crop and tree species best able to exploit a future high CO_2 world.

Regardless of the nature of changes in gene expression occurring in leaves, increased photosynthesis in plants grown with elevated CO_2 nearly always results in a greater supply of carbohydrates to fuel additional leaf and whole plant growth (discussed further). However, there are feedbacks at the organelle level that have the potential to dampen higher photosynthesis and modify changes in gene expression caused by elevated CO_2. For example, an overproduction of carbohydrates from higher photosynthesis sometimes results in an excessive accumulation of leaf starch that may negatively affect the internal organization of chloroplasts (Fig. 1).[6] Excessive accumulation of leaf starch occurs when carbohydrate production exceeds the plant's ability to transport, metabolize, or store available carbohydrates. In general, plants that receive adequate levels of water and nutrients have enough resources to build sufficient sinks and excessive accumulation of leaf starch is avoided. However, carbohydrates accumulate when plants are sink-limited as commonly occurs with low levels of nutrients and/or under water stress.[6]

WHOLE LEAF SCALE

Increasing levels of atmospheric CO_2 have been shown to affect leaf growth, development, anatomy, and longevity. Plants grown under high CO_2 often exhibit increased area per leaf,[7] which might indicate a higher leaf-level

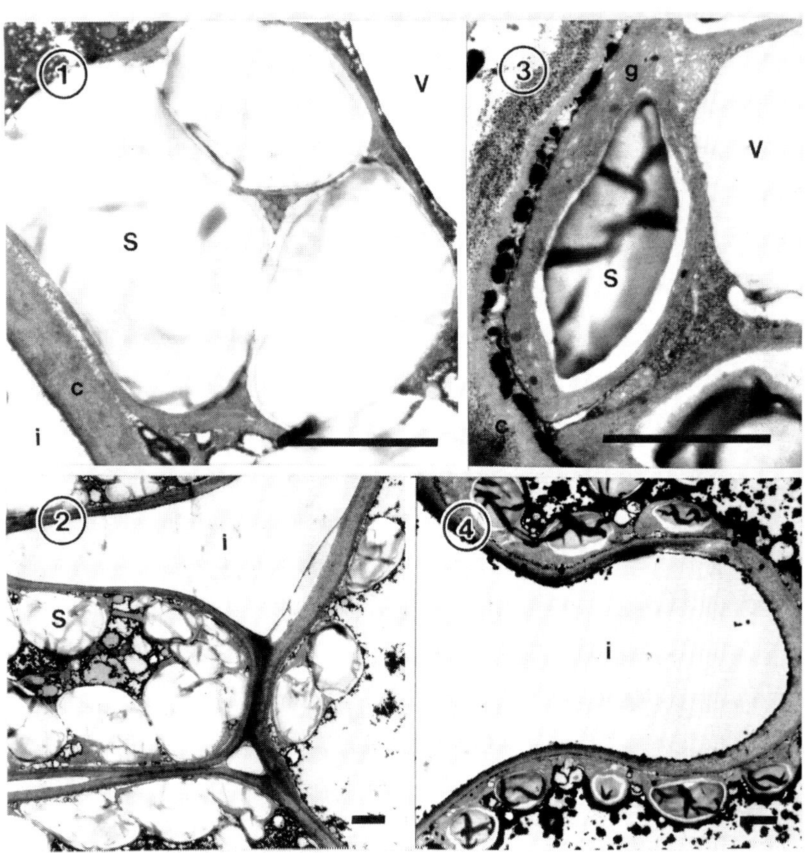

Fig. 1 Electron micrograph of mesophyll chloroplasts from plants grown for 12 months under low nitrogen (4 g m^{-2}), water stress (cyclically to –1.5 MPa predawn xylem pressure potential) and either elevated (720 µmol mol^{-1}; 1, 2) or ambient (365 µmol mo–^{1}; 3, 4) as described by Prichard et al.[6] Note the much smaller starch inclusions in the ambient than the elevated CO_2 treatment. S, starch grains; g, granum; c, cell wall, V, vacuole, I, intercellular space. Scale bar = 2 µm.

photosynthetic capacity. Leaf mesophyll and vascular tissue cross-sectional areas can increase leading to increases in leaf thickness,[7] as can an extra layer of palisade cells.[8] Changes in leaf size are more commonly caused by increased cell expansion rather than increased cell division. Leaf stomatal density has been shown to decrease with elevated CO_2 in some, but not all, cases.[7] Although decreased stomatal density and partial stomatal closure can increase WUE, it can also lessen the amount of CO_2 diffusing from the atmosphere into the leaf interior (thereby reducing photosynthesis and influencing gene expression).

Frequently, leaf chemistry also changes under high CO_2. Leaves can have lower nitrogen (N) concentration,[9–10] higher carbon to nitrogen ratios,[9–10] and (as mentioned earlier) higher levels of nonstructural carbohydrates.[6,9] These changes can affect interactions of leaves with pathogens and insect pests while still attached to the plant and decompositional processes following leaf fall.[11] Changes in epicuticular wax on high CO_2-grown leaves have also been observed.[12–13] These changes could affect water relations, susceptibility to pests and diseases, and surface properties that are important in chemical applications.

WHOLE PLANT SCALE

Plants grown under elevated CO_2 generally have more leaves per plant and higher total leaf area per plant,[7] which usually, but not always, results in greater leaf area index.[14–15] The increase in leaf number and leaf area, coupled with increases in leaf-level photosynthesis, generally leads to an increase in plant biomass, including leaf biomass. Summaries have consistently shown that biomass response to atmospheric CO_2 enrichment varies between plants with a C_3 (33–40% increase) versus a C_4 (10–15% increase) photosynthetic pathway.[16] This assumes that other resources (water and nutrients) are not limiting.

Plants will allocate energy and photosynthate to those organs associated with acquiring the most limiting resource. Thus, when CO_2 is elevated, plants tend to allocate more to belowground growth; roots often show the largest response to elevated CO_2.[17] So, in addition to C_3 versus C_4 differences, plants also vary in their response to CO_2 depending on phenology or growth habit. Plants with large storage organs for the extra carbohydrates produced (i.e., tubers or woody root systems) tend to have a greater response to elevated CO_2 and experience less photosynthetic down-regulation, than those without such organs (i.e., small, fibrous root systems).[18]

One benefit of high CO_2 is thought to be a decrease in drought stress. This would seem logical given larger root systems being able to supply more water coupled with a decrease in transpiration. However, it is important to note that while an increase in leaf-level photosynthesis generally indicates an increase in whole plant growth, a decrease in leaf-level transpiration does not necessarily translate to a reduction in whole plant water use. Plants grown under elevated CO_2 are generally larger (assuming adequate supply of other resources) and larger plants can use more water on a whole plant basis. Such feedbacks on leaf-level responses that are manifested at the whole plant level are further modified at the canopy scale.[19]

CANOPY AND ECOSYSTEM SCALE

Although molecular, cellular, and whole leaf response to CO_2 enrichment ultimately control a plant's C budget and growth, it is difficult to extrapolate from such fine-scale responses to an entire crop or forest canopy because of feedbacks that operate at these broader scales. For example, the relationship between stomatal conductance and transpiration commonly observed at the leaf or whole-plant level sometimes disappears at the canopy level. This is because, as transpiration decreases (due to partial stomatal closure induced by high CO_2 in and around leaves), the air in and directly above the canopy can become dryer, thereby steepening the water vapor diffusion gradient between the very wet leaf interior and canopy air (resulting in faster transpiration). This feedback, coupled with greater total leaf area index in CO_2 enriched canopies, likely explains the observation that stand-level transpiration is reduced far less than would be predicted based on leaf-level responses.

There is another potentially important canopy-scale feedback that is linked to partial closure of stomates induced by higher atmospheric CO_2 levels.[20–21] Reduced stomatal conductance and transpiration by individual leaves can slow canopy transpiration, thereby decreasing transpirational cooling and increasing the temperature of individual leaves in addition to the entire plant canopy (Fig. 2).[3] Higher canopy temperatures can drive faster transpiration and can modify the physiological responses of individual cells, leaves, and whole plants discussed earlier. Increased canopy temperatures caused by less cooling in CO_2-enriched environments, combined with increases in air temperatures caused by global warming, could be especially important for field crops. Wheat and rice yields, for example, commonly decrease by ~5% for each one degree Celsius increase in air temperature.[2]

Not all plants respond the same to elevated CO_2, therefore shifts in competitive interactions among plants growing in mixed stands, as is the case with unmanaged systems, are inevitable. Clearly, there will be winners and losers as atmospheric CO_2 concentration continues to rise, which could substantially complicate interpretation of leaf, cellular, and whole-plant responses to elevated CO_2. For instance, we grew an assemblage of species common to Southeastern U.S. longleaf pine savannahs under ambient and elevated CO_2 and predicted that broadleaf plants would have a greater response to CO_2 than would pines.[22] Total plant biomass and gross primary productivity (GPP = total

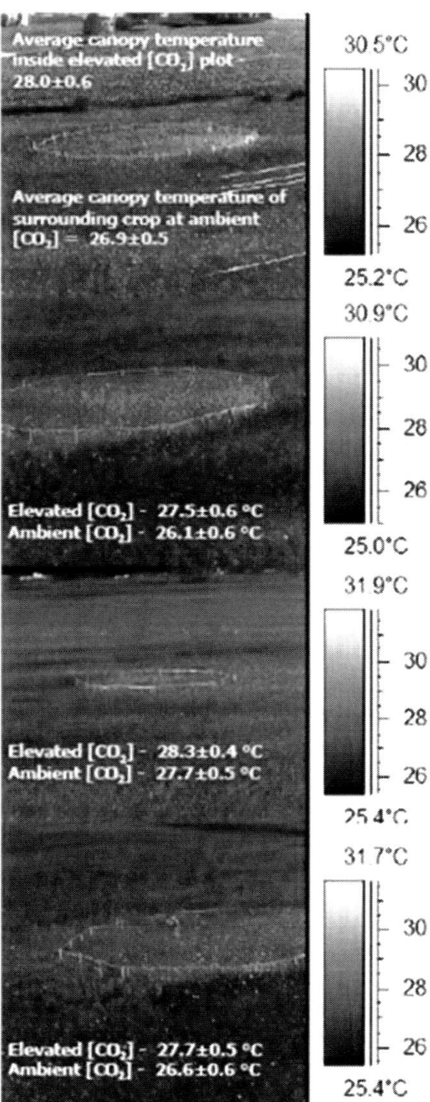

Fig. 2 Infrared image showing that maize growing inside an elevated [CO_2] plot was warmer than maize growing under ambient [CO_2] outside the plot at 15.30 h on 15 July 15, 2004. At that time, the average canopy temperature inside the four elevated [CO_2] plots at SoyFACE was 27.9°C + 0.2°C, significantly higher than the canopy temperature under ambient [CO_2] outside the plots (26.8°C + 0.3°C).
Source: This figure was reproduced with permission from the author and is copyrighted © by the Royal Society, London.[3]

photosynthesis or photosynthesis per unit land area) were greatly increased under high CO_2, and there were winners and losers, but not as we expected. The oaks had no response to twice ambient levels of CO_2 and the pines out-competed all other species. The pines had begun to reach canopy closure after only three years of CO_2 exposure and the understory species, wiregrass in particular, were being shaded out indicating that elevated CO_2 had substantially altered ecosystem structure. These sorts of ecosystem-level responses to elevated CO_2 could cancel out leaf-level responses in some cases or enhance those same responses in others.

Despite the canopy-scale feedbacks discussed earlier, most researchers have found stimulation of both GPP and net primary productivity (NPP = the amount of carbon incorporated into plant biomass or GPP-autotrophic respiration) following exposure to elevated level of atmospheric CO_2.[23–25]

CONCLUSION

Leaves represent the first contact between atmosphere and biosphere. An entire biological chain of events begins with atmospheric CO_2 entering leaves, being processed via photosynthesis into starch that ultimately feeds the planet. Rising levels of atmospheric CO_2 impact leaves at various scales which build from effects on the enzyme rubisco, to changes in leaf cells which can increase leaf size and alter leaf function. These changes usually lead to larger, leafier plants which increase plant yield and ecosystem productivity. Still, leaf responses to elevated atmospheric CO_2 need to be further elucidated if we are to accurately understand and predict the development and growth of crops under future environmental conditions. Future research will need to address the following:

1. How alterations in leaf structure are related to both cellular and higher-level growth processes
2. Whether increased cell division is driven by greater rates of cell expansion or by molecular cues
3. The role of ultrastructural, anatomical, and morphological leaf adjustments in photosynthesis
4. How CO_2-induced changes in leaf structure and leaf function will alter the degree of impact of multiple plant stresses (abiotic and biotic)
5. How CO_2-induced shifts in leaf quality and/or changes in leaf-surface features (e.g., epicuticular waxes, trichomes) will alter herbivore and pathogen attacks
6. How elevated CO_2 affects mechanisms of herbicide tolerance in plant;
7. How changes in leaf anatomy and physiology impact interactions with insect and disease pests
8. How changes in leaf chemistry impact decomposition and the residence time of leaf-derived carbon in the soil

REFERENCES

1. Keeling, C.D.; Whorf, T.P. Atmospheric CO_2 records from sites in the SIO air sampling network. In *Trends: A Compendium of Data on Global Change*; Carbon Dioxide Information Analysis Center: Oak Ridge National Laboratory, Oak Ridge, TN, 2001; 14–21 pp.

2. Pritchard, S.G.; Amthor, J.S. *Crops and Environmental Change*; Haworth Press: Binghamton, NY, 2005.
3. Leakey, A.D.B. Rising atmospheric carbon dioxide concentration and the future of C_4 crops for food and fuel. Proc. R. Soc., London, Ser. B, **2009**, *276*, 1333–2343.
4. Ahuja, I.; de Vos, R.C.H.; Bones, A.M.; Hall, R.D. Plant molecular stress responses face climate change. Trends Plant Sci. **2010**, *15*, 664–674.
5. Ainsworth, E.A.; Rogers, A. The response of photosynthesis and stomatal conductance to rising [CO_2]: Mechanisms and environmental interactions. Plant, Cell, Environ. **2007**, *30*, 258–270.
6. Pritchard, S.; Peterson, C.; Runion, G.B.; Prior, S.; Rogers, H. Atmospheric CO_2 concentration, N availability, and water status affect patterns of ergastic substance deposition in longleaf pine (*Pinus palustris* Mill.) foliage. Trees **1997**, *11*, 494–503.
7. Pritchard, S.G.; Rogers, H.H.; Prior, S.A.; Peterson, C.M. Elevated CO_2 and plant structure: A review. Glob. Chang. Biol. **1999**, *5*, 807–837.
8. Thomas, J.F.; Harvey, C.H. Leaf anatomy of four species grown under continuous CO_2 enrichment. Bot. Gaz. **1983**, *144*, 303–309.
9. Runion, G.B.; Entry, J.A.; Prior, S.A.; Mitchell, R.J.; Rogers, H.H. Tissue chemistry and carbon allocation in seedlings of *Pinus palustris* subjected to elevated atmospheric CO_2 and water stress. Tree Physiol. **1999**, *19*, 329–335.
10. O'Neill, E.G.; Norby, R.J. Litter quality and decomposition rates of foliar litter produced under CO_2 enrichment. In *Carbon Dioxide and Terrestrial Ecosystems*. Koch, G.W.; Mooney, H.A. Eds.; Academic Press: New York, 1996; 87–103 pp.
11. Torbert, H.A.; Prior, S.A.; Rogers, H.H.; Wood, C.W. Elevated atmospheric CO_2 effects on agro-ecosystems: Residue decomposition processes and soil C storage. Plant Soil **2000**, *224*, 59–73.
12. Graham, E.A.; Nobel, P.S. Long-term effects of a doubled atmospheric CO_2 concentration on the CAM species Agave deserti. J. Exp. Bot. **1996**, *47*, 61–69.
13. Prior, S.A.; Pritchard, S.G.; Runion, G.B.; Rogers, H.H.; Mitchell, R.J. Influence of atmospheric CO_2 enrichment, soil N, and water stress on needle surface wax formation in *Pinus palustris* (Pinaceae). Am. J. Bot. **1997**, *84*, 1070–1077.
14. Drake, B.G.; Gonzalez-Meler, M.A. More efficient plants: A consequence of rising atmospheric CO_2? Annu. Rev. Plant Physiol. Plant Mol. Biol. **1997**, *48*, 609–639.
15. Kimball, B.A.; Kobayashi, K.; Bindi, M. Responses of agricultural crops to free-air CO_2 enrichment. Adv. Agron. **2002**, *77*, 293–368.
16. Kimball, B.A. Carbon dioxide and agricultural yield: An assemblage and analysis of 430 prior observations. Agron. J. **1983**, *75*, 779–788.
17. Rogers, H.H.; Runion, G.B.; Krupa, S.V. Plant responses to atmospheric CO_2 enrichment with emphasis on roots and the rhizosphere. Environ. Pollut. **1994**, *83*, 155–189.
18. Runion, G.B.; Finegan, H.M.; Prior, S.A.; Rogers, H.H.; Gjerstad, D.H. Effects of elevated atmospheric CO_2 on non-native plants: Comparison of two important southeastern ornamentals. Environ. Control Biol. **2011**, *49(3)*, 107–117.
19. Runion, G.B.; Mitchell, R.J.; Green, T.H.; Prior, S.A.; Rogers, H.H.; Gjerstad, D.H. Longleaf pine photosynthetic response to soil resource availability and elevated atmospheric carbon dioxide. J. Environ. Qual. **1999**, *28*, 880–887.
20. Rogers, H.H.; Thomas, J.F.; Bingham, G.E. Response of agronomic and forest species to elevated atmospheric carbon dioxide. Science **1983**, *220*, 428–429.
21. Rogers, H.H.; Dahlman, R.C. Crop responses to CO_2 enrichment. Vegetatio **1993**, *104/105*, 117–131.
22. Runion, G.B.; Davis, M.A.; Pritchard, S.G.; Prior, S.A.; Mitchell, R.J.; Torbert, H.A.; Rogers, H.H.; Dute, R.R. Effects of elevated atmospheric carbon dioxide on biomass and carbon accumulation in a model regenerating longleaf pine community. J. Environ. Qual. **2006**, *35*, 1478–1486.
23. Ellsworth, D.S.; Oren, R.; Huang, C.; Phillips, N.; Hendrey, G.R. Leaf and canopy responses to elevated CO_2 in a pine forest under free-air CO_2 enrichment. Oecologia **1995**, *104*, 139–146.
24. DeLucia, E.H.; Hamilton, J.G.; Naidu, S.L.; Thomas, R.B.; Andrews, J.A.; Finzi, A.; Lavine, M.; Matamala, R.; Mohan, J.E.; Hendrey, G.R.; Schlesinger, W.H. Net primary production of a forest ecosystem with experimental CO_2 enrichment. Science **1999**, *284*, 1177–1179.
25. Witting, V.E.; Bernacchi, C.J.; Zhu, X.G.; Calfapietra, C.; Ceulemans, R.; Deangelis, P.; Gielen, B.; Miglietta, F.; Morgan, P.B.; Long, S.P. Gross primary production is stimulated for three *Populus* species grown under free-air CO_2 enrichment from planting through canopy closure. Glob. Chang. Biol. **2005**, *11*, 644–656.

Mangrove Forests

Ariel E. Lugo
International Institute of Tropical Forestry, Forest Service, U.S. Department of Agriculture (USDA-FS), Río Piedras, Puerto Rico
Ernesto Medina
International Institute of Tropical Forestry, Forest Service, U.S. Department of Agriculture (USDA-FS), Río Piedras, Puerto Rico, and Center for Ecology, Venezuelan Institute for Scientific Research, Caracas, Venezuela

Abstract
The mangrove environment is not globally homogeneous, but involves many environmental gradients to which mangrove species must adapt and overcome to maintain the familiar structure and physiognomy associated with the mangrove ecosystem. The stature of mangroves, measured by tree height, decreases along the following environmental gradients from low to high salinity, low to high wind speed, high to low air temperature, high to low nutrient availability, and high to low rainfall. Litterfall, an indirect measure of mangrove productivity, decreases along the same environmental gradients. Mangrove stature is low at the two extremes of the inundation period (hydroperiod) and peaks at intermediate levels of inundation. The main factors that control mangrove structure and function are the latitudinal temperature gradient, regional presence/absence of hurricanes, nutritional status of mangrove substrates, and local salinity gradients.

INTRODUCTION

Mangroves are woody plants that grow in saline soils. Tomlinson[1] considered 20 genera with 54 species as mangrove species of which 9 genera and 34 species constituted "true mangroves." He established five criteria to typify a true mangrove, including the ecophysiological capacity for excluding salinity. The other four criteria were the following: complete fidelity to the mangrove environment, a major role in the structure of the community and capacity to form pure stands, morphological specialization that adapts them to their environment, and taxonomic isolation. Tomlinson also listed 46 genera and 60 woody species as mangrove associates and was emphatic that the number of mangrove associates is a potentially large list of species, given the many habitats that converge with the mangrove environments:

The ecological literature seems incapable of being reduced to a simple set of rules to account for the diversity of vegetation types within the broad generic concept of mangal. Lack of uniformity is... a measure of the plasticity of mangroves and their ability to colonize such an enormous range of habitats.

Tomlinson, p. 5[1]

In 2005, mangrove forests covered about 15 to 17 million hectares worldwide.[2,3] They occur on low-latitude coastal zones where waters with different levels of salinity flood forests at different frequencies and depths. The mangrove environment is diverse, and we use the many environmental gradients under which mangroves grow to organize this review. At the extremes of any of the environmental gradients that we discuss, mangroves not only function differently, but also may appear to be exceptions to generalities. For example, some mangroves appear to grow in freshwater, whereas others appear never to flood. In both cases, the incursion of seawater or floods occurs but at very low frequencies that require long-term observation. In spite of the complexity of gradient space under which mangroves occur, they are all forested and tidal wetlands in estuarine environments. Mangroves have global importance because their carbon sequestration and dynamics are in the same order as the unaccounted global carbon sink.[4] The term "blue carbon" is used to depict the carbon sink function associated with sediment burial in coastal vegetation, particularly mangroves.[5] Mangroves are also important to the functioning of coastal ecosystems, and have economic and cultural importance to people.[6] However, between 1980 and 2005, there was a 20% to 30% loss in the global mangrove area.[2,3] Fortunately, mangroves can recover from deforestation if socioeconomic and environmental conditions are favorable.[7] We focus only on natural environmental gradients and ignore anthropogenic gradients, which are reviewed in Lugo et al.[8] and Cintrón and Schaeffer Novelli.[9]

Encyclopedia of Natural Resources DOI: 10.1081/E-ENRL-120047500

THE ECOPHYSIOLOGICAL CHALLENGE OF THE MANGROVE ENVIRONMENT

Growing on saline, periodically flooded, and low-oxygen environments imposes severe restrictions on plant growth. The effects of other environmental variables are exerted through salinity stress and oxygen supply at the root level. Salt exclusion from roots has high energy cost and effects on plant structure and nutrient uptake such that for plants in saline environments, the salt balance is more critical than the water balance. For example, restricted water uptake in halophytes prevents excess accumulation of salt in their tissues. Thus, in contrast to non-halophytes, high atmospheric evaporative demand (as a result of high temperature and irradiation) cannot be compensated in halophytes by high transpiration because of the resulting salt accumulation inside the plant. Restriction of freshwater supply to mangroves primarily affects salt concentration at the root level, thus affecting water and nutrient uptake. In the high-radiation environment where mangroves grow, maintenance of leaf temperature within physiological limits is strongly dependent on evaporative cooling provided by transpiration. However, because salinity restricts water uptake and transpiration, mangroves maintain leaf temperature and optimize water-use efficiency through variations in leaf inclination, leaf area, and succulence.[10]

The hypoxia factor in mangrove wetlands is counteracted mainly through structural devices allowing oxygen to reach waterlogged roots. Mangroves do not tolerate deep flooding for prolonged periods although in true mangroves, tolerance to periodic flooding is high. Finally, the gradient analysis that we undertake here is complex, particularly in the case of nutrients where actual gradients in the field are difficult to establish. Although there are certainly nutrient-rich (estuarine, high-runoff, high-rainfall) and nutrient-poor (calcareous islands in semiarid areas) sites, an actual nutrient availability gradient is observed only in connection with salinity gradients, as salinity interferes with nutrient uptake.

Temperature

Mangroves are considered a lowland tropical ecosystem, where normally tree species do not tolerate frost. However, mangrove forests occur over a wide range of temperature and a wide latitudinal range that extends from the equator to 30° and 38° north and south, respectively, for *Avicennia marina*.[11] Similar latitudinal range is observed for *Avicennia* spp. and *Laguncularia racemosa* in the Neotropics.[12] These high-latitude mangroves are clearly no longer tropical forests *sensu stricto*, as they may reproduce and regenerate in lowlands with frost, i.e., in warm temperate and temperate life zones.

Woody vegetation on coastal systems is completely replaced by herbaceous vegetation above 32° and 40° north and south, respectively.[12,13] In Florida, *Spartina alterniflora* salt marshes compete with mangroves for space at the coastal fringe, a competition that is mediated by freezing events,[14] with mangroves losing ground with increased freezing frequency or intensity or both.[15] *Spartina brasiliensis* and *S. alterniflora* co-occur with mangroves in the Neotropics,[16,17] but their competition outcome depends on a different ecological constraint, possibly depth of flooding and sediment erosion and deposition dynamics.

Mangroves cope with the latitudinal temperature gradient with a change of species. In Florida, for example, four mangrove species occur in the keys and Everglades, but as frost frequency increases northward, all species but *A. germinans* drop out.[18] In the Brazilian coast, two species of mangroves reach the same latitudinal limit (28°30′S), but only *A. schaueriana* grows as a large tree, *L. racemosa* grows stunted as a shrub.[12] Among mangrove species, *A. marina*, *A. schaueriana*, *A. germinans*, and *L. racemosa* are the most tolerant to lower temperatures. Duke[11] found that as the temperature decreases, *A. marina* undergoes significant phenological changes involving leaf production, flowering, and fruiting, such that with lower temperatures, growth rates and reproductive potential diminish. Distributional limits for this species coincide with trends toward zero reproductive success. For each 10°C increase, growth increased by a factor of two or three. Stuart et al.[13] found that at freezing temperatures, water deficits cause freeze-induced xylem failure (xylem embolism). Species with wider vessels experienced 60% to 100% loss of hydraulic conductivity after freezing and thawing under tension, whereas species with narrower vessels lost as little as 13% to 40% of hydraulic conductivity. They suggest that freeze-induced embolism may play a role in limiting latitudinal distribution of mangroves either through massive embolism or through constraints on water transport as a result of vessel size. For canopy leaves of *Rhizophora stylosa* at its northern limit in Japan (26°11′N), light-saturated maximum photosynthesis rate decreased with decreasing temperature reaching minimum values in January and maximum in June.[19]

Our observations of mangroves at high latitudes along the coasts of Florida, Brazil, and Australia show that trees become shrubby and leaves turn yellowish at the extremes. Multiple sprouting increases as well. However, in global-warming scenarios for the future, mangrove species are expected to move to higher latitudes in both hemispheres and establish closed-canopy mangrove forests where *Spartina* marshes grow today.

The effects of high insolation and its influence on photoinhibition confound the determination of high-temperature tolerance of mangroves. However, Biebl[20] found species differences in the temperature range of tolerance of four mangrove species in Puerto Rico: *Conocarpus erectus* −3°C to 54°C, *L. racemosa* −1°C to 51°C, *R. mangle* −3°C to 50°C, and *A. germinans* −3°C to 48°C. Notably, the high-temperature tolerance is higher than temperatures normally encountered by mangroves under natural conditions.

Rainfall

Mangroves occur from rain forest to very dry forest life zones, which cover a gradient from 500 to about 8000 mm annual rainfall. This wide rainfall gradient influences soil salinity and forest structure. In Mexico, for example, Méndez Alonzo et al.[21] found that average tree height, average tree dbh (diameter at breast height), and leaf mass per unit area of *Avicennia* trees increased with both precipitation (500 to 3000 mm/yr) and minimum annual temperature (−2°C to 14°C). High-rainfall areas also increase the likelihood of coupling mangroves to terrestrial nutrient sources through runoff. Those mangroves that receive both high rainfall and runoff are usually the most complex and productive mangroves of all, particularly those growing under riverine conditions.

Many mangroves, however, are isolated from significant terrestrial runoff and thus depend on rainfall for sustaining their primary productivity. In these cases, their development is proportional to rainfall. Those that receive terrestrial runoff (including groundwater) are less dependent on rainfall for sustaining primary productivity. Mangroves in dry life zones sometimes receive riverine runoff that originates in upland moist, wet, or rain forest life zones as happens with the mangroves at Tumbes in Peru[9] and those in the strait connecting Lake Maracaibo with the Caribbean Sea in western Venezuela, which receive runoff from Río Limón.[22,23] However, mangroves in dry life zones are generally exposed to drought and high soil salinity. *A. germinans* exhibits reduced photosynthetic rates when exposed to drought, irrespective of salinity,[24] although the reduction was stronger in plants grown at lower salinities. Coincidentally, plants under higher salinity have a higher osmotic potential in their leaves, which allows them to be more effective in water uptake as we discuss in the following text.

Salinity

Mangroves *sensu stricto* are halophytes with elevated salt concentration in their cells. They require a higher level of sodium (Na) than non-halophytes for optimum growth, achieved under controlled conditions at salinity equivalent to 25% of seawater.[25–27] The salinity gradient under which mangroves grow ranges from freshwater to about three times the salinity of seawater.[28] For a given species, such as *Rhizophora mangle*, mangrove height [28] and leaf size decreases,[29,30] whereas leaf thickness increases[31] with increasing salinity. Osmolality (concentration of osmotically active solutes) of leaf sap in mangroves increases with soil salinity.[32] Leaf area decreases linearly as osmolality increases with *R. mangle* appearing to be more sensitive than *A. germinans* and *L. racemosa.* Also, as salinity increases along a spatial gradient, mangrove species respond by forming monospecific zones according to their salinity tolerance.[33] The changes in species along a salinity gradient are usually accompanied by changes in photosynthetic rates and nutrient-use efficiency,[29] and when exposed to hypersaline conditions (>100‰), quick and massive mangrove mortalities ensue.[34,35]

Tolerance to high levels of salt in the vacuoles of mangroves requires the accumulation of "compatible solutes" in the cytoplasm to prevent dehydration of proteins. Those compatible solutes, which accumulate in the cytoplasm without toxic effects, counteract the osmotic effect of ions in the vacuoles. Among the most common are cyclitols such as mannitol in the species of *Aegiceras* and *Sonneratia*, pinitol in the species of *Bruguiera* and *Aegialitis*, methyl-muco-inositol in *Rhizophora* spp., nitrogen-containing glycinbetaine in *Avicennia* spp., and proline in the species of *Aegialitis* and *Xylocarpus*.[36,37] Mangroves accumulating betaines and proline as compatible solutes also have higher total nitrogen (N) concentration in their leaves than mangroves accumulating cyclitols.[22,29,32]

Although under natural conditions mangroves can grow in freshwater, at some point in their life cycle, they have to be exposed to seawater, as they are not effective competitors with non-halophytic species. Ball and Pidsley[38] showed that the distribution of *Sonneratia alba* and *S. lanceolata*, two species that grow at low salinities and freshwater, followed the temporal distribution of salt incursion nicely. Neither species grew under conditions where salinity was always absent, reaching just as far into the salinity gradient as was seasonally present. Biomass accumulation in *S. lanceolata* peaked at very low salinities and rapidly decreased with increasing salinity.[39] Mangrove trees grow to the largest sizes and biomass in estuaries with salinities well below sea water. Some species of mangroves such as *A. germinans* behave as euryhaline species with a wide tolerance to salinity. They tend to grow in basins where salinity can range through the whole range of tolerance of mangroves.[40] In contrast, species such as *R. mangle* behave as a stenohaline species with a narrower range of salinity tolerance. They occur mostly on fringes where salinities range narrowly and their range is from almost freshwater to salinities in the order of twice seawater.[41]

The salinity gradient in mangroves develops with distance from the ocean (higher-salinity inland in dry and moist environments and lower-salinity inland in wet and rainy environments), with distance from freshwater runoff (higher salinity toward the ocean), or seasonally depending on rainfall and runoff events or periodic droughts. Salinity gradients are sharper in arid coastlines where differences between seawater and inland hypersaline soils can span the range of tolerance of all mangroves. Interstitial soil water also exhibits a vertical salinity gradient from surface waters (usually lower salinity) to deep-soil water (salinity increases with depth).[42] However, freshwater discharges of aquifers can reverse the gradient and make soil water less saline than surface water, if the surface water is fed by tides.

The photosynthetic rate of mangrove species decreases with increasing salinity.[24,40,43,44,46] In *L. racemosa* and *A. germinans*, net photosynthesis and leaf conductance decrease in parallel as the salt concentration of the nutrient solution increases from 0‰ to 30‰ and 55‰, respectively.[43,45] In contrast to reports on other *Avicennia* species under natural conditions,[44,47] *A. germinans* grows well and shows higher photosynthetic rates in nutrient solutions without added salt than those grown at salinities above 10‰. Potassium (K) is the main ion accumulated in the leaves under those conditions.[45] Measurements of gas exchange during rainy and dry seasons showed that *A. germinans* and *C. erectus* maintain much higher photosynthetic rates and lower salinities of leaf sap during the rainy season.[40]

At intermediate to high salinities (30‰ to 55‰), fertilization with N or phosphorus (P) slightly increases photosynthetic rates of *Avicennia* and *Rhizophora*, but not over the levels of control trees,[48] i.e., fertilization is less effective in overcoming the effects of high salinity.

Mangrove associates are species that are either tolerant to salinity to a certain degree or that are present at those stages in their life cycle that are tolerant to high salt concentrations. *Pterocarpus officinalis* belongs to the first group. It is a mangrove associate throughout the Caribbean and northern South America in areas with high rainfall or surface runoff.[49,50] This species coexists with species of *Rhizophora* and *Laguncularia*, is able to restrict Na input to leaves, and appears to have a high affinity for K. The fern *Acrostichum aureum* is in the second group of mangrove associates, a species found throughout the tropics and that at times competes with mangrove establishment. The sporophytic phase of this fern has a salt tolerance similar to that of coexisting mangrove species,[51] but the gametophytic phase cannot survive under saline conditions; therefore, the distribution of the mangrove fern is restricted to areas with high rainfall or low salinities that allow the sexual reproduction of the species.

Hydroperiod

The importance of hydroperiod to mangrove structure and species zonation was first recognized by Watson,[52] who estimated the hydroperiod based on the number of tidal events that flooded particular areas of mangroves. Mangroves that grow over patches of coral reef or are located on off-coast overwash islands, or on fringes below low tide, are usually continuously flooded or experience a long hydroperiod. In contrast, some inland mangroves appear to always grow on dry land, except perhaps during the highest tidal events, storm tides, or excessive rainfall or runoff events; their hydroperiod is short. These extreme points in the hydroperiod gradient create numerous ecophysiological challenges to mangroves. Krauss et al.[53] found little effect of hydroperiod on gas exchange of seedlings and saplings of three mangrove species from south Florida, confirming studies that showed that the flooding regime affected mostly the maintenance of leaf area and biomass partitioning.[54]

Oxygen availability to roots can become limiting under long hydroperiods. Mangroves exposed to long hydroperiods or to abnormally high water depth develop adventitious roots on the water surface, which supplement oxygen supply to below-water parts. Water movement by tidal or runoff forces also mitigate low-oxygen conditions during chronic inundation. Low root respiration in mangroves[55] is another mitigating effect to low oxygen supply. In contrast to the long hydroperiod, mangroves with a short hydroperiod can experience excessive soil salinity or drought. Between these two extremes grow most of the mangroves with different degrees of soil oxygenation and salinity. Both long and short hydroperiods inhibit understory development and mangrove regeneration and reduce mangrove height.

As sea level rises, some coastal zones flood beyond the tolerance of mangroves. On the other hand, the saline wedge penetrates much further inland into estuaries and provides an opportunity for mangrove expansion. An example of how this phenomenon may proceed has been observed in one of the main tributaries of the Orinoco River, from which a large fraction of its freshwater supply was reduced due to the damming upriver. The consequence of the change in hydrology was that mangroves moved hundreds of kilometers inland.[56] In south Florida, mangroves expanded 3.3 km inland with a sea-level rise of 10 cm between 1940 and 1994.[57]

Hydrologic Energy

In general, mangroves occupy low-energy coastlines. In high-energy coastlines, mangroves grow behind sand dunes where waters are calm. However, fringe mangrove forests on different sections of low-energy coastlines face different levels of wave energy. This energy gradient is poorly studied but is now relevant to understanding mangrove responses to increased sea level[58] and tsunamis. Nevertheless, mangrove response to sea-level rise will depend not only on the amount of area with saline soils but also on air temperature (greater frost frequency at higher latitudes), competition with salt marshes[14] and other vegetation,[58] and the balance between the rate of sea-level rise, input of allochthonous sediments, and mangrove production of peat.[59]

Usually, mangrove peat and root systems are undercut by rising sea level, and it is common to see overturned mangroves in places where this process occurs. Mangrove stems and prop roots offer resistance to incoming tides and waves and in so doing, reduce the hydrologic energy dissipated on coastlines. Under high-energy conditions, prop root and stem density increase, allowing mangroves to perform this ecological service of coastline protection. However, this process occurs within a limited range of tolerance to high-wave or tidal energy. In the tropical Atlantic Brazilian coast, *Avicennia* is frequently at the fringe, and this species is less tolerant to the erosive force of tides, coastal currents, and strong winds.[42]

In locations affected by macrotides, where large rivers discharge, as in the Atlantic coast of Brazil north of the mouth of the Amazon river, mangroves face strong effects from erosion and sediment deposition. Batista et al.[60] used remote sensing to show that along the coast of Amapa, Brazil, during the period 1980–2003, large erosion rates caused the disappearance of 1.37 km^2/year of mangroves on one location, whereas in another location, progradation added 56 km^2 of mangrove areas to the shoreline. Such dramatic shifts in erosive and sedimentary forces maintain mangrove vegetation in a constant state of successional change.

When mangroves are massively killed by drought, hurricanes, or other disturbances, the organic peat on which they grow, collapses under the erosive power of tides and waves and because the production of organic matter by the forest ceases. This collapse of the forest floor causes the intrusion of seawater and converts the forest into a lagoon.[28,61] This sets succession back by many decades as it takes time for the mangroves to reverse a lagoon environment back to forest growing at the higher elevation afforded by the accumulation of peat and roots. McKee et al.[62] increased the rate of mangrove root accumulation, and therefore the rate of peat accretion by fertilizing mangroves in the Caribbean coast of Belize.

Mangroves are overtaken by tsunamis and are exposed to significant structural effects.[63,64] However, behind the mangroves, there is less destruction of property because of the energy dissipation involved in overcoming mangrove resistance to wave energy (search for mangroves and tsunamis in http://www.fao.org).

Nutrients

The levels of nutrient availability experienced by mangroves range from oligotrophic with extreme P limitation to highly eutrophic. Oligotrophy occurs in carbonate environments[65,66] or over acid peat soils.[67] The area covered by nutrient-limited mangroves is large and mostly in the wider Caribbean (including the Everglades of Florida) and many Pacific atolls. Nevertheless, mangroves in general are eutrophic systems and without nutrient limitation. Eutrophy occurs on alluvial floodplains and riverine fringes. Polluted coastlines also provide eutrophic conditions for mangrove growth. Within this generally favorable nutrient availability, conditions may vary with the type of water entering the mangroves (nutrient rich from land, nutrient poor from the ocean). For example, Chen and Twilley[68] found a gradient of mangrove structure and productivity from the mouth of the Shark River to inland sites in south Florida. The biotic gradient was not responding to salinity or sulfide concentration but to N and P concentrations in soil pore water. Fertile sites were dominated by *L. racemosa*, whereas *R. mangle* grew in the less fertile sites. These forests exhibited seasonal changes in photosynthetic rates in response to changes in air temperature and light intensity.[69]

Oligotrophy leads to the formation of dwarf mangroves, which are low-height trees (no more than 1 m) with normal leaf sizes and reduced leaf-turnover rates. Fertilization experiments with mangroves in Belize, Florida, and Panama illustrate some of the complexities associated with nutrient limitations in mangroves.[70–73] Dwarf *R. mangle* and *A. germinans* mangroves in carbonate environments always respond to P fertilization, but surrounding fringing *R. mangle* mangroves respond only to N fertilization, and mangroves under intermediate conditions responded to both N and P fertilization.[70] In a disturbed mangrove forest in the Indian River Lagoon, both *R. mangle* and *A. germinans* responded to N fertilization but not to P fertilization along a tree-height gradient.[71] Contrasting responses to nutrient fertilization by the same species under different environments differentially affected ecological process of the species.[72] Among the biotic responses affected by fertilization were changes in plant habit from stunted to larger-sized individuals, which were accomplished by increasing wood relative to leaf biomass and changes in leaf-specific area, nutrient uptake, and leaf herbivory.

Feller et al.[73] also observed changes in the cycling of nutrients at a stand level through leaf fall and within-stand cycling. They found that in P-limited environments, retranslocation (or resorption) of P by *R. mangle* is much higher (≈70%) than that of N (≈45%). Nitrogen fertilization did not change those percentages, but P fertilization decreased P resorption efficiency (<50%) but increased N resorption (≈70%). Studies of dwarf mangroves on P-deficient peats in Puerto Rico [67] and stunted mangroves under hypersaline conditions[29] confirmed larger resorption values for P than for N. It appears that under conditions of P limitation, resorption of P is greater than that of N and that a sufficient supply of P reduces its resorption levels below those of N.

In Panama, Lovelock et al.[74] found that fertilized dwarf mangroves responded to N and P fertilization by increasing hydraulic conductance sixfold by P and 2.5-fold by N. The response of the hydraulic conductivity, more than the photosynthetic response per unit leaf area, accounted for the increased in size of the fertilized mangroves.

Lovelock et al.[55] found that *R. mangle* root respiration per unit mass was low in Belize compared to temperate tree species at the same temperature. Root respiration did not differ significantly between zones (fringe vs. dwarf) and fertilization treatments (N or P), although rates were consistently higher after fertilization, particularly in dwarf mangroves. The fine roots fertilized with P responded to fertilization with increased P concentrations.

Wind

Coastal zones are usually windy mostly due to sea breezes, which favor gas exchange through their influence on gaseous gradients across leaf surfaces. Sea breezes ventilate the forest and moderate air temperatures. A significant fraction of mangroves occur in the hurricane belt, which results in periodic exposures to extreme wind events (velocities >100 km h^{-1} are common and can exceed 250 km h^{-1}). In the process, hurricanes dissipate high levels of energy over forests, as it happened when Hurricane Hugo dissipated about 210 J m^{-2} s^{-1} over the northeastern mangroves of Puerto Rico.[75] Cintrón and Schaeffer Novelli[76] found that for Neotropical mangroves, the maximum canopy height decreased with increasing latitude, where winds (Caribbean) and frost (Florida and Brazil) become significant factors affecting forest structure. The large tree sizes in the Pacific Island of Kosrae vs. Pohnpei,[77] or San Juan River mangroves in Venezuela[78] vs. mangroves in nearby Caribbean Islands are examples of wind effects on mangrove stature. Hurricane winds exert selective pressure on forests by periodically trimming or overturning taller trees, which are usually the ones most affected by wind energy.[75] As a result, hurricane-affected forests have lower stature and wind-sculptured canopies.

Redox

Redox gradients occur in mangroves in association with anoxic conditions in mud and the relatively high concentration of sulfate in seawater. Sulfur is the fourth-most abundant element in seawater after chlorine (Cl), Na, and magnesium (Mg). Under reduced conditions (no oxygen), sulfur is present as sulfide, which is toxic for mangroves and reduces photosynthesis and growth.[79] Mangroves mitigate sulfide accumulation around the roots by the transport of oxygen through aerenchyma cells.[80,81] The ecological significance of the redox gradient in soils is expressed by its effects on the rates of microbial decomposition processes as well as in the diversity of both microbial communities and metabolic pathways associated with increasingly reducing conditions with soil depth (Fig. 5.5 and Table 5.6 in Alongi).[82]

Other Organisms

In addition to mangrove trees, other groups of organisms respond to environmental gradients within the mangroves and in some cases, affect mangrove trees and ecosystem functioning. For example, gastropods respond to sediment metal concentration.[83] Crabs respond to tidal range, and epibionts increase with light availability but decrease with tidal energy. In the case of crabs, they act as ecosystem engineers, not only facilitating mangrove regeneration but also influencing soil aeration and transport of oxygen to anaerobic soil layers.[82]

A. germinans mangrove forests in Panama growing under different environmental conditions (Caribbean and Pacific coasts with different rainfall, hydroperiod, and soil salinity) differed in phenology, and invertebrate and bird composition.[84] Although the coastal zones shared 95% of the bird species, the mangroves only shared 34%, and each forest had a different feeding guild assemblage in the bird community. The results suggest that in spite of being the same mangrove species, the different environmental conditions along the environmental gradients ripple through the food chain and result in different community composition.

Integration and Spatial Scales of Mangroves

There is no mangrove model that considers all the gradients discussed here and uses the responses of mangrove organisms to these gradients to holistically explain mangrove structure and productivity (but see Lugo[33] for a zonation/succession diagram with many of the gradients included). Such a model would be extremely useful to mangrove conservation actions, including restoration and rehabilitation of mangrove sites. However, there have been several efforts to model or conceptualize mangrove functioning. Odum[85] developed trophic-level models of mangroves for south Florida, and Lugo et al.[86] used energy flow to simulate mangrove productivity and response to hurricanes. Cintrón et al.[28] developed a model of the effects of salinity on mangrove functioning, and Twilley et al.[87] modeled mangrove succession and applied it to restoration. Kangas[88] developed an energy theory for the landscape classification of wetlands, which dovetails with the stand classification of mangrove ecosystem types of Lugo and Snedaker.[89]

Thom[90–92] was the first to relate mangrove ecosystem structure and function with the geomorphology of coastlines. Twilley et al.[93] developed a spatial and functional hierarchy for assessing mangrove forests (Fig. 13.3 in Twilley et al.).[93] Their concept included the latitudinal or global distribution of mangroves at the top of the hierarchy. Within the latitudes, they included the environmental settings, which were based on Thom's geomorphological types. Geomorphologic conditions expose mangroves to different energy conditions such as direction and force of hydrological fluxes and origin and quality of waters interacting with the mangroves, i.e., coastal vs. inland waters. Inside these environmental settings, Twilley et al. included the ecological types of Lugo and Snedaker, which function in relation to topography and hydrology. Mangrove stands occur within any of the ecological types, and within a stand, both above and belowground processes take place. Twilley and Rivera Monroy[94] modified and refined their 1996 model and compiled what they called five ecogeomorphic models of nutrient biogeochemistry for mangrove wetlands. These models represent the state of understanding of the whole mangrove ecosystem functioning, taking into consideration the effects of multiple environmental gradients on these ecosystems.

CONCLUSION

Mangrove ecosystem function results in different rates of ecological processes and structural development depending on whether the mangroves are above or below the frost line, exposed or not to hurricanes, and in dry or rain forest climates. These overarching environmental extremes (temperature, rainfall, and wind disturbances) have controlling effects on how mangroves respond to environmental gradients and which species might predominate at the environmental extremes.

ACKNOWLEDGMENTS

This work was done in collaboration with the University of Puerto Rico. Mildred Alayón edited the manuscript. We received useful comments from Gilberto Cintrón, Frank Wadsworth, Blanca Ruiz, and two anonymous reviewers.

REFERENCES

1. Tomlinson, P.B. *The Botany of Mangroves*; Cambridge University Press: Cambridge, England, 1986.
2. Valiela, I.; Bowen, J.L.; York, J.K. Mangrove forests: one of the world's threatened major tropical environments. BioScience **2001,** *51*, 807–815.
3. FAO. *The World's Mangroves 1980–2005*; The Food and Agriculture Organization of the United Nations: Rome, Italy, 2007.
4. Bouillon, S.; Borges, A.V.; Castañeda-Moya, E.; Diele, K.; Dittmar, T.; Duke, N.C.; Kristensen, E.; Lee, S.Y.; Marchand, C.; Middleburg, J.J.; Rivera-Monroy, V.H.; Smith III, T.J.; Twilley, R.R. Mangrove production and carbon sinks: a revision of global budget estimates. Global Biogeochem. Cycles **2008,** *22*, GB2013. doi:2010.1029/2007GB003052.
5. Mcleod, E.L.; Chmura, G.L.; Bouillon, S.; Salm, R.; Björk, M.; Duarte, C.M.; Lovelock, C.E.; Schlesinger, W.; H. Silliman, B.R. A blueprint for blue carbon: toward an improved understanding of the role of vegetated coastal habitats in sequestering CO_2. Front. Ecol. Environ. **2011**, *9*, 552–560.
6. FAO. *Mangrove Forest Management Guidelines*; Food and Agriculture Organization of the United Nations: Rome, Italy, 1994.
7. Martinuzzi, S.; Gould, W.A.; Lugo, A.E.; Medina, E. Conversion and recovery of Puerto Rican mangroves: 200 years of change. Forest Ecol. Manag. **2009**, *257*, 75–84.
8. Lugo, A.E.; Cintrón, G.; Goenaga C. Mangrove ecosystems under stress. In *Stress Effects on Natural Ecosystems*; Barret, W.G.; Rosenberg, R., Eds.; John Wiley and Sons Limited: Sussex, England, 1981; 129–153.
9. Cintrón Molero, G.; Schaeffer Novelli, Y. Ecology and management of new world mangroves. In *Coastal Plant Communities of Latin America*; Seeliger, E., Ed.; Academic Press, Inc.: San Diego, CA, 1992; 233–258.
10. Ball, M.C.; Cowan, I.R.; Farquhar, G.D. Maintenance of leaf temperature and the optimization of carbon gain in relation to water loss in a tropical mangrove forest. Aust. J. Plant Physiol. **1988**, *15*, 262–276.
11. Duke, N.C. Phenological trends with latitude in the mangrove tree *Avicennia marina*. J. Ecol. **1990**, *78*, 113–133.
12. Schaeffer Novelli, Y.; Cintrón Molero, G.; Rothleder Adaime, R.; de Camargo, T.M. Variability of mangrove ecosystems along the Brazilian coast. Estuaries **1990**, *13*, 204–218.
13. Stuart, S.A.; Choat, B.; Martin, K.C.; Holbrook, N.M.; Ball, M.C. The role of freezing in setting the latitudinal limits of mangrove forests. New Phytol. **2007**, *173*, 576–583.
14. Kangas, P. C.; Lugo, A. E. The distribution of mangroves and saltmarshes in Florida. Trop. Ecol. **1990**, *31*, 32–39.
15. Lugo, A.E.; Patterson Zucca, C. The impact of low temperature stress on mangrove structure and growth. Trop. Ecol. **1977**, *18*, 149–161.
16. Braga, C.F.; Beasley C.R.; Isaac V.J. Effects of plant cover on the macrofauna of *Spartina* marshes in northern Brazil. Braz. Archi. Biol. Technol. **2009,** *52*, 1409–1420.
17. Colonnello, G. *Spartina alterniflora* Loisel, primer registro para Venezuela. Memorias Fundación La Salle de Ciencias Naturales **2001**, *59*, 29–33.
18. Odum, W.E.; McIvor, C.C. Mangroves. In *Ecosystems of Florida*; Myers, R.L.; Ewel, J.J., Eds.; University of Central Florida Press: Orlando, FL, 1990; 517–548.
19. Suwa, R.; Hagihara K.C. Seasonal changes in canopy photosynthesis and foliage respiration in a *Rhizophora stylosa* stand at the northern limit of its natural distribution. Wetlands Ecol. Manag. **2008**, *16*, 313–321.
20. Biebl, R. Temperaturresistenz tropischer Pflanzen auf Puerto Rico (Verglichen mit jener von Pflanzen der gemässigten Zone). Protoplasma **1965,** *59*, 133–156.
21. Méndez Alonzo, R.; López Portillo, J.; Rivera Monroy, V.H. Latitudinal variation in leaf and tree traits of the mangrove *Avicennia germinans* (Avicenniaceae) in the central region of the Gulf of Mexico. Biotropica **2008**, *40*, 449–456.
22. Medina, E.; Fonseca, H.; Barboza, F.; Francisco, M. Natural and man-induced changes in a tidal channel mangrove system under tropical semiarid climate at the entrance of the Maracaibo lake (Western Venezuela). Wetlands Ecol. Manag. **2001**, *9*, 243–253.
23. Barboza, F.; Barreto, M.B.; Figueroa, V.; Francisco, M.; González , A.; Lucena, M.; Mata, K.Y.; Narváez, E.; Ochoa, E.; Parra, L.; Romero, D.; Sánchez, J.; Soto, M.N.; Vera, A.J.; Villarreal, A.L.; Yabroudi, S.C.; Medina, E. Desarrollo estructural y relaciones nutricionales de un manglar ribereño bajo clima semi-árido. Ecotropic. **2006,** *19*, 13–29.

24. Sobrado, M.A. Drought effects on photosynthesis of the mangrove, *Avicennia germinans*, under contrasting salinities. Trees **1999**, *13*, 125–130.
25. Pannier, F. El efecto de distintas concentraciones salinas sobre el desarrollo de *Rhizophora mangle* L. Acta Científica Venezolana **1959**, *10*, 68–78.
26. Downton, W.J.S. Growth and osmotic relations of the mangrove *Avicennia marina,* as influenced by salinity. Aust. J. Plant Physiol. **1982**, *9*, 519–528.
27. Clough, B.F. Growth and salt balance of the mangroves *Avicennia marina* (Forsk.) Vierh. and *Rhizophora stylosa* Griff. in relation to salinity. Aust. J. Plant Physiol. **1984**, *11*, 419–430.
28. Cintrón, G.; Lugo, A. E.; Pool, D. J.; Morris, G. Mangroves of arid environments in Puerto Rico and adjacent islands. Biotropica **1978**, *10*, 110–121.
29. Lugo, A. E.; Medina, E.; Cuevas, E.; Cintron, G.; Laboy Nieves, E.N.; Novelli, Y.S. Ecophysiology of a mangrove forest in Jobos Bay, Puerto Rico. Caribbean J. Sci. **2007**, *43*, 200–219.
30. Duke, N. C. Morphological variation in the mangrove genus *Avicennia* in Australasia: systematic and ecological consideration. Aust. Syst. Bot. **1990b**, *3*, 221–239.
31. Camilleri, J. C.; Ribi, G. Leaf thickness of mangroves (*Rhizophora mangle*) growing in different salinities. Biotropica **1983**, *15*, 139–141.
32. Medina, E.; Francisco, M. Osmolality and δ13C of leaf tissues of mangrove species from environments of contrasting rainfall and salinity. Estuar. Coast. Shelf S. **1997**, *45*, 337–344.
33. Lugo, A. E. Mangrove ecosystems: Successional or steady state? Biotropica **1980**, *12* (supplement 2), 65–72.
34. Botero, L. Massive mangrove mortality on the Caribbean coast of Colombia. Vida Silvestre Neotrop. **1990**, *2*, 77–78.
35. Jiménez, J. A.; Lugo, A. E.; Cintrón, Tree mortality in mangrove forests. Biotropica **1985**, *17*, 177–185.
36. Stewart, G. R.; Popp, M. The ecophysiology of mangroves. In *Plant Life in Aquatic and Amphibious Habitats*; Crawford, R.M.M., Ed.; Blackwell Science Publishers: Oxford, 1987; 333–345.
37. Popp, M.; Polania, J. Compatible solutes in different organs of mangrove trees. Annales des Sci. Forest. **1989**, *46*(suppl.), 842–844.
38. Ball, M.C.; Pidsley, S.M. Growth responses to salinity in relation to distribution of two mangrove species, *Sonneratia alba* and *S. lanceolata*, in Northern Australia. Funct. Ecol. **1995**, *9*, 77–85.
39. Ball, M.C.; Sobrado, M.A. Ecophysiology of mangroves: challenges in linking physiological processes with patterns in forest structure. In *Physiological Plant Ecology*; Press, M.C.; Scholes, J.D.; Barker, M.G., Eds.; Blackwell Science: Oxford, England, 1999; 331–346.
40. Smith, J.A.C.; Popp, M.; Lüttge, U.; Cram, W.J.; Diaz, M.; Griffiths, H.; Lee, H.S.J.; Medina, E.; Schäfer, C.; Stimmel, K.-H.; Thonke, B. Water relations and gas exchange of mangroves. New Phytol. **1989**, *111*, 293–307.
41. Lugo, A.E. Fringe wetlands. In *Forested Wetlands*; Lugo, A.E.; Brinson, M.M.; Brown, S.; Eds.; Elsevier: Amsterdam, the Netherlands 1990; 143–169.
42. Mehlig, U.; Menezes, M.P.M.; Reise, A.; Schories, D.; Medina, E. Mangrove vegetation of the Caeté Estuary. In Mangrove dynamics and management in north Brazil; Saint-Paul, U; Schneider, H., Eds.; Springer: Berlin, 2010; 71–108.
43. Sobrado, M. A. Leaf characteristics and gas exchange of the mangrove *Laguncularia racemosa* as affected by salinity. Photosynthetica **2005**, *43*, 217–221.
44. Naidoo, G.; Tuffers, A.V.; von Willert, D.J. Changes in gas exchange and chlorophyll fluorescence characteristics of two mangroves and a mangrove associate in response to salinity in the natural environment. Trees **2002**, *16*, 140–146.
45. Suárez, N.; Medina, E. Influence of salinity on Na+ and K+ accumulation, and gas exchange in *Avicennia germinans*. Photosynthetica **2006**, *44*, 268–274.
46. Nandy (Datta), P.; Sauren Das, S.; Ghose, M.; Spooner-Hart, R. Effects of salinity on photosynthesis, leaf anatomy, ion accumulation and photosynthetic nitrogen use efficiency in five Indian mangroves. Wetlands Ecol. Manag. **2007**, *15*, 347–357.
47. Tuffers, A.; Naidoo, G.; von Willert, D. J. Low salinities adversely affect photosynthetic performance of the mangrove, *Avicennia marina.* Wetlands Ecol. Manag. **2001**, *9*, 235–242.
48. Lovelock, C. E.; Feller, I.C. Photosynthetic performance and resource utilization of two mangrove species coexisting in a hypersaline scrub forest. Oecologia **2003**, *134*, 455–462.
49. Medina, E.; Cuevas, E.; Lugo, A.E. Nutrient and salt relations of *Pterocarpus officinalis* L. in coastal wetlands of the Caribbean: assessment through leaf and soil analyses. Trees: Struct. Func. **2007**, *21*, 321–327.
50. Medina, E.; Francisco, M.; Quilice, A. Isotopic signatures and nutrient relations of plants inhabiting brackish wetlands in the northeastern coastal plain of Venezuela. Wetlands Ecol. Manag. **2008**, *16*, 51–64.
51. Medina, E.; Cuevas, E.; Popp, M.; Luga, A.E. Soil salinity, sun exposure, and growth of *Acrostichum aureum*, the mangrove fern. Botanical Gazette **1990**, *151*, 41–49.
52. Watson, J. G. Mangrove forests of the Malay Peninsula. Malayan Forest R. **1928**, *6*, 1–274.
53. Krauss, K.W.; Twilley, R.R.; Doyle, T.W.; Gardiner, E.S. Leaf gas exchange characteristics of three neotropical mangrove species in response to varying hydroperiod. Tree Physiol. **2006**, *26*, 959–968.
54. Pezeshki, S.R.; DeLaune, R.D.; Patrick Jr., W.H. Differential response of selected mangroves to soil flooding and salinity: gas exchange and biomass partitioning. Can. J. Forest Res. **1990**, *20*, 869–874.
55. Lovelock, C.E.; Ruess, R.W.; Feller I., C. Fine root respiration in the mangrove *Rhizophora mangle* over variation in forest stature and nutrient availability. Tree Physiol. **2006**, *26*, 1601–1606.
56. Colonnello, G.; Medina, E. Vegetation changes induced by dam construction in a tropical estuary: The case of the Mánamo River, Orinoco delta (Venezuela). Plant Ecol. **1998**, *139*, 145–154.
57. Ross, M.S.; Meeder, J.F.; Sah, J.P.; Ruiz, P.L.; Telesnicki, G.J. The southeast saline everglades revisited: 50 years of coastal vegetation change. J. Veg. Sci. **2000**, *11*, 101–112.
58. Doyle, T.W.; Krauss, K.W.; Conner, W.H.; From, A.S. Predicting the retreat and migration of tidal forests along the

northern Gulf of Mexico under sea-level rise. Forest Ecol. Manag. **2010**, *259*, 770–777.

59. Ellison, J.C.; Stoddart, D.R. Mangrove ecosystem collapse during predicted sea-level rise: Holocene analogues and implications. J. Coast. Res. **1991**, *7*, 151–165.
60. Batista, E.D.M.; Souza Filho, P.W.M.; Machado da Silveira, O.F.Avaliação de áreas deposicionais e erosivas em cabos lamosos da zona costeira Amazônica através da análise multitemporal de imagens de sensores remotos. Revista Brasileira de Geofísica **2009**, *27*. doi: 10.1590/S0102-261X2009000500007.
61. Cahoon, D.R.; Hensel, P.; Rybczyk, J.; McKee, K.L.;Proffitt, C.E.; Perez, B.C. Mass tree mortality leads to mangrove peat collapse at Bay Islands, Honduras after Hurricane Mitch. J. Ecol. **2003**, *91*, 1093–1105.
62. McKee, K.L.; Cahoon, D.R.; Feller, I.C. Caribbean mangroves adjust to rising sea level through biotic controls on change in soil elevation. Global Ecol. Biogeogr. **2007**, *16*, 545–556.
63. Baird, A.H.; Kerr, A.M. Landscape analysis and tsunami damage in Aceh: comment on Iverson and Prasad (2007). Landscape Ecology **2008**, *23*, 3–5.
64. Iverson, L.R.; Prasad, A.M. Modeling tsunami damage in Aceh: a reply. Landscape Ecol. **2008**, *23*, 7–10.
65. Feller, I.C. Effects of nutrient enrichment on growth and herbivory of dwarf red mangrove (*Rhizophora mangle*). Ecol. Monogr. **1995**, *65*, 477–505.
66. Cheeseman, J.M.; Lovelock, C.E. Photosynthetic characteristics of dwarf and fringe *Rhizophora mangle* L. in a Belizean mangrove. Plant, Cell Environ. **2004**, *27*, 769–780.
67. Medina, E.; Cuevas, E.; Lugo, A.E. Nutrient relations of dwarf *Rhizophora mangle* L. mangroves on peat in eastern Puerto Rico. Plant Ecol. **2010**, *207*, 13–24.
68. Chen, R.; Twilley, R.R. Patterns of mangrove forest structure and soil dynamics along the Shark River estuary, Flo. Estuaries **1999**, 22, 955–970.
69. Barr, J.G.; Jose, D.; Fuentes, J.D.; Engel, Vic.; Zieman, J.C. Physiological responses of red mangroves to the climate in the Florida Everglades. J. Geophys. Res. **2009,** *114*, G02008. doi:10.1029/2008JG000843.
70. Feller, I.C.; McKee, K.L.; Whigham, D.F.; O'Neill, J.P. Nitrogen vs. phosphorus limitation across an ecotonal gradient in a mangrove forest. Biogeochemistry **2002**, *62*, 145–175.
71. Feller, I.C.; Whigham, D.F.; McKee, K.L.; Nitrogen limitation of growth and nutrient dynamics in a disturbed mangrove forest, Indian River Lagoon, Florida. Oecologia **2003**, 134:405-414.
72. Feller, I.C.; Lovelock, C.E.; McKee, K.L. Lovelock, C.E. Nutrient addition differentially affects ecological processes of *Avicennia germinans* in nitrogen vs. phosphorus limited mangrove ecosystems. Ecosystems **2007**, 10, 347–359.
73. Feller, I.C.; Whigham, D.F.; O'Neill, J.P.; McKee, K.L. Effects of nutrient enrichment on within-stand cycling in a mangrove forest. Ecology **1999**, *80*, 2193–2205.
74. Lovelock, C.E.; Feller, I.C.; Mckee, K.L., The effect of nutrient enrichment on growth, photosynthesis and hydraulic conductance of dwarf mangroves in Panamá. Funct. Ecol. **2004**, *18*, 25–33.
75. Lugo, A.E. Visible and invisible effects of hurricanes on forest ecosystems: An international review. Aust. Ecol. **2008**, *33*, 368–398.
76. Cintrón, G.; Schaeffer Novelli, Y. Los manglares de la costa brasileña: revisión preliminar de la literatura. Oficina Regional de Ciencia y Tecnología de UNESCO para América Latina y el Caribe y Universidad Federal de Santa Catalina, Santa Catarina, 1981.
77. Allen, J.A.; Ewel, K.C.; Jack, J. Patterns of natural and anthropogenic disturbance of mangroves on the Pacific Island of Kosrae. Wetlands Ecol. Manag. **2001**, *9*, 279–289.
78. Twilley, R.R.; Medina, E. Forest dynamics and soil characteristics of mangrove plantations in the San Juan River estuary, Venezuela. In Annu. Lett.; Lugo, A.E., Ed.; International Institute of Tropical Forestry, USDA Forest Service: Río Piedras, PR, **1997**; 87–94.
79. Lin, G.H.; Sternberg; L.D.S.L. Effect of growth form, salinity, nutrient and sulfide on photosynthesis, carbon isotope discrimination and growth of red mangrove (*Rhizophora mangle* L.). Australian J. Plant Physiol. **1992**, *19*, 509–517.
80. McKee, K.L.; Mendelssohn, I.A.; Hester, M.K.A reexamination of pore water sulfide concentrations and redox potentials near the aerial roots of *Rhizophora mangle* and *Avicennia germinans*. Am. J. Bot. **1988**, *75*, 1352–1359.
81. McKee, K.L. Soil physicochemical patterns and mangrove species distribution - reciprocal effects? J. Ecol. **1993**, *81*, 477–487.
82. Alongi, D.M. *The Energetics of Mangrove Forests*; Springer: New York, NY, 2009.
83. Amin, B.; Ismail, A.; Arshad, A.; Yap, C.H.; Kamarudin, M.S. Gastropod assemblages as indicators of sediment metal contamination in mangroves. Water Air Soil Pollut. **2009**, *201*, 9–18.
84. Lefebvre, G.; Poulin, B. Bird communities in Panamanian black mangroves: potential effects of physical and biotic factors. J. Trop. Ecol. **1997**, *13*, 97–113.
85. Odum, W.E. Pathways of energy flow in a south Florida estuary; Sea Grant Technical Bulletin 7; University of Miami: Miami, FL, 1971.
86. Lugo, A.E.; Sell, M.; Snedaker, S.C. Mangrove ecosystem analysis. In *Systems Analysis and Simulation Ecology*; Patten, B.C., Ed.; Academic Press: New York, 1976; 13–145.
87. Twilley, R.R.; Rivera Monroy, V.H.; Chen, R.; Botero, L. Adapting an ecological mangrove model to simulate trajectories in restoration ecology. Mar. Pollut Bull. **1999**, *37*, 404–419.
88. Kangas, P.C. An energy theory of landscape for classifying wetlands. In *Forested Wetlands*; A.E.; Brinson, M.; Brown, S., Eds.; Elsevier: Amsterdam, 1990; 15–23.
89. Lugo, A.E.; Snedaker, S.C. The ecology of mangroves. Annu. Rev. Ecol. Syst. **1974**, *5*, 39–64.
90. Thom, B.G. Mangrove ecology and deltaic geomorphology, Tabasco, Mexico. J. Ecol. **1967**, *55*, 301–343.
91. Thom, B.G. Mangrove ecology from a geomorphic viewpoint. In *Proceedings of the International Symposium on Biology and Management of Mangroves*. Walsh, G; Snedaker, S.; Teas, H., Eds.; Institute of Food and Agricultural

Sciences, University of Florida: Gainesville, FL, 1974; 469–481.

92. Thom, B.G. 1982. Mangrove ecology-a geomorhological perspective. In *Mangrove Ecosystems in Australia: Structure, Function and Management*; Clough, B.F., Ed.; Australian Institute of Marine Sciences: Canberra, Australia, 1982; 3–17.
93. Twilley, R.R.; Snedaker, S.C.; Yáñez Arancibia, A.; Medina, E. Biodiversity and ecosystem processes in tropical estuaries: perspectives of mangrove ecosystems. In *Functional Role of Biodiversity: A Global Perspective*; Mooney, H.A.; Cushman, J.H.; Medina, E.; Sala, O.E. Schulza, E.-D., Eds.; John Wiley & Sons: New York, 1996; 327–370.
94. Twilley, R.R.; Rivera Monroy, V.H. Ecogeomorphic models of nutrient biogeochemistry for mangrove wetlands. In *Coastal Wetlands: An Integrated Approach.*; Perillo, G.M.E.; Wolanski, E.; Cahoon, D.R.; Ironson, M.M. Eds.; Elsevier: Amsterdam, 2009; 641–683.

Marshes: Salt and Brackish

Sally D. Hacker
Department of Integrative Biology, Oregon State University, Corvallis, Oregon, U.S.A.

Abstract

Salt and brackish marshes are highly productive habitats that occur at the intersection between the ocean and the land. They are the result of the interaction between vegetation, sediments, and the tides. Marsh plants capture sediments carried by rivers, thereby elevating the habitat above sea level. Salt and brackish marshes are exposed to contrasting salinities because they occupy different locations along an estuarine gradient. Coastal marshes provide many important services to humans including coastal protection, erosion control, and the maintenance of fisheries. However, humans have dramatically changed marshes for agriculture, development, and resource extraction; it is estimated that 50% of the coastal marshes in the U.S.A. are gone. In addition to direct human impact, marshes are highly susceptible to climate change via sea level rise, terrestrial flooding, and extreme disasters in the form of hurricanes and tsunamis. The future of salt and brackish marshes will depend on conservation, restoration, and climate.

INTRODUCTION

Salt and brackish marshes are vegetated coastal habitats that occur along the edges of estuaries, defined as places where rivers meet oceans. These highly productive habitats are dominated by terrestrial vegetation rooted in sediment and exposed to daily tidal inundation (Fig. 1). Coastal marshes are created when vegetation captures terrestrial sediments carried by rivers, thereby elevating the habitat above sea level. At high tide, a variety of estuarine animals including crabs, snails, and fish use the marsh for foraging and habitat protection. At low tide, terrestrial animals including insects, birds, rodents, and even ungulates are common visitors. Coastal marshes link marine and terrestrial ecosystems, making them unique and highly valuable.

Salt marshes differ from brackish marshes in their location within the estuary. Salt marshes are closer to the ocean and are thus flooded by nearly full strength seawater compared to brackish marshes, which occur upstream and are exposed to a mixture of seawater and freshwater. In reality, salt and brackish marshes are exposed to contrasting salinities because they occupy different locations along an estuarine gradient. Given the substantial influence of seawater on the physiology of terrestrial plants, the location of a particular marsh within an estuary has important consequences for the types of organisms that can live in these habitats, the types of interactions they experience, and their overall productivity.

Salt and brackish marshes, because they occur at the interface between terrestrial and marine environments, are important arbiters of critical ecosystem services, including coastal protection, maintenance of fisheries, and water purification.[1] The interface nature of coastal marshes also places them at great risk from human impact and climate change. Here, the basic information about the ecology of salt and brackish marshes including patterns of plant zonation, the engineering attributes of marsh plants, and the types of ecological interactions that shape them is provided. Then, the important ecosystem services of coastal marshes and the impacts of humans on these services is reviewed, and in conclusion, the conservation and restoration of these unique habitats is described.

BASIC ECOLOGY

Zonation Patterns of Salt and Brackish Marshes

Marsh plant distributions vary across estuarine gradients and also across intertidal zones, areas between the highest and lowest extent of the tides. As one moves up the intertidal, from the water's edge to the terrestrial border, there are distinct intertidal zones characterized by different plant assemblages and physical conditions. In salt marshes, the lowest zone is covered with seawater on a daily basis, making it stressful for most terrestrial plants. As a result, this zone is dominated by salt- and flood-tolerant plants, including grasses such as *Spartina* spp. and succulents such as *Salicornia* spp. In some cases, macroalgae may live attached to, or among, the stems of plants in this zone. In the middle zones of most salt marshes, tides cover the marsh surface less frequently, which can lead to highly variable salinity and oxygen conditions within the sediment. Plant species richness is typically greater in the middle marsh with mixtures of grasses (e.g., *Distichlis* spp.), rushes (e.g., *Juncus* spp.), succulents (e.g., *Salicornia* spp.), and forbs (e.g., *Atriplex* spp., *Plantago* spp.,

Encyclopedia of Natural Resources DOI: 10.1081/E-ENRL-120047521

Fig. 1 A salt marsh along the Pacific coast of North America.
Source: Photo by Sally D. Hacker.

Fig. 2 A brightly colored forb, *Potentilla pacifica*, amongst a background of the rush, *Juncus balticus*, in a Pacific coast salt marsh.
Source: Photo by Sally D. Hacker.

Triglochin spp., and *Potentilla* spp.; Fig. 2). In the high elevation zones of salt marshes, where tides rarely reach and freshwater runoff is common, less salt- and flood-tolerant grasses (e.g., *Deschampsia* spp.) and shrubs (e.g., *Iva* spp.) dominate.

In brackish marshes, both the tides and rivers influence the salinity of the water that floods these habitats. Although these marshes experience low salinity water (i.e., >15 g/kg NaCl), they may experience the same or even more flooding due to the influence of the river. Brackish marshes have greater overall biomass than salt marshes.[2] In the low marsh, low salinity but flood-tolerant sedges and rushes (e.g., *Carex* spp. and *Scirpus* spp.) dominate. In higher intertidal zones, tidal flooding might only occur a few days a year, with most water coming from rivers and terrestrial runoff. Here, freshwater species such as cattails (*Typha* spp.) can be common.

Coastal Marsh Plants as Ecosystem Engineers

Salt and brackish marshes are the result of the interaction between vegetation, sediments, and the tides. Marsh plants act as ecosystem engineers (i.e., organisms that create, modify, or maintain physical habitat such as trees, kelps, or beavers) by increasing the drag of water that moves across the surface of the marsh, causing sediments to settle out and be deposited on the marsh surface.[3] The roots of plants may contribute to this process by binding sediments and reducing erosion, but it is the upright stems of marsh plants that control sediment deposition.[4] Plants vary in their ability to accrete sediment depending on their morphology, density, and even stiffness.[5,6] With the deposition of sediment, plant growth is stimulated, creating a positive feedback between growth and deposition. The ability of a marsh to maintain a constant elevation with sea level is sensitive to plant productivity, sediment supply, and rates of sea-level rise.[7] As sea level rises with climate change, there are concerns about whether coastal marshes will be able to keep pace via sediment accretion.

Beyond accreting sediment, coastal marsh plants are masters at modifying the sediment chemistry via two main mechanisms.[8] First, they can passively shade the soil surface and reduce salt accumulation that occurs when seawater evaporates at the marsh surface (Fig. 3). Second, some marsh plants, particularly grasses, rushes, and sedges, can use specialized tissue called arenchyma to oxygenate their roots and rhizomes (underground stems) when exposed to flooding.

Species Interactions in Coastal Marshes

The distribution of marsh organisms across estuarine and elevation gradients is not only a product of salinity and tidal inundation but also of species interactions. Conceptual models[9,10] and numerous empirical studies in coastal marshes[2,11–13] show that species interactions can shift from mostly competitive (negative) under low stress and high productivity conditions (i.e., higher intertidal zones and more brackish marshes) to mostly facilitative (positive) under high stress and low productivity conditions (i.e., lower intertidal zones and more salty marshes). The main mechanisms behind the facilitative interactions were identified as those described earlier: shading and oxygenation of sediment by neighboring vegetation.[8] One well-studied example showed that positive interactions between the rush, *Juncus gerardii*, and a number of plants and their insect herbivores controlled the species richness of a New England marsh.[10]

Trophic interactions are also important to the distribution of organisms within coastal marshes. Plants are eaten by a variety of herbivores (e.g., snails, insects, rodents, and

Fig. 3 A high-salinity patch formed when vegetation was disturbed. The white on the surface is salt crystals.
Source: Photo by Sally D. Hacker.

ungulates), which, in turn, are fed on by carnivores (e.g., crabs, predatory insects, fish, and even turtles). For a long time, it was assumed that trophic interactions played a minor role in structuring marsh plant communities.[14] However, Silliman and colleagues[15,16] showed that so-called top-down effects of predators could dramatically affect the major marsh builder, the grass *Spartina alterniflora*. *Spartina* is indirectly killed by snails (*Littoraria irrorata*), which scrape the grass surface with their mouthparts and cause wounds that become infected by a fungus. Normally, snail populations are controlled by the predatory effects of the blue crab, *Callinectes sapidus*; however, recent disease and overharvesting of the crabs has released the snails to their destructive ways. Large diebacks of *Spartina* have occurred in some parts of the southeast U.S.A. due in part to the loss of snail predators.

HUMAN IMPORTANCE AND IMPACT

Coastal and estuarine ecosystems are some of the most heavily impacted systems on the Earth.[17] In the U.S.A. alone, nearly 50% of the population lives on the coast and this is expected rise to 70% by 2040.[18] Humans have drained, diked, and diverted water away from coastal marshes both for agriculture and development; it is estimated that 50% of the coastal marshes in the U.S.A. are gone.[19] In addition to direct human impact, marshes are highly susceptible to climate change via sea level rise, terrestrial flooding, and extreme disasters in the form of hurricanes and tsunamis. Thus, coastal marshes occur at the nexus that includes important ecosystem services, chronic and extreme climate effects, and the sustainability of human health and welfare. In the following text, some of the important ecosystem services, some of the ongoing human threats, and some of the attempts at restoration of coastal marshes are described.

Ecosystem Services of Coastal Marshes

Salt and brackish marshes, because they are the result of coastal and terrestrial processes, provide important and valuable ecosystem services to humans. Barbier et al.[1] reviewed these services in some detail and provided economic values, when available, from the literature. They identified six main services that included raw materials and food, coastal protection, erosion control, water purification, maintenance of fisheries, carbon sequestration, and tourism and recreation. One of the most understudied and likely undervalued services provided by coastal marshes is protection from waves and storm surge caused by large storms and hurricanes. Marshes are able to reduce the velocity, height, and duration of incoming waves by increasing the drag of water across a vegetated surface.[20] The value of this service will surely increase as more frequent and intense storms are projected to occur with climate change.

Salt marshes can also act as natural filters by purifying the water entering estuaries from watersheds.[21] This water often contains high nutrients and pollutants that can cause "dead zones" or extremely low oxygen conditions. When water passes over the vegetation in marshes, it traps sediment, stimulating plant growth and the uptake of nutrients. In some cases, coastal marshes have been used for wastewater treatment, saving significant amounts of money compared to conventional municipal treatment.[22] Finally, because coastal marshes are one of the most productive ecosystems on earth, they also are important for carbon sequestration, a role that may become more important as CO_2 continues to rise globally. Because marshes have extremely low oxygen soils, dead plant matter does not easily decompose and can be stored in the form of peat for hundreds of years.[21]

Human Impacts and Restoration

Despite the variety of ecosystem services provide by salt and brackish marshes, humans have had considerable negative impacts.[19,23] Loss of coastal marshes can be attributed to a number of factors that vary over local, regional, and global scales. At local scales, marshes have been converted for agriculture and development purposes through draining, diking, or diverting water away from the marsh. Some major cities, including Boston, San Francisco, and London, are built on filled or drained wetlands. Another local-scale impact from humans is that of non-native invasions. For example, *Spartina* spp. have been introduced into estuaries worldwide, transforming mudflats into marshes.[24,25] This reduces habitat for sea grass, oysters, fish, and foraging birds. At regional scales, extraction of groundwater, oil, and gas has caused subsidence, leading to the submergence and erosion of hundreds of square kilometers of salt marsh habitat in the Chesapeake Bay, San Francisco Bay, and Gulf of Mexico. Finally, at global scales, sea-level rise caused by warming temperatures threatens coastal marshes if they are unable to maintain their elevation above sea level through sediment accretion and/or if brackish marshes become more saline.

Generally, success at restoring coastal marshes once they are damaged is difficult and for the most part has only been attempted at local scales.[26,27] Restoration of old marshes that were converted for agricultural use typically involves reestablishing the tidal inundation regimes that were diverted or blocked. For example, the removal of dikes allows salt water to flow back into the fields, killing the vegetation, and reestablishing the connection with estuarine sediment, seeds, and animals. Slowly, coastal marsh vegetation can establish itself and start to build the marsh back to its former elevation. However, the restoration of marshes can be a very slow process, even with the intentional planting of vegetation, because the biological engineering of the habitat requires a certain density of plants that ameliorates the harsh physical conditions and promotes sediment accretion.[27] The establishment of marsh plants as ecosystem engineers in the restoration of marshes is particularly important to provide the foundation for the colonization of other species and thus the promotion of species diversity.[28] For this reason, if marshes are destroyed at very large spatial scales due to subsidence, erosion, or sea-level rise, their restoration will be virtually impossible without the reestablishment of the proper functional dynamics between vegetation and sediment.[29]

CONCLUSION

Salt and brackish marshes are the product of both terrestrial and ocean processes. This unique position makes them important arbiters of highly valued services such as the reduction of coastal vulnerability and the buffering of marine environments from nutrient loading and pollution. Yet, coastal marshes also are at the forefront of threats from humans and climate. As human populations disproportionately increase on the coast, and climate change threatens to alter sea level and the frequency of extreme storms, the sustainability of marshes will be an important indicator of the overall health of our coasts.

REFERENCES

1. Barbier, E.B.; Koch, E.W.; Silliman, B.R.; Hacker, S.D.; Wolanski, E.; Primavera, J.; Granek, E.F.; Polasky, S.; Aswani, S.; Cramer, L.A.; Stoms, D.M.; Kennedy, C.J.; Bael, D.; Kappel, C.V.; Perillo, G.M.; Reed, D.J. Coastal ecosystem-based management with nonlinear ecological functions and values. Science **2008**, *319* (5861), 321–323.
2. Crain, C.M.; Silliman, B.R.; Bertness, S.L.; Bertness, M.D. Physical and biotic drivers of plant distribution across estuarine salinity gradients. Ecology **2004**, *85* (9), 2539–2549.
3. Gutiérrez, J.L.; Jones, C.G.; Byers, J.E.; Arkema, K.; Berkenbusch, K.; Commito, J.A.; Duarte, C.M.; Hacker, S.D.; Hendriks, I.E.; Hogarth, P.J.; Lambrinos, J.G.; Palomo, G.; Wild, C. Physical ecosystem engineers and the functioning of estuaries and coasts. In *Treatise on Estuarine and Coastal Science*; Wolanski, E., McLusky, D.S., Eds.; Academic Press: Waltham, 2011, Vol. 7; 53–81.
4. Feagin, R.A.; Lozada-Bernard, S.M.; Ravens, T.M.; Möllerb, I.; Yeager, K.M.; Baird, A.H. Does vegetation prevent wave erosion of salt marsh edges? Proc. Nat. Acad. Sci. USA, **2009**, *106* (25), 10109–10113.
5. Bouma, T.J.; De Vries, M.B.; Low, E.; Peralta, G.; Tanczos, I.C.; Van de Koppel, J.; Herman, P.M.J. Trade-offs related to ecosystem-engineering: A case study on stiffness of emerging macrophytes. Ecology **2005**, *86* (8), 2187–2199.
6. Neumeier, U.; Amos, C.L. The influence of vegetation on turbulence and flow velocities in European salt-marshes. Sedimentology **2006**, *53* (2), 259–277.
7. Morris, J.T.; Sundareshwar, P.V.; Nietch, C.T.; Kjerfve, B.; Cahoon, D.R. Responses of coastal wetlands to rising sea level. Ecology **2002**, *83* (10), 2869–2877.
8. Hacker, S.D.; Bertness, M.D. Morphological and physiological consequences of a positive plant interaction. Ecology **1995**, *76* (7), 2165–2175.
9. Bertness, M.D.; Callaway, R. Positive interactions in communities. Trends Ecol. Evol. **1994**, *9* (5), 191–193.
10. Hacker, S.D.; Gaines, S.D. Some implications of direct positive interactions for community species diversity. Ecology **1997**, *78* (7), 1990–2003.
11. Bertness, M.D.; Hacker, S.D. Physical stress and positive associations among marsh plants. Am. Nat. **1994**, *144* (3), 363–372.
12. Hacker, S. D.; Bertness, M. D. Experimental evidence for factors maintaining plant species diversity in a New England salt marsh. Ecology **1999**, *80* (6), 2064–2073.
13. Guo, H.; Pennings, S.C. Mechanisms mediating plant distributions across estuarine landscapes in a low-latitude tidal estuary. Ecology **2012**, *93* (1), 90–100.
14. Teal, J.M. Energy flow in the salt marsh ecosystem of Georgia. Ecology **1962**, *43* (4), 614–624.
15. Silliman, B.R.; Zieman, J.C. Top-down control of *Spartina alterniflora* production by periwinkle grazing in a Virginia salt marsh. Ecology **2001**, *82* (10), 2830–2845.
16. Silliman, B.R.; Van de Koppel, J.; Bertness, M.D.; Stanton, L.E.; Mendelssohn, I.A. Drought, snails, and large-scale die-off of southern U.S. salt marshes. Science **2005**, *310* (5755), 1803–1806.
17. Halpern, B.S.; Walbridge, S.; Selkoe, K.A.; Kappel, C.V.; Micheli, F.; D'Agrosa, C.; Bruno, J.F.; Casey, K.S.; Ebert, C.; Fox, H.E.; Fujita, R.; Heinemann, D.; Lenihan, H.S.;

Madin, E.M.P.; Perry, M.T.; Selig, E.R.; Spalding, M.; Steneck, R.; Watson, R. A global map of human impact on marine ecosystems. Science **2008**, *319* (5865), 948–952.
18. Crossett, K.M.; Culliton, T.J.; Wiley, P.C.; Goodspeed, T.R. Population trends along the coastal United States: 1980–2008, National Oceanic and Atmospheric Administration. National Ocean Service, Management and Budget Office, Special Projects: Washington, DC, 2004.
19. Kennish, M .J. Coastal salt marsh systems in the U.S.: A review of anthropogenic impacts. J. Coastal Res. **2001**, *17* (3), 731–748.
20. Gedan, K.B.; Kirwan, M.L.; Wolanski, E.; Barbier, E.B.; Silliman, B.R. The present and future role of coastal wetland vegetation in protecting shorelines: Answering recent challenges to the paradigm. Clim. Change **2001**, *106* (1), 7–29.
21. Mitsch, W. J.; Gosselink, J.G.; Zhang, L.; Anderson, C.J. *Wetland Ecosystems*. Wiley: Hoboken, New Jersey, 2009.
22. Breaux, A.; Farber, S.; Day, J. Using natural coastal wetlands systems for wastewater treatment: an economic benefit analysis. J. Environ. Manag. **1995**, *44* (3), 285–291.
23. Bertness, M.; Silliman, B.R.; Jefferies, R. Salt marshes under siege: agricultural practices, land development and overharvesting of the seas explain complex ecological cascades that threaten our shorelines. Am. Sci. **2004**, *92* (1), 54–61.
24. Daehler, C.C.; Strong, D.R. Status, prediction and prevention of introduced cordgrass *Spartina* spp. invasions in Pacific estuaries, USA. Biol. Conserv. **1996**, *78* (1), 51–58.
25. Hacker, S.D.; Heimer, D.; Hellquist, C.E.; Reeder, T.G.; Reeves, B.; Riordan, T.J.; Dethier, M.N. A marine plant (*Spartina anglica*) invades widely varying habitats: Potential mechanisms of invasion and control. Biol. Invasions **2001**, *3* (2), 211–217.
26. Broome, S.W.; Seneca, E.D.; Woodhouse, W.W. Tidal salt marsh restoration. Aquat. Bot. **1988**, *32* (1), 1–22.
27. Zedler, J.B. Progress in wetland restoration ecology. Trends Ecol. Evol. **2000**, *15* (10), 402–407.
28. Halpern, B.S.; Silliman, B.R.; Olden, J.D.; Bruno, J.P.; Bertness, M.D. Incorporating positive interactions in aquatic restoration and conservation. Front. Ecol. Environ. **2007**, *5* (3), 153–160.
29. Moreno-Mateos, D.; Power, M.E.; Comin, F.A.; Yockteng, R. Structural and functional loss in restored wetland ecosystems. PLoS Biol. **2012**, *10* (1), e1001247, doi:10.1371/journal.pbio.1001247.

BIBLIOGRAPHY

1. Adam, P. *Saltmarsh Ecology*; Cambridge University Press: Cambridge, 1993.
2. Bertness, M.D. The ecology of a New England salt marsh. Am. Sci. **1992**, *80* (3), 260–268.
3. Bertness, M.D. *Atlantic Shorelines: Natural History and Ecology*; Princeton University Press: Princeton, 2007.
4. Keddy, P.A. *Wetland Ecology: Principles and Conservation*; Cambridge University Press: Cambridge, 2010.
5. Ranwell, D.S. *Ecology of Salt Marshes and Sand Dunes*; Chapman and Hall: London, 1972.
6. Silliman, B.R.; Grosholz, E.; Bertness, M.D. *Human Impacts on Salt Marshes: A Global Perspective*; University of California Press: Berkeley, 2009.

Net Ecosystem Production (NEP)

Mark E. Harmon
Oregon State University, Corvallis, Oregon, U.S.A.

Abstract

Net ecosystem production (NEP) refers to the carbon metabolism of ecosystems and represents the balance between photosynthesis and respiration. Hence, while NEP is related to the net gain or loss of ecosystem carbon, it does not include many processes either adding or removing carbon from ecosystems. Therefore other indices, such as Net Ecosystem Carbon Balance, are used to assess the net change in carbon stores of ecosystems. The temporal pattern of NEP is highly dependent on the amount and nature of carbon left by disturbances and a current topic of theoretical and empirical research.

INTRODUCTION

Net ecosystem production (NEP) quantifies the balance of ecosystem primary production minus ecosystem respiration. First used by Woodwell & Whittaker,[1] NEP provides an index of ecosystem metabolism and determines, in part, whether there is a net gain or loss of carbon. In some ways it is analogous to the energy balance of an organism. It also parallels the relationship between gross primary production, autotrophic respiration, and net primary production (NPP). The sign of NEP can be either positive or negative. When NEP is zero, the ecosystem is in carbon balance. NEP is expressed as the mass of carbon per area per time and can be expressed in terms of gross primary production (GPP) or NPP:

$$NEP = GPP - Re$$

$$NEP = NPP - Rh$$

where Re is ecosystem respiration and Rh is heterotrophic respiration resulting from the metabolism of consumers and decomposers. The two equations are equivalent because

$$GPP = NPP + Ra$$

and

$$Re = Ra + Rh$$

Here Ra is autotrophic respiration (i.e., respiration from primary producers).

NEP is parallel to NPP in that

$$NPP = GPP - Ra$$

Related terms include net ecosystem exchange (NEE), ΔC, and net ecosystem carbon balance (NECB). NEP is referenced to the ecosystem, whereas NEE is referenced to the atmosphere. This means that when NEP is positive NEE is negative and vice versa. As typically measured in terrestrial ecosystems (e.g., eddy covariance) NEE involves vertical exchanges of carbon dioxide with the atmosphere outside the ecosystem. While most consider NEE is solely due to photosynthesis and respiration, Chapin et al.[2] include carbon dioxide losses associated with combustion also; moreover they include flows in any direction. In contrast to NEE, NEP may not involve exchanges outside the ecosystem itself. For example, respired carbon could be absorbed by leaf stomata before leaving the ecosystem. ΔC is the net change in carbon stored in a fixed area over a time period, typically a year. In some cases NEP and ΔC have that same value, but conceptually differ in that carbon can enter and leave ecosystems by means other than photosynthesis or respiration.[2,3] This is most evident in aquatic systems with flowing water where carbon is constantly being imported or exported independent of the physiologic processes of photosynthesis and respiration.[3] For many terrestrial systems there can be extended periods when carbon losses via erosion, leaching, combustion, and harvest can be very low, and this gives the appearance that NEP and ΔC are the same.[4] However, over very long periods or broad expanses these other losses can be substantial. NECB was offered by Chapin et al.[2] to address this discrepancy and includes other processes that either add or remove carbon from ecosystems:

$$NECB = NEP - E - S - F - H - V$$

where erosion and particle transport (E), solution (S), fire (F), harvest (H), and volatiles (V) are some of the largely nonphysiologic processes influencing carbon balances. An alternative to specifying the processes moving carbon in and out of ecosystems is to specify the form of carbon moving, as was done by Chapin et al.[2] While many of these flows are out of the ecosystem, and hence the signs

Encyclopedia of Natural Resources DOI: 10.1081/E-ENRL-120047510

in the equation are negative, in some cases erosion and particle transport as well as solution transport can add carbon to ecosystems (particularly aquatic ecosystems). Net Biome Production (NBP) is used to indicate the overall carbon balance of an ecosystem[5] or landscape;[2] however, ecosystems need not achieve a carbon balance at the landscape or biome level and so a term that indicates the biological level that this biological property emerges (i.e., the ecosystem) is preferred by some.[4]

NEP is also related to the concepts of autotrophy and heterotrophy, particularly in aquatic ecosystems. A system that exhibits autotrophy is one in which NEP is positive, that GPP > Re. In contrast, a system that exhibits heterotrophy is one in which NEP is negative and GPP < Re. This is most useful in aquatic ecosystems because it indicates whether an ecosystem is fixing enough carbon or energy to support itself, or whether is it being subsidized by inputs from outside the system.

MEASUREMENT OF NEP

Until recently direct measurement of NEP or related terms was challenging. Traditional terrestrial methods involved first estimating NPP and then either measuring respiration losses or using a combination of decomposition rates and stores of detritus and soil to estimate Rh.[1] Accurately estimating NPP and Rh presents numerous methodological challenges, particularly belowground. As measuring GPP directly is difficult, most traditional estimates of NEP are done on an NPP basis. The advent of eddy covariance, a method that simultaneously measures carbon dioxide concentrations and vertical air movements, has allowed more direct estimates of NEP.[6] However, there are methodological issues with this technique as it tends to measure carbon exchanges better during the day than at night, the latter being when respiration is the primary process; moreover it assumes strictly vertical exchanges whereas advection (i.e., horizontal transport) is common at night. Both these limitations are likely to lead to an underestimation of Re, which could result in a positively biased estimate of NEP. Considerable effort has been made to eliminate these problems by locating measurements in areas with minimal topographic slope and using night time periods with sufficient turbulent mixing based on a minimum wind velocity.[7]

In flowing aquatic systems estimating NEP may not involve the measurement of carbon dioxide as in terrestrial systems. Instead, oxygen concentration and water velocity can indicate the rate at which photosynthesis versus respiration is occurring.[8] These net changes in oxygen can be converted to a net carbon balance. A confounding factor is the addition of oxygen-rich or -depleted waters from hyporheic or groundwater flows; the former would lead to an overestimate of GPP and the latter would overestimate Re.[9]

TERRESTRIAL VERSUS AQUATIC ECOSYSTEMS

There are significant differences in NEP between terrestrial and aquatic systems related to the transport of carbon, which influence how this variable is examined. In most terrestrial systems the input of carbon by processes other than photosynthesis is minimal. While losses of carbon via particle and solution transport occur in terrestrial ecosystems, for many locations they are small terms relative to photosynthesis and respiration. In aquatic ecosystems, particulate and solution transport can be very significant. This is one reason that NEP can often approximate a net carbon balance in many terrestrial ecosystems, but it often does not in aquatic ecosystems. While it is rare for terrestrial ecosystems to have a semipermanent negative NEP, in aquatic ecosystems with significant outside (allochthonous) carbon input it is possible for NEP to remain negative for extensive periods. The converse is also true, in that aquatic systems with primarily autochthonous input NEP may remain positive for extensive periods. This explains why in terrestrial ecosystems the focus of NEP studies is often on long-term temporal changes,[10] whereas in flowing aquatic ecosystems the focus is on broad-scale spatial changes.[11] Despite some aquatic ecosystems having a positive NEP, over broad spatial extents inland waters appear to have a negative value of NEP primarily due to the input of carbon from upland terrestrial ecosystems.[12]

TEMPORAL AND SPATIAL PATTERNS

In terrestrial ecosystems there are several classic temporal patterns of NEP at a local level (e.g., a small plot of land), which are distinguished by the amount of biological legacies initially present. In the case of primary succession on freshly created surfaces, there is no legacy carbon and NEP can remain positive for extensive periods, but gradually approaches zero (Fig. 1A). This is because Rh generally lags NPP, but with sufficient time the amount of carbon produced is theoretically equal to the amount dying and respiring. In the case of secondary succession, considerable legacy carbon is initially present and respiration losses can far exceed NPP. This means that NEP is initially negative after disturbance, but eventually becomes positive as legacy carbon respires away and NPP increases (Fig. 1B). With enough time NEP in secondary succession will approach zero as in primary succession. In terms of carbon accumulation, primary and secondary succession are quite different, with the former steadily accumulating carbon to a steady-state and the latter first declining and then increasing toward a steady-state value (Fig. 1C). Given that secondary succession can start with variable legacies (i.e., those left by clear-cut timber harvest differ from those left by fire or windstorms), there can be a considerable range in the magnitude and duration of the negative and positive phases of NEP during terrestrial succession. Although

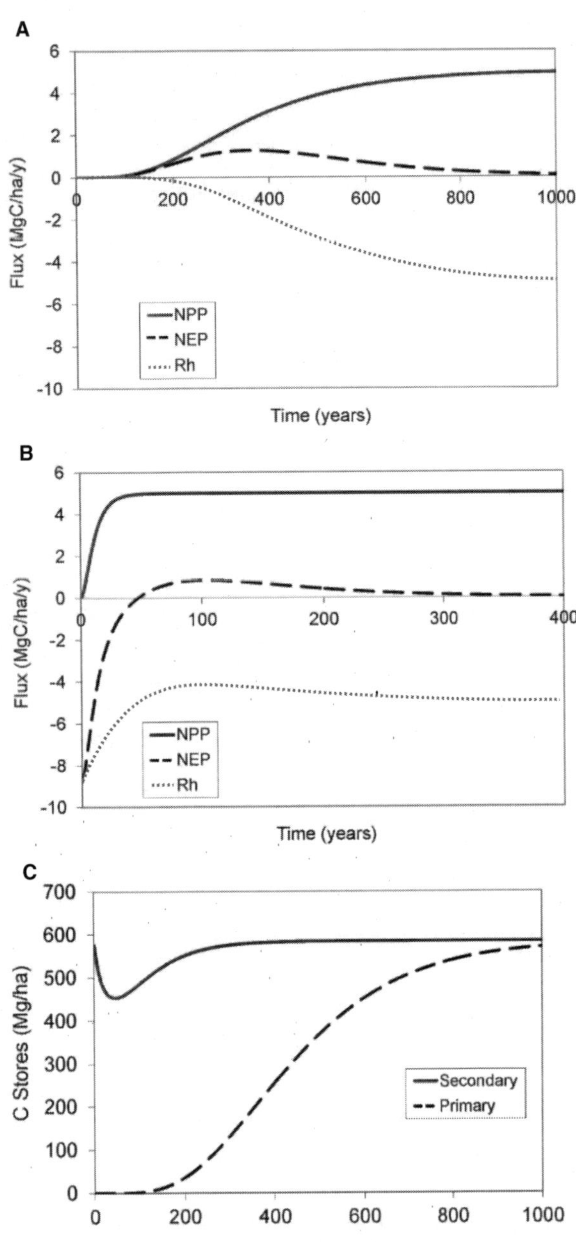

Fig. 1 Temporal trends in net primary production (NPP), heterotrophic respiration (Rh), net ecosystem production (NEP), and carbon stores predicted by the classic model. (**A**) Trend of fluxes for primary succession assuming the NPP maximum is reached in 800 years, (**B**) trend in fluxes for secondary succession assuming the NPP maximum is reached in 25 years and no removal of carbon by the disturbance, and (**C**) trends in total carbon stores.

many factors control NEP temporal dynamics, the shapes of the underlying NPP and Rh curves are a logical focus. Aside from the maximum level of NPP reached in an ecosystem, the time to reach that maximum is important in determining the NPP time curve. The primary control on the shape of the Rh time curve is the relative rate of respiration (i.e., rate constant) and the amount of biological legacy starting the succession. The effect of the NPP and Rh time curves on NEP varies with primary versus secondary succession (Fig. 2). In primary succession there is no or very little biological legacy, which means NPP is the primary controlling time curve. As the maximum NPP increases, so does the peak of NEP attained. More interesting is the effect of the time to reach the maximum NPP, with a decreasing, broader NEP peak, as the time to reach the maximum NPP increases (Fig. 2A). This is largely due to the fact that Rh has more time, so to speak, to catch up to NPP when the increase in NPP is slower. In secondary succession the Rh time curve appears to have the greatest influence on the temporal pattern of NEP. Reducing the biological legacy causes NEP to move from negative to positive faster and for the positive peak to be higher (Fig. 2B). In contrast increasing the time to reach maximum NPP has little effect on the time to move from negative to positive NEP, although it does increase the amount of carbon lost during the negative phase and the amount gained in the positive phase (Fig. 2C).

While NEP may approach zero with sufficient time in terrestrial ecosystems it is unlikely that it remains zero for an extended period of time. A more useful concept is that it averages zero, but given variation in climate, small scale disturbances, and other factors it is rarely exactly zero as described by classical theory.[13] Some have questioned whether NEP ever approaches zero in forests (Fig. 3A).[10] Given that NEP is not the same as NECB, NEP could remain positive for extended periods and carbon need not accumulate in terrestrial ecosystems; for example if there is a loss that is not associated with respiration such as erosion, significant solution losses, fire, or harvest (Fig. 3B). Because these processes remove carbon that has been produced, it is no longer available as substrates for respiration and cannot contribute to NEP; hence NPP > Rh. Without these losses an extended period of positive and constant NEP would mean ecosystem carbon stores would increase indefinitely in a straight line manner (Fig. 3C). This might be associated with forms of carbon that cannot be respired such as forms of soil organic matter.[14] However, classical studies of soil carbon increases over time using chronosequences indicate that while soil carbon accumulation can take long periods of time, the overall accumulation curve is not a straight line, but decreases in slope over time.[15,16] If there are major losses not associated with Rh, then it is possible that carbon accumulation is quite similar to the classical curve (Fig. 3C), albeit for different reasons than originally proposed. Additional research is required to measure long-term trends in NEP and to document the specific mechanisms responsible for extensive periods of constant, positive NEP.

Another aspect of temporal changes in NEP at the local level in terrestrial system being explored is whether there is only a single negative and positive phase after major disturbances. The classic theory assumes homogeneity of producers and the material being respired. In addition,

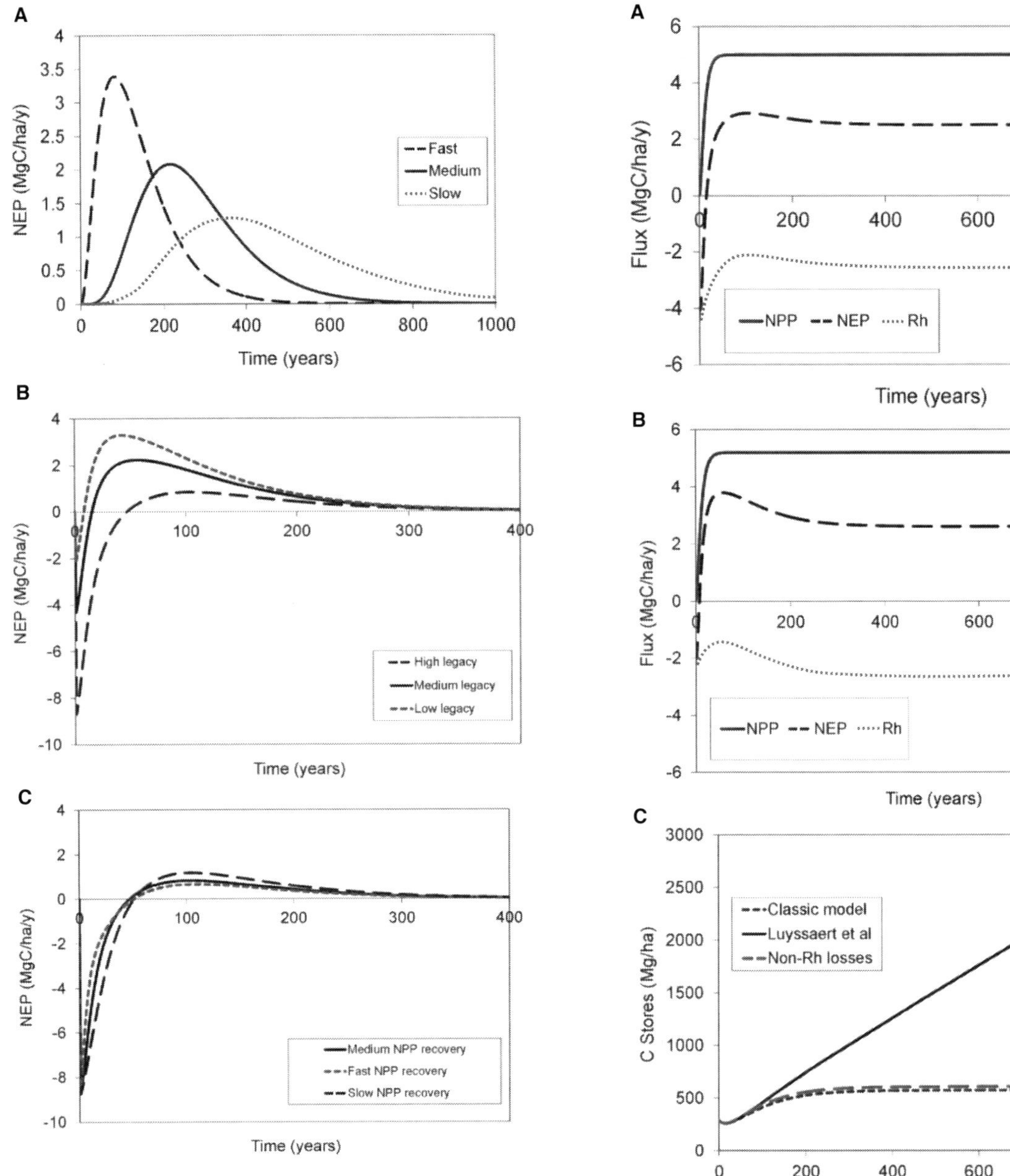

Fig. 2 Potential variability in trends in net primary production (NPP), heterotrophic respiration (Rh), and net ecosystem production (NEP) during primary and secondary succession. (**A**) Effect of varying time to reach maximum NPP in a primary succession. Fast, medium, and slow times to reach maximum NPP were 12, 25, and 50 years, respectively. (**B**) Effect of varying the size of the biological legacy in a secondary succession. Low, medium, and high legacies represented 25, 50, and 100% of the maximum legacy possible. (**C**) Effect of varying the time to reach maximum NPP in a secondary succession. Fast, medium, and slow times to reach maximum NPP were 12, 25, and 50 years, respectively, and the maximum possible legacy started the succession.

Fig. 3 The possible consequences of a semipermanent, positive net ecosystem production (NEP) on terrestrial carbon stores. (**A**) The time trend in net primary production (NPP), heterotrophic respiration (Rh), and NEP predicted by Luyssaert et al.[10] (**B**) An alternative explanation that invokes nonrespiration losses. (**C**) Implied carbon stores of (**A**) and (**B**) along with that predicted by the classic model.

temporal lags other than those associated with NPP are assumed to be minimal. In forest ecosystems in particular, forms of carbon have vastly different longevities in life and after they die. For example, leaves might remain alive for a year, whereas tree stems might remain alive for centuries. Some disturbances create forms of carbon, such as standing dead trees, which do not immediately begin to decompose. The heterogeneity of carbon forms and lags associated with respiration losses could lead to a situation in which NEP has multiple negative and positive phases (Fig. 4).

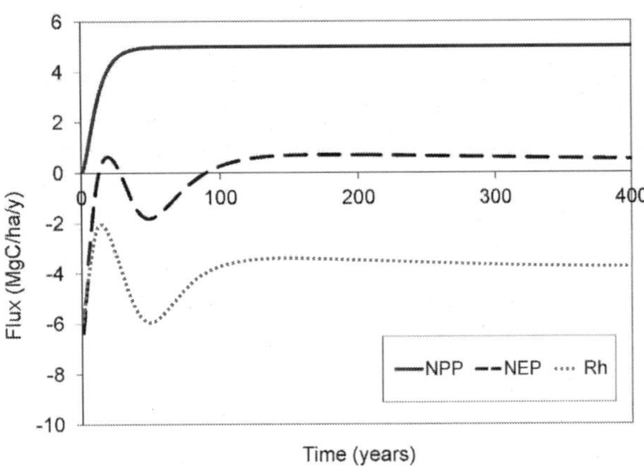

Fig. 4 A hypothetical situation in which net ecosystem production (NEP) can have multiple negative and positive phases. In this system net primary production (NPP) is allocated between live and dead parts with life spans that differ by an order of magnitude. Dead parts have a fast and slow phase of decomposition with a temporal lag before newly created dead parts can decompose at maximum rates. Dead plant parts are also assumed to eventually form soil carbon with a very slow rate of decomposition.

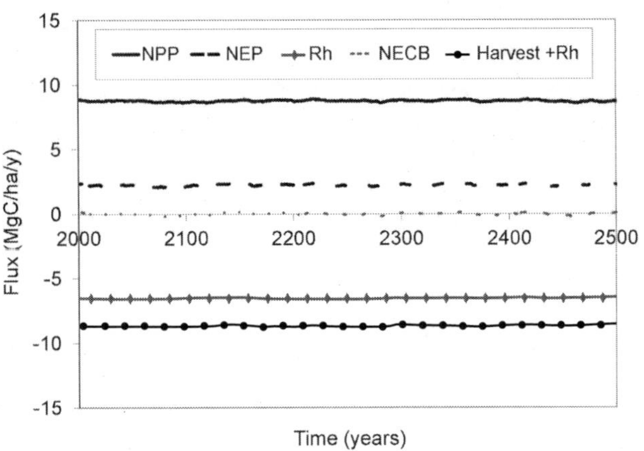

Fig. 5 Landscape level trends in net ecosystem production (NEP), net primary production (NPP), heterotrophic respiration (Rh), and NECB for a forest being clear-cut harvested every 50 years. The difference between NEP and Net Ecosystem Carbon Balance (NECB) is the amount of carbon harvested annually.

Determining if this is the case is important because if there are multiple negative phases, one cannot assume a priori that once NEP is positive ecosystems will accumulate carbon in the future. Most of the temporal trends of NEP described above for terrestrial ecosystems consider periods of decades to hundreds of years. Over longer periods of time there is evidence that NPP can decline, as elements such as phosphorus become depleted due to extensive weathering.[17] This suggests that without replenishment of fresh minerals for weathering that NEP might become negative, as carbon created by earlier periods of higher production is decomposed away. However, this is likely a very slow rate of carbon loss and may be difficult to measure over the short term. Temporal patterns of NEP in aquatic ecosystems appear to have not been studied to the degree they have in upland ecosystems. This may stem from the fact that distinct disturbances in aquatic systems are hard to identify;[18] one could argue that aquatic systems are continuously disturbed and this process maintains their productivity. Nonetheless, over very long periods, some aquatic systems may undergo a temporal pattern similar to terrestrial primary succession as these ecosystems move from an oligotrophic to a eutrophic condition with both production and decomposition increasing over time. In aquatic systems that receive large amount of their carbon in dissolved forms from adjoining upland systems NEP may remain negative regardless of system age. In contrast, those receiving particulate carbon inputs from adjacent uplands may have an extensive period of positive NEP if particulates are deposited in anaerobic sediments that greatly slow decomposition. For those with high rates of particulate carbon decomposition such as flowing waters, an extensive period of negative NEP may occur. It is also possible that riparian streams that typically have a negative NEP due to high allochthonous input have a period of positive NEP after disturbance of their associated forests. This might be caused by the simultaneous increase in light levels as well as temporary reduction in carbon inputs from the adjoining upland ecosystem. If so, then this would contrast from the typical initial negative phase observed in disturbed terrestrial ecosystems. When NEP is considered over a broader spatial extent than a local one (e.g., multiple areas with asynchronous timing of disturbances), the temporal pattern of distinct negative and positive phases of NEP become less prominent. This is because the negative phase of some locations is being offset to some degree by the positive phases elsewhere. In an area in which local ecosystems are perfectly counterbalancing each other, NEP would remain constant. While theoretically possible, NEP would more likely be relatively constant until the disturbance regime influencing the broad area was changed. An underappreciated feature of NEP is that at the local scale disturbances lead to major, systematic temporal trends in NEP, whereas at broader scales it is changes in the nature of the disturbance regime in terms of frequency and severity that cause NEP to change systematically over time.[19] Another feature is that local regions need not reach a period of zero NEP for a broader region to approach zero. As long as the timing of disturbances in local regions is asynchronous the modulating of effect of one location against another can cause NEP to approach zero. The presence of losses other than respiration in a broad area can cause NEP at this scale to remain positive for an extended period without the overall system gaining carbon (Fig. 5). Given these different temporal behaviors of NEP at different spatial scales, it is critical to clearly identify the spatial scale being considered when NEP is being examined.

ACKNOWLEDGMENTS

The author thanks Drs. Walter Dodds and Stanley Gregory for their perspectives on aquatic ecosystems. This work was supported by the Kaye and Ward Richardson Endowment.

REFERENCES

1. Woodwell, G.M.; Whittaker, R.H. Primary production in terrestrial communities. Am. Zool. **1968**, *8*, 19–30.
2. Chapin, F.S. III; Woodwell, G.M.; Randerson, J.T.; Rastetter, E.B.; Lovett, G.M.; Baldocchi, D.D.; Clark, D.A.; Harmon, M.E.; Schimel, D.S.; Valentini, R.; Wirth, C.; Aber, J.D.; Cole, J.J.; Goulden, M.L.; Harden, J.W.; Heimann, M.; Howarth, R.W.; Matson, P.A.; McGuire, A.D.; Melillo, J.M.; Mooney, H.A.; Neff, J.C.; Houghton, R.A.; Pace, M.L.; Ryan, M.G.; Running, S.W.; Sala, O.E.; Schlesinger, W.H.; Schulze. E.-D. Reconciling carbon-cycle concepts, terminology, and methods. Ecosystems **2006**, *9*, 1041–1050.
3. Lovett, G. M.; Cole, J.J.; Pace. M.L. Is net ecosystem production equal to ecosystem carbon storage? Ecosystems **2006**, *9*, 152–155.
4. Randerson, J.T.; Chapin, F.S. III; Harden, J.; Neff, J.C.; Harmon. M.E. Net ecosystem production: a comprehensive measure of net carbon accumulation by ecosystems. Ecol. Appl. **2002**, *12*, 937–947.
5. Schulze, E.-D.; Wirth, C.; Heimann, M. Climate change: Managing forests after Kyoto. Science **2000**, *289*, 2058–2059.
6. Goulden, M.L.; Munger, J.W.; Fan, S.-M.; Daube, B.C.; Wofsy. S.C. Measurements of carbon sequestration by long-term eddy covariance: methods and a critical evaluation of accuracy. Global Change Biol. **1996**, *2*, 169–182.
7. Paw, U,K.T.; Baldocchi, D.D.; Meyers, T.P.; Tilden, P.; Wilson, K.B. Kell B. Correction of eddy-covariance measurements incorporating both advective effects and density fluxes. Bound.-Lay. Meteorol. **2000**, *97*, 487–511.
8. Fellows, C.S.; Valet, H.M.; Dahm, C.N. Whole-stream metabolism in two montane streams: Contributions of the hyporheic zone. Limnol. Oceanogr. **2001**, *46*, 523–531.
9. Hall, R.O.; Tank, J.L. Correcting whole steam estimates of metabolism for groundwater inputs. Limnol. Oceanogr.: Methods **2005**, *3*, 222–229.
10. Luyssaert, S.; Schulze, E.-D.; Börner, A.; Knohl, A.; Hessenmöller, D.; Law, B.E.; Ciais, P.; Grace, J. Old-growth forests as global carbon sinks. Nature **2008**, *455*, 213–215.
11. Vannote, R.L.; Minshall, G.W.; Cummins, K.W.; Sedel, J.R.; Cushing, C.E. The river continuum concept. Can. J. Fish. Aquat. Sci. **1980**, *37*, 130–137.
12. Cole, J.J., Praire, Y.T.; Caraco, N.F.; McDowell, W.H.; Tranvik, L.J.; Striegl, R.G.; Duarte, C.M.; Kortelainnen, P.; Downing, J.A.; Middleburg, J.J.; Melack, J. Plumbing the global carbon cycle: Integrating inland waters into the terrestrial carbon budget. Ecosystems **2007**, *10*, 171–184.
13. Sierra C.A, Harmon, M.E.; Moreno, F.H.; Orrego, S.A.; del Valle, J.I. Spatial and temporal variation of net ecosystem production in a primary tropical forest in the Porce Region, Colombia. Global Change Biol. **2007**, *13*, 1–16
14. Schmidt, M.W.I.; Torn, M.S.; Abiven, S.; Dittmar, T.; Guggenberger, G.; Janssens, I.A.; Kleber, M.; Kögel-Knabner, I.; Lehmann, J.; Manning, D.A.C.; Nannipieri, P.; Rasse, D.P.; Weiner, S.; Trumbore, S.E. Persistence of soil organic matter as an ecosystem property. Nature **2011**, *478*, 49–56.
15. Olson, J.S. Rates of Succession and Soil Changes on Southern Lake Michigan Sand Dunes Botanical Gazette **1958**, *119*, 125–170.
16. Sollins, P.; Spycher, G.; Topik, C. Processes of soil organic matter accretion at a mudflow chronosequence, Mt. Shasta, California. Ecology **1983**, *64*, 1273–1282.
17. Wardle, D.A.; Walker, L.R.; Bardgett, R.D. Ecosystem properties and forest decline in contrasting long-term chronosequences. Science **2004**, *305*, 509–513.
18. Resh, V.H.; Brown, A.V.; Covich, A.P.; Gurtz, M.E.; Li, H.W.; Minshall, G.W.; Reice, S.R.; Sheldon, A.L.; Wallace, J.B.; Wissmar, R.C. The role of disturbance in stream ecology. J. North Am. Benthol. Soc. **1988**, *7*, 433–455.
19. Smithwick, E.A.H.; Harmon, M.E.; Domingo, J.B. Changing temporal patterns of forest carbon stores and net ecosystem carbon balance: The stand to landscape transformation. Landscape Ecol. **2007**, *22*, 77–94.

Nitrogen Fixation: Biological

Robert H. Burris
Department of Biochemistry, College of Agriculture and Life Sciences, University of Wisconsin, Madison, Wisconsin, U.S.A.

Abstract
For centuries farmers practiced mixed cropping of leguminous and non-leguminous plants without knowing the basis of the empirically observed benefit from the practice. In 1838, it was first recognized that leguminous plants could utilize nitrogen (N_2) from the air. This entry explores this process.

INTRODUCTION

For centuries farmers practiced mixed cropping of leguminous and non-leguminous plants without knowing the basis of the empirically observed benefit from the practice. The first clear recognition that leguminous plants could utilize nitrogen (N_2) from the air is attributed to Boussingault's report in 1838.[1] Skepticism remained until Hellriegel and Wilfarth in Germany in 1886 reported definitive experiments demonstrating that pea plants (*Pisum sativum*) in association with bacteria borne in their root nodules were capable of utilizing N_2.[1]

SYMBIOTIC AND FREE-LIVING NITROGEN FIXERS

The symbiotic process of N_2 fixation by leguminous plants plus associated root nodule bacteria (*Rhizobium* species) is of greatest practical importance to agriculture, but there also are a number of free-living bacteria capable of N_2 fixation. The first of these recognized was the anaerobic organism *Clostridium pasteurianum* as reported by Winogradsky in 1893.[2] In 1901 Beijerinck recorded that the aerobic bacterium *Azotobacter chroococcum* also fixed N_2.[2] These early observations were followed by practical applications, and the practice of inoculation of leguminous seeds with cultures of the rhizobia became widespread. The bacteria were grown commercially and farmers applied these bacteria to leguminous seeds at the time of planting. There is a specificity between the plants and bacteria, and this gave rise to the recognition of cross-inoculation groups, i.e., groups of plants all of which are nodulated by the same species of rhizobia. For example, *Rhizobium leguminosarum* infects garden peas, sweet peas (*Lathyrus odoratus*), common vetch (*Vicia faba*), hairy vetch (*Vicia villosa*), and lentil (*Lens culinaris*), whereas *Rhizobium japonicum* is rather specific for soybeans (*Glycine max*). Not all strains of root nodule bacteria that infect a plant are equally effective in N_2 fixation, and this is referred to as strain variation. Because of strain variation, commercial inoculants often are a mixture of effective strains. Inoculants are grown as large batches of effective organisms in liquid culture, and then these cultures are mixed on a peat base. The finely ground peat retains moisture, and the microorganisms remain viable for long periods. The peat culture is mixed with seeds before planting, and the viable organisms are in proximity when the seeds germinate and produce roots. The bacteria induce curling of root hairs and invade the plant through the root hairs. After invasion they induce the production of root nodules and proliferate in the nodules. The bacteria undergo morphological and metabolic changes in the nodule and are referred to as bacteroids. Bacteroids in the nodule are capable of fixing N_2 from the air.

Claims of fixation of N_2 by cultures of the rhizobia outside the host plant in general could not be verified. In 1975 three laboratory groups reported success in achieving such fixation, by furnishing succinic acid as the substrate for growth and by culturing the organisms under a very low pressure of oxygen.[3]

The rhizobia function at a low oxygen concentration in the nodule. The nitrogenase enzyme system is labile to oxygen, but there must be sufficient oxygen to support the generation of adenosine triphosphate (ATP). Hemoglobin in nodules achieves this necessary balance in the oxygen level, keeping it adequate to support the action of cytochrome oxidase which generates ATP.

There are well over 10,000 legume plant species that fix N_2, and about 200 non-legumes. Only one of these non-legumes has been reported to be nodulated by a *Rhizobium* species such as nodulate the legumes. This is *Parasponia andersonii*, a tree of the elm group. The other non-leguminous plants, such as the alder (e.g., *Alnus glutinosa*), fix nitrogen in association with actinomycetes. It proved extremely difficult to culture these actinomycetes, but since methodology to isolate and culture these organisms was developed in 1977,[3] information on them has expanded substantially. Some of the actinorhizal associations have found extensive use, particularly in tropical and low rainfall areas.

Encyclopedia of Natural Resources DOI: 10.1081/E-ENRL-120001554

BIOCHEMISTRY OF NITROGEN FIXATION

Early studies on biological nitrogen fixation emphasized practical applications of the process and the development of effective inoculants. Since 1925 there has been interest in the biochemistry of the N_2 fixation process. Wilson[1] utilized nodulated red clover plants (*Trifolium pratense*) to demonstrate that half-maximum fixation occurred at a pN_2 (partial pressure of N_2) of about 0.05 atmosphere (approximately 50 millibar). High levels of O_2 inhibited the fixation non-competitively. While growing clover plants under controlled atmospheres, Wilson[1] made the unexpected observation that fixation was inhibited by hydrogen gas (H_2). H_2 is a specific and competitive inhibitor of N_2 fixation. The H_2 inhibition of N_2 fixation results from an ordered, sequential process in which the H_2 must bind to the nitrogenase enzyme before the N_2.

There was speculation about the first or key roduct of N_2 fixation. Winogradsky[2] suggested that ammonia (NH_3) was the first product. Virtanen,[2] on the other hand, supported hydroxylamine as the key intermediate based on his claim that leguminous plants excreted aspartic acid, and that the hydroxylamine formed by fixation reacted with oxalacetic acid to form an oxime that then was converted to aspartic acid. Other investigators failed to confirm Virtanen's work. The use of the stable isotope ^{15}N as a tracer resolved the argument. ^{15}N-enriched N_2 was supplied to a culture of *Azotobacter vinelandii* fixing N_2 vigorously, and among the amino acids recovered from the hydrolysate of the cells, the highest enrichment in ^{15}N among the isolated amino acids was in glutamic acid[2] rather than in aspartic acid, which would have been supportive of Virtanen's hydroxylamine hypothesis. If NH_3 were the key product of fixation, it could logically be incorporated into glutamic acid by the action of known enzymes. Similar experiments were performed with species of bacteria from the genera *Clostridium*, *Chromatium*, *Chlorobium*, and *Rhodospirillum*, as well as the blue-green alga *Nostoc muscorum* and excised soybean nodules. In all cases the highest concentration of ^{15}N among the isolated amino acids was in glutamic acid.

CELL-FREE NITROGEN FIXATION

As the metabolism of nitrogenous compounds in intact microorganisms can be rather complex, it was important to verify the path of fixation in cell-free preparations where reactions can be defined more specifically. ^{15}N as a tracer furnished a sensitive tool for seeking reliable cell-free fixation. Carnahan and coworkers in 1960[3] developed a method for recovering active preparations consistently from *C. pasteurianum*. Such cell-free preparations supplied ^{15}N-enriched N_2 yielded NH_3 with over 50 atom % ^{15}N excess, a very convincing verification that NH_3 is the "key" intermediate in biological nitrogen fixation.

PURIFICATION OF NITROGENASE

Mortenson[3] demonstrated that the cell-free enzyme system consisted of two proteins. When the two proteins were purified, it turned out that one component was a molybdenum–iron (MoFe) protein and the other was an iron protein. In purification on a chromatographic column, the MoFe protein was eluted before the Fe protein, so the MoFe protein often was referred to as protein 1 or component 1, and the Fe protein as protein 2 or component 2. A number of other terminologies were used but abandoned. Descriptive terms are dinitrogenase for the MoFe protein, dinitrogenase reductase for the Fe protein, and nitrogenase for the complex of the two components. The primary function (it also has other functions) of the Fe protein is to transfer electrons to the MoFe protein, hence the designation dinitrogenase reductase.

REQUIREMENT FOR ADENOSINE TRIPHOSPHATE

In 1960 it was reported that ATP was inhibitory to N_2 fixation in cell-free preparations of nitrogenase, but in 1962 McNary[3] demonstrated that ATP is an absolute requirement for fixation by such preparations. The N–N bond is stable, so N_2 fixation is energy demanding whether it is accomplished chemically at high temperature and pressure, or whether it is accomplished by a biological system. A minimum of 16 ATPs are required for the reduction of one N_2 to $2NH_3$. Sixteen ATP are required under ideal conditions, and under most conditions in nature the requirement is 20 to 30 ATP.

SUBSTRATES OTHER THAN N_2

Nitrogenase is a versatile enzyme and is capable of reducing substrates other than N_2.[3] It was recognized early that H_2 is a by-product of the nitrogenase reaction. The next substrate to be recognized was nitrous oxide (N_2O) which is reduced to N_2. Cyanide and methyl isocyanide, and azide also can be reduced by the enzyme system. Nitrogenase can reduce acetylene to ethylene. Other substrates demonstrated include diazine, diazarine, cyanamide, and cyclopropene. Acetylene reduction to ethylene has become a widely used measure of nitrogenase activity. The substrate is easy to prepare, the ethylene formed as a product can be measured readily and accurately by gas chromatography, and the simple method has been applied to studies in the glasshouse and field.

CONTROL OF NITROGENASE

As nitrogen fixation is energy demanding, it is advantageous for the fixing organisms to turn off the system if they have access to a source of fixed nitrogen. Ludden[3] found that *Rhodospirillum rubrum* turned its nitrogenase off when supplied fixed nitrogen such as NH_3, or when it was placed in the dark and thus was deprived of light energy. When NH_3 was exhausted or when light was restored, it turned nitrogenase back on. Inactivation is achieved by ADP ribosylation of arginine 100 of one of the two equivalent subunits of dinitrogenase reductase, and reactivation is achieved by removal of this ADP-ribosyl group.

TERTIARY STRUCTURE OF NITROGENASE

The tertiary structure of the nitrogenase components was established in 1992.[2] This aids one in visualizing how the dinitrogenase and dinitrogenase reductase fit together and how the active Fe and MoFe sites are aligned.

ASSOCIATIVE NITROGEN FIXATION

The leguminous plants and their associated rhizobia are ofgreat practical importance in agriculture, but the non-leguminous actinorhizal systems often fix N_2 vigorously and occupy important niches in ecosystems. Certain non-legumes can benefit from association with bacteria that occupy their root surface or multiply within the plant without forming nodules.[3–4] In Brazil it has been demonstrated that sugarcane can derive nitrogen from fixation by *Acetobacter diazotrophicus* carried inside the cane. *A. diazotrophicus* is a remarkable bacterium, as it can grow in a 25% sucrose solution and fix N_2 at a pH as low as 3.0.

GENETICS OF THE NITROGENASE SYSTEMS

The genetics of biological N_2-fixing systems has occupied many investigators in recent years, as genetic manipulations may offer ways to improve the process or adapt it to new crops. Progress in our knowledge of the genetics of N_2-fixing organisms has led in recent years to a substantial reorganization of classification of these organisms.

REFERENCES

1. Wilson, P.W. *The Biochemistry of Symbiotic Nitrogen Fixation*; The University of Wisconsin Press: Madison, WI, 1940; 302 pp.
2. *Triplett, E.W. Prokaryotic Nitrogen Fixation*; Horizon Scientific Press: Wymondham, England, 2000; 800 pp.
3. Stacey, G.; Burris, R.H.; Evans, H.J., Eds.; *Biological Nitrogen Fixation*; Chapman and Hall: New York, 1992; 943 pp.
4. Postgate, J.R. *The Fundamentals of Nitrogen Fixation*; Cambridge University Press: Cambridge, 1982; 252 pp.

Ozone: Crop Yield and Quality Effects

Håkan Pleijel
Applied Environmental Science, Göteborg University, Göteborg, Sweden

Abstract
Air pollutants have been known for more than one hundred years to affect both yield and quality of agricultural crops. This is true for a wide range of pollutants including sulfur dioxide, fluorides, and nitrogen-containing pollutants such as nitrogen oxides (NO, NO_2) and ammonia (NH_3). Today, the main focus is on effects of the regional occurrence of elevated tropospheric ozone (O_3). In North America and Europe, where emissions of sulfur dioxide and fluorides have declined, the problem of ozone pollution has received the most attention. In contrast, in many developing countries emissions of a number of pollutants are increasing and there are strong indications of adverse effects on crops. However, these effects have been studied to a much lesser extent than in Europe, North America, Japan, and Australia.

DISCOVERY OF OZONE EFFECTS ON CROPS

Systematic studies of ozone effects on crop yield and quality started around 1950 in California, U.S.A.,[1] where the problems caused by ozone and other photochemical oxidants were first discovered. At that time, the studies were mainly of an observational nature, and they were based on visible injury appearing on the plant leaves. Controlled experiments in the laboratory began later; however, initially these yielded largely qualitative data which were not suitable as a basis to estimate the magnitude of potential effects in the field.

But, it should be noted here that in certain crops, such as spinach and tobacco, leaf injury is of large economic importance, apart from potential reductions in absolute yield. In some crops visible leaf injury, expressed as necrotic spots on the leaves, is the most pronounced effect by ozone and other pollutants, while in others a reduction in the leaf life span, the so-called premature senescence, is most important, as it reduces the length of the growth period. The latter occurs in wheat, in which characteristic leaf injury at moderate ozone exposure is generally lacking, while early senescence in response to ozone is pronounced.[2]

OPEN-TOP CHAMBERS

With the introduction of the open-top chamber (OTC) as an exposure system,[3] which permits semicontrolled exposure of plants to gaseous pollutants under ecologically realistic conditions in the field, an important step was taken towards a quantitative understanding of the impacts of ozone and other pollutants. An OTC is a transparent plastic cylinder through which air (ambient air, filtered air, or air enriched with various levels of pollutant gases) is ventilated using a fan. OTCs can be mounted in plots of a field grown crop and kept in place throughout the growing season. Although the system alters the microclimate of the plants to a certain extent,[4] the ecological realism is much larger than in a closed chamber system in the laboratory.

CROP LOSS ASSESSMENT PROGRAMS

In the U.S.A., a large experimental program, the National Crop Loss Assessment Network (NCLAN), involving the use of OTCs of crops, was executed during the 1980s. The NCLAN results showed that ambient levels of ozone had the potential to reduce the yield of a number of crops, including soy bean, wheat, and alfalfa.[5] In addition, economic estimates of crop losses for the U.S.A. indicated an annual loss for the farmers in the range of a few billion U.S. dollars.

A similar program, the European Open-top Chamber Programme (EOTC), was organized in Western Europe beginning in the second half of the 1980s until the early 1990s. Negative effects of ozone levels typical of wide areas of Europe were observed in beans and spring wheat.[6] In some, but not all, experiments with pastures the main effect of ozone was on the species composition, with clover being replaced by grasses. This effect acts to reduce the protein content and thus the fodder quality of the yield. Towards the end of the 1990s a European research program, Changing Climate and Potential Impacts on Potato Yield and Quality (CHIP), studied the effects of ozone on potato. It was concluded that current ozone levels in Europe can cause significant effects on potato, but that the sensitivity is less than in wheat.

Based on the experiences of both research programs in North America and Europe, it can be concluded that soybean, wheat, tomato, pulses, and watermelon can be

Encyclopedia of Natural Resources DOI: 10.1081/E-ENRL-120005563

classified as highly sensitive to ozone. Barley and certain fruits are rather insensitive, while potato, rapeseed, sugar beet, maize, and rice are intermediate with respect to ozone sensitivity. There exists a certain degree of intraspecific variation in ozone sensitivity, such as in different bean varieties, but a recent European study resulted in a rather uniform ozone response pattern for different cultivars of wheat.

"CRITICAL LEVELS" OF OZONE

The relationship between grain yield and ozone exposure was most consistent in spring wheat across cultivars and experimental sites, and, considering the pronounced sensitivity observed for this crop, the combined data were used as a basis to derive critical levels for ozone effects under the Convention on Long Range Transboundary Air Pollution of the United Nations Economic Commission for Europe (UNECE). For this purpose, the ozone exposure index AOT40 (the accumulated exposure over a concentration threshold of 40 nmol mol^{-1} ozone based on hourly averages) was used.[7]

OZONE UPTAKE

An important recent development in the field of crop loss assessments is the consideration of ozone uptake by the plant leaves, in place of a statistical index to characterize the concentration of ozone in the air surrounding the plants, such as the AOT40 exposure index. Key to this method is the quantification of ozone diffusion through the stomata of leaves.

The stomatal conductance of the plant leaves varies with a number of factors, such as soil and air moisture contents, temperature, solar radiation, phenology, and the levels of other pollutants, including carbon dioxide. Stomatal conductance models can now be used to establish relationships between yield loss and ozone uptake. An example is given in Fig. 1 showing the relationship between relative yield in two Swedish wheat cultivars and the calculated, cumulated uptake of ozone by flag leaves, taking into account an uptake rate threshold of 5 nmol m^{-2} s^{-1} (CUO^5).[8] This threshold provided the best correlation between effect and exposure, and it may represent the biochemical defense capacity of the plants. However, it remains to be shown that the two thresholds are directly related to each other.

EFFECTS OF OZONE ON PROTEIN CONTENTS

Much less attention has been paid to the influences of ozone on crop quality as compared to efforts made to understand the relationship with yield. The most discussed quality aspect has been the protein concentration or content, for instance, of wheat grains. The combined results of sixteen OTC experiments performed in four different European countries (Finland, Sweden, Denmark, and Switzerland) were used to show the relationship between grain yield and grain quality (Fig. 2), and grain yield and the off take of grain protein per unit ground area (Fig. 3), at different ozone or carbon dioxide levels.[9] From Fig. 2, it can be inferred that, relative to control (always the OTC treatment with non filtered air), the grain concentration of protein was higher and yield was lower in elevated ozone, which is the opposite result of situations with elevated carbon dioxide or in chambers where the ozone had been reduced by using charcoal filters. This situation could be attributed to a so-called growth dilution effect by elevated carbon dioxide and air filtration stimulating the carbohydrate accumulation relatively more than the uptake of nitrogen at a given nitrogen fertilizer application rate, while in situations of elevated ozone, carbohydrate accumulation was negatively affected more strongly than protein accumulation.

However, Fig. 3 shows that the protein yield per unit ground area was negatively affected by ozone and enhanced by elevated carbon dioxide, although the latter effect tended to saturate at a level determined by the availability of nitrogen in the soil.

CONCLUSION

There is strong evidence that current ozone concentrations over large areas in the industrialized world are high enough

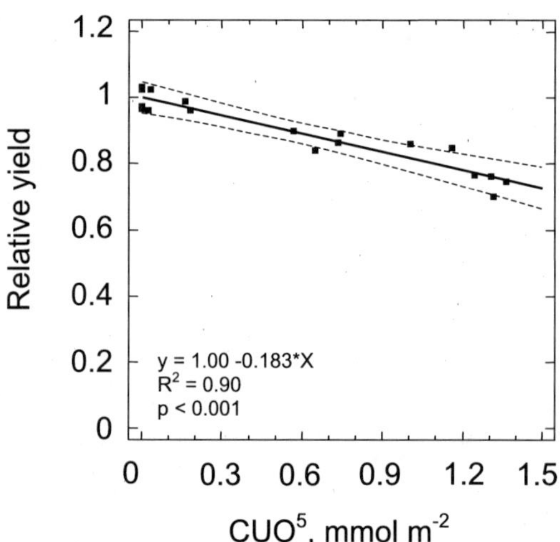

Fig. 1 Relationship between the relative yield of field grown spring wheat and the cumulative ozone uptake (CUO) by the flag leaves based on stomatal conductance modeling. An ozone uptake rate threshold of 5 nmol $m^{-2}s^{-1}$ was used, since this threshold resulted in the best correlation between relative yield and ozone uptake. Five OTC experiments performed in Sweden were included.

Source: Adapted from Danielsson, Pihl Karlsson, et al.[8]

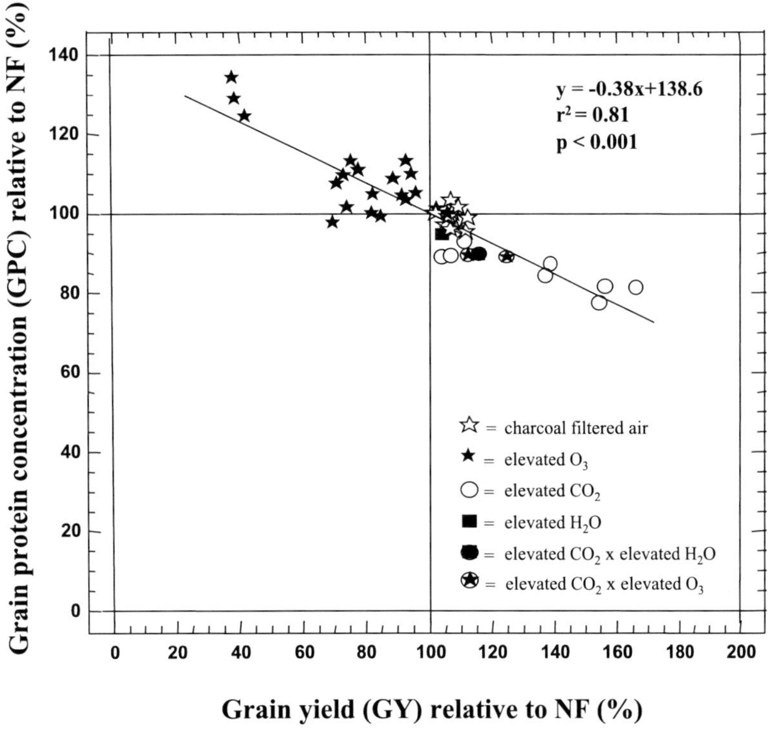

Fig. 2 Relationship between grain protein concentration and grain yield in field grown spring wheat based on data from OTC experiments performed in Sweden, Finland, Denmark, and Switzerland. The experiments included ozone treatments, carbon dioxide treatments, and a few irrigation treatments. Relative scales were used for both axes: The effects were related to the OTC treatment with non filtered air (NF) without extra ozone or carbon dioxide fumigation in each of the sixteen experiments included.
Source: Adopted from Pleijel, Mortenson, et al.[9]

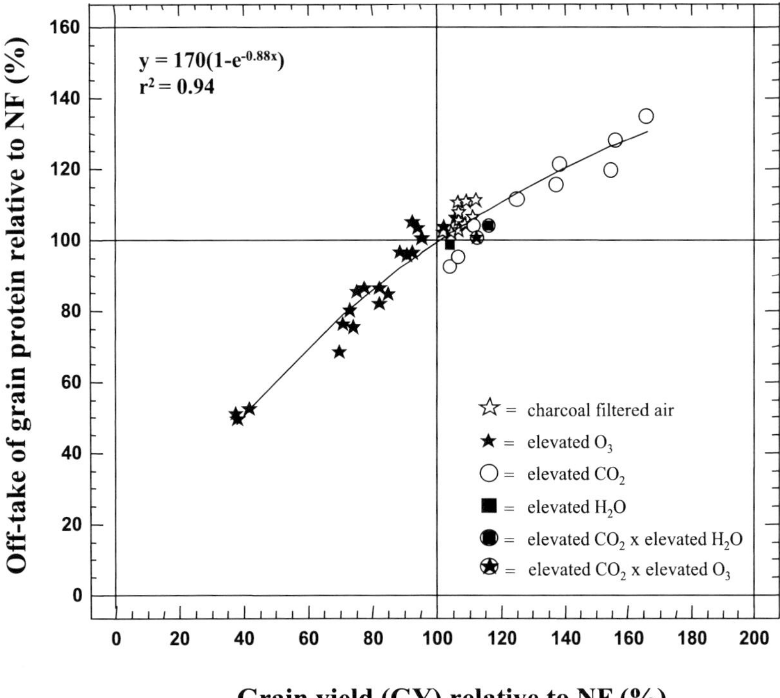

Fig. 3 Relationship between grain protein off take per unit ground area and grain yield in field grown spring wheat based on data from OTC experiments performed in Sweden, Finland, Denmark, and Switzerland. The experiments included ozone treatments, carbon dioxide treatments, and a few irrigation treatments. Relative scales were used for both axes: The effects were related to the OTC treatment with non-filtered air (NF) without ozone or carbon dioxide fumigation in each of the sixteen experiments included.
Source: Adopted from Pleijel, Mortenson, et al.[9]

to cause yield loss in several important agricultural crops, but it remains a challenge to quantify these effects exactly. New approaches based on the uptake of the pollutant by plants, rather than the concentration in the air surrounding the plant, are promising.

The protein concentration of the yield tends to increase with increasing ozone in some crops in connection with declining yield. On the other hand, the protein yield per unit ground area tends to decrease. In future research, the understanding of additional quality effects of air pollutant exposure should be given more attention.

In the developing world, such as many countries in Asia, there is a great risk for crop losses due to a number of air pollutants for which emission rates are increasing dramatically.[10] A loss in food production in these countries represents a much larger problem than in the industrialized world. More research should be devoted to this problem in the decades to come.

REFERENCES

1. Middleton, J.T.; Kendrick, J.B.; Schalm, H.W. Injury to herbaceous plants by smog or air pollution. *Plant Dis. Rep.* **1950**, *34*, 245–252.
2. Pleijel, H.; Ojanperä, K.; Danielsson, H.; Sild, E.; Gelang, J.; Wallin, G.; Skärby, L.; Selldén, G. Effects of ozone on leaf senescence in spring wheat—Possible consequences for grain yield. Phyton (Austria) **1997**, *37*, 227–232.
3. Heagle, A.S.; Body, D.; Heck, W.W. An open-top filed chamber to assess the impact of air pollution on plants. J. Environ. Qual. **1973**, *2*, 365–368.
4. Unsworth, M.H. Air pollution and vegetation, hypothesis, field exposure, and experiment. Proc. R. Soc. Edinb. **1991**, *97B*, 139–153.
5. Heck, W.W.; Taylor, O.C.; Tingey, D.T. *Assessment of Crop Loss from Air Pollutants*; Elsevier Applied Science: London, 1988.
6. Jäger, H.J.; Unsworth, M.H.; De Temmerman, L.; Mathy, P. *Effects of Air Pollution on Agricultural Crops in Europe. Results of the European Open-Top Chamber Project*; Air Pollution Research Report Series of the Environmental Research Programme of the Commission of the European Communities, Directorate-General for Science, Research and Development, 1992; 46.
7. Fuhrer, J.; Skärby, L.; Ashmore, M. Critical levels for ozone effects on vegetation in Europe. Environ. Pollut. **1997**, *97*, 91–106.
8. Danielsson, H.; Pihl Karlsson, G.; Karlsson, P.E.; Pleijel, H. Ozone uptake modeling and flux-response relationships—Assessments of ozone-induced yield loss in spring wheat. Atmos. Environ. **2003**, *37*, 475–485.
9. Pleijel, H.; Mortensen, L.; Fuhrer, J.; Ojanperä, K.; Danielsson, H. Grain protein accumulation in relation to grain yield of spring wheat (*Triticum aestivum* L.) grown in open-top chambers with different concentrations of ozone, carbon dioxide and water availability. Agric. Ecosyst. Environ. **1998**, *72*, 265–270.
10. Emberson, L.D.; Ashmore, M.R.; Murray, F.; Kuylenstierna, J.C.I.; Percy, K.E.; Izuta, T.; Zheng, Y.; Shimizu, H.; Sheu, B.H.; Liu, C.P.; Agrawal, M.; Wahid, A.; Abdel- Latif, N.M.; van Tienhoven, M.; de Bauer, L.I.; Domingos, M. Impacts of air pollutants on vegetation in developing countries. Water Air Soil Pollut. **2001**, *130*, 107–118.

Peatlands

Dale H. Vitt
Department of Plant Biology and Center for Ecology, Southern Illinois University, Carbondale, Illinois, U.S.A.
Paul Short
Canadian Sphagnum Peat Moss Association, St. Albert, Alberta, Canada

Abstract
Peatlands occupy only 3% of the global terrestrial surface, but hold 10% of the world's drinking water, and contain stored in peat about one-third of the Earth's soil carbon (about equal to the total CO_2 in the atmosphere) and 9–16% of the world's soil nitrogen. Peatlands act as watershed filters and serve as habitat for a number of unique suites of animal species. Fungi are common and abundant, and insects are diverse, but little studied. Plant diversity is low, but includes a number of rare and uncommon species. Peat has long been used as a soil amendment, and the harvesting of peatmoss is a small, but important industry where peatlands are abundant. The leading producer of commercial peat moss is Canada and the industry maintains a view of responsible environmental management that strives to restore peatlands to their before-harvest condition. Additionally, peatlands are managed for enhanced tree growth through drainage and fertilization in some countries and peat is harvested for electrical generation in a few northern countries. Peatlands contain a large, unstable stockpile of carbon that is sensitive to both climatic and anthropogenic disturbances. These disturbances have the potential to change the long-term peatland carbon sink into a source of carbon to the atmosphere, thus Wise Use, especially conservation, of our peatland resources should be a priority.

INTRODUCTION

Peatlands contain abundant natural resources and are ecosystems that over the long term have net primary production of organic matter exceeding losses from decomposition and export, and at millennial time scales have large stores of carbon.[1] As a result, peatlands develop a deep layer of organic soil, or peat, that is composed of at least 30%, but usually more than 90% organic matter. Deposits of peat accumulate in-place and contain a historical record of community change over time, as well as serve as proxies for past environmental and climatic changes,[2] and in some places hold a record of past human activities on the site. These records indicate that nearly all peatlands have initiated and developed over the past 10,000–12,000 years, and since the last glaciation. Globally, peatlands represent a major carbon stockpile, storing approximately one-third of the world's soil carbon (450 Pg; petagram = 1×10^{-15} g) and about 10% of the global drinkable water, yet cover only 3% of the world's terrestrial area.[3] As compared to ocean and lake sediments that also contain large reservoirs of stored carbon and are relatively unaffected by atmospheric processes, peatland ecosystems, and the carbon stored in the layers of peat, are extremely vulnerable to climate change and land-use changes. Additionally, peatlands are a sink for nitrogen and it has been estimated that boreal peatlands contain about 9–16% of the global pool of N.[4]

While all peatlands are wetlands, not all wetlands are peatlands, and the need to distinguish the difference is important to the recognition and management of peatland resources. In general, non-peat-forming wetlands are either treed or shrubby (swamps) or without woody vegetation (marshes). Both swamps and marshes have fluctuating water levels and relatively high amounts of nutrient inputs, hence organic matter produced in these systems decomposes rapidly and peat formation is limited. Peatlands (sometimes called "mires" in Europe) are usually classified as either fens or bogs (Fig. 1) depending on the source of the water. Fens receive waters from precipitation as well as from the surrounding uplands (i.e., geogenous water source), whereas bogs are somewhat raised above the surrounding area and receive water only from precipitation (i.e., ombrogenous). The ionic quality of these incoming waters greatly influences the structure and functioning of the ecosystem and dictates the amount and type of peat produced. Bogs and poor fens are dominated by the moss genus *Sphagnum* (Fig. 2), and it has been estimated that there is more carbon stored in *Sphagnum* than in any other plant genus, whereas rich fens have ground layers dominated by true mosses.[5]

Peatlands are estimated to have originally occupied 4,258,000 km^2, of which about 74% are in nontropical parts of the world. North America has 46%, Asia 35%, Europe 13%, South America 4%, Africa 1%, and Australia and Antarctica less than 1%. Peatlands are concentrated in the

Encyclopedia of Natural Resources DOI: 10.1081/E-ENRL-120047523

Fig. 1 Right side: A poor fen in the foreground and treed bog in the background. The fen dominated by sedges and *Sphagnum*. Left side: Continental bog of western Canada characterized by a dense tree layer of *Picea mariana*.
Source: Photos by Kimberli Scott.

Fig. 2 Ground cover of *Sphagnum lenense* with young leaves of *Rubus chamaemorus* (cloudberry) from the Yukon Territory, Canada.
Source: Photo by Dale Vitt.

boreal and arctic regions of the world where precipitation exceeds potential evapotranspiration. Countries with more than 50,000 km^2 of peatland are Russia (1,410,000 km^2), Canada (1,235,000 km^2), USA (625,000 km^2), Indonesia (270,000 km^2), Finland (96,000 km^2), Sweden (70,000 km^2), and Peru (50,000 km^2). Loss of natural peatlands to anthropogenic disturbance is variable and ranges from 52% of the natural peatlands being lost in Europe and 50% in nontropical Africa, 20% in nontropical South America, 8% in nontropical Asia, and 5% in North America. About 16% of nontropical peatlands have been lost to disturbance, with agriculture accounting for 50%, forestry 30%, peat extraction 10%, and the remainder from urbanization, inundation, and erosion. Restoration efforts, especially in peat harvested bogs of North America, have been developed to return some of these areas to functioning peatlands.[6]

SENSITIVITY TO DISTURBANCE

The large reservoirs of carbon and nitrogen found in peatlands have developed relatively recently over the past 10,000–12,000 years. Over this period of time it is estimated that peatlands accumulated carbon at a rate of about 17–20 $g/m^2/yr$ and that globally peat accumulation is about 40–70 Tg C/yr (teragram = 1×10^{-12} g). As a natural resource, peatlands are subject to both anthropogenic and natural disturbance impacts. In North America, particularly in the western Canadian provinces, wildfires account annually for the largest emissions of carbon, estimated at 6300 Gg C/yr, (gigagram = 1×10^{-9} g) while anthropogenic impacts that include flooding from hydroelectric construction (−100 Gg C/yr), peat harvesting (−140 Gg C/yr), and mining (−50 Gg C/y) account for −290 Gg C/yr. Considering that these western Canadian peatlands sequester ±8940 Gg C/yr and the results of permafrost melting contribute +160 Gg C/yr, the net estimate of annual carbon sequestration is +1320 Gg C/yr.[7]

Globally the conversion of peatlands to agriculture and forestry is estimated to contribute 100–200 Tg of C/yr to the atmosphere, and this combined with global peat extraction of approximately 15 Tg of C/yr would account for peatlands at the global scale to be a source rather than a sink for carbon.[8] However, these estimates are rapidly changing, with changes in forestry practices along with regional differences in anthropogenic activities and the application of appropriate responsible peatland management practices.

NATURAL RESOURCES OF PEATLANDS

Although peatlands have often been considered as wasteland, even cancers on the landscape (illustrated in *The Lord of the Rings,* "Meres of Dead Faces" NW of Mordor), they actually contain a wealth of natural resources, and perhaps more importantly carry out valuable ecosystem functions on the landscape. This view of peatlands as wastelands is exemplified by their use as prisons (e.g., the Gulags of the former Soviet Union and the German concentration camps such as Dachau); their use as landfills; and their use as military training grounds, with the latter allowing some peatlands to be preserved in areas of concentrated human occupation (such as western Europe).

Given the poor regard for these natural peatland resources, responsible peatland management has until recently been a challenge. *The Wise Use of Mires and Peatlands*, published in 2002[7] provided an excellent framework for decision making when considering the use of peatlands. More recently, the *Strategy for Responsible Peatland Management*, published by the International Peat Society in 2011,[9] has provided a comprehensive management construct for peatlands worldwide. It provides those involved in or responsible for peatland management with strategic objectives and actions for implementation that for the first time defines objectives and actions for the conservation, management, and rehabilitation of peatlands globally, based on the principles of Wise Use. According to the Strategy, peatland management includes organizing, controlling, regulating, and utilizing peatland and peat for specific purposes, which should be appropriate to the peatland type and use while respecting cultural, environmental, and socioeconomic conditions. The Strategy is applicable to all types of peatlands under every use, including nonuse, and it is directed to everyone responsible for or involved in the management of peatlands, or in the peat supply chain.

Water Resources

In areas where large portions of watersheds are covered by peatlands, these ecosystems provide important source areas for drinking water. For example, in Britain in Yorkshire, 45% of the public supply of water is from watersheds draining peatlands. Haworth Moor of *Wuthering Heights* fame was owned by a company providing drinking water. In the northern Andes, peatlands are source areas for drinking water for millions of people living in the large cities of Colombia and Peru. In western Canada's boreal forest, lake chemistry and functions are influenced by watersheds having large cover of peatlands. These peatland-dominated watersheds provide filtered waters to the downstream lakes, and in some cases acidifying the lakes.[10] The nutrient chemistry of these lakes is significantly different from lakes with only upland inputs.[11] Thus peatlands can provide a modifying influence on downstream ecosystems. The large dissolved organic carbon (DOC) component of peatlands, especially bogs, colors downstream lakes and streams. With warming and/or drying climatic conditions, coupled with increased atmospheric deposition and pasturing of animals, bogs can degrade releasing large amounts of DOC to downstream drainages. This is a loss of carbon from the long-term store found in these peatlands. This loss from agricultural uses has been reported as significant in the British Isles.

Soil Resources

Peat has been extracted from peatlands for centuries for a number of uses, foremost of which is currently an organic substrate in agriculture and gardening. Its uses include biodegradable pots for cultivation of seedlings, soil amendments for gardens, and growing media for both professional and amateur flower and vegetable growers. It also has been used as a substrate for mushroom growers and for tree nurseries. Among its unique properties for use as a soil amendment and as a growing medium are (1) excellent water-holding capacity—holding up to 20 times its weight in water and retaining water for exceptionally long periods of time; (2) a low pH that functions to reduce the microbial activity thus lessening the number of pathogens; (3) lowering the rate of decomposition and as a result the amendment remains in the soil medium longer. Additionally, peat is easy to harvest (Fig. 3), handle, and ship to market. Peat harvesting is a small, but important, industry in Canada ($10.3 \times 10^{-6} m^3$ harvested in 1999) and Germany ($9.5 \times 10^{-6} m^3$ in 1999).

In Canada the horticultural peat industry is a well-established although relatively recent commercial industrial use of peatlands. Peat for energy was the initial focus of peat use in Canada, but at present there is no industrial use of peat for energy production. Horticultural peat has become since the postwar era a major resource industry in rural areas of Canada, particularly in Quebec, and the Maritime Province of New Brunswick. Western Canada's Prairie Provinces (Alberta, Saskatchewan, and Manitoba) are areas of expansion for the industry and currently account for approximately 40% of the national production. The Canadian production principally services the horticultural industry of both Canada and the United States.

The Canadian industry, through policy and practice, is committed to the return of postharvest sites to functioning peatland ecosystems. Recognized throughout the world as leaders in restoration of peatlands is the Industrial Chair for Peatland Management and the Peatland Ecology Research Group at the University of Laval in Laval, Quebec. Beyond the applied level of peatland restoration the industry continues to actively promote responsible management of peatland resources throughout the provincial and federal agencies accountable for natural resource management. Peatlands (bogs and fens) need to be acknowledged as

Fig. 3 Vacuum harvesting of a drained peat field in Quebec, Canada.
Source: Photo by Dale Vitt.

natural biological resources and managed to ensure their environmental, social, and economic values are sustained. The key is responsible management that sets aside areas for protection and conservation and puts in place management practices that ensure the retention of ecosystem goods and services following development. The peat-harvesting industry has a commitment to support continued leading edge peatland research, application of science-based restoration techniques, and improved responsible peatland management in order to provide sustainable peatland management that recognizes the environmental, social, and economic values of peatland resources.

Historically, peat has also been used in a number of fascinating ways. During the 1800s and early 1900s, peat was used as litter in horse stables, especially for cavalry, railways, and transportation companies that stabled large numbers of horses. If fact, this use led to the rapid development of peat extraction in western Europe. It has been estimated that one French army with a 13,500-horse cavalry needed 22,000 tons of peat in one year.[8] *Sphagnum* peat was used extensively during the Napoleonic and the Franco-Prussian Wars, by the Japanese in the 1904–1905 war with Russia, and in World War I by both sides as a substitute for cotton in surgical dressings. In Britain, "War Work for Women" included collecting and processing *Sphagnum* in the extensive British open bogs. Aboriginal people in North America used *Sphagnum* moss for diapers and in the 1980s *Sphagnum* was used in the commercial production of feminine hygiene products. Peat also forms a component of filtering systems for water purification. The cation exchange properties of *Sphagnum* allow the exchange of hydrogen ions for all base cations, including cationic forms of heavy metals. The absorbent properties of peat allow not only the absorption of large quantities of water, but when dried to less than 8% moisture, peat becomes hydrophobic and absorbs large quantities of petroleum products; as such it has been used in oil spill cleanup and as floor mats in industries with petroleum spillage. The use of peat for building and insulation was widespread in Europe in the past. In Ireland, canal banks were lined with peat; in Finland it was used as a liner for roadways; and in Russia and Belarus it was pressed into sheets and used as insulation in buildings and industry.

Peat is also utilized for energy production. The use of peat for electrical generation is an important national strategic use of peat in Ireland, Finland, and Sweden. Currently, within the Nordic countries of the European Union, energy generation of peat is one of the most important uses of peat and will almost certainly remain important for countries with large peat reserves and in producing energy for isolated locations. The use of peat briquettes for burning in household stoves, furnaces, and fireplaces is also common in rural Ireland, but currently peat is mixed with *Miscanthus* biomass to form more energy-efficient renewable products. Bord da Mona reports that overall, about 4 megatons (=4 Tg) of peat are burned for energy generation each year, of which 1 megaton (=1 Tg) is used for local burning in households.

Biotic Resources

Since the early 1800s, and originating in Germany, peat has been used in balneology (baths). It is reported that owing to the biologically active substances in peat, such as humic acids, peat may influence the immune system and is effective against microorganisms. Peat is an important component in the distillation of Scotch whiskey. The special peaty flavor of Scotch whiskey is imparted by slowly drying the "green malt" over a smoldering peat fire.

Fens have long been cut for hay in both Canada and Eurasia. This "slough hay" is used for both fodder and straw for domestic animals. Also, drier peatlands have been long used as pasture. Wild plants are actively collected and eaten by local and indigenous people across the boreal zone. Cranberries (*Vaccinium oxycoccos* and *V. macrocarpon*) are commercially grown and processed for a number of juice products in eastern United States and Canada. In Fennoscandia and Quebec, Canada, the berries of cloudberry (*Rubus chamaemorus* (see Fig. 2) and lingonberry (*Vaccinium vitis-idaea*; Fig. 4) are distilled for liquors, as well as used in preserves and jellies. A large number of medicinal preparations are produced from sundews (*Drosera* spp.). First Nations peoples of Canada used plants from peatlands extensively for both food and medicinal preparations; for example, leaves of *Ledum groenlandicum* (Labrador Tea—Fig. 5) were boiled and drank for nausea.[5]

Drainage of peatlands to improve forest growth has long been practiced in Fennoscandia, Great Britain, and Russia. Current estimates are that 15×10^5 ha of northern peatlands has been drained for forestry use. In oceanic areas where peatlands are treeless, often nonindigenous tree species have been planted with subsequent fertilization, whereas in more continental areas that have stunted trees growing directly on the peat, the soils have been enhanced with drainage and fertilization. These practices, common in

Fig. 4 *Vaccinium vitis-idaea* (lingonberry) with red berries, growing on a mat of the reindeer lichen, *Cladina mitis*.
Source: Photo by Kimberli Scott.

Fig. 5 Flowering branch and characteristic leaves of *Ledum groenlandicum* (labrador tea).
Source: Photo by Kimberli Scott.

Europe and Asia, have not been implemented in any large scale in North America; however, black spruce (*Picea mariana*) is harvested from natural peatlands in eastern Canada.

Biodiversity

Peatlands are among the last large undisturbed ecosystems in the world and serve as habitat for many animals and birds. Although there are no vertebrate species that occur exclusively in peatlands, peatlands do serve as primary habitat for Caribou in North America, especially bogs with an abundance of reindeer lichens (mainly *Cladina mitis;* see Fig. 3). Both Moose and Black Bear utilize peatlands for food and shelter. In northern areas, peatland specialists include the Arctic Shrew, Northern Bog Lemming, and Southern Bog Lemming. Birds also occupy peatlands, but few are exclusive to these areas. Sand-hill Cranes nest in fens in western Canada; American Bittern, Wilson's Snipe, and Upland Sandpipers also use peatlands as major habitats in eastern Canada. Beavers, both in Europe and North America, may disturb peatlands but the lack of suitable substrate and food resources limit their activities to the edges. Surprisingly, little is known about the apparent rich invertebrate fauna of peatlands, where many undescribed species may occur. In the few studies completed the invertebrate diversity is extremely high, yet little studied. In comparison, such environmental indicators as testate amoebae have been well studied and have been extensively utilized in studies that reconstruct past environmental conditions of the local area.[12]

Carbon Resources

Over the long term, peatlands extract large amounts of carbon dioxide from the atmosphere and store it in deposits of peat. Presently, peatlands have stored about the same amount of carbon as there is in the atmosphere. Some believe that carbon extracted from the atmosphere by peatlands during the interglacials reduced atmospheric CO_2 concentrations and resulted in the development of the glacial periods. There is some evidence that global CO_2 concentrations are coupled to peatland initiation and global climate.[13] Although pristine peatlands have served as a sink for atmospheric CO_2, peatlands also are a source for methane (a potent greenhouse gas [GHG]) and also nitrous oxides. Wetlands, rice paddies, and animal livestock dominate global methane production, and emissions from peatlands are variable, but of less importance. While bogs produce little or no methane, fens may provide a rich source. Methane is a product of anaerobic decomposition and is often highest in wet habitats with relatively high rates of organic matter turnover (such as rich fens and eutrophic marshes). Nitrous oxide emissions from pristine peatlands are low.

Net GHGs from the horticultural peat harvesting process reveal that the entire life cycle of peat extraction emitted 0.54 Gg of CO_2 in 1999 rising to 0.89 Gg CO_2 in 2000. Seventy-one percent of the emissions were associated with peat decomposition, 15% from land use change, 10% from transportation to market, and 4% from processing.[14] Canadian peat horticultural emissions from all sources (0.89 Gg) represent 0.03% of all degraded peatlands (3 Pg) worldwide. Emissions are 0.006% of all total global net anthropogenic emissions (15.7 Pg). Within the Canadian context (given that the national total GHG in Canada at 771 Pg CO_2 in 2006) the peat industry represented 0.1% of total GHGs. Environmental Life Cycle Analysis such as this, has led both the Canadian Sphagnum Peat Moss Association and European Peat and Growing Medium Association to recognize that peat harvesting is a significant source of carbon emissions and peat harvesting companies are currently examining their emissions' footprint base and identifying opportunities to reduce their impacts.

PEATLAND RESTORATION

Following peat harvest, peat fields are bare and without vegetation. Decomposing peat produces greenhouse gases and these areas provide little in the form of suitable habitat or ecosystem services. Beginning in the late 1990s, however, techniques were developed (mainly in eastern Canada) to restore a viable living vegetative cover. Using the moss transfer technique developed through the research program at University of Laval's Peatland Ecology Group a *Sphagnum*-dominated plant cover has been reestablished within 3–5 years following restoration procedures, with biodiversity and hydrology approaching preharvest conditions. It is predicted that carbon sequestration should become a net sink within 15–20 years.[6]

Forestry practices result in a steady decrease in the carbon store due to increased aerobic conditions in the upper peat profile leading to increased decomposition. Even though greater initial tree growth of recently drained peatlands may increase carbon storage and exceed the losses from peat decomposition, longer term cumulative losses from the peat result in an increase in carbon emissions to the atmosphere.[15]

CONCLUSION

Peatlands occupy about 3% of the terrestrial surface of the world. These ecosystems are characterized by organic soils that contain about one-third of the world's soil carbon—about equal to the CO_2 in the atmosphere. This carbon store has accumulated gradually over the past 10,000–12,000 years and is extremely sensitive to land use changes and climatic change. Peatlands function as habitats for animals and provide valuable ecosystem services such as water storage and filtration. A variety of plant species that serve as food and medicine for local as well as aboriginal people are abundant in peatlands. *Sphagnum* moss has been used in a variety of commercial applications. Peat has been harvested for horticultural purposes and is a valuable soil amendment and peatlands have been drained for agricultural and forestry practices, releasing CO_2 to the atmosphere, but restoration of harvested peatlands currently provides a successful avenue for future sustainable management of this valuable resource.

REFERENCES

1. Yu, Z.; Loisel, J.; Brosseau, D.P.; Beilman, D.W.; Hunt, S.J. Global peatland dynamics since the Last Glacial Maximum. Geophys. Resear. Lett. **2010**, *37*, L13402.
2. Kuhry, P.; Nicholson, B.J.; Gignac, L.D.; Vitt, D.H.; Bayley, S.E. Development of *Sphagnum*-dominated peatlands in boreal continental Canada. Can. J. Bot. **1993**, *71*, 10–22.
3. Holden, J. Peatland hydrology and carbon release: Why small-scale process matters. Philos. T.: Math. Phys. Eng. Sci. **2005**, *363*, 2891–2913.
4. Limpens, J.; Heijmans, M.M.; Berendse, F. The peatland nitrogen cycle. In *Boreal Peatland Ecosystems*; Wieder, R.K., Vitt, D.H., Eds.; Springer Verlag: Berlin, 2006; 195–230.
5. Vitt, D.H. Peatlands: Ecosystems dominated by bryophytes. In *Bryophyte Biology*; Shaw, A.J., Goffinet, B. Eds.; Cambridge University Press: Cambridge, 2000; 312–343.
6. Rochefort, L.; Lode, E. Restoration of degraded boreal peatlands. In *Boreal Peatland Ecosystems*; Wieder, R.K., Vitt, D.H., Eds.; Springer Verlag: Berlin, 2006; 381–423.
7. Turetsky, M.R.; Wieder, R.K.; Halsey, L.; Vitt, D. Current disturbance and the diminishing peatland carbon sink. Geophys. Res. Lett. **2002**, *29* (11), 21-1–21-4, 10.1029/2001GLO14000.
8. Joosten, H.D.; Clarke, D. *Wise-use of Mires and Peatlands*; International Mire Conservation Group and International Peat Society: Helsinki, Finland, 2002.
9. Clarke, D.; Rieley, J., Eds.; *Strategy for Responsible Peatland Management*; International Peat Society: Jyäskylä, Finland, 2011.
10. Halsey, L.; Vitt, D.H.; Trew, D. Influence of peatlands on the acidity of lakes in northeastern Alberta, Canada. Water, Air Soil Pollut. **1997**, *96*, 17–38.
11. Prepas, E.E.; Planas, D.; Gibson, J.J.; Vitt, D.H.; Prowse, T.D.; Dinsmore, W.P.; Halsey, L.A.; McEachern, P.M.; Paquet, S.; Scrimgeour, G.J.; Tonn, W.M.; Paszkowski, C.A.; Wolfstein, K. Landscape variables influencing nutrients and phytoplankton communities in Boreal Plain lakes of northern Alberta: A comparison of wetland- and upland-dominated catchments. Can. J. Fish. Aquat. Sci. **2001**, *58*, 1286–1299.
12. Charman, D. *Peatlands and Environmental Change*; John Wiley & Sons Ltd.: Chicester, UK, 2002.
13. Yu, Z.; Campbell, I.D.; Campbell, C.; Vitt, D.H.; Bond, G.C.; Apps, M.C. Carbon sequestration in western Canadian peat highly sensitive to Holocene wet-dry climate cycles at millennial timescales. The Holocene **2003**, *13*, 801–808.
14. Cleary, J.; Roulet, N.T.; Moore, T.R. Greenhouse gas emissions from Canadian peat extraction, 1990-2000: A life-cycle analysis. Ambio **2005**, *34*, 456–461.
15. Laine, J.; Laiho, R.; Minkkinen, K.; Vasander, H. Forestry and boreal peatlands. In *Boreal Peatland Ecosystems*; Wieder, R.K.; Vitt, D.H. Eds.; Springer Verlag: Berlin, 2006; 331–358.

BIBLIOGRAPHY

1. Bauerochse, A.; Haßmann, H., Eds. *Peatlands Archaeological Sites - Archives of Nature – Nature Conservation – Wise Use*; Verlag Marie Leidorf: Rahden, 2003.
2. Crum, H. *A Focus on Peatlands and Peat Mosses*; University of Michigan Press: Ann Arbor, MI, 1992. Gore, A.J.P., Ed.; *Mires—Swamp, Bog, Fen and Moor*; General Studies; 4B – Regional Studies; Elsevier Publishing Co.: Amsterdam, 1983.
3. Rydin, H.; Jeglum, J. *The Biology of Peatlands*; Oxford University Press: Oxford, 2006. Wieder, R.K.; Vitt, D.H., Eds. *Boreal Peatland Ecosystems*; Springer Verlag: Berlin, 2006.

Plant Breeding: Global

Manjit S. Kang
Department of Plant Pathology, Kansas State University, Manhattan, Kansas, U.S.A.

Abstract

The world faces a major challenge of doubling food production within the next 25 years and tripling it in the next 40 years to feed and clothe a population of >9 billion. Scientists in the Consultative Group on International Agricultural Research (CGIAR) system, in collaboration with national breeding programs (centralized breeding) and with local farmers (participatory breeding), have helped develop modern varieties of major food staples (wheat, rice, maize, etc.). Increased crop production has helped reduce food prices and achieve a small measure of sustainable food security in developing countries, where farmers cannot afford varieties developed by private companies. Gene banks (e.g., CGIAR and Wheat Genetic and Genomic Resources Center at Kansas State University) provide genetic resources for future variety development.

INTRODUCTION

Global plant breeding generally refers to the development of improved, high yielding crop cultivars through organizations with international mandate, such as the Consultative Group on International Agricultural Research (CGIAR). The CGIAR consortium, which started with four centers in 1971 with the intended mission of achieving sustainable food security and reducing poverty in developing countries through scientific research and research-related activities in the fields of agriculture, forestry, fisheries, policy, and environment, now has 15 research centers around the globe.[1] Thirteen of these research centers are located in less developed countries. The CGIAR system is cosponsored by the World Bank, the Food and Agriculture Organization of the United Nations, the United Nations Development Program, and the United Nations Environment Program. Membership of CGIAR grew from 18 in 1971 to 64 in 2009, and it may grow further.[2] Prior to the establishment of the CGIAR system, private organizations such as the Rockefeller Foundation and the Ford Foundation supported international crop-breeding programs. The Green Revolution of the 1960s and 1970s was brought about by the deployment of high yielding cultivars of crops, such as wheat and rice, which were developed at international agricultural centers.

This entry has been divided into the following sections:

- World population and food needs
- The need for global plant breeding
- Accomplishments of global plant breeding
- Benefits of global plant breeding
- New directions of global plant breeding

WORLD POPULATION AND FOOD NEEDS

Someone has said that from the standpoint of food, the world must be regarded as one civilization and if there is not enough for everyone to eat, there will be no peace or prosperity in the world. Thus, it is imperative to produce food for citizens of all nations.

The world food problem is a complex and challenging issue. Human population has increased exponentially across centuries, whereas food production has increased only linearly. World population crossed the 7 billion mark in 2011 and it was pegged at 7.06 billion in mid-2012, developing countries accounting for 97% of this growth because of the dual effects of high birth rates and young populations.[3] World population is growing at an annual rate of 1.2% or 77 million annually.[3] Projections are that the world's human population will exceed 9 billion by 2050 (Fig. 1), and 80% of it is expected to be in less developed countries.[3,4] In 2010, about 925 million people, or 13.1% of the world population, remained chronically malnourished, with 98% of them living in less developed countries.[5]

Agriculture in less developed countries has to provide food for four times as many people per hectare of arable land than it does in developed countries, and it might have to support seven times as many per hectare by 2050.[4] In addition, global warming might have a more adverse effect on agricultural production, and the area of potential arable land in the lower latitudes where most less developed countries are located might decrease.[4]

Food production has to be doubled in the next 25 years and tripled in the next 40 years to feed people. Can it be achieved? There is optimism because cereal yields, like population, doubled during 1960–2000; although the

Encyclopedia of Natural Resources DOI: 10.1081/E-ENRL-120049196

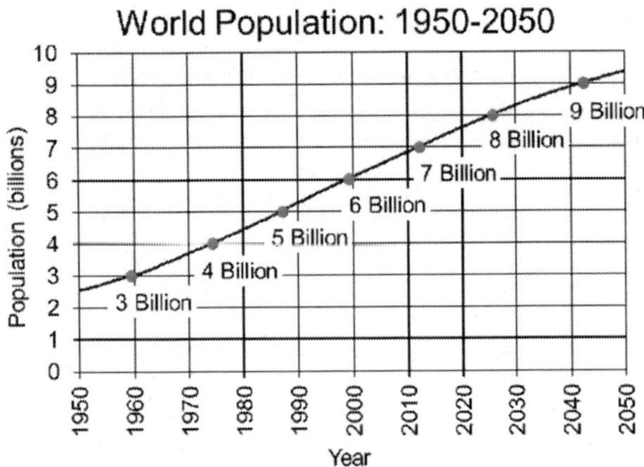

Fig. 1 The latest world population (actual or projected) for 1950–2050.
Source: Adapted from U.S. Census Bureau.[3]

average yields of root and tuber crops increased by only 36% and the average yields of pulse crops increased by only 29% during the same interval.[4]

THE NEED FOR GLOBAL PLANT BREEDING

There are two major reasons for a global plant-breeding network. First, in industrialized countries, private-sector plant breeding predominates in most of the major field crops.[6] Private industry employed about 70% of the field breeders for agronomic crops in the mid-1990s.[7] Because private companies must make profit, their products are expensive and farmers in poor countries cannot afford them. In developing countries, farmers and farm communities maintain and "develop" their own varieties, especially in regions that are least favorable for commercial crop production.[6] Thus, the CGIAR and other public organizations with international mandates cater to the needs of the farmers in less developed countries. Traditionally, CGIAR plant-breeding efforts have used a centralized global research approach, with CGIAR breeders collecting germplasm from various sources, evaluating the germplasm under carefully controlled experimental conditions, and making crosses among superior materials. The best progenies from selected crosses are then distributed as fixed or nearly fixed lines to collaborators in national agricultural research systems for testing.

Second, the universal presence of genotype (or cultivar) by environment interaction—that is, differential response of cultivars or genotypes to diverse environments—in crop breeding offers both challenges and opportunities for breeders.[8,9] Multienvironment testing helps characterize genotypes for their range of phenotypic expressions for different agronomic attributes across diverse environments. The CGIAR research centers conduct cultivar/nursery trials across countries, which can provide information on adaptation of test materials.[10] From an efficiency standpoint, most of the CGIAR breeders prefer genotypes/cultivars that exhibit stability of performance across regions/countries. Specifically adapted genotypes also are identified that can be used with advantage in specific regions/countries.

GLOBAL PLANT-BREEDING SYSTEM AND ITS ACCOMPLISHMENTS

The CGIAR research centers that are engaged in crop breeding are also known as Future Harvest Centers (FHC), as they provide improved germplasm for future cultivars and breeding methodologies. Their gene banks maintain an extensive collection of genetic resources including landraces, domesticated species, and sometimes wild relatives of domesticated species that could potentially serve as sources of useful traits.[11] The FHC scientists' strategic research generates information about basic plant biological processes, and also provides results leading to the development of novel breeding techniques and selection methods.[11] Germplasm enhancement or prebreeding is a major focus of CGIAR plant genetic improvement work. Prebreeding is essential when exotic (and by definition, unadapted) germplasm is used as a source of needed genetic diversity for elite but narrow-based breeding pools.[6] Some CGIAR centers only do prebreeding; for example, International Maize and Wheat Improvement Center (CIMMYT), International Center for Agricultural Research in the Dry Areas, International Crops Research Institute for the Semi-Arid Tropics, and International Rice Research Institute, do not develop their own cultivars or hybrids but provide improved germplasm worldwide for further breeding.

There are numerous accomplishments of CGIAR research centers. Modern varieties of wheat, rice, maize, barley, sorghum, pearl millet, potatoes, sweet potatoes, cassava, beans, and many other important crops developed by scientists from FHCs of the CGIAR, working in collaboration with colleagues in national breeding programs and more recently with farmers,[12] are grown on millions of hectares throughout the developing world.

BENEFITS OF GLOBAL PLANT BREEDING

During the past 50 years, modern varieties developed using improved germplasm from FHCs have effected significant gains in global food production and generated billions of dollars of benefits for producers and consumers, raising incomes and improving living standards of hundreds of millions of rural poor.[13] Increased crop production has helped reduce the prices of most major food staples, benefiting additional hundreds of millions of poor urban consumers. The FHCs have fostered synergistic activities among public and private national breeding programs and facilitated technology transfer.

NEW DIRECTIONS IN GLOBAL PLANT BREEDING

The centralized plant breeding (CPB) that was responsible for the Green Revolution has been successful in producing uniform, broadly adapted varieties. It is, however, believed that this approach did not benefit resource-poor farmers in marginal areas. The CPB system has a number of weaknesses. The most important among these is its limited success in producing locally adapted cultivars for many specialized environments, particularly for the most difficult environments where poverty is still rampant.

The CPB reduces crop diversity and is characterized by dependence on gene banks for the preservation of genetic resources. The CPB has other limitations as well; it works well in predictable, uniform environments, for example, when farmers can use/afford irrigation and fertilizers. In recent years, researchers in a number of FHCs and national breeding programs have employed a new approach known as participatory plant breeding (PPB).[11,14] Some nongovernmental organizations teach farmers how to use some of the techniques of professional breeding if they suit the farmers' capacities and needs.[11] The PPB involves close farmer–researcher collaboration to bring about genetic improvement within a crop.[15] The PPB is a noncentralized system of breeding that can help broaden the genetic base of crop germplasm and stabilize food production, which results from farmers' developing, identifying, and using locally developed crop varieties that are both acceptable and accessible to them. The PPB exploits genotype-by-environment interactions by identifying locally adapted varieties. The emergence of the PPB movement represents a response to a weakness of the traditional global approach to plant breeding.

A couple of important global initiatives are worth noting. International in scope and founded by Dr. Bikram S. Gill, the Wheat Genetic and Genomic Resources Center at Kansas State University maintains a gene bank of 2500 wheat accession as well as houses 2200 cytogenetic stocks developed by wheat scientists.[16,17] Dr. Norman E. Borlaug founded a new organization in 2005, which is now called Borlaug Global Rust Initiative (BGRI).[18] It contributes toward ensuring sustainable production of crops like wheat by organizing technical workshops on serious biotic threats from rust pathogens. Many countries have come together under the umbrella of BGRI to combat Ug99—a devastating strain of black stem rust.

In 2011, CIMMYT, in collaboration with the Government of India, established Borlaug Institute for South Asia (BISA), in India. The mission of BISA is to boost productivity of wheat and maize in South Asia in the face of climate change and a dwindling supply of natural resources.[19]

CONCLUSION

Global and local (e.g., PPB) breeding are complementary activities rather than substitutes. PPB helps especially resource-poor farmers who were not the beneficiaries of the Green Revolution of the 1960s and 1970s. Conventional breeding (centralized or PPB) has been highly successful in doubling or even tripling production of some cereals. Thus, the funding base of conventional plant breeding needs to be strengthened.

REFERENCES

1. Consultative Group on International Agricultural Research (CGIAR). 2012a; http://www.cgiar.org/cgiar-consortium/research-centers/ (accessed July 2012).
2. Consultative Group on International Agricultural Research (CGIAR). 2012b; http://www.cgiar.org/who-we-are/history-of-cgiar/members-chronology/ (accessed July 2012).
3. U.S. Census Bureau. 2012; http://www.census.gov/population/international/data/worldpop/graph_population.php (accessed September 2012).
4. Evans, L.T. Is crop improvement still needed?. In *Genetic and Productions Innovations in Field Crop Technology*; Kang, M.S., Ed.; Food Products Press—An Imprint of Haworth Press: Binghamton, NY, 2005.
5. World Hunger Education Service. 2012; http://www.world-hunger.org/articles/Learn/world%20hunger%20facts%20 2002.htm (accessed July 2012).
6. Duvick, D.N. Crop breeding in the twenty-first century. In *Genetic and Productions Innovations in Field Crop Technology*; Kang, M.S., Ed.; Food Products Press—An Imprint of Haworth Press: Binghamton, NY, 2005.
7. Frey, K.J. National plant breeding study I: human and financial resources devoted to plant breeding research and development in the United States in 1994. Iowa Agriculture and Home Economics Experiment Station, and Cooperative State Research, Education and Extension Service/USDA cooperating, No. 98, 1996.
8. Kang, M.S. Using genotype-by-environment interaction for crop cultivar development. Adv. Agron. **1998**, *62*, 199–252.
9. Yan, W.; Kang, M.S. GGE *Biplot Analysis: A Graphical Tool for Breeders, Geneticists, and Agronomists*; CRC Press: Boca Raton, FL, 2003.
10. Kang, M.S.; Pham, H.N. Simultaneous selection for high yielding and stable crop genotypes. Agron. J. **1991**, *83*, 161–165.
11. Bellon, M.R.; Morris, M.L. Linking global and local approaches to agricultural technology development: the role of participatory plant breeding research in the CGIAR. *CIMMYT Economics Working Paper 02–03*; CIMMYT: Mexico, DF, 2002.
12. Ceccarelli, S.; Grando, S.; Booth, R.H. International breeding programmes and resource-poor farmers: Crop improvement in difficult environments. In *Participatory Plant Breeding. Proceeding of a Workshop on Participatory Plant Breeding, 26–29* July 1995; Eyzaguirre, P., Iwanaga, M., Eds.; Wageningen, the Netherlands. IPGRI: Italy, 1996; 99–116.

13. Evenson, R.; Gollin, D., Eds.; *Impact of the CGIAR on International Crop Genetic Improvement*; CABI: Wallingford, U.K., 2001.
14. Ceccarelli, S.; Grando, S. Increasing the efficiency of breeding through farmer participation. In *Ethics and Equity in Conservation and Use of Genetic Resources for Sustainable Food Security*, pp 116–121. Proceeding of a workshop to develop guidelines for the CGIAR, 21–25 April 1997, Foz de Iguacu, Brazil. IPGRI, Rome, Italy, 1997.
15. Weltzien, E.; Smith, M.E.; Meitzner, L.S.; Sperling, L. Technical and institutional issues in participatory plant breeding from the perspective of formal plant breeding. An analysis of issues, results, and current experience. CGIAR Systemwide Program on Participatory Research and Gender Analysis for Technology Development and Institutional Innovation Working Document 3.
16. Wheat Genetic and Genomic Resources Center (WGGRC). 2012; http://www.k-state.edu/wgrc/ (accessed July 2012).
17. Gill, B.S.; Friebe, B.; Raupp, W.J.; Wilson, D.L.; Cox, T.S.; Sears, R.G.; Brown-Guedira, G.L.; Fritz, A.K. Wheat genetics resources center: The first 25 years. Adv. Agron. **2006**, *89*, 73–136.
18. Borlaug Global Rust Initiative. 2012; http://www.globalrust.org/traction/project/about (accessed July 2012).
19. Borlaug Institute for South Asia. 2012; http://www.cimmyt.org/en/about-us/partnerships/countries/doc_view/673-the-borlaug-institute-for-south-asia-bisa-a-brief-overview (accessed July 2012).

Population Genetics

Michael T. Clegg
Department of Ecology and Evolutionary Biology, University of California, Irvine, Irvine, California, U.S.A.

Abstract
This entry provides a brief outline of the theory of population genetics as it applies to changes in the composition of DNA molecules over time. Several excellent books provide a comprehensive introduction to population genetics theory, including works by Hartl and Clark and Hedrick.

INTRODUCTION

Genes are the fundamental hereditary units transmitted from parent to offspring in every generation. Population genetics is concerned with the statistical rules that govern the transmission of genes within collections of interbreeding individuals over time. Changes in the frequencies of genes within populations are the basis for evolutionary change, as well as the basis for the genetic improvement of plants and animals that provide human sustenance.

Population genetics employs mathematical models to describe trajectories of genetic change under various scenarios. These models are also used to design experiments to dissect the processes of evolution. The basic mathematical models of population genetics were developed long before it was established that DNA is the chemical basis of heredity. The challenge today is to analyze and learn from the rapidly accumulating body of genome sequence data. The basic models of population genetics are highly relevant to this enterprise because they are general, in the sense that change in gene frequency can be substituted with change in individual nucleotide frequency without any alteration in the underlying mathematics. This entry provides a brief outline of the theory of population genetics as it applies to changes in the composition of DNA molecules over time. Several excellent books provide a comprehensive introduction to population genetics theory, including works by Hartl and Clark and Hedrick.

EVOLUTIONARY FORCES THAT DRIVE DNA SEQUENCE CHANGE

Five forces determine genetic change.[1,2] These begin with mutation, defined as a permanent heritable change in a gene. A second force is recombination, which is a randomizing force that can also lead to permanent heritable changes in a gene. Genetic random drift, defined as the small fluctuations in gene frequencies caused by the sampling of individuals in reproduction, is third. The fourth force is migration, or the movement of genes among different populations. Fifth is natural selection, in which some individuals are more successful in transmitting their genes owing to superior adaptation to their environment and hence improved survival and reproduction. Natural selection is the force that leads to adaptive genetic change; it essentially sorts among mutations to eliminate those that are deleterious and increase the frequency of those that are advantageous.

MUTATION: THE SOURCE OF GENETIC NOVELTY

Gene or nucleotide sequences are subject to constant change. To begin, mutation takes many forms, each of which has some likelihood of occurring at each DNA replication cycle. From this it is apparent that any DNA molecule is subject to mutational erosion over time; this erosion process is a form of evolutionary change. DNA sequences that code for functionally important molecules are expected to be conserved because natural selection rejects most deleterious mutational changes in functionally important genes. Current genome sequencing efforts suggest that much of the DNA of organisms does not code for proteins or regulatory signals—these regions may be evolving relatively rapidly.

GENETIC RANDOM DRIFT: THE ROLE OF CHANCE IN EVOLUTIONARY CHANGE

Once a new mutation has arisen in a single DNA molecule (to create a new allelic form of a gene or a new allele), it may be transmitted to the next generation or it may be lost. Assuming the new mutation has no effect on fitness, the probability of transmission to the next generation is $1 - [1 - 1/2N]^{2N}$ in diploid organisms, where N is the population size. This result follows from the fact that a new mutation will initially occur in a single individual (frequency $= 1/2N$).

Encyclopedia of Natural Resources DOI: 10.1081/E-ENRL-120006084

The pool of gametes is assumed to be much larger than the N reproductive individuals in the population, so sampling with replacement is appropriate. (This expression is the probability of not being lost in the first generation, derived assuming binominal sampling.) In the vast majority of cases a new mutation will be lost through the vagaries of sampling within a generation or two. A very small proportion will drift to high frequency and be fixed (i.e., become the sole allele in the population or species).

An important result of mathematical genetics indicates that the probability of ultimate fixation of a new neutral mutation is $1/2N$ (the concept of neutrality is discussed later). Because there is a constant flux of new mutations, there will also be a constant rate of fixation of new allelic variants over the long term. In fact, the rate of fixation of new alleles is simply μ, the mutation rate. This result provides a formal justification for the molecular clock hypothesis, because it indicates that the rate of mutational change should be constant over time if the mutation rate is constant.

A second important theoretical result from population genetics demonstrates that the average time until the fixation of a new mutation is $T \sim 4N_e$, so for large N_e, many mutant alleles are expected to be in transition to fixation at any point in time. (N_e is a population size adjusted to take account of variation in reproductive output among different members of a breeding population.) For example, many plant species must have population sizes in the order of 10^6 to 10^7. The rate of mutation per nucleotide site per generation is approximately $\mu = 5 \times 10^{-9}$. If a typical gene is 10^3 nucleotide sites in length, the total probability of a new mutation in the gene in any single generation is 5×10^{-6}. For a species size ($2N$) of 10^6, five new mutations in this particular gene are expected in the total species per generation. Moreover, the average new mutation that is destined to be fixed will drift in the species of size 10^6 for 4,000,000 generations. The probability of drawing two different alleles of the gene in a population or species can be shown to be approximately $1 - 4N_e\mu/(1+4N_e\mu) = 19/21 = 0.90$ for this numerical example, so almost every pair of genes sampled is expected to be different. Thus, high levels of molecular polymorphism are expected in species. Current efforts to define all human single nucleotide polymorphisms (SNPs) verify these theoretical predictions.

Net—Savannas

THE NEUTRALITY HYPOTHESIS: AN INTERACTION BETWEEN MUTATION AND DRIFT

The theoretical results cited in the preceding discussion are based on the assumption that selection is absent. In the 1970s, Motoo Kimura[3] and associates advanced a theory called the neutrality hypothesis that claims that a large proportion of molecular evolutionary change is neutral to natural selection. This hypothesis was in part motivated by an important calculation by J. B. S. Haldane,[4] who showed that there is an upper bound to the rate of evolutionary change under natural selection. Because selective change requires differential survival and reproduction, the reproductive excess of the species defines this upper bound. Stated differently, some or all of the reproductive excess of a species is consumed by the process of selection. Empirical evidence on rates of molecular change appear to be too high to be consistent with this upper bound, especially for long-lived low-fecundity organisms like the human. As a consequence, it was postulated that a large proportion of molecular change is not driven by selection, but by random drift and mutation.

The test of the neutral theory is to ask how well observed data fit the theoretical calculations outlined earlier. Because the neutral theory provides precise mathematical predictions, the theory is testable; moreover, these tests provide a very useful context for identifying particular genes that are selected. This is because it is known precisely how neutral genes should behave, and departures from neutrality are likely to arise from selection. To date, available data do suggest that many polymorphisms are neutral or only very weakly selected. However, the statistical power to detect departures from neutrality increases with database size, which are growing very rapidly at present.

NATURAL SELECTION: THE ENGINE OF ADAPTIVE CHANGE

Natural selection operates at the level of phenotypic differences, so an allelic change must alter some aspect of phenotype to be perceived by selection. The phenotypic dimensions of organisms, and the ways these may be perceived by various environmental factors, are very complex. These range from subtle changes in rates of flow through a metabolic pathway, to resistance to disease, to the ability to withstand extreme stress environments, to aspects of behavior, and so on. The list is virtually endless. Moreover, each aspect of phenotype may be determined by a number of different genes and by the way these genes interact in development and metabolism. It is similarly difficult to identify the precise facets of an environment that may act as a selective agent. As a consequence, selection is often identified retrospectively. That is, the analytical tools of population genetics may support the inference that selection has affected a genetic locus based on the pattern of change through time. Such an analysis rarely tells the reasons why a particular change is adaptive.

A number of mathematical results that provide useful predictions about the statics and dynamics of selective genetic change are available from population genetic theory. We will mention just one result from this large body of work. The probability of fixation of a new mutant that is favored by selection (assuming the heterozygous genotype is intermediate in its fitness between the two homozygous genotypes) is $\sim 2s$, where s is the selective intensity favoring the new mutant allele. So if $s = 0.02$ (implying a

2% selective advantage), the favored allele will still be lost 96% of the time. This result demonstrates that most new favored mutations will still be lost owing to drift.

The average time to fixation of a new favored mutant is approximately $T \sim (2/s)\mathrm{Ln}(2N_e)$ generations (where Ln denotes the natural logarithmic function). Numerical iterations of this equation, compared to the neutral case, reveal that the time to fixation of selectively favored alleles is much shorter than for a neutral allele. Indeed, fixation of a selected allele can take place in orders of magnitude less time than a neutral allele. This calculation suggests that in contrast to the high levels of neutral polymorphism expected from the neutral theory outlined in the fore going, it may be relatively rare to find a selected gene in transition when one allele is favored over the other allele. The exception to this statement is the case where two or more alleles are favored, either because they lead to a higher fitness among heterozygotes or because their advantage is frequency-dependent (frequency dependent means that s is not constant, but instead depends on allele frequencies). Both of these situations are referred to as balanced polymorphism. Classic examples of a balanced polymorphism are often associated with disease resistance, where the resistance allele causes some other disadvantage or where there is a constant arms race between the disease agent and the target organism favoring new mutations at a resistance locus (a case of frequency dependence). An example of such a balanced polymorphism arises from the multiple major histocompatibility polymorphism (MHC) alleles in human populations.[5] In this case, disease resistance is strongly implicated because the MHC locus is involved in the recognition of potential disease agents.

RECOMBINATION AS BOTH RANDOMIZING AND CREATIVE FORCE

Recombination refers to the rearrangement of genes with respect to one another on chromosomes. Recombination is a randomizing force, but it is also much more. It is a creative force that operates at many levels within plant genomes.[6] Thus, molecular evidence has revealed that recombination is responsible for the generation of many gene novelties. For example, studies of different allelic sequences of a gene often show traces of past intragenic recombination events where new allelic forms have arisen, owing to the exchange of stretches of sequence between two different alleles. This reciprocal exchange can yield a new allele that is different from both parental forms.

THE COALESCENT: LOOKING BACKWARD IN TIME

An important theoretical construct from population genetics is known as the coalescent. Consider a set of DNA sequences that are drawn from a single locus within a species. The sample of sequences must have all descended from a common ancestor. The point where all alleles trace back to the most recent common ancestor is known as the coalescent. It is straightforward to estimate the genealogical history of the sample by using phylogeny estimation tools. If the nucleotide sequence diversity in the sample is the result of a drift/mutation process at equilibrium, then the arrival times of new mutation events are given by an exponential process.[7] This result provides a useful framework for testing sequence samples to see if they conform to a neutral process.

A particularly important statistic calculated from a sample of sequences is θ, a measure of the nucleotide sequence diversity. A different statistic that also measures nucleotide sequence diversity is π. (Both of these quantities are related to the probability of drawing two different nucleotides at a single site averaged over all sites examined.) If we calculate π and θ as an average over many independent replications of the evolutionary process that led to the observed sample, and if the underlying evolutionary process is the result of a drift/mutation process at equilibrium, then both θ and π should be approximately equal to $4N_e\mu$. Thus, the difference between π and θ should be zero under the drift/mutation assumption. A test statistic introduced by Tajima[8] based on a function of the difference between π and θ allows one to ask whether a data set conforms to the drift/mutation assumption by testing whether this difference is zero. If the difference is nonzero, then other forces such as selection or demographic change must be invoked. Many standard computer programs for sequence analysis implement these calculations (i.e., DnaSP).

Another statistic that features importantly in coalescence theory is T_{MARCA} (time to the most recent common ancestor of the sample). Mathematical calculations show that T_{MARCA} is also approximately equal to $4N_e$ when the drift/mutation assumption is satisfied. Finally, if μ is known it is possible to estimate N_e for the species or population from which the data came. Estimates of N_e from a few important crop plants vary over nearly an order of magnitude from 2×10^5 (wild barley) to 3×10^5 (pearl millet) to 18×10^5 (maize).[9] These data are derived from broad samples of the entire species so the estimates can be viewed as the best estimates of long-term effective population size for each species.

MIGRATION: THE SPATIAL DIMENSION OF GENETIC CHANGE

Every new mutation must have originated in a single individual at a specific location in space. This means that the spread of a mutation through a species is a spatial process, and it is possible to gain some insight into historical migration patterns through the analysis of geographic samples of gene sequences. Moreover, the coalescent framework has been refined to include spatial considerations, so migration rates can be quantified. Efforts to measure migration are still in very early stages, but contemporary data suggest that plants and their genes have been quite mobile over long periods of time.

There is great contemporary concern over the spread of so-called transgenes from agricultural plants into wild or weedy relatives through migration. This question has stimulated much work on the direct measurement of migration in plants. Current results suggest that gene migration rates may be higher than previously believed.[10]

CONCLUSION

Although population genetics is a mature science with a history of nearly 100 years, it is more relevant today than ever. The theoretical models of population genetics provide the conceptual and statistical tools to address important questions ranging from concerns about transgene escape to detection of the influence of selection on specific genes. As the body of genome sequence data expands, these tools will become ever more important in interpreting and mining this new resource.

REFERENCES

1. Hartl, D.L.; *Clark, A.G. Principles of Population Genetics*, 3rd Ed.; Sinauer Associates Inc.: Sunderland, MA, 1997; 542.
2. Hedrick, P.A. *Genetics of Populations*, 2nd Ed.; Jones and Barlett: Boston, MA, 2000; 553.
3. Kimura, M. *The Neutral Theory of Molecular Evolution*; Cambridge University Press: Cambridge, England, 1983; 367.
4. Haldane, J.B.S. The cost of natural selection. J. Genet. **1957**, *55*, 511–524.
5. Nei, M. DNA Polymorphism and Adaptive Evolution. In *Plant Population Genetics, Breeding and Genetic Resources*; Brown, A.H.D., Clegg, M.T., Kahler, A., Weir, B.W., Eds.; Sinauer Assoc.: Sunderland, MA, 1990; 128–142.
6. Clegg, M.T. The Role of Recombination in Plant Genome Evolution. In *Evolutionary Processes and Theory*; Waser, S.P., Ed.; Kluwer Academic Publishers: the Netherlands, 1999; 33–45.
7. Hudson, R.R. Gene genealogies and the coalescent process Oxf. Surv. Evol. Biol. **1990**, *7*, 1–44.
8. Tajima, F. Statistical method for testing the neutral mutation hypothesis by DNA polymorphism. Genetics **1989**, *123*, 585–595.
9. Cummings, M.P.; Clegg, M.T. Nucleotide sequence diversity at the alcohol dehydrogenase I locus in wild barley (*Hordeum vulgare* ssp. *spontaneum*): An evaluation of the background selection hypothesis. Proc. Natl. Acad. Sci. U. S. A. **1998**, *95*, 5637–5642.
10. Ellstrand, N. When transgenes wander, should we worry? Plant Physiol. **2001**, *125*, 1543–1545.

Protected Area Management

Tony Prato
Department of Agricultural and Applied Economics, University of Missouri, Columbia, Missouri, U.S.A.

Daniel B. Fagre
Northern Rocky Mountain Science Center, U.S. Geological Survey (USGS), West Glacier, Montana, U.S.A.

Abstract

Designated protected areas are diverse in scope and purpose and have expanded from Yellowstone National Park in the United States, the world's first national park, to 157,897 parks and protected areas distributed globally. Most are publicly owned and serve multiple needs that reflect regional or national cultures. With ever-increasing threats to the integrity of protected areas, managers are turning to flexible management practices such as scenario planning and adaptive management.

INTRODUCTION

The International Union for Conservation of Nature (IUCN) defines a protected area as a "clearly defined geographical space, recognized, dedicated and managed, through legal or other effective means, to achieve the long-term conservation of nature with associated ecosystem services and cultural values."[1] The Convention on Biological Diversity defines a protected area as a "geographically defined area which is designated or regulated and managed to achieve specific conservation objectives."[1] Protected areas encompass a broad range of ecosystems, including wetland, tropical or deciduous forest, alpine, savanna, and marine. Yellowstone National Park is the world's first public protected area. There has been increasing interest in transboundary protected areas that serve a dual purpose of protecting the landscape and generating cooperation between neighboring countries.

DESIGNATION, EXTENT, AND PURPOSES OF PROTECTED AREAS

The IUCN's World Commission on Protected Areas has established six protected area management categories: strict nature reserve, wilderness area, national park, natural monument or feature, habitat/species management area, protected landscape/seascape, and protected areas with sustainable use of natural resources.[2] Protected areas in the same category are managed based on similar objectives. For example, national parks are designated and managed to "(a) protect the ecological integrity of one or more ecosystems for present and future generations, (b) exclude exploitation or occupation inimical to the purposes of designation of the area, and (c) provide a foundation for spiritual, scientific, educational, recreational and visitor opportunities, all of which must be environmentally and culturally compatible."[3]

In addition to the IUCN categories, there are four other international designations of protected areas: Biosphere Reserve; World Heritage Site; Ramsar wetland; and marine. Biosphere Reserves serve three objectives: 1) conservation of landscapes, ecosystems, species, and genetic variation; 2) sustainable economic and human development; and 3) support for research, monitoring, education, and information exchange.[4] World Heritage Sites possess outstanding cultural, natural, or mixed (combination of cultural and natural) values that are protected for the benefit of all humanity through cooperative international arrangements.[5] The Convention on Wetlands[6] is an intergovernmental treaty whose member countries protect Wetlands of International Importance in their territories (Ramsar Convention on Wetlands undated). The United Nations Law of the Sea Treaty provides a legal framework for allocating rights to territorial seas and protecting marine-protected areas.[7]

In 2011, there were 157,897 protected areas in the world covering 24,236,479 km^2,[8] which amount to approximately 16% of the world's land surface area.[9] Locations of protected areas cover the gamut from densely populated urban areas to unpopulated wilderness areas. Protected areas range in size from a few square kilometers to thousands of square kilometers. Most protected areas are publicly owned and managed. The level of protection afforded protected areas depends on many factors, including adequate funding, diligent law enforcement, effective management practices, and the strength of citizen support.

Protected areas are established and managed for a wide range of social, economic, and ecological values, including preserving biodiversity, particularly for threatened and endangered species; contributing to regional conservation strategies; maintaining diversity of landscape or habitat and

Encyclopedia of Natural Resources DOI: 10.1081/E-ENRL-120048426

of associated species and ecosystems; ensuring the integrity and long-term maintenance of specified conservation targets; providing a range of ecosystem goods and services that support sustainable use of natural resources; providing research and educational opportunities; maintaining and protecting cultural sites and traditional values; providing opportunities for tourism and recreation; carbon storage; disaster mitigation, soil stabilization; water supply for human and agricultural uses; and wilderness preservation.[1,2]

NATURE AND TYPES OF PROTECTED AREAS

The establishment of public protected areas was motivated by setting aside novel or awe-inspiring resources for public enjoyment. Biodiversity or ecosystem integrity was not a relevant concern, and the size of many protected areas was the smallest necessary to protect unique features. With the passage of time, these areas became increasingly valuable for protecting species and accomplishing other societal goals that required relatively unaltered landscapes. Today protected areas are being considered in national strategic plans, such as carbon sequestration for mitigating global warming or the creation of continental conservation networks, purposes for which they were not created. Outside the United States, the concept of protected areas has expanded to include traditional uses deemed worthy of preservation. For that reason, historic agricultural practices are maintained in certain protected areas to continue the interaction between people and the landscape and increase public support for those areas. Some countries have established parks with the explicit aim of attracting tourists and increasing national income and employment (e.g., African wildlife parks). This demonstrates the important contributions of protected areas to local and regional economies and that the motivation for establishing protected areas has changed.

Publicly owned protected areas are managed by federal, state, and local governmental agencies. For example, at the federal level in the United States, the National Park Service manages the National Park System, the Forest Service manages the National Forest System, several federal agencies manage units in the National Wild and Scenic Rivers System and National Wilderness Preservation System, the National Oceanic and Atmospheric Administration manages Marine Protected Areas, and the U.S. Fish and Wildlife Service manages the National Wildlife Refuge System. Each federal agency operates under the authority of an organic act that dictates how certain types of protected areas should be managed. States also designate and manage protected areas. For instance, the Missouri Natural Areas System is an extensive statewide network of over 80 natural areas representing examples of the state's original natural landscape and outstanding biological and geological features.[10] Local park commissions manage urban parks, which are typically oriented towards outdoor sports, picnicking, hiking, swimming, and enjoyment of nature. Nongovernmental organizations purchase land and conservation easements to protect natural values and/or fish and wildlife habitat. For example, The Nature Conservancy works with landowners, communities, cooperatives, and businesses to establish local groups that protect land through the use of land trusts, conservation easements, private reserves, and incentives. Nature Conservancy lands, which are located primarily in the United States, constitute the world's largest system of private nature reserves covering 60,703 km^2.[11]

THREATS TO PROTECTED AREAS

As human populations, incomes, and leisure time increase, protected areas have come under greater pressure to serve more people and more diverse purposes often not envisioned when the protected areas were established. These human pressures vary greatly with country and the type of protected area but, in general, these dynamics have changed the role of protected area managers from caretakers to assertive advocates who manage their areas to provide benefits to local, regional, and national economies and minimize the impacts of threats to protected areas. Examples of such threats include rapid growth in visitation, changes in the mix of recreational and visitor uses (hunting, trapping, motorized/mechanized sports and transportation, hiking, camping, skiing, and air tours), development of roads and visitor facilities, spread of invasive species, residential and commercial development, legal resource extraction (logging, livestock grazing, farming, and energy development), illegal timber harvesting, wildlife poaching, air, water, and soil pollution, and climate change and variability.[12,13] Some threats, such as increases in visitation and invasive species, are internal to protected areas, whereas others such as residential and commercial development and climate change, are external. In terms of the nature of the threats, increases in visitation and invasive species exert pressure on natural resources of protected areas and make their preservation more difficult. Landscape fragmentation, caused by residential and commercial development adjacent to protected areas, and climate change reduce the ecological integrity of protected areas, particularly their ability to serve as refugia for threatened and endangered species. External threats to protected areas are typically more difficult to resolve than internal threats because protected area managers have limited legal authority and/or are unwilling to control events external to their protected areas.

MANAGEMENT OF PROTECTED AREAS

A major challenge facing protected area managers is how to manage those areas so as to achieve multiple management objectives. This challenge occurs for several reasons.

First, certain management objectives (e.g., resource protection and visitor satisfaction) are somewhat incompatible. Second, there is considerable uncertainty about the natural and/or socioeconomic processes influencing protected areas, how internal and external threats influence those processes, and how different management actions are likely to influence the achievement of management objectives. Third, some management objectives, such as minimizing adverse impacts on visitors and the local economy, are supported by one group of stakeholders, whereas other objectives, such as natural and cultural resource protection, are supported by another group of stakeholders. Fourth, many protected areas face declining financial and human resources, which makes it more difficult to develop and implement effective management plans.

Two methods can be used by protected area managers to evaluate and make management decisions under uncertainty, namely, scenario planning and adaptive management (AM). Scenario planning requires protected area managers to develop and evaluate decisions for all combinations of unknown future events (e.g., future climate scenarios) and management actions. The scenario planning approach used by the U.S. National Park Service to manage national parks for climate change requires park managers to: 1) choose the most probable climate scenario based on their subjective judgment; 2) determine the most preferred management action for that climate scenario; and 3) implement the most preferred management action.[14,15] A difficulty with scenario planning for climate change is that it requires protected area managers to choose the most probable climate scenario. This is problematic because scientists who develop climate scenarios have been unwilling or unable to assign probabilities to future climate scenarios. Thus, protected area managers need to develop management options for multiple scenarios.

AM postulates that "if human understanding of nature is imperfect, then human interactions with nature (e.g., management actions) should be experimental."[16] Kohm and Franklin state that "AM is the only logical approach under the circumstances of uncertainty and the continued accumulation of knowledge."[17] AM acknowledges that protected area managers cannot accurately predict the outcomes of management actions because of environmental, scientific, organizational, community, and political uncertainties.

AM can be either passive or active.[18] Application of passive AM to protected areas requires protected area managers to: 1) formulate a predictive model of how a protected area responds to alternative management actions; 2) select the most preferred management actions based on model predictions; 3) implement and monitor those management actions; 4) use monitoring results to revise the model; and 5) adjust management actions based on the revised model. Advantages of passive AM are that it is relatively simple to use and usually less expensive to implement than active AM. A disadvantage of passive AM is that it does not produce statistically reliable information about the impacts of management actions on protected areas because it does not utilize experimental methods.

Application of active AM to protected areas requires protected area managers to: 1) formulate hypotheses about how various management actions influence a particular feature of protected areas (e.g., biodiversity) when there is uncertainty about how future events (e.g., climate change) interact with management actions to influence that feature; 2) establish experiments to test hypotheses about the efficacy of management actions over time; 3) use test results to determine the best management action to implement in an upcoming management period; and 4) continue experiments and use experimental results to periodically retest the hypotheses to determine whether the best management action changes over time. Active AM typically involves major investments in research, monitoring, and modeling; has prerequisites that are not always satisfied; and is more complicated to implement than passive AM.[19–21]

CONCLUSION

Protected areas play increasingly important roles in our global society, including biodiversity conservation, providing ecosystem services, and supporting regional economies through recreational opportunities and tourism. These values of protected areas are reflected in the extensive distribution and diversity of protected areas and their increasing numbers. With a burgeoning global population and its demands on natural resources, protected areas will become even more critical to informed conservation planning and as scientific resources. Protected area managers must respond with vision and flexibility using new approaches such as scenario planning and AM.

ACKNOWLEDGMENT

Any use of trade, product, or firm names is for descriptive purposes only and does not imply endorsement by the U.S. Government.

REFERENCES

1. Dudley, N., Ed. *Guidelines for Applying Protected Area Management Categories*; IUCN: Gland, Switzerland, and Cambridge, UK, 2008.
2. IUCN. *Guidelines for Protected Area Management Categories*; IUCN Commission on National Parks and Protected Areas with assistance of the World Conservation Monitoring Centre: Gland, Switzerland, and Cambridge, UK, 1994.
3. Davey, A.G. *Protected Areas Categories And Management Objectives (Appendix 2)*; National System Planning for Protected Areas. IUCN: Gland, Switzerland, and Cambridge, UK, 1998.

4. UNESCO. Man and the Biosphere Program, http://www.unesco.org/new/en/natural-sciences/environment/ecological-sciences/man-and-biosphere-programme/ (accessed October 2012).
5. UNESCO. Our World Heritage, http://whc.unesco.org/uploads/activities/documents/activity-568-2.pdf (accessed October 2012).
6. Ramsar Convention on Wetlands. About the Ramsar Convention, http://www.ramsar.org/cda/en/ramsar-about-about-ramsar/main/ramsar/1-36%5E7687_4000_0__ (accessed October 10, 2012).
7. United Nations Law of the Sea Treaty. Law of the Sea Treaty (LOST) – Background, http://www.unlawoftheseatreaty.org/ (accessed October 2012).
8. IUCN and UNEP-WCMC. The World Database on Protected Areas (WDPA). Cambridge, UK, http://www.wdpa.org/ (accessed October 2012).
9. Coble, C.R.; Murray, E.G.; Rice, D.R. Earth Science 1987, 102.
10. Missouri Department of Natural Resources. About the Missouri State Park System, http://mostateparks.com/page/55047/about-missouri-state-park-system (accessed October 2012).
11. The Nature Conservancy. Public Lands Conservation, http://www.nature.org/about-us/private-lands-conservation/index.htm (accessed October 2012).
12. Prato, T.; Fagre, D. *National Parks and Protected Areas: Approaches for Balancing Social, Economic and Ecological Values*; Blackwell Publishers: Ames, Iowa. 2005.
13. Nolte, C.; Leverington, F.; Kettner, A.; Marr, M.; Nielsen, G.; Bastian, B.; Stoll-Kleemann, S.; Hockings, M. Protected Area Management Effectiveness Assessments in Europe: A Review of Application, Methods and Results; BfN-Skripten 271a. Bundesamt fur Naturschutz (BfN), Bonn, Germany, 2010.
14. Weeks, K.; Moore, P.; Welling, L. Climate change scenario planning: A tool for managing parks into uncertain futures. Park Sci. **2011**, *28*, 26–33.
15. Welling, L. National parks and protected areas management in a changing climate: Challenges and opportunities. Park Sci. **2011**, *28*, 6–9.
16. Lee, K.N. *Compass and Gyroscope: Integrating Science and Politics for the Environment*; Island Press: Washington, DC, USA. 1993.
17. Kohm, K.A.; Franklin, J.F. Introduction. In *Creating Forestry for the 21st Century: The Science of Ecosystem Management*; Kohm K.A.; Franklin J.F. Eds.; Island Press: Washington, DC, USA. 1997; 1–5.
18. Baron, J.S.; Gunderson, L.; Allen, C.D.; Fleishman, E.; McKenzie, D.; Meyerson, L.; Oropeza, J.; Stephenson, N. Options for national parks and reserves for adapting to climate change. Environ. Manag. **2009**, *44*, 1033–1042.
19. Prato, T. Bayesian adaptive management of ecosystems. Ecol. Model. **2005**, *183*, 147–156.
20. Prato, T. Accounting for risk and uncertainty in determining preferred strategies for adapting to future climate change. Mit. Adapt. Strateg. Global Change **2008**, *13*, 47–60.
21. Prato, T.; Fagre, D.B. Sustainable management of the Crown of the Continent Ecosystem. George Wright Forum **2010**, *27*, 77–93.

Protected Areas: Remote Sensing

Yeqiao Wang

Department of Natural Resources Science, University of Rhode Island, Kingston, Rhode Island, U.S.A.

Abstract

National parks, national forests, national seashores, natural reserves, wildlife refuges and sanctuaries, conservation areas, frontier lands, and marine protected areas are examples of protected lands and waters that serve as the fundamental building blocks of virtually all national and international conservation strategies. Protected areas are increasingly recognized as essential providers of ecosystem services and biological resources. Traditionally, management of protected areas is operated with field-oriented observation and mapping approaches. Repaid development of remote sensing science and technologies has profoundly influenced and contributed to the practice in understanding and management of protected areas from extended spatial, spectral, temporal, and thematic perspectives for which remote sensing has unique advantages. This entry provides a snapshot about remote sensing applications for inventory, monitoring, and management of protected areas.

INTRODUCTION

The World Commission on Protected Areas adopted a definition that describes a protected area as clearly defined geographical space, recognized, dedicated, and managed, through legal or other effective means, to achieve the long-term conservation of nature with associated ecosystem services and cultural values.[1] In general, protected areas include national parks, national forests, national seashores, all levels of natural reserves, wildlife refuges and sanctuaries, and designated areas for conservation of native biological diversity and natural and cultural heritage and significance. Protected areas also include some of the last frontiers that have unique landscape characteristics and ecosystem functions. Along the shoreline and over the ocean and sea, the International Union for the Conservation of Nature (IUCN) has defined marine protected areas (MPAs) as any area of intertidal or subtidal terrain, together with its overlying water and associated flora, fauna, historical, and cultural features, which has been reserved by law or other effective means to protect part or all of the enclosed environment.[2] As reported by the World Database on Protected Areas,[3] as of 2009, worldwide approximately 13% of the lands are designated as protected areas and about 0.8% waters along the shoreline and over the ocean are set as MPAs. In the United States, 14.81% of the terrestrial lands have been set as protected, and along the shoreline and over the ocean 24.75% of the terrestrial waters up to 12 nautical miles are set as MPAs. Protected lands and waters serve as the fundamental building blocks of virtually all national and international conservation strategies, supported by governments and international institutions. These provide the essential and best of efforts to protect the world's threatened species and are increasingly recognized as essential providers of ecosystem services and biological resources; key components in climate change mitigation strategies; and in some cases, are also vehicles for protecting threatened human communities or sites of great cultural and spiritual value.[1]

Humans have created protected areas over past millennia for a multitude of reasons.[4] The establishment of Yellowstone National Park in 1872 by the U.S. Congress ushered in the modern era of governmental protection of natural areas, which catalyzed a global movement.[5,6] However, even with implementation of a tremendous variety of monitoring programs and conservation planning efforts and achievements, species' population decline, biodiversity loss, extinction, system degradation, pathogen spread, and state change events are occurring at unprecedented rates.[7,8] The effects are augmented by continued changes in land use, invasive spread, alongside the direct, indirect, and interactive effects of climate change and disruption.[4] Protected areas become more important in serving as indicators of ecosystem conditions and functions either by their status and/or by comparison with unprotected adjacent areas. Protected lands are prized highly by the society, and yet different with diversified representative characteristics. Therefore inventory, monitoring, understanding, and management of protected areas present challenges.

Remote sensing refers to art, science, and technology for Earth system data acquisition through nonphysical contact sensors or sensor systems mounted on space-borne, airborne, and other types of platforms; data processing and interpretation from automated and visual analysis; information generation under computerized and conventional

Encyclopedia of Natural Resources DOI: 10.1081/E-ENRL-120049149

mapping facilities; and applications of generated data and information for societal benefits and needs. Remote sensing can provide comprehensive geospatial information to map and study protected areas in different spatial scales (e.g., high spatial resolution, large area coverage, etc.), different temporal frequencies (daily, weekly, monthly, annual observations, etc.), different spectral properties (visible, near infrared, microwave, etc.), and with spatial contexts (e.g., immediate adjacent areas of protected areas vs. a broader background of land and water bases). Remote sensing, in combination with field-based studies, has created new and exciting opportunities to meet monitoring needs to study protected areas.[9]

REMOTE SENSING OF CHANGING LANDSCAPE OF PROTECTED AREAS

Park units of the National Park Service (NPS) of the United States are important components in a system of reserves that protect biodiversity and other natural and cultural resources. To meet the NPS mission to manage resources so they are left "...unimpaired for the enjoyment of future generations," it is essential to know what resources occur in parks and to monitor the status and trends in the condition of key resource indicators. The NPS Inventory and Monitoring Program (I&M) was designed to provide the infrastructure and staff to identify critical environmental indicators, so-called "vital signs," and to implement long-term monitoring of natural resources in more than 270 parks that contain significant natural resources.[9] The 270+ parks are organized into 32 ecoregional networks (Fig. 1).[10]

The Park AnaLysis and Monitoring Support (PALMS) project, for example, was to enhance the quality of natural resource management in parks by better integrating the routine acquisition and analysis of NASA Earth System Science products and other data sources into NPS I&M.[10] The project focused on four sets of national parks to develop and demonstrate the approach. Those parks include the Sequoia-Kings Canyon and Yosemite National Parks (Sierra Nevada I&M Network), Yellowstone and Grand Teton National Parks (Greater Yellowstone I&M Network), Rocky Mountain National Parks (Rocky Mountain I&M Network), and a combination of Delaware Water Gap National Recreation Area and Upper Delaware Scenic and Recreational River (Eastern Rivers and Mountains I&M Network). The PALMS project developed a suite of indicators including measurements of weather and climate, stream health (water), land cover and land use, disturbances, primary production, and monitoring area.

Monitoring landscape dynamics of National Parks in the western United States by Landsat-based approaches is one of the examples.[11] The study is established by the approaches that translate ecologically based view of change into the spectral domain when the entire archive of TM spectral imagery is considered. A spectral index is used as a proxy for ecological attributes, and that index can be tracked as a time-series trajectory. The developed algorithms, collectively known as "LandTrendr," use simple statistical fitting rules to identify periods of consistent progression in the spectral trajectory (segments) and turning points (vertices) that separate those periods. The change-monitoring methods capture a wide range of processes affecting vegetation inside and outside of western national parks, such as decline/mortality processes; growth/recovery processes; combination of different processes, among others. The study concluded that even though national parks are protected from many forms of direct human intervention, their landscapes are anything but static. Vegetation is changing continuously in response to both endogenous and exogenous pressures, and monitoring these dynamic landscapes requires tools that capture a wide range of processes over large areas.

A study evaluated forest dynamics within and around the Olympic National Park using time-series Landsat observations.[12] The Olympic National Park (ONP) was established in the 1930s to protect the diversity of its ecosystems. It has some of the best examples of intact temperate rainforest in the Pacific Northwest and is home to endemic species. In 1976, it was designated by the UNESCO as an International Biosphere Reserve, and was declared a World Heritage Site in 1981. However, due to extensive logging in the early 20th century, most of the unprotected old growth forests outside the ONP and the Olympic National Forests were lost, and some of the private lands in the peninsular are among the most heavily logged in the United States. Viable protection of the unique biodiversity of the Olympic Peninsula requires continuous monitoring of forest dynamics. The time-series approach using remote sensing data allows reconstruction of disturbance history over the last decades by taking advantage of the temporal depth of the Landsat archive, and can be used to provide continuous monitoring as new satellite images are acquired.

Remote sensing has been used for monitoring wildlife habitat changes in Kejimkujik National Park and the National Historic Site in southern Nova Scotia of the Canadian Atlantic Coastal Uplands Natural Region.[13] The study addressed the major goals in protection and maintenance of ecological integrity through two key measures of "effective habitat amount" and "effective habitat connectivity." The study employed the Automated Multitemporal Updating by Signature Extension protocol, known otherwise as AMUSE, to generate multitemporal land cover maps from 1985, 1990, 1995, 2000 and 2005 Landsat remote sensing data. The resultant gradient maps showed areas of high and low effective habitat amount and connectivity throughout the park and greater park ecosystem for each species. Thresholds were then applied to identify species landscapes that were estimated to contain sufficient habitat. Statistics on the total amount and connectivity of effective

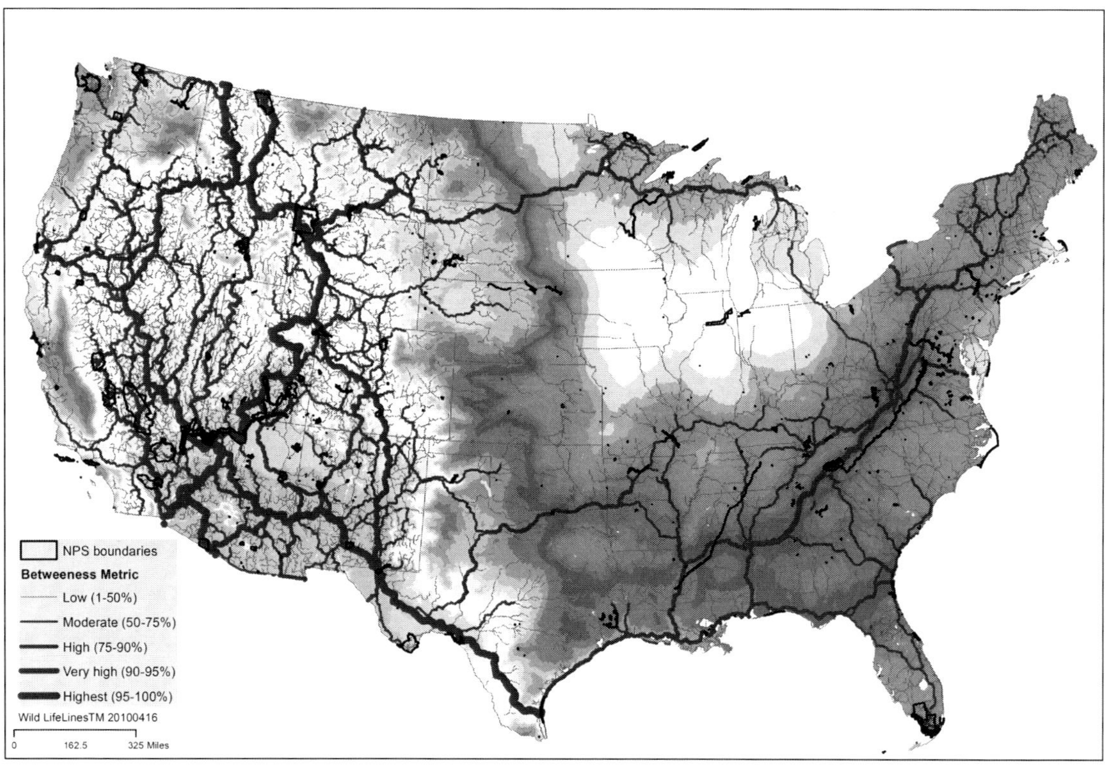

Fig. 1 Map showing connectivity of natural landscapes in the United States. The thickness of lines indicates magnitude of cumulative movement, assuming that animals avoid human-modified areas. The surface underneath the pathways depicts the averaged cost–distance surfaces, or the overall landscape connectivity surface. National Park Service units are outlined in black.
Source: Gross et al.[10]

habitat for Kejimkujik and the greater park ecosystem were then extracted for each 5-year time step and compared. The visual output from analyses provides powerful communication products on how land use can affect habitat patterns around the park. These communication products can support the park in engaging partner agencies and stakeholders in the development of collaborative conservation strategies that are vital in allowing Kejimkujik to manage toward achieving its long-term ecological integrity goals. Information is vital for sound decision making at all levels. Remote sensing and the related information technology have been playing and will continue to play critical roles in natural resource monitoring and management.

REMOTE SENSING FOR INVENTORY, MAPPING, AND CONSERVATION PLANNING OF PROTECTED AREAS

One particular advantage that remote sensing can provide for inventory and monitoring of protected areas is the information for understanding the past and current status, the changes occurred under different impacting factors and management practices, the trends of changes in comparison with those in the adjacent areas and implications of changes on ecosystem functions.[14,15] Field survey and *in situ* observations are essential to identify protected habitats through remote sensing. Almost every remote sensing exercise requires field survey to define habitats, to calibrate remote sensing imagery, and to evaluate the accuracy out of remote sensing outputs.[16] With GPS-guided positioning and field survey becoming a routine operation, challenges remain for incorporation of *in situ* measurements with remote sensing observations for quantitative analyses of protected habitats.

A study employed Landsat-7 ETM+ data and GPS collars to provide detailed information for conservation of Matschie's tree kangaroos in Papua New Guinea.[17] Matschie's tree kangaroos (*Dendrolagus matschiei*) are arboreal marsupials endemic to the Huon Peninsula in Papua New Guinea. Due primarily to increased hunting pressure and loss of habitat from agricultural expansion, *D. matschiei* are currently listed as endangered by the IUCN. The study concluded that *Dacrydium nidulum*-dominant forests are the most widespread forest throughout the study area and are also where tree kangaroos located. However, additional research has shown that these are not the only forest type that is used by the species. Clustered and independent movement locations indicate that animals do not utilize their habitat uniformly. These data provide vital information toward a better understanding of the habitat, the requirements of the animals, and the long-term conservation of the Matschie's tree kangaroo habitat. Aerial remote sensing from Wildlife Conservation Society's Flight Program has been used to support biodiversity conservation in Madagascar and Eastern and Southern Africa with a focus on the Albertine Rift.[18] Many sites in the Albertine Rift are protected as national parks, wildlife reserves, or forest reserves. The aerial imagery has been used to map threats to biodiversity, to develop land-use plans for protected area management, and to measure vegetation cover and dynamics.

MPAs are among critical components of protected waters. Important factors that affect the way plants and animals respond to MPAs include distribution of habitat types, level of connectivity to nearby fish habitats, wave exposure, depth distribution, prior level of resource extraction, regulations, and level of with regulations.[19] Conservation benefits are evident through increased habitat heterogeneity at the seascape level, increased abundance of threatened species and habitats, and maintenance of a full range of genotypes.[20] Remote sensing data that quantify spatial patterns in habitat type, oceanographic conditions, and benthic complexity can be integrated with in situ ecological data for design, evaluation, and monitoring of MPA networks to design, assess, and monitor MPAs.[21,22] Combining remote-sensing products with in situ ecological and physical data can support the development of a statistically robust monitoring program of living marine resources within and adjacent to marine protected areas.[23] Individual MPAs need to be networked in order to provide large-scale ecosystem benefits and to have the greatest chance of protecting all species, life stages, and ecological linkages if they encompass representative portions of all ecologically relevant habitat types in a replicated manner. High-resolution remote sensing data are capable of mapping physical and biological features of benthic habitat, such as monitoring of coral reef in the Hawaii Archipelago and near-shore protected areas in California and New England.[19]

REMOTE SENSING OF FRONTIER LANDS

Remote sensing has unique advantages in monitoring frontier lands, which are always in remote and difficult-to-reach locations and huge in area coverage. Different types of remote sensing data have been applied in the study of frontier lands, for example, using hyperspectral and radar data for monitoring of forests in the Amazon region [24–29] and Siberia,[30–32] and for hydrologic change detection in the lake-rich Arctic region.[33,34]

Representative case studies include satellite-observed endorheic lake dynamics across the Tibetan Plateau between 1976 and 2000; multisensor remote sensing of forest dynamics in Central Siberia; remote sensing and modeling for assessment of complex Amur tiger and Far Eastern leopard habitats in the Russian Far East; incorporating remotely sensed land surface properties to regional climate modeling; and InSAR for characterizing biophysical properties in protected tropical forests in Southeast Asia.

With a pronounced temperature rise, the Tibetan plateau is one of the world's most vulnerable areas to climate change. Tibetan warming has been accelerating since the 1950s, and the accelerated warming is expected to drive an array of complex physical and ecological changes in the region. Tibetan lakes in the endorheic basins serve as a sensitive indicator of regional climate and water cycle variability. The lakes are dynamic in their inundation area in response to climate and hydrological conditions. However, these lake dynamics at regional scales are not well understood due to inaccessibility and the inhospitable environment of this remote plateau, making satellite remote sensing the only feasible tool to detect lake dynamics across the plateau. A study monitors Tibetan lake changes using nearly a hundred Landsat scenes acquired around 1976 and 2000 and examined lake changes during ~25 years in endorheic basins across this broad plateau and analyzes the spatial patterns of the observed lake dynamics.[35]

The forested regions of Russian Siberia are vast and contain about a quarter of the world's forests that have not experienced harvesting. Monitoring the dynamics and mapping the structural parameters of the forest are important for understanding the causes and consequences of changes observed in these areas. Because of the inaccessibility and large extent of this forest, remote sensing data can play an important role in observing forest state and change. Ranson et al.[36] introduced a case study that used multisensor remote sensing data to monitor forest disturbances and to map aboveground biomass in Central Siberia. Radar images from the Shuttle Imaging Radar-C (SIR-C)/XSAR mission were used for forest biomass estimation in the Sayan Mountains. Radar images from the Japanese Earth Resources Satellite-1 (JERS-1), European Remote Sensing satellite-1 (ERS-1), and Canada's RADARSAT-1, and data from ETM+ on-board Landsat-7 were used to characterize forest disturbances from logging, fire, and insect damages.

Listed as endangered and critically endangered by the IUCN, respectively, fewer than 400 adult and subadult Amur (Siberian) tigers (*Panthera tigris altaica*) and only between 14 and 20 adult Amur (Far Eastern) leopards (*P. pardus orientalis*) exist in the wild in a topographically complex and biologically diverse landscape in the Russian Far East and in the Changbai mountain area in northeastern China near the Russia-China border. These endangered species face threats mostly from loss of habitat due to forest fire and logging to illegal poaching and competing prey with man. Remote sensing is a valuable tool for characterizing the vast habitat of wide-ranging endangered species, such as the Amur tiger, and the critically endangered Amur leopard. A study employed remote sensing for mapping, monitoring, and modeling tiger and leopard habitats related to vegetation and terrain.[37] The study helped target resources toward locations that are most important for ensuring a future in the wild for these species, as well as identify new areas that would be appropriate to connect good habitat, including that in established reserves.

In tropical forests in Southeast Asia, national parks have been established as a means of reducing deforestation in ecologically and economically favorable ways. Accurate estimates of forest biophysical properties in those national parks are important in assessing their current status and in supporting sustainable management of these protected lands. A study employed Landsat ETM+ and JERS-1 SAR images to estimate green and woody structures in a protected forest in northern Thailand.[29] Forest fractional cover and leaf area index are estimated from the ETM+ image using a linear unmixing model. A microwave/optical synergistic model has been modified to improve woody biomass estimation by removing leaf contribution from radar backscatter in tropical forests. Besides seasonal variations in different forest types, forest degradation by human disturbances is observed, which results in reduced green and woody biomass in forests closer to human settlements. The study demonstrates that a synergistic use of optical and SAR images could extract quantitative biophysical information of protected forests in tropical mountains.

CONCLUSIONS

Ecosystem indicators, whether process-based (e.g., productivity), pattern-based (e.g., land-use activities), or component-based (species populations), vary in space and time. A major limiting factor in comprehensive ecological models is lack of explanatory geospatial data.[4] The issues conspire against the ready, standardized integration of remote sensing into ecological research for protected area management. Remote sensing science is a universal tool for managers and researchers across many domains.[38] Remote sensing data and data products coupled with user-friendly data exploration, data management, analyses, and modeling tools in an accessible common platform, allow both scientists and practitioners a better understanding of how environmental impacts affect species populations and the ecosystem goods and services that sustain them. The lessons learned and recommendations put forward for remote sensing of protected areas include the following: "allocate sufficient time to develop a genuine science—management partnership," "communicate results in a management-relevant context," "confirm or embellish existing frameworks and processes," "plan for persistence and change," and "build on existing, widely used data analysis tools and software frameworks, even if they seem inefficient."[10]

Remote sensing is among the most fascinating frontiers of science and technology that are constantly improving. Protected areas are by no means uniform entities and have a wide range of management aims and are governed by many stakeholders. Advances in remote sensing have helped gather and share information about the protected areas at unprecedented rates and scales. There are many new and exciting applications of remotely sensed data that contribute to better informing management of protected areas. The achievements through applications of science and technologies, the challenges, the lessons learned, and recommendations in remote sensing of protected areas deserve special attention.

REFERENCES

1. Dudley, N., Ed.; *Guidelines for Applying Protected Area Management Categories*. IUCN: Gland, Switzerland, 2008; 86 pp.
2. Kelleher, G. *Guidelines for Marine Protected Areas*. IUCN: Gland, Switzerland and Cambridge, UK, 1999.
3. IUCN and UNEP-WCMC. *The World Database on Protected Areas (WDPA)*; UNEP-WCMC: Cambridge, UK, 2010.
4. Crabtree, R.; Sheldon, J. Monitoring and modeling environmental change in protected areas: integration of focal species populations and remote sensing (Ch. 21). In *Remote Sensing of Protected Lands*; Wang, Y., Ed.; CRC Press: Boca Raton, FL33487-2742, USA, 2012; 495–524 pp.
5. IUCN. Shaping a sustainable future. In *The IUCN Programme 2009–2012*; IUCN: Gland, Switzerland, 2008.
6. Heinen, J.; Hite K. Protected natural areas. In *Encyclopedia of Earth*; Cutler J. Cleveland. C.J., Ed.; Environmental Information Coalition, National Council for Science and the Environment: Washington, DC. 2007. http://www.eoearth.org/article/Protected_natural_areas (First published in the *Encyclopedia of Earth* November 29, 2007; Last revised November 29, 2007; Retrieved February 5, 2011)
7. Hoffmann, M.; Hilton-Taylor, C.; Angulo, A. et al. The impact of conservation on the status of the World's vertebrates. Science **2010**, *330* (6010), 1503–1509.
8. Pereira, H.M.; Leadley, P.W.; Proença, V., et al. Scenarios for global biodiversity in the 21st century. Science **2010**, *330* (6010), 1496–1501.
9. Fancy, S.G.; Gross, J.E.; Carter, S.L. Monitoring the condition of natural resources in U.S. National Parks. Environ. Monit. Asses. **2009**, *151* (1–4), 161–174.
10. Gross, J.E.; Hansen, A.J.; Goetz, S.J.; Theobald, D.M.; Melton, F.M.; Piekielek, N.B.; Nemani, R.R.; Remote Sensing for Inventory and Monitoring of U.S. National Parks (Ch. 2). In *Remote Sensing of Protected Lands*; Wang, Y., Ed.; CRC Press: Boca Raton, Florida, 2012; 29–56 pp.
11. Kennedy, R.E.; Yang, Z.; Cohen, W.B. Detecting trends in forest disturbance and recovery using yearly Landsat time series: 1. LandTrendr—Temporal segmentation algorithms. Remote Sensing Environ. **2010**, *114* (12), 2897–2910.
12. Huang, C.; Schleerweis, K.; Thomas, N.; Goward, S.N. Forest Dynamics within and around Olympic National Park Assessed Using Time Series Landsat Observations (Ch. 4). In *Remote Sensing of Protected Lands*; Wang, Y., Ed.; CRC Press: Boca Raton, Florida, 2012; 75–94 pp.

13. Zorn, P.; Ure, D.; Sharma, R.; O'Grady, S. Using earth observation to monitor species-specific habitat change in the Greater Kejimkujik National Park Region of Canada (Ch. 5). In *Remote Sensing of Protected Lands*; Wang, Y., Ed.; CRC Press: Boca Raton, Florida, 2012; 95–110 pp.
14. Hansen, A.J.; DeFries, R. Land use change around nature reserves: Implications for sustaining biodiversity. Ecol. Appl. **2007a**, *17* (4), 972–973.
15. Hansen, A.J.; DeFries, R. Ecological mechanisms linking protected areas to surrounding lands. Ecol. Appl. **2007b**, *17* (4), 974–988.
16. Wang, Y.; Traber, M.; Milestead, B.; Stevens, S. Terrestrial and submerged aquatic vegetation mapping in Fire Island National seashore using high spatial resolution remote sensing data. Mar. Geodesy **2007**, *30* (1), 77–95.
17. Stabach, J.A.; Dabek, L.; Jensen, R.; Wang, Y.Q. Discrimination of dominant forest types for Matschie's tree kangaroo conservation in Papua New Guinea using high-resolution remote sensing data. Int. J. Remote Sensing **2009**, *30* (2): 405–422.
18. Ayebare, S.; Moyer, D.; Plumptre, A.J.; Wang, Y. Remote sensing for biodiversity conservation of the Albertine Rift in Eastern Africa (Ch. 10). In *Remote Sensing of Protected Lands*; Wang, Y., Ed.; CRC Press: Boca Raton, Florida, 2012; 183–201 pp.
19. Friedlander, A.M.; Wedding, L.M.; Caselle, J.E.; Costa, B.M. Integration of remote sensing and *in situ* ecology for the design and evaluation of marine protected areas: examples from tropical and temperate ecosystems (Ch. 13). In *Remote Sensing of Protected Lands*; Wang, Y., Ed.; CRC Press: Boca Raton, FL, USA, 2012; 245–280 pp.
20. Edgar, G.J.; Russ, G.R.; Babcock, R.C. Marine protected areas. In *Marine Ecology*; Oxford University Press: South Melbourne, VIC, Australia, 2007; 533–555 pp.
21. Wedding, L.; Friedlander, A.M. Determining the influence of seascape structure on coral reef fishes in Hawaii using a geospatial approach. Mar. Geodesy **2008**, *31*, 246–266
22. Wedding, L.; Friedlander, A.; McGranaghan, M.; Yost, R.; Monaco, M.E. Using bathymetric Lidar to define nearshore benthic habitat complexity: implications for management of reef fish assemblages in Hawaii. Remote Sensing Environ. 2008, 112, 4159–4165.
23. Friedlander, A.M.; Brown, E.K.; Monaco, M.E. Coupling ecology and GIS to evaluate efficacy of màine protected areas in Hawaii. Ecol. Appl. **2007**, *17*, 715–730.
24. Sheng, Y.; Alsdorf, D. Automated ortho-rectification of Amazon basin-wide SAR mosaics using SRTM DEM data. IEEE Trans. Geosci. Remote Sensing **2005**, *43* (8), 1929–1940.
25. Arima, E.Y.; Walker, R.T.; Sales, M.; Souza, C., Jr.; Perz, S.G. The fragmentation of space in the Amazon basin: emergent road networks. Photogram. Eng. Remote Sensing **2008**, *74* (6), 699–709.
26. Walsh, S.J.; Messina, J.P.; Brown, D.G. Mapping & modeling land use/land cover dynamics in frontier settings. Photogram. Eng. Remote Sensing **2008**, *74* (6), 677–679.
27. Mena, C.F. Trajectories of land-use and land-cover in the northern Ecuadorian Amazon: temporal composition, spatial configuration, and probability of change. Photogram. Eng. Remote Sensing **2008**, *74* (6), 737–751.
28. Wang, C.; Qi, J.; Cochrane, M. Assessment of tropical forest degradation with canopy fractional cover from Landsat ETM+ and IKONOS imagery. Earth Interact. **2005**, *9* (22), 1–18.
29. Wang, C.; Qi J. Biophysical estimation in tropical forests using JERS-1 SAR and VNIR Imagery: II- aboveground woody biomass. Int. J. Remote Sensing **2008**, *29* (23), 6827–6849.
30. Sun, G.; Ranson, K.J.; Kharuk, V.I. Radiometric slope correction for forest biomass estimation from SAR data in Western Sayani mountains, Siberia. Remote Sensing Environ. **2002**, *79*, 279–287.
31. Bergen, K.M.; Zhao, T.; Kharuk, V.; Blam, Y.; Brown, D.G.; Peterson, L.K.; Miller, N. Changing regimes: forested land cover dynamics in central Siberia 1974–2001. Photogram. Eng. Remote Sensing **2008**, *74* (6), 787–798.
32. Kharuk, V.I.; Ranson, K.J.; Im, S.T. Siberian silkmoth outbreak pattern analysis based on SPOT VEGETATION data. Int. J. Remote Sensing **2009**, *30* (9), 2377–2388. Sensing Environ. *114*, 2897–2910.
33. Stow, D.A.; Hope, A. McGuire, D.; Verbyla, D.; Gamon, J.; Huemmrich, F.; Houston, S.; Racine, C.; Sturm, M.; Tape, K.; Hinzman, L.; Yoshikawa, K.; Tweedie, C.; Noyle, B.; Silapaswan, C.; Douglas, D.; Griffith, B.; Jia, G.; Epstein, H.; Walker, D.; Daeschner, S.; Petersen, A.; Zhou, L.; Myneni, R. Remote sensing of vegetation and land-cover change in Arctic tundra ecosystems. Remote Sensing Environ. **2004**, *89* (3), 281–308
34. Sheng, Y.; Shah, C.A.; Smith, L.C. Automated image registration for hydrologic change detection in the lake-rich arctic. IEEE Geosci. Remote Sensing Lett. **2008**, *5* (3), 414–418.
35. Sheng, Y., Li, J. Satellite-observed endorheic lake dynamics across the Tibetan plateau between circa 1976 and 2000 (Ch. 15). In *Remote Sensing of Protected Lands*; (Wang, Y., Ed.; CRC Press: Boca Raton, Florida, 2012; 305–319 pp.
36. Ranson, K.J.; Sun, G.; Kharuk, V.I.; Howl, J. Multisensor Remote Sensing of Forest Dynamics in Central Siberia (Ch. 16). In *Remote Sensing of Protected Lands*; Wang, Y., (Ed.; CRC Press, 2012; 321–377 pp.
37. Sherman, N.J.; Loboda, T.V.; Sun, G.; Shugart, H.H. Remote sensing and modeling for assessment of complex Amur (Siberian) Tiger and Amur (Far Eastern) Leopard Habitats in the Russian Far East (Ch. 17). In *Remote Sensing of Protected Lands*; Wang, Y., Ed.; CRC Press: Boca Raton, FL33487-2742, USA, 2012; 379–407 pp.
38. Kennedy, R.E.; Townsend, P.A.; Gross, J.E.; Cohen, W.B.; Bolstad, P.; Wang, Y.Q.; Adams, P. Remote sensing change detection and natural resource monitoring for managing natural landscapes. Remote Sensing Environ. **2009**, *113* (7), 1382–1396.

River Delta Processes and Shapes

Douglas A. Edmonds
Rebecca L. Caldwell
Department of Geological Sciences, Indiana University, Bloomington, Indiana, U.S.A.

Abstract

River deltas are arguably some of the most important environments on Earth; they are ecologically rich, usually contain hydrocarbons in the subsurface, and support ~10% to 25% of the world's population. River deltas form on shorelines where sediment can accumulate faster than it is taken away and are recognized by the presence of a distributary channel network or a bump that protrudes from the shoreline. Subaerial parts of river deltas grow by the processes of levee formation, river mouth bar deposition, avulsion, and organic and fine-sediment matter accumulation. The resulting balance of these processes, determined by upstream and downstream boundary conditions, sets the delta shape. Looking forward, the future of deltaic environments is grim under current estimates of relative sea-level rise. In order to protect these important environments, we must understand delta processes and shapes so that restoration schemes can be designed to take advantage of them.

INTRODUCTION

The flow of water and sediment creates environments of breath-taking beauty and complexity and few places exhibit this as clearly as river deltas. In deltas, the processes of water flow and sediment transport give rise to a seemingly ordered network of river channels that construct a semicircular platform from the sediments they carry (Fig. 1). These channels are the delta's artery system; they transport the water, sediment, and nutrients that promote delta growth and keep ecosystems healthy.

Beyond the aesthetic appeal of river deltas, they represent one of the most important environments on Earth. River deltas have rich, productive ecosystems, abundant coastal waterways for transport, and valuable hydrocarbon reservoirs in the subsurface. Humans have long recognized this importance. For example, many early civilizations emerged around river deltas,[1] and since then deltas have become home to 10–25% of the world's population.[2]

By their nature deltas are flat-lying landforms with little topographic relief. For instance, the average elevation of delta land is ~5 m above sea level.[3] Given their low elevations, they are vulnerable to flooding from sea-level rise and coastal subsidence.[4] To protect deltaic environments, we must understand how they function and how they will respond to environmental changes such as relative sea level change. Here, we focus on the processes and shapes that form the subaerial part of the delta, since that is where most humans live. This entry summarizes what is known about the subaerial part of river deltas by answering the following questions: 1) What is a river delta? 2) How and why do river deltas form? 3) What processes create river deltas? And 4) what determines the shape of a river delta? Finally, we conclude by considering the future of deltaic environments given the stress of changing sea-level, sediment fluxes, and subsidence rates.

WHAT IS A RIVER DELTA?

A river delta is an accumulation of sediment on a coastline adjacent to a river. Deltas can form on any coastline, and recognizing them on satellite images is not always straightforward, especially if the "sediment accumulation" of the delta is small or not easily visible. We define two geomorphic criteria for recognizing deltas. These criteria are chosen because they imply substantial sediment accumulation. First, deltas may have a noticeable depositional protrusion (or bump) that causes deviation from the mean shoreline position and is unambiguously linked to the river (see Fig. 1E for example). The minimum scale of the protrusion is not easy to define, but it should be wider (in a shore-perpendicular direction) than a few multiples of the river width to avoid confusing deltas with prograding river mouths or tie channels that have no delta.[5] The second criterion is clear evidence for a distributary channel network (Fig. 1A–D). Distributary channels are river channels that branch off from a parent channel and do not rejoin before intersecting the coast (Fig. 1B). By their nature, these channels facilitate sediment accumulation as they

Encyclopedia of Natural Resources DOI: 10.1081/E-ENRL-120047501

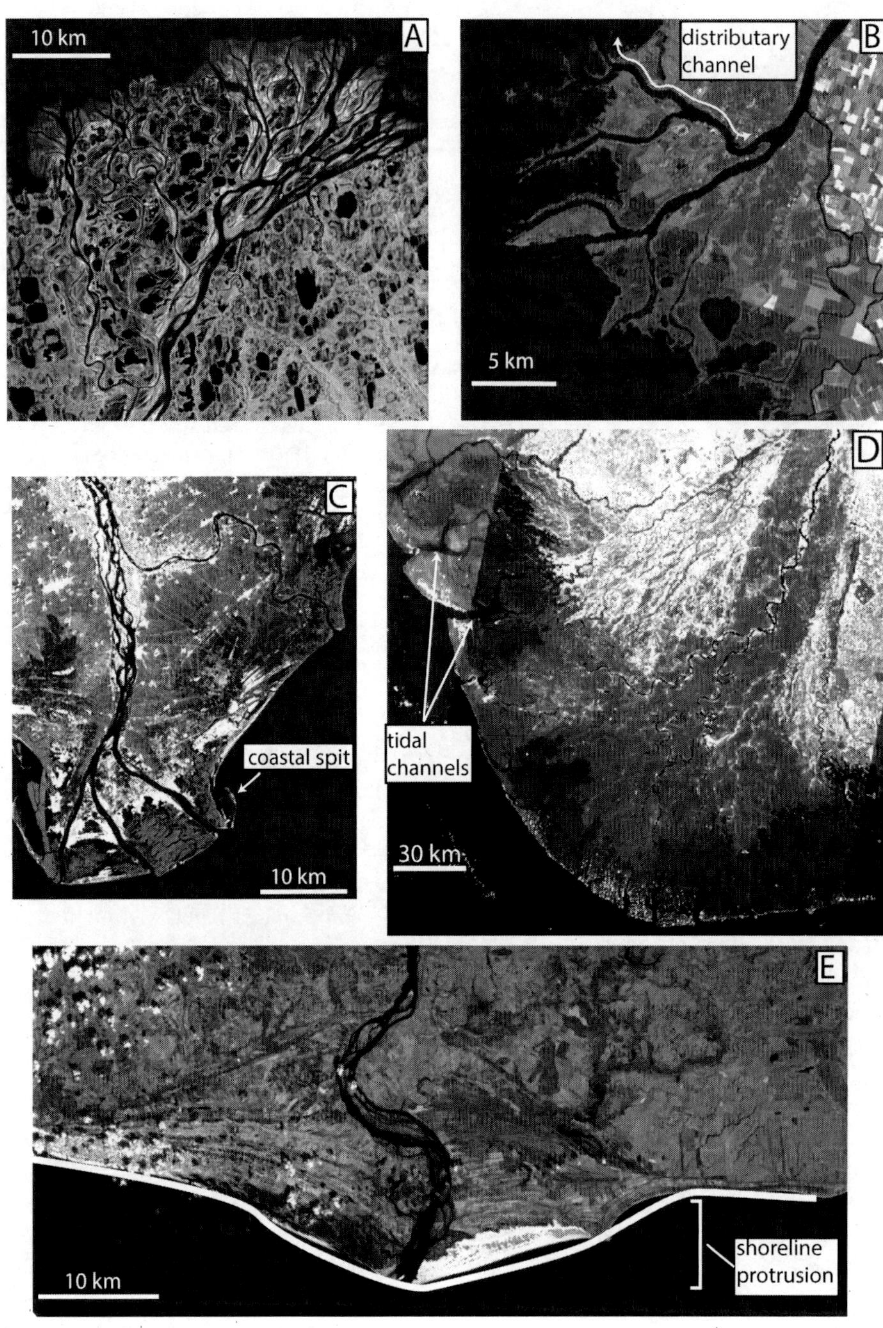

Fig. 1 Satellite imagery showing five examples of river deltas. Black represents water. (**A**) Colville River delta, Alaska, USA; (**B**) St. Clair River delta, Ontario, Canada; (**C**) Krishna River delta, India; (**D**) Niger River delta, Niger; and (**E**) São Francisco River delta, Brazil.
Source: Images are from Landsat 7 satellite obtained from the USGS through usgs.glovis.org.

distribute sediment over a broad area. Only one of these criteria is necessary for classifying a geomorphic feature as a delta, and often deltas exhibit both a protrusion and a distributary network (Fig. 1C).

HOW AND WHY DO RIVER DELTAS FORM?

A delta will form if sediment from a river or other coastal processes accumulates faster than is taken away by sediment compaction, and riverine and coastal erosive forces, while also accounting for a changing volume due to sea-level fluctuations and coastal subsidence. This is effectively a statement of conservation of mass for a control volume at the coastline. There are three main sources that supply sediments to that control volume. Rivers are the largest source of deltaic sediment; and on average, those rivers with larger drainage basins, higher relief, and in warmer climates carry more sediment.[4,6] It is somewhat unsurprising that the largest rivers in the world often have the largest river deltas.

But, interestingly, two rivers can have nearly the same sediment flux, such as the Mississippi River and the Amazon River, yet the Amazon delta has an area ten times that of the Mississippi delta.[3] Waves and tides, in some cases, can be a net sediment source to a delta,[7,8] especially during hurricanes when wind blows onshore. Finally, organic sediment created from the production and decay of plants can create substantial amounts of sediment, and this is a mechanism by which deltas aggrade their surfaces in response sea-level rise.[9] For example, in parts of the Mississippi delta, organic sediment makes up one-third of the total sediment volume.[10]

In principle, deltas should form at locations where these sediment sources are maximized, though it also depends on the size of the volume to be filled and the sediment sinks. The volume depends on the offshore depth and changes due to varying sea-level and coastal subsidence. Sediment sinks include fluvial bypass, which is the proportion of incoming sediment that is not deposited within the delta. In some cases, up to 40% of the river sediment brought to the delta is not deposited in the delta, but carried out to sea.[11] Other sediment sinks include sediment compaction, and waves and tides which can disperse river sediment, especially during storms.[8]

This conservation of mass framework helps us understand the origin of most of the world's modern deltas. Modern delta formation started around 7,000 years ago,[12] which is coincident with a deceleration of global sea-level rise. A decelerating global sea level rise reduced the volume to fill with sediment and tipped the balance in favor of delta formation.

WHAT PROCESSES CREATE RIVER DELTAS?

The natural processes that create deltas range in scale from microbial processes that influence sediment mobility to tectonic processes that influence sediment flux. Here, we focus on those mesoscale processes that construct the subaerial, geomorphic expression of the delta and its channel network. We acknowledge that even in this context many processes are not discussed and those processes may become more important depending on the delta and scale of observation.

The key hydrodynamic process that creates deltas is the turbulent expanding jet (Fig. 2). A turbulent jet forms at the river mouth as the river flow enters a standing body of water and transitions from confined to unconfined. Expansion due to loss of flow confinement and bed friction causes flow deceleration along the jet centerline. At the jet margins, there is strong lateral shear between the fast-moving currents in the jet and the relatively still-standing body of water, which produces turbulent eddies.[13] Flow expansion and deceleration within the jet ultimately cause sediment deposition, leading to embryonic delta formation.

The processes of levee and river mouth bar formation are linked to the dynamics of turbulent jets (Fig. 2). Subaqueous levees form when advection and diffusion transport sediment to the turbulent jet margins where it is deposited in the relatively still ambient water.[13,14] Concurrent with the process of levee formation, a river mouth bar forms at the location of maximum deceleration of velocity along the turbulent jet centerline.[15] River mouth bars are usually triangular in planview when formed only by fluvial processes,[16] but can be deflected in the presence of obliquely approaching waves.[17] Over time, the river mouth bar aggrades to the surface and flow bifurcates around it, creating two or more channels where there was originally one (Fig. 2).[15] In this way, the mouth bar sets the point of bifurcation, and relict mouth bars can be identified in the older, upstream parts of the delta (Fig. 2). As channels prograde past mouth bars, turbulent jets once again form at the river mouths and the process of levee growth and mouth bar deposition starts over.

Muddy, interdistributary plains form in bays between deltaic channels and often behind mouth bars (Fig. 2). These bays are usually shallow water bodies (1–4 m deep) and the water can be fresh or saline.[18] These interdistributary bays are sheltered from wave and tidal erosion during storms, allowing sediment brought in during floods to accumulate. As sediment slowly accumulates over time an interdistributary plain composed of coastal marshes will form.[18,19] These are also locations where vegetation is easily established allowing organic sediment to accumulate.

Deltas are also influenced by the process of channel avulsion. Avulsion occurs when flow is diverted out of an existing river channel and establishes a new channel on the adjacent floodplain (Fig. 3).[20,21] Deltaic avulsions are probable if there is a more attractive flow path down the delta, due to a steeper slope or superelevation of the channel. Superelevation occurs when channels perch themselves above their floodplain by sedimentation on the channel bed and levees. In some sense, avulsions are an inevitable outcome of delta growth. Differential progradation of the delta front will create more attractive, steeper paths on other parts of the delta.[22,23] Also, as deltas prograde their channels must aggrade lest their bed slopes decrease. Maintaining a constant slope requires aggradation which will lead to superelevation of the channel.

Once the conditions for avulsion are met, there are two main scales of deltaic avulsions that occur in different places within the delta system. Avulsions can occur upstream of the channel network, which results in creation of a new delta lobe. For these avulsions, the upstream extent of the backwater zone correlates with avulsion locations[24] because this is an area of higher sedimentation rates leading to superelevation.[25] Avulsions also occur within the channel network, creating new delta sublobes. This tends to occur where there is sustained overbank flow leading to time-dependent weakening of the levee.[26] In either case, once an avulsion occurs, it initiates the construction of new deltaic land as the processes of levee

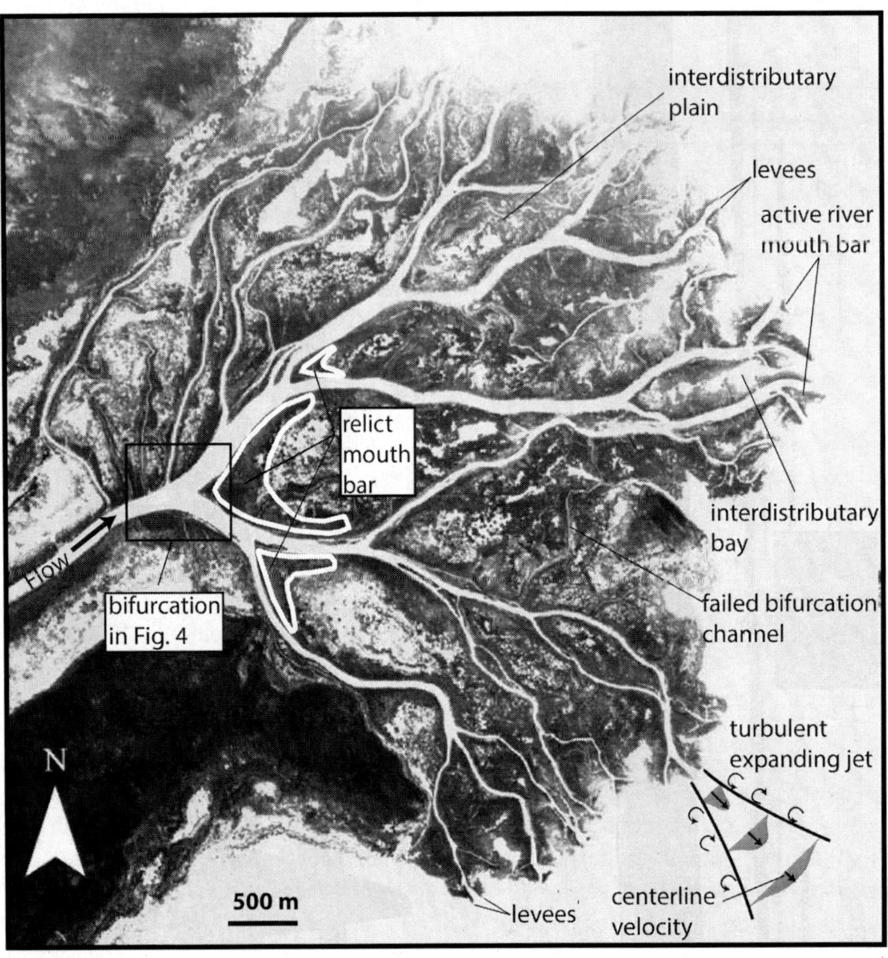

Fig. 2 Aerial image of the Mossy River delta, Saskatchewan, Canada taken in February 2009. Channel network and water are white due to snowfall. The turbulent jet depicted in the lower right is a cartoon, where the gray curves represent the horizontal velocity profile across the jet, and the curved arrows represent turbulence generation at jet margins. See text for a detailed discussion on the other features labeled in this image. Image copyright Digital Globe, Inc., obtained through National Geospatial Intelligence Agency commercial imagery program.

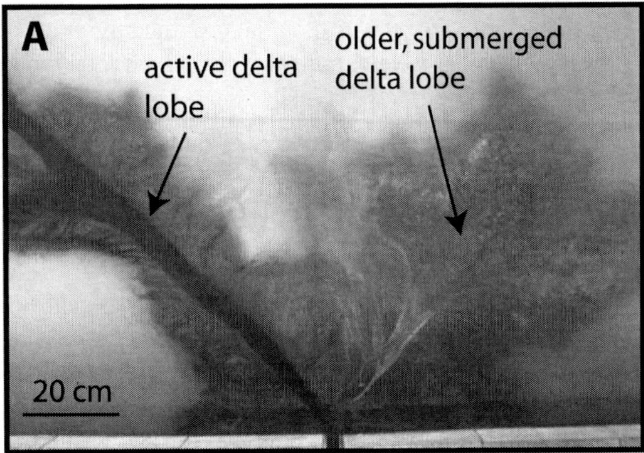

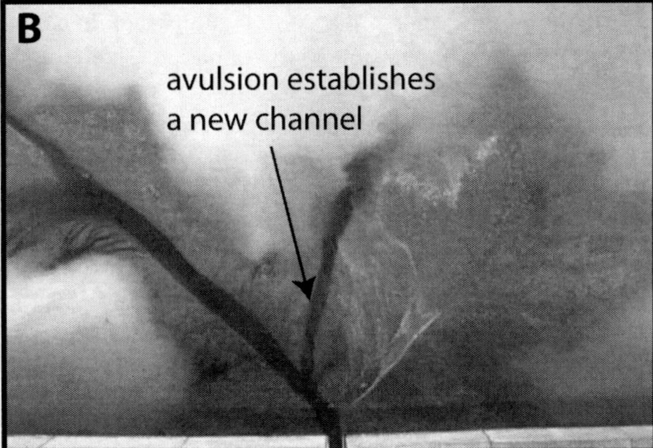

Fig. 3 An example of deltaic avulsion in an experimental delta created in the laboratory. Elapsed time between the images (**A**) and (**B**) is 80 minutes. The water is dyed so that channelized flow is visible.
Source: Figure modified from Edmonds et al.[26]

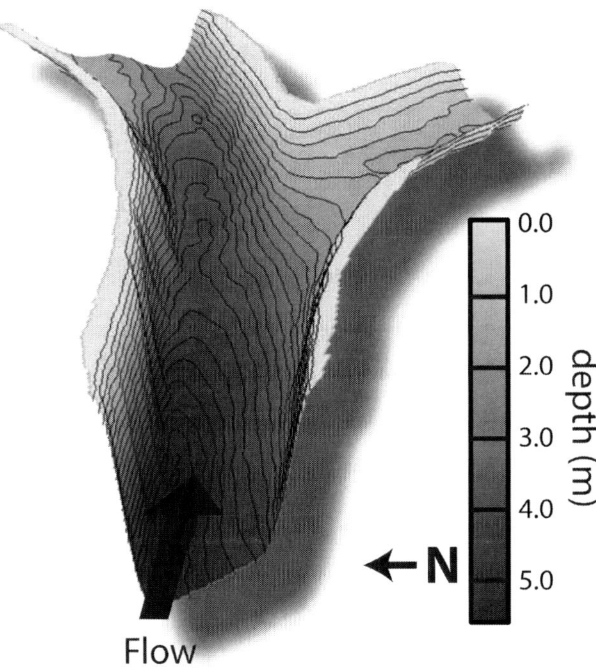

Fig. 4 Perspective view of the bathymetry of a deltaic bifurcation from the Mossy River delta (see Fig. 2 for location). The downstream bifurcate channels are asymmetric; the northern one is wider and deeper than the southern channel. The contour interval is 0.25 m and the vertical exaggeration is 50 times.

growth, mouth bar deposition, and interdistributary plain formation begin in new places.

As deltas prograde basinward through the deposition of mouth bars, levees, and interdistributary plains, they create channel networks. The basic unit of a channel network is the bifurcation, wherein one channel branches into two or more (Fig. 4). Bifurcations are created by channel splitting around mouth bars and avulsions. A global survey of modern deltas shows that on average bifurcations are asymmetric where one channel is ~70% wider than the other.[15] For example, the bifurcation at the head of the Mossy River delta (Fig. 2) is asymmetric—the northern channel is wider and deeper than the southern one (Fig. 4). Deltaic bifurcations can be stable where both channels receive water and sediment, or unstable where one channel captures all the flow. Stable bifurcations distribute water and sediment asymmetrically between their downstream bifurcate channels because this configuration is stable to perturbations, such as floods. Symmetric bifurcations, on the other hand, are not as common because if they are perturbed they return to the more stable asymmetric configuration.[27–30]

WHAT DETERMINES DELTA SHAPE?

We contend that delta morphology is intimately linked to the relative influences of processes operating on the delta. Those processes, in turn, are affected by the upstream and downstream boundary conditions of the system. Here we provide three generic measures of shape that capture first-order differences in delta morphology. We then describe how these shapes are linked to the different delta building processes, and finally how different boundary conditions affect the balance of delta building processes and thus morphology. We make the distinction between those boundary conditions originating from upstream (e.g., sediment type and flux) and downstream (e.g., marine processes such as waves and tides). This is only meant to serve as a guide to what controls delta shape because ultimately many different delta shapes exist depending on how these processes and boundary conditions interact.

Delta shape can be described by three attributes—bulk external shape, shoreline pattern, and channel network configuration. The bulk external shape of a delta, defined as the delta width divided by the length, ranges from wide flattened protrusions, to semicircles, to elongate shapes. An elongate delta, with a length longer than the width, preferentially constructs land further from the coastline compared to a semicircular delta. Delta shorelines range from smooth (Fig. 1D) to rough (Fig. 1B). Roughness can be defined by a sinuosity index, such as the ratio of the actual shoreline length to a spatially averaged shoreline length. For example, a sinuosity index greater than one represents a rough shoreline characterized by large-scale deviations from the average shoreline position. The configuration of a delta channel network has two end-member patterns: single-channel (Fig. 1E) to multichannel (Fig. 1A). Each channel can be a single-thread channel or a braided multithread channel, allowing for the possibility of a channel network where each channel is braided with multiple threads.[31]

A given combination of processes operating in a delta should create a distinct morphology. Here, we take a simple view and describe the delta morphology if a single process was dominant. For instance, deltas dominated by mouth bar construction are semicircular because the mouth bars create a well-developed distributary network that delivers sediment to a broad area. The shorelines are rough due to localized progradation at river mouths.[32] The channel networks are fractal in character because the channels reduce in size with successive bifurcations.[33,34] Deltas dominated by levee construction are usually elongate in shape, with smooth shorelines, and few channels.[32,35,36] This occurs because channels with well-developed levees inhibit overbank flow, and focus sedimentation at the river mouth creating localized progradation and an elongate form. Deltas dominated by avulsion have semicircular shapes due to frequent channel movement around the delta, but their shorelines have intermediate roughness, and single- to multi-channel networks. These end-member morphologies may be infrequently encountered in nature and in reality delta morphology depends on the relative influence of these processes and how they interact.

Predicting the morphology of a delta requires knowing the upstream and downstream boundary conditions, because those conditions determine the relative influence of each delta building process. Upstream boundary conditions discussed here include sediment size and total sediment input. Deltas constructed from coarse-grained sediments have steeper topset slopes and the additional shear stress creates channels that are more mobile because the banks are easily eroded. This creates channels that frequently avulse and in turn should limit mouth bar formation since channel mouths rarely stay in one position long. Morphologically, the result is a semicircular delta with a smooth shoreline. Deltas constructed from intermediate grain sizes have channels that are more stable and thus able to produce mouth bars. Mouth bar construction creates semicircular deltas with rougher shorelines. Deltas constructed from muddy sediments, on the other hand, tend to be elongate since mud is easily transported to the margins of a turbulent jet and preferentially deposited in levees.[32] An increase in sediment flux will increase the delta slope. This can trigger braiding in the distributary channels.[37] Higher sediment fluxes will create aggradation and frequent avulsions on the delta resulting in a semicircular planform with a smooth shoreline.[38] Additionally, bed load dominated systems should preferentially construct mouth bars,[15] and suspended load dominated systems should build levees since that material is more easily advected to the turbulent jet margins.[36]

Downstream boundary conditions that influence delta building processes include buoyancy forces, offshore bathymetry, waves, tides, and relative sea-level variations. Buoyancy forces and deep offshore bathymetry cause turbulent jets to detach from the bed, which prevents them from spreading as rapidly. These narrow turbulent jets, in turn, preferentially produce levees resulting in elongate delta shapes.[16,36] In general, strong wave energy creates smooth deltaic shorelines by redistributing sediment laterally via alongshore transport, and suppressing mouth bar formation (Fig. 1E).[39] The lateral distribution creates morphological features, such as the coastal spits of the Krishna delta (Fig. 1C). Additionally, waves have the effect of decreasing avulsion frequency because they decrease progradation rates and limit aggradation and superelevation potential.[40] On the contrary, certain kinds of waves can actually enhance delta growth[7] and influence mouth bar formation patterns[17] due to varying degrees of wave amplitude and angle to the shoreline. Tides, on the other hand, cause the remobilization of sediment, eroding mouth bars and maintaining abandoned distributary channels which results in delta morphologies characterized by dendritic channel networks with a large number of distributaries that widen basinward (Fig. 1D).[41,42] Tidal effects on bifurcation morphology and stability are not well constrained; in some cases, tides increase discharge asymmetry between bifurcate channels and in other cases they minimize it.[43,44] Little is also known about effects of sea-level rise on deltaic processes. Sea-level rise is hypothesized to limit mouth bar formation and increase avulsion frequency, since the rising elevation of the downstream boundary will drown mouth bars and force aggradation (and thus avulsion) upstream,[45] but this has never been tested.

WHAT IS THE FUTURE OF DELTAIC SYSTEMS?

The future of deltaic systems is bleak because global environmental changes have tipped conditions toward delta destruction, rather than creation. This is not altogether surprising; deltas occupy the shoreline where changes in upstream and downstream boundary conditions converge. For example, the Mississippi delta feels the combined effects of dam building and other engineering structures that have reduced the sediment flux of the Mississippi River by 50% compared to prehuman levels.[46] Alone this would be problematic, but it is further compounded by human withdrawal of subsurface natural resources (e.g., water and hydrocarbons) that has created accelerated sediment compaction, and by sea-level rise in the Gulf of Mexico at rates of 2 mm/year.[47] These observations suggest that there is not enough sediment reaching the Mississippi delta to offset these changes, and the delta will inevitably drown in the next 100 years.[47,48] Unfortunately, similar estimates have been made for the many of world's largest deltas, and the reality is that if nothing is done deltas could drown.[4]

In light of this, there is a broad call for developing restoration schemes that will save deltaic landforms.[11,49] Local restoration schemes cannot slow or reverse global trends of sea-level rise, but they can reverse some of the negative impacts of humans and restore the processes that effectively created the original delta. To start, this requires a predictive understanding of delta processes and shapes so that restoration schemes can be tuned for success. It is our contention that we know enough about delta processes to make scientifically defensible restoration decisions. Evidence of this includes recent advances in simulating delta growth (Fig. 5).[26,32,50] These tools simulate the complex interactions between delta processes and changing boundary conditions, and tests show that delta growth and shapes predicted by these methods are statistically similar to field-scale deltas.[33,34]

There are three promising areas of future research on deltaic systems. First, there are individual processes and boundary conditions left unexplored. For instance, there is little empirical data or theory that predicts how vegetation growth alters delta form. Second, we have only a minimal understanding of how delta processes interact to produce their morphology. We presented simple models for the effects of delta processes (i.e., mouth bar growth, avulsion, and levee growth) that operate in isolation. But, in most cases, many processes are interacting, possibly in nonlinear

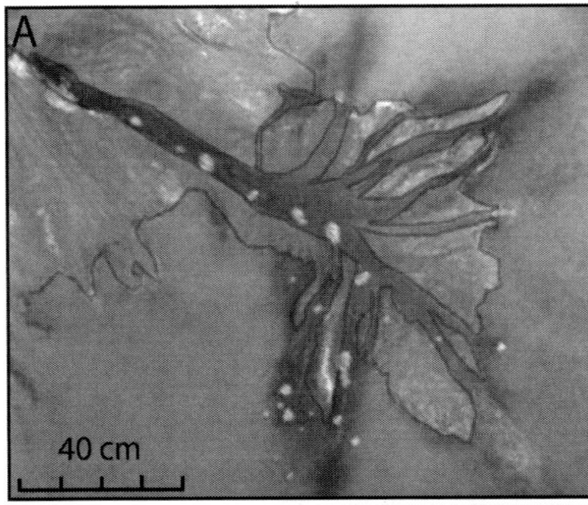

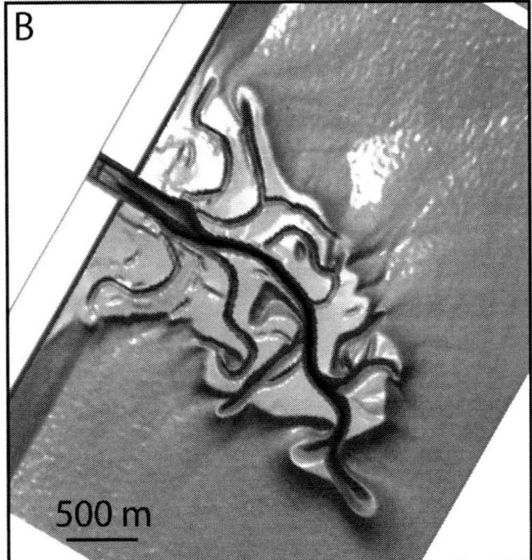

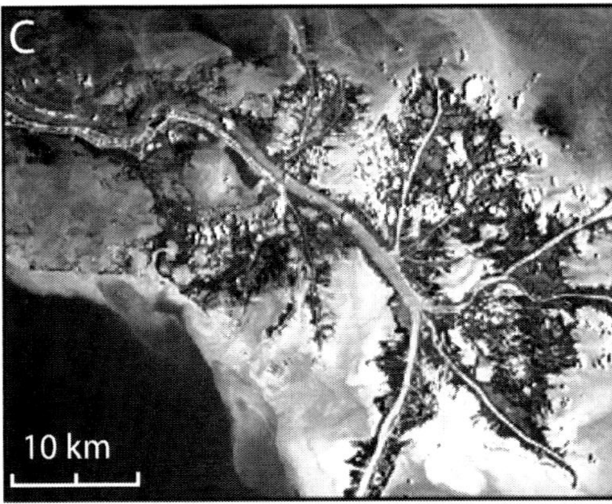

Fig. 5 Experimental (**A**) and numerical models (**B**) realistically simulate the basic morphology and shape of the Mississippi River delta (**C**). In the experimental delta, the water is dyed so that channelized flow is visible, the white dots on the surface are foam, and the gray line is a trace of the shoreline. For more details on experimental setup see Edmonds et al.[26] The numerical delta depicts bed elevation as shaded relief. For more details on numerical setup see Edmonds and Slingerland.[32] Image of the Mississippi delta is 2001 Advanced spaceborne thermal emission and reflection radiometer (ASTER) data (courtesy of U.S. Geological Survey National Center for Earth Resources Observation and Science, and National Aeronautics and Space Administration Landsat Project Science Office).

feedback loops. Experiments designed to understand these complex interactions and feedback loops will be important breakthroughs. Third, we must understand how deltas will respond to changing boundary conditions. Numerical and experimental models are valuable tools in this regard because they can simulate different scenarios of changing sea-level, coastal subsidence, and sediment flux.

REFERENCES

1. Day Jr, J.W.; Gunn, J.D.; Folan, W.J.; Yáñez-Arancibia, A.; Horton, B.P. Emergence of complex societies after sea level stabilized. Eos, Trans. Am. Geophys. Union **2007**, *88* (15), 169.
2. Ericson, J.P.; Vorosmarty, C.J.; Dingman, S.L.; Ward, L.G.; Meybeck, M. Effective sea-level rise and deltas: Causes of change and human dimension implications. Global Planet. Change **2006**, *50*, 63–82.
3. Syvitski, J.; Saito, Y. Morphodynamics of deltas under the influence of humans. Global Planet. Change **2007**, *57*, 261–282.
4. Syvitski, J.P.M.; Kettner, A.J.; Overeem, I.; Hutton, E.W.H.; Hannon, M.T.; Brakenridge, G.R.; Day, J.; Vorosmarty, C.; Saito, Y.; Giosan, L.; Nicholls, R.J. Sinking deltas due to human activities. Nat. Geosci **2009**, *2* (10), 681–686.
5. Rowland, J.C.; Dietrich, W.E.; Day, G.; Parker, G. Formation and maintenance of single-thread tie channels entering floodplain lakes: Observations from three diverse river systems. J. Geophys. Res.-Earth Surf. **2009**, *114*.
6. Orton, G.J.; Reading, H.G. Variability of deltaic processes in terms of sediment supply, with particular emphasis on grain size. Sedimentology **1993**, *40*, 475–512.
7. Ashton, A.D.; Giosan, L. Wave-angle control of delta evolution. Geophys. Res. Lett. **2011**, *38* (13), L13405.
8. Cahoon, D.R. A review of major storm impacts on coastal wetland elevations. Estuar. Coast. **2006**, *29* (6), 889–898.
9. Lorenzo-Trueba, J.; Voller, V.R.; Paola, C.; Twilley, R.R.; Bevington, A.E. Exploring the role of organic matter accumulation on delta evolution. J. Geophys. Res. **2011**, *117*, F00A02.
10. Wilson, C.A.; Allison, M.A. An equilibrium profile model for retreating marsh shorelines in southeast Louisiana. Estuar. Coast. Shelf Sci. **2008**, *80* (4), 483–494.
11. Paola, C.; Twilley, R.; Edmonds, D.; Kim, W.; Mohrig, D.; Parker, G.; Viparelli, E.; Voller, V. Natural processes in delta restoration: Application to the Mississippi delta. Annu. Rev. Mar. Sci **2011**, *3*, 3.1–3.25.
12. Stanley, D.J.; Warne, A.G. Worldwide initiation of holocene marine deltas by deceleration of sea-level rise. Science **1994**, *265* (5169), 228–231.
13. Rowland, J.; Stacey, M.; Dietrich, W. Turbulent characteristics of a shallow wall-bounded plane jet: Experimental implications for river mouth hydrodynamics. J. Fluid Mech. **2009**, *627*, 423–449.
14. Rowland, J.C.; Dietrich, W.E.; Stacey, M.T. Morphodynamics of subaqueous levee formation: Insights into river mouth morphologies arising from experiments. J. Geophys. Res. - Earth Surf. **2010**, *115* (F4), F04007.
15. Edmonds, D.A.; Slingerland, R.L. Mechanics of river mouth bar formation: Implications for the morphodynamics of delta distributary networks. J. Geophys. Res. **2007**, *112* (F2).

16. Wright, L.D. Sediment transport and deposition at river mouths: A synthesis. Geol. Soc. Am. Bull. **1977**, *88*, 857–868.
17. Nardin, W.; Fagherazzi, S. The effect of wind waves on the development of river mouth bars. Geophys. Res. Lett. **2012**, *39* (12), L12607.
18. Elliott, T. Interdistributary bay sequences and their genesis. Sedimentology **1974**, *21* (4), 611–622.
19. Tamura, T.; Saito, Y.; Nguyen, V.L.; Ta, T.K.O.; Bateman, M.D.; Matsumoto, D.; Yamashita, S. Origin and evolution of interdistributary delta plains; insights from Mekong River delta. Geology **2012**, *40* (4), 303–306.
20. Slingerland, R.; Smith, N.D. River avulsions and their deposits. Annu Rev. Earth Planet. Sci. **2004**, *32*, 257–285.
21. Hajek, E.A.; Wolinsky, M.A. Simplified process modeling of river avulsion and alluvial architecture: Connecting models and field data. Sediment. Geol. **2012**, *257*, 1–30.
22. Reitz, M.; Jerolmack, D.; Swenson, J. Flooding and flow path selection on alluvial fans and deltas. Geophys. Res. Lett. **2010**, *37* (6), L06401.
23. Reitz, M.D.; Jerolmack, D.J. Experimental alluvial fan evolution: Channel dynamics, slope controls, and shoreline growth. J. Geophys. Res. **2012**, *117* (F2), F02021.
24. Jerolmack, D.J.; Swenson, J. Scaling relationships and evolution of distributary networks on wave-influenced deltas. Geophys. Res. Lett. **2007**, *34*.
25. Chatanantavet, P.; Lamb, M.P.; Nittrouer, J.A. Backwater controls of avulsion location on deltas. Geophys. Res. Lett. **2012**, *39* (1), L01402.
26. Edmonds, D.A.; Hoyal, D.C.J.D.; Sheets, B.A.; Slingerland, R.L. Predicting delta avulsions: Implications for coastal wetland restoration. Geology **2009**, *37* (8), 759–762.
27. Edmonds, D.; Slingerland, R.; Best, J.; Parsons, D.; Smith, N. Response of river-dominated delta channel networks to permanent changes in river discharge. Geophys. Res. Lett. **2010**, *37* (12), L12404.
28. Edmonds, D.A.; Slingerland, R.L. Stability of delta distributary networks and their bifurcations. Water Resour. Res. **2008**, *44* (W09426).
29. Bolla Pittaluga, M.; Repetto, R.; Tubino, M. Channel bifurcations in braided rivers: Equilibrium configurations and stability. Water Resour. Res. **2003**, *39*, 10–46.
30. Miori, S.; Repetto, R.; Tubino, M. A one-dimensional model of bifurcations in gravel bed channels with erodible banks. Water Resour. Res. **2006**, 42, W11413.
31. Jerolmack, D.J.; Mohrig, D. Conditions for branching in depositional rivers. Geology **2007**, *35* (5), 463–466.
32. Edmonds, D.A.; Slingerland, R.L. Significant effect of sediment cohesion on delta morphology. Nat. Geosci. **2010**, *3* (2), 105–109.
33. Edmonds, D.A.; Paola, C.; Hoyal, D.C.J.D.; Sheets, B.A. Quantitative metrics that describe river deltas and their channel networks. J. Geophys. Res. **2011**, *116* (F4), F04022.
34. Wolinsky, M.; Edmonds, D.A.; Martin, J.M.; Paola, C. Delta allometry: Growth laws for river deltas. Geophys. Res. Lett. **2010**, *37* (21), L21403.
35. Kim, W.; Dai, A.; Muto, T.; Parker, G. Delta progradation driven by an advancing sediment source: Coupled theory and experiment describing the evolution of elongated deltas. Water Resour. Res. **2009**, *45*.
36. Falcini, F.; Jerolmack, D. A potential vorticity theory for the formation of elongate channels in river deltas and lakes. J. Geophys. Res. - Earth Surf. **2010**, *115*.
37. McPherson, J.G.; Shanmugam, G.; Moiola, R.J. Fan-deltas and braid deltas: Varieties of coarse-grained deltas. Geol. Soc. Am. Bull. **1987**, *99* (1), 331–340.
38. Sun, T.; Paola, C.; Parker, G.; Meakin, P. Fluvial fan deltas: Linking channel processes with large-scale morphodynamics. Water Resour. Res. **2002**, *38* (8), 1151.
39. Galloway, W.E. *Process Framework for Describing the Morphologic and Stratigraphic Evolution of Deltaic Depositional Systems*; Deltas: Models for Exploration. M.L. Broussard. Houston, TX, Houston Geological Society: 1975; 87–98.
40. Swenson, J.B. Relative importance of fluvial input and wave energy in controlling the timescale for distributary-channel avulsion. Geophys. Res. Lett. **2005**, *32*.
41. Dalrymple, R.W.; Choi, K. Morphologic and facies trends through the fluvial–marine transition in tide-dominated depositional systems: A schematic framework for environmental and sequence-stratigraphic interpretation. Earth-Sci. Rev. **2007**, *81* (3), 135–174.
42. Fagherazzi, S. Self-organization of tidal deltas. Proc. Nat. Acad. Sci. **2008**, *105* (48), 18692.
43. Buschman, F.A.; Hoitink, A.J.F.; van der Vegt, M.; Hoekstra, P. Subtidal flow division at a shallow tidal junction. Water Resour. Res. **2010**, *46* (12), W12515.
44. Sassi, M.G.; Hoitink, A.J.F.; de Brye, B.; Vermeulen, B.; Deleersnijder, E. Tidal impact on the division of river discharge over distributary channels in the Mahakam Delta. Ocean Dyn. **2011**, *61* (12), 2211–2228.
45. Jerolmack, D.J. Conceptual framework for assessing the response of delta channel networks to Holocene sea level rise. Quarter. Sci. Rev. **2009**, *28* (17–18), 1786–1800.
46. Meade, R.H.; Moody, J.A. Causes for the decline of suspended-sediment discharge in the Mississippi River system, 1940–2007. Hydrol. Process. **2010**, *24* (1), 35–49.
47. Blum, M.D.; Roberts, H.H. The Mississippi Delta region: Past, present, and future. Annu. Rev. Earth Planet. Sci. **2012**, *40*, 655–683.
48. Blum, M.D.; Roberts, H.H. Drowning of the Mississippi Delta due to insufficient sediment supply and global sea-level rise. Nat. Geosci. **2009**, *2* (7), 488–491.
49. Edmonds, D.A. Restoration sedimentology. Nat. Geosci. **2012**, *5* (11), 758–759.
50. Hoyal, D.; Sheets, B.A. Morphodynamic evolution of experimental cohesive deltas. J. Geophys. Res.-Earth Surf. **2009**, *114*.

Savannas and Grasslands

Niti B. Mishra
Kenneth R. Young
University of Texas at Austin, Austin, Texas, U.S.A.

Abstract

This entry discusses the ecological features of the extensive and important ecosystems of savannas and grasslands, which provide habitat to important species, regulate an important part of the Earth's nutrient cycles, and serve many purposes for people. The savannas are mostly tropical ecosystems wherein dominant woody and herbaceous plants coexist. In turn, grasslands are found in both tropical and temperate climates. Place-to-place changes in vegetation structure and land cover are caused by the interactions of plant available moisture, plant available nutrient, herbivore, and fire. Brief descriptions of these ecosystems are followed by a review of the significant ecological determinants of the structural and functional characteristics in savanna and grassland systems. There are a number of conceptual models that attempt to explain features of these ecosystems, including the unique coexistence of tree and grass life forms in savannas. Current threats to the sustainability of savanna and grassland systems worldwide arise as socio-economic processes interact with global environmental changes.

INTRODUCTION

Savannas and grasslands are important ecosystems spread over more than 20% of the Earth's terrestrial surface. While savannas are mostly tropical ecosystems that feature the unique co-existence of dominant woody and herbaceous life forms, grasslands are found in both tropical and temperate climates and are primarily dominated by grasses (Fig. 1).[1] The two biomes are distinct but share some similar ecological processes and biotic assemblages. Due to their large geographical extent, these ecosystems contribute to more than 30% of global net primary productivity and also significantly influence the global land-atmosphere energy balance and carbon and nutrient cycles.[2,3] Savannas and grasslands support most of the world's livestock; pasturelands and protected wildlife areas are the basis of socio-economic activities such as pastoralism, tourism, and conservation.[4,5] With a high variation in the ratio of woody to herbaceous life forms, the resulting spatial and structural heterogeneity of land cover is a defining characteristic of savanna systems. Due to their nutrient rich soil, temperate grasslands are productive, especially in North America where this biome houses a major crop and cereal-producing area. Changes in savanna and grassland occurrence and vegetation structure are primarily the result of the interaction of four factors: plant available moisture (PAM), plant available nutrient (PAN), grazing pressure, and fire. The complex interaction of these factors produces variations in vegetation structure at both local and regional scales. This high variation has made it difficult for studies to provide an overarching and simple definition of savannas; it furthermore has led many ecologists to consider savannas as ecologically unstable mixtures of woody and herbaceous life forms.

STRUCTURAL PROPERTIES AND GEOGRAPHIC DISTRIBUTION

Savannas have often been described in the ecological literature as landscapes with continuous grass cover and scattered trees/shrubs. However, the definition of the term "savanna" is debated given that the tree-grass ratio within these systems varies greatly from nearly treeless grasslands to areas with up to 80% woody canopy cover.[6] The woody cover component in savannas consists of trees/shrubs of evergreen or deciduous phenology and may include broad- or fine-leafed species. The herbaceous component can be either perennial or annual plant types of varying height. Since landscapes in various ecosystems fit this vegetation structural criterion, savannas in different parts of the world have been called rangelands, open forest, parkland, or simply grassland. The variation of the tree to grass ratio and its interpretation have also impacted the estimation of total area under savanna ecosystems. For example, while Atjay (1979) estimated it to be 22.5 million km^2, Alson (1983) found it to be 33.7 million km^2.

Based on climatic setting and disturbance regime, the world's savannas can be grouped into four broad categories:

- *Tropical and subtropical savannas* cover an eighth of the global terrestrial area (~1600 million ha) and represent the majority of savanna located in tropical and subtropical latitudes of Africa, Australia, South America,

Encyclopedia of Natural Resources DOI: 10.1081/E-ENRL-120047446

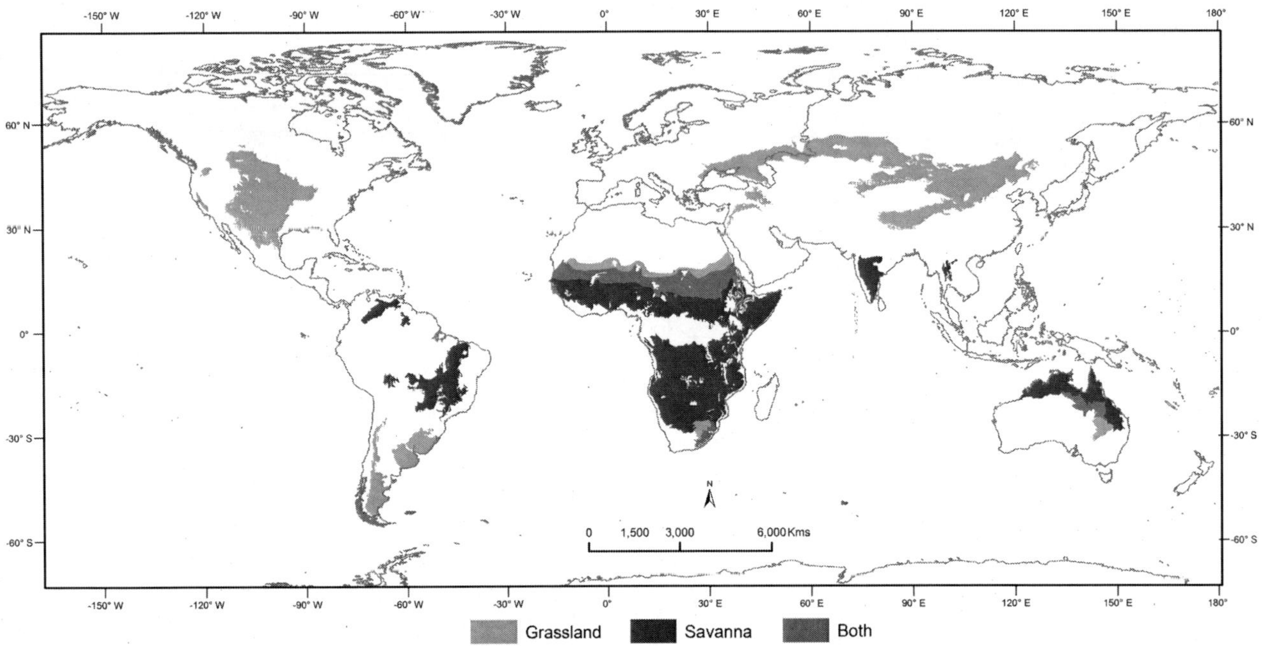

Fig. 1 World-wide distribution of natural savannas and grasslands.

India, and South East Asia (Fig. 1). African tropical savannas are scattered across a wide variety of soil conditions with annual rainfall ranging from 200 to 1800 mm; they form a semi-circle around the western central rainforest, bordered by deserts to the north and south.[7–9] In African savanna, fine-leaved woody vegetation mainly dominated by *Acacia* species occurs on nutrient-rich soils in low-lying semi-arid areas in Botswana, South Africa, and Namibia. On the contrary, broad-leaved savanna species (e.g., *Combretum* spp., *Lonchocarpus nelsii*, and *Terminalia* spp.) can be seen in the sub-humid interior plateau on infertile and highly weathered soils.[10] The Guinea-type savanna woodlands in humid climates replace moist semi-deciduous vegetation and form a transition to evergreen moist forests, whereas northern Sudan-type savannas with scattered *Acacia* trees form a transition with Sahara desert vegetation.[10,11] The Miombo woodland savanna covers an additional 2.7 million km^2 area across central and southern Africa and is characterized by the presence of deciduous species of *Brachystegia*, *Isoberlinia*, and *Fulbbernardia*, forming a discontinuous canopy with an average height of 10 m.[12] In South America, tropical savanna comprises the biologically rich Brazilian *Cerrado* and the Colombian and Venezuelan *Llanos*, together covering over 2 million km^2 in area. Rich in fauna, the *Cerrado* bioregion contains a gradient from completely open savannas to closed canopy forest along environmental shifts in rainfall, topography, and human influence. Besides producing most of the beef cattle of Brazil, with the implementation of soil improvement techniques and irrigation, *Cerrado* areas are now also important producers of soy and grains.[3,13] The Australian tropical savanna is contiguous with little topographic variation and covers about 12% of the total tropical savanna area.[14] In northern Australia, savanna open *Eucalyptus* spp. woodlands is the dominant land cover type. The woody canopy cover decreases from northern Australia toward the continent's interior, following the rainfall gradient.[15,16] In general, the woody vegetation in Australian tropical savannas has predominantly evergreen tree species that is different from African tropical savannas with mostly deciduous tree species.[17] Also, compared to the tropical savannas in other continents, Australian tropical savannas have lower population density and thus are ecologically less modified. However, due to increasing economic activities, particularly mining and agriculture, their future remains uncertain.[15,18]

- *Temperate grasslands* are known as prairies in North America (covering 5 million km^2), steppes in Eurasia, pampas in Argentina and Uruguay in South America, and the veld region in South Africa. These mid-latitude grasslands experience wet summers and dry winters; the climate can range from temperate to semi-arid to sub-humid. Compared to tropical savannas, temperate grasslands have lower biodiversity. Typical kinds of grasses include *Bouteloua dactyloides* (buffalo grass), *Lolium* (ryegrass), *Avena* (wild oats), and *Nassella pulchra* (purple needlegrass), which also are all part of the diet of herbivorous animals. Grass height depends on the amount of rainfall received from late spring to early summer, which ranges in total between 450 and 700 mm.[9] Often lightning from thunderstorms can ignite large grass fires in temperate grasslands, which keeps trees from overshadowing the grasses and acts to maximize biodiversity by creating landscape heterogeneity. Soils are deep and dark with fertile upper layers.

A longitudinal rainfall gradient can be used to help distinguish between tall- and short-grass grasslands in North America. The fauna in the temperate grasslands varies from herbivores such as bison, pronghorn, and deer (in North America), the Asian camel and various species of horses and saiga antelope (in Eurasia).

- Tropical *flooded savannas* are characterized by strong seasonality, that is, a dry period which causes severe water shortage and a wet period during which extensive flooding occurs.[19] Examples of flooded savannas include Zambezian flooded savanna of Mozambique, Okavango Delta, Botswana, and the Pantanal flooded savannas of Bolivia, Brazil, Paraguay, and some of the *Llanos* of Venezuela.[20–22]
- *Montane savannas* are high-altitude savannas, often of human origin due to burning, and include areas such as Angolan scarp savanna, Campos Rupestres montane savanna (Southeastern Brazil), degraded montane savanna of Ethiopia, and woodlands and high-altitude grasslands (e.g., páramo) of the Andes mountains.[9] At present, montane savannas are the most widespread vegetation type on the mountain chains running through Africa from Lesotho and South Africa to eastern Sudan. They are also abundant in India, Nepal (Terai), and Sri Lanka.

DETERMINANTS OF VEGETATION STRUCTURE

Findings from various studies suggest that the ecosystem structure and function in savannas and grasslands can be understood as the result of the complex interplay of four main ecological factors: PAM, PAN, fire, and herbivory (Fig. 2).

Plant Available Moisture/Nutrients

At regional to continental scales, PAM is the most important determinant of ecosystem structure and function in both savannas and grasslands. PAM can be quantified in terms of mean annual precipitation, number of days during which rainfall is greater than evapotranspiration, or the number of days during which water is available in the soil. In general, tree cover in savanna ecosystems increases with increasing rainfall; there is a corresponding decreasing in herbaceous cover (Fig. 3).[10,23] Like savannas, grasslands are water-limited ecosystems, and their locations are associated with places having dry periods that occur at least once a year. Grasslands are found in areas with mean annual precipitation ranging between 150 and 1200 mm. As temperature increases, the forest-grassland and desert-grassland boundaries occur at a higher amount of precipitation due to the increasing evaporation rate. For example, in North America where temperature isopleths run in an east-west direction and isohyets in a north-south direction, the grassland-forest boundary has a strong SE-NW orientation.[24] Unlike PAM, PAN acts as

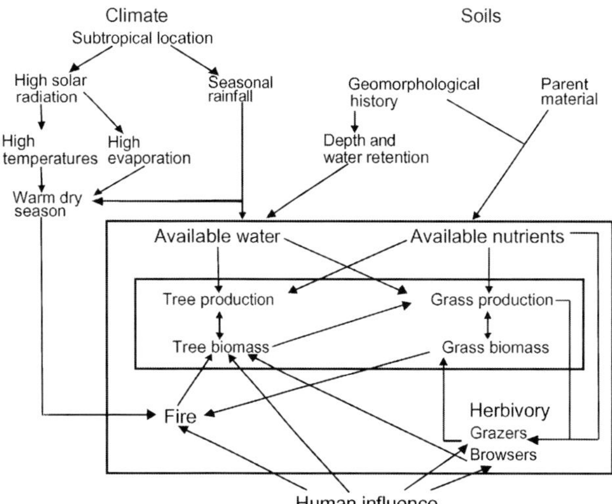

Fig. 2 Interaction between four important ecological determinants of savanna structure (i.e., PAN, PAM, fire, and herbivore). This interaction is realized at any given savanna or grassland location by climatic condition and soil type.
Source: Reproduced from House et al.,[42] with permission from Blackwell Publishing.

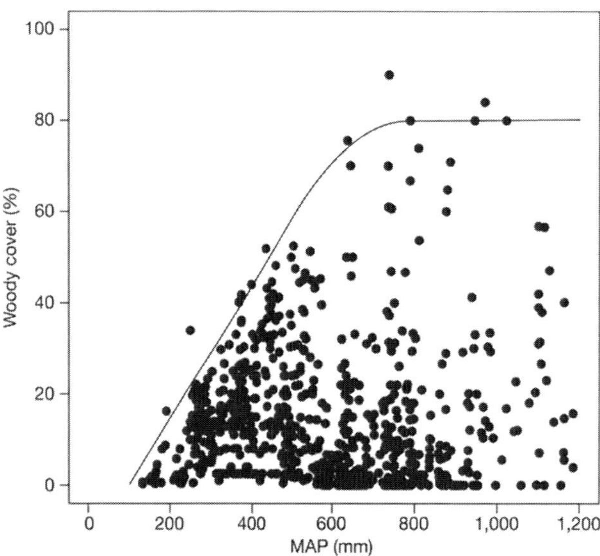

Fig. 3 Woody cover in African savannas as a function of mean annual precipitation (MAP). Data from 854 sites across Africa suggest that below 650 mm both herbaceous and woody life form coexist and above 650 mm MAP values woodlands can dominate. Below 101 mm MAP value trees are generally absent.
Source: Reproduced from Sankaran et al.,[10] with permission from Nature Publishing Group.

a determinant of ecosystem structure in savannas and grasslands at a much finer spatial scale. This is because nutrient availability is a function of not only the quality of soil parent material, but also its interaction with PAM that facilitates the release of nutrients via mineralization.[1] Plant growth in savannas takes place during high periods of PAM when nutrients are mineralized. In turn, this creates spatial changes in landscape mosaics and across topographic/edaphic gradients. The critical interaction of PAN and PAM has often been used to classify and compare savanna systems. Fine-scale changes in soil properties from site to site in a given landscape can thus result in different vegetation physiognomic patterns even within the same PAM regime.[7,15]

Fire

Fire is an important additional landscape scale determinant of the composition, ecology, and distribution of the savannas and grassland ecosystems; it controls the abundance of woody life forms below a precipitation determined boundary.[10] Savanna fires are predominantly surface fires that burn dry grasses and shrubs. These fires can spread rapidly due to the hot, windy, low-humidity conditions of the dry season. Early dry season fires are generally less intense than late dry season fires. The patch size of burned area can range from less than a hectare to several hundred square kilometers, creating a relatively fine-grained disturbance mosaic.[6] Since grasses and shrubs can grow rapidly post-fire, the same region can burn every year. Fire return intervals in savannas are commonly on the order of every one to three years and are further influenced by climatic factors and land-use practices.[25] With exceptions of sparsely populated areas where fire caused by lightening predominates, today most savanna fires are human-caused for reasons such as land clearing for farming, improving pasture for grazing, providing nutrients to crops, making fire breaks around human settlements, and fires due to accidents.[6,26,27]

In both savanna and grassland ecosystems, fire serves as a key ecological buffering mechanism that maintains woody plant densities.[28] The results of studies based on long-term fire exclusion plots in southern Africa and Australian savanna have shown that decreased fire frequencies can lead to bush encroachment, while increased fire frequencies can lead to conversion to grassland (Fig. 4).[29–31] Changing fire frequency has shifted the balance between savanna and forest vegetation cover over much of the seasonal tropics and recent studies also are attempting to understand how plant traits (e.g., savanna versus forest trees) determine the dynamics of savanna-forest boundary.[32,33] Besides fire frequency, the timing of fire in relation to the reproductive phenology can also limit or support plant reproduction, resulting in altered species composition in savanna and grassland vegetation communities.[34] In addition, fire activity in these ecosystems has also been found to directly affect bird and insect species richness and their behavioral responses. For example, studies in the northern territory of Australia found that grass removal by fire allowed greater access to fallen grass seed, resulting in higher numbers of granivores attracted to the burned areas. Insectivorous birds showed similar patterns, while the species preferring tall dense grass or understory vegetation were least abundant after fire events, which affect the spatial extent of recovering vegetation.[35]

Herbivores

The importance of herbivore as an ecological factor varies among savanna and grassland regions. The two main kinds of herbivores influencing vegetation structure in these ecosystems are large native mammals (often ungulates) and the comparatively overlooked but important invertebrates such as termites, ants, and grasshoppers. In African savannas, mammals, both wild and domesticated, exert an unparalleled control on the ecosystem structure by acting as grazers and browsers consuming vegetation, by

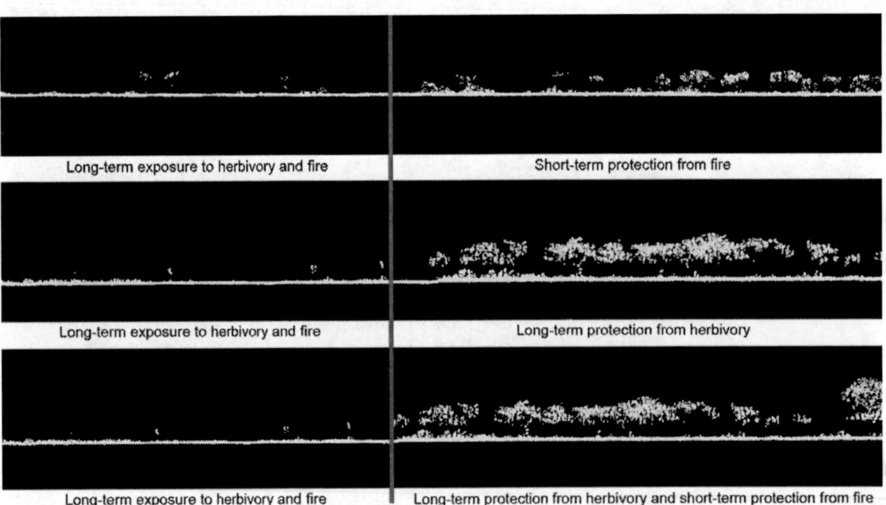

Fig. 4 Cross-profile comparisons of the herbivore and fire experimental treatments and the control landscape. Profiles display the light detection and ranging point cloud of both ground surface and vegetation canopy returns. The solid vertical line marks the boundary between control and treatment sites.
Source: Reproduced from Levick et al.,[31] with permission from Elsevier Publishing.

trampling soils and by destroying trees and shrubs.[36,37] Large herbivores can change vegetation structure and alter species composition in savanna ecosystems (Fig. 4). For example, in the African Serengeti system, after the eradication of rinderpest in the 1960s, the wildebeest population continued to increase until the 1980s, which transformed it in part into a woodland-dominated state due to the increased recruitment of sapling and adult trees.[38] There is an additional set of interactions with predators that control the behavior of some of the herbivores. Savanna structure and function are undoubtedly the result of all these interactive effects of herbivores and fire. More recent studies examining the impact of herbivore in southern African savanna found that areas under short-term (6 years) exclusion of herbivore contained 38–80% less bare ground compared to those sites that were exposed to herbivores. Additionally, areas under long-term exclusion from herbivores (>22 years) differed significantly in terms of their aboveground vegetation structure with up to 11-fold greater woody canopy cover in areas without herbivores.[39] Besides grazing, increased livestock pressure in farmlands in African and Australian savannas also impacted soil characteristics, resulting in losses of sensitive crust, soil compaction, erosion, and a related loss of nutrients.[40,41]

Although much smaller individually than the large and conspicuous mammalian grazers and browsers, insects are important herbivores in savanna systems, particularly in areas of low soil fertility. The key role of termites as ecosystem engineers in savannas is well established as they alter the mineral and organic composition of soil, and affect their hydrology, drainage, and infiltration rates as well as influencing decomposition and nutrient cycling.[1,42] Grasshoppers are important grazing insects in the majority of tropical savannas. In Nylsvley, South Africa, an average grasshopper biomass of 2.3 kg ha^{-1} was estimated to consume 130 kg ha^{-1} yr^{-1} biomass in *Burkea* savanna and 406 kg ha^{-1} yr^{-1} in *Acacia* savanna, values that represented 40% and 73% of the total herbivore consumption, respectively.[43] While there are site-specific studies highlighting the role of insect herbivores in savannas, in general there is a lack of landscape and regional scale studies describing their abundance and role in these ecosystems.

The spatiotemporal interaction and dynamics of the four determinants of savannas discussed above (i.e., PAM, PAN, fire, and herbivore) have traditionally been used to explain the ecosystem structure and function in savannas as well as the much-debated issue of woody plant encroachment. However, more recent studies based on results from experimental plots (with manually controlled disturbance regimes) and outputs from deterministic models of savanna dynamics have emphasized the role of increased atmospheric CO_2 concentration that could be responsible for increasing the woody cover in savannas.[44,45] Recent studies argue that increasing atmospheric CO_2 concentration can influence the growth rate of juvenile tree saplings that can grow faster to reach fire-resistant height. The resulting increased tree recruitment is causing the conversion of savannas into woodlands.[44,46,47]

PARADIGMS/MODELS OF TREE–GRASS INTERACTION IN SAVANNAS

The coexistence of woody and herbaceous life forms in the same ecosystem type has always puzzled ecologists. Over the last few decades of savanna ecology research, several conceptual models have been proposed to explain how trees and grasses can coexist in savanna systems. These models can broadly be categorized as either competition based or demographic based. Savanna structure, under competition-based models, is expected to be the result of spatial and temporal niche separation in resource (i.e., water and nutrients) acquisition; in these cases, fire and herbivore effects are only considered as modifiers.[48–50] Mosaic patterns of soils and of topography would lead to complicated vegetation patterns. Alternatively, demographic-based models assume that tree-grass coexistence occurs in savannas because climatic variability and disturbances act as bottlenecks to limit successful tree establishment. Proposed by Walter (1971), the *root niche separation model* considers water as the primary limiting factor for which trees and grasses compete in savannas. While grasses are rooted only in topsoil, trees have roots in both topsoil and subsoil with exclusive access to deeper water; this separation in theory could lead to the stable coexistence of the two. The *phenological niche separation model* proposes the wet season as a potential axis for niche separation between savanna trees and grasses.[1,42] Deciduous savanna trees achieve full leaf expansion early in the start of rainy season, whereas peak leaf areas of grasses are achieved only after several months.[7] In this scenario, trees have exclusive access to resources during both the early and the late rainy season while grasses are more competitive during the proper rainy season. Under a *balanced competition model*, coexistence in savannas arises because trees, being the superior competitor for light and for soil resources, become self-limiting at a biomass insufficient to exclude grasses, which are inferior competitors.[1] In the absence of external perturbations, this model predicts a threshold with increasing rainfall above which woody plants dominate and below which grasses replace trees. The *hydrologically driven competition-colonization model* emphasizes the tradeoffs between the competitive ability and the colonization potential of trees and grasses, which changes in response to fluctuations in soil water stress caused by inter-annual rainfall variability, and how that further changes in response to fluctuations in rainfall. Model predictions on the balance between trees and grasses are sensitive to the magnitude and variance of those inter-annual fluctuations.[51]

Demographic bottleneck models integrate the effects of multiple drivers (e.g., fire, herbivore, and climatic fluctuations) and rather than emphasizing the post-disturbance competitive interaction, they highlight the direct effect of these disturbances on germination, mortality, and the demographic transition of trees in savannas.[52–56] There exist alternative viewpoints among researchers using demographic bottleneck models in regards the degree of control of savanna structure and functioning. The transitional "disequilibrium" system approach, for example, considers disturbances such as fire and grazing as not only modifiers but also maintainers of the unique savanna structure by buffering the system against shifts to alternate states.[52] Alternatively, another viewpoint is that the non-equilibrium dynamics are driven by geographic variations in rainfall in arid regions,[56] while disequilibrium dynamics are driven by variations in fire intensity in more mesic areas.[56] In arid and semi-arid savannas, due to spatiotemporal variability in rainfall, tree recruitment is pulsed in time with occasional heavy rainfall events, provided that the mature trees live long enough to bridge these events. However in mesic savannas, where trees can dominate canopy cover, frequent high-intensity fires prevent the seedlings from escaping the flames and thereby control tree density.[56] Further, browsing by herbivores may prevent the seedlings from establishing as mature trees.[28,52,56]

Recently, demographic models have gained favor over competition-based models, which were found insufficient in explaining long-term tree-grass coexistence in the spatially explicit modeling of interactions in savannas using field data.[28] Although the different models discussed here have been supported by empirical evidence from different sites around the Earth, no single model has been able to provide a generic mechanism that provides a single explanation for all cases. Furthermore, observational data collected during various studies in savanna ecosystems suggest that the savanna structure and function results from the interaction of all the processes discussed above, implying that holistic explanations will be most robust. More recent studies examining the correlation among selected environmental variables and global savanna distribution emphasize that climatic variables do not necessarily have a deterministic effect on savanna distribution as divergences from the limits of savanna exit globally that cannot be explained by environmental determinants.[57,58] Studies using dynamic global vegetation models to resolve savanna distribution in different continents emphasize that although climatic determinants are important, extensive savannas in some high rainfall areas suggest a decoupling of climate and vegetation.[60]

BIODIVERSITY AND PRODUCTIVITY

The spatial patterns of biodiversity in grasslands and savannas vary widely, with many areas with these biomes having relatively high levels of species richness and types of organisms.[61] For example, Pampas grasslands of South America have more than 400 species of grasses and tall grass prairies have more than 250 plant species.[62] In contrast to these areas, some grasslands of Patagonia have fewer than 30 species of plants.[63] Considering the large area of grasslands, the overall faunal diversity is low with a total of 477 species of birds and 245 species of mammals (i.e., equivalent to 5% of the total number of species in each group).[64] Even within climatically similar grasslands, the pattern of diversity of large mammals is significantly different. For example, in Africa and South America there are approximately 200 total species of large mammals, but the contribution of different orders is very different. For example, in Africa, the order *Artiodactyla* (Bovidae) has 80 species, whereas in South America there are fewer than 20.[65] The tropical savannas of Africa represent the world's greatest diversity of ungulates (over 40 different species). Up to 16 grazing and browsing species of antelopes can coexist in a given area in African savannas. Besides the above-ground diversity in faunal life, below-ground organisms are also an important component in savanna and grassland ecosystems. Grasslands in general have been found to have more biomass and higher productivity below ground compared to above.[66] The concentration of organic matter below ground provides the substrate for diverse bacteria, fungal, and nematode groups. More than 100 species of fungi and more than 200 species of nematodes have been reported for the short-grass steppe and tall-grass prairie, respectively.[67] Tropical savanna and temperate grasslands account for 30% (i.e., 19.9 Gt yr^{-1}) and 9.1% (i.e., 5.6 Gt yr^{-1}) of the global net primary production (NPP), respectively, demonstrating that these are relatively productive ecosystems.[68] In tropical savanna systems, although the relative productivity of trees and grasses are highly variable, the NPP of grass layers is 2–3 times that of the tree NPP.

ECOLOGICAL CONCERNS AND SUSTAINABLE USE

Savannas and grasslands are among the oldest ecosystems used by humans. Anthropogenic usage of the resources in these ecosystems is increasing and humans have influenced the ecological factors affecting savanna vegetation by controlling fire and by using it for landscape management. With burning, humans have either modified or totally cleared natural vegetation in savannas and grasslands for agricultural purposes. The introduction of animal husbandry systems has resulted in changes in grazing and browsing pressure and has undoubtedly modified the competitive balance between woody and herbaceous life forms in savannas. Compared to savannas, grasslands are even more highly altered ecosystems, with less than half remaining in an intact natural condition. Intensive mechanized agriculture activities have replaced more than 40% of the global temperate grasslands and another 15% have been converted due to urbanization.[69] The remaining area is dominated by grasses, but is degraded and vulnerable to desertification.

Increasing the human exploitation of natural resources in both grasslands and savannas has resulted in a variety of environmental concerns such as decreasing vegetation cover and changing species composition, loss of biodiversity, increased soil erosion resulting in nutrient loss and decreased productivity, and shifts in water balance and availability.[4,7] As a result of changing land use and a resultant increase in grazing/browsing pressure and altered fire regime in southern African and Australian savannas, large areas are experiencing the encroachment of woody species. Presumably human influences shift the competitive edge toward woody plant species.[70] Weed infestation in certain tropical savannas such as the Northern Territory of Australia is posing huge environmental and economic threat as they are changing the savanna ecosystem structure and resultant fire regime.[71,72] While certain weed species are toxic (e.g., *Argemone ochroleuca*), others chock waterways or are so dense that animals cannot move through them.[73,74] Changing vegetation structure of this sort may limit the grazing potential for local people, and hence may be endangering the livelihood of dependent ranching and pastoral communities.[40,75] Large parts of natural vegetation in Brazil's *Cerrado* have been cleared and modified to not only produce 70% of the beef cattle of that country, but with the implementation of soil modification techniques and irrigation, this ecoregion is now also important for grain and soy. There are environmental risks associated with monoculture plantation establishment and with burning that affects carbon, soil acidity, and friability.[3,13] In North America, the utilization of *Prairie* grasslands for grazing, cattle ranching, extensive agriculture, and urbanization has modified most of the temperate grasslands. Ecological issues in *Prairies* include woody plant expansion, biodiversity loss due to monospecific crop agriculture, gully erosion, and decreased productivity.[73,74] Changing land use practices have also resulted in the invasion of non-native species, a problem encountered in savanna and grasslands worldwide.[76,77]

Besides human impacts, future conditions and sustainability in savanna and grassland systems will also be determined by impacts of global climate change, which may alter spatiotemporal patterns of rainfall and thus influence available moisture, nutrients, and land cover patterns. Studies predict that the increased concentration of atmospheric CO_2 will allow invasion and denser stands of C3 woody species in savannas as such species can potentially utilize higher CO_2 concentration more efficiently than C4 plants. Consequently, trees will have faster post-burn recovery rates and faster sapling growth, enabling them to dominate.[78–80]

CONCLUSION

The dynamics of the world's savannas and grasslands continue to be shaped by humans through changed ecological controls. Given the significant geographical extent of these ecosystems, any changes in the structural and functional properties of these ecosystems, either due to human influence or climatic variability, are likely to impact human well-being. Thus, increased knowledge about the interaction of ecological processes and resulting spatial patterns and their temporal dynamics will improve our understanding. This information is required for ecologically informed decision making in regards the future sustainability of these systems. Historically, savannas have been ignored in ecological research until recently when their carbon sequestration potential (a third that provided by tropical rainforests and temperate evergreen forests) and biodiversity importance became apparent. As a result, historical data on savannas are limited and most of modern savanna studies are short-term site studies involving only selected ecological aspects. Savanna research needs large-scale climate and bio-physical data along the broad environmental gradients of precipitation and water limitations. By managing fire activity in savannas, their productivity could be significantly increased. Savannas are suitable for agro-forestry and thus can contribute to carbon sequestration, greenhouse mitigation, and also provide more livelihood opportunities.

REFERENCES

1. Scholes, R.J.; Archer, S.R. Tree-grass interactions in savannas. Annu. Rev. Ecol. Syst. **1997**, *28*, 517–544.
2. Flanagan, L.B.; Wever, L.A.; Carlson, P.J. Seasonal and interannual variation in carbon dioxide exchange and carbon balance in a northern temperate grassland. Global Change Biol. **2002**, *8* (7), 599–615.
3. Mistry, J. Savannas. Prog. Phys. Geogr. **2000**, *24* (4), 601–608.
4. Adeel, Z.; de Kalbermatten, G.; Assessment, M.E. *Ecosystems and Human Well-Being: Desertification Synthesis*; Island Press, 2005.
5. Stott, P. Recent trends in the ecology and management of the world's savanna formations. Prog. Phys. Geogr. **1991**, *15* (1), 18–28.
6. Roy, D.P.; Boschetti, L.; Giglio, L. Remote sensing of global savanna fire occurence, extent and properties. In *Ecosystem Function in Savannas: Measurement and Modeling at Landscape to Global Scales*; Hill, M.J., Hanan, N.P., Eds.; CRC Press, 2011.
7. Scholes, R.J.; Walker, B.H. *Cambridge Studies in Applied Ecology and Resource Management; An African Savanna: Synthesis of the Nylsvley Study*; Cambridge University Press: Cambridge, England, 1993; xii+306 p.
8. Backeus, I. Distribution and vegetation dynamics of humid savannas in Africa and Asia. J. Veg. Sci. **1992**, *3* (3), 345–356.
9. Archibold, O.W. *Ecology of World Vegetation*; Chapman and Hall Ltd.: London, 1995; xi+510 p.
10. Sankaran, M.; Hanan, N.P.; Scholes, R.J.; Ratnam, J.; Augustine, D.J.; Cade, B.S.; Gignoux, J.; Higgins, S.I.; Le Roux, X.; Ludwig, F.; Ardo, J.; Banyikwa, F.; Bronn, A.; Bucini, G.; Caylor, K.K.; Coughenour, M.B.; Diouf1, A.; Ekaya, W.; Feral, C.J.; February, E.C.; Frost, P.G.H.; Hiernaux, P.; Hrabar, H.; Metzger, K.L.; Prins, H.H.T.;

Ringrose, S.; Sea, W.; Tews, J.; Worden, J.; Zambatis, N. Determinants of woody cover in African savannas. Nature **2005**, *438* (7069), 846–849.

11. Menaut, J.C.; Barbault, R.; Lavelle, P.; Lepage, M. African savannas: Biological systems of humification and mineralization; In *Ecology and Management of World's Savannas*; Tothill, J.C., Mott, J.J., Eds.; Australian Academy of Science: Canberra, 1985; 14–33.
12. Chidumayo, E.N. Climate and phenology of savanna vegetation in southern Africa. J. Veg. Sci. **2001**, *12* (3), 347–354.
13. Furley, P. Tropical savannas: Biomass, plant ecology, and the role of fire and soil on vegetation. Prog. Phys. Geogr. **2010**, *34* (4), 563–585.
14. Lehmann, C.E.; Prior, L.D.; Bowman, D.M. Decadal dynamics of tree cover in an Australian tropical savanna. Austral Ecol. **2009**, *34* (6), 601–612.
15. Williams, R.J.; Duff, G.A.; Bowman, D.M.J.S.; Cook, G.D. Variation in the composition and structure of tropical savannas as a function of rainfall and soil texture along a large-scale climatic gradient in the Northern Territory, Australia. J. Biogeogr. **1996**, *23* (6), 747–756.
16. Bowman, D.; Walsh, A.; Milne, D. Forest expansion and grassland contraction within a Eucalyptus savanna matrix between 1941 and 1994 at Litchfield National Park in the Australian monsoon tropics. Global Ecol. Biogeogr. **2001**, *10* (5), 535–548.
17. Bowman, D.; Prior, L. Turner Review No. 10. Why do evergreen trees dominate the Australian seasonal tropics? Aust. J. Bot. **2005**, *53* (5), 379–399.
18. Woinarski, J.; Braithwaite, R.W. Conservation foci for Australian birds and mammals. Search **1990**, *21* (65), 8.
19. Barreto, G.R.; Herrera, E.A. Foraging patterns of capybaras in a seasonally flooded savanna of Venezuela. J. Trop. Ecol. **1998**, *14* (1), 87–98.
20. Sarmiento, G.; Pinillos, M.; da Silva, M.P.; Acevedo, D. Effects of soil water regime and grazing on vegetation diversity and production in a hyperseasonal savanna in the Apure Llanos, Venezuela. J. Trop. Ecol. **2004**, *20* (2), 209–220.
21. Hoogesteijn, A.; Hoogesteijn, R. Cattle ranching and biodiversity conservation as allies in South America's flooded savannas. Great Plains Res. **2010**, *20* (1, Sp. Iss. SI), 37–50.
22. Mishra, N.B.; Crews, K.A.; Neuenschwander, A.L. Sensitivity of EVI-based harmonic regression to temporal resolution in the lower Okavango Delta. Int. J. Remote Sensing **2012**, *33* (24), 7703–7726.
23. Bucini, G.; Hanan, N.P. A continental-scale analysis of tree cover in African savannas. Global Ecol. Biogeogr. **2007**, *16* (5), 593–605.
24. Bailey, R.G.; Ropes, L. *Ecoregions: The Ecosystem Geography of the Oceans and Continents*; Springer: New York, 1998.
25. Van Wilgen, B.W.; Govender, N.; Biggs, H.C.; Ntsala, D.; Funda, X.N. Response of savanna fire regimes to changing fire-management policies in a large African National Park. Conserv. Biol. **2004**, *18* (6), 1533–1540.
26. Miranda, H.S.; Sato, M.N.; Neto, W.N.; Aires, F.S. Fires in the cerrado, the Brazilian savanna. In *Tropical Fire Ecology: Climate Change, Land Use, and Ecosystem Dynamics*; Springer-Verlag: Berlin, 2009; 427–450.
27. Gregory, N.C.; Sensenig, R.L.; Wilcove, D.S. Effects of controlled fire and livestock grazing on bird communities in East African savannas. Conserv. Biol. **2010**, *24* (6), 1606–1616.
28. Jeltsch, F.; Weber, G.E.; Grimm, V. Ecological buffering mechanisms in savannas: A unifying theory of long-term tree-grass coexistence. Plant Ecol. **2000**, *150* (1–2), 161–171.
29. Gillson, L.; Ekblom, A. Resilience and thresholds in savannas: Nitrogen and fire as drivers and responders of vegetation transition. Ecosystems **2009**, *12* (7), 1189–1203.
30. Hoffmann, W.A.; Schroeder, W.; Jackson, R.B. Positive feedbacks of fire, climate, and vegetation and the conversion of tropical savanna. Geophys. Res. Lett. **2002**, *29* (22), 4.
31. Levick, S.R.; Asner, G.P.; Kennedy-Bowdoin, T.; Knapp, D.E. The relative influence of fire and herbivory on savanna three-dimensional vegetation structure. Biol. Conserv. **2009**, *142* (8), 1693–1700.
32. Hoffmann, W.A.; Adasme, R.; Haridasan, M.; de Carvalho, M.T.; Geiger, E.L.; Pereira, M.A.; Gotsch, S.G.; Franco, A.C. Tree topkill, not mortality, governs the dynamics of savanna-forest boundaries under frequent fire in central Brazil. Ecology **2009**, *90* (5), 1326–1337.
33. Hoffmann, W.A.; Geiger, E.L.; Gotsch, S.G.; Rossatto, D.R.; Silva, L.C.; Lau, O.L.; Haridasan, M.; Franco, A.C. Ecological thresholds at the savanna-forest boundary: How plant traits, resources and fire govern the distribution of tropical biomes. Ecol. Lett. **2012**, *15* (7), 759–768.
34. Williams, R.J.; Myers, B.A.; Eamus, D.; Duff, G.A. Reproductive phenology of woody species in a north Australian tropical savanna. Biotropica **1999**, *31* (4), 626–636.
35. Woinarski, J.; Brock, C.; Fisher, A.; Milne, D.; Oliver, B. Response of birds and reptiles to fire regimes on pastoral land in the Victoria River District, Northern Territory. Rangeland J. **1999**, *21* (1), 24–38.
36. Hopcraft, J.G.C.; Olff, H.; Sinclair, A.R.E. Herbivores, resources and risks: Alternating regulation along primary environmental gradients in savannas. Trends Ecol. Evol. **2010**, *25* (2), 119–128.
37. Pringle, R.M.; Young, T.P.; Rubenstein, D.I.; McCauley, D.J. Herbivore-initiated interaction cascades and their modulation by productivity in an African savanna. Proc. Natl. Acad. Sci. U. S. A. **2007**, *104* (1), 193–197.
38. Sinclair, A.R.E. Serengeti past and present. In *Serengeti II: Dynamics, Management, and Conservation of an Ecosystem*; University of Chicago Press: Chicago, USA, 1995; 3–30.
39. Asner, G.P.; Levick, S.R.; Kennedy-Bowdoin, T.; Knapp, D.E.; Emerson, R.; Jacobson, J.; Colgan, M.S.; Martin, R.E. Large-scale impacts of herbivores on the structural diversity of African savannas. Proc. Natl. Acad. Sci. **2009**, *106* (12), 4947–4952.
40. Dougill, A.J.; Thomas, D.S.G.; Heathwaite, A.L. Environmental change in the kalahari: Integrated land degradation studies for nonequilibrium dryland environments. Ann. Assoc. Am. Geogr. **1999**, *89* (3), 420–442.
41. Beyer, S.; Kinnear, A.; Hutley, L.B.; McGuinnessa, K.; Gibb, K. Assessing the relationship between fire and grazing on soil characteristics and mite communities in a semi-arid savanna of northern Australia. Pedobiologia **2011**, *54* (3), 195–200.
42. House, J.I.; Archer, S.; Breshears, D.D.; Scholes, R.J. Conundrums in mixed woody-herbaceous plant systems. J. Biogeogr. **2003**, *30* (11), 1763–1777.

43. Gandar, M. *Ecological Notes and Annotated Checklist of the Grasshoppers (Orthoptera-Acridoidea) of the Savanna Ecosystem Project Study Area, Nylsvley*, SANSP Report 74; Cooperative Scientific Programmes: CSIR, Pretoria, South Africa, 1983.
44. Bond, W.J.; Midgley, G.F. Carbon dioxide and the uneasy interactions of trees and savannah grasses. Philos. Trans. Royal Soc. B Biol. Sci. **2012**, *367* (1588), 601–612.
45. Buitenwerf, R.; Bond, W.; Stevens, N.; Trollope, W.S.W. Increased tree densities in South African savannas: >50 years of data suggests CO_2 as a driver. Global Change Biol. **2012**, *18* (2), 675–684.
46. Bond, W.J.; Woodward, F.I.; Midgley, G.F. The global distribution of ecosystems in a world without fire. New Phytol. **2005**, *165* (2), 525–537.
47. Higgins, S.I.; Bond, W.J.; Trollope, W.S. Fire, resprouting and variability: A recipe for grass–tree coexistence in savanna. J. Ecol. **2000**, *88* (2), 213–229.
48. Walker, B.H.; Ludwig, D.; Holling, C.S.; Peterman, R.M. Stability of semi-arid savanna grazing systems. J. Ecol. **1981**, *69* (2), 473–498.
49. Belsky, A.J. Tree grass ratios in East-African savannas—A comparison of existing models. J. Biogeogr. **1990**, *17* (4–5), 483–489.
50. Eagleson, P.S.; Segarra, R.I. Water-limited equilibrium of savanna vegetation systems. Water Resour. Res. **1985**, *21* (10), 1483–1493.
51. Fernandez-Illescas, C.P.; Rodriguez-Iturbe, I. Hydrologically driven hierarchical competition-colonization models: The impact of interannual climate fluctuations. Ecol. Monogr. **2003**, *73* (2), 207–222.
52. Jeltsch, F.; Milton, S.J.; Dean, W.R.J.; van Rooyen, N. Tree spacing and coexistence in semiarid savannas. J. Ecol. **1996**, *84* (4), 583–595.
53. Holdo, R.M. Elephants, fire, and frost can determine community structure and composition in Kalahari woodlands. Ecol. Appl. **2007**, *17* (2), 558–568.
54. Zimmermann, J.; Higgins, S.I.; Grimm, V.; Hoffmann, J.; Linstädter, A. Grass mortality in semi-arid savanna: The role of fire, competition and self-shading. Perspect. Plant Ecol. Evol. Syst. **2010**, *12* (1), 1–8.
55. Sankaran, M.; Ratnam, J.; Hanan, N.P. Tree-grass coexistence in savannas revisited - insights from an examination of assumptions and mechanisms invoked in existing models. Ecol. Lett. **2004**, *7* (6), 480–490.
56. Higgins, S.I.; Bond, W.J.; Trollope, W.S.W. Fire, resprouting and variability: A recipe for grass-tree coexistence in savanna. J. Ecol. **2000**, *88* (2), 213–229.
57. Lehmann, C.E.; Archibald, S.A.; Hoffmann, W.A.; Bond, W.J. Deciphering the distribution of the savanna biome. New Phytol. **2011**, *191* (1), 197–209.
58. Hanan, N.P.; Lehmann, C.E. *Tree-Grass Interactions in Savannas: Paradigms, Contradictions, and Conceptual Models*; CRC Press: Boca Raton, FL, 2011.
59. Hill, M.J.; Hanan, N.P. *Ecosystem Function in Savannas: Measurement and Modeling at Landscape to Global Scales*; Taylor & Francis: U.S., 2010; 39.
60. Murphy, B.P.; Bowman, D.M. What controls the distribution of tropical forest and savanna? Ecol. Lett. **2012**, *15* (7), 748–758.
61. Sala, O.E. Temperate grasslands. In *Global Biodiversity in a Changing Environment*; Springer: New York, 2001; 121–137.
62. Freeman, C. The flora of Konza Prairie: A historical review and contemporary patterns. In *Grassland Dynamics: Long-Term Ecological Research in Tallgrass Prairie*; Oxford University Press: New York, 1998; 69–80.
63. Golluscio, R.A.; Sala, O.E. Plant functional types and ecological strategies in Patagonian forbs. J. Veg. Sci. **1993**, *4* (6), 839–846.
64. Groombridge, B. *Global Biodiversity: Status of the Earth's Living Resources*; Chapman & Hall: Cambridge, England, 1992.
65. Vrba, E.S. Mammal evolution in the African Neogene and a new look at the Great American Interchange. In *Biological Relationships between Africa and South America*; Yale University Press: New Heaven, USA, 1993; 393–432.
66. Sims, P.L.; Singh, J. The structure and function of ten western North American grasslands: III. Net primary production, turnover and efficiencies of energy capture and water use. J. Ecol. **1978**, *66* (2), 573–597.
67. Knapp, A.K.; Briggs, J.M.; Hartnett, D.C.; Collins, S.L. *Grassland Dynamics: Long-Term Ecological Research in Tallgrass Prairie*; Oxford University Press: New York, 1998.
68. Roy, J.; Mooney, H.A.; Saugier, B. *Terrestrial Global Productivity*; Academic Press: London, 2001.
69. Heidenreich, B. *What Are Global Temperate Grasslands Worth? A Case for Their Protection*; Temperate Grasslands Conservation Initiative: Vancouver, BC, Canada, 2009; 21.
70. Van Auken, O. Shrub invasions of North American semiarid grasslands. Annu. Rev. Ecol. Syst. **2000**, *31*, 197–215.
71. Silva, J.F. Biodiversity and stability in tropical savannas, In *Biodiversity and Savanna Ecosystem Processes*; Springer: New York, 1996; 161–171.
72. Setterfield, S.A.; Rossiter-Rachor, N.A.; Hutley, L.B.; Douglas, M.M.; Williams, R.J. Biodiversity Research: Turning up the heat: The impacts of *Andropogon gayanus* (gamba grass) invasion on fire behaviour in northern Australian savannas. Divers. Distrib. **2010**, *16* (5), 854–861.
73. Briggs, J.M.; Knapp, A.K.; Blair, J.M.; Heisler, J.L.; Lett, M.S.; Hoch, G.A.; McCarron, J.K. An ecosystem in transition: Causes and consequences of the conversion of mesic grassland to shrubland. Bioscience **2005**, *55* (3), 243–254.
74. Derksen, D.A.; Anderson, R.L.; Blackshaw, R.E.; Maxwell, B. Weed dynamics and management strategies for cropping systems in the northern Great Plains. Agron. J. **2002**, *94* (2), 174–185.
75. Sinden, J.; Jones, R.; Hester, S.; Odom, D.; Kalisch, C.; James, R.; Cacho, O. *The Economic Impact of Weeds in Australia*, Report to the CRC for Australian Weed Management, Technical Series; University of Adelaide Press: Australia, 2004; 8.
76. Archer, S.; Vavra, M.; Laycock, W.; Pieper, R.D. Woody plant encroachment into southwestern grasslands and savannas: Rates, patterns and proximate causes. In *Ecological Implications of Livestock Herbivory in the West*; Society for Range Management: Denver, USA, 1994.

77. Fensham, R.; Fairfax, R.; Archer, S. Rainfall, land use and woody vegetation cover change in semi-arid Australian savanna. J. Ecol. **2005**, *93* (3), 596–606.
78. Bond, W.J.; Midgley, G.F. A proposed CO_2-controlled mechanism of woody plant invasion in grasslands and savannas. Global Change Biol. **2000**, *6* (8), 865–869.
79. Tews, J.; Jeltsch, F. Modelling the impact of climate change on woody plant population dynamics in South African savanna. BMC Ecol. **2004**, *4* (1), 17.
80. Tews, J.; Esther, A.; Milton, S.J.; Jeltsch, F. Linking a population model with an ecosystem model: Assessing the impact of land use and climate change on savanna shrub cover dynamics. Ecol. Model. **2006**, *195* (3–4), 219–228.

Soil Carbon and Nitrogen (C/N) Cycling

Sylvie M. Brouder
Ronald F. Turco
Department of Agronomy, Purdue University, West Lafayette, Indiana, U.S.A.

Abstract
The demands on modern agriculture to provide ample food, feed, and energy coupled with anthropogenic (human)-driven global climate change has greatly increased the importance of understanding and managing the dynamics of soil carbon (C) and nitrogen (N) cycling (C/N cycling). The world's C and N supplies exist in a continuous and dynamic cycle, and key processes are constantly active in the soil beneath our feet. The amount of C in the top one meter of soil (the pedologic pool) is approximately 3.5 fold greater than that in the atmosphere; this C helps soil retain its structure and function including crop production and ecosystem stability. In contrast to C, the great majority of the Earth's N is not in soil but in the atmosphere. However, most of the forms useful to plants are found in the soil. Plant productivity in natural systems is often N limited; in agriculture, farmers routinely supplement soil N with fertilizers and manure and, thus, perturb C/N cycles. The cycles of C and N in soil are linked and are controlled by microorganisms. As a result, microorganisms and their activity have a significant impact on some of our greatest challenges including sustaining food production and water quality and reducing soil emission of the greenhouse gases that drive climate change. This entry provides a new overview of the forms and locations of C and N in soil, emphasizes the biological factors controlling pedologic C/N cycling, and addresses how today's anthropogenic activities stress the soil ecosystem, potentially impacting its stability.

INTRODUCTION

Carbon (C) and nitrogen (N) exist in soils in both inorganic and organic forms. Globally, only 780 petagrams (Pg; 10^{15} g) of C are in the atmosphere while the top one meter of soil is estimated to contain approximately 2700 Pg of C, 57% of which is soil organic matter (SOM).[1] In contrast, N in soil is less than one-hundredth of one percent of the N content of the atmosphere but, as with soil C, most soil N is in organic versus inorganic forms.[2] The inorganic C and N forms in soils include the gases carbon dioxide (CO_2) and dinitrogen (N_2) in the air space of soil pores and soluble forms in soil water such as carbonates from the chemical breakdown or weathering of rocks and minerals (e.g., silicates, limestones) and nitrate (NO_3^-) and ammonium (NH_4^+) from the breakdown or mineralization of SOM. C and N occur in all living tissues: C in sugars, starches, lipids, and a staggering array of complex molecules and N in nucleic acids (DNA and RNA) and in amino acids that are assembled into proteins. Both C and N are an integral component of SOM, which is formulated from plant and animal residues in various stages of decomposition including particulates of various size fractions, dissolved OM in soil water, and humus or stable OM as well as the living biomass of soil microorganisms (e.g., bacteria and fungi).[3] To a large degree, microorganisms control or drive global C and N (C/N) cycling. Even under relatively extreme environmental conditions, soil C and N are in a state of constant flux where microbiology transforms C and N among an array of molecules and drives C and N movement between the soil and the surrounding air and water. The goal of this entry is to summarize the state-of-the-art knowledge on the microbiology of soil C/N cycling and discuss how human activities including agriculture are causing large perturbations in soil C/N cycles for which outcomes are uncertain but anticipated to continue to contribute to global climate change.

BACTERIA AND FUNGI CONTROL SOIL C/N CYCLES

The proportion of soil C and N in the biomass of soil microorganisms is surprisingly small (1.2% and 2.6%;[4]), given the importance of soil microbiology in C/N cycling. Soil bacteria are typically 1 μm or less in length and live in tiny colonies (only 100 or so cells per colony) found inside of soil aggregates and are typically attached to clay and silt fractions. However, while the actual bacteria are small in physical size, soils are rich in cell number, microbial diversity, and, consequently, metabolic abilities. A surface soil can contain greater than 4000 different

Encyclopedia of Natural Resources DOI: 10.1081/E-ENRL-120047492

microorganism communities (i.e., genomes) and can support as many as 10^9 (one billion) bacteria in one gram.[5,6] Subsurface soils will tend to have lower population levels but will exceed 10^7 cells g^{-1}.[7] Because bacteria are small but soil surface area large, bacteria are found on less than 0.17% of the total surface area of SOM and less than 0.02% of the soil's mineral surface area.[8] Further, soil bacteria are largely sessile (not freely moving) and prefer to adhere to surfaces using a form of electrostatic interaction (London-van der Waals forces) followed by a hydrophobic interaction where the cells secrete a polymer that binds them to the surface.[9] This binding reaction attaches bacteria and soil particles together, helping to create the soil structure. Water is the critical connecting force in soil as it moves nutrients and oxygen from their sources to the sessile resident population (Fig. 1, Process [P] 1). The exact locations within soil colonized by bacteria reflect the availability of nutrients (typically embedded organic carbon) and a preference by bacteria for the small soil pores (between 0.25 and 6 μm diameter) associated with the microaggregate fractions.[10,11] Analysis has shown that greater than 80% of soil bacteria are located within the soil's microaggregate (2 and 53 μm in size) fraction.[12] Small pores offer an ideal location for bacteria as they allow for the passage of water, which carries nutrients and oxygen but blocks predators such as protozoa.

Fungi also contribute to soil C/N cycling, but fungi differ from bacteria in several important attributes. Soil fungi are resident on the outside of aggregates and are associated with larger organic materials and the coarse sand fractions.[13] More importantly, fungi are made mobile by extension of their hyphae and in doing so are the early colonizers and decomposers of fresh organic residues added to soil via natural and anthropogenic activities (Fig. 1, P 3). Fungi function by secreting a wide array of enzymes, which start many important soil processes. The combined abilities of residue colonization and enzyme secretion give fungi a critical role in initiating the C/N cycling processes in fresh residue additions to soil.

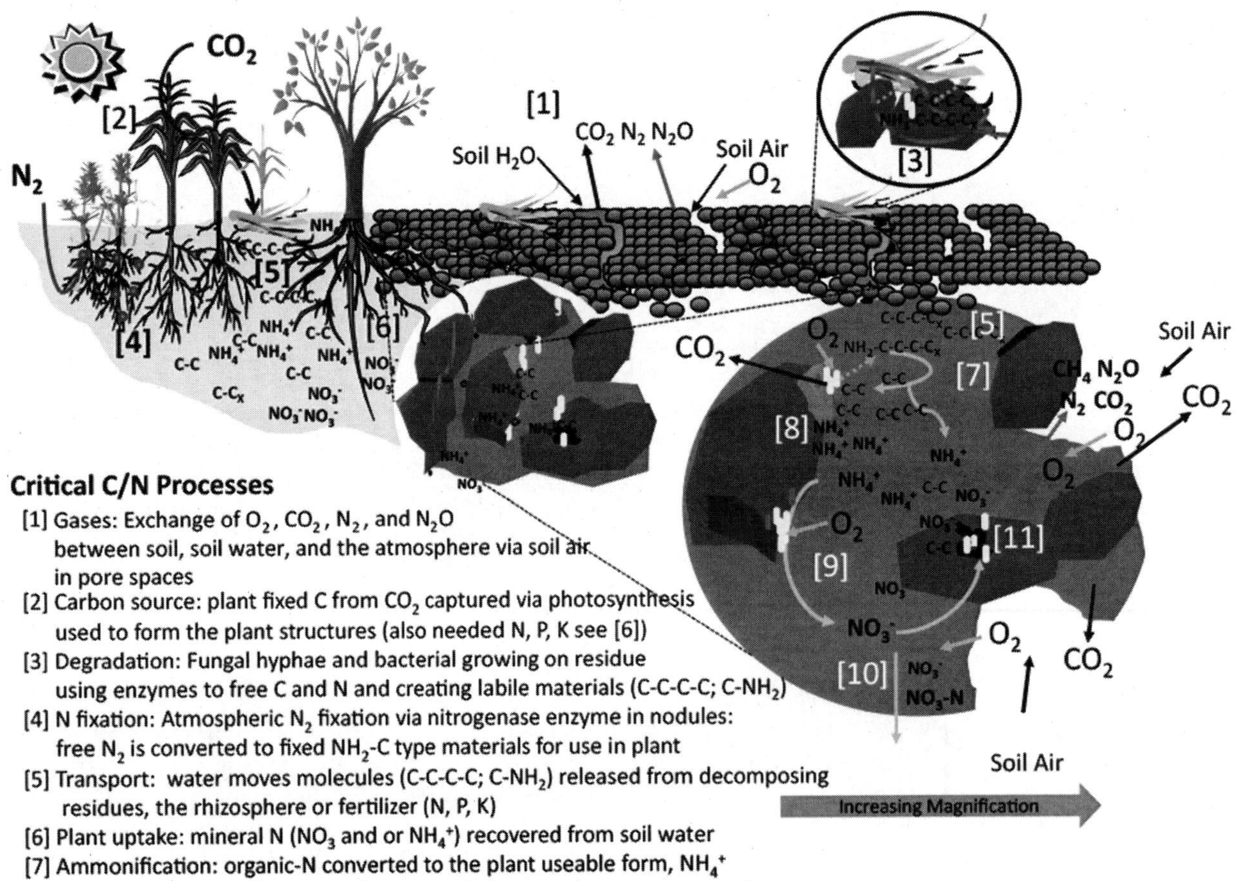

CO2
[2]
N2
[1]
Soil H2O
CO2 N2 N2O
Soil Air
O2
[3]
[4]
[5]
[6]
[7]
[8]
[9]
[10]
[11]
Soil Air
CO2
Increasing Magnification
Critical C/N Processes
[1] Gases: Exchange of O2, CO2, N2, and N2O between soil, soil water, and the atmosphere via soil air in pore spaces
[2] Carbon source: plant fixed C from CO2 captured via photosynthesis used to form the plant structures (also needed N, P, K see [6])
[3] Degradation: Fungal hyphae and bacterial growing on residue using enzymes to free C and N and creating labile materials (C-C-C-C; C-NH2)
[4] N fixation: Atmospheric N2 fixation via nitrogenase enzyme in nodules: free N2 is converted to fixed NH2-C type materials for use in plant
[5] Transport: water moves molecules (C-C-C-C; C-NH2) released from decomposing residues, the rhizosphere or fertilizer (N, P, K)
[6] Plant uptake: mineral N (NO3 and or NH4+) recovered from soil water
[7] Ammonification: organic-N converted to the plant useable form, NH4+
[8] Retention: NH4+ retained on soil's negatively charged cation exchange sites
[9] Nitrification: NH4+ from residues or fertilizer oxidized under aerobic conditions to the highly soluble NO3-
[10] Nitrate leaching: movement of NO3- with water
[11] Anaerobic processes: anaerobic microsite formed around C sources leading to denitrification & methanogenesis and the formation of N2, N2O and CH4

Fig. 1 Diagram of the major C/N cycle soil processes. Left side shows C/N cycling at the terrestrial scale; right side is magnified to highlight cycling within the soil matrix.

Continuous Soil C/N Cycling Requires Diversity in Organisms and Substrates

Soils and fresh additions of plant and animal materials present the resident microorganisms with a compositionally diverse array of organic substrates as potential energy and nutrient sources; genetic and associated functional diversity of the soil microbiology ensures ongoing soil C/N and other nutrient cycling in the ever-changing physical and chemical environments.[14] Simplistically, nutrient cycling reflects two sets of processes by which microorganisms meet their need for (i) energy and (ii) essential nutrients (e.g., C, N, phosphorus [P], sulfur) to maintain or increase their biomass (cell numbers). Because both bacteria and fungi must assimilate nutrients across a membrane, organic substrates must first be solubilized into physically smaller fractions. In organic residues of plant origin, the cell walls composed of lignin and cellulose form a recalcitrant physical barrier to degradation. Fungi are one of a few kinds of organisms that can secrete the enzymes necessary to break down cellulose to glucose and the only known organism that can completely degrade lignin via in-place oxidation. Thus, initial decomposition typically involves hyphal extension and enzyme excretion into newly introduced organic material to release labile substances (Fig. 1, P 3). This release process is general and creates a pool of soluble materials including physically smaller C and N molecules such as simple sugars, amino acids, and proteins that can be accessed by other bacteria, fungi, and growing plants. Labile materials not immediately assimilated by fungi may be washed out of decomposing tissue by rainfall; once these substrates reach the soil, they are transported into the soil matrix in soil water and diffuse to the resident microorganisms inside and outside of the soil aggregates (Fig. 1, P 5).

Under most conditions, once soluble organic N molecules are transported into the soil matrix, they rapidly enter a multistep process that sequentially creates the inorganic N forms of NH_4^+ and NO_3^-. Proteins and other organics rich in N are acted upon by bacteria that produce ammonification enzymes that free ammonia (NH_3) from C; in water, NH_3 dissolves to form NH_4^+ (Fig. 1, P 7). The NH_4–N can be assimilated by soil bacteria or plant roots (Fig. 1, P 6). When soil water concentrations of NH_4^+ are high, excess NH_4^+ may adsorb onto negatively charged soil surfaces where it is held until soil water NH_4^+ concentrations dissipate and the NH_4^+ diffuses back into solution (Fig. 1, P 8). Additionally, most soils are rich in chemoautotrophic bacteria. Whereas heterotrophic bacteria and fungi power their assimilation of organic C by the oxidation of organics, chemoautotrophic bacteria rely on the oxidation of inorganics (e.g., NH_3) and free CO_2 as their energy and C nutrition sources, respectively. Under aerobic conditions, chemoautotrophic (e.g., nitrifying) bacteria oxidize NH_4^+, consume CO_2, producing NO_3^- instead of CO_2 as an end product (Fig. 1, P 9). Nitrate is both water soluble and non-reactive with the negatively charged soil surface; thus, once in the NO_3–N form, N is not only available for assimilation by plants and microorganisms but also highly susceptible to loss from the soil system via the physical process of leaching with excess rainfall (Fig. 1, P 10).

It is important to note that different organisms and processes dominate when water from rainfall or flooding causes soils to become anoxic; these anaerobic processes have significant impacts on soil emission of greenhouse gases. When soil water content is high, macropores in soil are filled with water and O_2 consumption by aerobic bacteria, and plant root respiration (energy derived by electron transfer to O_2) exceeds the O_2 replacement rate from the soil atmosphere. As soil O_2 is depleted, the activity of anaerobic microorganisms that can use electron acceptors other than O_2 will increase. Provided adequate C substrate, many soil bacteria can use NO_3–N instead of O_2 as a terminal electron acceptor in the energy formation process with N_2O and N_2 as end products, a process called denitrification (Fig. 1, P 11). Alternatively, in fermentation, bacteria use the organic substrate itself to replace O_2 as the oxidizing agent; an important fermentation end product is methane (CH_4). Not only is denitrification an important loss pathway for soil N but products of both denitrification and fermentation have also been identified as potent greenhouse gases. The global-warming potentials on the individual molecule basis are 321- and 12-fold that of CO_2 for N_2O and CH_4, respectively (100-year timeframe).[15]

An additional subset of microorganisms is critical to the soil C/N cycle. While most soil microorganisms and plants cannot use the wealth of N_2 in the soil atmosphere, some have the ability to convert or "fix" the inert N_2 gas into ammonia NH_3. This biological N fixation (BNF) has three forms: it can be independent (free living), associative, or symbiotic with respect to higher plants. Free-living N_2-fixing bacteria are widely distributed and ubiquitous in the soil, but their contributions to terrestrial N is comparatively minor (<1kg ha^{-1} yr^{-1}) as these organisms lack easy access to C substrates for energy to drive the extremely energy-intensive fixation process. These bacteria fix N for their own metabolic needs but, when they die and decompose, the fixed N becomes available to plants. Some of these free-living bacteria live on the soil surface and can photosynthesize and have been estimated to add 10–38 kg N ha^{-1} yr^{-1}.[16] Much larger contributions come from soil microorganisms that can colonize root surfaces and invade plant tissues to gain more direct access to energy sources provided by plants. The largest contributions occur when soil microorganisms form true symbiotic relationships with plants such as occurs in legumes. Here, molecular dialogue between the soil microorganism and the plant root leads to the development of a specialized structure or nodule in the plant root to house the N_2 fixation process (Fig. 1, P 4). In the resulting symbiosis, plants and microorganisms achieve a relatively direct exchange of C for N. Symbiotic N fixation can contribute 24 to 250 kg N ha^{-1} in a growing season[17] and, when plant tissues die, this organic N is returned to the soil. Estimates of total annual contributions of BNF to fixed N in terrestrial ecosystems range from 107[18] to 195[19] teragrams (Tg; 10^{12} g) N yr^{-1}.

NATURAL VERSUS ANTHROPOGENIC CONTROLS OF C/N CYCLES

Among all the nutrient elements essential for soil microorganisms to grow, C and N are required in the largest quantities and are typically the most limiting in soils of natural systems; consequently, the general state of soil microbiological communities in their natural condition is one of repressed activity. Higher plants capture their C from atmospheric CO_2 with photosynthesis (Fig. 1, P 2) but, as with soil microorganisms, N must be acquired from soils (Fig. 1, P 6) and is required in higher quantities than any other soil-derived nutrient. Thus, N also constrains plant productivity in unmanaged ecosystems. Over the past several decades, human activities have increasingly perturbed pedologic C/N cycles by physically disturbing soils and/or by increasing quantities of C and/or N in labile form in the soil matrix thereby enhancing the overall activity of microorganisms and increasing opportunities for C and N loss from soils. Chief among anthropogenic activities known to disrupt previously stable cycles are (i) conversion of natural to managed systems with the goal of growing selected plant species (e.g., crops) with enhanced net primary productivity and (ii) the common agricultural practices of soil tillage and N fertilization.

In natural ecosystems, equilibrium exists between what happens when fresh organic residues are added and the genesis and decay of the resident soil humus. The rates of C/N cycling are modulated both by the physical environment and by the composition of added organic materials. Whereas labile substances such as simple sugars, amino acids, and proteins decompose rapidly, subsequent, slower rates of decomposition involve microorganisms attacking increasingly recalcitrant materials. Soil humus, the stable components of SOM that, by definition, bear no resemblance to the floral, faunal, or waste material of origin, is relatively resistant to biodegradation and includes materials derived from lignins, tannins, cutins, and so on.[20] In the absence of fresh residues, SOM serves as the major nutrient source for resident flora. However, as SOM is difficult to degrade and is N limited, the function of microorganisms is slow when they are constrained to this nutrient source. Without fresh additions, soil microorganisms are "starved" and waiting for the arrival of nutrient (C, N, P)-rich materials; as a general rule, new additions from an N-rich residue such as from an annual legume will decompose faster as compared to an equivalent mass of lignin-rich residues from tree branches and trunks.

Major physical determinants of cycling include the soil temperature and moisture regimes. Soil temperatures fluctuate daily and seasonally, especially at the soil surface. Cold or freezing temperatures can drastically slow or halt activity of microorganisms, often without killing them, while temperatures above 35°C not only halt activity but may also kill heat-sensitive organisms. Likewise, drought and low soil water status can restrict microbiology and desiccate microorganisms causing dormancy while soil flooding and anoxia slows growth of organisms that prefer aerobic conditions. However, niche differentiation is the hallmark of successful microbiological communities: heat-tolerant will proliferate at the expense of heat-sensitive soil microorganisms under high soil temperatures and, as discussed in the preceding text, anaerobic will take over from aerobic organisms with prolonged soil flooding.

Anthropogenic Factors Disrupting C/N Cycles: Land Conversion, Tillage, and N Fertilizers

Long-term stability in soil function is a direct result of centuries of buildup of SOM by soil microorganisms. However, the impacts of human activities related to agriculture can result in SOM degradation at rates that far outpace the comparatively slow but constant rates of SOM accumulation in natural systems. Ruddiman[21,22] dates anthropogenic soil losses of C as CO_2 to 8000 to 10,000 years ago with the beginnings of settled agriculture; soil C losses as CH_4 date to 5000 years ago and the introduction of paddy rice production. Recent estimates are that net anthropogenic emissions to the atmosphere total approximately 9.1 Pg C yr^{-1}, 82% of which comes from fossil fuel emissions; the remaining 1.6 Pg C yr^{-1} has been sourced to land use conversion (deforestation and biomass burning), agricultural cultivation, and associated erosion.[1] A meta-analysis of impacts of land use change on soil C showed that C stocks in the top 60 cm of soil decreased an average of 50% when soils were converted from native forests to row crop agriculture.[23] In contrast, converting forest to pasture did not negatively impact soil C stocks, but converting pastures into crops was as detrimental as direct conversion to crops from native forests. Conversion of cropland back to pasture or secondary forest restored some or all soil C, respectively.

Tillage, the mechanical disturbance of soil, is an agricultural practice that dates from the first settled agriculture.[24] It is a proven technology for conditioning the soil for agricultural crops including breaking up soil layers that are too hard and impede root exploration, mechanically controlling weeds that compete with crops for water and nutrients, incorporating nutrients, manure, and residue, controlling residue-borne pests and pathogens, and aerating the soil. Incorporation promotes rates of decomposition of residues of the soil surface by mechanically breaking down residues into small pieces, mixing residue fragments with soil, and enhancing the surface area of the residue directly in contact with the decomposer microorganisms in the soil. However, potential negative impacts are equally numerous.[25] Tilled soils that are not covered with a vigorously growing crop or even a thick mat of fallen residue are highly susceptible to erosion by wind and rainfall. Further, pulverization and compaction of the surface and underlying soil layers, respectively, can lead to reduced water infiltration and poor root development that reduce yields. The implementation of tillage on previously pastured and forested land typically results in dramatic losses of soil C.[23] The implementation of surface residue conservation and no-tillage crop management can restore some soil C but not under all conditions,[26] and the value of these practices as a universal solution remains uncertain.[27]

Unlike tillage, the application of large amounts of fertilizer N is a relatively recent practice but one with strong linkages to global climate change. The objective of fertilization in agriculture is to ensure crop productivity and is not limited by competition with soil microorganisms over inadequate mineral N supplies. Early agriculture could rely on only modest amounts of additional N from manures or legumes. Dramatic accelerations of soil C/N cycling in intensively managed agricultural systems are linked to the 1908 discovery of the Haber–Bosch process for high temperature and pressure conversion of N_2 to NH_3. Subsequent commercialization of the process revolutionized the production and thus availability of N fertilizers. At present, estimates of global N inputs into agricultural land are 213 Mt (10^6 t) N yr^{-1} of which 47% is fertilizer.[28] In contrast, livestock manures and BNF by leguminous crops account for only 16% and 28%, respectively, of annual global N inputs; cereals, primarily wheat, rice and maize, account for 55 Mt N yr^{-1} of fertilizer N. These fertilizer applications are associated with large increases in crop CO_2 capture but also with the acceleration of rates of C/N cycling including losses from soil. The negative impacts of N fertilizer on global N cycles are known to be substantial and range from eutrophication of aquatic systems to acid rain and ozone depletion.[29] However, implications of an altered N cycle for other biogeochemical cycles are much less certain and include a complex array of climate-relevant feedbacks.[30] Galloway and Gruber[30] identify reducing uncertainty in our understanding of coupled C/N cycle perturbations as essential to improving global climate change projections.

CONCLUSION

With reference to ecological function, Becking and Beijerinck concluded more than a century ago "everything is everywhere, but the environment selects."[31] Anthropogenic activities change the environment, and any form of agriculture should be viewed as a selective pressure on soil that will cause the equilibrium of the natural C/N cycle to shift. Soil microbiology, with its great diversity of organisms, will respond with strategies optimizing the use of both quantities and forms of C and N introduced by soil management. For agriculture, the ongoing challenge for the production of major staple crops that underpin global food security remains the design of management systems that appropriately balance the trade-offs between yield objectives, soil biology and long-term soil stability and sustainability of ecosystem function. The challenge is particularly complex as the nature and magnitude of trade-offs are site and soil specific. As we are managing soils to maximize crop yield, the soil biology is constantly responding to our interventions. For example, when reduced forms of N are applied, a subset of soil microorganisms will increase cell numbers and activity to oxidize the material and create the soluble forms such as NO_3–N. These soluble forms can be either leached out of the soil causing water pollution or transported to reduced locations where denitrifying microorganisms consume them and emit greenhouse gases. Conversely, when residues high in C but low in N are applied, soil microorganisms will tie up and use all forms of available nutrients, limiting the amount of N and P available to growing plants. Food security goals that seek to double current agricultural yields on all arable lands will require enhanced quantities of plant-available nutrients. More efficient nutrient management strategies to increase crop production while minimizing impacts on air and water quality and global climate change have been frequently identified as critical to achieving global food security. At present, much of our research investment has gone toward exploring genetic aspects of crop yield and resource (e.g., N and water)-use efficiencies. Research on functional roles of soil microorganisms and their responses to management has lagged and must be elevated to the prominence of plant genetics as the soil's resident microorganisms are the systems-level control point for many key processes governing use efficiency of soil resources.

REFERENCES

1. Lal, R. Sequestration of atmospheric CO_2 in global carbon pools. Energy Environ. Sci. **2008**, *1*, 86–100.
2. Tamm, C.O. *Nitrogen in Terrestrial Ecosystems: Questions of Productivity, Vegetational Changes, and Ecosystem Stability*; Springer-Verlag: New York, USA, 1991.
3. Stevenson, F.J. *Humus Chemistry: Genesis, Composition, Reactions*, 2nd Ed.; John Wiley & Sons: New York, USA. 1994.
4. Xu, X.; Thornton, P.E.; Post, W.M. A global analysis of soil microbial biomass carbon, nitrogen and phosphorus in terrestrial ecosystems. Global Ecol. Biogeogr. **2013**, *22*, 737–749.
5. Torsvik, V.; Goksoy, J.; Daae, F.L. High diverstiy in DNA of soil bacteria. Appl. Envir. Micro. **1990**, *56*, 782–787.
6. Torsvik, V.; Salte, K; Sorheim, R.; Goksoyr, J. Comparison of phenotypic diversity and DNA heterogeneity in a population of soil bacteria. Appl. Envir. Micro. **1990**, *56*, 776–781.
7. Konopka, A.E.; Turco, R.F. Biodegradation of organic compounds in vadose zone and aquifer sediments. Appl. Environ. Micro. **1991**, *57*, 2260–2268.

8. Hissett R.; Gray, T.R.G. Microsites and time changes in soil microbial ecology. In *The Role of Terrestrial and Aquatic Organisms in Decomposition Processes*; Anderson, J.M., MacFadyen, A. Eds.; Blackwell: Oxford, Great Britain, 1976; 23–39.
9. van Loosdrecht, M.C.M.; Lyklema, J.; Norde, W; Zehnder, A.J.B. Influence of interfaces on microbial activity. Microbiol. Rev. **1990**, *54*, 75–87.
10. Hattori, T.; Hattori, R. The physical environment in soil microbiology: An attempt to extend the principles of microbiology to soil microbiology. Crit. Rev. Microbiol. **1976**, *4*, 423–461.
11. Killham, K.; Amato, M.; Ladd, J.N. Effect of substrate location in soil and soil pore-water regime on carbon turnover. Soil Biol. Biochem. **1993**, *25*, 57–62.
12. Ranjard, L.; Poly, F.; Combrisson, J.; Richaume, A.; Gourbière, F.; Thioulouse, J.; Nazaret, S. Heterogeneous cell density and genetic structure of bacterial pools associated with various soil microenvironments as determined by enumeration and DNA fingerprinting approach (RISA). Microbial. Ecol. **2000**, *39*, 263–272.
13. Kandeler, E.; Tscherko, D.; Bruce, K.D.; Stemmer, M.; Hobbs, P.J.; Bardgett, R.D.; Amelung, W. Structure and function of the soil microbial community in microhabitats of a heavy metal polluted soil. Biol. Fert. Soils **2000**, *32*, 390–400.
14. Zak, D.R.; Tilman, D.; Paramenter, R.; Fischer, F.M.; Rice, C.; Vose, J.; Milchunas, D.G.; Martin, C.W. Plant production and soil microorganisms in late successional ecosystems: A continental-scale study. Ecology **1994**, *75*, 2333–2347.
15. Intergovernmental Panel on Climate Change. *Climate Change: The Physical Science Basis*; Cambridge Univ. Press: Cambridge, UK, 2007; 31–35.
16. Witty, J.F.; Keay, P.J.; Frogatt, P.J.; Dart, P.J. Alagal nitrogen fixation on temperate arable fields: The Broadbalk experiment. Plant Soil **1979**, *52*, 151–164.
17. Werner, D. Production and biological nitrogen fixation of tropical legumes. In *Nitrogen Fixation in Agriculture, Forestry, Ecology, and the Environment;* Werner, D., Newton, W.E., Eds; Springer: Dordrecht, the Netherlands, 2005; 1–13.
18. Galloway, J.N.; Dentener, F.J.; Capone, D.G.; Boyer, E.W.; Howarth, R.W.; Seitzinger, S.P; Asner, G.P.; Cleveland C.C; Green, P.A.; Holland, E.A.; Karl, D.M. Michaels, A.F.; Porter, J.H.; Townsend, A.R; Vorosmarty, C.J. Nitrogen cycles: Past, present and future. Biogeochemistry **2004**, *70*, 153–226.
19. Cleveland, C.C.; Townsend, A.R.; Fisher, H.; Howarth, R.W.; Hedin, L.O.; Perakis, S.S.; Latty, E.F.; Von Fischer, J.C.; Elseroad, A.; Wasson, M.F. Global patterns of terrestrial biological nitrogen (N2) fixation in natural ecosystems. Global Biogeochem. Cycles **1999**, *13*, 623–645.
20. Derenne, S.; Largeau, C. A review of some important families of refractory macromolecules: Composition, origin, and fate in soils and sediments. Soil Sci. **2001**, *166*, 833–847.
21. Ruddiman, W.F. *Plows, Plagues and Petroleum: How Humans Took Control of Climate*; Princeton Univ. Press: Princeton, USA, 2003.
22. Ruddiman, W.F. The anthropogenic greenhouse gas era began thousands of years ago. Clim. Change **2005**, *61*, 262–292.
23. Guo, L.B.; Gifford, R.M. Soil carbon stocks and land use change: A meta analysis. Global Change Biol. **2002**, *8*, 345–360.
24. Lal, R. The plow and agricultural sustainability. J. Sust. Agric. **2009**, *33*, 66–84.
25. Hobbs, P.R.; Sayre, K; Gupta, R. The role of conservation agriculture in sustainable agriculture. Phil. Trans. R. Soc. B. **2008**, *363*, 543–555.
26. Six, J.; Ogle, S.M.; Breidt, F.J.; Conant, R.T.; Mosier, A.R.; Paustian, K. The potential to mitigate global warming with no-tillage management is only realized when practiced in the long term. Global Change Biol. **2004**, *10*, 155–160.
27. Giller, K.E.; Witter, E.; Corbeels, M.; Tittonell, P. Conservation agriculture and smallholder farming in Africa: The heretics' view. Field Crops Res. **2009**, *114*, 23–34.
28. Connor, D.J.; Loomis, R.S.; Cassman, K.G. *Crop Ecology: Productivity and Management in Agricultural Systems,* 2nd Ed.; Cambridge University Press: New York, 2011; 195–261.
29. Galloway, J.N.; Aber, J.D.; Erisman, J.W.; Seitzinger, S.P.; Howarth, R.W.; Cowling, E.B. The nitrogen cascade. Bioscience **2003**, *53*, 341–356.
30. Gruber, N.; Galloway, J.N. An earth-system perspective of the global nitrogen cycle. Nature **2008**, *451* (17), 293–296.
31. De Wit, R.; T. Bouvier. Everything is everywhere, but, the environment selects; What did Baas Becking and Beijerinck really say? Environ. Microbiol. **2006**, *8*, 755–758.

Soil Degradation: Food Security

Michael A. Zoebisch
German Agency for International Cooperation, Tashkent, Uzbekistan
Eddy De Pauw
International Center for Agricultural Research in the Dry Areas (ICARDA), Aleppo, Syria

Abstract
Growing populations lead to an increasing demand for food. With decreasing per-capita areas of arable land, this leads to additional pressure and stresses on the limited land resources, especially the soils. As a consequence, large parts of the world are affected by soil degradation, and this has a direct effect on food security.

INTRODUCTION

Food security is commonly defined as the access by all people at all times to enough food for an active, healthy life. Extended concepts include food-quality aspects, cultural acceptability of the food, equitable distribution among the different social groups, and gender balance.[1–4] These concepts imply that, in addition to producing enough food, people must also have access to it. This opens the arena for the socioeconomic, cultural, policy, and political dimensions of food production and supply. Thus, food security—and consequently food insecurity—are multi-dimensional. It is not possible to separate clearly the biophysical and socioeconomic dimensions of food security, because they are closely interrelated and interdependent.[2] However, the most limiting of the natural resources needed for food production are soil and water. If these resources are depleting, land productivity—the key factor for crop production—declines, and the land is no longer capable of producing the biomass needed for direct and indirect human consumption. If soil degrades, its plant-life supporting functions are lost, together with its role in the hydrological cycle. Soil is a limited resource and, because of the long duration of soil-forming processes, can be considered as nonrenewable. The production of food and feed is directly linked to the productive capacity of the soil. The degree and extent of soil degradation and, consequently, its effect on food production are dependent on how the land is used.

UNDERNOURISHMENT—A MAJOR INDICATOR OF FOOD INSECURITY

The proportion of chronically undernourished people is a manifestation of the degree of food insecurity. Worldwide, more than 830 million people are undernourished; 800 million of these people live in the developing countries, i.e., about 18% of the population of these countries.[1] Table 1 shows that the most seriously affected region is Sub-Saharan Africa, with 33% of the population chronically undernourished. In Asia and the Pacific chronic undernourishment has decreased significantly over the past two decades.

Yields of the major food crops in many countries of these regions have risen remarkably. But in most cases the yield increases have not been able to keep pace with the needs of growing populations. Subregional trends, especially from Africa, China, South Asia, and Central America, indicate large yield declines due to soil degradation.[2,5,6] Countries of the less degraded temperate regions can substitute for the losses in other regions. However, in the affected areas, food prices will increase and hence the incidence of malnutrition. Densely populated countries and regions that depend solely on agriculture will be most affected. As input requirements—and their costs—grow with increasing degradation, these countries will probably not be able to maintain adequate levels of soil productivity without special efforts to stabilize the system.

VULNERABLE AREAS

Only about one tenth of the world's arable areas are not endangered by degradation. Most soils are vulnerable to degradation and have a low resilience to fully recover from stresses. The soils in the tropics and subtropics are more vulnerable to degradation than those in temperate areas, mainly due to the climatic conditions.[7] The total land area available for agricultural production is limited. Different types of land have different capabilities and limitations. Not all land is suitable for cultivation. The major natural limiting factors for soil productivity are steep slopes, shallow soils, low levels of natural fertility, poor soil drainage, sandy or stony soil, salinity, and sodicity.

Encyclopedia of Natural Resources DOI: 10.1081/E-ENRL-120001633

Table 1 Undernourishment in developing countries

Region/subregion	Undernourished population: Proportion of total population 1979/1981 (%)	Undernourished population: Proportion of total population 1995/1997 (%)	Number of people affected 1995/1997 (millions)
Total Developing World	*29*	*18*	*791.5*
Asia and Pacific	*32*	*17*	*525.5*
East Asia	29	14	176.8
Oceania	31	24	1.1
Southeast Asia	27	13	63.7
South Asia	38	23	283.9
Latin America and Caribbean	*13*	*11*	*53.4*
North America	5	6	5.1
Caribbean	19	31	9.3
Central America	20	17	5.6
South America	14	10	33.3
Near East and North Africa	*9*	*9*	*32.9*
Near East	10	12	27.5
North Africa	8	4	5.4
Sub-Saharan Africa	*37*	*33*	*179.6*
Central Africa	36	48	35.6
East Africa	35	42	77.9
Southern Africa	32	44	35
West Africa	40	16	31.1

Source: Adapted from Lal & Singh.[10]

Different agro-ecosystems are typically susceptible to different types of soil degradation.[6–8] In mountainous areas and steeplands, water erosion is the dominant form of soil degradation. For arid areas, both water and wind erosion and the loss of soil organic matter are typical. In humid areas, soil acidification and fertility decline are of importance. In irrigated areas, the hazards of soil salinization and waterlogging are of special significance. Knowledge of these hazards can help understand the limitations of the land and introduce appropriate land-use practices that reduce soil degradation and loss of soil productivity.

UNDERLYING CAUSES

Misuse of Land

The main causative factors of human-induced soil degradation, in a broad sense, are overuse and inappropriate management of agricultural land, deforestation and the removal of natural vegetation, and overgrazing as well as industrial activities that pollute the soil.[6] Worldwide, soil erosion by wind and water are the most important forms of soil degradation, accounting for between 70% and 90% of the total area affected by soil degradation.[5–6] High population pressure and the resultant increased need for higher land productivity is the main factor leading to inappropriate and exploitative land use.

In high-input systems, loss of soil quality due to degradation can be compensated to a certain extent by inputs, such as fertilizers and irrigation water. In low-input systems—usually practiced by resource-poor farmers (in poor countries)—soil quality contributes relatively more to agricultural productivity.[9] Therefore, soil degradation in these systems has more drastic and immediate effects on soil productivity and, hence, food production.

Pressure on the Land

When the land is not enough, i.e., when land scarcity becomes a limiting factor for food production, people usually resort to intensification of their land use and opening up new, often unsuitable, land for cultivation.[4,9] Both pathways will most certainly lead to soil degradation, if unsuitable land-use technologies are used. The majority of land users in the developing world do not have adequate access to appropriate land-use technologies and the financial means to invest in their land.[1] The area available for crop production therefore puts a definite limit to sustainable food production in many parts of the developing world. Alternative, i.e., nonagricultural means of income generation would enable the people to supplement their food requirements on the market and could relieve immediate pressure on the land. This would save soil and land resources.

Losses and Gains of Agricultural Land

Estimates show that more than 10 million hectares of agricultural land worldwide are lost per year, most of it due to soil degradation but also due to urbanization (i.e., between 2 and 4 million hectares).[4] Annual deforestation and clearing of savanna land accounts for about 20 million hectares, of which 16 million hectares are converted to cropland. Thus, there is an annual net gain of cropland of about 5–6 million hectares. Most of these converted lands can be considered marginal, i.e., less suitable for cropping. Loss of traditional sources of grazing and firewood, together with the ecological functions of forests, i.e., their role in regulating watershed hydrology, are factors inducing the degradation of soil.

MINIMUM PER-CAPITA CROPLAND REQUIREMENT

On a global scale, it is estimated that the minimum per-capita area of cropland required to produce an adequate quantity of food is around 0.1 ha.[2,5] This rough estimate is dependent on many different location-specific factors and circumstances, such as soil, terrain, climate, farming system, and land-use technology. However, it permits a suitable comparison at the overall global scale. Estimates predict that a large number of countries will have reached the 0.1 ha limit of per-capita cropland by the year 2025.[1,2,7] In the absence of more land for cultivation, future food needs of the people in these countries will have to depend on increased soil productivity through intensified land use. Resource-poor farmers with no option to expand their cultivated area usually cannot invest in necessary soil-conservation and soil-fertility maintenance measures. Over time, this leads to nutrient depletion and other forms of soil degradation that affect soil productivity.[4,9,10]

Although some developing countries have indeed increased their yields on a per-unit area basis significantly, these increases cannot compensate for the increased demands by growing populations. Table 2 illustrates past and future trends in per-capita available cropland in selected countries with high population growth and gives examples of yield-level trends. The table shows that, to a large extent, the gains in yields are offset by the decrease in available cropland to grow the required total quantities of crops to feed the population. In these countries, food insecurity can be directly related to the pressure on the land, or land scarcity.

The growing pressure on the land is not only a reflection of population growth and the direct needs for food, but also of changed diets and increasing cash needs (by the land users) that have to be met from land cultivation, putting additional stress on the soil resources.

ESTIMATING SOIL DEGRADATION ON A GLOBAL SCALE

Limitations of Methodology

The scientific relationships between different soil-degradation processes and soil productivity have been established mainly on field and plot level.[9,11] Although reliable data are available for a wide range of different agro-ecologies and farming systems, they cannot give a precise picture of degradation-dependent soil productivity for larger land units. The

Table 2 Trends of available per-capita cropland and average yield levels of main staple crops for selected countries

	Per-capita cropland (ha)						Average yield (kg/ha)			
Country	1961	1980	1990	2000	2015[a]	Crop	1961	1980	1990	2000[b]
Syria	1.40	0.62	0.46	0.34	0.17	Wheat	575	1,536	1,544	1,882
						Barley	461	1,312	310	381
Kenya	0.21	0.15	0.10	0.08	0.04	Wheat	1,090	2,156	1,864	1.400
						Maize	1,253	1,200	1,580	1,321
Pakistan	0.34	0.25	0.17	0.13	0.07	Wheat	822	1,568	1,825	2,492
						Rice	1.391	2.423	2,315	2.866
Bangladesh	0.17	0.10	0.09	0.07	0.05	Wheat	573	1,899	1,504	2.258
						Rice	1,700	2.019	2,566	2.851
Nepal	0.19	0.17	0.14	0.11	0.07	Wheat	1.227	1.199	1,415	1.820
						Rice	1.937	1.932	2,407	2.600
Zambia	1.52	0.92	0.65	0.57	0.28	Wheat	1.600	4.068	4,399	5.454
						Maize	882	1,688	1,432	1,431
Côte d'Ivoire	0.69	0.44	0.31	0.24	0.10	Rice	757	1.166	1,155	1.548
						Sorghum	667	583	575	352
Sudan	0.97	0.62	0.52	0.43	0.22	Rice	1.409	634	1,250	563
						Sorghum	970	712	428	634
Peru	0.20	0.20	0.17	0.14	0.10	Wheat	1.001	939	1,085	1.289
						Potato	5.287	7.196	7,881	10.818

[a]UN Projection, medium level.
[b]FAO Estimates.
Source: Adapted from Engelman & LeRoy,[2] and FAO.[12]

Table 3 Estimates of degradation and losses in soil productivity for different land-use types in the major regions

	Degraded areas									
	Agricultural land		Pasture land		Forests		Total degraded area		Loss in productivity[a]	
Region	(Mha)	(%)	(Mha)	(%)	(Mha)	(%)	(Mha)	(%)	Cropland (%)	Pasture (%)
Africa	121	65	243	31	130	19	494	30	25	6.6
Asia	206	38	197	20	344	27	747	27	12.8	3.6
South America	64	45	68	14	112	13	244	16	13.9	2.2
Central America	28	74	10	11	25	38	63	32	36.8	3.3
North America	63	26	29	11	4	1	96	9	8.8	1.8
Europe	72	25	54	35	92	26	218	27	7.9	5.6
Oceania	8	16	84	19	12	8	104	17	3.2	1.1
World	562	38	685	21	719	18	1,966	23	12.7	3.8

[a]Cumulative loss in productivity since 1945. Adjusted figures according to type and degree of degradation.
Source: Adapted from Scherr,[5] van Lynden & Oldeman,[6] and Oldeman.[13]

increasing complexity of the determining factors and processes on larger land units, the endless array of location-specific conditions, and the lack of adequate data make upscaling extremely difficult for country and regional levels. Estimates of the extent and severity of soil degradation—and their effects on soil productivity—on a country, regional, continental, and global level therefore bear a substantial degree of uncertainty. Most estimates use the methodology developed by the Global Assessment of Human-Induced Soil Degradation (GLASOD) Project.[5,6] The GLASOD estimates are based on expert assessment and are therefore largely subjective. However, the methodology has been widely adopted, and GLASOD estimates of soil degradation are accepted as the best estimates available. Global estimates are, therefore, only rough approximations and should not be taken literally.

Effects on Soil Productivity and Food Production

Table 3 shows estimates of degradation and cumulative losses in soil productivity on a continental level for different land-use types. The figures show that, overall, Africa and Central America are the most severely affected continents, both in terms of soil degradation and reduction of productivity. Sixty-five percent of African and 75% of Central American cropland are affected by soil degradation. The overall loss in productivity in these two regions over the last 50 years is estimated between 25% and 37%. On a continental basis, Africa still has land reserves, but large areas of the continent are marginal for crop production, and their food production is therefore most seriously affected by degradation. Europe also shows relatively high trends in degradation, but mainly of pastures and forests. Productivity losses generally are low, as for North America and Oceania (i.e., mainly Australia), and these can most probably be compensated for by improvements in technology and input supply.

A more detailed picture of trends in food productivity for the main food cereals in selected countries with high population growth rates and agriculture-based economies is given in Fig. 1. The overall past trends in the different countries show a consolidation of per-capita food production. However, the projections until 2025 indicate clear decreases. This suggests that long-term food security in these countries is at stake.

SOIL DEGRADATION AND DECLINE IN PRODUCTIVITY

Over the past decades, the cumulative loss of productivity of cropland has been estimated at about 13%. However, aggregate global food security does not appear to be under a significant threat.[2,5] There is no conclusive evidence that soil degradation has in the past affected global food security. We cannot conclude to such relationship because 1) the database on the extent and magnitude of various types of soil degradation is inadequate and 2) the impact of soil degradation on land productivity is very site-specific, often anecdotal and difficult to quantify at regional and global levels. In fact, globally, per-capita food production has increased by about 25% since 1961 when the first production surveys were conducted by Food and Agriculture Organization of the United Nations (FAO).[12] The yields per unit area of the major cereals (i.e., wheat, rice, maize) have steadily increased and are still rising.[1,12] This, however, does not imply that sufficient food is—and will be—available in all countries and regions. Food is not necessarily produced where it is most needed. A top priority should therefore be to improve inventories of land degradation at regional, national, and subnational levels.

However, the evidence is conclusive that soil degradation affects food security at the subnational and national levels in the developing countries by the gradual decline of the land's productive capacity and that this trend will continue in the future. We therefore have to assume that,

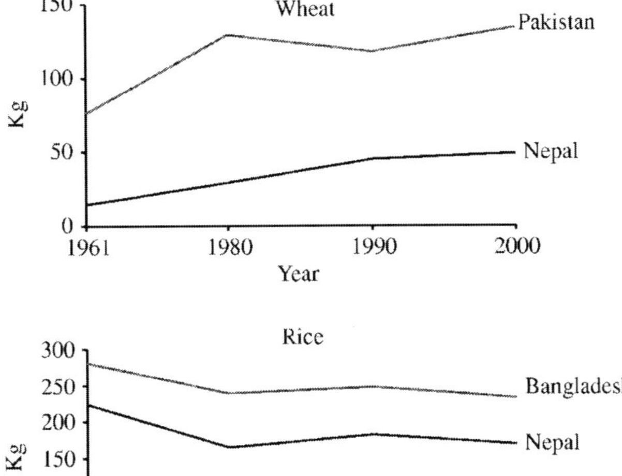

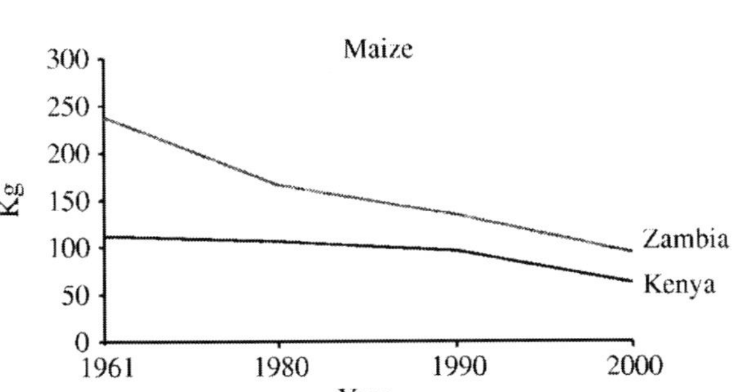

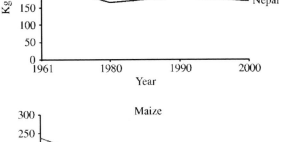

Fig. 1 Per-capita production of selected crops in countries with high population pressure.
Source: Adapted from Engelman & LeRoy,[2] and FAO.[12]

eventually, land degradation will constitute a serious threat to global food security by its particular impact on the developing countries. According to most scenarios, these countries are most vulnerable to degradation induced by increasing pressure on their land resources, the effects of climate change and their inability to finance programs to rehabilitate affected areas and prevent further degradation and decline of productivity.

REFERENCES

1. FAO. *The State of Food Insecurity in the World*; Food and Agriculture Organization of the United Nations (FAO): Rome, Italy, 1999.
2. Engelman, R.; LeRoy, P. *Conserving Land: Population and Sustainable Food Production*; Population and Environment Program; Population Action International: Washington, DC, U.S.A., 1995.
3. Saad, M.B. *Food Security for the Food-Insecure: New Challenges and Renewed Commitments*; Position Paper: Sustainable Agriculture; Centre for Development Studies, University College Dublin: Dublin, Ireland, 1999.
4. Kindall, H.W.; Pimentel, D. Constraints on the expansion of the global food supply. Ambio **1994**, *23* (3), 198–205.
5. Scherr, S.J. *Soil Degradation—A Threat to Developing- Country Food Security by 2020*? Discussion Paper 27; International Food Policy Research Institute: Washington, DC, 1999.
6. van Lynden, G.W.J.; Oldeman, L.R. *The Assessment of the Status of Human-Induced Soil Degradation in South and Southeast Asia*; International Soil Reference and Information Centre:Wageningen, the Netherlands, 1997.
7. Greenland, D.; Bowen, G.; Eswaran, H.; Rhoades, R.; Valentin, C. *Soil, Water, and Nutrient Management Research—A New Agenda*; IBSRAM Position Paper; International Board for Soil Research and Management: Bangkok, Thailand, 1994.
8. Lal, R. Soil management in the developing countries. Soil Science **2000**, *165* (1), 57–72.
9. Stocking, M.; Murnaghan, N. *Land Degradation—Guidelines for Field Assessment*; UNU/UNEP/PLEC Working Paper; Overseas Development Group, University of East Anglia: Norwich, UK, 2000.
10. Lal, R.; Singh, B.R. Effects of soil degradation on crop productivity in East Africa. J. Sustain. Agricul. **1998**, *13* (1), 15–36.
11. Stocking, M.; Benites, J. *Erosion-Induced Loss in Soil Productivity: Preparatory Papers and Country Report Analyses*; Food and Agriculture Organization of the United Nations: Rome, Italy, 1996.
12. FAO. FAOSTAT. Statistical Database of the Food and Agriculture Organization of the United Nations. Internet Web-Source http://www.fao.org (accessed January 2001), Rome, Italy, 2000.
13. Oldeman, L.R. *Soil Degradation: A Threat to Food Security*? Report 98-01; International Soil Reference and Information Centre (ISRIC): Wageningen, the Netherlands, 1998.

Soil Degradation: Global Assessment

Selim Kapur
University of Çukurova, Adana, Turkey
Erhan Akça
Adiyaman University, Adiyaman, Turkey

Abstract
Although soil provides food, fiber, and nutrients that are essential for humans, its value is rarely esteemed. Soils throughout human history were continuously mistreated, which led to the fall of civilizations. Agricultural and industrialization activities trigger erosion, salinity, and contamination all of which cause loss of soil quality to irrecoverable levels. Contemporary soil resource exhaustion has resulted in severe land degradation in many parts of the world; more than 25% of usable global land is degraded. Thus, combatting land degradation is humanity's major priority for global security.

INTRODUCTION

One of the major challenges that humanity will face in the coming decades is the need to increase food production to cope with the ever-increasing population, which is indeed the greatest threat for food/soil security.[1] The worldwide loss of 12 million hectares (ha) of agricultural land per year was reflected in the average annual loss up to €38 billion for EU25,[2] in spite of the strict laws for environmental protection enacted there. This is the unpalatable but important indicator of a steady decline in agricultural production and increase in soil damage, reflecting the "impaired productivity" in a quarter of the world's agricultural land.[3–5]

BACKGROUND

Soil can degrade without actually eroding. It can lose its nutrients and soil biota and can become damaged by waterlogging and compaction. Erosion is only the most visible part of degradation, where the forces of gravity, water flow, or wind actively remove soil particles.

Rather than taking the classical view that soil degradation was, is, and will remain an ongoing process, mainly found in countries of the developing world, this phenomenon should be seen as a worldwide process that occurs at different scales and different time frames in different regions. The causes of biophysical and chemical soil degradations are enhanced by socioeconomic interventions, which are the main anthropogenic components of this problem, together with agricultural mismanagement, overgrazing, deforestation, overexploitation, and pollution as reiterated by Lal,[6] UNEP/ISRIC,[7] Lal,[8] Eswaran & Reich,[9] Eswaran et al.,[10] Kapur et al.,[11] and Cangir et al.,[12] as the main reasons for erosion and chemical soil degradation.

Soil degradation, the threat to "soil security," is ubiquitous across the globe in its various forms and at varying magnitudes, depending on the specific demands of people and the inexorably increasing pressures on land. Europe provides many telling examples of the fragile nature of soil security and the destructive consequences of a wide range of soil degradation processes. Asia, Africa, and South and North America are not only partly affected by the nonresilient impacts of soil degradation but also experiencing more subtle destruction of soils through political developments, which seek to provide temporary relief and welfare in response to the demands of local populations.

Europe

The major problems concerning the soils of Europe are the loss of such resources owing to erosion, sealing, flooding, large mass movements as well as local and diffuse soil contamination, especially in industrial and urban areas, and soil acidification.[13,14] Salinity is a minor problem in some parts of Western and Eastern Europe, but with severe effects at the northern and western parts of the Caspian Sea (with low salinity)[15] mainly because of the shift to irrigated agriculture and destruction of the natural vegetation (Fig. 1).

Urbanization and construction of infrastructure at the expense of fertile land are widespread in Europe, particularly in the Benelux countries, France, Germany, and Switzerland, and such effects are most conspicuously destructive along the misused coasts of Spain, France, Italy, Greece, Turkey, Croatia, and Albania. The drastic increase in the rate

Encyclopedia of Natural Resources DOI: 10.1081/E-ENRL-120047486

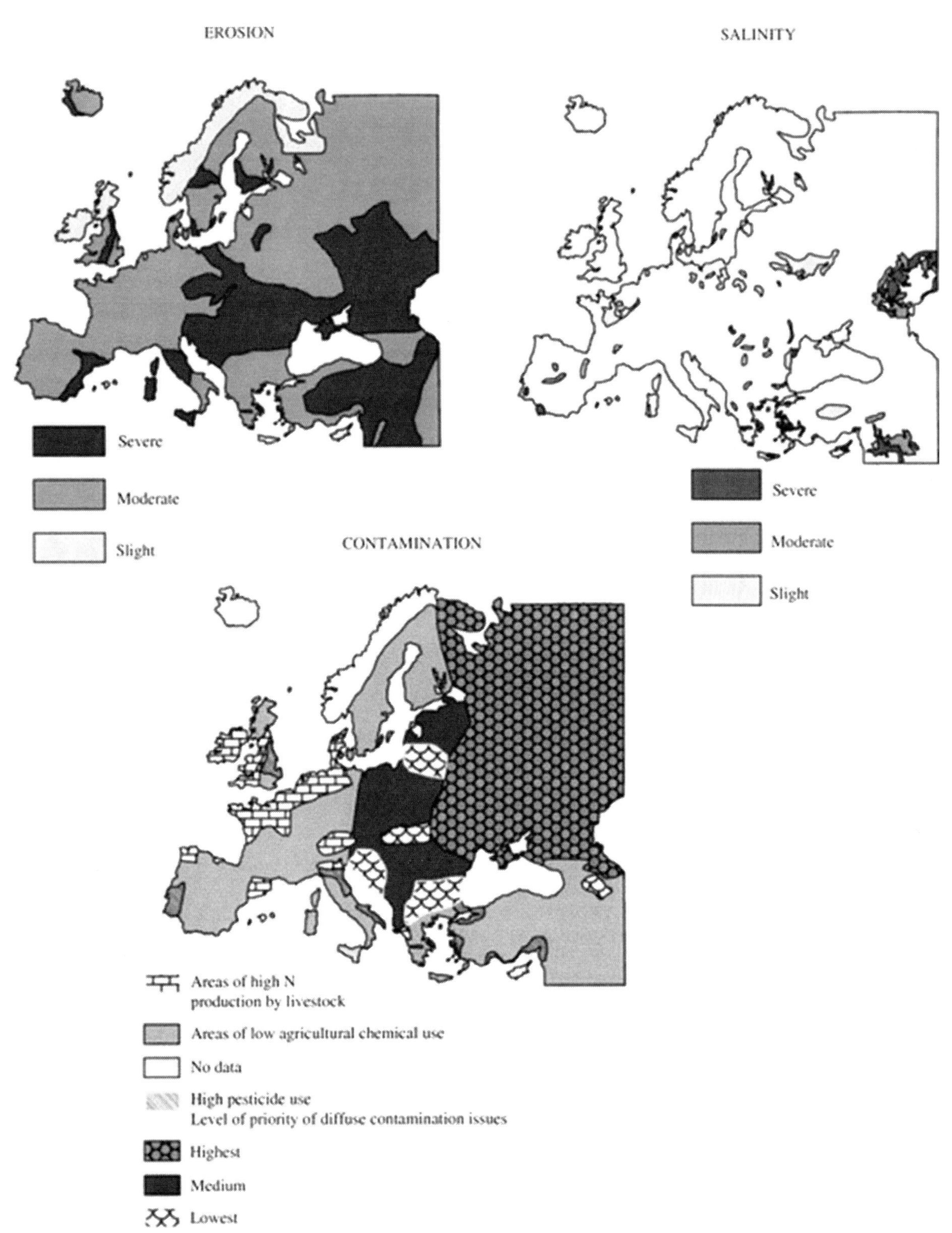

Fig. 1 Soil degradation in Europe.
Source: Adapted from UNEP/ISRIC,[7] Kobza,[16] USDA-NRCS,[17,18] Erol,[19] Kosmas et al.[20] Kharin et al.[21]

of urbanization since the 1980s is now expected to follow the Blue Plan, which seeks to create beneficial relationships among populations, natural resources, the main elements of the environment, and the major sectors of development in the Mediterranean Basin and to work for sustainable development in the Mediterranean region. The very appropriate term "industrial desertification" remains valid for the once degraded soils of East Europe, under the pressure of mining and heavy industry, as in Ukraine where such lands occupy 3% of the total land area of the country.[22]

There are three broad zones of "natural" erosion across Europe, including Iceland: (1) the southern zone (the Mediterranean countries); (2) a northern loess zone comprising the Baltic States and part of Russia; and (3) the eastern zone of Slovenia, Croatia, Bosnia-Herzegovina, Rumania, Bulgaria, Poland, Hungary, Slovakia, the Czech Republic, and Ukraine (Fig. 1). Seasonal rainfalls are responsible for severe erosion owing to overgrazing and the shift from traditional crops. Erosion in southern Europe is an ancient problem and still continues in many places, with marked on-site impacts and with significant decreases in soil productivity as a result of soil thinning. The northern zone of high-quality loess soils displays moderate effects of erosion with less intense precipitation on saturated soils. Local wind erosion on light textured soils is also responsible for the transportation of agricultural chemicals used in the intensive farming systems of the northern zone to adjacent water bodies, along with eroded sediments. The high erodibility of the soils of the eastern zone is exacerbated by the presence of large state-controlled farms that have introduced intensive agriculture at the expense of a decrease in the natural vegetation. Contaminated sediments are also present in this zone, particularly in the vicinity of former industrial operations/deserts, with high rates of erosion in Ukraine (33% of the total land area) and Russia (57% of the total land area), whose agricultural land has been subjected to strong water and wind erosion ever since the beginning of industrialization.

Localized zones of likely soil contamination through the activities of heavy industry are common in northwestern and central Europe as well as northern Italy, together with more scattered areas of known and likely soil contamination caused by the intensive use of agricultural chemicals. Sources of contamination are especially abundant in the "hot spots" associated with urban areas and industrial enclaves in the northwestern, southern, and central parts of the continent (Fig. 1). Acidification through deposition of windborne industrial effluents and aerosols has been a long-standing problem for the whole of Europe; however, this is not expected to increase much further, especially in western Europe, as a result of the successful implementation of emission-control policies over the recent years.[3]

The desertification of parts of Europe has been evident for some decades, and the parameters of the problem are now becoming clear, with current emphasis on monitoring of the environmentally sensitive areas[22,20] on selected sites, seeking quality indicators for (1) soil (particularly organic carbon; (2) vegetation; (3) climate; and (4) human management throughout the Mediterranean basin. Apart from the human factor, these indicators are inherent.

Asia

The most severe aspect of soil degradation on Asian lands has been desertification owing to the historical, climatic, and topographic character of this region as well as the political and population pressures created by the conflicts of the past 500 years or so. Salinization caused by the rapid drop in the level of the Aral Sea and the water-logging of rangelands in Central Asia owing to the destruction of the vegetation cover by overgrazing and cultivation provide the most striking examples of an extreme version of degradation—desertification caused by misuse of land. Soil salinity, the second colossal threat to the Asian environment, has occurred through the accumulation of soluble salts, mainly deposited from saline irrigation water or through mismanagement of available water resources, as in the drying Aral Sea and the Turan lowlands as well as the deterioration of the oases in Turkmenistan, with excessive abstraction of water in Central Asia (Fig. 2).[21]

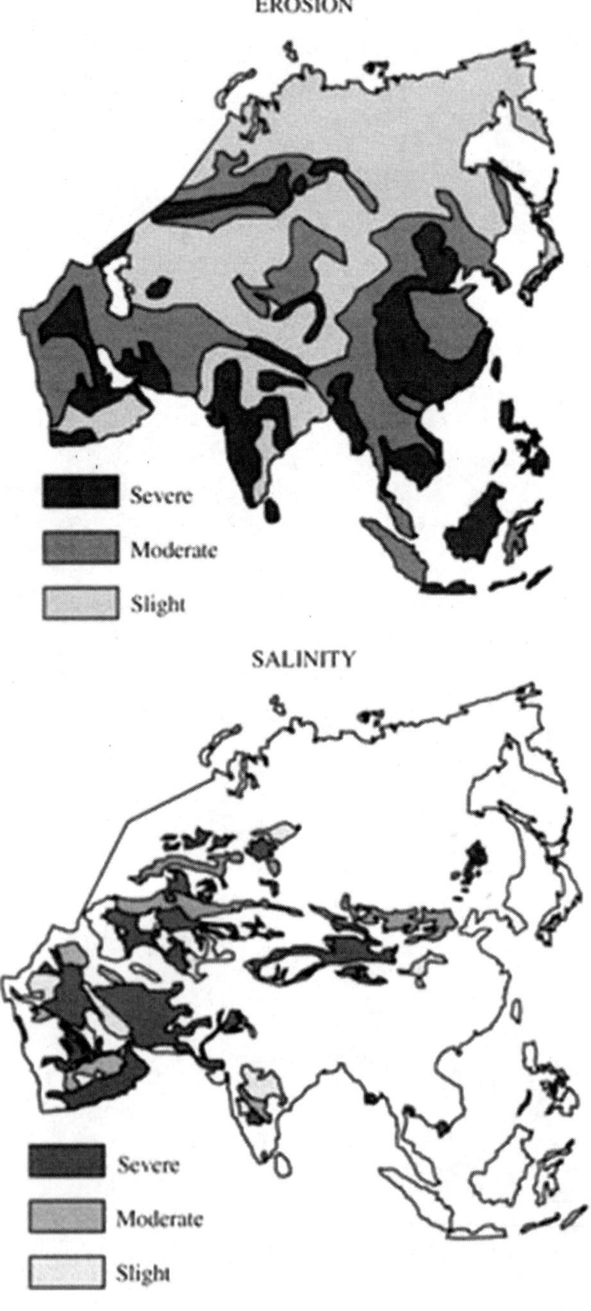

Fig. 2 Soil degradation in Asia.
Source: Adapted from UNEP/ISRIC,[7] Erol,[19] Kosmas et al.[20] Lowdermilk.[24]

The dry lands of the Middle East have been degrading, since the Sumerian epoch, with excessive irrigation causing severe salinity and erosion–siltation problems,[21] especially in Iraq, Syria, and Saudi Arabia. Iraq has been unique in the magnitude of the historically recorded build-up of salinity levels, with 4.81 million ha saline land, which is 74% of the total arable land surface (i.e., 90% of the land in the southern part of that country). The historical lands of Iran, Pakistan, Afghanistan, India, and China are also subject to ancient and ongoing soil degradation processes, which are subtle in some areas but evident and drastic in others (Fig. 2).

Africa

Africa's primary past and present concern has been the loss of soil security by nutrient depletion, that is, the decreasing NPK levels (kg/ha) along with micronutrients in cultivated soils following the exponential growth in population and the resulting starvation and migrations (Fig. 3). Intensification of land use to meet the increased food demands combined with the mismanagement of the land leads to the degradation of the continental soils. This poses the ultimate question of how the appropriate sustainable technologies that will permit increased productivity of soils can be identified. This problem is illustrated by the example of the Sudan, where nutrient depletion has steadily increased through more mechanized land preparation, planting, and threshing without the use of inorganic fertilizers, and legume rotations. Thus, aggregate yields have been falling as it became more difficult to expand the cultivated area without substantial public investments in infrastructure. This decline in yield has occurred at enormous rates in Vertisols of Sudan, with the mean annual sorghum yields decreasing from 1000 to 500 kg/ha and the wholesale price of sorghum increased 3.7% per month since 2007.[25] In Burkina Faso, the decreased infiltration and increased runoff causing erosion are further consequences of repeated cultivation. Thus, the technological measures to be identified for these two African examples must include development of water retention technologies in Burkina Faso, while polyculture/rotations with proper manuring and fertilization for cost-efficient provisions of N and P and preferably green manuring are all needed in Sudan to permit the balanced management of soil moisture, nutrients, and organic matter (and to enhance C-sequestration, a main goal for sustainability based on the earth sciences—to ensure the security of both the soil and global climate).[25,26]

Central and South America

Africa and Latin America have the highest proportion of degraded agricultural land. Water and wind erosion are the dominant soil degradation processes in Central and South America and have caused the loss of the topsoil at alarming rates because of the prevailing climatic and topographic conditions. Almost as important is the loss of nutrients from the Amazon basin (Fig. 4).[27] These effects are mainly attributable to deforestation and overgrazing, the former being responsible for the degradation of 576 million ha out of 1,000 million ha potential agricultural land. Another important factor has been the ever-increasing introduction of inappropriate agricultural practices derived from the so-called imported technology, which have not been properly adapted to indigenous land-use procedures. The traditional methods of permitting the land to recover naturally have been almost totally abandoned and replaced by unsuitable technological measures designed to maintain production levels (temporarily) and to overcome the loss of soil resilience, thus increasing chemical inputs.

The rapid industrialization/urbanization of the limited land resources in the Caribbean region has been expelling agricultural communities to remote and marginal regions that are at present rich in biodiversity and biomass—a major global C sink. Moreover, large-scale livestock herding of Central and South America is also a major threat to soil security and has been responsible for degrading 1 million km^2 of Argentinean, Bolivian, and Paraguayan pasturelands.

North America

The most prominent outcome of soil degradation (or more correctly desertification) in United States is exemplified by the accelerated dust storm episodes of the 1930s—the Dust Bowl years, marked by the "Black Blizzards," which were caused by persistent strong winds, droughts, and overuse of the soils. These resulted in the destruction of large tracts of farmland in the south and central United States. Recently, salinization has become an equally severe problem in the western part of the country (Fig. 5) through artificial elevation of water tables by extensive irrigation, with associated acute drainage problems. An area of about 10 million ha in the west of United States has been suffering from salinity-related reductions in yields, coupled with very high costs in both the Colorado River basin and the San Joaquin Valley.[28] Unfortunately, new irrigation technologies, such as the center pivot irrigation system (developed as an alternative to the conventional irrigation systems causing the salinity problems), have caused a decline of the water-table levels in areas north of Lubbock, Texas, by around 30–50 m, leading to a dramatic decrease in the thickness of the well-known Ogallala aquifer by 11% decrease between 2003 and 2011 only. In some areas, this has been followed by ground subsidence, which is an extreme form of soil structure degradation, that is, loss of the physical integrity of the soil.

Loss of topsoil, as a result of more than 200 years of intensive farming in United States, is estimated to vary between 25% and 75% and exceeds the upper limit in some

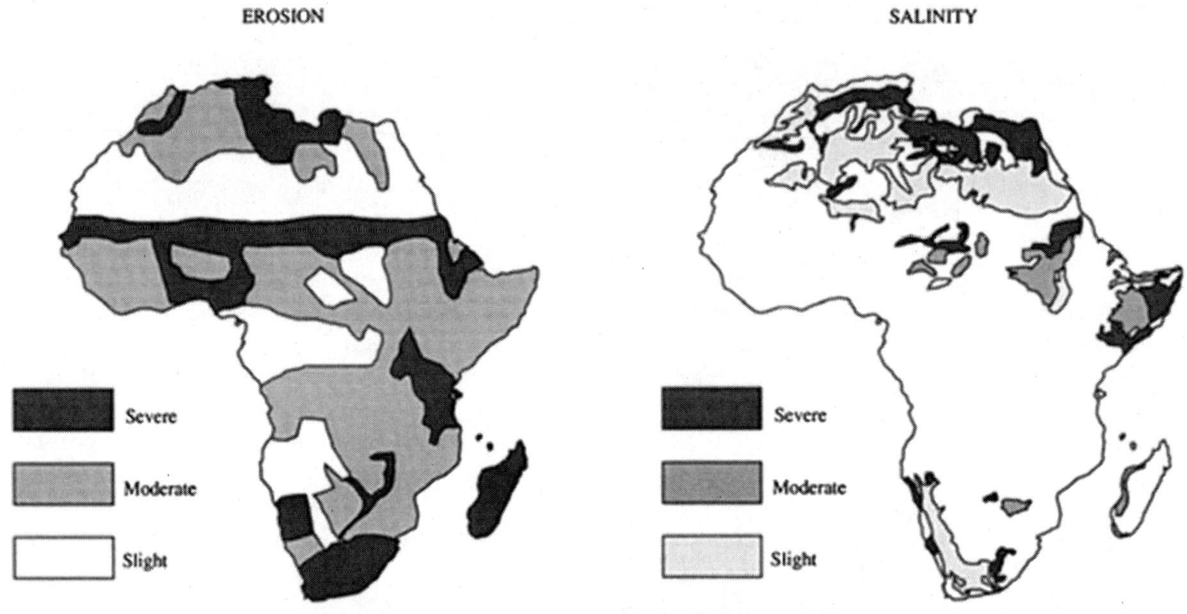

AVERAGE ANNUAL NUTRIENT DEPLETION (NPK)

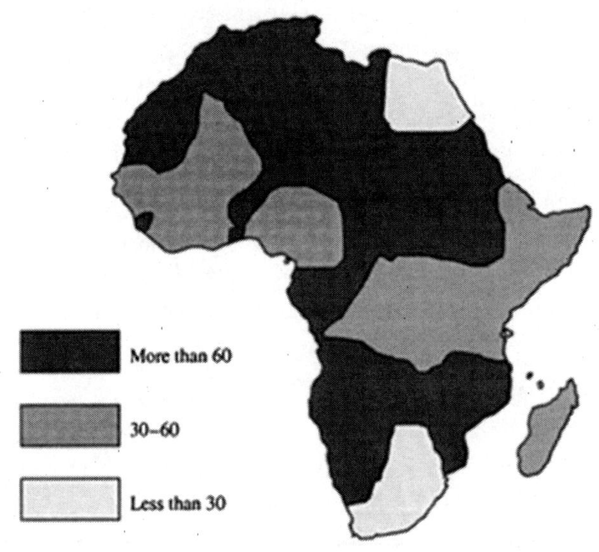

Kilogram of NPK per hectare

Fig. 3 Soil degradation in Africa.
Source: Adapted from UNEP/ISRIC,[7] Lowdermilk.[24]

parts of the country.[18,19] United States provides good examples of the difficulties involved in erosion control, with its large-scale intensive agriculture—deteriorating soil structure and increasing erosion of its susceptible soils. This problem could be overcome primarily by the strict introduction of the no-till system. No-till areas have increased from 4 million ha in 1989 to 25.3 million ha in 2004, and they are forecast to follow a linear extrapolation until 48 million ha out of the total 81 million ha of cultivated land will be attained.[29] Conservation farming is practiced in only about half of all U.S. agricultural land and on less than half of the country's most erodible cropland. Conservation farmers are encouraged to use only the basic types of organic fertilizers, such as animal and green manure together with compost, mulch farming, improved pasture management, and crop rotation, to conserve soil nutrients.

Canada is a large country where 68 million ha of available land is cultivated, which is only the 7.4% of the territorial of the country with an average farm size of 450 ha. It is reported that Canada has experienced annual soil

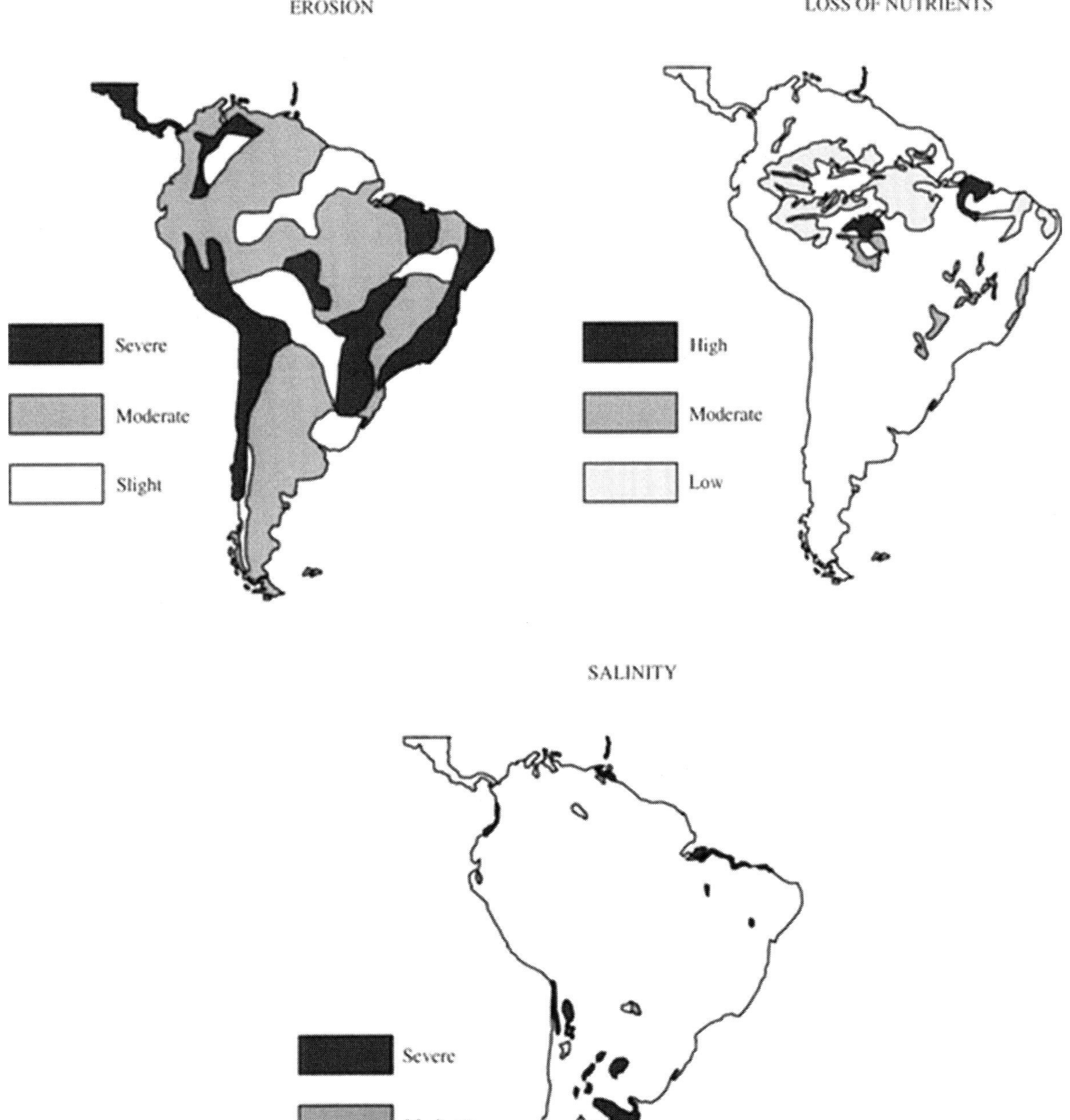

Fig. 4 Soil degradation in Central and South America.
Source: Adapted from UNEP/ISRIC,[7] Kharin et al.[21]

losses on the prairies, through wind and water erosions, that are similar to the Asian steppes, amounting, respectively, to 60 and 117 million tons. These annual rates are much higher than the rate of soil formation, resulting in an annual potential grain production loss of 4.6 million tons of wheat. With regard to primary soil salinity, during historic times, the prairies have experienced steady increases related partially to increasing groundwater levels. Major problems of secondary salinity are estimated to affect 2.2 million ha of land in Alberta, Saskatchewan, and parts of Manitoba, with an immense economic impact each year.

Australia

The Australian agricultural/soil resource base has been endangered, as the "business as usual" concept was adopted

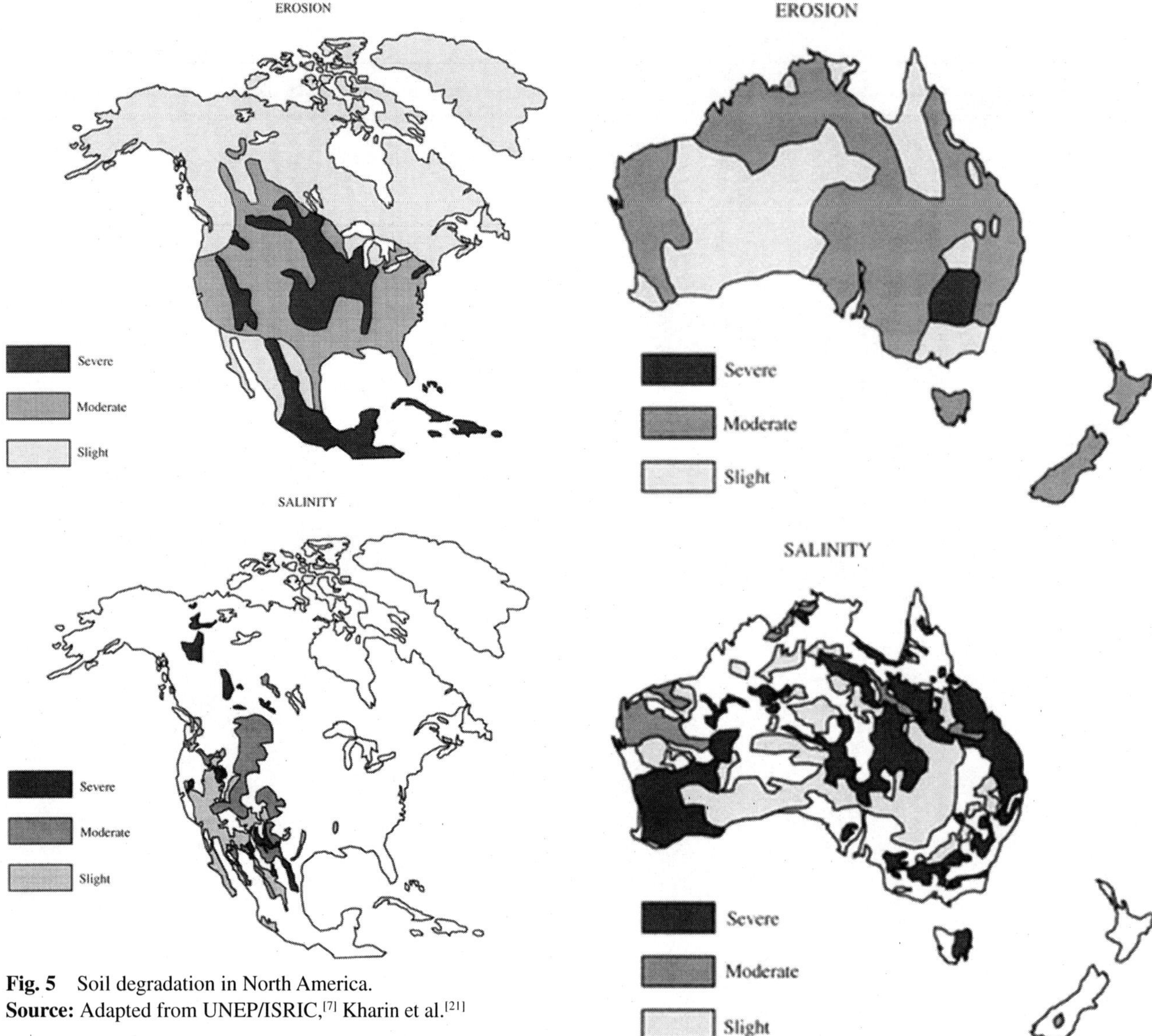

Fig. 5 Soil degradation in North America.
Source: Adapted from UNEP/ISRIC,[7] Kharin et al.[21]

Fig. 6 Soil degradation in Australia.
Source: Adapted from UNEP/ISRIC,[7] Kharin et al.[21]

on the continent to achieve temporary economic betterment. Identification of different types of soil degradation in Australia reveals that erosion has been the main component, primarily via dust storms, which are still a serious problem, especially where cropping practices do not include retention of cover and minimum tillage methods. Water erosion effects are also particularly severe in areas of summer rainfall and topographic extremities (Fig. 6). Although water is a scarce resource in Australia, about 14 million tons of soil is lost in Australia by water erosion. Remedial actions for this include the well-known measures of maintaining adequate cover and changing prevailing attitudes toward stock management, storage feed, redesign of watering sites, and management of riparian areas.

Part of the excess salinity in Australia is of primary origin and was retained in the subsoil by trees, which have now been cleared to create soil surfaces for cropping and pastures, allowing penetration of water to the saline subsoil, then followed by abstraction from the water table, thus leading to the ultimate disaster. About 30% of Australia's agricultural land is sodic, creating poor physical conditions and impeded productivity. This problem can only be alleviated by massive revegetation programs and by taking extra care of the water table and plant cover. Despite the introduction of costly conventional measures for reclamation, salinity levels continue to increase across Australia in the dry and irrigated soils. The dryland salinity in the continent affects about 5 million ha of farmland and is expanding at a rate of 3–5% per year.[30]

The retardation of organic matter levels also requires remediation measures, with economically justified fertilizer use strategies to be utilized throughout the continent. Moreover, overgrazing has resulted in the impoverishment of plant communities and loss of habitats as well as the decline in the chemical fertility of the soil by progressive depletion of organic matter in the topsoil, followed by deterioration in the soil structure.

Acidification caused by legume-based mixed farming plus use of ammonia-based fertilizers threatens 55 million ha of Australian land. Liming seems to be the most effective present remedy, but is costly, does not lead to rapid recovery, and is impractical for subsoil acidity. Thus, the precise remedies are yet to be developed for the conditions on this continent, utilizing careful, long-term monitoring and the experience of farmers to devise specific treatment and conservation procedures.

CONCLUSIONS

The state of soil degradation and its remedies as a multi-function–multi-impact approach have been identified through a Driving Force–Pressure–State–Impact–Response matrix by the European Environmental Agency[2] (Fig. 7) leading to sustainable land management (SLM)[31-33] measures to be taken for the future. SLM is concerned with more soil-friendly farming practices that minimize the erosion potential of soils, together with the adoption by landholders of property management planning procedures that involve community actions undertaken within several concerted action projects of the European Union, Food and Agriculture Organization, Global Environment Facility, United Nations Environment Programme, International Crops Research Institute for the Semi-Arid Tropics, Canadian International Development Agency, and World Bank.[34] Moreover, as Smyth & Dumanski[35] have stated, these combine socioeconomic principles with environmental concerns so that the production is enhanced, together with the reduction of its level of risk with the protection of natural resources, which would prevent degradation of soil and water quality to be successfully accepted by the farmer. The methods to be adopted for SLM via community actions include contour farming, terracing, vegetative barriers, and other land-use practices amalgamated with indigenous (traditional) technical knowledge (ITK)[33] as applied to farming and landscape preservation. The impetus to the use of ITK by scientists and local communities in creating new strategies for sustainable resource management was initiated in the United Nations Conference on Environment and Development held in Rio de Janeiro (Brazil) in 1992 and the need for raising awareness and combatting land degradation is man's priority issue for global social security.

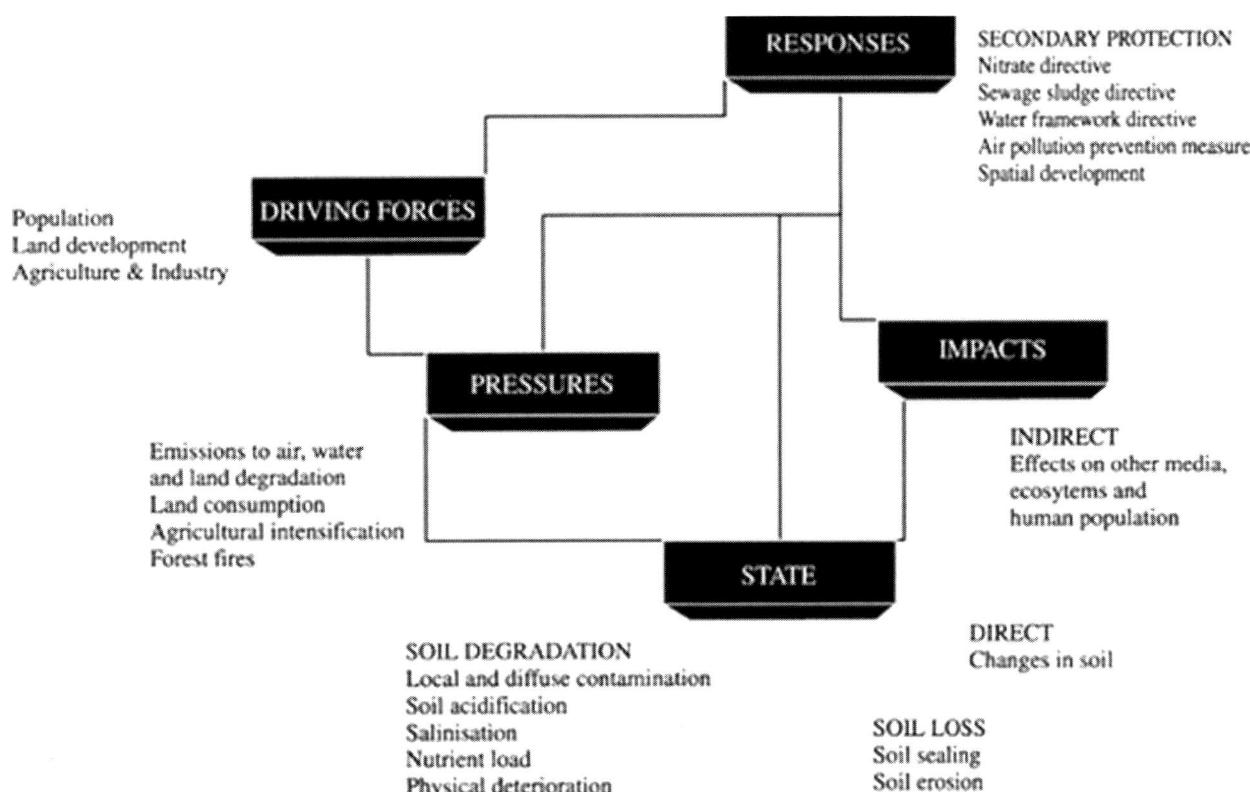

Fig. 7 Driving Force–Pressure–State–Impact–Response framework applied to soil.
Source: Adapted from World Bank.[31]

REFERENCES

1. International Food Policy Research Institute. The Economics of Desertification, Land Degradation, and Drought. Toward an Integrated Global Assessment. IFPRI Discussion Paper 01086, May 2011. Bonn, 188 p.
2. Toth, G.; Montanarella, L.; Rusco, E. *Threats to Soil Quality in Europe*; Institute for Environment and Sustainability, Land Management and Natural Hazards Unit: EUR 23438 EN. Milan, 2008; 162 p.
3. Amman, M.; Bertok, I.; Cabala, R.; Cofala, J.; Heyes, C.; Gyarfas, F.; Klimont, Z.; Schöpp, W.; Wagner, F. *A Further Emission Control Scenario for the Clean Air for Europe (CAFE) Programme*; International Institute for Applied Systems Analysis: Laxenburg, Austria, 2005.
4. Oldeman, L.R. The global extent of soil degradation. In *Soil Resilience and Sustainable Land Use*; CAB International: Oxon, U.K., 1994; 115 pp.
5. Steiner, K.G. *Causes of Soil Degradation and Development Approaches to Sustainable Soil Management*; Steiner, K.G., Ed.; GTZ. Weikersheim. Margraf Verlag: Filderstadt, Germany, 1996: 50 pp.
6. Lal, R. Forest soils and carbon sequestration. Forest Ecol. Manag. **2005**, *220*, 242–258 pp.
7. UNEP/ISRIC. In *GLASOD World Map of the Status of Human-Induced Soil Degradation, 1:10 M. 3 Sheets*; 2nd Revised Ed.; ISRIC: Wageningen, 1991.
8. Lal, R. Methods and guidelines for assessing sustainable use of soil and water resources in the tropics. In *N.21, SMSS Technical Monograph*; U.S. Agency for International Development: Washington, DC, 1994, 79 pp.
9. Eswaran, H.; Reich, P. Desertification: a global assessment and risks to sustainability. In *Proceedings of the 16th Int. Cong. Soil Science*; Montpellier, France, 1998; CD-ROM.
10. Eswaran, H.; Beinroth, F.H.; Reich, P. Global land resources and population supporting capacity. Am. J. Alternative Agric. **1999**, *14*, 129–136.
11. Kapur, S.; Atalay, P.İ.; Ernst, F.; Akça, E.; Yetiş, C.; İşler, F.; Öcal, A.D.; Üzel, İ.; Şafak, Ü. A review of the late quaternary history of Anatolia. In *1st International Symposium on Cilician Archaeology*; Durugönül, S., Ed.; Mersin University: Turkey, 1999; 253–272.
12. Cangir, C.; Kapur, S.; Boyraz, D.; Akça, E.; Eswaran, H. Land resource consumption in Turkey. J. Soil Water Conserv. **2000**, *3*, 253–259.
13. Millennium Ecosystem Assessment. *Ecosystems and Human Well-being: Desertification Synthesis*; World Resources Institute: Washington, DC, 2005.
14. Oldeman, L.R.; Hakkeling, R.T.A.; Sombroek, W.G. *World Map of the Status of Human-Induced Soil Degradation: An Explanatory Note*; International Soil Reference Center: Wageningen: the Netherlands, 1992; 34 pp.
15. Karpinsky, M.G. Aspects of the Caspian Sea benthic ecosystem. Mar. Pollut. Bull. **1992**, *24*, 3849–3862.
16. Kobza, J. Monitoring and contamination of soils in Slovakia in relation to their degradation. Proceedings of the 1st International Conference on Land Degradation. Kapur, S.; Akça, E.; Eswaran, H.; Kelling, G.; Vita-Finzi, C.; Mermut, A.R.; Öcal, A.D., Eds.; Adana, Turkey,1996; 10–14.
17. USDA-NRCS. *Risk of Human Induced Desertification, NRSC*; Soil Survey Division, World Soil Resources: 2003.
18. USDA-NRCS. *Global Desertification Vulnerability, NRSC*; Soil Survey Division, World Soil Resources: 2003.
19. Erol, O. Misuse of the Mediterranean coastal area in Turkey. In *Proceedings of the 1st International Conference on Land Degradation*, Kapur, S.; Akça, E.; Eswaran, H.; Kelling, G.; Vita-Finzi, C.; Mermut, A.R.; Öcal, A.D., Eds.; Adana: Turkey, 1999; 195–203.
20. Kosmas, C.; Ferrara, A.; Briassouli, H.; Imeson, A. Methodology for mapping environmentally sensitive areas (ESAs) to desertification. In *The MEDALUS Project. Mediterranean Desertification and Land Use Report*; Kosmas, C.; Kirkby, M.; Geeson, N., Eds.; European Commission: Belgium, 199; 31–47.
21. Kharin, N.G.; Tateishi, R.; Harahsheh, H. *Degradation of the Drylands of Asia*; Center for Environmental Remote Sensing Chiba Univ.: Japan, 1999; 81 pp.
22. Mnatsakanian, R. *Environmental Legacy of the Former Soviet Republics*; Center for Human Ecology, University of Edinburgh: Edinburgh, 1992.
23. Manfred, K.; Christian, P.; Markus, M. Strategic Environmental Assessment. Environmental Report. Central Europe Programme 2007–2013. Austrian Institute of Ecology: Wien. 44.
24. Lowdermilk, W.C. *Conquest of the Land through Seven Thousand Years. No. 99*; USDA, Soil Conservation Service, U.S. Government Printing Office: Washington, DC, 1997, 1–30.
25. Sanders, J.H.; Southgate, D.D.; Lee, J.G. *The Economics of Soil Degradation: Technological Change and Policy Alternatives, no. 22*; SMSS Technical Monograph; Department of Agricultural Economics, Purdue University: 1995; 8–11.
26. Croitoru, L.; Sarraf, M. *The Cost of Environmental Degradation*; Case Studies from the Middle East and North Africa. The World Bank Washington: DC 2010.
27. Lal, R. Monitoring soil erosion's impact on crop productivity. In *Soil Erosion Research Methods*; Lal, R., Ed.; Soil and Water Conservation Society: Ankeny, IA, 1988, 187–200.
28. Barrow, C.J. *Land Degradation*; Cambridge University Press: Cambridge, 1994.
29. Derpsch, R.; Friedrich, T.; Kassam, A.; Hongwen, L. Current status of adoption of no-till farming in the world and some of its main benefits. Int J Agric and Biol Eng. 1–25 pp.
30. State of the Environment 2011 Committee. *Australia State of the Environment 2011*; Independent report to the Australian Government Minister for Sustainability, Environment, Water, Population and Communities. Canberra: DSEWPaC, 2011
31. World Bank. Sustainable Land Management. Challenges, Opportunities, and Trade-offs. World Bank: Washington DC. 2006; 112P.
32. Eswaran, H.; Beinroth, F.; Reich, P. Biophysical considerations in developing resource management domains. In *Proceedings of the IBSRAM International Workshop on Resource Management Domains*, Kuala Lumpur, Aug. 26–29, 1996; Leslie, R.N., Ed.; The International Board for Soil Research and Management (IBSRAM): Thailand, 1996; 61–78.
33. Eswaran, H.; Berberoglu, S.; Cangir, C.; Boyraz, D.; Zucca, C.; Özevren, E.; Yazıcı, E.; Zdruli, P.; Dingil, M.; Dönmez, C.; Akça, E.; Çelik, I.; Watanabe, T.; Koca, Y.K.;

Montanarella, L.; Cherlet, M.; Kapur, S. The Anthroscape Approach in Sustainable Land Use. In *Sustainable Land Management: Learning from the Past for the Future*. Kapur, S.; Eswaran, H.; Blum, W.E.H. Eds.; Heidelberg, Springer: 2011; 1–50.

34. Zdruli, P.; Steduto, P.; Kapur, S.; Akça, E., Eds. Ecosystem-based assessment of soil degradation to facilitate land users' and land owners' prompt actions. Medcoastland Project Workshop Proceedings, June 2–7, 2003; Adana, Turkey, 2006.

35. Smyth, A.J.; Dumanski, J. A frame work for evaluating sustainable land management. Canadian J. soil Sci. **1995**, *75* (4), 401–406.

Soil: Erosion Assessment

John Boardman
Department of Geographical and Environmental Science, University of Cape Town, Cape Town, South Africa and Environmental Change Institute, University of Oxford, Oxford, U.K.

Abstract
Soil erosion is the loss of soil from the surface of the Earth. It is a two-stage process with detachment of soil particles preceding transport by the agency of water or wind. It is a natural process occurring at relatively low rates but which may be accelerated by human actions. Erosion is driven by both socioeconomic ("ultimate factors") and physical factors such as slope, soil, rainfall ("proximal factors"). Most erosion is associated with the formation of rills and gullies. Erosion has serious impacts on the fertility of soils, on water quality, and on reservoir storage capacity. Methods of assessing erosion include experimental plots, sediment yield in rivers, field monitoring, remote sensing, Cesium-137, historical analysis, expert opinion, and modeling. Combinations of these have often been used. Assessment of actual and potential erosion is necessary so that measures can be taken to prevent the loss of soil and limit the off-site impacts of erosion. A range of conservation measures have been used with varying degrees of success.

INTRODUCTION

Soil erosion is the loss of soil from the surface of the Earth (Fig. 1). It is a two-stage process with detachment of soil particles preceding transport by the agency of water or wind. It is a natural process occurring at relatively low rates, but which may be accelerated by human actions. Most erosion is associated with the formation of rills and gullies. Erosion has serious impacts on the fertility of soils, on water quality, and on reservoir storage capacity. Assessment of actual and potential erosion risk is necessary so that measures can be taken to prevent the loss of soil.

Agents of erosion are wind, water, and gravity (causing mass movement). Movement of people and animals may contribute, and soils will be more susceptible if weakened by weathering (wetting and drying, frost, etc.).

Erosion is an important contributor to global soil degradation. It is also a major component of desertification. Under semiarid conditions, vegetation damage typically precedes erosion. If climate variability is a major influence, then degradation may be cyclic, interspersed with periods of improved land quality. Frequently, because of population pressure, degraded land does not recover.

The major impact of soil erosion is on agricultural land, both arable and grazing. Other areas of concern are construction sites and protected natural areas (e.g., footpath erosion in National Parks).

This entry is mainly concerned with erosion by water which globally appears to affect twice the area of wind erosion.[1] Overgrazing as a cause of erosion is very widespread but is difficult to classify as it often leads to erosion by water and, in some areas, by wind.

The causes of erosion may be regarded as a) proximal and b) ultimate. Proximal causes relate to physical factors such as rainfall, runoff, wind velocity, gradient, and soil type. Ultimate causes relate to socioeconomic conditions that influence farmer decision-making, and therefore land-use and farming practices. Important farmer decisions concern what crops to plant, where to plant them and how to plant them. Such decisions are strongly influenced by farmer attitudes and traditions, community views and, especially in the Developed World, by government support for agriculture. Some such support may be regarded as a "perverse subsidy" in that it leads to unwanted and unforeseen environmental impacts including erosion and pollution; examples include European Union support for almonds in Spain,[2] land leveling in Norway[3] and olive oil production in Spain.[4] A major socioeconomic impact on eastern European landscapes was the collectivization of agriculture under Communism.[5,6]

The relationship between population growth, food production and exploitation of the soil is suggested by Montgomery (p. 84):[7]

"A fundamental model of agricultural development in which prosperity increases the capacity of the land to support people, allowing the population to expand to use the available land. Then, having eroded soils from marginal land, the population contracts rapidly before soil rebuilds in a period of low population density."

Major texts on erosion, its causes and consequences include those by Bennett,[8] Blaikie[9] Morgan,[10] Boardman and Poesen,[11] and Montgomery.[7]

Encyclopedia of Natural Resources DOI: 10.1081/E-ENRL-120047487

SCALE AND EXTENT OF EROSION

Rates of erosion on grassed or wooded areas are usually below 1 t/ha/yr. Rates of soil formation are very low—similar to natural rates of erosion. The oft-quoted USDA figure is 1 inch per 500 years which is about 0.6t/ha/yr.[7] Erosion in excess of this figure means the loss of, or decline in, a finite resource. Erosion rates on farmland in temperate areas occasionally exceed 100t/ha/yr. On the Loess Plateau of China, 7.2 Mha was reported to have erosion rates in excess of 100 t/ha/yr.[12]

Hotspots of global erosion are difficult to define and rank due to lack of, or unreliable data. However, we may suggest the following:[13]

- Loess plateau, the Yangtze basin and the southern hill country of China
- Ethiopia
- Swaziland and Lesotho
- The Andes
- South and East Asia
- Iceland
- Madagascar
- The Himalayas
- The west African Sahel
- Caribbean and Central America

There is very limited data for some of these areas particularly the Andes and Madagascar. If areas that have been severely affected in the past and to some extent are still vulnerable are to be included then the list would be extended:

- The Dust Bowl in the 1930s in the USA (parts of Colorado, Kansas, New Mexico, Texas and Oklahoma)[14] (Fig. 2)
- The Virgin Lands Project of the former Soviet Union[15]
- The Mediterranean basin[16]

Fig. 1 Severely eroded winter wheat fields, South Downs, UK, October 1987 (J. Boardman).

IMPACT OF EROSION

Impacts of erosion are the following:

1. On-site, which includes loss of crop, seeds, and fertilizer and thinning of the soil layer thus directly affecting present and future productivity. The amount of yield reduction is disputed with recent work suggesting a figure of 4% per 10 cm of soil loss.[17] In extreme cases, land may become unworkable or unprofitable because of soil loss and nutrient depletion, as in the case of badlands (Fig. 3). Montgomery discusses many cases where land is abandoned and farmers move on e.g., 18th century tobacco growing in Virginia, Georgia, and the Carolinas or the Amazon lowlands today (p. 118).[7] This process may result in exploitation of less suitable land or land at greater risk of erosion e.g., Rome (p. 62).[7]
2. Off-site, where people, water/air, or property not directly responsible for erosion are affected. Runoff from agricultural land generally transports sediment particles and attached phosphorous and pesticides. Impacts include dam sedimentation,[18] pollution of water courses,[19] and property damage by muddy floods.[20]

Fig. 2 Dust storm approaching Stratford, Texas Dust bowl surveying in Texas, April 18, 1935. Image ID: theb1365, Historic C&GS Collection, NOAA George E. Marsh Album.

Fig. 3 Badlands, northern Spain (J. Boardman).

Costs of erosion are underresearched. This is partly because it is difficult to put a value on soil. As soils thin crop yields decline or they are artificially enhanced by the use of fertilizers. This is a cost which less developed societies cannot afford. Serious soil depletion results in abandonment and migration (see above). Costs resulting from off-site impacts are easier to estimate; for example, water cleaning costs in England and Wales are estimated at several hundred million pounds per year for the removal of nitrates, phosphates and pesticides, some of which are associated with runoff and erosion events.[21] The costs of muddy floods in central Belgium range from 16 to 172 million euros per year.[22] Fine-particle sediments (mainly of agricultural origin) affect fish spawning in gravel-bedded rivers.[23,24] Sedimentation of ports is recorded from the Mediterranean and from the eastern states of the United States (p. 139).[7]

Impacts may also be seen as either immediate or long-term. The former includes the formation of gullies, the burial of crops, and the flooding of houses; the latter refers to the gradual thinning of soils or sedimentation of reservoirs. Unfortunately the long-term effects are easy to overlook. Societies in the Middle East, Rome, Greece, MesoAmerica, etc. have all been seriously affected by soil loss.[7]

EROSION ASSESSMENT

Erosion assessment may describe the "actual" risk of erosion: the Global Assessment of Soil Degradation project (GLASOD) is an example of this type of assessment applied globally.[1] Similarly, Canada has published maps of the risk of wind and water erosion at the provincial level.[25] New Zealand has combined, on single maps at 1: 250,000 scale, the actual and "potential" risk of erosion.[26] Potential in this case refers to a change of land use to pastoral or arable. Maps from the CORINE project show actual and potential soil erosion in southern Europe; the latter is defined as the worst possible situation that might be reached.[27]

Erosion assessment may focus on changes through time in the same area. Trimble[28] has measured and modeled changes in erosion and deposition rates in the Coon Creek catchment in Wisconsin since the beginning of European settler farming in the 1850s. This study relates erosion to land use change, management, and conservation techniques, as well as to the capacity of rivers to transport eroded materials. It therefore links on-site and downstream effects through time and space. Erosion rates since Neolithic clearance have been modeled on the loessic soils of the South Downs, UK. This assessment is based on changing land use, climate, soils, and farming practices (Fig. 4).[29] Assessment

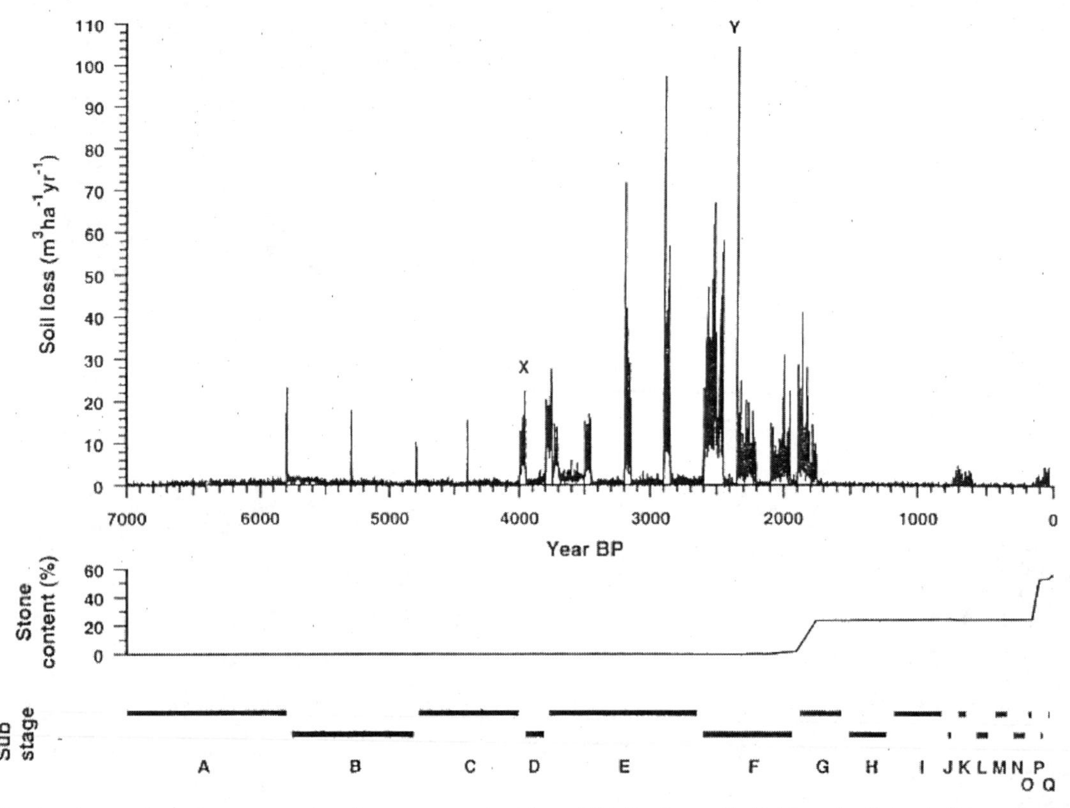

Fig. 4 Simulated annual erosion rates and surface stone content for a "thin" soil profile (1.22 m depth to chalk) on the South Downs, southern England
Source: Adapted from Favis-Mortlock et al.[29]

Soil—Soil

of erosion and its impacts may be combined in complex scenarios of change involving population pressure e.g., Hurni for Ethiopia.[30]

Modeling constitutes an important approach to erosion assessment. This is for several reasons: many areas lack the data that allow direct estimates of erosion; modeling may be based on scenarios of future land use or climate and therefore estimates may be made of erosion under assumed conditions; and, application of a model may be far less expensive and time-consuming than field surveys.

The spatial scale of modeling may vary from a few m^2 to thousands of km^2. The global change and terrestrial ecosystems (GCTEs) model evaluation exercise divides models into field, catchment, and landscape, based on spatial scale.[31] Resolution of erosion models is determined by grid size, that is, km^2 grid size used in CORINE for southern Europe and the 250 m grid used in the maps of seasonal erosion risk in France.

Assessments are frequently expressed in map form.[32] Landscape features indicative of erosion may be mapped as a record of an actual event or series of events. In arable landscapes features may be temporary; thus rills, ephemeral gullies, and fans are usually plowed out annually e.g., erosional features produced as a result of a snowmelt event in May 1993 in Slovakia.[33] In semiarid, grazed landscapes erosional features may become permanent, e.g., gullies and badlands. Assessments should distinguish features active at the present time from those that are inactive and represent erosion under former conditions.

Several methods of erosion assessment are commonly used and Table 1[34] and will be reviewed briefly.

1. Small experimental plots in the laboratory or in the field; these may operate with natural or simulated rainfall[35] (Figs. 5 and 6). They are assumed to represent larger parts of the landscape. Results are frequently extrapolated to fields or catchments and are used as input to models. Extrapolation from small plots to regional, national, or continental size areas is not acceptable.[36]
2. Sediment yield of rivers represents the net loss of soil from a catchment.[37] Erosion rates on fields may be very different owing to storage of soils between the field and the river. Highest rates of sediment yields from a large river are from the Yellow River, China, with an average silt content of 38 kg m^3.[12] Rivers of east and south Asia deliver about 67% of the sediment reaching the world's oceans; rates are highest in this region owing to tectonic uplift and steep slopes, high rainfall, deforestation, and population pressure resulting in intensive farming activities.[38] Alternatively, the sediment yield from river basins may be estimated by measurement of the volume of sediment trapped in a still water body such as a lake, reservoir, or pond. A date for the construction of a reservoir is often known. Sediment including bedload will be trapped. Losses in water passing through the dam may be estimated (Trap Efficiency). The advantage of this method is the possibility of medium-term records of >100 years.[39] Sediment yield to three small dams in the Karoo, South Africa, is estimated at 115, 357 and 654 t km^{-2} yr^{-1} for a period of 65 years. These rates are adjusted to take into account trap efficiency and the contrasts are due to differences in land use within the catchments.[40] Similar approaches have been used in Australia[41] and in Ethiopia.[42]
3. Field monitoring is defined as "field-based measurement of erosional and/or depositional forms over a significant area (e.g., >10 km^2) and for a period of >2 years."[43] Regular measurements of volumes of soil loss from rills and gullies or deposited in fans are made. Notable examples are the monitoring of

Table 1 Assessment methods: their merits and limitations

Assessment method	Merits	Limitations
Experimental plots in lab or field	Control over erosion factors; ease of measurement of runoff and soil	Small size: difficult to extrapolate to larger areas
Sediment yield of rivers	Ease of measurement; may cover >100 years if reservoir sediments used	Measures total soil lost from watershed, not from specific areas
Field monitoring	Records soil lost from rills and gullies; covers large areas; inexpensive	Gives very approximate amounts of soil lost
Remote sensing	Covers large areas and may cover long time-span	Limited availability for many areas; may not be suitable for small scale features
Cesium-137	Gives 55-year average	Many unknowns: unreliable without careful calibration; expensive and time-consuming
Stratigraphy and pedology	Deposited soil may contain artifacts; suitable for long-time-period studies	Scarcity of suitable sites; low precision of some dating methods
Expert opinion	May reveal unrecorded information	Subjectivity: difficult to compare different areas
Models	Objectivity; ease of use even by "non-experts"; numerical output	Availability of data; complexity of process descriptions; frequent lack of validation

Fig. 5 Experimental plot (flume), University of Toronto (J. Boardman).

Fig. 6 Experimental plots, South-Limbourg, the Netherlands (J. Boardman).

17 localities in England and Wales during 1982–1986,[44] a 10-year program on the South Downs, England during 1982–1991,[45] monitoring of ephemeral gullies in Belgium during 1982–1993[46] and a long-term monitoring scheme in the Swiss midlands.[47,48]

Simple techniques of measurement and assessment of erosion suitable for Less Developed Countries include pedestal height, depth of armor layer, plant/tree root and fence post exposure, tree mounds and build up of sediment behind barriers and in drains.[49] Many studies have used erosion pins to assess rates of erosion over time (Fig. 7). These are particularly suited to bare degraded land such as badlands. Rates may be related to topographic position, presence of vegetation, weathering and rainfall/wind factors[50] (Fig. 8).

4. Remote sensing is used increasingly in erosion assessment. For rill and gully erosion, it may be used as a locational technique to allow detailed field measurements to be made at selected sites. Gullies and badlands may be identified on black and white air photographs of 1:20,000 scale; in the Karoo of South Africa comparisons have been made between the extent of gullying in 1945 and 1980.[51] Also in South Africa, in a pioneering

Fig. 7 Erosion pin site in the Karoo, South Africa (J. Boardman).

study, Talbot mapped gullying in the Swartberg resulting from land conversion to wheat farming[52] (Fig. 9). More recently, the extent of Icelandic land degradation has been assessed using Landsat imagery.[53]

5. Cesium-137, derived from weapons testing since 1954, is used as a tracer to assess amounts of erosion and deposition. Annual average rates for the last ca. 55 years are estimated by measurement of the amount and distribution of cesium in comparison with an uneroded reference site.[54] Studies have been carried out in Australia, Canada, China, the Netherlands, Poland, Thailand, United Kingdom, and United States. However, results are controversial with particular issues regarding the original assumed regular distribution of cesium across the landscape via the assumed homogenous spatial distribution of rainfall; and the assumption that cesium "sticks" to fine soil particles in a predictable way. These assumptions have recently been strongly challenged and erosion rates obtained using this technique now appear unreliable.[55]
6. Total amounts of soil loss since a historical baseline such as woodland clearance or the beginning of European settlement have been assessed by comparison of existing soil profiles with uneroded ones. Evans estimates historical losses of topsoil of 150–250 mm in lowland England and Wales, much of which is now stored as alluvium and colluvium.[56] Sedimentation above marker horizons in reservoirs may be used to compute sediment yield for periods of time.[40] Pedological and stratigraphic approaches have also been used to assess the effect of past extreme events, for example, the major storms and floods in Germany in the early 14th century.[57]
7. Expert opinion has been used to assess erosion on a regional or global scale. GLASOD assesses erosion at a global scale (10 M).[1] A total of 12 degradation types are recognized under the broad headings of water erosion, wind erosion, chemical deterioration and physical deterioration. The "degree" of degradation is assessed for each mapping unit, as is its "relative extent." A combination of degree and relative extent gives an assessment of "severity" which is expressed in "severity classes" and mapped in four colors. However,

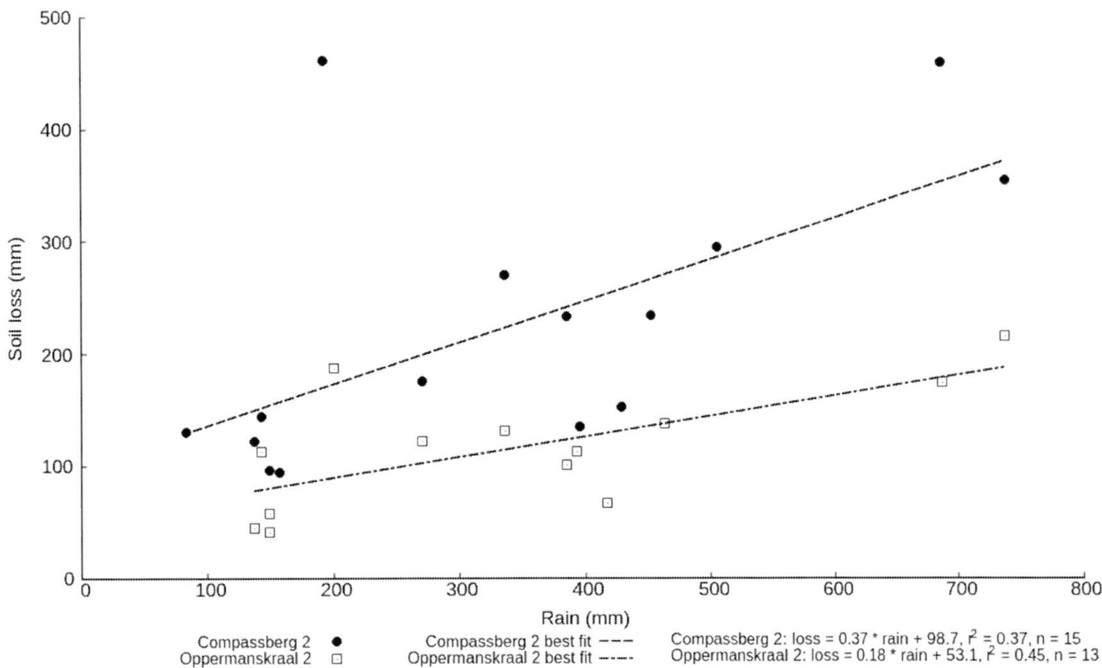

Fig. 8 Average rates of erosion for eroding pins at two sites in the Karoo, South Africa plotted against rainfall amount for each measurement period (J. Boardman).

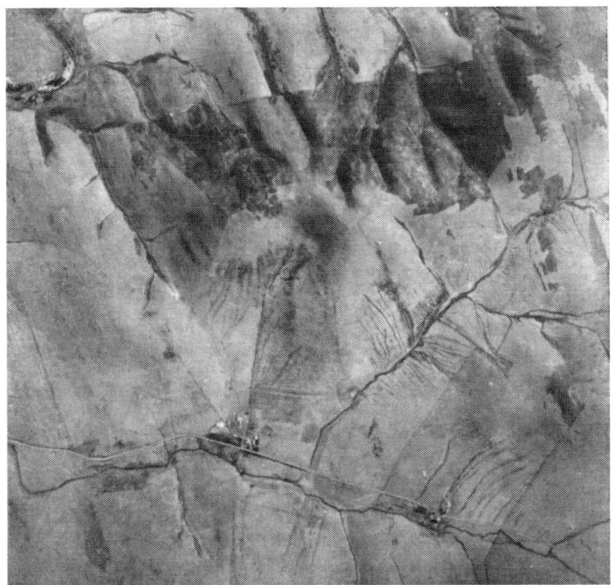

Fig. 9 Severely gullied hill country ca. 8 km northwest of Durbanville, South Africa. Gullies are on ploughed land. Photography March 1938: the land had been abandoned by 1944.
Source: Adapted from Talbot.[52]

there is evidence that GLASOD is a very imperfect predictive tool. Van-Camp et al.[58] are particularly critical and suggest that the GLASOD approach should be abandoned: comparison with maps and data for Spain show that assessment of water erosion was poor.[59] GLASOD has also been assessed by other authors.[60] At a more local and detailed level, expert opinion appears more reliable: see for example a recent South African assessment.[61]

8. Assessment of erosion has frequently been based on the application of models. The Universal Soil Loss Equation (USLE) is most widely used because of its simplicity and low-level data requirements.[62] Recent developments in modeling have concentrated on simulation of erosional processes and computerization. Models may be used to estimate average, long-term erosion rates under specified conditions (USLE), or to assess the effect of a particular rainfall/runoff event e.g., EUROSEM.[63] Erosion models have also been used to predict wind erosion, and to assess chemical losses in solution or attached to soil particles (for reviews, see Morgan).[64]

CONTROLLING SOIL EROSION AND OFF-SITE IMPACTS

Controlling soil erosion is based on two principles, protecting the soil from runoff or wind by providing vegetative protection, or reducing the slope and thereby decreasing water velocity. Erosion generally takes place on bare soil (Fig. 1), and therefore the protection of a crop or crop debris at vulnerable times of the year is important. Most radical solutions such as minimum tillage and conservation tillage aim to have some degree of vegetation cover throughout the year, and therefore drill the new crop through the remnants of the old thus avoiding the bare ground inherent in tillage systems involving the plough.

Large areas of the same crop give rise to the risk of runoff at the same time. Thus the break-up of large fields into

Fig. 10 (**A**) Old abandoned terraces and recently rebuilt terraces, Mallorca, Spain (J. Boardman); (**B**) Abandoned terraces, central Corsica, France (J. Boardman); (**C**) Abandoned and degraded terraces, Eritrea (J. Boardman).

strips of different crops can reduce the risk. Valley-bottom zones of concentrated flow may be grassed (grass waterways) to prevent erosion. Buffer strips of grass are widely used around arable fields in order to reduce runoff volumes and loss of soil from fields. Evans and Boardman show how return to grass of a small area interrupts valley-bottom runoff from large areas of winter cereals thus preventing flooding of houses.[65]

The desire to reduce slope gradients has traditionally been addressed by the construction of terraces. An added advantage is that a terrace allows for cultivation of slopes that were too steep for agriculture. Terrace systems are known from the Middle Bronze Age on Crete[66] and from the Incas of Peru. Many systems around the Mediterranean are now abandoned and erosion associated with their post-abandonment state has been much researched.[67] Maintenance of terraces is therefore vital if they are to prevent erosion (Fig. 10).

It is doubtful if any one conservation technique will be effective on its own. Most successful schemes involve several approaches. For example, the Melsterbeek Catchment soil and water conservation scheme in Flanders uses grass waterways, buffer strips, and retention ponds (Fig. 11). Successful schemes, such as this, must involve the local population of land owners and managers which generally requires financial support and technical assistance.[68]

In the Palouse wheat growing area of Washington State high erosion rates have persisted for over a 100 years and are particularly associated with the practice of summer fallow. The costs of reducing erosion by conservation practices are well illustrated:

"Applying various measures of conservation treatment to the land will affect the economy of the basin. But erosion rates can be reduced by 40% in the low and high precipitation zones and 60% in the intermediate precipitation zone—without decreasing farm income. The erosion rate can be reduced 80% through maximum levels of conservation practices and retirement of 35,000 acres of steep, erodible land. Achieving this level of reduction would cost

Fig. 11 Mitigation measures: grass buffers along field edges, the Melsterbeek catchment, Flanders (Belgium) (Karel Vandaele, Samenwerking Land & Water).

more than $29 million in reduced productivity and increased operating costs."[69]

Soil erosion is often best addressed in a wide framework of natural resource improvements. The issue of sustaining the livelihoods of the farmers as well as introducing conservation measures is critical. Examples of successful approaches from West Africa, New Zealand, southern Brazil, northwest India, northern Cameroon, and Kenya are discussed by El Swaify et al.[70] There is also the well-known case of the Machakos District, Kenya.[71] Hudson discusses reasons for failure of soil conservation projects.[72]

Previously quoted extremely high erosion rates for the Loess Plateau, China, and associated high sediment yields for the Yellow River, appear to have been substantially reduced in recent years. Water–soil conservation practices are credited with 40% reduction in sediment loads, sediment trapping by major reservoirs with 30% and most of the remaining is due to precipitation decrease.[73]

CONCLUSION

The assessment of erosion is necessary in order to understand the scale and extent of the problem. Many assessment methods have been used including experimental plots, sediment yield in rivers, field monitoring, remote sensing, Cesium-137, historical analysis, expert opinion, and modeling. All have advantages and disadvantages and may not be appropriate for particular circumstances. Assessment will include an evaluation of the type of erosion processes and their relative importance. The design of mitigation measures will depend on the outcome of the assessment. Mitigation measures must take into account the driving forces of erosion particularly its socioeconomic context and the need to support local peoples in recommending appropriate solutions.

REFERENCES

1. Oldeman, L.R.; Hakkeling, R.T.A.; Sombroek, W.G. *World Map of the Status of Human-Induced Soil Degradation: An Explanatory Note*; ed. ISCRIC and UNEP, 1990; p. 32.
2. Faulkner, H. Gully erosion associated with the expansion of unterraced almond cultivation in the coastal Sierra de Lujar, S. Spain. Land Degrad. Rehabil. **1995**, *6* (1), 115–127.
3. Lundekvam, H.E.; Romstad, E.; Oygarden, L. Agricultural policies in Norway and effects on soil erosion. Environ. Sci. Policy **2003**, *6* (1), 57–68.
4. de Graaff, J.; Eppink, L.A.A.J. Olive oil production and soil conservation in southern Spain, in relation to EU subsidy policies. Land Use Policy **1999**, *40*, 259–267.
5. Stankoviansky, M. Geomorphic effect of surface runoff in the Myjava Hills, Slovakia. Zietshrift fur Geomorphologie **1997**, *110*, 207–217.
6. Boardman, J.; Poesen, J.; Evans, R. Socio-economic factors in soil erosion and conservation. Environ. Sci. Policy **2003**, *6* (1), 1–6.
7. Montgomery, D.R. *Dirt: The Erosion of Civilizations*; University of California Press, Berkeley, 2007.
8. Bennett, H.H. *Soil Conservation*; McGraw Hill: New York, 1939.
9. Blaikie, P. *The Political Economy of Soil Erosion in Developing Countries*: Longman, 1985.
10. Morgan, R.P.C. *Soil Erosion and Conservation;* Third Edition, Blackwell: Oxford, 2005.
11. Boardman, J.: Poesen, J. *Soil Erosion in Europe*; John Wiley: Chichester, 2006.
12. Wen Dazhong. Soil erosion and conservation in China. In *World Soil Erosion and Conservation*, Pimentel, D., Ed.; Cambridge University Press: Cambridge, 1993: 63–85.
13. Boardman, J. Soil erosion science: reflections on the limitations of current approaches. Catena **2006**, *68*, 73–86.
14. Worster, D. *Dust Bowl: The Southern Plains in the 1930s*; Oxford University Press: New York, 1979.
15. McCauley, M. *Khrushchev and the Development of Soviet Agriculture*; The Virgin Lands Program 1953–1964, Holmes & Meier Publishers, Inc.: New York, 1976.
16. Faulkner, H., Hill A. Forests, soils and the threat of desertification. In *The Mediterranean: Environment and Society*, King, R., Proudfoot, L., Smith, B. Eds.; Arnold: London, 1997; 252–272.
17. Bakker, M.M.; Govers, G.; Rounsevell, M.D.A. The crop productivity-erosion relationship: an analysis based on experimental work. Catena **2004**, *57*, 55–76.
18. Renwick, W.H. Continental-scale reservoir sedimentation patterns in the United States. In *Erosion and Sediment Yield: Global and Regional Perspectives*; Walling, D.E., Webb, B.W. Eds., Proceedings of the Exeter Symposium, July 1996. IAHS Publication: 236, 1996, Wallingford, 513–522.
19. Evans, R. Pesticide run off into English rivers – a big problem for farmers. Pesticides News **2009**, *85*, 12–15.
20. Boardman, J.; Verstraeten, G.; Bielders, C. Muddy floods. In *Soil Erosion in Europe;* Boardman, J., Poesen. J. Eds.; John Wiley: Chichester, 2006, 743–755.
21. Evans, R. Soil erosion and land use: towards a sustainable policy, In *Soil Erosion and Land Use: Towards a Sustainable Policy*; Cambridge Environmental Initiative, Professional Seminar Series 7, White Horse Press: 1995; 14–26.

22. Evrard, O.; Bielders, C.; Vandaele, K.; van Wesemael, B. Effectiveness of erosion mitigation measures to prevent muddy floods in central Belgium, off-site impacts and potential control measures. Catena **2007**, *70*, 443–454.
23. Theurer, F.D.; Harrod, T.R.; Theurer, M. *Sedimentation and Salmonids in England and Wales*; Technical Report, Environment Agency: England and Wales, 1998.
24. Birt, K. Runoff from potato farms blamed for fish kills on Canadian Island. J. Soil Water Conserv. **2007**, *62* (6), 136A.
25. Manitoba. *Water Erosion Risk Map 1:1,000,000, Agriculture Canada*; Publication 5259/B: Ottawa, Canada, 1988.
26. Prickett, R.C. Sheet 23 Oamaru: Erosion Map of New Zealand, 1:250,000, National Water and Soil Organisation, Wellington, New Zealand, 1984.
27. CORINE. *Soil Erosion Risk and Important Land Resources in the Southern Regions of the European Community*; EUR 1323: Luxembourg, 1992.
28. Trimble, S.W. Decreased rates of alluvial sediment storage in the Coon Creek Basin, Wisconsin, 1975–1993. Science **1999**, *285*, 1244–1246.
29. Favis-Mortlock, D.T., Boardman, J., Bell, M. Modelling long-term anthropogenic erosion of a loess cover, South Downs, UK. Holocene **1997**, *7* (1), 79–89.
30. Hurni, H. Land degradation, famine, and land resource scenarios in Ethiopia. In *World Soil Erosion and Conservation,* Pimentel, D., Ed.; Cambridge University Press: Cambridge, 1993; 27–61.
31. Boardman, J.; Favis-Mortlock, D.T. Modelling soil erosion by water. In *Modelling Soil Erosion by Water*, Boardman, J., Favis-Mortlock, D.T. Eds.; NATO ASI Series; Springer: Berlin, 1998; 3–6.
32. Morgan, R.P.C. Chapter 4, Erosion hazard assessment. In *Soil Erosion and Conservation*; Third Edition, Blackwell: Oxford, 2005.
33. Stankoviansky, M. Geomorphic effect of surface runoff in the Myjava Hills, Slovakia. Zeitshrift fur Geomorphologie **1997**, *110*, 207–217.
34. Boardman, J. Soil erosion: the challenge of assessing variation through space and time. In *Geomorphological Variations*, Goudie, A.S., Kalvoda, J. Eds., Nakladatelsti P3K: Prague, 2007; 205–220.
35. Mutchler, C.K.; Murphree, C.E.; McGregor, K.C. Laboratory and field plots for erosion studies. In *Soil Erosion Research Methods*, Lal, R., Ed., Soil and Water Conservation Society: Ankeny, Iowa, 1988.
36. Boardman, J. An average soil erosion rate for Europe: myth or reality? J. Soil Water Conserv. **1998**, *53* (1), 46–50.
37. Walling, D.E. Measuring sediment yield from river basins. In *Soil Erosion Research Methods*; Lal, R. Ed., Soil and Water Conservation Society: Ankeny, IA 1988, 9–73.
38. Milliman, J.D.; Meade, R.H. World-Wide delivery of river sediment to the oceans. J. Geol. **1983**, *91*, 751–762.
39. Verstraeten, G.; Bazzoffi, P.; Lajczak, A.; Radoane, M.; Rey, F.; Poesen J.; de Vente, J. Reservoir and pond sedimentation in Europe. In *Soil Erosion in Europe,* Boardman, J., Poesen. J. Eds.; John Wiley: Chichester, 2006, 759–774.
40. Boardman, J.; Foster, I.; Rowntree, K.; Mighall, T.; Gates, J. Environmental stress and landscape recovery in a semi-arid area, the Karoo, South Africa. Scott. Geogr. J. **2010**, *126* (2), 64–75.
41. Erskine, W.D.; Mahmoudzadeh, A.; Myers, C. Land use effects on sediment yields and soil loss rates in small basins of Triassic sandstone near Sydney, NSW, Australia. Catena, **2002**, *49* (4), 271–287.
42. Haregeweyn, N.; Poesen, J.; Nyssen, J.; De Wit, J.; Haile, M.; Govers, G.; Deckers, S. Reservoirs in Tigray (northern Ethiopia): characteristics and sediment deposition problems. Land Degrad. Dev. **2006**, *17*, 211–230.
43. Boardman, J. Soil erosion by water: problems and prospects for research. In *Advances in Hillslope Processes*, Anderson, M.G. and Brooks, S.M. Eds.; Volume 1, Wiley: 1996; 489–505.
44. Evans, R. Extent, frequency and rates of rilling of arable land in localities in England and Wales. In *Farm Land Erosion: In Temperate Plains Environment and Hills*, Wicherek, S. Ed., Elsevier: Amsterdam, 1993; 177–190.
45. Boardman, J. Soil erosion and flooding on the eastern South Downs, southern England, 1976–2001. Trans. Inst. Br. Geogr. **2003**, *28*, 176–196.
46. Poesen, J. Gully typology and gully control measures in the European loess belt. In *Farm Land Erosion: In Temperate Plains Environment and Hills*, Wicherek, S. Ed.; Elsevier: Amsterdam, 1993, 221–239.
47. Prasuhn, V. Soil erosion in the Swiss midlands: results of a 10-year field survey. Geomorphology **2011**, *126*, 32–41.
48. Prasuhn, V. On-farm effects of tillage and crops on soil erosion measured over 10 years in Switzerland. Soil & Tillage Research **2012**, *120*, 137–146.
49. Stocking M.A.; Murnaghan, N. *Handbook for the Field Assessment of Land Degradation*; Earthscan: 2001, London.
50. Haigh, M.J. The use of erosion pins in the study of slope evolution. In *Shorter Technical Methods (11), Technical Bulletin 18*, Br. Geomorphol. Res. Group, 1977, 31–49.
51. Keay-Bright, J.; Boardman, J. Evidence from field based studies of rates of erosion on degraded land in the central Karoo, South Africa. Geomorphology **2009**, *103*, 455–465.
52. Talbot, W.J. *Swartland and Sandveld*; Oxford University Press: Cape Town, 1947.
53. Arnalds, O.; Borarinsdottir, E.F.; Metusalemsson, S.; Jonsson, A.; Gretarsson, E.; Arnason, A. Soil Erosion in Iceland, Soil Conservation Service and the Agricultural Research Institute, Iceland, 2001.
54. Quine, T.A.; Walling, D.E. Assessing recent rates of soil loss from areas of arable cultivation in the UK. In *Farm Land Erosion: In Temperate Plains Environment and Hills*, Wicherek, S., Ed., Elsevier: Amsterdam, 1993; 357–371.
55. Parsons, A.J.; Foster, I.D.L. What can we learn about soil erosion from the use of ^{137}Cs? Earth-Sci. Rev. **2011**, *108*, 101–113.
56. Evans, R. Soil erosion: its impact on the English and Welsh landscape since woodland clearance. In *Soil Erosion on Agricultural Land,* Boardman, J., Foster, I.D.L., Dearing, J.A., Eds., Wiley: Chichester, 1990; 231–254.
57. Bork, H-R. Soil erosion during the past millennium in Central Europe and its significance within the geomorphodynamics of the Holocene. Catena **1989**, *15*, 121–131.
58. Van-Camp, L.; Bujarrabal, B.; Gentile, A-R.; Jones, R.J.A.; Montarnarella, L.; Olazabal, C.; Selvaradjou, S-K. Reports of the Technical Working Groups Established under the Thematic Strategy for Soil Protection. EUR 21319 EN/2, 2004, Office for Official Publications of the European Communities, Luxembourg.

59. Sanchez, J.; Recatala, L.; Colomer, J.C.; Ano, C. Assessment of soil erosion at national level: a comparative analysis for Spain using several existing maps. In *Ecosystems and Sustainable Development III*. Villacampa, Y., Brevia, C.A., Uso, J.L. Eds.; Advances in Ecological Sciences 10, WITT Press: Southampton, UK, 2001; 249–258.
60. Sonneveld, B.G.J.S.; Dent, D.L. How good is GLASOD? J. Environ. Manag. **2009**, *90*, 274–283.
61. Hoffman, T., Ashwell, A. *Nature Divided: Land Degradation in South Africa*; University of Cape Town Press: 2001.
62. Wischmeier, W.H.; Smith, D.D. *Predicting Rainfall Erosion Losses, Agricultural Research Service Handbook 537*; U.S. Department of Agriculture: Washington, DC, 1978.
63. Morgan, R.P.C.; Quinton, J.N.; Smith, R.E.; Govers, G.; Poesen, J.W.A.; Auerswald, K.; Chischi, G.; Torri, D.; Styczen, M.E. The European Soil Erosion Model (EUROSEM): a dynamic approach for predicting sediment transport from fields and small catchments. Earth Surf. Process. Landforms **1998**, *23*, 527–544.
64. Morgan, R.P.C. *Soil Erosion and Conservation*; Third Edition, Blackwell: Oxford, 2005; ch. 6.
65. Evans, R.; Boardman, J. The curtailment of muddy floods in the Sompting catchment, South Downs, West Sussex, southern England. Soil Use Manag. **2003**, *19*, 223–231.
66. Grove, A.T.; Rackham, O. *The Nature of Mediterranean Europe: An Ecological History*; Yale University Press: New Haven, 2001; ch. 6 Cultivation terraces.
67. Lesschen, J.P.; Cammeraat, L.H.; Nieman, T. Erosion and terrace failure due to agricultural land abandonment in a semi-arid environment. Earth Surf. Process. Landforms **2008**, *33*, 1574–1584.
68. Boardman, J.; Vandaele, K. Soil erosion, muddy floods and the need for institutional memory. Area **2010**, *42*, 502–513.
69. USDA. *Erosion in the Palouse: A Summary of the Palouse River Basin Study*, United States Department of Agriculture, Soil Conservation Service, Forest Service, Economics, Statistics, and Cooperative Service, Washington, 1979.
70. El-Swaify, S.A. with contributors. Sustaining the Global Farm - Strategic Issues, Principles, and Approaches. International Soil Conservation Organisation (ISCO), and the Department of Agronomy and Soil Science, University of Hawaii at Manoa, Hololulu, USA, 1999.
71. Tiffin, M.; Mortimore, M.; Gichuki, F. *More People, Less Erosion: Environmental Recovery in Kenya;* Wiley: Chichester, 1994.
72. Hudson, N.W. *A Study of the Reasons for Success or Failure of Soil Conservation Projects*. FAO Soil Bull. FAO, Rome, 1991; 64.
73. Peng, P.; Chen, S.; Dong, P. Temporal variation of sediment load in the Yellow River basin, China, and its impacts on the lower reaches and the river delta. Catena **2010**, *83*, 135–147.

Soil: Evaporation

William P. Kustas
Hydrology and Remote Sensing Lab, Agricultural Research Service, U.S. Department of Agriculture (USDA-ARS), Beltsville, Maryland, U.S.A.

Abstract

Soil evaporation can significantly influence energy flux partitioning of partially vegetated surfaces, ultimately affecting plant transpiration. While important, quantification of soil evaporation, separately from canopy transpiration, is challenging. Techniques for measuring soil evaporation exist and continually improve. The large variability in soil water content requires that there be careful thought to the design of soil evaporation measurements in the field. Numerical models for simulating soil evaporation have been developed and are shown to be fairly robust. However, the required inputs for defining model parameters often limit their application. For many operational applications where detailed soils and ancillary weather data are unavailable or where daily evaporation values are only needed, some of the analytical models described may provide the necessary level of accuracy. Moreover, in the application of weather forecast and hydrologic models, the use of simplified approaches is necessitated by the computational requirements or the lack of adequate data or both for defining more complex numerical model inputs. For large area estimation, the use of remotely sensed soil moisture and surface temperature offer the greatest potential for operational applications.

INTRODUCTION

Soil evaporation not only determines partitioning of available energy between sensible and latent heat flux for bare soil surfaces but can also significantly influence energy flux partitioning of partially vegetated surfaces. This latter effect occurs via the impact of soil evaporation on the resulting surface soil moisture and temperature. These, in turn, strongly influence the microclimate in partially vegetated canopies, indirectly affecting plant transpiration.[1] Over a growing season, soil evaporation can be a significant fraction of total water loss for agricultural crops.[2] On a seasonal basis in semiarid and arid regions, soil evaporation can significantly alter the relative fraction of runoff to rainfall, which in turn has a major impact on the available water for plants.[3] In deserts, in spite of its small magnitude, soil evaporation can introduce significant errors in meteorological forecasting if neglected.[4]

The measurement of soil evaporation at field scale is typically obtained using standard micrometeorological techniques, namely Bowen ratio and eddy covariance methods. Traditionally, due to fetch and measurement requirements, under partial canopy cover conditions, these techniques are not able to partition the total evapotranspiration into its soil evaporation and plant transpiration components. Recently, a novel procedure for partitioning evapotranspiration through utilizing the measured high-frequency time series of carbon dioxide and water vapor concentrations has been developed and tested.[5] This approach relies upon the simple assumption that contributions to the time series of carbon dioxide and water vapor concentrations derived from stomatal processes (i.e., photosynthesis and transpiration), and nonstomatal processes (i.e., respiration and direct evaporation) separately conform to flux-variance similarity. Vegetation water-use efficiency is the only parameter needed to perform the partitioning. Further work is needed to evaluate the utility of this technique with eddy covariance data collected over a variety of land cover and climate conditions.

Soil evaporation in partial canopy cover conditions varies spatially depending primarily on soil water distribution, canopy shading, and under-canopy wind patterns. These effects are magnified in row crops and under various irrigation techniques (e.g., drip irrigation). Soil evaporation can be measured using microlysimeters,[6] chambers,[7] time-domain reflectometers (TDRs),[8] a combination of microlysimetry and TDR,[9] micro-Bowen ratio systems,[3,10] or heat pulse probes.[11] Given the high spatial variability in the driving forces under partial canopy cover conditions, these point-based measurements are difficult to extrapolate to the field scale. Therefore, models have been developed to estimate the contribution of soil evaporation to the total evapotranspiration process.

Measurement methods are described, and models of varying degrees of complexity are reviewed, focusing primarily on relatively simple analytical models, some of which provide daily estimates and can be implemented operationally. The potential application of models using remote sensing data for large-scale estimation is also briefly discussed.

Encyclopedia of Natural Resources DOI: 10.1081/E-ENRL-120049129

METHODOLOGIES

Measurement Methods

Microlysimeters

Microlysimeters have been widely used to measure evaporation from the soil surface of irrigated crops.[8,12,13] Typically, an undisturbed soil sample (a representative vertical section of the soil profile) is inserted into a small cylinder open at the top. The microlysimeter is inserted back into the soil with its upper edge level with the soil surface and weighed either periodically or continuously. Changes in weight reflect an evaporative flux. To eliminate vertical heat conduction through the microlysimeter cylinder and minimize horizontal heat flux in the deeper layers of the sample, poly(vinyl chloride) (PVC) has been found to be the most suitable material. The microlysimeter's dimensions are typically a diameter of ~8 cm and a depth of 7–10 cm.[14–16] Theoretically, the microlysimeters provide absolute reference for soil evaporation, as long as their soil and the heat balance are similar to the surrounding area.

Chambers

Chambers are used to directly measure the flux of gases between the soil surface and the atmosphere by enclosing a volume and measuring all flux into and out of the volume.[17] Reviews of chamber designs and calculations of fluxes based on chamber methods can be found in Livingston and Hutchinson[18] and Hutchinson and Livingston.[19]

With infrared gas analyzers (IRGAs) becoming increasingly common, they are widely considered to be the method of choice today for chamber-based soil respiration and evaporation measurements.[20] Chambers can be used in either of two modes to calculate fluxes:[18] (1) in steady-state mode, the flux is calculated from the concentration difference between the air flowing at a known rate through the chamber inlet and outlet after the chamber headspace air has come to equilibrium concentration of carbon dioxide; (2) in the non-steady-state mode, the flux is calculated from the rate of increasing concentration in the chamber headspace of known volume shortly after the chamber is put over the soil.

In both modes, air is circulated between a small chamber that is placed on the soil and an IRGA. Typically, a soil chamber of ca. 1 L volume is placed on a PVC collar of about 80 cm^2 area. This collar is inserted about 2 cm into the soil and secured to prevent movement when the chamber is placed on it. When the chamber is placed on the collar, circulation of air between the chamber and the external IRGA is induced by a pump, and the water vapor concentration is measured.[21]

Soil water balance

Soil evaporation (E) can be extracted from the water balance equation, provided that all other components are known:

$$E = I + P - R + F - \Delta S \tag{1}$$

where I is irrigation, P is precipitation, R is runon or runoff, F is deep soil water flux (percolation), and ΔS is change in soil water storage. For an experimental field site, irrigation and precipitation can be easily monitored, and runoff and runon may be controlled to near-zero amounts by diking. Deep soil water flux errors can be reliably estimated in several ways, with the most important being the monitoring or measuring of the soil water content well below the root zone. The change in soil water storage can be determined fairly accurately with profile measurements of soil water content over multiple depths at the beginning and end of a defined time period.

There are many soil water content sensors, all of which work by measuring a surrogate property that is empirically or theoretically related to the soil water content. A recent comparative review by Evett et al.[22] concluded that soil water content is best determined using the neutron probe, gravimetric sampling, and conventional TDR[23] methods as compared to bore hole capacitance methods. Of the three optimal methods in the study, TDR is the only methodology capable of providing automated continuous measurements. However, other continuous measurement sensors such as frequency-domain reflectometers (FDRs),[24] and time-domain transmission (TDT)[25] sensors are also emerging as options for long-term installations, given appropriate calibration.

Micro-Bowen ratio systems

The Bowen ratio energy balance (BREB) approach is one of the simplest and most practical methods of estimating water vapor flux[10,26] and has thus been used extensively under a wide range of conditions providing robust estimates.[27] Use of the BREB concept[28] enables solving the energy balance equation by measuring simple gradients of air temperature and vapor pressure in the near-surface layer above the evaporating surface. The BREB equation is

$$LE = \frac{Rn - G}{1 + B_o} \tag{2}$$

in which LE is the latent heat flux, Rn is the net radiation, G is the soil heat flux, and B_o the Bowen ratio, which is found from measurements of temperature and vapor pressure at two heights within the constant flux layer.[29] Assuming equal transfer coefficients for heat and vapor, the Bowen ratio is defined as

$$B_o = \frac{H}{LE} = \gamma \frac{\partial T}{\partial e} \tag{3}$$

where H is the sensible heat flux, γ is the psychrometric constant, T is the air temperature, and e is the water vapor pressure.

Application of the Bowen ratio concept to measure bare soil surface evaporation was suggested[3,10] by measuring temperature and vapor pressure close to the soil surface (e.g., at 1 and 6 cm). Compared to microlysimeters, the micro-Bowen ratio (MBR) system yielded good results over bare soil.[10] The potential of the MBR approach was demonstrated by successfully measuring soil evaporation within a maize field.[30] To date, testing of the MBR is very limited, and further research is required to examine the performance of the technique under various environmental and agronomic conditions.

Heat pulse probes

A novel approach for measuring soil evaporation has been recently proposed, based on the soil sensible heat balance.[11,31] In this approach, a sensible heat balance is used to determine the amount of latent heat involved in the vaporization of soil water following Gardner and Hanks:[32]

$$LE = (H_0 - H_1) - \Delta S \tag{4}$$

where H_0 and H_1 are soil sensible heat fluxes at depths 0 and 1, respectively; ΔS is the change in soil sensible heat storage between depths 0 and 1; L is the latent heat of vaporization; and E is evaporation.

Typically, three-needle sensors like those described by Ren et al.[33] are used, which are spaced 6 mm apart, in parallel. Temperature is measured by all three needles, and the central needle also contains a resistance heater for producing a slight pulse of heat required for the heat pulse method. At a given time interval (2–4 hours), a heat pulse is executed, and the corresponding rise in temperature at the outer sensor needles is recorded. Soil volumetric heat capacity and thermal diffusivity are then determined from the heat input and temperature response following the procedures described by Knight and Kluitenberg[34] and Bristow et al.,[35] respectively. Having measurements of soil temperature, thermal conductivity, and volumetric heat, evaporation is estimated from Eq. 4.

Numerical Models

Numerous mechanistic/numerical models of heat and mass flows exist and are primarily based on the theory of Philip and de Vries.[36] However, they continue to be refined through improved parameterization of the moisture and heat transport through the soil profile.[37–39] Some of these mechanistic models have been used recently to explore the utility of bulk transfer approaches used in weather forecasting models[40] and in soil–vegetation–atmosphere models,[41] computing field to regional-scale fluxes. These bulk transport approaches are commonly called the "alpha" and "beta" methods defined by the following expressions

$$LE_s = \frac{\rho C_p}{\gamma}\left(\frac{a e_*(T_s) - e_A}{R_A}\right) \tag{5}$$

and

$$LE_g = \frac{\rho C_p}{\gamma}\beta\left(\frac{e_*(T_s) - e_A}{R_A}\right) \tag{6}$$

where ρ is the air density (~1 kg/m^3), C_p the heat capacity of air (~1000 J/kg/K), γ the psychrometric constant (~65 Pa), $e_*(T_S)$ the saturated vapor pressure (Pa) at soil temperature T_S(K), e_A the vapor pressure (Pa) at some reference level in the atmosphere, and R_A is the resistance (s/m) to vapor transport from the surface usually defined from surface layer similarity theory.[42] For a, several different formulations exist[43,44] with one of the first relating a to the thermodynamic relationship for relative humidity in the soil pore space, h_R[45]

$$h_R = \exp\left(\frac{\psi g}{R_V T_S}\right) \tag{7}$$

where ψ is the soil matric potential (m), g the acceleration of gravity (9.8 m/s^2), and R_V the gas constant for water vapor (461.5 J/kg/K). From Eq. 6, β can be defined as a ratio of aeodynamic and soil resistance to vapor transport from the soil layer to the surface, R_S, namely,

$$\beta = \frac{R_A}{R_A + R_S} \tag{8}$$

Both modeling[44] and observational results[46,47] indicate that more reliable results are obtained using the beta method. In fact, Ye and Pielke[44] formulate an expression similar to Camillo and Gurney,[48] which combines Eqs. 5 and (6) and provides more reliable evaporation rates for a wider range of conditions,

$$LE_s = \frac{\rho C_p}{\gamma}\beta\left(\frac{a e_*(T_s) - e_A}{R_A}\right) \tag{9}$$

Unfortunately, there is no consensus concerning the depth in the soil profile to consider in defining the a and β terms. In particular, studies evaluating the soil resistance term R_S use a range of soil moisture depths: 0–1/2 cm,[48] 0–1 cm,[7] 0–2 cm,[39,46] and 0–5 cm.[39] The study conducted by Chanzy and Bruckler[38] appears to be one of the few studies that attempts to determine the most useful soil moisture depth for modeling soil evaporation by considering the penetration depth of passive microwave sensors of varying wavelengths. Using field data and numerical simulations with a mechanistic model, they find that the 0–5 cm depth, which can be provided by the L-band microwave frequency, appears to be the most adequate frequency for evaluating soil evaporation.

Besides the depth of the soil layer to consider, the equations relating soil moisture to R_S have ranged from linear to exponential (Table 1). Furthermore, observations and numerical models have shown that varies significantly throughout the day and that its magnitude is also affected by climatic conditions.[38,39]

These are not the only complicating factors that make the use of such a bulk resistance approach somewhat tenuous. From detailed observations of soil moisture changes and water movement, Jackson et al.[49] found the soil water flux in the 0–9 cm depth to be very dynamic with fluxes at all depths continually changing in magnitude and sometimes direction over the course of a day. These phenomena observed by Jackson et al.[49] are owing in part to a process that occurs during soil drying where dry surface soil layer forms, significantly affecting the vapor transport through the profile.[50] Yamanaka et al.[51] recently developed and verified, using wind tunnel data, a simple energy balance model in which the soil moisture available for evaporation is defined using the depth of the evaporating/drying front in the soil. This approach removes the ambiguity of defining the thickness of the soil layer and resulting moisture available for evaporation. However, the depth of the evaporating surface is not generally known *a priori*, nor can it be measured in field conditions; hence, this approach at present is limited to exploring the effects of evaporating front on R_S type formulations.

Analytical Models

To reduce the effect of temporal varying, R_S, Chanzy and Bruckler[38] developed a simple analytically based daily LE_S (E_D) model using simulations from their mechanistic model for different soil texture, moisture, and climatic conditions as quantified by potential evaporation (E_{PD}), as given by Penman.[52] The analytical daily model requires midday 0–5 cm soil moisture θ, daily potential evaporation, and daily average wind speed (U_D). The simple model has the following form:

Table 1 Bulk soil resistance formulations, R_S, from previous studies

R_S formula (s/m)	Value of coefficients	Soil type	Depth (cm)	References
$R_S = a\left(\frac{\theta_S}{\theta}\right)^n + b$	$a = 3.5$	Loam[a]	0–1/2	[21]
	$b = 33.5$			
	$n = 2.3$			
$R_S = a(\theta_S - \theta) + b$	$a = 4{,}140$	Loam[b]	0–1/2	[48]
	$b = -805$			
$R_S = R_{SMIN} \exp[a(\theta_{MIN} - \theta)]$	$R_{SMIN} = 10$ s/m	Fine sandy loam	0–1	[7]
	$\theta_{MIN} = 15\%$			
	$a = 0.3563$			
$R_S = \frac{a(\theta_S - \theta)^n}{2.3\times10^{-4}(T_s/273.16)^{1.75}}$	$a = 2.16 \times 10^2$	Loam	0–2	[46]
	$n = 10$			
	$\theta_S = 0.49$			
	$a = 8.32 \times 10^5$	Sand	0–2	[46]
	$n = 16.16$			
	$\theta_S = 0.392$			
$R_S = a\theta + b$	$a = -73{,}420 - -51{,}650$	Sand	0–2	[39]
	$b = 1{,}940 - 3{,}900$			
$R_S = \exp\left(b - a\left(\frac{\theta}{\theta_S}\right)\right)$	$a = 4.3$	Silty clay loam[a]	0–5	[41]
	$b = 8.2$			
	$a = 5.9$	Gravelly sandy loam	0–5	[66]
	$b = 8.5$			

[a]Soil type for the data from Jackson et al.[49] was determined from texture-dependent soil hydraulic conductivity and matric potential equations of Clapp and Hornberger evaluated by Camillo and Gurney.[48]

[b]Soil type was determined from texture-dependent soil hydraulic conductivity and matric potential equations of Clapp and Hornberger evaluated by Sellers et al.[41]

$$\frac{E_D}{E_{PD}} = \left[\frac{\exp(A\theta + B)}{1+\exp(A\theta + B)}\right]C + (1-C) \tag{10a}$$

$$A = a + 5\max(3 - E_{PD}, 0) \tag{10b}$$

$$B = b - 5(-0.025b - 0.05)\max(3 - E_{PD}, 0) + \alpha(U_D - 3) \tag{10c}$$

$$C = 0.90 - 0.05c(U_D - 3) \tag{10d}$$

where the coefficients a, b, and c depend on soil texture (Table 1) and were derived from their detailed mechanistic model simulations for loam, silty clay loam, and clay soils.[38] In Fig. 1, a plot of Eq. 10a is given for two soil types, loam (Fig. 1A) and silty clay loam (Fig. 1B) under two climatic conditions, namely, a relatively low evaporative demand condition with U_D = 1 m/s and E_{PD} = 2 mm/d and high demand U_D = 5 m/s and E_{PD} = 10 mm/d. Notice the transition from E_D/E_{PD} ~ 1 to E_D/E_{PD} < 1 as a function of θ varies not only with the soil texture, but also with the evaporative demand. The simplicity of such a scheme outlined in Eq. 10 needs further testing for different soil textures and under a wider range of climatic conditions.

The ratio E_D/E_{PD} as a function of θ illustrated in Fig. 1 also depicts the effect of the two "drying stages" typically used to describe soil evaporation.[42] The "first stage" (S_1) of drying is under the condition where water is available in the near-surface soil to meet atmospheric demand, i.e., E_D/E_{PD}~1. In the "second stage" (S_2) of drying, the water availability or θ falls below a certain threshold where the soil evaporation is no longer controlled by the evaporative demand, namely, E_D/E_{PD} <1. Under S_2, several studies find that a simple formulation can be derived by assuming that the time change in θ is governed by desorption, namely, as isothermal diffusion with negligible gravity effects from a semi-infinite uniform medium. This leads to the rate of evaporation for S_2 being approximated by[42,53]

$$E_D = 0.5 D_E t^{-1/2} \tag{11}$$

where the desorptivity D_E (mm/d$^{1/2}$) is assumed to be a constant for a particular soil type and t is the time (in days) from the start of S_2. Although both numerical models and observations indicate that the soil evaporation is certainly a more complicated process than the simple analytical expression given by Eq. 11, a number of field studies[3,54–58] have shown that for S_2 conditions, reliable daily values can be obtained using Eq. 11. In many of these studies for determining D_E, the integral of Eq. 11 is used, which yields the cumulative evaporation as a function of $t^{1/2}$

$$\sum E_D = D_E (t - t_O)^{1/2} \tag{12}$$

where t_O is the number of days where E_D/E_{PD}~1 or is in S_1. In practice, observations of ΣE_D are plotted vs. $(t - t_O)^{1/2}$ and in many cases the choice of the starting point of S_2 is $t_O \approx 0$ or immediately after the soil is saturated. As shown by Campbell and Norman,[58] the course of evaporation rate for three drying experiments (see Fig. 9.6 in Campbell and Norman[58]) indicates that for a loam soil, t_O depends on the evaporative demand or E_{PD} with t_O ~ 2 days when E_{PD} is high vs. t_O~5 days when E_{PD} is low. On the other hand, for a sandy soil, there is almost an immediate change from S_1 to S_2 conditions with $t_O \approx 1$ day. As suggested by the analysis of Jackson et al.[56] and as stated more explicitly by Brutsaert and Chen,[59] the value of t_O can significantly influence the value computed for D_E. Jackson et al.[56] also show that for the same soil type, the value of D_E has a seasonal dependency (ranging from 4 to 8 mm/d$^{1/2}$) most likely related to the evaporative demand, which they correlate to daily average soil temperature (see Fig. 2 in Jackson, Idso, et al.[56]). Values of D_E from the various studies are listed in Table 2. Brutsaert and Chen[59] modified Eq. 11 for deriving D_E by rewriting in terms of a "time-shifted" variable $T = t - t_O$ and expressing it in the form

$$E_D^{-2} = \left(\frac{2}{D_E}\right)^2 T \tag{13}$$

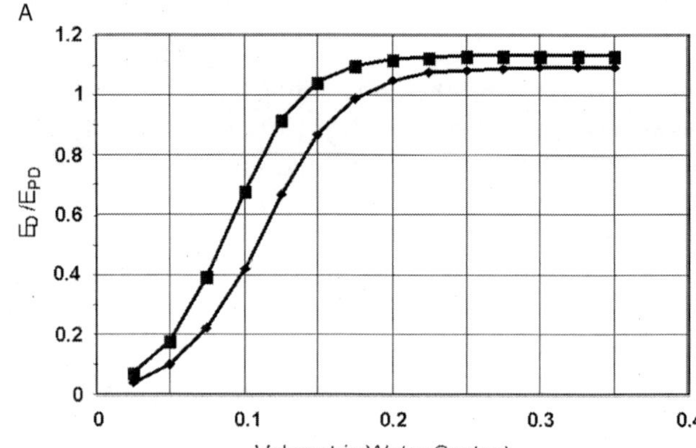

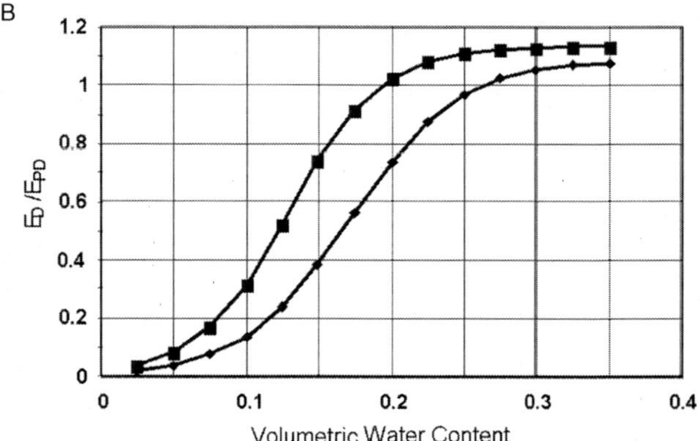

Fig. 1 A plot of E_D/E_{PD}, estimated using Eq. 10 from Chanzy and Bluckler[38] vs. volumetric water content for (**A**) loam and (**B**) silty clay loam soil under two evaporative demand conditions: U_D = 1 m/s and E_{PD} = 2 mm/d (squares) and U_D = 5 m/s and E_{PD} = 10 mm/d (diamonds).

Table 2 Values of desorptivity, D_E, evaluated from various experimental sites

Desorptivity D_E (mm/d$^{1/2}$)	Soil type	References
4.96–4.30	Sand	[54]
5.08	Clay loam	[55]
4.04	Loam	[55]
3.5	Clay	[55]
~4 to ~8[a]	Loam[b]	[56]
5.8	Clay loam	[57]
4.95[c]	Silty clay loam[d]	[59]
2.11	Gravelly sandy loam	[3]

[a]The magnitude of D_E was found to have a seasonal dependency.
[b]Soil type was determined from texture-dependent soil hydraulic conductivity and matric potential equations of Clapp and Hornberger evaluated by Camillo and Gurney.[48]
[c]This value was evaluated for a vegetated surface.
[d]Soil type was determined from texture-dependent soil hydraulic conductivity and matric potential equations of Clapp and Hornberger evaluated by Sellers et al.[41]

where D_E and t_O will come from the slope and intercept (see Fig. 1 in Brutsaert and Chen[59]). It follows that ΣE_D under S_2 will start at $T = T_O$ and not at $T = 0$, so that Eq. 12 is rewritten as

$$\sum E_D = D_E\left(T^{1/2} - T_O^{1/2}\right) \tag{14}$$

They evaluated the effect on the derived D_E using this technique with the data from Black et al.[54] The value of D_E using Eqs. 13 and 14 was estimated to be approximately 3.3 mm/d$^{1/2}$, which is smaller than D_E values reported by Black et al.,[54] namely, 4.3–5 mm/d$^{1/2}$. However, Brutsaert and Chen[59] show that this technique yields a better linear fit to the data points that were actually under S_2 conditions.

Equations 13 and 14 were used with the September 1973 data set from Jackson et al.[56] and compared to using Eq. 12 with $t_O = 0$. The plot of Eq. 12 with the regression line in Fig. 2 yields D_E~10 mm/d$^{1/2}$, which is significantly larger than any previous estimates (Table 2). Moreover, it is obvious from the figure that Eq. 12 should not be applied with $t_O = 0$, as this relationship is not linear over the whole drying processes. With Eq. 13, applied to the data, t_O is estimated to be approximately 4.3 days, and thus a linear relationship should start at the shifted time scale $T = t - 4.3$; this means ΣE_D should start on day 5 or $T_O \approx 5 - 4.3$ (Fig. 3A). With Eq. 14, a more realistic $D_E \approx 4.6$ mm/d$^{1/2}$ is estimated for the linear portion of daily evaporation following the S_2 condition (Fig. 3B).

While this approach is relatively easy to implement operationally, D_E will likely depend on climatic factors as well as soil textural properties. However, it might be feasible

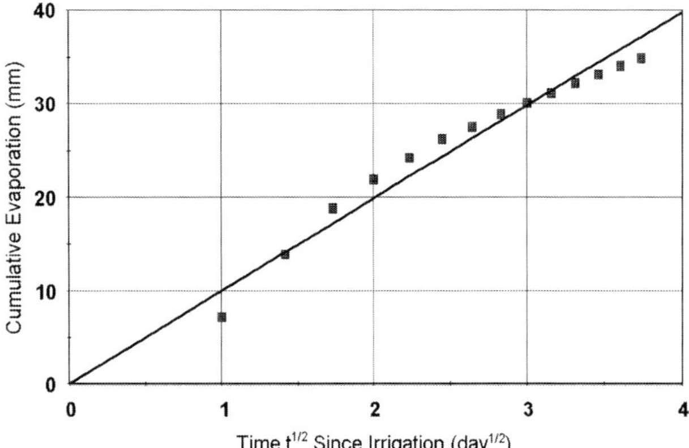

Fig. 2 The desorptivity D_E (mm/d$^{1/2}$) estimated from a least squares linear fit to the data from Jackson et al.[56] assuming $t_O = 0$ (i.e., stage-two drying occurs immediately after irrigation/ precipitation).

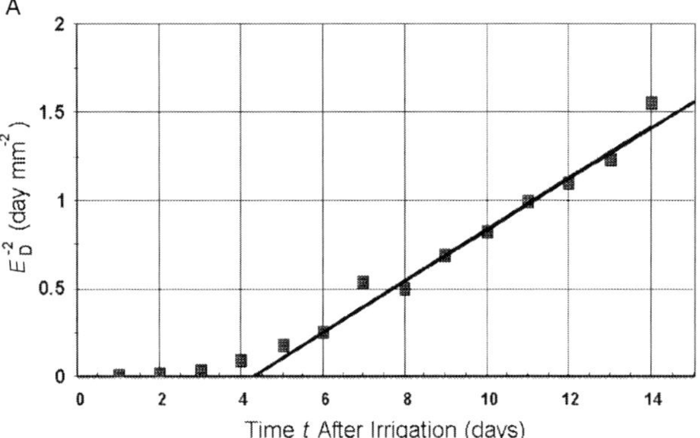

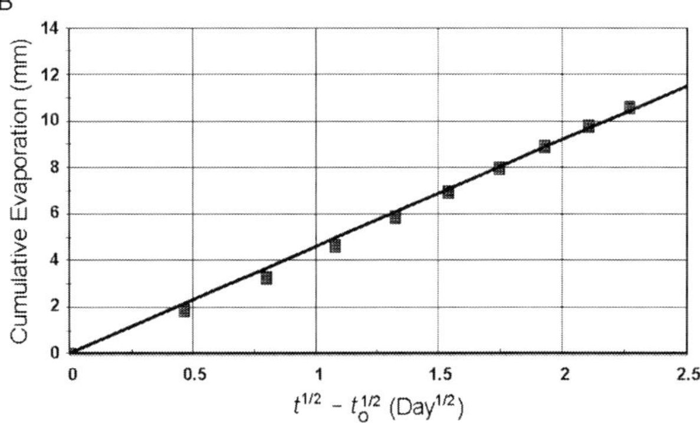

Fig. 3 Estimation of (**A**) D_E and t_O with the data from Fig. 2 using Eq. 13 from Brutsaert and Chen[59] and (**B**) the resulting cumulative evaporation ΣE_D curve under second stage drying using Eq. 14.

Soil—Soil

to describe the main climate/seasonal effect on D_E from soil temperature observations.[56] These might come from weather station observations or possibly from multitemporal remote sensing observations of surface temperature.

The difficulty in developing a formulation for R_S, which correctly describes the water vapor transfer process in the soil, was recognized much earlier by Fuchs and Tanner[60] and Tanner and Fuchs.[61] They proposed a combination method instead that involves atmospheric surface layer observations and remotely sensed surface temperature, T_{RS}. Starting with the energy balance equation

$$R_N = H + G + LE \tag{15}$$

where R_N is the net radiation, H the sensible heat flux, LE the latent heat flux and G the soil heat flux all in W/m^2, and assuming the resistance to heat and water vapor transfer are similar yielding,

$$H = \rho C_P \left(\frac{T_{RS} - T_A}{R_A} \right) \tag{16}$$

$$LE = \frac{\rho C_P}{\gamma} \left(\frac{e_{RS} - e_A}{R_A} \right) \tag{17}$$

an equation of the following form can be derived

$$LE = \left(\frac{\gamma + \Delta}{\Delta} \right) LE_P - \frac{\rho C_P}{\Delta} \left(\frac{e_*(T_{RS}) - e_A}{R_A} \right) \tag{18}$$

where

$$LE_P = \rho C_P \left(\frac{e_*(T_A) - e_A}{R_A(\Delta + \gamma)} \right) + \Delta \left(\frac{R_N - G}{\Delta + \gamma} \right) \tag{19}$$

The difference $e_*(T_A) - e_A$ is commonly called the saturation vapor pressure deficit, and the value of soil surface vapor pressure e_{RS} is equal to $h_{RS} e_*(T_{RS})$ where h_{RS} is the soil surface relative humidity. Substituting Eq. 19 into Eq. 18 yields

$$LE = R_N - G - \frac{\rho C_P}{\Delta} \left(\frac{e_*(T_{RS}) - e_*(T_A)}{R_A} \right) \tag{20}$$

This equation has the advantage over the above bulk resistance formulations using R_S in that there are no assumptions made concerning the saturation deficit at or near the soil surface or how to define h_{RS}. Instead, this effect is accounted for by T_{RS} because as the soil dries, T_{RS} increases and hence $e_*(T_{RS})$, which generally results in the last term on the right-hand side of Eq. 20 to increase, thus causing LE to decrease. In a related approach,[62] the magnitude of LE is simply computed as a residual in the energy balance equation, Eq. 19, namely,

$$LE = R_N - G - \rho C_P \left(\frac{T_{RS} - T_A}{R_A} \right) \tag{21}$$

Particularly crucial in the application of either Eq. 20 or 21 is a reliable estimate of R_A and T_{RS}. Issues involved in correcting radiometric temperature observations for surface emissivity, viewing angle effects, and other factors are summarized in Norman and Becker.[63] The aerodynamic resistance R_A is typically expressed in terms of Monin–Obukhov similarity theory[42]

$$R_A = \frac{\left\{ \ln\left[\frac{z - d_O}{z_{OM}} \right] - \psi_M \right\} \left\{ \ln\left[\frac{z - d_O}{z_{OS}} \right] - \psi_S \right\}}{k^2 U} \tag{22}$$

where z is the observation height in the surface layer (typically 2–10 m), d_O the displacement height, z_{OM} the momentum roughness length, z_{OS} the roughness length for scalars (i.e., heat and water vapor), k (~0.4) von Karman's constant, ψ_M the stability correction function for momentum, and ψ_S the stability correction function for scalars. Both d_O and z_{OM} are dependent on the height and density of the roughness obstacles and can be considered a constant for a given surface, while the magnitude of z_{OS} can vary for a given bare soil surface as it is also a function of the surface friction velocity.[42] Experimental evidence suggests that existing theory with possible modification to some of the "constants" can still be used to determine z_{OS} providing acceptable estimates of H for bare soil surfaces. However, application of Eq. 21 in partial canopy cover conditions has not been successful in general because z_{OS} is not well defined in Eq. 22, exhibiting large scatter with the existing theory.[64]

For this reason, estimating soil evaporation from partially vegetated surfaces using T_{RS} invariably has to involve "two-source" approaches whereby the energy exchanges from the soil and vegetated components are explicitly treated.[65] Similarly, when using remotely sensed soil moisture for vegetated surfaces, a two-source modeling framework needs to be applied.[66] In these two-source approaches, there is the added complication of determining aerodynamic resistances between soil and vegetated surfaces and the canopy air space. Schematically, the resistance network and corresponding flux components for two-source models is shown in Fig. 4. An advantage with the two-source formulation of Norman et al.[65] is that R_S is not actually needed for computing LE_S as it is solved as a residual. Nevertheless, the formulations in such parameterizations that are used (such as the aerodynamic resistance formulations) are likely to strongly influence LE_S values.[1] Yet this two-source formulation is found to be fairly robust in separating soil and canopy contributions to evapotranspiration.[67]

CONCLUSIONS

Techniques for measuring soil evaporation, separately from plant transpiration, exist and continually improve. Given the high variability in soil water content often exhibited in the field, the exact position of the instrumentation may have a large effect on the measurement. This is magnified in row crops, where the variability is structured, rather than random. In row crops, soil evaporation is not only dependent

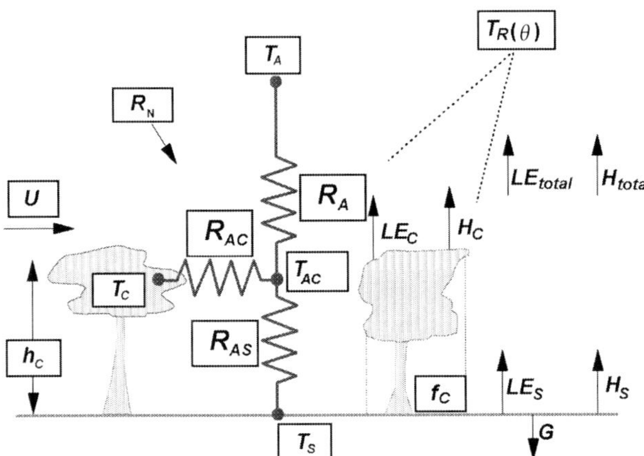

Fig. 4 A schematic diagram illustrating the resistance network for the two-source approach where the subscript c refers to the vegetated canopy and s refers to the soil surface. The symbol $T_R(\theta)$ refers to a radiometric temperature observation at a viewing angle θ, T_{AC} is the model-derived within-canopy air space temperature, h_C is the canopy height, f_C is the fractional vegetation/canopy cover, and R_{AC} and R_{AS} are the aerodynamic resistances to sensible heat flux from the canopy and soil surface, respectively. The main meteorological inputs required for the model are also illustrated, namely wind speed, U, air temperature, T_A, and the net radiation, R_N.
Source: Adapted from Norman et al.[65]

on the soil properties, but also local climatic conditions, and the timing of irrigation/precipitation. Under low-to-moderate wind speed conditions (~1 to 4 m s^{-1}), row orientation relative to the wind direction also affects the amount of soil evaporation.[68] This large variability requires that there be careful thought to the design of soil evaporation measurements in the field. The methodology proposed by Scanlon and Kustas[5] using eddy covariance measurements suffers less from the sampling issue but requires *a priori* knowledge of the vegetation water-use efficiency, which will vary with vegetation type and condition.

For many landscapes having partial vegetative cover, the contribution of soil evaporation to the total evapotranspiration flux cannot be ignored, particularly with regard to the influence of surface soil moisture and temperature on the microclimate in the canopy air space (Fig. 4). Numerical models for simulating soil evaporation have been developed and are shown to be fairly robust. However, the required inputs for defining model parameters often limit their application to field sites having detailed soil profile information and that are well instrumented with ancillary weather data.

For many operational applications where detailed soil and ancillary weather data are unavailable or where daily evaporation values are only needed, some of the analytical models described may provide the necessary level of accuracy. Moreover, in the application of weather forecast and hydrologic models, the use of simplified approaches is necessitated by the computational requirements or the lack of adequate data or both for defining more complex numerical model parameters and variables.

For large area estimation, the use of remotely sensed soil moisture and surface temperature offer the greatest potential for operational applications. Development of modeling schemes that can incorporate this remote sensing information and readily apply it on a regional scale basis has been proposed and shows promise.[66,69–71]

REFERENCES

1. Kustas, W.P.; Norman, J.M. Evaluation of soil and vegetation heat flux predictions using a simple two-source model with radiometric temperatures for partial canopy cover. Agric. Forest Meteorol. **1999**, *94*, 13–25.
2. Tanner, C.B.; Jury, W.A. Estimating evaporation and transpiration from a row crop during incomplete cover. Agron. J. **1976**, *68*, 239–242.
3. Wallace, J.S.; Holwill, C.J. Soil evaporation from tiger-bush in south-west Niger. J. Hydrol. **1997**, *188–189*, 426–442.
4. Agam, N.; Berliner, P. R.; Zangvil, A.; Ben-Dor, E. Soil water evaporation during the dry season in an arid zone. *J. Geophys. Res. Atmos.* **2004**, *109*, D16103, doi:16110.11029/12004JD004802.
5. Scanlon, T.M.; Kustas, W.P. Partitioning carbon dioxide and water vapor fluxes using correlation analysis. Agric. Forest Meteorol. **2010**, *150*, 89–99.
6. Allen, S.J. Measurement and estimation of evaporation from soil under sparse barley crops in northern Syria. Agric. Forest Meteorol. **1990**, *49*, 291–309.
7. Van de griend, A.A.; Owe, M. Bare soil surface resistance to evaporation by vapor diffusion under semiarid conditions. Water Resour. Res. **1994**, *30*, 181–188.
8. Plauborg, F. Evaporation from bare soil in a temperate humid climate - Measurement using micro-lysimeters and time domain reflectrometry. Agric. Forest Meteorol. **1995**, *76*, 1–17.
9. Baker, J.M.; Spaans, E.J. Measuring water exchange between soil and atmosphere with TDR-Microlysimetry. Soil Sci. **1994**, *158*, 22–30.
10. Ashktorab, H.; Pruitt, W.O.; Pawu, K.T.; George, W.V. Energy-balance determinations close to the soil surface using a micro-Bowen ratio system. Agric. Forest Meteorol. **1989**, *46*, 259–274.
11. Heitman, J.L.; Horton, R.; Sauer, T.J.; Desutter, T.M. Sensible heat observations reveal soil-water evaporation dynamics. J. Hydrometeorol. **2008a**, *9*, 165–171.
12. Shawcroft, R.W.; Gardner, H.R. Direct evaporation from soil under a row crop canopy. Agric. Meteorol. **1983**, *28*, 229–238.
13. Lascano, R.J.; Van bavel, C.H.M. Simulation and measurement of evaporation from a bare soil. Soil Sci. Soc. Am. J. **1986**, *50*, 1127–1133.
14. Boast, C.W.; Robertson, T.M. A "micro-lysimeter" method for determining evaporation from bare-soil: description and laboratory evaluation. Soil Sci. Soc. Am. J. **1982**, *46*, 689–696.
15. Walker, G.K. Measurement of evaporation from soil beneath crop canopies. Can. J. Soil Sci. **1983**, *63*, 137–141.

16. Evett, S.R.; Warrick, A.W.; Matthias, A.D. Wall material and capping effects on microlysimeter temperatures and evaporation. Soil Sci. Soc. Am. J. **1995**, *59*, 329–336.
17. Denmead, O.T.; Dunin, F.X.; Wong, S.C.; Greenwood, E.A.N. Measuring water use efficiency of Eucalyptus trees with chambers and micrometeorological techniques. J. Hydrol. **1993**, *150*, 649–664.
18. Livingston, G.P.; Hutchinson, G.L. Enclosure-based measurement of trace gas exchange: Applications and sources of error. In *Biogenic Trace Gases: Measuring Emissions from Soil and Water*; Matson, P.A., Harriss, R.C., Eds.; Blackwell Scientific Publications: Oxford, 1995; pp. 14–51.
19. Hutchinson, G.L.; Livingston, G.P. Gas flux. In *Methods of Soil Analysis: Part 1, Physical Methods*; Dane, J.H., Topp, G.C., Eds.; Soil Science Society of America: Madison, WI: 2002.
20. Davidson, E.A.; Savage, K.; Verchot, L.V.; Navarro, R. Minimizing artifacts and biases in chamber-based measurements of soil respiration. Agric. Forest Meteorol. **2002**, *113*, 21–37.
21. Norman, J.M.; Kucharik, C.J.; Gower, S.T.; Baldocchi, D.D.; Crill, P.M.; Rayment, M.; Savage, K.; Striegl, R.G. A comparison of six methods for measuring soil-surface carbon dioxide fluxes. J. Geophys. Res.-Atmospheres **1997**, *102*, 28771–28777.
22. Evett, S.R.; Schwartz, R.C.; Casanova, J.J.; Heng, L.K. Soil water sensing for water balance, ET and WUE. Agric. Water Manag. **2012**, *104*, 1–9.
23. Topp, G.C.; Davis, H.L.; Annan, A.P. Electromagnetic determination of soil water content: measurements in coaxial transmission lines. Water Resour. Res. **1980**, *16*, 574–582.
24. Logsdon, S.D.; Green, T.R.; Seyfried, M.S.; Evett, S.R.; Bonta, J. Hydra probe and twelve-wire probe comparisons in fluids and soil cores. Soil Sci. Soc.Am. J. **2010**, *74*, 5–12.
25. Miralles-Crespo, J.; Van Lersel, M.W. A calibrated time domain transmissometry soil moisture sensor can be used for precise automated irrigation of container-grown plants. Hort. Science **2011**, *46*, 889–894.
26. Allen, R.G.; Pereira, L.S.; Howell, T.A.; Jensen, M.E. Evapotranspiration information reporting: I. Factors governing measurement accuracy. Agric. Water Manag. **2011**, *98*, 899–920.
27. Farahani, H.J.; Howell, T.A.; Shuttleworth, W.J.; Bausch, W.C. Evapotranspiration: Progress in measurement and modeling in agriculture. Trans. Am. Soc. Agric. Biol. Engineers **2007**, *50*, 1627–1638.
28. Bowen, I.S. The ratio of heat losses by conduction and by evaporation from any water surface. Phys. Rev. **1926**, *27*, 779–787.
29. Monteith, J.L.; Unsworth, M.H. *Principles of Environmental Physics*, 3rd ed.; Elsevier/Academic Press: 2008.
30. Zeggaf, A.T.; Takeuchi, S.; Dehghanisanij, H.; Anyoji, H.; Yano, T. A Bowen ratio technique for partitioning energy fluxes between maize transpiration and soil surface evaporation. Agronomy J. **2008**, *100*, 988–996.
31. Heitman, J.L.; Xiao, X.; Horton, R.; Sauer, T.J. Sensible heat measurements indicating depth and magnitude of subsurface soil water evaporation. Water Resour. Res. **2008b**, *44*.
32. Gardner, H.R.; Hanks, R.J. Evaluation of the evaporation zone in soil by measurement of heat flux. Soil Sci. Soc. Am. Proc. **1966**, *32*, 326–328.
33. Ren, T.S.; Ochsner, T.E.; Horton, R. Development of thermo-time domain reflectometry for vadose zone measurements. Vadose Zone J. **2003**, *2*, 544–551.
34. Knight, J.H.; Kluitenberg, G.J. Simplified computational approach for dual-probe heat-pulse method. Soil Sci. Soc. Am. Journal. *68*, 447–449.
35. Bristow, K.L.; Kluitenberg, G.J.; Horton, R. Measurement of soil thermal-properties with a dual-probe heat-pulse technique. Soil Sci. Soc. Am. J. **1994**, *58*, 1288–1294.
36. Philip, J.R.; De vries, D.A. Moisture movement in porous materials under temperature gradients. Trans. Am. Geophys. Union **1957**, *38* (2), 222–232.
37. Camillo, P.J.; Gurney, R.J.; Schmugge, T.J. A soil and atmospheric boundary layer model for evapotranspiration and soil moisture studies. Water Resour. Res. **1983**, *19*, 371–380.
38. Chanzy, A.; Bruckler, L. Significance of soil surface moisture with respect to daily bare soil evaporation. Water Resour. Res. **1993**, *29*, 1113–1125.
39. Daamen, C.C.; Simmonds, L.P. Measurement of evaporation from bare soil and its estimation using surface resistance. Water Resour. Res. **1996**, *32*, 1393–1402.
40. Noilhan, J.; Planton, S. A simple parameterization of land surface processes for meteorological models. Mon. Weather Rev. **1989**, *117*, 536–549.
41. Sellers, P.J.; Heiser, M.D.; Hall, F.G. Relations between surface conductance and spectral vegetation indices at intermediate (100 m^2 to 15 km^2) length scales. J. Geophys. Res. **1992**, *97*, 19033–19059.
42. Brutsaert, W. *Evaporation into the Atmosphere: Theory, History and Applications*; D. Reidel Publishing Co.: Boston, 1982.
43. Mahfouf, J.F.; Noilhan, J. Comparative study of various formulations of evaporation from bare soil using in situ data. J. Appl. Meteorol. **1991**, *30*, 1354–1365.
44. Ye, Z.; Pielke, R.A. Atmospheric parameterization of evaporation from non-plant-covered surfaces. J. Appl. Meteorol. **1993**, *32*, 1248–1258.
45. Philip, J.R. 1957, Evaporation, and moisture and heat fields in the soil. Journal of Meteorology. *14*, 354–366.
46. Kondo, J.; Saigusa, N.; Sato, T. A parameterization of evaporation from bare soil surfaces. J. Appl. Meteorol. **1990**, *29*, 385–389.
47. Cahill, A.T.; Parlange, M.B.; Jackson, T.J.; O'Neill, P.; Schmugge, T.J. Evaporation from nonvegetated surfaces: Surface aridity methods and passive microwave remote sensing. J. Appl. Meteorol. **1999**, *38*.
48. Camillo, P.J.; Gurney, R.J. A resistance parameter for bare-soil evaporation models. Soil Sci. **1986**, *141*, 95–104.
49. Jackson, R.D.; Kimball, B.A.; Reginato, R.J.; Nakayama, F.S. Diurnal soil water evaporation: Time-depth-flux patterns. Soil Sci. Soc. Am. Proc. **1973**, *37*, 505–509.
50. Hillel, D. *Applications of Soil Physics*; Academic Press: San Diego, 1980.
51. Yamanaka, T.; Takeda, A.; Sugita, F. A modified surface-resistance approach for representing bare-soil evaporation: Wind tunnel experiments under various atmospheric conditions. Water Resour. Res. **1997**, *33*, 2117–2128.
52. Penman, H.L. Natural evaporation from open water, bare soil and grass, Proc. R. Soc. Lond. **1948**, *A193*, 120–146.

53. Gardner, W.R. Solution of the flow equation for the drying of soils and other porous media. Soil Sci. Soc. Am. Proc. **1959**, *23*, 183–187.
54. Black, R.A.; Gardner, W.R.; Thurtell, G.W. The prediction of evaporation, drainage, and soil water storage for a bare soil. Soil Sci. Soc. Am. Proc. **1969**, *33*, 655–660.
55. Ritchie, J.T. Model for predicting evaporation from a row crop with incomplete cover. Water Resour. Res. **1972**, *8*, 1204–1213.
56. Jackson, R.D.; Idso, S.B.; Reginato, R.J. Calculation of evaporation rates during the transition from energy-limiting to soil-limiting phases using albedo data. Water Resour. Res. **1976**, *12*, 23–26.
57. Parlange, M.B.; Katul, G.G.; Cuenca, R.H.; Kavvas, M.L.; Nielsen, D.R.; Mata, M. Physical basis for a time series model of soil water content. Water Resour. Res. **1992**, *28*, 2437–2446.
58. Campbell, G.S.; Norman, J.M. *An Introduction to Environmental Biophysics*; Springer-Verlag: New York, 1998.
59. Brutsaert, W.; Chen, D. Desorption and the two stages of drying of natural tallgrass prairie. Water Resour. Res. **1995**, *31*, 1305–1313.
60. Fuchs, M.; Tanner, C.B. Evaporation from a drying soil. J. Appl. Meteorol. **1967**, *6*, 852–857.
61. Tanner, C.B.; Fuchs, M. Evaporation from unsaturated surfaces: a generalized combination method. J. Geophys. Res. **1968**, *73*, 1299–1304.
62. Jackson, R.D.; Reginato, R.J.; Idso, S.B. Wheat canopy temperature: A practical tool for evaluating water requirements. Water Resour. Res. **1977**, *13*, 651–656.
63. Norman, J.M.; Becker, F. Terminology in thermal infrared remote sensing of natural surfaces. Agric. Forest Meteorol. **1995**, *77*, 153–166.
64. Verhoef, A.; De bruin, H.A.R.; Van den hurk, B.J.J.M. Some practical notes on the parameter kB-1 for sparse vegetation. J. Appl. Meteorol. **1997**, *36*, 560–572.
65. Norman, J.M.; Kustas, W.P.; Humes, K.S. A two-source approach for estimating soil and vegetation energy fluxes in observations of directional radiometric surface temperature. Agric. Forest Meteorol. **1995**, *77*, 263–293.
66. Kustas, W.P.; Zhan, X.; Schmugge, T.J. Combining optical and microwave remote sensing for mapping energy fluxes in a semiarid watershed. Remote Sensing Environ. **1998**, *64*, 116–131.
67. Kustas, W.P.; Anderson, M.C. Advances in thermal infrared remote sensing for land surface modeling. Agric. Forest Meteorol. **2009**, *149*, 2071–2081.
68. Agam, N.; Evett, S.R.; Tolk, J.A.; Kustas, W.P.; Colaizzi, P.D.; Alfieri, J.G.; Mckee, L.G.; Copeland, K.S.; Howell, T.A.; Chávez, J.L. Evaporative loss from irrigated interrows in a highly advective semi-arid agricultural area. Adv. Water Resour. **2012**, http://dx.doi.org/10.1016/j.advwatres.2012.07.010
69. Mecikalski, J.R.; Diak, G.R.; Anderson, M.C.; Norman, J.M. Estimating fluxes on continental scales using remotely sensed data in an atmospheric-land exchange model. J. Appl. Meteorol. **1999**, *38*, 1352–1369.
70. Kustas, W.P.; Jackson, T.J.; French, A.N.; Macpherson, J.I. Verification of patch- and regional-scale energy balance estimates derived from microwave and optical remote sensing during SGP97. J. Hydrometeorol. **2001**, *2*, 254–273.
71. Anderson, M.C.; Kustas, W.P.; Norman, J.M.; Hain, C.R.; Mecikalski, J.R.; Schultz, L.; Gonzalez-Dugo, M.P.; Cammalleri, C.; D'urso, G.; Pimstein, A.; Gao, F. Mapping daily evapotranspiration at field to continental scales using geostationary and polar orbiting satellite imagery. Hydrol. Earth Syst. Sci. **2011**, *15*, 223–239.

Soil: Fauna

Mary C. Savin
Department of Crop, Soil, and Environmental Sciences, University of Arkansas, Fayetteville, Arkansas, U.S.A.

Abstract
Soil fauna are diverse, abundant, and critical to ecosystem functions. Soil fauna account for almost one-quarter of living species and facilitate important ecological interactions that contribute to decomposition, nutrient cycling, and stabilization of terrestrial ecosystems. Soil fauna are composed of diverse groups of organisms that range across orders of magnitude in size, with many less than a few centimeters or even a few millimeters. These organisms function under different abiotic conditions, and despite their small size, contribute to characteristics and processes observable at field and landscape scales. Future research efforts that build upon past discoveries should enhance understanding of how soil fauna connect aboveground and belowground ecology for proper management and sustainability of ecosystems and ecosystem services.

INTRODUCTION

Soil fauna are diverse, abundant, and critical to ecosystem functions. Soil fauna account for almost 25% of living species.[1] Although they compose a group of organisms that has been ignored historically when it comes to land management, there is growing recognition and understanding of the roles soil fauna play in ecosystems. Appreciation for which organisms are present in soil, their habitats, community diversity, and contributions to ecosystem functions, stability and sustainability is increasing with accumulating evidence from scientific research. Further development of our understanding of soil ecology is essential because soil fauna contribute directly and indirectly to properties and processes at the scale of the organism and its habitat with outcomes observed up to field and landscape levels. Future research efforts that build upon past discoveries should enhance our understanding of how soil fauna connect aboveground and belowground ecology for the proper management and sustainability of ecosystems and ecosystem services.[2]

SOIL FAUNA—ORGANISMS

Soil fauna, protists, and animals include organisms that span orders of magnitude in size (from μm to cm to m). Generally, soil fauna are eukaryotic, aerobic chemoheterotrophs, but they occupy many niches in soil food webs.[3] Organisms are commonly distributed into groupings which can range from low to high taxonomic resolution. One common differentiation is classification by size. Sized-based groups are microfauna, mesofauna, macrofauna, and megafauna. (See soil microbiology and ecology textbooks, e.g., references,[4–7] for more detailed descriptions of the groups.)

Because categorization based on size is an empirical separation, the upper size cut-off value may differ depending on the reference in the literature that one is searching. Microfauna may be considered up to 100–200 μm in length and include protozoa and nematodes. Nematodes (phylum Nematoda) may also be designated as mesofauna in the literature. Protozoa (kingdom Protista) are often grouped into flagellates, amoeba, and ciliates. Flagellates and amoebae tend to be smaller and numerous while ciliates are larger relatively and less numerous. Protozoa consume bacteria, protozoa, and other organisms. Abundances range from $10–10^5$/g soil.[8] (The diverse range of numbers for soil faunal abundances is related to the heterogeneity of soil conditions.) Nematodes, roundworms, are often grouped by into trophic groups: bacterivores, fungivores, omnivores, predators, and plant parasites. Microfauna, and also the mesofauna enchytraeids, are considered to reside in water films and water-filled pores inside aggregates.[9–10]

Mesofauna range from 100 (or 200) to 2000 (sometimes considered up to 10,000) μm. At this larger size range, mesofauna are considered to be in the interaggregate pore space, which is likely to be air-filled, rather than the intra-aggregate pore space expected to be occupied by microfauna.[11] This category includes microarthropods such as springtails (subclass Collembola) and mites (subclass Acari), and the enchytraeids (family Enchytraeidae) (potworms). Microarthropod abundances may range from $10–10^5/m^2$ (reviewed in Coleman et al.).[5] Enchytraeidae are particularly important in temperate and boreal forests (reviewed in Huhta et al.)[12] at abundances ranging from 10^3 to greater than $10^5/m^2$. Enchytraeidae consume fine plant litter covered in fungi and bacteria. Collembola are likely omnivorous, although they are often considered fungivorous, and mites feed on fungi, plant detritus, or are predaceous.[5] Mesofauna are generally regarded as

Encyclopedia of Natural Resources DOI: 10.1081/E-ENRL-120047488

important consumers in litter transformation, especially important given that about 90% of terrestrial organic inputs are processed through detrital food webs.[5]

Macrofauna are generally considered to be greater than 2 mm, up to a few cm, and some species of earthworms are meters long. Macrofauna are often cited as including ants (widely distributed, family Formicidae), earthworms (Phylum Annelida, class Oligochaeta), termites (tropical and some temperate, Termitidae in the Order Isoptera), and macroarthropods (phylum Arthropoda). Earthworm abundances range from less than 10 to greater than $1000/m^2$ in temperate and tropical ecosystems.[13] When present, ants and termites can be so abundant that researchers often do not enumerate abundances. Megafauna sometimes include earthworms and include vertebrates such as moles, gophers, and prairie dogs which can be important in many ecosystems.[14] Macrofauna and megafauna can be ecosystem engineers. These organisms are too large to exist within the existing pores; in soil, they move soil particles, modify soil pore structure, and create structures such as aggregates.[15–16] In general, ecosystem engineers modify their physical environment and resources for others.[14–17] Further, they alter the environment distinctively from abiotic processes.[14,17]

Much of the biological activity in soil occurs in a small volume relative to whole which poses logistical challenges in representative sampling. However, these "hotspots" or small volumes of soil where conditions are conducive for high activity can be placed in important functional domains.[16] These domains, or spheres of influence regulating microbial activity, are often associated with the activity of larger organisms such as plant roots and macrofauna, and include examples such as the rhizosphere (zone of influence around a root), drilosphere (zone of influence around an earthworm burrow), and termitosphere (zone of influence under termite activity). Macrofauna create conditions conducive for high rates of activity by directly changing the "architecture" of soil at scales relevant to smaller organisms,[18] and the accumulation of their direct and indirect activities changes properties and processes in fields. Macrofauna may reside in the surface litter layer (e.g., epigeic earthworms), soil (e.g., endogeic earthworms), and/or move across soil-litter boundaries (e.g., anecic earthworms). Macrofauna are also involved in important symbiotic relationships in soil, such as protozoan gut communities in termites, fungus gardens of termites, and microbes in the external rumen of earthworms.[7]

SOIL FAUNA—DIVERSITY AND BIOMASS

The enormous diversity of fauna means that not all groups have been identified and described. Fauna grouped by size may be placed into broad taxonomic categories that use trophic status for differentiation. Beyond taxonomic classifications, soil fauna are often placed into functional groups. Characterizing communities based on functional groups or higher level taxonomic groups may provide stronger statistical power[19] than trying to identify organisms at finer resolution. However, species may feed across trophic levels,[20] and thus taxonomic trophic groups may not be most appropriate level of resolution for understanding the functional importance of soil fauna. Salamon et al.,[21] for example, found that different species of Collembola respond to increases in fungal compared to bacterial biomass. Although there is one food web aboveground, there are both bacterial-based and fungal-based food webs operating belowground.

Despite the difficulty of determining the appropriate level of resolution to understand soil faunal diversity, diversity is an important consideration for processes, ecosystem functions, and productivity. There remains much that is not well known about the diversity of belowground communities. Regardless, it is safe to state that both local and global diversity of soil fauna is huge (reviewed in Coleman).[3] Estimates for numbers of species can span hundreds in a few hundred square meters.[22] However, relatively few taxa may dominate soil communities within a land use. In the southeastern United States, rank–abundance relationships suggest that a few taxa dominated in four land uses, with most abundant taxa belonging to different types of beetles in cultivated and pasture sites, centipedes in hardwood forest, and millipedes in pine stand forest.[23] A large portion of biomass in tundra, boreal forest, and temperate forest soil (high latitude ecosystems) was accounted for in enchytraeids, while earthworms dominated in tropical forests and temperate grassland soils.[24] Mites, Collembola, and Enchytraeidae are abundant in acidic forests with thick organic layers.[20] Despite the diversity of organisms, there are a few predominant organisms in ecosystems in many studies, that is, many rare, and few cosmopolitan taxa.

A recent survey of soil animals (i.e., not including protozoa) in 11 locations around the global using molecular analysis of 18S rDNA sequences resulted in a dominance of nematodes and microarthropods.[25] Almost 96% of operational taxonomic units (OUTs) (surrogate for species defined by 99% sequence similarity) were detected in one location only, indicating high endemism and a lack of cosmopolitan species.[25] Predominance of nematodes (grasslands) and microarthropods (forests) was related to pH with nematodes positively correlated and microarthropods negatively correlated with pH. In fact, several environmental variables were related to abundances of different groups; microarthropods predominate in forests with lower pH, root biomass, mean annual temperature, inorganic N, and higher C:N, litter, and moisture.[25] Further, on a global scale the soils from ecosystems with higher aboveground plant diversity had lower belowground faunal diversity.[25] Diversity matrices have been less informative than multivariate analyses.[25–26] Thus, it may be that diversity per se is not the best approach to assessing soil faunal communities, rather multivariate approaches relating abundance and functional structure to environmental variables may be informative.

While it is generally postulated that high biodiversity levels are positive for ecosystem health, stability, and function, there may be much functional redundancy among soil fauna. Impacts of fauna depend in part on characteristics of the ecosystem. In a field experiment investigating the contributions of soil fauna to litter decomposition, fauna of all size classes increased decomposition of litter in its home environment when the plant community was in late succession and contained more recalcitrant compounds.[27] In early succession, when litter contained easily degraded compounds, the presence of an indigenous community was less important. Often there are changes in taxonomic groups within a system, but little change in diversity. Thus, it may be that functional structure of soil fauna is more important than taxonomic structure in ecosystems. Responses to elevated CO_2 concentrations were related to trophic structure among soil biota in a meta-analysis.[28] Abundances of detritivores increased while there were no changes at higher trophic levels. So, change in the functional characteristics of species composition is important. In a microcosm study, Cole et al.[29] found that an increasing density of microarthropods increased nitrate concentrations, but increased richness decreased N mineralization and nitrification. The authors suggest that increased predation pressure with increased richness decreased the microbivore grazing of remaining microarthropods. Thus, they concluded it was not a change in richness per se that was important but a change that altered functional group status that was important for functioning.

The study by Cole et al.[29] also brings to light another question in understanding the impacts of soil fauna on ecosystems. In addition to the extent of diversity, be it functional or taxonomic, among ecosystems, there is also a question of whether total abundance of soil fauna or diversity is more important. In another global meta-analysis of seven biomes, the authors estimated biomass of five major faunal groups. Despite the challenges in computing estimates for biomass that resulted in low to medium confidence for several of the estimates from several biomes, faunal biomass was found to be 40–80% of animal biomass in different biomes except desert.[24] However, soil faunal biomass (with the exception of deserts at <0.02% of microbial biomass) averaged 2% (1.5–3.6%) of microbial biomass.[24] Microbial biomass is 2–5% soil organic carbon, which is usually 1–5% of soil by volume. Despite composing a minor amount of the biomass in soil, soil fauna are major players in many important functions in terrestrial ecosystems.

SOIL FAUNA—ROLES IN ECOSYSTEM FUNCTIONS, STABILITY, AND SUSTAINABILITY

Lavelle et al.[2] argue that invertebrates are a "resource that needs to be properly managed to enhance ecosystem services" since fauna mediate soil formation, nutrient cycling, primary production, and regulation services. Soil organisms alter soil aggregation through the production of particular compounds, the physical breakdown of organic materials, and the movement of soil particles. Changing soil pore and aggregate structure impacts water infiltration or run-off, and distribution and water storage within the soil profile. Water content affects the habitat, diffusion of gases and compounds, and activity and access of soil organisms, which has implications for the rates and extent of nutrient cycling. Grazing consumes microbes, changes community structure and activity, and releases nutrients. Soil fauna may consume microorganisms, other fauna, and plant detritus. Thus, soil fauna impact nutrient cycling directly through incorporation and comminution of plant detrital materials, and indirectly through consumption, movement, selection, and activation of microbial communities. These processes impact sequestration of carbon (and other nutrients) in soil and gaseous releases which affect climate. The activities of soil fauna impact root production, and enhanced nutrient cycling can increase aboveground plant biomass.[30]

While the negative impacts of pests and pathogens are well studied and appreciated because of the economic damage they can cause, the positive impacts of the activities of soil fauna on plant growth and other ecosystem functions are often less appreciated and intentionally utilized. Ingham et al.[30] demonstrated the positive influence of free-living nematodes on plant biomass by inoculating soil with bacteria and bacterial-feeding nematodes. About 30% N taken up by plants can come from mineralization as a result of grazing of fauna on soil microorganisms.[31] The presence of earthworms has been shown to decrease plant parasitic nematode infection of rice by more than 80%, allowing plants to maintain biomass production similar to uninfected plants.[32] In a recent study as part of the biodiversity experiment in Jena, Germany, use of the insecticide chlorpyrifos decreased biomass of two functional plant groups (forbs and grass), indicating that negative effects from reductions in Collembola abundance, whose grazing of microorganisms contributes to nutrient cycling, outweighed benefits that might have been gained from reductions in herbivore populations.[33]

Soil fauna are an integral component of belowground ecology which can alter aboveground ecology, but these fauna also respond to aboveground management. Fauna may respond to or emit plant signaling compounds (reviewed in Kardol and Wardle).[34] Re-establishing mixed forests in coniferous stands changed *Collembola* species richness and diversity, highlighting how aboveground management can affect belowground ecology.[35] In a study comparing conventional practices to organic wheat farming, organic management promoted both above- and belowground interactions with positive effects on soil fauna compared to a conventional system using mineral fertilizers and herbicides.[36] Organic management improved soil quality by enhancing generalist predators (top-down control);

however, the system was also enhanced through the addition of resources (bottom-up control).[36]

Subject to much debate is whether terrestrial systems and soil food webs are controlled by bottom-up or top-down forces. Invertebrate numbers and populations respond to or covary with soil properties, such as soil organic matter (OM), temperature, and electrical conductivity.[26] However, there are conflicting data about which properties and the degree of importance of properties that control soil fauna. Mites correlated positively with high soil OM.[27] Soil pH is an important controlling variable. For example, enchytraeids are abundant in low pH soil.[20] Soils are often considered to be controlled through bottom-up controls. However, research results do not necessarily clarify the debate. Although microorganisms are the primary agents of decomposition in soil, microbes do not always respond to resource inputs. This may be in part because fauna are consuming biomass, and thus preventing measurable increases in microbial production. In a temperate deciduous forest in Germany, while particular organisms responded to resource input, the structure of the food web was not clearly related to bottom-up control.[21] Earthworms changed habitat for other organisms, reducing effects of resource inputs.[21] Increased predation pressure reduces N mineralization, presumably through cascading effects that reduce grazing of microorganisms.[29,36] These results indicate that top-down control of food web dynamics is also important in understanding how functions emerge in systems. Taken together, there remain many gaps in the data for many groups of fauna to adequately assess what conditions lead to the greatest diversity. It is often hypothesized that with little disturbance and much competition, diversity will be limited, and under too much stress, diversity will be limited. However, the heterogeneity of soil, patchy community structures, and the activities of ecosystem engineers may all complicate these relationships and lead to high diversity of fauna in soil ecosystems.[37]

In addition to diverse community composition, inherent spatial and temporal variability in soil properties and activities and interactions of organisms poses challenges in accurate assessments. In a study investigating the literature from 1940 to 1992, spatial variability of mesofauna and macrofauna was analyzed.[19] While common species showed lower variation than rare species, the coefficient of variation for common species was often 100% with much of the variability attributed to random error.[19] An important step in making cross-study comparisons is to standardize field collection methods to assess soil faunal communities.[38] Abundances and diversity are often greatest at the soil surface. While organisms can be found at depths below 30 cm, Callaham et al.[23] discontinued sampling in four land uses in the southeastern United States at the 15–30 cm depths for their faunal community assessments because fauna were found at the 0–15 cm soil depth. Abundances often need to be transformed to fit parametric statistical analyses. Distributions can be difficult to sample representatively because they are patchy at best within an ecosystem and variable in response to soil properties and to disturbances.[39,40] Further, distributions and activity are temporally variable. For example, enchytraeid activity is closely related to rainfall.[41]

Soil fauna are responsive to changes in their environment. This responsiveness, which occurs in part because the soil is their habitat, makes soil fauna useful indicator organisms of disturbance, recovery, and soil quality.[42] Nematodes, for example, have been used as indicators of ecosystem recovery and function, maturity, and the soil food web.[43,44] Earthworms are utilized as indicators of pollution, including heavy metals. However, their diversity in niche separation may make it difficult to interpret responses. Currently, we remain limited not only by scale of resolution in sampling and identification, but by what we do not know about the biology and ecology of unidentified and even identified organisms. Barbercheck et al.[26] suggested separating Collembola by habitat type for effectiveness of using Collembola as indicators. However, relying on one indicator or group of indicators to describe effects of disturbance among ecosystems is not advised.[19,26]

In restoration ecology, it has been proposed that to restore ecosystems, one must integrate the belowground with the aboveground.[34] Organisms are useful as indicators because they integrate cumulative effects of chemistry and physics. However, there are different types of disturbances which occur at different frequencies, intensities, durations, and extent of spatial coverage. Greater diversity of fauna was measured in the less disturbed hardwood forest site, followed by the pine stand, followed by pastures, and the least diversity occurred in cultivated soil in four land uses in southeastern United States.[23] Native earthworms were present in fields and forests, but most commonly collected from forests.[23] Introduced earthworms were most often collected from cultivated fields and pastures. Introduced species are well known for drastically altering characteristics and functions of ecosystems, especially if they become invasive. Introduction of non-native earthworms has resulted in change in location and cycling of OM in forests, such that ecosystems are no longer sustainable (northern U.S. forests).[45] It is frequently expected that introduced organisms will be more common in disturbed areas and native species more common in undisturbed locales.[46] However, although reduced in abundance, native species are not always eliminated from an ecosystem as a result of a disturbance. To what extent native and introduced organisms co-occur and alter ecosystem function still needs research.

CONCLUSION

Although there is growing recognition of their integral effects on soil functions, soil fauna have been largely ignored in development of land management plans. There is an accumulating body of knowledge of the importance of these organisms in driving ecosystem functions. Beyond the small size of many organisms, soil fauna may not have received deserved levels of research attention because there is general agreement that most important drivers of nutrient cycling are microorganisms. However, soil fauna interact with microbes, influence community structure and dispersal, change the physical environment, and facilitate decomposition and nutrient cycling. Soil fauna may be the link that is necessary to connect microorganisms to fields and landscapes, that is, the scale at which processes are observed.

REFERENCES

1. Decaëns, T.; Jiménez, J.J.; Gioia, C.; Measey, G.J; Lavelle, P. The values of soil animals for conservation biology. Eur. J. Soil Biol. **2006**, *42*, S23–S38.
2. Lavelle, P.; Decaëns, T.; Aubert, M.; Barot, S.; Blouin, M.; Bureau, F.; Margerie, P.; Mora, P.; Rossi, J.P. Soil invertebrates and ecosystem services. Eur. J. Soil Biol. **2006**, *42*, S3–S15.
3. Coleman, D.C. From peds to paradoxes: Linkages between soil biota and their influences on ecological processes. Soil Biol. Biochem. **2008**, *40*, 271–289.
4. Amador, J.A.; Görres, J.H. Chapter 8 Fauna. In *Principles and Applications of Soil Microbiology 2nd Edition*; Sylvia, D.M.; Fuhrmann, J.J.; Hartel, P.G.; Zuberer, D.A., Eds.; Pearson Education Inc.: Upper Saddle River, NJ, 2005; 181–200.
5. Coleman, D.C.; Crossley, Jr., D.A.; Hendrix, P.F. *Fundamentals of Soil Ecology 2nd Edition*; Elsevier Academic Press: Burlington, MA, 2004.
6. Coleman, D.C.; Wall, D.H. Fauna: The engine for microbial activity and transport. In *Soil Microbiology, Ecology, and Biochemistry 3rd Edition*; Paul, E.A., Ed.; Elsevier Academic Press: Burlington, MA. 2007; 163–191.
7. Lavelle, P.; Spain, A.V. *Soil Ecology*; Kluwer Academic Publishers: Dordrecht, the Netherlands, 2001.
8. Clarholm, M. Protozoan grazing of bacteria in soil – impact and importance. Microb. Ecol. **1981**, *7*, 343–350.
9. Bouwman, L.A.; Zwart, K.B. The ecology of bacterivorous protozoans and nematodes in arable soil. Agric. Ecosyst. Environ. **1994**, *51*, 145–160.
10. Görres, J.H.; Savin, M.C.; Neher, D.A.; Weicht, T.H.; Amador. J.A. Grazing in a porous environment: 1. The effect of pore structure on C and N mineralization. Plant Soil **1999**, *212*, 75–83.
11. Lavelle, P. Faunal activities and soil processes: Adaptive strategies that determine ecosystem function. Adv. Ecol. Res. **1997**, *27*, 95–132.
12. Huhta, V.; Persson, T.; Setälä, H. Functional implications of soil fauna diversity in boreal forests. Appl. Soil Ecol. **1998**, *10*, 277–288.
13. Curry, J.P. Factors affecting earthworm abundances in soils. In *Earthworm Ecology*; Edwards, C.A. Ed; CRC Press: Boca Raton, 2000; 37–64.
14. Reichman, O.J.; Seabloom, E.W. The roles of pocket gophers as subterranean ecosystem engineers. Trends Ecol. Evol. **2002**, *17*, 44–49.
15. Lavelle, P.; Barois, I.; Martion, A.; Zaidi, Z.; Schaefer, R. Management of earthworm populations in agro-ecosystems: A possible way to maintain soil quality? In *Ecology of Arable Land*; Clarholm, M.; Bergström, L. Eds.; Kluwer Academic Publishers: Dordrecht, the Netherlands, 1989; 109–122.
16. Brown, G.G.; Barois, I.; Lavelle, P. Regulations of soil organic matter dynamics and microbial activity in the drilosphere and the role of interactions with other edaphic functional domains. Eur. J. Soil Biol. **2000**, *36*, 177–198.
17. Jones, C.G.; Lawton, J.H; Shachak, M. Organisms as ecosystem engineers. Oikos **1994**, *69*, 373–386.
18. Görres, J.H.; Savin, M.C.; Amador, J.A. Soil micropore structure and carbon mineralization in burrows and casts of anecic earthworms (*Lumbricus terrestris*). Soil Biol. Biochem. **2001**, *33*, 1881–1887.
19. Ekschmitt, K. Population assessments of soil fauna: General criteria for the planning of sampling schemes. Appl. Soil Ecol. **1998**, *9*, 439–445.
20. Scheu, S.; Falca, M. The soil food web of two beech forests (*Fagus sylvatica*) of contrasting humus type: stable isotope analysis of a macro- and a mesofauna-dominated community. Oecologia **2000**, *123*, 285–286.
21. Salamon, J.A.; Alphei, J.; Ruf, A.; Schaefer, M.; Scheu, S.; Schnieder, K.; Sührig, A.; Maraun, M. Transitory dynamic effects in the soil invertebrate community in a temperate deciduous forest: Effects of resource quality. Soil Biol. Biochem. **2006**, *38*, 209–221.
22. Hansen, R.A. Effects of habitat complexity and composition on a diverse litter microarthropods assemblage. Ecology **2000**, *81*, 1120–1132.
23. Callaham, Jr., M.A.; Richter, Jr, D.D.; Coleman, D.C.; Hofmockel, M. Long-term land-use effects on soil invertebrate communities in Southern Piedmont soils, USA. Eur. J. Soil Biol. **2006**, *42*, S150–S156.
24. Fierer, N.; Strickland, M.S.; Liptzin, D.; Bradford, M.A.; Cleveland, C.C. Global patterns in belowground communities. Ecol. Lett. **2009**, *12*, 1238–1249.
25. Wu, T.; Ayres, E.; Bardgett, R.D; Wall, D. H.; Garey, J.R. Molecular study of worldwide distribution and diversity of soil animals. PLOS 2011, doi: 10.1073/pnas.1103824108.
26. Barbercheck, M.E.; Neher, D.A.; Anas, O.; El-Allaf, S.M.; Weicht, T.R. Response of soil invertebrates to disturbance across three resource regions in North Carolina. Environ. Monit. Assess. **2009**, *152*, 283–298.
27. Milcu, A.; Manning, P. All size classes of soil fauna and litter quality control the acceleration of litter decay in its home environment. Oikos **2011**, *120*, 1366–1370.
28. Blankinship, J.C.; Niklaus, P.A.; Hungate, B.A. A meta-analysis of responses of soil biota to global change. Oecologia **2011**, *165*, 553–565.
29. Cole, L.; Dromph, K.M.; Boaglio, V.; Bardgett, R.D. Effect of density and species richness of soil mesofauna on nutrient mineralisation and plant growth. Biol. Fertil. Soil **2004**, *39*, 337–343.
30. Ingham, R.E.; Trofymow, J.A.; Ingham, E.R.; Coleman, D.C. Interactions of bacteria, fungi, and their nematode grazers: Effects on nutrient cycling and plant growth. Ecol. Monogr. **1985**, *55*, 119–140.
31. Verhoef, H.A.; Brussaard, L. Decomposition and nitrogen mineralization in natural and agro-ecosystems: The contribution of soil animals. Biogeochemistry **1990**, *11*, 175–211.

32. Blouin, M.; Zuily-Fodil, Y.; Pham-Thi, A.-T.; Laffray, D.; Reversat, G.; Pando, A.; Tondoh, J.; Lavelle, P. Below-ground organism activities affect plant aboveground phenotype, inducing plant tolerance to parasites. Ecol. Lett. **2005**, *8*, 202–208.
33. Eisenhauer, N.; Sabais, A.C.W.; Schonert, F.; Scheu, S. Soil arthropods beneficially rather than detrimentally impact plant performance in experimental grasslands systems of different diversity. Soil Biol. Biochem. **2010**, *42*, 1418–1424.
34. Kardol, P.; Wardle, D.A. How understanding aboveground-belowground linkages can assist restoration ecology. Trends Ecol. Evol. **2010**, *25*, 670–679.
35. Chauvet, M.; Titsch, D.; Zaytsev, A.S.; Wolters, V. Changes in soil faunal assemblages during conversion from pure to mixed forest stands. For. Ecol. Manag. **2011**, *262*, 317–324.
36. Birkhofer, K.; Bezemer, T.M.; Bloem, J.; Bonkowski, M.; Christensen, S.; Dubois, D.; Ekelund, F.; Fließbach, A.; Gunst, L. Hedlund, K.; Mäder, P.; Mikola, J.; Robin, C.; Setälä, H.; Tatin-Froux, F.; Van der Putten, W.H.; Scheu, S. Long-term organic farming fosters below and aboveground biota: Implications for soil quality, biological control and productivity. Soil Biol. Biochem. **2008**, *40*, 2297–2308.
37. Decaëns, T. Macroecological patterns in soil communities. Global Ecol. Biogeograph. **2010**, *19*, 287–302.
38. Römbke, J.; Sousa, J.P.; Schouten, T.; Riepert, F. Monitoring of soil organisms: A set of standardized field methods proposed by ISO. Eur. J. Soil Biol. **2006**, *42*, S61–S64.
39. Jiménez, J.-J.; Decaëns, T.; Amézquita, E.; Rao, I.; Thomas, R.J. and Lavelle, P. Short-range spatial variability of soil physico-chemical variables related to earthworm clustering in a neotropical gallery forest. Soil Biol. Biochem. **2011**, *43*, 1071–1080.
40. Rossi, J.P.; Huerta, E.; Fragoso, C.; Lavelle, P. Soil properties inside earthworm patches and gaps in a tropical grassland (la Mancha, Veracruz, Mexico). Eur. J. Soil Biol. **2006**, *42*, S284–S288.
41. Nieminen, J.K. Enchytraeid population dynamics: Resource limitation and size-dependent mortality. Ecol. Monit. **2009**, *220*, 1425–1430.
42. Wardle, D.A.; Yeates, G.W.; Watson, R.N.; Nicholson, K.S. The detritus food-web and the diversity of soil fauna as indicators of disturbance regimes in agro-ecosystems. Plant Soil **1995**, *170*, 35–43.
43. Bongers, T. The maturity index: An ecological measure of environmental disturbance based on nematode species composition. Oecologia **1990**, *83*, 14–19.
44. Neher, D.A. Role of nematodes in soil health and their use as indicators. J. Nematol. **2001**, *33*, 161–168.
45. Bohlen, P.J.; Groffman, P.M.; Fahey, T.J.; Fisk, M.C.; Suarez, E.; Pelletier, D.M.; Fahey, R.T. Ecosystem consequences of exotic earthworm invasion of north temperate forests. Ecosystems **2004**, *7*, 1–12.
46. Hendrix, P.F.; Baker, G.H.; Callaham, Jr., M.A.; Damoff, G.A.; Fragoso, C.; Gonzalez, G.; James, S.W.; Lachnicht, S.L.; Winsome, T.; Zou, X. Invasion of exotic earthworms into ecosystems inhabited by native earthworms. Biol. Invasions **2006**, *8*, 1287–1300.

Soil: Fertility and Nutrient Management

John L. Havlin
Department of Soil Science, North Carolina State University, Raleigh, North Carolina, U.S.A.

Abstract

Future population growth will drive increased food and fiber production per unit of land. Increasing crop productivity increases soil nutrient removal and the importance of replenishing soil fertility through effective and efficient nutrient management. Native soil nutrient supply depends on the soils' ability to buffer nutrient loss through crop removal. Mineralization of soil organic fractions provides limited supplies of N, S, and micronutrients, while mineral dissolution and surface exchange reactions resupply P, K, Ca, Mg, and micronutrients. Nutrient mobility in soil influences ion transport to plant roots, evaluation of nutrient availability to plants, and ultimately nutrient management decisions. Effective nutrient management requires quantifying crop nutrient requirement and the nutrient supplying capacity of the soil through soil testing. Once these are determined, the nutrient management plan includes determination of the optimum nutrient rate, appropriate nutrient source(s), most effective nutrient placement method(s), and nutrient application timing. Management decisions will vary depending on the specific nutrient. Effective management of soil nutrients will ensure crop productivity meets consumption demand, while minimizing impacts of nutrient use on the environment.

INTRODUCTION

Nutrients are essential to support life and are acquired through diverse food sources, ultimately supplied by the soil. While plants are a major direct human food source, animals used as human food also obtain nutrients from a variety of plants (forage and grains) in their diets. As plants are consumed or removed from a field, the nutrients are also removed. With continued crop removal over time, soil nutrients are depleted and plant productivity declines. As food demand increases with increasing population growth, a greater agricultural output demand requires substantial replacement of soil nutrients through addition of inorganic fertilizers and organic wastes. Therefore, to optimize nutrient availability to crops and minimize environmental risk of nutrient use, it is essential to understand nutrient reactions and processes in soils (*soil fertility*), and to efficiently manage nutrient inputs (*nutrient management*) to ensure adequate soil nutrient supply to support life. Without effective management of soil nutrients, 30–40% more land area will be needed today to produce the same crop yields depending on the crop, soil, and climatic region. Therefore, crop producers must take advantage of technologies that increase productivity, as well as those that minimize productivity loss.

SOIL FERTILITY

Soil fertility represents the diverse interactions between physical, chemical, and biological soil processes that control plant nutrient supply. Understanding these processes and interactions is critical to optimizing nutrient availability to crops. Maximum crop production depends on the growing season environment and the producers' skill to minimize yield limiting factors using appropriate management technologies. Crop productivity is maximized when the *most* limiting factor to yield potential is corrected first, the second most limiting factor next, and so on (*Law of the Minimum*). For example, the correct nutrient supply may not result in the highest yield potential if plant-available water is the *most* limiting factor. Thus, minimizing water as a yield-limiting factor must be accomplished before the crop can fully respond to management of any other factor.

Essential Plant Nutrients

Seventeen elements are considered essential plant nutrients since they are involved in metabolic functions such that the plant cannot complete its life cycle without these elements (Table 1). The most abundant nonmineral nutrients (C, H, and O) are obtained from CO_2 and H_2O and converted into carbohydrates that form amino acids, proteins, nucleic acids, and other compounds through photosynthesis. The supply of CO_2 has increased from ~310 to 390 ppm CO_2 since 1960. The supply of H_2O rarely limits photosynthesis directly but reduces plant growth when plant-available H_2O is limited. The remaining 14 macronutrients and micronutrients are classified on their relative abundance in

Encyclopedia of Natural Resources DOI: 10.1081/E-ENRL-120047493

Table 1 Essential plant nutrients and their relative and average concentration

	Nutrient		Concentration in plants[a]	
Classification	**Name**	**Symbol**	**Relative**	**Average**
Macronutrients	Hydrogen	H	60,000,000	6%
	Carbon	C	40,000,000	45%
	Oxygen	O	30,000,000	45%
	Nitrogen	N	1,000,000	1.5%
	Potassium	K	250,000	1.0%
	Calcium	Ca	125,000	0.5%
	Magnesium	Mg	80,000	0.2%
	Phosphorus	P	60,000	0.2%
	Sulfur	S	30,000	0.2%
Micronutrients	Chloride	Cl	3000	100 ppm
	Iron	Fe	2000	100 ppm
	Boron	B	2000	20 ppm
	Manganese	Mn	1000	50 ppm
	Zinc	Zn	300	20 ppm
	Copper	Cu	100	6 ppm
	Nickel	Ni	2	<1 ppm
	Molybdenum	Mo	1	<1 ppm

[a]Concentration expressed on a dry matter weight basis.

plants. The N concentration in most plants is greater than the other 13 nutrients combined. Although nutrient content varies greatly between plant species, most plants also require substantial amounts of Ca and K (Table 1). Plants absorb many nonessential elements present in the soil solution. For example, Al^{3+} in plants can be high when soils contain relatively large amounts of soluble Al (low pH soil). In some severely acid soils, Al toxicity can reduce plant growth and yield. In high-pH soils, Na^{+} can be elevated in soil solution and/or on the exchange complex, increasing Na^{+} in the plant, which can also be detrimental to plant growth.

Many soil, climate, and management factors influence plant nutrient availability. When nutrient concentration is *deficient* enough to severely reduce plant growth and yield, distinct visual deficiency symptoms appear (Fig. 1). Extreme deficiencies can result in plant death. Visual symptoms may not appear when a nutrient is marginally deficient (*hidden hunger*), but plant yields may still be reduced. Plant nutrient concentration above which plant growth or yield is not increased is the *critical level* or *range*. Critical nutrient ranges vary among plants but always occur in the transition between nutrient deficiency and sufficiency (Fig. 1). *Sufficiency* or *luxury consumption* is the concentration range where added nutrient does not increase yield but increases nutrient concentration. Nutrient concentration is considered *excessive* or *toxic* when plant growth and yield are reduced.

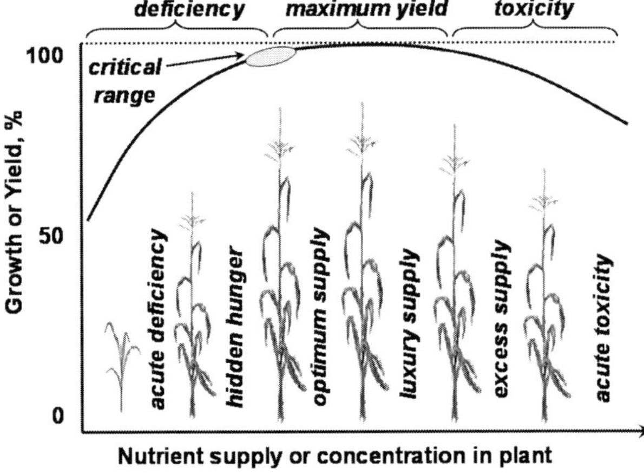

Fig. 1 Optimum nutrient supply is essential for maximizing plant growth and nutrient use. Nutrient addition beyond critical nutrient range does not increase plant growth.

Soil Processes Supplying Plant-Available Nutrients in Soil

After plants absorb nutrients from soil, nutrients are exported in harvested plant materials or returned to soil in plant residues left in the field. These nutrients are subject to the same biological and chemical reactions as nutrients added as fertilizers and organic wastes. Although this "cycle" varies among nutrients, understanding nutrient dynamics in the soil–plant–atmosphere system is essential to successful nutrient management.

As plants absorb nutrients from soil solution, reactions occur to resupply the soil solution (Fig. 2). These reactions are important to nutrient availability; however, the dominant reaction varies between nutrients. For example, biological processes are more important for N and S availability, and surface exchange reactions are important for Ca, Mg, and K supply to roots.

Cation and anion exchanges

Exchange of cations and anions on surfaces of clay minerals, inorganic compounds, organic matter (OM), and roots is one of the most important soil chemical properties and greatly influences nutrient availability (Fig. 2). Ions adsorbed on mineral and organic surfaces are reversibly exchanged with other ions in the solution. *Cation exchange capacity* (CEC) represents the quantity of negative (–) charge available to attract cations in solution and the *anion exchange capacity* (AEC) represents positive (+) charge attracting anions in a solution. In most soils CEC > AEC.

The source of (+) or (–) surface charges found in clay minerals is comprised of permanent charge unaffected by solution pH, and on (+) and (–) charges located on clay mineral edges and soil OM (Fig. 3).

At low pH more (+) charge exists due to higher H^+ ions on mineral edges and OM; as pH increases, H^+ concentration decreases, which increases (–) edge or surface charge. At pH >7 all H^+ on edges are neutralized, which maximizes (–) pH-dependent charge (Fig. 3). Soils with greater clay and OM contents have a higher CEC than sandy, low OM soils (Table 2).

CEC influences nutrient availability, as most adsorbed cations are essential plant nutrients, except for Al^{3+} in acid soils and Na^+ in alkaline pH soils. Cations are adsorbed to CEC with different strengths, which influences the relative ease of desorption. The *lyotropic series*, or relative strength of adsorption, is Al^{3+} (H^+) > Ca^{2+} > Mg^{2+} > K^+ = NH_4^+ > Na^+. Cations with greater charge are more strongly adsorbed. For cations of similar charge, adsorption strength is determined by the size of the hydrated cation; greater hydrated cation radii (Mg^{2+}) reduces adsorption because they cannot get as close to the exchange surface as smaller cations (Ca^{2+}).

Base saturation (BS) represents the percentage of CEC occupied by bases [Ca^{2+}, Mg^{2+}, K^+, (Na^+)], which increases with increasing soil pH (Fig. 4). The availability of Ca^{2+}, Mg^{2+}, and K^+ increases with increasing %BS. For example, 80% BS provides essential nutrients (cations) to plants easier than 40% BS. At pH 5 and 6 most soils have ~50% and 80% BS, respectively.

Soil solution anions are adsorbed to (+) charged sites on the surface of clay minerals and soil OM. AEC increases with decreasing soil pH (Fig. 3). The adsorption strength is $H_2PO_4^-$ > SO_4^{2-} >> NO_3^- = Cl^-. In most soils $H_2PO_4^-$ is the primary anion adsorbed, although in severely acid soils SO_4^{2-} adsorption represents a major source of plant-available S.

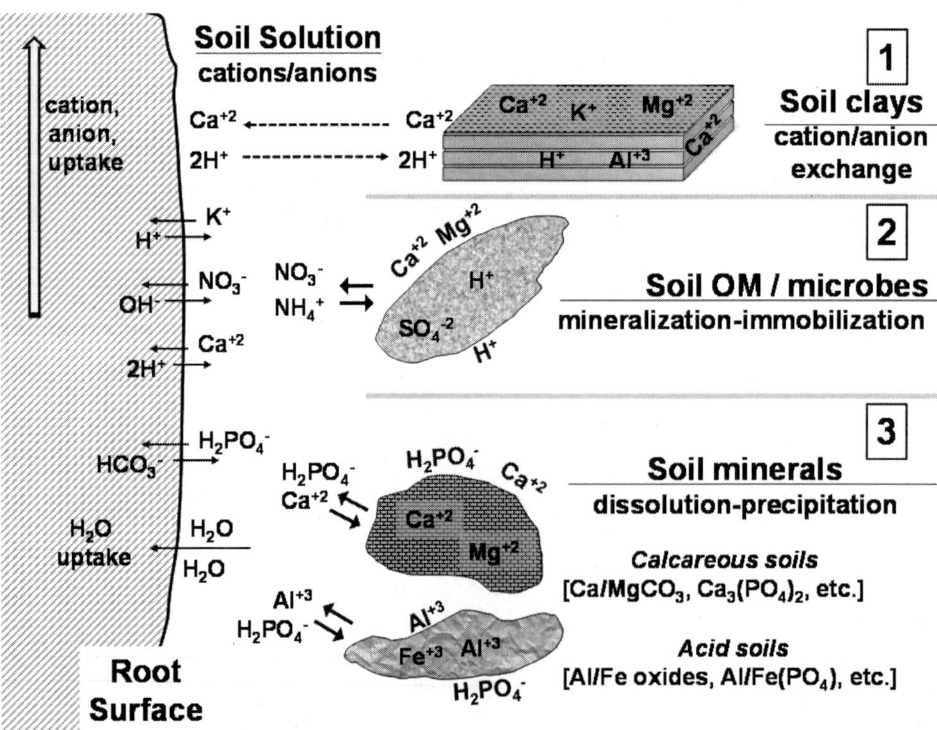

Fig. 2 Primary processes of buffering nutrients in soil solution absorbed by plant roots. Clay minerals (cation/anion exchange),[1] organic matter (mineralization),[2] and mineral compounds (dissolution)[3] resupply nutrients to soil solution as they are removed from solution by plant uptake.

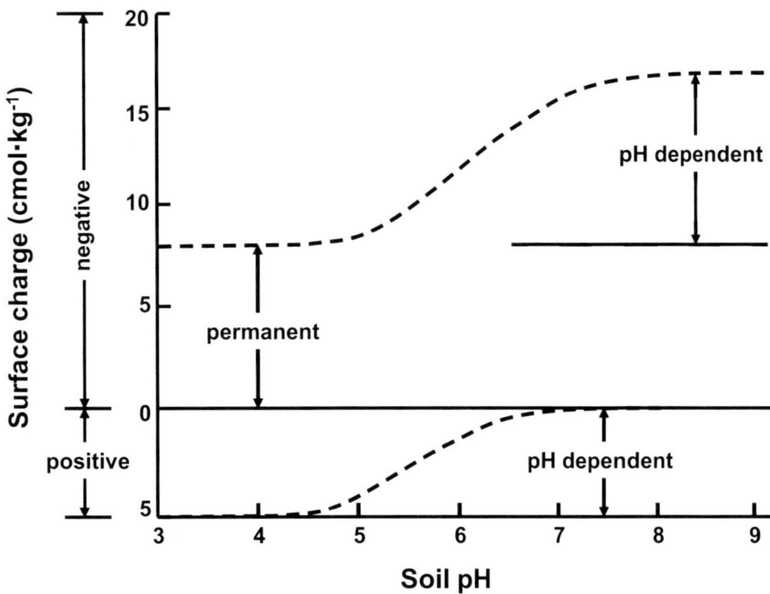

Fig. 3 Permanent and pH-dependent charge associated with clay minerals. As soil pH increases (+), charge decreases and (−) charge increases, increasing cation exchange capacity.

Table 2 Typical range in cation exchange capacity for moderately weathered (Mollisol) and highly weathered (Ultisol) soils

	Mollisol	Ultisol
Soil textural class	(cmol kg^{-1})	
Sands (light colored)	3–5	~1
Sands (dark colored)	10–20	1–3
Loams	10–15	1.5–5
Silt loams	15–25	2–6
Clay and clay loams	20–50	3–5
Organic soils	50–100	20–40

Soil Acidity and Alkalinity. Acid soils usually occur in regions where annual precipitation >600–800 mm. Natural soil acidification is enhanced with increasing rainfall since rain pH ≤ 5.7, depending on pollutants such as SO_2, NO_2, and others which lower rainfall pH. Management factors increasing soil acidity include (1) use of acid forming fertilizers, (2) crop removal and/or leaching of cations decreasing %BS, and (3) decomposition of organic residues. Optimum soil pH depends on the crop and ranges between 4.5 and 7.5. Neutralizing soil acidity by liming to ≥pH 5.5 will reduce exchangeable Al^{3+} to ~20% of the CEC and will usually prevent Al toxicity-related yield loss.

Calcareous soils contain the solid mineral $CaCO_3$, exhibit soil pH ≥7.2, and occur in regions of <500 mm annual precipitation. In areas where annual rainfall is >800 mm, the acidifying effect of rain increases depth to $CaCO_3$ to below the root zone. If $CaCO_3$ exists, all of it must be dissolved and neutralized before soil pH could decrease. In most situations, reducing soil pH by neutralizing $CaCO_3$ is too costly. High soil pH in calcareous soils reduces availability of several micronutrients (e.g., Fe^{3+}, Zn^{2+}).

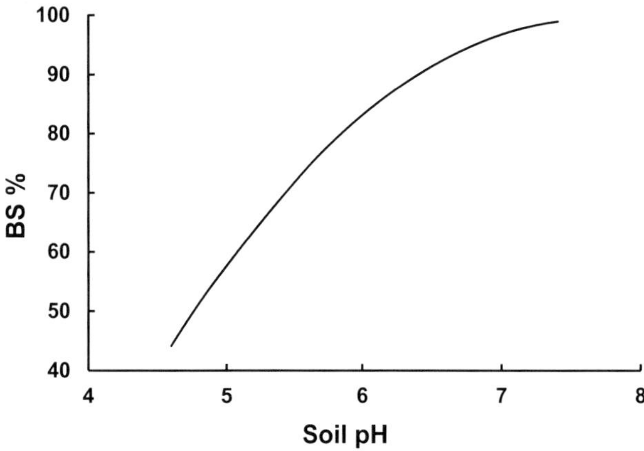

Fig. 4 Influence of soil pH on base saturation% (BS%) in mineral soils.

Table 3 Classification and properties of salt-affected soils

Classification	EC_{se} (dS m^{-1})[a]	Soil pH	ESP%[a]	Physical condition
Saline	>4	<8.5	<15	Normal
Sodic	<4	>8.5	>15	Poor
Saline-sodic	>4	<8.5	>15	Normal

[a]EC_{se} represents electrical conductivity of soil water removed from a sample of saturated soil. Increasing salt concentration increases EC_{se}. ESP% (exchangeable Na percent) represents % of cation exchange capacity occupied with Na^+.

Excessive salts in alkaline soils can reduce plant growth and yield depending on crop sensitivity. In arid and semiarid regions as water evaporates, salts are deposited near the soil surface to form saline and/or sodic soils, and are characterized by their electrical conductivity (EC_{se}), soil pH, and exchangeable Na% (ESP) (Table 3). In saline soils soluble salts interfere with plant growth, although salt tolerance varies with plant species. In sodic soils (ESP > 15%), soil aggregates disperse reducing infiltration and soil water permeability, while excess Na can also create nutritional disorders. Saline-sodic soils exhibit both high salt and Na content.

Soil organic matter

Soil OM represents organic materials in various stages of decay (Fig. 5). The most important fraction to soil fertility is the stable OM that is relatively resistant to further decomposition (Fig. 5). Soil humus, the largest component of soil OM, is relatively resistant to microbial degradation; but is essential for maintaining optimum soil physical conditions (aggregation, infiltration, aeration, etc.) important for plant growth. Fresh plant and animal residues undergo rapid decomposition. The microbial biomass represents soil microorganisms responsible for mineralization and immobilization processes that influence availability of N, P, S, and other nutrients. Residue decomposition and stable OM formation depend on climate, soil type, and soil and crop management practices. Soil OM greatly influences various soil processes and properties that influence nutrient availability (Table 4). For example, increasing soil OM increases soil aggregation and decreases bulk density, which improves water infiltration, air exchange, root

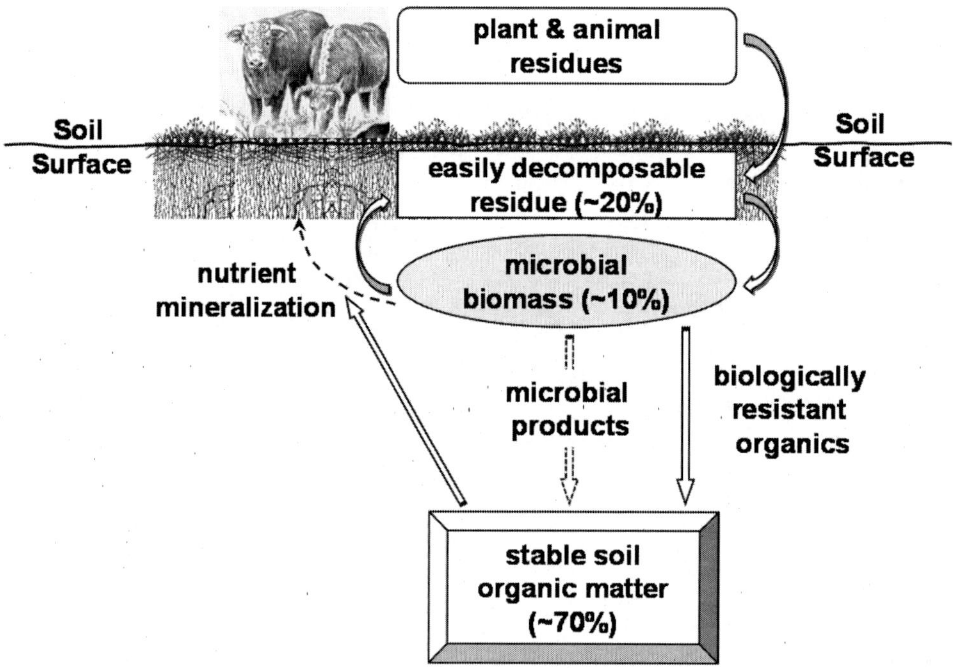

Fig. 5 Pathway of degradation of plant and animal residues to stable soil organic matter. Mineralization of organic compounds produces plant-available nutrients.

Table 4 Characteristics of soil OM and associated effects on soil and plants

Property	Effect on soil	Effect on plant
Soil color	Imparts dark color	Darker color causes soils to warm faster; >surface soil temperature advances germination and seedling growth (depending on residue cover)
H_2O holding capacity	Holds ~20 times weight in H_2O	>H_2O holding capacity, >plant-available H_2O esp. in sandy soils
OM–clay interaction	Cements soil particles into aggregates	Improve soil structure & porosity, enhances gas exchange, infiltration, root proliferation in soil
Ion exchange / buffering	Increases CEC and AEC 20–70%	>Nutrient retention and availability
Mineralization / immobilization	Nutrient cycling	Increases nutrient availability, retains / conserves nutrients
Chelation	OM forms stable metal (M^{n+}) complexes	Enhance micronutrient availability
Solubility	Insoluble humus–clay complexes, many soluble low MW organic compounds	OM-bound nutrients less likely to leach

Abbreviations: ACE, anion exchange capacity; CEC, cation exchange capacity; MW, molecular weight; OM, organic matter.

proliferation, and crop productivity. Soil OM levels depend on soil and crop management. If management is changed, a new OM level is eventually attained that may be lower or higher than the previous level. Management systems should be adopted that sustain or increase soil OM and productivity. Proper nutrient management will produce high yields, which will increase the quantity of residue and soil OM.

Mineral solubility in soils

Mineral solubility refers to the concentration of ions in solution supported or maintained by solid phase mineral(s). There are many soil minerals influencing nutrient concentrations in soil solution. For example, in acid soils, $FePO_4{\cdot}2H_2O$ influences P availability by

$$FePO_4{\cdot}2H_2O + H_2O \leftrightharpoons H_2PO_4^- + H^+ + Fe(OH)_3$$

The K_{sp} for this reaction is

$$K_{sp} = (H_2PO_4^-)(H^+)$$

As $H_2PO_4^-$ decreases with P uptake, strengite dissolves to resupply or maintain solution $H_2PO_4^-$ concentration. This reaction also shows that as H^+ increases (decreasing pH), $H_2PO_4^-$ decreases. Therefore, specific P minerals present in soil and the concentration of solution P supported by these minerals are dependent on solution pH. Solubility relationships are particularly important for plant availability of P and many of the micronutrients.

Buffer capacity

As nutrient concentration in the soil solution decreases by plant uptake, nutrients in soil solution are replenished from exchange surfaces, soil OM, or dissolution of soil minerals (Fig. 2). Nutrient supply to plants depends on nutrient concentration in solution and on *buffering capacity* (BC) or the ability of the soil to resupply nutrients to solution (Fig. 6). For example, as plant roots absorb K^+, adsorbed K^+ on the CEC is desorbed to resupply solution K^+. Similarly, $H_2PO_4^-$ adsorbed to the AEC will desorb, while P-bearing minerals may also dissolve to supply solution $H_2PO_4^-$ (Fig. 2). Both processes, desorption from exchange sites and dissolution of minerals, comprise soil BC.

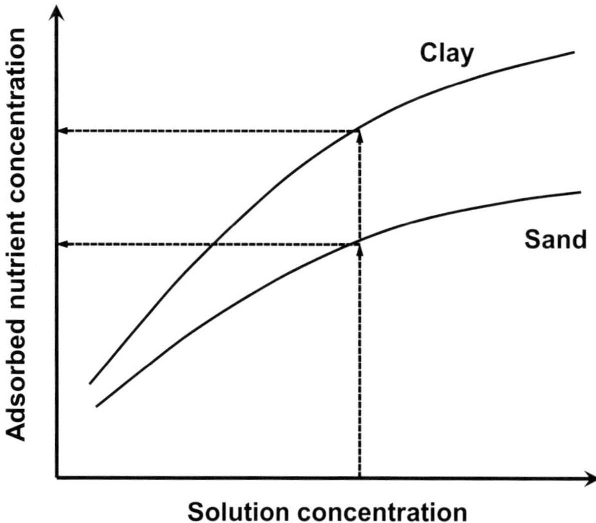

Fig. 6 Influence of soil texture on the quantity of adsorbed and solution nutrient concentrations. Buffer capacity is the ratio of adsorbed to solution nutrient concentration. To maintain the same solution concentration in both soils, the clay soil should have a greater adsorbed nutrient concentration than the sand soil. As solution concentration decreases with the nutrient uptake, the clay soil will have a greater supply of adsorbed nutrients to resupply the soil solution.

Nutrient Transport from Soil to Roots

Nutrient absorption by plant roots requires contact between the nutrient ion and the root surface (Fig. 2). Nutrients reach the root surface by *root interception*, *mass flow*, and *diffusion*. Although each is important, mass flow and diffusion dominate nutrient transport (Table 5).

Root interception represents ion exchange through physical contact between the root and exchange surfaces. Ions such as H^+ adsorbed to root surfaces may exchange with ions adsorbed to mineral and OM surfaces through direct contact of both surfaces. As roots and associated mycorrhiza develop, more ions adsorbed to soil surfaces are contacted by increasing root mass. The quantity of nutrients in direct contact with plant roots is the amount in a soil volume equal to the root volume (~2% of soil volume).

Mass flow occurs when ions in soil solution are transported to the root by transpiration of water by the plant, water evaporation at the soil surface, and percolation of water in the soil profile. The quantity of nutrients transported by mass flow is determined by the rate of water flow and the nutrient concentration in the soil water. Decreasing soil moisture decreases water transport to the root surface.

Diffusion occurs when an ion moves from an area of high to low concentration. As roots absorb nutrients from the soil solution, concentration at the root surface is lower than that in solution not influenced by the root. Therefore, increasing nutrient concentration gradient increases diffusion rate toward the root. A high plant requirement for a nutrient results in a large concentration gradient, favoring a high diffusion rate. Most of the P and K move to the root by diffusion (Table 5).

The importance of diffusion and mass flow in supplying ions to the root surface depends on the ability of the solid phase of the soil to replenish or buffer the soil solution. Ion concentrations are influenced by the types of clay minerals and OM in the soil and the distribution of cations and anions on CEC or AEC.

Mass flow and diffusion processes are also important in nutrient management. Soils that exhibit low diffusion rates because of high BC, low soil moisture, or high clay content may require application of immobile nutrients near the roots to maximize nutrient availability and plant uptake.

Table 5 Significance of root interception, mass flow, and diffusion in nutrient transport to roots. For some nutrients, the quantity transported can exceed that required by the crop

Nutrient	Nutrients required ($kg\ ha^{-1}$) for 10 ton ha^{-1} of corn	Percentage supplied by: Root interception	Mass flow	Diffusion
N	200	1	99	0
P	40	2	4	94
K	180	2	20	78
Ca	45	120	440	0
Mg	50	27	280	0
S	22	4	94	2
Cu	0.11	8	400	0
Zn	0.36	25	30	45
B	0.22	8	350	0
Fe	2.2	8	40	52
Mn	9.3	25	130	0
Mo	0.01	8	200	0

Note: Contribution of diffusion was estimated by the difference between total nutrient needs and amounts supplied by interception and mass flow. If interception + mass flow is >100%, then diffusion = 0.

Nutrient Mobility in Soil

Nutrient mobility in soil influences ion transport to plant roots, evaluation of nutrient availability to plants, and ultimately nutrient management decisions. Nutrient mobility varies between ions, where NO_3^-, SO_4^{2-}, Cl^-, and $H_3BO_3^{\circ}$ are not strongly attracted to exchange sites and are soluble in soils so they can readily move through the root zone with water. As a result mobile nutrients within the whole soil volume occupied by the plant root system are available for transport to the root in percolating and transpirational water (Fig. 7). The relative mobility of each nutrient will depend on soil pH, temperature, moisture, soil texture, type of clay, and OM content.

Immobile nutrients interact with mineral and OM surfaces, are less soluble, and do not readily move throughout the root zone (Fig. 7). While classified as immobile nutrients in soil, some are more mobile than others. Generally, NH_4^+, K^+, Ca^{2+}, and Mg^{+2} are more soluble and mobile than the micronutrient cations, and much more mobile than $H_2PO_4^-/HPO_4^{2-}$ and MoO_4^{2-}. As these nutrients are relatively immobile in soil, plant roots access them from a small soil volume surrounding individual roots. Plants create a small zone around the root that has very low concentration of these immobile nutrients due to plant uptake,

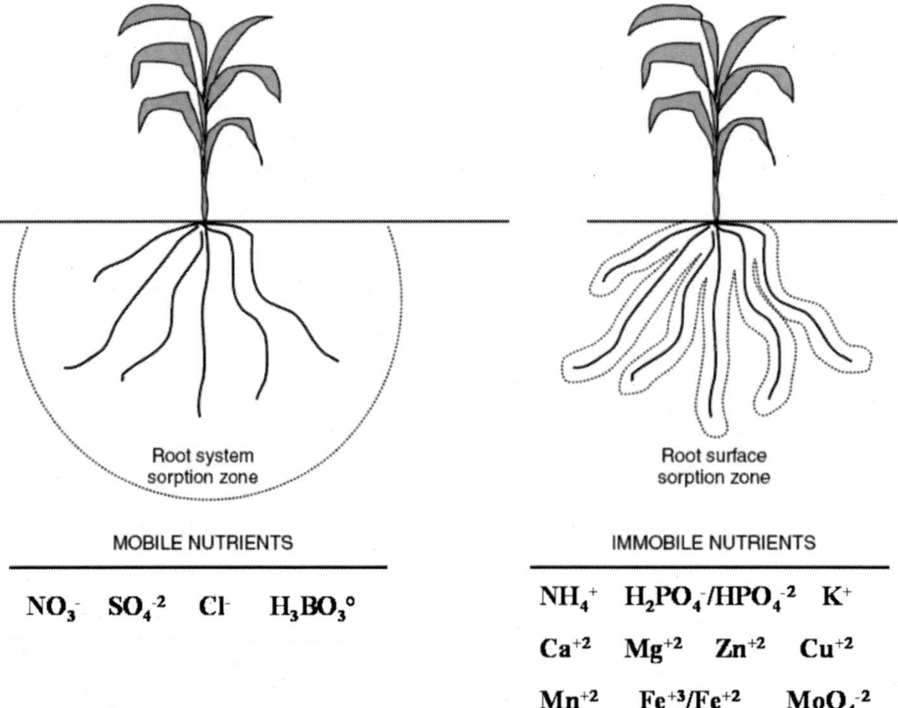

Fig. 7 Relationship between nutrient mobility and nutrient extraction zones. Plants obtain mobile nutrients from the whole soil volume occupied by plants roots. In contrast, plants obtain immobile nutrients from the small soil volume immediately surrounding the plant root.

causing a concentration gradient encouraging nutrient diffusion. If the soil has a high buffer capacity for an immobile nutrient, then the solution can be replenished and diffusion continues. With a low buffer capacity, solution concentration (and diffusion) ultimately decreases, causing a nutrient deficiency.

Understanding nutrient mobility in soils is essential to managing nutrient applications to maximize plant growth and recovery of applied nutrients by the plant. For example, mobile nutrients can be broadcast or band applied with fairly similar results because of their mobility in soil. However, P is generally placed in concentrated bands because it is generally immobile in soil.

NUTRIENT MANAGEMENT

Efficient nutrient management programs supply plant nutrients in adequate quantities to sustain maximum plant growth and yield while minimizing environmental impacts of nutrient use. Substantial economic and environmental consequences occur when nutrients limit plant productivity or when nutrients are applied in excess of plant requirement.

The essential requirements of an effective nutrient management plan includes (1) assessment of crop nutrient requirement, (2) evaluation of nutrient supplying capacity of the soil, (3) quantify optimum nutrient rate, (4) identify appropriate and available nutrient source(s), (5) determine most efficient nutrient placement method, and (6) schedule nutrient applications to meet plant nutrient demand. Management decisions will vary depending on the specific nutrient and crop.

Assessing Plant Nutrient Requirement

The quantity of nutrients required by plants depends on the nutrient (Table 1), plant characteristics (specific crop, yield level, variety, or hybrid), growing season environmental factors (moisture, temperature, etc.), soil characteristics (soil properties, soil fertility, and landscape position), and soil and crop management. Management practices that enhance crop productivity will increase nutrient requirements.

Evaluation of Nutrient Supplying Capacity of the Soil

The *soil testing–nutrient recommendation system* is comprised of four consecutive steps: (1) collect a representative soil sample from the field, (2) determine the quantity of plant-available nutrient in the soil sample (soil test), (3) interpret the soil test results (Fig. 8), and (4) estimate the quantity of nutrient required by the crop. The greatest potential for error is in collecting a soil sample that accurately represents the field sampling area. Great care is

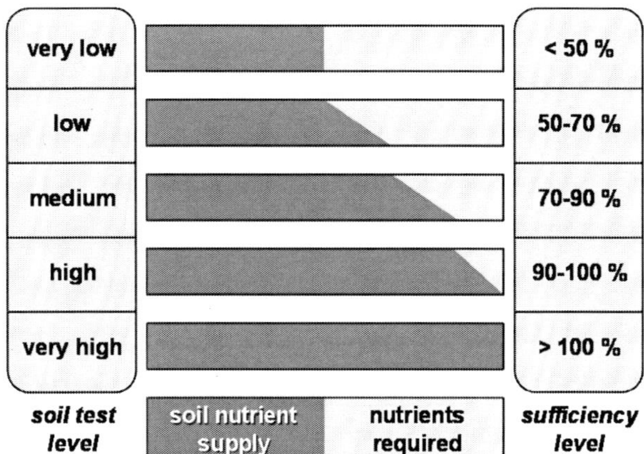

Fig. 8 As soil test levels increase, the capacity of the soil to provide sufficient plant nutrients increases, reducing the quantity of nutrients added to meet crop requirement.

required in identifying the sampling area from which a composited sample is sent to a laboratory for analysis. Selected parameters used in separating sampling areas within a field include differences in topography, soil type, crop productivity, and past management.

Soil testing is essential for determining the relative availability of plant nutrients in soil. Soil tests extract part of the total nutrient content in soil that is related to (but not equal to) the quantity of nutrients removed by plants. Thus, the soil test level represents an *index* of nutrient availability representing the relative nutrient sufficiency of a soil (Fig. 8).

Quantify Optimum Nutrient Rate

Soil test interpretation for purposes of making nutrient recommendations is influenced by the mobility of the nutrient (Fig. 7). With mobile nutrients, crop yield is proportional to total quantity of nutrient present in the root zone, because of minimal interaction with soil constituents (Fig. 2). For NO_3^-, SO_4^{2-}, and Cl^-, a 0.5–1.0 m profile sample is important for accurately assessing mobile nutrient availability (Fig. 7). For N, recommendations are usually based on yield goal, where N required to produce each unit of yield is known (e.g., 0.3 kg N·kg^{1-} grain). This concept is also evident when additional in-season N is recommended because better-than-average growing conditions increased the yield potential above initial estimates provided before planting. Additional factors affecting optimum N rate include estimates of N mineralized during the growing season from previous legume crops and organic N amendments (manures, biosolids, etc.). With immobile nutrients (e.g., $H_2PO_4^-$, K^+, Zn^{2+}), crop yield potential is limited by the quantity of nutrient available at the soil–root interface. Generally, solution concentrations of immobile nutrients are low, and replenishment occurs through exchange, mineralization, and mineral solubility reactions (Fig. 2). If nutrient uptake demand exceeds the soil's capacity to replenish solution nutrients, then plant growth and yield

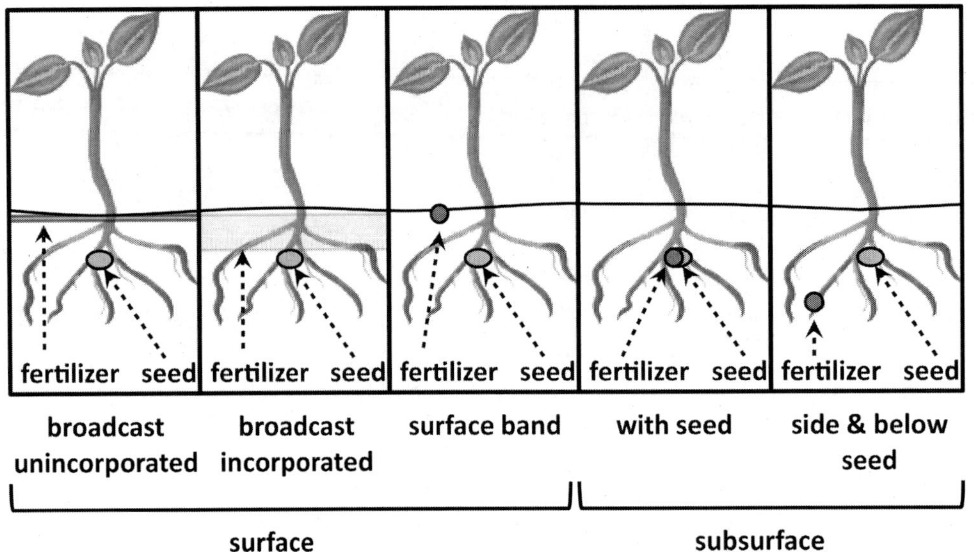

Fig. 9 Options for application of nutrients at or below the soil surface.

will be limited. Immobile nutrient recommendations are based on sufficiency levels determined through soil testing (Fig. 8). Soil tests for immobile nutrients provide an index of nutrient availability that is generally independent of environment.

Identify Nutrient Source(s)

The primary criteria for selecting a nutrient source are availability and cost. For example, if organic nutrient sources (manures, biosolids, etc.) are available they should be utilized if the cost per unit of plant-available nutrients is comparable to inorganic fertilizers. There are important benefits of using organic sources beyond their nutrient value, which should also be considered (Table 4). Nutrient content of organic materials must be quantified to determine the application rate needed to meet crop nutrient requirement in the year of application.

In most cases there are few differences in crop response between inorganic fertilizer sources. Although most plants prefer a mixture of NH_4^+ and NO_3^-, most NH_4^+ or NH_4^+-forming N sources (urea-based products) readily dissolve and nitrify to form NO_3^-. Under conditions of high denitrification, volatilization, and/or nitrification potential, numerous additives or coatings (solid fertilizers) are available to reduce losses of applied NH_4^+ and NO_3^-. When applying nutrients with the seed, nutrient sources with a low salt index are important in reducing salt damage to germinating seeds and seedlings.

Nutrient Placement Methods

Nutrient placement decisions involve knowledge of crop and soil characteristics, whose interactions determine nutrient availability. Although numerous placement methods have been developed, the primary goal is to ensure optimum nutrient availability from plant emergence to maturity. Vigorous seedling growth (i.e., no early growth stress) is essential to optimize plant growth and yield. Merely applying nutrients does not ensure that they will be absorbed by the plant.

Fertilizer placement options generally involve surface or subsurface applications before, at, or after planting. Placement practices depend on the crop and crop rotation, degree of deficiency or soil-test level, nutrient mobility in the soil, degree of acceptable soil disturbance, and equipment availability.

Preplant application

Preplant applications are important to provide adequate nutrient supply during germination and early plant growth stages. Broadcast nutrients are applied uniformly on the soil surface over the entire area before planting, and can be incorporated by tillage (Fig. 9). In no-till there is limited opportunity for incorporation; thus, broadcast N applications may reduce N recovery by the crop due to enhanced immobilization, denitrification, and volatilization of applied N. Crop recovery of nutrients can be increased by placement below the soil surface where soil moisture might be more favorable for nutrient uptake (Fig. 9). Surface band applied fertilizers can be effective before planting (Fig. 9); however, if not incorporated, dry surface soil conditions can reduce nutrient uptake, especially with immobile nutrients. Surface band N applications can improve N availability compared with broadcast application in some soils and cropping systems due to reduced interaction with surface plant residues.

Animal waste and other organic nutrient sources are commonly preplant applied, usually as a broadcast or subsurface band placement. Broadcast applications, especially unincorporated, are subject to greater denitrification and volatilization N losses and surface runoff losses of all nutrients.

At planting application

Fertilizer placement can occur at numerous locations near the seed (2–8 cm below and/or to side of seed), depending on equipment and crop (Fig. 9). Fertilizer application with the seed is a subsurface band used to enhance early seedling vigor, especially in cold, wet soils. Usually, low nutrient rates are applied to avoid germination or seedling damage. Fertilizers can be surface applied at planting in bands directly over or to the side of the row (Fig. 9).

After planting application

Broadcast application after planting is commonly referred to as topdressing. Topdress N applications are common on permanent or close seeded crops (e.g., turf, small grains, pastures, etc.); however, N immobilization with high surface residue reduces recovery of topdress N. Topdressed P and K are not as effective as preplant applications. Sidedress N application is common in row crops (e.g., corn, grain sorghum, etc.) as either a surface or subsurface band. Sidedress applications increase flexibility since applications can be made almost any time equipment is operated without damage to the crop. Sidedress placement is particularly suited for spatially variable N application using remote sensors mounted on the applicator to guide application rate. Subsurface sidedress applications too close to the plant can cause damage by either root pruning or nutrient toxicity. Sidedress application of immobile nutrients (e.g., P and K) is not recommended because most crops need P and K early in the season.

Timing of Nutrient Applications

Nutrients are needed in the greatest quantities during periods of maximum plant growth. Thus, nutrient management plans are designed to ensure adequate nutrient supply before the exponential growth period. In this case, N should be applied preplant or split applied before the stem extension phase. All of the immobile nutrients, including P and K, should be applied before planting. Knowledge of crop-uptake patterns facilitates improved management for maximum productivity and recovery of applied nutrients.

Variable Nutrient Management

Variable or *site-specific* nutrient management can improve nutrient use efficiency by distributing nutrients based on spatial variation in yield potential, soil test levels, and other spatially variable factors that influence soil nutrient availability and crop nutrient demand. Preseason and in-season assessments of spatial information are two variable nutrient management approaches used to guide nutrient application decisions.

CONCLUSION

Meeting food needs for a growing population requires increased food production on the same or less agricultural land area. Land managers must adopt economically viable technologies that maintain, enhance, or protect the productive capacity of our soil resources to ensure future food and fiber supplies. While organic nutrient sources are important to meeting the nutritional needs of higher-yielding cropping systems, inorganic fertilizer nutrients will remain the predominant nutrient source. The challenge to the agricultural community is to ensure maximum recovery of applied nutrients, regardless of source, through use of diverse soil, crop, water, nutrient, and other input management technologies to maximize plant productivity. Accomplishing this will significantly reduce nutrient losses to the environment.

BIBLIOGRAPHY

1. Barker, A.V.; Pilbeam, D.J. *Handbook of Plant Nutrition*; CRC: Boca Raton, FL, 2007.
2. Grant, C.A.; Milkha, M.S. *Integrated Nutrient Management for Sustainable Crop Production*; The Hanworth Press: NY, 2008.
3. Havlin, J.L.; Tisdale, S.L.; Nelson, W.L.; Beaton, J.D. *Soil Fertility and Fertilizers: An Introduction to Nutrient Management*. 8th Edition. Pearson Inc.: Upper Saddle River, NJ, 2014; 516 p.
4. Rengel, Z.; Ed. *Handbook of Soil Acidity*; Marcel Dekker: New York, NY, 2003.
5. Sparks, D.L. *Environmental Soil Chemistry*; Academic Press: 2003.

Soil: Organic Matter

R. Lal
Carbon Management and Sequestration Center, The Ohio State University, Columbus, Ohio, U.S.A.

Abstract
Soil organic matter (SOM) is essential to maintaining soil quality and providing numerous ecosystem services. Its magnitude in soil depends on climate, soil properties, terrain characteristics, vegetation, and landscape position. Its concentration is more in soils of cooler and humid than warmer and dry climates. There are numerous cost-effective and nondestructive methods of determination of SOM. Its magnitude and dynamics can be modeled and related to several index properties. Managing SOM is essential for advancing food security, mitigating climate change, improving water quality, and enhancing biodiversity.

INTRODUCTION

The origin and importance of soil organic matter (SOM) were appropriately summarized in 1973 by Allison:[1] "Soil organic matter has over the centuries been considered by many as an elixir of life. Ever since the dawn of history, some eight thousand or more years ago, man has appreciated the fact that dark soils, commonly found in the river valleys and broad level plains, are usually productive soils. He also realized at a very early date that color and productivity are commonly associated with organic matter derived chiefly from decaying plant materials." The SOM comprises the sum of all organic substances in soil, and primarily consists of heterogeneous substances of plant, animal, and microbial origin at various stages of decomposition. Its decomposition products include (1) soluble organic compounds such as sugars, proteins, and other metabolites; (2) amorphous organic compounds such as humic acids, fats, waxes, oils, lignin, and polyuronides; and (3) organomineral complexes, which involve hybrid compounds of organic molecules attached to clay particles through polyvalent cations such as Al^{3+}, Ca^{2+}, and Mg^{2+}.[2]

Two principal components of SOM are highly active humus and relatively inert charcoal. Humus is a dark brown or black amorphous material. Being a colloidal substance, it is characterized by a large surface area, high charge density, and high affinity for clay. Charcoal is widely present in soils of the fire-dependent ecosystems (i.e., savanna, steppe, grasslands). The charcoal, also called black carbon (C), is of pyrogenic origin. It comprises of lightly charred organic matter, soot particles rich in graphite, and polycondensed aromatic groups. Highly fertile soils of the Amazon, *terra preta do Indios*, are enriched by addition of charcoal.

SOIL ORGANIC MATTER DYNAMICS

In natural ecosystems, SOM is in dynamic equilibrium with its environment. Its amount in the soil is stabilized when the rate of input equals that of the output. The input of SOM is mainly through residues of plants grown in situ and from substances brought in by alluvial, aeolian, and colluvial processes. The output of SOM is through decomposition or oxidation, erosion, and leaching. The rate of decomposition depends on temperature and moisture regimes, and is more in warmer and wetter than in cooler and drier climates. Soil erosion, by water and wind, preferentially removes SOM because it is lighter (lower density) than mineral fraction, and is concentrated in the vicinity of the soil surface where erosional processes are the most effective. The soluble and colloidal components of SOM are leached into the subsoil and eventually into aquatic ecosystems along with the percolating water. Removal of crop residues for alterative uses, such as traditional or modern biofuels (cellulosic ethanol) and livestock feed, reduces input of biomass and decreases the soil organic C pool. Thus, there is no such thing as a free biofuel from crop residues.

Natural factors affecting the SOM concentration include climate, soil type, vegetation, hydrology, landscape position, and soil biodiversity. Principal climatic parameters are precipitation amount and distribution, temperature and its seasonal and diurnal variation, annual and seasonal water budget, and the duration of the growing season as determined by the degree days and hydrologic balance. Predominant soil properties that affect SOM include texture, clay minerals, pH, ionic composition and concentration, and soil depth. Internal drainage of the soil profile, affecting soil moisture regime, is also important. Landscape affects SOM content through the position, drainage density, and slope shape, and aspect. Slope aspect

Encyclopedia of Natural Resources DOI: 10.1081/E-ENRL-120047483

determines soil temperature, slope position determines moisture regime, and slope shape determines the sedimentation and depositional processes. Analyses of these factors indicate that once SOM has been severely depleted by natural or anthropogenic factors, it is difficult to restore in tropical agroecosystems, coarse-textured soils, drought-prone environments, and nutrient-deficit (N, P, S) conditions. Vegetation, and its residues, influences SOM content through the relative concentration of lignin, cellulose, polyphenols, and other recalcitrant compounds. Root system, its depth distribution and turnover are also important to SOM stock in the profile, especially in the subsoil horizons. Roots contribute more to refractory SOM (relatively stable soil C pool) than the same amount of above-ground residue-derived C.[3] Presence of stabilizing elements (especially N, P, S) is important to converting biomass C into humus.[4] In drier climates, SOM stock is favorably influenced by the hydraulic lift of deep-rooted shrubs.[5]

ASSESSMENT OF SOIL ORGANIC MATTER

There is a long history of determination of SOM concentration in the laboratories.[6] Concentration of SOM in the laboratory is determined by either the wet combustion[7] or dry combustion method.[8,9] There are numerous modern techniques of measuring SOM stock in situ,[10] and rapidly and in a cost-effective manner.[11,12] There are also molecular-level methods for monitoring SOM concentration in relation to climate change,[13] including nuclear magnetic resonance techniques and SOM biomarker techniques.[14–19] Several practical indicators of changes in SOM concentration have been developed.[20,21] While concentration of SOM in the root zone (measured as % or g/kg) has been the unit of assessment for agricultural and forestry land uses, the assessment at regional, watershed, national, and global scales is done in units of Mg/ha, Tg, or Pg.

MANAGEMENT OF SOIL ORGANIC MATTER

The quantity and quality of SOM affect numerous soil processes and ecosystem services of significance at local, regional, and global scales (Fig. 1). The SOM concentration of 2–3% dry soil weight is needed to improving the use of efficiency of inputs and enhancing agronomic production.[22] It is essential to advance global food security[23]

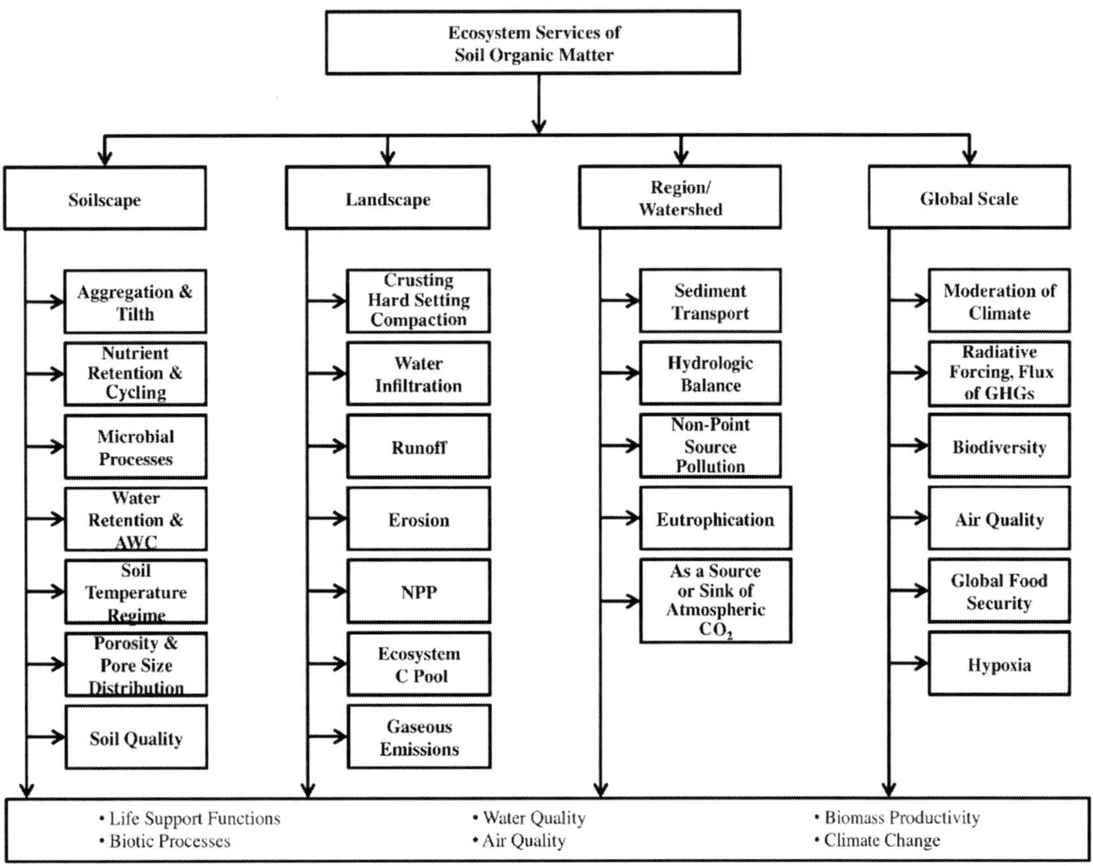

Fig. 1 Ecosystem services provided by soil organic matter.
Abbreviations: AWC, available water capacity; GHGs, greenhouse gases (e.g., CO_2, CH_4, N_2O); NPP, net primary productivity.

and create a positive C budget by reducing losses and increasing inputs of biomass-C. Maintenance of SOM at above the critical level enhances physical, chemical, and biological processes, which are strong determinants of the overall soil quality (Table 1, Fig. 2). The processes are enhanced by adoption of soil-specific practices. Some recommended management practices in agroecosystems include conservation tillage, mulch farming, cover cropping, integrated nutrient management including use of manures, biofertilizers, and biochar; agroforestry; controlled grazing; and improved pasture. Restoration of wetlands, erosion control, and afforestation of degraded and desertified lands are important options of enhancing the SOM concentration and pool. In 1938, William Albrecht[24] emphasized that "soil organic matter is one of our most precious resources; its unwise exploitation has been devastating; and it must be given its proper rank in any conservation policy as one of the major factors affecting the level of crop production in the future." It was in a similar context, in view of the rising energy costs in late 1970s and the suggestion about the use of crop residues for biofuels, when Hans Jenny[25] stated in 1980 that "I am arguing against indiscriminate conversion of biomass and organic wastes to fuels. The humus capital, which is substantial, deserves being maintained because good soils are a national asset."

CONCLUSION

Quantity and quality of SOM and its maintenance above the threshold/critical level have been the foundation of past civilizations, and are the bases on which the future of mankind depends. Its concentration in the root zone determines numerous ecosystem services including food security, climate change mitigation and adaptation, biodiversity, and water resources. It is the global public good, whose importance to human well-being and ecosystem functions can never be overstated. While strengthening and promoting long-term basic and applied research on processes and practices of its management and assessment at different scales, policy instruments must be in place at national and international levels for its enhancement and sustainable management.

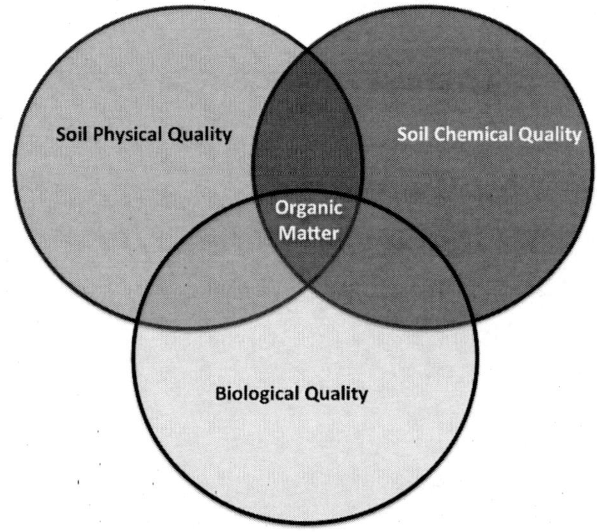

Fig. 2 Interactive physical, chemical, and biological processes moderated by soil organic matter as a strong determinant of soil quality.

Table 1 Soil organic matter as a determinant of specific components of soil quality

Soil physical quality	Soil chemical quality	Soil biological quality
1. Soil structure and tilth	1. Soil reaction (pH)	1. Food for soil organisms
2. Porosity and port size distribution; biopores	2. Cation exchange capacity	2. Habitat for biota
3. Plant's available water holding capacity	3. Elemental toxicity (Al^{3+}, Fe^{3+}, Mn^{3+})	3. Microbial biomass carbon
4. Soil erodibility	4. Electrical conductance, salt concentration	4. Products of decomposition
5. Infiltration, percolation, run off, and leaching	5. Plant available nutrients (i.e., N, P, K, Cu, Zn)	5. Soil respiration rate
6. Crusting, compaction, and hard setting	6. Nutrient/elemental cycling	6. Methanogenesis
7. Soil temperature regime and thermal conductivity	7. Chemical transformations	7. Nitrification/denitrification
8. Gaseous diffusion (diffusion coefficient)	8. Nutrient diffusion	8. Soil C sink capacity
9. Surface area	9. Organomineral complexes	9. Gaseous composition of soil air
10. Soil strength	10. Redox potential	10. Activity and species diversity of soil flora and fauna

REFERENCES

1. Allison, F.E. *Soil Organic Matter and Its Role in Crop Production*; Amsterdam: Elsevier, 1973.
2. Schnitzer, M. Soil organic matter – the next 75 years. Soil Sci. **1991**, *151*, 41–58.
3. Kätterer, T.; Bolinder, M.A; Andrén, O.; Kirchmann, H.; Menichetti, L. Roots contribute more to refractory soil organic matter than above-ground crop residues, as revealed by a long-term field experiment. Agr. Ecosyst. Environ. **2011**, *141*, 184–192.
4. Kirkby, C.A.; Kirkegaard, J.A.; Richardson, A.E.; Wade, L.J.; Blanchard, C.; Batten, G. Stable soil organic matter: A comparison of C:N:P:S ratios in Australian and other world soils. Geoderma **2011**, *163*, 197–208.
5. Armas, C.; Kim J.H.; Bleby, T.M.; Jackson, R.B. The effect of hydraulic lift on organic matter decomposition, soil nitrogen cycling, and nitrogen acquisition by a grass species. Oecologia **2012**, *168* (1), 11–22. doi:10.1007/s00442-01102065-2 (accessed August 2011).
6. Manlay, R.J.; Feller, C.; Swift M.J. Historical evolution of soil organic matter concepts and their relationships with the fertility and sustainability of cropping systems. Agric. Ecosyst. Env. **2007**, *119*, 217–233.
7. Allsion, F.E. Wet combustion apparatus and procedure for organic and inorganic carbon in soil. Soil Sci. Soc. Am. Proc. **1960**, *24*, 36–40.
8. Nelson, D.W.; Sommers, L.E. Total carbon, organic carbon and organic matter. In *Methods of Soil Analysis, Part 2. Chemical and Microbial Properties*. Amer. Soc. Agron. No. 9 (Part 2). Agronomy Series. Page, R.L.; Miller, R.H.; Keeney, D.R.; Eds.; ASA/SSSA: Madison WI, 1982; 539–580.
9. Tiesson, H.; Moir, J.O. Total and organic carbon. In *Soil Sampling and Methods of Analysis*; Carter, M.R. Ed; Lewis Publishers: Boca Raton, FL, 1993; 187–199.
10. Wielopolski, L.; Chatterjee, A.; Mitra, S.; Lal, R. In-situ determination of soil carbon pool by inelastic neutron scattering. Geoderma **2011**, *160* (3–4), 394–399, doi:1016/j/Geoderma.2010.10.009 (accessed August 2011).
11. Chatterjee, A.; Lal, R.; Wielopolski, L.; Martin, M.Z.; Ebinger, M.H. Evaluation of different soil carbon determination methods. Crit. Rev. Plant Sci. **2009**, *28*, 164–178.
12. Lal, R; Cerri, C.; Bernoux, M; Etchevers, J.; Cerri, E. *The Potential of Soils of Latin America to Sequester Carbon and Mitigate Climate Change*; The Hawthorn Press: West Hazelton, PA 2006; 554.
13. Feng, X.; Simpson, M.J. Molecular-level methods for monitoring soil organic matter responses to global climate change. J. Environ. Monit. **2011**, *13*, 1246–1254.
14. Braun, S.; Agren, G.I.; Christensen, B.T.; Jensen, L.S. Measuring and modeling continuous quality distributions of soil organic matter. Biogeosciences **2010**, *7*, 27–41.
15. Parton, W.J.; Schimel, D.S.; Cole, C.V.; Ojima, D.S. Analysis of factors controlling soil organic matter levels in Great Plains grasslands. Soil Sci. Soc. Am. J. **1987**, *51*, 1173–1179.
16. Nye, P.H.; Greenland, D.J. The soil under shifting cultivation. Commonwealth Bureau Soils Tech. Commun. **1960**, *51*, 156.
17. Viaud, V.; Angers, D.A.; Walter, C. Toward landscape-scale modeling of soil organic matter dynamics in agroecosystems. Soil Sci. Soc. Amer. J. **2010**, *74*, 1847–1860.
18. Neill, C. Impacts of crop residue management on soil organic matter stocks: a modeling study. Ecol. Model. **2011**, *222*, 2751–2760.
19. Kutsch, W.L.; Bahn. M.; Heinemeyer, A. *Soil Carbon Dynamics: An Integrated Methodology*; Cambridge University Press: Cambridge, U.K. 2009; 286.
20. Sohi, S.P.; Yates, H.C.; Gaunt, J.L. Testing a practical indicator for changing oil organic matter. Soil Use Manag. **2010**, *26*, 108–117.
21. Plante, A.F.; Fernández, J.M.; Haddix, M.L.; Steinweg, J.M.; Conant, R.T. Biological, chemical and thermal indices of soil organic matter stability in four grassland soils. Soil Biol. Biochem. **2011**, *43*, 1051–1058.
22. Lal, R. Beyond Copenhagen: Mitigating climate change and achieving food security through soil carbon sequestration. Food Secur. **2010**, *2*, 169–177.
23. Magdoff, F.; Van Es, H. *Building Soils for Better Crops: A Sustainable Soil Management*; SARE/USDA: Beltsville, MD, 2009; 294.
24. Albrecht, W. Loss of soil organic matter and its restoration. In *Soils and Men*; *USDA Yearbook of Agriculture*, U.S. Government Printing Office: Washington, D.C., 1938; 347–360.
25. Jenny, H. Alcohol or humus. Science **1980**, *209*, 444.

Soil: Organic Matter and Available Water Capacity

T. G. Huntington
U.S. Geological Survey (USGS), Augusta, Maine, U.S.A.

Abstract

Available water capacity (AWC) is defined as the amount of water (cm^3 water/100 cm^3 soil) retained in the soil between "field capacity" (FC) and the "permanent wilting point" (PWP). FC and PWP are defined as the volumetric fraction of water in the soil at soil water potentials of 10 to 33 kPa and 1500 kPa, respectively. One of the paradigms of soil science is that AWC is positively related to soil organic matter (SOM) because SOM raises FC more than PWP. SOM enhances soil water retention because of its hydrophilic nature and its positive influence on soil structure. It improves soil structure by enhancing aggregate formation, thereby increasing porosity in the range of pore sizes that retain plant-available water, and enhancing infiltration and water retention throughout the rooting zone. Understanding the sensitivity of AWC to SOM is important for assessments of how environmental changes that influence SOM content could affect AWC and sustainability of agricultural productivity.

INTRODUCTION

Available water capacity (AWC) is defined as the amount of water (cm^3 water/100 cm^3 soil) retained in the soil between "field capacity" (FC) and the "permanent wilting point" (PWP).[1] FC and PWP are defined as the volumetric fraction of water in the soil at soil water potentials of 10 to 33 kPa and 1500 kPa, respectively. One of the paradigms of soil science is that AWC is positively related to soil organic matter (SOM) because SOM raises FC more than PWP[2–4] (Fig. 1). SOM enhances soil water retention because of its hydrophilic nature and its positive influence on soil structure.[5,6] Increasing SOM increases soil aggregate formation and aggregate stability[7] (Fig. 2), thereby increasing porosity in the range of pore sizes that retain plant-available water and enhancing infiltration and water retention throughout the rooting zone. When SOM decreases, soil aggregation and aggregate stability decrease and bulk density increases.[8] These changes in physical properties result in lower infiltration rates and higher susceptibility to erosion.[9,10] This entry reviews the literature on the sensitivity of AWC to soil organic carbon (SOC) and discusses the environmental implications of changes in SOM and AWC. This entry supersedes and updates a 2003 review[11] with a description of the recent literature on the sensitivity of AWC to SOM and discusses the environmental implications of changes in SOM and AWC.

SENSITIVITY OF AWC TO SOM

Many studies have determined that AWC decreases as SOM decreases, but considerable variation in this relation has also been reported. A majority of studies have used regression analysis to quantify the sensitivity of AWC to SOM. Other estimates of the relation between SOM and AWC have been obtained from changes in water retention characteristics over time for a specific soil where different management, fertilization, or erosion histories have resulted in changes in SOM. To facilitate comparison among estimates of the relation between AWC and SOM in this review, sensitivity was defined as the average change in AWC (cm^3 water/100 cm^3 soil) predicted for a 5% relative decrease in SOM where the initial SOM concentration was in the range of 2% to 4% by weight. If texture was included in the published regression, it was held constant at 65% silt, 20% clay, and 15% sand. If bulk density was included in the regression, it was assumed that bulk density decreased with increasing SOM following the regression equation:

$$\text{Bulk density} = 1.723 - 0.212(\text{OC})^{0.5} - 0.0006(\text{WC15})^2$$

where OC = percent soil organic carbon and WC15 = water content at 1500 kPa.[12] WC15 for a soil with the texture of 65% silt, 20% clay, and 15% sand was assumed to be 11.3 cm^3 water/100 cm^3 soil. Where data or regression analysis required transformation of soil organic carbon (SOC) to SOM, it was assumed that SOM = 1.724 * SOC.

Reported sensitivities fell in the following ranges (0 to +0.27),[13–15] (0 to –1),[16–21] (–1 to –2),[22–25] and (<–2)[26–28] (cm^3 water/100 cm^3 soil) per 5% relative decrease in SOM concentration. The average value for the sensitivity was –1.1 (+/– 0.34 SE) cm^3 water/100 cm^3 soil per 5% relative decrease in SOM concentration. Out of the 15 studies, 3 found a negative or no relation between SOM and AWC. Regression models that

Encyclopedia of Natural Resources DOI: 10.1081/E-ENRL-120049130

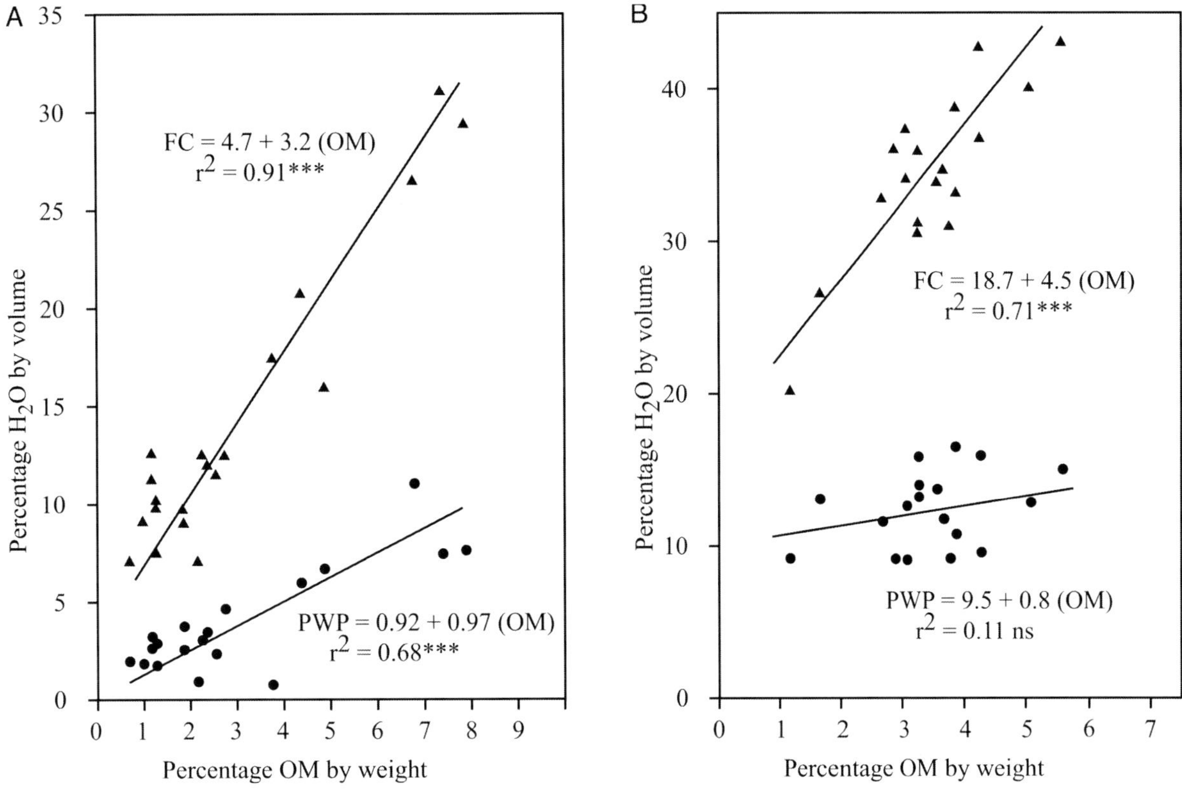

Fig. 1 Water content at field capacity (FC) and permanent wilting point (PWP) for sandy (**A**) and silt loam (**B**) textured soils. *** indicates that the relationship is statistically significant with p-value <0.0001 and ns indicates that the relationship was not statistically significant with p-value >0.05.
Source: Reproduced from the original in Hudson[3] with permission.

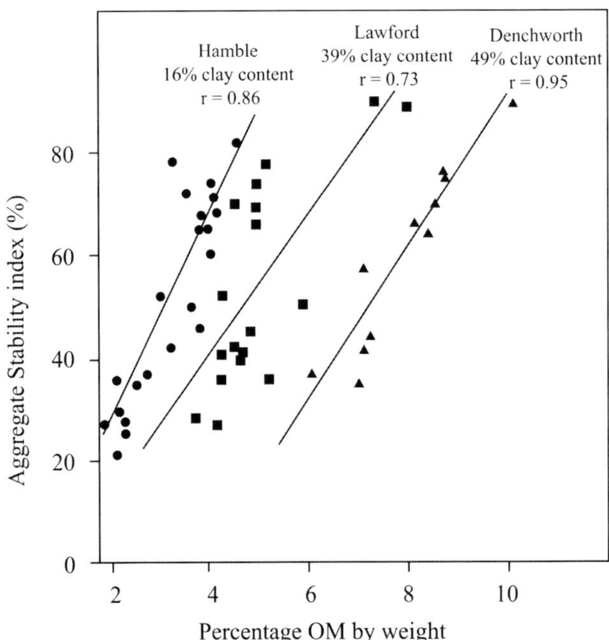

Fig. 2 Relationship between aggregate stability and soil organic matter content for samples from the Hamble, Lawford, and Denchworth soil series from southern England.
Source: Redrawn from Haynes and Beare.[7]

indicated no relationship between SOM and AWC typically found that SOM resulted in equivalent increases in water retention at both FC and PWP with the result that AWC did not change.

In addition, two large studies have used the U.S. National Soil Inventory database to study the relationship between SOM and AWC where AWC was based on reported values for FC and PWP.[29,30] One study[29] showed a positive relationship between SOM and plant AWC that was dependent on soil texture and reported that FC was more sensitive to SOM than PWP. In another study,[30] a complex relation was reported such that a 1% increase in SOM causes a 2% to >5% increase in soil AWC depending on the soil texture. Both studies concluded that coarse-textured soils had a higher sensitivity to SOM than fine-textured soils and that sensitivity decreased as SOM content increased. These two studies[29,30] are consistent with the findings of most of the other cited studies that have found that increases in SOM result in proportionately larger increases in AWC in medium- to coarse-textured soils compared with finer-textured soils. Most studies found that aggregate stability has been shown to increase with increasing SOM at similar rates across a wide range of clay contents[7] (Fig. 2). However, in that study[7] it was also shown that with increasing clay content, soils require a higher SOM content to maintain a given aggregate stability value (Fig. 2). As SOM levels decrease, as occurs following chronic high rates of erosion, the effect of incremental decreases in SOM on AWC becomes progressively smaller. In heavily eroded soils, clay contents become increasingly more important than SOM in influencing hydraulic properties.[10]

The physical and chemical properties of SOM could influence how SOM affects soil water retention. Differences in SOM quality may arise from differences in inputs, such as natural plant resides, manure, or sludge, or they may result from different soil drainage conditions. These and other factors such as mineralogy, exchangeable cations, and surface chemical properties that can influence the formation and stability of aggregates and mesopores complicate the simple relation between AWC and SOM.[6]

There are several other possible explanations for the high variability in the reported sensitivity of AWC to SOM. Differences in methods used to measure water retention at both FC and PWP could introduce bias. Both *in situ* and laboratory methods have been used for the determination of water retention at FC and PWP. Laboratory methods can involve everything from intact cores to air-dried, sieved, rewetted, and recompacted soil samples.[1] Any pretreatment that changes soil porosity and soil aggregation will affect estimated water retention. There is no agreed-upon standard water potential for the estimation of field capacity in the laboratory. Different methods have been used to determine SOC and SOM. Differences in texture, mineralogy, bulk density, coarse fragment volume, and quality of the SOM of the soils used in the regression analyses could mask the effects of SOM. Some reports involve controlled regressions where variables such as texture are held constant. It may not be surprising that in uncontrolled regressions, the effects of SOM are obscured.[3,6]

Soil water retention data from long-term plots, such as classic crop rotations, fertility experiments, and no-till versus conventional tillage experiments, may be more appropriate for this type of sensitivity analysis. In these types of experiments, SOM changes measurably over time, but other variables are essentially held constant. However, there have been few reports of this kind. At the Sanborn Field in Missouri, continuous cropping for over a century resulted in large decreases in SOM and soil productivity.[31] Losses of SOM at the Sanborn Field could have increased erosion and loss of topsoil that likely decreased AWC.[31] However, in another study at the Sanborn Field,[15] no significant differences in water retention among treatments were observed for soil cores collected between 2.5 and 10 cm depth. A study in Denmark reported that after 90 years of manure additions, AWC was substantially higher on the amended plots than on unfertilized reference plots.[28] In a 32-year continuous corn experiment, cropping treatments that resulted in lower SOM (removal of corn stover and tillage) reduced SOM by 8% and 25% and decreased AWC by 8% and 13%, respectively.[32] In a shorter-term study, the addition of manures to grasses, cereals, and sugar beets for 7 or more years increased AWC at four sites with different soils in the United Kingdom.[33]

ENVIRONMENTAL IMPLICATIONS

Environmental changes such as land-use conversions, changes in agricultural management practices, climate change, and changes in acidic deposition could affect SOM contents and, consequently, AWC. Land-use conversions from grasslands or forests to cultivated agriculture[34] or intensification of agriculture[35] are likely to result in net losses of SOM, but improvements in agricultural management practices have the potential to substantially enhance SOM contents and increase AWC.[36] Historical increases in acidic deposition followed by decreases in recent decades have been associated with concomitant increases and decreases in SOM in forest soils.[37,38]

It is now generally acknowledged that global mean annual temperature will likely increase in the range 1.4°C to 5.8°C during the 21st century as atmospheric carbon dioxide (CO_2) concentrations increase.[39] However, there is substantial uncertainty in how SOM levels will respond to the associated soil warming and elevated CO_2 concentrations. Some modeling studies[40–42] and empirical studies where land use and management were held constant[43–48] are consistent with declines in SOM following increases in temperature, but other environmental factors cannot be ruled out. Other studies suggest only no or small declines in SOM with increasing temperature or increases in SOM if precipitation increases along with increasing temperature.[49–52] In one study[53] it was concluded that temperature affects SOM polymerization and adsorption and desorption to mineral surfaces as well as the production and conformation of microbial enzymes. They argued that the net effect of these processes on the availability and degradability of SOM will determine how SOM will be affected by increasing soil temperatures. In another recent study,[54] it was reported that warming accelerates decomposition of decades-old carbon in forest soils. In yet another recent study,[55] elevated CO_2 was shown to have a priming effect on the decomposition of older SOM and to accelerate the decomposition of new detrital inputs. Complex relations have been reported between SOM and temperature and CO_2 that suggest both increases and decreases in SOM,[53,56,57] with the result that there is no consensus on the overall effect of these environmental changes on SOM in soils.

In areas where climate change results in a hotter and drier environment with more variability and extreme events of drought and heavy rainfall, as has been predicted for some regions,[58,59] soils will be subjected to more influence from a number of processes that are conducive to structural degradation.[60] Soil aggregate stability and soil structural integrity are likely to be adversely affected by losses in SOM that would accompany drying conditions. Structural degradation is associated with a loss in AWC and crop water use efficiency and increasing susceptibility to erosion. Decreases in SOM content would have adverse effects

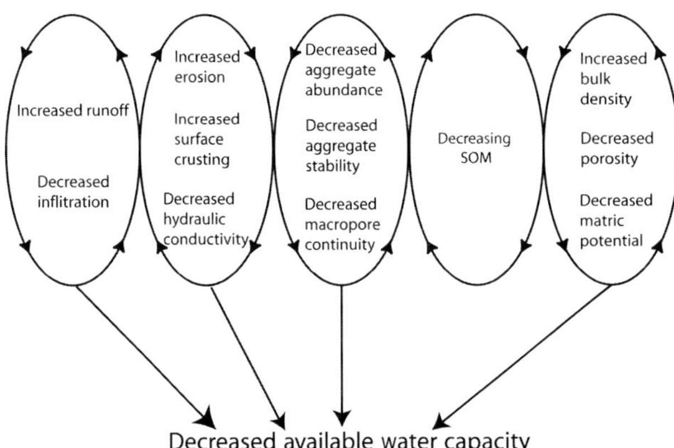

Fig. 3 A conceptual model of the major effects of decreasing soil organic matter on soil physical properties that interact through the loops shown to decrease available water capacity in cultivated, eroding agricultural fields.

on soil chemical and physical properties because of the importance of SOM to soil structure and hydrologic properties. SOM is important for the maintenance of good soil structure through its positive influence on soil aggregate formation and stability[7] (Fig. 2). Aggregation increases soil porosity and provides for improved water retention within the range of soil water potential where water is available to plants. It has been recognized that decreases in SOC could reduce AWC and soil fertility with resulting adverse consequences on soil health and productivity.[61,62]

Reductions in SOM in the surface soil result in increased bulk density, decreased porosity, and increases in matric potential that, in turn, result in decreased AWC (Fig. 3). Decreases in SOM also result in decreases in soil aggregation, aggregate stability,[63,7] and macropore continuity, thus causing increases in surface crusting and decreases in hydraulic conductivity (Fig. 3). These changes in physical properties lead to reduced rates of water infiltration and increases in runoff and erosion.[10] A climate warming-induced intensification of the hydrologic cycle could increase runoff and erosion in some regions.[64] Decreased infiltration and increased runoff result in less precipitation retained throughout the soil rooting zone so that less precipitation becomes available for plant growth. SOM exerts a strong influence on AWC through these interacting feedback loops that affect many soil physical properties (Fig. 3).

CONCLUSIONS

Despite considerable uncertainty in the relation between AWC and SOM, most studies are consistent with the classical conceptual understanding that these soil properties are positively related. SOM promotes soil aggregate formation and the development of soil porosity that enhances infiltration and retention of water in a range that is available to plants. Environmental changes that result in increases in SOM will increase AWC. On the other hand, environmental changes that result in decrease in SOM will decrease AWC with resulting adverse consequences for the sustainability of agricultural productivity in some regions. There is a need to improve the quantitative understanding of the sensitivity of AWC to SOM to reduce the current level of uncertainty and to provide resource managers with better decision support systems.

ACKNOWLEDGMENTS

Rattan Lal, Harry Vereecken, Alan Olness, and Jeffery Kern provided useful insights and references on the sensitivity of AWC to SOM. Miko Kirschbaum and Gregory Lawrence provided valuable insights that substantially improved this revised entry. Funding for this entry was provided by U.S. Geological Survey Climate and Land Use Change Research and Development Program.

REFERENCES

1. Romano, N.; Santini, A. Measurement of water retention: Field methods. In *Methods of Soil Analysis, Part 4*; Dane, J. H., Topp, G. C., Eds., Soil Sci. Soc. Am. Madison, WI, **2002**, 721–738.
2. Jenny, H. *The Soil Resource*; Springer-Verlag: New York, 1980.
3. Hudson, B. D. Soil organic matter and available water capacity. J. Soil Water Conserv. **1994**, *49*, 189–194.
4. Emerson, W.W. Water retention, organic carbon and soil texture. Aust. J. Soil Res. **1995**, *33*, 241–251.
5. Stevenson, F.J. *Humus Chemistry*; John Wiley and Sons: New York, N.Y., 1982.
6. Kay, B.D. Soil structure and organic carbon: A review. In *Soil Processes and the Carbon Cycle*; Lal, R., Ed.; CRC Press: Boca Raton, 1997; 169–197.
7. Haynes, R.J.; Beare, M.H. Aggregation and organic matter storage in meso-thermal, humid soils. In *Structure and Organic Matter in Agricultural Soils*; Carter, M.R., Stewart, B.A., Eds.; CRC Lewis: Boca Raton, 1996.
8. Haynes, R.J.; Naidu, R. Influence of lime, fertilizer and manure applications on soil organic matter content and soil physical conditions: A review. Nutrient Cycling Agroecosyst. **1998**, *51*, 139–153.
9. Ascough, J.C. II; Baffaut, C.; Nearing, M.A.; Liu, B.Y. The WEPP watershed model: I. Hydrology and erosion. Trans. ASAE **1997**, *40*, 921–933.
10. Rhoton, F.E.; Lindbo, D.L. A soil depth approach to soil quality assessment. J. Soil Water Cons. **1997**, *52*, 66–72.
11. Huntington, T.G. Available water capacity and soil organic matter. In *Encyclopedia of Soil Science*; Lal, R., Ed.; Marcel Dekker: New York, 2003; 1–5 pp, DOI: 10.1081/E-ESS 120018496.
12. Manrique, L.A.; Jones, C. Bulk density of soils in relation to soil physical and chemical properties. Soil Sci. Soc. Am. J. **1991**, *55*, 476–481.
13. Veereecken, H.; Maes, J.; Feyen, J.; Darius, P. Estimating the soil moisture retention characteristics from texture, bulk density and carbon content. Soil Sci. **1989**, *148*, 389–403.

14. Lal, R. Physical properties and moisture retention characteristics of some Nigerian soils. Geoderma **1979**, *21*, 209–223.
15. Anderson, S.H.; Gantzer, C.J.; Brown, J.R. Soil physical properties after 100 years of continuous cultivation. J. Soil Water Cons. **1990**, *45*, 117–121
16. Gupta, S.C.; Larson, W.E. Estimating soil water retention characteristics from particle size distribution, organic matter content and bulk density. Water Resour. Res. **1979**, *15*, 1633–1635.
17. Thomasson, A.J.; Carter, A.D. Current and future uses of the UK soil water retention dataset. In *Indirect Methods for Estimating the Hydraulic Properties of Unsaturated Soils*, van Genuchten, T.M., Leij, F.J., Lund, L.J., Eds.; Proceedings of the International Workshop on Indirect Methods for Estimating the Hydraulic Properties of Unsaturated Soil, Riverside, California, USA, 1992; 355–358.
18. Salter, P.J.; Berry, G.; Williams, J.B. The influence of texture on the moisture characteristics of soils, III: Quantitative relationships between particle size, composition, and available water capacity. J. Soil Sci. **1966**, *17*, 93–98.
19. Pidgeon, J.D. The measurement and prediction of available water capacity of ferrallitic soils in Uganda. J. Soil Sci. **1972**, *23*, 431–444.
20. Schroo, H. Soil productivity factors of the soils of the Zandry formation in Surinam. De Surinaamse Landbouw **1976**, *24*, 68–84.
21. Karlen, D.L.; Wollenhaupt, N.C.; Erbach, D.C.; Berry, N.C.; Swain, J.B.; Eash, N.S.; Jordahl, J.L. Crop residue effects on soil quality following 10 years of no-till corn. Soil Tillage Res. **1994**, *31*, 149–167.
22. Rawls, W.L.; Brakensiek, D.L.; Saxton, K.E. Estimation of soil water properties. Trans. ASAE **1982**, *25*, 1316–1320.
23. Hollis, J.M.; Jones, R.J.A.; Palmer, R.C. The effects of organic matter and particle size on the water retention properties of some soils in the West Midlands of England. Geoderma **1977**, *17*, 225–231.
24. Tran-Vinh-An; Nguba, H. Contribution a l'etude de l'eau utile de qelques sols dur Zaire. Sols Africana **1971**, *16*, 91–103.
25. Hamblin, A.P.; Davies, D.B. Influence of organic matter on the physical properties of some east Anglian soils of high silt content. J. Soil Sci. **1977**, *28*, 11–22.
26. Salter, P.J.; Haworth, F. The available water capacity of a sandy loam soil, II. the effects of farm yard manure and different primary cultivations. J. Soil Sci. **1961**, *12*, 335–342.
27. Schertz, D.L.; Moldenhauer, W.C.; Livingston, S.J.; Weesies, G.A.; Hintz, E.A. Effect of past soil erosion on crop productivity in Indiana. J. Soil Water Conserv. **1989**, *44*, 604–608.
28. Schjonning, P.; Christensen, B.T.; Carstensen, B. Physical and chemical properties of a sandy loam receiving animal manure, mineral fertilizer or no fertilizer for 90 years. Eur. J. Soil Sci. **1994**, *45*, 257–268.
29. Rawls, W.J.; Pachepsky, Y.A.; Ritchie, J.C.; Sobecki, T.M.; Bloodworth, H. Effect of soil organic carbon on soil water retention. Geoderma **2003**, *116*, 61–76.
30. Olness, A.; Archer, D. Effect of organic carbon on available water in soil. Soil Sci. **2005,** *170*, 90–101.
31. Mitchell, C.C.; Weserman, R.L.; Brown, J.R.; Peck, T.R. Overview of long-term agronomic research. Agron. J. **1991**, *83*, 24–29.
32. Moebius-Clune, B.N.; van Es, H.M.; Idowu, O.J.; Schindelbeck , R.R.; Moebius-Clune, D.J.; Wolfe, D.W.; Abawi, G.S.; Thies, J.E.; Gugino, B.K.; Lucey, R. Long-term effects of harvesting maize stover and tillage on soil quality. Soil Sci. Soc. Amer. J. **2008,** *72*, 960–969.
33. Bhogal, A.; Nicholson, F.A.; Chambers, B.J. Organic carbon additions: Effects on soil bio-physical and physico-chemical properties. Eur. J. Soil Sci. **2009**, *60*, 276–286.
34. Paul, E.A.; Paustian, K.; Elliott, E.T.; Cole, C.V. *Soil Organic Matter in Temperate Agroecosystems: Long-Term Experiments in North America*; CRC Press: Boca Raton, 1997.
35. Mattson, K.G.; Swank, W.T.; Waide, J.B. Decomposition of woody debris in a regenerating, clear-cut forest in the southern Appalachian. Can. J. For. Res. **1987**, *17*, 712–721.
36. Rosenberg, N.J.; Izaurralde, R.C.; Malone, E.L. *Carbon Sequestration in Soils: Science, Monitoring and Beyond: Conference Proceedings: St. Michaels Workshop, Dec.* Battell Press: Columbus, OH; 1999.
37. Oulehle, F.; Evans, C.D.; Hofmeister, J; Krejci, R.; Tahovska, K.;Persson, T.; Cudlin, P.; Hruska, J. Major changes in forest carbon and nitrogen cycling caused by declining sulphur deposition. Global Change Biol. **2011**, *17*, 3115–3129.
38. Lawrence, G.B.; Shortle, W.C.; David, M.B.; Smith, K.T.; Warby, R.A.F.; Lapenis, A.J. Early indications of soil recovery from acidic deposition in U.S. red spruce forests. Soil Sci. Soc. Amer. J. **2012**, *76*, 1407–1417.
39. Houghton, J.T.; Ding, Y.; Griggs, D.C.; Noguer, M.; van der Linden, P.J.; Dai, X.; Maskell, K.; Johnson, C.A. *Climate Change 2001: The Scientific Basis*; Cambridge University Press: Cambridge, UK, 2001.
40. Rounsevelle, M.D.A.; Evans, S.P.; Bullock, P. Climatic change and agricultural soils: Impacts and adaptation. Clim. Change **1999**, *43*, 683–709.
41. Jenkinson, D.S.; Adams, D.E.; Wild, A. Model estimates of CO_2 emissions from soil in response to global warming. Nature **1991**, *351*, 304–306.
42. Cox, P.M.; Betts, R.A.; Jones, C.D.; Spall, S.A.; Totterdel, I.J. Acceleration of global warming due to carbon-cycle feedbacks in a coupled climate model. Nature **2000**, *408*, 184–187.
43. Carter, J.O.; Howden, S.M. *Global Change Impacts on the Terrestrial Carbon Cycle*, Working Paper Series 99/10, Report to the Australian Greenhouse Office, CSIRO Wildlife and Ecology: Canberra, Australia, **1999**.
44. Bellamy, P.H.; Loveland, P.J.; Bradley, R.I.; Lark, R.M; Kirk G.J.D. Carbon losses from all soils across England and Wales. Nature **2005**, *437*, 245–248.
45. Senthilkumar, S.; Basso, B.; Kravchenko, A.N.; Robertson, G.P. Contemporary evidence of soil carbon loss in the U.S. corn belt. Soil Sci. Soc. Amer. J. **2009**, *73*, 2078–2086.
46. Nafziger, E.D.; Dunker, R.E. Soil organic carbon trends over 100 years in the morrow plots. Agron. J. **2011**, *103*, 261–267.
47. Khan, S.A.; Mulvaney, R.L.; Ellsworth, T.R.; Boast, C.W. The myth of nitrogen fertilization for soil carbon sequestration. J. Environ. Qual. **2007**, *36*, 1821–1832.
48. Fantappié, M.; L'Abate, G.; Costantini, E.A.C. The influence of climate change on the soil organic carbon content in Italy from 1961 to 2008. Geomorphology **2011**, *135*, 343–352.

49. Smith, T.M.; Weishampel, J.F.; Shugart, H.H.; Bonan, G.B. The response of terrestrial c storage to climate change - modeling c-dynamics at varying temporal and spatial scales. Water Air Soil Pollut. **1992**, *64*, 307–326.
50. Schimel, D.S.; Braswell, B.H.; Holland, B.A.; McKeown, R.; Ojima, D.S.; Painter, T.H.; Parton, W.J.; Townsend, A.R. Climatic, edaphic, and biotic controls over the storage and turnover of carbon in soils. Global Biogeochem. Cycles **1994**, *8*, 279–293.
51. Torn, M.S.; Lapenis, A.G.; Timofeev, A.; Fischer, M.L.; Babikov, B.V.; Harden, J.W. Organic carbon and carbon isotopes in modern and 100-year-old-soil archives of the Russian steppe. Global Change Biol. **2002**, *8*, 941–953.
52. Johnston, A.E.; Poulton, P.R.; Coleman, K. Soil organic matter. Its importance in sustainable agriculture and carbon dioxide fluxes. Adv. Agronomy, **2009**, *101*, 1–57.
53. Conant, R.T.; Ryan, M.G.; Ågren, G.I.; Birge, H.E.; Davidson, E.A.; Eliasson, P.E.; Evans, S.E.; Frey, S.D.; Giardina, C.P.; Hopkins, F.M.; Hyvönen, R.; Kirschbaum, M.U.F.; Lavallee, J.M.; Leifeld, J.; Parton, W.J.; Megan Steinweg, J.; Wallenstein, M.D.; Martin-Wetterstedt, J.Å.; Bradford, M.A., Temperature and soil organic matter decomposition rates – synthesis of current knowledge and a way forward. Global Change Biol. **2011**, *17*, 3392–3404.
54. Hopkins, F.M.; Torn, M.S.; Trumbore, S.E. Warming accelerates decomposition of decades-old carbon in forest soils. Proc. Nat. Acad. Sci. **2012**, Early Edition, http://www.pnas.org/cgi/doi/10.1073/pnas.1120603109
55. Phillips, R.P.; Meier, I.C.; Bernhardt, E.S.; Grandy, A.S.; Wickings, K.; Finzi, A.C. Roots and fungi accelerate carbon and nitrogen cycling in forests exposed to elevated CO_2. Ecol. Lett. **2012**, *15*, 1042–1049.
56. Smith, P.; Fang, C.; Dawson, J.J.C.; Moncrieff, J.B.; Donald, L.S. *Impact of Global Warming on Soil Organic Carbon, Advances in Agronomy*; Academic Press: 2008; pp. 1–43.
57. Kirschbaum, M.U.F. The temperature dependence of organic matter decomposition: Seasonal temperature variations turn a sharp short-term temperature response into a more moderate annually averaged response. Global Change Biol. **2010**, *16*, 2117–2129.
58. Meehl, G.A.; Stocker, T.F.; Collins, W.D.; Friedlingstein, P.; Gaye, A.T.; Gregory, J.M.; Kitoh, A.; Knutti, R.; Murphy, J.M.; Noda, A.; Raper, S.C.B.; Watterson, I.G.; Weaver, A.J.; Zhao, Z.-C. Global climate projections. In *Climate Change: The Physical Science Basis. Contribution of Working Group I to the Fourth Assessment Report of the Intergovernmental Panel on Climate Change;* Solomon, S., Qin, D., Manning, M., Chen, Z., Marquis, M., Averyt, K.B., Tignor, M., Miller, H.L., Eds.; Cambridge University Press: Cambridge, United Kingdom, 2007; pp. 747–845.
59. Tebaldi, C.; Hayhoe, K.; Arblaster, J.M.; Meehl, G.A., Going to the extremes: An intercomparison of model-simulated historical and future changes in extreme events. Clim. Change **2006**, *79*, 185–221.
60. Chan, K.Y. Soil attributes and soil processes in response to the climate change. In *Soil Health and Climate Change*. Singh, B.P., Allen, D.E., Dalal, R.C. Eds.; Springer: Berlin, Soil Biol. 2011; *29*, 49–67.
61. McCarty, J.J.; Canziani, O.F.; Leary, N.A.; Dikken, D.J.; White, K.S. *IPCC Climate Change 2001: Impacts, Adaptation & Vulnerability;* Cambridge University Press, Cambridge UK. 2001.
62. Allen, D.E.; Singh, B.P.; Dalal, R.C.; Cowie, A.L.; Chan, K.Y. Soil health indicators under climate change: A review of current knowledge. In *Soil Health and Climate Change*, Singh, B.P., Allen, D.E., Dalal, R.C. Eds.; Springer: Berlin, Soil Biol. 2011; *29*, 25–55.
63. Pimentel, D.; Harvey, C.; Resosudarmo, P.; Sinclair, K.; Kurtz, D.; McNair, M.; Crist, S.; Shpritz, I.; Fitton, L.; Saffouri, R.; Blair, R. Environmental and economic costs of soil erosion and conservation benefits. Science **1995**, *267*, 1117–1123.
64. Huntington, T.G. Climate warming-induced intensification of the hydrologic cycle: A review of the published record and assessment of the potential impacts on agriculture. Advances in Agronomy **2010**, *109*, 1–53.

Soil: Spatial Variability

Josef H. Görres
Department of Plant and Soil Science, University of Vermont, Burlington, Vermont, U.S.A.

Abstract
Spatial variability occurs at all scales—from micrometer scale (1–10 μm) relevant to microbial processes to continental scale (1000 km) relevant to the distribution of soil orders and climate change. The order of magnitude of the coefficient of variation (CV) of soil properties ranges from a few percent to greater than 100%. Jenny's soil forming factors, climate, topography, organisms, parent material, and time, represent the fabric of the natural variability of soil that is modified by anthropogenic properties resulting in a complex distribution of soil characteristics. Very small scale soil property variations may affect biological and chemical processes and may even explain soil ecological puzzles such as the high level of functional redundancy and diversity of soil biota. Spatial and temporal variations in soil properties may affect the spatial distribution of important soil processes. The spatial analysis of these variations is often done using semivariance analysis and mapping with kriging, an unbiased, optimal interpolation technique developed for estimating ore deposits from limited field data. Variable rate agriculture, in which GPS technology is used to map crop yield, uses spatial analysis to predict spatially explicit fertilizer applications to improve yield and lower environmental impacts. Spatial analysis is also used in environmental management, environmental modeling, and a multitude of research investigations. All applications need to consider the patterns and scale of variations in order to obtain accurate estimates of the variability of soil properties and processes.

INTRODUCTION

All soil properties, whether they be physical, biological, or chemical in nature, show considerable multi-scale variability, ranging from microbial scales (1–1000 μm) to regional (10–100 kilometers) and even continental scales (thousands of kilometers). Variability at all scales is high. The coefficients of variation (CVs—ratio of the standard deviation and mean) for individual properties may range from several percent to several hundred percent. Warrick[1] classified variations of physical soil properties with CVs less than 15%, between 15% and 50%, and greater than 50% as low, medium, and high, respectively. Bulk density and porosity were classed as low, whereas electrical conductivity, hydraulic conductivity, and solute concentrations were classified as high variability. For biological properties, similar ranges are reported in the literature. For example, CVs for phosphatase activity were reported as 39%.[2] For forest soils, the CV for nematode abundance was of the order of 100% as compared to moisture and organic matter with CVs of 10–40%.[3] Where earthworms are present, the CV of water percolation may vary from 10% to 100%.[4,5] To make matters more complex, some soil properties and their variability change seasonally[3] or indeed on a daily basis.

Soil scientists engaged in soil mapping, classification, and morphology may have to integrate over several scales to identify soil taxa. Soil taxonomy recognizes that transformations and translocations of soil materials connect soil horizons in depth. Evaluating the three-dimensional variation of soil properties in combination at the pedon-scale (meter-scale) may thus produce thousands of unique profiles, each representing a soil series, the finest taxonomic resolution in soil classification systems and the pedology analog of species in Linnaean taxonomy.

Both natural processes and human activity contribute to the spatial variability encountered in soils. Jenny's equation of soil formation is the most commonly used model of pedogenesis that attempts to understand differences among soils in terms of five spatially variable factors. These five factors are responsible for landscape to continental-scale variations in soil properties.[6] The mosaic of different soil series created by the superposition of Jenny's factors may be regarded as the fabric of soil variability which is further modified by land management.

Soil biological processes create very much smaller, microscale variations in properties other than those considered in soil classification. These variations can be tracked to particular origins.[7] For example, it has long been known that soil biota are more numerous in soil surrounding roots[8] and burrows of soil animals[9] than in the bulk soil. Microscale heterogeneity may cause the spatial separation of key nutrient cycling transformations that may drive many below-ground ecological processes.[10,11]

Any record of the spatial distribution of properties is a snapshot in time. Soil processes are strongly dependent on

Encyclopedia of Natural Resources DOI: 10.1081/E-ENRL-120047495

meteorological and climatic forces and thus respond to temporal variations in temperature, precipitation, wind speed, insolation, and relative humidity. Denitrification, the conversion of nitrate into a suite of nitrogen gases, for example, depends on the distribution of carbon at the microscale. However, denitrification is mainly an anaerobic process that is strongest under saturated conditions. Large rainfall events thus cause an increase in denitrification activity. Nitrification, on the other hand, is an aerobic process and thus occurs when soils are drying. As biochemical processes often occur at microsites, the diffusion of biochemical reaction products further links space and time.[11]

Soil variability has practical consequences for experimental designs, farming, and the implementation and design of best management practices. For example, precision farming employs variable rate technologies that strive to apply agrochemicals to soils based on georeferenced spatial variations of field fertility parameters. This may save farm resources and improve the environmental quality by reducing inputs of fertilizers and pesticides. Yet, environmental and economic impacts of variable rate agriculture may depend on many factors and it is not clear whether this technology always leads to a reduction in application rate on the field scale.[12] In environmental application, land managers and soil evaluators often face questions associated with landscape-scale variations, such as how site-specific soil evaluations need to be when determining soil suitability for land use projects.

What all applications that account for the spatial variability of soils have in common is that the scale of variation ought to be a focus. The scale germane to an application will have to be isolated to design an appropriate sampling strategy and determine the volume (sample support) of an individual soil sample.[13] Soil variability represents interesting challenges to soil scientists, making knowledge of spatial distribution of soil properties and processes paramount to current practice in both fundamental and applied pedology.

FACETS OF SPATIAL VARIABILITY IN SOILS

Soils, Soil Properties, Soil Processes, and Time

When discussing the spatial variability of soils, a distinction has to be made between the variability of soils, i.e., among taxonomic units, and soil properties, a measurable quantity. Pedologists classify terrestrial and subaqueous soils based on the vertical distribution of matter and its properties within a representative soil volume called a pedon.[14] Each soil series thus has a unique sequence of horizons with specific diagnostic properties that are intertwined with spatially and temporally variable processes within and outside of the soil. Soil survey maps are quintessentially two-dimensional as the nature of a soil series is given by the relationship and interpretation of properties in the vertical as well as the horizontal dimensions of its profile. Soil classification relies on the spatial, three-dimensional arrangement of measurable and relatively stable properties within a soil profile.[14]

However, stability and equilibrium are not one of the hallmarks of many soil properties. Spatial distributions change over time. Daily variations in soil moisture and temperature may engender rapid changes in food web dynamics and biochemical transformation rates responsible for much of the below-ground decomposition and nutrient cycling. Variations also occur at seasonal and inter-annual scales. Even longer term environmental variations relevant to climate and vegetation affect soil development in a way that may have tangible effects on soil loss.[15] Among these decade-long cycles, space weather, in particular the 22-year cycles of the solar magnetosphere and the concomitant variation in cosmic ray fluxes, may be responsible for droughts such as the Dustbowl episode in Southern U.S.A. in 1930.[16] While devastating for farmers, these drought episodes are responsible for a spatial redistribution of top soil materials. However, even at a single point in time, snapshots of soils present observers with sufficiently puzzling spatial variability.

The importance of the dependence of variability on time is best illustrated by how changes in soil properties may affect the variability of soil processes. It is well known that small differences in soil moisture content may translate into large differences in soil processes. The affected processes are physico-chemical and biological in nature. Moisture saturation at the center of aggregates in an otherwise dry soil may result in considerable denitrification rates.[17] Also, only small differences in moisture or the presence and location of macro-pores in soils may result in large differences in water fluxes through the soil.

Factors of Soil Formation: Intrinsic Sources of Spatial Variability

Early soil scientists were intrigued by the differences in soils that they encountered in their travels. Dukochaev (1846–1903), Hilgard (1833–1916), and Jenny (1899–1992) formulated theories in which independent factors could explain geographic variations in soils.[18] These factors are intrinsic or natural sources of variability. In an attempt to create a more quantitative foundation for soil science, Jenny's equation[6] expressed the magnitude of soil-forming processes in terms of five variables or factors: climate, parent material, topography, organisms, and time.

$$\text{Process} = F(\text{climate, topography, parent material, organisms, time}) \quad (1)$$

The five factors act on all soil-forming processes with each factor imbuing different qualities to the soil. Overlaying these qualities across the landscape differentiates

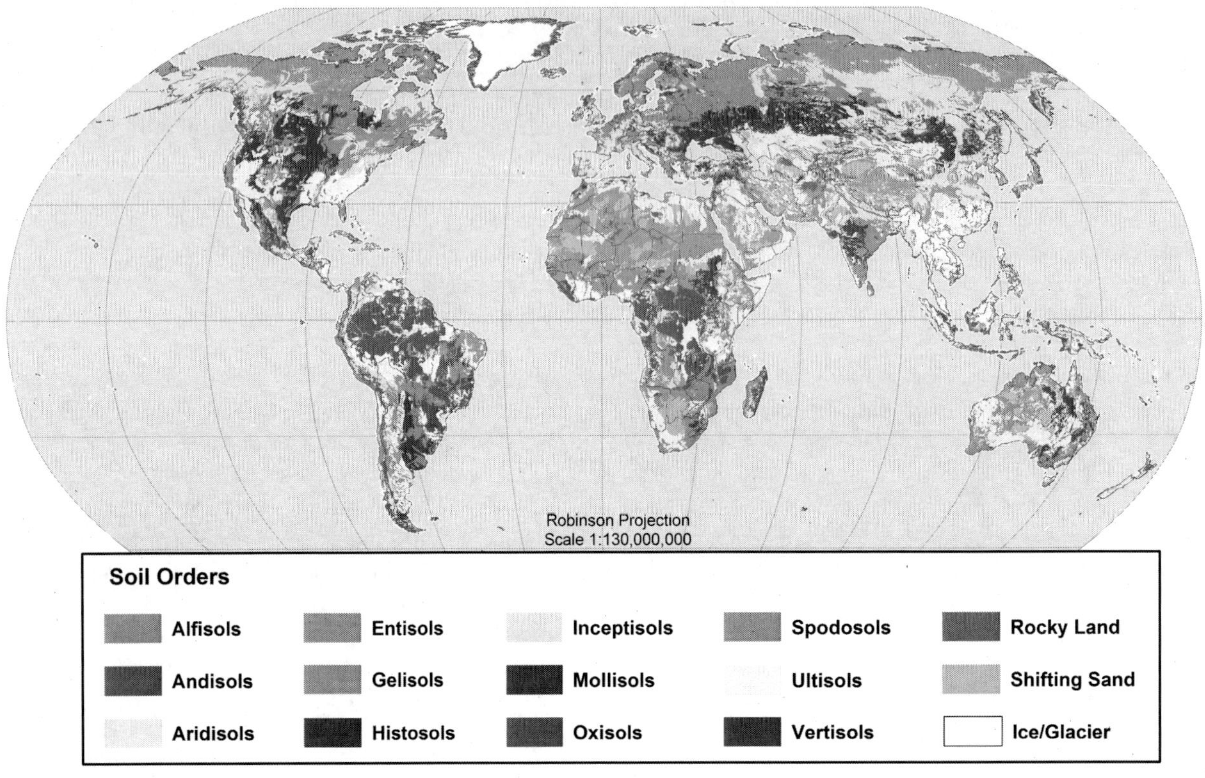

Fig. 1 The planetary distribution of soil orders across the land area. Distinct geographic patterns of soil orders emerge based on climate, topography, organisms, parent material, and time.
Source: Reproduced with permission from Dr. Paul F. Reich, Geographer–World Soil Resources, USDA Natural Resources Conservation Service.

among soil series, the spatial variability considered by soil taxonomy. At the continental scale, large-scale variations in climate, vegetation, and parent material give rise to characteristic patterns of soil orders, the coarsest soil taxonomic scale (Fig. 1). For example, most spodosols are found in northern, temperate climes with boreal, coniferous forests, where precipitation is high and recent glaciation has left deposits of sands leading to distinct patterns of reduced and oxidized horizon environments (Fig. 2A). In contrast, inceptisols have not developed very distinctive horizons due to a lack of time that the other factors have acted upon them. In Fig. 2B, an agricultural inceptisol is shown with a horizon altered by plough action. This young soil (only thousands of years old) developed after the Laurentide ice sheet of the last glaciation melted and receded. In this soil, the combination of two parent materials, a fine silt mantle overlying glacial sand and gravel deposits, results in characteristic reduction-oxidation horizons that dominate the spatial variability in the depth of the soil. The fine silt material may retain water long enough above the gravel deposit to reduce iron oxides resulting in a horizon where the distinctive orange color of iron oxides is absent.

The five factors do not act independently, and Jenny's equation can thus not be understood as a linear algebraic equation. Toposequences are a good example of the interrelationship between soil-forming factors at the landscape scale. A special toposequence called a hydrosequence is shown in Fig. 3. The landscape position along this hydrosequence determines not only elevation and, thus, the distance to ground water and the drainage class of the soil. It also determines vegetation, decomposition, and production rates of the ecosystems as the land surface slopes from an upland area into a wetland. Together, these factors are responsible for the variation of soils along the toposequence.

Action of *Homo sapiens*: Extrinsic Factors of Variability

Superimposed on the patchwork of intrinsic variation are extrinsic spatial patterns introduced by land management.

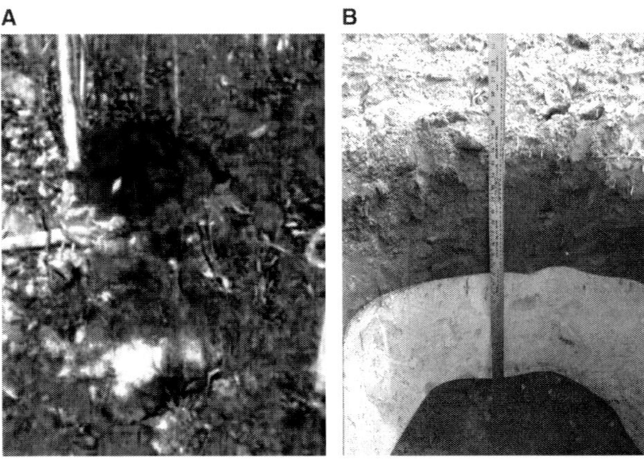

Fig. 2 (**A**) Spodosol at Hubbard Brook, NH. Distinct horizonation with a "bleached" E horizon above a characteristic dark area B_{hs} horizon, indicating the translocation of iron and humus from the E and O horizons. Provided by Dr. Joel Tilley of the Agricultural and Environmental Testing Lab at the University of Vermont. (**B**) Inceptisol at the Peckham Farm, Kingston, RI. A silt cap deposited over coarse sand gravel deposits created conditions that favor reduction of iron oxides within the silt horizon resulting in a white, colorless layer and a streak of white iron oxides at the contact between the silt and the sand and gravel.

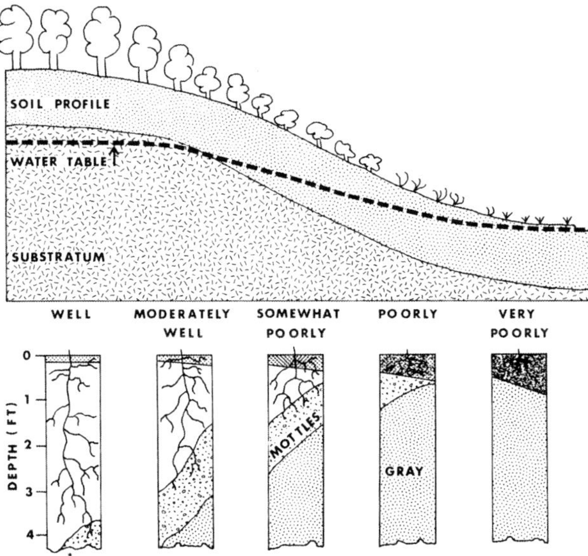

Fig. 3 Soil profiles along a toposequence vary by elevation or more importantly with the distance to the water table.
Source: Rhode Island Agricultural Research Station 1988. Wright and Sautter. Soils of the RI Landscape, reproduced with permission of the Director, Rhode Island Agricultural Research Station.

At the landscape scale, the mosaic of agricultural fields (Fig. 4) may impose an additional variability of soils. In addition, within each field, management such as tillage and irrigation may introduce further variability at the field scale. Notably, extrinsic patterns may persist for many years even when the practice causing the initial variation is abandoned. For example, ridge and furrow tillage in the nineteenth century may still have affected the crop yield pattern in 1911 at Rothamsted.[19]

Fig. 4 Landscape-scale field-planting pattern, an extrinsic variation contributing to the spatial variability of soils.

Apart from the desired patterns of fertilizer application and planting, the use of machinery causes compaction which can also persist for many years.[20] Likewise, the hoof fall of large grazing animals may affect soil compaction in pastures with characteristic spatial variations that depend on grazing pattern, paddock geography, and animal stocking rate. Dung deposition patterns in the pasture add to the variability of pasture soils[21] and render the distribution of nutrient cycling processes uncertain.

These examples suggest that land management creates additional variability at the field scale. However, it can also be argued that land conversions may make soils more homogeneous. For example, the conversion from a heterogeneous land use mosaic to large fields that support monocultures such as corn or sunflowers would smooth variations at the landscape scale while increasing heterogeneity at the field scale.

Variation at the Microscale

The patterns of soil heterogeneity observed by soil surveys, i.e., those that are caused by soil-forming factors, essentially integrate soil properties over the pedon that has a volume of several meters cubed. Yet, the diversity of processes at much smaller scale is apparent even when observing differences at the soil surface. The pattern of vegetation may reveal small-scale variations in soil properties, a fact that is used in plant indicator systems.[22] Equally obvious at the soil surface are the patterns of excavation mounts of burrowing animals like ants and earthworms, which may coexist within centimeters of each other but have quite divergent microbial communities.[23]

Very small scale spatial variability is also regarded as the key to better understanding why soil faunal communities maintain high levels of diversity with much apparent

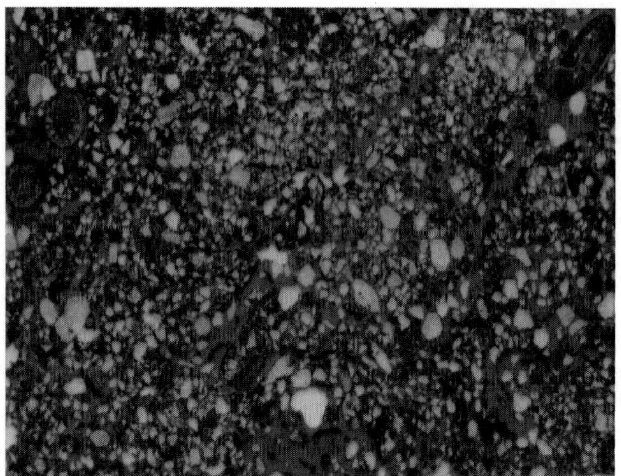

Fig. 5 Micrograph of a thin section of sandy loam core showing the small-scale spatial variation in pore sizes. Pores are segmented in a light color. The network shows how the physical pore habitat varies even at the millimeter scale. The vertical extent of the image is 1 cm.

functional redundancy.[24] Pore networks comprise a diversity of pore neck and chamber diameters that allow access only to those organisms that fit into the pores. The variability of pore diameter at the μm-scale is depicted in Fig. 5, where the light-colored segmented areas represent the pore network of a sandy loam collected in Rhode Island. The physical pore structure interacts with biologically active "spheres," where soil organisms exert their influence on the soil and its microbial and faunal communities.[7] Drilosphere and rhizosphere are the soils surrounding earthworm burrows and plant roots, respectively. The enrichment of the drilosphere with detrital resources welded to the burrow walls by earthworms and modification of the soil pore structure[25] causes a large increase in biological activity.[9,26] Similarly, resource enrichment can cause increases in microbial populations in the rhizosphere.[8] Here, local photosynthate exudations prime soil biological and chemical activity.

Hattori and Hattori[27] introduced the concept of microhabitats of protozoa and microorganisms that are located within soil aggregates, which Beare et al.[7] called the "aggregatusphere." Now, with molecular techniques, it is possible to establish the spatial variability of microorganisms within single aggregates revealing great diversity at the millimeter scale.[28]

The spatial distribution of pore spaces (also called the "porosphere") and its partitioning into air and water-filled pores may select for specific soil organisms. Moisture controls the level of competition and predator–prey relationships that occur among soil organisms. Naturally, the relative fraction of air-filled and water-filled pores changes as soils dry and wet up, and thus, the mosaic of habitat of water and air-based soil organisms changes too. The concept of habitable pore space[29,30] encapsulates this variability at scales of pore diameters ranging from tens to hundreds of micrometers. At these scales, soil pores are microhabitats for protozoa[27] and nematodes.[25,31] However, there are different interpretations as to what happens to these habitable spaces when drying and wetting cycles rapidly alter their habitability for the aerial and the aquatic communities that inhabit soil pores.[32]

The idea that microsites can significantly affect bulk soil processes suggests that measurement may be scale dependent. Parkin[10] found that a soil core (100 cm^3-scale) had very high denitrification activity. A single, small woodchip was isolated as the source of the high denitrification rate measured for the core. Parkin's experiment demands that sampling support be considered in designs that are to estimate a soil property or process. Similarly, the moisture distribution within aggregates should be considered when examining denitrification in an otherwise dry soil.[17]

Concepts of Variability

Variability is often conceptualized as the superposition of intrinsic and extrinsic factors, some of which were discussed earlier. However, there are other ways to look at soil variability. One such approach distinguishes between stochastic variations and deterministic variations that when added together are responsible for the variation in soil properties.

Deterministic variations are those that can be predicted from soil processes, and they may be spatially dependent at various scales. Spatial dependence is introduced by the translocation of matter and energy from one location to another or by the extent of large parent material deposits or even by vegetation patterns. Stochastic variations are those that cannot be predicted and are much like white noise and thus not spatially dependent. For spatial interpretations, the deterministic variables may have to be removed so that random variations can be properly characterized.

Webster[33] argued that with sufficient knowledge about all soil processes, all variability is deterministic and that only incomplete knowledge and the sheer complexity of interacting soil processes may make variations in soil appear random. Notwithstanding philosophical argumentation about random and deterministic variations, apparent spatial dependence or independence may require different methods of estimating a central tendency and the variation about it.

One measure that is often used to describe variability is the *CV*, which measures the variation as standard deviation, σ, standardized by the mean, μ:

$$CV = \frac{\sigma}{\mu}$$

The *CV*, however, requires that sample points are independent and may be falsely applied when there is any degree of spatial dependence. This casts some doubt on *CV*s estimated at different scales as soil properties may have different levels of spatial dependence at different scales.

Sampling and Quantifying Variability

Quantification of the variability of soil properties is more complex than it may appear. Even when designing the sampling method, important decisions need to be made. Apart from the classical equation for the number of samples that need to be taken to determine the mean of a sample distribution at a given confidence level and accuracy, the physical size of the sample support (the physical volume or mass of a sample specimen) becomes important as it will determine the smallest variations that can be resolved.

Sometimes, the nature of the assay will determine the sample support. For enzyme activity assays, only a couple of grams of soil are necessary. To get reliable enzyme activity information, several soil samples need to be homogenized and several replicates may need to be analyzed. The purpose of homogenization and analytical replication is to smooth small-scale variations and increase the certainty of estimation at a given scale. For measurements of bulk density, cores are taken that are typically 5–10 cm in diameter and 10 cm long. Variations that are contained within that volume are averaged out, or smoothed. Large-scale examples of spatial averaging come from remote sensing where luminosity is averaged over the area of pixels that are tens to thousands of meters in length. Selection of the satellite and sensor allows matching the scale of resolution with the research or management need. Spatial averaging is an essential tool not only for improving the reliability of data, but also focusing on the scale of the relevant variation.

Given a sufficient number of samples that were randomly collected, first-order statistics can describe soil variability in terms of statistical moments of the sample distribution: mean, variance, skewness, and kurtosis. This assumes that measured soil properties are independent of each other at the sample locations and for the scale represented by the sample support.

Many soil properties are spatially dependent. First-order, or classical, estimates of mean and variance can become biased when spatial relationships, so-called second-order properties, are not taken into account. A second-order measure commonly used in soil science is the semivariogram, although other methods such as power spectra derived from Fourier harmonic decomposition have also been used.[19] A semivariogram measures one half of the covariance between soil property values of sample pair classes that are classed by pair-separation distances called lags. The further samples are apart, the greater their semivariance and the less related they are in space unless there are repeating patterns. Theoretically, the semivariance is limited to a maximum value that is reached when the lag exceeds the range of spatial dependence. To parameterize the semivariogram, an analytical semivariance function is fitted that best describes the variation. For spatially dependent samples, the semivariance functions increase as lag increases until the variance reaches a constant value called the "sill" (Fig. 5). The lag value at which the sill is reached is called "range" and gives the maximum distance up to which the measured soil property is spatially correlated. Thus, for truly random variations, the sill variance is reached at lag 0. The semivariance parameter at lag 0 is also called "nugget variance." It serves as a good aggregate estimator of error due to sampling uncertainty, unresolved variations, and analytical errors.

However, estimates of second-order properties should be carried out with some care because just as there are rules for sampling spatially independent sample sets, there are rules for measuring the variability of spatially dependent samples. The Nyquist sampling theorem states that any variation potentially resolved by the sampling support needs to be sampled with a frequency of greater than twice the wavelength or characteristic length of the variation. If the rule is not obeyed, the signal cannot be adequately reconstructed or worse the signal is misrepresented by an *alias* signal. For soil science, this means that under-sampling a deterministic variation can result in a biased estimate of the mean and variance and potentially a misrepresentation of a pattern. An example for this is the soil moisture distribution across a plowed field (Fig. 6). The solid line in Fig. 7 shows the original, exhaustive 230-point sample set for soil moisture. The sample interval was 10 cm. The stippled line is based on 32 data points randomly selected from the original. This represents an average sampling interval of just under the wavelength, associated with a ridge and furrow pattern. The range of the original is 50 cm, but 90 cm for the under-sampled variation. In addition to the overestimation of the range distance, the sill variance is overestimated by 30%. This error is not captured by the nugget variance, which usually is a good

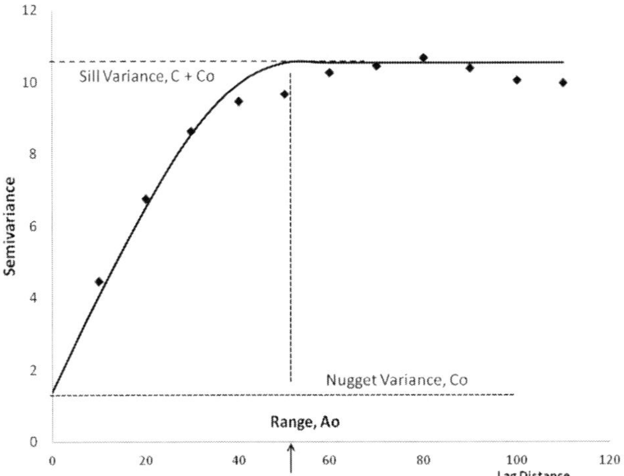

Fig. 6 Diagram of a semivariance function. There are three spatial parameters most often estimated with semivariance analyses. The distance, A_o, where the data is no longer spatially dependent is called range. The maximum semivariance, $C + C_o$, is called sill and comprises the variance due to spatial variations, C, plus the nugget variance, C_o, which is an aggregate measure of uncertainty due to analytical and sampling errors.

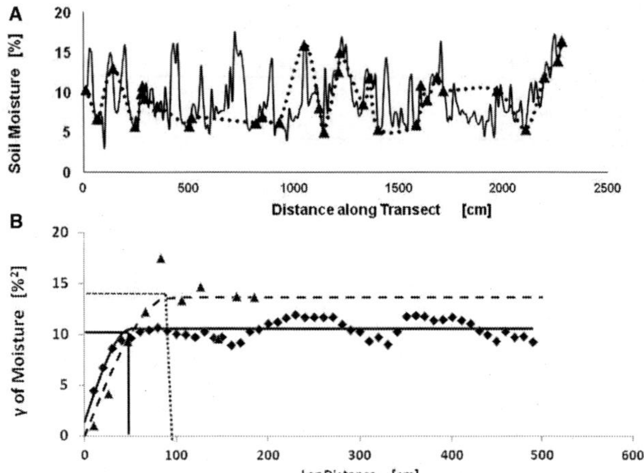

Fig. 7 (**A**) Exhaustive trace of soil moisture variation across a plowed field (solid line) with a randomly selected subset of data points (stippled line and triangular markers) at an average spacing just less of the wavelength of the ridge-furrow pattern (Gorres, unpublished data). This is a violation of the Nyquist sampling theorem. The sampled variation mimics or aliases a variation with much longer spatial characteristics. (**B**) The semivariance function of the more exhaustively sampled moisture profile (solid line and diamond markers) has a spatially correlated range of 50 cm while the semivariance of the aliased variation (triangular markers and broken line) is 90 cm long overestimating both the correlated range and the sill variance. Note that the nugget variance of the aliased signal is zero, showing that when signals are under-sampled, the nugget variance is not a good indicator of the estimation error made.

estimator of spatial estimation errors, but is zero for the under-sampled moisture variation in this case.

For complete sample sets where sample points abut each other, the Nyquist theorem does not need to be invoked as the sample support would smooth out or integrate over any small unresolved variations. This kind of smoothing can be accomplished by taking several samples from each location, followed by combining and homogenizing the samples. Prior to analysis satellite technology may also provide unabridged sample sets, where small-scale variations in luminosity are integrated over the pixel support. Apart from several specialized measurements (e.g., light detection and ranging [LIDAR]), environmental satellite imagery is of low resolution and suitable only for large-scale investigations. Much progress needs to be made before remotely sensed data can replace *in situ* sampling for the analysis of some properties.

While it is clearly important to obtain as many samples as possible to satisfy needs for accuracy and improved inferential power, the cost of sampling and analysis often competes with the requirements of the central limit theorem and the Nyquist sampling theorem. Cost may prohibit high-frequency sampling across an area of interest. As a result a bias toward long wavelength, variations may develop, exposing the estimates to aliasing of patterns. To include and improve the estimates at shorter scales, random sampling schemes have been devised that include nests of closely spaced variations.[26] In their study of physical and biological properties, 25 × 25 m plots were subdivided into twenty-five 5 × 5 m plots. A random core sample was taken from each plot. However, at five locations nine samples were collected in 3 × 3 sample nests. The nested samples were 10 cm apart. In this way, variations at the meter and centimeter scale could be resolved.[34]

Kriging

To estimate regional means of spatially correlated soil properties, the semivariance values are used in a spatial interpolation method called kriging. Kriging gives unbiased means and minimizes the error of estimation of regional parameters. The method was developed as a way to estimate the yield of ore deposits. In geological investigations, depositional patterns are large scale and aliasing may not be as much of a problem as in soils. Thus, the true error of the estimation of regional means with kriging depends strongly on the quality of the data collected and the care given to considering the spatial scale of patterns, even though kriging will mathematically minimize estimation error.

Some Applications of Spatial Estimation Methods

Spatial variations in soil properties are particularly important when the agricultural or environmental management require targeted measures. For example, nutrients need to be managed differently on a sandy soil than a clay soil. Sometimes environmental risks can be better assessed when a spatially explicit distribution of soil properties are available. There is a wide range of applications that use the spatial variability of soils in order to optimize estimates of properties or assess environmental risk.

One of the foremost commercial applications of the analysis of spatial patterns of soil properties is in variable rate agriculture, also known as precision agriculture. Improved productivity has been attributed to applying fertilizers at rates tailored to the previous year's yields or soil fertility status measured and spatially referenced using GPS. In field areas where fertility is low, more fertilizer is applied than in areas where it is high. Not only does this approach improve productivity and economic return, but it can also reduce nutrient loading to ground and surface water.

Losses of nutrients from fields are not only a risk to water quality but they also represent valuable farm resources. Spatial variations of soil properties are thus considered in farm nutrient management plans. Soil management tools such as the N leaching index and P index are used to estimate the risk of nutrient losses. In the U.S.A., many states assign the nitrate leaching index based on the hydrologic group of the soil, a landscape scale measure of infiltration potential, as well as the regional rainfall patterns.

At much larger scales, the spatial variability of soil alkalinity has been incorporated into approaches to regulating acid air pollution. The critical load concept is central to this. The critical load of acid deposition is reached when observable changes in the ecosystem occur, i.e., when the acid challenge can no longer be buffered by the ecosystem. The critical load concept relates the atmospheric deposition of SOx and NOx to the rate of base cation release from soil minerals during weathering as well as other factors. Weathering is a soil-forming process and thus climate, parent material, organisms, and topography impose a spatial pattern, which has to be accounted for when mapping critical loads of acid deposition for a region. The typical resolution of critical load maps is one to several kilometers. Once critical loads have been determined, exceedance of the critical load for a map unit can be calculated and emissions can be regulated.

Increasingly, variations in soil properties are being used in combination with soil process models such as the Leaching Estimation and Chemistry Model (LEACHM) or Nitrogen Loss and Environmental Assessment Package (NLEAP) in pollution risk assessment. For some applications, it is sufficient to treat the distribution of soil properties as independent. In this case, the means and measures of uncertainty of a particular soil process can be simulated by randomly selecting soil properties from their respective cumulative distribution functions as inputs to a soil process model. The repetition of these simulations gives a distribution of values for the soil process from which statistical moments can be derived. For example, LEACHN, a submodel of LEACHM that estimates nitrogen and carbon dynamics, was parameterized using soil properties randomly selected from cumulative distribution functions of measured soil organic matter, bulk density, texture, and field capacity to simulate the mean of NO_3-N in water extracted at a hypothetical production well.[35] This approach also allowed an estimate of the probability that the water pumped at the well exceeded the EPA drinking water standard of 10 mg NO_3-N/L under different land use scenarios.

Watershed-scale model estimates of runoff and erosion have been improved using spatially explicit representations of the landscape. These models use remotely sensed data or spatially referenced soil survey data to establish a spatial context rather than assuming random distributions of soil property values across the landscape. For example, the incorporation of soil information into a watershed model improved the estimate of storm hydrographs by accounting for delays in runoff contributions to stream flow.[36]

CONCLUSIONS

Soils exhibit high levels of variability. CVs vary from less than 10% to greater than 100%. The magnitude of the CV varies with the scale of measurement. Classical statistics can parameterize spatial distributions of spatial variations in terms of first-order statistical moments when the variations are spatially independent. However, more often than not, variations are spatially dependent making it necessary to evaluate a spatial correlation between points. Any analysis of spatial heterogeneity that does not consider the properties of spatial patterns in the design of measurement and monitoring schemes will likely produce erroneous information.

REFERENCES

1. Warrick, A.W. Spatial variability. In *Environmental Soil Physics*; Hillel, D., Ed.; Academic Press: New York, 1998; 655–676.
2. Amador, J.A.; Glucksman, A.; Lyons, J.; Görres, J.H. Spatial distribution of phosphatase activity within a riparian forest. Soil Sci. **1997**, *162* (11), 808–825.
3. Görres, J.H.; DiChiaro, M.; Lyons, J.; Amador, J.A. Spatial and temporal patterns of soil biological activity in a forest and an old field. Soil Biol. Biochem. **1998**, *30* (2), 219–230.
4. Subler, S.; Baranski, C.A.; Edwards, C.M. Earthworm additions increased short term nitrogen availability and leaching in two grain-crop agro ecosystems. Soil Biol. Biochem. **1997**, *29* (3), 413–421.
5. Dominguez, J.; Bohlen, P.J.; Parmelee, R.W. Earthworms increase nitrogen leaching to greater soil depths in row crop agroecosystems. Ecosystems **2004**, *7* (6), 672–685.
6. Jenny, H. *Factors of Soil Formation: A System of Quantitative Pedology*; McGraw Hill Book Company: New York, NY, USA, 1941; 281 pp.
7. Beare, M.H.; Coleman, D.C.; Crossley, D.A.; Hendrix, P.F.; Odum, E.P. A hierarchical approach to evaluating the significance of soil biodiversity to biochemical cycling. Plant Soil **1995**, *170* (1), 5–22.
8. Rouatt, J.W.; Katznelson, H.; Payne, T.M.B. Statistical evaluation of the rhizosphere effect. Soil Sci. Soc. Am. J. **1960**, *24* (4), 271–273.
9. Savin, M.C.; Görres, J.H.; Amador, J.A. Dynamics of soil microbial and micro faunal communities as influenced by macropores, litter incorporation, and the anecic earthworms *Lumbricus terrestris* (L.). Soil Sci. Soc. Am. J. **2004**, *68* (1), 116–124.
10. Parkin, T.B. Soil microsites as a source of denitrification variability. Soil Sci. Soc. Am. J. **1987**, *51* (5), 1194–1199.
11. Drury, C.F.; Voroney, R.P.; Beauchamp, E.G. Availability of NH_4^+-N to microorganisms and the soil internal N-cycle. Soil Biol. Biochem. **1991**, *23* (2), 165–169.
12. Batte, M.T. Factors influencing the profitability of precision farming systems. Soil Water Conserv. **2000**, *55* (1), 12–18.
13. Parkin, T.B.; Starr, J.L.; Meisinger, J.J. Influence of sample size on measurement of soil denitrification. Soil Sci. Soc. Am. J. **1987**, *51* (6), 1492–1501.
14. Soil Survey Staff. *Soil Taxonomy: A Basic Guide for Soil Classification for Making and Interpreting Soil Surveys*, 2nd Ed.; USDA-NRCS: Washington, DC, 1999.

15. Nearing, M.A.; Pruski, F.F.; O'Neal, M.R. Expected impacts of climate change on soil erosion rates. Soil Water Conserv. **2005**, *59* (1), 43–50.
16. Kallenrode, M.-B. Droughts in the Western U.S. In *Space Physics: An Introduction to Plasmas and Particles in the Heliosphere Magnetospheres*; Springer: Berlin, Germany, 2004; 397–398.
17. Sexstone, A.J.; Revsbech, N.P.; Parkin, T.B.; Tiedje, J.M. Direct measurement of oxygen profiles and denitrification rates in soil aggregates. Soil Sci. Soc. Am. J. **1985**, *49* (3), 645–651.
18. Fanning, D.S.; Fanning, M.C.B. The factors of soil formation—Overview. In *Soil Morphology, Genesis and Classification*; John Wiley and Sons: New York, USA, 1989; 287–290.
19. McBratney, A.B.; Webster, R. Detection of a ridge and furrow pattern by spectral analysis of crop yield. Int. Stat. Rev. **1981**, *49* (1), 45–52.
20. Raghavan, G.S.V.; McKyes, E.; Chassé, M.; Merineau, F. Development of compaction patterns due to machinery operation in an orchard soil. Canad. J. Plant Sci. **1976**, *56* (3), 505–509.
21. Afzal, M.; Adams, W.A. Heterogeneity of soil mineral nitrogen in pasture grazed by cattle. Soil Sci. Soc. Am. J. **1991**, *56* (4), 1160–1166.
22. Ellenberg, H.; Weber, H.E.; Duell, R.; Wirth, V.; Werner, W.; Paulissen, D. Indicator values of plants in Central Europe. Scripta Geobotanica **1992**, *18*, 160–166.
23. Amador, J.A.; Görres, J.H. Microbiological characterization of the structures built by earthworms and ants in an agricultural field. Soil Biol. Biochem. **2007**, *39* (8), 2070–2077.
24. Ettema, C.H.; Wardle, D.A. Spatial soil ecology. Trends Ecol. Evol. **2002**, *17* (4), 177–183.
25. Görres, J.H.; Amador, J.A. Partitioning of habitable pore space in earthworm burrows. J. Nematol. **2010**, *42* (1) 68–72.
26. Görres, J. H.; Savin, M.; Amador, J.A. Dynamics of carbon and nitrogen mineralization, microbial biomass, and nematode abundance within and outside the burrow walls of anecic earthworms (*Lumbricus terrestris*). Soil Sci. **1997**, *162* (9), 666–671.
27. Hattori, T.; Hattori, R. The physical environment in soil microbiology: An attempt to extend principles of microbiology to soil microorganisms. CRC Crit. Rev. Microbiol. **1976**, *4* (4), 423–461.
28. Mummey, D.; Holben, W.; Six, J.; Stahl, P. Spatial stratification of soil bacterial populations in aggregates of diverse soils. Microb. Ecol. **2006**, *51* (3), 404–411.
29. Elliott, E.T.; Anderson, R.V.; Coleman, D.C.; Cole, C. V. Habitable pore space and microbial trophic interactions. Oikos **1980**, *35* (3), 327–335.
30. Hassink, J; Bowmann, L.A.; Zwart, K.B. Relationship between habitable pore space, soil biota and mineralization rates in grassland soils. Soil Biol. Biochem. **1993**, *25* (1), 47–55.
31. Neher, D.A.; Weicht, T.R.; Savin, M.C.; Görres, J.H.; Amador, J.A. Grazing in a porous environment. 2. Nematode community structure. Plant Soil **1999**, *212* (1), 85–99.
32. Görres, J.H.; Savin, M.C.; Neher, D.A.; Weicht, T.R.; Görres, J.H. Grazing in a porous environment: 1. The effect of soil pore structure on C and N mineralization. Plant Soil **1999**, *212* (1), 75–83.
33. Webster, R. Is soil variation random? Geoderma **2000**, *97* (3–4), 149–163.
34. Amador, J.A.; Wang, Y.; Savin, M.C.; Görres, J.H. Small-scale variability of physical and biological soil properties in Kingston, Rhode Island. Geoderma **2000**, *98* (1), 83–94.
35. Görres, J.H.; Gold, A. Incorporating spatial variability into GIS to estimate nitrate leaching at the aquifer scale. J. Environ. Qual. **1996**, *25* (3), 491–498.
36. Zhu, A.X.; Mackay, D.S. Effects of spatial detail of soil information on watershed modeling. J. Hydrol. **2001**, *248* (1), 54–77.

Soil: Taxonomy

Mark H. Stolt
Department of Natural Resources Science, University of Rhode Island, Kingston, Rhode Island, U.S.A.

Abstract
Soil, like any natural resource, needs to be classified in order to best utilize and conserve the resource. As long as humans have lived in communities, soils have been classified. In most cases, the classification approach is strongly influenced by the concept that the most important use of soils is for agriculture. Taxonomic classifications of the soil resource, however, can be applied to any number of use and management purposes when the classification is based on the physical, chemical, and morphological properties of the soils. As such, use and management assessment of soil resources are made through soil surveys based on taxonomic soil classifications. In this entry, I discuss the development and architecture of the soil taxonomic system used in the United States for the past 50 years—Soil Taxonomy. The discussion includes references and comparisons to other national classification systems, and the international soil classification system known as the World Reference Base. Recent advances in soil classification and current issues are also discussed.

INTRODUCTION

One could argue that of all the natural resources—soil, water, and air are the only essential resources. Soil, like any natural resource, needs to be classified in order to best utilize and conserve the resource. Fanning and Fanning[1] suggested fairly simple reasons for classification: to provide a means to organize our knowledge; and to provide a system to retrieve the information for any number of purposes. Soils have been classified for almost as long as humans have lived in communities.[2] One of the primary reasons was to assess the land for agriculture; those with the best soils for agriculture were taxed the greatest amount because they owned the best land. Since such classification approaches were not designed around the natural physical, chemical, and morphological properties of the soils (taxonomic properties), but based on their productivity or other use assessments, these soil classification approaches are termed interpretive classification systems.[1]

Agriculture is also the root of most soil taxonomic systems. For example, in Soil Taxonomy (the current system used in the United States and many other areas of the world), in deciding which properties to use as criteria for classifying the soil, the property with the most effect on agriculture is the one that is chosen.[3] Although throughout history agriculture has driven the design of a soil classification system, there are many other uses of soil taxonomy. This is clear in the title of the U.S. soil classification system; although typically referred to as just "Soil Taxonomy," the complete title is "Soil Taxonomy: A Basic System for Making and Interpreting a Soil Survey." In every U.S. soil survey there is a list of use and management interpretations for each soil taxa. These include many nonagronomic interpretations such as use of the soil for construction materials, recreational purposes, supporting roads and buildings, waste disposal treatment, constructing earthen dams, silvaculture, etc. Thus, the ultimate purpose of soil taxonomy is to provide a means to assess the potential of the soil resource to serve any number of purposes through soil surveys. In this entry, we will focus on taxonomic soil classification systems.

SOIL TAXONOMY OVER THE LAST CENTURY

Soil science and taxonomy, as we know them today, are little more than a century old. Prior to the late 1800s, soils were considered a surface mantle of loose and weathered rock.[4] V.V. Dokuchaiev, a Russian geologist working mostly between 1870 and the turn of the 20th century, articulated our current concept of soil: a natural body that has formed over time through a series of physical, chemical, and biological processes acting upon a given soil parent material. These processes are governed by the landscape setting, climate, and soil biology. Jenny[5] called these the five state factors of soil formation: climate, organisms, relief, parent materials, and time. Thus, soils with similar parent materials and biology that have formed under similar climates and landscape settings, will have similar properties, and thus taxonomic classification.

Of the taxonomic soil classification systems prior to the turn of the 20th century, Dokuchaiev's system was the most influential. His works were translated to German by Glinka[6] and later translated from German to English by C.F. Marbut.[1,7–9] Marbut considered many of the ideas and concepts of soils at that time relative to mapping

Encyclopedia of Natural Resources DOI: 10.1081/E-ENRL-120047494

soils and developed a classification system for the United States that was published in its final form in 1935.[10] Followers of Dokuchaiev, especially Sibertsev,[11] further developed Dokuchaiev's system that later became the foundation of the soil classification system that was used in the United States between 1938 and 1965.[1] The 1938 Yearbook system[12] was based primarily on Dokuchaiev's concepts of soil orders and were named Zonal, Intrazonal, and Azonal soils in the 1938 Yearbook.[13] Schaetzl and Anderson[14] called the 1938 Yearbook the first serious attempt of classifying soils in the United States. Zonal soils essentially matched the concept of "normal soils" described by Dokuchaiev and more fully articulated by Marbut:[15] soils that have developed in better drained conditions over enough time to be considered to have reached maturity under current climatic conditions. Azonal soils were the young soils developing in areas such as shallow to bedrock, floodplains, or aridic sands. Intrazonal soils were controlled by wetness or chemical composition such as salts. Examples and summaries of the 1938 Yearbook system can be found in Fanning and Fanning[1] and Schaetzl and Anderson.[14]

Although the 1938 Yearbook was used for many years in the United States, serious flaws were widely recognized in the system. There were six categories in the hierarchy, but the suborder (second) and family (fourth) categories went essentially unused.[16] Likewise, the series-level category failed in the hierarchical scheme; since it would not always match with a higher-level taxa. Some soils went unrecognized, such as some of those formed from volcanic ash or shrink-swell clays.[1] Another issue was that classification was based on unaltered (virgin) soils, thus a soil that had been cultivated or eroded was classified into a taxa that represented what it would have been like in a virgin form.[14] Because of these issues a more quantitative and less subjective system was needed.

In the early 1950s, Guy Smith began to work toward a new classification system.[17] Over the next decade, six versions (labeled approximations) of the system were circulated among soil scientists mostly in the United States.[18] The 7th approximation,[19] was circulated worldwide at the World Soil Congress in 1960. At first there was little acceptance of the new system, but Simonson[20] noted regarding his paper published in Science "evidence of growing interest was indicated by requests for a few thousand reprints of my 1962 paper." By 1965, the system was adopted for USDA-SCS soil surveys and a decade later published as Soil Taxonomy.[3]

A list of guidelines for the development of an effective soil classification system was given within the 7th Approximation[19] and later revised for inclusion in Soil Taxonomy.[3] In general, the guidelines suggested that the system be flexible enough that soils that were not currently identified could easily be added; the system could be used worldwide; the properties used to classify the soil will be measured (not necessarily by instruments); the focal point of the properties measured are soil genesis related; if a choice needs to be made between which of two properties to use in the classification criteria, the one with the most affect on plant growth be used; and properties that are easily changed by anthropogenic activities such as plowing, be only used at the lowest of levels when classifying the soil.

THE SOIL THAT WE CLASSIFY

Our concept of soil depends on our background and experiences.[21] For example, someone with a potted plant may say the plant is supported by soil; and many would agree. A soil scientist would argue that this concept of soil is analogous to saying a cup of water taken from the lake shore is a lake. No one would agree with that! Knox[22] reviewed the various bodies of soil materials that could be considered soil. These ranged from individual particles, like a sand grain, to landscape-level bodies that could be mapped out at a 1:24000 scale. That pot of soil materials described above would be considered a "hand specimen" by Knox.[22] In this entry, we will use the USDA-NRCS[23] definition of soil: a collection of natural bodies; composed of mineral and organic materials; that supports growth of plants out-of-doors or show evidence of soil forming (pedogeneic) processes. Within this context, our collection natural bodies are the series of horizons that vary with depth and over an area of the landscape on the order of 1 to 10 m^2. Such a three-dimensional area of soil is called a pedon.[19,21,24] Most soils are described using a 2-dimensional view of a pedon, called a soil profile (Fig. 1). In most current soil taxonomic systems, a pedon is the smallest soil unit that is classified.

SOIL TAXONOMY INTO THE 21ST CENTURY

Soil Taxonomy has six hierarchical categories: order, suborder, great group, subgroup, family, and series (Table 1). The first five categories are set up as a key so that every soil can be classified, the last taxa in the key is the default. In most cases, classification to the order, great group, or subgroup level in mineral soils is predicated on the presence or absence of diagnostic horizons. Those diagnostic horizons forming at the soil surface are called epipedons. Subsurface diagnostic horizons are typically in the position of the B horizons in the profile. The standard for the epipedon is the mollic because soils with mollic epipedons typically are excellent for agriculture. Most of the soils in the Great Plains in the United States and the Steppe regions in Russia have mollic epipedons. These are the soils that Dokuchaiev studied in Russia and formulated his early concepts of soils. Mollic epipedons are thick (typically 25 cm or more), dark in color, rich in organic matter, and rich in basic cations (base saturation >50%). Umbric epipedons are similar to the mollic except more acidic, having a lower base status. Histic epipedons are typically 20–40 cm thick and dominated by organic soil materials. There are also folistic

Fig. 1 (**A**) A Typic Epiaquod in Rhode Island. The tape is in inches. The spodic diagnostic subsurface horizon varies in thickness and depth. Near the tape the spodic horizon occurs from 10 to 20 inches. The soil is saturated to the surface for parts of year, thus the aquic suborder. The water is held, or perched, above a restrictive layer (epi) starting at 36″. There are no other diagnostic features (Typic). (**B**) An Umbric Endoaquult in Pennsylvania. The soil has an umbric epipedon, an argillic diagnostic subsurface horizon, and a base saturation <35%. The soil is saturated throughout the profile and to the soil surface (Endoaqu) for extended periods of the year. The small marks on the tape are 10 cm increments. The large marks are in feet. (**C**) Fluvaquentic Eutrudept on a Pennsylvania floodplain. The tape is in cm. Fluvaquentic indicates the soil is saturated during the year for a considerable amount of time above 60 cm and there is buried soil organic carbon (note the buried A horizon between 80 and 90 cm). The soil has a cambic diagnostic subsurface horizon (20 to 80 cm) with a base saturation >60% (Eutr). The climate is humid, thus a Udept. (**D**) Calcic Haplustalf in Texas. The small increments on the tape are 10 cm apart. The larger marks are one-foot apart. The soil has calcic and argillic diagnostic subsurface horizons. The base saturation is greater than 35% and the moisture regime is semiarid (ustic).

Table 1 The architecture of Soil Taxonomy

There are six categories from the highest (least detail) to lowest (most detail) level: order, suborder, great group, subgroup, family, and series. The names for classes in the order, suborder, great group and subgroup categories are all coined terms using Greek, Latin, French, or similar roots. The soil orders are listed in the order they key out, a limited number of examples are provided for the other categories.

Order: Names of the twelve orders consist of three or four syllables and every name ends in the suffix "sol."

Order name	Formative element	Derivation
Entisol	ent	Nonsense syllable, think of recent
Vertisol	ert	L. *Verto*, turn, think of invert
Inceptisol	ept	L. *inceptum*, think of inception
Aridisol	id	L. *aridus*, dry, think of arid
Mollisol	oll	L. *mollis*, soft, think of mollify
Spodosol	od	Gk. *Spodos*, wood ash, think of Podzol*
Alfisol	alf	Nonsense syllable, think of Pedalfer**
Ultisol	ult	L. *ultimus*, last, think of ultimate
Oxisol	ox	F. *oxide*, think of oxide
Andisol	and	Gk. *Andes*, think andesite
Histosol	ist	Gk. *histos*, tissue, think of Histology
Gelisol	el	L. *gelare*, to freeze, think of gel

*Podzol-classic term used for soils with spodic horizons; **Pedalfer-one of the two taxa in Marbut's (1935) highest category.

Suborder: The name of every suborder is a two-syllable term consisting of a prefix syllable with some specific connotation plus the formative element from the order to which the suborder belongs. The examples are for Ultisols, Inceptisols, and Vertisols, respectively.

Formative element	Connotation	Example
Aqu	Aquic moisture regime	Aquult
Ud	Udic moisture regime	Udept
Ust	Ustic moisture regime	Ustert

Great group: The name of each great group is constructed by adding a second prefix element with a specific connotation to the two-syllable term which is the name of the suborder to which the great group belongs.

Formative element	Connotation	Example
Umbra	An umbric epipedon	Umbraquult
Eutr	High base status	Eutrudept
Calci	A calcic horizon	Calciustert

Subgroup: Names of subgroups are binomial. An adjective (sometimes two) is used to modify the name of the great group to give the name of each subgroup within that great group.

Adjective	Connotation	Example
Plinthic	Presence of plinthite	Plinthic Umbraquult
Arenic	Thick sandy surface	Arenic Eutrudept
Lithic	Shallow to rock	Lithic Calciustert

(*Continued*)

Table 1 (*Continued*) The architecture of Soil Taxonomy

Family: A sequence of descriptive adjectives is used in family names. The common examples are used to describe particle-size classes, mineralogy classes, activity classes, and soil temperature classes.

An example is Coarse-loamy, mixed, active, mesic, Typic Dystrudept. Coarse-loamy (particle-size class), mixed (mineralogy class), active (activity class), mesic (soil temperature class), Aquic Dystrudept (subgroup).

Series: is a family that falls within a given range of characteristics. Series names are typically derived from towns or localities that are near where the soils are found (i.e., Newport). The range in characteristics are related to morphological properties such as thickness of the B horizons, soil colors, rock fragment contents, soil textures, etc. The range in characteristics are not found in Soil Taxonomy but are accessed online: http://soils.usda.gov/technical/classification/osd/index.html

(unsaturated histic epipedons), mellanic (associated with volcanic ash), plaggen (heavily manured for centuries), and anthropic (P rich from long-term human occupation) epipedons. In general, the ochric epipedon is the default, meaning soils that do not meet the criteria for the other six epipedons have an ochric epipedon (very new soils may have no epipedon).

In Soil Taxonomy, there are 18 diagnostic subsurface horizons. Only the more common are briefly discussed here so that the reader can understand the concept and follow the examples in Table 1. Most diagnostic subsurface horizons are related to the accumulation of soil constituents in the B horizon position of the soil profile. The accumulations may be of silicate clay (argillic), sesquioxides bound to organic matter (spodic), calcium carbonate (calcic), gypsum (gypsic), and salts more soluble than gypsum (salic). Certain diagnostic subsurface horizons are dense or cemented by the constituent that has accumulated such as petrocalcic, petrogypsic, or a duripan (cemented by silica that has reprecipitated in the B horizon). Younger soils may not have had enough time to form any of the diagnostic horizons previously mentioned but have B horizons recognized by a simple change in color. These "color" B horizons typically meet the criteria for a cambic diagnostic horizon.

In keying out soils to the order level, Gelisols are the first possibility. The important criterion is the presence of permafrost (generally within a meter of the soil surface). Histosols key out next, these soil are dominated by organic soil materials (at least 40 cm) in the upper 80 cm. Spodosols follow, having spodic diagnostic horizons. Andisols have andic soil properties (high exchange capacity and water holding capacity). Oxisols are extremely weathered and dominated by Fe and Al oxides (oxic horizon). Vertisols are dominated by shrink-swell clays that results in a self mixing soil. Aridsols may have a number of diagnostic horizons such as argillic, calcic, cambic as long as they are in aridic (dry) moisture regimes. Ultisols have argillic horizons and low base saturation (<35%). Mollisols have mollic epipedons and a base saturation throughout the profile greater than 50%. These soils may have calcic or argillic horizons. Alfisols almost always have argillic horizons, but unlike Ultisols their base saturation is >35%. Inceptisols are weakly developed soils typically with cambic horizons. Entisols are the least developed, characterized by a lack of diagnostic horizons or epipedons other than ochric, and key out last.

Because soil moisture (or lack of) has a dramatic effect on soil morphology, suborders typically bring the wetness attributes or soil moisture states related to climate (soil moisture regimes) into the classification (see examples in Table 1 and Fig. 1). These moisture regimes include aquic (wet), udic (humid), ustic (semiarid), and xeric (dry parts of the year and moist other parts).

There were only a few major changes to Soil Taxonomy over the years. In 1975, there were 10 orders; that changed to 12 with the addition of Gelisols (soils with permafrost) and Andisols (soils formed in volcanic ash type materials). The definition of soil was changed in the 2nd edition.[25] Previously by definition, a soil was required to be able to support rooted plants.[3] In the new edition, soils, as long as they showed evidence of pedogenesis, were considered soils even if they did not support rooted plants. This departure from the agricultural roots of Soil Taxonomy was designed to accommodate soils in the arctic that were permanently covered with snow or ice[26,27] or soils that were permanently under water (subaqueous soils).[28]

In 1983, field versions of Soil Taxonomy were first published. The field versions essentially just included what was needed to key the soil out to the family level, much of the supplemental materials of the classification system was left out. These versions were called the *Keys to Soil Taxonomy*. As we learned more and more about the soils, changes in the taxonomy were introduced in revised versions of *Keys to Soil Taxonomy*. For example, taxa were included in the 11th edition of *Keys to Soil Taxonomy*[23] to accommodate subaqueous soils.[28] Although by design Soil Taxonomy was meant to be flexible enough so that new soils could be added when deemed necessary,[3] some argue that over time it might be too big and cumbersome to be useful any longer. For example, the first *Keys to Soil Taxonomy*[29] was a 4″ × 9″ paperback that easily fit in your back pocket for use in the field (<250 pages). By 1996, the

7th edition of *Keys to Soil Taxonomy* in the same 4″ × 9″ format was well over 600 pages.

One area of soils that continues to be difficult to classify in Soil Taxonomy are the anthropogenic soils (i.e., soils developed in materials moved by man).[30] In an attempt to resolve this issue, an International Committee on Anthropogenic Soils (ICOMANTH) was formed with the charges defining appropriate classes in Soil Taxonomy for soils that have their major properties derived from human activities. The first activities of this committee were reported in the 1st circular letter dated 1995 (http://clic.cses.vt.edu/icomanth/03-AS_Circulars.pdf). Efforts of the committee resulted in new horizon designations for soils with restrictive properties because of the presence of a continuous man-made material such as cement (master horizon designation M) and horizons containing human artifacts such as glass, brick, or cement (subordinate horizon designation u). These changes appeared in the 10th edition of *Keys to Soil Taxonomy*.[31] Suggestions for classifying such soils have primarily focused on subgroups such as Spolic, Urbic, or Garbic[1,32] or a separate order with Soil Taxonomy—Anthrosols.[33] Efforts toward classifying anthropogenic soils continue through ICOMANTH such that at the present time there are seven circular letters describing proposals, changes, and related activities of the committee (http://clic.cses.vt.edu/icomanth/circlet.htm).

OTHER NATIONAL AND INTERNATIONAL CLASSIFICATION SYSTEMS

There are numerous other taxonomic soil classification systems. Buol et al.[34] reviewed a number of these including the Russian, French, Belgian, British, Canadian, Australian, Brazilian, and Chinese. Many of these systems take elements directly or indirectly from Soil Taxonomy.[25] For example, at the order level the Australian system uses "Organosols" instead of Histosols.[35] Seven of the diagnostic horizons or features in the Chinese system are taken directly from Soil Taxonomy.[36] The Chinese and Australian systems have soil orders for anthropogenic soils.[35,36]

The World Reference Base[37] is the only true international soil classification system. This system, known as the WRB, was endorsed and adopted by the International Union of Soil Sciences (IUSS) as the classification system for soil correlation and international communication. The intention of the WRB was not to provide an alternative for any national soil classification system, but to serve as a soil classification translator for communication at an international level.[37] During the development of the WRB, a focused effort was put forth to include as much nomenclature from Soil Taxonomy,[25] and other major national soil classification systems, as possible.

The WRB is a two-tiered system. At the highest category are 32 Reference Soil Groups. A number of these are essentially the same as the soil orders in Soil Taxonomy; examples of these include Vertisols, Histosols, Podzols (Spodosols), Cryosols (Gelisols), and Chernozems (Mollisols). Like Soil Taxonomy,[25] diagnostic horizons are used as criteria in many of the taxa. Although their definitions are not identical, a number of the WRB diagnostic horizons are conceptually the same as those in Soil Taxonomy. Some examples in include mollic, ochric, histic, gypsic, calcic, and cambic.

The second tier of the WRB is a series of qualifiers that are added as a prefix or suffix to the reference soil groups to identify in greater detail properties of the particular soil. Each reference soil group is assigned a list of qualifiers that may be used. In the WRB[37] an example is given for a Cryosol soil reference group: "Histic Turbic Cryosol (Reductaquic, Dystric)." Histic and Turbic are the prefix qualifiers indicating that the Cryosol has a histic epipedon and evidence of cryoturbation (mixing by freeze-thaw), respectively. Reductaquic and Dystric are the suffixes indicating the soil is both saturated and reduced, and has a base saturation less than 50%, respectively. Unlike Soil Taxonomy,[25] climatic parameters are not applied in the classification of soils in the WRB and the system is not designed for mapping soils at resource-level scales.

SUMMARY AND CONCLUSIONS

Soil is one of the basic and essential natural resources. Thus, some form of soil classification has been around since humans formed communities. The early systems were primarily based on a use assessment of the soils for a single purpose such as agriculture productivity (i.e., low yields, moderate yields, or high yields). These systems are often referred to as interpretive soil classifications. Taxonomic classification of soils is based on the natural physical, chemical, and morphological properties of the soil and can provide many use and management interpretations. Application of taxonomic classification for soil resources is through the use of soil surveys. The best example of a taxonomic system for mapping and assessment of the soil resource is Soil Taxonomy. This is a hierarchical system that is set up like a key so that every soil can be classified. There are six categories in the system with order the highest (least amount of detail) and series the lowest (greatest amount of detail). Although Soil Taxonomy is used worldwide, many countries have their own soil classification system. There is also a truly international system called the World Reference Base (WRB). The WRB is designed as an international translator for systems such as Soil Taxonomy and was not meant to provide classifications for resource soil mapping.

Current issues with Soil Taxonomy appear to be related to the lack of significant taxa for anthropogenic soils, its field applications, and the rapidly expanding number of taxa. Issues with field applications center around the need for considerable laboratory data to correctly classify a soil. A possible resolution is to develop guidelines based on reasonable assumptions that can assist the user in applying Soil Taxonomy in the field without any laboratory data. As we learn more and more about soils, the number of taxa in Soil Taxonomy continues to grow. Currently, proposals are being considered to add taxa to accommodate the classification of some anthropogenic soils. Some users feel that continued additions leaves the classification system too cumbersome to key through to find the correct taxa. Whether this is a problem or not is debatable, but correcting such a problem may require reformatting the hierarchical system. An approach to resolve this issue that is currently being considered is new classification system using some of the format of the WRB and some of Soil Taxonomy. As the world's population continues to grow resource managers will turn more and more to soils information to best utilize this essential resource. As such, developing an understanding of soil taxonomy will continue to be a need for natural resource users and managers.

REFERENCES

1. Fanning, D.S.; Fanning, M.C.B. *Soil Morphology, Genesis, and Classification*; John Wiley and Sons: New York, NY, 1989.
2. Simonson, R.W. Soil classification in the United States. Science **1962**, *137*, 1027–1034.
3. Soil Survey Staff. *Soil Taxonomy: A Basic System of Soil Classification for Making and Interpreting Soil Surveys, USDA-NRCS Agricultural Handbook 436*; U.S. Government Printing Office: Washington, DC, 1975.
4. Simonson, R.W. Historical aspects of soil survey and soil classification. Part 1. 1899–1910. Soil Survey Horizons **1986a**, *27* (1), 3–11.
5. Jenny, H. *Factors of Soil Formation*. McGraw Hill Book Co. Inc.: New York, NY, 1941.
6. Glinka, K.D. *Die Typen der Bodenbildung: Ihre Kalssification und Geographische Berbreitung*. Gebruder Borntrager: Berlin, 1914.
7. Marbut, C.F. *The Great Soil Groups of the World and Their Development. Translated from Glinka, 1914*. Edwards Brothers: Ann Arbor, MI, 1927.
8. Smith, G.D. Historical development of soil taxonomy-background. In *Pedogenesis and Soil Taxonomy*. Wilding, L.P., Smeck, N.E.; Hall, G.F.; Eds.; Elsevier, Amsterdam, 1983, 23–49 pp.
9. Simonson, R.W. Historical aspects of soil survey and soil classification. Part 2. 1911–1920. Soil Surv. Horizons **1986b**, *27* (2), 3–9.
10. Marbut, C.F. Soils of the United States. In *Atlas of American Agriculture*; Baker, O.E., Eds.; USDA, U.S. Government Printing Office: Washington, D.C., 1935; 12–15 pp.
11. Sibertsev, N.M. *Russian Soil Investigations. Israel Programs for Science Translations Jerusalem. 1966. Translated from Russian by N. Kaner*; U.S. Department of Commerce: Springfield, VA, 1901.
12. Baldwin, M., Kellogg, C.E.; Thorp, J. Soil classification. In *Yearbook of Agriculture: Soils and Men*; USDA, U.S. Government Printing Office: Washington, DC, 1938; 979–1001 pp.
13. Simonson, R.W. Historical aspects of soil survey and soil classification. Part 4. 1931–1940. Soil Survey Horizons **1986c**, *27* (4), 3–10.
14. Schaetzl, R.; Anderson, S. *Soils: Genesis and Geomorphology*; Cambridge University Press: New York, NY, 2005.
15. Marbut, C.F. A scheme of soil classification. In *Proceeding of the 1st International Congress of Soil Science*; Washington, DC., 1928; 1–31 pp.
16. Simonson, R.W. Historical aspects of soil survey and soil classification. Part 5. 1941–1950. Soil Survey Horizons **1987a**, *28* (1), 1–8.
17. Cline, M. *Soil Classification in the United States. Agronomy Mimeo 79–12*; Cornell University: Ithaca, NY, 1979.
18. Simonson, R.W. Historical aspects of soil survey and soil classification. Part 6. 1951–1960. Soil Survey Horizons **1987b**, *28* (2), 39–46.
19. Soil Survey Staff. *Soil Classification, a Comprehensive System—7th Approximation*; USDA. U.S. Government Printing Office: Washington, D.C, 1960.
20. Simonson, R.W. Historical aspects of soil survey and soil classification. Part 7. 1961–1970. Soil Survey Horizons **1987c**, *28* (3), 77–84.
21. Simonson, R.W. Concept of soil. Adv. Agronomy. **1968**, *20*, 1–47.
22. Knox, E.G. Soil individuals and soil classification. Soil Sci. Soc. Am. Proc. **1965**, *29*, 79–84.
23. Soil Survey Staff. *Keys to Soil Taxonomy*, 11th Ed.; United States Department of Agriculture Natural Resources Conservation Service. Government Printing Office: Washington, D.C, 2010.
24. Simonson, R.W.; Gardner, D.R. The concepts and functions of the pedon. Transactions of the 7th International Congrees of Soil Science. 1960; 129–131.
25. Soil Survey Staff. *Soil Taxonomy: A Basic System of Soil Classification for Making and Interpreting Soil Surveys*, 2nd Edn.; USDA-NRCS Agricultural Handbook 436, U.S. Government Printing Office: Washington, DC, 1999.
26. Bockheim, J.G. Soil development rates in the Transantarctic Mountains. Geoderma **1990**, *47*, 59–77.
27. Bockheim, J.G. Properties and classification of cold desert soils from Antarctica. Soil Science Soc. Am. J. **1997**, *64*, 224–231.
28. Stolt, M.H.; Rabenhorst, M.C. Subaqueous soils. In *Handbook of Soil Science*, 2nd Ed.; Huang, P.M.; Li, Y.; Sumner, M.E., Eds.; CRC Press: Boca Raton, FL, 2011; 36.1–36.14 pp.
29. Soil Survey Staff. *Keys to Soil Taxonomy*; United States Department of Agriculture Natural Resources Conservation Service, Government Printing Office: Washington, D.C, 1983.
30. Kimble, J.M., Aherns, R.; Bryant, R. *Classification, Correlation, and Management of Anthropogenic Soils*; USDA-NRCS, National Soil Survey Center: Lincoln, NE, 1999.
31. Soil Survey Staff. *Keys to Soil Taxonomy*, 10th Ed.; United States Department of Agriculture Natural Resources Conservation Service, Government Printing Office: Washington, D.C, 2006.

32. Pouyat, R.V.; Ellland, W.R. The investigation and classification of humanly-modified soils in the Baltimore ecosystem study. In *Classification, Correlation, and Management of Anthropogenic Soils*. Kimble, J.M., Aherns, R., Bryant, R., Eds.; USDA-NRCS, National Soil Survey Center: Lincoln, NE, 1999; 141–154 pp.
33. Bryant, R.B.; Galbraith, J.M. Incorporating anthropogenic processes in soil classification. In *Soil Classification: A Global Desk Reference*; Eswaren, H.; Rice, T.; Aherns, R.; Stewart, B.A., Eds.; CRC Press: Boca Raton, FL, 2003; 57–65 pp.
34. Buol, S.W.; Southard, R.J.; Graham, R.C.; McDaniel, P.A. *Soil Genesis and Classification*, 5th Ed.; Iowa State Press: Ames, Iowa, 2003.
35. Isbell, R.F. *The Australian Soil Classification System. Australian Soil and Land Survey Handbook*; CSIRO Publishing: Collingwood, Australia, 1998.
36. Gong, Z.; Zhang, G.; Luo, G. The Anthrosols in Chinese soil taxonomy. In *Classification, Correlation, and Management of Anthropogenic Soils*; Kimble, J.M., Aherns, R., Bryant, R., Eds.; USDA-NRCS, National Soil Survey Center: Lincoln, NE, 2003; 40–51 pp.
37. Food and Agriculture Organization of the United Nations. *World Reference Base for Soil Resources*; FAO: Rome, 2006.

Sustainability and Sustainable Development

Alan D. Hecht
Office of Research and Development, U.S. Environmental Protection Agency (EPA), Washington, District of Columbia, U.S.A.

Joseph Fiksel
Center for Resilience, The Ohio State University, Columbus, and Office of Research and Development, U.S. Environmental Protection Agency (EPA), Cincinnati, Ohio, U.S.A.

Abstract

Sustainability is generally understood to mean a state or condition that allows for the fulfillment of economic and social needs without compromising the natural resources and environmental quality that are the foundation of human health, safety, security, and economic well-being. Sustainable development is both a *goal* of achieving this state for the benefit of current and future generations, and a *process* whereby innovative tools, models, and approaches are adopted to advance economic prosperity while minimizing adverse global impacts. Many case studies of business and government show that adopting sustainable practices can both enhance economic growth and protect the environment and social well-being. The need for sustainable approaches to business and economic development is made more urgent today by increased rates of degradation and depletion of natural resources and by growing threats to human health and the environment. These stresses will only increase as the global population grows from 7 billion people today to 9 billion by 2050. The practice of sustainable development is essential for reaching the next level of environmental protection, social development, and economic progress. This entry identifies some key origins of sustainability in the United States and around the world and concludes with eight steps toward achieving sustainable development.

INTRODUCTION

Sustainable development was defined in a 1987 United Nations report as "economic and social development that meets the needs of the present without compromising the ability of future generations to meet their own needs."[1] The concept of "sustainability" has been widely characterized as resting on three pillars: social well-being, economic prosperity, and environmental protection (Fig. 1).

As described in a 2011 report from the U.S. National Research Council, sustainability can be viewed as both a *goal* and a *process* that improves the economy, the environment, and society for the benefit of current and future generations.

As a *goal*, sustainability is aimed at meeting society's basic economic and social needs without undermining the natural resource base and environmental quality that are necessary for continuing to meet these needs in the future. Sustainability also describes the *process* whereby innovative tools, models, and approaches are applied in order to meet that goal, striving to advance economic prosperity while protecting natural resources. From the perspective of business and finance, sustainability as both a goal and process aims to create value for shareholders and society while efficiently using resources and minimizing adverse effects on the environment.

In the United States, the concept of sustainable development has roots that are over a century old. President Theodore Roosevelt relied on Gifford Pinchot, Roosevelt's chief forester and the "father of American conservation," for advice to develop public land policies in the late 1890s. Pinchot believed that forests should be made to serve the nation's future as well as its present. He was an early advocate of sustainable development, writing in his 1910 book, *The Fight for Conservation,* "the central thing for which conservation stands is to make this country the best possible place to live in, both for us and for our descendants."[2]

Nearly fifty years later, the National Policy and Environmental Act established in 1969 the national goal of creating and maintaining "conditions under which [humans] and nature can exist in productive harmony, and fulfill the social, economic and other requirements of present and future generations of Americans."[3]

Sustainability has been investigated worldwide, as the Club of Rome composed of European economists and scientists published "The Limits to Growth" in 1972.[4] This controversial report highlighted the negative impact of economic growth on the environment and served as a signal warning of the need for strategic planning in economic growth and environmental protection. The concept of sustainability was developed at the 1992 Earth Summit in Rio

Encyclopedia of Natural Resources DOI: 10.1081/E-ENRL-120047481

Fig. 1 The three pillars of sustainability.

de Janeiro, where 179 nations endorsed the principles of Agenda 21, which declared, "The right to development must be fulfilled so as to equitably meet developmental and environmental needs of present and future generations."[5]

Twenty years after the Earth Summit, corporate, government, and non-governmental leaders converged again in Rio de Janeiro in June 2012 to examine and reaffirm commitments to global sustainability. The meeting provided a timely opportunity to further develop a common understanding of sustainability and how it can improve the lives of people in all countries around the world.

Today sustainability is incorporated in many government goals and policy objectives. The European Union has been promoting sustainability since 2000, when it formulated the Lisbon Strategy with the goal of achieving the most competitive and dynamic knowledge-based economy in the world. The European Council has stated, "Clear and stable objectives for sustainable development would present significant economic opportunities and the potential to unleash a new wave of technology innovation, generating growth and employment."[6]

Sustainability has been advanced in the United States largely through Presidential Executive Orders such as 13423 under George W. Bush and 13514 under Barack Obama. Both orders set specific targets and goals for how the U.S. government should manage its own facilities. Federal procurement policies also reflect sustainability goals, their requirements for purchasing green and energy-efficient products.[7]

The United States has recently sought to promote sustainability internationally through the first presidential "Policy Directive on Global Development," whose aim is to focus on sustainable development outcomes that "place a premium on broad-based economic growth, democratic governance, game-changing innovations, and sustainable systems for meeting basic human needs."[8]

Around the world, sustainable business practices are advocated by such varied organizations as the Global Reporting Initiative, the CERES coalition, and the United Nations Environmental Programme. At the state and local level, an association of over 1220 local governments from 70 different countries, known as ICLEI—Local Governments for Sustainability provides technical consultation, training, and information services to implement sustainable development.[9] Many cities in the United States have adopted sustainability goals and a number of states have established governing bodies responsible for advancing sustainable practices.[10]

A SENSE OF URGENCY

Global Footprint Network's *Living Planet Report 2010* notes that humanity's ecological footprint has more than doubled since 1966.[11] This footprint represents the pressures on land and water area caused by the industrial and lifestyle patterns of human population. A challenge for global society is to "decouple" this global footprint from continued economic growth.

World population is expected to reach 9 billion people by 2050.[12] Almost all of the increase will take place in the developing nations, where hundreds of millions of people are seeking more secure access to food, clothing, and shelter, as well as sanitation, education, health care, energy, communication, and consumer goods. Coupled with the increasingly high levels of resource consumption in developed nations, this will place significant stress on global ecosystems. Ironically, the pressures on natural resources in developing nations could impede industrialization and further aggravate income gaps between rich and poor.

In addition to its impact on global warming, the inexorable growth of the global economy is contributing to a number of disturbing trends: sea level is rising, available fresh water and arable land are growing scarce, forests are disappearing, and biodiversity is threatened by changing natural habitats. The impacts of population growth and economic development are detailed in the 2005 Millennium Ecosystem Assessment, which found that 15 of 24 important global ecosystem services are being degraded or used unsustainably.[13]

Accomplishing sustainable development in an increasingly interconnected global economy will require a holistic "systems approach" to problem solving, as illustrated in Fig. 2. The challenge of sustainability is to enable continued economic growth while reducing the rates of both resource use and waste generation. For example, companies that import components and feedstocks from overseas must be mindful of the potential environmental and social impacts in their supply chains. Considering the direct and indirect effects of sustainability policies and strategies will help avoid unintended adverse consequences.

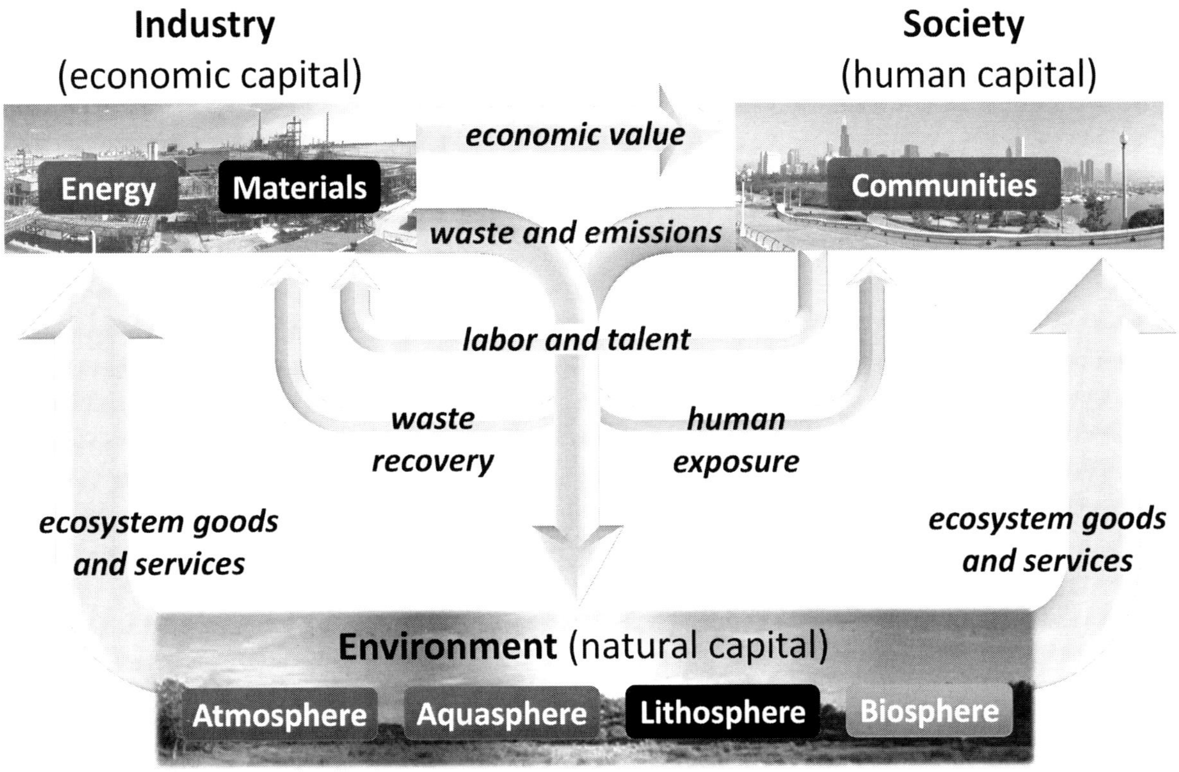

Fig. 2 Systems view of sustainability.
Source: Center for Resilience, The Ohio State University.

OPPORTUNITIES FOR POSITIVE CHANGE

Sustainable practices can enhance and support economic growth. Many businesses have recognized the importance of sustainable practices to support a growing economy. Reducing the use of resources typically leads to reduced purchasing and operating costs, so companies generally view sustainable development as enhancing their bottom line while reducing their environmental footprint. In addition, sustainability enhances shareholder value by improving brand image, reputation, and stakeholder relationships. According to the Dow Jones Sustainability Indexes: "Sustainability is a business approach that creates long-term shareholder value by embracing opportunities and managing risks that derive from economic, environmental and social developments."[14]

Sustainability strategies have been incorporated into the management objectives of major companies around the world.[15] A 2009 study published in the *Harvard Business Review* concluded that "sustainability is a mother lode of organizational and technological innovations that yield both bottom-line and top-line returns" and that "there is no alternative to sustainable development."[16]

The World Business Council for Sustainable Development (WBCSD) has developed an ambitious agenda to assist global industries in moving toward sustainable growth. The WBCSD Vision 2050 report coined the phrase "green race," and outlines a "pathway that will require fundamental changes in governance structures, economic frameworks, business and human behavior." The report argues that these changes are "necessary, feasible and offer tremendous business opportunities for companies that turn sustainability into strategy."[17]

Financial industries, including banking, investment, and insurance have also recognized the importance of sustainability and protection of natural resources. For example, the Equator Principles, launched in Washington in June 2003, require that banks assess the social and environmental impacts of projects that they finance. The principles were initially adopted by ten global financial institutions: ABN AMRO Bank, N.V., Barclays plc, Citi, Crédit Lyonnais, Credit Suisse First Boston, HVB Group, Rabobank Group, The Royal Bank of Scotland, WestLB AG, and Westpac Banking Corporation.

Influenced by advocacy groups such as the Rainforest Action Network, Citigroup has gone beyond the Equator Principles by committing to refuse funding for projects that could result in illegal logging, other environmental damage, or harm to indigenous people.[18] A recent report from the non-profit organization CERES lays out a road map for sustainability based on assessing the company's baseline environmental and social performance, analyzing corporate management and accountability structures and systems, and conducting a materiality analysis of risks and opportunities.[19]

While these corporate and institutional commitments are valuable, scientific and technological innovation will be critical elements in achieving genuine progress in sustainability. The required scientific knowledge to apply a systems approach will come from many disciplines—including chemistry, biology, and physics as well as economic and social sciences. New approaches such as "transdisciplinary" research and "sustainability science" have been introduced to describe the fusion of multiple disciplines to generate new knowledge, which is important for advancing sustainable innovation.[20]

There are many excellent examples of innovations that dramatically improve the sustainability of products and processes. For example, traditional wood adhesives often contain formaldehyde, which is toxic to humans. Researchers have been able to develop new adhesives from soy flour, which are environmentally friendly, stronger, and cost-competitive. These new adhesives have replaced more than 50% of formaldehyde-based adhesives, and earned a Presidential Green Chemistry Challenge Award.[21]

Finally, it is important to assure the continued resilience of environmental, economic, and social systems in the face of turbulent change. *Resilience* is defined as the capacity not only to absorb disturbances, but also to adapt and reorganize in order to maintain or improve functional capacity.[22] In order to cope with unpredictable events, whether caused by humans or nature, infrastructure and supply chain systems must be designed for long-term resilience. One example is harnessing ecosystem services to provide "green infrastructure" that helps to protect community-based water resource systems from storms, floods, and other climate disruptions.[23]

EIGHT STEPS TO SUSTAINABILITY

Sustainability represents the next level of environmental protection.[24] In previous centuries, the traditional approaches to environmental issues focused on land conservation and risk mitigation. However, the 21st century brings a new set of environmental challenges that are more complex and globally interconnected. These new challenges demand new, more integrated approaches that account for the linkages among environmental resources, national security, human well-being, and economic competitiveness. In addition to regulatory approaches, a variety of strategies to achieve both sustainability and resilience, including public-private collaboration, integrated systems thinking, technological innovation, and adaptive management will be needed.

Effective working toward global sustainability incorporates these eight steps.

Take the Long View

Sustainability requires long-term thinking, not only in natural resource management and urban development but also in corporate strategic planning. Taking a long view leads us to consider emerging pressures on ecosystems at different scales of resolution, and how the built environment and ecosystems can be managed in a synergistic way. For industrial innovators, the long view involves thinking about new business models in which material and energy resources can be used more effectively, while wastes and hazardous residuals can be reduced or eliminated.

Understand the System Dynamics

At all levels of society, humans participate in many different complex systems. As a growing population and consumption increase pressures on the ecosystem and the world becomes more tightly connected, small perturbations or changes in any one system can quickly cascade to other systems. Traditional industrial and engineering systems were not created with a systems view and may therefore be vulnerable to unexpected disruptions, such as natural disasters, industrial accidents, sabotage, or terrorism. By understanding system vulnerabilities and leverage points, we can develop cost-effective and resilient management strategies.

Define Sustainability Goals

President Kennedy set a goal for the United States to land on the moon. This goal was achieved because it was fully supported with the needed resources and polices. Making sustainable development an explicit policy goal at all levels of government, as well as the private sector, can send a clear message about the need for breakthrough innovation and creative regulatory and compliance approaches that serve the collective interests of the public, business, and government. A commitment to sustainable development can be a driver for a new generation of technology and policy innovation that restores U.S. leadership in the global economy.

Use Effective Tools

In order to practice sustainable development, advanced tools and approaches are necessary. Science and technology are key underlying contributors, along with innovative environmental and economic policies. New instrumentation, data handling, and methodological capabilities have expanded our understanding of the environment and of how complex biological, chemical, and human systems interact. A growing body of economic and environmental assessment tools is available to support planners and decision makers at all levels of government and industry.

Find the Right Collaborators

Around the world, collaboration and partnerships among stakeholders are crucial to achieving solutions that are less polarized, more economically viable, and focused on balancing short- and long-term goals. Collaboration has begun among government, industry, and non-governmental organizations, and has already yielded many sustainable innovations.

Lead by Example

The profound changes needed to achieve sustainability will require confidence and bold leadership. The most persuasive approach for overcoming uncertainty and hesitation is leading by example—demonstrating that these changes are both realistic and beneficial. Government can help by establishing framework conditions and incentives that encourage innovation, and by initiating collaborative pilot projects that bridge knowledge gaps and overcome inertia.

Measure and Track Progress

The use of sustainability indicators is essential for an integrated systems approach to the multifaceted challenges of sustainability. Indicators can help managers and policy makers anticipate and assess key trends, provide early warning of potential disruptions, quantify progress toward sustainability goals, and support decision making about complex trade-offs. A broad spectrum of available indicators is available that captures the economic, environmental, and social attributes of the systems that we strive to manage.

Learn and Adapt

The path to sustainability cannot be planned precisely due to the enormous complexity and uncertainty inherent in global political, economic, social, and natural systems. To navigate this path successfully, we cannot cling to preconceived notions or prescriptive solutions. We must be prepared to learn from experience, rethink our assumptions, and continuously adapt to change. Taking a resilient approach in the short term will enable sustainability in the long term.

CONCLUSION

The current approach to energy use and resource management is not sustainable. Moving from the status quo toward sustainability requires new approaches to problem solving that take an entire system into account, rather than isolated components. The challenge of achieving sustainability is urgent since unsustainable behavior—whether it is economic, social, or environmental—can lead to national and global crises. For example, when withdrawal rates from freshwater aquifers exceed recharge rates, water scarcity is the eventual result. Conversely, sustainable practices can preserve and protect natural capital and enhance economic growth. The challenge of realizing sustainability goals raises a number of practical questions: "What kind of government policies and business strategies can advance us toward sustainability?" "How can business and government work together to advance sustainable practices?" and "What can citizens do to advance the concept of sustainability?" Making sustainability operational is not something a government or business can do alone. A sustainable and thriving economy will require the convergence of four major elements: (1) regulations and policies, (2) advances in science and technology, (3) green business practices, and (4) public support and participation. A global commitment to accelerate movement toward sustainability is essential to safeguard social well-being, shared prosperity, and our natural environment. There is really no alternative.

ACKNOWLEDGMENTS

The authors are grateful to Edward Fallon (Senior Environmental Employment participant at EPA) and Meadow Anderson (American Association for Advancement of Science Fellow at EPA) for their helpful edits and comments.

DISCLAIMER

Views expressed in this entry are those of the authors and do not necessarily reflect the views or policies of the U.S. EPA. Mention of trade names or commercial products does not constitute EPA endorsement or recommendations for use.

REFERENCES

1. World Commission on the Environment and Development. *Our Common Future*; Oxford University Press: New York, 1987. http://www.un.org/documents/ga/res/42/ares42-187.htm (accessed September 2011).
2. Pinchot, Gifford. *The Fight for Conservation*; Doubleday, Page & Company: New York. 1910. (accessed October 2011).
3. National Environmental Policy Act (NEPA). http://ceq.hss.doe.gov/nepa/regs/nepa/nepaeqia.htm.
4. Meadows, D.H.; Meadows, D.K.; Randers, R.: Behrens, W.W. *The Limits to Growth*; Universe Books: New York, 1972.
5. United Nations. Division for Sustainable Development. Agenda 21. http://www.un.org/esa/dsd/agenda21/res_agenda21_00.shtml (accessed October 2011).

6. Larsson A.; Azar, C.; Sterner, T.D.; Strömberg, D.B.; Andersson, B. *Technology and Policy for Sustainable Development*; Centre for Environment and Sustainability at Chalmers University of Technology and the Göteborg University: Goteborg, 2002.
7. Bush, G.W. Executive Order 13423—Strengthening Federal Environmental, Energy, and Transportation Management. 2007. http://edocket.access.gpo.gov/2007/pdf/07-374.pdf and Obama, B.H. Executive Order 13514—Federal Leadership in Environmental, Energy, and Economic Performance, 2009. http://edocket.access.gpo.gov/2009/pdf/e9-24518.pdf (accessed October 2011).
8. ICLEI - Local Governments for Sustainability. http://www.iclei.org/index.php?id=about (accessed March 2012).
9. Obama, B.H. Fact Sheet: U.S. Global Development Policy. 2010. http://www.whitehouse.gov/the-press-office/2010/09/22/fact-sheet-us-global-development-policy (accessed March, February 2012).
10. Engel, K.H.; Miller, M.L. State Governance: Leadership on Climate Change (Arizona Legal Studies Discussion Paper No. 07-37). In *Agenda for a Sustainable America*; Dernbach, J.C., Ed.; Environmental Law Institute: Washington, 2009; 441 pp. http://papers.ssrn.com/sol3/papers.cfm?abstract_id=1081314 (accessed October 2011).
11. WWF, Global Footprint Network, and ZSL. Living Planet Report 2010: Biodiversity, Biocapacity, and Development. http://www.footprintnetwork.org/press/LPR2010.pdf (accessed October 2011).
12. UN. World Population to 2300. http://www.un.org/esa/population/publications/longrange2/WorldPop2300final.pdf (accessed March 2012).
13. Millennium Ecosystem Assessment. *Guide to the Millennium Assessment Reports*. Island Press: Washington, 2005. http://www.maweb.org/en/index.aspx (accessed October 2011).
14. Dow Jones Sustainability Indexes, in Collaboration with SAM. Corporate Sustainability. http://www.sustainability-index.com/07_htmle/sustainability/corpsustainability.html. (accessed October 2011).
15. Esty, D.; Winston, A. *Green to Gold: How Smart Companies Use Environmental Strategy to Innovate, Create Value, and Build Competitive Advantage*; Yale University Press: New Haven, 2006.
16. Nidumolu, R.; Prahalad, C.K.; Rangaswami, M.R. Why sustainability is now the key driver of innovation. Harvard Bus Rev. **2009**, *87*, 57–64.
17. World Business Council for Sustainable Development. *Vision 2020: The New Agenda for Business*; http://www.wbcsd.org/web/projects/BZrole/Vision2050-FullReport_Final.pdf (accessed October 2011).
18. Swiss Re Center for Global Dialogue. *Capitalizing on Natural Resources*. New Dynamics in Financial Markets. 9th International Sustainability Leadership Symposium 10/11 September 2008. Swiss Re Centre for Global Dialogue: Rüschlikon, Switzerland. 2008. http://www.sustainability-zurich.org/cm_data/Report_08_V8_final.pdf (accessed October 2011).
19. Ceres. The 21st Century Corporation : The Ceres Roadmap for Sustainability. http://www.ceres.org/resources/reports/ceres-roadmap-to-sustainability-2010 (accessed October 2011).
20. Kates, R.W.; T.M. Parris, T.M. Long-term trends and a sustainability transition. Proc. Nat. Acad. Sci. **2003**, *100* (14), 8062–8067.
21. U.S. EPA. The Presidential Green Chemistry Challenge. http://www.epa.gov/greenchemistry/pubs/pgcc/presgcc.html (accessed October 2011).
22. Fiksel, J. Sustainability and resilience: towards a systems approach. Sustainability: Sci. Pract. Policy **2006**, *2* (2), 1–8.
23. U.S. EPA. Green Infrastructure. http://cfpub.epa.gov/npdes/greeninfrastructure/gisupport.cfm. (accessed October 2011)
24. Hecht A. The next level of environmental protection: business strategies and government policies converging on sustainability. Sustain. Dev. Law Policy **2007**, *7*, 19–25.

Sustainable Agriculture: Social Aspects

Frederick H. Buttel
Department of Rural Sociology, University of Wisconsin, Madison, Madison, Wisconsin, U.S.A.

Abstract

Agriculture—conventional and sustainable, "traditional" and "modern"—is intrinsically social (or sociotechnical) in nature. The social character of agriculture is particularly apparent, particularly widely recognized, and particularly crucial in the case of sustainable agriculture. Not only must any reasonable definition of sustainable agriculture include both social and economic as well as technological and ecological components, the roots of unsustainability and the pathways to sustainability are fundamentally social processes. This entry will provide an overview of the most fundamental social dimensions of sustainable agriculture.

THE DEFINITION OF SUSTAINABLE AGRICULTURE

It is widely agreed that sustainable agriculture is intrinsically a joint social and ecological construct. Thus, for example, Ikerd[1] stresses the "anthropocentric" as well as "ecocentric" nature of agricultural sustainability, noting that the essence of sustainable agriculture is that "we (in sustainable agriculture) are concerned about sustaining agriculture for the benefit of humans, both now and into the indefinite future." Sustainable agriculture is an agriculture that is "ecologically sound, economically viable, and socially responsible."[2] Allen similarly emphasizes that a conception of sustainable agriculture that fails to recognize the role of people, social actors, social institutions, and social movements is a limited one.[3]

Sustainable agriculture and organic farming are very closely related but are not exactly coterminous. Organic farming systems tend to be highly sustainable systems according to the ecological and social components of the definition of sustainable agriculture, and are often quite economically viable as well.[4,5] Many in the sustainable agriculture community, however, feel strongly that sustainability improvements in mainstream or conventional agriculture are as important as the expansion of organic farming and organic agro-food systems.

STRUCTURAL CAUSES OF THE LACK OF SUSTAINABILITY

Agricultural sustainability is not only a vision of the long-term goals for agriculture, which imply measuring sticks for or indicators of its achievement.[6] Sustainability also implies a critique of or a set of concerns about "conventional" agriculture—that mainstream agriculture has shortcomings in terms of environmental soundness, economic soundness, and social justice—and suggests that there have been trends over time that jeopardize the achievement of these three ends.

The historical U.S. pattern of abundant land, limited and/or relatively expensive rural labor, and a strong tendency toward overproduction and low commodity prices has encouraged a capital-intensive, chemical-intensive monocultural system with a very high degree of enterprise and spatial homogeneity (or specialization). Commodity, trade, and public research policies have historically reinforced these tendencies to monoculture, specialization, and heavy reliance on chemicals, leading to an agriculture that has significant shortcomings on all of the criteria implied in the definition of sustainable agriculture.[7]

SUSTAINABILITY AND THE CORPORATE-INDUSTRIAL AGRICULTURAL ECONOMY

Agricultural sustainability cannot be considered apart from the globalized and highly concentrated corporate economy within which sustainable agriculture, and agriculture as a whole, is enmeshed. Corporate concentration on input provision, food processing, and retailing limit the choices available to sustainable and other farmers. Corporate control over agriculture is increasing through processes such as the disappearance of "open markets" for agricultural products; the extension of vertical integration and contractual relationships; and the ever closer alignment of multinational chemical and seed companies on one hand, and food processors, manufacturers, and global trading companies on the other.[8] There is evidence that

Encyclopedia of Natural Resources DOI: 10.1081/E-ENRL-120012948

the increased corporate domination of agro-food systems has negative implications for achieving agricultural sustainability. Farmer contractees, for example, are typically directed contractually by "integrators" to use particular practices which, more often than not, are unsustainable.

It should be stressed, however, that corporations—often small ones, but also large firms—have been major agents in extending the scope of organic food production and organic marketing. There are two major nationwide organic chain grocery stores, and a number of corporations have been very successful in producing organic products such as potatoes, grapes, fresh fruits and vegetables, and so on. The corporate organization of the organic food industry represents a major dilemma for proponents of agricultural sustainability. Highly successful sustainable agriculture ideas will be attractive to private corporations, which will often be successful in diffusing products and practices. At the same time, corporate agricultural sustainability practices may undermine certain aspects of the sustainability agenda (e.g., large-scale monocultural production of organic potatoes in the Northern Intermountain states is associated with a considerable threat of soil erosion and loss of biodiversity).

There is considerable debate in the sustainable agriculture community as to whether the best indicator of the growth of sustainable agriculture is the national market share of organic food—much of which is due to the activities of very large corporations such as Whole Foods—or whether the extent of sustainable agriculture is most accurately measured by the prevalence of community-supported agriculture (CSA), farmers markets, local co-ops, community gardens, and direct marketing of food. Those with the strongest commitments to sustainable agriculture are also most likely to see food system localization and the re-creation of more local "foodsheds" as the heart and soul of a genuine and enduring sustainable agriculture.[9]

SUSTAINABILITY AS A SOCIAL MOVEMENT

Agricultural sustainability is as much a social movement as it is a set of technologies and production practices. Social definitions of sustainability—that is, the concerns, views, ideologies, and agendas of movement participants—are as or more important in shaping what sustainability is as ecological or biological indicators such as soil erosion rates, levels of soil organic matter, biodiversity, levels of runoff, and so on. What has put sustainability on the national and global agenda is primarily the organization and activities of three major types of groups: farmer-driven sustainable agriculture organizations and movements, consumer-driven or consumer-oriented sustainable agriculture initiatives (such as many CSA farms and most food co-ops and local food councils), and national and global environmental groups. Environmental groups have arguably done the most to increase the visibility and persuasiveness of the overall notion of sustainability, and of the particular concept of agricultural sustainability, though farmer-oriented groups do not always agree with environmentalists' agendas for agricultural sustainability.

Each of the three main types of sustainable agriculture organizations, however, agrees on the main components of the movement agenda: the need for more public research on sustainable practices, the need for public policy incentives for sustainability, the desirability of family farming, and the imperative to improve the environmental performance of agriculture. Increasingly, this consensus has extended to issues such as opposition to major trade liberalization agreements (the World Trade Organization and NAFTA) and opposition to genetic engineering of crops and foods.[10]

PUBLIC POLICY AND THE FUTURE OF AGRICULTURAL SUSTAINABILITY

Three necessary but insufficient conditions for the continued advance of sustainable agriculture are the development of improved sustainable technology; the presence of a vigorous and dynamic sustainability movement; and the development of a public policy environment that reduces the public policy disincentives to an environmentally sound, socially responsible, and economically viable agriculture. Of these three conditions, the public policy environment of agricultural sustainability is arguably the most important over the long term, even though the macro-public-policy environment is difficult to redirect, and there are areas of public policy disagreement among farmers, consumers, and environmental actors in the sustainable agriculture community. For example, some sustainability advocates, particularly those in the environmental community (and some farm groups in that community), believe that the federal (and global) levels of action are most important or efficacious, whereas others believe that at this time the local (community or regional) arena is where advocates can most easily make a difference. Nonetheless, it is apparent that over the long term sustainable agriculture cannot advance far in a public policy environment that involves the strong disincentives to sustainability that currently prevail. There is a need for more active federal, state, and local regulation of both the on-site and the off-site impacts of agriculture; for ending the commodity-driven pattern of federal agricultural policy that shovels the lion's share of subsidies in the direction of large, monocultural producers of overproduced commodities; and for redirection of the public agricultural research agenda. Some of the most innovative public policy ideas in sustainable agriculture are being developed in Europe. Green payments, taxes on pesticides and fertilizers, and the embracement of a "multifunctionality" approach to government agricultural policy are particularly promising policy instruments.[11] Each of these policy instruments would support sustainability and yield other benefits (reduced government outlays, rural development, reduced greenhouse gas emissions). Still, there are no technocratic shortcuts to sustainability. Sustainability is, and must remain, as much a social movement as it is a set of practices and measuring sticks of agro-food system performance.

REFERENCES

1. Ikerd, J.E. Assessing the health of ecosystems: A socioeconomic perspective. Paper Presented at the First International Ecosystem Health and Medicine Symposium, Ottawa, Canada, 1994. (http://www.ssu.missouri.edu/ faculty/jikerd/papers/Otta-ssp.htm), p. 2 (accessed October 2002).
2. Ikerd, J.E. New farmers for a new century. Paper Presented at the Youth in Agriculture Conference. Ulvik, Norway, February, 2000. (http://www.ssu.missouri.edu/faculty/ jikerd/papers/Newfarmer1.htm), p. 3 (accessed October 2002).
3. Allen, P. Connecting the Social and Ecological in Sustainable Agriculture. In *Food for the Future: Conditions and Contradictions of Sustainability*; Allen, P., Ed.; Wiley: New York, 1993; 1–16.
4. Flora, C.B.; Francis, C.; King, L., Eds. *Sustainable Agriculture in Temperate Zones*; Wiley-Interscience: New York, 1990.
5. *Interactions between Agroecosystems and Rural Communities*; Flora, C.B., Ed.; CRC Press: Boca Raton, FL, 2001.
6. Airstars, G.A., Ed. *A Life Cycle Approach to Sustainable Agriculture Indicators: Proceedings. Center for Sustainable Systems*; University of Michigan: Ann Arbor, MI, 1999.
7. Buttel, F.H. The sociology of agricultural sustainability: Some observations on the future of sustainable agriculture. Agric. Ecosyst. Environ. **1993**, *46*, 175–186.
8. Heffernan, W.D. *Consolidation in the Food and Agriculture System*; National Farmers Union: Denver, CO, 1999. (http://www.nfu.org/images/heffernan_1999.pdf) (accessed October 2002).
9. Kloppenburg, J., Jr.; Hendrickson, J.; Stevenson, G.W. Coming into the foodshed. Agric. Human Values **1996**, *13*, 33–42.
10. Kirschenmann, F. Questioning Biotechnology's Claims and Imagining Alternatives. In *Of Frankenfoods and Golden Rice;* Buttel, F., Goodman, R., Eds.; Wisconsin Academy of Sciences, Arts, and Letters: Madison, WI, 2002; 35–61.
11. Organisation for Economic Cooperation and Development (OECD). *Multidimensionality: A Framework for Analysis*; OECD: Paris, 1998.

Transpiration

Thomas R. Sinclair
Crop Genetics and Environmental Research Unit, U.S. Department of Agriculture (USDA), University of Florida, Gainesville, Florida, U.S.A.

Abstract
Nearly all water evaporated from vegetated surfaces to the atmosphere originates from leaves. Water is vaporized from cell walls inside leaves and diffuses from the leaf interior to the bulk atmosphere around plants. This process is called transpiration, and this entry discusses the regulation of and methods to estimate transpiration rates. While transpiration involves basically the vaporization of water and the diffusion of the vapor into the bulk atmosphere, transpiration is complex because the water vapor must move from the leaf interior, through pores in the leaf epidermis called stomata, and finally into the atmosphere. Stomata are under active control so that transpiration rates are dynamic and rapidly respond to the environment. An understanding of the influence of stomata regulation on both carbon dioxide and water vapor flux density leads to various approaches to calculate plant canopy transpiration rates.

BACKGROUND

Carbon dioxide (CO_2) assimilation in photosynthesis was greatly facilitated, and thereby allowing rapid plant growth, when plants invaded the Earth's land masses. No longer was it necessary for CO_2 to diffuse at very slow rates through the water surrounding aquatic plant life, but rather CO_2 could be absorbed directly from the atmosphere into individual cells. While photosynthesis rates and plant growth were enhanced substantially by allowing direct exposure of photosynthetic cells to the atmosphere, a potentially fatal consequence was that water could be evaporated at very high rates from exposed cell surfaces.

The problem of rapid evaporation from cell surfaces can be solved by having either an especially effective plant structure to rapidly transport large quantities of water in order to replenish each cell with water, or mechanisms to substantially inhibit water loss (and also CO_2 assimilation rates) from cell surfaces. While evolution "experimented" with each of these approaches, an innovative third, anatomical solution dominates. The photosynthetic cells are packaged inside thin broad tissues, i.e., leaves, that can throttle water vapor diffusion between the cells inside leaves and the atmosphere outside the leaves (Fig. 1).

The exterior of leaf epidermal cells is coated with a cuticle containing waxy materials that effectively block water loss directly from epidermal cells to the atmosphere. Consequently, virtually all water lost by leaves must move through stomatal pores that are scattered in the leaf epidermis (Fig. 2). Stomata regulate gas diffusion through the epidermis by adjustments in the dimension of the stomatal pore. The apertures of stomatal pores adjust to maintain a fairly stable CO_2 concentration inside leaves. When leaf photosynthetic rates are high, stomata adjust to increase the size of the pore aperture for rapid diffusion of CO_2 into leaves. On the other hand, when photosynthetic rates are low, aperture size decreases. Since water vapor diffuses through the same stomatal aperture, adjustments to accommodate photosynthesis cause changes in transpiration rate.

STOMATA

Stomata are embedded in the leaf epidermis and are formed by a pair of cells called guard cells. The guard cells in monocots plants, as shown in Fig. 2, tend to resemble barbells with bulbous structures at each end. The guard cells of dicot plants shown in Fig. 3, have a kidney shape. In both cases, the pair of guard cells is attached to each other at both ends. Swelling of the bulbous end of guard cells in monocots causes the cylindrical midsections to move apart increasing the aperture of the pore. The entire kidney-shaped cells of dicots swell and, as a result of a specialized cell wall structure bordering the pore, the aperture of the pore increases. Conversely, a decrease in the size of guard cells in both cases results in a decrease in pore aperture.

The shrinking and swelling of guard cells are under active control by plants as a result of changes in the concentrations of specific compounds in guard cells and neighboring cells. An increase in solute concentration in guard cells causes water to flow into the cells so that guard cells swell. There appears to be two mechanisms in guard cells that result in changes in solute concentration.[1] One mechanism is based on potassium-malate transport and appears

Encyclopedia of Natural Resources DOI: 10.1081/E-ENRL-120010277

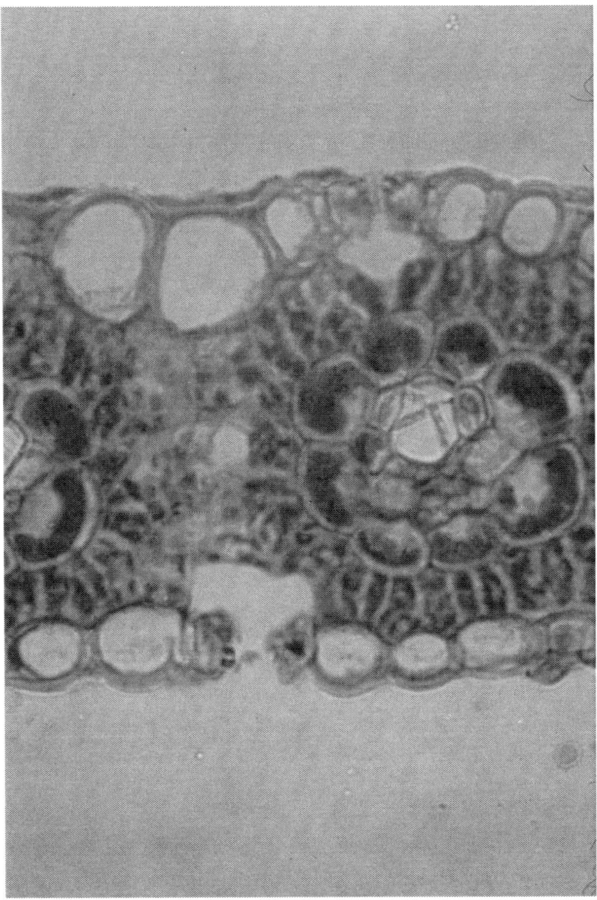

Fig. 1 Cross-section of monocot leaf showing stomata in epidermis of both sides of the leaf.

to be particularly involved with stomata opening in response to light at sunrise. The second mechanism involves sucrose accumulation in guard cells and is associated with maintenance of stomata aperture during the day. Both mechanisms are potentially sensitive to changes in the CO_2 environment of the leaf interior.[1] The sucrose mechanism seems especially sensitive to leaf photosynthetic rates so that stomata aperture can be fine tuned throughout the day as the photosynthetic rate of the leaf responds to changing environmental conditions. Of course, changes in aperture to match CO_2 diffusion to leaf photosynthetic capacity also results in changes in transpiration rate.

LEAF TRANSPIRATION RATE

Leaf transpiration involves the diffusion of water vapor through stomatal pores, so transpiration rate is dependent on conductance of vapor through and above the pores, and on the gradient of water vapor across the pores. Not surprisingly, conductance of an individual stomatal pore (g_p, $cm^3 sec^{-1}$) is directly dependent on the dimensions of the pore aperture and can be expressed quantitatively[2] as

$$g_p \approx \pi abD / d \tag{1}$$

where a is the semilength of the major axis (cm), b, the semilength of the minor axis (cm), d, the depth of the pore (cm), and D, the molecular diffusion coefficient of water vapor ($0.24\,cm^2 sec^{-1}$ at 20°C).

The semilength and depth of the pore usually remain fairly stable, so the main variable influencing g_p is the semilength of the minor axis of the pore.

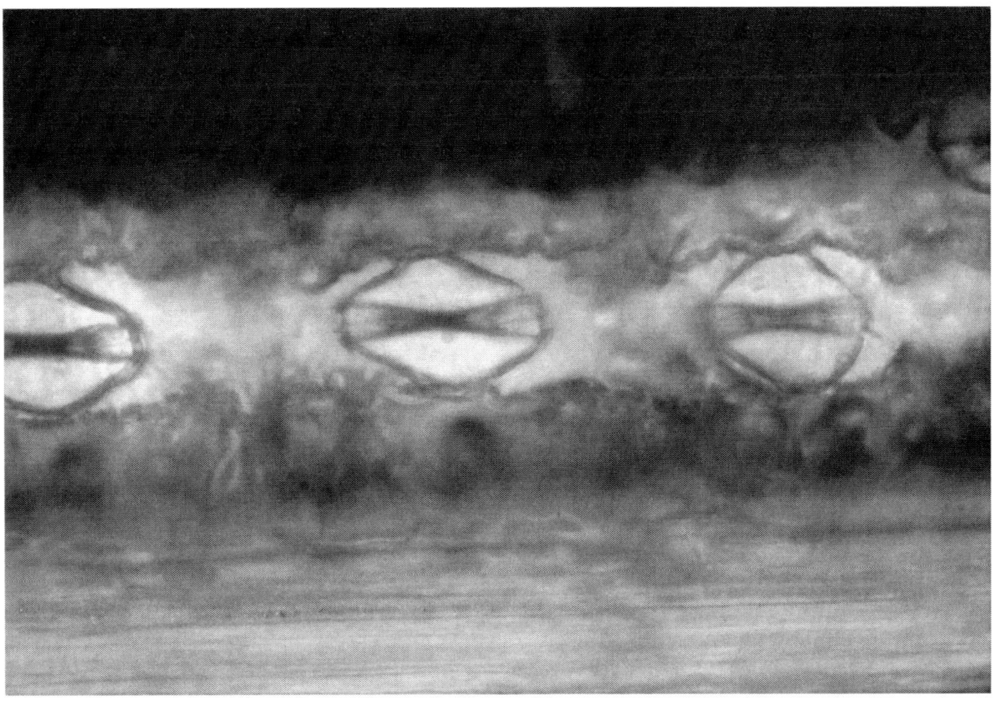

Fig. 2 Photograph showing stomata in the leaf surface of sorghum.

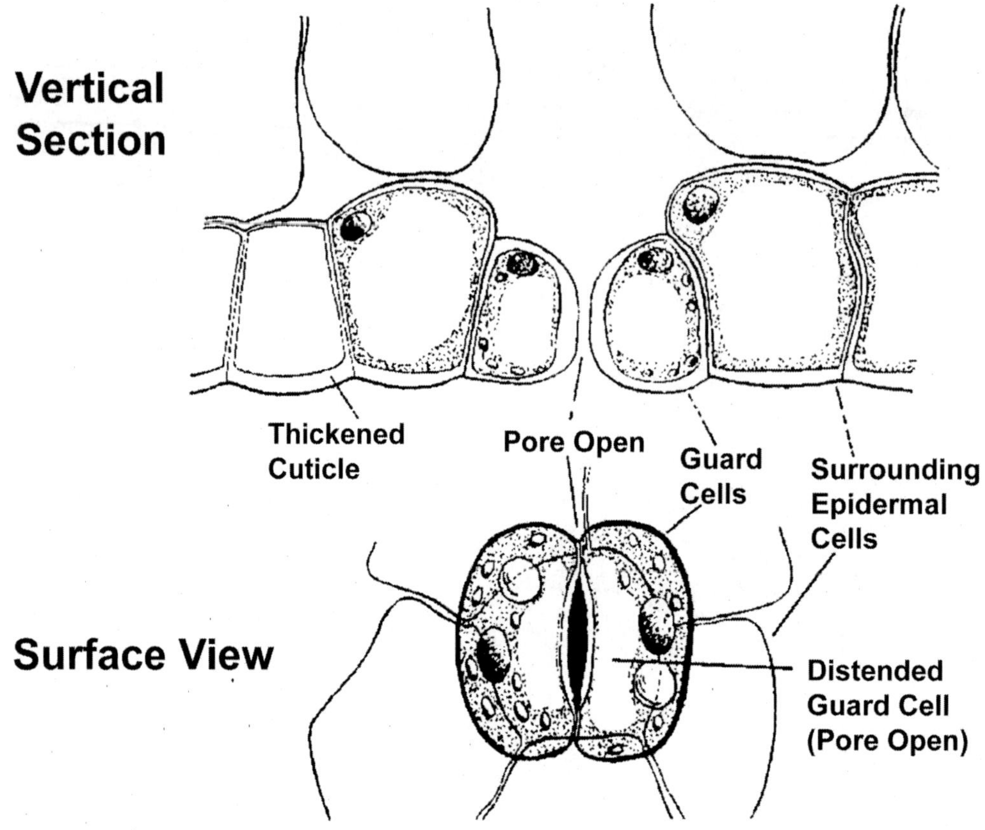

Fig. 3 Drawing of cross-section and surface view of a dicot stomata.
Source: Adapted from Sinclair & Gardner.[10]

The overall stomata conductance of the leaf (g_s, cm sec^{-1}) is calculated by incorporating into Eq. (1) stomatal density (n, stomata cm^{-2}) and the influence of "end effects" exterior to the leaf.[2] Consequently,

$$g_s \approx \pi abD / \left(d + b/2\ln\left(4a/b\right)\right) \tag{2}$$

An estimate of maximum g_s can be calculated for fully open stomata ($b \approx 3 \times 10^{-4}$ cm) by assuming that $a = 8 \times 10^{-4}$ cm, $d = 10 \times 10^{-4}$ cm, and $n = 10 \times 10^3$ stomata cm^{-2}. Equation 2 gives 1.3 cm sec^{-1} for a leaf with stomata on only one side, and 2.5 cm sec^{-1} for a leaf with stomata on two sides. Of course, when the stomatal pore is closed ($b = 0$), Eq. (2) gives a conductance of zero.

In addition to the restriction on water vapor diffusion resulting from stomatal conductance, there is also a limitation on water loss resulting from the aerodynamic boundary layer around leaves (g_{bl}, cm sec^{-1}). The value of g_{bl} is dependent on wind speed and leaf dimensions. In the case of a 100 cm sec^{-1} wind speed and 8 cm wide leaf, the value of g_{bl} is approximately 2.5 cm sec^{-1}.[3] Consequently, in this case, the value of g_{bl} is equal to that of gs.

As g_s and g_{bl} are in series, the inverse of the two conductances are added together to calculate leaf transpiration rate, Tr_L (g cm^{-2} sec^{-1}). The combined conductance is multiplied by the vapor pressure difference between the interior of the leaf, which is calculated as the saturated vapor pressure at leaf temperature and atmospheric vapor pressure (P_a).[3]

$$Tr_L = \varepsilon\left(g_s g_{bl}/\left(g_s + g_{bl}\right)\right)\left(P_L^* - P_a\right)/H_v \tag{3}$$

where ε is the molecular weight of water (18 g mol^{-1}) and H_v the heat of vaporization (44 kJ mol^{-1} at 25°C).

CANOPY TRANSPIRATION RATE

In principle, the transpiration of a leaf canopy (Tr_C) can be calculated by summing Tr_L for all the individual leaves in the canopy. This is a formidable task, however, because values of g_s, g_{bl}, and leaf temperature are required for each individual leaf. Consequently, "summary" expressions have been developed in an effort to express the transpiration rates of the entire canopy.

One of the more popular approaches relies on predictions of the energy balance of the leaf canopy. The Penman–Monteith equation[4] gives an explicit solution for canopy transpiration rate based on the energy balance of the entire canopy. The difficulty with this approach is that it requires an estimate of the canopy boundary layer conductance and the "canopy conductance" for the entire canopy, which must be appropriately weighted to represent vapor transfer through all the stomata distributed in the canopy. There is really no independent method for measuring canopy conductance, and estimates are obtained only by back-solving the energy balance equation. In practice, the Penman–Monteith is often applied as an empirical equation where boundary layer and canopy conductances are estimated from empirical functions.

A recent innovation has been developed to calculate Tr_C based on the water use efficiency of leaf canopies. Water use efficiency has been a topic of research since at least 1699[5] giving it a longer history of study than virtually any other plant trait. Roughly 100 years ago, there was a particularly intensive period of study in both Europe and the United States on plant water use efficiency, culminating in the classic investigations by Briggs and Shantz.[6,7] In an analysis of much of the data from this period, deWit[8] found a highly stable relationship within each species between accumulated plant mass and cumulative transpiration over a wide range of conditions when normalized by "atmospheric demand" (Fig. 4).

Tanner and Sinclair[9] extended the analysis of deWit[8] by deriving a mechanistic expression for transpirational water use efficiency of canopies. Their derivation defined a specific transpirational water use efficiency coefficient (k, Pa) for each crop species dependent on the photosynthetic pathway and the biochemical composition of the plant products. Their estimates of k for maize, wheat, and soybean were 12 Pa, 5 Pa, and 4 Pa, respectively. Then, for each species, water use efficiency based on accumulated plant mass (M) was stable when expressed in the following equation:

$$M / \mathrm{Tr}_C = k / \left(P_a^* - P_a\right) \tag{4}$$

Tanner and Sinclair[9] approximated for daily transpiration by assuming this value was 75% of the maximum atmospheric vapor pressure deficit calculated at the daily maximum temperature.

A solution for TrC is obtained directly by the rearrangement of Eq. (4):

$$\mathrm{Tr}_C = M\left(P_a^* - P_a\right)/k \tag{5}$$

This equation can be readily used based on estimates of M, which in turn can be calculated based on the interception of solar radiation and the radiation use efficiency of the canopy. Consequently, Eq. (5) is consistent with the

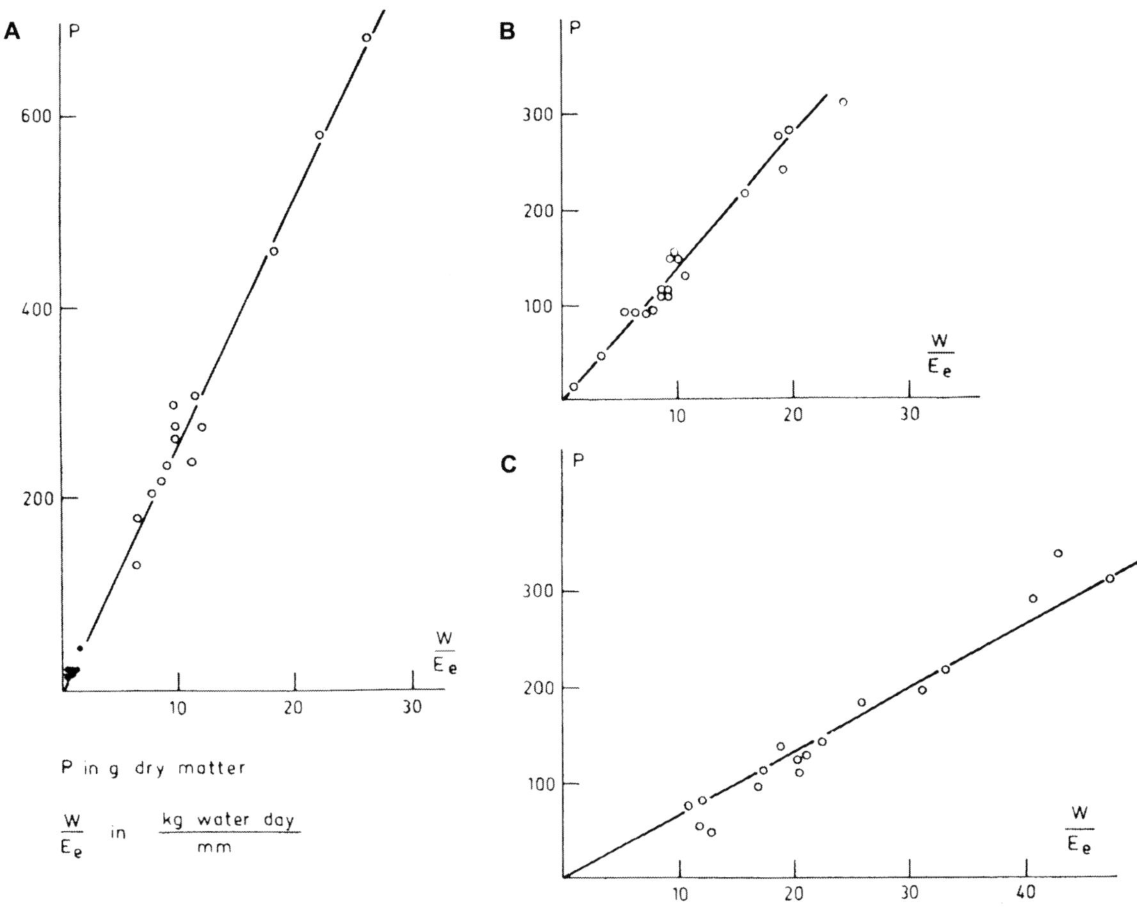

Fig. 4 Plot of crop growth (P) against transpiration water loss (W) divided by pan evaporation for (**A**) sorghum, (**B**) wheat, and (**C**) alfalfa.
Source: Adapted from deWit.[8]

energy balance approach in that it is sensitive to the amount of solar radiation intercepted by the canopy.

Not only is Eq. (5) easier in principle to implement that an energy balance approach, but it is conceptually much more compatible with the understanding of stomata regulation of transpiration. That is, Eq. (5) is calculated from a direct dependence of Tr_C on CO_2 assimilation, which is consistent with the fact that stomata are regulated for CO_2 assimilation and the rate of water loss from leaves is simply a consequence of this process. Equation 5 has been used very effectively to calculate transpiration rates and evaluate limitations of water availability on crop yields.

REFERENCES

1. Talbott, L.D.; Zeiger, E. The role of sucrose in guard cell osmoregulation. J. Exp. Bot. **1998**, *49*, 329–337.
2. Parlange, J.; Waggoner, P.E. Stomatal dimensions and resistance to diffusion. Plant Physiol. **1970**, *46*, 337–342.
3. Sinclair, T.R. Theoretical considerations in the description of evaporation and transpiration. In *Irrigation of Agricultural Crops*; Stewart, B.A., Nielsen, D.R., Eds.; American Society of Agronomy: Madison, WI, 1990; 343–361.
4. Monteith, J.L. Evaporation and environment. In *The State and Movement of Water in Living Organisms, XIX Symp. Soc. Exp. Biol.* Academic Press: NY, 1965; 205–234.
5. Woodward, J. Some thoughts and experiments concerning vegetation. Philos. Trans. R. Soc. London **1699**, *21*, 193–227.
6. Briggs, L.J.; Shantz, H.L. The water requirement of plants. II. A review of the literature. U.S. Dep. Agric. Bur. Plant Ind. Bull. **1913**, *285*, 1–96.
7. Briggs, L.J.; Shantz, H.L. The water requirement of plants as influenced by environment. Proc. Second Pan-American Science Congress, Washington, DC, 1917; 95–107.
8. deWit, C.B. *Transpiration and Crop Yields*; Institute of Biological and Chemical Research on Field Crops and Herbage: Wageningen, the Netherlands, 1958; Vol. 64.6, 1–88.
9. Tanner, C.B.; Sinclair, T.R. Efficient water use in crop production: research or re-search? In *Limitations to Efficient Water Use in Crop Production*; Taylor, H.M., Jordan, W.R., Sinclair, T.R., Eds.; American Society of Agronomy: Madison, WI, 1983; 1–27.
10. Sinclair, T.R.; Gardner, F.R. *Principles of Ecology in Plant Production*; CAB International: Wallingford, UK, 1998; 1 pp.

Transpiration: Water Use Efficiency

Miranda Y. Mortlock
School of Agriculture and Food Sciences, University of Queensland, Brisbane, Queensland, Australia

Abstract

The balance between carbon assimilation (net photosynthetic production) and the throughput of water by transpiration (resource use in terms of water) results in a benefit–cost ratio of interest to eco-physiologists and crop physiologists, known as water use efficiency. The differences in concentration of CO_2 and water vapor between the intercellular surfaces of the leaf mesophyll and the atmosphere drive the fluxes of carbon dioxide and water through the plant. Hot dry environments provide conditions of high evaporative demand. CO_2 concentrations are low in the atmosphere, and this gas diffuses through the stomata, which need to be open to allow gas exchange. There is a need under most environments to conserve water, and under drought stress stomata close which conserves water. Water use efficiency is an expression of the benefit–cost ratio for a plant and integrates the physiology of photosynthesis and plant water relations over a particular growth period or cropping season.

DEVELOPMENT OF THE CONCEPT OF WATER USE EFFICIENCY

Early last century, Briggs and Shantz studied the water requirements (WRs) of crops by weighing containers and working out WR per plant; however, they did not use an area basis.[1] Water requirement is the inverse of water use efficiency. Viets confined his definition of water use efficiency to the ratio of plant production to ET measured on the same area.[2] Tanner and Sinclair summarized some early studies and defined water use efficiency as the biomass of water accumulated per unit of water transpired and evaporated per unit crop area.[1] Biomass is expressed as total yield or economic yield. To compare species with different chemical composition (protein vs. carbohydrate products), grams of glucose *equivalents* are used. In irrigation studies, water use efficiencies may be referring to a broader definition of water, with efficiencies comparing situations where soil water drainage, surface run-off, or soil evaporation are considered. These types of water use efficiency are not included in this discussion.

WATER USE EFFICIENCY ON AN EVAPOTRANSPIRATION BASIS

Water use can be defined per unit of evapotranspiration (ET), and under field conditions, this is more practical than the narrower definition using just transpiration. Measures of ET integrate both soil and crop factors for the season, which confound the respective efficiencies of the plant and soil evaporation. Timing and frequency of irrigations and rain, the soil type and plant or mulch cover can affect soil evaporation. The transpiration part of ET use is a measure of crop performance.

WATER USE EFFICIENCY ON A TRANSPIRATION BASIS

Another definition of water use efficiency is in terms of transpired water only. Measuring the transpiration component is hard to do in practice, as it is difficult to prevent soil evaporation. Deep soil drainage also needs to be measured or prevented. It is only possible to measure transpiration on an experimental basis with the use of weighing lysimeters. Typically, the lysimeter is a large pot in the greenhouse and may weigh up to 80 kg. The lysimeter is weighed frequently over the crop season, and known quantities of water are added which is, both costly and limits the practical size of the trial. When extended to field studies (lysimeters within a growing crop), a limited number of comparisons are made due to the setup cost and expense of rainout facilities at sites.

TRANSPIRATION EFFICIENCY UNITS

Water use efficiency can be expressed as $g\,kg^{-1}$ of water transpired. Typical values may be 1.6–2.4 $g\,kg^{-1}$ for sunflower or around 9 $g\,kg^{-1}$ for sorghum. It can also be expressed by using a molar scale. Transpiration may be typically 1–5 $mmol\,m^{-2}\,sec^{-1}$ and photosynthetic rates for C_3 20–25 $\mu mol\,m^{-2}\,sec^{-1}$ or 40 $\mu mol\,m^{-2}\,sec^{-1}$ for C_4 plants. Hence, C_4 plants have higher transpiration efficiencies than C_3 plants.

Encyclopedia of Natural Resources DOI: 10.1081/E-ENRL-120010279

TRANSPIRATION EFFICIENCY AND COMPARISON ACROSS SEASONS

As the evaporative demand of the atmosphere will vary from place to place and from season to season, transpiration efficiency (TE) from particular trials are adjusted for the vapor pressure deficit (VPD) of the atmosphere:

$$Y/T = k/(e^* - e_i)$$

where Y is the yield, T, the transpiration, k, the transpiration coefficient, e^*, the saturated vapor pressure, and e_i, the vapor pressure of the atmosphere.[1] The coefficient (k) is estimated as 9 kPa for sorghum.

Typically, a mean value of VPD for hours of daylight over the stress period season can be used. The mean daily value of the 9.00 hr VPD and VPD at the time of maximum temperature can been used to compute a seasonal VPD over the stress period.[3]

INSTANTANEOUS TRANSPIRATION EFFICIENCY MEASURED AT AN INDIVIDUAL LEAF LEVEL

Transpiration and photosynthesis measured by using a canopy gas exchange system are used to compute an instantaneous measurement of TE at the leaf level. This rarely correlates with TE computed for a season as there are many processes integrated within the plant alone and over the cropping season.

Water use efficiency is the molar ratio of CO_2 uptake (A) to transpiration (E) and can be written as

$$A / E = (c_a - c_i) / 1.6\Delta_W$$

where c_a is external and c_i is the internal partial pressure of CO_2, respectively. Δ_w is the leaf-air VPD.[4]

This shows that the internal partial pressure of CO_2 is linked to water use efficiency.

PHYSIOLOGY OF TRANSPIRATION EFFICIENCY

C_3 and C_4 plants vary in TE as the carboxylation pathways give rise to different efficiencies.[5-7] C_4 plants, such as maize and sorghum, have higher transpiration efficiencies than C_3 plants. Legumes have a lower TE than cereal crops due to the metabolic cost of symbiotic N fixation. Variations in TE for wheat correlate with carbon isotope discrimination.[8] Carbon isotope discrimination (Δ) as determined on dried leaf tissue of C_3 plants has shown to be correlated to TE. This is useful, but is not necessarily practical as carbon isotope discrimination is a costly measurement and therefore not ideal as a selection criterion. Progress has been made in C_3 crops, towards obtaining a selection index for high TE lines. In peanut, the carbon isotope discrimination was linearly related to the specific leaf area of the leaves. Thus, a surrogate for Δ is available and has been used to breed high TE lines.[9]

ENVIRONMENTAL INFLUENCES ON TRANSPIRATION EFFICIENCY

Water Deficit Effect

Under moderate water deficits, TE increased in grain sorghum.[3,10] The leaf area index is irreversibly reduced under stress, cell density is maintained but cell enlargement is irreversibly affected.[11] Some plants may be able to adjust osmotically which may contribute to resistance to water deficits.

Rising CO_2

As the CO_2 concentration doubles, transpiration will decline and photosynthesis will increase.[12]

TRANSPIRATION EFFICIENCY IN DROUGHT RESEARCH

If this trait can be identified as significant and suitable selection criteria can be developed, more efficient crops will result through the incorporation of high TE lines in breeding programs. Benefits of high TE lines, anticipated from crop simulation modeling of grain sorghum, suggest a 10% increase in yield from moderate to good environments.[13] Genetic differences in TE appear to be detectable in a range of crops. The combination of C_4 productivity in marginal environments along with improved drought efficiency may have useful economic benefits in the future.

REFERENCES

1. Tanner, C.B.; Sinclair, T.R. Efficient water use in crop production: research or re-search? In *Limitations to Efficient Water Use in Crop Production*; ASA-CSSASSSA: Madison, WI, 1983.
2. Viets, F.G., Jr. Fertiliser and the efficient use of water. Adv. Agron. **1962**, *14*, 223–264.
3. Mortlock, M.Y.; Hammer, G.L. Genotype and water limitation effects on transpiration efficiency in sorghum. In *Water Use in Crop Production*; Kirkham, M.B., Ed.; Food Products Press, The Haworth Press: Binghamton, New York, 1999; 265–286.
4. Griffiths, H. Carbon isotope discrimination. In *Photosynthesis and a Changing Environment: A Field and Laboratory Manual*; Hall, D.O., Sturlock, O., Bolhàr-Nordenkampf, R., Leegood, R.C., Long, S.P., Eds.; Chapman & Hall: London, 1993.
5. Farquhar, G.D.; O'Leary, M.H.; Berry, J.A. On the relationship between carbon isotope discrimination and the intercellular carbon dioxide concentration in leaves. Aust. J. Plant Physiol. **1982**, *9*, 121–137.

6. Farquhar, G.D. On the nature of carbon isotope discrimination in C4 species. Aust. J. Plant Physiol. **1983**, *10*, 205–226.
7. Hubick, K.T.; Hammer, G.L.; Farquhar, G.D.; Wade, L.; von Caemmerer, S.; Henderson, S. Carbon isotope discrimination varies genetically in C_4 species. Plant Physiol. **1990**, *91*, 534–537.
8. Farquhar, G.D.; Richards, R.A. Isotopic composition of plant carbon correlates with water use efficiency in wheat genotypes. Aust. J. Plant Physiol. **1984**, *11*, 539–552.
9. Wright, G.C.; Hubick, K.T.; Farquhar, G.D. Discrimination in carbon isotopes of leaves correlates with water-use efficiency of field grown peanut cultivars. Aust. J. Plant Physiol. **1988**, *15*, 815–825.
10. Donatelli, M.; Hammer, G.L.; Vanderlip, R.L. Genotype and water limitation effects on phenology, growth, and transpiration efficiency in grain sorghum. Crop Sci. **1992**, *32* (3), 781–786.
11. Hsiao, T.C.; Bradford, K.J. Physiological consequences of cellular water deficits. In *Limitations to Efficient Water Use in Crop Production*; ASA-CSSA-SSSA: Madison, WI, 1983.
12. Nobel, P.S. Leaves and fluxes. In *Physicochemical and Environmental Plant Physiology*; Academic Press: New York, 1991.
13. Hammer, G.L.; Butler, D.; Muchow, R.C.; Meinke, H. Integrating physiological understanding and plant breeding via crop modelling and optimisation. In *Plant Adaptation and Crop Improvement*; Cooper, M., Hammer, G.L., Eds.; CAB International: Wallingford, U.K., ICRI-AT, IRRI, 1996.

Urban Environments: Remote Sensing

Xiaojun Yang
Department of Geography, Florida State University, Tallahassee, Florida, U.S.A.

Abstract

The application of remote sensing in the urban environment has received more attention than ever, mainly due to the capability of this modern technology in developing the information base that is essential for the planning and management of global cities now containing more than half of the 7 billion people of our planet. This topical entry provides a brief overview on the utilities of remote sensing for the urban environment. While remote sensing has predominately been used for urban feature extraction, we have witnessed a steady and strengthening extension of the applicability beyond this conventional domain and into areas of urban socioeconomic and environmental analysis and predictive modeling of urbanization over the past two decades. This entry covers these aspects, along with some areas that need further research.

INTRODUCTION

The concentration of large population in urban areas continues to be one of the major global changes happening in both developing and developed counties alike.[1] While intensifying urban growth and development have often been considered an indication of the vitality of a regional economy, they have rarely been well planned, thus provoking concerns over the degradation of our environment and ecosystems.[2] Monitoring urban growth and landscape changes is critical both to those who study urban dynamics and those who must manage resources and provide services in the urban environment.[3]

Assessment of urban growth and landscape changes involves the procedure of inventorying and mapping that can help derive a reliable information base.[4] Urban and landscape patterns are observable and therefore can be mapped through ground surveys or remote sensing. While ground surveys are largely localized by nature, remote sensing makes direct observations across a large area of the land surface, thus allowing urban landscape to be mapped in a timely and cost-effective mode. Nevertheless, the urban environment, because of its complex and dynamic landscape types, challenges the applicability and robustness of remote sensing and geospatial technologies.[5] Encouragingly, recent innovations in data, technologies, and theories in the wider arena of remote sensing and geospatial technologies have provided us with invaluable opportunities to advance the studies on the urban environment.

This topical entry will present a brief overview on the utilities of remote sensing for the urban environment (Fig. 1). While an overwhelming share of the existing work has targeted the use of remote sensing to derive various urban physical features, we have witnessed a steady and strengthening expansion of the applicability beyond this domain and into areas of urban socioeconomic and environmental analysis and predictive modeling of urbanization over the past two decades.[6] The sections provide an overview on each of the major areas of research and discuss several issues that need further research.

REMOTE SENSING AND URBAN ANALYSIS

Remote Sensing Systems for Urban Areas

Remote sensor data used for the urban environment should meet certain conditions in terms of spatial, spectral, radiometric, and temporal characteristics.[5] There is a wide variety of passive and active remote sensing systems acquiring data with various resolutions that can be useful for urban analysis. Medium-resolution remote sensor data have been used to examine large-dimensional urban phenomena or processes since early 1970s when Landsat-1 was successfully launched.[3] Images acquired by the U.S. Landsat program and French SPOT satellites, along with those by the Indian remote sensing satellites, the NASA's Terra satellite, and the China–Brazil Earth resources satellites, are the principal sources of data under this category. With the launch of IKONOS, the world's first commercial, high-resolution imaging satellite, on September 24, 1999, very high spatial resolution satellite imagery became available, which allowed a detailed work concerning the urban environment. Images under this category include those acquired by QuickBird (since 2001), WorldView-1 (since 2007), GeoEye-1 (since 2008), WorldView-2 (since 2009), Pleiades-1 (since 2011), and SPOT-6 (since 2012). Independent of weather conditions, active remote sensing systems, such as airborne or space-borne radar, can be particularly useful for applications such as housing damage

Encyclopedia of Natural Resources DOI: 10.1081/E-ENRL-120049151

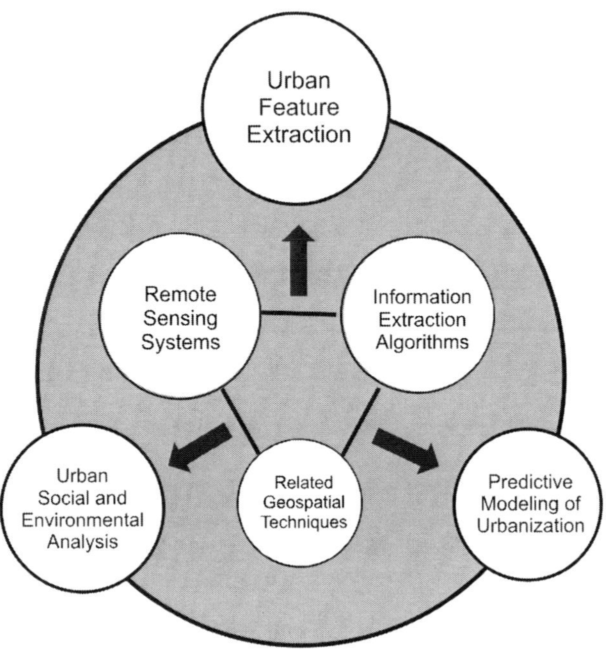

Fig. 1 Remote sensing of urban environments: an overview.

assessment or ground deformation estimation in connection with some disastrous events in urban areas.[7] Another active sensor system, similar in some respects to radar, is lidar (light detection and ranging), which can be used to derive height information useful for reconstructing three-dimensional city models[8] or detailed land use mapping.[9] Last, image fusion techniques allow the combined use of images varying in spatial, spectral, temporal, or radiometric resolution in support of various urban applications.

Algorithms and Techniques for Urban Attribute Extraction

The urban environment is characterized by the presence of heterogeneous surface covers with large interpixel and intrapixel spectral variations, thus challenging the applicability and robustness of conventional image-processing algorithms and techniques. Largely built upon parametric statistics, conventional pattern classifiers generally work well for medium-resolution scenes covering spectrally homogeneous areas, but not in heterogeneous regions, such as urban areas, or when scenes contain severe noises due to the increase of image spatial resolution. Developing improved image processing algorithms and techniques for working with different types of remote sensor data has, therefore, become a very active research area in urban remote sensing. For years, various strategies have been developed to improve urban mapping performance at the per-pixel, subpixel, or object level.[10–12] While the pattern recognition approach at the per-pixel level has been widely adopted over a long period of time, the one at the subpixel level allows the decomposition of each pixel into independent end members or pure materials to map urban subpixel composition. This new approach can be particularly useful for working with medium-resolution images. While the approach at the per-pixel or subpixel level is heavily built upon the use of image spectral features, the object-oriented approach allows the combined use of spectral and spatial characteristics. This latest approach aims to create homogeneous image objects in terms of spectral and shape elements, and these objects are then analyzed using a traditional classification algorithm (e.g., nearest-neighbor, minimum distance, maximum likelihood) or a knowledge-based approach.

Urban Socioeconomic Analysis

Applying remote sensing in urban socioeconomic analysis has been an expanding research area in urban remote sensing. There are two major types of analysis. The first type of centers on linking socioeconomic data to land change data derived from remote sensing in order to identify the drivers of landscape changes.[13] This type of analysis usually needs to integrate image analysis with geographic information system (GIS) and spatial statistical analysis. The other type focuses on the development of indicators of urban socioeconomic status by a combined use of remote sensing and census or field-survey data.[14] For example, defining and measuring urban sprawl, which is considered an important issue for urban and regional planning, involves not only urban spatial characteristics derived from remote sensing but also GIS-based socioeconomic conditions such as population density and transportation.[15] Another example is the estimation of small area population; this type of information is critical for decision making by both public and private sectors but is only available for one date per decade. Remote sensing can provide an alternative to derive reliable population estimation in a timely and cost-effective fashion.[16]

Urban Environmental Analysis

Although urban areas are quite small relative to the global land cover, they significantly alter hydrology, biodiversity, biogeochemistry, and climate at local, regional, and global scales.[2] Understanding environmental consequences of urbanization is a critical concern to both the planning and global change science communities. Urban environmental analysis can help understand the status, trends, and threats in urban areas so that appropriate management actions can be planned and implemented. This is a research area in which remote sensing can play a critical role. For example, remote sensing can be used to derive the spatial distribution of impervious surfaces in urban areas that can be linked with water quality indicators.[17–19] Remote sensing of land surface temperature can help characterize the urban head island phenomenon that can be further linked with human health problems.[20,21]

Remote sensing can be also used to estimate gross primary production in urban areas that can be associated with various settlement densities; such knowledge can help understand the net carbon exchange between land and the atmosphere due to urban development.[22] And remote sensing can be used to characterize biodiversity in urban areas that can help assess how urbanization affects biodiversity.[23]

Urban Growth and Landscape Change Modeling

A group of important activities in urban studies are to understand urban dynamics and to assess future urban growth impacts on the environment. There are two major types of models that can be used to support such activities: analytical models that are useful to explain urban expansion and evolving patterns as well as dynamic models that can be used to predict future urban growth and landscape changes. The group of dynamic models can be particularly valuable because of their predictive power, which can be used to imagine, test, and assess the spatial consequences of urban growth under specific socioeconomic and environmental conditions. The role of remote sensing is indispensible in the entire model development process from model conceptualization to implementation that includes input data preparation, model calibration, and model validation.[24]

CONCLUSION AND FURTHER RESEARCH

This topical entry provides an overview on the utilities of remote sensing for the urban environment and discusses the major areas of research in urban remote sensing. Although many exciting progresses have been made in the field of urban remote sensing, there are several major conceptual or technical areas that deserve further attention. First, although the development of urban remote sensing has been largely technology driven, urban remote sensing professionals should be equipped with not only solid technical skills but also essentials of intellectual knowledge on the urban environment that can help better plan and implement an urban remote sensing project. Second, while current literature overwhelmingly focuses on the issue of image spatial resolution, recent studies suggest the importance of image spectral characteristics for urban feature mapping.[25] Continuing research is needed to help acquire good and sufficient *in situ* data for building comprehensive spectral libraries of different urban features. Third, an emerging research effort is needed to balance the different needs by remote sensing and urban planning communities. Finally, more efforts are needed to develop innovative data models used for representing dynamic processes, to identify improved methods and techniques that can be used to deal with data incompatibility in terms of parameter measuring and sampling schemes, and to develop more realistic dynamic models that can be used to support various urban and regional planning activities.

REFERENCES

1. Foley, J.A.; DeFries, R.; Asner, G.P.; Barford, C.; Bonan, G.; Carpenter, S.R.; Chapin, F.S.; Coe, M.T.; Daily, G.C.; Gibbs, H.K.; Helkowski, J.H.; Holloway, T.; Howard, E.A.; Kucharik, CJ.; Monfreda, C.; Patz, JA.; Prentice, IC.; Ramankutty, N.; Snyder, PK. Global consequences of land use. Science **2005**, *309* (5734), 570–574.
2. Grimm, N.B.; Faeth, S.H.; Golubiewski, N.E.; Redman, C.L.; Wu, J.;Bai, X.; Briggs, J.M. Global change and ecology of cities. Science **2008**, *319* (5864), 756–760.
3. Yang, X. Satellite monitoring of urban spatial growth in the Atlanta metropolitan region. Photogram. Eng. Remote Sensing **2002**, *68* (7), 725–734.
4. Yang, X. Remote sensing and GIS for urban analysis: An introduction. Photogram. Eng. Remote Sensing **2003**, *69* (8), 937, 939.
5. Jensen, J.R.; Cowen, D.C. Remote sensing of urban/suburban infrastructure and socio-economic attributes. Photogram. Eng. Remote Sensing **1999**, *65* (5), 611–622.
6. Yang, X. What is urban remote sensing? In *Urban Remote Sensing: Modeling, Synthesis and Modeling in the Urban Environment*; Yang, X., Ed.; Wiley-Blackwell: Chichester, UK, 2011a; 3–11 pp.
7. Dell'Aqua, F.; Gamba, P.; Polli, D. Very-high-resolution spaceborne synthetic aperture radar and urban areas: looking into details of a complex environment. In *Urban Remote Sensing: Modeling, Synthesis and Modeling in the Urban Environment*; Yang, X., Ed.; Wiley-Blackwell: Chichester, UK, 2011; 63–73 pp.
8. Li, J.; Guan, H. 3D building reconstruction from airborne lidar point clouds fused with aerial imagery. In *Urban Remote Sensing: Modeling, Synthesis and Modeling in the Urban Environment*; Yang, X., Ed.; Wiley-Blackwell: Chichester, UK, 2011; 75–91 pp.
9. Meng, X.; Currit, N.; Wang, L.; Yang, X. Detect residential buildings from lidar and aerial photographs through object-oriented land-use classification. Photogram. Eng. Remote Sensing **2012**, *78* (1), 35–44.
10. Yang, X. Parameterizing support vector machines for land cover classification. Photogram. Eng. Remote Sensing **2011b**, *77* (1), 27–37.
11. Verhoeye, J.; Wulf, R.D. Land cover mapping at sub-pixel scales using linear optimization techniques. Remote Sensing Environ. **2002**, *79* (1), 96–104.
12. Walker, J.S.; Briggs, J.M. An object-oriented approach to urban forest mapping in Phoenix. Photogram. Eng. Remote Sensing **2007**, *73* (5), 577–583.
13. Lo, C.P.; Yang, X. Drivers of land-use/land-cover changes and dynamic modeling for the Atlanta, Georgia metropolitan area. Photogram. Eng. Remote Sensing **2002**, *68* (10), 1073–1082.
14. Yu, D.L.; Wu, C.S. Incorporating remote sensing information in modeling house values: A regression tree approach. Photogram. Eng. Remote Sensing **2006**, *72* (2), 129–138.
15. Frenkel, A.; Orienstein, D. A pluralistic approach to defining and measuring urban sprawl. In *Urban Remote*

Sensing: Modeling, Synthesis and Modeling in the Urban Environment; Yang, X., Ed.; Wiley-Blackwell: Chichester, UK, 2011; 165–181 pp.
16. Wang, L.; Cardenas, J.S. Small area population estimation with high-resolution remote sensing and lidar. In *Urban Remote Sensing: Modeling, Synthesis and Modeling in the Urban Environment*; Yang, X., Ed.; Wiley-Blackwell: Chichester, UK, 2011; 183–193 pp.
17. Ridd, M.K. Exploring a VIS (vegetation-impervious surface-soil) model for urban ecosystem analysis through remote sensing: comparative anatomy for cities? Int. J. Remote Sensing **1995**, *16* (12), 2165–2185.
18. Wu, C.; Murray, A.T. Estimating impervious surface distribution by spectral mixture analysis. Remote Sensing of Environment **2003**, *84* (4), 493–505.
19. Yang, X. Estimating landscape imperviousness index from satellite imagery. IEEE Geosci. Remote Sensing Lett. **2006**, *3* (1), 6–9.
20. Lo, C.P.; Quattrochi, D.A.; Luvall, J.C. Application of high-resolution thermal infrared remote sensing and GIS to assess the urban heat island effect. Int. J. Remote Sensing **1997**, *18* (2), 287–304.
21. Lo, C.P. Quattrochi, D.A. Land-use and land-cover change, urban heat island phenomenon, and health implications: A remote sensing approach. Photogram. Eng. Remote Sensing **2003**, *69* (8), 1053–1063.
22. Buyantuyev, A.; Wu, J. Urbanization alters spatiotemporal patterns of ecosystem primary production: a case study of the Phoenix metropolitan region, USA. J. Arid Environ. **2009**, *73*, 512–520.
23. Hedblom, M.; Söderström, B. Landscape effects on birds in urban woodlands: an analysis of 34 Swedish cities. J. Biogeography **2010**, *37* (7), 1302–1316.
24. Yang, X.; Lo, C.P. Modeling urban growth and landscape change for Atlanta metropolitan region. Int./ J. Geographical Info. Sci. **2003**, *17* (5), 463–488.
25. Herold, M.; Roberts, D.A.; Gardner, M.E.; Dennison, P.E. Spectrometry for urban area remote sensing - Development and analysis of a spectral library from 350 to 2400 nm. Remote Sensing Environ. **2004**, *91*, 304–319.

BIBLIOGRAPHY

1. Donnay, J.P.; Barnsley, M.J.; Longley, P.A., Eds.; *Remote Sensing and Urban Analysis*; Taylor & Francis: London, UK, 2001.
2. Jensen, R.R.; Gatrell, J.D.; McLean, D., Eds.; *Geo-Spatial Technologies in Urban Environments: Policy, Practice, and Pixels (2nd)*; Springer: Berlin, Germany, 2007.
3. Mesev, V., Ed.; *Remotely-Sensed Cities*; CRC Press: Boca Raton, Florida, 2003.
4. Netzband, M.; Stefanov, W.L.; Redman, C. L., Eds. *Applied Remote Sensing for Urban Planning, Governance and Sustainability*; Springer: Berlin, Germany, 2007.
5. Rashed, T.; Jürgens, C., Eds.; *Remote Sensing of Urban and Suburban Areas*; Springer: Berlin, Germany, 2010.
6. Ridd, M.K.; Hipple, J.D., Eds.; *Manual of Remote Sensing, Volume 5: Remote Sensing of Human Settlements*; ASPRS: Maryland, U.S.A., 2006.
7. Weng, Q., Ed.; *Remote Sensing of Impervious Surfaces*; CRC Press: Boca Raton, Florida, 2008.
8. Weng, Q.; Quattrochi, D., Eds.; *Urban Remote Sensing*; CRC Press: Boca Raton, Florida, 2007.
9. Yang, X., Ed.; *Urban Remote Sensing: Monitoring, Synthesis and Modeling in the Urban Environment*; Wiley-Blackwell: Chichester, UK, 2011.

Vernal Pool

Peter W. C. Paton
Department of Natural Resources Science, University of Rhode Island, Kingston, Rhode Island, U.S.A.

Abstract
Vernal pools are small, shallow wetlands that usually dry annually. They often are inundated by snowmelt and spring rains and are frequently dry by the end of summer, although some pools do not become dry every year. These wetlands occur globally and are widespread in North America along the Pacific Coast, in the glaciated Northeast and Midwest and in parts of the Southeast. The hydroperiod (days with surface inundation) and hydroregime (timing of inundation) are the primary factors determining the flora and fauna occupying these pools. The geomorphological setting of vernal pools is variable, with pools along the Pacific Coast perched above impervious substrates, while pools occur across a broader spectrum of geological settings in the eastern North America. Biologists are just beginning to understand the complex food webs and life histories of the flora and fauna of these biodiversity hotspots. In some parts of the world, vernal pools are well known for their flora, invertebrates, and vertebrates. Due to habitat loss and other anthropogenic impacts, several endangered species occur in vernal pools, thus this habitat type is a major conservation concern for biologists.

INTRODUCTION

Broadly defined, vernal pools are generally hydrologically isolated from open water bodies, temporary to semipermanent wetlands that form in small, shallow permanent basins during the coolest part of the year, and dry annually or during drought years.[1–3] In the United States, these ephemeral wetlands are found in Pacific Coast states, in the North, and Northeast,[1] however they can be found globally.[4,5] They are called vernal pools because in many areas, particularly Mediterranean climates such as California, these wetlands are inundated with water in the spring; vernal is from the latin word "vernalis" which means "pertaining to spring" and "pool" is often used for a small water body.[3] A somewhat more restrictive definition for vernal pools in California is "precipitation-filled seasonal wetlands inundated during periods when temperature is sufficient for plant growth, followed by a brief waterlogged-terrestrial stage and culminating in extreme desiccating soil conditions of extended duration."[4] In the widely used Cowardin wetland classification system,[6] vernal pools with vegetation are classified as seasonally flooded emergent wetlands, while ponds with no vegetation are open water or unconsolidated bottom habitats in the palustrine system, although some ponds would be better classified as intermittently flooded or semi-permanently flooded. Other terms that biologists have used for this wetland type include spring pools, ephemeral wetlands, semipermanent pond, fishless ponds, seasonal astatic waters, vernal ponds, vernal marshes, buffalo wallows, seasonal pools, seasonal ponds, vleis, or temporary waters.[2,4]

Four phases have been identified in the annual cycle of vernal pools: (i) *desiccated phase*: pond substrate is completely dry and desert-like, which often occurs in the summer or fall; (ii) *preinundation phase*: initial precipitation makes the soil moist, allowing germination of flora and hatching of some vertebrates and invertebrates; (iii) *pool phase*: the pond basin is completely full of water from precipitation; and (iv) *drying phase*: evapotranspiration directly or indirectly causes the pond and soil substrate to dry.[4] To classify vernal pools, biologists suggest that one needs to understand three elements: first, the *water source* to fill pools, which may be rain-fed or groundwater or a combintion; second, the *hydroperiod*, which is the duration of inundation and the waterlogged phase; and third, the *hydroregime*, which is the timing of inundation.[1,4] The geomorphologic setting also helps one to understand the function of vernal pools.[7]

Vernal pools primarily occur in confined, permanent basins with no permanent inflow or outflow outlet; thus, there is no continuous surface water connection with permanently inundated water bodies. Vernal pools in Mediterranean climates are usually found in small, shallow basins, where surface area usually covers 50–5000 m^2,[8] and depths ranging from 0.1 to 1 m.[9,10] In California, vernal pools occur in small depressions that are underlain by an impervious substrate that limits drainage.[11,12] Therefore, they tend to be clustered on the landscape because they are associated with specific geological formations and soils.[13] In the glaciated, northeastern North America, vernal pools are found in a broad array of geomorphic settings.[7] The surface area of pond basins in the glaciated Northeast

Encyclopedia of Natural Resources DOI: 10.1081/E-ENRL-120047524

averages about 0.1 ha in size, but ranges from 0.01 to 2.25 ha.[2] Pond basins typically are often about 1 m depth at maximum water depth, with a range from 0.2 to >3.8 m deep.[2]

Hydroperiod, the number of days with surface inundation by water, is a function of climate and water balance, where changes in surface water in a pool is equal to the sum of inputs from precipitation, groundwater, and surface water, minus losses from evapotranspiration.[10] Water storage will vary as a function of climatic conditions and hydrogeologic settings, with considerable variation among pond basins. In ponds in the glaciated Northeast, the hydroperiod for most vernal pools averages about 190–270 days inundation annually and can range from 70 days to permanently flooded in some wet years.[14] One set of vernal pools in California has average hydroperiods of 115 days,[15] while the hydroperiod at other pools ranged from 10 to 200 days.[16,17] Precipitation is the primary water input for most vernal pools, with weekly precipitation accounting for over 50% of the variation in water depth at one site in Massachusetts.[18] However, in certain geomorphic settings, vernal pools can be fed by groundwater.[7,10,15]

Vernal pools are unique ecosystems that include many regionally endemic (e.g., in California) or obligate species (e.g., glaciated Northeast).[1] However, many genera and species that are found in vernal pools are distributed across the continent. Vernal pools are important for local, regional, and national biodiversity because they provide habitat for unique flora,[19] invertebrates,[2,20] and vertebrates.[21] In particular, pool-breeding amphibians have their greatest reproductive success in fishless vernal pools, although they will use other aquatic habitats for breeding.[22]

FLORA

Hydroperiod and hydroregime, coupled with soil and canopy cover, are probably the most influential factors that affect plant community composition within vernal pools.[4,8] One of the unique attributes of vernal pools is the flora adapted to these ephemeral wetlands, particularly in California. Vernal pools are among the last ecosystems in California dominated by native flora.[13] Along the Pacific Coast, over 100 plant taxa commonly occur in vernal pools, with over half endemic to California and the other species dispersed from Washington to Baja California. Although vernal pools have a global distribution, California is the only region that evolved an extensive flora endemic to vernal pools.

At least 15 taxa are federally listed as either endangered (10 species) or threatened (5 species) and 37 other species are either state-listed or state species of concern.[23] In the glaciated Northeast, there are at least 20 at-risk plant taxa that occur in vernal pools, with 1 federally listed species (Northern bulrush *Scirpus ancistrochaetus*) and 4 threatened species.[24]

Plants occurring in vernal pools are generally dispersed along a hydrological gradient, with obligate wetlands plants found in deeper parts of a pond and facultative wetland plants occupying the shallow edges of vernal pools.[24] Vegetation has been used successfully in the Northeast to predict the hydroperiod of vernal pools. Researchers in Rhode Island were able to predict the hydroperiod class of 72% of 65 ponds they studied, but many ponds had no vascular vegetation which made it impossible to predict the hydroperiod of those ponds.[25]

INVERTEBRATES

Vernal pools with different hydroperiods and hydroregimes usually are occupied by different community assemblages, which can include hundreds of species.[2,20] Invertebrate community structure and food webs within vernal pools can be complex, with invertebrate species occupying a range of trophic levels. Lower trophic levels include decomposers, detritivores, and photosynthesizers; intermediate levels include grazers, shredders, filter feeders, and deposit feeders; while upper tropic levels include predators and parasites.[2] Two primary ecological forces structuring vernal pool communities are competition and predation. Predatory dragonfly nymphs and diving-beetle larve can affect the composition of the entire community, including vertebrates breeding in the pools.[20]

Fairy shrimp (Order Anostraca) rank among the most charismatic invertebrates that are closely linked to vernal pools. In the northeastern United States, most species are the genus *Eubranchipus*, which are filter feeders that eat algae, zooplankton and bacteria; several species in the region are rare or endangered. On the Pacific Coast, several species of endemic fairy shrimp are among the flagship species leading to conservation efforts for this imperiled wetland because several are either federally listed as endangered (four species: conservancy, longhorn, san diego, and riverside fairy shrimp) or threatened (one species: vernal pool fairy shrimp).[26] These species are threatened by habitat loss and the introduction of alien species, such as predatory fish (*Gambusia affinis*).[27]

VERTEBRATES

Among the vertebrates that use vernal pools, pool-breeding amphibians are the taxa that many biologists associate with vernal pools. In California, at least two species of amphibians (California Tiger Salamander (*Ambystoma californiense*) and Western Spadefoot Toad (*Spea hammondii*)) regularly use vernal pools, but these two species also use other types of wetlands.[19] In the glaciated Northeast, at least 10 species of salamanders and three species of frogs use vernal pools as their primary breeding habitat, and one other species of salamander and 14 species of frogs use vernal pools in addition to other types of wetlands for breeding.[22] Pool-breeding amphibians have complex life histories, with aquatic larvae developing through metamorphosis in vernal pools, while juveniles and adults spend most of their life in the terrestrial realm surrounding vernal pools. Vernal pool characteristics that are important in determining amphibian use of the pool include hydroperiod, canopy cover over the pool, potential predator composition, and vegetation composition in area surrounding the pond.[22]

There are additional vertebrates that often use vernal pools for various aspects of their life history. Although reptiles, birds, and mammal may not often breed in vernal pools, they often use these ephemeral wetlands as foraging and resting habitat.[25,28] In the glaciated Northeast, of the 24 species of turtles in the region, at least 64% obtain food, shelter, or other resources from vernal pools, with at least three species considered as vernal pool specialists.[25] Of the 43 snakes in the Northeast, over 50% use vernal pools. Over 80% of the 232 birds in the region occasionally use vernal pools. In the western United States, vernal pools provide important habitat for resident and migratory birds, with many species of waterfowl and shorebirds often using these wetlands as foraging habitat.[29]

CONCLUSION

In North America, vernal pools are abundant on the Pacific Coast, in the glaciated Northeast, and in parts of the Southeast. Due to their hydroperiod and hydroregime, they provide important habitat for many species of regionally endemic invertebrates and vertebrates that are adapted to this unique type of wetland. Many species of endangered plants, invertebrates, and vertebrates occur in vernal pools, thus vernal pools are a global conservation concern for biologists. Therefore, land planners and biologists have developed a variety of strategies to conserve vernal pools.[30] However, for many species, it is just as important to protect habitat surrounding the vernal pool, as some species may disperse hundreds of meters from vernal pools during other parts of their annual cycle.[31]

REFERENCES

1. Zedler, P.H. Vernal pools and the concept of isolated wetlands. Wetlands **2003**, *2* (3), 597–607.
2. Colburn, E.A. *Vernal pools: Natural History and Conservation*; McDonald and Woodward Publishing Co.: Blacksburg, Virginia, 2004.
3. Calhoun, A.J.K.; deMaynadier, P.G. *Science and Conservation of Vernal Pools in Northeastern North America*; CRC Press: Boca Raton, Florida. 2008.
4. Keeley, J.E.; Zedler. P.H. Characterization and global distribution of vernal pools. In *Ecology, Conservation, And Management of Vernal Pool Ecosystems – Proceedings from a 1996 Conference*; Witham, C.W., Bauder, E.T., Belk, D., Ferren W.R., Jr., R. Ornduff, R, Eds.; California Native Plant Society: Sacramento, CA, 1998; 1–14.
5. Vanschoenwinkel, B.; Hulsmans, A.; De Roeck, E.; De Vries, C; Seaman, M; Brendonck, L. Community structure in temporary freshwater pools: disentangling the effects of habitat size and hydroregime. Freshwater Biol. **2009**, *54* (7), 1476–1500.
6. Cowardin, L.M.; Carter, V.; Golet, F.C.; LaRoe, E.T. *Classification of Wetlands and Deepwater Habitats of the United States*. U.S. Fish and Wildlife Service, Washington, D.C. 1979.
7. Rheinhardt, R.D.; Hollands, G.G. Classification of vernal pools: geomorphic setting and distribution. In *Science and Conservation of Vernal Pools in Northeastern North America*; Calhoun, A.J.K., deMaynadier, P.G. Eds.; CRC Press: Boca Raton, FL, 2008; 31–54.
8. Mitsch, W.J.; Gosselink, J.G. *Wetlands*; 3rd Ed. John Wiley and Sons Inc: New York, 2000.
9. Hanes, T.; Stromberg, L. Hydrology of vernal pools on non-volcanic soils in the Sacramento Valley. In *Ecology, Conservation, and Management of Vernal Pool Ecosystems*; Witham, C.W., Bauder, E.T., Belk, D., Ferren, W.R. Jr, Ornduf, R. Eds.; California Native Plant Society: Sacramento, CA; 1998; 38–49.
10. Brooks, R.T.; Hayashi, M. Depth–area–volume and hydroperiod relationships of ephemeral (vernal) forest pools in southern New England. Wetlands **2002**, *22* (2), 247–255.
11. Leibowitz, S.G.; Brooks, R.T. Hydrology and landscape connectivity of vernal pools. In *Science and Conservation of Vernal Pools in Northeastern North America*; Calhoun, A.J.K., deMaynadier, P.G., Eds.; CRC Press: Boca Raton, FL; 2008; 31–53.
12. Holland, R.F. The vegetation of vernal pools: a survey. In *Vernal Pools: Their Ecology and Conservation*; Jain, S. Ed.; Institute of Ecology Publication No. 9. University of California: Davis, CA; 1976; 11–15.
13. Rains, M.C.; Fogg, G.E.; Harter, T.; Dahlgren, R.A.; Williamson, R.J. The role of perched aquifers in hydrological connectivity and biogeochemical processes in vernal pool landscapes, Central Valley, California. Hydrol. Proces. **2006**, *20* (5), 1157–1175.
14. Smith, D.W.; Verrill, W.L. Vernal pool–soil–landform relationships in the Central Valley, California. In *Ecology, Conservation, and Management of Vernal Pool Ecosystems*, Witham, C.W., Bauder, E.T., Belk, D., Ferren, W.R. Jr, Ornduf, R. Eds.; California Native Plant Society: Sacramento, CA. 1998; 15–23.
15. Skidds, D.E.; Golet, F.C. Estimating hydroperiod suitability for breeding amphibians in southern Rhode Island seasonal forest ponds. Wetland Ecol. Manag. **2005**, *13* (3), 349–366.
16. Marty, J.T. Effects of cattle grazing on diversity in ephemeral wetlands. Conserv. Biol. **2005**, *19* (5), 1626–1632.
17. Black, C.; Zedler, P.H. An overview of 15 years of vernal pool restoration and construction activities in San Diego County, California. In *Ecology, Conservation, and Management of Vernal Pool Ecosystems*, Witham, C.W., Bauder, E.T., Belk, D., Ferren, W.R. Jr, Ornduf, R. Eds.; California Native Plant Society: Sacramento, CA; 1998; 195–205.

18. Brooks, R.T. Weather-related effects on woodland vernal pool hydrology and hydroperiod. Wetlands **2004**, *24* (1), 104–114.
19. Zedler, P.H. The ecology of southern California vernal pools: A community profile. Biology Report 85, U.S. Fish and Wildlife Service: Washington, D.C. 1987.
20. Colburn, E.A.; Weeks, S.C.; Reed, S.K. Diversity and ecology of vernal pool invertebrates. In *Science and Conservation of Vernal Pools in Northeastern North America*; Calhoun, A.J.K., deMaynadier, P.G., Eds.; CRC Press: Boca Raton, FL, 2008; 105–126.
21. Mitchell, J.C.; Paton, P.W.C.; Raithel, C.J. The importance of vernal pools to reptiles, birds, and amphibians. In *Science and Conservation of Vernal Pools in Northeastern North America*; Calhoun, A.J.K.; deMaynadier, P.G., Eds.; CRC Press: Boca Raton, FL, 2008;169–192.
22. Semlitsch, R.D.; Skelly, D.K. Ecology and conservation of pond-breeding amphibians. In *Science and Conservation of Vernal Pools in Northeastern North America*; Calhoun, A.J.K., deMaynadier, P.G., Eds.; CRC Press: Boca Raton, FL, 2008; 127–148.
23. Lazar, K.A. Characterization of rare plant species in the vernal pools of California. MS thesis, Plant Biology Graduate Group, University of California: Davis, CA, 2006.
24. Cutka, A.; Rawinski, T.J. Flora of Northeastern vernal pools. In *Science and Conservation of Vernal Pools in Northeastern North America*; Calhoun, A.J.K.; deMaynadier, P.G., Eds.; CRC Press: Boca Raton, FL, 2008; 71–104.
25. Mitchell, J. Using plants as indicators of hydroperiod class and amphibian habitat suitability in Rhode Island seasonal pond. M.S. thesis. University of Rhode Island: Kingston, RI, 2005. http://www.fws.gov/endangered (accessed October 2011).
26. Leyse, K.E.; Lawler, S.P.; Strange, T. Effects of an alien fish, *Gambusia affinis*, on an endemic California fairy shrimp, *Linderiella occidentalis*: implications for conservation of diversity in fishless waters. Biol. Conserv. **2003**, *118* (1), 57–65.
27. Paton, P.W.C. A review of vertebrate community composition in seasonal forest pools of the northeastern United States. Wetlands Ecol. Manag. **2005**, *13* (3), 235–246.
28. Silveira, J.G. Avian uses of vernal pools and implications for conservation practice. In *Ecology, Conservation, and Management of Vernal Pool Ecosystems*; Witham, C.W., Bauder, E.T., Belk, D., Ferren, W.R. Jr, Ornduf, R. Eds.; California Native Plant Society: Sacramento, CA, 1998; 195–205.
29. Calhoun, A.J.K; Reilly, P. Conserving vernal pool habitat through community based conservation. In *Science and Conservation of Vernal Pools in Northeastern North America*; Calhoun, A.J.K.; deMaynadier, P.G., Eds.; CRC Press: Boca Raton, FL, 2008; 319–341.
30. deMaynadier, P.G.; Houlahan, J.E. Conserving vernal pool amphibians in managed forests. In *Science and Conservation of Vernal Pools in Northeastern North America*; Calhoun, A.J.K.; deMaynadier, P.G., Eds.; CRC Press: Boca Raton, FL, 2008; 253–280.

Water Deficits: Development

Kadambot H. M. Siddique
Helen Bramley
The UWA Institute of Agriculture, University of Western Australia, Crawley, Western Australia, Australia

Abstract

Water deficit affects plant growth and productivity and is the single most important factor limiting crop yields worldwide. To breed crops better able to cope with limited water availability we need to know how water moves through plants and mechanisms that control water transport. This entry provides a summary of our current understanding of these processes. It briefly describes the way in which water flows through plants and soils due to gradients in water potential. Mechanisms that improve the retention of water within the plant and ways that the plant can manipulate the flow of water are also described. An emphasis is given to crop plants because improvement in maintaining the hydration of crops is important for efficient agricultural water use and greater productivity.

INTRODUCTION

Despite water being the Earth's most abundant compound, water deficit is the single most important factor limiting crop yields worldwide. Of the Earth's total water supply less than 1% exists in freshwater aquifers, rivers, and lakes. Agriculture consumes more than two-thirds of this available fresh water worldwide. The increasing global demands on both food and water resources require more efficient agricultural water use to maintain, or even improve production. To enable more efficient and sustainable use of our water resources, a greater understanding of plant water use is needed, especially the way in which water deficits develop and affect plant growth and productivity.

THE SOIL, PLANT, AND ATMOSPHERE CONTINUUM

In plants, water deficits develop because of water loss from leaves and insufficient resupply.[1] Water evaporates through open stomata during the uptake of CO_2 from the atmosphere for photosynthesis. This transpired water is replaced by water taken up by the roots and transported to the shoots and leaves. Water absorbed from the soil by roots has to travel a tortuous pathway, crossing cells via the symplasm and cell walls (apoplast), and flowing longitudinally through xylem vessels to reach the sites of evaporation within the leaves.[2] This pathway forms a hydraulic continuum between the soil, plant, and atmosphere. The rate of water flow through this continuum depends on the resistance of the pathway to water flow and the size of the force driving the water flow.

Water moves passively down a chemical potential gradient, which is the free energy associated with water and is more generally called water potential by plant scientists. Hence, gradients in water potential can drive water flow through the soil–plant–atmosphere continuum. The unsaturated atmosphere has an extremely negative water potential, whereas a soil saturated with fresh water has a water potential close to zero. As the soil dries out or if the water is salty, the water potential decreases. Accordingly, a plant must lower its water potential in order to absorb water from the soil.

There are two mechanisms that generate the water potential gradients within plants to induce soil water uptake and drive water upward to the leaves. In the first mechanism, solutes are released into the root xylem, thereby lowering the xylem water potential and inducing root water uptake.[3] As more water is drawn into the xylem a positive pressure develops that pushes water upward. Woody species tend to have low pressures but some herbaceous crop species can generate root pressures up to 300 kPa,[4] which would be sufficient to push water to the top of the plant. However, the high volumes of water that flow through a plant when it is transpiring dilute the xylem sap so that root pressure is negligible during the day.

The second and dominant mechanism generating water potential gradients within plants is due to transpiration. As water evaporates from the interstitial tissues of leaves, surface tension is generated at the air–water interface. Surface tension places the mass of water behind the air–water interface under negative pressure, which is transmitted throughout the liquid continuum due to the cohesion of water molecules,[5] drawing water into the roots and upward to the leaves. As more water evaporates, the greater the

Encyclopedia of Natural Resources DOI: 10.1081/E-ENRL-120049220

surface tension, the more negative the hydrostatic pressure and the greater the pulling force. Hydrogen bonds between water molecules give water its unique tensile strength that can withstand the high tensions generated in transpiring plants. Tensions within the xylem are typically −1 to −2 MPa, but can be as extreme as −10 MPa.[6]

Although water columns in the xylem do not break under the tensions generated by transpiration, the water column can be disrupted by cavitation.[6] Cavitation is generally thought to occur by air-seeding through intervessel pit membranes.[6] This is particularly problematical in dry and saline soils, and hot dry atmospheric conditions, when tensions in the plant are at their largest. Once cavitated, the vessel quickly embolizes and water flow is obstructed. Embolisms can occur in herbaceous species at fairly high water potentials in comparison to woody species, but the reason behind this is not known. However, cavitation may not be as detrimental to crop plants compared with woody species because embolized vessels of herbaceous species are possibly refilled overnight by root pressure, but many trees rely on new growth in the spring to produce water-filled vessels.

Large-water-potential gradients are needed to drive water uptake from the soil if the soil is dry or saline, or if the soil–root interface and the plant tissue through which water flows form significant resistances to water flow. The difference in water potential between the plant and the soil and progressive changes in soil– and plant–water potential as the soil dries out are presented schematically (Fig. 1).

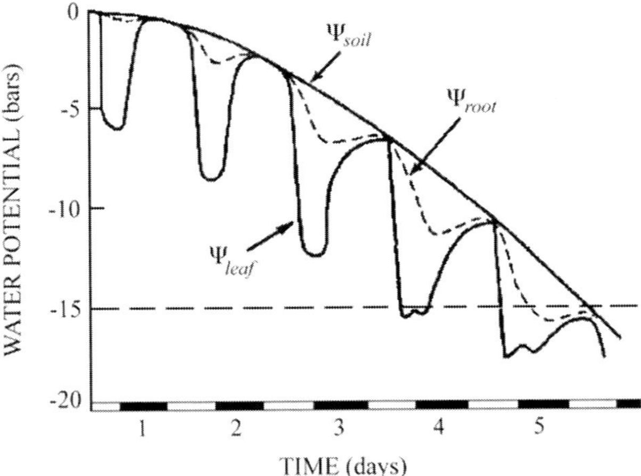

Fig. 1 Schematic representation of changes in leaf–water potential (ψ_{leaf}), root–water potential (ψ_{root}), and soil–water potential (ψ_{soil}) as transpiration proceeds during a drying cycle. The same evaporative conditions are considered to prevail each day; the horizontal dashed line indicates the value of ψ_{leaf} at which wilting occurs.

Source: From Slatyer, R.O. 1967. *Plant–Water Relations*, Academic Press, New York.

WATER STATUS OF CROP PLANTS

The water status of a crop plant is usually defined in terms of its relative water content or water potential. Relative water content is the amount of water in a leaf relative to that at full hydration and standardized by unit dry weight. The net water potential of a plant comprises of two principal components: osmotic pressure and hydrostatic pressure. A gravitational component may also be included, but this is only relevant for tall trees. Osmotic pressure is defined by the concentration of solutes in solution, whereas hydrostatic pressure results from the pressure exerted by water on its surroundings. Within plant cells, water is drawn into the cell due to the concentration of solutes in the cytoplasm and tonoplast. As more water accumulates, the cell volume resists expansion because of the rigidity of cell walls, and a positive hydrostatic pressure develops within the cell called turgor pressure.

Maintenance of leaf hydration and turgor is paramount to maintaining growth and stomatal opening. Plants with higher relative water contents at lower leaf water potentials are generally more resistant to drought, but typically have slow growth rates. Many crop plants are known to accumulate solutes in leaves and roots to maintain turgor in a process called osmotic adjustment. The elasticity of plant cells also determines their capacity for water retention and relative changes in turgor with changes in water content.[7] Plant cells with high volumetric elastic modulus (low elasticity) hold less water at full turgor and for any given change in water content the changes in turgor are larger.

Leaf expansion is one of the first processes that is affected by water deficit and is inhibited at higher leaf water potentials than photosynthesis or metabolism, causing a change in the sink for photoassimilates.[8] Maintenance of plant water status is also important for reproductive processes and fruit or grain growth and development. The grains of legumes are isolated from the water relations of the maternal plant, such that seed-coat turgor is maintained despite changes in the water potential of leaves, stems, and pod walls.[9] The water relations of cereal grains have been shown to decouple from the maternal plant during the grain-filling stage, about two weeks after flowering, which is when the water content of the grain is at its highest.[10] Whether hydraulic isolation occurs because of xylem discontinuity into the grain, formation of an apoplastic barrier, or some other process, the mechanism may ensure continued seed filling and assimilate redistribution, even when the water potential of the environment has reduced the current availability of assimilates from the leaves.

MECHANISMS FOR CONTROL OF WATER TRANSPORT

The relative magnitude of water supply and water loss, as well as the water-holding capacity determine the degree to which plants are water deficient. However, because plants are living organisms, it is important to realize they do not solely provide a pathway for the movement of water from the soil to the atmosphere. In addition to mechanisms for osmotic regulation, plants can also effectively control the flow of water according to their water status through stomatal regulation, changes in anatomy and aquaporin activity.

Water loss from plant communities is usually termed evapotranspiration, because of the combined soil evaporation and plant transpiration. Evapotranspiration from nonstressed crops is largely governed by meteorological conditions. The plant, in contrast to the soil, exercises control over water loss through its stomata and cuticle, and also through its ability to reduce the radiation load through active and/or passive changes in leaf orientation. Because these controls are located at the leaf surface, the internal water status of the plant is closely coupled to that of the soil and less closely coupled to the atmosphere.[1]

A high cuticular resistance is associated with either thicker cuticles or with cuticles enriched with hydrophobic materials. Although the plant cannot flexibly control this, cuticular resistance has been found to decrease with increasing temperature and relative humidity of the atmosphere. Thicker or waxier cuticles may develop in response to drought, which in addition to reducing cuticular transpiration can also increase light reflectance and reduce radiation load.

The density of stomata on the leaf surface as well as stomata apertures determine stomatal conductance to water vapor, CO_2, and other gases. Stomatal density is determined during leaf development, but stomatal aperture is controlled by the turgor pressure of two flanking guard cells, which enables the plant to reversibly modify gas exchange with the atmosphere, with a time scale of only a few minutes. This short-term regulation may both optimize the dry matter production and offer some protection from xylem embolism or other detrimental effects of very low leaf water potential.[11] There is wide variation between species in the critical threshold level in leaf–water potential or relative water content when stomata close. Stomata of crops such as lupin close before any detectable changes in leaf water potential occur in response to depletion of soil water, whereas the stomata of other crops such as wheat gradually close with decreasing soil moisture content and leaf water potential.[12] These differences are, in part, related to differing sensitivity to chemical signals originating in dehydrating roots and conveyed toward the stomatal complexes that can control stomatal apertures.

Crop canopy photosynthesis, transpiration, or leaf extension growth may show midday depressions caused by high evaporative demand if the soil water supply is depleted below a threshold of ca. 50% of total extractable soil water. The exact point of the threshold water content is difficult to distinguish and is usually extrapolated from measurements taken when evapotranspiration is clearly reduced by drought.[13] Because stomata act as regulators for CO_2 exchange, water deficits sufficient to close stomata must also depress photosynthesis. Photosynthesis declines initially as a result of stomatal closure, but prolonged and severe water stress can lead to depression of chloroplast and enzyme activity and to nonstomatal effects on photosynthesis.[1] In shoot tissues, when water deficits limit the supply of CO_2 for photosynthesis, a number of reversible mechanisms downregulate the efficiency of excitation energy transfer to photosystem II.

The rate of water transport through the plant in the liquid phase is constrained by the hydraulic resistance of the plant tissue. Tracheids and xylem vessels have relatively low resistances in comparison with living tissues. Anatomical changes, such as suberin deposition in cell walls, may occur in roots and leaves in response to soil water deficit, which help conserve water. However, these changes are irreversible, slow, and growth dependent, and increase the resistance to water flow. It is now generally accepted that changes in aquaporin activity can regulate rapid and reversible changes in root or leaf hydraulic resistance.[2] Aquaporins are membrane-embedded proteins that form water-conducting channels. By increasing the abundance or opening the water channels in roots, the plant can decrease the resistance to water flow to match transpiration demands, minimize the drop in water potential needed to drive water flow and aid in the maintenance of leaf water status. Conversely, a decrease in aquaporin activity may help conserve water in adverse conditions.

MEASUREMENT OF PLANT–WATER RELATIONS

Any study of plant–water relations requires accurate methods for measuring them. Indirect methods that have been used to characterize or infer leaf water status include visible wilting, color change, leaf rolling, leaf temperature, and leaf thickness. Pressure chambers and psychrometers directly, but destructively, measure water potential, although new hygrometers can be clamped directly to stems and leaves. Infrared gas analyzers are traditionally used to measure gas exchange. Acoustic emission sensors detect cavitation events and pressure probes can directly measure turgor of plant cells, although cell pressure probes are impractical for field measurements. Infrared thermography and remote sensing are rapidly being adopted to monitor canopy temperature as a surrogate for stomatal conductance in crop phenotyping programs. New sensors such as magnetic probes, which detect relative changes in leaf hydration, and modified time-domain reflectance probes, which detect stem water content, are being used for irrigation scheduling and increasing sustainable water use.

EFFECTS OF WATER DEFICITS

Long-term severe water deficits can have a negative effect on crop growth and yield, and may even result in senescence of the plant. It has been suggested that drought resistance of crops may be improved via breeding for specific phenological, morphological, physiological, or biochemical characteristics that improve yields in water-limited environments. These characteristics include early vigor, transpiration efficiency, stomatal control, abscisic acid accumulation, proline accumulation, osmotic adjustment, carbon remobilization, rapid grain growth, changes in hydraulic conductance, membrane stability, and lethal water potential.[14] It is, however, important to note that water deficits do not always lead to detrimental effects on yield. Mild or early water deficits may in fact increase yields or at least have no effect on yield, especially in indeterminate species.

CONCLUSION

Water deficit is an inevitable consequence of life on dry land for plants. A variety of mechanisms have evolved to control plant water status, particularly the regulation of water loss, turgor maintenance, and rate of water transport. Many of these mechanisms are being selected by breeding programs to increase crop productivity and drought resistance under water-limited conditions. With increasing global atmospheric CO_2, regional changes in climate are expected. These include warmer and drier conditions, changes to length of growing season, precipitation, and evaporation; all of which will affect plant–water relations.[15] Changes in land use and regional climate brought about by anthropogenic incursions mean that crop productivity in more marginal rainfall areas will be of increasing importance in world agriculture. Expansion to these areas requires the ability to identify and breed for characteristics that better enable crop plants to cope with limited water availability.

REFERENCES

1. Begg, J.E.; Turner, N.C. Crop water deficits. Adv. Agron. **1976**, *28*, 161–217.
2. Bramley, H.; Turner, D.W.; Tyerman, S.D.; Turner, N.C. Water flow in the roots of crop species: the influence of root structure, aquaporin activity and waterlogging. Adv. Agron. **2007**, *96*, 133–196.
3. Zholkevich, V.N. Root pressure. In *Plant Roots: The Hidden Half*; Waisel, Y.; Eshel, A.; Kafkafi, U., Eds.; Marcel Dekker Inc.: New York, 1991; pp. 589–603.
4. Steudle, E.; Jeschke, W.D. Water transport in barley roots. Planta. **1983**, *158*, 237–248.
5. Dixon, H.H. *Transpiration and the Ascent of Sap in Plants*. 1914, Macmillan: London.
6. Tyree, M.T.; Sperry, J.S. Vulnerability of xylem to cavitation and embolism. Ann. Rev. Plant Physiol. Plant. Mol. Biol. **1989**, *40*, 19–36.
7. Cosgrove, D.J. In defence of the cell volumetric elastic modulus. Plant Cell. Environ. **1988**, *11*, 67–69.
8. Boyer, J.S. Leaf enlargement and metabolic rates in corn, soybean and sunflower at various leaf water potentials. Plant Physiol. **1970**, *43*, 1056–1062.
9. Shackel, K.A.; Turner, N.C. Seed coat cell turgor in chickpea is independent of changes in plant and pod water potential. J. Exp. Bot. **2000**, *51* (346), 895–900.
10. Barlow, E.W.R.; Lee, J.W.; Munns, R.; Smart, M.G. Water relations of the developing wheat grain. Aust. J. Plant Physiol. **1980**, *7*, 519–525.
11. Smith, J.A.C.; Griffiths, H. Integrating Plant Water Deficits from Cell to Community. In *Water Deficits: Plant Responses From Cell to Community*; Smith, J.A.C.; Griffiths, H., Eds.; BIOS Scientific Publishers Ltd.: Oxford, UK, 1993; 1–4.
12. Henson, I.E.; Jensen, C.R.; Turner, N.C. Leaf gas exchange and water relations of lupins and wheat. I. Shoot responses to soil water deficits. Aust. J. Plant Physiol. **1989**, *16*, 401–403.
13. Ritchie, J.T. Water dynamics in the soil–plant–atmosphere system. Plant Soil **1981**, *58*, 81–96.
14. Turner, N.C. Further progress in crop water relations. Adv. Agron. **1997**, *58*, 293–338.
15. Jarvis, P.G. Global Change and Plant Water Relations. In *Water Transport in Plants Under Climatic Stress*; Borghetti, M.; Grace, J.; Raschi, A., Eds.; Cambridge University Press: Cambridge, UK, 1993, 1–13.

Weeds: Seed Banks and Seed Dormancy

Lynn Fandrich
Oregon State University, Corvallis, Oregon, U.S.A.

Abstract

The relationship between seed dormancy and success of a plant as an agricultural weed is significant. The existence of large populations of weed seed with varying degrees and states of dormancy is the basis for the annual weed problem. Dormancy allows a weed seed to avoid germination when conditions are not appropriate to complete its life cycle and prolongs population survival in the form of a soil seed bank.

INTRODUCTION

The relationship between seed dormancy and success of a plant as an agricultural weed is significant. Holm et al.[1] estimated that there are about 250 significant weeds in world agriculture. If the top 40 weeds classified as "serious" are selected by ranking them according to the number of countries in which they are considered "serious" weeds, it can be shown, without exception, that each one of the 40 species has dormancy. If weed seed were forever dormant, or simply nondormant, the complexities of weed management would be greatly reduced. However, weed seeds vary widely with respect to degree, duration, and source of dormancy. The existence of large populations of weed seed with varying degrees and states of dormancy is the basis for the annual weed problem. Dormancy allows a weed seed to avoid germination when conditions are not appropriate to complete its life cycle and prolongs population survival in the form of a soil seed bank.

CHARACTERISTICS OF THE SOIL SEED BANK

All of the viable seed present on the surface and in the soil is usually described as the soil seed bank. The number of seeds in the seed bank is determined by the rate of input in the seed rain, minus the losses resulting from disease, predation, and germination. The seed bank consists of new seeds recently shed by a parent plant, as well as older seeds that have persisted in the soil for several years. All of these seeds will vary in depth of burial and state of dormancy. Only a small proportion of seeds, varying between and among species, germinate when conditions are favorable. It is generally assumed that annual species contribute up to 95% of the seeds in the agricultural seed bank and only 2 to 10% of weed seeds emerge as seedlings each year.

Because most weeds are capable of producing a high number of seeds, especially compared to cultivated species, a few uncontrolled plants can rapidly increase the number of seeds in the soil seed bank. Data collected from plants grown in monoculture showed that wild oat (*Avena fatua*) produced 372–623 seeds,[2] common lambsquarters (*Chenopodium album*) produced 75,600–150,400 seeds,[3] and Palmer amaranth (*Amaranthus palmeri*) produced 200,000 to 600,000 seeds per plant.[4] It should be noted that biotic (inter- and intraspecific competition) and abiotic (climate) factors affect seed production, and most plants in agricultural situations do not often approach these high seed production values.

Large numbers of seeds are dispersed both horizontally and vertically in the soil. Plowing, chiseling, disking, and other forms of tillage are major ways that seeds are moved and covered with soil. After a number of years of soil disturbance, a relatively stable vertical distribution of seeds is reached in the tillage zone. Early estimates of seed numbers in agricultural soils exceeded 100,000 seeds per hectare in the tillage layer. Samples of soil on well managed farms in Minnesota averaged 16 million seeds per hectare, and as many as 45 million per hectare have been counted.

PERIODICITY AND INTERMITTENT GERMINATION

Most weed species exhibit annual periodicity in germination and emergence restricted to certain times of the year. This behavior was part of the original evidence for annual cycles in the relief and imposition of seed dormancy. Observations on seed germination and seedling emergence from soil samples placed in the greenhouse revealed that there was a typical periodicity, as well as a period of maximum germination for more than 100 weed species.[5] Exhumed seeds of corn spurry (*Spergula arvensis*) showed clear seasonal changes in dormancy during three successive years of germination tests.[6] Often, seeds are conditionally dormant, i.e., the appropriate conditions for germination

Encyclopedia of Natural Resources DOI: 10.1081/E-ENRL-120020236

are not present; seeds of winter annual species will not germinate in late spring, and spring annual species will not germinate in late fall.

WEED SEEDS AND DEPTH OF BURIAL

Seed germination and emergence of seedlings from various depths below the soil surface depend on the degree of dormancy present and conditions of the immediate environment. However, dormancy-breaking stimuli are most effectively encountered and germination is rapid when seeds remain at or near the soil surface. There is an overall agreement that the number of seedlings and rate of seedling emergence decrease proportionately with increases in the depths of seed burial. Examination of seeds recovered from 12 cm deep showed that approximately 85% were completely dormant.[7] The results of these investigations also revealed a strong association between seed mass and maximum depth of emergence, with the heaviest seeds emerging from the greatest depths.

PERSISTENCE OF WEED SEEDS IN SOIL

It is quite common for seeds to be dormant when they are fully mature on the parent plant and for dormancy to be lost after they are shed. The term "after-ripening" has frequently been used to denote the interaction between environment and seed over time that leads to a loss of dormancy.[8] The length of the after-ripening period varies between and among species and by seed location on the plant. Most weed species show one or two periods of high percent germination in seed produced during any given year, and the percent germination in subsequent years can be quite low. The mean length of the dormancy period in populations drives turnover in the seed bank. Grime[9] makes a distinction between "transient" (complete turnover in less than one year) and "persistent" seed banks (some seed remains in the bank longer than one year). The term "aging" is sometimes used incorrectly to describe after-ripening. Aging reduces seed viability and seedling vigor.

ENVIRONMENTAL FACTORS THAT AFFECT SEED DORMANCY

Temperature

Temperature is the major environmental factor to cause a change in seed dormancy. Temperatures may break dormancy but also send seeds into dormancy when other environmental conditions are limited. Seeds of summer and winter annual weeds may be conditionally dormant due to temperature influences. Experimentally, the effect of chilling is tested through exposure of imbibed seeds to low temperatures, usually around 0 to 1°C, for a substantial period. Seeds of bur buttercup (*Ranunculus testiculatis*) germinated only at very cold incubation temperatures, and maximum germination occurred at 5°C.[10]

Seeds of winter annual species must be exposed to high summer temperatures for several months to germinate at fall temperatures. If lack of moisture prevents seeds of obligate winter annuals from germinating in the fall, low winter temperatures may induce dormancy. Consequently, viable seeds that fail to germinate in the fall cannot germinate in spring because they are dormant. Seeds of the winter annual Japanese brome (*Bromus japonicus*) were forced into dormancy with winter temperatures and would not germinate during the spring and summer, despite favorable germination conditions.[11]

Under natural conditions in the soil, seeds are subjected to fluctuating temperatures. It has been observed that random fluctuations in temperature, in and above a low temperature, are as effective in promoting germination as exposure to constant low temperatures. It has also been found that seeds which remain dormant at constant temperatures may be induced to germinate by a diurnal fluctuation of temperatures. Thompson and Grime[12] investigated the effect of fluctuating temperatures on the germination of 112 species of herbaceous plants. Forty-six of the species examined were found to have their germination stimulated by temperature fluctuations in the light.

Light

Exposure to light breaks dormancy in many weed species, but there may be situations in which light has no effect or even inhibits germination. Baskin and Baskin[13] observed that among 142 species, germination of 107 was promoted by light, 32 were unaffected by the difference in light and dark conditions, and 3 were inhibited by light. Wesson and Wareing[14] concluded that "under natural conditions in the field, the germination of buried seeds following a disturbance of the soil is completely dependent upon exposure of seed to light." Even a short exposure or "light break" was sufficient to significantly increase the total number of seedlings emerging from soil.

CONCLUSION

Understanding the processes of dormancy loss and germination is essential to the development of good weed management programs. Currently, most management programs aim to control emerged weed populations and affect the quantity of seed returned to the soil, but it should be recognized that almost all agronomic practices also affect weed seed dormancy and germination through manipulation of the soil environment. Because many agricultural weeds exhibit periodicity of emergence, tillage treatments, planting dates, and other cultural practices can be altered to follow peak germination times and maximize depletion of the seed bank. A natural population of viable weed seeds in the field declined at a rate of approximately 50% per year with frequent cultivation.[15] A review of the utilization of specific agronomic practices to manipulate weed seed dormancy and germination requirements for weed management is available from Dyer[16] and others.[17,18]

REFERENCES

1. Holm, L.; Pancho, J.V.; Herberger, J.P.; Plucknett, D.L. *A Geographical Atlas of World Weeds*; John Wiley and Sons: New York, 1979.
2. O'Donnell, C.C.; Adkins, S.W. Wild oat and climate change: The effect of CO_2 concentration, temperature, and water deficit on the growth and development of wild oat in monoculture. Weed Sci. **2001**, *49*, 694–702.
3. Colquhoun, J.; Stoltenberg, D.E.; Binning, L.K.; Boerboom, C.M. Phenology of common lambsquarters growth parameters. Weed Sci. **2001**, *49*, 177–183.
4. Keeley, P.E.; Carter, C.H.; Thullen, R.J. Influence of planting date on growth of Palmer amaranth (*Amaranthus palmeri*). Weed Sci. **1987**, *35*, 199–204.
5. Brenchley, W.E.; Warrington, K. The weed seed population of arable soil. I. Numerical estimation of viable seeds and observations on their natural dormancy. J. Ecol. **1930**, *18*, 235–272.
6. Bouwmeester, H.J.; Karssen, C.M. The effect of environmental conditions on the annual dormancy pattern of seeds of *Spergula arvensis*. Can. J. Bot. **1993**, *71*, 64–73.
7. Roberts, H.A. Studies on the weeds of vegetable crops. II. Effect of six years cropping on the weed seeds in the soil. J. Ecol. **1962**, *50*, 803–813.
8. Simpson, G.M. *Seed Dormancy in Grasses*; Cambridge University Press: Cambridge, England, 1990.
9. Grime, J.P. The role of seed dormancy in vegetation dynamics. Ann. Appl. Biol. **1981**, *98*, 555–558.
10. Young, J.A.; Martens, E.; West, N.E. Germination of bur buttercup seeds. J. Range Manag. **1992**, *45*, 358– 362.
11. Baskin, J.M.; Baskin, C.C. Ecology of germination and flowering in the weedy winter annual grass *Bromus japonicus*. J. Range Manag. **1981**, *34*, 369–372.
12. Thompson, K.; Grime, J.P. A comparative study of germination responses to diurnally-fluctuating temperatures. J. Appl. Ecol. **1983**, *20*, 141–156.
13. Baskin, C.C.; Baskin, J.M. Germination ecophysiology of herbaceous plant species in a temperate region. Am. J. Bot. **1988**, *75*, 286–305.
14. Wesson, G.; Wareing, P.F. The role of light in the germination of naturally occurring populations of buried weed seeds. J. Exp. Bot. **1969**, *20*, 402–413.
15. Burnside, O.C.; Wilson, R.G.; Wicks, G.A.; Roeth, F.W.; Moomaw, R.S. Weed seed decline and buildup in soils under various corn management systems across Nebraska. Agron. J. **1986**, *78*, 451–454.
16. Dyer, W.E. Exploiting weed seed dormancy and germination requirements through agronomic practices. Weed Sci. **1995**, *43*, 498–503.
17. Forcella, F. Real-time assessment of seed dormancy and seedling growth for weed management. Seed Sci. Res. **1998**, *8*, 201–209.
18. Mulugeta, D.; Stoltenberg, D.E. Weed and seedbank management with integrated methods as influenced by tillage. Weed Sci. **1997**, *45*, 706–715.

Wetlands

Ralph W. Tiner
National Wetlands Inventory (NWI), U.S. Fish and Wildlife Service, Hadley, Massachusetts, U.S.A.

Abstract
Wetland is a universal term used to describe a variety of flooded or saturated environments. Given that wetlands include a diverse assemblage of ecosystems, classification schemes have been developed to separate and describe these different systems and to group similar habitats. Wetlands provide a number of functions that are considered valuable to society. The purpose of this entry is to provide an understanding of what wetlands are, how they vary globally, and their extent as determined by various inventories.

INTRODUCTION

Wetland is a universal term used to describe the collection of flooded or saturated environments that have been referred to as marshes, swamps, bogs, fens, salinas, pocosins, mangroves, wet meadows, sumplands, salt flats, varzea forests, igapo forests, bottomlands, sedgelands, moors, mires, potholes, sloughs, mangals, palm oases, playas, muskegs, and other regional and local names. It has been defined as a basis for inventorying these natural resources, for conducting scientific studies, and, in some countries, for regulating uses of these areas. Given that wetlands include a diverse assemblage of ecosystems, classification schemes have been developed to separate and describe these different systems and to group similar habitats. Wetlands provide a number of functions that are considered valuable to society (e.g., surface water storage to minimize flood damages, sediment retention and nutrient transformation to improve water quality, shoreline stabilization, streamflow maintenance, and provision of vital habitat for fish, shellfish, wildlife, and plants that yield food and fiber for people). Because of these values and the widespread recognition of wetlands as important natural resources, numerous wetland definitions and classification systems have been developed to inventory these resources around the globe. The purpose of this entry is to provide readers with an understanding of what wetlands are (wetland definition), how they vary globally (wetland types), and their extent as determined by various inventories. This entry should serve as a starting point for learning about wetlands, with the listed references being sources of more detailed information.

WETLAND DEFINITIONS

Wetlands are aquatic to semiaquatic ecosystems where permanent or periodic inundation or prolonged waterlogging creates conditions favoring the establishment of aquatic life. Wetlands are often located between land and water and have, therefore, been referred to as ecotones (i.e., transitional communities). However, many wetlands are not ecotones between land and water, since they are not associated with a river, lake, estuary, or stream.[1] Wetlands may derive water from many sources, including groundwater, river overflow, surface water runoff, precipitation, snowmelt, tides, melting permafrost, and seepage from impoundments or irrigation projects.

While the term "wetland" has many definitions, all definitions have common elements (see Table 1; some definitions even include deepwater habitats.). The presence of water in wetlands may be permanent or temporary. Their water may be salty or fresh. Wetlands may be natural habitats or artificially created. They range from shallow water environments to temporarily wet (i.e., flooded or saturated) areas. All are wet long enough and often enough to, at least, periodically support hydrophytic vegetation and other aquatic life (including anaerobic microbes), to create hydric soils or substrates, and to activate biogeochemical processes associated with wet environments.

WETLAND TYPES

Differences in climate, soils, vegetation, hydrology, water chemistry, nutrient availability, and other factors have led to the formation of a multitude of wetland types around the globe. In general, wetlands are characterized by their hydrology (e.g., tidal vs. nontidal, inundation vs. soil saturation, frequency and duration of wetness), the presence or absence of vegetation (vegetated vs. nonvegetated), the type of vegetation (forested or treed, shrub, emergent, or aquatic bed), and soil type (e.g., organic vs. inorganic, peatland vs. nonpeatland). Table 2 presents brief descriptions of some North American types and Fig. 1 shows examples of vegetated wetlands.

Encyclopedia of Natural Resources DOI: 10.1081/E-ENRL-120015639

Table 1 Examples of wetland definitions used for inventories

Country/organization	Wetland definition (source)
International/Ramsar	"areas of marsh, fen, peatland, or water, whether natural or artificial, permanent or temporary, with water that is static or flowing, fresh, brackish, or salt, including areas of marine water the depth of which at low tide does not exceed 6 m.... ...may incorporate riparian and coastal zone adjacent to wetlands, and islands or bodies of marine water deeper than 6 m at low tide lying within the wetlands."[2]
Australia	"areas of seasonally, intermittently, or permanently waterlogged soils or inundated land, whether natural or artificial, fresh or saline, e.g., waterlogged soils, ponds, billabongs, lakes, swamps, tidal flats, estuaries, rivers, and their tributaries.[12]
Canada	"land that is saturated with water long enough to promote wetland or aquatic processes as indicated by poorly drained soils, hydrophytic vegetation, and various kinds of biological activity which are adapted to a wet environment."[3]
U.S.	"lands transitional between terrestrial and aquatic systems where the water table is usually at or near the surface or the land is covered by shallow water." Wetland attributes include hydrophytic vegetation, undrained hydric soil, or saturated or flooded substrates.[4]

Table 2 Brief nontechnical descriptions of some wetland types in North America

Wetland type	General description
Marsh	Herb-dominated wetland with standing water through all or most of the year, often with organic (muck) soils
Tidal marsh	Herb-dominated wetland subject to periodic tidal flooding
Salt marsh	Herb-dominated wetland occurring on saline soils, typically in estuaries and interior arid regions
Swamp	Wetland dominated by woody vegetation and usually wet for extended periods during the growing season
Mangrove swamp (Mangal)	Tidal swamp dominated by mangrove species
Peatland, mire, moor, or muskeg	Peat-dominated wetland
Bog	Nutrient-poor peatland, typically characterized by ericaceous shrubs, other woody species, and peat mosses
Fen	More or less nutrient-rich peatland, often represented by sedges and/or calciphilous herbs and woody species
Wet meadow	Herb-dominated wetland that may be seasonally flooded or saturated for extended periods, often with mineral hydric soils
Bottomland	Riverside or streamside wetland, usually on floodplain
Flatwood	Forested wetland with poorly drained mineral hydric soils located on broad flat terrain of interstream divides, common on coastal plains and glaciolacustrine plains
Farmed wetland	Wetland cultivated for rice, cranberries, sugar cane, mints, or other crops

These types may be defined differently in other regions.

Various countries have devised classification systems for describing differences among their wetlands and for categorizing wetlands for natural resources inventories. Scientists have created systems to organize certain wetlands into meaningful groups for analysis and management (e.g., peatland classifications; Tiner[1]). In 1998, the Ramsar Convention Bureau published a multinational classification system to provide consistency for inventorying wetlands and designating wetlands of international importance.[2] This system includes 11 types of marine or coastal wetlands (i.e., shallow water and intertidal habitats; permanent shallow marine waters; marine subtidal aquatic beds; coral reefs; rocky marine shores; sand; shingle or pebble shores; estuarine waters; intertidal mud, sand, or salt flats; intertidal marshes; intertidal forests; coastal brackish/saline lagoons; coastal freshwater lagoons). This system also includes 19 inland wetland types (i.e., permanently flooded aquatic habitats to intermittently flooded sites are represented: permanent inland deltas; permanent rivers/streams/creeks; seasonal, intermittent, or irregular rivers/streams/creeks; permanent freshwater lakes; seasonal or intermittent freshwater lakes; seasonal or intermittent saline/brackish/alkaline lakes and flats; permanent saline/brackish/alkaline marshes and pools; seasonal or intermittent saline/brackish/alkaline marshes and pools; permanent freshwater marshes and pools; seasonal or intermittent freshwater marshes and pools; nonforested peatlands; alpine wetlands including meadows and temporary

Fig. 1 Some examples of North American wetlands (top to bottom, left to right): tidal salt marsh, inland marsh, pothole marsh, wet meadow/shrub swamp, northern peatland, bottomland swamp, hardwood swamp, and flatwood wetland. **Source:** Photos courtesy of U.S. Fish and Wildlife Service.

snowmelt waters; tundra wetlands; shrubby-dominated wetlands; freshwater tree-dominated wetlands on inorganic soils; forested peatlands; freshwater springs and oases; geothermal wetlands; and subterranean karst and cave hydrological systems). Last, nine man-made wetland types (aquaculture ponds; ponds; irrigated land including rice paddies; seasonally flooded agricultural land; salt exploitation sites; water storage areas including impoundments generally more than 8 ha; excavations; wastewater treatment areas; and canals and drainage channels) are also included in the system.

In North America, the Canadian and United States wetland classification systems were developed by government agencies interested in wetland conservation and management. The Canadian system emphasizes wetland origin (class), form, and vegetation in describing the wetland types.[3] Five wetland classes are recognized: bog, fen, marsh, swamp, and shallow water. Within each class, different forms and types are characterized. Eight general vegetation types are defined by the presence or absence of vegetation (treed, shrub, forb, graminoid, moss, lichen, aquatic bed, and nonvegetated). These types may be subdivided into other types (e.g., treed into coniferous or deciduous, shrub into tall, low, and mixed, graminoids into grass, reed, tall rush, low rush, and sedge, aquatic bed into floating and submerged). The U.S. Fish and Wildlife Service's wetland and deepwater habitat classification[4] is the official federal system used for mapping wetlands and for reporting the status and trends of wetlands in the U.S. The features separating wetlands include general ecological and physical factors and specific features such as vegetation, soil/substrate composition, hydrology, water chemistry, and human alterations. Classification follows a hierarchical approach with five main levels designated: ecological system (marine, estuarine, lacustrine, riverine, and palustrine), subsystem, class (vegetated: forested, scrub-shrub, emergent, and aquatic bed; nonvegetated: unconsolidated shore, rocky shore, streambed, and reef), subclass, and modifiers. The modifiers are used to describe a wetland's hydrology (water regime), pH and salinity (water chemistry), soils, and the influence of humans and beaver (special modifiers). Common types include estuarine intertidal emergent wetlands (e.g., salt and brackish marshes), estuarine intertidal unconsolidated shore (e.g., tidal flats and beaches), palustrine emergent wetlands (e.g., marshes, fens, and wet meadows), palustrine forested wetlands, and palustrine scrub-shrub wetlands (e.g., shrub bogs and shrub swamps).

A hydrogeomorphic approach (HGM) to wetland classification has also been developing in the U.S.[5] The HGM system emphasizes abiotic features important for assessing wetland functions. Seven hydromorphic classes are identified: riverine, depressional, slope, mineral soil flats, organic soil flats, lacustrine fringe, and estuarine fringe. The U.S. Fish and Wildlife Service has adapted the HGM approach to provide additional modifiers to its classification system on a pilot basis. These HGM-type descriptors include landscape position (i.e., lotic, lentic, terrene, estuarine, and marine), landform (i.e., slope, basin, interfluve, floodplain, flat, island, and fringe), and water flow path (i.e., inflow, outflow, throughflow, bidirectional flow, isolated, and paludified).[6] These descriptors provide the required information to aid the evaluation of functions of wetlands across watersheds and large geographic areas.

EXTENT OF WETLANDS

Comprehensive wetland inventories do not exist in most countries. There are many inconsistencies among the inventories (e.g., different levels of effort, focus on particular types, and artificial wetlands such as rice paddies are often not included in wetland inventories).[7] Consequently, comparative analysis is fraught with problems. Nonetheless, Table 3 provides some perspective on the extent of wetlands in many regions. Most of the data came from a series of reports produced for the Bureau of the Ramsar Wetlands Convention.[8] Globally, estimates for wetlands range from about 750 million ha[7] to about 1.5 billion ha. Ten countries have over 2 million ha of peatlands alone, with Canada leading at nearly 130 million ha (represents about 18% of the country) followed by the former U.S.S.R. at 83 million ha.[9,10] About a third of Finland is covered by peatlands (10 million ha). The Pantanal of South America, perhaps the largest wetland in the world, reportedly covers about 200,000 km² (or 2 million ha) during the wet season.[11]

Table 3 Estimates of the current extent of wetlands in different regions of the world

Region/country	Wetland extent (ha)	Source
Africa	121,321,683–124,686,189	[13]
Asia	211,501,790–224,117,790	[14]
Central America		
Mexico	3,318,500 (very incomplete)	[15]
Europe		
Eastern	225,849,930	[16]
Western	28,821,979	[17]
Middle East	7,434,790	[18]
Neotropics	414,996,613	[19]
North America		
Canada	127,199,000–150,000,000	[20]
U.S.	114,544,800	[1]
Oceania	35,748,853	[21]
South America		
Tropical region	200,000,000	[22]

REFERENCES

1. Tiner, R.W. *Wetland Indicators: A Guide to Wetland Identification, Delineation, Classification, and Mapping*; Lewis Publishers, CRC Press: Boca Raton, FL, 1999; 392 pp.
2. Ramsar Convention Bureau. Information Sheet on Ramsar Wetlands: Gland, Switzerland, 1998.
3. National Wetlands Working Group. *Wetlands of Canada*; Ecological Land Classification Series No. 21, Land Conservation Branch, Canadian Wildlife Service; Environment Canada: Ottawa, Ont., 1988; 452 pp.
4. Cowardin, L.M.; Carter, V.; Golet, F.C.; LaRoe, E.T. *Classification of Wetlands and Deepwater Habitats of the United States*; FWS/OBS-79/31; U.S. Department of the Interior, Fish and Wildlife Service: Washington, DC, 1979; 131 pp.
5. Brinson, M.M. *A Hydrogeomorphic Classification for Wetlands*; Wetlands Research Program Tech. Rep. WRP-DE-4; U.S. Army Waterways Expt. Station: Vicksburg, MS, 1993; 103 pp.
6. Tiner, R.W. *Keys to Waterbody Type and Hydrogeomorphic-Type Wetland Descriptors for U.S. Waters and Wetlands, Operational Draft*; U.S. Department of the Interior, Fish and Wildlife Service; Northeast Region: Hadley, MA, 2000; 20 pp.
7. Finlayson, C.M.; Davidson, N.C. Global review of wetland resources and priorities for wetland inventory: summary report. In *Global Review of Wetland Resources and Priorities for Wetland Inventory*; Finlayson, C.M., Spiers, A.G., Eds.; Supervising Scientist Report 144; Canberra, Australia, 1999; 9 pp.
8. Finlayson, C.M., Spiers, A.G., Eds.; *Global Review of Wetland Resources and Priorities for Wetland Inventory*; Supervising Scientist Report 144; Canberra, Australia, 1999.
9. Taylor, J.A. Peatlands of the British Isles. In *Mires: Swamp, Bog, Fen, and Moor; Regional Studies*; Gore, A.J.P., Ed.; Elsevier: Amsterdam, the Netherlands; 1–46.
10. Botch, M.S.; Massing, V.V. Mire ecosystems in the U.S.S.R. In *Mires: Swamp, Bog, Fen, and Moor; Regional Studies*; Gore, A.J.P., Ed.; Elsevier: Amsterdam, the Netherlands; 95–152.
11. Swarts, F.A., Ed. *The Pantanal of Brazil, Bolivia, and Paraguay*; Waterland Research Institute; Hudson MacArthur Publishers: Gouldsboro, PA, 2000; 287 pp.
12. Wetland Advisory Committee. *The Status of Wetlands Reserves in System Six*; Report of the Wetland Advisory Committee to the Environmental Protection Authority: Australia, 1977.
13. Stevenson, N.; Frazier, S. Review of wetland inventory information in Africa. In *Global Review of Wetland Resources and Priorities for Wetland Inventory*; Finlayson, C.M., Spiers, A.G., Eds.; Supervising Scientist Report 144; Canberra, Australia, 1999; 94 pp.
14. Watkins, D.; Parish, F. Review of wetland inventory information in Asia. In *Global Review of Wetland Resources and Priorities for Wetland Inventory*; Finlayson, C.M., Spiers, A.G., Eds.; Supervising Scientist Report 144; Canberra, Australia, 1999; 26 pp.
15. Olmstead, I. Wetlands of Mexico. In *Wetlands of the World: Inventory, Ecology, and Management Volume I*; Whigham, D.F., Dykyjova, D., Heiny, S., Eds.; Kluwer Academic Publishers: Dordrecht, the Netherlands, 1993; 637–677.
16. Stevenson, N.; Frazier, S. Review of wetland inventory information in Eastern Europe. In *Global Review of Wetland Resources and Priorities for Wetland Inventory*; Finlayson, C.M., Spiers, A.G., Eds.; Supervising Scientist Report 144; Canberra, Australia, 1999; 53 pp.
17. Stevenson, N.; Frazier, S. Review of wetland inventory information in Western Europe. In *Global Review of Wetland Resources and Priorities for Wetland Inventory*; Finlayson, C.M., Spiers, A.G., Eds.; Supervising Scientist Report 144; Canberra, Australia, 1999; 57 pp.
18. Frazier, S.; Stevenson, N. Review of wetland inventory information in the Middle East. In *Global Review of Wetland Resources and Priorities for Wetland Inventory*; Finlayson, C.M., Spiers, A.G., Eds.; Supervising Scientist Report 144; Canberra, Australia, 1999; 19 pp.
19. Davidson, I.; Vanderkam, R.; Padilla, M. Review of wetland inventory information in the Neotropics. In *Global Review of Wetland Resources and Priorities for Wetland Inventory*; Finlayson, C.M., Spiers, A.G., Eds.; Supervising Scientist Report 144; Canberra, Australia, 1999; 35 pp.
20. Davidson, I.; Vanderkam, R.; Padilla, M. Review of wetland inventory information in the North America. In *Global Review of Wetland Resources and Priorities for Wetland Inventory*; Finlayson, C.M., Spiers, A.G., Eds.; Supervising Scientist Report 144; Canberra, Australia, 1999; 35 pp.
21. Watkins, D. Review of wetland inventory information in Oceania. In *Global Review of Wetland Resources and Priorities for Wetland Inventory*; Finlayson, C.M., Spiers, A.G., Eds.; Supervising Scientist Report 144; Canberra, Australia, 1999; 26 pp.
22. Junk, W.J. Wetlands of tropical South America. In *Wetlands of the World: Inventory, Ecology, and Management Volume I*; Whigham, D.F., Dykyjova, D., Heiny, S., Eds.; Kluwer Academic Publishers: Dordrecht, the Netherlands, 1993; 679–739.

Wetlands: Biodiversity

Jean-Claude Lefeuvre
Laboratory of the Evolution of Natural and Modified Systems, University of Rennes, Rennes, France
Virginie Bouchard
School of Natural Resources, The Ohio State University, Columbus, Ohio, U.S.A.

Abstract
Wetlands are highly complex ecosystems subjected to a variety of hydrologic regimes, climatic conditions, soil formation processes, and geomorphologic settings. They include swamps, bogs, marshes, mire, fens, salt marshes, mangroves, and other types of ecosystems saturated by water during all or part of the growing season. They are found on every continent, except Antarctica, and in various climes, from the tropics to the tundra. The extent of the world's wetland is generally estimated to be from 7 to 9 million km^2 or about 4–6% of the Earth. The variety of seasonal and perennial wetlands provides environmental conditions for highly distinctive fauna and flora.

ROLE OF HYDROLOGY

Wetlands differ significantly in their water source and seasonal hydrologic regime. Hydrological patterns (i.e., flooding frequency, duration and hydroperiod) influence physical and chemical characteristics (e.g., salinity, oxygen and other gas diffusion rates, reduction–oxidation potential, nutrient solubility) of a wetland. In return, these internal parameters and processes control flora and fauna distribution as well as ecosystem functions. Plants, animals and microbes are often oriented in predictable ways along the hydrological gradient (Fig. 1). Conversely, the biotic component affects the hydrology by eventually modifying flow or water level in a wetland.[1,2] Species also influence nutrient cycles and other ecosystem functions.[3]

A LANDSCAPE PERSPECTIVE

Although the hydrology is part of the ecological signature of an individual wetland, wetlands are considered as neither aquatic nor terrestrial systems. They have characteristics from both systems and are defined as ecotones placed under this dual influence.[4] Because wetlands are located at the interface of multiple systems, they assure vital functions (e.g., wildlife habitats) beneficial at the landscape level. Reduction of wetland area often reduces biodiversity in the landscape.[2,5] Increases in biodiversity occur when wetlands are created or restored in a disturbed landscape.[6]

WETLAND FLORA

Wetland plants are adapted to a variety of stressful abiotic conditions (e.g., immersion, wave abrasion, water level fluctuation, low oxygen conditions). Identical adaptations to common environmental features have led taxonomically distinct species to sometimes look similar in terms of morphology, life cycle and life forms.[7] Traditionally, wetland plants have been classified into groups of different life forms, primarily in relation with hydrological conditions. Helophytes are defined as plant species with over-wintering buds in water or in the submerged bottom.[8] They are differentiated from hydrophytes in that their vegetative organs are partially raised above water level.[8] Hejny's classification is based on relatively stable vegetative features that determine the ability of wetland plants to survive two unfavorable conditions, cold and drought.[7] This classification uses the types of photosynthetic organs present in both the growth and flowering phases. Other classifications include both life form and growth form. When a species has a range of growth form, it is classified under the form showing the greatest achievement of its potential.[7]

Plant communities in wetlands can be more or less homogeneous, mosaic-like, or distributed along a gradient resulting in a clear zonation of species. A gradient exists if one or several habitat parameters change gradually in space. This phenomenon is common in fresh water marshes that present a gradient of water depth and water saturated soils (Fig. 1). Such a gradient is often accompanied by differences in peat accumulation that is influenced by waves or currents. General principles of the zonation of aquatic plants have been largely described.[7,9,10] Littoral vegetation can belong to several types of communities, which derived from the general principle that, from deeper water to the shore, we may expect successively submergent, floating, and emergent macrophytes. The most important habitat factor is water depth, depending on slope and peat accumulation.[10] Other factors may be poor irradiance caused by high turbidity or exposure to waves or flow.[7,10]

Encyclopedia of Natural Resources DOI: 10.1081/E-ENRL-120001727

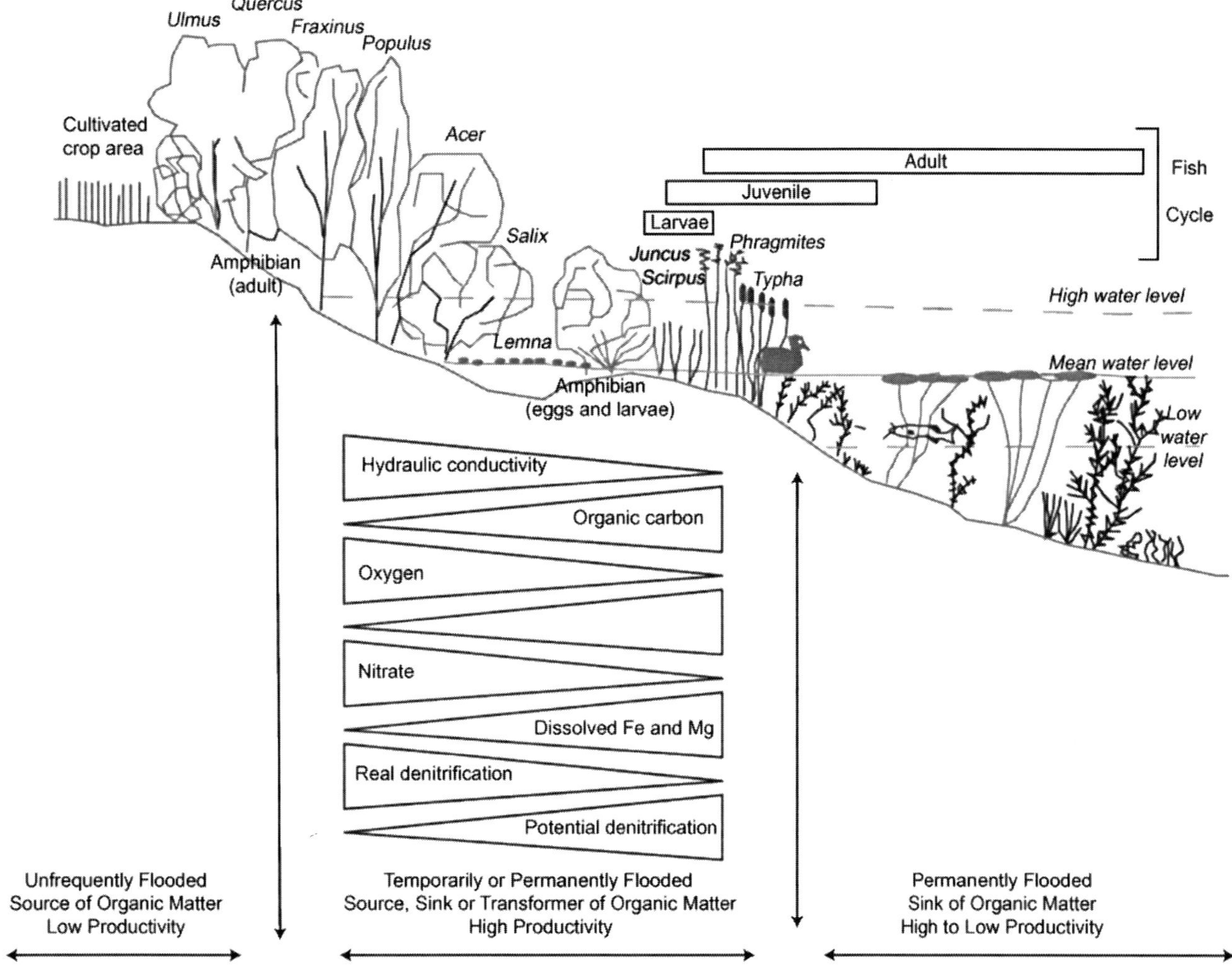

Fig. 1 Species distribution along the hydrological gradient in a freshwater marsh.

Riparian ecosystems are found along streams and rivers that occasionally flood beyond confined channels or where riparian sites are created by channel meandering in the stream network. Riparian or bottomland hardwood forests contain unique tree species that are flood tolerant. Species distribution is associated with floodplain topography, flooding frequency, and flooding duration.[11,12] In South-Eastern U.S. bottomland, seasonally flooded forests are colonized by *Platanus occidentalis* (sycamore), *Ulmus americana* (American elm), *Populus deltoides* (cottonwood) and are flooded between 2% and 25% of the growing season (Fig. 1). Other species such as *Fraxinus pennsylvanica* (green ash), *Celtis laevigate* (sugarberry) and *Carya aquatica* (water hickory) colonized bottomlands that are flooded by less than 2% of the growing season.[11] Freshwater marshes are dominated with emergent macrophytes rooted in the bottom with aerial leaves (i.e., helophytes). Species such as *Typha* (cattail), *Phragmites* (reed grass) and *Scirpus* (bulrush) are often clonal. A plant community is usually organized in sequence of patches that are dominated by one species. The second plant groups are the rooted plants with floating leaves (*Nymphaeid*). Lotus (*Nelumbo*) and water lilies (*Nymphea*) have very identical morphology (i.e., similar leaves and flowers) but a genetic analysis showed that lotus is more closely related with plane-tree than with water lilies.[10] Submerged plants include elodeids (i.e., cauline species whose whole life cycle can be completed below the water surface or where only the flowers are emergent) and isoetids (i.e., species growing on the bottom whose whole life cycle can be completed without contact with the surface). Submerged species include species such as coontail (*Ceratophyllum demersum*) and water milfoil (*Myriophyllum* spp.). Plant species found in salt marshes are called halophytes (i.e., plants which complete their cycle in saline environments). A saltmarsh can be divided into low, middle, and upper marsh, according to flooding frequency and duration. Each zone is dominated by different plant species according to their tolerance to saline immersion.[10]

A dominant competitive species—often a clonal species—can modify the theoretical zonation. Change in water chemistry (i.e., eutrophication) or hydrology may favor a particular species over the natural plant community. Highly competitive species are often invasive and aggressive in displacing native species. The expansion of *Phragmites australis* into tidal wetlands of North America causes a reduction in biodiversity as many native species of plants are replaced by a more cosmopolitan species.[13] In riparian ecosystems, biodiversity is usually higher in the intermediate zones, whereas it is lower upstream and downstream (Fig. 2). The percentage of exotic species is low in upstream areas, but can represent up to 40 in downstream zones (Fig. 1).[14]

WETLAND FAUNA

Diversity of vertebrate and invertebrate species is the result of a community composed of resident and transient species, which use the space differently and at various times of day and year. The density and variety of animal populations at a particular wetland site are also explained by climatic events that affect geographic areas on a large scale. For example, the population of waterfowl during winter is largely dependent on climate variations in the northern part of the continents.[15]

Resident species are often dependent on the type of vegetation. Animal communities are generally distributed along a zonation pattern, parallel to the plant communities, which is driven by the hydrological gradient. A few species depend entirely on a single plant species for their survival. One beetle (*Donacia* spp.) may depend on reeds, at least during its larval stage where another beetle (*Galerucella* spp.) uses only water lilies as its habitat and diet. For many other residents, their habitats extend to several plant communities during their life span. It is generally the case for many vertebrates such as amphibians, rodents, passerines, and waterfowl.

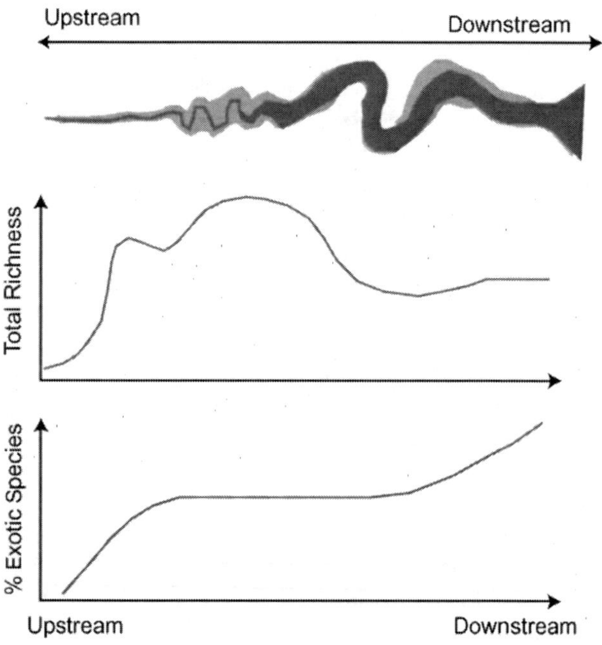

Fig. 2 Total richness and invasive species distribution along a stream longitudinal gradient.
Source: Adapted from Planty-Tabacchi.[14]

Many amphibian species depend on wetland or riparian zones for reproduction and larval stage. Likewise, wet meadows are necessary as a reproduction zone and nursery for a number of freshwater fishes. About 220 animals and 600 plant species are threatened by a serious reduction of wetlands in California, and the state's high rate of wetland loss (91% since the 1780s) is partly responsible.[15] Many waterfowl species are sensitive to areas of reduction, patch size and distribution, wetland density, and proximity to other wetlands.[16] When the Marais Poitevin (France), one of the principal wintering and passages sites for waterfowl in the Western Europe, underwent agricultural intensification in the 1980s, the population of ducks and waders declined tremendously. This decline was partly due to a 50% reduction of wet meadows between 1970 and 1995, primarily caused by the conversion to arable farmlands.[17] Thus, maintenance of biodiversity depends on the existence of inter connections between wetlands, and between aquatic and terrestrial ecosystems. In fact, some authors have pointed out that an increase in biodiversity occurs when wetlands are created or restored in a disturbed landscape.[6,18]

CONCLUSIONS

Despite the importance of wetlands, conservation efforts have ignored them for a long time. It is urgent to conserve the existing wetlands, and also to restore and create wetland ecosystems. A wide range of local, state, federal, and private programs are available to support the national policies of wetland "No Net Loss" in the U.S., and around the world. From a biodiversity perspective, on-going wetland protection policies may not be working because restored or created wetlands are often very different from natural wetlands.[19] Created wetlands often result in the exchange of one type of wetland for another, and result in the loss of biodiversity and functions at the landscape level.[19] We know now that it takes more than water to restore a wetland,[20] even if an important place should be given to the ability of self-design of wetland ecosystems.[21]

REFERENCES

1. Naiman, R.J.; Johnston, C.A.; Kelley, J.C. Alteration of North American streams by beaver. Bioscience **1988**, *38*, 753–762.
2. National Research Council. *Wetlands: Characteristics and Boundaries*; National Academy Press: Washington, DC, 1995; 306 pp.
3. Naiman, R.J.; Pinay, G.; Johnston, C.A.; Pastor, J. Beaver influences on the long-term biogeochemical characteristics of boreal forest drainage networks. Ecology **1994**, *75*, 905–921.

4. Hansen, A.J.; Di Castri, F.; Eds.; Landscape boundaries. In *Consequences for Biotic Diversity and Ecological Flows; Ecological Studies*; Springer: Berlin, 1983; 452 pp.
5. Gibbs, J.P. Wetland loss and biodiversity conservation. Conser. Biol. **2000**, *14*, 314–317.
6. Weller, M.W. *Freshwater Marshes*, 3rd Ed.; University of Minnesota Press: Minneapolis, U.S., 1994; 192.
7. Westlake, D.F.; Kvet, J.; Szczepanski, A.; Eds.; *The Primary Ecology of Wetlands*; Cambridge University Press: 1998; 568 pp.
8. Raunkiaer, C. *The Fife Forms of Plants and Statistical Plant Geography*; Oxford University Press: Oxford, England, 1934; 632 pp.
9. Grevilliot, F.; Krebs, L.; Muller, S. Comparative importance and interference of hydrological conditions and soil nutrient gradients in floristic biodiversity in flood meadows. Biodivers. Conserv. **1998**, *7*, 1495–1520.
10. Mitsch, W.J.; Gosselink, J.G. *Wetlands*, 3rd Ed.; Wiley: New York, 2000; 920 pp.
11. Clark, J.R.; Benforado, J., Eds.; *Wetlands of Bottomland Hardwood* Forest; Elsevier: Amsterdam, the Netherlands, 1981; 401 pp.
12. Keogh, T.M.; Keddy, P.A.; Fraser, L.H. Patterns of tree species richness in forested wetlands. Wetlands **1999**, *19*, 639–647.
13. Chambers, R.M.; Meyerson, L.A.; Saltonstall, K. Expansion of *Phragmites australis* into tidal wetlands of North America. Aquat. Bot. **1999**, *64*, 261–273.
14. Planty-Tabacchi, A.M. Invasions Des Corridors Riverains Fluviaux Par Des Espéces Végétales D'origine Étrangére. These University Paul Sabatier Toulouse III, France, 1993; 177 pp.
15. Hudson, W.E., Ed.; *Landscape Linkages and Biodiversity*; Island Press: Washington, DC, 1991.
16. Leibowitz, S.C.; Abbruzzese, B.; Adams, P.R.; Hughes, L.E.; Frish, J. *A Synoptic Approach to Cumulative Impact Assessment: A Proposed Methodology*, 1992; 138 pp.
17. Duncan, P.; Hewison, A.J.M.; Houte, S.; Rosoux, R.; Tournebize, T.; Dubs, F.; Burel, F.; Bretagnolle, V. Long-term changes in agricultural practices and wildfowling in an internationally important wetland, and their effects on the guild of wintering ducks. J. Appl. Ecol. **1999**, *36*, 11–23.
18. Hickman, S. Improvement of habitat quality for resting and migrating birds at the Des Plaines River wetlands demonstration project. Ecol. Engng. **1994**, *3*, 485–494.
19. Whigham, D.F. Ecological issues related to wetland preservation, restoration, creation and assessment. Sci. Total Environ. **1999**, *240*, 31–40.
20. Zedler, J.B. Progress in wetland restoration ecology. Trends Ecol. Evol. **2000**, *15*(10), 402–407.
21. Mitsch, W.J.; Wu, X.; Nairn, R.W.; Weihe, P.E.; Wang, N.; Deal, R.; Boucher, C.E. Creating and restoring wetlands: a whole ecosystem experiment in self-design. BioScience **1998**, *48*, 1019–1030.

Wetlands: Classification

Arnold G. van der Valk
Ecology, Evolution and Organismal Biology, Iowa State University, Ames, Iowa, U.S.A.

Abstract

Today it is widely accepted that wetlands have three main characteristics: (1) regular inundation or saturated soils, (2) hydric plant species, and (3) hydric soils. Wetland classification systems began in the United States in the 1890s in order to inventory wetlands that could be reclaimed (i.e., drained). Later, American classification systems were developed to inventory wetlands as waterfowl habitat and subsequently for multiple legal, conservation, and scientific purposes. Although the main goal of classification systems originally was to provide a framework for inventories, the most recent classification systems, the hydrogeomorphic systems, are primarily intended to identify wetlands with similar functions and to provide a framework for quantifying their functions. Many countries, besides the United States, have developed wetland classification systems that are tailored to local conditions, including Canada, Australia, and South Africa. Initially, wetland classifications were horizontal in structure with no subdivisions of the various types of wetlands recognized. Later, they became hierarchical in structure with multiple subdivisions. The Ramsar Convention Bureau, Switzerland, has developed a classification system that is intended to be worldwide in scope. This bureau, however, uses a less formal definition of what is a wetland, "areas of marsh, fen, peatland, or water," than that used in most national systems. Nevertheless, it provides a useful framework for the rapid assessment of wetland types. New classification systems currently being developed, like the new South African system, suggest that more sophisticated wetland classification systems can be developed that could become the basis of a new international wetland classification system. Before a new international system can be developed, a universally accepted definition of what constitutes a wetland is needed.

INTRODUCTION

There are many kinds of wetlands around the world that differ because of their geographic location (tropical, temperate, arctic, etc.), geomorphological setting (channel, coastal or lake fringe, basin, slope, etc.), water chemistry (freshwater, brackish water, salt water, etc.), soils (mineral, organic), and vegetation (herbaceous, shrub, forested, mosses, etc.). In many parts of the world, wetlands are utilized by local people, for example, for grazing cattle and other domestic animals in sub-Saharan Africa and crop cultivation in Asia, and this can result in significant impacts on their size, hydrology, flora, and fauna. Although most wetlands are natural features of the landscape, there are also wetlands that were purposefully constructed or that developed in association with dams and irrigation systems. Nevertheless, all these wetlands have three things in common: (1) They are shallowly flooded or have saturated soils; (2) their soils are developed to some extent under anaerobic conditions, and consequently have characteristics that differ from those of adjacent terrestrial soils; and (3) their flora and fauna are different from those of adjacent terrestrial ecosystems. Wetlands, however, are often highly dynamic systems: their hydrology and vegetation can change from year to year; and because they can occur in multiple states, this complicates their classification.

The need to classify wetlands has grown over the years, as the perceived societal and scientific value of wetlands has increased. Wetlands historically were often viewed as nuisances or even dangerous places that harbored a variety of human diseases. Wetland classification and inventorying were initially driven by the need to identify wetlands to drain. As waterfowl began to decline in the 20th century, the importance of wetlands as habitat for waterfowl stimulated early efforts to classify them. Waterfowl interests wanted to preserve those wetlands that were important as breeding or overwintering habitat to ensure the survival of recreationally important waterfowl species. By the late 20th century, wetlands came to be valued not only as wildlife habitat but for their other services to society, such as pollutant removal, flood water storage, groundwater recharge, etc. In some places, this resulted in legal protection being given to wetlands in order to preserve these wetland services. Legal protection required that all wetlands in a jurisdiction be identified and classified.

In order to classify wetlands, we first have to define what a wetland is. In 1995, a Committee on Characterization of Wetlands[1] in the United States provided a formal definition of a wetland: "A wetland is an ecosystem that depends on constant or recurrent, shallow inundation or saturation at or near the surface of the substrate. The minimum essential characteristics of a wetland are recurrent,

Encyclopedia of Natural Resources DOI: 10.1081/E-ENRL-120047526

sustained inundation or saturation at or near the surface and the presence of physical, chemical, and biological features reflective of recurrent, sustained inundation or saturation. Common diagnostic features of wetlands are hydric soils and hydrophytic vegetation. These features will be present except where specific physicochemical, biotic, or anthropogenic factors have removed them or prevented their development." This definition or its minor variants is now universally accepted by wetland ecologists. In short, wetlands are what they are because of their hydrology, and their hydrology creates physical and chemical conditions that produce distinct soils and habitats that can be exploited only by subsets of the local flora and fauna that are adapted to these conditions. Among the important fauna found in wetlands are waterfowl, capybaras, water buffalo, alligators, and many kinds of fish. Among the most important plants are cattails (*Typha* spp.), water lilies (*Nymphaea* spp.), pondweeds (*Potamogeton* spp.), mangroves (e.g., *Avicennia* spp.), papyrus (*Cyperus papyrus*), and common reed (*Phragmites* spp.)

There are two ways in which the term wetland is commonly used: (1) an entire landscape feature (basin, fringe, flat, channel) that qualifies as a wetland because of its hydrology, plants, and soils; for example, the Everglades or Okavango Delta; (2) a distinguishable portion or area of such a landscape feature, usually because of its dominant vegetation, for example, a sawgrass flat or wet prairie within the Everglades. In a wetland classification, it is important to distinguish between these two uses of the term wetland. They represent different scales at which a classification can be done. Landscape features are often not homogeneous units with regard to hydrology, soils, or vegetation, but they can be delimited, mapped, measured, and counted. Distinguishable portions or areas of landscape features are homogeneous units with regard to hydrology, water chemistry, vegetation, soils, or some other feature. This distinction was sometimes ignored in some older classification systems. Most wetlands when classified as a landscape feature contain a complex of plant communities that reflect local topographic gradients (i.e., water depths, soils, water chemistry), related plant distribution patterns, and secondary patterns caused by alterations of the original landscape by plants and animals, for example, tree islands in the Everglades. Consequently, classifications based solely on landscape features collect very different information than those based on homogeneous distinguishable areas. Modern classification systems often combine both approaches with higher units in the system based on landscape features (rivers, lakes, depressions, flats, etc.) and subunits based on homogenous distinguishable areas.

Ecologists in Europe and North America began describing wetlands in some detail in the 19th century.[2] They distinguished between coastal/saline wetlands, interior/freshwater wetlands, and peat bogs; and they recognized many subtypes of these on the basis of their vegetation, for example, reed swamp, carr, forest swamp, fen, etc. Early ecologists distinguished between wetlands (marshes, swamps, tidal marshes) and peatlands (mires, bogs, fens) and treated them as different kinds of entities. Today, this is no longer the case, and peatlands are considered to be wetlands with organic soils.

Although plant ecologists and geographers in Europe in the 19th century had described a variety of different types of wetland types, one of the first attempts to develop a national wetland classification occurred in the United States in the 1890s, and the first national wetland inventories were conducted in 1906 and 1922. These inventories were carried out by federal government agencies and were done using questionnaires, soil survey and other maps, drainage information, and various kinds of state reports. These early inventories used very crude classification systems (e.g., the 1922 survey recognized only five wetland types). Their main purpose was to identify wetlands for "reclamation," that is, drainage. Limited and crude as these classification systems and inventories were, they did provide the first information about the state of wetlands in the United States.[2] The next generation of wetland classification systems developed in the United States were more positively disposed to wetlands and initially focused on evaluating and conserving wetlands as habitat for waterfowl. But as wetlands became more visible in public policy and research arenas in the 1970s, the need arose for more general wetland classification systems that could be used for multiple purposes (conservation, restoration, legal, management, research) and especially for implementing national wetland inventories and wetland protection policies. Wetland classification systems in the United States since the 1950s will be reviewed first and then selected contemporary classifications that were developed outside the United States. Only those classification systems that were adopted or proposed by a national agency or international organization will be considered.

AMERICAN CLASSIFICATION SYSTEMS

The designers of any wetland classification to be used in inventorying wetlands have to deal with a number of major issues. What will be the "unit" classified? How will data on wetlands be collected? How accurate do the data have to be? Who will use the data generated? In what form will it be stored? How will inventory data be made available/distributed/retrieved? In the United States, the first two national classification systems developed in the last half of the 20th century were intended to be used as a basis for a national wetland inventory (NWI). Consequently, the practical realities (people, time, money) of collecting data at a large geographic scale were major factors in their design. The practical realities of collecting data and distributing them, however, changed dramatically from the 1950s to the late 1970s. The advent of easily available, high-resolution aerial photography and satellite imagery; inexpensive computers; geographic information system software for collecting, storing, and analyzing spatial data; high speed graphics printers; the Internet; etc. has had an impact on the design of classification systems and on the distribution of inventory data collected using them. This can be seen through the comparison of the *Classification of Wetlands of the United States*[3] commonly referred to as the Martin et al. (1953) system and its replacement *Classification of Wetlands and Deepwater Habitats of the United States*[4] commonly referred to as the Cowardin et al. (1979) system.

Martin et al. (1953) developed a national wetland classification system to inventory waterfowl habitat. This classification system was horizontal in that its 20 wetland types, which covered all inland and coastal wetlands in the conterminous United States, had no subdivisions or subtypes. One of the main limitations faced by its developers was the small number of staff available to collect data in the field and their limited expertise in wetland ecology. Wetland types (e.g., inland fresh water meadows, inland shallow fresh marshes, inland deep marshes, and inland open fresh water) were vegetation types largely defined by certain characteristic species and water depth: features easy to observe in the field. This classification system was not consistent in the kinds of categories used, and it mixed vegetation types like inland shallow fresh marshes (distinguishable units) with landscape unit like bogs and sounds and bays. It also put wetlands from different parts of the country with little in common (Cypress-Tupelo Swamps from the southeastern and Black Spruce forests from northeastern United States) into the same category, namely, wooded swamps. The use of water depth to distinguish some freshwater wetland types also caused problems because this was highly dependent on the time of year a wetland was visited. The uneven nature of the categories reflected the main purpose of this classification system to find out how much waterfowl habitat of various types still existed in the United States. It emphasized types of wetlands used by waterfowl (six types of inland freshwater wetlands) and plays down those that were not considered important to waterfowl (e.g., bogs). This classification was used to do a detailed inventory of American wetlands, and the results of this inventory were published in 1956 in *Wetlands of the United States. Their Extent and Their Value to Waterfowl and Other Wildlife*.[5] The results of this inventory were only available in this Fish and Wildlife report, commonly referred to as Circular 39. A number of regional wetland classifications systems were developed in the United States in the 1960s and 1970s that dealt with many of the shortcomings of the Martin et al. (1953) system.[2]

By the time that the next American NWI was developed 20 years later, advances in technology, especially the availability of high quality aerial photography and computers to store and retrieve spatial information, greatly influenced the development of the *Classification of Wetlands and Deepwater Habitats of the United States*.[4] The more widespread interest in wetlands in the 1970s is reflected in the makeup of the committee that developed this new classification system. All the authors of the Martin et al. system worked for the U.S. Fish and Wildlife Service. Although the U.S. Fish and Wildlife Service was still heavily involved, the authors of the new classification also came from other federal agencies (U.S. Geological Survey, National Oceanographic and Atmospheric Administration) as well as academia. One of the main differences between the new classification and its predecessor is that it is a hierarchical system. Much like the hierarchical system used to classify plant and animal species, it is possible to select and examine several levels of aggregation. Overall, the developers of the new classification system had many major objectives for it: (1) to build the system using consistent and uniform terminology that could be applied to any wetland in the United States; (2) to be able to group ecologically similar wetlands together at all levels of the classification hierarchy; (3) to make it readily applicable to mapping all the wetlands of the United States from aerial photographs as part of a NWI. The system has three levels in its hierarchy: system, subsystem, and class. Systems and subsystems are based on hydrology (Fig. 1). There are five systems: marine, estuarine, riverine, lacustrine, and palustrine. Except for the palustrine, each system has two or more subsystems. In turn, each system or subsystem has multiple classes. Classes are mostly types of plant communities (aquatic bed, moss-lichen wetland, forested wetland, etc.) or various kinds of non-vegetated substrates (bedrock, rubble, sand, etc.) that can be identified on aerial photographs. Classes can further be subdivided into subclasses and dominance types. While classes and subclasses are based primarily on dominant life form or easily recognized substrate features for unvegetated areas, dominance types are described by their most abundant plant species or sessile macro invertebrate species (e.g., oysters). In addition, water regime, soil, water chemistry, and special modifiers can also be used to characterize wetlands. Special modifiers are used to describe human-made wetlands (excavated, impounded, diked, farmed, etc.). In short, the Cowardin et al. system is a classification in which the basic units are distinguishable areas on aerial photographs. This classification system has been used successfully by the NWI in the United States for over 30 years, and the NWI, which is part of the Fish and Wildlife Service, periodically issues status and trend reports for the country that document changes in different kinds of wetlands. NWI maps are now distributed on the World Wide Web (http://www.fws.gov/wetlands/).

Although it has served its primary purpose well, the Cowardin et al. (1979) wetland classification system has proven to be inadequate for other purposes, especially the evaluation of wetland functions and values. In 1993, another wetland classification system was published by Mark Brinson[6] for this purpose that he called *A Hydrogeomorphic*

Classification for Wetlands. It is most commonly known as the HGM system. Unlike the Cowardin et al. system, the basic unit in the HGM system is a landscape unit. The HGM system identifies wetland types believed to have similar functions. The position of the wetland in the landscape and the source(s) of water and direction of flow into and out of the wetland are the major classificatory variables. In the HGM system, seven basic HGM classes of wetlands are recognized: depressional, riverine, slope, lacustrine, mineral soil flats, organic soil flats, and estuarine (tidal). The HGM system deals only with wetlands as defined by the Committee on Characterization of Wetlands. It excludes

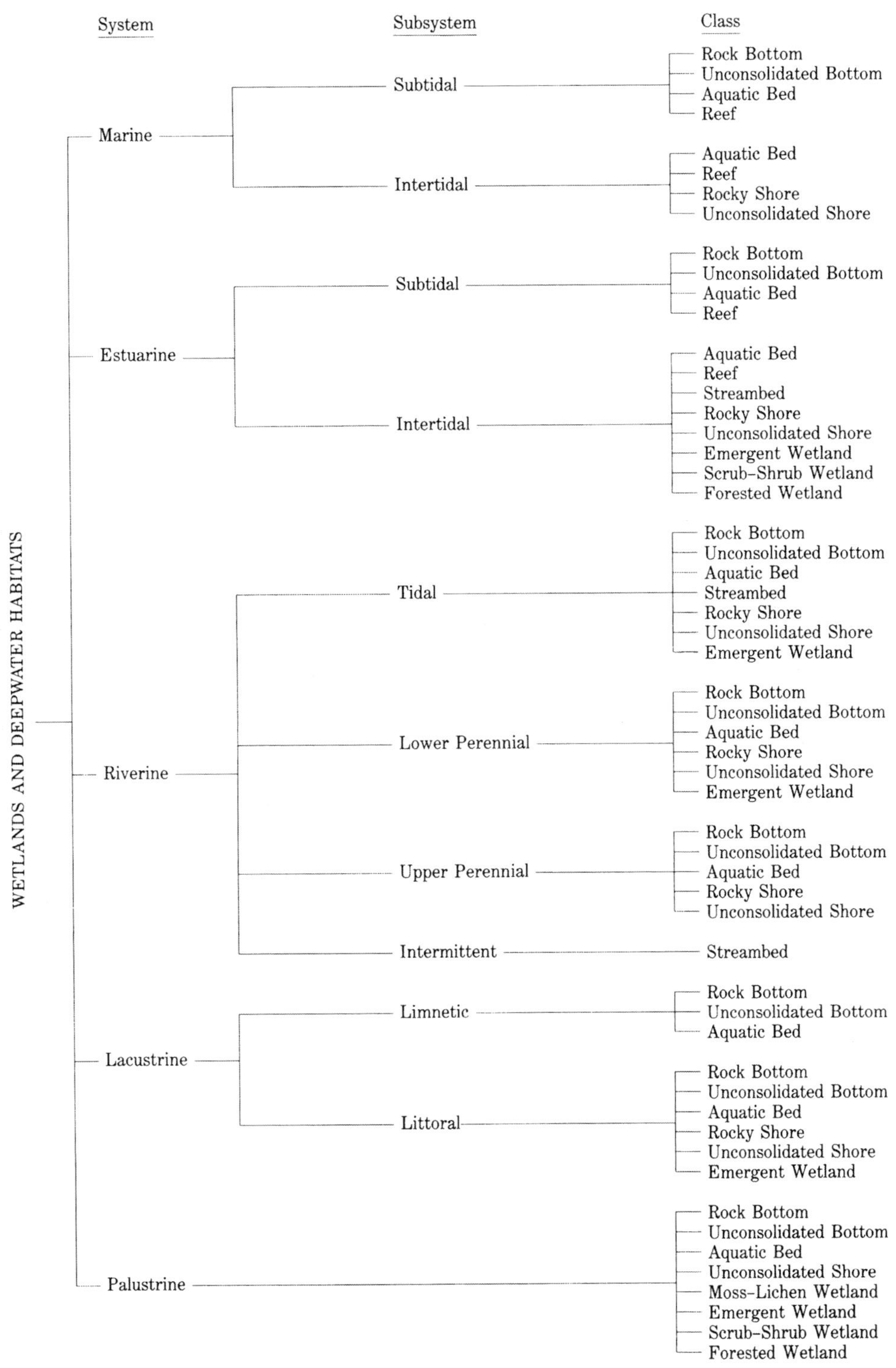

Fig. 1 The classification hierarchy of the Cowardin et al. (1979) American wetland classification system.
Source: Cowardin et al.[4]

the deepwater systems that are also covered in the Cowardin et al. classification system. Various characteristics of a wetland and its watershed can then be used to assess its functions (flood peak attenuation, sediment retention, water storage, nutrient retention, habitat value, carbon sequestration, etc.). A series of regional manuals have been developed to do functional assessments using the HGM system. These can be downloaded from the U.S. Army Corps of Engineers website (http://el.erdc.usace.army.mil/wetlands/wlpubs.html).

OTHER CLASSIFICATION SYSTEMS

Besides those in the United States, wetland classifications have been developed in many other countries.[2,7] For example, Canada developed a national classification system that was built on a variety of provincial and regional systems proposed in the 1970s and 1980s. *Wetlands of Canada*,[8] which was published in 1988, describes the Canadian system in detail. Like the Cowardin et al. system in the USA, the Canadian system is hierarchical, but its levels (class, form, and type) are completely different. Its classes are landscape units. These are widespread wetlands that are distinguished primarily by their dominant vegetation and soils (bog, fen, swamp, marsh, and shallow open water). Forms are based on landscape position, surface morphology of the wetland, its sources of water, presence of permafrost, etc. It is an open-ended category and more forms can be added as needed. Types are based on the physiognomy of the dominant plant species (trees, shrubs, graminoid, moss, lichen, etc.) of the wetland's vegetation. Subforms and phases are also recognized based on species composition, depth of water, water chemistry, etc. The lower level units are homogeneous, distinguishable units. Thus, the Canadian system includes both landscape and distinguishable area units.

In Australia, Semeniuk and Semeniuk (1995) developed a geomorphic classification system for non-coastal wetlands that has much in common with the HGM system.[9] Like the HGM system, it is built around various landforms (basins, channels, flats, slopes, and highlands) and hydrology. They, however, emphasized water longevity (permanent inundation, seasonal inundation, intermittent inundation, and seasonal waterlogged) rather than source and direction of water flow. When landform and length of inundation/saturated are combined, they recognized 13 different wetland types. Likewise, Australia has developed its own classification system for inventory purposes.[10] It recognizes six general wetland types: lakes, swamps, lands subject to inundation, river and creek channels, tidal flats, and coastal water bodies (estuaries, lagoons). Each major type is subdivided into multiple subtypes, mostly based on length of inundation or stream flow for inland wetlands and patterns of tidal inundation for tidal flats.

South Africa has spent a great deal of time and effort developing a classification system that is suitable for the diverse conditions found in southern Africa. As in the United States, a number of regional wetland classifications had developed, but they were inadequate for a NWI. In 2006, the *National Wetland Inventory: Development of a Wetland Classification System for South Africa* was published[11] and an update, *Further Development of a Proposed National Wetland Classification System for South Africa*, was published in 2009.[12] There is also a user's manual, *Classification System for Wetlands and Other Aquatic Ecosystems in South Africa*.[13] Like the American Cowardin et al. system, the South African system is hierarchical, but, unlike it, the South African system incorporates an HGM approach at some levels. It uses primary discriminators to classify wetland (HGM) units in the first three levels of the hierarchy (Fig. 2). At level 1, the highest level, three basic types of wetland systems are recognized: marine, estuarine, and inland. At level 2, bioregional differences (e.g., Namaqua Inshore Bioregion off the western coast) are the main discriminators rather than wetland types. Level 3 is based on subsystems of each of the level 1 systems. For estuarine systems, connectivity (permanently open, temporarily open) to the ocean is used to delimit subsystems and for the inland system, four landscape units (slope, valley floor, plain, and bench) are recognized. There are no marine subsystems. At level 4, HGM units are used to classify wetlands. These HGM units are based on landform; sources and movement of water into, through, and out of the wetland; and hydrodynamics (direction and strength of water flow), as per Brinson's (1993) HGM system in the United States. Secondary discriminators are used at level 5 to subdivide the HGM units at level 4. These include tidal regimes for estuarine and marine wetlands and water regimes for inland wetlands. Level 6 includes various descriptors that can be used to further characterize the wetland. There are six classes of descriptors: geology, natural vs. artificial, vegetation type, substratum, salinity, and pH. The South African wetland classification system is, in effect, a fusion of the two contemporary American classification systems, Cowardin et al.[4] and Brinson's HGM[6] with HGM units nested within broader (system) units. Like the Cowardin et al. system, the South African system includes not only inland and coastal wetlands but also deepwater habitats.

During the 1980s and 1990s, wetland inventories were done in many countries or regions around the world.[7,14] Some of these consisted primarily of a list of nationally important wetlands with a brief description of their major features. In spite of being an important source of information about wetlands globally, these various efforts did not use a standardized method or classification system, and thus the results are uneven and sometimes not comparable. The Ramsar Convention Bureau has spearheaded efforts to develop a truly international classification system.[15] The main goal of the Ramsar Convention was to protect wetland habitats important to migratory water birds. Countries that are signatories to the Ramsar Convention agree to designate

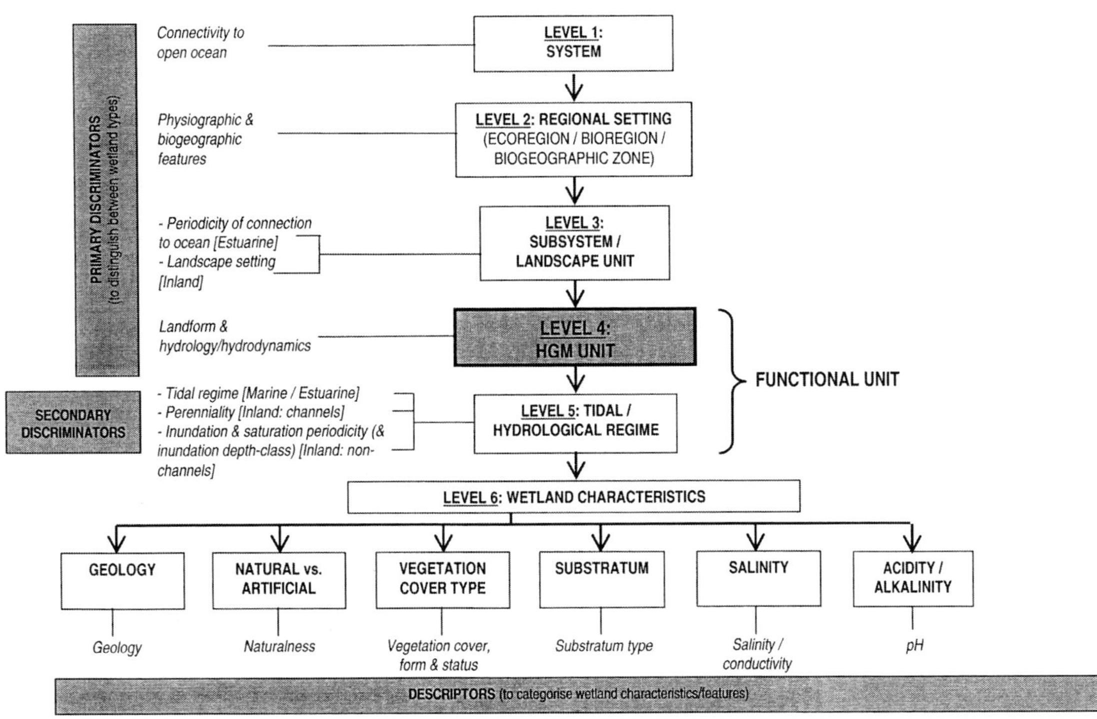

Fig. 2 The structure of the proposed South African wetland classification system. Primary discriminators are used in levels 1–3 to classify the hydrogeomorphic (HGM) units in level 4. Secondary discriminators in level 5 are used to subdivide the HGM units in level 4. In level 6, there are various descriptors that can be used to further characterize the HGM units in level 4.
Source: SANBI.[12]

one wetland as a wetland of international importance. These are widely known as Ramsar Sites. Each signatory government (160 as of 2011) is also expected to implement a plan to promote the conservation and wise use, as far as possible, of its wetlands. This includes an inventory of its wetlands.

The Ramsar Convention's approach to wetlands is both practical and flexible. Almost anything that is shallowly flooded is considered to be a wetland, regardless of its origin or current use. It has its own definition of wetlands:[15] "... areas of marsh, fen, peatland or water, whether natural or artificial, permanent or temporary, with water that is static or flowing, fresh, brackish or salt, including areas of marine water the depth of which at low tide does not exceed six metres." Although the Ramsar definition of a wetland recognizes that hydrology (flooding/saturation) is the defining feature of a wetland, it differs from the Committee on Characterization of Wetlands' definition in that it does not explicitly mention hydric soils or plants. Like the Cowardin et al. and South African systems, it also includes deepwater marine habitats. The Ramsar classification system (Table 1) recognizes six major wetland types: marine, estuarine, lacustrine, riverine, palustrine, and human made.[15] The Ramsar Wetland Classification system is designed to enable all countries to carry out a rapid inventory of the wetlands, especially of wetlands that are nominated as Ramsar sites. The current version (2011) of the system recognizes 42 wetland types grouped into three broad classes (marine and coastal wetlands, inland wetlands, and human-made wetlands). Although at first sight, it appears to be a horizontal classification much like the Martin et al. (1953) American classification, it can also be viewed as a hierarchical system as demonstrated by Scott and Jones (1995).[7] In fact, the Ramsar system closely parallels the Cowardin et al. (1979) American classification system, except that in the Ramsar classification human-made wetlands are at the same level in the hierarchy as hydrologic systems while they are only descriptors of classes in the Cowardin et al. system This reflects the difference between extant wetlands in much of the United States, which are generally less impacted and exploited by people than their counterparts in other parts of the word, particularly in Europe, northern Africa, and most of southern Asia. Like the Martin et al. system, however, it is a mixed classification system with both landscape units (coastal brackish/saline lagoons) and homogeneous distinguishable units (intertidal forested wetlands).

CONCLUSION

Although a number of successful national and international wetland classifications have been developed, we do not have a universally accepted international wetland classification system. Current national and international wetland classifications often use different definitions of wetlands, terminology to describe them, and methods for delineating

Table 1 Ramsar classification system and codes for wetland types

Marine/Coastal Wetlands

A – Permanent shallow marine waters in most cases less than 6 m deep at low tide; includes sea bays and straits.

B – Marine subtidal aquatic beds; includes kelp beds, sea-grass beds, tropical marine meadows.

C – Coral reefs.

D – Rocky marine shores; includes rocky offshore islands, sea cliffs.

E – Sand, shingle or pebble shores; includes sand bars, spits and sandy islets; includes dune systems and humid dune slacks.

F – Estuarine waters; permanent water of estuaries and estuarine systems of deltas.

G – Intertidal mud, sand or salt flats.

H – Intertidal marshes; includes salt marshes, salt meadows, saltings, raised salt marshes; includes tidal brackish and freshwater marshes.

I – Intertidal forested wetlands; includes mangrove swamps, nipah swamps and tidal freshwater swamp forests.

J – Coastal brackish/saline lagoons; brackish to saline lagoons with at least one relatively narrow connection to the sea.

K – Coastal freshwater lagoons; includes freshwater delta lagoons.

Zk(a) – Karst and other subterranean hydrological systems, marine/coastal.

Inland Wetlands

L – Permanent inland deltas.

M – Permanent rivers/streams/creeks; includes waterfalls.

N – Seasonal/intermittent/irregular rivers/streams/creeks.

O – Permanent freshwater lakes (over 8 ha); includes large oxbow lakes.

P – Seasonal/intermittent freshwater lakes (over 8 ha); includes floodplain lakes.

Q – Permanent saline/brackish/alkaline lakes.

R – Seasonal/intermittent saline/brackish/alkaline lakes and flats.

Sp – Permanent saline/brackish/alkaline marshes/pools.

Ss – Seasonal/intermittent saline/brackish/alkaline marshes/pools.

Tp – Permanent freshwater marshes/pools; ponds (below 8 ha), marshes and swamps on inorganic soils; with emergent vegetation water-logged for at least most of the growing season.

Ts – Seasonal/intermittent freshwater marshes/pools on inorganic soils; includes sloughs, potholes, seasonally flooded meadows, sedge marshes.

U – Non-forested peatlands; includes shrub or open bogs, swamps, fens.

Va – Alpine wetlands; includes alpine meadows, temporary waters from snowmelt.

Vt – Tundra wetlands; includes tundra pools, temporary waters from snowmelt.

W – Shrub-dominated wetlands; shrub swamps, shrub-dominated freshwater marshes, shrub carr, alder thicket on inorganic soils.

Xf – Freshwater, tree-dominated wetlands; includes freshwater swamp forests, seasonally flooded forests, wooded swamps on inorganic soils.

Xp – Forested peatlands; peat swamp forests.

Y – Freshwater springs; oases.

Zg – Geothermal wetlands.

Zk(b) – Karst and other subterranean hydrological systems, inland.

Human-Made Wetlands

1 – Aquaculture (e.g., fish/shrimp) ponds.

2 – Ponds; includes farm ponds, stock ponds, small tanks; (generally below 8 ha).

3 – Irrigated land; includes irrigation channels and rice fields.

4 – Seasonally flooded agricultural land (including intensively managed or grazed wet meadow or pasture).

5 – Salt exploitation sites; salt pans, salines, etc.

6 – Water storage areas; reservoirs/barrages/dams/impoundments (generally over 8 ha).

(*Continued*)

Table 1 (***Continued***) Ramsar classification system and codes for wetland types

7 – Excavations; gravel/brick/clay pits; borrow pits, mining pools.

8 – Wastewater treatment areas; sewage farms, settling ponds, oxidation basins, etc.

9 – Canals and drainage channels, ditches.

Zk(c) – Karst and other subterranean hydrological systems, human-made.

Note: "Floodplain" is a broad term used to refer to one or more wetland types, which may include examples from the R, Ss, Ts, W, Xf, Xp, or other wetland types. Some examples of floodplain wetlands are seasonally inundated grassland (including natural wet meadows), shrub lands, woodlands, and forests. Floodplain wetlands are not listed as a specific wetland type herein.
The categories are intended to provide only a very broad framework to aid in the rapid identification of main wetland habitats represented at each site.
Source: Ramsar Convention Secretariat.[15]

them. Consequently, some wetland classification systems do not include aquatic systems that are recognized as wetlands in others.[13,14] Inconsistencies in the treatment of man-made wetlands are a particular problem. To date, the Ramsar classification system is the only international classification that is being used by multiple countries, mostly those that do not have a national system. Although very useful for classifying wetlands at a coarse scale, the Ramsar classification, in its current form, is too crude to be used for detailed national inventories of wetlands like that of the U.S. NWI.

Advances in aerial photography and satellite-born sensors as well as in the acquisition, processing, and storage of spatial data have greatly reduced the time and cost of collecting wetland inventory data. Nevertheless, the need to develop and implement a global inventory of wetlands is hampered by the lack of a universally accepted classification system. The United States historically has led the way in the development of wetland classification systems and in carrying out a detailed NWI, but new developments in other parts of the world like South Africa suggest that more sophisticated wetland classification systems can be developed that could become the basis of a new international wetland classification system. Before such a new international system can be developed, however, a universally accepted definition of what constitutes a wetland needs to be adopted.

REFERENCES

1. Commission on Characteristics of Wetlands. *Characteristics and Boundaries*; National Academy Press: Washington, DC, USA, 1995.
2. Tiner, R.W. *Wetland Indicators: A Guide to Wetland Identification, Delineation, Classification, and Mapping*; Lewis Publishers: Boca Raton, FL, USA, 1999.
3. Martin, A.C.; Hotchkiss, N.; Uhler, F.M.; Bourn, W.S. Classification of Wetlands of the United States. *Fish and Wildlife Service Spec. Sci. Rep., Wildlife No. 20*; U.S. Department of the Interior Fish and Wildlife Service: Washington, DC, USA, 1953; 1–14.
4. Cowardin, L.W.; Carter, Virginia; Golet, F.C.; LaRoe, E.T. Classification of wetlands and deepwater habitats of the United States. *Fish and Wildlife Service FWS/OBS-79/31*; U.S. Department of the Interior Fish and Wildlife Service: Washington, DC, USA, 1979; 1–131.
5. Shaw, S.P.; Fredine, C.G. Wetlands of the United States. *Their Extent and Their Value to Waterfowl and Other Wildlife*. U.S. Fish and Wildlife Service Circular 39; U.S. Department of the Interior Fish and Wildlife Service: Washington, DC, USA, 1956; 1–67.
6. Brinson, M.M. *A Hydrogeomorphic Classification of Wetlands*. U.S. Army Corps of Engineers Wetlands Research Program Tech. Rept. WRP-DE-4; U.S. Army Corps of Engineers Waterways Experiment Station: Vicksburg, MS, USA, 1993; 1–92.
7. Finlayson, C.M.; van der Valk, A.G., Eds. *Classification and Inventory of the World's Wetlands*; Kluwer Academic Press; Dordrecht, 1995.
8. National Wetlands Working Group. Wetlands of Canada. Ecological Land Classification No. 24. Sustainable Development Branch Canadian Wildlife Service, Environment Canada; Ottawa, Canada.
9. Semeniuk, C.A.; Semeniuk, V.A. Geomorphic approach to global classification for inland wetlands. In *Classification and Inventory of the World's Wetlands*; Finlayson, C.M.; van der Valk, A.G., Eds.; Kluwer Academic Press; Dordrecht, 1995; 103–124. 1988.
10. Paijamans, K.; Galloway, R.W.; Faith, D.P.; Fleming, P.M.; Haantjens, H.A.; Heyligers, P.C.; Kalma, J.D.; Loffler, E. Aspects of Australian Wetlands. Division of Water and land Resources Tech. Paper No. 44, Commonwealth Scientific and Industrial Research Organization: Melbourne, Australia, 1985, 1–71.
11. Ewart-Smith, J.L.; Ollis, D.J.; Malan, H.L. National Wetland Inventory: Development of a Wetland Classification System. Water Research Commission WRC Report No. KV174/06: Water Research Commission: Pretoria, South Africa, 2006.
12. SANBI. Further Development of a Proposed National Wetland Classification System for South Africa. Primary Project Report. South African National Biodiversity Institute: Pretoria, South Africa, 2009.
13. Ollis D.J.; Snaddon, C.D.; Job, N.M.; Mbona, N. Classification System for wetlands and other aquatic ecosystems in South Africa. SANBI Biodiversity Series 22. South African National Biodiversity Institute: Pretoria, South Africa, 2013.
14. Finlayson, C.M.; Davidson, N.C.; Spiers, A.G.; Stevenson, N.J. Global wetland inventory–current status and future priorities. Mar. Freshwater Res. **1999**, *50*, 717–727.
15. Ramsar Convention Secretariat, The Ramsar Convention Manual: A Guide to the Convention on Wetlands (Ramsar, Iran, 1971), 5th ed. Ramsar Convention Secretariat; Gland, Switzerland, 2011.

Wetlands: Coastal, InSAR Mapping

Zhong Lu
Cascades Volcano Observatory, U.S. Geological Survey (USGS), Vancouver, Washington, U.S.A.
Jinwoo Kim
The Ohio State University, Columbus, Ohio, U.S.A.
C. K. Shum
Division of Geodetic Science, School of Earth Sciences, The Ohio State University, Columbus, Ohio, U.S.A.

Abstract

Interferometric synthetic aperture radar (InSAR) analysis allows the mapping of coastal wetlands under all weather conditions. InSAR intensity and coherence images are used to characterize wetland types, extents, conditions, and their changes while InSAR phase images can map water level changes to infer water dynamics over coastal wetlands at an unprecedented spatial resolution between two data acquisition times. InSAR analysis of fully polarized radar imagery has the potential to image vegetation structure and characteristics of coastal wetlands.

INTRODUCTION

Coastal wetlands include hydrologic and other processes that are fundamental to understanding ecological and climatic changes.[1,2] Characterizing wetland types and changes is critical for monitoring human, climatic, and other effects on wetland systems including their restoration. Measuring changes in water level over wetlands, and consequently changes in water storage capacity, provides a governing parameter in hydrologic models and is required for comprehensive assessment of flood hazards.[3] With frequent coverage over wide areas, satellite sensors can provide accurate measurements of coastal wetland characteristics and underneath-water storage.

A synthetic aperture radar (SAR) is an advanced radar system that utilizes image-processing techniques to synthesize a large virtual antenna, which provides much higher spatial resolution than is practical using a real-aperture radar.[4] A SAR system transmits electromagnetic waves at a wavelength that can range from a few millimeters to tens of centimeters. Because a SAR actively transmits and receives signals backscattered from the target area and the radar signals with long wavelengths are mostly unaffected by weather clouds, a SAR can operate effectively during day and night under most weather conditions. Through SAR processing, both the intensity and phase of the radar signal backscattered from each ground resolution element (typically meters to submeters in size) can be calculated and combined to form a complex-valued SAR image that represents the radar reflectivity of the ground surface.[4] The intensity (or strength) of the SAR image is determined primarily by the terrain slope, surface roughness, and dielectric constant. The phase of the complex-valued SAR image is related to the apparent distance from the satellite to a ground resolution element as well as the interaction between radar waves and scatterers within a resolution element of the imaged area.

Interferometric SAR (InSAR) utilizes the phase components of two coregistered SAR images of the same area acquired from similar vantage points to extract the landscape topography and patterns of surface change, including ground displacement.[5,6] The spatial separation between two SAR antennas or two vantage points of the same SAR antenna used to acquire the two SAR images is called the InSAR baseline. Two antennas can be displaced in the along-track direction to produce an along-track InSAR image if the two SAR images are acquired with a short time lag.[5,7] Along-track InSAR can be used to measure the velocity of targets moving toward or away from the radar to determine sea ice drift, ocean currents, river flows, and ocean wave parameters. Along-track InSAR applications in the past have been limited to technology demonstration in experimental studies from airborne SAR observations. Two antennas can be displaced in the cross-track direction to produce cross-track InSAR. For cross-track InSAR, the two antennas can be mounted on a single platform for simultaneous, single-pass InSAR observation, which is the usual implementation for airborne systems and spaceborne systems (e.g., Shuttle Radar Topography Mission) for generating high-resolution, precise digital elevation models

Encyclopedia of Natural Resources DOI: 10.1081/E-ENRL-120049157

(DEMs) over large regions.[8] Alternatively, InSAR images can be formed by using a single antenna on an airborne or spaceborne platform in nearly identical repeating flight lines or orbits for repeat-pass InSAR.[9,6] In this case, even though successive observations of the target area are separated in time, the observations will be highly correlative if the backscattering properties of the surface have not changed in the interim. In this way, InSAR is capable of measuring surface displacement with subcentimeter vertical precision for X-band and C-band sensors (λ = 2–8 cm) or few-centimeter precision for L-band sensors (λ = 15–30 cm), in both cases at a spatial horizontal resolution of tens of meters over an image swath (width) of a few tens of kilometers up to ~100 km. This is the typical implementation for past, present, and future spaceborne SAR sensors (Table 1). The required wavelength to monitor the earth surface varies among applications. Among SAR systems with similar repeat intervals, spatial resolutions, and imaging geometries, long-wavelength (such as L-band) SAR is generally a better choice for studying coastal wetlands due to the penetration of the long-wavelength SAR signal into vegetation to capture canopy characteristics and reveal the water surface beneath.

Interactions of radar waves with wetlands can be complicated.[10] Over flooded vegetation, the radar backscattering consists of contributions from the interactions of radar waves with the canopy surface, canopy volume, and water surface. On the basis of a canopy backscattering model for continuous tree canopies,[11] the total radar backscattering over wetland can be approximated as the incoherent summation of contributions from a) canopy surface backscattering, b) canopy volume backscattering that includes backscattering from multiple-path interactions of canopy–water, and c) double-bounce trunk–water backscattering (i.e., the radar signal is initially reflected away from the sensor by the water's surface, toward a tree bole or other vertical structure and is then directly reflected toward the sensor). The relative contributions from surface backscattering, volume backscattering, and double-bounce backscattering are controlled primarily by radar parameters including wavelength, polarization, incidence angle, and spatial resolution as well as wetland characteristics including vegetation type (and structure), vegetation leaf on/off condition, canopy closure, and other environmental factors.[10,12]

InSAR STUDIES OF COASTAL WETLANDS

Typical InSAR processing includes precise registration of an interferometric SAR image pair, a common Doppler/bandwidth filter to improve InSAR coherence, interferogram generation, removal of the curved earth phase trend, adaptive filtering, phase unwrapping, precision estimation of interferometric baseline, generation of a surface displacement (deformation) image (or a DEM map), estimation of interferometric correlation, and rectification of interferometric products.[6,13–15] Using a single pair of SAR images as

Table 1 Satellite SAR sensors capable of InSAR mapping

Mission	Agency	Period of operation[1]	Orbit repeat cycle (days)	Band/frequency (GHz)	Wave-length (cm)	Incidence angle (°) at swath center	Resolution (m)
Seasat	NASA	06/1978 to 10/1978	17	L-band/1.275	23.5	23	25
ERS-1	ESA	07/1991 to 03/2000	3, 168, and 35[1]	C-band/5.3	5.66	23	30
JERS-1	JAXA	02/1992 to 10/1998	44	L-band/1.275	23.5	39	20
ERS-2	ESA	04/1995 to 07/2011	35	C-band/5.3	5.66	23	30
Radarsat-1	CSA	11/1995 to 2013	24	C-band/5.3	5.66	10 to 60	10 to100
Envisat	ESA	03/2002 to 04/2012	35, 30[2]	C-band/5.331	5.63	15 to 45	20 to 100
ALOS	JAXA	01/2006 to 05/2011	46	L-band/1.270	23.6	8 to 60	10 to 100
TerraSAR-X	DLR	6/2007 to present	11	X-band/9.65	3.1	20 to 55	1 to 16
Radarsat-2	CSA	12/2007 to present	24	C-band/5.405	5.55	10 to 60	3 to 100
COSMO-SkyMed	ASI	6/2007 to present	1, 4, 5, 7, 8, 9, 12 and 16[3]	X-band/9.6	3.1	20 to 60	1 to 100

Abbreviations: ALOS, Advanced Land Observing Satellite; ASI, Agenzia Spaziale Italiana (Italian Space Agency); COSMO-SkyMed, COnstellation of small Satellites for the Mediterranean basin Observation; CSA, Canadian Space Agency; DLR, Deutsches Zentrum für Luft- und Raumfahrt e.V. (German Space Agency); Envisat, Environmental Satellite; ERS, European Remote Sensing Satellite; ESA, European Space Agency; JAXA, Japanese Aerospace Exploration Agency; JERS, Japanese Earth Resources Satellite; NASA, National Aeronautics and Space Administration.

[1]To accomplish various mission objectives, the ERS-1 repeat cycle was 3 days from July 25, 1991, to April 1, 1992, and from December 23, 1993, to April 9, 1994; 168 days from April 10, 1994, to March 20, 1995; and 35 days at other times.

[2]ENVISAT repeat cycle was 35 days from March, 2002 to October, 2010, and 30 days from November, 2010 to April 8, 2012.

[3]A constellation of 4 satellites, each of which has a repeat cycle of 16 days, can collectively produce repeat-pass InSAR images at intervals of 1 day, 4 days, 5 days, 7 days, 8 days, 9 days, and 12 days, respectively.

input, a typical InSAR processing chain outputs two SAR intensity images, a displacement or DEM map, and an interferometric correlation map.

When more than two scenes of SAR images are available over a study area, multiple-temporal InSAR images can be produced. Multi-interferogram InSAR processing should then be employed to improve the accuracy of displacement measurement (or other InSAR products).[16–20] The goal of multi-interferogram InSAR processing is to characterize the spatial and temporal behaviors of the displacement signal and various artifacts and noise sources (atmospheric delay anomalies including radar frequency-dependent ionosphere refraction and non-dispersive troposphere delay of the radar signals, orbit errors, DEM-induced artifacts) in individual interferograms and then to remove the artifacts and anomalies to retrieve time-series displacement measurements at the SAR pixel level.

SAR Intensity Image

In its simplest form, a SAR intensity image can be regarded as a thematic layer containing information about specific surface characteristics and is particularly sensitive to terrain slope, surface roughness, and the target's dielectric constant. Surface roughness refers to the SAR wavelength-scale variation in the surface relief. Dielectric constant is an electric property of material that influences radar return strength and is controlled primarily by moisture content of the imaged surface. SAR backscattering intensity images have been used in characterizing types, conditions, and flooding of wetlands[21–25] (Fig. 1).

InSAR Coherence Image

An InSAR coherence image is a cross correlation product derived from two coregistered complex-valued (both intensity and phase components) SAR images.[26,27] It depicts changes in backscattering characteristics on the scale of radar wavelength. Loss of InSAR coherence is often referred to as decorrelation. Decorrelation over wetlands can be caused by the combined effects of a) thermal decorrelation caused by uncorrelated noise sources in radar instruments, b) geometric decorrelation resulting from imaging a target from different look angles, c) volume decorrelation caused by volume backscattering effects, and d) temporal decorrelation due to environmental changes over time.[28–30,12] On the one hand, decorrelation renders an InSAR image useless for measuring surface displacement or constructing a topographic map. On the other hand, the pattern of decorrelation within an image can indicate surface modifications caused by flooding, wildfire, or other processes.

Geometric decorrelation increases as the perpendicular baseline length increases until a critical length is reached at which InSAR coherence is lost.[26] For surface backscattering,

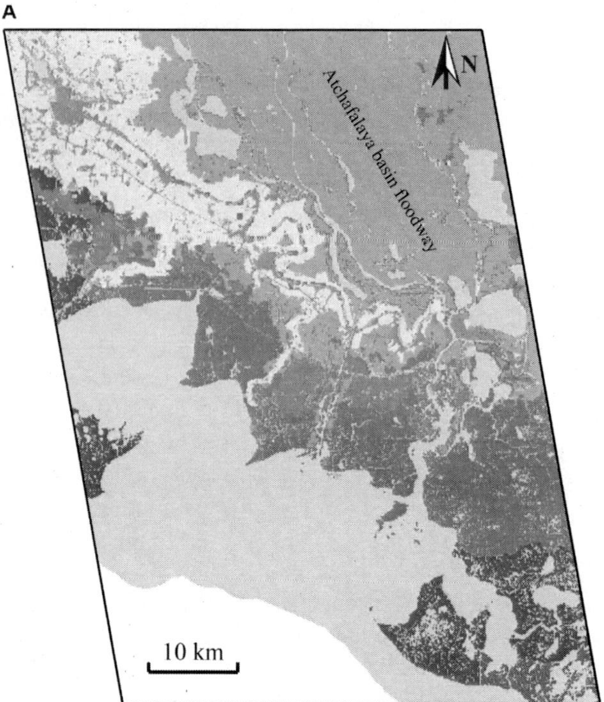

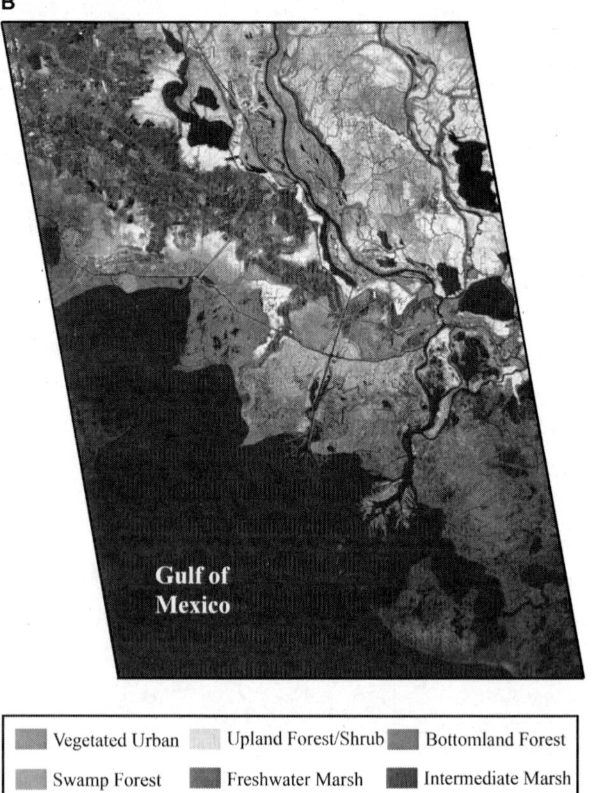

Fig. 1 **(A)** Thematic map showing major land cover classes of coastal forests in southeastern Louisiana, U.S.A. **(B)** L-band (wavelength of ~24 m) SAR intensity image produced from Advanced Land Observing Satellite (ALOS) Phased Array type L-band Synthetic Aperture Radar (PALSAR) image acquired on January 12, 2007.

most of the effect of baseline geometry on the measurement of interferometric coherence can be removed by the common spectral band filtering.[31] Volume backscattering describes multiple scattering of the radar wave within a distributed volume over wetlands and can be significantly affected by the InSAR baseline geometry configuration. As a result, volume decorrelation is most often coupled with geometric decorrelation and is a complex function of vegetation canopy structure. Generally, the contribution of volume backscattering is controlled by the proportion of transmitted signal that penetrates the surface and the relative two-way attenuation from the surface to the volume element and back to the sensor.[30] Because both surface backscattering and volume backscattering consume and attenuate the transmitted radar signal, they determine the proportion of radar signal that is available to produce double-bounce backscattering. Hence, surface backscattering and volume backscattering over wetlands combine to lower InSAR coherence and reduce double-bounce backscattering that is utilized to sense water surface beneath wetlands.

Temporal decorrelation describes any event that changes the physical orientation, composition, or scattering characteristics and spatial distribution of scatterers within an imaged volume. Over wetlands, these decorrelations are primarily caused by wind changing the leaf orientations, moisture condensation and rain changing the dielectric constant, and flooding changing the dielectric and roughness of canopy background, and seasonal phenology, as well as anthropogenic activities such as cultivation and timber harvesting.[30] Temporal decorrelation is the net effect of changes in radar backscattering and therefore depends on the stability of the scatterers, the canopy penetration depth of the transmitted pulse, and the response to the changing conditions with respect to the SAR wavelength.

DEM

A precise DEM can produce a very important data set for characterizing and monitoring coastal wetlands. The ideal SAR configuration for DEM production is a single-pass (simultaneous) two-antenna system.[8] However, repeat-pass single-antenna InSAR also can be used to produce useful DEMs.[32] Either technique is advantageous in areas where the photogrammetric approach to DEM generation is hindered by persistent clouds or other factors.[33] There are many sources of error in DEM construction from repeat-pass SAR images, e.g., inaccurate determination of the InSAR baseline, atmospheric delay anomalies, possible surface displacement due to groundwater movement, or other geological processes during the time interval spanned by the images, etc. To generate a high-quality DEM, these errors must be identified and corrected using a multi-interferogram approach.[33] A data fusion technique can be used to combine DEMs from several interferograms with different spatial resolution, coherence, and vertical accuracy to generate the final DEM product.

For vegetated terrains, InSAR analysis based on fully polarized SAR images (i.e., radar signals are transmitted and received with both vertical and horizontal polarizations) is often employed to estimate vegetation characteristics and underlying topography.[34–36] Polarization signatures of the vegetation canopy, the bulk volume of vegetation, and the ground are different and can be separated using polarimetric analysis. An optimization procedure can be employed to maximize interferometric coherence between two polarimetric radar images, thereby reducing the effect of baseline and temporal decorrelation in the interferogram. Then, using a coherent target decomposition approach that separates distinctive backscattering returns from the canopy top, the bulk volume of vegetation, and the ground surface, one can derive height differences between physical scatterers with differing scattering characteristics.[34] Physical radar backscattering models for different vegetation types including wetlands can be developed to calculate the canopy height, the bare earth topography, and other parameters based on measurements from polarimetric InSAR images.

InSAR Displacement Image

An InSAR displacement image is derived from phase components of two SAR images. SAR is a side-looking sensor; so an InSAR displacement image depicts surface displacements in the SAR line-of-sight (LOS) direction, which include both vertical and horizontal motion. Typical look angles for satellite-borne SARs are less than 45° from vertical; so LOS displacements in InSAR displacement images are more sensitive to vertical motion (uplift or subsidence) than horizontal motion.

When double-bounce backscattering dominates the returning radar signal, a repeat-pass InSAR image can be sufficiently coherent to allow the measurement of water level changes from the interferometric phase values—a significant research front on coastal wetland studies with SAR. Alsdorf et al.[37,38] discovered that interferometric analysis of L-band (wavelength of ~24 cm) SAR imagery can yield centimeter-scale measurements of water level changes throughout inundated floodplain vegetation over the Amazon. Their work confirmed that scattering elements for L-band radar consist primarily of the water surface and vegetation trunks, which allows double-bounce backscattering returns. Later, Wdowinski et al.[39,40] applied L-band images to study water level changes over the Everglades in Florida. Using C-band SAR images, Lu et al.[41] found that the generated InSAR images maintained adequate coherence to measure phase change over swamp forests in southeastern Louisiana [Fig. 2(**A**) and Fig. 2(**B**)]. This finding was unexpected because the radar signal with wavelengths shorter than L-band, such as C-band (wavelength of ~5.7 cm), was thought to backscatter from the upper canopy of swamp forests rather than the underlying water surface, and a double-bounce travel path could only occur over inundated macrophytes and small shrubs.[42,43] Later, detailed studies

A

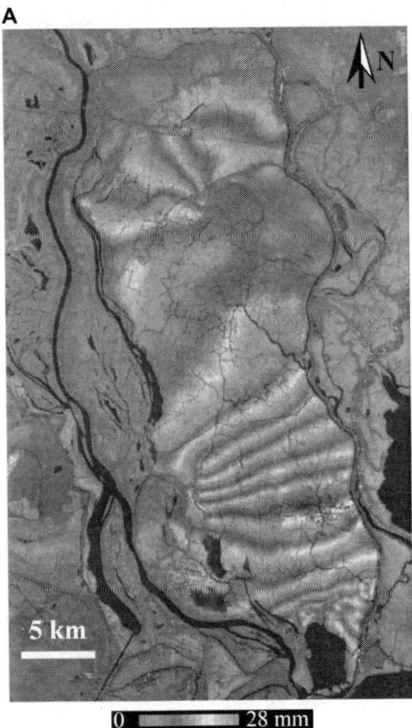

B

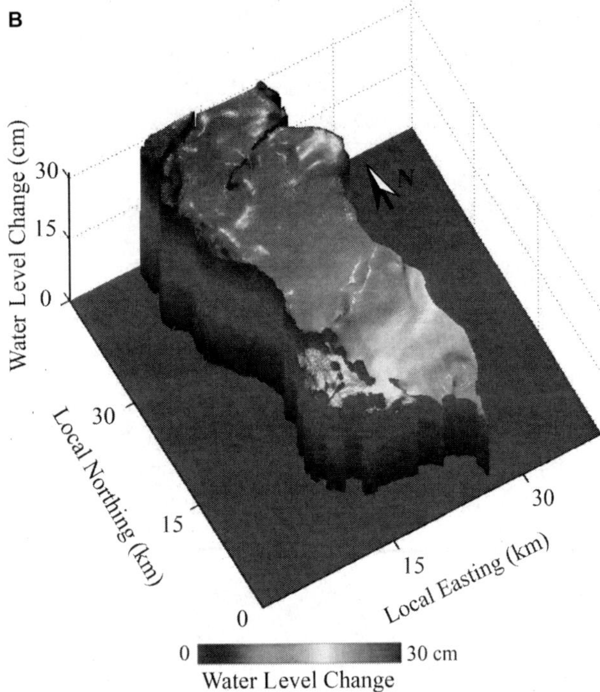

Fig. 2 **(A)** C-band (wavelength of ~5.7 cm) interferometric synthetic aperture radar (InSAR) image produced from RADARSAT-1 images, showing heterogeneous water level changes in swamp forests in southeastern Louisiana, U.S.A between May 22 and June 15, 2003. The interferometric phase image is draped over the radar intensity image. Each fringe (full-color cycle) represents a 2.8 cm LOS range change or 3.1 cm vertical change in water level. **(B)** Three-dimensional view of water volume changes derived from the InSAR image combined with radar altimeter and water level gauge measurements. Presumably, the uneven distribution of water level changes results from dynamic hydrologic effects in a shallow wetland with variable vegetation cover.

of C-band and X-band SAR images confirmed that the dominant radar backscattering mechanism for selected coastal wetlands is double-bounce backscattering, implying that even C-band and X-band InSAR images can be used to estimate water level changes over wetlands.[12,44–46]

InSAR studies of coastal wetlands suggest that water level changes can be dynamic and spatially heterogeneous and cannot be represented by readings from sparsely distributed gauge stations. However, InSAR phase measurements are relative[15] and are often disconnected by structures and other barriers in wetlands.[12] This makes it difficult to obtain absolute water level measurements over coastal wetlands from InSAR phase measurements alone. To address this issue, Kim et al.[47] and Lu et al.[48] successfully combined radar altimetry and InSAR to estimate absolute water level changes in the swamp forests of the Atchafalaya basin, Louisiana, and showed agreement between InSAR-estimated and altimeter-observed wetland water level changes over the smaller Helmand River wetland, Afghanistan, respectively. Therefore, high-resolution InSAR images, calibrated by satellite radar altimeter or other gauge measurements, can provide precise estimates of absolute water level changes over coastal wetlands.

CONCLUSION

InSAR analysis can provide timely observations of vegetation characteristics and water level changes over coastal wetlands at high spatial resolution. Future InSAR analysis of fully polarimetric radar images will offer the capability of imaging 3-D structure of coastal wetlands on a large scale for improved characterization and management purposes. With more operational satellite radar platforms available for timely data acquisitions, InSAR will continue to provide solutions to many scientific questions related to the monitoring of coastal wetlands and is poised to become a critical tool in wetland restoration.

ACKNOWLEDGMENTS

Radarsat-1 SAR images and Advanced Land Observing Satellite (ALOS) Phased Array type L-band Synthetic Aperture Radar (PALSAR) images are copyright © 2003 Canadian Space Agency and 2007 Japan Aerospace Exploration Agency/Ministry of Economy, Trade and Industry (METI), respectively. All SAR images were provided by the Alaska Satellite Facility. The authors thank O. Kwoun, E. Ramsey, and R. Rykhus for help and contribution to this project. Funding from USGS and NASA is acknowledged. The Ohio State University component of this work is partially supported by grants from USGS (No. G10AC00628) and by the Chinese Academy of Sciences/SAFEA International Partnership Program for Creative Research Teams (KZZD-EW-TZ-05).

REFERENCES

1. Prigent, C.; Matthews, E.; Aires, E.; Rossow, W. Remote sensing of global wetland dynamics with multiple satellite datasets. Geophys. Res. Lett. **2001**, *28*, 4631–4634.
2. Alsdorf, D.; Rodríguez, E.; Lettenmaier, D. Measuring surface water from space. Rev. Geophys. **2007**, *45*, RG2002, doi: 10.1029/2006RG000197.
3. Coe, M. A linked global model of terrestrial hydrologic processes: Simulation of the modern rivers, lakes, and wetlands. J. Geophys. Res. **1998**, *103*, 8885–8899.
4. Curlander, J.C.; McDonough, R.N. *Synthetic Aperture Radar: Systems and Signal Processing*, 1st Ed.; Wiley series in remote sensing and image processing; Publisher: Wiley-Interscience, 1991.
5. Goldstein, R.; Zebker, H. Interferometric radar measurements of ocean surface currents. Nature **1987**, *328*, 707–709.
6. Massonnet, D.; Feigl, K. Radar interferometry and its application to changes in the Earth's surface. Rev. Geophys. **1998**, *36*, 441–500.
7. Romeiser, R.; Thompson, D. Numerical study on the along-track interferometric radar imaging mechanism of oceanic surface currents. IEEE Trans. Geosci. Remote Sensing **2000**, *38*, 446–458.
8. Farr, T.G.; Rosen, P.A.; Caro, E.; Crippen, R.; Duren, R.; Hensley, S.; Kobrick, M.; Paller, M.; Rodriguez, E.; Roth, L.; Seal, D.; Shaffer, S.; Shimada, J.; Umland, J.; Werner, M.; Oskin, M.; Burbank, D.; Alsdorf, D. The shuttle radar topography mission. Rev. Geophys. **2007**, *45*, RG2004, doi: 10.1029/2005RG000183.
9. Gray, L.; Farris-Manning, P. Repeat-pass interferometry with airborne synthetic aperture radar. IEEE Trans. Geosci. Remote Sensing **1993**, *31*, 180–191.
10. Ramsey III, E.W. Radar remote sensing of wetland. In *Remote Sensing Change Detection*; Lunetta, R.S., Elvidge, C.D. Eds.; Ann Arbor Press: Chelsea, Michigan, 1999; 211–243.
11. Sun, G. *Radar Backscattering Modeling of Coniferous Forest Canopies* Ph.D. Dissertation; The University of California at Santa Barbara; California, 1990; 121 p.
12. Lu, Z.; Kwoun, O. Radarsat-1 and ERS interferometric analysis over southeastern coastal Louisiana: implication for mapping water-level changes beneath swamp forests. IEEE Trans. Geosci. Remote Sensing **2008**, *46*, 2167–2184.
13. Bamler, R.; Hart, P. Synthetic aperture radar interferometry. Inverse Probl. **1998**, *14*, R1–R54.
14. Rosen, P.; Hensley, S.; Joughin, I.R.; Li, F.K; Madsen, S.N.; Rodriguez, E.; Goldstein, R.M. Synthetic aperture radar interferometry. Proc. IEEE **2000**, *88*, 333–380.
15. Lu, Z. InSAR imaging of volcanic deformation over cloud-prone areas – Aleutian Islands. Photogram. Eng. Remote Sensing **2007**, *73*, 245–257.
16. Ferretti, A.; Prati, C.; Rocca, F. Permanent scatterers in SAR interferometry. IEEE Trans. Geosci. Remote Sensing **2001**, *39*, 8–20.
17. Ferretti, A.; Savio, G.; Barzaghi, R.; Borghi, A.; Musazzi, S.; Novali, F.; Prati, C.; Rocca, F. Submillimeter accuracy of InSAR time series: Experimental validation. IEEE Trans. Geosci. Remote Sensing **2007**, *45*, 1142–1153.
18. Berardino, P.; Fornaro, G.; Lanari, R.; Sansosti, E. A new algorithm for surface deformation monitoring based on small baseline differential SAR interferograms. IEEE Trans. Geosci. Remote Sens. **2002**, *40*, 2375–2383.
19. Hooper, A.; Segall, P.; Zebker, H. Persistent scatterer interferometric synthetic aperture radar for crustal deformation analysis, with application to Volcán Alcedo, Galápagos. J. Geophys. Res. **2007**, *112*, B07407, doi: 10.1029/2006 JB004763.
20. Rocca, F. Modeling interferogram stacks. IEEE Trans. Geosci. Remote Sensing **2007**, *45*, 3289–3299.
21. Bourgeau-Chavez, L.L.; Smith, K.B.; Brunzell, S.M.; Kasischke, E.S.; Romanowicz, E.A.; Richardson, C.J. Remote monitoring of regional inundation patterns and hydroperiod in the greater Everglades using synthetic aperture radar. Wetlands **2005**, *25*, 176–191.
22. Kasischke, E.S.; Smith, K.B.; Bourgeau-Chavez, L.L.; Romanowicz, E.A.; Brunzell, S.M.; Richardson, C.J. Effects of seasonal hydrologic patterns in south Florida wetlands on radar backscatter measured from ERS-2 SAR imagery. Remote Sensing Environ. **2003**, *88*, 423–441.
23. Kiage, L.M.; Walker N.D.; Balasubramanian, S.; Babin, A.; Barras, J. Applications of Radarsat-1 synthetic aperture radar imagery to assess hurricane-related flooding of coastal Louisiana. Int. J. Remote Sensing **2005**, *26*, 5359–5380.
24. Rykhus, R.; Lu, Z. *Hurricane Katrina Flooding and Possible Oil Slicks Mapped with Satellite Imagery. USGS Circular 1306*; Science and the Storms - The USGS Response to the Hurricanes of 2005: 2007; 50–53.
25. Kwoun, O.; Lu, Z. Multi-temporal RADARSAT-1 and ERS backscattering signatures of coastal wetlands at southeastern Louisiana. Photogram. Eng. Remote Sensing **2009**, *75*, 607–617.
26. Zebker, H.; Villasenor, J. Decorrelation in interferometric radar echoes. IEEE Trans. Geosci. Remote Sensing **1992**, *30*, 950–959.
27. Lu, Z.; Freymueller, J. Synthetic aperture radar interferometry coherence analysis over Katmai volcano group, Alaska. J. Geophys. Res. **1998**, *103*, 29887–29894.
28. Hagberg, J.; Ulander, L.; Askne, J. Repeat-pass SAR interferometry over forested terrain. IEEE Trans. Geosci. Remote Sensing. **1995**, *33*, 331–340.
29. Wegmüller, U.; Werner, C. Retrieval of vegetation parameters with SAR interferometry. IEEE Trans. Geosci. Remote Sensing **1997**, *35* (1), 18–24.
30. Ramsey III, E.W.; Lu, Z.; Rangoonwala, A.; Rykhus, R. Multiple baseline radar interferometry applied to coastal land cover classification and change analyses. GISci. Remote Sensing **2006**, *43*, 283–309.
31. Gatelli, F.; Guarnieri, A.M.; Parizzi, F.; Pasquali, P.; Prati, C.; Rocca, F. The wavenumber shift in SAR interferometry. IEEE Trans. Geosci. Remote Sensing **1994**, *32*, 855–865.
32. Lu, Z.; Jung, H.S.; Zhang, L.; Lee, W.J.; Lee, C.W.; Dzurisin, D. DEM generation from satellite InSAR. In *Advances in Mapping from Aerospace Imagery: Techniques and Applications*, Yang, X., Li, J., Eds.; CRC Press: 2013; 119–144.
33. Lu, Z.; Fielding, E.; Patrick, M.; Trautwein, C. Estimating lava volume by precision combination of multiple baseline spaceborne and airborne interferometric synthetic aperture radar: The 1997 eruption of Okmok Volcano, Alaska. IEEE Trans. Geosci. Remote Sensing **2003**, *41*, 1428–1436.

34. Cloude S.; Papathanassiou, K. Polarimetric SAR interferometry. IEEE Trans. Geosci. Remote Sensing **1998**, *36*, 1551–1565.
35. Treuhaft, R.; Siqueira, P. Vertical structure of vegetated land surfaces from interferometric and polarimetric radar. Radio Sci. **2000**, *35*, 141–177.
36. Touzi, R.; Boerner, W.; Lee, J.; Lueneburg, E. A review of polarimetry in the context of synthetic aperture radar: concepts and information extraction. Can. J. Remote Sensing **2004**, *30*, 380–407.
37. Alsdorf, D.; Melack, J.; Dunne, T; Mertes, L.; Hess, L.; Smith, L. Interferometric radar measurements of water level changes on the Amazon floodplain. Nature **2000**, *404*, 174–177.
38. Alsdorf, D.; Birkett, C.; Dunne, T.; Melack, J.; Hess, L. Water level changes in a large Amazon lake measured with spaceborne radar interferometry and altimetry. Geophys. Res. Lett. **2001**, *28*, 2671–2674.
39. Wdowinski, S.; Amelung, F.; Miralles-Wilhelm, F.; Dixon, T.; Carande, R. Space-based measurements of sheet-flow characteristics in the Everglades wetland, Florida. Geophys. Res. Lett. **2004**, *31*, L15503, doi: 10.1029/2004GL020383.
40. Wdowinski, S.; Kim, S.; Amelung, F.; Dixon, T.; Miralles-Wilhelm, F.; Sonenshein, R. Space-based detection of wetlands surface water level changes from L-band SAR interferometry. Remote Sensing Environ. **2008**, *112* (3), 681–696.
41. Lu, Z.; Crane, M.; Kwoun, O.; Wells, C.; Swarzenski, C.; Rykhus, R. C-band radar observes water level change in swamp forests. EOS **2005**, *86* (14), 141–144.
42. Hess, L.; Melack, J.; Filoso, S.; Wang, Y. Delineation of inundated area and vegetation along the Amazon floodplain with SIR-C synthetic aperture radar. IEEE Trans. Geosci. Remote Sensing **1995**, *33*, 896–904.
43. Wang, Y.; Hess, L.; Filoso, S.; Melack, S. Understanding the radar backscattering from flooded and nonflooded Amazon forests: Results from canopy backscatter modeling. Remote Sensing Environ. **1995**, *54*, 324–332.
44. Lu, Z.; Kwoun, O. Interferometric synthetic aperture radar (InSAR) study of coastal wetlands over southeastern Louisiana. In *Remote Sensing of Wetlands*, Wang, Y.Q. Ed.; CRC Press; 2009: 25–60.
45. Hong, S.-H.; Wdowinski, S.; Kim, S.-W. Evaluation of TerraSAR-X observations for wetland InSAR application. IEEE Geosci. Remote Sensing **2010a**, *48*, 864–873.
46. Hong, S.-H.; Wdowinski, S.; Kim, S.-W. Space-based multi-temporal monitoring of wetland water levels: Case study of WCA1 in the Everglades. Remote Sensing Environ. **2010b**, doi: 10.1016/j.rse.2010.05.019.
47. Kim, J.W.; Lu, Z.; Lee, H.K.; Shum, C.K.; Swarzenski, C.M.; Doyle, T.W. Integrated analysis of PALSAR/Radarsat-1 InSAR and ENVISAT altimeter for mapping of absolute water level changes in louisiana wetland. Remote Sensing Environ. **2009**, *113*, 2356–2365.
48. Lu, Z.; Kim, J.W.; Lee, H.K.; Shum, C.K.; Duan, J.; Ibaraki, M.; Akyilmaz, O.; Read, C. Helmand river hydrologic studies using ALOS PALSAR InSAR and ENVISAT altimetry, Mar. Geodesy **2009**, *32* (3), 320–333, doi: 10.1080/01490410903094833.

Wetlands: Economic Value

Gayatri Acharya
Environmental Economist, World Bank Institute (WBI), Washington, District of Columbia, U.S.A.

Abstract
Benefits of some wetlands will always be difficult to quantify and measure. However, various wetland goods and services are extremely valuable and, in many cases, a measurable economic value can be obtained. The economic value of ecological functions in sustaining the livelihood and cultures of human societies cannot be disregarded in development and conservation policy since these values allow us to make informed decisions about tradeoffs and help in the conservation of natural resources to enhance human welfare. Valuation studies are necessary to fully capture the economic and ecological linkages that make wetlands valuable.

INTRODUCTION

Wetlands have been centers of human civilization throughout history. For example, the Mekong and Red River deltas of Vietnam together comprise 16% of the country but support 42% of the population.[1] Yet, wetlands have been historically transformed or converted and their capacity diminished in some manner by the impacts of land use changes. OECD[2] noted that "The world may have lost 50% of the wetlands that existed since 1900; whilst much of this occurred in the northern countries during the first 50 yr of the century, increasing pressure for conversion to alternative land use has been put on tropical and subtropical wetlands since the 1950s. By 1985, it was estimated that 56–65% of the available wetland had been drained for intensive agriculture in Europe and North America; the figures for tropical and subtropical regions were 27% for Asia, 6% for South America, and 2% for Africa, making a total of 26% worldwide."

The conversion of multiple use ecosystems to single land uses may signify a welfare loss through reduced access to, or availability of, valuable goods and services. Whether or not such conversions are merited can be assessed only with sufficient information on the relative efficiency of converting or conserving the wetland ecosystem vis-à-vis alternative land uses. Economic values are a measure of how much people are willing to pay for, and how much better or worse off they would consider themselves to be as a result of changes in the supply of different goods and services. Valuation provides a means of quantifying the benefits that people receive from wetlands. The net benefits of investing in land uses that are compatible with wetlands conservation, relative to those economic activities, which contribute to wetlands degradation, can then be assessed. Values generated through economic analysis are limited however by a) their inability to capture intrinsic values associated with wetlands and b) the level of understanding, both economists and natural scientists have with regard to the level of use of goods and services, and the variation in their availability with fluxes in the natural system. Furthermore, economic values measure preferences and welfare impacts associated with changes in the natural system or the availability of a service or product; they do not, in themselves, suggest anything about inter- or intra-generational distributional effects of conservation or development investments.

Wetland goods and services are also particularly difficult to value. This is because: a) many goods are not marketed but traded or consumed directly; b) wetland services, such as water quality or groundwater recharge, often occur in areas away from the physical location of the wetland and may not be easily attributable to the wetland; c) many wetlands are transboundary resources and data on the use and consumption of goods and services are difficult to obtain; and d) many wetlands are public property. For these and other related reasons, the economic benefits generated by wetlands, and the economic costs associated with wetlands degradation or loss, are frequently unknown and omitted in project or policy analysis. As a result, the potential of wetlands to be used as contributors to economic growth, income generating activities, and as sources of goods and services, has been underestimated in many parts of the world, resulting in the loss of valuable species, services, and livelihoods.

ECONOMIC VALUES OF WETLANDS

Wetlands cover an estimated 6% of the world's land surface and may be of various types such as marshes, floodplains, freshwater ponds, lakes, swamps, etc. While these ecosystems are associated with a diverse and complex

Encyclopedia of Natural Resources DOI: 10.1081/E-ENRL-120001699

array of direct and indirect uses, the precise functions and benefits of a specific wetland can be determined only after careful study of the hydrological, physical, and biological characteristics of the site, over a period of time.[3] Since the economic value of an ecosystem service (or good) is determined by the contribution it makes to maintaining the present level of human well being, it is calculated as a measure of a change in well being, or *social welfare*. Therefore, valuing a good or service requires us to study the change in a person's welfare due to a change in the availability of the resource. Economic value expressed in currency units is the monetary expression of the tradeoff between identified alternative land uses (see Freeman,[4] among others, for more on valuation theory).

To fully assess these tradeoffs, the ultimate goal of valuation should be to assess the total economic value (TEV) of a system. Total economic value is defined as the sum of use value + existence value where use value is comprised of a) direct use value (consumptive or nonconsumptive); b) indirect use value; and c) option value. Table 1 defines each of these values while Fig. 1 shows how direct and indirect benefits can be identified in a wetland system. While valuation attempts to take account of all the components of the total economic benefit of wetlands, it is generally constrained by data availability and, therefore, partial valuation studies are more common (see Barbier, Adams, et al.[7] for more on the differences in total and partial valuation).

Direct uses include the use of the wetland for water supply and harvesting of wetland products such as fish and plant resources. Commonly recognized values of wetlands include those captured from the use of raw materials and physical products that are generally used in economic activities such as fisheries, water supply, and agriculture.

Table 1 Types of benefits value by economic valuation techniques

Direct benefits: These include the raw materials and physical products that are used directly for production, consumption, and sale including those providing energy, shelter, foods, water supply, transport, and recreation.

Indirect benefits: These include ecological functions which maintain, protect, and support natural and human systems through services such as maintenance of water quality, flow and storage, flood control and storm protection, nutrient retention, and micro-climate stabilization, and other productive and consumptive activities.

Option benefits: These refer to the premium placed on maintaining a pool of wetlands species and genetic resources for future uses such as for leisure or in commercial, industrial, agricultural and pharmaceutical applications and water-based developments, some of which may not be used or known in the present but may be used in the future.

Existence benefits: The intrinsic value of wetlands species and areas regardless of their current or future use possibilities, such as cultural, aesthetic, heritage and bequest significance.

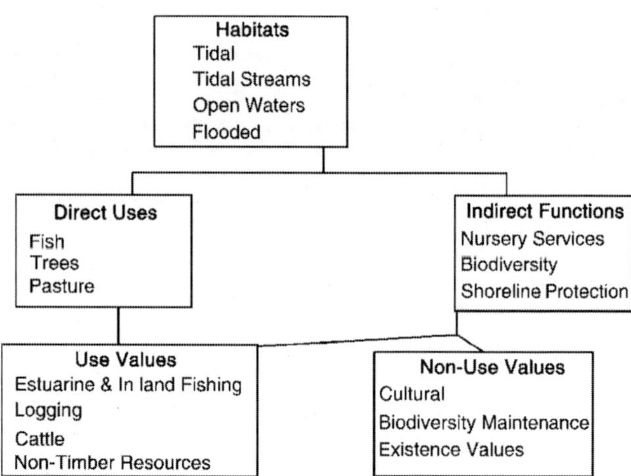

Fig. 1 Identifying direct and indirect benefits in a wetland system.

Where wild foods, grasses, bird life, and other naturally occurring species are exploited for significant economic returns, such uses are also easily identifiable and may be valued in existing markets.

Direct uses however represent only a small proportion of the total value of wetlands, which generate economic benefits in excess of just physical products. *Indirect benefits* are derived from environmental functions that wetlands may provide, depending on the type of wetland, soil, and water characteristics, and associated biotic influences.[5] Wetland functions are defined as a process or series of processes that take place within a wetland, including habitat and hydrological and water quality functions. Such functions have an economic value if they contribute to economic activity and enhance human welfare.

Valuing Wetlands Goods—Techniques and Examples

A combination of the available techniques used to value environmental goods and services can be used to approximate a partial or total value of any one ecosystem, taking care that we can identify the various components of the system and have adequate economic and ecological data to do so. The following methods have been used in valuing wetland resources:

- Market prices: to estimate economic values for ecosystem products or services that are bought and sold in markets.
- Change in productivity: to estimate economic values for ecosystem products or services that contribute to the production of marketed goods or services.
- Travel cost: to estimate economic values associated with ecosystems/sites used for recreation based on the assumption that the value of a site is reflected in how much people are willing to pay to travel to visit the site.

Table 2 Using contingent valuation and travel cost methods to assess the recreational value of Lake Nakuru, Kenya

Lake Nakura National Park is an important international tourist destination in Kenya. Fees currently charged to enter the park underestimate the total value that tourists place on the wetland and its component species, especially flamingos. A travel cost survey of visitors elicited information about length of stay, travel costs, place of origin, and visitation rates, distinguishing between resident and nonresident tourists. The contingent valuation survey used in this study asked visitors what their personal total costs of travel were; how much they would be willing to increase their expenditures to visit the park; how much they would contribute to a fund to clean-up and control urban pollution affecting the park; and how much they would contribute to a project to conserve flamingos. The survey also asked respondents for the minimum reduction in trip costs that they would be willing to accept should there be no flamingos. Results suggests that the annual recreational value of wildlife viewing in Lake Nakura National Park was between U.S. $7.5 and $15 million and over a third of this value was accounted for by the willingness to pay for viewing flamingos.

Source: Adapted from Navrud & Mungatana.[6]

Table 3 Valuing the Hadeija Nguru Wetlands, Nigeria

In the Hadeija–Jama are floodplain region in northern Nigeria, more than one half of the wetlands have already been lost to drought and upstream dams. Ecosystem valuation has been used in this area to weigh the costs and benefits of development projects that would divert some more water away from the floodplain for irrigated agriculture in upstream areas. The net benefits of such a diversion are estimated at U.S. $29 per hectare. In comparison, the floodplain, under the present flooding regime, provides U.S. $167 per hectare in benefit to a wide range of local people engaged in farming, fishing, grazing livestock, or gathering fuelwood and other wild products—benefits that would be greatly diminished by the project.

A study of the groundwater recharge function of the wetlands confirms furthermore that the wetlands play an important role by maintaining groundwater recharge in the floodplain. Groundwater recharge supports irrigated agricultural production in the floodplain. Irrigated agriculture using water from the shallow groundwater aquifer has a value of 36,308 Naira (U.S. $413) per hectare for the study area. A value of at least 2863 Naira or U.S. $32.5 per farmer per dry season or U.S. $62 per hectare is attributable to the present rate of groundwater recharge. In terms of maintaining water supply resources, the value of the recharge function is 1,146,588 Naira or U.S. $13,029 per day for the wetlands.

Source: Adapted from Barbier, et al.[7] Acharya & Barbier,[8] Acharya.[9]

Table 4 Economic values of wetlands: some examples

Valuation approach	Good or service valued	Reference	Value	Location
Revealed willingness to pay (WTP)	Commercial fishing and trapping	[18]	$486.25 per acre	Louisiana, U.S.
Revealed WTP	Freshwater forested wetlands	[23]	$78,000 per acre	Maryland, U.S.
Mitigation costs	Freshwater emergent wetlands		$49,000 per acre	
Contingent valuation, replacement cost	Nitrogen abatement	[21]	U.S. $59/kgN reduction capacity	Gotland, Sweden
Loss in productivity, market prices	Agriculture, forestry fishing from floodplain;	[7]	N109 (U.S. $15)/$10^3\,m^3$ N381 (U.S. $51)/h	Hadejia–Nguru
Production function, contingent behavior	Groundwater recharge function (drinking water supply)	[9]	U.S. $13,029 per day for the wetlands	Wetlands, Nigeria
Production function	Groundwater recharge function (agriculture)	[8]	$62/h	
Market prices	Artisanal fisheries, aquarium fish, sea cucumbers, shells, tourism	[24]	U.S. $2.14 million/yr	Kisite Marine National Park and Mpunguti Marine National Reserve, Kenya
Preventive expenditure	Coastal zone/shoreline protection	[25]	U.S. $603,248/yr	Mahe, Seychelles

- Hedonic pricing: to estimate economic values for ecosystem or environmental services that directly affect market prices of some other good, e.g., housing prices may reflect the value of local environmental attributes or amenities.
- Contingent valuation method: to estimate economic values for products and ecosystem services. This can be used to estimate nonuse values by asking people to directly state their willingness to pay for specific environmental services, based on hypothetical scenarios or contingent markets.
- Contingent choice, contingent ranking, contingent behavior: to estimate economic values for products and for ecosystem services by asking people to make tradeoffs between sets of ecosystem or environmental services or characteristics.
- Benefit transfer: to infer economic values by transferring benefit estimates from localized studies to similar areas under similar market conditions.
- Damage cost avoided, replacement cost, and substitute costs: to estimate economic values based on costs of avoided damages resulting from lost or diminished ecosystem services, replacing ecosystem services, or providing substitute services.

Examples of how these techniques have been applied to elicit values of various wetland benefits are given in Tables 2 and 3 and in Table 4. These studies illustrate the range of values associated with different types of goods and services, wetlands provide.

In some parts of the U.S., the benefits of protecting natural watersheds to assure safe and plentiful drinking water supplies are being recognized. The city of New York recently found that it could avoid spending U.S. $6–8 billion in constructing new water treatment plants by protecting the upstate watershed that has traditionally provided these purification services. Based on this assessment, the city invested U.S. $1.5 billion in protecting its watershed by buying land around reservoirs and providing other incentives for land use.[10] This decision ensures that in addition to getting a relatively cheap treatment system for its water supply, the city also maintains the watershed for recreation, wildlife habitat, and other ecological benefits derived from a healthy ecosystem.

CONCLUSIONS

Some caveats to valuation also apply. In a number of recent papers,[11-13] researchers have attempted to value the world's ecosystems, some by extrapolating localized studies of specific resource losses to capture the impact of a global loss of such resources. Even if local values are calculated using the best possible data, extrapolating these values to all wetland resources on the planet creates problem not least because there is wide variation in the types of wetlands and their component products. Heimlich et al.[14] presents the results of over 30 studies carried out on wetlands. After accounting for geographical variation and other differences across the studies by calculating the present value over a 50 yr period discounted at 6% for all the studies and then converting these values to an annual basis, they find significant variation among the different studies. For example, general recreational values range from $105 to $9859 per acre; waterfowl hunting from $108 to $3101 per acre; and recreational fishing from $95 to $28,845 per acre. Table 4 gives a few examples of various studies and types of goods or services valued. As the figures show, there is a wide variation among the types of wetlands valued and the values attributable to them. Great caution must therefore be exercised when extrapolating localized studies to other geographical areas. Resource values are influenced by local consumptive and productive use patterns, market conditions, and the institutional arrangements affecting the use of the resources. The good news is that there are now hundreds of studies of wetland values (Bardecki[15] for an extensive review) that we can refer to for some guidance on the range of values associated with wetlands.[16-22]

Benefits of some wetlands will always be difficult to quantify and measure primarily because the required scientific, technical, or economic data is difficult to obtain and also that certain intrinsic values are not measurable by existing economic valuation methods. However, as the studies reported in this paper suggest, various wetland goods and services are extremely valuable and in many cases, a measurable economic value can be obtained. The economic value of ecological functions in sustaining the livelihood and cultures of human societies clearly cannot be disregarded in development and conservation policy since these values allow us to make informed decisions about tradeoffs and help in the conservation of natural resources to enhance human welfare. Hence, valuation studies need to be carried out wherever possible and with collaboration between economists and natural scientists to fully capture the economic and ecological linkages that make wetlands valuable.

REFERENCES

1. Dugan, P.J. *Wetland Conservation: A Review of Current Issues and Required Action*; IUCN: Gland, Switzerland, 1990.
2. OECD/IUCN. *Guidelines for Aid Agencies for Improved Conservation in Sustainable Use of Tropical and Subtropical Wetlands*; OECD: Paris, 1996.
3. Mitsch, W.J.; Gosselink, J.G. *Wetlands*, 2nd Ed.; Van Nostrand Reinhold: New York, 1993.
4. Freeman, A.M. *The Measurement of Environmental and Resource Values: Theory and Methods*; Resources for the Future: Washington, DC, 1993.
5. Maltby, E. *Waterlogged Wealth*; Earthscan: London, 1986.
6. Navrud, S.; Mungatana, E. Environmental valuation in developing countries: the recreation value of wildlife viewing. Ecol. Econ. **1994**, *11*, 135–151.

7. Barbier, E.B.; Adams, W.; Kimmage, K. Economic valuation of wetland benefits. In *The Hadejia–Nguru Wetlands*; Hollis, Ed.; IUCN: Gland, 1993.
8. Acharya, G.; Barbier, E.B. Valuing groundwater recharge through agricultural production in the Hadejia–Nguru wetlands in Northern Nigeria. Agric. Econ. **2000**, *22*, 247–259.
9. Acharya, G. Valuing the hidden hydrological services of wetland ecosystems. Ecol. Econ. **2000**, *35*, 63–74.
10. National Research Council, (NRC). Committee to review the New York city watershed management strategy. In *Watershed Management for Potable Water Supply: Assessing the New York City Strategy*; National Academy Press: Washington, DC, 2000.
11. Costanza, R.; d'Arge, R.; de Groot, R.; Farber, S.; Grasso, M.; Hannon, B.; Limburg, K.; Naeem, S.; O'Neill, R.V.; Paruelo, J.; Raskin, R.G.; Sutton, P.; van den Belt, M. The value of the world's ecosystem services and natural capital. Nature **1997**, *387*, 253–260.
12. Pimentel, D.; Wilson, C.; McCullum, C.; Huang, R.; Dwen, P.; Flack, J.; Tran, Q.; Saltman, T.; Cliff, B. Economic and environmental benefits of biodiversity. BioScience **1997**, *47*(11).
13. Ehrlich, P.R.; Ehrlich, A.H. *Betrayal of Science and Reason: How Anti-Environmental Rhetoric Threatens Our Future*; Putnam: New York, 1996.
14. Heimlich, R.E.; Wiebe, K.D.; Claassen, R.; Gadsby, D.; House, R.M. *Wetlands and Agriculture: Private Interests and Public Benefits*; Agriculture Economic Report No. 765; Resource Economics Division, Economic Research Service, USDA, 1998.
15. Bardecki, M. Wetlands and economics: an annotated review of the literature, 1988–1998 with special reference to the wetlands of the Great Lakes. In *Report Prepared for Environment Canada—Ontario Region*; http://www.on.ec.gc.ca/glimr/data/wetland-valuation/intro.html.; 1998.
16. Hammack, J.; Brown, G.M. *Waterfowl and Wetlands: Toward Bioeconomic Analysis*; Resources for the Future: Washington, DC, 1974.
17. Lynne, G.D.; Conroy, P.; Prochaska, F.J. Economic valuation of marsh areas for marine production processes. J. Environ. Econo. and Manag. **1981**, *8*, 175–186.
18. Farber, S.; Costanza, R. The economic value of wetland systems. J. of Environ. Manag. **1987**, *24*, 41–51.
19. Bell, F. The economic valuation of saltwater marshes supporting marine recreational fishing in south-eastern United States. Ecol. Econ. **1997**, *21*, 243–254.
20. Bergstrom, J.C.; Stoll, J.R.; Titre, J.P.; Wright, V.L. Economic value of wetlands based recreation. Ecol. Econ. **1990**, *2*, 129–147.
21. Gren, I.M.; Folke, C.; Turner, K.; Bateman, I. Primary and secondary values of wetland ecosystems. Environ. Resour. Econ. **1994**, *4*(1), 55–74.
22. Morrison, M.D.; Bennett, J.W.; Blamey, R.K. *Valuing Improved Wetland Quality Using Choice Modelling*; Choice Modelling Research Report No. 6; University College, The University of New South Wales: Canberra, 1998.
23. Bohlen, C.; King, D. *Towards a Sustainable Coastal Watershed: The Chesapeake Experiment*; Proceedings of a conference, CRC Publication No. 149; Nelson, S., Hiss, P., Eds.; Chesapeake Research Consortium: Edgewater, MD, 1995.
24. Emerton, L.; Tessema, Y. *Economic Constraints to the Management of Marine Protected Areas: The Case of Kisite Marine National Park and Mpunguti Marine National Reserve, Kenya*; IUCN—The World Conservation Union: 2001.
25. Emerton, L. *Economic Tools for Valuing Wetlands in Eastern Africa*; IUCN—The World Conservation Union: 1998.

Wetlands: Ecosystems

Sherri DeFauw
New England Plant, Soil and Water Laboratory, Agricultural Research Service, U.S. Department of Agriculture (USDA-ARS), University of Maine, Orono, Maine, U.S.A.

Abstract
Wetlands perform key roles in the global hydrologic cycle. This entry profiles several of the most rapidly disappearing wetland ecosystems in North America, in terms of properties and processes.

INTRODUCTION

Wetlands perform key roles in the global hydrologic cycle. These transitional ecosystems vary considerably in their capacity to store and subsequently redistribute water to adjacent surface water systems, groundwater, the atmosphere, or some combination of these. Saturation in the root zone or water standing at or above the soil surface is key to defining a wetland. When oxygen levels in waterlogged soils decline below 1%, anaerobic (or reducing) conditions prevail. Most, but not all, wetland soils exhibit redoximorphic features formed by the reduction, translocation, and oxidation of iron (Fe) and manganese (Mn) compounds; the three basic kinds of redoximorphic features include redox concentrations, redox depletions, and reduced matrix.[1] Microbial transformations in flooded soils also impact other biogeochemical cycles (C, N, P, S) at various spatial and temporal scales. Several of the most rapidly disappearing wetland ecosystems in North America are profiled here, in terms of properties and processes.

HYDROLOGIC CONSIDERATIONS

Wetland water volume and source of water are heavily influenced by landscape position, climate, soil properties, and geology. Wetlands may be surface flow dominated, precipitation dominated, or groundwater discharge dominated systems (Fig. 1). Surface flow dominated wetlands include riparian swamps and fringe marshes. In unregulated settings (i.e., no dams or diversions) these ecosystems are subject to large hydrologic fluxes, and vary the most in terms of soil development, sediment loads, and nutrient exchanges. Precipitation-dominated wetlands (e.g., prairie potholes and bogs) reside in landscape depressions and typically have a relatively impermeable complex of clay and/or peat layers that retard infiltration (or recharge) and also impede groundwater discharge (or inflow). Groundwater dominated wetlands (e.g., fens and seeps) may form in riverine settings, at slope breaks, or in areas where abrupt to rather subtle changes in substrate porosity occur. Groundwater contributions to wetlands are complex, dynamic, and rather poorly understood.[2]

Frequency and duration of flooding, and the long-term amplitude of water level fluctuations in a landscape are the three most important hydrologic parameters that "shape" the aerial extent of a wetland complex as well as determine the relative abundance of four intergrading wetland settings (i.e., swamps, wet meadows, marshes, and aquatic ecosystems.[3] Wetland hydrodynamics control soil redox conditions. The hydroperiod–redox linkage, in turn, controls plant macronutrient concentrations (N, P, K, Ca, Mg, S), micronutrient availability (B, Cu, Fe, Mn, Mo, Zn, Cl, Co), pH, organic matter accumulation, decomposition, and influences plant zonation. Oftentimes, the zonation of plants (as determined by competition and/or physiological tolerances) provides key insights into the hydrodynamics and biogeochemistry of an area.[3]

ECOSYSTEM PROFILES

Four of the most rapidly disappearing wetland ecosystems in North America are summarized here in terms of key properties and processes. These profiles include comments on geographical extent, geomorphology, soils, hydrodynamics, biogeochemistry, vegetation structure, and/or indicator species, as well as recent estimates of ecosystem losses and ecological significance.

Riparian Swamps

Bottomland hardwood forests once dominated the river floodplains of the eastern, southern, and central United States. In the Mississippi Alluvial Plain, an estimated 8.6 million hectare area of bottomland hardwoods has been reduced to 2 million forested hectares remaining.[4] Although their true extents are not well-documented (due to difficulties in determining upland edges), these hydrologically open, linear landscape features have been logged, drained, and converted to other uses (predominantly agriculture) at

Encyclopedia of Natural Resources DOI: 10.1081/E-ENRL-120010127

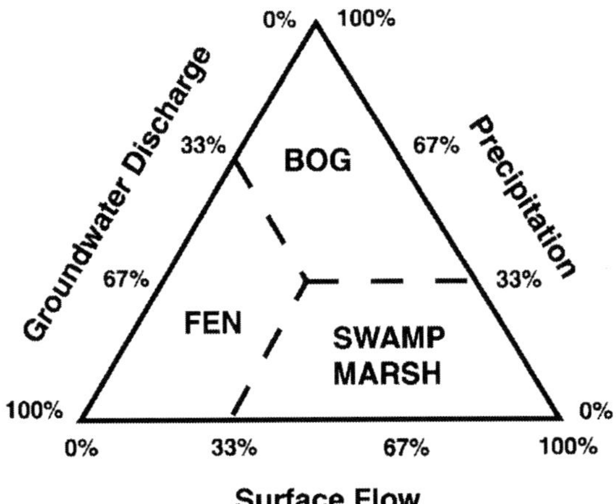

Fig. 1 The relative contributions of groundwater discharge, precipitation, and surface flow determine main wetland types. Swamps and marshes are distinguished by the frequency and duration of surface flows.
Source: Modified from Keddy.[3]

alarming rates. Between 1940 and 1980, bottomland hardwood forests were cleared at a rate of 67,000 ha yr^{-1}.[5]

Riparian wetlands are unique, vegetative zonal expressions of both short- and long-term fluvial processes. A widely used classification scheme relates flooding conditions (frequency and flood duration during the growing season) with zonal associations of hardwood species.[6] Zone II (intermittently exposed "swamp"—bald cypress and water tupelo usually dominate the canopy) and Zone III (semipermanently flooded "lower hardwood wetlands"—overcup oak and water hickory are common) are accepted as wetlands by most; however, Zones IV (seasonally flooded "medium hardwood wetlands" that include laurel oak, green ash, and sweetgum) and V (temporarily flooded "higher hardwood wetlands"—typically an oak–hickory association with loblolly pine) have been controversial when dealing with wetland management issues.[7] Typically the complexity of floodplain microtopography abates smooth transitions from one zone to the next, with unaltered floodplain levees oftentimes exhibiting the highest plant diversity.[8]

Riparian wetlands process large influxes of energy and materials from upstream watersheds and lateral runoff from agroecosystems. As a result of these inputs, combined with decomposition of resident biomass, the organic matter content of these alluvial soils usually ranges from 2% to 5%.[8] Clay-rich bottomland tracts have higher concentrations of N and P as well as higher base saturations (i.e., Ca, Mg, K, Na) compared to upland areas. Watersheds dominated by riparian ecosystems export large amounts of organic C in dissolved and particulate forms.[9]

Prairie Potholes

Prairie potholes comprise a regional wetland mosaic that includes parts of the glaciated terrains of the Dakotas, Iowa, Minnesota, Montana, and Canada. Originally, this region encompassed 8 million hectares of wetland prior to drainage for agriculture; an estimated 4 million hectares remained at the close of the 1980s.[10] These landscape depressions are the most important production habitat in North America for most waterfowl.[11]

Most prairie potholes are seasonally flooded wetlands dependent on snowmelt, rainfall, and groundwater. Four main hydrological groupings are recognized: ephemeral, intermittent, semi-permanent, and permanent.[11] Water level fluctuations are as high as 2–3 m in some settings. Measures of soil hydraulic conductivity demonstrate that groundwater flow is relatively slow (0.025–2.5 m yr^{-1}); therefore, the potholes are hydrologically isolated from each other in the short-term.[12]

Long-term ecological studies at the Cottonwood Lake Study Area, NPWRC-USGS[11] have revealed the intricacies of several prairie pothole phases. During periods of drought, marsh soils, sediments, and seed banks are exposed (i.e., dry marsh phase). Seed banks in natural sites have 3000–7000 seeds/m^2. A mixture of annuals (usually the dominant group) and emergent macrophyte species germinate on the exposed mudflats and a wet meadow develops. As water levels increase, the annuals decline and emergent macrophytes rapidly recolonize (i.e., regenerating marsh). If the flooding is consistently shallow, emergent macrophytes will eventually dominate the entire pothole. Sustained deep water flooding results in extensive declines in emergent macrophytes (i.e., degenerating marsh), and intensive grazing by muskrats may culminate in a lake marsh phase as submersed macrophytes become established. When water levels recede, emergent macrophytes re-establish. The rich plant communities that develop in these dynamic marshland complexes are also controlled by two additional environmental gradients, namely, salinity and anthropogenic disturbances (involving conversion to agriculture and extensive irrigation well pumping).[11]

Northern Peatlands

Deep peat deposits, in the United States, occur primarily in Alaska, Michigan, Minnesota, and Wisconsin, and are scattered throughout the glaciated northeast and northwest, as well as mountaintops of the Appalachians. The most extensive peatland system, in North America, is the Hudson Bay lowlands of Canada that occupies an estimated 32 million hectares.[13] The Alaskan and Canadian peatlands are relatively undisturbed, and the least threatened by developmental pressures. Elsewhere, peatlands have either been converted for agricultural use (including forestry) or mined for fuel and horticultural materials.

Bogs and fens are the two major types of peatlands that occupy old lake basins or cloak the landscape. The most influential, interdependent physical factors shaping these ecosystems include: 1) water level stability; 2) fertility;

3) frequency of fire; and 4) grazing intensity.[3] Northern bogs are dominated by oligotrophic *Sphagnum* moss species, and may be open, shrubby, or forested tracts. These predominantly rainfed (ombrogenous) systems have low water flow (with a water table typically 40–60 cm below the peat surface), are extremely low in nutrients (especially poor in basic cations), and accumulate acidic peats (pH 4.0–4.5).[14] Fens are affected by mineral-bearing soil waters (groundwater and/or surface water flows), and possess water levels at or near the peat surface. Fens may be subdivided into three hydrologic types: soligenous (heavily influenced by flowing surface water); topogenous (largely influenced by stagnant groundwater); or limnogenous (adjacent to lakes and ponds).[14] These minerogenous ecosystems range from acidic (pH 4.5) to basic (pH 8.0); vegetation varies from open, sedge-dominated settings to shrubby, birch-willow dominated associations to forested, black spruce-tamarack tracts. Nutrient availability gradients do not necessarily coincide with the ombrogenous–minerogenous gradient; recent investigations indicate higher P availability in more ombrogenous peatlands, and greater N availability in more minerogenous peatlands.[15]

Northern peatlands represent an important, long-term carbon sink, with an estimated 455 Pg (1 Petagram = 10^{15} g) stored worldwide.[16] An estimated 220 Pg of C is currently stored in North American peatlands, compared to about 20 Pg in storage during the last glacial maximum.[17] High latitude peatlands also release about 60% of the methane generated by natural wetlands.[18] In addition, sponge-like living *Sphagnum* carpets facilitate permanently wet conditions, and the high cation exchange capacity of cells retains nutrients and serves to acidify the local environment.[19]

Pocosins

The Pocosins region of the Atlantic coastal plain extends from Virginia to the Georgia–Florida border.[20] These nonalluvial, evergreen shrub wetlands are especially prevalent in North Carolina; in fact, pocosins once covered close to 1 million hectare in this state.[21] Derived from an Algonquin Indian word for "swamp-on-a-hill," pocosins are located on broad, flat plateaus and sustained by waterlogged, acidic, nutrient-poor sandy, or peaty soils usually far removed from large streams. Wetland losses are high, with 300,000 hectare drained for agriculture and forestry uses between 1962 and 1979.[22]

Pocosins are characterized by a dense, ericaceous shrub layer; an open canopy of pond pine may be present or absent.[21] A typical low pocosin ecosystem [less than 1.5 m (5 ft) tall] includes swamp cyrilla (or titi), fetterbush, bayberry, inkberry, sweetbay, laurel-leaf greenbrier, and sparsely distributed, stunted pond pine.[22] Pocosin soils may be either organic (with a deep peat layer—e.g., Typic Medisaprist) or mineral (usually including a water restrictive spodic horizon—e.g., Typic Endoaquod). As peat depth decreases, the stature of the vegetation increases. High pocosin [with shrubby vegetation 1.5–3.0 m (5–10 ft) tall and canopy trees approximately 5 m (16 ft) in height] usually occurs on peat deposits of 1.5 m (5 ft) or less in thickness or on wet sands.[20] The major natural disturbance to these wetlands is periodic burning (with a fire frequency of about 15–50 years).

Pocosin surface and subsurface waters are similar to northern ombrogenous bogs, but are more acidic with higher concentrations of sodium, sulfate, and chloride ions.[23] Carbon : Phosphorus (C : P) ratios increase sharply during the growing season; phosphorus availability limits plant growth and probably plays a crucial role in controlling nutrient export. Undisturbed pocosins export organic N and inorganic phosphate in soil water.[23]

CONCLUSIONS

Despite existing preservation policies, U.S. wetland conversions are anticipated to continue at a rate of 290,000–450,000 acres (117,408–182,186 ha) annually.[24] It is widely known that wetlands are the product of many environmental factors acting simultaneously; perturbations in one realm (e.g., hydrology) not only impact local wetland properties and processes, but also have consequences in linked ecosystems as well. Wetlands are major reducing systems of the biosphere, transforming nutrients and metals, and regulating key exchanges between terrestrial and aquatic environments.

ACKNOWLEDGMENTS

William J. Rogers, B. A. Howell, and Van Brahana commented on an earlier draft; their reviews are genuinely appreciated.

REFERENCES

1. Soil survey staff. *Keys to Soil Taxonomy*, 8th Ed.; USDA-NRCS: Washington, DC, 1998.
2. http://water.usgs.gov/nwsum/WSP2425/hydrology.html (accessed July 2002).
3. Keddy, P.A. *Wetland Ecology: Principles and Conservation*; Cambridge University Press: Cambridge, 2000.
4. Llewellyn, D.W.; Shaffer, G.P.; Craig, N.J.; Creasman, L.; Pashley, D.; Swan, M.; Brown, C. A decision support system for prioritizing restoration sites on the Mississippi River alluvial plain. Conserv. Biol. **1996**, *10*(5), 1446–1455.
5. Kent, D.M. *Applied Wetlands Science and Technology*, 2nd Ed.; Lewis Publishers: Boca Raton, FL, 2001.
6. Clark, J.R., Benforado, J., Eds.; *Wetlands of Bottomland Hardwood Forests*; Elsevier: Amsterdam, 1981.
7. Mitsch, W.J.; Gosselink, J.G. *Wetlands*; Van Nostrand Reinhold Company: New York, 1986.

8. Wharton, C.H.; Kitchens, W.M.; Pendleton, E.C.; Sipe, T.W. *The Ecology of Bottomland Hardwood Swamps of the Southeast: A Community Profile*; U.S. Fish and Wildlife Service, Biological Services Program FWS/OBS-81/37: Washington, DC, 1982; 1–133.
9. Brinson, M.M.; Swift, B.L.; Plantico, R.C.; Barclay, J.S. *Riparian Ecosystems: Their Ecology and Status*; U.S. Fish and Wildlife Service, Biological Services Program FWS/OBS-81/17: Washington, DC, 1981; 1–151.
10. Leitch, J.A. Politicoeconomic overview of prairie potholes. In *Northern Prairie Wetlands*; van der Valk, A.G., Ed.; Iowa State University Press: Ames, 1989; 2–14.
11. http://www.npwrc.usgs.gov/clsa (accessed July 2002).
12. Winter, T.C.; Rosenberry, D.O. The interaction of ground water with prairie pothole wetlands in the cottonwood lake area, East-Central North Dakota, 1979–1990. Wetlands **1995**, *15*(3), 193–211.
13. Wickware, G.M.; Rubec, C.D.A. *Ecoregions of Ontario*, Ecological Land Classification Series No. 26; Environment Canada, Sustainable Development Branch: Ottawa, Ontario, 1989.
14. http://www.devonian.ualberta.ca/peatland/peatinfo.htm (accessed July 2002).
15. Bridgham, S.D.; Updegraff, K.; Pastor, J. Carbon, nitrogen, and phosphorus mineralization in Northern Wetlands. Ecology **1998**, *79*, 1545–1561.
16. Gorham, E. Northern peatlands: role in the carbon cycle and probable responses to climatic warming. Ecol. Appl. **1991**, *1*(2), 182–195.
17. Halsey, L.A.; Vitt, D.H.; Gignac, L.D. *Sphagnum* dominated peatlands in North America since the last glacial maximum: their occurrence and extent. The Bryologist **2000**, *103*, 334–352.
18. Cicerone, R.J.; Ormland, R.S. Biogeochemical aspects of atmospheric methane. Global Biogeochem. Cycles **1988**, *2*(2), 299–327.
19. van Breeman, N. How *Sphagnum* bogs down other plants. Trends Ecol. Evol. **1995**, *10*(7), 270–275.
20. Sharitz, R.R.; Gresham, C.A. Pocosins and Carolina bays. In *Southern Forested Wetlands: Ecology and Management*; Messina, M.G., Conner, W.H., Eds.; Lewis Publishers: Boca Raton, FL, 1998; 343–389.
21. 21. Richardson, C.J. Pocosins: an ecological perspective. Wetlands **1991**, *11*(S), 335–354.
22. Richardson, C.J.; Evans, R.; Carr, D. Pocosins: an ecosystem in transition. In *Pocosin Wetlands: An Integrated Analysis of Coastal Plain Freshwater Bogs in North Carolina*; Richardson, C.J., Ed.; Hutchinson Ross Publishing Company: Stroudsburg, PA, 1981; 3–19.
23. Wallbridge, M.R.; Richardson, C.J. Water quality of pocosins and associated wetlands of the Carolina coastal plain. Wetlands **1991**, *11*(S), 417–439.
24. Schultink, G.; van Vliet, R. *Wetland Identification and Protection: North American and European Policy Perspectives*; Agricultural Experiment Station Project 1536; Michigan State University, Department of Resource Development: East Lansing, MI, 1997.

Wetlands: Freshwater

Brij Gopal
Centre for Inland Waters in South Asia, Jaipur, India

Abstract
Freshwater wetlands are habitats with vegetation especially adapted to prolonged water saturation or shallow flooding of the soils. They occur in all climates, from sea level to more than 5000 m altitude. Wetlands are usually highly productive, support high biodiversity, and provide numerous direct and indirect benefits to humans, which include flood regulation, groundwater recharge, and water quality improvement. Wetlands are threatened by all kinds of human activities, especially those impacting upon their hydrology. All countries of the world are now making efforts to conserve them under the aegis of the Ramsar Convention.

INTRODUCTION

Wetland is the term used since the late 1960s for a very large diversity of habitats that were previously known by common terms such as bog, fen, mire, marsh, swamp, and mangrove and hundreds of local names in different countries. Wetlands occur in all climates, across a wide range of latitudes, and from sea level to more than 5000 m altitude (as in the Himalaya). The unifying characteristic of these areas on the Earth's surface is that the land remains either water-saturated or under shallow water for a period ranging from several weeks to the whole year and is generally covered with vegetation that differs from that in the adjacent areas (see also[1]). Wetlands are sometimes transitional in character between deepwater and terrestrial habitats, and are often located between them. These habitats may depend for their water entirely upon the precipitation (bogs and prairie potholes) or on the surface water from the rivers (riparian/floodplain wetlands), lakes (littoral zones), or the oceans (coastal wetlands). Wetlands also develop where the ground water is discharged naturally onto the surface (fens). Wetlands are usually grouped into freshwater and marine wetlands or into inland and coastal wetlands. The inland wetlands, occurring above the mean sea level, also include those developing in saline waters (salt lakes) besides those experiencing estuarine/brackish water conditions (lagoons and backwaters).

KINDS OF WETLANDS

There are many ways of categorizing wetlands on the basis of diverse characteristics such as hydrology, geomorphology, nutrient status, production, salinity, and so on. Freshwater wetlands can be readily grouped into bogs, fens, marshes, swamps, and shallow waters. The bogs and fens occur in temperate climates and accumulate peat, whereas marshes are dominated by herbaceous plants (mostly grasses and sedges), swamps are dominated by trees and shrubs, and shallow waters may have more of submerged and floating plants. *Taxodium* swamps in southeast United States, *Melaleuca* swamps of Malaysia and Lower Mekong basin, *Myristica* swamps of southwest India, peat swamp forests (usually dominated by dipterocarps) of Indonesia and Maputaland (South Africa), and the floodplain forests of the Amazonia are examples of important swamps. Well-known marshes include the reed and papyrus marshes of Okavango, Mesopotamian marshes of Iraq, reed marshes of the Danube River delta, papyrus marshes of Egypt, Camargue marshes of southern France, and the extensive Everglades of Florida (U.S.A.). Wetlands also develop in depressions that are flooded seasonally for only a few months.

Humans have also created wetlands around the world, especially in arid and semi-arid regions, by constructing small and large reservoirs and dugout ponds to store water for various human uses. Another kind of wetlands includes the extensive paddy fields and aquaculture ponds (or fish ponds), which are modified by humans from the naturally occurring wetlands for specific production systems. Among the human-made wetlands, mention may also be made of the 200 ha Putrajaya wetlands (Malaysia) that were constructed in late 1990s to treat storm water and wastewater from the new capital township, Putrajaya.[2]

WETLAND CHARACTERISTICS

Wetlands can be distinguished from lakes with deep water by the dominant community of primary producers. The planktonic organisms dominate the deep water lakes, whereas macrophytes (including macroalgae, bryophytes, and vascular plants) are usually the dominant producers in wetlands. Wetlands generally support high densities and diversity of fauna, particularly birds, fish, and

Encyclopedia of Natural Resources DOI: 10.1081/E-ENRL-120048162

macro-invertebrates. Another characteristic of most freshwater wetlands is the hydric soils, which develop under prolonged waterlogging or submergence that eliminates oxygen from the soil. The hypoxic (reduced supply of oxygen) to anoxic (total absence of O_2) conditions in the soil affect the root respiration and consequently the plant growth. Wetland plants have a variety of morphological, anatomical, and physiological adaptations against hypoxia. These adaptations include the presence of air spaces within their roots, stems, and leaves that allow the storage and transport of gases from the atmosphere to the roots. Wetland plants are particularly tolerant to many toxic substances, such as heavy metals, which are either precipitated on the surface of their roots or stored in isolated areas of their roots and stems. In the absence of oxygen, anaerobic micro-organisms mediate several chemical transformations that often result in the reduction of C, S, N, Fe, Mn, and other substances to their toxic forms. The reduction of carbon results in the production of methane (commonly known as marsh gas), whereas the reduction of nitrates (denitrification) results in harmful ammonia or loss of nitrogen to the atmosphere. Nitrogen deficiency in wetlands is often explained by the occurrence of insectivorous plants such as *Drosera* and *Utricularia* in wetlands. The reduction, translocation, and oxidation of iron (Fe) and manganese (Mn) compounds in the soil impart characteristic redoximorphic features to many wetland soils.

Besides climate and geomorphic features of the area, the hydrological regimes (i.e., the duration, depth, and annual amplitude of changes as well as the frequency and timing of flooding) primarily determine all wetland characteristics. (For more on wetland hydrology, see [1].) Flooding regimes control soil chemistry, pH and nutrient availability, microbial decomposition and organic matter accumulation, and distribution and zonation of biota. Nutrients, salinity, and disturbance are other important modifiers that affect wetland structure and function.[3] Peat (partly decomposed organic matter) accumulation occurs in both temperate and tropical climates. In northern temperate regions, peat is formed largely by moss (*Sphagnum* species), whereas in the Amazon basin and tropical Southeast Asia, peat develops from woody material in swamp forests.

WETLAND BIODIVERSITY

Wetlands are usually very rich in biodiversity.[4] Often, their biodiversity is underestimated because one or two species of plants may dominate large areas (e.g., Lake Neusiedler See in Austria with almost exclusive occurrence of *Phragmites australis*) and similarly large populations of one animal species may appear dominant. However, many wetlands are extraordinarily rich in species of higher plants alone: 1150 species in the Everglades (U.S.A.), 672 species in Teici (Latavia), 650 species in Kushiro (Japan), 873 species in Dongdong Tinghu (China), 1256 species in Okavango (Botswana), and approximately 900 species of only trees in the whitewater floodplains of Amazon basin. The faunal diversity is exceptionally large considering the fact that wetland animals not only include the residents in the wetland proper and regular migrants from deepwater habitats, terrestrial uplands, and other wetlands but also occasional visitors and those indirectly dependent on wetland biota.[5] On a regional scale, the beta-diversity of wetlands is quite large as 15–20% of all biota of the region occurs in wetlands.

WETLAND PRODUCTIVITY

Wetlands are usually considered to be highly productive systems. However, this is true only for warmer climates because primary production is greatly constrained by low temperatures in northern temperate regions. Peat bogs are usually very low in production (0.5 $g/m^2/yr$), whereas marshes and swamps (such as Amazonian swamp forests and Indonesian peat swamp forests) have very high production – as high as or even higher than that of rainforests. Nutrient availability is also a major constraint to wetland production. Wetlands are accordingly classified as oligotrophic or eutrophic depending on their nutrient status.

WETLAND USE AND HUMAN IMPACTS

Humans have extensively used, impacted, and transformed freshwater wetlands in all parts of the world. Wetlands have provided, and are still used as a valuable source of food, fuel, fodder, and fiber. Many wetland plants have been a source of medicine and various chemicals. In the valleys of Tigris–Euphrates and Indus, wetlands formed the cornerstone of human civilizations that grew there. In most of South and Southeast Asia, wetlands were closely integrated into the social and cultural life of the people, and were even considered sacred. In temperate regions, although peat bogs were greatly despised, peat formed a major source of fuel and reeds (*Phragmites*) were extensively used for thatching the cottages. However, with growing agricultural and urban development, wetlands were drained and reclaimed on a large scale. Forested wetlands (swamps) were also cleared. Marshes and shallow water bodies were converted to paddy fields and fish ponds. Within the United States, more than half of the wetlands were lost to drainage during 1780–1980, and more than 90% of wetlands in New Zealand were lost within a few decades after colonization started in the mid-1800s. Wetland losses were relatively small in Asia where they were converted rapidly and until recently, still lower in Africa and South America where developmental and human pressure had remained rather low.

VALUES AND ECOSYSTEM SERVICES

After large areas of wetlands were lost worldwide, people started realizing their importance as critical habitats for waterfowl in the early 1950s. By the 1970s, wetlands were recognized as ecosystems and researchers discovered their important functions and values to human beings. Various studies demonstrated their major role in regulating water quality, controlling erosion, supporting fisheries and wildlife, and above all in the hydrological cycle. The Millennium Ecosystem Assessment[6] brought into focus the term "ecosystem services" for the benefits derived by the humans from the functioning of ecosystems. Wetlands also offer a wide range of ecosystem services. They provide freshwater and a variety of plant and animal resources, regulate floods and droughts, climate and soil erosion, support high biodiversity and formation of soils, and have many cultural, recreational, and aesthetic benefits to different communities. However, it is important to note that wetlands differ greatly in their ability to provide various ecosystem services and no wetland can provide all of them.

THREATS TO WETLANDS

Wetlands in general and freshwater wetlands in particular, are among the most threatened ecosystems as humans continue to alter, directly and indirectly, the hydrological regimes. Surface water flows are extensively diverted and impounded for irrigation, hydropower, and other human uses and thereby affect the downstream wetlands. The abstraction of both surface and ground water, drainage, and the wastewater discharge also alter the hydrological regimes directly. Indirect alterations are caused by land use changes throughout the wetland catchments. Discharge of domestic wastewater and industrial effluents into wetlands directly or through lakes and rivers, seriously affect the biota, particularly the benthic and microbial communities, and a number of ecosystem processes.

Another common threat to freshwater wetlands throughout the world has been the introduction of exotic invasive species, mostly plants. Water hyacinth tops the list in all tropical countries and is becoming a problem in its native Brazil as well. *Lythrum salicaria, Melaleuca quinquenervia, Typha agrustifolia, Phragmites australis, Mimosa pigra*, and *Arundo donax* are among the invasives that are causing large changes in wetland plant and animal communities and their functioning in different parts of the world.

Both natural and human-induced fires are common threats to wetlands, although fire has also been used as a management tool in some areas. For example, the Southeast Asian peat swamp forests are now affected by increased frequency of fire—largely caused by drainage, land use changes, and long dry spells. These fires not only cause the loss of wetlands with high biodiversity but also have an impact on wildlife, human health, and economy.

WETLANDS AND CLIMATE CHANGE

Wetlands play an important role in regulating climate change. On one hand, they contribute significantly to carbon sequestration through their high primary production and low decomposition rates, and on the other, they are an important source of the greenhouse gas, methane. It is projected that the draining of peatlands and fire in peat swamps will add significantly to the climate change processes. At the same time, wetlands themselves are directly threatened by climate change as the melting of glaciers and the increased variability in precipitation regimes will impinge upon the hydrological regimes of wetlands in different parts of the world. Climate change is also projected to favor the spread of invasive species.

CONSERVATION AND RESTORATION

During the past few decades, there have been many efforts at national and international level to conserve wetlands and not only protect them against degradation but also to restore them. The Ramsar Convention on Wetlands of International Importance, which was agreed upon in 1971 and came into force in 1980, is the only global convention dealing with an ecosystem type. It has served as the single most important instrument for wetland conservation. It requires the contracting parties to not only designate important wetlands within their country but also to include wetland conservation in their national land-use planning so as to promote the wise use of wetlands, to establish wetland nature reserves, and to promote training in the fields of wetland research and management. As of October 12, 2013, 168 nations had designated 2165 wetlands (including marine and saline wetlands) covering a total area of 205,830,125 ha under the Ramsar Convention.

Wetland restoration requires the restoration of hydrological regimes, reduction in nutrient loadings and wastewater discharges, and protection against other threats to the specific wetland. Ramsar Convention also promotes wetland restoration through an analysis of change in their ecological character, and along with its approach for wise use of wetlands.

REFERENCES

1. Mitsch, W.J.; Gosselink, J.G. *Wetlands*; 4th Ed.; John Wiley: New York, 2007; 582 pp.
2. Shutes, R.B.E. Artificial wetlands and water quality improvement. Environ. Int. **2001**, *26*, 441–447.
3. Keddy, P.A. *Wetland Ecology*, 2nd Ed.; Cambridge University Press: Cambridge, 2010; 497 pp.
4. Gopal, B. Biodiversity in wetlands. In *The Wetlands Handbook*; Maltby, E., Barker, T., Eds.; Blackwell Science: Oxford, U.K, 2009; 65–95.
5. Gopal, B.; Junk, W.J. Biodiversity in wetlands: An introduction. In *Biodiversity in Wetlands: Assessment, Function and Conservation*, Vol. 1.; Gopal, B., Junk, W.J., Davis, J.A., Eds.; Backhuys Publishers: Leiden, 2000; 1–10.
6. Millennium Ecosystem Assessment. *Ecosystems and Human Well-Being: Wetlands and Water Synthesis*. World Resources Institute: Washington, DC, 2005; 68 pp.

BIBLIOGRAPHY

1. Brinson, M.M.; Swift, B.L.; Plantico, R.C.; Barclay, J.S. Riparian Ecosystems: Their Ecology and Status; U.S. Fish and Wildlife Service, Biological Services Program FWS/OBS-81/17: Washington, DC, 1981; 1–151.
2. Fraser, L.H.; Keddy, P.A. *The World's Largest Wetlands*: Ecology and Conservation. Cambridge University Press: New York, 2005; 488.
3. Grobler, R.; Moning, **C.;** Sliva, J.; Bredenkamp, G.; Grundling, P.-L. Subsistence farming and conservation constraints in coastal peat swamp forests of the Kosi Bay Lake system, Maputaland, South Africa. *Géocarrefour* 2004, 79/4, [online] URL: http://geocarrefour.revues.org/842.
4. Lähteenoja, O.; Ruokolainen, K.; Schulman, L.; Alvarez, J. Amazonian floodplains harbour minerotrophic and ombrotrophic peatlands. Catena **2009**, *79*, 140–145.
5. Lähteenoja, O.; Rojas Reátegui, Y.; Räsänen, M.; Del Castillo Torres, D.; Oinonen, M.; Page, S. The large Amazonian peatland carbon sink in the subsiding Pastaza-Marañón foreland basin, Peru. Global Change Biol. **2012**, *18*, 164–178.
6. LePage, B.A. *Wetlands: Integrating Multidisciplinary Concepts*; Springer: New York, 2011; 261 p.
7. Lewis, W.M. *Wetlands: Characteristics and Boundaries*; National Academies Press: Washington, DC, 1995; 307 p.
8. Schöngart, J.; Wittmann, F.; Worbes, M. Biomass and net primary production of Central Amazonian floodplain forests. In *Amazonian Floodplain Forests: Ecophysiology, Biodiversity, and Sustainable Management*; Junk, W.J.; Piedade, M.T.F.; Wittmann, F.; Schöngart, J., Parolin, P., Eds.; Springer: New York, 2010; 347–388.
9. Wittmann, F.; Schöngart, J.; Montero, J.C.; Motzer, T.; Junk, W.J.; Piedade, M.T.F.; Queiroz, H.L.; Worbes, M. Tree species composition and diversity gradients in white-water forest across the Amazon basin. J. Biogeogr. **2006**, *33*, 1334–1347.

WEBSITE

1. http://www.ramsar.org/.

Wetlands: Remote Sensing

Niti B. Mishra
University of Texas at Austin, Austin, Texas, U.S.A.

Abstract
Wetlands are ecologically and economically significant ecosystems that require monitoring and management due to increasing threats from both anthropogenic pressure and climatic variability. Remote sensing provides timely, up-to-date, relatively accurate, and cost-effective information on wetlands. This entry provides an overview of remote sensing data and the methods, challenges, and application in wetland ecosystems. Data acquired from different remote sensors such as multispectral and hyperspectral imagers, radar, and lidar systems have been utilized for wetland mapping and change detection, characterizing biophysical and biochemical properties of wetland vegetation and its temporal dynamics at a range of spatiotemporal scales. The increasing number of Earth observation sensors and their spatial, spectral, and temporal resolution is enabling more accurate detection of wetland extent, biological productivity, habitat quality, and overall ecological dynamics. Integration of remotely sensed data acquired at multiple scales combined with advancements in image analysis techniques is overcoming data-specific and methodological challenges in wetland remote sensing.

INTRODUCTION

Wetlands are vital ecosystems for planetary and human health as they provide essential ecological services such as purification of water, cycling of nutrients, and recharge of ground water and also provide wetland products such as fish, rice, and some forest resources.[1,2] Wetlands are often rich in biodiversity and are important sites for biogeochemical processes such as the production of methane and nitrous oxide as well as fixation of carbon and nitrogen.[1,3,4] Historically, wetlands were considered a hindrance to development and were modified following dredge and fill operations, pollutant runoff, eutrophication, and fragmentation by roads.[5] The key role of wetlands in sustaining ecological health has only been recognized recently prompting their monitoring and conservation.[5,6] Over the last four decades, remote sensing has served as an important and powerful tool for mapping, monitoring, and characterization of wetland ecosystems. All types of wetlands including inland freshwater marshes, coastal tidal marshes, mangrove ecosystems, and forested wetlands or swaps have been studied using remotely sensed data available at different spatial, spectral, and temporal resolutions. This entry provides an overview of the data, methods, application areas, and challenges of remote sensing of wetlands.

TYPES OF REMOTELY SENSED DATA USED IN WETLAND STUDIES

Aerial Photography and Multispectral Remote Sensing

The earliest remote sensing applications in wetlands utilized color and color-infrared aerial photos, and orthophoto quads acquired over wetlands for applications such as delineating wetland boundaries,[7–9] mapping wetland vegetation,[10–12] classifying wetland types, and assessing vegetation growth and water quality.[13,14] Although aerial photographs provided the high spatial resolution important for characterizing wetland heterogeneity, most of these studies acknowledged the need for improved spectral resolution for enhanced characterization of spatially heterogeneous and spectrally complex wetlands. Furthermore, aerial photography-based studies were site specific and had limited spatial coverage. The availability of multispectral satellite sensors (e.g., Landsat MSS, TM, ETM+, SPOT, IRS) allowed overcoming these limitations by providing larger swath area and repetitive coverage and enabled monitoring at landscape-to-regional scales. The Landsat 1, –2, and –3 had multispectral scanner (MSS) sensors with four spectral bands that were used to study different aspects of wetland ecology in a cost-effective manner.[15–19] Launched in the

Encyclopedia of Natural Resources DOI: 10.1081/E-ENRL-120049156

year 1982, the Landsat Thematic Mapper (TM) represented significant improvement in radiometric and spatial resolution over MSS and was used by several studies for wetland mapping and monitoring.[20–25] The Landsat TM band combination of 2, 4 and 5 is effective for wetland detection; band 5 has the ability to distinguish vegetation and soil moisture levels and band 1 is utilized for determining water quality in wetlands.[23,26,27] Unlike Landsat MSS or TM sensors, Systeme Pour l'Observation de la Terre (SPOT) was the first satellite to have pointable optics that provided increased capabilities for capturing any given area. SPOT High Resolution Visible (HRV) sensor has green, red, and near infrared spectral bands at 20 m spatial resolution while the SPOT-4 sensor launched in 1998 also included a middle infrared band which was intensively utilized in wetland studies.[28–32] Indian Remote Sensing (IRS) satellites equipped with Linear Imaging Self-Scanning Sensor (LISS-II, III) providing multispectral images at medium spatial resolution (i.e., 20–30 m) have also been utilized for wetland studies at landscape scale.[33–36] Multispectral sensors with very high temporal revisit capability (e.g., Moderate Resolution Imaging Spectroradiometer [MODIS], Advanced Very High Resolution Radiometer [AVHRR]) have been utilized for studying temporally dynamic wetland systems (e.g., estuaries, deltas) to characterize the vegetation phenology and rapid vegetation succession.[37,38] However, in spite of high temporal resolution and large swath area, these data sets have not been preferred for wetland studies primarily due to coarse spatial resolution.[26]

High Spatial Resolution Remote Sensing

Most wetlands display high spatial heterogeneity, patchiness, and considerable functional diversity that may require spatial details higher than those offered by medium spatial resolution imagery (Fig. 1).

The availability of high spatial resolution sensors (e.g., IKONOS, Quickbird, GeoEye with spatial resolution <5 m) has enabled more detailed characterization of wetland properties such as identifying multiple classes of wetland vegetation.[39,40] Multitemporal high spatial resolution images have been found useful because many wetland species have overlapping spectral reflectance properties and are spectrally distinguishable at a particular time of year, likely due to differences in biomass and pigments and the rate at which change occurs throughout the growing

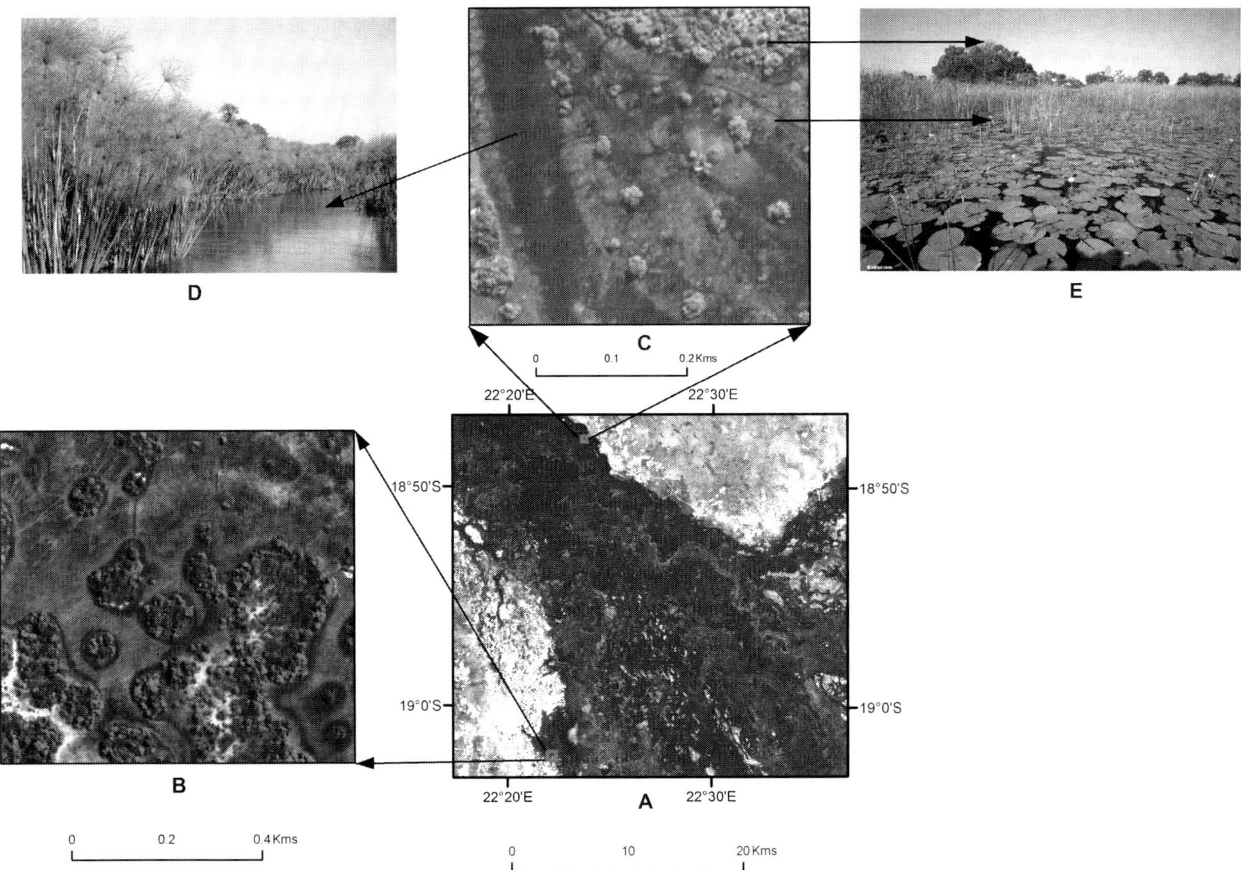

Fig. 1 Spatial heterogeneity and structural diversity of vegetation and land cover in the Okavango Delta, northern Botswana. (**A**) represents part of the delta observed from Landsat TM imagery (RGB:421). (**B**) and (**C**) represent two subset areas as observed from high spatial resolution Quickbird imagery (RGB:421). (**D**) and (**E**) are photographs acquired in the field.

season.[41] High spatial resolution imagery in wetlands shows high intraclass spectral variability due to which the Object-Based Image Analysis (OBIA) approach has been adopted for producing higher characterization accuracy.[42,43]

Hyperspectral Remote Sensing

Imaging spectroscopy or hyperspectral remote sensing instruments with hundreds of narrow continuous spectral bands between 400 and 2500 nm wavelength offer high spectral sensitivity and have enabled species-level discrimination of wetland vegetation, in particular mapping invasive vegetation species which was not possible with multispectral sensors. Important hyperspectral sensors used in wetland studies include the Airborne Visible/Infrared Imaging Spectrometer (AVIRIS), HyMAP, Compact Airborne Imaging Spectrometer (CASI), and Hyperspectral Digital Imagery Collection Experiment (HYDICE). While all of these are airborne sensors, spaceborne hyperspectral sensors (e.g., EO-1 Hyperion) have a comparatively lower signal-to-noise ratio making them less effective for discriminating vegetation species in wetlands.

Radar Remote Sensing

Active remote sensors such as radar backscatter provide different information than optical sensors, and radar systems can acquire images at any time of day and almost under any weather conditions. L-band radar data generally provides good distinction between flooded and non-flooded forest and between forest and marsh vegetation. Radar imagery has been used in wetland locations ranging from tropics to boreal regions.[44,45] Radar imaging applications in forested wetlands have found that longer wavelength had greater penetration in forest canopies, and C-band and L-band sensors could clearly differentiate between flooded and non-flooded forests compared to optical Landsat TM sensor, which was unable to detect water under canopy.[46]

WETLAND APPLICATIONS OF REMOTELY SENSED DATA

Wetland Inventory/Mapping and Change Detection

Remotely sensed data has been utilized for identifying, describing, and mapping the distribution of wetlands at a range of scales from local to regional to global.[47,48] Remote sensing-based wetland mapping has been cost effective and has minimized the need for extensive fieldwork in often inaccessible or logistically challenging wetland areas or both. At the global level, wetland inventory and mapping based on remotely sensed data have been carried out by notable studies such as the International Geosphere-Biosphere Programme Data and Information System (IGBP-DIS) Wetland Data Initiative,[49] the global wetland inventory promoted through the Ramsar Convention in partnership with international, national, and local organizations,[27] and the incorporation of information from individual projects based on existing and new Earth observation data sources.[50] At the regional and national scales, both government and non-governmental organizations worldwide have been using remotely sensed data for wetland inventory. For the last three decades in the United States, the Fish and Wildlife Service has been using aerial photos and multispectral images at high and medium spatial resolution to support the National Wetland Inventory.[51]

Over the past decades, the dramatic decrease and loss of wetland areas have prompted studies to quantify changes in wetlands using multitemporal remotely sensed data. Both short-term (seasonal) and long-term (>1 year) changes in wetland systems have been studied using remote sensing. Short-term changes in wetland extents are often in response to the seasonal rainfall pattern or flooding cycles, and its characterization is important for understanding wetland productivity, the source–sink distribution for nutrient and greenhouse gases, and associated wildlife habitat and hazard risks. Remote sensing-based studies have examined these seasonal changes in different regions such as wetlands in Australia,[15] Poyang Lake, China,[52,53] Okavango Delta, Botswana,[54,55] and in Brazilian wetlands.[44] Contrary to seasonal dynamics, long-term changes in wetlands are driven by anthropogenic activities as well as climatic variability influencing rainfall patterns or changes in sea level that are mainly impacting coastal wetlands. Satellite images have been utilized in several notable studies for quantifying long-term changes in wetlands at the local-to-regional scales.[26,29,56–59] Additionally, wetland monitoring and change detection programs at the national level have also been initiated by different countries. In the United States, the Coastal Change Analysis Program (C-CAP) uses multitemporal remotely sensed data to provide inventories of coastal intertidal areas, wetlands, and adjacent uplands that is updated every 5 years.

Remote Sensing of Wetland Vegetation

Wetland vegetation is a vital component of the wetland ecosystem that also serves as an indicator for early signs of any physical or chemical degradation. Different types of remotely sensed data have been used for biophysical and biochemical characterization of wetland vegetation. Earlier studies found high spatial resolution of aerial photos suitable for detailed mapping of wetland vegetation but were restricted by both limited swath and spectral dimensionality of these data sets.[60] The availability of consistent multispectral images (e.g., Landsat, SPOT) enabled improved characterization of vegetation types, density, vigor, and moisture content in wetland vegetation, nevertheless with significant challenges and limitations. In wetlands, commonly available multispectral sensors (Landsat TM, ETM+, SPOT) have been successfully utilized for mapping plant functional types where species are grouped based on traits such as life-form, photosynthetic pathway, and leaf longevity.[61–63] However, the spectral dimensionality of multispectral sensors (i.e., <10 spectral bands) has been found insufficient for discriminating individual vegetation species in wetlands. The availability of hundreds of spectral bands with field spectroradiometers and airborne hyperspectral sensors (e.g., AVIRIS, HyMap, CASI) has made it possible to discriminate and classify individual wetland vegetation species (Fig. 2). Studies based on the analysis of field spectroradiometer-acquired canopy-level spectral reflectance of individual wetland vegetation species found that (i) a single species in different phenological stages showed significant variation in its spectral reflectance and (ii) the difference in the spectral reflectance and absorption features of different wetland vegetation species could be exploited to discriminate them, and these characteristics could be scaled from canopy to airborne remote sensing platforms.[41,64,65] Most studies using either field spectroradiometer or hyperspectral images have found that wetland vegetation has the greatest variation in the near infrared and red-edge regions.[66–68] Using wave bands in these spectral regions, studies have successfully mapped wetland vegetation species[41,64,68–70]

Estimating Biophysical and Biochemical Parameters of Wetland Vegetation

The most important biophysical and biochemical properties that characterize wetland vegetation are biomass and chlorophyll concentration and leaf water content. Estimating wetland biomass is necessary for understanding productivity, carbon cycle, and nutrient allocation. For biomass estimation, researchers have examined the correlation between *in situ*-derived biomass estimates with individual reflectance bands as well as various vegetation indices (VIs) derived from multispectral images.[71–74] These studies obtained modest-to-poor results depending upon landscape complexity and species composition and concluded that (i) growing season is the optimal time for biomass estimation in wetlands using remotely sensed data and (ii) near infrared region has the potential for vegetation biomass estimation in wetlands. Although saturation of VIs in densely vegetated areas has posed limitations to VI-based biomass estimation,[75] the development of techniques such as the band depth analysis approach has resolved this saturation issue.[76]

Leaf area index (LAI) is an important biophysical variable for quantifying evapotranspiration, photosynthesis, and primary productivity in terrestrial ecosystems. While research efforts on estimating LAI from remote sensing has intensively focused on forest and agriculture systems, similar

Fig. 2 Wetland vegetation species map produced from PROBE-1 airborne hyperspectral imagery by combining *in situ* observations with field spectral libraries.
Source: Reproduced from Zomer et al.[106] with permission from Elsevier Publishing.

studies for wetlands have only been attempted in forest wetlands and mangrove wetlands. Combining Landsat TM- and SPOT XS-derived normalized difference vegetation index (NDVI) with a model based on gap fraction analysis, Green et al.[77] produced a thematic map of LAI for three wetland species in a mangrove forest in West Indies with high accuracy (88%). Another study in a degraded mangrove forest of Mexico found a strong linear relationship between field-derived LAI and high-resolution IKONOS imagery-derived NDVI and sample ratio.[78,79] Hyperspectral remote sensing applications in various terrestrial ecosystems have shown that the narrow band indices can significantly improve biophysical characterization of vegetation. However, very few studies in wetlands have exploited hyperspectral data sets for quantifying biophysical variables such as LAI,[80] and therefore more studies are required.

Plant leaf and canopy water content is an important indicator of the physiological status of plants. There has been rapid growth in studies to access vegetation water content mainly based on laboratory and field-based spectra. However, most of these studies are not conducted in wetlands because remote sensing–based studies in wetlands have mainly focused on discriminating and mapping rather than estimating plant physiognomic properties. Therefore such studies in wetlands are needed that can help quantify the integrated condition of wetland plants and can identify plant water stress across a range of scales.[81]

Remote Sensing of Submerged Aquatic Vegetation

In wetlands, submerged aquatic vegetation (SAV) plays a major role in ecological functioning by providing habitat for many aquatic species, dissipating disturbance, enriching sediments, and improving water quality.[82] The main challenge for remote sensing of SAV is to isolate the weakened plant signal from interference by the water column, the bottom, and the atmospheric effects. Earlier studies showed that in shallow water, SAV density dominates a pixel's reflectance but as the depth increases, dominance of reflectance shifts to water column components.[83] Hence, studies have incorporated bathymetric information to reduce the effect of water column variation.[84,85] Due to the spatial complexity of SAV, high spatial resolution aerial photos have been found useful for monitoring the status and trend of SAV.[86,87] Studies using medium as well as high spatial resolution multispectral imagery have mapped both the various types of SAV and their biomass achieving moderate accuracy.[88–92] The spectral and spatial capabilities offered by airborne hyperspectral images have enabled detailed mapping of SAV types, habitat, and ecological conditions.[69,93] Hyperspectral sensors have been used in combination with LiDAR, field spectroscopy, and ancillary details to identify SAV areas in healthy or degraded conditions for conservation prioritization.

CHALLENGES AND METHODS USED IN WETLAND REMOTE SENSING

The physical complexity, spatial heterogeneity, and diversity of life-forms in wetlands pose a significant challenge for their remote characterization at various spatial and temporal scales. Imagery from traditional multispectral sensors is subject to limitations of spatial and spectral resolution compared to the narrow vegetation units found in wetlands.[42] In wetlands, the biophysical and biochemical estimates derived from VIs obtained from broadband sensors can produce unstable results due to variation in underlying soil type, canopy, and leaf properties.[81] Despite effective performance of hyperspectral sensors in discriminating wetland species, there are still significant challenges because (i) reflectance from different vegetation species could be similar and (ii) within species, spectral variability is often high because of age difference, miroclimate, soil and topographic difference. To address these complexities and challenges, researchers are evaluating different data sources and developing improved wetland-specific methods for using newly available and advanced remote sensing data. Recent studies have utilized advanced machine-learning classifiers such as decision trees,[94] artificial neural networks,[95] and fuzzy hybrid classifiers.[96] These techniques have proved useful especially when combined with ancillary information such as topography, soil type, and water level and often provided improved mapping accuracy than more traditionally used parametric classifiers (e.g., maximum likelihood).[97]

Due to the narrow patch size of wetland vegetation, commonly used multispectral sensors (e.g., Landsat, SPOT) often contain pixels with more than one vegetation or land cover type (mixed pixels). Improving upon the limitations posed by per-pixel analysis approach for such mixed pixels, studies have utilized soft classifiers such as Spectral Mixture Analysis (SMA). SMA models a pixels spectral response as a linear/non-liner proportionally weighted combination of known endmember (pure pixel) spectra and provides subpixel abundance estimates. A modified version of SMA known as the Multiple Endmember SMA (MESMA) allows both number and type of endmembers to vary on a per-pixel basis thus producing more realistic subpixel fractional estimates. Both SMA and MESMA have been successfully utilized for wetland characterization with multispectral[26,98] and hyperspectral images.[64,99]

Contrary to the classical per-pixel analysis approach, more recent studies in wetlands have adopted the OBIA approach. OBIA segments the image into objects representing group of pixels according to desired scale, shape, and compactness criteria followed by classification of objects. The OBIA approach offers several advantages over per-pixel analysis such as absence of the "salt-and-pepper" effect in classification results, ability to utilize shape, contextual and textural features, and treatment of landscapes as consisting of relatively homogeneous objects that increases the signal-to-noise ratio of classification input. Important multispectral satellite sensors with which the OBIA approach has been utilized for wetland studies include Landsat TM, ETM,[61,100,101] and SPOT-5.[102] For tackling the physical complexity and heterogeneity of wetlands and their improved characterization, studies have integrated remotely sensed data from multiple sensors (e.g., optical, radar, lidar) and at multiple spatial scales. Airborne lidar has been integrated with multispectral and hyperspectral images to map SAV.[103,104] Additionally, using only optical sensors, bisensor monitoring of temporally dynamic wetlands by integrating medium spatial resolution imagery (e.g., Landsat) with high temporal resolution imagery (e.g., MODIS) has also been suggested.[105]

CONCLUSION

The progress in both remote sensing data and analysis methods is offering advanced capabilities and new frontiers for better characterizing the spatial heterogeneity and biophysical complexity of wetlands. Nevertheless, there are several major challenges in remote sensing of wetlands that need to be addressed. Additional research is needed to develop more robust classification techniques considering wetland complexity and utilizing commonly available multispectral images. For the optimum utilization of hyperspectral data, there is an increased need for building comprehensive spectral libraries of wetland vegetation types that could be used to calibrate and validate remotely derived results. Furthermore, the fundamental understanding of the relationship between reflectance and wetland vegetation canopy shape, density, and effects of background needs to be studied in detail. Hyperspectral instruments are considered to hold the future for remote sensing of wetlands as they can identify both species and their quality/conditions in wetlands. Ideally, remote sensing of wetland ecosystems requires high spatial, spectral, and temporal resolutions. However, like all scientific endeavors, in remote sensing, the limited number of photons arriving at a sensor requires trade-offs among bandwidth, pixel size, and noise. More recent studies in wetlands are integrating remotely sensed data at multiple spatial and temporal scales to overcome these challenges. In the future, remote sensing–based characterization of wetlands could lead to the development of an early warning system capable of detecting any subtle changes in wetlands and identifying healthy versus disturbed areas.

REFERENCES

1. Mitsch, W.J.; Gosselink, J.G. The value of wetlands: Importance of scale and landscape setting. Ecol. Econ. **2000**, *35* (1), 25–33.
2. Houlahan, J.E.; Keddy, P.A.; Makkay, K.; Findlay, C.S. The effects of adjacent land use on wetland species richness and community composition. Wetlands **2006**, *26* (1), 79–96.
3. Liu, D.S.; Xia, F. Assessing object-based classification: advantages and limitations. Remote Sensing Lett. **2010**, *1* (4), 187–194.
4. Hunt, R.J.; Walker, J.F.; Krabbenhoft, D.P. Characterizing hydrology and the importance of ground-water discharge in natural and constructed wetlands. Wetlands **1999**, *19* (2), 458–472.
5. Gibbs, J.P. Wetland loss and biodiversity conservation. Conserv. Biol. **2000**, *14* (1), 314–317.
6. Dudgeon, D.; Arthington, A.H.; Gessner, M.O.; Kawabata, Z.-I.; Knowler, D.J.; Lévêque, C.; Naiman, R.J.; Freshwater biodiversity: importance, threats, status and conservation challenges. Biol. Rev. **2006**, *81* (2), 163–182.
7. Stewart, W.R.; Carter, V.; Brooks, P.D. Inland (non-tidal) wetland mapping. Photogram. Eng. Remote Sensing **1980**, *46* (5), 617–628.
8. Carter, V.; Garrett, M.K.; Shima, L.; Gammon, P. The great dismal swamp – Management of a hydrologic resource with aid of remote sensing. Water Resour. Bull. **1977**, *13* (1), 1–12.
9. Mead, R.A.; Gammon, P.T. Mapping wetlands using orthophotoquads and 35-mm aerial photo. Photogram. Eng. Remote Sensing **1981**, *47* (5), 649–652.
10. Shima, L.J.; Anderson, R.R.; Carter, V.P. The use of aerial color infrared photography in mapping the vegetation of a freshwater marsh. Chesapeake Sci. **1976**, *17* (2), 74–85.
11. Lovvorn, J.R.; Kirkpatrick, C.M. Analysis of freshwater wetland vegetation with large-scale color infrared aerial photography. J. Wildl. Manag. **1982**, *46* (1), 61–70.
12. Scarpace, F.L.; Quirk, B.K.; Kiefer, R.W.; Wynn, S.L. Wetland mapping from digitized aerial photography. Photogram. Eng. Remote Sensing **1981**, *47* (6), 829–838.
13. Welch, R.; Remillard, M.M.; Slack, R.B. Remote sensing and geigraphic information system techniques for aquatic resource evaluation. Photogram. Eng. Remote Sensing **1988**, *54* (2), 177–185.
14. Martyn, R.D.; Noble, R.L.; Bettoli, P.W.; Maggio, R.C. Mapping aquatic weeds with aerial color infrared photography and evaluating their control by grass carp. J. Aquat. Plant Manag. **1986**, *24*, 46–56.
15. Johnston, R.M.; Barson, M.M. Remote sensing of Australian wetlands - An evaluation of Landsat TM data for inventory and classification. Aust. J. Mar. Freshwater Res. **1993**, *44* (2), 235–252.
16. Ackleson, S.G.; Klemas, V.; McKim, H.L.; Merry, C.J. A comparison of SPOT simulator data with Landsat MSS imagery for delineating water masses in Delaware Bay, Broadkill River, and adjacent wetlands. Photogram. Eng. Remote Sensing **1985**, *51* (8), 1123–1129.
17. Jensen, J.R.; Christensen, E.J.; Sharitz, R. Nontidal wetland mapping in South Carolina using airborne multispectral scanner data. Remote Sensing Environ. **1984**, *16* (1), 1–12.
18. Lee, C.T.; Marsh, S.E. The use of archival Landsat MSS and ancillary data in a GIS environment to map historical change in an urban riparian habitat. Photogram. Eng. Remote Sensing **1995**, *61* (8), 999–1008.
19. Wickware, G.M.; Howarth, P.J. Change detection in the Peace—Athabasca delta using digital Landsat data. Remote Sensing Environ. **1981**, *11* (1), 9–25.

20. Ackleson, S.G.; Klemas, V. Remote sensing of submerged aquatic vegetation in lower chesapeake bay: A comparison of Landsat MSS to TM imagery. Remote Sensing Environ. **1987**, *22* (2), 235–248.
21. Ramsey, E.W.; Laine, S.C. Comparison of Landsat thematic mapper and high resolution photography to identify change in complex coastal wetlands. J. Coast. Res. **1997**, *13* (2), 281–292.
22. Zainal, A.; Dalby, D.; Robinson, I. Monitoring marine ecological changes on the east coast of Bahrain with Landsat TM. Photogram. Eng. Remote Sensing **1993**, *59* (3), 415–421.
23. Jensen, J.R.; Hodgson, M.E.; Christensen, E.; Mackey, H.E.; Tinney, L.R.; Sharitz, R.R. Remote sensing in inland wetlands - A multispectral approach. Photogram. Eng. Remote Sensing **1986**, *52* (1), 87–100.
24. Dronova, I.; Gong, P.; Clinton, N.E.; Wang, L.; Fu, W.; Qi, S.; Liu, Y. Landscape analysis of wetland plant functional types: The effects of image segmentation scale, vegetation classes and classification methods. Remote Sensing Environ. **2012**, *127*, 357–369.
25. Mishra, N.B.; Crews, K.A.; Neuenschwander, A.L. Sensitivity of EVI-based harmonic regression to temporal resolution in the lower Okavango Delta. Int. J. Remote Sensing **2012**, *33* (24), 7703–7726.
26. Ozesmi, S.L.; Bauer, M.E. Satellite remote sensing of wetlands. Wetlands Ecol. Manag. **2002**, *10* (5), 381–402.
27. Davidson, N.C.; Finlayson, C.M. Earth Observation for wetland inventory, assessment and monitoring. Aquat. Conserv.-Mar. Freshwater Ecosyst. **2007**, *17* (3), 219–228.
28. McCarthy, T.S.; Franey, N.J.; Ellery, W.N.; Ellery, K. The use of SPOT imagery in the study of environmental processes of the Okavango Delta, Botswana. S. Afr. J. Sci. **1993**, *89* (9), 432–436.
29. Houhoulis, P.F.; Michener, W.K. Detecting wetland change: A rule-based approach using NWI and SPOT-XS data. Photogram. Eng. Remote Sensing **2000**, *66* (2), 205–211.
30. Narumalani, S.; Jensen, J.R.; Burkhalter, S.; Mackey, H.E. Aquatic macrophyte modeling using GIS and logistic multiple regression. Photogram. Eng. Remote Sensing **1997**, *63* (1), 41–49.
31. Harvey, K.; Hill, G. Vegetation mapping of a tropical freshwater swamp in the Northern Territory, Australia: a comparison of aerial photography, Landsat TM and SPOT satellite imagery. Int. J. Remote Sensing **2001**, *22* (15), 2911–2925.
32. Lee, J.K.; Park, R.A.; Mausel, P.W. Application of geoprocessing and simulation modeling to estimate impacts of sea level rise on the northeast coast of Florida. Photogram. Eng. Remote Sensing **1992**, *58* (11), 1579–1586.
33. Kushwaha, S.; Dwivedi, R.; Rao, B. Evaluation of various digital image processing techniques for detection of coastal wetlands using ERS-1 SAR data. Int. J. Remote Sensing **2000**, *21* (3), 565–579.
34. Chopra, R.; Verma, V.; Sharma, P. Mapping, monitoring and conservation of Harike wetland ecosystem, Punjab, India, through remote sensing. Int. J. Remote Sensing **2001**, *22* (1), 89–98.
35. Shanmugam, P.; Ahn, Y.-H.; Sanjeevi, S.A. comparison of the classification of wetland characteristics by linear spectral mixture modelling and traditional hard classifiers on multispectral remotely sensed imagery in southern India. Ecol. Model. **2006**, *194* (4), 379–394.
36. Prasad, S.N.; Ramachandra, T.V.; Ahalya, N.; Sengupta, T.; Kumar, A.; Tiwari, A.K.; Vijayan, V.S.; Vijayan, L. Conservation of wetlands of India - a review. Trop. Ecol. **2002**, *43* (1), 173–186.
37. Zhao, B.; Yan, Y.; Guo, H.; He, M.; Gu, Y.; Li, B. Monitoring rapid vegetation succession in estuarine wetland using time series MODIS-based indicators: An application in the Yangtze River Delta area. Ecol. Indicators **2009**, *9* (2), 346–356.
38. Yan, Y.-E.; Ouyang, Z.-T.; Guo, H.-Q.; Jin, S.-S.; Zhao, B. Detecting the spatiotemporal changes of tidal flood in the estuarine wetland by using MODIS time series data. J. Hydrol. **2010**, *384* (1), 156–163.
39. Rutchey, K.; Vilchek, L. Air photointerpretation and satellite imagery analysis techniques for mapping cattail coverage in a northern Everglades impoundment. Photogram. Eng. Remote Sensing **1999**, *65* (2), 185–191.
40. Sawaya, K.E.; Olmanson, L.G.; Heinert, N.J.; Brezonik, P.L.; Bauer, M.E. Extending satellite remote sensing to local scales: land and water resource monitoring using high-resolution imagery. Remote Sensing Environ. **2003**, *88* (1–2), 144–156.
41. Schmidt, K.S.; Skidmore, A.K. Spectral discrimination of vegetation types in a coastal wetland. Remote Sensing Environ. **2003**, *85* (1), 92–108.
42. Gilmore, M.S.; Wilson, E.H.; Barrett, N.; Civco, D.L.; Prisloe, S.; Hurd, J.D.; Chadwick, C. Integrating multitemporal spectral and structural information to map wetland vegetation in a lower Connecticut River tidal marsh. Remote Sensing Environ. **2008**, *112* (11), 4048–4060.
43. Laba, M.; Blair, B.; Downs, R.; Monger, B.; Philpot, W.; Smith, S.; Sullivan, P.; Baveye, P.C. Use of textural measurements to map invasive wetland plants in the Hudson River National Estuarine Research Reserve with IKONOS satellite imagery. Remote Sensing Environ. **2010**, *114* (4), 876–886.
44. Hess, L.L.; Melack, J.M.; Novo, E.M.L.M.; Barbosa, C.C.F.; Gastil, M. Dual-season mapping of wetland inundation and vegetation for the central Amazon basin. Remote Sensing Environ **2003**, *87* (4), 404–428.
45. Hess, L.L.; Melack, J.M.; Simonett, D.S. Radar detection of flooding beneath the forest canopy: a review. Int. J. Remote Sensing **1990**, *11* (7), 1313–1325.
46. Townsend, P.A.; Walsh, S.J. Remote sensing of forested wetlands: application of multitemporal and multispectral satellite imagery to determine plant community composition and structure in southeastern USA. Plant Ecol. **2001**, *157* (2), 129A–149A.
47. Lehner, B.; Doll, P. Development and validation of a global database of lakes, reservoirs and wetlands. J. Hydrol. **2004**, *296* (1–4), 1–22.
48. Rebelo, L.M.; Finlayson, C.M.; Nagabhatla, N. Remote sensing and GIS for wetland inventory, mapping and change analysis. J. Environ. Manag. **2009**, *90* (7), 2144–2153.
49. Darras, S. A first step towards identifying a global delineation of wetlands. In *GBP-DIS Wetland Data Initiative*; IGBP-DIS Office: Toulouse, France, 1999.

50. Rosenqvist, A.; Finlayson, C.M.; Lowry, J.; Taylor, D. The potential of long-wavelength satellite-borne radar to support implementation of the Ramsar Wetlands Convention. Aquat. Conserv.-Mar. Freshwater Ecosyst. **2007**, *17* (3), 229–244.
51. Klemas, V. Remote sensing of emergent and submerged wetlands: an overview. Int. J. Remote Sensing **2013**, *34* (18), 6286–6320.
52. Dronova, I.; Gong, P.; Wang, L. Object-based analysis and change detection of major wetland cover types and their classification uncertainty during the low water period at Poyang Lake, China. Remote Sensing Environ. **2011**, *115* (12), 3220–3236.
53. Hui, F.M.; Xu, B.; Huang, H.B.; Yu, Q.; Gong, P. Modelling spatial-temporal change of Poyang Lake using multitemporal Landsat imagery. Int. J. Remote Sensing **2008**, *29* (20), 5767–5784.
54. McCarthy, T.S.; Cooper, G.R.J.; Tyson, P.D.; Ellery, W.N. Seasonal flooding in the Okavango Delta, Botswana - recent history and future prospects. S. Afr. J. Sci. **2000**, *96* (1), 25–33.
55. Neuenschwander, A.L.; Crawford, M.M.; Ringrose, S. Monitoring of seasonal flooding in the Okavango Delta using EO-1 data. *Igarss: IEEE International Geoscience and Remote Sensing Symposium and 24th Canadian Symposium on Remote Sensing*, Proceedings: Remote Sensing: Integrating Our View of the Planet, New York: IEEE. 2002, Vols. 1–6, 3124–3126.
56. Gong, P.; Niu, Z.; Cheng, X.; Zhao, K.; Zhou, D.; Guo, J.; Liang, L.; Wang, X.; Li, D.; Huang, H.; Wang, Y.; Wang, K.; Li, W.; Wang, X.; Ying, Q.; Yang, Z.; Ye, Y.; Li, Z.; Zhuang, D.; Chi, Y.; Zhou, H.; Yan, J. China's wetland change (1990–2000) determined by remote sensing. Sci. China Earth Sci. **2010**, *53* (7), 1036–1042.
57. Shuman, C.S.; Ambrose, R.F. A comparison of remote sensing and ground-based methods for monitoring wetland restoration success. Restoration Ecol. **2003**, *11* (3), 325–333.
58. Nielsen, E.M.; Prince, S.D.; Koeln, G.T. Wetland change mapping for the U.S. mid-Atlantic region using an outlier detection technique. Remote Sensing Environ. **2008**, *112* (11), 4061–4074.
59. Munyati, C. Wetland change detection on the Kafue Flats, Zambia, by classification of a multitemporal remote sensing image dataset. Int. J. Remote Sensing **2000**, *21* (9), 1787–1806.
60. Howland, W.G. Multispectral aerial photography for wetland vegetation mapping. Photogram. Eng. Remote Sensing **1980**, *46* (1), 87–99.
61. Dronova, I.; Gong, P.; Clinton, N.E.; Wang, Li.; Fu, W.; Qi, S.; Liu, Y. Landscape analysis of wetland plant functional types: The effects of image segmentation scale, vegetation classes and classification methods. Remote Sensing Environ. **2012**, *127*, 357–369.
62. Khanna, S.; Santos, M.J.; Ustin, S.L.; Haverkamp, P.J. An integrated approach to a biophysiologically based classification of floating aquatic macrophytes. Int. J. Remote Sensing **2011**, *32* (4), 1067–1094.
63. Wang, L.; Dronova, I.; Gong, P.; Yang, W.; Li, Y.; Liu, Q. A new time series vegetation–water index of phenological–hydrological trait across species and functional types for Poyang Lake wetland ecosystem. Remote Sensing Environ. **2012**, *125* (0), 49–63.
64. Rosso, P.H.; Ustin, S.L.; Hastings, A. Mapping marshland vegetation of San Francisco Bay, California, using hyperspectral data. Int. J. Remote Sensing **2005**, *26* (23), 5169–5191.
65. Fyfe, S.K. Spatial and temporal variation in spectral reflectance: Are seagrass species spectrally distinct? Limnol. Oceanogr. **2003**, *48* (1), 464–479.
66. Thenkabail, P.S.; Enclona, E.A.; Ashton, M.S.; Meer, B.V.D. Accuracy assessments of hyperspectral waveband performance for vegetation analysis applications. Remote Sensing Environ. **2004**, *91* (3–4), 354–376.
67. Asner, G.P. Biophysical and biochemical sources of variability in canopy reflectance. Remote Sensing Environ. **1998**, *64* (3), 234–253.
68. Vaiphasa, C.; Ongsomwang, S.; Vaiphasa, T.; Skidmore, A.K. Tropical mangrove species discrimination using hyperspectral data: A laboratory study. Estuarine Coast. Shelf Sci. **2005**, *65* (1–2), 371–379.
69. Belluco, E.; Camuffo, M.; Ferrari, S.; Modenese, L.; Silvestri. S.; Marani, A.; Marani, M. Mapping salt-marsh vegetation by multispectral and hyperspectral remote sensing. Remote Sensing Environ. **2006**, *105* (1), 54–67.
70. Pengra, B.W.; Johnston, C.A.; Loveland, T.R. Mapping an invasive plant, *Phragmites australis*, in coastal wetlands using the EO-1 Hyperion hyperspectral sensor. Remote Sensing Environ. **2007**, *108* (1), 74–81.
71. Tan, Q.; Shao, Y.; Yang, S.; Wei, Q. *Wetland Vegetation Biomass Estimation Using Landsat-7 ETM+ data.* In Geoscience and Remote Sensing Symposium, 2003. IGARSS'03. Proceedings. 2003 IEEE International; 2003.
72. Ren-dong, L.; Ji-yuan, L. Wetland vegetation biomass estimation and mapping from Landsat ETM data: A case study of Poyang Lake. J. Geogr. Sci. **2002**, *12* (1), 35–41.
73. Ramsey, E.W.; Jensen, J.R. Remote sensing of mangrove wetlands: relating canopy spectra to site-specific data. Photogram. Eng. Remote Sensing **1996**, *62* (8), 939–948.
74. Moreau, S.; Bosseno, R.; Gu, X.F.; Baret, F. Assessing the biomass dynamics of Andean bofedal and totora high-protein wetland grasses from NOAA/AVHRR. Remote Sensing Environ. **2003**, *85* (4), 516–529.
75. Thenkabail, P.S.; Smith, R.B.; De Pauw, E. Hyperspectral vegetation indices and their relationships with agricultural crop characteristics. Remote Sensing Environ. **2000**, *71* (2), 158–182.
76. Mutanga, O.; Skidmore, A.K. Narrow band vegetation indices overcome the saturation problem in biomass estimation. International J. Remote Sensing **2004**, *25* (19), 3999–4014.
77. Green, E.P.; Mumby, P.J.; Edwards, A.J.; Clark, C.D.; Ellis, A.C. Estimating leaf area index of mangroves from satellite data. Aquat. Bot. **1997**, *58* (1), 11–19.
78. Kovacs, J.M.; Flores-Verdugo, F.; Wang, J.F.; Aspden, L.P. Estimating leaf area index of a degraded mangrove forest using high spatial resolution satellite data. Aquat. Bot. **2004**, *80* (1), 13–22.
79. Kovacs, J.M.; Wang, J.F.; Flores-Verdugo, F. Mapping mangrove leaf area index at the species level using IKONOS and LAI-2000 sensors for the Agua Brava Lagoon,

Mexican Pacific. Estuarine Coast. Shelf Sci. **2005**, *62* (1–2), 377–384.
80. Darvishzadeh, R.; Skidmore, A.; Atzberger, C.; Wieren, S.V. Estimation of vegetation LAI from hyperspectral reflectance data: Effects of soil type and plant architecture. Int. J. Appl. Earth Observation Geoinfo. **2008**, *10* (3), 358–373.
81. Adam, E.; Mutanga, O.; Rugege, D. Multispectral and hyperspectral remote sensing for identification and mapping of wetland vegetation: A review. Wetlands Ecol. Manag. **2010**, *18* (3), 281–296.
82. Hurley, L.M. *Submerged Aquatic Vegetation. Habitat Requirements for Chesapeake Bay Living Resources*, 2nd edn.; Chesapeake Research Consortium, Inc.: Solomons, MD, 1991; 2–1.
83. Silva, T.S.; Costa, M.P.; Melack, J.M.; Novo, E.M.L.M. Remote sensing of aquatic vegetation: theory and applications. Environ. Monitor. Assess. **2008**, *140* (1–3), 131–145.
84. Dierssen, H.M.; Zimmerman, R.C.; Leathers, R.A.; Downes, T.V.; Davis, C.O. *Ocean Color Remote Sensing of Seagrass and Bathymetry in the Bahamas Banks by High-Resolution Airborne Imagery;* DTIC Document: 2003.
85. Heege, T.; Bogner, A.; Pinnel, N. Mapping of submerged aquatic vegetation with a physically based process chain. In *Remote Sensing*; International Society for Optics and Photonics: 2004.
86. Ferguson, R.L.; Wood, L.; Graham, D. Monitoring spatial change in seagrass habitat with aerial photography. Photogram. Eng. Remote Sensing **1993**, *59* (6), 1033–1038.
87. Finkbeiner, M.; Stevenson, B.; Seaman, R. *Guidance for Benthic Habitat Mapping: An Aerial Photographic Approach;* 2001.
88. Wabnitz, C.C.; Andréfouët, S.; Torres-Pulliza, D.; Müller-Karger, F.E.; Kramer, P.A. Regional-scale seagrass habitat mapping in the Wider Caribbean region using Landsat sensors: Applications to conservation and ecology. Remote Sensing Environ. **2008**, *112* (8), 3455–3467.
89. Gullström, M.; Lundén, B.; Bodin, M.; Kangwe, J.; Öhman, M.C.; Mtolera, M.S.P.; Björk, M. Assessment of changes in the seagrass-dominated submerged vegetation of tropical Chwaka Bay (Zanzibar) using satellite remote sensing. Estuarine, Coast. Shelf Sci. **2006**, *67* (3), 399–408.
90. Nobi, E.; Thangaradjou, T. Evaluation of the spatial changes in seagrass cover in the lagoons of Lakshadweep islands, India, using IRS LISS III satellite images. Geocarto Int. **2012**, *27* (8), 647–660.
91. Mishra, D.; Narumalani, S.; Rundquist, D.; Lawson, M. Benthic habitat mapping in tropical marine environments using QuickBird multispectral data. Photogrammetric Eng. Remote Sensing **2006**, *72* (9), 1037.
92. Wolter, P.T.; Johnston, C.A.; Niemi, G.J. Mapping submergent aquatic vegetation in the U.S. Great Lakes using Quickbird satellite data. Int. J. Remote Sensing **2005**, *26* (23), 5255–5274.
93. Williams, D.J.; Rybicki, N.B.; Lombana, A.V.; O'Brien, T.M.; Gomez, R.B. Preliminary investigation of submerged aquatic vegetation mapping using hyperspectral remote sensing. Environ. Monitor. Assess. **2003**, *81* (1–3), 383–392.
94. Xu, M.; Watanachaturaporn, P.; Varshney, P.K.; Arora, M.J. Decision tree regression for soft classification of remote sensing data. Remote Sensing Environ. **2005**, *97* (3), 322–336.
95. Berberoglu, S.; Lloyd, C.D.; Atkinson, P.M.; Curren, P. The integration of spectral and textural information using neural networks for land cover mapping in the Mediterranean. Comput. Geosci. **2000**, *26* (4), 385–396.
96. Sha, Z.; Bai, Y.; Xie, Y. et al. Using a hybrid fuzzy classifier (HFC) to map typical grassland vegetation in Xilin River Basin, Inner Mongolia, China. Int. J. Remote Sensing **2008**, *29* (8), 2317–2337.
97. Xie, Y.; Sha, Z.; Yu, M. Remote sensing imagery in vegetation mapping: A review. J. Plant Ecol. **2008**, *1* (1), 9–23.
98. Mertes, L.A.; Daniel, D.L.; Melack, J.M.; Nelson, B.; Martinelli, L.A.; Forsberg, B.R. Spatial patterns of hydrology, geomorphology, and vegetation on the floodplain of the Amazon River in Brazil from a remote sensing perspective. Geomorphology **1995**, *13* (1), 215–232.
99. Li, L.; Ustin, S.L.; Lay, M. Application of multiple endmember spectral mixture analysis (MESMA) to AVIRIS imagery for coastal salt marsh mapping: A case study in China Camp, CA, USA. Int. J. Remote Sensing **2005**, *26* (23), 5193–5207.
100. Grenier, M.; Demers, A.-M.; Labrecque, S.; Benoit, M.; Fournire, R.A.; Drolet, B. An object-based method to map wetland using RADARSAT-1 and Landsat ETM images: Test case on two sites in Quebec, Canada. Can. J. Remote Sensing **2007**, *33*, S28–S45.
101. Frohn, R.C.; Autrey, B.C.; Lane, C.R.; Reif, M. Segmentation and object-oriented classification of wetlands in a karst Florida landscape using multi-season Landsat-7 ETM+ imagery. Int. J. Remote Sensing **2011**, *32* (5), 1471–1489.
102. Powers, R.P.; Hay, G.J.; Chen, G. How wetland type and area differ through scale: A GEOBIA case study in Alberta's Boreal Plains. Remote Sensing Environ. **2012**, *117*, 135–145.
103. Yang, J.; Artigas, F.J.; Wang, J. Mapping salt marsh vegetation by integrating hyperspectral and LiDAR remote sensing. Remote Sensing Coast. Environ. CRC: Boca Raton, Florida, **2010**, 173–190.
104. Chust, G.; Galparsoro, I.; Borja, Á.; Javire, F.; Adolfo, U. Coastal and estuarine habitat mapping, using LIDAR height and intensity and multi-spectral imagery. Estuarine, Coast. Shelf Sci. **2008**, *78* (4), 633–643.
105. Michishita, R.; Gong, P.; Xu, B. Spectral mixture analysis for bi-sensor wetland mapping using Landsat TM and Terra MODIS data. Int. J. Remote Sensing **2012**, *33* (11), 3373–3401.
106. Zomer, R.J.; Trabucco, A.; Ustin, S.L. Building spectral libraries for wetlands land cover classification and hyperspectral remote sensing. J. Environ. Manag. **2009**, *90*, 2170–2177.

Wetlands: Tidal

William H. Conner
Baruch Institute of Coastal Ecology and Forest Science, Clemson University, Georgetown, South Carolina, U.S.A.
Ken W. Krauss
National Wetlands Research Center, U.S. Geological Survey (USGS), Lafayette, Louisiana, U.S.A.
Andrew H. Baldwin
Department of Environmental Science & Technology, University of Maryland, College Park, Maryland, U.S.A.
Stephen Hutchinson
Baruch Institute of Coastal Ecology and Forest Science, Clemson University, Georgetown, South Carolina, U.S.A.

Abstract

Tidal wetlands are some of the most dynamic areas of the Earth and are found at the interface between the land and sea. Salinity, regular tidal flooding, and infrequent catastrophic flooding due to storm events result in complex interactions among biotic and abiotic factors. The complexity of these interactions, along with the uncertainty of where one draws the line between tidal and nontidal, makes characterizing tidal wetlands a difficult task. The three primary types of tidal wetlands are tidal marshes, mangroves, and freshwater forested wetlands. Tidal marshes are dominated by herbaceous plants and are generally found at middle to high latitudes of both hemispheres. Mangrove forests dominate tropical coastlines around the world while tidal freshwater forests are global in distribution. All three wetland types are highly productive ecosystems, supporting abundant and diverse faunal communities. Unfortunately, these wetlands are subject to alteration and loss from both natural and anthropogenic causes.

INTRODUCTION

Tidal wetlands occur in some of the most dynamic areas of the Earth and are positioned at the interface between land and sea where two of the most powerful forces acting on the planet's waters collide. Light and heat from solar energy evaporate water and move it to the atmosphere, while gravity from the moon and sun propels the ebbing and flowing of tides, resulting in low and high water events on diurnal time scales. Vegetation, also fueled by solar energy, plays a crucial role in stabilizing coastal boundaries. At the mouths of the rivers, edges of embayments, and lagoons, these two great forces interact in complex ways to form the hydrodynamic framework for the development of tidal wetlands.

Tidal wetlands reduce the energy of tides, storms, and floods through complex interactions with biotic and abiotic constituents. At low tide, most energy dissipation is a result of linear force being converted to helical motion and undulating bottom structure, and friction with the channel surface. As the banks are topped at high tide, vegetation plays an increasingly important role in energy dissipation. Sedimentation and deposition of nutrient laden materials are important in maintaining this same vegetation, as is the salinity profile determined by the relative contribution of riverine and oceanic water. Animals can also play an important role through such means as oyster bar development, benthic burrows, beaver dams, and excavations. The complexity of the interactions makes characterizing tidal wetlands a difficult task. In this entry, we briefly review the three primary tidal wetland types that occur globally, including tidal marshes, mangroves, and freshwater forested wetlands.

TIDAL MARSHES

Tidal marshes (Fig. 1) are coastal wetlands dominated by herbaceous plants, unlike tidal freshwater forested wetlands and tropical tidal forests (mangroves). Marshes occur along coastlines at middle to high latitudes of both the northern and southern hemispheres and are less common at tropical latitudes, where mangrove trees are the dominant form of intertidal wetland vegetation.[1] Contrary to popular belief, not all tidal marshes are saline. Rather, this ecosystem spans salinity regimes ranging from fresh to concentrations above that of seawater (Fig. 2) and includes tidal freshwater (<0.5 ppt salinity), oligohaline (0.5–5.0 ppt salinity), brackish (5.0–18 ppt salinity), and salt (>18 ppt salinity) marshes.[2–4] There is a great deal of uncertainty in estimates of the area of these wetlands depending upon where one draws the line between tidal and nontidal areas. Worldwide, there are an estimated 12×10^6 ha of tidal marshes with nearly 25% of them in the United States.[5]

Encyclopedia of Natural Resources DOI: 10.1081/E-ENRL-120047505

Fig. 1 Tidal marshes are dominated by herbaceous vegetation comprising a wide range of species. In Chesapeake Bay on the U.S. Atlantic Coast, both tidal freshwater (**A**) and brackish (**B**) marshes are abundant. Tidal marshes are less abundant in much of Europe, but include reed-dominated tidal freshwater marshes along the Elbe River on Germany's Atlantic coast (**C**), and salt marshes dominated by species including common cordgrass (*Spartina anglica* C.E. Hubbard) and sea aster (*Aster tripolium* [Jacq.] Dobrocz.) along the North Sea (**D**).
Source: Photo by A.H. Baldwin, University of Maryland.

Soil and Biogeochemistry

Tidal marshes in sheltered estuaries and deltas receive inputs of mineral riverine sediment, while in marine settings they may receive considerable inputs of marine sediment.[6,7] Due to high primary productivity, most tidal marshes accumulate organic matter from roots, stems, and leaves. Organic matter content of tidal marsh soils may be low in areas receiving abundant mineral sediment, such as salt marshes of the Yangtze River in China[8] or high (>80% organic matter) where there is little mineral sediment input such as the floating marshes of the Louisiana delta in the United States.[9] Organic matter deposition is an important contribution to vertical accretion of the marsh surface in many tidal marshes, helping them to maintain elevation under rising sea-level conditions. For example, across 76 tidal freshwater marshes in North America and Europe organic matter was responsible for 62% of marsh vertical accretion, with the remainder coming from mineral matter.[10]

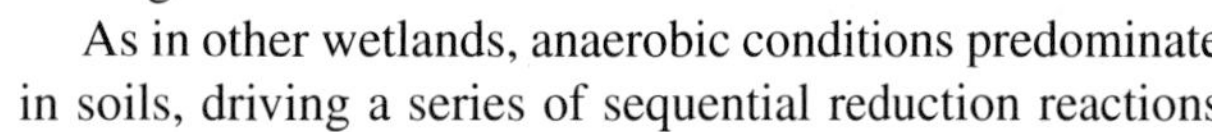

As in other wetlands, anaerobic conditions predominate in soils, driving a series of sequential reduction reactions as soil oxidation–reduction (redox) potential decreases.[11] In tidal freshwater marshes, methanogenesis is generally the terminal redox reaction, but when sulfate is present due to ocean water intrusion, for example, in saline marshes, sulfate reduction is likely to predominate because it is energetically more favorable than methanogenesis.[12,13] One implication of these redox processes is that increases in sea level may result in sulfate intrusion into tidal freshwater marshes, resulting in a shift from methanogenesis to sulfate reduction, accelerating decomposition of organic matter, and potential loss of wetlands due to reduced vertical accretion.[14]

Vegetation and Fauna

Because they span a range of inundation frequencies and salinity regimes, tidal marshes support spatially diverse assemblages of plant and animal communities. In general, the number of plant species in a given wetland increases as elevation increases (i.e., inundation decreases) and salinity decreases.[15] However, due in part to mixing of freshwater

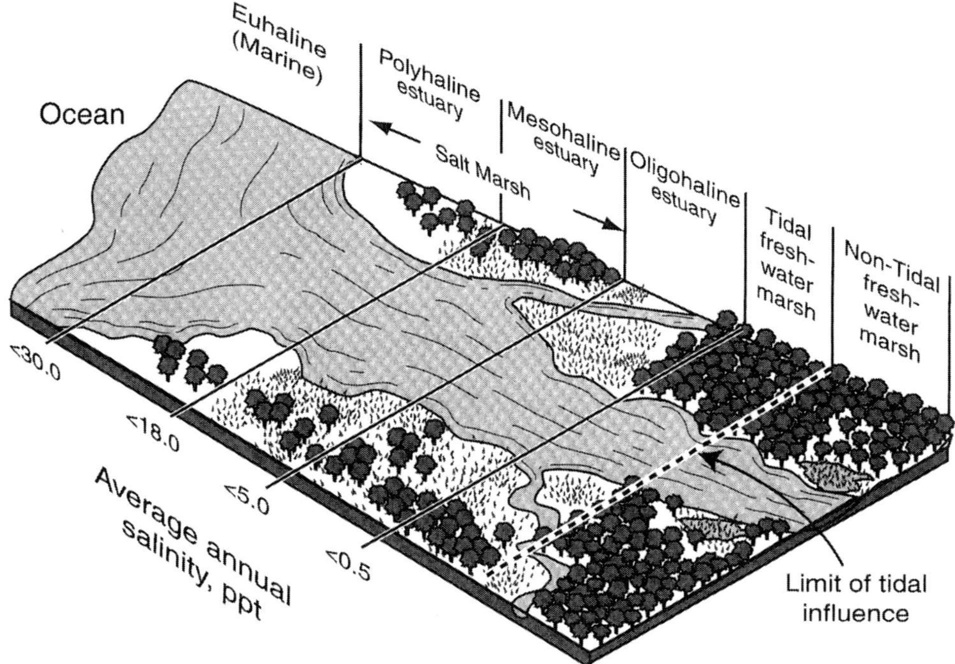

Fig. 2 The salinity of tidal marshes varies across estuaries. Most large estuaries contain a tidal freshwater zone of varying width that experiences tidal fluctuation but little or no intrusion of saline water. The salinity of estuarine water has an overriding influence on the species composition of plants and animals present and on biogeochemical processes.
Source: © John Wiley & Sons, Inc. 2009 as included in *Wetland Ecosystems*. Reprinted with permission.

and brackish species, the transition between the tidal freshwater and brackish zones of estuaries may contain the most species-rich tidal marshes.[16] Among the most globally widespread taxa of tidal marsh plants are common reed (*Phragmites australis* [Cav.] Trin. ex Steud.), bullrushes (*Schoenoplectus* spp.), and cordgrasses (*Spartina* spp.).[17–19] Across inundation gradients, vegetation often forms distinct bands of differing plant communities (horizontal zonation), with more distinct zonation occurring in salt marshes than in tidal freshwater marshes.[3] The distribution of plant species across estuarine salinity gradients is driven by a tradeoff between competitive ability and salinity stress tolerance. Salt-tolerant species can grow in tidal freshwater marshes, but only in the absence of competition from neighboring freshwater species, while freshwater species die or exhibit reductions in growth due to salt stress when transplanted to saline marshes.[19,20] While many of the dominant species of tidal marshes are clonal perennials, seed dispersal and seed banks are important sources of propagules for the maintenance or regeneration of vegetation in both tidal freshwater and salt marshes.[21–23] In some tidal freshwater marshes, annual species comprise half or more of the species richness and biomass production.[24,25] In addition to vascular plants, benthic and epiphytic microalgae are important but often overlooked primary producers in marshes across estuarine salinity gradients.[3,26]

The high primary production of tidal marshes supports abundant and diverse faunal communities, including aquatic and terrestrial invertebrates, fish, birds, mammals, reptiles, and (in tidal freshwater zones) amphibians.[3] Fish use marsh edges and, to a lesser extent, marsh interiors as nursery habitats,[27] and fish and invertebrate species composition differs between salinity regimes.[3,28] Terrestrial invertebrate communities (spiders, beetles, isopods, and others) also vary along salinity gradients in tidal marshes.[29]

Ecosystem Processes

Primary production from vascular plants, benthic algae, and phytoplankton in tidal marshes and associated creeks can be high.[6] Aboveground primary production in tidal freshwater marsh emergent plants varies considerably between vegetation types and species, but can be above 2,000 g m^{-2} yr^{-1}, and total production including belowground may reach 8,500 g m^{-2} yr^{-1}.[30] Primary production in emergent salt marsh plants is also commonly in the range of 1,000–2,000 g m^{-2} yr^{-1} aboveground and may exceed 8,000 g m^{-2} yr^{-1} including belowground.[6]

The high primary production of tidal marshes results in abundant litter and detritus that are important in internal carbon (C) and nutrient cycling and export of materials supporting food webs in adjacent waters. In general, decomposition rates of plant material are faster in tidal freshwater wetlands than in salt and brackish tidal wetlands because the freshwater plant species have relatively lower C:N ratios.[3] However, decomposition rates may differ considerably between species within a wetland of a particular salinity regime, and even within the same individual plant due to differences in lignin content, for example between stems and leaves.[31] Much of the organic matter and nutrients flow out of tidal marshes into adjacent marshes, with estimates of the amount of primary production exported varying considerably and depending on factors such as geomorphology, hydrology, and weather.[6]

Trends, Restoration, and Management

Tidal marshes are subject to alteration and loss from both natural and anthropogenic causes. Among the most important threats to tidal marshes are global climate change, coastal eutrophication, and invasive species, as well as continued direct physical impacts on marshes via draining, dredging, or filling. Acceleration in the rate of sea-level rise may exceed the ability of marshes to accrete vertically, resulting in their conversion to open water systems,[32] although vegetation allows some marshes to maintain elevation under conditions of rising sea level.[33] Higher atmospheric concentration of CO_2 favor C_3 over C_4 species,[34] but nitrogen (N) addition in combination with added CO_2 promotes encroachment of C_4 plants, negating increases in community biomass that would otherwise occur due to CO_2 alone.[35] Excess nutrients alter plant community composition in both tidal freshwater and saline marshes.[36,37] On the Atlantic and Gulf of Mexico coasts of North America, Eurasian genotypes of common reed are displacing native marsh plant communities in both tidal freshwater and brackish marshes,[38] although some faunal communities are not strongly impacted.[39,40] Similarly, the dominant species at lower elevation in U.S. Atlantic and Gulf coast salt marshes, smooth cordgrass (*Spartina alterniflora* Loisel.), has been introduced in estuaries around the globe, including the U.S. west coast and the Yangtze River delta in China.[41,42] In addition to these impacts, salt marsh dieback is a recent phenomenon that is not fully understood but may be linked to drought and changes in soil chemistry.[43]

Restoration and management of tidal marshes have offset some of their losses and changes. Provided suitable tidal hydrology, vegetation appears to regenerate fairly quickly (5–10 years) in restored salt marshes, although development of soil carbon and nitrogen pools may require much longer time frames.[44,45] In tidal freshwater marshes, seed dispersal may rapidly establish a diverse plant community and seed banks in restored sites, but in an urban environment constraints on vegetation development (altered hydrology, nutrients, invasive species) may prevent establishment of vegetation similar to that seen in marshes within rural environments.[23,46–48] Control of invasive plants may help restore native marsh vegetation, at least temporarily, but may have little impact on invertebrate or fish communities that are not altered by the presence of the invader.[49] Well-intentioned efforts to divert water from the Mississippi River to introduce sediments, nutrients, and fresh water to prevent marsh loss may have negative impacts on resilience to storms due to excess nitrogen, which inhibits root growth and thus may weaken soil strength.[50] Over small areas, sediment can be sprayed onto the surface to maintain surface elevation in rapidly subsiding wetlands.[51]

TROPICAL TIDAL FORESTS

This tidal habitat type is dominated by a group of plant species known as mangroves.[52] With some exception, mangroves are mostly woody, facultative halophytes sensu[53] that dominate tropical coastlines throughout the world due to a number of strong, and unique, life history characteristics (Fig. 3).[54] While mangroves are also found in subtropical and some warm temperate environments (e.g., New Zealand's North Island, northern Gulf of Mexico), they readily outcompete tidal salt marshes in the tropics,[55] differentiating them from marsh distribution more in terms of climate than in their similar hydrodynamic requirements.[56] Mangrove distribution includes 123 countries comprising between 13,776,000–15,236,100 ha worldwide,[57,58] and they provide critical goods and services to many inhabitants of low-lying, underdeveloped nations.[59] Mangrove forests are threatened by development on nearly every coastline they exist.[60,61]

Soil and Biogeochemistry

While the marine intertidal environment can be quite harsh to vegetation, process transformations of nutrients and energy from anaerobic soils—despite growing in stressful conditions—define mangrove ecosystems and enable high degrees of productivity. In fact, Alongi [p. 108][62] states that "highly evolved and energetically efficient plant-soil-microbe relations are a major factor in explaining why mangroves are highly productive in harsh tropical environments." Paradoxically, there are no "normal"

Fig. 3 Mangrove forest at low tide along the Shark River in Everglades National Park, Florida, U.S.A.
Source: Photo by K.W. Krauss, U.S. Geological Survey.

mangrove soils; pedogenic properties range from coralline to highly organic, acidic to alkaline (pH 5.8–8.5), and anoxic to suboxic (redox, –200 to +300 mv).[62] The type of vegetation can also influence the biogeochemistry of mangrove soils;[63–65] the tight connections between plant and soil in mangroves provide a model for the entire field of ecology.[62]

Twilley[66] described the fluxes of organic matter and nutrients in south Florida mangrove ecosystems as a balance among all sources and sinks (Fig. 4); whether a particular mangrove wetland is considered a source, sink, or conveyor of specific nutrients relates strongly to the process of material exchange between wetland and estuary and in situ transformations in lieu of geophysical processes in the estuary.[66] For example, N in the form of NO_3 and NH_4 is taken up by fine roots within the soil during material flux through the wetland; by one account, leaving only about 4–12% as the fraction of N buried within the mangrove soil.[67] Denitrification (efflux of N_2O and N_2 to the atmosphere) and dissolved fluxes of organic and inorganic N can represent an important conduit for N loss from mangroves. However, where growth limitations have been observed, studies have often implicated phosphorus (P) availability.[68] Feller[69] discovered that fertilizing stunted mangroves in Belize with N and P produced a P-only response, both in growth and in altering the relative sclerophylly of leaves.

Carbon fluxes from mangroves are large when modeled on a global perspective (Table 1); net export of dissolved inorganic C from all mangrove wetlands ranges from 112–160 Tg C year^{-1}.[62,70] Mangrove soils can be incredibly proficient at sequestering and storing atmospheric C,[71] rating high among tropical and coastal ecosystems in C burial rates (226 g C m^{-2} year^{-1}).[72] The volume of litterfall from mangrove ecosystems that can serve as a source of organic C to the soil ranges globally from 1.3–18.7 t ha^{-1} yr^{-1} and is directly related to standing forest structural attributes (Fig. 5). While productivity is important, the true potential for C sequestration within mangrove soils lies in the high relative rates of root growth balanced against low rates of decomposition in low oxygen soils.[73] Root growth has been documented to be as high as 525 g m^{-2} yr^{-1} from biogenic Caribbean atolls, contributing directly to a layer of peat 10 m deep and 7000–8000 years old.[74]

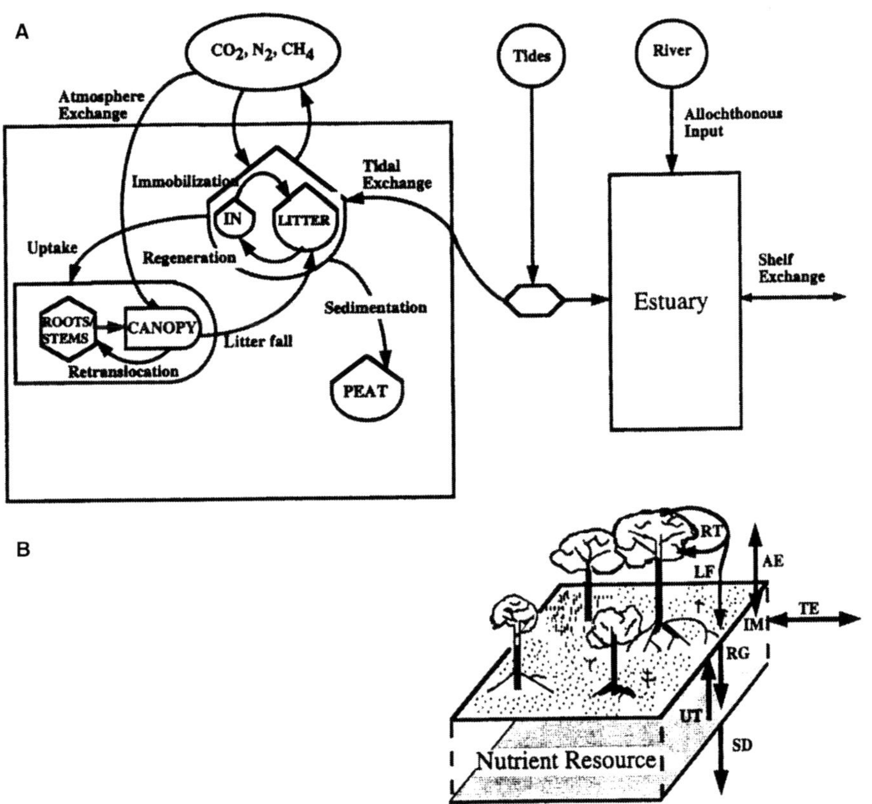

Fig. 4 (**A**) The suite of organic matter and nutrient fluxes between mangrove soils and the remainder of the estuary. (**B**) Conceptual overview of spatial linkages among ecological processes in mangroves, with application to all tidal wetlands being reviewed herein.[66]
Source: © Lewis Publishers 1998 as included in Southern Forested Wetland Ecology and Management. Reprinted with permission.
Abbreviations: AE, atmosphere exchange; IM, immobilization; LF, litterfall; RT, retranslocation; RG, regeneration; SD, sedimentation; TE, tidal exchange; UT, uptake.

Table 1 Summary of material carbon fluxes (Tg C $year^{-1}$) from mangrove wetlands globally, assuming an area of 160,000 ha[62]

Vegetation		Tidal fluxes	
GPP	+735	POC	−29
R_c	−423	DOC	−14
NPP	+309	R_{water}	−35
Wood	+67		
Litter	+68	**Total Respiration**	−500
Root	+174		
		Export	−160
Soil Fluxes			
Burial	+29		
R_s	−42		

Abbreviations: GPP, gross primary productivity; NPP, net primary productivity; R, respiration; total respiration, $R_{canopy} + R_{soil} + R_{water}$; '+', C import/sequestration; '−', C export/loss.

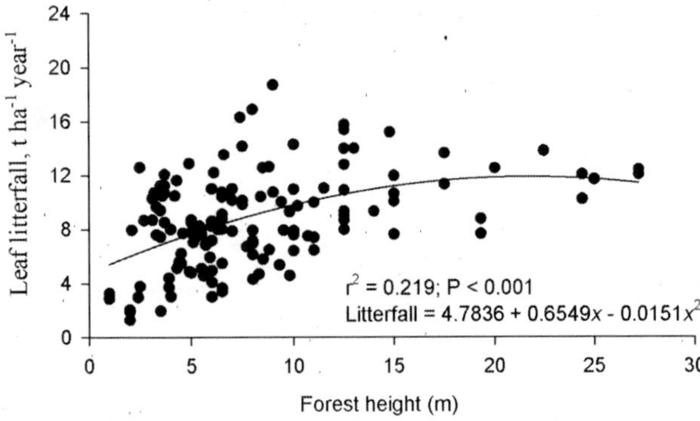

Fig. 5 Mangrove litterfall (mt ha^{-1} yr^{-1}) as a function of mean forest height (m) for both Atlantic East Pacific (AEP) and Indo West Pacific (IWP) biogeographic regions (sensu)[75] from studies reported in the literature.

Source: Adapted from Saenger & Snedaker,[128] Day et al.[129] Steinke & Charles,[130] Jardel et al.,[131] Hardiwinoto et al.,[132] Lu & Lin,[133] Flores-Verdugo et al.,[134] Schaeffer-Novelli et al.,[135] Flores-Verdugo et al.,[136] Sukardjo & Yamada,[137] Flores-Verdugo et al.,[138] Singh et al.,[139] Clarke,[140] Day et al.,[141] Twilley et al.,[142] Lugo et al.[143]

Vegetation and Fauna

Globally, there are nearly 73 species and/or hybrids of mangroves,[75] and forests range in height from sub-meter, stunted mangroves in the Caribbean[69] to tall forests in the 35 m range in places such as Micronesia where stress gradients and tropical storm disturbances are low.[76] From an ecophysiological perspective, mangroves stand out among tropical forest vegetation and tolerate at least five environmental regimes that would lead to rapid mortality in many plants: (1) mangroves are facultative halophytes that often flourish in high salinity, in fact, optimal growth rates vary by species but range from 5% to 75% full-strength seawater;[54] (2) mangrove leaves absorb high, persistent light levels inherent to the tropics without undergoing photoinhibition;[77] (3) mangroves often occur in low nutrient environments[78] and are especially efficient at nutrient resorption prior to leaf senescence;[79,80] (4) mangroves have high levels of water use efficiency both at the individual leaf[81] and tree levels[82] to ensure less water demand against osmotic gradients; and (5) mangroves tolerate tidal flooding quite well.[83,84]

Mangrove faunal assemblages are much more diverse than vegetation assemblages, and include crustaceans, fish, insects, spiders, reptiles, and fur-bearing mammals. Mangroves provide critical habitat for economically important species, such as the mud crab (*Scylla serrata*), which can affect local economies considerably.[85] However, of the faunal guilds present, most research has focused on the effects that insects and crustaceans have on litter and propagule decomposition, recycling of organic matter and nutrients, and structuring of forests. Insects can affect trees in several ways; Robertson and Duke[86] found that among 14 tree species surveyed in an Australian mangrove wetland, 8–100% of the leaves were damaged, accounting for 0.3–35% of total leaf area lost. Individual tree photosynthetic capacity and ecosystem nutrient cycling are impacted by such patterns of herbivory,[87] which itself can be influenced by different available nutrient regimes. Likewise, insect herbivory transcends life form in mangroves to include stems[69] and propagules[88] along with leaves.

Research has also focused on the effects that crabs have on influencing which mangrove tree species regenerate, and potentially which tree species are allowed to dominate the overstory.[89] Globally, over 54% of all reproductive propagules or seeds that fall to the floor of a mangrove-forest are consumed by crabs.[90] This predation is highly species specific, with mangroves producing palatable propagules (e.g., *Avicennia* spp.), sometimes harboring greater predation pressure from crabs than those maintaining less palatable propagules or seeds (e.g., *Xylocarpus* spp., *Rhizophora* spp.). Crabs often have primacy over many environmental variables in determining regeneration potential of a mangrove forest, especially considering how tolerant mangrove seedlings are to a diverse array of other stressors which often prove more detrimental to seedlings in other types of tropical forests.[91]

Ecosystem Processes

Because of the high rates of productivity and potential for organic matter export (see Table 1), a central paradigm was proposed ("Outwelling Hypothesis"),[92] implicating tidal wetlands (incl. mangroves) in enabling high productivity in adjacent coastal environments. Enhanced material export is thought to promote aquatic primary productivity, trophic exchange between autotrophic and heterotrophic communities, and support, for example, local fisheries. A rate of C export of 64–333 g C m^{-2} yr^{-1} from mangroves (*low*;[93]

high,[94]) echoes scaled rates of 112–160 Tg C yr^{-1} previously referenced to suggest strong potential for supporting energetic transformations in adjacent nearshore environments. However, Alongi[62] provides the caveat that "the number of factors (influencing export from mangroves) and their nature is such that each system is unique; some mangroves export nutrients and some do not" (p. 130).

Mangroves also provide important protection for coastal communities: in 2004, mangroves and other coastal vegetation protected human communities in India during the Asian tsunami[95] and suppressed storm surge at a rate of 9.4 cm per linear km inland during a hurricane in south Florida.[96] The capacity for storm surge suppression through mangroves may be even higher. Further, mangroves respond to sea-level rise actively along many coastlines through sediment trapping, enhanced root growth, or both.[97] The potential for mangrove migration inland as sea level rises is also strong, but many mangrove communities occur in prime real estate areas and thus become "squeezed" by development. This considerably diminishes the capacity of mangroves to protect coastlines, filter nutrients, and support high levels of coastal productivity.

Trends, Restoration, and Management

Impacts to mangroves include land clearing, hydrological alteration, over-harvesting both of trees and fauna, coastal eutrophication, pruning of canopies, sea-level rise, road construction, and maricultural operations. Overall, mangrove loss has been tremendous globally, and continues at a rate of 1–2% per year, equaling or surpassing loss of coral reefs or tropical rain forests.[61] Yet, the potential ecological services that mangroves provide, including strong recent interest in their potential for C sequestration,[72] have prompted efforts to restore mangroves. Indeed, treatises on the silvics of mangroves have been around for over a century,[98] culminating from a classic overview for Malaysia.[83] However, such attention to tree regeneration has taken the focus off of hydrological restoration, which must precede any efforts to restore mangrove communities through planting in areas where they have been destroyed.[99] Timber production, fisheries and wildlife enhancement, mitigation and legislative compliance, social enrichment, and restoration of ecosystem services provide the necessary impetus for mangrove restoration,[98] and nurseries have been established to support wide-scale plantings in some areas (Fig. 6). Restoration (or rehabilitation, sensu)[100] of mangroves is certainly more than tree planting; where successful, appropriate attention is made to re-establishing tidal and dispersal connections within degraded former wetlands.[99]

Management of most mangroves involves passive approaches, from allowing natural forest dynamics to providing larger protected areas that include mangroves (e.g., Everglades National Park [USA], Sungei Buloh Wetland Reserve [Singapore]). However, only about 6.9% of mangrove forests are in protected areas globally.[58] The aim of

Fig. 6 Mangrove nursery in Bali, Indonesia established to make effective use of natural tides for mass propagation of mangroves to support forestry operations and ecological restoration projects.
Source: Photo credit by K.W. Krauss, U.S. Geological Survey.

management of these unique systems "should be to maintain the use of mangroves as a renewable resource, providing fisheries and forestry products and possessing an inherent amenity value based on their geomorphological, recreational and scientific characteristics" (p. 275).[98] Often, management of ecological services can be melded with forestry operations to provide needed revenue for land purveyors. This type of management is perhaps most advanced in Malaysia and Thailand, and benefits greatly by encouraging local community involvement. Yet, mangrove timber management per se dates to the late 1700s in the Sundarbans of India and Bangladesh.[101] While scientific discoveries have been largely ignored when establishing management strategies,[102] some managers have made use of discoveries, specifically related to zonation of forests along specific inundation regimes to establish harvest and developmental zones.[103] Where mangroves are undergoing reduced growth or succumbing to sea-level rise, harvest and human activity is sometimes discouraged (e.g.);[104] however, the ultimate responsibility for establishing official public use, harvest, or exploitation guidelines lies with a diverse array of community, regional, and/or national governments overlaying the global distribution of mangroves, complicating their sustainable management.

TEMPERATE TIDAL FRESHWATER FORESTS

While global in distribution, nearly all studies on tidal freshwater forests originate from the southeastern United States, where at least 200,000 ha are generally found at the outlets of coastal rivers with low gradient and low topographic relief at or near sea level.[105] The actual extent of tidal freshwater forests in other parts of the world is not well documented. These forests are unique in hydrologic and salinity characteristics. They are flooded and drained regularly by freshwater overflow attributed to local high tides, and they are also prone to saltwater influx during low river flows (as occurs in drought years) or high storm tides (usually during hurricanes). Tide patterns vary across the range of tidal forests from diurnal, semidiurnal, or mixed and are of different amplitudes and range depending upon the location.[105] These forests are found in the upper estuaries of U.S. Atlantic and Gulf coast river systems where there is sufficient freshwater flow to maintain salinities of less than 0.5 ppt,[105] but insufficient flow to dampen upstream tidal movement.[106] Hydrology within these forests can be difficult to interpret because of river discharge, tidal stage, local precipitation, evapotranspiration, groundwater, and prevailing winds.[107] River discharge and tidal stage vary on a seasonal basis and have potential implications for wetland saltwater intrusion. These forests are subject to regular flooding from <1 m in the western Gulf of Mexico (Apalachicola Bay in Florida to southern tip of Texas) to between 2 and 3 m on the Atlantic coast (southern North Carolina to northeast Florida).[105]

Soil and Biogeochemistry

There is very little published research on tidal freshwater forest soils, but those that have included the entire tidal forest range have confirmed a wide range of soil characteristics.[108,109] For example, an analysis of soil profiles from the Suwannee River found soils representing 7 orders and 18 taxonomic subgroups, ranging from upland in appearance to deep mucks.[109] Factors such as microtopography, local climate, elevation, proximity to river mouth, vegetative cover, and physiographic origin lead to highly variable soil conditions that are difficult to generalize.[110] Generally, soils of tidal freshwater forests tend to be highly organic, a result of suppressed decomposition under anaerobic conditions and moderate to high plant production.[111]

Biogeochemistry of tidal freshwater forests is highly complex because of being influenced by coastal tides.[110] They also receive water at different frequencies from river floods, saltwater surges, and groundwater sources, with the quantity and timing of each source varying seasonally and annually. Because of anaerobic conditions and high organic matter content, tidal freshwater forest soils have the potential to retain nutrients and other material through sedimentation, plant uptake/detritus storage, sorption, and microbial immobilization, and can promote N loss through denitrification.[110] The accumulation of silts and clays contributes to the retention of P and other nutrients,[6] while high organic matter content contributes to high cation exchange capacity and the retention of heavy metals and other potential pollutants.[106] Similar to tidal marshes undergoing salinization, mineralization of soil organic matter may be promoted by greater sulfate availability (associated with saltwater),[14] resulting in potential losses of soil surface elevation as freshwater forests become more degraded (but see[112]).

Vegetation and Fauna

Canopy tree richness tends to be low in tidal freshwater forests. Dominant tree species are bald cypress (*Taxodium distichum* (L.) Rich.), water tupelo (*Nyssa aquatica* L.), and swamp tupelo (*Nyssa biflora* Walter) in lower elevation areas and ash (*Fraxinus* spp.), red maple (*Acer rubrum* L.), sweetgum (*Liquidambar styraciflua* L.), American hornbeam (*Carpinus caroliniana* Walter), and sweetbay (*Magnolia virginiana* L.) in higher elevation areas. These trees are sensitive to saltwater intrusion, however, with bald cypress generally being the most salt tolerant, although its growth is reduced considerably at mean annual salinity concentrations above 2 ppt.[113,114] Understory trees, shrub, and herb layers are generally sparse and low in diversity because of dense canopy and frequent flooding in the upper reaches of the river. However, as salinity levels increase, tree canopy decreases, resulting in more extensive and species-rich subcanopies and herbaceous layers.[5] Common species vary from river system to river system (see for examples[109,115,116]).

Low salinity tidal freshwater forests likely serve as reservoirs for abundant infaunal populations,[117] with oligochaetes, especially from the taxonomic families Tubificidae and Lumbriculidae, being dominant.[113] Faunal communities in tidal freshwater forests are likely affected by both hydrological patterns and salinity levels, but very little is known about faunal populations of these areas. While structural differences related to woody versus herbaceous vegetation were expected to have a significant impact on infauna and epifauna, researchers in the Cape Fear River, North Carolina, found salinity variations to have a stronger direct impact on fauna.[113]

Ecosystem Processes

There are few estimates of primary productivity for tidal freshwater forests, but these forests are expected to benefit from tidal subsidies of nutrients and energy just as marshes do.[5] There is a broad range in productivity values since forests near their downstream limit are likely to be stunted by higher salinities and flooding frequency. Upstream areas may be more productive due to nutrient subsidies from flood waters and reduced saltwater intrusion. In South Carolina, litterfall in downstream areas was 88–118 g m^{-2} yr^{-1} compared to 563–686 g m^{-2} yr^{-1} in upstream forests.[118] Similar trends have been found in Louisiana where litterfall productivity of tidal forests was 170 g m^{-2} yr^{-1} versus 700 g m^{-2} yr^{-1} for freshwater forests without any salinity impact.[119]

Nutrient dynamics in tidal freshwater forests can be variable and change along the estuarine gradient.[5] Tidal export of detritus matter is highly significant, but little data are available for these systems. The export of C, N, and P from forested wetlands contributes to the productivity of the lower estuary.[120] The major pulse of materials to the lower estuary coincides with the time of high detrital formation and the arrival of migrant species entering the estuary for growth and spawning purposes.

Trends, Restoration, and Management

When settlers arrived in the United States, rivers were the main means of travel and trade in coastal areas.[121] The adjacent tidal freshwater forests provided material for construction activities, trade, and fuel as well as land for agricultural practices. The result was a much reduced expanse of forested area. In South Carolina, 40,000 ha of tidal forests were cleared for rice production, most of which today still remains as marsh or pond.[122] Extensive logging between 1890 and 1925 resulted in nearly every virgin stand of bald cypress disappearing. In Louisiana, oil and gas field canals, shipping channels, and pipelines crisscross the coastal zone, creating conduits for saltwater intrusion into freshwater swamps. Signs of saltwater stress and forest dieback are so severe along these channels that there has been die-off of trees resulting in "ghost forests."[105] Unfortunately, we have no records of how much tidal freshwater forest existed in the past. The actual extent of tidal freshwater forests in the southeastern United States today is conservatively estimated to be over 200,000 ha.[123]

Eustatic sea-level rise and land subsidence have resulted in widespread hydrological changes in many freshwater forested wetlands.[124–126] The most widespread change is increased flooding depth and duration, followed by more prevalent and pervasive events of saltwater intrusion.[127] The frequency and severity of droughts and hurricanes are major natural factors that influence the extent and concentration of saltwater distribution that contributes to forest dieback in the coastal zone. Projected sea-level rise and changing climate are expected to accelerate the process and extent of saltwater intrusion into coastal freshwater forested wetlands, further impacting these habitats and restoration efforts.[105]

There is an overall paucity of large-scale restoration efforts in tidal freshwater forests. Most efforts have been of small-scale and represent a hodgepodge of funding sources over the past 25 years. Restoration efforts may benefit from being tied to large engineering designs such as freshwater or sediment diversions, which could provide continuous freshwater for mitigating persistent salinity incursion. While tidal bald cypress plantings may avoid stress in any given year without such measures, site elevation relative to mean sea-level makes plantings susceptible to salt-induced mortality during droughts or hurricanes. The physical difficulties associated with propagating good trees, planting seedlings on a large scale, and monitoring requires a dedicated multi-institutional approach.

ACKNOWLEDGMENTS

We thank Drs. Michael J. Osland and Alex T. Chow for helpful comments on this manuscript. Technical contribution No. 5997 of the Clemson University Experiment Station. This entry is based upon work supported by the USGS Climate and Land Use Change Research and Development Program and upon work supported by NIFA/USDA, under project number SC-1700424. Any use of trade, product, or firm names is for descriptive purposes only and does not constitute endorsement by the U.S. Government.

REFERENCES

1. Sharitz, R.R.; Pennings, S.C. Development of wetland plant communities. In *Ecology of Freshwater and Estuarine Wetlands*; Batzer, D.P., Sharitz, R.R., Eds.; University of California Press: Berkeley, 2006; 177–241.
2. Gosselink, J.G. *The Ecology of Delta Marshes of Coastal Louisiana: A Community Profile*; U.S. Fish and Wildlife Service, Department of the Interior: Washington, D.C.; 1984.
3. Odum, W.E. Comparative ecology of tidal fresh-water and salt marshes. Annu. Rev. Ecol. Syst. **1988**, *19*, 147–176.
4. Barendregt, A.; Whigham, D.F.; Baldwin, A.H., Eds. *Tidal Freshwater Wetlands*; Backhuys: Leiden, the Netherlands, 2009.
5. Mitsch, W.J.; Gosselink, J.G.; Anderson, C.J.; Zhang, L. *Wetland Ecosystems*; John Wiley and Sons, Inc.: Hoboken, NJ, 2009.
6. Mitsch, W.J.; Gosselink, J.G. *Wetlands*, 3rd Ed.; John Wiley & Sons: New York, 2000.
7. Kolka, R.K.; Thompson, J.A. Wetland geomorphology, soils, and formative processes. In *Ecology of Freshwater and Estuarine Wetlands*; Batzer D.P., Sharitz, R.R., Eds.; University of California Press: Berkeley, 2006; 7–42.
8. Zhou, J.L.; Wu, Y.; Kang, Q.S.; Zhang, J. Spatial variations of carbon, nitrogen, phosphorous and sulfur in the salt marsh sediments of the Yangtze Estuary in China. Estuar. Coast. Shelf Sci. **2007**, *71*, 47–59.
9. Sasser, C.E.; Gosselink, J.G.; Holm, G.O.; Visser, J.M. Tidal freshwater wetlands of the Mississippi River deltas. In *Tidal Freshwater Wetlands*; Barendregt, A., Whigham, D.F.; Baldwin, A.H., Eds.; Backhuys: Leiden: the Netherlands, 2009; 167–178.
10. Neubauer, S.C. Contributions of mineral and organic components to tidal freshwater marsh accretion. Estuar. Coast. Shelf Sci. **2008**, *78*, 78–88.
11. Megonigal, J.P.; Hines, M.E.; Visscher, P.T. Anaerobic metabolism: Linkages to trace gases and aerobic processes. In *Biogeochemistry;* Schlesinger, W.H., Ed.; Elsevier-Pergamon: Oxford, UK, 2004; 317–424.
12. Howarth, R.W.; Teal, J.M. Sulfate reduction in a New England salt marsh. Limnol. Oceanogr. **1979**, *24*, 999–1013.
13. Neubauer, S.C.; Givler, K.; Valentine, S.K.; Megonigal, J.P. Seasonal patterns and plant-mediated controls of subsurface wetland biogeochemistry. Ecology **2005**, *86*, 3334–3344.
14. Weston, N.B.; Vile, M.A.; Neubauer, S.C.; Velinsky, D.J. Accelerated microbial organic matter mineralization following salt-water intrusion into tidal freshwater marsh soils. Biogeochemistry **2011**, *102*, 135–151.
15. Engels, J.G.; Jensen, K. Patterns of wetland plant diversity along estuarine stress gradients of the Elbe (Germany) and Connecticut (USA) Rivers. Estuar. Coast. **2009**, *2*, 301–311.

16. Sharpe, P.J.; Baldwin, A.H. Patterns of wetland plant species richness across estuarine gradients of Chesapeake Bay. Wetlands **2009**, *29*, 225–235.
17. Yang, S.L. Trapping effect of tidal marsh vegetation on suspended sediment, Yangtze Delta. Estuar. Coast. Shelf Sci. **1998**, *47*, 227–233.
18. Byrd, K.B.; Kelly, M. Salt marsh vegetation response to edaphic and topographic changes from upland sedimentation in a Pacific estuary. Wetlands **2006**, *26*, 813–829.
19. Engels, J.G.; Jensen, K. Role of biotic interactions and physical factors in determining the distribution of marsh species along an estuarine salinity gradient. Oikos **2010**, *119*, 679–685.
20. Crain, C.M.; Silliman, B.R.; Bertness, S.L. Physical and biotic drivers of plant distribution across estuarine salinity gradients. Ecology **2004**, *85*, 2539–2549.
21. Huiskes, A.H.L.; Koutstaal, B.P.; Herman, P.M.J.; Beeftink, W.G.; Markusse, M.M.; De Munck, W. Seed dispersal of halophytes in tidal salt marshes. J. Ecol. **1995**, *83*, 559–567.
22. Leck, M.A.; Simpson, R.L. Ten-year seed bank and vegetation dynamics of a tidal freshwater marsh. Am. J. Bot. **1995**, *82*, 1547–1557.
23. Neff, K.P.; Baldwin, A.H. Seed dispersal into wetlands: techniques and results for a restored tidal freshwater marsh. Wetlands **2005**, *25*, 392–404.
24. Whigham, D.F.; Simpson, R.L. Annual variation in biomass and production of a tidal freshwater wetland and comparison with other wetland systems. Va. J. Sci. **1992**, *43*, 5–14.
25. Baldwin, A.H.; Egnotovich, M.S.; Clarke, E. Hydrologic change and vegetation of tidal freshwater marshes: Field, greenhouse, and seed-bank experiments. Wetlands **2001**, *21*, 519–531.
26. Rizzo, W.M.; Wetzel, R.L. Intertidal and shoal benthic community metabolism in a temperate estuary: studies of spatial and temporal scales of variability. Estuaries **1985**, *8*, 342–351.
27. Minello, T.J.; Able, K.W.; Weinstein, M.P.; Hays, C.G. Salt marshes as nurseries for nekton: testing hypotheses on density, growth and survival through meta-analysis. Mar. Ecol.-Prog. Ser. **2003**, *246*, 39–59.
28. Mathieson, S.; Cattrijsse, A.; Costa, M.J.; Drake, P.; Elliot, M.; Gardner, J.; Marchand, J. Fish assemblages of European tidal marshes: a comparison based on species, families and functional guilds. Mar. Ecol.-Prog. Ser. **2000**, *204*, 225–242.
29. Desender, K.; Maelfait, J.P. Diversity and conservation of terrestrial arthropods in tidal marshes along the River Schelde: a gradient analysis. Biol. Conserv. **1999**, *8*, 221–229.
30. Whigham, D.F. Primary production in tidal freshwater wetlands. In *Tidal Freshwater Wetlands*; Barendregt, A.; Whigham, D.F.; Baldwin, A.H., Eds.; Backhuys: Leiden, the Netherlands, 2009; 115–122.
31. Findlay, S.E.G.; Nieder, W.C.; Ciparis, S. Carbon flows, nutrient cycling, and food webs in tidal freshwater wetlands. In *Tidal Freshwater Wetlands*; eds. Barendregt, A., Whigham, D.F.; Baldwin, A.H., Eds.; Backhuys: Leiden, the Netherlands, 2009; 137–144.
32. Craft, C.; Clough, J.; Ehman, J.; Joye, S.; Park, R.; Pennings, S.; Guo, H.; Machmuller, M. Forecasting the effects of accelerated sea-level rise on tidal marsh ecosystem services. Front. Ecol. Environ. **2009**, *7*, 73–78.
33. Morris, J.T.; Sundareshwar, P.V.; Nietch, C.T.; Kjerfve, B.; Cahoon, D.R. Responses of coastal wetlands to rising sea level. Ecology **2002**, *83*, 2869–2877.
34. Erickson, J.E.; Megonigal, J.P.; Peresta, G.; Drake, B.G. Salinity and sea level mediate elevated CO_2 effects on C_3–C_4 plant interactions and tissue nitrogen in a Chesapeake Bay tidal wetland. Global Change Biol. **2007**, *13*, 202–215.
35. Langley, J.A.; Megonigal, J.P. Ecosystem response to elevated CO_2 levels limited by nitrogen-induced plant species shift. Nature **2010**, *466*, 96–99.
36. Levine, J.M.; Brewer, J.S.; Bertness, M.D. Nutrients, competition, and plant zonation in a New England salt marsh. J. Ecol. **1998**, *86*, 285–292.
37. Baldwin, A.H. Nitrogen and phosphorus differentially affect annual and perennial plants in tidal freshwater and oligohaline wetlands. Estuar. Coast. **2013**, *36,* 547–558.
38. Chambers, R.M.; Meyerson, L.A.; Saltonstall, K. Expansion of *Phragmites australis* into tidal wetlands of North America. Aquat. Bot. **1999**, *64*, 261–273.
39. Fell, P.E.; Weissbach, S.P.; Jones, D.A.; Fallon, M.A.; Zeppieri, J.A.; Faison, E.K.; Lennon, K.A.; Newberry, K.T.; Reddington, L.K. Does invasion of oligohaline tidal marshes by reed grass, *Phragmites australis* (Cav.) Trin. ex Steud., affect the availability of prey resources for the mummichog, *Fundulus heteroclitus* L.? J. Exp. Mar. Biol. Ecol. **1998**, *222*, 59–77.
40. Posey, M.H.; Alphin, T.D.; Meyer, D.L.; Johnson, J.M. Benthic communities of common reed *Phragmites australis* and marsh cordgrass *Spartina alterniflora* marshes in Chesapeake Bay. Mar. Ecol.-Prog. Ser. **2003**, *261*, 51–61.
41. Ayres, D.R.; Smith, D.L.; Zaremba, K.; Klohr, S.; Strong, D.R. Spread of exotic cordgrasses and hybrids (*Spartina* sp.) in the tidal marshes of San Francisco Bay, California, USA. Biol. Invasions **2004**, *6*, 221–231.
42. Xiao, D.R.; Zhang, L.Q.; Zhu, Z.C. The range expansion patterns of *Spartina alterniflora* on salt marshes in the Yangtze Estuary, China. Estuar. Coast. Shelf Sci. **2010**, *88*, 99–104.
43. Alber, M.; Swenson, E.M.; Adamowicz, S.C.; Mendelssohn, I.A. Salt marsh dieback: an overview of recent events in the U.S. Estuarine Coast. Shelf Sci. **2008**, *80*, 1–11.
44. Craft, C., J. Reader, J.; Sacco, N.; Broome, S.W. Twenty-five years of ecosystem development of constructed *Spartina alterniflora* (Loisel) marshes. Ecol. Appl. **1999**, *9*, 1405–1419.
45. Eertman, R.H.M.; Kornman, B.A.; Stikvoort, E.; Verbeek, H. Restoration of the Sieperda tidal marsh in the Scheldt estuary, the Netherlands. Restor. Ecol. **2002**, *1*, 438–449.
46. Leck, M.A. Seed-bank and vegetation development in a created tidal freshwater wetland on the Delaware River, Trenton, New Jersey, USA. Wetlands **2003**, *23*, 310–343.
47. Baldwin, A.H. Restoring complex vegetation in urban settings: the case of tidal freshwater marshes. Urban Ecosyst. **2004**, *7*, 137.
48. Neff, K.P.; Rusello, K.; Baldwin, A.H. Rapid seed bank development in restored tidal freshwater wetlands. Restor. Ecol. **2009**, *17*, 539–548.

49. Warren, R.S.; Fell, P.E.; Grimsby, J.L.; Buck, E.L.; Rilling, G.C.; Fertik, R.A. Rates, patterns, and impacts of *Phragmites australis* expansion and effects of experimental *Phragmites* control on vegetation, macroinvertebrates, and fish within tidelands of the lower Connecticut River. Estuaries **2001**, *24*, 90–107.
50. Kearney, M.S.; Riter, J.C.A.; Turner, R.E. Freshwater river diversions for marsh restoration in Louisiana: twenty-six years of changing vegetative cover and marsh area. Geophy. Res. Lett. **2011**, *38*, doi:10.1029/2011GL047847.
51. Ford, M.A.; Cahoon, D.R.; Lynch, J.C. Restoring marsh elevation in a rapidly-subsiding salt marsh by thin-layer deposition of dredge material. Ecol. Eng. **1999**, *12*, 189–205.
52. Tomlinson, P.B. *The Botany of Mangroves*; Cambridge University Press: Cambridge, UK, 1986.
53. Ball, M.C. Ecophysiology of mangroves. Trees **1988**, *2*, 129–142.
54. Krauss, K.W.; Lovelock, C.E.; McKee, K.L.; López-Hoffmanc, L.; Ewe, S.M.L.; Sousa, W.P. Environmental drivers in mangrove establishment and early development: A review. Aquat. Bot. **2008**, *89*, 105–127.
55. Saintilan, N.; Rogers, K.; McKee, K.L. Salt marsh-mangrove interactions in Australasia and the Americas. In *Coastal Wetlands: An Integrated Ecosystem Approach*; Gerardo, M.E.P., Wolanski, E., Cahoon, D.R., Brinson, M.M., Eds.; Elsevier: 2009; 855–883.
56. Friess, D. A.; Krauss, K.W.; Horstman, E.M.; Balke, T.; Bouma, T.J.; Galli, D.; Webb, E.L. Are all intertidal wetlands naturally created equal? Bottlenecks, thresholds and knowledge gaps to mangrove and saltmarsh ecosystems. Biol. Rev. **2012**, *87*, 346–366.
57. Spalding, M; Kainuma, M.; Collins, L. *World Atlas of Mangroves*; Earthscan: London, 2010.
58. Giri, C.; Ochieng, E.; Tieszen, L.L.; Zhu, Z.; Singh, A.; Loveland, T.; Masek, J.; Duke, N. Status and distribution of mangrove forests of the world using earth observation satellite data. Global Ecol. Biogeogr. **2011**, *20*, 154–159.
59. Ewel, K.C.; Twilley, R.R.; Ong, J.E. Different kinds of mangrove forests provide different goods and services. Global Ecol. Biogeogr. Lett. **1998**, *7*, 83–94.
60. Valiela, I.; Bowen, J.L.; York, J.K. Mangrove forests: one of the world's threatened major tropical environments. BioScience **2001**, *51*, 807–815.
61. Duke, N.C. Mangrove floristics and biogeography. In *Tropical Mangrove Ecosystems*; Robertson, A.I.; Alongi, D.M., Eds.; American Geophysical Union: Washington, 1992; 63–100.
62. Alongi, D.M. *The Energetics of Mangrove Forests*; Springer: New York, 2009.
63. McKee, K.L. Soil physicochemical patterns and mangrove species distribution – reciprocal effects? J. Ecol. **1993**, *81*, 477–487.
64. Alongi, D.M.; Tirendi, F.; Dixon, P.; Trott, L.A.; Brunskill, G.J. Mineralization of organic matter in intertidal sediments of a tropical semi-enclosed delta. Estuar. Coast. Shelf Sci. **1999**, *48*, 451–467.
65. Gleason, S.M.; Ewel, K.C.; Hue, N. Soil redox conditions and plant-soil relationships in a Micronesian mangrove forest. Estuar. Coast. Shelf Sci. **2003**, *56*, 1065–1074.
66. Twilley, R.R. Mangrove wetlands. In *Southern Forested Wetlands Ecology and Management*; Messina, M.G.; Conner, W.H., Eds.; Lewis Publishers: Boca Raton, FL, 1998; 445–473.
67. Alongi, D.M.; Trott, L.A.; Wattayakorn, G.; Clough, B. Below-ground nitrogen cycling in relation to net canopy production in mangrove forests of southern Thailand. Mar. Biol. **2002**, *140*, 855–864.
68. Sherman, R.E.; Fahey, T.J.; Howarth, R.W. Soil-plant interactions in a neotropical mangrove forest: iron, phosphorus and sulfur dynamics. Oecologia **1998**, *115*, 553–563.
69. Feller, I.C. Effects of nutrient enrichment on growth and herbivory of dwarf red mangrove (*Rhizophora mangle*). Ecol. Monogr. **1995**, *65*, 477–505.
70. Bouillon, S.; Borges, A.V.; Castañeda-Moya, E.; Diele, K.; Dittmar, T.; Duke, N.C.; Kristensen, E.; Lee, S.Y.; Marchland, C.; Middelburg, J.J.; Rivera-Monroy, V.H.; Smith III, T.J.; Twilley, R.R. Mangrove production and carbon sinks: a revision of global budget estimates. Global Biogeochem. Cycle. **2007**, *22*, GB2013.
71. Donato, D.C.; Kauffman, J.B.; Murdiyarso, D.; Kurnianto, S.; Stidham, M.; Kanninen, M. Mangroves among the most carbon-rich forests in the tropics. Nat. Geosci. **2011**, *4*, 293–297.
72. McLeod, E.; Chmura, G.L.; Bouillon, S.; Salm, R.; Björk, M.; Duarte, C.M.; Lovelock, C.E.; Schlesinger, W.H.; Silliman, B.R. A blueprint for blue carbon: toward an improved understanding of the role of vegetated coastal habitats in sequestering CO_2. Front. Ecolo Environ. **2011**, *9*, 552–560.
73. Middleton, B.A.; McKee, K.L. Degradation of mangrove tissues and implications for peat formation in Belizean island forests. J. Ecol. **2001**, *89*, 818–828.
74. McKee, K.L.; Cahoon, D.R.; Feller, I.C. Caribbean mangroves adjust to rising sea level through biotic controls on change in soil elevation. Global Ecol. Biogeogr. **2007**, *16*, 545–556.
75. Duke, N.C.; Ball, M.C.; Ellison, J.C. Factors influencing biodiversity and distributional gradients in mangroves. Global Ecol. Biogeogr. Lett. **1998**, *7*, 27–47.
76. Ewel, K.C.; Bourgeois, J.A.; Cole, T.G.; Zheng, S. Variation in environmental characteristics and vegetation in high-rainfall mangrove forests, Kosrae, Micronesia. Global Ecol. Biogeogr. Lett. **1998**, *7*, 49–56.
77. Cheeseman, J.M. The analysis of photosynthetic performance in leaves under field conditions: a case study using *Bruguiera* mangroves. Photosynth. Res. **1991**, *29*, 11–22.
78. Reef, R.; Feller, I.C.; Lovelock, C.E. Nutrition of mangroves. Tree Physiol. **2010**, *30*, 1148–1160.
79. Feller, I.C.; Whigham, D.F.; O'Neill, J.P.; McKee, K.L. Effects of nutrient enrichment on within-stand cycling in a mangrove forest. Ecology **1999**, *80*, 2193–2205.
80. Lovelock, C.E.; Feller, I.C.; Ball, M.C. Testing the growth rate vs. geochemical hypothesis for latitudinal variation in plant nutrients. Ecol. Lett. **2007**, *10*, 1154–1163.
81. Andrews, T.J.; Muller, G.J. Photosynthetic gas exchange of the mangrove, *Rhizophora stylosa* Griff., in its natural environment. Oecologia **1985**, *65*, 449–455.
82. Krauss, K.W.; Young, P.J.; Chambers, J.L.; Doyle, T.W.; Twilley, R.R. Sap flow characteristics of neotropical mangroves in flooded and drained soils. Tree Physiol. **2007**, *27*, 775–783.

83. Watson, J.G. Mangrove forests of the Malay Peninsula. Malayan Forest R. **1928**, *6*, 1–275.
84. Chapman, V.J. Mangrove Vegetation; J. Cramer; Vaduz, 1976.
85. Naylor, R.; Drew, M. Valuing mangrove resources in Kosrae, Micronesia. Environ. Dev. Econ. **1998**, *3*, 471–490.
86. Robertson, A.I.; Duke, N.C. Insect herbivory on mangrove leaves in North Queensland. Aust. J. Ecol. **1987**, *12*, 1–7.
87. Hogarth, P.J. *The Biology of Mangroves*; Oxford University Press: Oxford, 1999.
88. Farnsworth, E.J.; Ellison, A.M. Global patterns of pre-dispersal propagule predation in mangrove forests. Biotropica **1997**, *29*, 318–330.
89. Smith, T.J., III. Seed predation in relation to dominance and distribution in mangrove forests. Ecology **1987**, *68*, 266–273.
90. Allen, J.A.; Krauss, K.W.; Hauff, R.D. Factors limiting the intertidal distribution of the mangrove species *Xylocarpus granatum*. Oecologia **2003**, *135*, 110–121.
91. Lindquist, E.S.; Krauss, K.W.; Green, P.T.; O'Dowd, D.J.; Sherman, P.M.; Smith, III, T.J. Land crabs as key drivers in tropical coastal forest recruitment. Biol. Rev. **2009**, *84*, 203–223.
92. Odum, E.P. The status of three ecosystem-level hypotheses regarding salt marsh estuaries: tidal subsidy, outwelling, and detritus-based food chains. In *Estuarine Perspectives*; Kennedy, V.S., Ed.; Academic Press: New York, 1980; 485–495.
93. Twilley, R.R. The exchange of organic carbon in basin mangrove forests in a southwest Florida estuary. Estuar. Coast. Shelf Sci. **1985**, *20*, 543–557.
94. Alongi, D.M. *Coastal Ecosystem Processes*; CRC Press, Boca Raton, 1998.
95. Danielsen, F.; Sorensen, M.K.; Olwig, M.F.; Selvam, V.; Parish, F.; Burgess, N.D.; Hiraishi, T.; Karunagaran, V.M.; Rasmussen, M.S.; Hansen, L.B.; Quarto, A.; Suryadiputra, N. The Asian tsunami: a protective role for coastal vegetation. Science **2005**, *310*, 643.
96. Krauss, K.W.; Doyle, T.W.; Doyle, T.J.; Swarzenski, C.M.; From, A.S.; Day, R.H.; Conner, W.H. Water level observations in mangrove swamps during two hurricanes in Florida. Wetlands **2009**, *29*, 142–149.
97. McKee, K.L. Biophysical controls on accretion and elevation change in Caribbean mangrove ecosystems. Estuar. Coast. Shelf Sci. **2011**, *91*, 475–483.
98. Saenger, P. *Mangrove Ecology, Silviculture and Conservation*; Kluwer Academic Publishers: Dordrecht, 2002.
99. Lewis, R.R. Ecological engineering for successful management and restoration of mangrove forests. Ecol. Eng. **2005**, *24*, 403–418.
100. Field, C.D. Rehabilitation of mangrove ecosystems: an overview. Mar. Poll. Bull. **1998**, *37*, 383–392.
101. Chowdhury, R.A.; Ahmed, I. History of forest management. In *Mangroves of the Sundarbans*, Volume 2; Hussain, Z., Acharya, G., Eds.; IUCN Wetlands Program: Switzerland, 1994; 155–180.
102. Kairo, J.G.; Dahdouh-Guebas, F.; Bosire, J.; Koedam, N. Restoration and management of mangrove ecosystems–a lesson for and from the East African region. S. Afr. J. Bot. **2001**, *67*, 383–389.
103. Aksornkoae, S. Scientific mangrove management in Thailand. In *Tropical Forestry in the 21st Century*; Aksornkoae, S., Puangchit, L., Thaiutsa, B., Eds.; Kasetsart University: Bangkok, 1997; 118–126.
104. Krauss, K.W.; Cahoon, D.R.; Allen, J.A.; Ewel, K.C.; Lynch, J.C.; Cormier, N. Surface elevation change and susceptibility of different mangrove zones to sea-level rise on Pacific high islands of Micronesia. Ecosystems **2010**, *13*, 129–143.
105. Doyle, T.W.; O'Neil, C.P.; Melder, P.V.; et al. Tidal freshwater swamps of the Southeastern United States: effects of land use, sea-level rise and climate change. In *Ecology of Tidal Freshwater Forested Wetlands of the Southeastern United States*; Conner, W.H., Doyle, T.W., Krauss, K.W., Eds.; Springer: Dordrecht, 2007; 1–28.
106. Simpson, R.L.; Good, R.E.; Leck, M.A.; Whigham, D.F. The ecology of freshwater tidal wetlands. BioScience **1983**, *33* (4), 255–259.
107. Anderson, C.J.; Lockaby, B.G. Soils and biogeochemistry of tidal freshwater forested wetlands. In *Ecology of Tidal Freshwater Forested Wetlands of the Southeastern United States*; Conner, W.H., Doyle, T.W., Krauss, K.W., Eds.; Springer: Dordrecht, 2007; 65–88.
108. Doumlele, D.G.; Fowler, K.; Silberhorn, G.M. Vegetative community structure of tidal freshwater swamp in Virginia. Wetlands **1984**, *4*, 129–145.
109. Light, H.M.; Darst, M.R.; Lewis, L.J.; Howell, D.A. Hydrology, vegetation, and soils of riverine and tidal forests of the Lower Suwannee River, Florida, and potential impacts of flow reductions. Professional Paper 1656A. U.S. Geological Survey, Tallahassee, FL, 2002.
110. Anderson, C.J.; Lockaby, B.G. Seasonal patterns of river connectivity and saltwater intrusion in tidal forested freshwater wetlands. River Res. Applic. **2012**, *28*, 814–826.
111. Wharton, C.H.; Kitchens, W.M.; Pendleton, E.C.; Sipe, T.W. The ecology of bottomland hardwood swamps of the southeast: a community profile, FWS/OBS-81/37; U.S. Fish and Wildlife Service, Biological Services Program: Washington, DC, 1982.
112. Hackney, C.T.; Posey, M.; Leonard, L.L.; Alphin, T.; Avery, G.B. Monitoring the effects of a potential increased tidal range in the Cape Fear River ecosystem due to deepening Wilmington Harbor, North Carolina. Year 1: June1, 2003-May 31, 2004. U.S. Army Corps of Engineers, Wilmington District (Contract No. DACW 54-00-R-0008), Wilmington, NC, 2005.
113. Hackney, C.T.; Avery, G.B.; Leonard, L.A.; Posey, M.; Alphin, T. Biological, chemical, and physical characteristics of tidal freshwater swamp forests of the Lower Cape Fear River/Estuary, North Carolina. In *Ecology of Tidal Freshwater Forested Wetlands of the Southeastern United States*; Conner, W.H., Doyle, T.W., Krauss, K.W., Eds.; Springer: Dordrecht, 2007; 183–221.http://onlinelibrary.wiley.com/doi/10.1002/rra.1489/abstract (accessed August 2011).
114. Krauss, K.W.; Duberstein, J.A.; Doyle, T.W.; Conner, W.H.; Day, R.H.; Inabinette, L.W.; Whitbeck, J.L. Site condition, structure, and growth of bald cypress along tidal/non-tidal salinity gradients. Wetlands **2009**, *29*, 505–519.

115. Rheinhardt, R. A multivariate analysis of vegetation patterns in tidal freshwater swamps of lower Chesapeake Bay, USA. Bull. Torrey Bot. Club **1992**, *119*, 192–207.
116. Duberstein, J.; Kitchens, W. Community composition of select areas of tidal freshwater forest along the Savannah River. In *Ecology of Tidal Freshwater Forested Wetlands of the Southeastern United States*; Conner, W.H., Doyle, T.W., Krauss, K.W., Eds.; Springer: Dordrecht, 2007; 321–348.
117. Rozas, L.P.; Hackney, C.T. Use of oligohaline marshes by fishes and macrofaunal crustaceans in North Carolina. Estuaries **1984**, *7*, 213–224.
118. Cormier, N.; Krauss, K.W.; Conner, W.H. Periodicity in stem growth and litterfall in tidal freshwater forested wetlands: influence of salinity and drought on nitrogen recycling. Estuar. Coasts **2013**, *36*, 533–546.
119. Effler, R.S.; Shaffer, G.P.; Hoeppner, S.S.; Goyer, R.A. Ecology of the Maurepas swamp: effects of salinity, nutrients, and insect defoliation. In *Ecology of Tidal Freshwater Forested Wetlands of the Southeastern United States*; Conner, W.H., Doyle, T.W., Krauss, K.W., Eds.; Springer: Dordrecht, 2007; 349–384.
120. Day, J.W. Jr.; Butler, T.J.; Conner, W.H. Productivity and nutrient export studies in a cypress swamp and lake system in Louisiana. In *Estuarine Processes*: Vol.2; Wiley, M., Ed.; Academic Press: New York, 1977; 255–269.
121. Rodgers, G.C. Jr. *The History of Georgetown County, South Carolina*; The University of South Carolina Press: Columbia, SC, 1970.
122. Gresham, C.A.; Hook, D.D. Rice fields of South Carolina: A resource inventory and management policy evaluation. Coast. Zone Manage. J. **1982**, *9*, 183–203.
123. Field, D.W.; Reyer, A.J.; Genovese, P.V.; Shearer, B.D. *Coastal Wetlands of The United States: An Accounting of a Valuable National Resource*; Office of Oceanography and Marine Assessment, National Ocean Service, National Oceanic and Atmospheric Administration: Rockville, MD, 1991.
124. Brinson, M.M.; Bradshaw, H.D.; Jones, M.N. Transitions in forested wetlands along gradients of salinity and hydroperiod. J. Elisha Mitchell Sci. Soc. **1985**, *101*, 76–94.
125. Conner, W.H.; Day, J.W. Jr. Rising water levels in coastal Louisiana: implications for two forested wetland areas in Louisiana. J. Coast. Res. **1988**, *4*, 589–596.
126. Pezeshki, S.R.; DeLaune, R.D.; Patrick, W.H. Jr. Flooding and saltwater intrusion: potential effects on survival and productivity of wetland forests along the U.S. Gulf Coast. For. Ecol. Manage. **1990**, *33* (34), 287–301.
127. Shaffer, G.P.; Wood, W.B.; Hoeppner, S.S.; Perkins, T.E.; Zoller, J.; Kandalepas, D. Degradation of bald cypress – water tupelo swamp to marsh and open water in Southeastern Louisiana, USA: an irreversible trajectory? J. Coast. Res. **2009**, *54*, 152–165.
128. Saenger, P.; Snedaker, S.C. Pantropical trends in mangrove above-ground biomass and annual litterfall. Oecologia **1993**, *96*, 293–299.
129. Day, J.W.; Day, R.H.; Barreiro, M.T.; Ley-Lou, F.; Madden, C.J. Primary production in the Laguna de Terminos, a tropical estuary in the Southern Gulf of Mexico. Oceanol. Acta **1982**, *SP*, 269–276.
130. Steinke, T.D.; Charles, L.M. Litter production by mangroves. I: Mgeni Estuary. S. Afr. J. Bot. **1986**, *52*, 552–558.
131. Jardel, E.J.; Saldaña A.A.; Barreiro, M.T. Contribución al conocimiento de la ecología de los manglares de la Laguna de Términos, Campeche, México. Ciencias Marinas **1987**, *13*, 1–22.
132. Hardiwinoto, S., Nakasuga, T.; Igarashi, T. Litter production and decomposition of a mangrove forest at Ohura Bay, Okinawa. Res. Bull. Coll. Exp. Forests (Hokkaido Univ.) **1989**, *46*, 577–594.
133. Lu, C.; Lin, P. Studies on litter fall and decomposition of *Bruguiera sexangula* (Lour.) Poir. community on Hainan Island, China. Bull. Mar. Sci. **1990**, *47*, 139–148.
134. Flores-Verdugo, F., González-Farías, F.; Ramírez-Flores, O.; Amezcua-Linares, F.; Yáñez-Arancibia, A.; Alvarez-Rubio, M.; Day, J.W. Mangrove ecology, aquatic primary productivity, and fish community dynamics in the Teacapán-Agua Brava lagoon-estuarine system (Mexican Pacific). Estuaries **1990**, *13*, 219–230.
135. Schaeffer-Novelli, Y.; Mesquita, H. de S.L.; Cintrón-Molero, G. The Cananéia lagoon estuarine system, São Paulo, Brazil. Estuaries **1990**, *13*, 193–203.
136. Flores-Verdugo, F., González-Farías, F.; Zamorano, D.S.; et al. Mangrove ecosystems of the Pacific Coast of Mexico: distribution, structure, litterfall, and detritus dynamics. In *Coastal Plant Communities of Latin America*; Seeliger, U., Ed.; Academic Press: San Diego, CA, 1992; 269–288.
137. Sukardjo, S.; Yamada, I. Biomass and productivity of a *Rhizophora mucronata* Lamarck plantation in Tritih, Central Java, Indonesia. For. Ecol. Manage. **1992**, *49*, 195–209.
138. Flores-Verdugo, F.; González-Farías, F.; Zaragoza-Araujo, U. Ecological parameters of the mangroves of semi-arid regions of Mexico important for ecosystem management. In *Towards the Rationale Use of High Salinity Tolerant Plants*, Volume 1; Leith, H, Al Masoom, A., Eds.; Kluwer Academic Publishers: the Netherlands, 1993; 123–132.
139. Singh, V.P.; Garge, A.; Mall, L.P. Study of biomass, litter fall and litter decomposition in managed and unmanaged mangrove forests of Andaman Islands. In *Towards the Rationale Use of High Salinity Tolerant Plants*, Volume 1; Leith, H, Al Masoom, A., Eds.; Kluwer Academic Publishers, the Netherlands, 1993; 149–154.
140. Clarke, P.J. Baseline studies of temperate mangrove growth and reproduction; demographic and litterfall measures of leafing and flowering. Aust. J. Bot. **1994**, *42*, 37–48.
141. Day, J.W., Coronado-Molina, C.; Vera-Herrera, F.R.; Twilley, R.; Rivera-Monroy, V.H.; Alvarez-Guillen, H.; Day, R.; Conner, W. A 7-year record of above-ground net primary production in a southeastern Mexican mangrove forest. Aquat. Bot. **1996**, *55*, 39–60.
142. Twilley, R.R.; Pozo, M.; Garcia, V.H.; Rivera-Monroy, V.H.; Zambrano, R.; Bodero, A. Litter dynamics in riverine mangrove forests in the Guayas River estuary, Ecuador. Oecologia **1997**, *111*, 109–122.
143. Lugo, A.E.; Medina, E.; Cuevas, E.; Cintron, G.; Nieves, E.N.L.; Novelli, Y.S. Ecophysiology of a mangrove forest in Jobos Bay, Puerto Rico. Caribb. J. Sci. **2007**, *43*, 200–219.

BIBLIOGRAPHY

Tidal Marshes

1. Adam, P. *Saltmarsh Ecology*; Cambridge University Press; Cambridge, 1990.
2. Barendregt, A.; Whigham, D.F.; Baldwin, A.H., Eds. *Tidal Freshwater Wetlands*; Backhuys: the Netherlands, 2009.
3. Mitsch, W.J.; Gosselink, J.G.; Anderson, C.J.; Zhang, L. *Wetland Ecosystems*; John Wiley and Sons: New Jersey, 2009.
4. Perillo, G.M.E.; Wolanski, E., Cahoon, D.R.; Brinson, M.M., Eds. *Coastal Wetlands: An Integrated Ecosystem Approach*; Elsevier: the Netherlands, 2009.

Tropical Tidal Forests

1. Alongi, D.M. *The Energetics of Mangrove Forests*; Springer: New York, 2009.
2. Dahdouh-Guebas, F.; Koedam, N. Mangrove ecology: applications in forestry and coastal zone management. Aquatic Botany 2008, Volume 89, Issue 2 (Special Issue).
3. Field, C.B.; Whittaker, R.J. Biodiversity and function of mangrove ecosystems. Global Ecol. Biogeogr. Lett. **1998**, *7* (1), (Special Issue).
4. Hogarth, P.J. *The Biology of Mangroves*; Oxford University Press: Oxford, 1999.
5. Robertson, A.I.; Alongi, D.M., Eds. *Tropical Mangrove Ecosystems*; American Geophysical Union: 1994.
6. Saenger, P. *Mangrove Ecology, Silviculture and Conservation*; Kluwer Academic Publishers: 2002.
7. Spalding, M; Kainuma, M.; Collins, L. *World Atlas of Mangroves*; Earthscan: London, 2010.
8. Tomlinson, P.B. *The Botany of Mangroves*; Cambridge University Press: Cambridge, 1986.

Temperate Tidal Forests

1. Conner, W.H., Doyle, T.W., Krauss, K.W., Eds. *Ecology of Tidal Freshwater Forested Wetlands of the Southeastern United States*; Springer: Dordrecht, 2007.

Index

Volume I: Pages 1–588; *Volume II:* Pages 589–1088.
Note: Page references in roman refer to Volume I and page references in italics refer to Volume II.

Volume I: Pages 1–588; *Volume II:* Pages 589–1088.
Note: Page references in roman refer to Volume I and page references in italics refer to Volume II.

Volume I: Pages 1–588; *Volume II:* Pages 589–1088.
Note: Page references in roman refer to Volume I and page references in italics refer to Volume II.

Volume I: Pages 1–588; *Volume II:* Pages 589–1088.

Note: Page references in roman refer to Volume I and page references in italics refer to Volume II.

Volume I: Pages 1–588; *Volume II:* Pages 589–1088.
Note: Page references in roman refer to Volume I and page references in italics refer to Volume II.

Volume I: Pages 1–588; *Volume II:* Pages 589–1088.
Note: Page references in roman refer to Volume I and page references in italics refer to Volume II.

Volume I: Pages 1–588; *Volume II:* Pages 589–1088.
Note: Page references in roman refer to Volume I and page references in italics refer to Volume II.

Volume I: Pages 1–588; *Volume II:* Pages 589–1088.
Note: Page references in roman refer to Volume I and page references in italics refer to Volume II.

Volume I: Pages 1–588; *Volume II:* Pages 589–1088.
Note: Page references in roman refer to Volume I and page references in italics refer to Volume II.

Volume I: Pages 1–588; *Volume II:* Pages 589–1088.
Note: Page references in roman refer to Volume I and page references in italics refer to Volume II.

Volume I: Pages 1–588; *Volume II:* Pages 589–1088.
Note: Page references in roman refer to Volume I and page references in italics refer to Volume II.

Volume I: Pages 1–588; *Volume II:* Pages 589–1088.
Note: Page references in roman refer to Volume I and page references in italics refer to Volume II.

Volume I: Pages 1–588; *Volume II:* Pages 589–1088.
Note: Page references in roman refer to Volume I and page references in italics refer to Volume II.

Volume I: Pages 1–588; *Volume II:* Pages 589–1088.
Note: Page references in roman refer to Volume I and page references in italics refer to Volume II.

Volume I: Pages 1–588; *Volume II:* Pages 589–1088.
Note: Page references in roman refer to Volume I and page references in italics refer to Volume II.

Volume I: Pages 1–588; *Volume II:* Pages 589–1088.
Note: Page references in roman refer to Volume I and page references in italics refer to Volume II.